ENCYCLOPEDIA OF POLYMER SCIENCE AND ENGINEERING

VOLUME 10

Molecular Weight Determination
to
Pentadiene Polymers

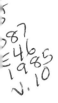

ENCYCLOPEDIA OF POLYMER SCIENCE AND ENGINEERING

VOLUME 10

**Molecular Weight Determination
to
Pentadiene Polymers**

A WILEY-INTERSCIENCE PUBLICATION
John Wiley & Sons
NEW YORK · CHICHESTER · BRISBANE · TORONTO · SINGAPORE

Library of Congress Cataloging in Publication Data:
Main entry under title:

Encyclopedia of polymer science and engineering.

   Rev. ed. of: Encyclopedia of polymer science and
technology. 1964–
   "A Wiley-Interscience publication."
   Includes bibliographies.
   1. Polymers and polymerization—Dictionaries.
I. Mark, H. F. (Herman Francis), 1895–
II. Kroschwitz, Jacqueline I.   III. Encyclopedia
of polymer science and technology.

TP1087.E46 1985   668.9   84-19713
ISBN 0-471-80942-X (v. 10)

Printed in the United States of America

In tribute to his towering role in the development of polymer science, Volume 10 of the *Encyclopedia of Polymer Science and Engineering* is dedicated to

PAUL JOHN FLORY

PUBLICATIONS OF P. J. FLORY

1. "The Photochemical Decomposition of Nitric Oxide" *J. Am. Chem. Soc.* **57**, 2641 (1935) P. J. Flory and H. L. Johnston

2. "Predissociation of the Oxygen Molecule" *J. Chem. Phys.* **4**, 23 (1936) P. J. Flory

3. "Molecular Size Distribution in Linear Condensation Polymers" *J. Am. Chem. Soc.* **58**, 1877 (1936) P. J. Flory

4. "The Mechanism of Vinyl Polymerizations" *J. Am. Chem Soc.* **59**, 241 (1937) P. J. Flory

5. "Kinetics of Condensation Polymerization: The Reaction of Ethylene Glycol with Succinic Acid" *J. Am. Chem. Soc.* **59**, 466 (1937) P. J. Flory

6. "The Heat of Combustion and Structure of Cuprene" *J. Am. Chem. Soc.* **59**, 1149 (1937) P. J. Flory

7. "Intramolecular Reaction Between Neighboring Substituents in Vinyl Polymers" *J. Am. Chem. Soc.* **61**, 1518 (1939) P. J. Flory

8. "Kinetics of Polyesterification: A Study of the Effects of Molecular Weight and Viscosity on Reaction Rate" *J. Am. Chem. Soc.* **61**, 3334 (1939) P. J. Flory

9. "Viscosities of Linear Polyesters. An Exact Relationship Between Viscosity and Chain Length" *J. Am. Chem. Soc.* **62**, 1057 (1940) P. J. Flory

10. "Molecular Size Distribution in Ethylene Oxide Polymers" *J. Am. Chem. Soc.* **62**, 1561 (1940) P. J. Flory

11. "Kinetics of the Degradation of Polyesters by Alcohols" *J. Am. Chem. Soc.* **62**, 2255 (1940) P. J. Flory

12. "A Comparison of Esterification and Ester Interchange Kinetics" *J. Am. Chem. Soc.* **62**, 2261 (1940) P. J. Flory

13. "Viscosities of Polyester Solutions and the Staudinger Equation" *J. Am. Chem. Soc.* **62**, 3032 (1940) P. J. Flory and P. B. Stickney

14. "Molecular Size Distribution in Three Dimensional Polymers. I. Gelation"

J. Am. Chem. Soc. **63,** 3083
(1941)
P. J. Flory

15. "Molecular Size Distribution in
Three Dimensional Polymers.
II. Trifunctional Branching
Units"
J. Am. Chem. Soc. **63,** 3091
(1941)
P. J. Flory

16. "Molecular Size Distribution in
Three Dimensional Polymers.
III. Tetrafunctional Branching
Units"
J. Am. Chem. Soc. **63,** 3096
(1941)
P. J. Flory

17. "Thermodynamics of High
Polymer Solutions"
J. Chem. Phys. **9,** 660 (1941)
P. J. Flory

18. "Thermodynamics of High
Polymer Solutions"
J. Chem. Phys. **10,** 51 (1942)
P. J. Flory

19. "Constitution of Three-
Dimensional Polymers and the
Theory of Gelation"
J. Phys. Chem. **46,** 132 (1942)
P. J. Flory

20. "Viscosities of Polyester
Solutions"
J. Phys. Chem. **46,** 870 (1942)
P. J. Flory

21. "Statistics of Intramolecular
Aldol Condensations in
Unsaturated Ketone Polymers"
J. Am. Chem. Soc. **64,** 177
(1942)
P. J. Flory

22. "Random Reorganization of
Molecular Weight Distribution
in Linear Condensation
Polymers"
J. Am. Chem. Soc. **64,** 2205
(1942)
P. J. Flory

23. "Molecular Weights and

Intrinsic Viscosities of
Polyisobutylenes"
J. Am. Chem. Soc. **65,** 372
(1943)
P. J. Flory

24. "Statistical Mechanics of Cross-
Linked Polymer Networks. I.
Rubberlike Elasticity"
J. Chem. Phys. **11,** 512 (1943)
P. J. Flory and J. Rehner, Jr.

25. "Statistical Mechanics of Cross-
Linked Polymer Networks. II.
Swelling"
J. Chem. Phys. **11,** 521 (1943)
P. J. Flory and J. Rehner, Jr.

26. "Statistical Theory of Chain
Configuration and Physical
Properties of High Polymers"
Ann. N.Y. Acad. Sci. **44,** 419
(1943)
P. J. Flory and J. Rehner, Jr.

27. "Thermodynamics of
Heterogeneous Polymer
Solutions"
J. Chem. Phys. **12,** 114 (1944)
P. J. Flory

28. "Network Structure and the
Elastic Properties of Vulcanized
Rubber"
Chem. Rev. **35,** 51 (1944)
P. J. Flory

29. "Effect of Deformation on the
Swelling Capacity of Rubber"
J. Chem. Phys. **12,** 412 (1944)
P. J. Flory and J. Rehner, Jr.

30. "Thermodynamics of
Heterogeneous Polymers and
Their Solutions"
J. Chem. Phys. **12,** 425 (1944)
P. J. Flory

31. "Tensile Strength in Relation to
Molecular Weight of High
Polymers"
J. Am. Chem. Soc. **67,** 2048
(1945)
P. J. Flory

32. "Thermodynamics of Dilute
Solutions of High Polymers"

J. Am. Chem. Soc. **72,** 5052
(1950)
P. J. Flory

64. "Molecular Configuration and
Thermodynamic Parameters
from Intrinsic Viscosities"
J. Polym. Sci. **5,** 745 (1950)
P. J. Flory and T. G. Fox, Jr.

65. "Treatment of Osmotic and
Light Scattering Data for
Dilute Solutions"
J. Am. Chem. Soc. **73,** 285
(1951)
T. G. Fox, Jr., P. J. Flory,
and A. M. Bueche

66. "Further Studies on the Melt
Viscosity of Polyisobutylene"
J. Phys. Chem. **55,** 221 (1951)
T. G. Fox, Jr. and P. J. Flory

67. "The Effect of Rate of Shear on
the Viscosity of Dilute
Solutions of Polyisobutylene"
J. Am. Chem. Soc. **73,** 1901
(1951)
T. G. Fox, Jr., J. C. Fox, and
P. J. Flory

68. "Treatment of Intrinsic
Viscosities"
J. Am. Chem. Soc. **73,** 1904
(1951)
P. J. Flory and T. G. Fox, Jr.

69. "Intrinsic Viscosity–
Temperature Relationships for
Polyisobutylene in Various
Solvents"
J. Am. Chem. Soc. **73,** 1909
(1951)
T. G. Fox, Jr. and P. J. Flory

70. "Intrinsic Viscosity
Relationships for Polystyrene"
J. Am. Chem. Soc. **73,** 1915
(1951)
T. G. Fox, Jr. and P. J. Flory

71. "Crystallization in High
Polymers. VII. Heat of
Fusion of Poly-(N,N'-
sebacoylpiperazine) and its
Interaction with Diluents"

J. Am. Chem. Soc. **73,** 2532
(1951)
P. J. Flory, L. Mandelkern,
and H. K. Hall

72. "Melting and Glassy State
Transitions in Cellulose Esters
and their Mixtures with
Diluents"
J. Am. Chem. Soc. **73,** 3206
(1951)
L. Mandelkern and
P. J. Flory

73. "The Dependence of the
Diffusion Coefficient on
Concentration in Dilute
Polymer Solutions"
J. Chem. Phys. **19,** 984 (1951)
L. Mandelkern and
P. J. Flory

74. "Statistical Thermodynamics of
Rubber Elasticity"
J. Chem. Phys. **19,** 1435
(1951)
F. T. Wall and P. J. Flory

75. "Thermodynamics of High
Polymer Solutions"
Annu. Rev. Phys. Chem. **II,**
383 (1951)
P. J. Flory and
W. R. Krigbaum

76. "Molecular Dimensions of
Natural Rubber and Gutta
Percha"
J. Am. Chem. Soc. **74,** 195
(1952)
H. L. Wagner and P. J. Flory

77. "The Frictional Coefficient for
Flexible Chain Molecules in
Dilute Solution"
J. Chem. Phys. **20,** 212 (1952)
L. Mandelkern and
P. J. Flory

78. "Molecular Dimensions of
Cellulose Triesters"
J. Am. Chem. Soc. **74,** 2517
(1952)
L. Mandelkern and
P. J. Flory

79. "Molecular Size Distribution in Three Dimensional Polymers. VI. Branched Polymers Containing A-R-B$_{f-1}$ Type Units"
    J. Am. Chem. Soc. **74,** 2718 (1952)
    P. J. Flory

80. "Statistical Mechanics of Dilute Polymer Solutions. III. Ternary Mixtures of Two Polymers and a Solvent"
    J. Chem. Phys. **20,** 873 (1952)
    W. R. Krigbaum and P. J. Flory

81. "Molecular Dimensions of Polydimethylsiloxanes"
    J. Am. Chem. Soc. **74,** 3364 (1952)
    P. J. Flory, L. Mandelkern, J. B. Kinsinger, and W. B. Schultz

82. "Heats of Fusion of Aliphatic Polyesters"
    J. Am. Chem. Soc. **74,** 3949 (1952)
    L. Mandelkern, R. R. Garrett, and P. J. Flory

83. "Sedimentation Behavior of Flexible Chain Molecules: Polyisobutylene"
    J. Chem. Phys. **20,** 1392 (1952)
    L. Mandelkern, W. R. Krigbaum, H. A. Scheraga, and P. J. Flory

84. "Molecular Weight Dependence of Intrinsic Viscosity of Polymer Solutions"
    J. Polym. Sci. **9,** 381 (1952)
    W. R. Krigbaum, L. Mandelkern, and P. J. Flory

85. "Phase Equilibria in Polymer–Solvent Systems"
    J. Am. Chem. Soc. **74,** 4760 (1952)
    A. R. Shultz and P. J. Flory

86. "Treatment of Osmotic Pressure Data"
    J. Polym. Sci. **9,** 503 (1952)
    W. R. Krigbaum and P. J. Flory

87. "Macromolecular Extension and Hydrodynamic Parameters in Intrinsic Viscosity"
    J. Polym. Sci. **10,** 121 (1953)
    S. Newman and P. J. Flory

88. "Molecular Configuration of Polyelectrolytes"
    J. Chem. Phys. **21,** 162 (1953)
    P. J. Flory

89. "The Dependence of the Intrinsic Viscosity Sodium Polyacrylate on Salt Concentration"
    J. Chem. Phys. **21,** 164 (1953)
    P. J. Flory, W. R. Krigbaum, and W. B. Schultz

90. "Statistical Mechanics of Dilute Polymer Solutions. IV. Variation of the Osmotic Second Coefficient with Molecular Weight"
    J. Am. Chem. Soc. **75,** 1775 (1953)
    W. R. Krigbaum and P. J. Flory

91. "Molecular Weight Dependence of the Intrinsic Viscosity of Polymer Solutions. II"
    J. Polym. Sci. **11,** 37 (1953)
    W. R. Krigbaum and P. J. Flory

92. *Principles of Polymer Chemistry*
    Cornell University Press, Ithaca, N.Y., 1953
    P. J. Flory

93. "Phase Equilibria in Polymer–Solvent Systems. II. Thermodynamic Interaction Parameters from Critical Miscibility Data"
    J. Am. Chem. Soc. **75,** 3888 (1953)
    A. R. Shultz and P. J. Flory

94. "Statistical Mechanics of Dilute Polymer Solutions. V. Evaluation of Thermodynamic Interaction Parameters from Dilute Solution Measurements"
   J. Am. Chem. Soc. **75,** 5254 (1953)
   W. R. Krigbaum and P. J. Flory

95. "Phase Equilibria in Polymer–Solvent Systems. III. Three-component Systems"
   J. Am. Chem. Soc. **75,** 5681 (1953)
   A. R. Shultz and P. J. Flory

96. "The Configuration and Properties of Polymer Molecules in Dilute Solution"
   Proc. Int. Conf. Theoretical Phys., Japan, 381 (Sept. 1953)
   P. J. Flory

97. "Thermodynamics of Crystallization in High Polymers. Cellulose Trinitrate"
   J. Polym. Sci. **12,** 97 (1954)
   P. J. Flory, R. R. Garrett, S. Newman, and L. Mandelkern

98. "Crystallization Kinetics in High Polymers. I. Bulk Polymers"
   J. Appl. Phys. **25,** 830 (1954)
   L. Mandelkern, F. A. Quinn, Jr., and P. J. Flory

99. "Intrinsic Viscosities of Polyelectrolytes. Poly(Acrylic Acid)"
   J. Phys. Chem. **58,** 653 (1954)
   P. J. Flory and J. E. Osterheld

100. "The Glass Temperature and Related Properties of Polystyrene. Influence of Molecular Weight"
   J. Polym. Sci. **14,** 315 (1954)
   T. G. Fox and P. J. Flory

101. "Molecular Dimensions in Relation to Intrinsic Viscosities"
   J. Polym. Sci. **14,** 451 (1954)
   S. Newman, W. R. Krigbaum, C. Laugier, and P. J. Flory

102. "Polymer Chain Dimensions in Mixed-Solvent Media"
   J. Polym. Sci. **15,** 231 (1955)
   A. R. Shultz and P. J. Flory

103. "Molecular Configuration of Gelatin"
   J. Polym. Sci. **16,** 383 (1955)
   E. V. Gouinlock, Jr., P. J. Flory, and H. A. Scheraga

104. "Theory of Crystallization in Copolymers"
   Trans. Faraday Soc. **51,** 848 (1955)
   P. J. Flory

105. "Mechanism of Crystallization in Polymers"
   J. Polym. Sci. **18,** 592 (1955)
   P. J. Flory and A. D. McIntyre

106. "Melting and Glass Transitions in Polymers"
   Suppl. La Ricera Scientifica, 3 (1955)
   P. J. Flory

107. "Statistical Thermodynamics of Semi-Flexible Chain Molecules"
   Proc. R. Soc. London Ser. A **234,** 60 (1956)
   P. J. Flory

108. "Phase Equilibria in Solutions of Rod-Like Particles"
   Proc. R. Soc. London Ser. A **234,** 73 (1956)
   P. J. Flory

109. "Evidence for a Reversible First-Order Phase Transition in Collagen–Diluent Mixtures"
   Nature **177,** 176 (1956)
   R. R. Garrett and P. J. Flory

110. "Role of Crystallization in Polymers and Proteins"
   Science **124,** 53 (1956)
   P. J. Flory

(1963)
J. E. Mark and P. J. Flory

143. "Thermodynamic Stability of Solution-Crystallized Polyethylene"
Trans. Faraday Soc. **59,** 1906 (1963)
J. B. Jackson, P. J. Flory, and R. Chiang

144. "Macromolecules in the Solid State"
in P. Leurgenes, ed., *Lecture in Materials Science*
W. A. Benjamin, New York, 1963, p. 27
P. J. Flory

145. "Crystallization and Melting of Copolymers of Polymethylene"
Polymer **4,** 221 (1963)
M. J. Richardson, P. J. Flory, and J. B. Jackson

146. "Shear Modulus in Relation to Crystallinity in Polymethylene and its Copolymers"
Polymer **4,** 237 (1963)
J. B. Jackson, P. J. Flory, R. Chiang, and M. J. Richardson

147. "Melting Points of Linear-Chain Homologs. The Normal Paraffin Hydrocarbons"
J. Am. Chem. Soc. **85,** 3548 (1963)
P. J. Flory and A. Vrij

148. "The Crystallization of Polymethylene Copolymers; Morphology"
Polymer **5,** 159 (1964)
J. B. Jackson and P. J. Flory

149. "Configuration of the Poly-(Dimethylsiloxane) Chain. I. The Temperature Coefficient of the Unperturbed Extension"
J. Am. Chem. Soc. **86,** 138 (1964)
J. E. Mark and P. J. Flory

150. "Configuration of the Poly-(Dimethylsiloxane) Chain. II.

Unperturbed Dimensions and Specific Solvent Effects"
J. Am. Chem. Soc. **86,** 141 (1964)
V. Crescenzi and P. J. Flory

151. "Configuration of the Poly-(Dimethylsiloxane) Chain. III. Correlation of Theory and Experiment"
J. Am. Chem. Soc. **86,** 146 (1964)
P. J. Flory, V. Crescenzi, and J. E. Mark

152. "The Configuration of the Polyoxymethylene Chain"
Makromol. Chem. **75,** 11 (1964)
P. J. Flory and J. E. Mark

153. "Understanding Unruly Molecules"
Chemistry **37,** 6 (1964)
P. J. Flory

154. "Mean-Square Moments of Chain Molecules"
Proc. Nat. Acad. Sci. U.S. **51,** 1060 (1964)
P. J. Flory

155. "Statistical Thermodynamics of Chain Molecule Liquids. I. An Equation of State for Normal Paraffin Hydrocarbons"
J. Am. Chem. Soc. **86,** 3507 (1964)
P. J. Flory, R. A. Orwoll, and A. Vrij

156. "Statistical Thermodynamics of Chain Molecule Liquids. II. Liquid Mixtures of Normal Paraffin Hydrocarbons"
J. Am. Chem. Soc. **86,** 3515 (1964)
P. J. Flory, R. A. Orwoll, and A. Vrij

157. "Thermodynamic Properties of Nonpolar Mixtures of Small Molecules"
J. Am. Chem. Soc. **86,** 3563 (1964)

P. J. Flory, J. E. Mark, and A. Abe

174. "Effect of Volume Exclusion on the Dimensions of Polymer Chains"
J. Chem. Phys. **44,** 2243 (1966)
P. J. Flory and S. Fisk

175. "On the Interpretation of Nuclear Magnetic Resonance Spectra of Stereoregular Polymers"
J. Am. Chem. Soc. **88,** 873 (1966)
P. J. Flory and J. D. Baldeschwieler

176. "Treatment of Liquid–Liquid Phase Equilibria. Hydrocarbon–Perfluorocarbon Mixtures"
J. Am. Chem. Soc. **88,** 2887 (1966)
A. Abe and P. J. Flory

177. "Macrocyclization Equilibrium Constants and the Statistical Configuration of Poly(dimethylsiloxane) Chains"
J. Am. Chem. Soc. **88,** 3209 (1966)
P. J. Flory and J. A. Semlyen

178. "Analysis of the Unperturbed Dimensions of the Linear Polyphosphate Chain"
Trans. Faraday Soc. **62,** 2622 (1966)
J. A. Semlyen and P. J. Flory

179. "Dipole Moments of Chain Molecules. I. Oligomers and Polymers of Oxyethylene"
J. Am. Chem. Soc. **88,** 3702 (1966)
J. E. Mark and P. J. Flory

180. "Treatment of the Effect of Excluded Volume and Deduction of Unperturbed Dimensions of Polymer Chains. Configurational Parameters for Cellulose Derivatives"

Makromol. Chem. **98,** 128 (1966)
P. J. Flory

181. "Stress–Strain Isotherms for Poly-(Dimethylsiloxane) Networks"
J. Appl. Phys. **37,** 4635 (1966)
J. E. Mark and P. J. Flory

182. "Configurational Statistics of Chain Molecules"
Proceedings of the Robert A. Welch Foundation Conferences on Chemical Research, X. Polymers, Houston, Texas, Nov. 21–23, 1966
P. J. Flory

183. "Conformational Energy Estimates for Statistically Coiling Polypeptide Chains"
J. Mol. Biol. **23,** 47 (1967)
D. A. Brant, W. G. Miller, and P. J. Flory

184. "Random Coil Configurations of Polypeptide Copolymers"
J. Mol. Biol. **23,** 67 (1967)
W. G. Miller, D. A. Brant, and P. J. Flory

185. "Configurational Statistics of Polyamide Chains"
J. Polym. Sci. Part A-2 **5,** 399 (1967)
P. J. Flory and A. D. Williams

186. "Configurational Statistics of Poly-(Ethylene Terephthalate) Chains"
J. Polym. Sci. Part A-2 **5,** 417 (1967)
A. D. Williams and P. J. Flory

187. "Stereochemical Equilibrium in Chain Molecules"
J. Am. Chem. Soc. **89,** 1798 (1967)
P. J. Flory

188. "Conformational Energy and Configurational Statistics of

Poly-L-Proline"
Proc. Nat. Acad. Sci. U.S. **58,**
52 (1967)
P. R. Schimmel and
P. J. Flory

189. "Epimerization of 2,4-
Diphenylpentane, an Oligomer
of Polystyrene"
J. Am. Chem. Soc. **89,** 4807
(1967)
A. D. Williams,
J. I. Brauman, N. J. Nelson,
and P. J. Flory

190. "Analysis of the Skeletal
Configuration of Crystalline
Hen Egg-White Lysozyme"
Proc. Nat. Acad. Sci. U.S. **58,**
428 (1967)
D. A. Brant and
P. R. Schimmel
(communicated by P. J. Flory)

191. "Optical Anisotropy of Chain
Molecules. Theory of
Depolarization of Scattered
Light with Application to
n-Alkanes"
J. Chem. Phys. **47,** 1999
(1967)
R. L. Jernigan and
P. J. Flory

192. "Configurational Statistics of
Polypeptide Chains"
*International Symposium on
Conformation of Biopolymers,*
University of Madras, India,
Jan. 16–17, 1965; in G. N.
Ramachandran, ed.,
Conformation of Biopolymers,
Vol. 1, Academic Press, Inc.,
London, 1967, pp. 339–363
P. J. Flory

193. "Dipole Moments in Relation to
Configuration of Polypeptide
Chains"
J. Am. Chem. Soc. **89,** 6807
(1967)
P. J. Flory and
P. R. Schimmel

194. "Equation of State Parameters
for Normal-Alkanes:
Correlation with Chain Length"
J. Am. Chem. Soc. **89,** 6814
(1967)
R. A. Orwoll and P. J. Flory

195. "Thermodynamic Properties of
Binary Mixtures of *n*-Alkanes"
J. Am. Chem. Soc. **89,** 6822
(1967)
R. A. Orwoll and P. J. Flory

196. "Analysis of Thermodynamic
Excess Properties of Liquid
Mixtures of Argon and
Krypton"
Trans. Faraday Soc. **64,** 1188
(1968)
H. Höcker and P. J. Flory

197. "Rayleigh Scattering by Real
Chain Molecules"
J. Am. Chem. Soc. **90,** 3128
(1968)
P. J. Flory and
R. L. Jernigan

198. "Conformational Energies and
Configurational Statistics of
Copolypeptides Containing
L-Proline"
J. Mol. Biol. **34,** 105 (1968)
P. R. Schimmel and
P. J. Flory

199. "Thermodynamics of Mixing of
n-Alkanes with
Polyisobutylene"
Macromolecules **1,** 279 (1968)
P. J. Flory, J. L. Ellenson,
and B. E. Eichinger

200. "Determination of the Equation
of State of Polyisobutylene"
Macromolecules **1,** 285 (1968)
B. E. Eichinger and
P. J. Flory

201. "Thermodynamics of Mixing
Polymethylene and
Polyisobutylene"
Macromolecules **1,** 287 (1968)
P. J. Flory, B. E. Eichinger,
and R. A. Orwoll

P. J. Flory and H. Höcker

234. "The Thermodynamics of Polystyrene Solutions. II. Polystyrene and Ethylbenzene"
Trans. Faraday Soc. **67**, 2270 (1971)

H. Höcker and P. J. Flory

235. "The Thermodynamics of Polystyrene Solutions. III. Polystyrene and Cyclohexane"
Trans. Faraday Soc. **67**, 2275 (1971)

H. Höcker, H. Shih, and P. J. Flory

236. "Configuration-Dependent Properties of Polymer Chains"
Pure Appl. Chem. **26**, 309 (1971)

P. J. Flory

237. "Spatial Configurations of Polynucleotide Chains. I. Steric Interactions in Polyribonucleotides: A Virtual Bond Model"
Biopolymers **11**, 1 (1972)

W. K. Olson and P. J. Flory

238. "Spatial Configuration of Polynucleotide Chains. II. Conformational Energies and the Average Dimensions of Polyribonucleotides"
Biopolymers **11**, 25 (1972)

W. K. Olson and P. J. Flory

239. "Spatial Configurations of Polynucleotide Chains. III. Polydeoxyribonucleotides"
Biopolymers **11**, 57 (1972)

W. K. Olson and P. J. Flory

240. "Theory of Optical Anisotropy of Chain Molecules"
J. Chem. Phys. **56**, 862 (1972)

P. J. Flory

241. "Depolarized Rayleigh Scattering and the Mean-Square Optical Anisotropies of N-Alkanes in Solution"
J. Chem. Soc. Faraday Trans. II **68**, 1098 (1972)

G. D. Patterson and P. J. Flory

242. "Optical Anisotropies of Polyoxyethylene Oligomers"
J. Chem. Soc. Faraday Trans. II **68**, 1111 (1972)

G. D. Patterson and P. J. Flory

243. "Optical Anisotropy of Polypeptide Chains"
Biopolymers **11**, 1527 (1972)

R. T. Ingwall and P. J. Flory

244. "Stress-Optical Behavior of Polymethylene and Poly(dimethylsiloxane)"
Macromolecules **5**, 550 (1972)

M. H. Liberman, Y. Abe, and P. J. Flory

245. "Molecular Configuration and States of Aggregation of Biopolymers"
Polymerization in Biological Systems, Ciba Foundation 7 (1972)

P. J. Flory

246. "Equation-of-State Parameters for Poly(dimethylsiloxane)"
Macromolecules **5**, 758 (1972)

H. Shih and P. J. Flory

247. "Thermodynamics of Solutions of Poly(dimethylsiloxane) in Benzene, Cyclohexane, and Chlorobenzene"
Macromolecules **5**, 761 (1972)

P. J. Flory and H. Shih

248. "Stereochemical Equilibrium in 2,4,6-Trichloro-n-Heptane with Applications to Poly(Vinyl Chloride)"
J. Chem. Soc. Faraday Trans. II **69**, 632 (1973)

P. J. Flory and C. J. Pickles

249. "The Temperature Coefficient of the Unperturbed Dimensions of Polyoxyethylene"
Macromolecules **6**, 300 (1973)

J. E. Mark and P. J. Flory

250. "Conformational Energies,

Stereoregularity, and the Role of Nonstaggered Conformations in Polymer Chains"
J. Polym. Sci. Polym. Phys. Ed. **11,** 621 (1973)
P. J. Flory

251. "Moments of the End-to-End Vector of a Chain Molecule, Its Persistence and Distribution"
Proc. Nat. Acad. Sci. **70,** 1819 (1973)
P. J. Flory

252. "Depolarized Light Scattering by Amides and Peptides"
Biopolymers **12,** 1123 (1973)
R. T. Ingwall and P. J. Flory

253. "Kerr Constants of Amides and Peptides"
Biopolymers **12,** 1137 (1973)
R. T. Ingwall, E. A. Czurylo, and P. J. Flory

254. "Molecular Configuration in Bulk Polymers"
International Symposium on Macromolecules, Helsinki, Finland, July 2–7 1972; *Pure Appl. Chem. Macromol. Chem.* **8,** 1 (1972); *Rubber Chem. Technol.* **48**(3), 513 (July–Aug. 1975)
P. J. Flory

255. "Macromolecules Vis-à-Vis the Traditions of Chemistry"
164th National Meeting of ACS, New York, Aug. 27–Sept. 1, 1972; *J. Chem. Education* **50,** 732 (1973)
P. J. Flory

256. "The Challenge to Macromolecular Science"
Dedication of the Midland Macromolecular Institute, Sept. 28, 1972; *Int. J. Polym. Mater.* **2,** 265 (1973); *Angew. Chem.* **13,** 97 (1974)
P. J. Flory

257. "Optical Anisotropy of Polyisobutylene-Strain Birefringence"
J. Polym. Sci. Polym. Phys. Ed. **12,** 187 (1974)
M. H. Liberman,
L. C. DeBolt, and
P. J. Flory

258. "The Elastic Properties of Elastin"
Biopolymers **13,** 677 (1974)
C. A. J. Hoeve and
P. J. Flory

259. "Molecular Concepts in Polymer Science"
Priestley Medal Address, ACS Meeting, Los Angeles, Calif., Apr. 1, 1974; *C&E News,* 36 (Apr. 8, 1974)
P. J. Flory

260. "The Science of Molecules"
Priestley Address, Selinsgrove, Pa, Apr. 25, 1974; *C&E News,* 23 (July 29, 1974)
P. J. Flory

261. "The Interpretation of Viscosity–Temperature Coefficients for Polyoxyethylene Chains in a Thermodynamically Good Solvent"
Macromolecules **7,** 325 (1974)
S. Bluestone, J. E. Mark, and
P. J. Flory

262. "Foundations of Rotational Isomeric State Theory and General Methods for Generating Configurational Averages"
Macromolecules **7,** 381 (1974)
P. J. Flory

263. "Configurational Statistics of Vinyl Polymer Chains"
J. Am. Chem. Soc. **96,** 5015 (1974)
P. J. Flory,
P. R. Sundararajan, and
L. C. DeBolt

264. "Configurational

Characteristics of Poly(methyl methacrylate)"
J. Am. Chem. Soc. **96,** 5025 (1974)
P. R. Sundararajan and P. J. Flory

265. "Introductory Lecture, General Discussion on Gels and Gelling Processes"
Faraday Disc. Chem. Soc. **57,** 7 (1974)
P. J. Flory

266. "Moments and Distribution Functions for Polymer Chains of Finite Length. I. Theory"
J. Chem. Phys. **61,** 5358 (1974)
P. J. Flory and D. Y. Yoon

267. "Moments and Distribution Functions for Polymer Chains of Finite Length. II. Polymethylene"
J. Chem. Phys. **61,** 5366 (1974)
D. Y. Yoon and P. J. Flory

268. "The Elastic Free Energy and the Elastic Equation of State: Elongation and Swelling of Polydimethylsiloxane Networks"
J. Polym. Sci. Polym. Phys. Ed. **13,** 683 (1975)
P. J. Flory and Y.-I. Tatara

269. "Spatial Configuration of Macromolecular Chains"
Les Prix Nobel en 1974, reprinted in the following:
Science **188,** 1268 (1975);
Brit. Polym. J. **8,** 1 (1976); *J. Polym. Sci. Polym. Symp.* **54,** 19 (1976)
P. J. Flory

270. "Small Angle Neutron and X-ray Scattering by Poly(methyl methacrylate) Chains"
Polymer **16,** 645 (1975)
D. Y. Yoon and P. J. Flory

271. "Conformational Energy and Configurational Statistics of Polypropylene"
Macromolecules **8,** 765 (1975)
U. W. Suter and P. J. Flory

272. "Conformational Characteristics of Polystyrene"
Macromolecules **8,** 776 (1975)
D. Y. Yoon, P. R. Sundararajan, and P. J. Flory

273. "Conformational Characteristics of Poly(methyl acrylate)"
Macromolecules **8,** 784 (1975)
D. Y. Yoon, U. W. Suter, P. R. Sundararajan, and P. J. Flory

274. "Moments and Distribution Functions for Poly(dimethylsiloxane) Chains of Finite Length"
Macromolecules **9,** 33 (1976)
P. J. Flory and V. W. C. Chang

275. "Moments and Distribution Functions for Polypeptide Chains. Poly-L-alanine"
Macromolecules **9,** 41 (1976)
J. C. Conrad and P. J. Flory

276. "Small-Angle X-ray and Neutron Scattering by Polymethylene, Polyoxyethylene, and Polystyrene Chains"
Macromolecules **9,** 294 (1976)
D. Y. Yoon and P. J. Flory

277. "II. Small-Angle Neutron and X-Ray Scattering by Poly(methyl methacrylate) Chains"
Macromolecules **9,** 299 (1976)
D. Y. Yoon and P. J. Flory

278. "Persistence Vectors and Higher Moment Tensors of Polyoxyalkanes"
J. Polym. Sci. Polym. Phys. Ed. **14,** 1337 (1976)
A. Abe, J. W. Kennedy, and

P. J. Flory

279. "Persistence Vectors for Polypropylene, Polystyrene, and Poly(methyl methacrylate) Chains"
J. Polym. Sci. Polym. Phys. Ed. **14,** 1425 (1976)
D. Y. Yoon and P. J. Flory

280. "Theoretical Predictions on the Configurations of Polymer Chains in the Amorphous State"
J. Macromol. Sci. Phys. Ed. **B12**(1), 1 (1976)
P. J. Flory

281. "Macrocyclization Equilibria. 1. Theory"
J. Am. Chem. Soc. **98,** 5733 (1976)
P. J. Flory, U. W. Suter, and M. Mutter

282. "Macrocyclization Equilibria. 2. Poly(dimethylsiloxane)"
J. Am. Chem. Soc. **98,** 5740 (1976)
U. W. Suter, M. Mutter, and P. J. Flory

283. "Macrocyclization Equilibria. 3. Poly(6-aminocaproamide)"
J. Am. Chem. Soc. **98,** 5745 (1976)
M. Mutter, U. W. Suter, and P. J. Flory

284. "Statistical Thermodynamics of Random Networks"
Proc. R. Soc. London Ser. A **351,** 351 (1976)
P. J. Flory

285. "Analysis of Nuclear Magnetic Resonance Spectra of Protons in Predominantly Isotactic Polystyrene"
Macromolecules **10,** 562 (1977)
D. Y. Yoon and P. J. Flory

286. "Small-angle Neutron Scattering by Semicrystalline Polyethylene"

Polymer **18,** 509 (1977)
D. Y. Yoon and P. J. Flory

287. "Theory of Elasticity of Polymer Networks. The Effect of Local Constraints on Junctions"
J. Chem. Phys. **66,** 5720 (1977); *Rubber Chem. Technol.* **52**(1), 110 (1979)
P. J. Flory

288. "Concept and Innovation in Polymer Science"
Perkin Medal Lecture, New York, Feb. 18, 1977; *Chem. Ind.,* 369 (May 21, 1977)
P. J. Flory

289. "Separation of Collision-Induced from Intrinsic Molecular Depolarized Rayleigh Scattering. Optical Anisotropy of the C—Cl Bond"
J. Chem. Soc. Faraday Trans. II **73,** 1505 (1977)
C. W. Carlson and P. J. Flory

290. "Optical Anisotropy of Polystyrene and Its Low Molecular Analogues"
J. Chem. Soc. Faraday Trans. II **73,** 1521 (1977)
U. W. Suter and P. J. Flory

291. "Optical Anisotropies of *para*-Halogenated Polystyrenes and Realted Molecules"
J. Chem. Soc. Faraday Trans. II **73,** 1538 (1977)
E. Saiz, U. W. Suter and P. J. Flory

292. "Dipole Moments of Poly(*p*-chlorostyrene) Chains"
Macromolecules **10,** 967 (1977)
E. Saiz, J. E. Mark, and P. J. Flory

293. "The Molecular Theory of Rubber Elasticity"
in E. M. Pearce and J. R. Schaefgen, eds., *Contemporary Topics in*

Polymer Science, Vol. 2, Plenum Press, New York, 1977, pp. 1–18.

294. "Statistical Thermodynamics of Macromolecular Liquids and Solutions"
Ber. Bunsenges. Phys. Chem. **81,** 885 (1977)
P. J. Flory

295. "Molecular Morphology in Semicrystalline Polymers"
Nature **272,** 226 (1978)
P. J. Flory and D. Y. Yoon

296. "Rubber Elasticity in the Range of Small Uniaxial Tensions and Compressions. Results for Poly(dimethylsiloxane)"
J. Polym. Sci. Polym. Phys. Ed. **16,** 1115 (1978)
B. Erman and P. J. Flory

297. "Theory of Elasticity of Polymer Networks. II. The Effect of Geometric Constraints on Junctions"
J. Chem. Phys. **68,** 5363 (1978)
B. Erman and P. J. Flory

298. "Chemistry, Macromolecules, and the Needs of Man"
Pure Appl. Chem. **50,** 255 (1978)
P. J. Flory

299. "Small-Angle Neutron Scattering by *n*-Alkane Chains"
J. Chem. Phys. **69**(6), 2536 (1978)
D. Y. Yoon and P. J. Flory

300. "Statistical Thermodynamics of Mixtures of Rodlike Particles. 1. Theory for Polydisperse Systems"
Macromolecules **11,** 1119 (1978)
P. J. Flory and A. Abe

301. "Statistical Thermodynamics of Mixtures of Rodlike Particles. 2. Ternary Systems"
Macromolecules **11,** 1122 (1978)
A. Abe and P. J. Flory

302. "Statistical Thermodynamics of Mixtures of Rodlike Particles. 3. The Most Probable Distribution"
Macromolecules **11,** 1126 (1978)
P. J. Flory and R. S. Frost

303. "Statistical Thermodynamics of Mixtures of Rodlike Particles. 4. The Poisson Distribution"
Macromolecules **11,** 1134 (1978)
R. S. Frost and P. J. Flory

304. "Statistical Thermodynamics of Mixtures of Rodlike Particles. 5. Mixtures with Random Coils"
Macromolecules **11,** 1138 (1978)
P. J. Flory

305. "Statistical Thermodynamics of Mixtures of Rodlike Particles. 6. Rods Connected by Flexible Joints"
Macromolecules **11,** 1141 (1978)
P. J. Flory

306. "Spatial Configuration of Natural and Synthetic Macromolecules"
The Third Philip Morris Science Symposium (1978)
P. J. Flory

307. "The Elastic Free Energy of Dilation of a Network"
Macromolecules **12,** 119 (1979)
P. J. Flory

308. "Relationship of Stress to Uniaxial Strain in Crosslinked Poly(Dimethylsiloxane) over the Full Range from Large Compressions to High Elongations"
J. Polym. Sci. Polym. Phys. Ed. **17,** 1845 (1979)

H. Pak and P. J. Flory
309. "Molecular Theory of Rubber Elasticity"
Polymer **20,** 1317 (1979)
P. J. Flory

310. "Theory of Systems of Rodlike Particles. I. Athermal Systems"
Mol. Cryst. Liq. Cryst. **54,** 289 (1979)
P. J. Flory and G. Ronca

311. "Theory of Systems of Rodlike Particles. II. Thermotropic Systems with Orientation-dependent Interactions"
Mol. Cryst. Liq. Cryst. **54,** 311 (1979)
P. J. Flory and G. Ronca

312. "Introductory Lecture: Levels of Order in Amorphous Polymers"
Faraday Disc. Chem. Soc. **68,** 14 (1979)
P. J. Flory

313. "Molecular Morphology in Semicrystalline Polymers"
Faraday Disc. Chem. Soc. **68,** 288 (1979)
D. Y. Yoon and P. J. Flory

314. "Crystalline and Mesomorphic Phases in Polymers"
Ferroelectrics **30,** 1 (1980)
P. J. Flory

315. "Molecular Morphology in Amorphous and Glass Polymers"
J. Non-Crystalline Solids **42,** 117 (1980)
P. J. Flory

316. "Molecular Structure, Conformation and Properties of Macromolecules"
Pure Appl. Chem. **52,** 241 (1980)
P. J. Flory

317. "Structural Geometry and Torsional Potentials in *p*-Phenylene Polyamides and Polyesters"
Macromolecules **13,** 479

(1980)
J. P. Hummel and P. J. Flory
318. "Moments of the End-to-End Vectors for *p*-Phenylene Polyamides and Polyesters"
Macromolecules **13,** 484 (1980)
B. Erman, P. J. Flory, and J. P. Hummel

319. "Interphases of Chain Molecules: Monolayers and Lipid Bilayer Membranes"
Proc. Nat. Acad. Sci. U.S. **77**(6), 3115 (1980)
K. A. Dill and P. J. Flory

320. "Elastic Modulus and Degree of Cross-Linking of Poly(ethyl acrylate) Networks"
Macromolecules **13,** 1554 (1980)
B. Erman, W. Wagner, and P. J. Flory

321. "The Phase Equilibria in Thermotropic Liquid Crystalline Systems"
J. Chem. Phys. **73**(12), 6327 (1980)
M. Warner and P. J. Flory

322. "Molecular Organization in Micelles and Vesicles"
Proc. Nat. Acad. Sci. U.S. **78**(2), 676 (1981)
K. A. Dill and P. J. Flory

323. "Structural Regularity and Crystallinity in Macromolecules"
in F. Ciardelli and P. Guisti, eds., *Structural Order in Polymers,* Pergamon Press, Oxford, UK, 1981
P. J. Flory

324. "Statistical Thermodynamics of Mixtures of Semirigid Macromolecules: Chains with Rodlike Sequences at Fixed Locations"
Macromolecules **13,** 954 (1981)

R. R. Matheson, Jr. and
P. J. Flory

325. "Intermediate Angle Scattering
Functions and Local Chain
Configurations of
Semicrystalline and Amorphous
Polymers"
Polym. Bull. **4,** 693 (1981)
D. Y. Yoon and P. J. Flory

326. "Direction of the Dipole
Moment in the Ester Group"
J. Phys. Chem. **85,** 3211
(1981)
E. Saiz, J. P. Hummel,
P. J. Flory, and M. Plavsic

327. "Optical Anisotropies of
Aliphatic Esters"
J. Phys. Chem. **85,** 3215
(1981)
P. J. Flory, E. Saiz,
B. Erman, P. A. Irvine, and
J. P. Hummel

328. "Elastic Activity of Imperfect
Networks"
Macromolecules **15,** 99 (1982)
P. J. Flory

329. "Optical Anisotropies of Model
Analogues of Polycarbonates"
Macromolecules **15,** 664
(1982)
B. Erman, D. C. Marvin,
P. A. Irvine and P. J. Flory

330. "Optical Anisotropy of the
Polycarbonate of
Diphenylolpropane"
Macromolecules **15,** 670
(1982)
B. Erman, D. Wu,
P. A. Irvine, D. C. Marvin,
and P. J. Flory

331. "Theory of Elasticity of
Polymer Networks. 3"
Macromolecules **15,** 800
(1982)
P. J. Flory and B. Erman

332. "Relationships between Stress,
Strain, and Molecular
Constitution of Polymer
Networks. Comparison of
Theory with Experiments"
Macromolecules **15,** 806
(1982)
B. Erman and P. J. Flory

333. "Treatment of Disordered and
Ordered Systems of Polymer
Chains by Lattice Methods"
Proc. Nat. Acad. Sci. U.S. **79,**
4510 (1982)
P. J. Flory

334. "Molecular Theories of Liquid
Crystals"
in A. Ciferri, ed., *Polymer
Liquid Crystals,* Academic
Press, Inc., New York, 1982,
Chapt. 4
P. J. Flory

335. "The Science of
Macromolecules"
*National Research Council,
Outlook for Science and
Technology:—The Next Five
Years,* W.H. Freeman and
Co., San Francisco, Calif.
1982
P. J. Flory and Committee

336. "Optical Anisotropies of
Aromatic Esters and of
Oligomers of Poly(*p*-
oxybenzoate)"
J. Phys. Chem. **87,** 2929
(1983)
P. A. Irvine, B. Erman, and
P. J. Flory

337. "Conformational
Characteristics of
Polyisobutylene"
Macromolecules **16,** 1317
(1983)
U. W. Suter, E. Saiz, and
P. J. Flory

338. "Configuration of the
Polyisobutylene Chain
According to Neutron and
X-ray Scattering"
Macromolecules **16,** 1328
(1983)

H. Hayashi, P. J. Flory, and
G. D. Wignal

339. "Configuration of the
Polyisobutylene Chain in Bulk
and in Solution According to
Elastic Neutron Scattering"
Europhysica **120B,** 408 (1983)
H. Hayashi and P. J. Flory

340. "Theory of Strain Birefringence
of Amorphous Polymer
Networks"
Macromolecules **16,** 1601
(1983)
B. Erman and P. J. Flory

341. "Experimental Results
Relating Stress and
Birefringence to Strain in
Poly(dimethylsiloxane)
Networks. Comparisons with
Theory"
Macromolecules **16,** 1607
(1983)
B. Erman and P. J. Flory

342. "Theoretical Basis for Liquid
Crystallinity in Polymers"
*Recent Advances in Liquid
Crystalline Polymers,
Proceedings of European
Science Foundation Sixth
Polymer Workshop,* Lyngby,
Denmark, 1983
P. J. Flory

343. "Conformations of
Macromolecules in Condensed
Phases"
Pure Appl. Chem. **56**(3), 305
(1984)
P. J. Flory

344. "Silicone Networks with
Junctions of High Functionality
and the Theory of Rubber
Elasticity"
*J. Polym. Sci. Polym. Phys.
Ed.* **22,** 49 (1984)
P. J. Flory and B. Erman

345. "The Equation-of-State Theory
of Mixtures and the Polymer–
Solvent Interaction Parameter"
Polym. Commun. **25,** 132
(1984)
B. Erman and P. J. Flory

346. "Molecular Theory of Liquid
Crystals"
Adv. Polym. Sci. **59,** 1 (1984)
P. J. Flory

347. "The Interphase in Lamellar
Semicrystalline Polymers"
Macromolecules **17,** 862
(1984)
P. J. Flory, D. Y. Yoon, and
K. A. Dill

348. "Chain Packing at Polymer
Interactions"
Macromolecules **17,** 868
(1984)
D. Y. Yoon and P. J. Flory

349. "Liquid Crystalline Transitions
in Homologous *p*-Phenylenes
and their Mixtures. I.
Experimental Results"
*J. Chem. Soc. Faraday Trans.
I* **80,** 1795 (1984)
P. A. Irvine, D. Wu, and
P. J. Flory

350. "Liquid Crystalline Transitions
in Homologous *p*-Phenylenes
and their Mixtures. II.
Theoretical Treatment"
*J. Chem. Soc. Faraday Trans.
I* **80,** 1807 (1984)
P. J. Flory and P. A. Irvine

351. "Liquid Crystalline Transitions
in Homologous *p*-Phenylenes
and their Mixtures. III.
Relation of Orientation-
Dependent Interactions to
Optical Anisotropies"
*J. Chem. Soc. Faraday Trans.
I* **80,** 1821 (1984)
P. A. Irvine and P. J. Flory

352. "Phase Equilibria in Liquid
Crystalline Systems. Part I.
Synthesis and Liquid
Crystalline Properties of
Oligomers of the *p*-Oxybenzoate
Series"

Ber. Bunsenges. Phys. Chem.
88, 524 (1984)
M. Ballauff, D. Wu, and
P. J. Flory

353. "Phase Equilibria in Liquid
Crystalline Systems. Part II.
Theory and Interpretation of
Experimental Results"
Ber. Bunsenges. Phys. Chem.
88, 537 (1984)
M. Ballauff and P. J. Flory

354. "Helical Conformations of
Isotactic Poly(methyl
methacrylate). Energies
Computed with Bond Angle
Relaxation"
Polym. Commun. **25,** 258
(1984)
M. Vacatello and P. J. Flory

355. "Statistical Thermodynamics of
Semirigid Macromolecules:
Chains with Interconvertible
Rodlike and Random-Coil
Sequences in Equilibrium"
J. Phys. Chem. **88,** 6606
(1984)
P. J. Flory and
R. R. Matheson, Jr.

356. "Molecular Theory of Rubber
Elasticity"
Soc. Polym. Sci. Jpn. (1985);
Polym. J. **17,** 1 (1985)
P. J. Flory

357. "Network Topology and the
Theory of Rubber Elasticity"
Brit. Polym. J. **17**(2) (1985)
P. J. Flory

358. "Conformational Statistics of
Poly(methyl methacrylate)"

Macromolecules **19,** 405
(1986)
M. Vacatello and P. J. Flory

359. "Critical Phenomena and
Transitions in Swollen Polymer
Networks and in Linear
Macromolecules"
Macromolecules **19,** 2342
(1986)
B. Erman and P. J. Flory

360. "Development of Concepts in
Polymer Science: A Half
Century in Retrospect"
Makromol. Chem. J.
Makromol. Symp. **1,** 5 (1986)
P. J. Flory

361. "Optical Anisotropics of
Alkylcyanobicyclohexyls and
Related Compounds"
J. Chem. Soc. Faraday Trans.
82(1), 3367 (1986)
P. Navard and P. J. Flory

362. "Optical Anisotropies of
Alkylcyanobiphenyls,
Alkoxycyanobiphenyls and
Related Compounds"
J. Chem. Soc. Faraday Trans.
82(1), 3381 (1986)
P. Navard and P. J. Flory

363. "Chain Molecules at Interfaces"
J. Polym. Sci. Polym. Symp.
Ed. **75** (1987)
P. J. Flory

364. "The Elastic Behavior of Cis-
1,4-Polybutadiene"
Macromolecules **20,** 351
(1987)
P. J. Flory and
R. W. Brotzman

CONTENTS

EDITORIAL STAFF FOR VOLUME 10

Executive Editor: JACQUELINE I. KROSCHWITZ
Editorial Supervisor: TERRY ANN KREMER
Editors: ANNA KLINGSBERG, ROSE MARIE PICCININNI, ANDREA SALVATORE, ELENA MANNARINO

CONTRIBUTORS TO VOLUME 10

E. T. Adams, Jr., *Texas A&M University, College Station, Texas,* Osmometry

Gary A. Baum, *The Institute for Paper Chemistry, Appleton, Wisconsin,* Paper

David L. Beach, *Chevron Chemical Company, Kingwood, Texas,* Olefin polymers

Norbert M. Bikales, *National Science Foundation, Washington, D.C.,* Nomenclature

John K. Borchardt, *Shell Development Company, Houston, Texas,* Oil-field applications

F. A. Bovey, *AT&T Bell Laboratories, Murray Hill, New Jersey,* Nuclear magnetic resonance

Aaron L. Brody, *Schotland Business Research, Inc., Princeton, New Jersey,* Flexible materials and Rigid containers under Packaging materials

Mukerrem Cakmak, *University of Akron, Akron, Ohio,* Orientation and Orientation processes

Charles E. Carraher, *Florida Atlantic University, Boca Raton, Florida,* Organometallic polymers

Francesco Ciardelli, *Università di Pisa, Pisa, Italy,* Optically active polymers

Anthony R. Cooper, *Lockheed Missiles & Space Company, Inc., Palo Alto, California,* Molecular weight determination

D. K. Dandge, *New Mexico Institute of Mining and Technology, Socorro, New Mexico,* Nitroso polymers

L. G. Donaruma, *University of Alabama in Huntsville, Huntsville, Alabama,* Nitroso polymers

Arthur Drelich, *Chicopee, Dayton, New Jersey,* Survey under Nonwoven fabrics

M. P. Dreyfuss, *Michigan Molecular Institute, Midland, Michigan,* Oxetane polymers

P. Dreyfuss, *Michigan Molecular Institute, Midland, Michigan,* Oxetane polymers

Hardev S. Dugal, *The Institute of Paper Chemistry, Appleton, Wisconsin,* Paper

Dwight B. Easty, *The Institute of Paper Chemistry, Appleton, Wisconsin,* Paper

Allan H. Fawcett, *The Queen's University of Belfast, United Kingdom,* Olefin–sulfur dioxide copolymers

Lewis J. Fetters, *Exxon Research and Engineering Company, Annadale, New Jersey,* Monodisperse polymers

†Paul J. Flory, *Stanford University, Stanford, California,* Networks

J. Neil Henderson, *Consultant, Hudson, Ohio,* Pentadiene polymers

Jack D. Hultman, *The Institute for Paper Chemistry, Appleton, Wisconsin,* Paper

L. W. Jelinski, *AT&T Bell Laboratories, Murray Hill, New Jersey,* Nuclear magnetic resonance

Martin B. Jones, *University of North Dakota, Grand Forks, North Dakota,* Oxidative polymerization

Stuart M. Kaback, *Exxon Research and Engineering Company, Linden, New Jersey,* Patent information

Yury V. Kissin, *Mobil Chemical Company, Edison, New Jersey,* Olefin polymers

Peter Kovacic, *University of Wisconsin, Milwaukee, Wisconsin,* Oxidative polymerization

L. M. Landoll, *Hercules, Inc., Oxford, Georgia,* Olefin fibers

Earl W. Malcolm, *The Institute of Paper Chemistry, Appleton, Wisconsin,* Paper

Carl D. Marotta, *Innovation Technology, Inc., Warrington, Pennsylvania,* Medical devices under Packaging materials

N. J. Mills, *University of Birmingham, Birmingham, United Kingdom,* Optical properties

Virgil Percec, *Case Western Reserve University, Cleveland, Ohio,* Oligomers

Anton Peterlin, *Consultant, Bethesda, Maryland,* Morphology

Charles U. Pittman, *Mississippi State University, Mississippi State, Mississippi,* Organometallic polymers

Coleen Pugh, *Case Western Reserve University, Cleveland, Ohio,* Oligomers

John R. Reynolds, *University of Texas-Arlington, Arlington, Texas,* Organometallic polymers

John D. Sinkey, *The Institute for Paper Chemistry, Appleton, Wisconsin,* Paper

Ronald L. Smorada, *E.I. du Pont de Nemours & Co., Inc., Wilmington, Delaware,* Spunbonded under Nonwoven Fabrics

Gavin G. Spence, *Hercules, Inc., Wilmington, Delaware,* Paper additives and resins

Howard Starkweather, *E.I. du Pont de Nemours & Co., Inc., Wilmington, Delaware,* Olefin–carbon monoxide copolymers

M. C. Throckmorton, *Consultant, Akron, Ohio,* Pentadiene polymers

---

†Deceased.

D. B. Todd, *Baker Perkins, Inc., Saginaw, Michigan,* Pelletizing

James L. White, *University of Akron, Akron, Ohio,* Orientation and Orientation processes

G. D. Wignall, *Oak Ridge National Laboratories, Oak Ridge, Tennessee,* Neutron scattering

CONVERSION FACTORS, ABBREVIATIONS, AND UNIT SYMBOLS

SI Units (Adopted 1960)

A new system of measurement, the International System of Units (abbreviated SI), is being implemented throughout the world. This system is a modernized version of the MKSA (meter, kilogram, second, ampere) system, and its details are published and controlled by an international treaty organization (The International Bureau of Weights and Measures) (1).

SI units are divided into three classes:

Base Units

length	meter[†] (m)
mass[‡]	kilogram (kg)
time	second (s)
electric current	ampere (A)
thermodynamic temperature[§]	kelvin (K)
amount of substance	mole (mol)
luminous intensity	candela (cd)

Supplementary Units

plane angle	radian (rad)
solid angle	steradian (sr)

[†]The spellings "metre" and "litre" are preferred by ASTM; however, "-er" is used in the *Encyclopedia*.

[‡]"Weight" is the commonly used term for "mass."

[§]Wide use is made of "Celsius temperature" (*t*) defined by

$$t = T - T_0$$

where T is the thermodynamic temperature, expressed in kelvins, and $T_0 = 273.15$ K by definition. A temperature interval may be expressed in degrees Celsius as well as in kelvins.

Derived Units and Other Acceptable Units

These units are formed by combining base units, supplementary units, and other derived units (2–4). Those derived units having special names and symbols are marked with an asterisk in the list below:

Quantity	Unit	Symbol	Acceptable equivalent
*absorbed dose	gray	Gy	J/kg
acceleration	meter per second squared	m/s^2	
*activity (of ionizing radiation source)	becquerel	Bq	1/s
area	square kilometer	km^2	
	square hectometer	hm^2	ha (hectare)
	square meter	m^2	
*capacitance	farad	F	C/V
concentration (of amount of substance)	mole per cubic meter	mol/m^3	
*conductance	siemens	S	A/V
current density	ampere per square meter	A/m^2	
density, mass density	kilogram per cubic meter	kg/m^3	g/L; mg/cm^3
dipole moment (quantity)	coulomb meter	C·m	
*electric charge, quantity of electricity	coulomb	C	A·s
electric charge density	coulomb per cubic meter	C/m^3	
electric field strength	volt per meter	V/m	
electric flux density	coulomb per square meter	C/m^2	
*electric potential, potential difference, electromotive force	volt	V	W/A
*electric resistance	ohm	Ω	V/A
*energy, work, quantity of heat	megajoule	MJ	
	kilojoule	kJ	
	joule	J	N·m
	electronvolt[†]	eV[†]	
	kilowatt hour[†]	kW·h[†]	
energy density	joule per cubic meter	J/m^3	
*force	kilonewton	kN	
	newton	N	kg·m/s^2

[†]This non-SI unit is recognized by the CIPM as having to be retained because of practical importance or use in specialized fields (1).

Quantity	Unit	Symbol	Acceptable equivalent
*frequency	megahertz	MHz	
	hertz	Hz	1/s
heat capacity, entropy	joule per kelvin	J/K	
heat capacity (specific), specific entropy	joule per kilogram kelvin	J/(kg·K)	
heat transfer coefficient	watt per square meter kelvin	W/(m²·K)	
*illuminance	lux	lx	lm/m²
*inductance	henry	H	Wb/A
linear density	kilogram per meter	kg/m	
luminance	candela per square meter	cd/m²	
*luminous flux	lumen	lm	cd·sr
magnetic field strength	ampere per meter	A/m	
*magnetic flux	weber	Wb	V·s
*magnetic flux density	tesla	T	Wb/m²
molar energy	joule per mole	J/mol	
molar entropy, molar heat capacity	joule per mole kelvin	J/(mol·K)	
moment of force, torque	newton meter	N·m	
momentum	kilogram meter per second	kg·m/s	
permeability	henry per meter	H/m	
permittivity	farad per meter	F/m	
*power, heat flow rate, radiant flux	kilowatt	kW	
	watt	W	J/s
power density, heat flux density, irradiance	watt per square meter	W/m²	
*pressure, stress	megapascal	MPa	
	kilopascal	kPa	
	pascal	Pa	N/m²
sound level	decibel	dB	
specific energy	joule per kilogram	J/kg	
specific volume	cubic meter per kilogram	m³/kg	
surface tension	newton per meter	N/m	
thermal conductivity	watt per meter kelvin	W/(m·K)	
velocity	meter per second	m/s	
	kilometer per hour	km/h	
viscosity, dynamic	pascal second	Pa·s	
	millipascal second	mPa·s	
viscosity, kinematic	square meter per second	m²/s	
	square millimeter per second	mm²/s	

Quantity	Unit	Symbol	*Acceptable equivalent*
volume	cubic meter	m^3	
	cubic decimeter	dm^3	L(liter) (5)
	cubic centimeter	cm^3	mL
wave number	1 per meter	m^{-1}	
	1 per centimeter	cm^{-1}	

In addition, there are 16 prefixes used to indicate order of magnitude, as follows:

Multiplication factor	*Prefix*	*Symbol*	*Note*
10^{18}	exa	E	[a]Although hecto, deka, deci, and centi
10^{15}	peta	P	are SI prefixes, their use should be
10^{12}	tera	T	avoided except for SI unit-multiples
10^9	giga	G	for area and volume and
10^6	mega	M	nontechnical use of centimeter, as
10^3	kilo	k	for body and clothing
10^2	hecto	h[a]	measurement.
10	deka	da[a]	
10^{-1}	deci	d[a]	
10^{-2}	centi	c[a]	
10^{-3}	milli	m	
10^{-6}	micro	μ	
10^{-9}	nano	n	
10^{-12}	pico	p	
10^{-15}	femto	f	
10^{-18}	atto	a	

For a complete description of SI and its use, the reader is referred to ASTM E 380 (4).

A representative list of conversion factors from non-SI to SI units is presented herewith. Factors are given to four significant figures. Exact relationships are followed by a dagger. A more complete list is given in ASTM E 380–84 (4) and ANSI Z 210.1–1976 (6).

Conversion Factors to SI Units

To convert from	*To*	*Multiply by*
acre	square meter (m^2)	4.047×10^3
angstrom	meter (m)	$1.0 \times 10^{-10\dagger}$
are	square meter (m^2)	$1.0 \times 10^{2\dagger}$
astronomical unit	meter (m)	1.496×10^{11}
atmosphere	pascal (Pa)	1.013×10^5
bar	pascal (Pa)	$1.0 \times 10^{5\dagger}$
barn	square meter (m^2)	$1.0 \times 10^{-28\dagger}$

†Exact.

To convert from	To	Multiply by
barrel (42 U.S. liquid gallons)	cubic meter (m³)	0.1590
Bohr magneton (μ_β)	J/T	9.274×10^{-24}
Btu (International Table)	joule (J)	1.055×10^3
Btu (mean)	joule (J)	1.056×10^3
Btu (thermochemical)	joule (J)	1.054×10^3
bushel	cubic meter (m³)	3.524×10^{-2}
calorie (International Table)	joule (J)	4.187
calorie (mean)	joule (J)	4.1908
calorie (thermochemical)	joule (J)	4.184†
centipoise	pascal second (Pa·s)	$1.0 \times 10^{-3†}$
centistokes	square millimeter per second (mm²/s)	1.0†
cfm (cubic foot per minute)	cubic meter per second (m³/s)	4.72×10^{-4}
cubic inch	cubic meter (m³)	1.639×10^{-5}
cubic foot	cubic meter (m³)	2.832×10^{-2}
cubic yard	cubic meter (m³)	0.7646
curie	becquerel (Bq)	$3.70 \times 10^{10†}$
debye	coulomb·meter (C·m)	3.336×10^{-30}
degree (angle)	radian (rad)	1.745×10^{-2}
denier (international)	kilogram per meter (kg/m)	1.111×10^{-7}
	tex‡	0.1111
dram (apothecaries')	kilogram (kg)	3.888×10^{-3}
dram (avoirdupois)	kilogram (kg)	1.772×10^{-3}
dram (U.S. fluid)	cubic meter (m³)	3.697×10^{-6}
dyne	newton (N)	$1.0 \times 10^{-5†}$
dyne/cm	newton per meter (N/m)	$1.0 \times 10^{-3†}$
electron volt	joule (J)	1.602×10^{-19}
erg	joule (J)	$1.0 \times 10^{-7†}$
fathom	meter (m)	1.829
fluid ounce (U.S.)	cubic meter (m³)	2.957×10^{-5}
foot	meter (m)	0.3048†
footcandle	lux (lx)	10.76
furlong	meter (m)	2.012×10^{-2}
gal	meter per second squared (m/s²)	$1.0 \times 10^{-2†}$
gallon (U.S. dry)	cubic meter (m³)	4.405×10^{-3}
gallon (U.S. liquid)	cubic meter (m³)	3.785×10^{-3}
gallon per minute (gpm)	cubic meter per second (m³/s)	6.308×10^{-5}
	cubic meter per hour (m³/h)	0.2271
gauss	tesla (T)	1.0×10^{-4}
gilbert	ampere (A)	0.7958
gill (U.S.)	cubic meter (m³)	1.183×10^{-4}
grad	radian	1.571×10^{-2}
grain	kilogram (kg)	6.480×10^{-5}

†Exact.
‡See footnote on p. xxxviii.

To convert from	To	Multiply by
gram-force per denier	newton per tex (N/tex)	8.826×10^{-2}
hectare	square meter (m²)	$1.0 \times 10^{4\dagger}$
horsepower (550 ft·lbf/s)	watt (W)	7.457×10^{2}
horsepower (boiler)	watt (W)	9.810×10^{3}
horsepower (electric)	watt (W)	$7.46 \times 10^{2\dagger}$
hundredweight (long)	kilogram (kg)	50.80
hundredweight (short)	kilogram (kg)	45.36
inch	meter (m)	$2.54 \times 10^{-2\dagger}$
inch of mercury (32°F)	pascal (Pa)	3.386×10^{3}
inch of water (39.2°F)	pascal (Pa)	2.491×10^{2}
kilogram-force	newton (N)	9.807
kilowatt hour	megajoule (MJ)	$3.6^{\dagger}$
kip	newton (N)	4.48×10^{3}
knot (international)	meter per second (m/s)	0.5144
lambert	candela per square meter (cd/m²)	3.183×10^{3}
league (British nautical)	meter (m)	5.559×10^{3}
league (statute)	meter (m)	4.828×10^{3}
light year	meter (m)	9.461×10^{15}
liter (for fluids only)	cubic meter (m³)	$1.0 \times 10^{-3\dagger}$
maxwell	weber (Wb)	$1.0 \times 10^{-8\dagger}$
micron	meter (m)	$1.0 \times 10^{-6\dagger}$
mil	meter (m)	$2.54 \times 10^{-5\dagger}$
mile (statute)	meter (m)	1.609×10^{3}
mile (U.S. nautical)	meter (m)	$1.852 \times 10^{3\dagger}$
mile per hour	meter per second (m/s)	0.4470
millibar	pascal (Pa)	1.0×10^{2}
millimeter of mercury (0°C)	pascal (Pa)	$1.333 \times 10^{2\dagger}$
minute (angular)	radian	2.909×10^{-4}
myriagram	kilogram (kg)	10
myriameter	kilometer (km)	10
oersted	ampere per meter (A/m)	79.58
ounce (avoirdupois)	kilogram (kg)	2.835×10^{-2}
ounce (troy)	kilogram (kg)	3.110×10^{-2}
ounce (U.S. fluid)	cubic meter (m³)	2.957×10^{5}
ounce-force	newton (N)	0.2780
peck (U.S.)	cubic meter (m³)	8.810×10^{-3}
pennyweight	kilogram (kg)	1.555×10^{-3}
pint (U.S. dry)	cubic meter (m³)	5.506×10^{-4}
pint (U.S. liquid)	cubic meter (m³)	4.732×10^{-4}
poise (absolute viscosity)	pascal second (Pa·s)	$0.10^{\dagger}$
pound (avoirdupois)	kilogram (kg)	0.4536
pound (troy)	kilogram (kg)	0.3732
poundal	newton (N)	0.1383
pound-force	newton (N)	4.448

†Exact.

To convert from	To	Multiply by
pound-force per square inch (psi)	pascal (Pa)	6.895×10^3
quart (U.S. dry)	cubic meter (m³)	1.101×10^{-3}
quart (U.S. liquid)	cubic meter (m³)	9.464×10^{-4}
quintal	kilogram (kg)	$1.0 \times 10^{2\dagger}$
rad	gray (Gy)	$1.0 \times 10^{-2\dagger}$
rod	meter (m)	5.029
roentgen	coulomb per kilogram (C/kg)	2.58×10^{-4}
second (angle)	radian (rad)	4.848×10^{-6}
section	square meter (m²)	2.590×10^6
slug	kilogram (kg)	14.59
spherical candle power	lumen (lm)	12.57
square inch	square meter (m²)	6.452×10^{-4}
square foot	square meter (m²)	9.290×10^{-2}
square mile	square meter (m²)	2.590×10^6
square yard	square meter (m²)	0.8361
stere	cubic meter (m³)	$1.0^\dagger$
stokes (kinematic viscosity)	square meter per second (m²/s)	$1.0 \times 10^{-4\dagger}$
tex	kilogram per meter (kg/m)	$1.0 \times 10^{-6\dagger}$
ton (long, 2240 pounds)	kilogram (kg)	1.016×10^3
ton (metric)	kilogram (kg)	$1.0 \times 10^{3\dagger}$
ton (short, 2000 pounds)	kilogram (kg)	9.072×10^2
torr	pascal (Pa)	1.333×10^2
unit pole	weber (Wb)	1.257×10^{-7}
yard	meter (m)	$0.9144^\dagger$

Abbreviations and Unit Symbols

Following is a list of commonly used abbreviations and unit symbols appropriate for use in the *Encyclopedia*. In general they agree with those listed in *American National Standard Abbreviations for Use on Drawings and in Text (ANSI Y1.1)* (6) and *American National Standard Letter Symbols for Units in Science and Technology (ANSI Y10)* (6). Also included is a list of acronyms for a number of private and government organizations as well as common industrial solvents, polymers, and other chemicals.

Rules for Writing Unit Symbols (4):

1. Unit symbols should be printed in upright letters (roman) regardless of the type style used in the surrounding text.

2. Unit symbols are unaltered in the plural.

3. Unit symbols are not followed by a period except when used as the end of a sentence.

4. Letter unit symbols are generally written in lowercase (eg, cd for candela) unless the unit name has been derived from a proper name, in which case the first letter of the symbol is capitalized (W, Pa). Prefix and unit symbols retain their prescribed form regardless of the surrounding typography.

†Exact.

5. In the complete expression for a quantity, a space should be left between the numerical value and the unit symbol. For example, write 2.37 lm, *not* 2.37lm, and 35 mm, *not* 35mm. When the quantity is used in an adjectival sense, a hyphen is often used, for example, 35-mm film. *Exception:* No space is left between the numerical value and the symbols for degree, minute, and second of plane angle, and degree Celsius.

6. No space is used between the prefix and unit symbols (eg, kg).

7. Symbols, not abbreviations, should be used for units. For example, use "A," not "amp," for ampere.

8. When multiplying unit symbols, use a raised dot:

$$\text{N·m for newton meter}$$

In the case of W·h, the dot may be omitted, thus:

$$\text{Wh}$$

An exception to this practice is made for computer printouts, automatic typewriter work, etc, where the raised dot is not possible, and a dot on the line may be used.

9. When dividing unit symbols use one of the following forms:

$$\text{m/s } or \text{ m·s}^{-1} or \ \frac{m}{s}$$

In no case should more than one slash be used in the same expression unless parentheses are inserted to avoid ambiguity. For example, write:

$$\text{J/(mol·K) } or \text{ J·mol}^{-1} \cdot \text{K}^{-1} or \text{ (J/mol)/K}$$

but *not*

$$\text{J/mol/K}$$

10. Do not mix symbols and unit names in the same expression. Write:

$$\text{joules per kilogram } or \text{ J/kg } or \text{ J·kg}^{-1}$$

but *not*

$$\text{joules/kilogram } nor \text{ joules/kg } nor \text{ joules·kg}^{-1}$$

Abbreviations and Units

A	ampere	*ac-*	alicyclic
A	anion (eg, H*A*); mass number	ACGIH	American Conference of Governmental Industrial Hygienists
a	atto (prefix for 10^{-18})		
AATCC	American Association of Textile Chemists and Colorists	ACS	American Chemical Society
		AGA	American Gas Association
		Ah	ampere hour
ABS	acrylonitrile–butadiene– styrene	AIChE	American Institute of Chemical Engineers
abs	absolute	AIME	American Institute of Mining, Metallurgical, and Petroleum Engineers
ac	alternating current, *n.*		
a-c	alternating current, *adj.*		

AIP	American Institute of Physics	bid	twice daily
AISI	American Iron and Steel Institute	Boc	t-butyloxycarbonyl
alc	alcohol(ic)	BOD	biochemical (biological) oxygen demand
Alk	alkyl	bp	boiling point
alk	alkaline (not alkali)	Bq	becquerel
-alt-	alternating as in alternating copolymer	C	coulomb
amt	amount	°C	degree Celsius
amu	atomic mass unit	C-	denoting attachment to carbon
ANSI	American National Standards Institute	C_M	chain-transfer constant for monomer
AO	atomic orbital	C_P	chain-transfer constant for polymer
AOAC	Association of Official Analytical Chemists	C_S	chain-transfer constant for solvent
AOCS	American Oil Chemists' Society	c	centi (prefix for 10^{-2})
APHA	American Public Health Association	c	critical
		ca	circa (approximately)
API	American Petroleum Institute	cd	candela; current density; circular dichroism
aq	aqueous	CFR	Code of Federal Regulations
Ar	aryl		
ar-	aromatic	cgs	centimeter-gram-second
as-	asymmetric(al)	CI	Color Index
ASH-RAE	American Society of Heating, Refrigerating, and Air Conditioning Engineers	cis-	isomer in which substituted groups are on same side of double bond between C atoms
ASM	American Society for Metals	cl	carload
ASME	American Society of Mechanical Engineers	cm	centimeter
		cmil	circular mil
ASTM	American Society for Testing and Materials	cmpd	compound
		CNRS	Centre National de la Recherche Scientifique
at no.	atomic number	CNS	central nervous system
at wt	atomic weight	-co-	copolymerized with
av(g)	average	CoA	coenzyme A
AWS	American Welding Society	COC	Cleveland open cup
b	bonding orbital	COD	chemical oxygen demand
bbl	barrel	coml	commercial(ly)
bcc	body-centered cubic	conc	concentration
bct	body-centered tetragonal	cp	chemically pure
Bé	Baumé	cph	close-packed hexagonal
BET	Brunauer-Emmett-Teller (adsorption equation)	CPSC	Consumer Product Safety Commission
		cryst	crystalline

cub	cubic	eng	engineering
D	Debye	EPA	Environmental Protection Agency
D-	denoting configurational relationship	epr	electron paramagnetic resonance
d	differential operator		
d-	*dextro-*, dextrorotatory	ϵ	dielectric constant (unitless)
da	deka (prefix for 10^1)	eq.	equation
dB	decibel	esca	electron-spectroscopy for chemical analysis
dc	direct current, *n.*		
d-c	direct current, *adj.*	esp	especially
dec	decompose	esr	electron-spin resonance
detd	determined	est(d)	estimate(d)
detn	determination	estn	estimation
dia	diameter	esu	electrostatic unit
dil	dilute	η	viscosity
dl-; DL-	racemic	$[\eta]$	intrinsic viscosity
DMA	dimethylacetamide	η_{inh}	inherent viscosity
DMF	dimethylformamide	η_r	relative viscosity
DMG	dimethyl glyoxime	η_{red}	reduced viscosity
DMSO	dimethyl sulfoxide	η_{sp}	specific viscosity
DOD	Department of Defense	exp	experiment, experimental
DOE	Department of Energy	ext(d)	extract(ed)
DOT	Department of Transportation	F	farad (capacitance)
		F	faraday (96,487 C); free energy
DP	degree of polymerization		
dp	dew point	f	femto (prefix for 10^{-15})
DPH	diamond pyramid hardness	FAO	Food and Agriculture Organization (United Nations)
DS	degree of substitution		
dsc	differential scanning calorimetry		
		fcc	face-centered cubic
dstl(d)	distill(ed)	FDA	Food and Drug Administration
dta	differential thermal analysis		
		FEA	Federal Energy Administration
E	Young's modulus		
(E)-	entgegen; opposed	FHSA	Federal Hazardous Substances Act
e	polarity factor in Alfrey-Price equation		
		fob	free on board
e^-	electron	fp	freezing point
ECU	electrochemical unit	FPC	Federal Power Commission
ed.	edited, edition, editor	FRB	Federal Reserve Board
ED	effective dose	frz	freezing
EDTA	ethylenediaminetetraacetic acid	G	giga (prefix for 10^9)
		G	gravitational constant $=$ 6.67×10^{11} N·m²/kg²; Gibb's free energy
em	electron microscopy		
emf	electromotive force		
emu	electronmagnetic unit	g	gram
en	ethylene diamine	(g)	gas, only as in $H_2O(g)$

g	gravitational acceleration	ir	infrared
-g-	graft as in graft copolymer	IRLG	Interagency Regulatory
gc	gas chromatography		Liaison Group
gem-	geminal	ISO	International Organization
glc	gas-liquid chromatography		for Standardization
g-mol	gram-molecular weight	IU	International Unit
wt;		IUPAC	International Union of Pure
gmw			and Applied Chemistry
GNP	gross national product	IV	iodine value
gpc	gel-permeation	iv	intravenous
	chromatography	J	joule
GRAS	Generally Recognized as	K	kelvin
	Safe	K	equilibrium constant
grd	ground	k	kilo (prefix for 10^3)
Gy	gray	k	reaction rate constant
H	henry	kg	kilogram
H	enthalpy	L	denoting configurational
h	hour; hecto (prefix for 10^2)		relationship
ha	hectare	L	liter (for fluids only) (5)
HB	Brinell hardness number	*l*-	*levo*-, levorotatory
Hb	hemoglobin	(l)	liquid, only as in $NH_3(l)$
hcp	hexagonal close-packed	LC_{50}	conc lethal to 50% of the
hex	hexagonal		animals tested
HK	Knoop hardness number	LCAO	linear combination of
hplc	high-pressure liquid		atomic orbitals
	chromatography	LCD	liquid crystal display
HRC	Rockwell hardness (C scale)	lcl	less than carload lots
HV	Vickers hardness number	LD_{50}	dose lethal to 50% of the
hyd	hydrated, hydrous		animals tested
hyg	hygroscopic	LED	light-emitting diode
Hz	hertz	liq	liquid
^{i}(eg, Pr^i)	iso (eg, isopropyl)	lm	lumen
i-	inactive (eg,	ln	logarithm (natural)
	i-methionine)	LNG	liquefied natural gas
IACS	International Annealed	log	logarithm (common)
	Copper Standard	LPG	liquefied petroleum gas
ibp	initial boiling point	ltl	less than truckload lots
IC	inhibitory concentration	lx	lux
ICC	Interstate Commerce	M	mega (prefix for 10^6); metal
	Commission		(as in MA)
ICT	International Critical Table	M	molar; actual mass
ID	inside diameter; infective	$\overline{M}_w$	weight-average mol wt
	dose	$\overline{M}_n$	number-average mol wt
ip	intraperitoneal	$\overline{M}_v$	viscosity-average mol wt
IPS	iron pipe size	m	meter; milli (prefix for
IPTS	International Practical		10^{-3})
	Temperature Scale (NBS)	m	molal

m-	meta	neg	negative
max	maximum	NEMA	National Electrical
MCA	Chemical Manufacturers'		Manufacturer's
	Association (was		Association
	Manufacturing Chemists	NF	*National Formulary*
	Association)	NIH	National Institutes of
MEK	methyl ethyl ketone		Health
meq	milliequivalent	NIOSH	National Institute of
mfd	manufactured		Occupational Safety and
mfg	manufacturing		Health
mfr	manufacturer	nmr	nuclear magnetic resonance
MIBC	methyl isobutyl carbinol	NND	New and Nonofficial Drugs
MIBK	methyl isobutyl ketone		(AMA)
MIC	minimum inhibiting	no.	number
	concentration	NOI(BN)	not otherwise indexed (by
min	minute; minimum		name)
mL	milliliter	NOS	not otherwise specified
MLD	minimum lethal dose	nqr	nuclear quadruple
MO	molecular orbital		resonance
mo	month	NRC	Nuclear Regulatory
mol	mole		Commission; National
mol wt	molecular weight		Research Council
mp	melting point	NRI	New Ring Index
MR	molar refraction	NSF	National Science
ms	mass spectrum		Foundation
mxt	mixture	NTA	nitrilotriacetic acid
μ	micro (prefix for 10^{-6})	NTP	normal temperature and
N	newton (force)		pressure (25°C and 101.3
N	normal (concentration);		kPa or 1 atm)
	neutron number	NTSB	National Transportation
N-	denoting attachment to		Safety Board
	nitrogen	*O*-	denoting attachment to
n (as	index of refraction (for 20°C		oxygen
n_{D}^{20})	and sodium light)	*o*-	ortho
ⁿ (as	normal (straight-chain	OD	outside diameter
Buⁿ),	structure)	ω	frequency
n-		OPEC	Organization of Petroleum
n	neutron		Exporting Countries
n	nano (prefix for 10^9)	OSHA	Occupational Safety and
na	not available		Health Administration
NAS	National Academy of	owf	on weight of fiber
	Sciences	Ω	ohm
NASA	National Aeronautics and	P	peta (prefix for 10^{15})
	Space Administration	p	pico (prefix for 10^{-12})
nat	natural	*p*-	para
NBS	National Bureau of	*p*	proton
	Standards	p.	page

Pa	pascal (pressure)	RI	Ring Index
PAN	polyacrylonitrile	rms	root-mean square
pd	potential difference	rpm	rotations per minute
PE	polyethylene	rps	revolutions per second
pH	negative logarithm of the effective hydrogen ion concentration	RT	room temperature
		^{s} (eg, Bu^s); *sec-*	secondary (eg, secondary butyl)
phr	parts per hundred of resin (rubber)	S	siemens
π	osmotic pressure	*(S)-*	sinister (counterclockwise configuration)
p-i-n	positive-intrinsic-negative		
pmr	proton magnetic resonance	*S*	entropy
p-n	positive-negative	*S-*	denoting attachment to sulfur
po	per os (oral)		
POP	polyoxypropylene	*s-*	symmetric(al)
pos	positive	s	second
PP	polypropylene	(s)	solid, only as in $H_2O(s)$
pp.	pages	SAE	Society of Automotive Engineers
ppb	parts per billion (10^9)		
pph	parts per hundred	SAMPE	Society for the Advancement of Material and Process Engineering
ppm	parts per million (10^6)		
ppmv	parts per million by volume		
ppmwt	parts per million by weight	SAN	styrene–acrylonitrile
PPO	poly(phenyl oxide)	sat(d)	saturate(d)
ppt(d)	precipitate(d)	satn	saturation
pptn	precipitation	SBR	styrene–butadiene– rubber
Pr (no.)	foreign prototype (number)		
PS	polystyrene	sc	subcutaneous
pt	point; part	SCF	self-consistent field; standard cubic feet
PVC	poly(vinyl chloride)		
pwd	powder	Sch	Schultz number
py	pyridine	SFs	Saybolt Furol seconds
Q	reactivity of monomer in Alfrey-Price equation	SI	Le Système International d'Unités (International System of Units)
qv	quod vide (which see)		
R	univalent hydrocarbon radical	sl sol	slightly soluble
		SMC	sheet molding compound
(R)-	rectus (clockwise configuration)	sol	soluble
		soln	solution
r	precision of data	soly	solubility
rad	radian; radius	sp	specific; species
rds	rate determining step	SPE	Society of Plastics Engineers
Ref.	reference		
rf	radio frequency, *n.*	sp gr	specific gravity
r-f	radio frequency, *adj.*	SPI	Society of the Plastics Industry
rh	relative humidity		
ρ, d	density	sr	steradian

std	standard	TOC	Tagliabue (Tag) open cup
STP	standard temperature and pressure (0°C and 101.3 kPa)	*trans-*	isomer in which substituted groups are on opposite sides of double bond between C atoms
sub	sublime(s)		
SUs	Saybolt Universal seconds	TSCA	Toxic Substance Control Act
syn	synthetic	TWA	time-weighted average
ᵗ (eg, Buᵗ), *t-,* *tert-*	tertiary (eg, tertiary-butyl)	Twad	Twaddell
		UL	Underwriters' Laboratory
		USDA	United States Department of Agriculture
T	tera (prefix for 10^{12}); tesla (magnetic flux density)	USP	*United States Pharmacopeia*
		uv	ultraviolet
t	metric ton (tonne); temperature	V	volt (emf)
		var	variable
TAPPI	Technical Association of the Pulp and Paper Industry	*vic-*	vicinal
		vol	volume (not volatile)
τ	relaxation	vs	versus
TCC	Tagliabue (Tag) closed cup	v sol	very soluble
tex	tex (linear density)	W	watt
T_g	glass-transition temperature	Wb	weber
tga	thermogravimetric analysis	Wh	watt hour
θ	adjective of condition, solvent, temperature	WHO	World Health Organization (United Nations)
THF	tetrahydrofuran	wk	week
tlc	thin layer chromatography	yr	year
TLV	threshold limit value	(Z)-	zusammen; together; atomic number
TMS	tetramethylsilane		

Non-SI (Unacceptable and Obsolete) Units		*Use*
Å	angstrom	nm
at	atmosphere, technical	Pa
atm	atmosphere, standard	Pa
b	barn	cm^2
bar⁺	bar	Pa
bbl	barrel	m^3
bhp	brake horsepower	W
Btu	British thermal unit	J
bu	bushel	m^3; L
cal	calorie	J
cfm	cubic foot per minute	m^3/s
Ci	curie	Bq
cSt	centistokes	mm^2/s
c/s	cycle per second	Hz
cu	cubic	exponential form
D	debye	C·m

⁺Do not use bar (10^5Pa) or millibar (10^2Pa) because they are not SI units, and are accepted internationally only for a limited time in special fields because of existing usage.

Non-SI (Unacceptable and Obsolete) Units		*Use*
den	denier	tex
dr	dram	kg
dyn	dyne	N
dyn/cm	dyne per centimeter	mN/m
erg	erg	J
eu	entropy unit	J/K
°F	degree Fahrenheit	°C; K
fc	footcandle	lx
fl	footlambert	lx
fl oz	fluid ounce	m^3; L
ft	foot	m
ft·lbf	foot pound-force	J
gf/den	gram-force per denier	N/tex
G	gauss	T
Gal	gal	m/s^2
gal	gallon	m^3; L
Gb	gilbert	A
gpm	gallon per minute	(m^3/s); (m^3/h)
gr	grain	kg
hp	horsepower	W
ihp	indicated horsepower	W
in.	inch	m
in. Hg	inch of mercury	Pa
in. H_2O	inch of water	Pa
in.-lbf	inch pound-force	J
kcal	kilogram-calorie	J
kgf	kilogram-force	N
kilo	for kilogram	kg
L	lambert	lx
lb	pound	kg
lbf	pound-force	N
mho	mho	S
mi	mile	m
MM	million	M
mm Hg	millimeter of mercury	Pa
mμ	millimicron	nm
mph	mile per hour	km/h
μ	micron	μm
Oe	oersted	A/m
oz	ounce	kg
ozf	ounce-force	N
η	poise	Pa·s
P	poise	Pa·s
ph	phot	lx
psi	pound-force per square inch	Pa

Non-SI (Unacceptable and Obsolete) Units		*Use*
psia	pound-force per square inch absolute	Pa
psig	pound-force per square inch gauge	Pa
qt	quart	m^3; L
°R	degree Rankine	K
rd	rad	Gy
sb	stilb	lx
SCF	standard cubic foot	m^3
sq	square	exponential form
thm	therm	J
yd	yard	m

BIBLIOGRAPHY

1. The International Bureau of Weights and Measures, BIPM (Parc de Saint-Cloud, France) is described on page 22 of Ref. 4. This bureau operates under the exclusive supervision of the International Committee of Weights and Measures (CIPM).
2. *Metric Editorial Guide (ANMC-78-1)* 3rd ed., American National Metric Council, 5410 Grosvenor Lane, Bethesda, Md. 20814, 1981.
3. *SI Units and Recommendations for the Use of Their Multiples and of Certain Other Units (ISO 1000–1981),* American National Standards Institute, 1430 Broadway, New York, N.Y. 10018, 1981.
4. Based on *ASTM E 380–84 (Standard for Metric Practice),* American Society for Testing and Materials, 1916 Race Street, Philadelphia, Pa. 19103, 1984.
5. *Fed. Regist.,* Dec. 10, 1976 (41 FR 36414).
6. For ANSI address, see Ref. 3.

R. P. LUKENS
American Society for Testing and Materials

M

Continued

MOLECULAR WEIGHT DETERMINATION

This article discusses the determination of molecular weight averages and molecular weight distributions. The more important methods are described in greater detail in separate articles (see CHROMATOGRAPHY; FRACTIONATION; CHARACTERIZATION OF POLYMERS).

Molecular Weight Averages and Distributions

Most polymeric materials are comprised of mixtures of molecules of various sizes. This distribution of molecular weights is caused by the statistical nature of the polymerization process (1). A complete description of the molecular weight distribution of a polymer is necessary in order to understand its physical, rheological, and mechanical properties.

Certain techniques for molecular weight determination are only capable of yielding one of the molecular weight averages of the distribution. These averages are defined in terms of the molecular weight M_i and the number of moles n_i or the weight w_i of the component molecules. The molecular weight averages are defined by equations 1–4.

Number-average molecular weight:

$$\overline{M}_n = \frac{\Sigma n_i M_i}{\Sigma n_i} = \frac{\Sigma w_i}{\Sigma w_i/M_i} \tag{1}$$

Weight-average molecular weight:

$$\overline{M}_w = \frac{\Sigma n_i M_i^2}{\Sigma n_i M_i} = \frac{\Sigma w_i M_i}{\Sigma w_i} \tag{2}$$

z-Average molecular weight:

$$\overline{M}_z = \frac{\Sigma n_i M_i^3}{\Sigma n_i M_i^2} = \frac{\Sigma w_i M_i^2}{\Sigma w_i M_i} \tag{3}$$

(z + 1)-Average molecular weight:

$$\overline{M}_{z+1} = \frac{\Sigma n_i M_i^4}{\Sigma n_i M_i^3} = \frac{\Sigma w_i M_i^3}{\Sigma w_i M_i^2} \tag{4}$$

Viscosity-average molecular weight:

$$\overline{M}_v = \left[\frac{\Sigma n_i M_i^{1+a}}{\Sigma n_i M_i}\right]^{1/a} = \left[\frac{\Sigma w_i M_i^a}{\Sigma w_i}\right]^{1/a} \tag{5}$$

Equation 5 is an important practical molecular weight average derived from viscometry (qv). In order to calculate this average, the exponent a of the Mark-Houwink relationship relating intrinsic viscosity $[\eta]$ to molecular weight must be known.

$$[\eta] = KM^a \tag{6}$$

The value of a lies between 0.5 and 1.0 for random coils, depending on the solvent employed to determine the relationship of the intrinsic viscosity to molecular weight. For rigid rod molecules, the value of a is expected to be ~1.8. With these limits for a, it may be seen that $\overline{M}_v$ is always larger than $\overline{M}_n$, but can equal $\overline{M}_w$ when the upper limit of a is reached for random coils.

If molecular weight is considered a continuous variable, the molecular weight distribution may be described by a set of moments μ_r given by the integrals

$$\mu_r = \int_0^\infty M^r f(M)dM$$

where $r = 0, 1, 2, 3$, etc, $f(M)$ is the number-density distribution, and $f(M)dM$ is the number of moles of molecules with molecular weight between M and $(M + dM)$.

The average molecular weights are defined as

$$\overline{M}_n = \frac{\mu_1}{\mu_0}$$

$$\overline{M}_w = \frac{\mu_2}{\mu_1}$$

$$\overline{M}_z = \frac{\mu_3}{\mu_2}$$

$$\overline{M}_{z+1} = \frac{\mu_4}{\mu_3}$$

Molecular weight averages can be determined from these moments (2).

Molecular Weight Distribution Functions

Various mathematical functions have been employed to describe the distribution of molecular weights. Some of the more common functions are shown in Table 1 in terms of the mole fraction X.

Width of Molecular Weight Distributions. The width of the Gaussian distribution function may be expressed in terms of the standard deviation of the

Table 1. Molecular Weight Distribution Functions

Name	Function	Comments
Gaussian	$X(M) = \dfrac{1}{\sigma_n(2\pi)^{1/2}} \exp\left[\dfrac{(M - M_m)^2}{2\sigma_n^2}\right]$	M_m = median value equal to $\overline{M}_n$
log-normal	$X(M) = \dfrac{1}{\sigma_n(2\pi)^{1/2}} \exp\left[\dfrac{(\ln M - \ln M_m)^2}{2\sigma_n^2}\right]$	M_m = geometric mean
Poisson[a]	$X(M) = \dfrac{\nu^{M-1}}{\Gamma(M)} \exp(-\nu)$	$\nu = \overline{M}_n - 1$
Flory-Schulz[b]	$X(M) = \dfrac{\beta^{k+1} M^{k-1} \overline{M}_n}{\Gamma(k+1)} \exp(-\beta M)$	k = degree of coupling $\beta = \dfrac{k}{\overline{M}_n}$

[a] $\Gamma(M) = \gamma$-function.
[b] $\Gamma(k + 1) = \gamma$-function.

mole fraction MWD function σ_n or the mass fraction MWD function σ_w:

$$\sigma_n = (\overline{M}_w \overline{M}_n - \overline{M}_n)^{0.5}$$
$$\sigma_w = (\overline{M}_z \overline{M}_w - \overline{M}_w)^{0.5}$$

The standard deviation is an absolute measure of the width for the Gaussian function only. Widths of molecular weight distribution for other functions have to be calculated for each case from the distribution function itself.

The relationships between molecular weight averages and parameters in the molecular weight distribution functions are as follows:

Gaussian distribution function
    mole-fraction distribution function   median value = $\overline{M}_n$
    mass-fraction distribution function   median value = $\overline{M}_w$
Log-normal distribution function
    mass-fraction distribution function   median value = $\overline{M}_m$

$$\overline{M}_n = \overline{M}_m \exp[(\sigma_w)^2/2]$$
$$\overline{M}_w = \overline{M}_m \exp[3(\sigma_w)^2/2]$$
$$\overline{M}_z = \overline{M}_m \exp[5(\sigma_w)^2/2]$$

which leads to

$$\exp(\sigma_w)^2 = \frac{\overline{M}_w}{\overline{M}_n} = \frac{\overline{M}_z}{\overline{M}_w}$$

Thus the ratios of two adjacent averages are constant.

Poisson Distribution Function. The ratio of the weight-average to number-average molecular weight is given by

$$\frac{\overline{M}_w}{\overline{M}_n} = 1 + \left(\frac{1}{\overline{M}_n}\right) - \left(\frac{1}{\overline{M}_n}\right)^2$$

and therefore depends only on $\overline{M}_n$.

Flory-Schulz Distribution Function. The molecular weight averages for this function are related by

$$\frac{\overline{M}_n}{k} = \frac{\overline{M}_w}{k+1} = \frac{\overline{M}_z}{k+2}$$

where k is the coupling constant, defined as the number of independently growing chains required to form one dead chain.

Polydispersity. Traditionally, polydispersity Q has been defined as

$$Q = \overline{M}_w/\overline{M}_n = U + 1$$

where U, the molecular inhomogeneity, has a numerical value of one less than Q. The width of molecular weight distributions increases with increasing Q and U. Except for the Gaussian distribution function, the standard deviation is only a relative measure of distribution width (3,4).

Determination of Molecular Weight

Number-average Molecular Weight

End-group Analysis. In end-group analysis the concentration of an end group is measured by a suitable technique, and then, from the known structure of the polymer, the value for $\overline{M}_n$ may be calculated (5–9). In condensation polymers one end of the polymer should have a specific group that can be titrated by appropriate means. Polyesters (8), polyamides (8), and polyurethanes (10) have been subjected to this analysis. Polyethers have been analyzed by hydroxyl-group titrations similar to those used for polyesters or by reaction with excess phenyl isocyanate followed by reaction with excess di-*n*-butylamine; the latter was back-titrated with perchloric acid (11). Nonaqueous titrations (12) have been used to characterize polysulfones with $\overline{M}_n$ values of up to 25,000. Spectrophotometric analyses have also been widely utilized (9), notably uv–visible, infrared techniques, and nmr (13). Examples include the determination of phenolic end groups in polymers by uv spectroscopy (14) and potentiometric titration (15). Infrared spectroscopy has been used (16) to estimate carboxyl groups in the presence of carbonyl groups by reaction of the former with SF_4 (16); the resulting thionyl halides were quantitatively measured. Nmr spectroscopy (17) has been used to characterize $\overline{M}_n$ of hydroxyl end-capped polystyrenes. Reaction of the single proton on the hydroxyl group yields a trimethylsilyl group containing nine protons. The enhanced sensitivity allows determination of $\overline{M}_n$ values of up to 80,000. A similar method has been reported for hydroxy-terminated polybutadienes (18), polyester polyurethanes (19), and epoxides (20).

In certain cases extremely high sensitivity may be attained with radioactive labeling (21). Using radiolabeled (^{35}S) bisulfite initiator, the number-average molecular weight of Teflon samples was determined in the range of 389,000–8,900,000 (22). This is a particularly useful technique for insoluble polymers.

Measurement of Colligative Properties. Colligative properties are those that depend on the number of species present rather than on their kind. From thermodynamic arguments it may be shown that for very dilute ideal solutions

$$\ln a_1 = -X_2$$

where a_1 is the activity of the solvent in a dilute ideal solution and X_2 is the mole fraction of solute. From this relationship the solute molecular weight may be calculated if the weight fraction w_2 is known.

$$X_2 = \frac{n_2}{n_1} = \frac{w_2 M_1}{M_2 w_1}$$

This equation demonstrates that the activity, measurable by several methods, is proportional to the number of solute molecules. Thus in the case of polydisperse solutes, the number-average molecular weight is the average determined. At higher concentrations and molecular weights, the solutions become nonideal, and higher powers of the solute concentration are introduced into the equations. Measurements of several solute concentrations are required, and appropriate extrapolation techniques must be used.

Lowering of Vapor Pressure. The partial vapor pressure p_1 of solvent 1 over a solution is lower than the vapor pressure over the pure solvent p_1^0. This is expressed by Raoult's law:

$$p_1 = X_1 p_1^0$$

where X_1 is the mole fraction of the solvent.

For a binary solution containing a mole fraction X_2 of solute then,

$$X_2 = \frac{p_1^0 - p_1}{p_1^0} = \frac{\Delta p_1}{p_1^0} \tag{7}$$

For a dilute solution,

$$X_2 = \frac{n_2}{n_1} = \frac{w_2 M_1}{M_2 w_1} \tag{8}$$

Combining equations 7 and 8 yields

$$M_2 = w_2 \frac{M_1}{w_1} \frac{p_1^0}{\Delta p_1}$$

The unknown molecular weight of solute M_2 may be calculated from the lowering of the vapor pressure caused by the addition of w_2 grams of solute to form a binary solution; $\overline{M}_n$ values of up to 1000 may be determined (23–26).

A variation of this technique, the isopiestic method (27–29), was devised to avoid the difficulty of accurately determining the small difference in vapor pressure caused by the addition of solute. In this method (30) two limbs of a container joined by a common vapor space are filled, one with a reference solution of known weight concentration w_s and molecular weight M_s, and the other with the unknown solution of known weight concentration w_2 and unknown molar concentration. The apparatus is immersed in a constant-temperature bath controlled to

±0.001°C; the absolute temperature is not important. Distillation of the solvent continues until the vapor pressures above the solutions in each limb are equal and thus the solutions contain equal mole fractions of solute. At equilibrium, which may require several weeks, the final volumes V_s and V_2 of the known and unknown solutions are read from the calibrations on each limb. Assuming ideal solution behavior, the unknown molecular weight is calculated from

$$\overline{M}_n = \frac{w_2 M_s V_s}{V_2 w_s}$$

$\overline{M}_n$ values of up to 20,000 have been determined by this method.

Ebulliometry. Ebulliometry (23,31–36) is another technique for determining the depression of the solvent activity by the solute. In this case the elevation of the boiling point is determined. The boiling-point elevation ΔT_b is measured with sensitive thermocouples or matched thermistors in a Wheatstone bridge. The molecular weight $\overline{M}_n$ is calculated from

$$\overline{M}_n = \frac{K_b c}{\Delta T_b}$$

where c is the concentration of solute in g/1000 g of solvent and

$$K_b = \frac{RT_b^2 M}{1000 \, \Delta H_v} \tag{9}$$

is the molal ebullioscopic constant. M is the molecular weight of the solvent and T_b its boiling point; ΔH_v is the molar latent heat of vaporization of the solvent.

The ebullioscopic constant may be evaluated directly from equation 9 and thus provides an absolute method. However, the value for K_b is often determined by using a high purity solute of known molecular weight.

Currently, a sensitivity of 1×10^{-5}°C or 2.4×10^{-6}°C is attainable using a 160-junction thermocouple (37) or thermistors (38), respectively. The upper limit of molecular weight, which may be determined by this method, also depends on the solvent and solution nonideality. With careful measurements and good ebulliometer design, $\overline{M}_n$ values of up to 100,000 may be determined on isotactic polypropylene (39). The apparatus employed is shown in Figure 1. This technique has been applied to aqueous solutions (40). An ebulliometer of simple design having a temperature sensitivity of 50×10^{-6}°C has recently been reported (41). A rotating ebulliometer attempts to overcome the superheating problem (42). Errors in the methods for treating ebullioscopic data to obtain molecular weight have been discussed (43).

Cryoscopy. The freezing point of a solution is depressed below that of the pure solvent by an amount proportional to the mole fraction of solute. The value for $\overline{M}_n$ is obtained from

$$\overline{M}_n = \frac{K_f c}{\Delta T}$$

where c is the concentration of solute in g/1000 g of solvent and

$$K_f = \frac{RT_m^2 M}{1000 \, \Delta H_{fus}}$$

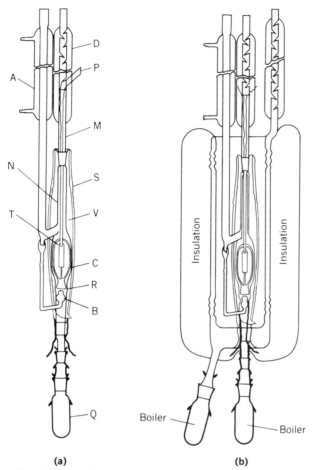

(a)   (b)

Fig. 1. (a) Ebulliometer design for molecular weight determination of up to $\overline{M}_n =$ 100,000. A, Ebulliometer cooler; B, boiler; C, Cottrell pump; D, adiabatic jacket cooler; M, thermopile; N, upper thermopile weldings; P, thermopile terminals connected with detector; Q, adiabatic jacket boiler; R, platinum resistance; T, lower thermopile weldings; and V, vacuum jacket circulated by solvent vapor. (b) Complete ebulliometer with adiabatic jackets and insulating apparatus (39). Courtesy of Huethig and Wepf Verlag.

is the molal cryoscopic constant; M is the molecular weight of the solvent and T_m its melting point; ΔH_{fus} is the molar latent heat of fusion of the solvent.

Some materials, such as camphor, have very large cryoscopic constants and may be used to increase the sensitivity of the method. The method is used to determine $\overline{M}_n$ of polymers (44–48), such as polyesters (49) and ethylene–vinyl acetate copolymers (50).

A novel variation of this technique (51) involves depression of the first-order, nematic–isotropic melting transition of N-(p-ethoxybenzylidene)-p-n-butylaniline. Polystyrene and poly(ethylene oxide) are soluble in both phases, and $\overline{M}_n$ values of up to 10^6 have been studied.

Vapor-pressure Osmometry. When a drop of solution is exposed to pure solvent vapor, the solvent vapor condenses onto the droplet, because the solute lowers the vapor pressure of the drop (52–55). The heat of condensation causes the temperature of the drop to rise, until theoretically the solution vapor pressure is increased to equal that of the solvent. In practice this does not occur, but a state of equilibrium is reached, where the heat losses from the droplet are matched by the heat of condensation. The temperature increase is proportional to the number of moles of solute present.

The design of a commercial instrument is given in Figure 2 (56). The temperature stability needed is ~0.001°C. In another design, by Wescan Instruments, Inc., small platinum mesh screens ensure that a constant amount of solvent or solution is present on the thermistor. The resistance value ΔR, which is proportional to the temperature difference, is measured with a bridge circuit in all instruments. A calibration curve of ΔR vs molality m is determined by employing a solute of known molecular weight. After several minutes ΔR reaches a constant value, the calibration curve is often highly linear, but may not go through the zero point (55).

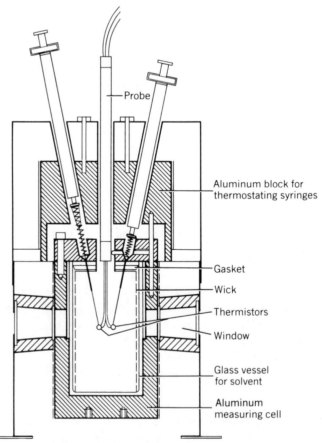

Fig. 2.   Vapor-pressure osmometer apparatus (56). Courtesy of Dr. -Ing. Herbert Knauer GmbH.

$$\Delta R = a + bm$$

The number-average molecular weight of the unknown sample may then be calculated from equation 10,

$$\overline{M}_n = \frac{bc}{\Delta R - a} \qquad (10)$$

where c is the concentration of solute in g/1000 g of solvent. In older osmometers the calibration curve depended on the material used to perform the calibration. For the Wescan instrument and another research vapor-pressure osmometer (57), both based on the same design (58), the calibration constant was independent of the material. These instruments have determined polystyrene $\overline{M}_n$ values of 400,000 (59), 100,000 (60), and 100,000 (57); polyolefins have been characterized at 140°C (61). For higher molecular weight solutes, the virial coefficients become important. In this case $\Delta R/c$ must be plotted against c in order to extrapolate the $\Delta R/c$ value to infinite dilution. When the plot is not linear, it is necessary to plot $(\Delta R/c)^{1/2}$ vs c to obtain a linear extrapolation to infinite dilution.

Membrane Osmometry. Membrane osmometry (qv) is a well-established technique (62–68); commercial instrumentation is available. The main problem is membrane permeation by lower molecular weight species (69). Improvements in membranes (qv) with well-defined, low molecular weight cutoffs can be expected as a consequence of the renewed interest in membrane-separation techniques (70,71). Membrane osmometry has been applied to a wide variety of polymers, including operation at high temperatures for polymers difficult to dissolve. The upper limit for molecular weight determination is ca 1×10^6. The lower limit depends on the molecular weight distribution, ie, the details of the composition of the low molecular weight tail. In most cases the concentration dependence of osmotic pressure π must be determined and π/c extrapolated to zero concentration by plotting π/c vs c, or $(\pi/c)^{1/2}$ vs c.

Viscosity-average Molecular Weight

The viscosity of dilute polymer solutions may be related to the molecular weight of the polymer by the appropriate calibration (see VISCOMETRY). The polymer is usually separated into narrow molecular weight distribution fractions, which are characterized by absolute molecular weight methods. The molecular weight is related to the intrinsic viscosity $[\eta]$ by the Mark-Houwink relationship (eq. 6).

The intrinsic viscosity is obtained by plotting the reduced or inherent viscosity of a series of polymer solutions of various concentrations against the solution concentration. Extrapolation to zero concentration yields the intrinsic viscosity. The technique has found widespread application (72–75), and compilations of Mark-Houwink constants are available (76).

If the Mark-Houwink constants are determined using narrow molecular weight distribution materials, the average molecular weight obtained from equation 6 is the viscosity-average molecular weight $\overline{M}_v$. The range of molecular weights that may be characterized is very large. Special precautions are required for extremely high molecular weight samples to avoid degradation during the solution process or the measurement (77).

Weight-average Molecular Weight

Light Scattering. Light scattering (qv) is a technique (78–80) widely used to characterize polymeric materials. Commercial instrumentation is available to measure the light-scattering properties of polymer solutions and the refractive index increment (81). (This instrumentation includes the Brice Phoenix Photometer from C. N. Wood Manufacturing Company and the Sofica Light Scattering Photometer as well as the Chromatix from LDC/Milton Roy. Malvern Instruments, Ltd. also produces similar equipment.) Lasers are used as light sources in order to measure scattering at small angles to the beam. This is important since the data obtained must be extrapolated to zero concentration and zero-scattering angle using Zimm plots (82). The weight-average molecular weight is determined by this technique, and the molecular weights accessible range from a few hundred to several million. The solutions employed must be free of dust and are usually subjected to filtration or centrifugation. Solutions of high molecular weight, degradable polymers are the most difficult to clarify. The solvent is selected to give a reasonable difference in refractive index with the solute.

Copolymers with homogeneous distribution of monomer units in the polymer chain may be analyzed by the same methods as homopolymers. If the composition varies as a function of molecular weight, the analysis is more complicated (83). Terpolymers have also been considered (84).

Dynamic light scattering (85) has been used to determine the number-average molecular weight (86) and molecular weight distribution of polymers in solution (87,88).

Pulse-induced critical scattering (89,90), although not utilized as a method for molecular weight characterization, may be a sensitive procedure for detecting small differences in molecular weight distribution.

X-ray (91) and neutron scattering (qv) (92) may also be used to determine weight-average molecular weights. The main drawback is the greater expense and the lack of suitable facilities.

Ultracentrifuge. The ultracentrifuge is used to determine molecular weights (93). Sedimentation velocity employs a sufficiently high centrifugal field so that the sedimentation rate may be measured. The sedimentation coefficient S is empirically corrected for concentration and pressure effects to give S_0, the sedimentation coefficient at zero concentration. The method also requires the measurement of the diffusion coefficient at infinite dilution D_0 (see ULTRACENTRIFUGATION). The molecular weight is calculated from the Svedberg equation

$$\frac{S_0}{D_0} = \frac{M(1 - \nu\rho)}{RT}$$

where ν is the partial specific volume of the solute and ρ the density of the solution.

For a polydisperse solute the correct averages for S and D are combined in the Svedberg equation to yield a well-defined average molecular weight. If D is obtained from intensity-fluctuation spectroscopy (94) and combined with the weight-average S value, the weight-average molecular weight is obtained. The application of the sedimentation velocity method to calculate molecular weight distribution is extremely complex, but has been successful in special studies (95).

In the sedimentation equilibrium method, a lower centrifugal field is maintained for a period of time in such a way that sedimentation is balanced by diffusion and an equilibrium distribution of polymer is established in the cell. Although $\overline{M}_w$ and $\overline{M}_z$ are easily determined, the length of time of the experiment is a disadvantage. In contrast to light scattering, this method is not affected by dust particles, and no calibration is needed. The molecular weight distribution may be obtained from the sedimentation velocity data, but not without mathematical difficulties (96) or requiring additional data at other rotor speeds (97,98). An improved detection system for the analytical ultracentrifuge has been reported (99,100). A new 8-cell rotor that allows increased throughput of samples has been described (101).

Improved data analysis has led to the simultaneous determination of the sedimentation coefficient and diffusion coefficient, and a calculation of the molecular weight from the Svedberg equation (102). The Svedberg method is capable of determining molecular weights of up to 40×10^6 (77) on an absolute basis. Sedimentation equilibrium in a density gradient has been used to determine the compositional analysis of copolymers, eg, butadiene–styrene (103).

Higher Molecular Weight Averages

Classical techniques often cannot be used for higher molecular weight averages ($\overline{M}_z$, $\overline{M}_{z+1}$, etc). These are usually measured by determining the molecular weight distribution by fractionation (qv) or chromatography (qv) and back-calculating the averages from the distribution.

Determination of Molecular Weight Distribution

Fractionation.   Polymers may be separated on the basis of molecular weight by precipitation or phase-separation techniques (104,105). The polymer that is originally in solution separates on the basis of molecular weight when the thermodynamic quality of the solvents is changed, ie, by changing the solvent composition or temperature. The principal techniques are fractional precipitation and coacervate extraction, both solution-based (Fig. 3). The former is based on the removal of material from the polymer-rich precipitated phase and the latter on the removal of fractions from the polymer-lean solution phase. Computer simulations and experimental results have shown the coacervate extraction technique (107) to yield narrower molecular weight distribution fractions than the fractional precipitation method (108–110).

To obtain the molecular weight distribution, the fractions of known mass must be characterized by a molecular weight technique. The integral and differential molecular weight distribution may be evaluated using a method (111) that assumes no particular molecular weight distribution function for the fraction or treatments that assume a particular model for the molecular weight distribution of the fractions (112). Fractionation may also be achieved by crystallization from dilute solution (113).

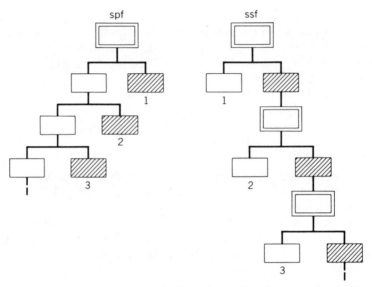

Fig. 3. Successive precipitation (spf) and solution fractionation (ssf) (coacervate extraction): ▢, mother solution; ▢, polymer-rich phase; and ▨, polymer-lean phase. Numbers denote fraction numbers (106). Courtesy of Society of Chemical Industry.

In column fractionation (114,115) the polymer is precipitated onto an inert support, which is placed at the top of a packed column (116). A solvent mixture of increasing solvent power is pumped through the column; a temperature gradient is often maintained. This is known as Baker-Williams fractionation (117). This technicue is applicable to all amorphous homopolymers and crystalline homopolymers above the melting point. For copolymers and more complex compositions, the same technique may be employed, but the analysis is considerably more difficult.

Chromatography. *Gel-permeation (Size-exclusion) Chromatography.* Gel-permeation chromatography (gpc) has been in use since the 1960s and has been a most important development in molecular weight determination (118–122). For this technique the polymer must be soluble, and calibration with suitable standards is required. The method is extremely efficient; it requires 0.5–2 h per sample and only a few milligrams of material. Commercial instrumentation capable of operating up to 150°C is available from Millipore Corp. Detection is normally accomplished by differential refractometry, although uv and ir absorption spectrometry have also been employed. The chromatographic columns are packed with spherical- or irregular-shaped porous beads. The packing may be a cross-linked polymer, a porous glass, or a silica material. The pore size distribution determines the separation range of the columns. Typical calibration curves (123) are shown in Figure 4. By combining columns in series, a separation range may be achieved covering the molecular weight distribution of the sample. Mathematical treatments of this chromatogram and calibration curve to calculate the molecular weight distribution are given in Ref. 124. Broadening by the chromatographic process must be calibrated for and corrections applied to produce true molecular weight distributions. The design and use of a continuous, visco-

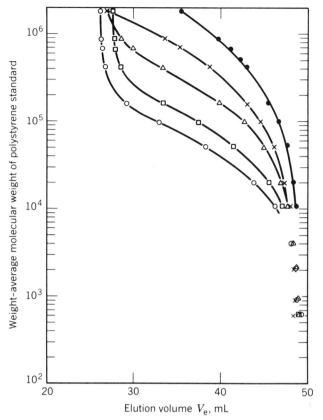

Fig. 4. Gpc calibration curves for Corning porous glasses (CPG) with various pore sizes: ○, CPG 10–240; □, CPG 10–370; △, CPG 10–700; ×, CPG 10–1250; and ●, CPG 10–2000 (123). Courtesy of John Wiley & Sons, Inc.

metric detector (125,126) has further advanced the application of the gpc technique, and a commercial viscometric detector is on the market from Viscotek Corp. (127,128).

Direct determination of the polymer molecular weight in the detector cell after separation by gpc is possible by low angle laser-light scattering (lalls). The technique (129–131) is applicable to the dilute polymer solutions emerging from the chromatographic column; commercial equipment is available, eg, the Chromatix KMX6 and LSD100 from LDC/Milton Roy. This method is used for on-line molecular weight measurement and branching characterization of water-soluble polymers (132,133). Shear degradation during the chromatographic process may be a problem (134), but can be detected by viscometry and lalls.

Other detectors that are capable of determining the polymer concentration in the effluent are based on dielectric constant (135) and density (136–142). Mass can be determined directly with a piezoelectric quartz sensor having a sensitivity of 10^{-10} g (143). A universal mass detector based on the formation of droplets in a nebulizer gas has been developed by Applied Chromatographic Systems, Ltd. and by Varex Corp. and applied to gpc (144).

A direct molecular weight method (145) employs the lalls detector, but uses sedimentation velocity instead of gpc to separate the polymers.

High Performance Liquid Chromatography and Supercritical Fluid Chromatography. For lower molecular weights, high performance liquid chromatography (hplc) and supercritical fluid chromatography (sfc) offer higher resolution than gel-permeation chromatography. The oligomers are resolved into individual peaks, and molecular weight averages are calculated (146). Hplc is more common and has been applied to phenol–formaldehyde resins (147), polyurethanes (148), polyesters (149), and polycarbonates (150). Supercritical chromatography can resolve higher oligomers than hplc. Polystyrene oligomers of up to $n = 50$ may be resolved by sfc; the process requires 18 h (151). The introduction of two commercial sfc units by Lee Scientific and Suprex Corp. should rapidly increase the application of sfc to characterize oligomeric species.

Thin-layer Chromatography. Thin-layer chromatography (tlc) is widely applicable to polymer characterization and has been used to characterize molecular weight distribution and compositional distribution (152–154). Cellulose nitrate has been fractionated by this method on the basis of molecular weight or nitrogen content (155).

Phase-distribution Chromatography. A chromatographic fractionation is achieved in this method (156) by partitioning a solute between a solvent flowing through a column, packed with a support coated with a thin layer of high molecular weight, noncross-linked polymer. The temperature of operation is below the θ-temperature of the solute, and separation efficiency increases sharply with decreasing temperature. A fully automated chromatograph has been designed (157), and applications have been reported (158–160).

Field-flow Fractionation. Field-flow fractionation employs a one-phase chromatographic system (161). Separation occurs in a thin channel containing a single moving fluid. The field applied across the channel may be selected on the basis of the solute. Possible fields include sedimentation, cross-flow, concentration, dielectric, thermal, and magnetic.

Thermal fields have been used for the separation of polymers in organic solvents (162). For aqueous systems a cross-flow of solvent has been used to separate polyelectrolytes (163). Extremely high molecular weights of up to 10^{12} can be determined (164); because no abrupt changes occur in the forces imposed on the polymer molecules, no degradation takes place. Sedimentation fields were investigated (165), and small samples were quickly fractionated.

Electrophoresis. Electrophoresis (qv) is widely practiced on biopolymers (166,167) for the determination of molecular weights of proteins and their subunits. Typically, a polyacrylamide bed is used to separate or characterize these solutes under the influence of an electric field, on the basis of charge and molecular size. The proteins are frequently denatured with sodium dodecyl sulfonate (SDS), and these subunit–SDS complexes are separated on the basis of molecular size. A densitometer employing a soft laser scanning device is commercially available to automate the data-reduction process (168).

Other Techniques. Atomization of polymer solutions to form single molecules deposited in a solvent-free state on a surface has been demonstrated for a wide variety of polymers (169). The electron microscope is used at 20,000–30,000× after platinum shadowing to determine size and size distribution (170). The scanning transmission electron microscope is employed for biological macromolecules (171) (see ELECTRON MICROSCOPY).

BIBLIOGRAPHY

"Molecular Weight Determination" in *EPST* 1st ed., Vol. 9, pp. 182–193, by Manfred J. R. Cantow, Airco Central Research Laboratories, and Julian F. Johnson, Chevron Research Company.

1. L. H. Peebles, *Molecular Weight Distributions in Polymers,* Wiley-Interscience, New York, 1971.
2. K. W. Min, *J. Appl. Polym. Sci.* **22,** 589 (1978).
3. C. W. Pyun, *J. Polym. Sci. Polym. Phys. Ed.* **17,** 2111 (1979).
4. *Ibid.,* **24,** 229 (1986).
5. N. C. Billingham, *Molar Mass Measurements in Polymer Science,* Halsted Press, a division of John Wiley & Sons, Inc., New York, 1977, Chapt. 9, p. 234.
6. G. F. Price in P. W. Allen, ed., *Techniques of Polymer Characterization,* Butterworth & Co. (Publishers) Ltd., London, 1959, Chapt. VII.
7. S. R. Rafikov, S. A. Pavlova, and I. I. Tverdokhlebora, *Determination of Molecular Weights and Polydispersity of High Polymers,* Israel Programme for Scientific Translation, Jerusalem, Israel, 1964, Chapt. 8.
8. R. G. Garmon in P. E. Slade, Jr., ed., *Polymer Molecular Weights,* Marcel Dekker, Inc., New York, 1975, Chapt. 3.
9. J. Urbanski, N. Czerwinski, K. Janicka, F. Majewska, and H. Zowall, *Handbook of Analysis of Synthetic Polymers and Plastics,* Wiley-Interscience, New York, 1977.
10. D. J. David and H. B. Staley, *Analytical Chemistry of Polyurethanes,* Wiley-Interscience, New York, 1959.
11. D. H. Reed, F. E. Critchfield, and D. K. Elder, *Anal. Chem.* **35,** 571 (1963).
12. A. J. Wnuk, T. F. Davidson, and J. E. McGrath, *J. Appl. Polym. Sci. Appl. Polym. Symp.* **34,** 89 (1978).
13. J. R. Ebdon in L. S. Bark and N. S. Allen, eds., *Analysis of Polymer Systems,* Applied Science Publishers, London, 1982, p. 21.
14. E. Shchori and J. E. McGrath, *J. Appl. Polym. Sci. Appl. Polym. Symp.* **34,** 103 (1978).
15. A. J. Wnuk, T. F. Davidson, and J. E. McGrath, *J. Appl. Polym. Sci. Appl. Polym. Symp.* **34,** 89 (1978).
16. J. F. Heacock, *J. Appl. Polym. Sci.* **7,** 2319 (1963).
17. A. D. Edwards and M. J. R. Loadman, Malaysian Rubber Producers Research Association, unpublished data, 1976.
18. G. Fages and Q. T. Pham, *Makromol. Chem.* **180,** 2435 (1979).
19. F. W. Yeager and J. W. Becker, *Anal. Chem.* **49,** 722 (1977).
20. W. B. Moniz and C. F. Poranski, Jr. in R. S. Bauer, ed., *Epoxy Resin Chemistry, ACS Symp. Ser.* **114,** American Chemical Society, Washington, D.C., 1979.
21. K. L. Berry and J. H. Peterson, *J. Am. Chem. Soc.* **73,** 5195 (1951).
22. R. C. Dobau, A. C. Knight, J. H. Peterson, and C. A. Sperati, *130th Meeting of the American Chemical Society,* Atlantic City, N.J., Sept. 1956.
23. D. F. Rushman in Ref. 6, p. 113.
24. I. M. Kolthoff and P. J. Elving, eds., *Treatise on Analytical Chemistry, Theory and Practice,* Vol. 7, Wiley-Interscience, New York, Part 1, p. 4446.
25. G. Radakoff, *Z. Chem.* **1,** 135 (1961).
26. R. U. Bonnar, M. Dimbat, and F. H. Stross, *Number Average Molecular Weights,* Wiley-Interscience, New York, 1958, p. 275.
27. Ref. 23, p. 114.
28. Ref. 24, p. 4448.
29. Ref. 26, p. 268.
30. R. L. Parette, *J. Polym. Sci.* **15,** 450 (1955).
31. Ref. 26, p. 113.
32. Ref. 24, p. 4450.

33. M. Ezrin in M. McIntyre, ed., *Characterization of Macromolecular Structure, Publication 1573,* National Academy of Science, Washington, D.C., 1968, p. 3.
34. C. A. Glover in Ref. 8, Part I, p. 105.
35. C. A. Glover in M. Ezrin, ed., *Polymer Molecular Weight Methods, Adv. Chem. Ser.* **125,** American Chemical Society, Washington, D.C., p. 1.
36. G. Davison in Ref. 13, p. 209.
37. C. A. Glover, C. N. Reilly, and F. W. McLafferty, eds., *Advances in Analytical Chemistry and Instrumentation,* Vol. 5, Wiley-Interscience, New York, 1966, p.1.
38. E. Zichy, *SCI Monogr.* **17,** 122 (1963).
39. P. Parrini and M. S. Vacanti, *Die Makromol. Chem.* **175,** 935 (1974).
40. W. De Oliveira, *Differential Ebulliometry,* Ph.D. thesis, Clarkson College of Technology, Potsdam, N.Y., 1975.
41. W. De Oliveira and G. Francisco, *Chem. Biomed. Environ. Instrum.* **10,** 189 (1980).
42. J.-T. Chen, H. Sotobayashi, and F. Asmussen, *Colloid Polym. Sci.* **259,** 1202 (1981).
43. J. Melsheimer and H. Sotobayashi, *Die Makromol. Chem.* **179,** 2913 (1978).
44. Ref. 34, p. 124.
45. E. J. Newitt and V. Kokle, *J. Polym. Sci. Part A-2* **4,** 705 (1966).
46. Ref. 23, p. 125.
47. Ref. 26, p. 17.
48. Ref. 7, p. 255.
49. L. Carbonnel, R. Guieu, C. Ponge, and J. C. Rosso, *J. Chim. Phys. Physicochim. Biol.* **70,** 1400 (1973).
50. A. Ya Ryasnyanskaya, S. L. Lyubimova, V. V. Kalashnikov, R. A. Terteryan, V. N. Mosastyrskii, and B. V. Gryaznov, *Neftepererab. Neftekhim. Moscow,* 69 (1973).
51. B. Kronberg and D. Patterson, *Macromolecules* **12,** 916 (1979).
52. A. V. Hill, *Proc. R. Soc. London Ser. A* **127,** 9 (1930).
53. C. A. Glover in Ref. 8, p. 142.
54. Ref. 5, p. 81.
55. M. J. R. Cantow, R. S. Porter, and J. F. Johnson, *J. Polym. Sci. Part A* **2,** 2547 (1964).
56. H. Knauer, Berlin, unpublished data.
57. K. Kamide, T. Terakuwa, and H. Uchiki, *Makromol. Chem.* **177,** 1447 (1976).
58. R. E. Dohner, A. H. Wachter, and W. Simon, *Helv. Chim. Acta* **50,** 2193 (1967).
59. A. H. Wachter and W. Simon, *Anal. Chem.* **41,** 90 (1969).
60. D. E. Burge, *J. Appl. Polym. Sci.* **24,** 293 (1979).
61. F. M. Mirabella, *J. Appl. Poly. Sci.* **25,** 1775 (1980).
62. Ref. 26, p. 191.
63. Ref. 5, p. 44.
64. Ref. 6, p. 68.
65. H. Coll and F. H. Stross in Ref. 35, p. 10.
66. R. D. Ulrich in Ref. 34, p. 9.
67. W. R. Krigbaum and R.-J. Roe in Ref. 24, p. 4461.
68. D. E. Burge, *Am. Lab.,* 41 (June 1977).
69. H. Coll, *J. Polym. Sci. Macromol. Rev.* **5,** 541 (1971).
70. "Polymeric Separation Media" in A. R. Cooper, ed., *Polymer Science and Technology,* Vol. 16, Plenum Press, New York, 1982.
71. "Ultrafiltration Membranes and Applications" in Ref. 70, Vol. 13, 1981.
72. P. F. Onyon in Ref. 6, Chapt. VI.
73. D. K. Carpenter and L. Westerman in Ref. 8, Part II, p. 441.
74. M. Van Oene in Ref. 35, p. 353.
75. Ref. 5, p. 172.
76. M. Kurata, Y. Tsunashima, M. Iwama, and K. Kamada in J. Brandrup and E. H. Immergut, eds., *Polymer Handbook,* 2 ed., Wiley-Interscience, New York, 1975, Chapt. IV, p. 1. Third edition to be published in late 1987.
77. B. Appelt and G. Meyerhoff, *Macromolecules* **13,** 657 (1980).
78. M. B. Huglin, ed., *Light Scattering from Polymer Solutions,* Academic Press, Inc., Orlando, Fla., 1972.
79. M. Kerker, *The Scattering of Light,* Academic Press, Inc., Orlando, Fla., 1969.

80. H. Yamakawa, *Modern Theory of Polymer Solutions,* Harper & Row, Publishers, Inc., New York, 1971.
81. H. Utiyama in Ref. 78.
82. B. H. Zimm, *J. Chem. Phys.* **16,** 1093 (1948).
83. E. F. Cassassa and G. C. Berry in Ref. 8, Part I, p. 161.
84. H. Kambe, Y. Kambe, and C. Honda, *Polymer* **14,** 460 (1973).
85. B. Chu, *Laser Light Scattering,* Academic Press, Inc., Orlando, Fla., 1974.
86. J. C. Selser, *Macromolecules* **12,** 909 (1979).
87. Q. Ying, B. Chu, R. Qian, J. Bao, J. Zhang, and C. Xu, *Polymer* **26,** 1401 (1985).
88. B. Chu, Q. Ying, C. Wu, J. R. Ford, and M. S. Dhadal, *Polymer* **26,** 1408 (1985).
89. M. Gordon, J. Goldsbrough, B. W. Ready, and K. W. Derham in J. H. S. Green and R. Dietz, eds., *Industrial Polymers: Characterization by Molecular Weight,* Transcripta Books, London, 1973, p. 45.
90. H. Galina, M. Gordon, P. Irvine, and L. A. Kleintjens, *Pure Appl. Chem.* **54,** 365 (1982).
91. D. Glotter and O. Kratky, *Small Angle X-Ray Scattering,* Academic Press Ltd., London, 1982.
92. R. W. Richards in J. V. Dawkins, ed., *Developments in Polymer Characterization-5,* Elsevier Applied Science Publishers, Ltd., Barking, UK, 1986.
93. H. Fujita, *Mathematical Theory of Sedimentation Analysis,* Academic Press, Inc., Orlando, Fla., 1962.
94. D. N. Pussey in Ref. 89, p. 26.
95. R. Dietz in Ref. 89, p. 19.
96. D. A. Lee, *J. Polym. Sci. Part A-2* **8,** 1039 (1970).
97. T. G. Scholte, *J. Polym. Sci. Part A-2* **6,** 11 (1968).
98. T. G. Scholte, *Eur. Polym. J.* **6,** 51 (1970).
99. J. Flossdorf, H. Schillig, and K. P. Schindler, *Makromol. Chem.* **179,** 1617 (1978).
100. J. Flossdorf, *Makromol. Chem.* **181,** 715 (1980).
101. N. Machtle and U. Klodwig, *Makromol. Chem.* **180,** 2507 (1979).
102. R. Wohlschiess, K. F. Elgert, and H-J. Cantow, *Angew. Makromol. Chem.* **74,** 323 (1978).
103. C. J. Stacy, *J. Appl. Polym. Sci.* **21,** 2231 (1977).
104. M. J. R. Cantow, ed., *Polymer Fractionation,* Marcel Dekker, Inc., New York, 1967.
105. L. H. Tung, ed., *Fractionation of Synthetic Polymers,* Marcel Dekker, Inc., New York, 1977.
106. K. Kamide, Y. Miyazaki, and T. Abe, *Br. Polym. J.* **13,** 168 (1981).
107. K. Kamide and Y. Miyazaki, *Polym. J.* **12,** 153 (1980).
108. K. Kamide in Ref. 104, p. 103.
109. K. Kamide, Y. Miyazaki, and T. Abe, *Makromol. Chem.* **117,** 485 (1976).
110. K. Kamide, Y. Miyazaki, and T. Abe, *Polym. J.* **9,** 395 (1977).
111. G. V. Schulz, *Z Phys. Chem. Abt. B* **46,** 137 (1940).
112. K. Kamide, T. Ogawa, M. Sanada, and M. Matsumoto, *Kobunshi Kagaku* **25,** 440 (1968).
113. R. Koningsveld and A. J. Pennings, *Recl. Trav. Chim. Pays-Bas Belg.* **83,** 552 (1964).
114. J. H. Elliot in Ref. 104, p. 67.
115. E. M. Barrall, II, J. F. Johnson, and A. R. Cooper in Ref. 105, p. 267.
116. V. Desreux and M. C. Spiegels, *Bull. Soc. Chim. Belg.* **59,** 476 (1950).
117. C. A. Baker and R. J. P. Williams, *J. Chem. Soc.,* 2352 (1956).
118. K. H. Altgelt and L. Segal, eds., *Gel Permeation Chromatography,* Marcel Dekker, Inc., New York, 1971.
119. W. W. Yau, J. J. Kirkland, and D. D. Bly, eds., *Modern Size Exclusion Chromatography,* Wiley-Interscience, New York, 1979.
120. J. Janca, ed., *Steric Exclusion Liquid Chromatography of Polymers,* Marcel Dekker, Inc., New York, 1984.
121. A. R. Cooper in Ref. 92, p. 131.
122. A. R. Cooper in Ref. 13, p. 243.
123. A. R. Cooper, A. R. Bruzzone, J. H. Cain, and E. M. Barrall, II, *J. Appl. Polym. Sci.* **15,** 571 (1971).
124. J. Janca in J. C. Giddings, E. Grushka, J. Cazes, and P. R. Brown, eds., *Advances in Chromatography,* Vol. 19, Marcel Dekker, Inc., New York, 1981, p. 37.
125. A. C. Ouano, *J. Polym. Sci. Part A-1* **10,** 2169 (1972).
126. A. C. Ouano, D. L. Horne, and A. R. Gregges, *J. Polym. Sci. Polym. Phys. Ed.* **12,** 307 (1974).

127. M. A. Haney, *J. Appl. Polym. Sci.* **30**, 3023 (1985).
128. *Ibid.,* p. 3037.
129. W. Kaye, *Anal. Chem.* **45**, 221 (1973).
130. A. C. Ouano and W. Kaye, *J. Polym. Sci.* **12**, 1151 (1974).
131. A. C. Ouano, *J. Chromatogr.* **118**, 303 (1976).
132. A. E. Hamielec and H. Meyer in Ref. 92, p. 95.
133. D. J. Nagy, *J. Polym. Sci. Polym. Lett. Ed.* **24**, 87 (1986).
134. H. G. Barth and F. J. Carlin, Jr., *J. Liquid Chromatogr.* **7**, 1717 (1984).
135. R. A. Sanford, R. K. Bade, and E. N. Fuller, *Am. Lab.*, 99 (Mar. 1983).
136. J. Francois, M. Jacot, Z. Grubisic-Gallot, and H. Benoit, *J. Appl. Polym. Sci.* **22**, 1159 (1978).
137. D. Sarazin, J. LeMoigne, and J. Francois, *J. Appl. Polym. Sci.* **22**, 1377 (1978).
138. H. Leopold and B. Trathnigg, *Die Angew. Makromol. Chem.* **68**, 185 (1978).
139. B. Trathnigg, *Monatsh. Chem.* **109**, 467 (1978).
140. B. Trathnigg, *Die Angew. Makromol. Chem.* **89**, 65 (1980).
141. *Ibid.,* p. 73.
142. B. Trathnigg, *Makromol. Chem. Rapid Commun.* **1**, 569 (1980).
143. W. W. Schulz and W. H. King, Jr., *J. Chromatogr. Sci.* **11**, 343 (1973).
144. H. S. Huang and H. G. Barth, *Antec '85, Conference Proceedings of the SPI 43rd Ann. Tech. Conf.,* Washington, D.C., 1985, p. 277.
145. G. Holzworth, L. Soni, and D. N. Schulz, *Macromolecules* **19**, 422 (1986).
146. S. Mori, *J. Chromatogr.* **156**, 111 (1978).
147. A. Sebenik and S. Lapanje, *J. Chromatogr.* **106**, 454 (1975).
148. P. McFadyen, *J. Chromatogr.* **123**, 468 (1976).
149. L. M. Zaborsky, II, *Anal. Chem.* **49**, 1166 (1977).
150. C. Bailly, D. Daoust, R. Legras, J. P. Mercier, and M. de Valck, *Polymer* **27**, 776 (1986).
151. E. Klesper and W. Hartmann, *J. Polym. Sci. Polym. Lett. Ed.* **15**, 9 (1977).
152. B. G. Belenkii and E. S. Gankina, *J. Chromatogr. Chromatogr. Rev.* **141**, 13 (1977).
153. H. Inagaki and T. Tanaka in J. V. Dawkins, ed., *Developments in Polymer Characterization-3,* Applied Science Publishers, London, 1982, p. 1.
154. H. Inagaki in Ref. 105, p. 649.
155. K. Kamide, T. Okada, T. Terakawa, and K. Kaneo, *Polym. J.* **10**, 547 (1978).
156. R. H. Casper and G. V. Schulz, *Sep. Sci.* **6**, 321 (1971).
157. G. S. Greschner, *Makromol. Chem.* **180**, 2551 (1979).
158. *Ibid.,* **181**, 1435 (1980).
159. *Ibid.,* **182**, 2845 (1981).
160. G. S. Greschner, *Advances in Polymer Science,* Vol. 73, Springer-Verlag, Berlin, 1985, p. 1.
161. J. C. Giddings, *Pure Appl. Chem.* **51**, 1459 (1979).
162. J. C. Giddings, M. N. Myers, G. C. Lin, and M. Martin, *J. Chromatogr.* **142**, 23 (1977).
163. J. C. Giddings, G. C. Lin, and M. N. Myers, *J. Liquid Chromatogr.* **1**, 1 (1978).
164. J. C. Giddings, *J. Chromatogr.* **125**, 3 (1976).
165. J. J. Kirkland, W. W. Yau, and W. A. Doerner, *Anal. Chem.* **52**, 1944 (1980).
166. P. C. Allen, E. A. Hill, and A. M. Stokes, *Plasma Proteins, Analytical and Preparative Techniques,* Blackwell Scientific Publications, Oxford, UK, 1978, p. 7.
167. T. J. Mantle, *Techniques in Protein and Enzyme Biochemistry, B105b,* 1978, p. 1.
168. J. A. Zeineh, M. M. Zeineh, and R. A. Zeineh, *Am. Lab.,* 124 (June 1986).
169. D. V. Quayle, *Br. Polym. J.* **1**, 15 (1969).
170. G. Koszterszitzs and G. V. Schulz, *Makromol. Chem.* **178**, 2437 (1977).
171. R. Freeman and K. R. Leonard, *J. Microsc.* **122**, 275 (1980).

General References

Refs. 1, 6, 7, 26, 33, 35, 78, 85, 89, 104, and 118 are good general references.
N. C. Billingham, *Molar Mass Measurements in Polymer Science,* Halsted Press, a division of John Wiley & Sons, Inc., New York, 1977.
P. E. Slade, Jr., ed., *Polymer Molecular Weights,* Marcel Dekker, Inc., New York, 1975, Parts I and II.
L. S. Bark and N. S. Allen, eds., *Analysis of Polymer Systems,* Applied Science Publishers, London, 1982.

J. V. Dawkins, ed., *Developments in Polymer Characterization 1–5,* Elsevier Applied Science Publishers, Ltd., Barking, UK, 1986.
L. H. Tung, ed., *Fractionation of Synthetic Polymers,* Marcel Dekker, Inc., New York, 1977.
W. W. Yau, J. J. Kirkland, and D. D. Bly, eds., *Modern Size Exclusion Chromatography,* Wiley-Interscience, New York, 1979.
J. Janca, ed., *Steric Exclusion Liquid Chromatography of Polymers,* Marcel Dekker, Inc., New York, 1984.

ANTHONY R. COOPER
Lockheed Missiles & Space Company, Inc.

MOLECULAR WEIGHT DISTRIBUTION.   See FRACTIONATION.

MONODISPERSE POLYMERS

The preparation of homopolymers and block copolymers (qv) with near-monodisperse (uniform) molecular weight distributions is of both academic and commercial value. The absence, in certain cases, of a spontaneous termination step during propagation allows for the preparation of linear and star-shaped polymers. This facilitates studies in block polymer morphology, chain dynamics, diffusion (both tracer and self-diffusion), unperturbed chain dimensions, and rheology. These investigations have benefited from the narrow distribution of chain lengths present in the samples. The materials evaluated have been prepared by anionic polymerization of dienes and by hydrogenation of these polydienes.

Before monodisperse polymers were prepared, investigations were conducted on the effect of the relative rates of initiation and propagation on the molecular weight distribution in nonterminating addition polymerizations (1). Flory was the first to treat such systems in a quantitative fashion (2). Basing his treatment on the anionic polymerization of ethylene oxide initiated by sodium alkoxides, he derived relations which predicted a very narrow Poisson distribution for such polymers. The crucial criteria were based on the assumptions that there is no termination or transfer during chain growth, that the initiation event is at least as rapid as the subsequent propagation process, and that all monomers have equal probability of reacting with perfect mixing maintained throughout the reaction. In certain systems, all of these requirements can be met experimentally.

The molecular weights in Flory's derivation are defined by:

$$P_j = e^{-x}x^{j-1}/(j-1)!$$
$$W_j = [x/(x+1)]je^{-x}x^{j-2}/(j-1)!$$

where P_j and W_j are the number and weight fraction, respectively, of polymers with length j, and where x denotes the number of monomers reacted per initiator molecule. The ratio of the weight-average ($\bar{x}_w$) to number-average ($\bar{x}_n$) chain length is given by:

$$\bar{x}_w/\bar{x}_n = 1 + (x_n - 1)/x_n^2$$

which at high values of x reduces to:

$$\bar{x}_w/\bar{x}_n = 1 + 1/x_n$$

Thus when x equals 100, the molecular weight distribution, given by $\bar{x}_w/\bar{x}_n$, is 1.01. Such a system is virtually monodisperse.

The usual experimental uncertainty ($\pm 2\%$) in the measurements of number- and weight-average molecular weights does not allow the confirmation of such a narrow distribution. An $\bar{x}_w/\bar{x}_n$ ratio of near unity is, in fact, not definitive proof that a polymer is near-monodisperse. For instance, a sample with equal parts by weight of three monodisperse fractions with molecular weights of 4×10^5, 5×10^5, and 6×10^5, respectively, exhibits an $\bar{x}_w/\bar{x}_n$ ratio of 1.03. Therefore, measurements of at least three moments, M_n, M_w, and M_z, are needed to gain a truly quantitative description of the molecular weight distribution. To this end, size exclusion chromatography (qv), properly executed, provides the most accurate insight into polymer polydispersity, column-broadening effects not withstanding. From an operational standpoint, polymers with ratios of <1.1 of the various moments in the distribution are considered near-monodisperse.

Polymers with near-monodisperse molecular weight distribution have been prepared by anionic polymerizations involving monofunctional (from organolithiums) or difunctional chains (via electron-transfer species, such as sodium naphthalene) (3–8). In an early claim, the implied presence of polystyrenes with monomodal distributions, prepared via sodium naphthalene, was in error (see ANIONIC POLYMERIZATION) (9).

In any consideration of molecular weight distributions resulting from termination-free systems, the potential influence of the relative rates of initiation and propagation must be recognized (2). Quantitative relations have been developed between the molecular weight distribution and the ratio of propagation-to-initiation rates (10–12). Even in the extreme case where $k_p/k_i = 10^6$, the resultant x_w/x_n ratio does not exceed ~ 1.4. Termination-free polymerizations, in which the active center has two forms in dynamic equilibrium, with each exhibiting its own propagation rate, have been examined (13,14). This approach has been used to develop a method for dissecting k_p into the contributions from the various ionic structures of the propagating center in polar media from an analysis of the resultant molecular weight distribution (15). Any desired polydispersity in molecular weights is obtained by controlling the addition of initiator during a nonterminating polymerization (16).

Premature termination of active centers can influence molecular weight distributions. In anionic polymerizations, the following possibilities must be distinguished: partial deactivation of initiator by impurities; termination of chain and active centers by impurities; monomer, solvent, or bond rearrangements; and chain-transfer reactions.

The partial elimination of a monofunctional initiator does not affect the molecular weight distribution but does influence the resultant molecular weight since predicted weight is based on the monomer–initiator stoichiometry. The same applies to the situation where difunctional growing chains are formed, eg, from sodium naphthalene, as long as the termination reaction involves only the initiator species. However, if some of the active centers of the newly formed difunctional growing chains are prematurely terminated, a bimodal distribution

Table 1. Polymers with Narrow Molecular Weight Distributions

Polymer and initiator	Solvent	$\overline{M}_w/\overline{M}_n$	Refs.[a]
Polystyrene[b]			
sodium naphthalene	THF	1.22–1.62	3
sodium naphthalene	THF	1.1	4
sodium naphthalene	THF	1.04–1.19	5
sodium naphthalene	THF	1.61	36
sodium naphthalene	THF	1.04	6
sodium naphthalene	THF	1.06–1.09	7
sodium naphthalene	THF	1.07–1.56	8
sodium naphthalene	THF	1.04–1.39	22
tetra(α-methylstyrene) disodium	THF	1.05–1.09	37
α-phenylethylpotassium	THF	1.05	38
cumylpotassium	THF	1.1	39
sodium biphenyl	THF	1.09	40,41
C_2H_5Li or n-C_4H_9Li	benzene	1.05–1.12	42
n-C_4H_9Li	benzene[c]	1.03–1.09	43,44
n-C_4H_9Li	THF	1.09–1.41	45
n-C_4H_9Li	benzene/THF	1.1	46
sec-C_4H_9Li	THF	1.12, 1.17[d]	47
Polyvinyltoluene			
sodium naphthalene	THF	1.25	48
Poly-(α-methylstyrene)[b]			
sodium naphthalene	THF	1.0–1.03	49,50
tetra(α-methylstyrene) disodium	THF	1.05–1.06	51
n-C_4H_9Li	THF	1.05	52
n-C_4H_9Li + additives[e]	THF	1.00–1.02	53
Poly(p-tert-butylstyrene)			
sec-C_4H_9Li	benzene	1.04	54
Poly(methyl methacrylate)[b]			
n-butyllithium	pyridine/toluene (30/70 v/v)	<1.1	55
sodium biphenyl	THF	1.03–1.08	56
lithium biphenyl	dimethoxyethane	1.1	41
cumylcesium	THF	<1.1	57,58
1,4-Polybutadiene[b]			
n-C_4H_9Li	n-hexane	1.1	59
n-, *sec*-, and *tert*-C_4H_9Li	cyclohexane	1.10–1.20	60
n-C_2H_9Li	hexane	1.09–1.43	61,62
sec-C_4H_9Li	cyclohexane[f]	1.02–1.10	63
1,4-Polyisoprene[b]			
n-C_4H_9Li	n-hexane	1.05–1.18	64
n-C_4H_9Li	n-heptane	1.1	65
n-C_4H_9Li	cyclohexane	1.1	66
sec-C_4H_9Li	n-hexane	1.1	67
n-, *sec*-, and *tert*-C_4H_9	cyclohexane	1.13–1.35	60
sec-C_4H_9Li	cyclohexane	1.05	68
sec-C_4H_9Li	cyclohexane[f]	1.02–1.1	69
Polyacrylonitrile			
sodium triethyltriisopropoxyaluminate	dimethylformamide	1.1	70
Poly(tert-butyl crotonate)			
2-methylbutyllithium	THF	1.01	71
Poly(1-vinylpyrene)			
cumylpotassium	THF	1.14–1.3	72

Table 1. (*Continued*)

Polymer and initiator	Solvent	$\overline{M}_w/\overline{M}_n$	Refs.[a]
Poly(n-butyl isocyanate)			
n-C₄H₉Li	toluene	1.1	73,74
fluorenylsodium	toluene	1.1	73,74
Poly(propylene sulfide)			
sodium naphthalene	THF	<1.1	75–77
n-C₄H₉Li	THF	<1.1	78
Poly-β-propiolactone(α-methyl and α-n-propyl)			
tetrahexylammonium benzoate	THF	~1.1	79
Polydimethylsiloxane			
n-C₄H₉Li or LiOH	THF, o-xylene	<1.1	80
sec-C₄H₉Li	THF, o-xylene	1.02–1.27[g]	81
lithioacetal	benzene/THF	~1.15	82
Polytetrahydrofuran[b]			
trialkyl oxonium ions, alkyl trifluorosulfonates, acyl or alkyl halides/ silica salts	THF	~1.1	83–85
Poly[2-(4-vinylphenyl)ethoxy-trimethylsilane][h]			
lithium naphthalene	THF	<1.1	86
Poly(2-methylpentadiene)[i]			
sec-C₄H₉Li	cyclohexane	<1.1	87

[a] These references give syntheses and characterization.
[b] Commercial standards available from The Goodyear Tire and Rubber Co., Polymer Laboratories, Pressure Chemicals, and Toya Soda Manufacturing Co., Ltd.
[c] A small amount of THF or triethylamine was added to accelerate initiation.
[d] $\overline{M}_z/\overline{M}_w$ values for polystyrene of $\overline{M}_w = 43.7 \times 10^6$ and 27.3×10^6, respectively.
[e] LiBr, LiOH, and LiOC₄H₉.
[f] Polydienes prepared by the addition of amine promoters.
[g] Polydispersity increases with molecular weight.
[h] Acid hydrolysis yields poly[2-(4-vinylphenyl)ethanol].
[i] Hydrogenation yields atactic, in the Bernoullian sense, polypropylene (87).

forms from the mixture of mono- and difunctional chains (17–19). The presence of this type of termination explains the erroneous Mark-Houwink-Sakurada relation for polystyrene in Ref. 9.

Although chain transfer (qv) is known to occur in certain anionic polymerizations, there has been no extensive determination of transfer constants as in free-radical polymerizations. The solvent liquid ammonia acts as a transfer agent in the polymerization of styrene initiated by potassium amide (20). The solvent toluene ($pK_a = 35$) serves as a transfer agent in the sodium-initiated polymerization of butadiene (21). Both reactions are examples of metallation where a labile proton on the solvent is replaced by an alkali metal; the resultant species is capable of initiating chain growth. The role of toluene as a transfer agent in anionic polymerizations was studied for the sodium counterion (22,23) and for lithium (24). Toluene is used as a chain-transfer agent in the commercial preparation of low molecular weight (10^3–10^4) polybutadienes of various microstructures (25). As expected, the transfer rate is increased in the presence of polar solvents, such as diamines or ethers.

Spontaneous termination occurs in some anionic polymerizations, involving styrene or dienes, by the elimination of NaH or LiH (26–33). However, these reactions are minimized by hydrocarbon solvents or by moderate temperatures (30°C) and are thus not competitive with propagation. Either polar solvents or higher reaction temperatures (>60°C) increase the frequency of such chain-end deactivations.

Monomers rarely act as transfer agents in anionic systems; an example is the n-butyllithium-initiated polymerization of 9-vinylanthracene in THF (33–35). Here, the occurrence of transfer to monomer supposedly explained why oligomers having 4–12 units were obtained even at 100% conversion.

Table 1 lists polymers having a narrow distribution of chain lengths.

With the exception of oxonium ion-based polymerization of tetrahydrofuran, the polymers given in Table 1 are prepared by an anionic process. Carbocationic systems exhibit side reactions which, from a rate standpoint, are competitive with the propagation event. Thus monomers such as isobutylene are not suitable for the preparation of near-monodisperse chains. Such materials have not been prepared even when it was claimed that isobutylene can form so-called living polymers (qv) through the use of BCl_3 complexed with organic tertiary esters (88) (see also LIVING POLYMER SYSTEMS in the Supplement).

BIBLIOGRAPHY

"Monodisperse Polymers" in *EPST* 1st ed., Vol. 9, pp. 194–203, by H. K. Livingston, Wayne State University.

1. H. Dostal and M. Mark, *Z. Phys. Chem. Abt. B* **29,** 299 (1935); *Trans. Faraday Soc.* **32,** 54 (1936).
2. P. J. Flory, *J. Am. Chem. Soc.* **62,** 1561 (1940).
3. G. Meyerhof, *Z. Elektrochem.* **61,** 1245 (1957); *Z. Phys. Chem.* **23,** 100 (1960).
4. H. J. Cantow, *Makromol. Chem.* **30,** 169 (1959); G. Meyhoff and H. J. Cantow, *J. Polym. Sci.* **34,** 503 (1959).
5. H. W. McCormick, *J. Polym. Sci.* **36,** 341 (1959).
6. C. Stretch and G. Allen, *Polymer* **2,** 151 (1961).
7. F. Wenger, *Makromol. Chem.* **64,** 151 (1963).
8. M. Morton, R. Milkovich, D. McIntyre, and J. L. Bradley, *J. Polym. Sci. Part A* **1,** 443 (1963).
9. R. Waack, A. Rembaum, J. D. Coombes, and M. Szwarc, *J. Am. Chem. Soc.* **79,** 2026 (1957).
10. M. Magat, *J. Chem. Phys.* **47,** 841 (1950).
11. L. Gold, *J. Chem. Phys.* **28,** 91 (1958).
12. S. E. Bresler, A. A. Kovotkov, M. I. Moevitski, and I. Ya. Poddubnyi, *Rubber Chem. Technol.* **23,** 669 (1960).
13. B. D. Coleman and T. G. Fox, *J. Am. Chem. Soc.* **85,** 1241 (1963).
14. R. V. Figini, *Makromol. Chem.* **71,** 193 (1964).
15. G. V. Schulz, *Ber. der Bunsenges.* **78,** 1064 (1974).
16. A. Eisenberg and D. A. McQuarrie, *J. Polym. Sci. Polym. Chem. Ed.* **4,** 737 (1966).
17. R. V. Figini, *Makromol. Chem.* **44,** 497 (1961).
18. T. A. Orofino and F. Wenger, *J. Chem. Phys.* **35,** 532 (1961).
19. B. D. Coleman, F. Gornick, and G. Weiss, *J. Chem. Phys.* **39,** 3233 (1963).
20. W. C. E. Higginson and N. S. Wooding, *J. Chem. Soc.,* 760 (1952).
21. R. E. Robertson and L. Marion, *Can. J. Res. Sect. B* **26,** 657 (1948).
22. F. M. Bower and H. W. McCormick, *J. Polym. Sci. Part A* **1,** 1749 (1963).
23. B. W. Brooks, *Chem. Commun.,* 68 (1967).

24. A. Gatzke, *J. Polym. Sci. Polym. Chem. Ed.* **7**, 2281 (1969).
25. A. Luxton, *Rubber Rev.* **54**, 596 (1981).
26. M. Levy, M. Szwarc, S. Bywater, and D. J. Worsfold, *Polymer* **1**, 515 (1960).
27. G. Spach, M. Levy, and M. Szwarc, *J. Chem. Soc.,* 355 (1962).
28. T. A. Antkowiak, *Polym. Prepr. Am. Chem. Soc. Div. Polym. Chem.* **12**(2), 393 (1971).
29. W. Nentwig and H. Sinn, *Makromol. Chem. Rapid Commun.* **1**, 59 (1980).
30. B. J. Schmitt, *Makromol. Chem.* **156**, 243 (1972).
31. D. Margerison and V. A. Nysa, *J. Chem. Soc. C,* 3065 (1968).
32. M. D. Glasse, *Prog. Polym. Sci.* **9**, 133 (1983).
33. J. Comyn and M. D. Glasse, *J. Polym. Sci. Polym. Chem. Ed.* **21**, 209, 227 (1983).
34. A. Eisenberg and A. Rembaum, *J. Polym. Sci. Part B* **2**, 157 (1964).
35. R. H. Michal and W. P. Baker, *J. Polym. Sci. Part B* **2**, 163 (1964).
36. T. Lyssy, *Helv. Chim. Acta* **42**, 2245 (1959).
37. J. M. G. Cowie, D. J. Worsfold, and S. Bywater, *Trans. Faraday Soc.* **57**, 705 (1961).
38. S. P. S. Yen, *Makromol. Chem.* **81**, 152 (1965).
39. J. Herz, M. Hert, and C. Straszielle, *Makromol. Chem.* **160**, 213 (1960).
40. F. Wenger, *Makromol. Chem.* **36**, 200 (1960).
41. G. M. Guzman and A. Bello, *Makromol. Chem.* **107**, 46 (1967).
42. M. Morton, A. A. Rembaum, and J. L. Hall, *J. Polym. Sci. Part A* **1**, 461 (1963).
43. T. A. Altares, Jr., D. P. Wyman, and V. R. Allen, *J. Polym. Sci. Part A* **2**, 4533 (1964).
44. L. J. Fetters and M. Morton in W. J. Bailey, ed., *Macromolecular Syntheses,* John Wiley & Sons, Inc., New York, 1972, p. 77.
45. S. Onogi, T. Masuda, and K. Kitagawa, *Macromolecules* **3**, 117 (1970).
46. T. Masuda, Y. Ohta, and S. Onogi, *Macromolecules* **4**, 763 (1971).
47. D. McIntyre, L. J. Fetters, and E. Slagowski, *Science* **176**, 1041 (1972); *Macromolecules* **7**, 394 (1974).
48. F. M. Bower and H. W. McCormick, *J. Polym. Sci. Part A* **1**, 1749 (1963).
49. H. W. McCormick, *J. Polym. Sci.* **41**, 327 (1959).
50. A. F. Sirianni, D. J. Worsfold, and S. Bywater, *Trans. Faraday Soc.* **55**, 2124 (1959).
51. F. Wenger, *J. Am. Chem. Soc.* **82**, 4281 (1960); *Makromol. Chem.* **37**, 143 (1960).
52. Roestamjah, L. A. Wall, R. E. Florin, M. H. Aldrich, and L. J. Fetters, *J. Polym. Sci. Polym. Phys. Ed.* **13**, 1783 (1975).
53. T. Kato, K. Miyaso, I. Noda, T. Fujimoto, and M. Nagasawa, *Macromolecules* **3**, 777 (1970).
54. L. J. Fetters, E. Firer, and M. Dafauti, *Macromolecules* **10**, 1200 (1977).
55. B. C. Anderson, G. D. Andrews, P. Arthur, Jr., H. W. Jacobson, L. R. Melby, A. J. Playtis, and W. H. Sharkey, *Macromolecules* **14**, 1599 (1981).
56. A. Roig, J. E. Figueruelo, and E. Llano, *J. Polym. Sci. Part B* **3**, 171 (1965).
57. G. Lohr and G. V. Schulz, *Makromol. Chem.* **172**, 137 (1973).
58. A. H. E. Müeller, H. Höcker, and G. V. Schulz, *Macromolecules* **10**, 1086 (1977).
59. J. F. Meier, Ph.D. dissertation, University of Akron, Akron, Ohio, 1963.
60. H. L. Hsieh and O. F. McKinney, *J. Polym. Sci. Part A* **4**, 843 (1966).
61. H. E. Adams, K. Farhat, and B. L. Johnson, *Ind. Eng. Chem. Prod. Res. Dev.* **5**, 126 (1966).
62. B. L. Johnson, H. E. Adams, F. C. Weissert, and K. Farhat, *Proceedings of the 5th International Rubber Conference,* 1968, p. 29.
63. Z. Xu, N. Hadjichristidis, J. M. Carella, and L. J. Fetters, *Macromolecules* **16**, 925 (1983).
64. M. Morton, E. E. Bostick, and R. G. Clarke, *J. Polym. Sci. Part A* **1**, 475 (1963).
65. W. H. Beattie and C. Booth, *J. Appl. Polym. Sci.* **7**, 507 (1963).
66. N. Calderon and K. W. Scott, *J. Polym. Sci. Part A* **3**, 551 (1965).
67. N. Nemoto, M. Moriwaki, H. Odani, and M. Kurata, *Macromolecules* **4**, 215 (1971).
68. L. J. Fetters and M. Morton, *Macromolecules* **7**, 552 (1974).
69. J. W. Mays, N. Hadjichristidis, and L. J. Fetters, *Macromolecules* **17**, 2723 (1984).
70. R. Chiang, J. H. Rhodes, and A. R. Evans, *J. Polym. Sci. Polym. Chem. Ed.* **4**, 3089 (1966).
71. T. Kitano, T. Fujimoto, and M. Nagasawa, *Macromolecules* **7**, 719 (1974).
72. J. J. O'Malley, J. F. Yanus, and J. M. Pearson, *Macromolecules* **5**, 158 (1972).
73. L. J. Fetters and H. Yu, *Macromolecules* **4**, 384 (1971).
74. A. J. Bur and L. J. Fetters, *Macromolecules* **6**, 874 (1973).
75. S. Boileau, G. Champetier, and P. Sigwalt, *Makromol. Chem.* **69**, 180 (1963).

76. R. S. Nevin and E. M. Pearce, *J. Polym. Sci. Part B* **3,** 491 (1965).
77. P. Sigwalt, *Chim. Ind. Milan* **96,** 909 (1966).
78. M. Morton, R. F. Kammereck, and L. J. Fetters, *Macromolecules* **4,** 11 (1971); *Br. Polym. J.* **3,** 120 (1971).
79. J. Cornibert, R. H. Marchessault, A. E. Allegrezza, and R. W. Lenz, *Macromolecules* **6,** 676 (1973).
80. C. L. Lee, C. L. Frye, and O. K. Johannson, *Polym. Prepr. Am. Chem. Soc. Div. Polym. Chem.* **10,** 1361 (1969).
81. J. G. Zilliox, J. E. L. Roovers, and S. Bywater, *Macromolecules* **8,** 573 (1975).
82. P. M. Lefebvre, R. Jerome, and P. Teyssie, *Macromolecules* **10,** 871 (1977).
83. P. Dreyfuss and M. P. Dreyfuss, *Adv. Polym. Sci.* **4,** 528 (1967).
84. D. H. Richards, S. B. Kingston, and T. Souel, *Polymer* **19,** 68 (1978).
85. T. G. Crocher and R. E. Wetton, *Polymer* **17,** 205 (1976).
86. A. Hirao, K. Yamaguchi, K. Takenaka, K. Suzuki, S. Nakahama, and N. Yamazaki, *Makromol. Chem. Rapid Commun.* **3,** 941 (1982).
87. Z. Xu, J. Mays, X. Chen, N. Hadjichristidis, F. C. Schilling, H. E. Bair, D. S. Pearson, and L. J. Fetters, *Macromolecules* **18,** 2560 (1985).
88. R. Faust and J. P. Kennedy, *Polym. Bull.* **15,** 317 (1986).

LEWIS J. FETTERS
Exxon Research and Engineering Company

MONOMER

A monomer is defined as a compound consisting of molecules each of which can provide one or more constitutional units of a polymer (or oligomer) (1). For example, the monomer molecule styrene ($\bigcirc$—CH=CH$_2$) provides the constituent unit —CH—CH$_2$—. The monomer molecule CH$_2$=CH—CH=CH$_2$ can

provide the constitutional units

$$-CH_2CH=CHCH_2- \quad \text{or} \quad -CHCH_2-$$
$$\underset{\displaystyle CH=CH_2}{|}$$

The monomer molecule CH$_2$N$_2$ can provide the constitutional unit —CH$_2$—.

A monomer molecule of δ-valerolactone can provide the constitutional unit

$$-NH(CH_2)_4C-$$
$$\overset{\displaystyle O}{\overset{\displaystyle \|}{}}$$

By a condensation process, the monomer molecules hexamethylene diamine $H_2N(CH_2)_6NH_2$ and adipoyl chloride $ClCO(CH_2)_4COCl$ can lead to the regular polymer molecule $-\!\!\left[NH(CH_2)_6NHCO(CH_2)_4CO\right]_n$ which contains the constitutional units

$$-NH(CH_2)_6NH- \quad \text{and} \quad -\overset{\overset{\displaystyle O}{\|}}{C}(CH_2)_4\overset{\overset{\displaystyle O}{\|}}{C}-$$

A monomeric unit or "mer" is the largest constitutional unit contributed by a single monomer molecule in a polymerization process.

BIBLIOGRAPHY

1. "Basic Definitions of Terms Relating to Polymers," IUPAC, *Pure Appl. Chem.* **40,** 475 (1974).

MONOMER REACTIVITY RATIOS.   See COPOLYMERIZATION.

MOONEY VALUE.   See RUBBER COMPOUNDING AND PROCESSING.

MORPHOLOGY

The morphology of polymers is concerned with the shape, arrangement, and function of crystals alone or embedded in the solid. Polymers are either truly homogeneous amorphous or heterogeneous semicrystalline solids. Even single crystals obtained from dilute solution have a lower density than the ideal crystal lattice, thus inviting the notion of a finite amorphous component. Completely crystalline polymers are obtainable only from crystalline oligomers by polymerization in the solid state.

The emphasis on technical applications reduces the importance of the morphology of single crystals. Of technical importance are molded materials and highly oriented polymers which, because of the low density ρ of polymers, yield in one direction much higher specific elastic modulus E/ρ and strength σ_b/ρ than metals.

The morphology of polymeric solids differs from that of low molecular weight substances, such as metals, salts, and molecular crystals, ie, of solids made of small molecules. The macromolecule is a long thread of monomers held together in the chain direction by strong covalent forces. The van der Waals forces between adjacent uncharged macromolecules are at least 100 times smaller. Such an enormous anisotropy of the force field dominating the properties of the amorphous and crystalline components is in striking contrast to the almost isotropic force field in atomic or molecular solids. Hence the experience of the polymer chemist does not agree with that of the metallurgist. Although it is relatively easy to separate macromolecules held together by weak van der Waals forces, it is 100 times more difficult to break the macromolecular chain held together by the much stronger covalent forces. Hence the morphology of the separation plane and its bridging by taut tie molecules (TTM), crystalline bridges, or microfibrils is an indication of the nature of the structural elements of the sample and their mutual arrangement. The plastic deformation accompanying fracture modifies the morphology observed in the separation plane (1). At room temperature and a moderate draw rate, the break surfaces of the lamellar sample first transform in the microfibrillar structure. Even at liquid nitrogen temperature, the sorption of N_2 and O_2 at unit activity by the strained sample modifies the fracture surfaces so much that they cannot be used reliably for the determination of original polymer morphology.

Under special conditions, the completely amorphous material may partially crystallize and be transformed into the semicrystalline polymer. The coexistence of crystalline and amorphous components is observed in all synthetic polymers that are not obtained by polymerization in the solid state of crystalline monomers or cyclic oligomers. The amorphous state is often called the liquid state, although it may differ quite appreciably from a normal low molecular weight liquid. The dominance of the long, randomly coiled, linear macromolecules with the enormous anisotropy of their force field and the presence of crystals strongly affects the behavior of the chains in the amorphous component. Virtually all amorphous chain sections have one end (cilia) or both ends (folds, two different crystals connecting tie molecules) fixed in the crystal lattice. Hence the properties of the polymer chains in the transition region between the crystals and the amorphous layers are substantially different from those in truly amorphous material without any crystals or strongly sorbing solids.

In the absence of crystals, the material is fully amorphous like a true rubber or glass. These two states are separated by the glass-transition temperature T_g. Chain mobility is pronounced in the rubbery state but almost absent in the glassy state. This influences the time dependence of the eventual morphology, although the conformations of the macromolecules in both states are rather similar. In some polymer melts, the rate of crystallization at any temperature is so low compared to the normally applied cooling rate and diffusion of the heat to the surrounding that they can be quenched to the glassy amorphous state. Such a polymer crystallizes if it is annealed above T_g (2) or mixed with solvent (3) or if the chains are made parallel by drawing (4). To prevent crystallization in readily crystallizable polyethylene (PE), small droplets (dia below 1 μm) must be quenched to liquid nitrogen temperature (5–13). Search by electron microscopy (em) for precrystalline order did not reveal any crystals in such truly amorphous material (14–16). Investigation of poly(ethylene terephthalate) (PET) quenched in ice water detected 7.5-nm spherical nodulae (Fig. 1), which upon heating above T_g showed enough mobility to align themselves into the two-dimensional lamellarlike ribbons (17). Very likely, PET at room temperature contains some small, defective,

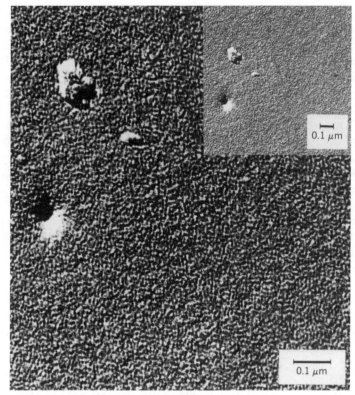

Fig. 1.   Electron micrograph of shadowed amorphous PET quenched in ice water showing balls (7.5–10 nm), which are interpreted as nodulae, crystal precursors, or defective crystals. Insert: minor magnification (17). Courtesy of Marcel Dekker, Inc.

and hence not easily detectable crystals, which are then seen as nodulae. Truly amorphous, rapidly cooled, linear polyethylene (LPE) below T_g does not show any nodulae (18).

The PE chain adapts very slowly to changes in the temperature of the melt (19–21). The relaxation time is nearly proportional to the third power of molecular weight, as expected for all parallel chain stems in the crystals (22), if the same consideration is applied to nodulae. Furthermore, up to 50 K above the melting point, the value of the thermal-expansion coefficient of the melt is larger than that at higher temperatures (23). This may indicate the existence of an additional phase that finally disappears at the temperature of the inflexion point of the density. A similar persistence of some order above the melting point to 175°C was observed by birefringence of melted cross-linked drawn PE (24).

Crystal Models.   The unexpected lamellar shape of polymer crystals with chains perpendicular to the lamellar surface observed (25) in solid gutta percha did not advance the study of polymer morphology. Rapid progress was made after the publication in 1957 (26) of chain folding on the lamellar surface of the synthetic LPE (Fig. 2).

Thin lamellae with long chains nearly perpendicular to their wide surfaces have been known for many years (27). At nearly the same time, bending was

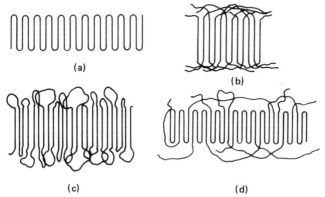

Fig. 2. Chain arrangement in a PE single crystal: (**a**) regular folds; (**b**) perfect switchboard model; (**c**) loose loops with adjacent reentry; and (**d**) mixture of (**a**), (**b**), and (**c**). The picture is not precise because the density of the amorphous layers is higher than that of the crystals.

suggested to account for the drawing process in fibers (28). In 1943 a perpendicular periodic striation of iodine-treated cellulose fibers was observed by em (28); the repeating distance L was less than 100 nm, which could not be explained (Fig. 3). Similar results were published in 1954 (30). The steric possibility of the folded

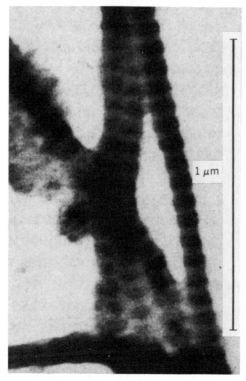

Fig. 3. Electron micograph of microfibrils of iodine-treated cellulose fiber showing the axial periodicity (29).

arrangement was known from the fact that cyclic paraffins (31) crystallize in the same orthorhombic lattice as linear paraffins. The regular fold contains about five CH_2 groups with increased conformational energy above that of all gauche conformations since the chains are not exactly in the minima of gauche conformations. The surface energy of such a fold-containing plane is thus higher than that of other crystallographic planes.

In the fringed micelle model (Fig. 4) (32), it is assumed that bundles of macromolecules form crystals as long as the parallelism of the chains in the bundle is not too disturbed by separation and inclusion of part of the chain in other crystals independent of the first one. Since there is no uniformity in the crystal length, this model cannot explain the existence of L in small-angle x-ray scattering (saxs). If each densely packed chain in the laterally extended crystal enters the amorphous layer after a single passage, the density of the quasi-amorphous phase with random-chain conformation is at least twice the density of the crystals with parallel chains. These two deviations from the experimental facts generally exclude the micelle model. In the extreme switchboard model, all the chains in the lamella pass to another crystal, whereas in the model of fibrillar material, TTMs connect the blocks and a substantial part of the rest of the chains folds back. The former model is too dense in the amorphous layers, but the latter is free of this deficiency.

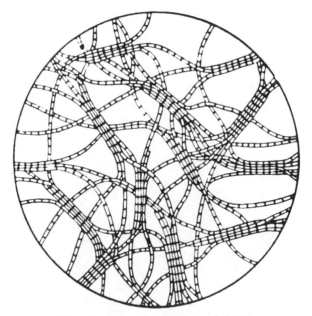

Fig. 4.   Fringed micelle model (32).

Composite Structure of the Semicrystalline Solid.   In semicrystalline solids, the polymer chains in the amorphous part are intimately connected with the crystals and cannot act independently. The crystalline and the amorphous components depend on each other in a way that is different from the situation in the nonpolymeric composites. The chain mobility is limited by the crystals or other solids where the chains are anchored. This causes a volume change during uniaxial extension or compression in total disagreement with the infinite rubber model in which the volume remains constant with deformation. The inflow of amorphous molecules from the sides tries to achieve constancy. In a polymeric semicrystalline solid the amorphous chains cannot move as far in the long and thin amorphous layers between the crystals. Consequently, the deformation of the semicrystalline solid polymer has to include in the elastic modulus of the amorphous component both moduli of rubber (33), the large bulk modulus K that considers the volume changes, and the small shear modulus G without any volume change.

The polymer solid is usually described as a two- or three-component system. In the former case, the crystalline and amorphous phases are treated as the two components. In the latter case, the amorphous phase is divided into two parts: the ideal amorphous and the interface layer that has chain mobility, density, and heat content between the ideal crystal and the ideal amorphous value.

The crystallinity is determined from the x-ray scattering as the ratio of the crystalline scattering and the total scattering of the sample (34). In practice, this tedious operation is rarely performed. The crystallinity is deduced from the density or fusion of the sample (see CRYSTALLINITY DETERMINATION).

The two-phase model gives a survey over the fraction of crystals in the sample by the ideal volume or mass crystallinity

$$\alpha_v = (\rho - \rho_a)/(\rho_c - \rho_a) \tag{1}$$
$$\alpha_m = (\rho_c/\rho)\alpha_v$$

where ρ, ρ_c, and ρ_a are the density of the sample, the ideal crystal lattice, and the ideal amorphous component, respectively. Actually, the crystals contain defects. The smaller the longer period L and the lateral extension of crystal, the more the folds laterally expand the unit cell (35–37). Both effects yield $\rho'_c < \rho_c$. Regular tight folds yield in PE a calculated density of 1 g/cm^3 which is identical with that of the ideal crystal. Hence single crystals with such folds have a crystallinity $\alpha_m = 1$. Density experiments usually yield a crystallinity $\alpha_m = 0.88$. In a two-component system, this excludes tight regular folding, but requires long cilia if the crystals have the ideal density or crystal defects that would reduce $\rho'_c < \rho_c = 1$ g/cm^3 (38–42).

The defective ordering of the chains in the crystal lattice may be studied in terms of the influence of the quadrupole moment of D = ^{2}H on an nmr spectrum that is limited to the immediate neighborhood of the ^{2}H nucleus (43–45). The method yields much better information if applied to oriented samples, eg, oriented single-crystal mats.

The chains in the amorphous component are partially aligned at the crystal surfaces, thus yielding $\rho'_a > \rho_a$. Both corrections change α_v and α_m into the actual α'_v and α'_m as shown in Figure 5. Since neither ρ'_c nor ρ'_a are usually known, ideal crystallinity is always quoted, which sometimes agrees with the actual case (47–49). Hence the separation of the polymer sample into ideal crystalline and amorphous components is in many respects more a mathematical maneuver than a realistic procedure. The situation is particularly critical in polymer single crystals, which have a lower density than ideal crystals, a great many defects in

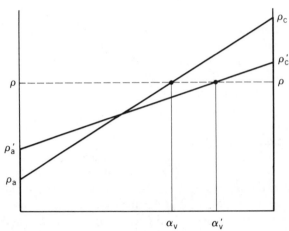

Fig. 5. Volume crystallinity of a two-component model: α_v with ideal density of both components and α_v' with $\rho_c' < \rho_c$ and $\rho_a' > \rho_a$, taking into account the depression of ρ_c by crystal defects and the increase of ρ_a by partial parallelism of chains in the amorphous component close to the crystal–amorphous interface (46). Courtesy of Marcel Dekker, Inc.

the crystal lattice, and two, much-disputed, amorphous surface layers where the loops are accommodated. The situation is similar in drawn or extruded samples when, as draw ratio increases, the separation of the amorphous and crystalline components gradually becomes less precise and finally disappears completely.

The same applies to the interpretation of the melting enthalpy h; if it is assumed that the amorphous component is already in the liquid state, $\Delta h_a = 0$. In this case, the mass crystallinity $\alpha_m = \Delta h/\Delta h_c$ (49) is the ratio of the measured melting enthalpy Δh of the sample and the ideal melting enthalpy of the crystals Δh_c. The situation is similar for the crystallinity deduced from density and by thermal methods. Any observed difference indicates the deviation from ideality of the crystalline or amorphous component (50). Up to $M_w = 850{,}000$, the crystallinity of quenched LPE samples derived from the density agrees approximately with that from the heat of fusion. Above this M_w, the ideal crystallinity derived from the density (0.5) is much larger than that from the heat of fusion (0.3). In this range, the change from spherulites to randomly oriented stacks of lamellae can be observed; this may eventually explain the observed differences (51–53).

The separation into two phases is partially supported by etching experiments. Fuming nitric acid and other aggressive chemicals, which usually destroy the sample by oxidizing and cutting the chains, first attack the accessible amorphous layers which are rapidly etched away (54–61). The crystals are more resistant and hence are removed more slowly than the amorphous layers. By measuring the time dependence of the weight loss of such samples, an increase in weight caused by the oxidation can be noticed. This first stage is followed very soon by a steady decrease. After some time, the rapid initial drop in weight attributed to the removal of the exposed amorphous areas changes to the slower rate corresponding to the attack on the crystalline core of the lamellae. More insight is obtained by determining the molecular weight distribution of the crystalline debris (55–57) that yields different locations of folds in the lamellar structure. Some are very deeply buried by the overlaying quasi-amorphous layers.

They therefore reduce the density of chains which must be accommodated in higher layers and thus make possible the gradual approach to true amorphicity.

In spite of the simplicity of the two-component system, care has to be taken, particularly in staining experiments, not to damage the sample (62). If the etching or staining molecules (or their precursors and products) are not able to diffuse into the narrow quasi-amorphous layers, the sample resists etching or staining of these layers nearly as strongly as if they were crystalline. Such a case seems to occur in single crystals with a suspected mosaic structure and in highly drawn or extruded samples. The more reliable em studies indicate that the latter have a fully developed microfibrillar structure, ie, thick, rugged amorphous layers between the crystal blocks in each microfibril, as well as extremely thin quasi-amorphous or partially crystalline layers between the microfibrils made from extended intermicrofibrillar TTM (63). The latter kind is so narrow or so well-ordered that it cannot be penetrated by staining. The amorphous layers, although much less permeable than in the undrawn sample (64), admit some staining. This can be seen in electron micrographs of drawn samples treated with uranium salts. The laterally adjacent crystal blocks of the microfibrils look like lamellae that are an artifact of the staining preparation of the sample for em.

A little more realistic than the two-component model is that with three components: crystalline, ideal amorphous, and a transition layer between these two. The density of the crystalline component may be reduced by crystal defects. The transition layer between the crystal and the ideal amorphous component may extend over the whole noncrystalline layer between the crystal cores. Additional information is needed to determine the extent of the transition layer if its density gradient is not known.

Crystallization. Chains crystallize if they are sufficiently regular and not too hampered in the liquid state by entanglements. Hence a normally polydisperse medium molecular weight ($<100,000$) highly regular PE without side chains crystallizes very easily. The crystals are of very special nature. The extension L in the chain direction is only a few nanometers, whereas the chain may be 1000-nm long (26). The thickness D of the crystal lattice is less than L, which also includes the thickness of the quasi-amorphous surface layer between consecutive crystals, $L = D + \ell$. Therefore, during crystallization the macromolecule can emerge from the surface by folding back and either adjacently or randomly reentering the growing face of the same crystal, or by proceeding to the growing face of another close-by crystal (65). In crystallization from dilute solution, there is no close-by crystal and the last possibility is not available. The resulting absence of tie molecules (TM) may substantially reduce the thickness $\ell/2$ of the amorphous layer of such isolated single crystals (66,67).

The chain ends (cilia), the folds, the connecting TM, and any physically adsorbed macromolecules form the amorphous layer between the crystalline cores of superimposed crystalline lamellae. As long as the stems in the crystal are perpendicular to the crystal–amorphous interface, the number N_a of chains per unit area of the perfectly amorphous component must be only $\rho_a/2\rho_c$ of the number N_c in the fully crystalline component (68). Hence the finite difference $N_c - N_a$ (proportional to $1 - \rho_a/2\rho_c > 0.5$) must fold back before the area of perfectly amorphous polymer is reached. The later this occurs, the longer the amorphous surface layer deviates from true amorphicity. Without any folding, as proposed in the extreme switchboard model without reentry (69–72), no realistic amorphous layer can be constructed since the local density is too high, much higher than in the crystals. This limitation (73) is less strict for a nonzero angle θ between the chain direction in the crystal lattice and the interface normal; it disappears completely when $\theta = \sin^{-1} (1 - \rho_a/2\rho_c) = 35°$.

Although the thickness of the crystalline core D and the quasi-amorphous layer ℓ are determined by the crystallization conditions, there are no limits to the lateral dimensions of the crystals. The result of this peculiar anisotropy of crystal dimensions is extremely thin, folded chain lamellae. In the solidification from melt, the lamellae are initially arranged in stacks of parallel lamellae. In a gel formed from dilute solution, the crystals or crystal stacks are widely separated and only occasionally connected to form a coherent network (Fig. 6).

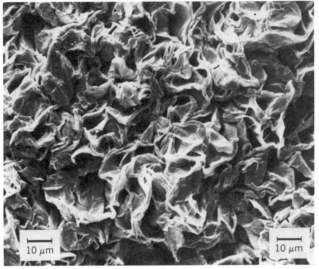

Fig. 6.   Electron micrograph of gel obtained from dilute solution of PE (74). Courtesy of *British Polymer Journal.*

Irregularly spaced side chains that also may be a different kind are partially or totally prevented from inclusion in the crystal lattice of the main chain (75). Thus only those sections of the main chain situated between two consecutive noncrystallizable side chains sufficiently close to or larger than D are included in the crystal. The excluded side chains, together with the short sections of the main chain, form the amorphous component. The ideal mass or volume amorphicity ($1 - \alpha_v$ or $1 - \alpha_m$, respectively) thus increases with branching. In some cases, the regular side chains crystallize, although the irregular main chain remains amorphous (76). This is particularly true in comb-shaped macromolecules with equal and sufficiently long regular side chains which crystallize even if, as a consequence of atacticity, the main chain does not.

If crystallization requires rearrangement of short but bulky regular side chains in identical positions, for instance in isotactic polystyrene (i-PS), it proceeds extremely slowly (77). Thus freshly quenched i-PS seems to be amorphous and only after a rather long time becomes semicrystalline. Atactic polystyrene does not crystallize at all.

The entanglements in the liquid state may hamper crystallization (70). However, the reptation model and simple sliding motion through the entanglements permit crystals to be formed even from entangled chains in the melt (78). With increasing temperature, the irregular extent of the sliding-through entanglements makes the average chain distance between consecutive entanglements a poor parameter for characterization of crystallizability. Since the number of entanglements per molecule increases linearly with molecular weight, crystallization becomes more and more difficult. Very high molecular weight material is so entangled that the crystallization is drastically reduced (79). The same applies to dissolution and melting where many clumps are persistently entangled and unaffected by the usual solvents or heat. Hence these materials cannot be processed in conventional ways. Large pieces may be formed by sintering (80–84), which coalesces the persistent agglomerates of the long macromolecules. Their density differs from that of crystals or amorphous material. If they are very large, they are even visible to the naked eye.

The mechanical, transport, and degradation properties of the polymeric solids depend to a large extent on the amorphous component. In any direction, both components alternate since the crystals are lamellae only a few nanometers thick. The crystals have an extremely large elastic modulus in the chain direction, ca 300 GPa (43.5×10^6 psi) in zigzag PE and 42 GPa (6×10^6 psi) in helical isotactic polypropylene (i-PP), and a much smaller one, ~5 GPa (725×10^3 psi) perpendicular to it. At room temperature, the elastic modulus E of the amorphous layers increases with crystallinity α_m (85) since the ideal amorphous state develops slowly with increasing distance from the crystal surface. For $\alpha_m = 0.7$, it is about 0.5 GPa (72,500 psi) in LPE. It is almost 10 times lower in branched PE with lower $\alpha_m < 0.5$ (86).

The amorphous component in polymers contains numerous TM connecting one crystal to another. If they are not taut, they do not substantially increase the resistance of the sample to deformation. In many respects, TTM act like crystalline bridges connecting the lamellae with the amorphous layer. Their presence substantially increases the small elastic modulus E of the amorphous layer. Hence the highly drawn or extruded fibers and films with a large fraction of TTM or crystalline bridges in the draw direction are axially much tougher and stronger than undrawn material with very few, if any, TTM. Since the TTM do not affect the perpendicular properties of drawn polymers, almost the same conditions exist in the transverse direction as in the undrawn material, although the substantial reduction in the thickness of the amorphous layers in this direction reduces their influence. However, since there are few lateral TTM, it is easy to separate microfibrils. The elastic lateral elongation soon gets replaced by the irreversible formation of longitudinal voids between microfibrils.

The crystalline bridges connecting the blocks in the microfibril make these blocks coherent for x-ray scattering (87). Interpretation of such a scattering coherence as crystal extension in the draw direction reveals, in drawn or extruded samples, needlelike crystals whose length increases almost linearly with the draw ratio (88,89). As soon as the crystalline bridges of TTM melt at the annealing temperature, each free TTM exerts a contracting force that may be measured or directly observed in shrinkage.

Experimental Methods

The static mechanical properties E and σ_b are usually measured in a tensile field. Both depend on the rate of force application and the measurement time, but also on the time the sample has relaxed after the last treatment (90). The microhardness measurements with an indenter (91) determine the dimensions of the depression a certain load has made on the surface of the specimen. The method is rapid and measures the plastic resistance and its anisotropy on the surface of the sample.

The mechanical modulus of wave propagation may be used to detect the mechanical anisotropy of the sample at the frequency of the wave. Based on the two-component model, it permits the calculation of the degree of crystallinity and the arrangement and orientation of the crystalline and amorphous components. At zero frequency, the static values are measured (92). At much higher frequency, ultrasonic waves are used since the sensors are not affected by unavoidable vibrations of the surroundings. Similar information is extracted from the measurement of the low frequency mechanical and high frequency dielectric tensor at different temperatures, although in both cases the responding elements are different making the comparison difficult (93,94). Before investigation of the sample, the surface layer is removed by radiation (95), ion bombardment in plasma (96–98), or chemicals, eg, fuming nitric acid (54–61). The amorphous regions are more affected (99); etching agents must be used with care.

Microscopy. *Optical microscopy (om)* is the simplest means of examination. Frequently, considerable morphological detail may be observed without elaborate specimen preparation. The polarizing om is used extensively for spherulites (100–104). The high birefringence of polymer crystals produces striking colored patterns. Upon melting, the birefringence usually disappears. However, some orientation and order may remain for a long time in the liquid state far above the melting point (23,24). The interference microscope uses a reference beam that is recombined with the beam passing through the sample. It may detect very small changes of a few nanometers in single crystals (105,106), which are also manifested in changes-of-color patterns (107). Phase contrast and dark-field illumination enhance the contrast in specimens of nearly the same refractive index as their surrounding areas. In om the lateral resolution is limited to approximately the wave length of the light used. Hence the crystal must extend laterally to cover 1 μm in order to be seen and measured (see also LIGHT MICROSCOPY).

Acoustic microscopy of polymers is based on local elastic modulus and density. The former cannot be directly assessed by other types of microscopy. In a scanning acoustical microscope with 395-MHz frequency, the intrinsic structure of isotactic polypropylene (i-PP) spherulites is visible growing on a glass slide at low supercooling (108).

Electron Microscopy (em). High resolution transmission em (tem) (109) has led to the most significant discoveries in polymer morphology. The resolution can be as low as a few nanometers; 2 nm usually suffices. Transmission em compliments electron diffraction (qv). Wide- and small-angle electron scattering (waes, saes) give information on the crystallographic and molecular scale directly related to x-ray studies. The diffraction spots may be used for dark-field em images which show bright areas that contribute to the chosen diffraction spots. The defocuscontrast technique permits the observation of crystals and amorphous component of a sample without staining that may affect the sample (110,111). Noise distortion may be interpreted as density or order fluctuation (112).

Denser crystals have a higher electron density and hence absorb the electron beam more readily than lower density amorphous components. This effect is enhanced by diffraction scattering and reversed by staining the accessible amorphous layer with OsO_4 (113), RuO_4 (114,115), permanganate (116,117), or uranyl acetate (62), which work satisfactorily in PE, i-PP, and other polyolefins. In the case of nylon-6 and nylon-6,6, a solution of 2% phosphotungstic acid and 2% benzyl alcohol in water (118) seems to give adequate contrast.

Since most polymer single crystals are small, em is ideally suited for their investigation. Its main disadvantage is the rapid degradation of polymer crystals in the intense electron beam needed for normal viewing. The damage may occur by chain scission (119) which shortens chain sections, eg, in polytetrafluoroethylene (PTFE) (120), or results in volatile monomers, eg, from polyoxymethylene (POM) (121,122); cross-linking prevails in PE (95,123–128), poly(4-methyl-1-pentene) (P4MP), isotactic PS, i-PP, polychlorotrifluoroethylene (PCTFE), and polyamides (129–132). As a consequence of the gradual destruction of the crystalline order, the diffraction-contrast distinction between lattice orientation in different parts of the crystal, the intensity of Moire patterns, and dislocation networks are reduced. This limits magnification for dark-field and diffraction-contrast electron microscopy (qv).

Since unevenness in the fold surfaces favors cross-linking (133), such effects occur frequently in pyramidal crystals collapsed on the substrate, but are less likely in pyramids with smooth surfaces. Hence the latter tolerate a dose up to 2400 C/m^2, whereas the former are destroyed at 100 C/m^2 (134). The more stable pyramidal crystals of PE transform irreversibly from the orthorhombic to the hexagonal structure at 800 C/m^2. Independent confirmation of this peculiar effect of increased resistivity to radiation has not been provided.

Increasing the voltage of the em from 100 kV to 1–2 MV limits the electron bombardment damage to ca 10–20% (135). However, the contrast is less. Channel-plate image intensifiers permit scanning of crystalline specimens without serious damage to the lattice (136). Replicating techniques eliminate degradation. Although the fine-structural information is not obtained, other characteristics can be gleaned. Moreover the metal or carbon replica yields an excellent sample for the saes technique (137).

Scanning em (sem) is widely used where the high magnification of the normal em is not needed and om is too low in resolution (138). The advantages of sem are large depth of focus and a simple specimen preparation of fibers, bulk polymers, and biological samples. Resolution, however, is limited to about 20.0 nm. This may introduce serious errors without the results of the higher resolution tem.

Scanning tem (stem) has advantages in the observation of diffraction-contrast features and diffraction patterns from radiation-sensitive crystalline polymers. More than one reflection results in increased dark-field image intensity and resolution and visibility of the detailed arrangement of lamellae (139).

"Decoration" with a thin coating of gold or other heavy atom reveals the roughness of the surface which provides nuclei for the crystallization of the impinging atoms or ions (140). The swelling of the decorated lamellar surfaces shows

a large mobility of the quasi-amorphous chains of the sample (141,142). If, instead of heavy atoms, a crystallizable polymer or oligomer is used, the crystal overgrowth shows the direction of the surface molecules of the test crystal or sample because the decorating molecules are forming crystals with a parallel chain axis (parallel epitaxial growth). The lateral extension of the decorating polymer crystals is perpendicular to the chain folds of the polymer crystal on which they grow by depositing chains (secondary and tertiary nucleation) epitaxially (143,144) in excellent agreement with single-crystal fracture experiments (145). In decorated PE crystals grown slowly from dilute solution, the parallelism of folds and deposition planes in each sector reveals a substantial fold regularity (regimes I and II of crystallization (172)).

Scattering.   *Wide-angle x-ray scattering (waxs)* was the first method of mass-crystallinity determination. The ratio of intensity of crystal scattering divided by the total scattering intensity yields the true mass crystallinity (34,146). Usually the ratio of one intense crystalline peak and amorphous background is measured and correlated with the truly measured x-ray crystallinity which serves as calibration (see SCATTERING; X-RAY DIFFRACTION).

Peak positions and intensities are used to determine unit cell parameters and complete crystal structures in the usual manner. The tangential intensity distribution of a scattering peak may give a view on the orientation of the crystals, and the radial width some information about the size of crystallites and the relative amounts of order and disorder in the samples. Most interpretations are based on the two-component system (crystalline and amorphous); the crystalline phase has thermal or first-order and paracrystalline or second-order defects (147,148). A survey of the core structure of single crystals without the folds and the amorphous surface layer can be obtained from the optical transform of the waxs Patterson pattern (149).

Small-angle x-ray scattering (saxs) yields information about the thickness L and distribution of the lamellae (150,151). With the help of synchrotron radiation, the exposure time is reduced by a factor of 10^4 (152–154). This permits the observation of rapid changes of chain arrangement in crystals during annealing and drawing. The limited number of properly ordered lamellae of constant thickness in a single stack seems to be responsible for the absence of higher orders (155,156). Higher orders (up to 5 in PE) are seen, however, in well-oriented, single-crystal mats precipitated from dilute solutions if all the crystallization has occurred at the same temperature (157). The limiting number of maxima is given by the paracrystallinity of D and ℓ layers. The best fit with experimental data of PE crystals grown from 0.1% solution in xylene at 80°C was obtained with $L = 12.5$ nm and $\ell = 1.2$ nm (158). The width of the transition layer may also be estimated from saxs (158–163). Other electron-density fluctuations can lead to similar patterns (147,148).

The deformation of the crystals and amorphous component, and their relaxation may be deduced from the dynamic saxs instrument (164–166) that, by repetition of the experiment, permits the investigation of the sample at chosen phases of the imposed strain or stress cycle. The combination with dynamic light scattering or birefringence permits the determination of the orientation and relaxation of the crystalline and amorphous components.

Small-angle neutron scattering (sans) permits evaluation of the size and shape of labelled macromolecules (167–171). In contrast to light scattering of solutions, extrapolation to infinite dilution is not required. Both molecular weight and R (gyration radius) have been calculated in concentrated solution (170) or more importantly, in concentrated blends of deuterated and nondeuterated components of the same polymer (168). It was demonstrated that the gyration radius R is not changed substantially by solidification. This observation affected the kinetic theory of crystallization. In contrast with former formulations (172), random folds and sequences averaging only three adjacent reentries are recognized (173) (see CRYSTALLIZATION KINETICS in the Supplement). The constancy of R upon solidification requires some folding, interrupted by longer folds or tie molecules (174,175). It further yields the solidification model of crystallization (176–181) where the sections of the chain are incorporated with the smallest translational motion at the face of the growing lamellae. According to this model, the molecular coil is pulled in the crystal in such a way that only the crystallizing sections of length D are straightened and included in the growing crystal without any concern for the rest of the macromolecule. With such minimum displacement, the molecule may participate in adjacent crystals. Since the growth is not uniform and not all adjacent lamellae advance at the same time and place, this implies that the number of connecting TM is smaller (182) than expected from the model of completely random reentry with the immediate availability of needed growth faces in all lamellae found later in the solid (65).

The replacement of hydrogen by deuterium changes the thermodynamic properties of the material much more than expected (182–185), and sans is no longer completely reliable. The decrease in the ideal melting temperature T_m° of the infinite lattice by 6 K after replacement of all PE protons with deuterons affects the nucleation and growth of crystals that both depend on supercooling $\Delta T = T_m^\circ - T$, where T is the temperature at which the phase change occurs. As a consequence, the molecules of one kind partially aggregate, yielding a much higher apparent molecular weight which may influence all subsequent evaluations. Quenched samples seem to be relatively free of this effect (186–188) (see also NEUTRON SCATTERING).

Small-angle visible light scattering (sals) measures the spatial correlation of the refractive indexes (189,190) and is useful for the investigation of spherulites in thin films. The larger the scattering pattern, the smaller the average spherulite. Even with isotropic spheres, a cloverleaf pattern is observed (191,192). A two-dimensional position-sensitive detector rapidly detects the intensity in the scattering pattern; it utilizes a vidicon detector and a microprocessor (193). Upon deformation, the symmetrical cloverleaf pattern is compressed in the draw direction and extended perpendicular to it. The spherulite film must be thin enough to avoid severe multiple-scattering effects.

The laser speckle interferometer uses the speckle pattern created by the random interference of laser light scattered from various parts of the object (194,195). The change in the speckle pattern contains information about changes occurring, eg, during drawing (196).

Spectroscopy. Spectroscopic methods (197–202) permit the determination of conformational statistics of chain elements in crystals and amorphous layers and in the transition layers. These methods can be used for an independent determination of crystallinity and amorphicity ratio which may differ from that obtained from the density or heat of fusion (47–50,52,53). Raman spectroscopy (qv) offers more lines than infrared (ir). Spectroscopy may solve or indicate a solution of special problems involving chain folds, adjacent reentry, chain neighbors, and so on. No spectroscopic evidence was obtained for adjacent reentry (203–205), although the fine structure of the partially deuterated PE seems to indicate sharp folding (206–208).

Raman spectroscopy is used routinely to observe the accordian-type longitudinal oscillation of the straight-chain sections in crystals (209–221). By assuming a higher axial elastic modulus, $E_c = 344$ GPa (50×10^6 psi), than the experimentally determined 290 GPa (42×10^6) (222), a length of the apparent oscillating rod is obtained which may be equated with the thickness L of the crystals, although the straight section is only D long. Actually, the length obtained is closer to D than to L (223). A completely independent oscillation of straight sections and hence the use of observed linear accordian model (LAM) spectra for an analysis of the distribution of such sections is only possible with lateral and longitudinal decoupling of oscillators. The former decoupling is a consequence of the large difference between the forces in and perpendicular to the chain direction. The latter decoupling, however, is only achieved by rapid damping of the longitudinal oscillation in the amorphous layers between the crystals as shown in Figure 7 (224,225).

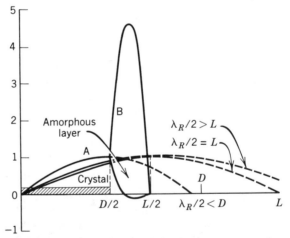

Fig. 7. Amplitude of first LAM oscillation in crystalline and amorphous layers if the wave in the amorphous component is (A) damped or (B) undamped. It is assumed that $E_c/E_a = 29$, $\rho_c/\rho_a = 1/0.853$, $L = 20.0$ nm, $D = 160$ nm, and damping length $\ell^* = 1.0$ nm. For comparison, the unrealistic oscillation of the whole distance $L = \lambda_R/2$ of the lamellae is also shown. The thus determined $E_R = \rho_c(c\lambda_R\nu')^2$ is too large in the last and the undamped case because in both cases λ_R is larger than 20 (224).

Experiments on highly oriented, drawn PE (226) have shown that any irregular change of trans to gauche conformation, ie, any crystal defect, acts almost as a free end of the oscillating rod (224). Thus the shape of the corresponding scattering or absorption line may give a rough survey of the length distribution of straight sections D in the sample. Nearly the same information may be obtained from the inelastic neutron spectra of lattice oscillations as from LAM oscillations (227–229).

Differential Scanning Calorimetry (dsc). This widely used method is rapid and simple, and yields Δh and the temperature peaks associated with the melting temperature of the lamellae depressed by the ratio of the surface free energy σ_e and the crystal thickness D.

$$T_m = T_m^\circ(1 - 2\sigma_e/\rho_c D\Delta h_c) = T_m^\circ(1 - A/D) \tag{2}$$

Partial recrystallization of the melt during the slow heating may hide endothermic effects of melting thinner crystals; superheating must be avoided.

Other Methods. The wettability and anisotropic thermal conductivity of single chains depend on the angle with the chain direction. Hence they may be used to determine the chain orientation on the surface and the interior of the polymer (230,231). In highly oriented samples, thermal conductivity may serve for the detection of TTM (232).

Thermal expansion at low temperature permits the detection of the semicrystalline bridges across the amorphous layers between the crystals (233–237). The method works well in highly oriented samples where the chains in the bridges and in the crystals have the same axial orientation. Since polymers of practical value do not change chain conformation at such a low temperature, the fraction of oriented chains can be measured. If these chains are in a zigzag, extended trans conformation, as in PE crystals and TTM, they yield a negative thermal-expansion coefficient in the direction the chains are predominantly oriented.

The most sensitive method for the study of the amorphous component in polymer solids is based on the transport of small molecules (238–240). The laterally compressed amorphous areas in highly drawn or extruded semicrystalline samples sorb less and more slowly transmit the penetrant molecules than completely relaxed amorphous material. Even in the undeformed semicrystalline solid, the change of crystallinity α_v influences the properties of the amorphous component, since the chains on the surface of the crystals behave differently from those further away from the boundary. On the crystal surfaces, the emerging amorphous chains are practically immobilized and hence do not permit any transport (241).

Crystals from Dilute Solution

Nucleation of Single Crystals. First, a nucleus is formed which grows and yields the lamellae. In the presence of a large number of primary nuclei, small crystals tend to develop (242). Fewer nuclei give large, uniform crystals. The origin of nuclei in crystallization from dilute solution is uncertain. Because crystallization may occur at much higher temperature than is theoretically considered possible by homogeneous nuclei, it must be concluded that invisible heterogeneous nuclei, most probably fragments of the original crystals or small dust particles, exist in solution above the so-called dissolution temperature.

In self-seeding (243–245), crystals are produced by cooling the solution to a temperature at which crystallization is terminated in a few minutes. The suspension is slowly heated (10 K/h), until the crystals dissolve, evidenced by the disappearance of turbidity. The solution, still containing a few folded-chain blocks nearly twice as thick as the dissolved lamellae, is cooled to the desired crystallization temperature. The crystals are formed simultaneously on the microscopic undissolved nuclei. Thus more accurate and meaningful data can be obtained from kinetic and structural viewpoints (Fig. 8) (246). This method was used for obtaining isothermally solution-grown crystals of LPE, which seemingly obeyed the same rules developed for crystallization from the melt (247).

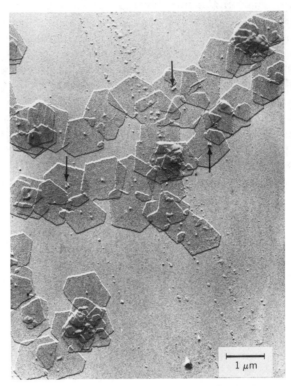

Fig. 8.   Self-seeded single crystals (246).

Single-crystal Growth.   Subsequent to primary nucleation, the crystal grows into a two-dimensional lamella by epitaxial deposition of chains (172). The crystal grows from the edge of the nucleus with the same structure of unit cell. The same applies to the heterogeneous nucleus, even if the unit cells of the nucleus and those of the deposited chains differ (248). The long-chain linear molecules adopt a folded configuration in the crystals.

Folding profoundly influences the developing geometry of the lamellar crystals. Well-defined, lateral growth-shapes reflect the symmetry of the unit cell. As indicated in Figure 9, the lateral growth faces in the lozengelike crystal of PE are {110} faces, whereas in the truncated type, the longer faces are {110} and the shorter faces are {100} (Fig. 10). In the case of the hexagonal unit cell of polyoxymethylene (POM), the simplest lamellar crystal is in the shape of a hexagon (Fig. 11).

Crystals of the type shown in Figures 9 and 10 must be grown under specific conditions. At low temperature of crystallization, true lozenges with only diagonal

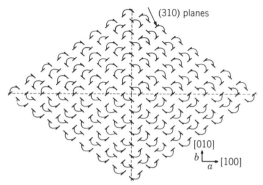

Fig. 9. Regular folds with adjacent reentry showing the "sectorization" of a single crystal of PE (249). Courtesy of *Journal of Applied Physics.*

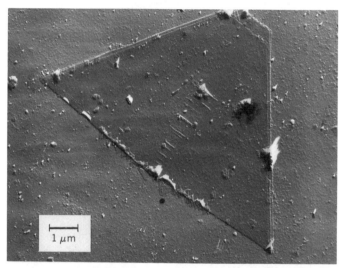

Fig. 10. Trapezoidal PE single crystal obtained from truncated pyramidal shape of lamella as formed after the pyramid broke along a diagonal (249). Courtesy of *Journal of Applied Physics.*

faces are formed. With increasing temperature, the {100} faces develop and at sufficiently high temperatures can even predominate (251). Increasing the concentration above 0.1% at any temperature always leads to an increase in complexity of the crystal habit and generally tends to enhance the truncation effect with predominance of {100} faces.

Excellent proof that folding occurs parallel to the outer-boundary planes of the single crystal is given by the appearance of microfibrils at certain angles of the fracture. No microfibrils are seen if the fracture plane is parallel to the outer boundary, whereas a maximum number is observed if the fracture is perpendicular to this direction (Fig. 12). In the former case, the fracture plane separates the chains, whereas in the latter case, both sides of the crack are bridged by a great many chains which are not easy to break. In the case of PE, the fold directions in each sector are parallel to the outer limit of the crystals.

Fig. 11. Hexagonal crystal of POM (250).

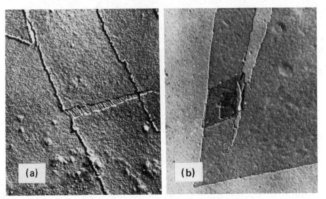

Fig. 12. Accidentally broken single crystals of PE showing cracks approximately parallel and perpendicular to the outer surface. In (**a**) the break plane cuts many chains which in response yield numerous microfibrils bridging the crack. In (**b**) no or too few of such chains cross the crack and microfibrils are not visible. The fracture stops at the sector boundary (145).

    This effect causes severe problems in the case of i-PP, which crystallizes in lath-shaped crystals. Many microfibrils cross a crack perpendicular to the long axis, but none are observed if the crack is parallel to it (252,253). It must therefore be concluded that the fold direction is parallel to the long and not to the short axis. A similar situation seems to occur with barium poly(L-glutamate) (254). The decoration technique with crystallizable aliphatic paraffins (143) was employed in a study of the actual growth pattern which erratically changes in the interior of the sample, with a preference of parallelity with the long axis of the lath.

    The space requirement of the regular folds accounts for the hollow pyramidal morphology of polymer crystals (249). The question of the regularity of the folds is still very much under debate. Fold planes exist and the faces of developing crystals are formed by depositing successive layers of folded-chain molecules along the fold planes. Depending on the number of growth faces in the crystal, a number of different orientations of fold planes are possible within the same crystal (Figs. 9 and 10). This "sectorization" affects its physical properties and divides the single crystal in four sectors in the case of PE (simple pyramidal crystal) and P4MP, but the truncated, single crystal of PE and the hexagonal crystal of POM into six sectors (122). Since the folds expand the crystal core (35–37), the planes containing the straight sections and the folds are more separated than the same planes without any folds. These crystallographic planes are therefore slightly bent, although the resolution of the em is not sufficiently high to show this effect directly.

Fold geometry is most important in crystal growth. In general, subsequent lamellar layers are added by a spiral-growth mechanism originating from a screw dislocation (255,256). This growth probably originates from a wedge-shaped edge of a lamella presenting a nucleus to which new material can be added (257). However, it is not a new secondary nucleation step, as might take place in the addition of initially separated layers in nonpolymeric materials, but it is directly akin to screw-dislocation growth in such substances as low molecular weight paraffins. Nevertheless, a significant difference is evident in that there need not be a lattice register between the layers in polymers, only contact between fold surfaces. In fact, this pertinent information on the nature of the fold surface can be obtained (258) by examination of regularity (splayed layers) or absence of rotation (contacting layers) of many spiral layers (Fig. 13).

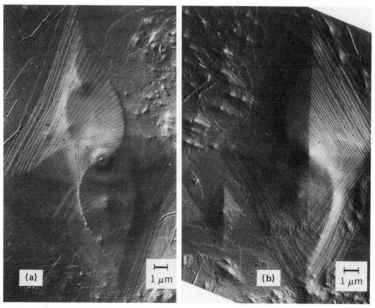

Fig. 13. Spiral growth in PE single crystals: (**a**) rotated and (**b**) unrotated lamellae (258). Courtesy of Steinkopff-Verlag Darmstadt.

Irregular Lateral-growth Habit.   In some cases, the lamella does not exhibit simple geometrical shapes. The periphery of the single crystal is rounded or serrated on such a fine scale that it is difficult to ascertain the occurrence of crystallographic microfaceting. The polymers from whose solutions single crystals are grown have irregular shape and may have similar energy levels of different lateral plans and folds that depend on concentration and temperature of the solution and on the molecular weight of the polymer. Modes of crystallization or folding are easily changed (259–263). With irregular chains, such effects are the consequence of low tacticity, eg, in polyacrylonitrile (PAN) (264–267), polychlorotrifluoroethylene (PCTFE) (268–270), and poly(vinyl chloride) (271); side chains in branched PE (272,273); random side groups in chlorinated, brominated, or sulfochlorinated PE (274–284); and random copolymers of ethylene with propylene or 1-butene (285,286). In view of the irregular nature of the polymer chain with random or atactic side groups or large atoms, the irregular lateral habit can be attributed to the inherent irregularities of the chain, which influences the packing in the crystal lattice and the folds on the surface of lamella. The same effect is also observed in the extremely slow crystallization of i-PS from a good solvent (287,288), whereas the much more rapid crystallization from a solution in a poor solvent yields simple hexagonal crystals (288).

The formation of curved polymer crystals from dilute solution has been reported for PE (289), nylon-6,6 (290), orthorhombic poly(1-butene) (291), poly(vinylidene fluoride) (PVF$_2$) (292), P4MP (293), POM (250), and PCTFE (270). The curvature increases with intensive supercooling, and finally a scroll (one-plane curvature) or a hollow bowl (two-dimensional curvature) is formed. The radius of such systems can be as small as a few μm (Fig. 14). The main reason

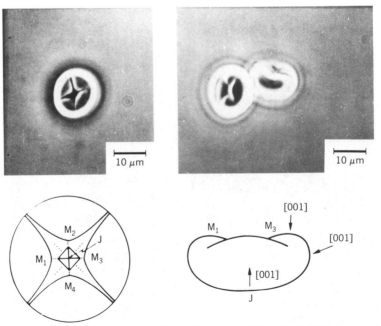

Fig. 14.   Spherical growth of single crystals of P4MP (293).

for the curved habit is the slightly conical sector shape which contributes to the overall curvature of the lamellae. The more multisectored the growth, the larger the successive increases in inclination at the periphery of the crystallizing layer. Very likely the stress caused by bulky folds also bends the single lamellae (294).

Single-crystal Mats. These are obtained by precipitation of single crystals from dilute solution (Fig. 15) (295). From highly dilute solutions, individual lamellae are formed with few connections. In such a case, the mat is extremely brittle with little lateral cohesion. Drying in a vacuum oven does not assure complete removal of solvent. This may be achieved by replacing the original solvent with more volatile liquids and keeping the system below the temperature of the solution point. If an oriented mat is required, a slight pressure is applied during filtering the solvent (296). Such a mat cannot be drawn, but it may be extruded yielding in PE the highest axial E and σ_b (297). Annealing removes much of the brittleness since the interpenetration of the crystalline stems increases their extension in the chain direction and produces enough lateral contact (295).

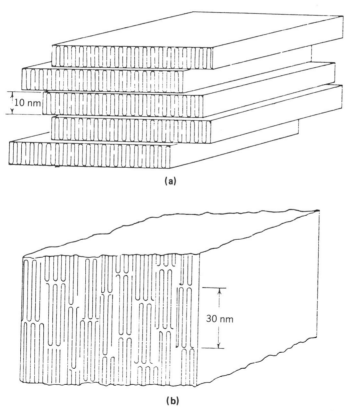

Fig. 15. Oriented single-crystal mats of PE (**a**) as grown and (**b**) after annealing (295). Courtesy of *Journal of Applied Physics*.

Lamellae. The thickness L of single lamella and lamellae-constituting spirals tends to be uniform at any particular crystallization temperature and solvent, of which it is a sensitive function. This is indicated by both em and saxs of oriented single-crystal mats where many orders of diffraction usually are obtained when the crystals are dried (157). The LAM-determined fold period L_R (with the high E_c) depends on crystallization temperature T for PE from dilute solution in various solvents (Fig. 16) (298). Any temperature changes during growth are reflected in the thickness D or L (242,296,299). Abrupt changes cause small steps to appear. The effect may be seen in different decoration density.

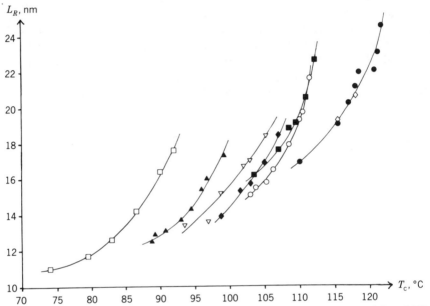

Fig. 16. Experimental values of LAM-determined saxs long period L_R of PE as a function of the crystallization temperature T in different solvents. □, Xylene; ▲, octane; ▽, dodecane; ◆, hexadecane; ○, ethyl esters; ■, hexyl acetate; ◇, tetradecanol; and ●, dodecanol (298). Courtesy of *Journal of Materials Science*.

Very soon after the discovery of the chain-folded lamellae, several studies on fractionated material were conducted (242,286,300–303). Generally, L is found to be nearly independent of molecular weight if, as in crystals from solution and melt, the folded-chain lamellae are favored and the extended chain length is sufficiently larger than L. Working with oligomers of urethane (304), nylon-6 (305), low molecular weight POE (301) and paraffins (306), respectively, it was shown that L, as determined by saxs, increases linearly with the degree of polymerization, ie, molecular length, up to a definite value and then remains constant. The intersection point of the two straight lines may be interpreted as the onset of folding.

If in a monodisperse oligomer of poly(ethylene oxide) (PEO) the temperature of crystallization and hence the supercooling is changed, an abrupt change is obtained in the saxs long period as soon as L is an integer divisor of the extended chain length (307). The superfluous cilium $\delta\ell < L(T)$ seems not to be included in the lattice (308,309)

$$\delta\ell = \ell_{\text{chain}} - nL(T) \tag{3}$$

with $n = 1, 2, 3$, etc. With increasing T (reduced supercooling), the L multiplier n plunges to $n - 1$. This abrupt change was directly observed on single crystals of monodisperse PEO with attached uncrystallizable PS segments (31). The transition from the less stable, once-folded chain crystal to the more stable, extended-chain crystal without any folds was described (311). As expected, the growth of the extended chain crystal, $n = 1$, is much slower than that of the folded one, $n > 1$, since it takes much longer to deposit the longer straight section. In the unfractionated polymer of PEO, the long period of saxs seems to be a linear function of the gyration radius (312).

In nylon polymers, hydrogen bonds are required in the growth plane. This overrides the long-period demands that are valid in uniform PE. The location of hydrogen bonds in planes containing the folds determines the permitted values L of lamellae. From saxs data on nylon-6,6 (313–315), regular folds are obtained from dilute solution, but a low crystallinity α_m has the maximum value 0.7 that depends on the assumed density of the amorphous material (316). The lower density material cannot be accommodated on the surface of the lamellae since the folds most likely are supposed to be tight and regular. Hence it may be a consequence of crystal defects. The simple two-component model with ideal density of the crystals and amorphous material cannot be applied.

Attachment of Heavy Atoms to Folds. Initially, Cl and Br atoms or ions do not enter the crystal structure (39,317). They react with amorphous chains up to the interface with the crystals, ie, with the loops and cilia on the surface of the crystals. Thus they make the lamellar surfaces impermeable. Since they are heavier than the replaced H atoms, they shift the LAM frequency to smaller values (318). Additional details are seen with ir, such as the formation of double bonds and the attachment of the second Cl or Br atom on the same monomer (319). At a later stage, the attacking atoms enter the crystals primarily from the side faces which are not chlorinated or brominated (320). The crystalline core is thus divided into the penetrated part and the part at a distance from the side faces. The attached atoms are too large to be included in the crystal lattice of an eventually recrystallized solid (39).

The mosaic structure of single crystals aroused a prolonged controversy. The crystals show a variation of the c-axis orientation because the Moire lines of two superimposed single crystals are not smooth, and the radial width of waxs and waes peaks, even after experimental corrections, remains finite. The minimum extension of blocks of parallel chains is ca 30 nm (321). The variation in the orientation leads to the concept of mosaic blocks (322–327), which are separated by disordered regions in such a way that the scattering of radiation from a single block is completely independent from the rest. However, some contend that the variation of the c-axis orientation and the observed striations (328,329) in the single crystal results from the nonuniform bending on the substrate (288,329–335). If the lamella is free, it shows no variation of the inclination of chains and has no domains with different orientation, which would show up in dark-field em as regions of different darkness, ie, a checkerboard appearance, in contrast with em observation of perfectly free crystals. However, in a pyramidal habit, the coherent (hk0) x-ray scattering persists only for as long as the normal to the chain can be traced inside the crystal. With $D = 15.0$ nm and the angle $\theta = 26°$, the shortest extension of the "mosaic" crystal $D/\tan\theta = 30.6$ nm, which nearly corresponds to the dimension claimed by the proponents of the mosaic blocks.

Computer simulation of PE single crystals was recently performed in order to obtain the elastic constants of the perfect crystal (336), the boundary properties of twins and stacking faults (337), and the structure of dislocations (338). Only short-range interactions between closely adjacent atoms were considered. The agreement with rather sparse experimental data may be called satisfactory for such attempts.

In polydisperse samples, crystallization results in chain fractionation. The longest molecules form the primary nucleus and crystallize first by secondary and tertiary nucleation (286,339). The same applies to row nuclei formed in strained solution (340). The effect may be partially derived from the theory of crystallization from dilute solution (341). The polymer still in solution has a lower average molecular weight than the formed crystal. In the case of PE, the fractionation occurs at a molecular weight corresponding to about three passages through the lamella (342). During crystallization three straight sections of the polymer chain may be simultaneously deposited on the growing face. If the uncrystallized remainder of the chain or the whole macromolecule is shorter, the deposition may occur reluctantly or not at all.

Pyramidal Morphology.   A number of observations lead to the conclusion (328,343) that PE crystals in suspension are not in a planar but a pyramidal form. This is applicable equally to other polymers crystallizing in the lamellar habit. In most cases, analogous features are observed. The most obvious and frequent morphological feature appearing on dry crystals is a central pleat and sometimes, as in the case of a truncated crystal broken in half across the b axis, a trapezoid-shaped structure (Fig. 10). Corrugations that can be related to the special type of pyramidal crystal often appear (Fig. 17) and do not disappear for up to several months after removal of the solvent. Possible modes of the collapse of pyramidal crystals with a tilt in the center of corrugated appearance are shown in Figure 18.

Fig. 17.   Corrugated lozenge of PE dried on substrate (344). Courtesy of *Philosophical Magazine.*

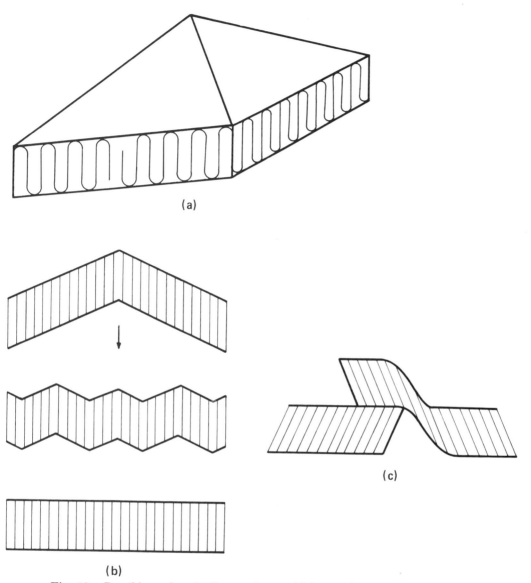

Fig. 18. Possible modes of collapse of pyramidal crystals: (**a**) as formed from dilute solution; (**b**) with plastic deformation only; and (**c**) with a break along a diagonal. The chains are vertical in case (**b**) and inclined to the substrate in (**c**) (345).

Some types of PE pyramids with {110} and {100} faces as seen by om are shown in Figure 19. The existence of the pyramid habit implies that the fold surface is regular and uniform. Such a uniformity could hardly be expected in the deposition of highly mobile macromolecules during crystallization. Hence it was suggested that during chain deposition there was, in fact, a considerable difference in the length of straight sections (347). The initially irregular surface was smoothed out by longitudinal diffusion after growth. Although this suggestion has led to some controversy as to the true nature of the fold surface and the

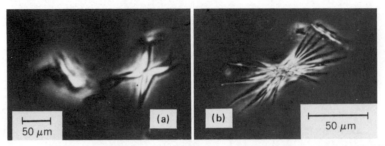

Fig. 19. Optical micrograph of (**a**) single and (**b**) multilayer pyramidal PE single crystals floating in suspension (346).

crystallization theories, it could explain the regular corrugation by a displacement of adjacent folds by one or two C—C distances in a positive or negative direction along the chain. In this case, a hollow pyramid or a chair-shaped crystal is formed, depending on the particular direction of the displacement.

Twinning.   Growth twins, in which two identical crystals grow with a specific mirror relationship to each other, are frequently found in polymer crystallization (348,349). In complexity, they are intermediate between dendrites and lozenges. They may play a very important role in the development of grosser morphological structures, but this point has not yet been investigated. Planar twinning has been mainly recorded in PE (189,350–354) in association with spiral growth of lamellae (Fig. 20) by deformation of the single crystals (355,356) or with growth-determined mirror planes (352,357,358). The nature of the twinning depends on the folds (352,359). Direct microscopic examination of the fine struc-

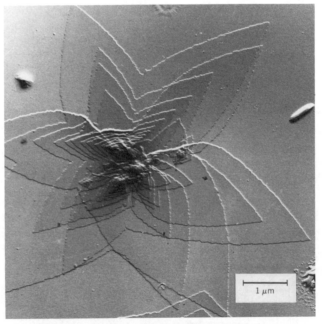

Fig. 20.   Rosette-twinned PE single crystals.

ture of the twin lamellae in PE (348) using diffraction contrast strongly suggests mirroring on the {110} planes. In addition, the mirroring may occur on different planes in the same crystal, thus producing multiple twins.

Twins of self-seeded, single crystals grown from dilute PEO solution have been investigated (360,361). Since pure PEO is difficult to use for growing single crystals for prolonged observation (362,363), monodisperse PEO molecules were attached to an uncrystallizable block of PS. Thus crystal growth may be better controlled and the resulting crystals are preserved and multilayer development is suppressed. However, large amounts of uncrystallizable PS heavily influence the crystallization and thickening of poly(ethylene oxide) (PEO) crystals. Twin boundaries may be observed by using replicas or dark-field em. All the multiple twins including 14 different crystal habits can be derived from a rational combination of two basic (hk0) twin modes, namely {100} and {120} twins, ruling out the higher energy {010} mode. The only observable twins are those in which the boundary plane is a mirror plane with respect to the monoclinic unit cell of PEO. This twinning rule is tentatively related to the helical conformation of the chain.

Dendrites are treelike crystals that resemble simpler single crystals with many reentrant growth faces on their main growth face (Fig. 21). They can develop many branches and often numerous spiral overgrowths. The formation of the serrated edge may be a type of secondary crystallization (296). It would take longer for the molecules to deposit evenly in one fold plane than to begin a new crystal face. Secondary and even tertiary branches are formed. Some of these branches overlap the primary lamellae. With optical microscopy, the three-dimensional nature of the structures can be seen to resemble the early stages of a spherulite (364). The overlap of planes where microsectorization or microserration has taken place can easily provide a step for nucleation of spiral growth. The higher the molecular weight of the polymer, the greater the chances of the crystals being dendritic.

Interlocked single crystals and crystal halves (308) may arise from interlocked, seeding nuclei that are mutually oriented by some interaction (Fig. 22). A purely mechanical fit must exist between the growth faces in close contact or

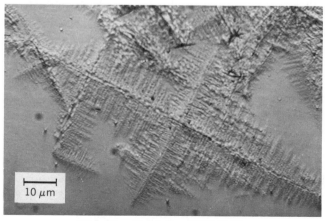

Fig. 21. Dendrite of PE (346).

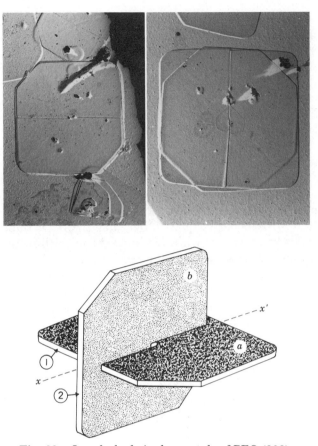

Fig. 22.   Interlocked single crystals of PEO (308).

some epitaxial interaction between the fold surfaces of the parent lamella. The split or abutting edge of the daughter crystal may play the decisive role in such an orientation.

A special case of twinning occurs in isotactic polypropylene (i-PP). It forms complex twinned dendrites called quadrites (365–367) consisting of a cross-hatched pattern of 12.5-nm thick rectangular lamellae with axes forming an angle of ~80° between branches (Fig. 23). The macromolecular stems are in the plane of the two branches at the same crystallographic angle of 80°40'. The c axis in the parent branch is oriented parallel to the a axis of the daughter branch and vice versa. This peculiar arrangement is caused by the fact that the α-unit cell of i-PP has nearly the same dimensions in the c direction as the γ-unit cell in the a direction (367). Hence on the lateral {010} face of the growing lathlike α-phase crystal, a small patch of γ-phase crystal is deposited epitaxially at ca 80°. A rudimentary branch forms with a long axis making a dihedral angle of 80°40' with that of the parent. Subsequent deposition of α-phase material on the branch continues until another patch of γ-phase forms on the growth front. This is a special type of epitaxial overgrowth ruled by the electric field of the substrate and sensed by depositing chains regardless of the direction of the backbone.

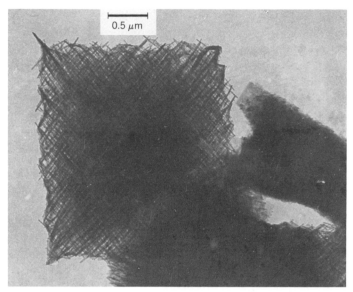

Fig. 23. Quadrite of isotactic polypropylene (365).

Multilayers. Multilayered crystals very likely result from spiral growth of single lamellae (Fig. 13). Crystals with such sharp fold surfaces would not be expected to exhibit any rotation; they should fit into a good lattice register. This has been demonstrated (368,369) by dislocation networks of low molecular weight fractions of PE (Fig. 24) on such double crystals. The mechanism (370), origins (371), and direct observation (372) of dislocation networks have been reported.

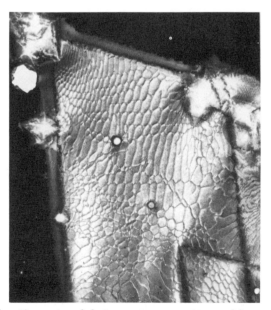

Fig. 24. Dislocation network between two superimposed low molecular weight PE single crystals (369).

Annealing. Dry, single-crystal annealing has been studied mainly by em and saxs, and density and birefringence measurements. Annealing experiments have been performed on isolated crystals on substrates or on sedimented cakes. The precise effect of the substrate is largely unexplored (373). Using the Moire technique (374), em experiments on bilayered single crystals on a carbon substrate show that the dislocation density increases by a factor of about 10^3 after annealing at 95°C; this may be due to chain sliding. For dry, single-crystal mats, saxs at RT does not indicate any change in L by annealing at 90°C (375).

Although annealing usually increases L, the occasionally observed drop during the annealing of single-crystal mats may be attributed to chain tilting resulting from a different foldpacking characteristic at that temperature. At high growth temperature, the single crystals have a more oblique pyramidal structure (289). Such changes in tilting do not involve a change in the length of the straight section of macromolecules in the crystal lattice but reduce the observed L.

The L of single-crystal mats increases with time t_A and temperature T_A of annealing (376); the macroscopic dimensions of the sample do not change (377). By observing the waxs, saxs, birefringence, and density data of single-crystal mats of PE, it was shown that the linear increase of L with the log t_A proceeds faster with higher T_A, provided T_A is above the temperature of the usual crystallization from dilute solution (378). In the beginning of the annealing, disordering takes place as seen from the reduction of birefringence and waxs, indicating a partial melting. Other authors (93,379) find a limiting L; extensive work was performed by other investigators (380–382). The general picture of $L(t_A)$ is given in Figure 25. The absolute increase of L is higher at higher T_A. The intensity of the saxs scattering first drops and, after passing through a minimum, remains constant or rises rather slowly. The occasional maximum of L as function of time seems to be a consequence of irreversible initial melting of the thinnest crystal

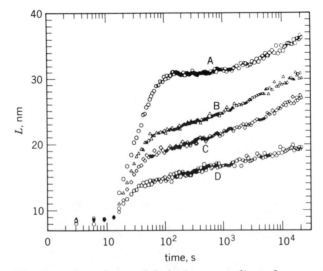

Fig. 25. The time dependence of L during annealing of a quenched PE single-crystal mat at A, 124°C; B, 120°C; C, 116°C; and D, 110°C (383). Courtesy of Steinkopff-Verlag Darmstadt.

L_{min} that yields an increased amorphous layer between the remaining thicker crystals and thus increases the local long period from the triplet L, L_{min}, L to the doublet $L + L_{min}/2, L + L_{min}/2$. Hence the local long period is higher than the average L at that time. If such contributions to the average L are high enough, they give rise to maximum L.

These experiments have led to the conclusion that thickening of the dry lamellae can take place by self-diffusion of the chains along their backbones by collective jumps, which yields the dependence of L on $\log t_A$ (22,377,384,385) or by melting of thinner crystals and recrystallization with a larger new fold period characteristic of the particular temperature (67,380,386–389). The melting effect yields a constant maximum L that is a function of T_A. No such limit exists for the sliding motion of the chains. This mechanism yields a high dependence of the annealing effects on the molecular weight of the polymer which reduces the slope of L vs $\log t_A$ by M_w^3. The resulting small slope at high molecular weight leads to an apparently finite limit of L_{max}. The best fit with experimental data on a single molecular weight (382) is obtained by assuming a simultaneous melting of the thinner crystals and the L growth by sliding motion of the chains (383) with a computer simulation of the lamellar thickening.

The L growth has been explained (390) by gradual doubling of crystal thickness, ie, by attaching one lamella to another and simultaneously destroying the surface layer. A more reasonable mechanism (391) for the supposedly final doubling of L of nylon-6 requires the unfolding of each second loop, which demands a doubling of the energy for sliding motion while the slope of L, $\log t_A$ becomes independent of molecular weight .

Dissolution. Dissolution by the same solvent occurs if the temperature is above the original formation temperature of the crystals. Dissolution is actually melting with the added effects of polymer–solvent interaction. The penetration by the solvent proceeds from the lateral surfaces of the crystal and some internal crystal defects (Fig. 26a). Holes are formed inside the crystal since the areas around the defects are dissolved. The undissolved parts of the original crystal serve as epitaxial nuclei for the recrystallization with larger thickness of dissolved macromolecules. Gradually the interior of the crystal disappears completely, leaving a picture-frame morphology (Fig. 26b). The increase in thickness during such an annealing experiment is a consequence of dissolution and epitaxial recrystallization (392).

The presence of a small amount of solvent (393), which is not sufficient for the dissolution of the whole sample at any annealing temperature, enhances the rapidity of the saxs long-period increase to a new constant value. The original crystals are gradually dissolved. The solute recrystallizes extremely rapidly and epitaxially on the remaining crystals, according to the new supercooling $\Delta T = T_s - T_A$. This is smaller than the original supercooling $T_s - T_p$ during which the crystals were precipitated from solution since T_A is higher than T_p, the temperature of crystallization; the equilibrium solubility temperature T_s is the same, neglecting the influence of concentration. At any temperature of annealing, T_s is much smaller than that of a dry single-crystal mat which, depending on thickness, melts below $T_m^\circ = 142°C$. Hence at any temperature the supercooling of the crystals in contact with the solvent is less and L is larger than for the dry samples. Since the epitaxial recrystallization is rapid, the new long period soon

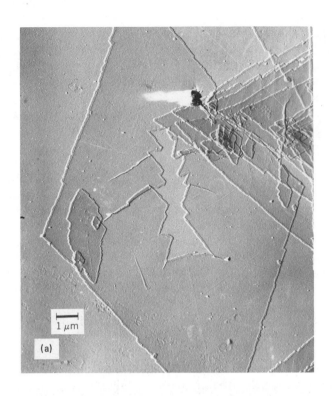

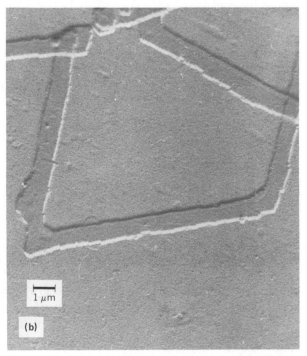

Fig. 26. Single crystals of PE grown at 90°C and subsequently annealed at 90°C in xylene: (**a**) intermediate state showing dissolution of the interior and recrystallized thicker edges and (**b**) final picture-frame morphology where only the thickened edges remain (392). Courtesy of *Journal of Applied Physics*.

reaches a fixed value that hardly changes with longer annealing time. In drying a single-crystal mat, it is difficult to remove all the solvent, which explains the wide fluctuation of the reported results in the annealing of supposedly dry crystal mats (93,379–383).

The behavior of branched PE depends on the frequency and character of the branches. Regular truncated lamellae have been grown from methyl- and ethyl-branched PE by a self-nucleating technique (275). Electron diffraction (394) confirmed the increase in the a-axis spacing and indicated a continuous change in fold length with annealing, supercooling, and solvent power, very much in the same manner as with folded-chain linear homopolymers. This implies that chain folding is not affected by short branches which can be included in the crystal lattice. Melting point and density (395–399) are considerably lowered by the introduction of branches, whereas the so-called amorphous fraction is increased (397,400–402). This increase is far more than would appear to be accommodated in a defective lattice. If, however, the side chains are large and bulky, they are rejected from the crystal lattice. If they are long and regular enough, they may crystallize by themselves, even if the main chain does not crystallize because the many excluded branch points make it too irregular (76).

Effect of Substrate. Nucleation of polymer crystals on solid surfaces affects orientation. In epitaxy, the electrical field of the unit cell of the substrate rules the mode of deposition. The common deposition of lamellae (kebab) on the row nucleus (shish) in shish kebab seems to be a consequence of parallel arrangement of the lattice of the substrate and of the overgrowth as demanded in normal crystallization. The same applies to the basic concept of crystallization by secondary and tertiary nucleation where any new layer of straight macromolecular sections is deposited on the already existing surface of the crystal. If the deposition is on the folds containing surfaces, two types of epitaxy are observed. In the more frequent fold-plane epitaxy (403–409), the chains are arranged parallel to the fold surface, whereas in fold-surface epitaxy (410–412), the chains are perpendicular to it.

Many homopolymers crystallize in thin needles with the longest dimension parallel to the {001} face of cleaved KCl crystal and, consequently, with the chain direction parallel to the ⟨110⟩ direction of the substrate (272,390,413–416). Hence the needles are lamellae-oriented perpendicular to the alkali halide surface.

A fold-plane epitaxy is observed in polymer decorated with short-chain molecules (143,144), where the crystallizing chains tend to be parallel to the folds of the decorated sample. In single crystals, the decoration chains are parallel to the fold planes and the decorating crystals, growing by additional deposition of short chains, are perpendicular to the folds on which they originate, demonstrating "sectorization." Their directional variation may be associated with the spread of local fold planes that, according to the currently prevailing crystallization theories, may deviate from absolute parallelism to the macroscopically observed growth face of the decorated lamella.

Crystals from Concentrated Solution and Melt

The most common mode of crystallization from melt and concentrated solution is the formation of spherulites (101,417–421). With very few primary nuclei, eg, at slight supercooling with complete exclusion of heterogeneous nuclei, the spherulites may grow to large dimensions before they impinge on each other. Extensive supercooling, quenching, or dense heterogeneous nucleation produce many primary nuclei. The spherulites are small and poorly developed. Only randomly oriented stacks of parallel lamellae can be observed (53). High molecular weight slows the crystallization since the melt viscosity increases with molecular weight to the 3.4th power. A monodisperse high molecular weight sample yields only randomly oriented stacks of lamellae of different thickness (422).

Depending on the heat removal from the sample, the temperature of crystallization may differ from that of the environment (423). Extensive supercooling requires small samples in which the heat removal is faster than the time to complete the crystallization. For easily crystallizable PE, small droplets of the melt are used with a diameter of a few μm (9–13) (423) in order to obtain a supercooling of nearly 70 K (T_{cryst} = 75°C) instead of the normally achieved ΔT < 30 K with T_{cryst} > 115°C.

Crystallization increases density and reduces the volume. In molds, the crystallization starts at the surfaces where the melt is in contact with the cooler parts. Crystallization may be normal although columnar crystallization is preferred adjacent to the cold surface. The interior crystallizes subsequently with an appreciable reduction in volume. As a result, the tensile stress in the center of the molded specimen and the compressive stress on the surface influence the mechanical properties (424).

Solid polymer crystallizes from the melt or concentrated solution in compact multilayer polyhedral crystals called hedrites or axialites, similar to interlocked single or oval crystals (Fig. 27). They seem to be related to the development of spherulites (425), although they are not spherulitic.

Fig. 27. PE hedrite (axialite) crystallized from concentrated (>1%) xylene solution (346).

Spherulites are also found in minerals and low molecular weight compounds grown from viscous impure melts. They seem to be formed as a consequence of certain conditions and are present in practically all polymers. Spherulites are easily identified under the polarizing om as fibrous arrays of ribbonlike crystallites emanating with spherical symmetry from a central point and increasing linearly in numbers with increasing distance. Spherulites are clearly observed (426) and identified by their maltese cross of birefringent extinction parallel and perpendicular to the direction of polarization (Fig. 28).

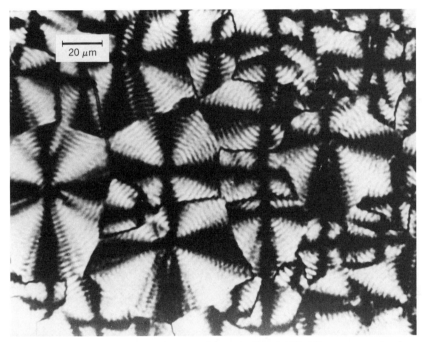

Fig. 28. Spherulitic structure of melt-crystallized PE. The concentric rings are a consequence of the helical twisting of single lamellae. Courtesy of F. Khoury.

Most ribbonlike lamellae solidify extremely slowly from the melt (Fig. 29) in a quasi-pyramidal shape (428–431). Kinking does not change the orientation and the length of the crystalline sector of the chains in the spherulite, but does influence the mechanical properties. At any strain, the crystals have to be deformed by irreversible chain sliding because the lamellae cannot change their arrangement by sliding past one another.

The optical pattern of spherulites can be represented in terms of the arrangements of crystals along the spherulite radius (100). If these crystals are aligned parallel to each other, their appearance is uniform and fibrous. Helically twisted lamellae have various extinction effects in the form of concentric rings (Fig. 28). In PE, the spacing of the rings decreases with increased supercooling (418). In contrast, high tension favors straight lamellae in shish kebabs of a mold (432). Occasionally, at very slow crystallization, the maltese cross is distorted into a zigzag line (433) which seems to be the consequence of large-ring spacing.

The fine structure observed by em fits into this model. The growth of the

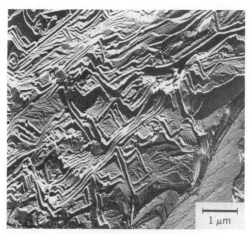

Fig. 29. Sharp bending of lamellae of PE solid crystallized at 130.4°C for 27 days indicating their pyramidal shape (427). Courtesy of Royal Society of Chemistry.

Fig. 30. The two-dimensionally grown spherulite of isotactic PS showing fibrous sheafs. Courtesy of F. P. Price.

spherulite emanates from the sheaflike arrangements of lamellae (Fig. 30). Partially noncrystalline branching of lamellae-stacks loosely fills the space of the spherulite with nearly radial branches. Such spherulites are porous and have large amorphous spaces. At the high temperature of crystallization, crystallinity is improved by repairing some defects and increasing the thickness. The impure crystals appear later, usually during cooling (434).

The existence of pure and impure crystals was detected by saxs or em (435) and is supported by dsc of spherulitic samples (240). Direct determination of their morphology is extremely difficult since it is not possible to distinguish between similar components in primary and secondary crystallization. The components can be partially removed by selective solvents revealing the existence of TM or TTM with overlying microfibrils (436). Such intercrystalline links are observed in PE crystallized from concentrated solutions in a paraffin that was later removed by dissolution in xylene. Similar structures have been observed on plastically deformed acetal homopolymers (437) and P4MP (438,439).

The ribbonlike growth of lamellae seems to be a consequence of the segregation of the impurities surrounding the lamellae, such as low molecular weight, atactic, or branched polymers. The separation proceeds much in the same way as in the zone-refining method of metals (440–443). If the diffusion constant D_i of the impurities in the melt is low enough compared with the growth rate G_t of the ribbon tips, impurity-rich layers form at the developing crystal faces and do not diffuse away quickly enough to prevent ribbonlike growth. The ratio $\delta = D_i/G_t$ with the dimension of length is a measure of the separation of layers, ie, the width of the developing radial ribbons. Since these ribbons can be expected to be highly temperature dependent, systematic variation of D_i and G_t yields spectacular changes in ribbon width over several orders of magnitude, ie, from ca 10 nm to several μm. A much coarser structure develops at slight supercooling in i-PP, for example, where D_i is larger and G_t smaller than in PE. This picture was recently modified and improved (444). It was questioned by some (445), who by varying δ by slow crystallization could not explain their data.

The measurement of the molecular weight of the formed crystals and the remaining solution or melt shows some fractionation during crystallization similar to that occurring in dilute solution (446).

The one-dimensional concept of paracrystallinity (148,322–325,347) in polymers applies to the saxs of a stack of parallel lamellae obtained by spherulitic growth or to the saxs of crystal blocks in microfibrils obtained by drawing or extrusion. The maximum number N of cooperating lamellae in such a scattering stack depends on the thickness fluctuation g of single-scattering units (447):

$$gN^{1/2} = a^* \approx 0.2 \qquad (4)$$

This empirical relationship yields a maximum $N = 4$ of cooperating lamellae, if their thickness of 20 nm on the average fluctuates by 2.0 nm, ie, $g = 0.1$. The same principle may be applied to waxs of a single polymer crystal where the average square root of the paracrystalline displacement increases with distance, whereas that of the thermal defects remains independent (448–450).

Relationship of Single Crystals Grown from Solution to Melt-grown Crystals. Folded-chain lamellae in the form of single crystals cannot be obtained from the melt in their original shape. At the high temperature of the melt, the thickness of lamellae grows extremely rapidly. In solution, the slow nucleation affects the growth time so much that the resulting lamellae are too thick and uneven. Only recently (247) was it possible to extrapolate data to the start of solidification, $t = 0$, and to prove by LAM that v' is equal in crystals obtained from dilute solution (298) and melt (234) at the same temperature and at the same supercooling. All previous extrapolations to $t = 0$ (451) turned out to be erroneous (231).

The results shown in Figure 31 confirm supercooling as the main parameter for crystal thickness. Since the thickness of the amorphous layer is of minor

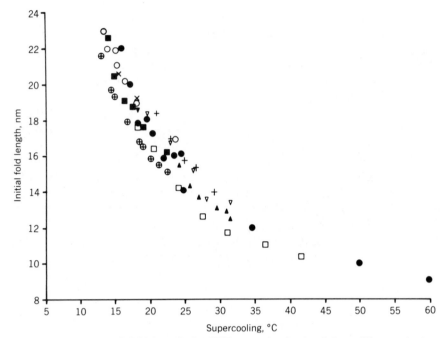

Fig. 31. The initial fold length L of PE crystals obtained from dilute solution and melt as a function of supercooling. The data are corrected for the different temperatures of crystallization. Solvents: $\square$, xylene; $\blacksquare$, hexyl acetate; $\oplus$, ethyl esters; $\bigcirc$, dodecanol; $\triangledown$, dodecane; $\blacktriangle$, octane; $\times$, tetradecanol; $+$, hexadecane; and $\bullet$, melt-crystallized (247). Courtesy of *Journal of Materials Science.*

importance as soon as it is sufficiently larger than the damping length ℓ^* (Fig. 7), the thickness D of the crystalline core is derived from ν'. The long period L may still differ since sans shows that, in contrast with adjacent reentry in isolated crystals from dilute solutions, a random reentry prevails in crystals from the melt which have numerous neighbors (66). This may influence the thickness $\ell/2$ of the surface layers as it is demonstrated in the annealing of single-crystal mats as shown below.

The time dependence of the crystallization from the melt is usually described by the Avrami equation (452), which is derived for the crystallization of low molecular weight material,

$$\alpha/\alpha_\infty = 1 - \exp\left(-kt^n\right) \qquad (5)$$

where α/α_∞ is the relative, α the actual, and α_∞ the maximum achievable crystallinity; n is the dimension of crystal nucleation and growth, k a crystal growth constant (Fig. 32). The equation was described even earlier (453) and later but independently (454). The equation is inadequate for polymers since there are multiple impingements of the spherulites (456) and changes in density of the crystals and melt (457,458). The formula best describes the initial volume growth of unimpinging spherulites which later differs appreciably from the growth of the crystal phase.

Solidification is sometimes described by two Avrami equations correspond-

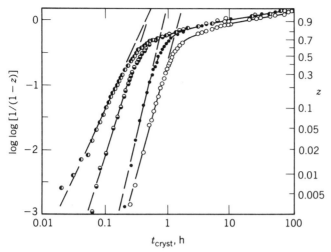

Fig. 32. Log log $(1 - z)$ with $z = \alpha/\alpha_\infty$ of solidifying four samples of PE at $T = 126°C$ over log t. For each sample, two straight lines are obtained as expected from two Avrami equations, one corresponding to the rapid primary crystallization and the other to the slow crystallization of the first rejected melt. The solidification seems to be completed in 100 h. Molecular weight = ●, 10^5; ○, 0.9×10^5; ◑, 1.25×10^5; and ◐, 1.25×10^5 (455). Courtesy of *Journal of Applied Physics*.

ing to primary crystallization until the radial growth of spherulites is completed, and secondary crystallization involving the solidification of the first rejected remainder of the sample having different density (459); the constant k can be a function of time (460).

Consideration of nucleation and crystal growth gives equation 6 which includes the time dependence of relative crystallinity

$$\alpha/\alpha_\infty = 1 - (C_0 + 1)/[C_0 + \exp(C_1 t)] \qquad (6)$$

The initial nucleation rate $k_0 = C_1/(1 + C_0)$ and the relative growth rate $k_0 C_0$, derived from the long linear range if $[(1 - \alpha)/\alpha_\infty]^{-1} d(\alpha/\alpha_\infty)/dt$ is plotted over α/α_∞, are obtained from the parameters C_0 and C_1 (461–463). At the end of crystallization, the impurities with a different density change and crystallization temperature contribute most of the density increase. Hence they are not considered in equations 5 and 6.

Quenching of the melt results in heterothermal nucleation and the parameters of the Avrami equation have to be changed. This also applies to initially isothermal solidification which, during subsequent cooling to room temperature, completes the epitaxial crystallization on the existing crystals. The corrected terms for LPE below 124°C (464) are not constant; in i-PP they are constant from 110 to 136°C (465). The exponent n in equation 5 is between 2.8 and 3.3, which shows a gradual change of crystallization mode with time. Some similar, more fundamental, although less handy, corrections may be applied (466,467).

Extended-chain Crystals.   Under high pressure, the polymers, especially PE, may crystallize from the melt in the form of so-called extended-chain crystals whose average thickness of the lamellae of 150 nm is much larger than the usual L of 20 nm and comparable to the extended macromolecule length (468–471).

Such crystals usually are not made of fully extended chains, but contain folds which are less numerous the thicker the crystals and the lower the molecular weight. The name "extended-chain crystals" is misleading. In polytetrafluoroethylene, crystal thickness is 3000 nm (468). A fracture surface of LPE crystallized at 500 kPa (5 atm) is shown in Figure 33. Even at 101 kPa (1 atm), PE crystallized slowly from the melt exhibits extended-chain crystals up to 12,000 mol wt (472). Polydisperse polymers give evidence of fractionation. The lower molecular weight component crystallizes in extended-chain form and the higher molecular weight component in the folded-chain form. Crystallization of PE under high pressure and temperature favors a phase transition (473); nearly all crystals are of the extended-chain type but with different thicknesses.

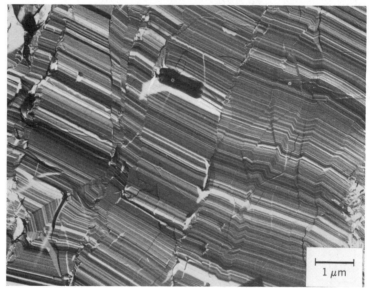

1 µm

Fig. 33.   Fracture surface of material containing extended-chain crystals of PE (428). Courtesy of D. C. Bassett.

A hexagonal lattice in PE samples (Fig. 34) crystallizes under high pressure and elevated temperature as shown by waxes (470,473,474). Depending on molecular weight, the phase boundaries are a function of applied pressure and temperature (Fig. 35). The new phase is characterized by extremely weak lateral forces, which explains the high L and its fast growth (473–476). Under normal pressure the hexagonal lattice transforms easily in the orthorhombic lattice. In uniaxial drawing the deformation of such crystals by chain slip does not go to the end of long straight sections and hence prevents the transformation into microfibrils.

Special instruments were built for the investigation of PE and other polymeric material (477): high pressure isotherms (478), phase diagrams (479), annealing (480), waxes (481), ir (482), Raman (483), Raman and Brillouin (484), and thermoluminescence after irradiation (485,486). No LAM frequency could be de-

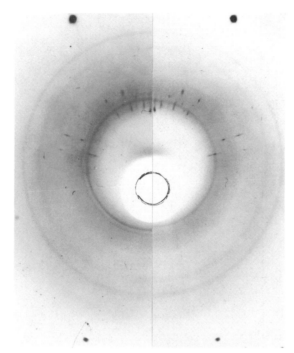

Fig. 34.   Waxs of extended-chain phase of PE showing the hexagonal unit cell (474). Courtesy of Butterworth & Co. (Publishers) Ltd.

tected (487) in spite of former claims (488). The axial E is quite conventional (489), and the fracture is brittle at moderate molecular weight.

A metastable monoclinic phase of PE in extended-chain crystals is found when samples crystallized at very slight supercooling are subsequently cooled to room temperature (468,480). However, a similar stress-induced monoclinic phase of PE, which disappears upon heating, shows up in deformed bulk material (490–492), spherulites (493), and single crystals (494).

Melting (51) is usually measured by dsc which may show many endothermic peaks. The thinnest crystals melt first. This melt crystallizes epitaxially on thicker crystals not yet melted. The exothermic recrystallization peak may be hidden by the superimposed endothermic crystallization since the difference of both effects is observed. This depends on the competition of the rate of heating with that of melting and recrystallization. The large number of primary nuclei, ie, of unmelted crystals, speeds up the crystallization (495). Very rapid increase of temperature delays melting, which is strongly influenced by superheating, and reduces the crystal thickness growth. In some cases, however, the peaks in dsc indeed yield the same crystal thickness distribution as saxs (496).

Annealing behavior of crystals obtained from the melt (111,497) is similar to that of single-crystal mats (377–381), and the same theoretical treatment is applicable. The shortening of the time needed for saxs by the introduction of synchrotron radiation facilitates application of this technique (498). Since the growth of L by sliding chain motion depends on the temperature and not on

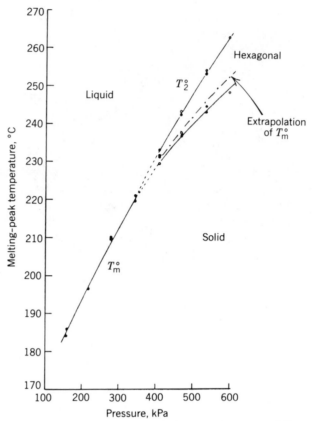

Fig. 35. Phase diagram T_m, P of PE solid (417). To convert kPa to atm, divide by 100. Courtesy of D. C. Bassett.

supercooling, it plays an important role in the solidification from the melt at high temperature (497) and a lesser role in the crystallization from solution at lower temperature.

Branched polymers crystallize from concentrated solution and from the melt in the same manner as from dilute solution. Short branches may be included in the lattice by expansion of the unit cell (499,500). With larger side groups most of the branching points are in the noncrystalline layers on the lamella surfaces (501–504).

Increased frequency of branching results in a narrow molecular weight domain where spherulitic growth occurs. As shown by sals and melting in dsc, the remaining temperatures and molecular weight yield only randomly oriented stacks of lamellae indicating the beginning of spherulitic growth, which soon stops as a consequence of numerous excluded branching points (504–506).

Cross-linked Polymers. Since the cross-links tend to be excluded from the crystal lattice, a sufficient length of segments between consecutive cross-links is needed for crystallization. As a consequence of the numerous excluded cross-links, the lamellae are thinner than in a normal LPE, full of defects, and with less contrast between the crystals and the quasi-amorphous layers. The spherulites are smaller, about 20–30 μm, as in the case of highly branched material (507,508). A grosser structure, found earlier, is an artifact caused by crystallization of shorter sections of the cross-linked material obtained by oxidizing the solid (62,117,509).

Lamella Structure

Simple spatial considerations require that each macromolecule occupy a larger volume in the amorphous, rather than in the crystalline region. If the chain in the amorphous region has no preference for one direction, ie, if its orientation is completely randomized, the number of amorphous chains per cross-section is halved. This applies equally to macromolecules crystallizing from dilute solution or from the melt (68). Thus with the empirical knowledge that chain folding occurs, two models for the individual crystals from the melt were proposed. In one, the folds are sharp (510) and the macromolecule reenters the lattice adjacent to the emerging chain, and in the other, the folds are irregular (68) and any reentrance is at random (switchboard model). No experimental data are in complete agreement with either theory. The switchboard model, without some nearby reentry in the same crystal, yields too high a density of the amorphous layers where the chains are perpendicular to the surface (511); the adjacent reentry model without the tie molecules and cilia does not yield a thick-enough amorphous layer (512–516).

Sharp-fold Surface.   Much of the evidence for the sharp-folded model comes from density and direct em observation of individual crystals grown from dilute solution. Thus the sectorization, nonplanar pyramids, and truncated habits seen in Figure 10 could all be explained quantitatively by assuming that the C—C distance along the main chain and the sharp folds obtained by regularization after chain deposition were staggered by half (339). Examination of internal sector boundaries (517,518) and direct observation of the slightly different lattice spacing in each section (319), due to the presence or absence of folds in crystallographically equivalent planes, can be accounted for by envisaging the folds to be part of the lattice itself, ie, sharp.

The most convincing evidence for an ordered fold surface is the observation of dislocation networks between bilayered crystals (368–372), which can be accounted for only on the basis of a specific interaction between the lattices of both crystals. Such an interaction could not be expected with disordered surface layers. It may be the result of the folds fitting together crystallographically. Such an order may also be established by some cilia of one crystal penetrating the other to establish the relative orientation needed for the undisturbed area of the dislocation network (370,371).

Further support for sharp-fold surfaces is found in epitaxial overgrowth of crystals of one polymer on themselves, on crystals of polymers with similar lattices, and on nonpolymers. All overgrowth crystals have the same orientation, implying at least some form of lattice matching through a fold surface with a crystallographic regularity similar to the body lattice (402,431).

Disordered-fold Surfaces.   Evidence for fold surfaces with various degrees of disorder has been obtained by a number of methods. Decoration of fold surfaces with gold particles measures the nucleation density of the surface and strongly suggests a model with some sharp folds interrupted by cilia and some loose loops and irregular reentries. The idea of adsorption of macromolecules on the surface may be supported by some fracture experiments on POM single crystals (519), where a missing narrow ribbon on the surface of one side is found on the other side of the crack, as if during fracture it were taken away by the side with better adherence.

The best evidence for a disordered-fold surface is obtained from PE samples treated with fuming nitric acid, which oxidizes and degrades the chains and produces low molecular weight water-soluble debris (45–62). Evidence from such experiments strongly suggests an amorphous surface layer with different loop depth. The crystalline interior consists of straight sections with some persisting folds (58).

The thickness of the amorphous area may be obtained from the sample density after annealing (Fig. 36)

$$\rho = \rho_c - (\rho_c - \rho_a)\ell/L \qquad (7)$$

yielding a straight line plotted vs $1/L$ (378). The weight defect per unit area of the lamella $(\rho_c - \rho_a)\ell$ seems to increase with higher T_A but is completely independent of lamella thickness L. In single-crystal mats, the initial position of ρ is higher with $1/L$, which on annealing at T_A drops below the straight line corresponding to T_A; after a short while, it moves according to equation 7. The mat first adjusts to annealing by partial melting and then recrystallizes with a constant $(\rho_c - \rho_a)\ell$ weight defect which does not change with t_A. This observation

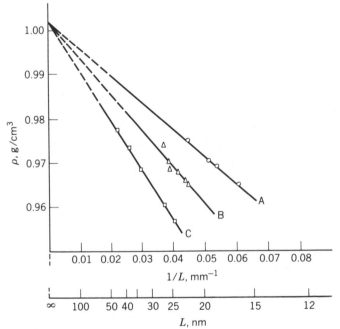

Fig. 36. Density of annealed single-crystal mats over the inverse long period: $\rho_c = 1.002$ g/cm³, $\rho = \rho_c - (\rho_c - \rho_a)\ell/L = \rho_c - A/L$ at A, 120°C; B, 125°C; and C, 130°C. Drawn according to Ref. 378 and corrected.

leads to the conclusion that the surface area has less weight defect in crystals obtained from dilute solution than in annealed single-crystal mats and solidified melts. A similar conclusion follows from sans (66), since in dilute solution the chains have to reenter the same crystal, while in solidification from the concentrated solution or melt and in annealing they may also enter adjacent crystals (73).

The density defects can also be detected through the presence of an amorphous halo in waxs. In principle, the existence of an amorphous phase on the surface of the crystals and the defects within the crystal lattice cannot be distinguished. Strong evidence has been presented for the existence of an amorphous layer (490). However, alternatives such as an "intermediate state of order" (520,521) and a three-dimensional defect network (522) generated from chain ends in the lattice have been suggested.

Theoretical Aspects. Before 1957 there was little need to take account of detailed morphology in describing crystallization, since the fringed-micelle concept of polymer structure did not require it. With the discovery of folded-chain crystals from solution and their identification in melt-crystallized polymers, it became necessary to consider their mode of formation. In particular, it was imperative to explain the dependence of the saxs long period L on crystallization and annealing temperature. Actually, any theory of crystallization is necessarily based on the thickness D and not on the spacing L of the crystals. Hence in the following discussion L is replaced by D.

Different theories of polymer crystallization have been developed: one is based on equilibrium consideration of chains in an ideal infinite lattice and the other on kinetic consideration of chain deposition. According to the thermodynamic equilibrium theory (523–527), the straight sections of the chains in the lattice assume a length determined by the minimum of total free-energy density. This theory does not deal with nucleation and crystal growth rate. The metastable minimum of not fully extended chains is highly temperature dependent. In PE, it exists only up to 110°C. The thickness increase by longitudinal chain motion by annealing below ca 110°C and the dependence of D on supercooling seem to be hardly compatible with the thermodynamic theory in the present formulation.

In a different equilibrium approach (528), it was assumed that the amorphous chains in the melt or solution form a bundle, ie, precursors of the crystalline state. The bundles cannot be discovered if they are formed only in the immediate vicinity of the growing face. In the true melt or in solutions, no such bundles were detected.

The kinetic theories (529–533) specifically require reentrant folding, assuming that thin lamellae must be formed. Nucleation theory is used to calculate the rate of formation of new crystals and their subsequent growth by additional deposition of ribbons of regularly folded molecules in a repeating sequence on the growing face of the crystal. The thickness of the secondary and tertiary nuclei on growing crystal faces is about one-half the extension of the primary nucleus. In some crystals, morphological evidence of this is found; the small bumps in the center of some lamellae may be these nuclei (Fig. 8) (534).

The crystal thickness D is the sum of two terms: one small and nearly constant, and the other large and proportional to $(T\Delta T)^{-1}$, where T is the slowly varying absolute temperature of crystallization and ΔT, the rapidly varying supercooling. The interplay between adsorbed chain deposition and dissolving avoids in small, almost constant terms, the D-infinity at ca 100 K supercooling. Here, according to any kinetic theory, D ought to go to infinity if this interplay is not properly included. The macromolecule is first adsorbed at the surface of the growing crystal and only subsequently enters the growing face (172,173).

In crystallization from the melt and solution, complete agreement between the most recent kinetic theories (173) and the experimental L is claimed (247). Since the theories are concerned with the crystallized chains with no effort wasted on the thickness $\ell/2$ of each of the two amorphous layers on the crystal surface, agreement is expected with $D = \alpha_v L$ and not with L. The sans-derived identity of the gyration radius R of the macromolecule in the melt and the crystalline solid (169) gave rise to the solidification model (176–182). The dependence of L on R in PE (19,21,267,535–537) and PEO (182,312) is mainly explained by chain stiffness in the liquid state. The obvious discrepancies between the kinetic theory and experiment originate from the difficulty of the basic postulates of the theoretical model in describing the exact actual mechanism of crystal growth (538–540).

The interface (541) influences the chain-reentry problem (65). In a lattice model with all adjacent lamellae able to accept parts of the crystallizing chain, the transition from the crystal to the isotropic amorphous state requires 70% of the chains to reenter the same lamella from which they emerge. However, fewer than 20% enter the crystal adjacently; most reenter at nearby sites that are not adjacent to the emergent stem. The tilting of chains in the crystal reduces the thickness of the interface layers, the fraction of reentry in the same crystal, and the fraction of adjacent reentries (73).

The interphase in which order persists is calculated (65) to be 1.0–1.2-nm thick in agreement with the two-phase experiment on single crystals (158). The interfacial free energy, 5.5–7 $\mu J/cm^2$ (55–70 erg/cm^2), calculated from the persistence of order beyond the crystal boundary, is compared with the observed depression of the melting point by the finite thickness of PE lamellae. The experimental values for lower molecular weight are so close to the theoretical estimate that the persistence of order beyond the face of the crystal (542,543) must be the main source of the comparatively large interfacial free energy between 9 $\mu J/cm^2$ (low mol wt) and 30 $\mu J/cm^2$ (high mol wt) (90–300 erg/cm^2). The experimental value for high molecular weight is so much higher than the theoretical value that it must be attributed to departures from equilibrium caused by the rapidity of crystallization.

With the same model, it was found that 60% of the chains have to choose adjacent reentry in order to make space for the true amorphous conformation of the remaining chains (544). When tight folding is destabilized the adjacent reentry becomes less prevalent. Nevertheless, in the examples considered, the combined adjacent-and next-nearest-neighbor reentry never fall below ~50%. A simple model consisting of random walk on a cubic lattice between two absorbing barriers gives a drastic reduction in the probability of adjacent reentry that seems to correspond better to the experimental data.

Oriented Polymers

Most semicrystalline polymers, particularly those produced commercially, are partially oriented; that is, their chains have an overall alignment which may impart to the bulk polymer certain advantageous properties, eg, increased me-chanical strength or dielectric polarizability. Molecular orientation, whether aris-ing from crystallization under stress or deformation of a solidified polymer, or in naturally occurring oriented crystalline polymers like cellulose or keratin, is always associated with an orientational morphology (545).

The chain orientation in a semicrystalline polymer is against the equilib-rium, which in large domains requires complete randomness. The sample is stable only at sufficiently low temperatures that extend chain- and crystal-relaxation times longer than the time of the experiment which may extend over a few years. The oriented material is not absolutely stable. Increasing the temperature in-creases the mobility of the macromolecules and crystals; the chains move per-ceptibly toward their equilibrium position.

Orientation, even if it is complete in easily oriented crystals, for instance in hard elastomers, does not yield by itself a high axial elastic modulus E and strength σ_b in the orientation direction because these quantities are modified by the weak amorphous component that more or less regularly alternates with the axially strong crystals. In hard elastomers without sufficient TTM and with most amorphous layers being voids, the high chain orientation does not yield a high E and σ_b material. Crossing of the amorphous layers by TTM or crystalline bridges or replacing the small shear modulus of the amorphous layer by the much higher bulk modulus yields a high E and σ_b since this stiffens the amorphous layer and increases fracture resistance. In the extreme case of nearly perfect chain orien-tation, the smaller and stronger amorphous layers disrupt the continuity of the crystal lattice as little as the isolated crystal defects.

Crystallization from a strained melt or solution yields partly oriented chains. Flowing liquids are strained in the flow direction by a positive longitudinal velocity gradient, as exists in jet flow, around the eddies in turbulence, in the Taylor four-cylinder flow, in molding, in acoustic fields, and others. With increas-ing gradient, a shish-kebab structure gradually prevails (539,546–548), ie, a very long linear row nucleus (shish) in the gradient direction with spaced, folded-chain lamellae (kebab) that are to be epitaxially grown on the shish. An extremely low gradient is needed as shown in sonicated solutions (Fig. 37). A survey of these methods is given in Refs. 550, 551.

The shish is formed first from the longest macromolecules, which involves a pronounced fractionation. The distance between subsequent kebabs seems to measure the length between subsequent chain entanglements (552,553). The chains are almost extended in the shish, but some folds are revealed by fuming nitric acid etching (554), and the shish itself may have a shish-kebab structure. A similar structure is found in a solidified melt-spinning jet (548,555). At low filament speed, the spherulitic structure prevails; at higher uptake speed, the shish-kebab structure gradually takes over (Fig. 38). The coexistence of shish kebabs and spherulites also was shown in solids obtained from the polymer melt sheared between two plates (556). Very likely the spherulites, characteristic for crystallization from a liquid at rest, were formed after the shishes accommodated the initial shear stress so that the remaining melt was no longer strained.

Under appropriate conditions, fiber can be continuously produced from a flowing melt or solution for many days (557). The shish kebab is dried and ex-truded at high temperature (558), usually 170°C, that is far above the melting point. Addition of a lubricant, eg, aluminum stearate, facilitates extrusion (559).

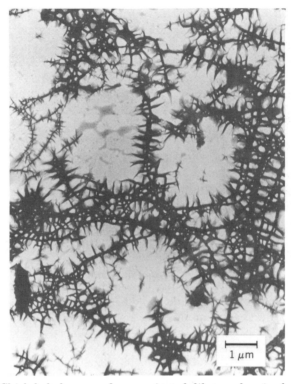

Fig. 37.   Shish kebab grown from sonicated dilute nylon-4 solution (549).

During extrusion the kebabs seem to be pulled into the shish morphology, increasing the distance between the kebab crystals (560,561). The maximum elastic modulus, up to 150 GPa (21.75 × 10⁶ psi) for high molecular weight LPE (562), is obtained when all the kebab has disappeared, ie, has been transformed in the shish. The extremely high maximum draw ratio of the shish kebab seems to equal the ratio of the kebab to shish.

If the flow is stopped immediately after the formation of the row nuclei, the solute or melt crystallizes epitaxially on the oriented row nuclei. Lamellae perpendicular to the flow direction fill the space completely. In order to conserve the volume under unidirectional stress, the deformation of the laterally extended amorphous layers between such lamellae requires a much larger lateral displacement of the amorphous sections of the molecules than that in the narrow microfibrils. Since such a displacement is not possible, the specific volume of the amorphous layers is increased. Deformation is also opposed by the bulk modulus K and not only by the much smaller shear modulus G of the rubbery state. Such lamellar structure shows an extremely high elastic modulus in spite of being formed from lamellae with few axial TTM (563–565).

A much lower orientational effect from nonuniform flow is obtained in polymer injection molding (qv) or solidification between compressed plates where the flow has a substantially lower gradient. The cooling gradient has a special effect, ie, nucleation on the colder mold surfaces favors columnar crystallization (566,567). The numerous primary nuclei at the surface start with lamellar growth which,

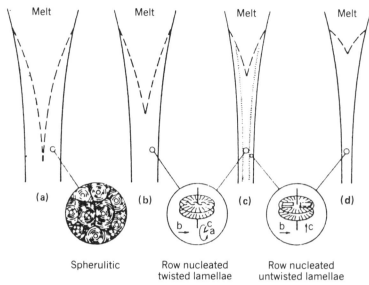

Spherulitic   Row nucleated   Row nucleated
                      twisted lamellae   untwisted lamellae

Fig. 38. Shish-kebab development in fiber spinning at steadily increasing speed:
(**a**) very low take-up velocity; (**b**) low take-up velocity; (**c**) medium take-up velocity; and
(**d**) high take-up velocity (548).

due to the mutual impediment, soon results in growth mainly perpendicular to
the mold surface. The nuclei in the interior of the sample with a much smaller
or no velocity gradient yield normal spherulites. If, however, at solidification the
fluid in the interior of the mold is still under stress, solidification occurs primarily
in the same way as shish kebabs are formed in the strained solution (400,556,568).

Uniaxial Deformation of Solidified Polymers.   Highly oriented material is
obtained by drawing or extrusion below the melting point (63,569–573). The first
method requires a sample with sufficient longitudinal cohesion by which it resists
the uniaxial elongation force in the stretching device. Samples with insufficient
longitudinal cohesion may only be extruded because the transformation from the
original to the final fibrous structure and the drawing of this structure are achieved
by transverse compressive forces generated when the sample is forced through
a conical die many times smaller than the original cross section of the billet. In
order to accelerate the rather slow extrusion, pressure extrusion can be combined
with a uniaxial tension on the extrudate (572).

Extrusion is substantially improved by the split-billet technique. The cylindrical billet is cut in half and the sample is inserted (573). This permits the
drawing of incoherent samples into thin sheets, for instance, dried gels and oriented single-crystal mats. Repeated extrusion gives unusually high draw ratios
(up to 300) and an extremely high elastic modulus $E = 225$ GPa (32.6×10^6 psi)
(297). At such a high E, the chain is almost fully extended or a high concentration
of TTM is present. If the crystals are assigned a static E just below the high
frequency value derived with LAM, $E_{cLAM} = 290$ GPa (42.0×10^6 psi) (222),
about three-fourths of chains in the crystal lattice are required to serve as TTM
in each amorphous layer. The amorphous layers hardly exist as an independent
structural element since they are almost completely bridged by the crystal lattice.

Hence it would be more realistic to regard the extruded fiber as a single crystal with numerous defects which are responsible for the lower density and elastic modulus compared with the ideal crystal.

The temperature of drawing T_d usually determines the observed L which seems to be independent of L of the undrawn sample (574–577). Drawing becomes easier at higher temperature because viscosity decreases and relaxation increases. A good compromise must be reached in order to draw or extrude the material to the highest possible E. Experiments on high T_d of i-PP in a long oven (578) have shown an unexpected effect of the draw rate $\lambda^{\cdot}$ on L giving a straight plot of L vs log $\lambda^{\cdot}$. The largest L occurs with very slow drawing and the smallest L with very rapid drawing. Since during drawing the sample is maintained for longer times at high T_d, a lower $\lambda^{\cdot}$ increases the long period L. Before and after drawing, the change in L is so slow that it may be completely neglected, giving the impression that drawing by itself accelerates the chain sliding and partial melting needed for a change in L.

Deformation of Bulk Polymer.   Plastic deformation by drawing or extrusion is the best method to obtain highly oriented samples from unoriented solid material. According to the microfibrillar model of fibrous structure and its formation (Fig. 39), the wide, thin lamellae are destroyed and the sheared-off or melted and recrystallized blocks are incorporated in highly aligned, long and narrow microfibrils (63). The lamellae are a few nanometers thick and 1000-nm wide, whereas the microfibrils are about 10–20-nm wide and 1000-nm long. The long persistence in deformed spherulitic samples of the helical twist of lamellae indicates that the angle between the direction of plastic deformation and the lamella influences the deformation (579). If this angle is close to zero, a peculiar destructive mode shifts the folded-chain blocks from one lamella to the next and slowly incorporates them in microfibrils (580).

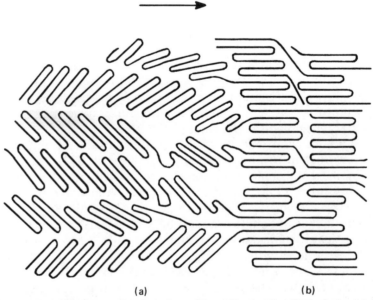

(a)   (b)

Fig. 39.   Transformation by drawing of lamellae with tilted chain (**a**) in the microfibrillar (**b**) structure (345).

During degradation, partial but not complete quasi-melting is indicated by sans (581). The dependence of L on the supercooling derived from the drawing temperature indicates that the temperature of quasi-melting agrees with T_d. It does not agree with the thermodynamic melting point at $\Delta F = 0$ yielding $T_m = \Delta H/\Delta S$. The quasi-melt is highly oriented and thus has a small entropy. In this respect it differs from the normal melt, which consists of completely unoriented chains. The quasi-melt consists of irregularly wider-spaced crystal chains that demand a higher heat content without a substantial increase in entropy. Since the temperature remains the same, a part of the quasi-melting energy can be transferred to the element next to be quasi-melted; the available energy soon is higher than the input energy. The solidification of such a quasi-melt is extremely rapid and may happen in the small region between the dwindling lamella and the emerging microfibrils. It must occur at T_d.

Single-crystal Drawing. To elucidate the structural aspects of oriented polymers, the deformation process of single crystals was investigated. These crystals are the strongest constituent of semicrystalline polymers. The initial stages of the deformation of PE single crystals were homogeneous up to ca 15% elongation, accomplished by twinning or phase changes (582). If the crystalline chains firmly adhere to the strained, deforming substrate, whole lamellar segments tilt and slip toward the draw direction. Free crystals are transformed into microfibrils without any detectable chain tilt and slip (583). The orientation of the chains in the crystal lattice and in the amorphous layers between the crystalline blocks is established during the formation of microfibrils and does not improve substantially during later drawing.

If T_d is very low, eg, liquid nitrogen temperature, the crystals resist chain shearing; deformation proceeds as destruction of lamella by gradual separation of folded-chain blocks. The microfibril consists of a group of fully extended chains. From time to time, such microfibrils tear off a large piece of folded chain from the lamella. Heating to T_g or above destroys such a strained piece, separating it into much smaller blocks and producing microfibrils of the usual type (584).

Hard elastomers are obtained by drawing or extrusion (585,586) of the melt in a sharp temperature gradient (587) or by moderate drawing of a swollen polymer (588). After annealing, these elastomers display a high partially reversible axial elongation up to 700%, a small elastic modulus ($\sim$0.1 MPa) (14.5 psi), and numerous small voids which increase the permeability of the strained film. As long as the holes are not penetrated by the measuring liquid, the bulk density decreases inversely with elongation. Perpendicular to the extrusion direction, the material behaves like a normal semicrystalline polymer. Locally bonded, highly oriented parallel lamellae, which bend easily upon extension, yield a small E, in spite of the high chain orientation in the crystals, and create many holes (585,586). A small fraction of TTM is essential for these properties. The orientation itself does not affect this fraction.

Drawing of the Fibrous Structure. Annealing of highly drawn LPE produces bundles of microfibrils called fibrils (Fig. 40). The drawing of the fibrous structure (590) based on longitudinal motion of fibrils instead of the displacement of microfibrils requires less motion (Fig. 41). The shear forces on each strongly displaced fibril result in a weaker shear motion of microfibrils; the fibrils are more skewed, but the bulk-draw ratio is not increased. The nearly linear increases of the TTM fraction per amorphous area yields a similar increase in the axial elastic

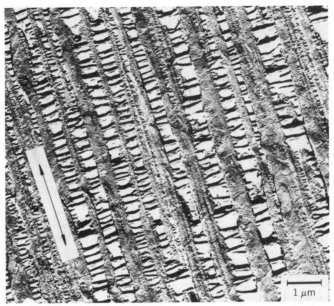

Fig. 40.   Annealing for 2 h at 120°C of the fibrous structure of linear PE drawn at 60°C to λ = 13.5 showing the fibrils (589). Courtesy of Marcel Dekker, Inc.

modulus (Fig. 42). The forces on each microfibril are shear forces that do not affect the structure of the axially strong microfibril, particularly the longitudinal alternation of amorphous layers and crystalline blocks. However, these forces displace the folds and straight sections of the chains in the crystalline blocks to such an extent that the axial projection of the differences in the electron density of the microfibrils continuously decreases. This effect levels off and finally completely cancels the meridional maximum of the saxs; no long period may be deduced any more from it (591). If annealing effects are excluded, the long period L of the microfibril does not change during the drawing of the fibrous structure.

   The time dependence of the axial elastic modulus E after drawing or extrusion shows a rapid increase (Fig. 43); E stabilizes in a few hours (90). This effect may be due to slow crystallization of nearly extended TTM. The high surface-to-volume ratio so depresses T_m that crystallization is prevented during the short drawing at $T_d > 25$°C or during annealing at the higher $T_A > T_d$. The new crystalline bridges increase the axial elastic modulus of the amorphous layers and impart coherence to the blocks they connect. The coherence length is now larger than D or L (89,591,592).

   The depression of the crystallization or melting point T_m of a finite crystal, compared to T_m° of an infinite crystal, can be explained by thermodynamics. For a parallelepiped, with the length D parallel to the chains and the thicknesses d and w perpendicular to the chains, the depression is

$$T_m = T_m^\circ[1 - (2/\rho_c \Delta h_c)(\sigma_D/D + \sigma_d/d + \sigma_w/w)] \tag{8}$$

where Δh_c is the melting enthalpy of the crystal and σ_D is the corresponding surface energy on the two faces perpendicular to D. The surface energy is the difference between the binding energy of the crystal chain on one side and the

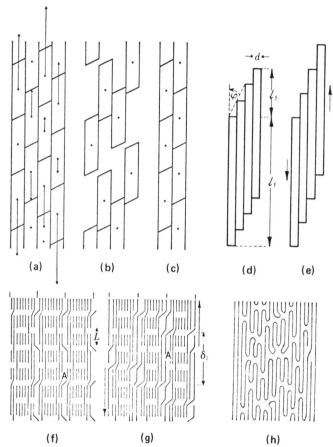

Fig. 41. Drawing of microfibrillar structure: (**a**)–(**c**) the displacement of the fibrils; (**d**)–(**g**) the shear deformation of the fibrils that increases the TTM content; and (**h**) the deformation of the microfibril leading to the gradual disappearance of the meridional saxs maximum. For simplification, the loops are regular, the reentry is adjacent, and the fibrils are axially reduced; the full length of all TTM is on the outside of the microfibrils (590).

environment chain on the other side of the boundary. It strongly depends on the environment, which may be the amorphous phase as in a melt-for-melt crystallization or a solvent in precipitation from a dilute solution. In most cases of thin lamellae, d and w are large compared with D, and σ_d and σ_w are small compared with σ_D; therefore,

$$T_{\mathrm{m}} = T_{\mathrm{m}}^{\circ}(1 - 2\sigma_D/\rho_c\Delta h_c D) \tag{9}$$

The measurement of T_{m} and D yields σ_D if T_{m}°, ρ_c, and Δh_c are known.

The situation is the opposite in drawn or extruded PE samples with a modest TTM content bridging the amorphous layer between the crystal blocks. These bridges are incorporated or even crystallized in the folded-chain blocks in such a way that σ_D approaches zero. However, d and w are so small in single- or double-layer bridges that a substantial depression is reached. For one-molecular-layer bridges of n chains, the depression is $(1 + 1/n) \times 0.24 \times T_{\mathrm{m}}^{\circ} = (1 + 1/n) \times 99.5$ K. It starts with 199 K for the first chain and drops to 149 K for the

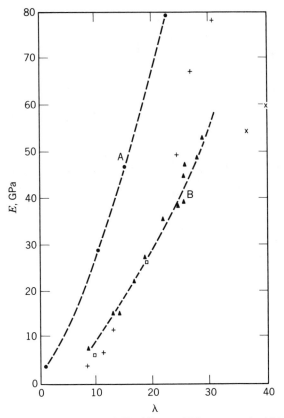

Fig. 42. Dependence of the axial E of linear PE on λ at A, 196°C and B, RT (46). Courtesy of Marcel Dekker, Inc.

second chain. Even for the highest $n = \infty$, the corresponding T_m is only slightly above RT.

High molecular weight samples (>300,000) cannot be drawn to a high draw ratio (593). The drawn sample breaks as soon as the forces on the microfibril are larger than the microfibril can sustain. The stretching forces on the microfibril are transmitted mainly by the intermicrofibrillar and interfibrillar TTM that connect the microfibril with its surroundings; the strength of the microfibril is based on the constant initial fraction of intramicrofibrillar TTM. Hence the number of the former TTM must be reduced in order to obtain a high draw ratio and thus a high E. Large draw ratios of high molecular weight LPE can be achieved only by extrusion of a single-crystal mat, gel, or shish-kebab structure obtained from a strained, dilute solution. In all three cases, the connections between the microfibrils are drastically reduced.

The gel formed from a supercooled, very dilute solution consists of loosely packed single crystals or stacks of parallel crystals (Fig. 6). Each stack is almost independent from the other gel particles, with very few molecules moving from one stack to another. Such a material produces a high λ and consequently a high axial E up to 150 GPa (21.75 × 10^6 psi), and the strength of the drawn sample increases. Experiments (594), as shown in Figure 44, give

$$\sigma_b = KM^{0.48}E^{0.79} \tag{10}$$

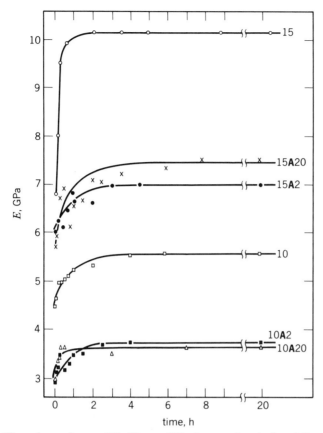

Fig. 43. Time dependence of E of linear PE after mechanical and thermal treatment (90).

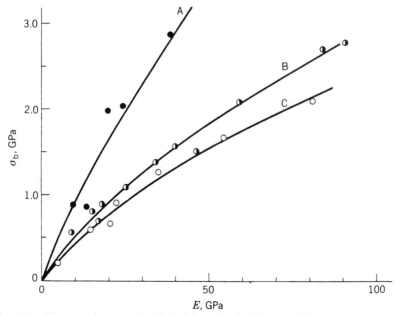

Fig. 44. The axial strength of highly extruded linear PE as a function of E and molecular weight. A, 4×10^6; B, 1.1×10^6; and C, 0.8×10^6 (594).

with σ_b in GPa, $K = 3$, M in Mdaltons, and E in 100 GPa. For high strength, a high molecular weight sample is drawn to a high draw ratio λ, yielding a high E.

If, however, in certain angular ranges of drawing, it is easier to orient the lamellae into blocks than to destroy them, a mixture of microfibrils and oriented lamellae is obtained, as, for example, with nylon-6 gels. The drawn sample contained the highly aligned blocks in the well-organized microfibrils as well as chains perpendicular to the draw direction that nullified the beneficial effects of gel-drawing (595). The strong hydrogen bonds hold the crystal together, resisting the destructive forces in drawing or extrusion. Instead of tearing off blocks and incorporating them in newly formed microfibrils, the lamellae are oriented. Since the chains in the oriented lamellae are perpendicular to the draw direction, they act with the weak van der Waals forces or with the larger hydrogen bond forces between adjacent chains instead of with the 100-times stronger covalent forces in the chain direction. Hence the axial elastic modulus and strength of the extruded gel are greatly depressed.

Annealing of the drawn samples increases the distinction between crystalline and amorphous components and restores the properties of the amorphous component almost to those in undrawn samples, ie, a large portion of TTM is removed. The annealed samples clearly show by em and saxs the alternation of crystalline and amorphous regions. The mechanical and transport properties of the amorphous layers are first restored to the sample before drawing. They change further with the time at RT after annealing. The sample remains completely free or exhibits fixed ends and attains a stationary value after a few hours (Fig. 43). The sample has the tendency to approach this situation before annealing, when it possessed higher elastic modulus and lower diffusivity (596). In most experiments not directed toward time effects, these effects decay quickly and are missed if the measurements are performed some hours or days after the last mechanical or thermal treatment.

BIBLIOGRAPHY

"Morphology" in *EPST* 1st ed., Vol. 9, pp. 204–274, by P. Ingram and A. Peterlin, Research Triangle Institute.

1. J. P. Berry, *J. Polym. Sci.* **50**, 107, 313 (1961).
2. R. Lam and P. H. Geil, *Science* **205**, 1388 (1979).
3. R. P. Sheldon and P. R. Blakey, *Nature London* **195**, 172 (1962).
4. J. M. Schulz, *Bull. Am. Phys. Soc.* **255**(3), 249 (1980).
5. P. H. Hendra, H. P. Jobic, and K. Holland-Moritz, *J. Polym. Sci. Part B* **13**, 193, 365 (1975).
6. J. Breedon-Jones, S. Barenberg, and P. H. Geil, *J. Macromol. Sci. Part B* **15**, 329 (1978).
7. R. Lam and P. H. Geil, *Polym. Bull.* **1**, 127 (1978).
8. J. Breedon-Jones, S. Barenberg, and P. H. Geil, *Polymer* **20**, 903 (1979).
9. R. Lam and P. H. Geil, *J. Macromol. Sci. Part B* **20**, 37 (1981).
10. P. J. Barham, R. A. Chivers, D. A. Jarvis, J. Martinez-Salazar, and A. Keller, *J. Polym. Sci. Part B* **19**, 539 (1981).
11. R. A. Chivers, J. Martinez-Salazar, and A. Keller, *J. Polym. Sci. Polym. Phys. Ed.* **20**, 1717 (1982).
12. P. J. Barham, D. A. Jarvis, and A. Keller, *J. Polym. Sci. Polym. Phys. Ed.* **20**, 1743 (1982).
13. J. Martinez-Salazar, P. J. Barham, and A. Keller, *J. Polym. Sci. Polym. Phys. Ed.* **22**, 1085 (1984).

14. P. J. Flory, *Faraday Discuss. R. Soc. Chem.* **68**, 14 (1979).
15. D. R. Uhlmann, *Faraday Discuss. R. Soc. Chem.* **68**, 87 (1979).
16. E. L. Thomas and E. J. Roche, *Polymer* **20**, 1413 (1979).
17. G. S. Y. Yeh and P. H. Geil, *J. Macromol. Sci. Part B* **1**, 235 (1967); *J. Mater. Sci.* **2**, 457 (1967).
18. S. Lee, H. Miyaji, and P. H. Geil, *J. Macromol. Sci. Part B* **22**, 489 (1983).
19. J. Rault and E. Robelin, *Polym. Bull.* **2**, 393 (1980).
20. J. Rault, *Bull. Am. Phys. Soc.* **27**(3), 258 (1982).
21. E. S. Hsiue, R. E. Robertson, and G. S. Y. Yeh, *J. Macromol. Sci. Part B* **22**, 305 (1983).
22. A. Peterlin, *Polymer* **6**, 25 (1965).
23. J. K. Krüger, L. Peetz, W. Wildner, and M. Pietralla, *Polymer* **21**, 620 (1980).
24. S. B. Clough, *J. Macromol. Sci. Part B* **4**, 199 (1970).
25. K. H. Storcks, *J. Am. Chem. Soc.* **60**, 1753 (1938).
26. A. Keller, *Philos. Mag.* **2**, 1171 (1957).
27. J. J. Trillat and H. Motz, *Ann. Phys.* **4**, 273 (1935).
28. H. Mark, *Ergeb. Tech. Roentgenkd.* **4**, 75 (1934).
29. K. Hess and H. Kiessig, *Naturwissenschaften* **31**, 171 (1943).
30. K. Hess and H. Mahl, *Naturwissenschaften* **41**, 86 (1954).
31. A. Mueller, *Helv. Chim. Acta* **16**, 155 (1933).
32. K. Herrmann, O. Gerngross, and W. Abitz, *Z. Phys. Chem. Abt. B* **10**, 1371 (1930).
33. L. R. G. Treloar, *The Physics of Rubber Elasticity,* Clarendon Press, Oxford, UK, 1975.
34. J. L. Matthews, H. S. Peiser, and R. B. Richards, *Acta Crystallogr.* **2**, 85 (1949).
35. R. K. Eby, *J. Res. Nat. Bur. Stand. Sect. A* **68**, 268 (1964).
36. G. T. Davis, R. K. Eby, and J. P. Colson, *J. Appl. Phys.* **41**, 4316 (1970).
37. G. T. Davis, J. J. Weeks, G. M. Martin, and R. K. Eby, *J. Appl. Phys.* **45**, 4175 (1975).
38. T. Kawai and A. Keller, *Philos. Mag.* **8**, 1203 (1963).
39. I. R. Harrison and E. Baer, *J. Polym. Sci. Part A-2* **9**, 1305 (1971).
40. I. R. Harrison and T. Juska, *J. Polym. Sci. Polym. Phys. Ed.* **17**, 491 (1979).
41. G. Zerbi and M. Gussoni, *Polymer* **21**, 1129 (1980).
42. W. Fronk and W. Wilke, *Colloid Polym. Sci.* **260**, 1107 (1982).
43. R. Hentschel, J. Schlitter, H. Sillescu, and H. W. Spiess, *J. Chem. Phys.* **68**, 56 (1978).
44. R. Hentschel, H. Sillescu, and H. W. Spiess, *Polymer* **22**, 1516 (1981).
45. H. W. Spiess in I. M. Ward, ed., *Development in Oriented Polymers,* Vol. 1, Applied Science Publishers, London, 1981.
46. A. Peterlin in L. E. Murr, ed., *Industrial Materials in Science and Engineering,* Marcel Dekker, Inc., New York, 1984, p. 145.
47. E. W. Fischer, H. Goddar, and G. F. Schmidt, *Makromol. Chem.* **118**, 144 (1968).
48. D. J. Blundell, D. R. Beckett, and P. H. Willcocks, *Polymer* **22**, 704 (1981).
49. M. J. Richardson, *Plast. Rubber Mater. Appl.* **1**, 182 (1976).
50. J. C. Hser and S. H. Carr, *Polym. Eng. Sci.* **19**, 436 (1979).
51. B. Wunderlich, *Macromolecular Physics,* Vols. 1–3, Academic Press, Inc., Orlando, Fla., 1973, 1976, 1980.
52. L. Mandelkern, A. L. Allou, Jr., and M. Gopaian, *J. Phys. Chem.* **72**, 309 (1968).
53. J. Maxfield and L. Mandelkern, *Macromolecules* **10**, 1141 (1977).
54. R. P. Palmer and A. J. Cobbold, *Makromol. Chem.* **74**, 174 (1962).
55. A. Keller and S. Sawada, *Makromol. Chem.* **74**, 190 (1962).
56. A. Peterlin and G. Meinel, *J. Polym. Sci. Part B* **3**, 1059 (1965).
57. D. J. Blundell, A. Keller, and T. Connor, *J. Polym. Sci. Part A-2* **5**, 991 (1967).
58. J. T. Williams, A. Keller, and I. M. Ward, *J. Polym. Sci. Part A-2* **6**, 1613, 1621 (1968).
59. M. E. Cagiao, D. R. Rueda, and F. J. Balta-Calleja, *Makromol. Chem.* **182**, 2705 (1981).
60. H. Tanaka, *Colloid Polym. Sci.* **260**, 1101 (1982).
61. H. J. Grimm and E. L. Thomas, *Polymer* **26**, 27, 38 (1985).
62. G. Kanig, *Colloid Polym. Sci.* **251**, 782 (1973); *Proc. Colloid Polym. Sci.* **57**, 176 (1975).
63. A. Peterlin, *J. Mater. Sci.* **6**, 490 (1971).
64. A. Peterlin, J. L. Williams, and V. Stannett, *J. Polym. Sci. Part A-2* **5**, 957 (1967).
65. P. J. Flory, D. Y. Yoon, and K. A. Dill, *Macromolecules* **17**, 972 (1984).
66. D. M. Sadler and A. Keller, *Macromolecules* **10**, 1128 (1977).
67. E. W. Fischer, *Colloid Polym. Sci.* **218**, 976 (1967).
68. P. J. Flory, *J. Am. Chem. Soc.* **84**, 2857 (1962).

69. D. Y. Yoon and P. J. Flory, *Polymer* **18**, 509 (1977).
70. P. J. Flory and D. Y. Yoon, *Nature London* **272**, 226 (1978).
71. D. Y. Yoon and P. J. Flory, *Faraday Discuss. R. Soc. Chem.* **68**, 288 (1969).
72. D. Y. Yoon and P. J. Flory, *Polym. Bull.* **4**, 693 (1981).
73. D. Y. Yoon and P. J. Flory, *Macromolecules* **17**, 878 (1984).
74. P. J. Lemstra and P. Smith, *Br. Polym. J.* **12**, 220 (1980).
75. P. J. Flory, *Trans. Faraday Soc.* **51**, 848 (1955).
76. N. A. Plate and V. P. Shibaev, *J. Polym. Sci. Macromol. Rev.* **8**, 117 (1974).
77. J. N. Hay, *J. Polym. Sci. Part A* **3**, 433 (1965).
78. F. A. DiMarzio, C. M. Guttman, and J. D. Hoffman, *Faraday Discuss. R. Soc. Chem.* **68**, 210 (1969).
79. E. Ergos, J. G. Fatou, and L. Mandelkern, *Macromolecules* **5**, 147 (1972).
80. J. F. Lontz, "Sintering and Plastic Deformation" in L. J. Bonis and H. H. Hauser, eds., *Fundamental Phenomena in Material Sciences,* Vol. 1, Plenum Press, New York, 1963.
81. G. W. Haldin and I. L. Kamel, *Polym. Eng. Sci.* **17**, 21 (1977).
82. R. W. Truss, K. S. Hahn, J. T. Wallace, and P. H. Geil, *Polym. Eng. Sci.* **20**, 747 (1980); *J. Macromol. Sci. Part B* **19**, 313 (1981).
83. K. S. Han, J. P. Wallace, R. W. Truss, and P. H. Geil, *J. Macromol. Sci. Part B* **19**, 313 (1981).
84. R. J. Crawford and D. W. Paul, *J. Mater. Sci.* **17**, 2261, 2267 (1982).
85. R. H. Boyd, *Polym. Eng. Sci.* **19**, 1010 (1979).
86. J. Brandrup and E. H. Immergut, *Polymer Handbook,* John Wiley & Sons, Inc., New York, 1975.
87. R. Bonart and R. Hosemann, *Makromol. Chem.* **39**, 105 (1960).
88. J. Clements, R. Jakeways, and I. M. Ward, *Polymer* **19**, 639 (1978).
89. A. G. Gibson, G. R. Davies, and I. M. Ward, *Polymer* **19**, 683 (1978).
90. F. DeCandia, V. Vittoria, and A. Peterlin, *J. Polym. Sci. Polym. Phys. Ed.* **23**, 1217 (1985).
91. F. J. Balta-Calleja, *Microhardness Relating to Crystalline Polymers, Adv. Polym. Sci.* **66**, Springer-Verlag, New York, 1985, p. 117.
92. M. Takayanagi, *Kautsch. Gummi Kunstst.* **17**, 164 (1964).
93. M. Takayanagi, S. Uemura, and S. Minami, *J. Polym. Sci. Part C* **5**, 113 (1964).
94. J. D. Ferry, *Viscoelastic Properties of Polymers,* John Wiley & Sons, Inc., New York, 1980.
95. M. Dole and K. Katsuura, *J. Polym. Sci. Part B* **3**, 467 (1965).
96. M. Amand, R. E. Cohen, and R. F. Badour, *Polymer* **22**, 361 (1981).
97. B. Neppert, B. Heise, and H. G. Kilian, *Colloid Polym. Sci.* **261**, 577 (1983).
98. B. Catoire, P. Bourioty, O. Demuth, A. Baszkin, and M. Cheviur, *Polymer* **24**, 766 (1984).
99. A. Keller, *J. Polym. Sci.* **17**, 291, 351 (1955); **39**, 151 (1959).
100. A. Keller and J. R. S. Waring, *J. Polym. Sci.* **39**, 447 (1959).
101. A. Keller in R. H. Doremus, B. W. Roberts, and D. W. Turnbull, eds., *Growth and Perfection of Crystals,* John Wiley & Sons, Inc., New York, 1958, p. 499.
102. H. D. Keith and F. J. Padden, *J. Polym. Sci.* **39**, 101, 123 (1959).
103. F. K. Price, *J. Polym. Sci.* **37**, 71 (1959); **39**, 139 (1959).
104. G. Ungar and A. Keller, *Polymer* **21**, 1273 (1980).
105. P. Sullivan and B. Wunderlich, *SPE Trans.* **4**, 113 (1964).
106. P. Prentia and S. Hashkemi, *J. Mater. Sci.* **19**, 518 (1984).
107. B. Wunderlich and P. Sullivan, *J. Polym. Sci.* **56**, 19 (1962).
108. P. A. Tucker and R. A. Wilson, *J. Polym. Sci. Part B* **18**, 97 (1980).
109. D. R. Clarke, *J. Mater. Sci.* **8**, 279 (1973).
110. J. Petermann, M. Miles, and H. Gleiter, *J. Macromol. Sci. Part B* **12**, 393 (1976).
111. M. Miles and J. Petermann, *J. Macromol. Sci. Part B* **16**, 243 (1979).
112. E. J. Roche and E. L. Thomas, *Polymer* **22**, 333 (1981).
113. H. Kato, *J. Electron Microsc.* **14**, 220 (1965); *Polym. Eng. Sci.* **7**, 38 (1967).
114. J. S. Trent, J. I. Scheinbeim, and P. R. Couchman, *Macromolecules* **16**, 589 (1983); *J. Polym. Sci. Part B* **19**, 315 (1981).
115. A. Vitali and E. Montani, *Polymer* **21**, 1220 (1980).
116. R. H. Olley, *Colloid Polym. Sci.* **255**, 1005 (1977).
117. R. H. Olley, A. M. Hodge, and D. C. Bassett, *J. Polym. Sci. Polym. Phys. Ed.* **17**, 627 (1979).
118. J. Martinez-Salazar and C. G. Camson, *J. Mater. Sci. Lett.* **3**, 693 (1984).
119. G. Ungar, *J. Mater. Sci.* **16**, 2635 (1981).

120. H. Vanni and J. F. Rabolt, *J. Polym. Sci. Polym. Phys. Ed.* **18**, 587 (1980).
121. J. V. Pascale, D. B. Herrmann, and R. J. Miner, *Mod. Plast.* **41**, 239 (1963).
122. D. C. Bassett, *Philos. Mag.* **10**, 595 (1964).
123. O. Yoda and I. Kuriyama, *J. Mater. Sci.* **14**, 1733 (1979).
124. G. Ungar, D. T. Grubb, and A. Keller, *Polymer* **21**, 1284 (1980).
125. O. Yoda, I. Kuriyama, and A. Odajima, *J. Mater. Sci. Lett.* **1**, 451 (1982).
126. S. K. Batheja, E. H. Andrews, and R. J. Young, *J. Polym. Sci. Polym. Phys. Ed.* **21**, 523 (1983).
127. S. K. Batheja, *J. Macromol. Sci. Part B* **22**, 159 (1983).
128. G. S. Y. Yeh, C. J. Chen, and D. C. Boose, *Colloid Polym. Sci.* **263**, 109 (1985).
129. A. N. Shaitanova, K. D. Pismannik, T. V. Bazhbenk-Melikova, A. S. Goikman, N. P. Matsibora, and N. A. Myalov, *Vysokomol. Soedin. Ser. B* **16**, 783 (1974).
130. A. J. Hughes and P. J. Bell, *J. Polym. Sci. Polym. Phys. Ed.* **16**, 201 (1978).
131. *Ibid.*, p. 215.
132. M. S. Ellison, S. H. Zeronian, and Y. Fujiwara, *J. Mater. Sci.* **19**, 82 (1984).
133. G. N. Patel and A. Keller, *J. Polym. Sci. Polym. Phys. Ed.* **13**, 303 (1975).
134. S. Giorgio and R. Kern, *J. Polym. Sci. Polym. Phys. Ed.* **22**, 1931 (1984).
135. L. E. Thomas, C. J. Humphrey, W. R. Duff, and D. T. Grubb, *Radiat. Eff.* **3**, 89 (1970).
136. L. E. Thomas and S. Danyluk, *J. Phys. E* **4**, 843 (1971).
137. A. Peterlin and K. Sakaoku in G. Goldfinger, ed., *Clean Surfaces,* Marcel Dekker, Inc., New York, 1970, p. 1.
138. A. V. Creve, *J. Microsc.* **100**, 247 (1974).
139. E. S. Sherman, W. W. Adams, and E. L. Thomas, *J. Mater. Sci.* **16**, 1 (1981).
140. G. A. Bassett, D. A. Blundell, and A. Keller, *J. Macromol. Sci. Part B* **1**, 161 (1967).
141. D. J. Blundell and A. Keller, *J. Macromol. Sci. Part B* **7**, 213 (1973).
142. A. Keller and D. M. Sadler, *J. Macromol. Sci. Part B* **7**, 261 (1973).
143. J. C. Wittmann and B. Lotz, *Makromol. Chem. Rapid Commun.* **3**, 733 (1982).
144. J. C. Wittmann and B. Lotz, *J. Polym. Sci. Polym. Phys. Ed.* **23**, 205 (1985).
145. P. H. Lindenmeyer, *J. Polym. Sci. Part C* **1**, 5 (1963).
146. W. J. Ruland, *J. Appl. Crystallogr.* **4**, 170 (1977).
147. A. Guinier and G. Fournet, *Small Angle Scattering of X-Rays,* John Wiley & Sons, Inc., New York, 1955.
148. R. Hosemann and S. N. Bagchi, *Direct Analysis of Diffraction by Matter,* North-Holland Publishers, Amsterdam, Netherlands, 1962.
149. A. Kawaguchi, *Polymer* **22**, 753 (1981).
150. W. O. Statton in B. Ke, ed., *Newer Methods of Polymer Characterization,* John Wiley & Sons, Inc., New York, 1964, p. 231.
151. P. H. Geil, *J. Polym. Sci. Part C* **13**, 149 (1966).
152. J. H. Koch, J. Bordas, E. Scholz, and H. C. Brocker, *Polym. Bull.* **1**, 709 (1979).
153. H. G. Zachmann and G. Elsner, *Third IUPAC Conference,* Mainz, FRG, Pergamon Press, London, 1979, p. 1314.
154. P. Forgacs, M. A. Sheromov, B. P. Tolochko, N. A. Mezentsev, and V. F. Pindurin, *J. Polym. Sci. Polym. Phys. Ed.* **18**, 2155 (1980).
155. B. Crist, *J. Polym. Sci. Polym. Phys. Ed.* **11**, 635 (1973).
156. B. Crist and N. Morosoff, *J. Polym. Sci. Polym. Phys. Ed.* **11**, 1023 (1973).
157. Ref. 15, p. 86.
158. W. D. Varnell, I. R. Harrison, and S. J. Kozimski, *J. Polym. Sci. Polym. Phys. Ed.* **19**, 1237 (1981).
159. D. Yu. Tsvankin, *Polym. Sci. USSR* **6**, 2304 (1964).
160. U. Ruland, *J. Appl. Crystallogr.* **4**, 70 (1971).
161. C. G. Vonk, *J. Appl. Crystallogr.* **6**, 81 (1973).
162. A. Todo, T. Hashimoto, and H. Kawai, *J. Appl. Crystallogr.* **11**, 558 (1978).
163. U. Siemann and W. Ruland, *Colloid Polym. Sci.* **260**, 999 (1982).
164. H. Kawai, T. Ito, D. A. Keedy, and R. S. Stein, *J. Polym. Sci. Part B* **2**, 1075 (1964).
165. S. Suehiro, T. Yamada, H. Inagaki, and H. Kawai, *Polym. J.* Jpn. **10**, 315 (1978).
166. P. Young, T. Kyu, J. S. Lin, and R. S. Stein, *J. Polym. Sci. Polym. Phys. Ed.* **21**, 881 (1983).
167. R. G. Kirste, W. A. Kruse, and J. Schelten, *Makromol. Chem.* **162**, 299 (1973).
168. G. D. Wignall, D. G. Ballard, and J. Schelten, *Eur. Polym. J.* **10**, 861 (1974).

169. G. Lieser, E. W. Fischer, and K. Ibel, *J. Polym. Sci. Part B* **13**, 39 (1975).
170. C. E. Williams, M. Nierlich, J. P. Cotton, G. Jannink, F. Boue, M. Daoud, B. Farnou, C. Picot, P. G. deGennes, M. Rinando, M. Moan, and C. Wolff, *J. Polym. Sci. Part B* **17**, 379 (1979).
171. L. H. Sperling, *Polym. Eng. Sci.* **24**, 1 (1984).
172. J. D. Hoffman, G. T. Davies, and J. I. Lauritzen, Jr. in N. B. Hannay, ed., *Solid State Chemistry,* Plenum Press, New York, 1975, p. 497.
173. J. D. Hoffman, *Polymer* **24**, 3 (1983).
174. F. A. DiMarzio and C. M. Guttman, *Polymer* **21**, 733 (1980).
175. J. M. Grenet, *Polymer* **21**, 1385 (1980).
176. E. W. Fischer, *Pure Appl. Chem.* **50**, 1319 (1978).
177. P. Calvert, *J. Polym. Sci. Polym. Phys. Ed.* **17**, 1341 (1979).
178. M. Stamm, E. W. Fischer, M. Dettenmaier, and P. Convert, *Faraday Discuss. R. Soc. Chem.* **68**, 263 (1979).
179. E. W. Fischer, M. Stamm, M. Dettenmaier, and P. Herchenroder, *Am. Chem. Soc. Prepr.* **20**, 219 (1979).
180. M. Dettenmaier, E. W. Fischer, and M. Stamm, *Colloid Polym. Sci.* **258**, 343 (1980); J. Schelten and M. Stamm, *Macromolecules* **14**, 818 (1981).
181. E. W. Fischer in H. Benoit and P. Rempp, eds., *IUPAC Macromolecules,* Pergamon Press, London, 1982, p. 191.
182. E. W. Fischer, K. Hahn, J. Kugler, U. Struth, R. Born, and M. Stamm, *J. Polym. Sci. Polym. Phys. Ed.* **22**, 1491 (1985).
183. J. M. Peterson, *J. Appl. Phys.* **39**, 11 (1968).
184. F. C. Stehling, E. Ergos, and L. Mandelkern, *Macromolecules* **4**, 672 (1971).
185. W. Korschak, N. M. Kozireva, G. N. Menchikova, A. V. Gorshkova, E. A. Grigoryan, and F. S. Dyachkovskij, *Polym. Sci. USSR* **21**, 159 (1979).
186. J. Schelten, D. G. H. Ballard, G. D. Wignall, G. Longman, and W. Schmatz, *Polymer* **17**, 751 (1976).
187. J. Schelten, G. D. Wignall, D. G. H. Ballard, and G. W. Longman, *Polymer* **18**, 1111 (1977).
188. J. Schelten, A. Zinken, and D. G. H. Ballard, *Colloid Polym. Sci.* **259**, 260 (1981).
189. R. S. Stein in R. W. Lenz and R. S. Stein, eds., *Structure and Properties of Polymer Films,* Plenum Press, New York, 1973, p. 1.
190. R. S. Stein and M. B. Rhodes, *J. Appl. Phys.* **31**, 1873 (1960).
191. G. H. Meeten and P. Navard, *J. Polym. Sci. Polym. Phys. Ed.* **22**, 2159 (1984).
192. J. V. Champion, A. Killy, and G. H. Meeten, *J. Polym. Sci. Polym. Phys. Ed.* **23**, 1467 (1985).
193. R. J. Tabor, R. S. Stein, and M. B. Long, *J. Polym. Sci. Polym. Phys. Ed.* **20**, 2041 (1982).
194. A. E. Ames in J. C. Danty, ed., *Laser Speckle and Related Phenomena,* Springer Publishing Co., Berlin, 1975, p. 203.
195. Y. Y. Hung in R. K. Erf, ed., *Speckle Metrology,* Academic Press, Inc., Orlando, Fla., 1978, p. 203.
196. J. Holoubek and B. Sedlaček, *J. Macromol. Sci. Part B* **23**, 143 (1984).
197. P. D. Vasko and J. L. Koenig, *J. Macromol. Sci. Part B* **6**, 117 (1972).
198. P. D. Frayer, J. L. Koenig, and J. B. Lando, *J. Macromol. Sci. Part B* **6**, 129 (1972).
199. P. J. Hendra in K. J. Ivin, ed., *Structural Studies of Macromolecules by Spectroscopic Methods,* John Wiley & Sons, Inc., New York, 1976, p. 73.
200. G. Capaccio, I. M. Ward, M. A. Wilding, and G. W. Longman, *J. Macromol. Sci. Part B* **15**, 381 (1978).
201. J. Maxfield, R. Stein, and M. C. Chen, *J. Polym. Sci. Polym. Phys. Ed.* **16**, 37 (1978).
202. G. Hasetti, F. Cabassi, and G. Zerbi, *Polymer* **21**, 143 (1980).
203. P. C. Painter, M. Watzel, and J. L. Koenig, *Polymer* **18**, 1169 (1977).
204. J. N. Hay, *Polymer* **22**, 718 (1981).
205. J. N. Hay and P. A. Fitzgerald, *Polymer* **22**, 1003 (1981).
206. M. I. Bank and S. Krimm, *J. Polym. Sci. Part A-2* **7**, 1785 (1969).
207. S. Krimm and T. S. Cheam, *Faraday Discuss. R. Soc. Chem.* **68**, 244 (1979).
208. X. Jing and S. Krimm, *J. Polym. Sci. Polym. Phys. Ed.* **20**, 1155 (1982).
209. W. L. Peticolas, G. W. Hibler, J. L. Lippert, A. Peterlin, and H. G. Olf, *Appl. Phys. Lett.* **18**, 87 (1971).
210. A. Peterlin, H. G. Olf, W. L. Peticolas, G. W. Hibler, and J. L. Lippert, *J. Polym. Sci. Part B* **9**, 583 (1971).

211. H. G. Olf, A. Peterlin, and W. L. Peticolas, *J. Polym. Sci. Polym. Phys. Ed.* **12**, 359 (1974).
212. J. L. Koenig and D. L. Tabb, *J. Macromol. Sci. Part B* **9**, 141 (1974).
213. J. V. Fraser, P. J. Hendra, M. E. A. Cudby, and H. A. Willis, *J. Mater. Sci.* **9**, 1270 (1975).
214. M. J. Folkes, A. Keller, J. Stejny, P. L. Goggin, G. V. Fraser, and P. J. Hendra, *Colloid Polym. Sci.* **253**, 3554 (1975).
215. P. J. Hendra, E. P. Marsden, M. E. A. Cudby, and H. A. Willis, *Makromol. Chem.* **167**, 2442 (1975).
216. J. D. Lugosz, G. V. Fraser, D. Grubb, A. Keller, and J. A. Odell, *Polymer* **17**, 471 (1976).
217. G. J. Farell and A. Keller, *J. Mater. Sci.* **12**, 966 (1977).
218. R. G. Snyder, S. J. Krause, and J. R. Scherer, *J. Polym. Sci. Polym. Phys. Ed.* **16**, 1593 (1978).
219. R. G. Snyder and J. R. Scherer, *J. Polym. Sci. Polym. Phys. Ed.* **18**, 421 (1980).
220. R. G. Snyder, J. A. Scherer, and A. Peterlin, *Macromolecules* **14**, 177 (1981).
221. B. Fanconi and J. F. Rabolt, *J. Polym. Sci. Polym. Phys. Ed.* **23**, 1201 (1985).
222. G. R. Strobl and R. E. Eckel, *Colloid Polym. Sci.* **258**, 570 (1980).
223. I. M. Ward, *Faraday Discuss. R. Soc. Chem.* **68**, 478 (1979).
224. A. Peterlin, *J. Polym. Sci. Polym. Phys. Ed.* **20**, 2329 (1982).
225. A. Peterlin, *Croat. Chem. Acta* **60**, 103 (1987).
226. A. Peterlin and R. G. Snyder, *J. Polym. Sci. Polym. Phys. Ed.* **19**, 1727 (1981).
227. V. LaGarde, H. Prask, and S. Trevino, *Faraday Discuss. Chem. Soc.* **48**, 15 (1969).
228. M. Sakamoto, M. Izumi, N. Masaki, H. Motohashi, N. Minakawa, K. Doji, I. Kuriyama, O. Yoda, N. Tamura, and A. Odajima, *J. Polym. Sci. Part B* **11**, 377 (1973).
229. J. White in Ref. 199, p. 41.
230. U. W. Gedde and J. F. Jansson, *Polymer* **24**, 1521, 1532 (1983).
231. J. Martinez-Salazar, P. J. Barham, and A. Keller, *J. Mater. Sci.* **20**, 1616 (1985).
232. R. S. Porter, personal communication.
233. C. L. Choy, F. C. Chen, and E. L. Ong, *Polymer* **20**, 191 (1979).
234. F. C. Chen, C. L. Choy, and K. Young, *J. Polym. Sci. Polym. Phys. Ed.* **18**, 2313 (1980); **19**, 335 (1981).
235. F. C. Chen, C. L. Choy, S. P. Wong, and K. Young, *J. Polym. Sci. Polym. Phys. Ed.* **19**, 971 (1981).
236. G. K. White and C. L. Choy, *J. Polym. Sci. Polym. Phys. Ed.* **22**, 835 (1984).
237. J. Rault, *J. Macromol. Sci. Part B* **15**, 567 (1978).
238. A. Peterlin in H. Hopfenberg, ed., *Permeability of Plastic Fibers and Coatings,* Plenum Press, New York, 1974, p. 9.
239. A. Peterlin, *Makromol. Chem. Suppl.* **3**, 215 (1979).
240. A. Peterlin in J. C. Seferis and P. S. Theocaris, eds., *Interrelations Between Processing Structures and Properties of Polymeric Materials,* Elsevier Science Publishing Co., Amsterdam, the Netherlands, 1984, p. 372.
241. A. Peterlin and F. L. McCrackin, *J. Polym. Sci. Polym. Phys. Ed.* **19**, 1003 (1981).
242. V. F. Holland and P. H. Lindenmeyer, *J. Polym. Sci.* **57**, 589 (1962).
243. D. J. Blundell, A. Keller, and A. J. Kovacs, *J. Polym. Sci. Part B* **4**, 481 (1966).
244. D. J. Blundell and A. Keller, *J. Macromol. Sci. Part B* **2**, 301 (1968).
245. S. H. Carr, A. Keller, and E. Baer, *J. Polym. Sci. Part A-2* **8**, 1467 (1970).
246. A. Keller and F. M. Willmouth, *J. Polym. Sci. Part A-2* **8**, 1443 (1970).
247. P. J. Barham, R. A. Chivers, A. Keller, J. Martinez-Salazar, and S. J. Organ, *J. Mater. Sci.* **20**, 1625 (1985).
248. J. C. Wittmann, A. M. Hodge, and B. Lotz, *J. Polym. Sci. Polym. Phys. Ed.* **21**, 2495 (1983).
249. D. H. Reneker and P. H. Geil, *J. Appl. Phys.* **31**, 1916 (1960).
250. F. Khoury and J. D. Barnes, *J. Res. Nat. Bur. Stand. Sect. A* **78**, 95 (1974).
251. T. Kawai and A. Keller, *Philos. Mag.* **11**, 1165 (1965).
252. J. A. Sauer, D. R. Morrow, and G. C. Richardson, *J. Appl. Phys.* **36**, 3017 (1965).
253. P. Cerra, D. R. Morrow, and J. A. Sauer, *J. Macromol. Sci. Part B* **3**, 33 (1969).
254. H. D. Keith, G. Giannoni, and F. J. Paden, Jr., *Biopolymers* **7**, 775 (1969).
255. F. C. Frank, *Faraday Discuss. Chem. Soc.* **5**, 48 (1949).
256. H. B. van der Heijde, *J. Polym. Sci. Part A-2* **5**, 225 (1967).
257. A. Keller, *Philos. Mag.* **6**, 329 (1961).
258. A. Keller, *Colloid Polym. Sci.* **219**, 118 (1967).
259. A. K. Patel and B. L. Farmer, *Polymer* **21**, 153 (1980).

260. G. C. Alfonso, S. Falchetti, and E. Pedemonte, *Europhys. Conf. Abstr.* **6g**, 49 (1982).
261. D. M. Sadler, *Polymer* **24**, 140 (1983).
262. E. Passaglia and F. Khoury, *Polymer* **25**, 631 (1984).
263. D. M. Sadler and G. H. Gilmer, *Polymer* **25**, 1447 (1984).
264. V. F. Holland, S. B. Mitchell, W. L. Hunter, and P. H. Lindenmeyer, *J. Polym. Sci.* **62**, 141 (1962).
265. J. J. Klement and P. H. Geil, *J. Polym. Sci. Part A-2* **6**, 1381 (1968).
266. G. N. Patel and R. D. Patel, *J. Polym. Sci. Part A-2* **8**, 47 (1970).
267. G. Hinrichsen and H. Orth, *Colloid Polym. Sci.* **247**, 844 (1971).
268. V. A. Kargin, N. F. Bakeev, and L. Li-shen, *Vysokomol. Soedin.* **3**, 1100 (1961).
269. Y. Myamoto, C. Nakafuku, and T. Takemura, *Polym. J. Jpn.* **3**, 122 (1972).
270. J. D. Barnes and F. Khoury, *J. Res. Nat. Bur. Stand. Sect. A* **78**, 363 (1974).
271. A. Nakajima and S. Hayashi, *Colloid Polym. Sci.* **228**, 12 (1969).
272. R. Eppe, E. W. Fischer, and H. A. Stuart, *J. Polym. Sci.* **34**, 721 (1959).
273. P. H. Geil, *J. Polym. Sci.* **51**, S10 (1961).
274. H. G. Kilian and E. W. Fischer, *Colloid Polym. Sci.* **211**, 40 (1966).
275. P. J. Holdsworth and A. Keller, *J. Polym. Sci. Part B* **5**, 605 (1967).
276. R. J. Roe, H. F. Cole, and D. R. Morrow in K. D. Pae, F. R. Morrow, and Y. Chen, eds., *Advances in Polymer Science and Engineering,* Plenum Press, New York, 1972, p. 27.
277. J. S. Mijovic and J. A. Koutsky, *Polym. Plast. Technol. Eng.* **9**, 139 (1977).
278. T. Tagawa and J. Mori, *Electron Microsc.* **27**, 267 (1978).
279. A. A. Popov, N. N. Blinov, B. E. Krisyuk, S. G. Karpova, A. N. Preverov, and G. Ye. Zaikov, *Polym. Sci. USSR* **23**, 1665 (1981).
280. M. Kojima and H. Satake, *J. Polym. Sci. Polym. Phys. Ed.* **20**, 2153 (1982).
281. M. K. Jain and A. S. Akhiraman, *J. Mater. Sci.* **18**, 179 (1983).
282. R. T. Y. Trilla, J. M. Perina, and J. G. Fatou, *Polymer* **15**, 803 (1983).
283. J. Martinez-Salazar, A. Keller, M. E. Cagiao, D. R. Rueda, and F. J. Balta-Calleja, *Colloid Polym. Sci.* **261**, 412 (1983).
284. N. Ya. Rapoport and G. E. Zaikov, *Eur. Polym. J. Jpn.* **20**, 409 (1984).
285. J. Martinez-Salazar, E. J. Cabaras, D. R. Rueda, M. E. Cagiao, and F. J. Balta-Calleja, *Polym. Bull.* **12**, 269 (1984).
286. T. Kawai, K. Ujihara, and H. Maeda, *Makromol. Chem.* **132**, 87 (1970).
287. P. Blais and R. St. J. Manley, *J. Polym. Sci. Part A-2* **4**, 1022 (1966).
288. H. D. Keith, R. G. Vadimsky, and F. J. Padden, Jr., *J. Polym. Sci. Part A-2* **8**, 1687 (1970).
289. D. C. Bassett, F. C. Frank, and A. Keller, *Philos. Mag.* **8**, 1753 (1963).
290. P. H. Geil, *J. Polym. Sci.* **44**, 449 (1960).
291. V. F. Holland and R. L. Miller, *J. Appl. Phys.* **35**, 3241 (1964).
292. K. Sakaoku and A. Peterlin, *J. Macromol. Sci. Part B* **1**, 401 (1967).
293. F. Khoury and J. D. Barnes, *J. Res. Nat. Bur. Stand. Sect. A* **76**, 225 (1972).
294. R. B. Williamson and W. F. Busse, *J. Appl. Phys.* **38**, 4187 (1967).
295. W. O. Statton, *J. Appl. Phys.* **38**, 4149 (1967).
296. D. C. Bassett and A. Keller, *Philos. Mag.* **7**, 1553 (1962).
297. T. Kanamoto, A. Tsurata, K. Tanaka, M. Takeda, and R. S. Porter, *Polym. J. Jpn.* **15**, 327 (1983).
298. S. J. Organ and A. Keller, *J. Mater. Sci.* **20**, 1602 (1985).
299. B. Wunderlich, E. A. James, and Tsao-Wen Shu, *J. Polym. Sci. Part A* **2**, 2759 (1964).
300. A. Keller and A. O'Connor, *Polymer* **1**, 163 (1960).
301. F. J. Balta-Calleja and A. Keller, *J. Polym. Sci. Part A* **2**, 2151, 2171 (1964).
302. G. M. Stack, L. Mandelkern, and I. G. Voigt-Martin, *Polym. Bull.* **8**, 421 (1982); *Macromolecules* **17**, 321 (1984).
303. I. G. Voigt-Martin and L. Mandelkern, *J. Polym. Sci. Polym. Phys. Ed.* **22**, 1901 (1984).
304. W. Kern, J. Davidovits, K. J. Rauterkus, and G. F. Schmidt, *Makromol. Chem.* **43**, 106 (1961).
305. H. Zahn and W. Pieper, *Colloid Polym. Sci.* **180**, 97 (1962).
306. G. Ungar, J. Stejny, A. Keller, I. Bidd, and M. C. Whiting, *Science* **229**, 444 (1985).
307. J. P. Arlie, P. Spegt, and A. Skoulios, *Makromol. Chem.* **99**, 160 (1966); **104**, 212 (1967).
308. B. Lotz, A. J. Kovacs, and J. C. Wittmann, *J. Polym. Sci. Polym. Phys. Ed.* **13**, 909 (1975).
309. D. M. Sadler and A. Keller, *Polymer* **17**, 37 (1976).

310. A. J. Kovacs, A. Gonthier, and C. Straupe, *J. Polym. Sci. Part C* **50**, 283 (1975).
311. *Ibid.*, **59**, 31 (1977).
312. J. Kugler, U. Struth, R. Born, E. W. Fischer, and K. Hahn, *J. Polym. Sci. Polym. Phys. Ed.*, to be published.
313. E. D. T. Atkins, A. Keller, and D. M. Sadler, *J. Polym. Sci. Part A-2* **10**, 863 (1972).
314. P. Dreyfus and A. Keller, *J. Polym. Sci. Polym. Phys. Ed.* **11**, 201 (1973).
315. B. Hinrichsen, *Makromol. Chem.* **166**, 291 (1973).
316. H. W. Starkweather and R. E. Moynihan, *J. Polym. Sci.* **22**, 363 (1956).
317. N. A. Arroyo and A. Hiltner, *J. Appl. Polym. Sci.* **23**, 1473 (1979).
318. J. Runt and I. R. Harrison, *J. Polym. Sci. Part B* **18**, 83 (1980).
319. M. E. Guiluz, H. Ishida, and A. Hiltner, *J. Polym. Sci. Polym. Phys. Ed.* **17**, 893 (1979).
320. B. H. Chang, A. Siegmann, and A. Hiltner, *J. Polym. Sci. Polym. Phys. Ed.* **22**, 255 (1982).
321. E. W. Fischer and R. Lorenz, *Colloid Polym. Sci.* **189**, 97 (1962).
322. R. Hosemann, *Polymer* **3**, 349 (1962).
323. H. Čačkovic, R. Hosemann, and W. Wilke, *Colloid Polym. Sci.* **234**, 1000 (1968).
324. R. Hosemann, *CRC* **1**, 351 (1972).
325. R. Hosemann, *Endeavour* **32**, 99 (1973).
326. J. Haase, R. Hosemann, and H. Čačkovic, *Polymers* **18**, 743 (1977).
327. R. Hosemann, P. H. Lindenmeyer, and G. S. Yeh, *J. Macromol. Sci. Part B* **15**, 17 (1978).
328. D. C. Bassett, F. C. Frank, and A. Keller, *Nature* **184**, 810 (1959).
329. A. Boudet, *J. Mater. Sci.* **19**, 2989 (1984).
330. E. L. Thomas, S. L. Sass, and E. J. Kramer, *J. Polym. Sci. Polym. Phys. Ed.* **12**, 1015 (1974).
331. I. R. Harrison and J. Runt, *J. Polym. Sci. Polym. Phys. Ed.* **14**, 317 (1976).
332. I. R. Harrison, A. Keller, D. M. Sadler, and E. L. Thomas, *Polymer* **17**, 736 (1976).
333. J. R. White, *J. Polym. Sci. Polym. Phys. Ed.* **16**, 387 (1978).
334. C. Vonk, *J. Appl. Crystallogr.* **11**, 541 (1978).
335. D. L. Dorset, *J. Polym. Sci. Polym. Phys. Ed.* **17**, 1797 (1979).
336. D. J. Bacon and N. A. Geary, *J. Mater. Sci.* **18**, 853 (1983).
337. *Ibid.*, p. 864.
338. D. J. Bacon and K. Tharmalingan, *J. Mater. Sci.* **18**, 884 (1983).
339. A. Peterlin and G. Meinel, *J. Appl. Phys.* **35**, 817 (1964).
340. A. J. J. Pennings, J. M. M. A. van der Mark, and A. M. Kiel, *Colloid Polym. Sci.* **237**, 336 (1970).
341. I. C. Sanchez and E. A. DiMarzio, *J. Res. Nat. Bur. Stand. Sect. A* **76**, 213 (1972).
342. A. Mehta and B. Wunderlich, *J. Polym. Sci. Polym. Phys. Ed.* **12**, 255 (1974); *Colloid Polym. Sci.* **253**, 193 (1975).
343. W. D. Niegisch and P. R. Swan, *J. Appl. Phys.* **31**, 1906 (1960).
344. D. C. Bassett, F. C. Frank, and A. Keller, *Philos. Mag.* **8**, 1739 (1963).
345. A. Peterlin, *J. Polym. Sci. Part C* **9**, 61 (1965).
346. D. C. Bassett, A. Keller, and S. Mitsuhashi, *J. Polym. Sci. Part A* **1**, 762 (1963).
347. R. Hosemann, W. Wilke, and F. J. Balta-Calleja, *Acta Crystallogr.* **21**, 118 (1966).
348. I. Heber, *Colloid Polym. Sci.* **191**, 30 (1963).
349. M. Kojima, *J. Polym. Sci. Part A-2* **5**, 615 (1967).
350. A. Keller and A. O'Connor, *Discuss. Faraday Soc.* **25**, 114 (1958).
351. C. Sella and J. J. Trillat, *C. R.* **248**, 410 (1959).
352. R. D. Burbank, *Bell Syst. Tech. J.* **39**, 1627 (1960).
353. D. J. Blundell and A. Keller, *J. Macromol. Sci. Part B* **2**, 337 (1968).
354. A. Keller, *J. Macromol. Sci. Part B* **2**, 337 (1968).
355. H. Kiho, A. Peterlin, and P. H. Geil, *J. Appl. Phys.* **35**, 1599 (1964).
356. P. Allan, E. B. Crelli, and M. Bevis, *Philos. Mag.* **27**, 127 (1973).
357. J. R. Knox, *J. Polym. Sci. Part C* **18**, 69 (1967).
358. R. J. Young, R. Dulniak, D. N. Batchelder, and D. Bloor, *J. Polym. Sci. Polym. Phys. Ed.* **17**, 1325 (1979).
359. M. Bevis, *Colloid Polym. Sci.* **256**, 234 (1978).
360. A. J. Kovacs, B. Lotz, and A. Keller, *J. Macromol. Sci. Part B* **3**, 385 (1969).
361. A. J. Kovacs and A. Gonthier, *Colloid Polym. Sci.* **250**, 530 (1972).
362. B. Lotz, A. J. Kovacs, G. A. Bassett, and A. Keller, *Colloid Polym. Sci.* **209**, 115 (1966).
363. J. C. Wittman and A. J. Kovacs, *Ber. Bunsenges. Phys. Chem.* **74**, 901 (1970).

364. B. Wunderlich and P. Sullivan, *J. Polym. Sci.* **61,** 195 (1962).
365. F. Khoury, *J. Res. Nat. Bur. Stand. Sect. A* **70,** 29 (1966).
366. A. J. Lovinger, *J. Polym. Sci. Polym. Phys. Ed.* **21,** 97 (1983).
367. H. D. Keith and F. J. Padden, Jr., *J. Appl. Phys.* **44,** 1217 (1973).
368. V. F. Holland and P. H. Lindenmeyer, *J. Appl. Phys.* **36,** 3049 (1965).
369. P. H. Lindenmeyer, *J. Polym. Sci. Part C* **15,** 109 (1967).
370. N. Niinomi, K. Abe, and M. Takayanagi, *J. Macromol. Sci. Part B* **2,** 649 (1968).
371. D. M. Sadler and A. Keller, *Colloid Polym. Sci.* **239,** 641 (1970).
372. J. Peterman and H. Gleiter, *Philos. Mag.* **25,** 813 (1972).
373. C. A. Garber and P. H. Geil, *Makromol. Chem.* **113,** 246 (1968).
374. V. F. Holland, *J. Appl. Phys.* **35,** 1351, 3235 (1964).
375. T. J. Weaver and I. R. Harrison, *Polymer* **22,** 1590 (1981).
376. J. H. Magill, *J. Polym. Sci. Part B* **20,** 7 (1982).
377. N. Hirai, Y. Yamashita, T. Mitsuhata, and Y. Tamura, *Rep. Res. Lab. Surf. Sci. Okayama Univ.* **2,** 1 (1961).
378. E. W. Fischer and G. F. Schmidt, *Angew. Chem.* **74,** 551 (1962).
379. M. Takayanagi and F. Nagatoshi, *Mem. Fac. Eng. Kyushu Univ.* **24,** 33 (1965).
380. T. Kawai, *Colloid Polym. Sci.* **201,** 104 (1965).
381. A. Nakajima, S. Hayashi, and H. Nishimura, *Colloid Polym. Sci.* **229,** 107 (1967).
382. A. Kawaguchi, T. Ichida, S. Murakami, and K. Katayama, *Colloid Polym. Sci.* **262,** 597 (1984).
383. T. Ichida, M. Tsuji, S. Murakami, A. Kawaguchi, and K. Katayama, *Colloid Polym. Sci.* **263,** 293 (1985).
384. A. Peterlin, *J. Polym. Sci. Part B* **1,** 279 (1963).
385. J. D. Hoffman and J. J. Weeks, *J. Chem. Phys.* **42,** 4301 (1965).
386. L. Mandelkern and A. L. Allou, *J. Polym. Sci. Part B* **4,** 447 (1966).
387. L. Mandelkern, *Polym. Eng. Sci.* **7,** 232 (1967).
388. D. C. Bassett and A. M. Hodge, *Proc. R. Soc. London Ser. A* **359,** 121 (1978).
389. I. G. Voigt-Martin, L. Mandelkern, and E. W. Fischer, *J. Polym. Sci. Polym. Phys. Ed.* **18,** 2347 (1980).
390. T. Asaki, Y. Miyamoto, H. Miyaji, and K. Assi, *Polymer* **23,** 773 (1982).
391. P. Dreyfus and A. Keller, *J. Polym. Sci. Part B* **8,** 253 (1970).
392. V. F. Holland, *J. Appl. Phys.* **35,** 59 (1964).
393. W. O. Statton and P. H. Geil, *J. Appl. Polym. Sci.* **3,** 357 (1960).
394. P. R. Swan, *J. Polym. Sci.* **56,** 409 (1962).
395. M. J. Richardson, P. J. Flory, and J. B. Jackson, *Polymer* **4,** 221 (1963).
396. J. B. Jackson and P. J. Flory, *Polymer* **5,** 159 (1964).
397. C. H. Baker and L. Mandelkern, *Polymer* **7,** 7, 71 (1966).
398. J. P. Colson and R. K. Eby, *J. Appl. Phys.* **37,** 3511 (1966).
399. P. J. Balta-Calleja and A. Schoenfeld, *Faserforsch. Textiltech.* **18,** 170 (1967).
400. P. J. Flory, *Trans. Faraday Soc.* **51,** 848 (1955).
401. H. Baur, *Colloid Polym. Sci.* **203,** 97 (1965).
402. W. R. Krigbaum and I. Uematsu, *J. Polym. Sci. Part A* **3,** 2915 (1965).
403. J. Willems and E. W. Fischer, *Discuss. Faraday Soc.* **25,** 204 (1957).
404. J. Willems and I. Willems, *Experientia* **13,** 465 (1957).
405. E. W. Fischer in B. Ke, ed., *Newer Methods of Polymer Characterization,* Vol. 6, Wiley-Interscience, New York, 1964, p. 136.
406. S. Wellinghoff, E. Rybnikar, and E. Baer, *J. Macromol. Sci. Part B* **10,** 1 (1974).
407. S. Wu, *J. Macromol. Sci. Part B* **10,** 1 (1974).
408. S. E. Rickert and E. Baer, *J. Polym. Sci. Polym. Phys. Ed.* **16,** 895 (1978).
409. P. Lovinger, *Macromolecules* **14,** 322 (1981).
410. J. A. Koutsky, A. G. Walton, and E. Baer, *J. Polym. Sci. Part A-2* **4,** 611 (1966); *Part B* **5,** 177, 185 (1967).
411. S. E. Rickert and E. Baer, *J. Appl. Phys.* **47,** 304 (1976).
412. C. M. Balik and A. J. Hopfinger, *J. Appl. Crystallogr.* **13,** 999 (1980).
413. E. W. Fischer, *Colloid Polym. Sci.* **159,** 108 (1958).
414. E. W. Fischer and J. Willems, *Makromol. Chem.* **99,** 85 (1966).
415. J. Willems, *Discuss. Faraday Soc.* **25,** 111 (1957); *Experientia* **23,** 409 (1967).

416. F. Lovinger, *Polym. Prepr. Am. Chem. Soc. Div. Polym. Chem.* **21**, 253 (1980).
417. D. C. Bassett, *Principles of Polymer Morphology*, Cambridge University Press, Cambridge, UK, 1981.
418. R. L. Miller, I. R. Raff, and M. W. Bonk, eds., *Crystalline Olefin Polymers*, John Wiley & Sons, Inc., New York, 1965.
419. A. Galeski and E. Piotrowska, *J. Polym. Sci. Polym. Phys. Ed.* **21**, 1299 (1983).
420. A. Peterlin in Ref. 276, p. 1.
421. T. W. Haas and P. H. MacRae, *Polym. Eng. Sci.* **9**, 227 (1969).
422. I. G. Voigt-Martin and L. Mandelkern, *J. Polym. Sci. Polym. Phys. Ed.* **19**, 1769 (1981).
423. H. D. Keith and T. C. Loomis, *J. Polym. Sci. Polym. Phys. Ed.* **22**, 295 (1984).
424. L. D. Coxon and J. R. White, *J. Mater. Sci.* **14**, 1114 (1979).
425. W. D. Niegisch, *J. Polym. Sci.* **40**, 203 (1959).
426. H. Staudinger and R. Signer, *Z. Kristallogr. Kristallgeom. Kristallphysik. Kristallchemie* **70**, 193 (1929).
427. D. C. Bassett, A. M. Hodge, and R. H. Olley, *Faraday Discuss. R. Soc. Chem.* **68**, 218 (1979).
428. D. C. Bassett, *Developments in Crystalline Polymers,* Applied Science Publishers, London, 1982.
429. W. Rose and C. Meurer, *J. Mater. Sci.* **16**, 883 (1981).
430. D. C. Bassett, *Polymer* **19**, 469 (1978).
431. D. T. Grubb and A. Keller, *J. Polym. Sci. Polym. Phys. Ed.* **18**, 207 (1980).
432. A. Keller and M. J. Machin, *J. Macromol. Sci. Part B* **1**, 41 (1967).
433. A. Keller, *Nature* **169**, 913 (1952).
434. R. G. Strobl, T. Engelke, H. Meier, and G. Urban, *Colloid Polym. Sci.* **260**, 394 (1982).
435. J. Martinez-Salazar and F. J. Balta-Calleja, *Polym. Bull.* **3**, 7 (1980).
436. H. D. Keith, F. J. Padden, and R. G. Vadimsky, *Science* **150**, 1026 (1965); *J. Polym. Sci. Part A-2* **4**, 267 (1966); *J. Appl. Phys.* **37**, 4027 (1966).
437. E. S. Clark, *Soc. Plast. Eng. Tech. Pap.* **23**, 46 (July 1967).
438. Y. Hase and P. H. Geil, *Polymer J. Jpn.* **2**, 560, 581 (1971).
439. F. Rybnikar and P. H. Geil, *J. Macromol. Sci. Part B* **7**, 1 (1973).
440. H. D. Keith and F. J. Padden, Jr., *J. Appl. Phys.* **34**, 2409 (1963); **35**, 1270, 1286 (1964).
441. H. A. Davis, *J. Polym. Sci. Part A-2* **4**, 267 (1966).
442. H. D. Keith, F. J. Padden, Jr., and R. G. Vadimsky, *J. Appl. Phys.* **42**, 4585 (1971).
443. M. E. Cagiao and F. J. Balta-Calleja, *J. Macromol. Sci. Part B* **21**, 559 (1982).
444. P. D. Calvert, *J. Polym. Sci. Part B* **21**, 467 (1983).
445. D. C. Bassett and M. Hodge, *Proc. R. Soc. London Part A* **377**, 65 (1981).
446. F. R. Anderson, *J. Polym. Sci. Part C* **8**, 275 (1965).
447. W. Wilke, *Colloid Polym. Sci.* **258**, 360 (1980); **259**, 577 (1981).
448. R. Bonart, *Colloid Polym. Sci.* **194**, 97 (1964).
449. R. Hosemann and W. Wilke, *Makromol. Chem.* **118**, 230 (1968).
450. W. Wilke, W. Vogel, and R. Hosemann, *Colloid Polym. Sci.* **237**, 317 (1970).
451. P. J. Barham, R. A. Chivers, J. Martinez-Salazar, and A. Keller, *J. Polym. Sci. Part B* **19**, 539 (1981); *J. Polym. Sci. Polym. Phys. Ed.* **20**, 1717 (1982); P. J. Barham, D. A. Jarvis, and A. Keller, *J. Polym. Sci. Polym. Phys. Ed.* **20**, 1733 (1982); J. Martinez-Salazar, P. J. Barham, and A. Keller, *J. Polym. Sci. Polym. Phys. Ed.* **22**, 1085 (1984).
452. M. Avrami, *J. Chem. Phys.* **7**, 1103 (1939); **8**, 212 (1940).
453. A. N. Kolmogoroff, *Izv. Akad. Nauk Kaz. SSR Ser. Mat.* **1**, 355 (1937).
454. U. R. Evans, *Trans. Faraday Soc.* **41**, 365 (1945).
455. J. Rabesiaka and A. J. Kovacs, *J. Appl. Phys.* **32**, 2314 (1961).
456. J. N. Hay and Z. J. Przekop, *J. Polym. Sci. Polym. Phys. Ed.* **16**, 81 (1978).
457. *Ibid.,* **19**, 951 (1979).
458. D. Grenier and R. E. Prudhomme, *J. Polym. Sci. Polym. Phys. Ed.* **18**, 1655 (1980).
459. J. Rohleder and H. A. Stuart, *Makromol. Chem.* **41**, 110 (1960).
460. W. Dietz, *Colloid Polym. Sci.* **259**, 413 (1981).
461. A. Ya. Malkin, V. P. Begishev, I. A. Keapin, and S. A. Bolgov, *Polym. Eng. Sci.* **24**, 1396 (1984).
462. A. Ya. Malkin, V. P. Begishev, I. A. Keapin, and Z. S. Andrianova, *Polym. Eng. Sci.* **24**, 1402 (1984).
463. A. Ya. Malkin, V. P. Begishev, and I. A. Kipin, *Polymer* **24**, 81 (1983).
464. T. Ozawa, *Polymer* **12**, 150 (1971).

465. M. Eder and A. Wlochovicz, *Polymer* **24**, 1593 (1983).

466. M. R. Kamal and E. Chou, *Polym. Eng. Sci.* **23**, 27 (1983).

467. V. P. Megishev, I. A. Kipin, and A. Ya. Malkin, *Vysokomol. Soedin. Ser. B* **24**, 656 (1982).

468. C. J. Speerschneider and C. H. Li, *J. Appl. Phys.* **33**, 1871 (1962).

469. B. Wunderlich and T. Arakawa, *J. Polym. Sci. Part A* **2**, 3697 (1964).

470. P. H. Geil, F. R. Anderson, B. Wunderlich, and T. Arakawa, *J. Polym. Sci. Part A* **2**, 3707 (1964).

471. P. J. Holdsworth and A. Keller, *J. Macromol. Sci. Part B* **1**, 595 (1967).

472. F. R. Anderson, *J. Polym. Sci. Part C* **3**, 123 (1963); *Part B* **3**, 721 (1965).

473. D. C. Bassett and B. Turner, *Nature London Phys. Sci.* **240**, 146 (1972); *Philos. Mag.* **29**, 925 (1974).

474. D. C. Bassett, *Polymer* **17**, 460 (1976).

475. D. C. Bassett, S. Block, and G. J. Piermarini, *J. Appl. Phys.* **45**, 4146 (1974).

476. I. Poddubny, G. K. Elyashevich, V. G. Baranov, and S. Frenkel, *Polym. Eng. Sci.* **20**, 206 (1980).

477. U. Leute and W. Dollhopf, *Colloid Polym. Sci.* **258**, 353 (1980).

478. K. Matsushige and T. Takemura, *J. Polym. Sci. Polym. Phys. Ed.* **16**, 921 (1978).

479. R. K. Jain and R. Simha, *J. Polym. Sci. Part B* **17**, 33 (1979).

480. T. Asaki, *J. Polym. Sci. Polym. Phys. Ed.* **22**, 175 (1984).

481. Y. Maeda, H. Kanetsuma, K. Nagata, K. Matsushige, and T. Takemura, *J. Polym. Sci. Polym. Phys. Ed.* **19**, 1313 (1981).

482. Y. Maeda, H. Kanetsuma, K. Tagashira, and T. Takemura, *J. Polym. Sci. Polym. Phys. Ed.* **19**, 1325 (1981).

483. T. S. Hsu, *J. Polym. Sci. Polym. Phys. Ed.* **18**, 2379 (1980).

484. L. Wunder, *Macromolecules* **14**, 1024 (1981).

485. T. Hashimoto, T. Sakai, and S. Miyata, *J. Polym. Sci. Polym. Phys. Ed.* **16**, 1965 (1978).

486. T. Hashimoto, K. Ogita, S. Uemoto, and T. Sakai, *J. Polym. Sci. Polym. Phys. Ed.* **21**, 1347 (1983).

487. N. E. Schlotter and J. F. Rabolt, *Macromolecules* **17**, 1581 (1984).

488. J. F. Rabolt and C. H. Wang, *Macromolecules* **16**, 1698 (1983).

489. G. E. Attenburry and D. C. Bassett, *J. Mater. Sci.* **14**, 2679 (1979).

490. P. W. Teare and D. R. Holmes, *J. Polym. Sci.* **24**, 497 (1957).

491. K. Tanaka, T. Seto, T. Haa, *J. Phys. Soc. Jpn.* **17**, 873 (1962); *Rep. Prog. Polym. Phys. Jpn.* **6**, 293 (1963).

492. A. Turner-Jones, *J. Polym. Sci.* **62**, S53 (1962).

493. P. Ingram and A. Peterlin, *J. Polym. Sci. Part B* **2**, 739 (1964).

494. A. Wlochovicz and M. Eder, *Polymer* **25**, 1268 (1984).

495. K. A. Narh, J. A. Odell, A. Keller, and G. V. Fraser, *J. Mater. Sci.* **15**, 2001 (1980).

496. G. V. Fraser, A. Keller, and J. A. Odell, *J. Appl. Polym. Sci.* **22**, 279 (1978).

497. J. J. Weeks, *J. Res. Nat. Bur. Stand. Sect. A* **67**, 441 (1963).

498. D. T. Grubb, J. J. H. Lin, M. Caffrey, and D. H. Bilderbach, *J. Polym. Sci. Polym. Phys. Ed.* **22**, 367 (1984).

499. J. J. Mucigrosse and P. J. Phillips, *Coat. Plast. Prepr.* **38**, 424 (1978).

500. S. R. Barnes, *Polymer* **21**, 723 (1980).

501. R. Kuhn and H. Kroner, *Colloid Polym. Sci.* **260**, 1083 (1982).

502. G. I. Asbach and W. Wilke, *Colloid Polym. Sci.* **260**, 113 (1982).

503. T. Paluda, *Polymer* **23**, 1300 (1982).

504. P. J. Flory and A. Vrij, *J. Am. Chem. Soc.* **85**, 534 (1963).

505. J. N. Hay and M. Wiles, *Makromol. Chem.* **178**, 623 (1977).

506. I. G. Voigt-Martin, *J. Polym. Sci. Polym. Phys. Ed.* **18**, 1513 (1980).

507. L. Mandelkern, M. Glotin, R. A. Benson, *Macromolecules* **14**, 22 (1981).

508. M. Glotin and L. Mandelkern, *Macromolecules* **14**, 1394 (1981).

509. J. L. Pezzetti, N. J. Capiati, and E. M. Valles, *J. Polym. Sci. Part B* **22**, 401 (1984).

510. A. Keller, *Polymer* **3**, 393 (1962).

511. *Faraday Discuss. R. Soc. Chem.* **68** (1979).

512. A. Peterlin, *J. Macromol. Sci. B* **3**, 19 (1969).

513. H. G. Zachmann and A. Peterlin, *J. Macromol. Sci. Part B* **3**, 495 (1969).

514. A. Peterlin and H. G. Zachmann, *J. Polym. Sci. Part C* **34**, 11 (1971).

515. W. M. Ewers, H. G. Zachmann, and A. Peterlin, *J. Macromol. Sci. Part B* **6**, 695 (1972).

516. A. Peterlin, *J. Coll. Interface Sci.* **43**, 255 (1973).

517. D. C. Bassett, *Philos. Mag.* **12,** 907 (1965).

518. K. Kiss, S. H. Carr, A. G. Walton, and E. Baer, *J. Polym. Sci. Ser. B* **5,** 1087 (1967).

519. J. Breedon-Jones and P. H. Geil, *J. Res. Nat. Bur. Stand. Sect. A* **79,** 609 (1975).

520. E. Segerman and P. G. Stern, *Nature* **210,** 1258 (1966).

521. P. G. Stern, *Colloid Polym. Sci.* **215,** 140 (1967).

522. P. Predecki and W. O. Statton, *J. Appl. Phys.* **37,** 4053 (1966).

523. E. W. Fischer, *Z. Naturforsch* **14a,** 584 (1959).

524. A. Peterlin and E. W. Fischer, *Z. Phys.* **159,** 272 (1960).

525. E. W. Fischer, *Ann. N.Y. Acad. Sci.* **89,** 620 (1961).

526. A. Peterlin, E. W. Fischer, and C. Reinhold, *J. Chem. Phys.* **37,** 1403 (1962); *J. Polym. Sci.* **62,** S59 (1962).

527. A. Peterlin and C. Reinhold, *J. Polym. Sci. Part A* **3,** 2801 (1965).

528. G. Allegra, *J. Chem. Phys.* **66,** 5453 (1977).

529. J. I. Lauritzen and J. D. Hoffman, *J. Res. Nat. Bur. Stand. Sect. A* **64,** 73 (1960).

530. J. D. Hoffman, *SPE Trans.* **4,** 315 (1964).

531. F. C. Frank and H. P. Tosi, *Proc. R. Soc. London Ser. A* **263,** 323 (1961).

532. F. P. Price, *J. Chem. Phys.* **35,** 1884 (1961).

533. F. P. Price, *J. Polym. Sci.* **42,** 49 (1960); *SPE Trans.* **4,** 151 (1964).

534. G. F. Bahr, F. B. Johnson, and E. Zeitler, *Lab. Invest.* **14,** 377 (1965).

535. D. Aitken, M. Glotin, P. J. Hendra, H. Jobic, and E. Marsden, *J. Polym. Sci. Part B* **14,** 619 (1976).

536. D. Aitken, D. J. Cutler, M. Glotin, P. J. Hendra, M. E. Cudby, and H. A. Willis, *Polymer* **20,** 1465 (1979).

537. E. Robelin, F. Rousseaux, M. Lemonnier, and J. Rault, *J. Phys. Paris* **41,** 1469 (1980).

538. J. J. Point, *Macromolecules* **12,** 770 (1979).

539. J. J. Point and A. J. Kovacs, *Macromolecules* **13,** 399 (1980).

540. G. Kanig, *Colloid Polym. Sci.* **261,** 993 (1983).

541. K. A. Dill, *Faraday Discuss. R. Soc. Chem.* **68,** 106 (1979).

542. L. Mandelkern, *Faraday Discuss. R. Soc. Chem.* **68,** 310 (1979).

543. R. Kitamaru and F. Horii, *Adv. Polym. Sci.* **26,** Springer-Verlag, New York, 1980, p. 137.

544. M. L. Mansfield, *Macromolecules* **16,** 914 (1983).

545. A. J. Pennings and A. M. Keil, *Colloid Polym. Sci.* **205,** 160 (1965).

546. A. J. Pennings, "Crystal Growth," *International IUPAC Conference,* Boston, Pergamon Press, London, 1966, p. 389.

547. R. Salovey and M. Y. Hellman, *J. Polym. Sci. Part B* **5,** 647 (1967).

548. J. K. Dees and J. E. Spruiell, *J. Appl. Polym. Sci.* **18,** 1053 (1974).

549. K. Sakaoku, H. G. Clark, and A. Peterlin, *J. Polym. Sci. Part A-2* **6,** 1035 (1968).

550. R. L. Miller, ed., *Flow-induced Crystallization in Polymer Systems,* Gordon & Breach Science Publishers, Inc., New York, 1979.

551. A. J. McHugh, *Polym. Eng. Sci.* **22,** 15 (1982).

552. J. C. M. Torfs, G. O. R. Aberda van Eckenstein, and A. J. Pennings, *Eur. Polym. J.* **17,** 157 (1981).

553. B. Wunderlich, C. M. Cormier, A. Keller, and M. J. Machin, *J. Macromol. Sci. Part B* **1,** 93 (1967).

554. F. M. Willmouth, A. Keller, I. M. Ward, and T. Williams, *J. Polym. Sci. Part A-2* **6,** 1627 (1968).

555. K. Katayama, T. Amano, and K. Nakamura, *Colloid Polym. Sci.* **226,** 125 (1968).

556. W. George and P. Tucker, *Polym. Eng. Sci.* **15,** 451 (1975).

557. J. C. M. Torfs, J. Smook, and A. J. Pennings, *J. Appl. Polym. Sci.* **28,** 57, 77 (1983).

558. B. Kalb and A. J. Pennings, *Polymer* **21,** 3 (1980).

559. J. Smook and A. J. Pennings, *Polym. Bull.* **9,** 75 (1983).

560. A. J. Pennings, R. Langeveen, R. S. de Vries, *Colloid Polym. Sci.* **255,** 532 (1977).

561. M. J. Hill, P. J. Barham, and A. Keller, *Colloid Polym. Sci.* **258,** 1023 (1980); M. J. Hill and A. Keller, *Colloid Polym. Sci.* **259,** 335 (1981); M. J. Hill, P. J. Barham, and A. Keller, *Colloid Polym. Sci.* **261,** 721 (1983); P. J. van Hutten, C. E. Koning, and A. J. Pennings, *Colloid Polym. Sci.* **262,** 521 (1984).

562. A. J. Pennings, J. Smook, J. de Boer, S. Gogolevski, and P. F. van Hutten, *Pure Appl. Chem.* **55,** 777 (1983).

563. J. A. Odell, D. T. Grubb, and A. Keller, *Polymer* **19,** 617 (1978).

564. J. A. Odell, A. Keller, and M. J. Miles, *Colloid Polym. Sci.* **262,** 683 (1984).
565. Z. Bashir, J. A. Odell, and A. Keller, *J. Mater. Sci.* **20,** (1985).
566. E. Jenckel, E. Teege, and W. Hinrichs, *Colloid Polym. Sci.* **129,** 19 (1952).
567. D. R. Fitchmun and S. Newman, *J. Polym. Sci. Part A-2* **8,** 1545 (1970).
568. M. J. Hill and A. Keller, *J. Macromol. Sci. Part B* **3,** 153 (1969).
569. W. G. Perkins and R. S. Porter, *J. Mater. Sci.* **12,** 2355 (1977).
570. A. Cifferi and I. M. Ward, eds., *Ultrahigh Modulus Polymers,* Applied Science Publishers, London, UK, 1979.
571. A. E. Zachariades and R. S. Porter, eds., *The Strength and Stiffness of Polymers,* Marcel Dekker, Inc., New York, 1983.
572. T. Shimada, A. E. Zachariades, M. P. C. Watts, and R. S. Porter, *J. Appl. Polym. Sci.* **26,** 1309 (1981).
573. P. D. Griswold, A. E. Zachariades, and R. S. Porter, *J. Polym. Sci. Part A-2* **6,** 205 (1968).
574. H. Hendus, *Colloid Polym. Sci.* **165,** 32 (1959).
575. R. Corneliussen and A. Peterlin, *Makromol. Chem.* **105,** 192 (1967).
576. J. F. Balta-Calleja, *J. Macromol. Sci. Part B* **4,** 519 (1970).
577. J. F. Balta-Calleja, A. Peterlin, and B. Crist, *J. Polym. Sci. Part A-2* **10,** 1749 (1972).
578. A. Garton, R. F. Stepaniak, D. J. Carlsson, and D. M. Wilkes, *J. Polym. Sci. Part B* **16,** 587 (1978).
579. A. O. Muzzi and D. Hansen, *Textile Res. J.* **41,** 436 (1970).
580. A. Peterlin, *Polym. Eng. Sci.* **17,** 183 ((1977).
581. G. D. Wignall and W. Wu, *Polym. Comment.* **24,** 354 (1983).
582. H. Kiho, A. Peterlin, and P. H. Geil, *J. Polym. Sci. Part B* **3,** 157, 257, 263 (1965).
583. J. Petermann and H. Gleiter, *J. Polym. Sci. Polym. Phys. Ed.* **11,** 359 (1973).
584. J. Petermann, W. Kluge, and H. Gleiter, *J. Polym. Sci. Polym. Phys. Ed.* **17,** 1943 (1979).
585. S. L. Cannon, G. B. McKenna, and W. O. Statton, *Macromol. Rev.* **11,** 209 (1976).
586. G. K. Elyashevich, *Adv. Polym. Sci.* **43,** Springer-Verlag, New York, 1982, p. 205.
587. R. G. Quinn and B. S. Sprague, *J. Polym. Sci. Part A-2* **8,** 1971 (1970).
588. J. L. Williams and A. Peterlin, *Makromol. Chem.* **135,** 41 (1970).
589. A. Peterlin and K. Sakaoku in G. Goldfinger, ed., *Clean Surfaces,* Marcel Dekker, Inc., New York, 1970, p. 1.
590. A. Peterlin, *Colloid Polym. Sci.* **253,** 809 (1975).
591. F. J. Balta-Calleja and A. Peterlin, *J. Mater. Sci.* **4,** 722 (1969).
592. G. Capaccio and I. M. Ward, *J. Polym. Sci. Polym. Phys. Ed.* **19,** 667 (1981).
593. G. Meinel and A. Peterlin, *Eur. Polym. J.* **7,** 657 (1971).
594. P. Smith and P. J. Lemstra, *J. Polym. Sci. Polym. Phys. Ed.* **19,** 1007 (1981).
595. H. H. Chuah and R. S. Porter, *Polymer* **27,** 1022 (1986).
596. F. DeCandia, R. Russo, V. Vittoria, and A. Peterlin, *J. Polym. Sci. Polym. Phys. Ed.* **18,** 2083 (1980).

ANTON PETERLIN
Consultant

N

NAPHTHENATES. See Drying oils.

NATURAL FIBERS. See Cotton; Fibers, vegetable; Silk; Wool.

NATURAL GUMS. See Gums, industrial; Polysaccharides.

NATURAL POLYMERS. See Biopolymers.

NATURAL RESINS. See Resins, natural.

NATURAL RUBBER. See Rubber, natural.

NEOPRENE. See Chloroprene polymers.

NETWORKS

This article deals with the theoretical description of elastomeric, polymeric networks in terms of their molecular-chain configurations (see also Cross-linking; Elasticity; Gels).

Polymeric networks, ie, systems of interconnected macromolecular chains, exhibit the property of high extensibility coupled with the capacity for full recovery that is implied by the term "rubber elasticity." Although the constitutional units comprising various networks may differ widely, their mechanical properties have much in common; this is especially evident in their stress–strain relationships.

The junctions that confer recoverability on the network ordinarily are provided by covalent bonds. In some instances, however, physical combinations of chains, mediated, for example, by crystallites, may serve the same function as the permanent chemical interlinkages. Thus rubber elasticity assumes importance beyond the narrow limits by which it is commonly designated. It is operative in the swelling of polymeric networks, the deformation of semicrystalline polymers, and the viscoelastic behavior of linear polymers under flow in the liquid or amorphous state, and is essential to the functions of elastic proteins and muscle. The theory of rubber elasticity is centrally important to much of polymer science.

The density of junctions or cross-linkages in elastomeric networks is generally low, typically 0.05–0.2 mol/kg. Yet, the degree of interlinking places them far beyond the gel point. It follows that imperfections in the form of chains attached to the network at one end are usually of minor consequence. To characterize the constitution of the network in a way that comprehends network imperfections, the cycle rank of the network defined below turns out to be the appropriate measure of connectivity.

A feature of networks, second in importance only to the degree of connectivity, is the copious interpenetration of the chains comprising them. Chains emanating from a given junction of the network are embedded in a maze of other chains and junctions. Most of the latter are only remotely topologically related to the junction considered. This fact, often overlooked, has an important bearing on the properties of elastomeric networks. It precludes their treatment in terms of a lattice, albeit a disordered one, consisting of junctions related to their spatial neighbors by connecting chains (see also INTERPENETRATING POLYMER NETWORKS).

Molecular Theory

The basic premise of the molecular theory of elasticity in networks asserts that the stress in a typical strained network originates within the molecular chains of the structure; contributions from interactions between the chains are negligible. This premise finds direct support in elasticity measurements on polymeric networks. The temperature coefficient of the stress at fixed strain and its constancy with dilution are especially significant in this connection (1–4).

Even more pervasive confirmation is provided by experiments showing that configurations of polymer chains are unperturbed by their neighbors in amorphous polymers. Neutron scattering (qv) has been particularly decisive in demonstrating the absence of appreciable perturbations in the configurations of polymer chains when interspersed with other randomly configured polymer molecules of the same kind (5,6). The (free) energy of interaction between neighboring chains must therefore be sensibly independent of their configurations. It follows that the intermolecular energy should not be significantly affected by the changes in configurations of the chains of a network induced by deformation (7). The stored elastic free energy, which is central to the theory of rubber elasticity, therefore comprises the sum of contributions of the individual network chains. Other contributions, such as those often postulated to arise from interchain interactions, may be ignored according to the stated premise and the compelling evidence in its support.

The principal task of theory is to establish the relationship between the macroscopic strain and the distortion of the distribution of configurations of the network chains (7,8). It was assumed originally that the locations of the network junctions may be considered to be affine in the macroscopic strain, from which it followed that the distribution of end-to-end vectors of the chains, ie, the chain vectors, should likewise be affine in the strain (9–14). The theory of James and Guth appeared at first to corroborate this conjecture, inasmuch as it showed the mean locations of junctions in a phantom network of Gaussian chains to be affine in the strain (15). Nearly 30 years elapsed before the important distinction between the distribution of mean chain vectors and their instantaneous (or time-averaged) distribution was recognized (7). Modern theory of rubber elasticity is an outgrowth of the recognition that the actual distribution of chain vectors in a network of Gaussian chains is, in general, nonaffine in the strain.

The Isolated Chain

The chains in elastomeric networks typically consist of 100–1000 skeletal bonds. For free chains of this length, the function describing the distribution of end-to-end chain vectors $\mathbf{r}$ is Gaussian in good approximation (16), ie, the distribution is well represented by

$$W(\mathbf{r}) = (\tfrac{3}{2}\pi\langle r^2\rangle_0)^{3/2} \exp\left[-(\tfrac{3}{2}\langle r^2\rangle_0)r^2\right] \tag{1}$$

where $\langle r^2\rangle_0$ is the mean-square magnitude of $\mathbf{r}$ for the free chain averaged over all configurations. It follows that the free energy of the chain is given as a function of its displacement length $r \equiv |\mathbf{r}|$ by

$$A(\mathbf{r}) = \text{const} - kT \ln W(\mathbf{r}) = A^0(T) + (3kT/2\langle r^2\rangle_0)r^2 \tag{2}$$

The magnitude of the average retractive force exerted by the chain at fixed r, obtained by differentiation of equation 2, is

$$\bar{f} = 3kT\langle r^2\rangle_0^{-1}r \tag{3}$$

It is directed along the chain vector. Proportionality between the average force and the displacement length of the chain follows directly from equation 1, as is obvious. Conversely, primary assertion of equation 3 would lead to equation 1.

According to the premise enunciated above, the elastic properties of a network of Gaussian chains must follow from these relationships.

Network Structure and Topology

A polymer network may be characterized by the number μ_J of its junctions, their functionality ϕ (or average functionality $\bar{\phi}$), and by the number ν_{ends} of chain ends (7). The number of chains in the network, including those with only one end attached, is

$$\nu = \tfrac{1}{2}(\mu_J\phi + \nu_{\text{ends}}) \tag{4}$$

The effective number ν_e of chains is less than ν as a result of the imperfections caused by free chain ends. For a perfect network where $\nu_{\text{ends}} = 0$,

$$\nu_e = \nu = \mu_J\phi/2 \tag{5}$$

A quantity that characterizes the network with greater generality, regardless of the nature of its imperfections, is the cycle rank ξ or number of independent circuits it contains (7,17). It may be defined alternatively as the minimum number of scissions required to reduce the network to a spanning tree, ie, a unified structure comprising all of the chains and containing no closed circuits or loops. This quantity will be used in due course to characterize the elastic response of the network. It suffices to observe that in a perfect network, ξ is given by the difference between the number of chains and the number of junctions of functionality $\phi \geq 3$, that is,

$$\xi = \nu - \mu_J = \nu(1 - 2/\phi) \tag{6}$$

See equation 5.

A prominent and important feature that is characteristic of polymeric networks is the copious interpenetration of chains and junctions (7,18). The region of space pervaded by a given chain is shared with many other chains and junctions. The domain roughly demarcated by the junctions that are topologically first neighbors of a given junction is occupied by many other junctions. This is illustrated in Figure 1 for a tetrafunctional network.

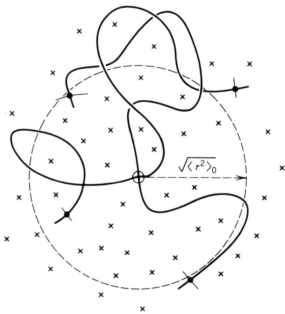

Fig. 1. x, Spatial neighbor junctions surrounding a given junction and ●, its four topological neighbors in a tetrafunctional network.

The average number Γ of junctions within the region of radius $\langle r^2 \rangle_0^{1/2}$ offers a quantitative measure of the degree of interpenetration. It is given by

$$\Gamma = (4\pi/3)\langle r^2 \rangle_0^{3/2}(\mu_J/V^0) \tag{7}$$

where V^0 is the volume of the network in its state of reference (see below). Since $\langle r^2 \rangle_0$ increases linearly with the length of a chain, it must increase linearly with μ_J^{-1}. Hence, Γ is inversely proportional to the square root of the degree of interlinking. For typical elastomeric networks, Γ is in the range 25–100. Clearly, the chains and junctions are profusely interspersed. The first tier of topological neighbors is located well beyond the nearest spatial neighbors (Fig. 1). It is to be noted also that the shortest topological pathway from a given junction to one of its nearest neighbors in space may span many chains.

Affine Networks

The high degree of interpenetration in elastomeric networks and the fact that each junction is located in an environment dominated by chains and junctions whose structural relation to the junction considered is remote lends credence to the assumption universally adopted in the earliest theories of rubber elasticity (9–14) that the positions of the junctions are approximately affine in the macroscopic strain (13). It follows at once from the premise introduced above that the elastic free energy of the network is the sum of expressions, eg, equation 2, for each chain of the network. The sum $\Sigma_{i=1}^{\nu} r_i^2$ is required over all chains. According to the assumption that the transformation of chain vectors is affine in the displacement gradient tensor $\boldsymbol{\lambda}$ that defines the macroscopic strain, this sum is just $\nu \langle r^2 \rangle_0 (\lambda_x^2 + \lambda_y^2 + \lambda_z^2)$, where λ_x, λ_y, and λ_z are the principal extension ratios measured relative to the dimensions of the specimen when isotropic and at the volume V^0, such that the mean-square magnitude of the chain vectors matches the value $\langle r^2 \rangle_0$ for unperturbed chains. Adding the term $-\mu_J kT \ln V$ for the dispersion of the junctions over the prevailing volume V and expressing the free energy relative to the state of reference in which $\lambda_x = \lambda_y = \lambda_z = 1$ and $V = V^0$, the following is obtained (14,19):

$$\Delta A_{\text{aff}} = (\nu/2)kT(\lambda_x^2 + \lambda_y^2 + \lambda_z^2 - 3) - \mu_J kT \ln (V/V^0) \tag{8}$$

The stress is obtained as a function of strain by differentiation of equation 8. For uniaxial elongation parallel to the x axis, $\lambda_x = \lambda = L/L^0$ and $\lambda_y = \lambda_z = (V/V_0\lambda)^{1/2}$, and the force of retraction for the affine network is

$$\begin{aligned} f_{\text{aff}} &= (\partial \Delta A_{\text{el}}/\partial L)_{T,V} = (\partial \Delta A_{\text{el}}/\partial \lambda)_{T,V}/L^0 \\ &= (\nu kT/L^0)(\lambda - V/V^0\lambda^2) \tag{9} \\ &= (\nu kT/L_{i,V})(V/V^0)^{2/3}(\alpha - \alpha^{-2}) \tag{10} \end{aligned}$$

where $\alpha = L/L_{i,V} = \lambda(V/V^0)^{-1/3}$ is the extension ratio relative to the length $L_{i,V} = L^0(V/V^0)^{1/3}$ of the unstretched (isotropic) specimen at the volume V prevailing in the elongated state. Equations 9 and 10 are traditionally identified as alternative stress–strain relations for Gaussian networks.

Phantom Networks

The theory of James and Guth addresses networks of Gaussian chains whose only action is to deliver contractile forces (proportional to their displacement lengths r) at the junctions to which they are attached (15). The chains have no other material properties; they may pass through one another freely and they are not subject to the volume exclusion requirements of real molecular systems. Being free of constraints by neighboring chains, the junctions of the phantom network thus described undergo displacements that are affected only by their connections to the network and not at all by their immediate surroundings (7).

Without prior assumptions concerning the disposition of the junctions in a Gaussian phantom network, James and Guth showed: (1) that their mean positions in this hypothetical network are affine in the strain, (2) that their fluctuations around these mean positions are Gaussian, and (3) that these fluctuations should be independent of the strain (15). The fluctuations of the junctions are substantial. The mean-squared magnitude of the fluctuations in the chain vectors caused by them is given by (20,21)

$$\langle (\Delta r)^2 \rangle = (2/\phi)\langle r^2 \rangle_0 \tag{11}$$

The corresponding measure of the dispersion in the magnitudes of the mean vectors is (7)

$$\langle \bar{r}^2 \rangle = (1 - 2/\phi)\langle r^2 \rangle_0 \tag{12}$$

Thus for a tetrafunctional phantom network, the fluctuations account for half of $\langle r^2 \rangle_0$ for the free chain.

It follows directly from deductions (1) and (3) that the instantaneous positions must be nonaffine in the strain. The distribution of junctions is the convolution of their mean positions affine in the strain with their fluctuations, which are invariant with the strain. The distribution of chain vectors in the phantom network, although a function of the strain, is not therefore affine in the strain.

This theory leads to an elastic free energy of the same form as the first term in equation 8, but with a smaller coefficient (15). For a tetrafunctional network, $\nu/2$ should be replaced by $\nu/4$ in the adaptation of that equation to a phantom network. Additionally, the second term of equation 8 disappears. As was shown subsequently (7), the elastic free energy for a phantom network with junctions of any functionality is given with complete generality by

$$\Delta A_{\mathrm{ph}} = (\xi/2)kT(I_1 - 3) \tag{13}$$

where ξ is the cycle rank (see above) and I_1 is the first strain invariant defined by

$$I_1 = \lambda_x^2 + \lambda_y^2 + \lambda_z^2 \tag{14}$$

The form of the dependence of the retraction force on strain under uniaxial deformation is the same as given by equations 9 or 10. The number ν of chains is replaced by ξ. Hence, the retractive force is

$$f_{\mathrm{ph}} = (\xi kT/L_{i,V})(V/V^0)^{2/3}(\alpha - \alpha^{-2}) \tag{15a}$$

$$= (\xi kT/L^0)(V/V^0)^{1/3}(\alpha - \alpha^{-2}) \tag{15b}$$

For a perfect tetrafunctional network $\xi = \nu/2$, as follows from equation 6. Hence, the predicted retractive force in this case is half that for the affine network. This difference reflects the fact that only the mean vectors **r** are altered by the strain; the fluctuations, which in a tetrafunctional network account for half of $\langle r^2 \rangle_0$, are unaffected by strain.

Affine and Phantom Network Theories vs Experiments

It follows from equation 10 and likewise from equation 15 that

$$[\partial \ln (f/T)/\partial T]_{L,V} = -(\tfrac{2}{3})d \ln V^0/dT = -d \ln \langle r^2 \rangle_0/dT \tag{16}$$

Values of $d \ln \langle r^2 \rangle_0/dT$ determined from stress–temperature coefficients using this relationship are not appreciably affected by swelling of the network with a diluent (1–4). They are in agreement with results of measurements conducted on dilute solutions of the linear polymer (3,22). These findings lend assurance that the primary premise of the molecular theories of rubber elasticity is valid.

The experimental relationship of stress to strain is strikingly at variance with the traditional theories previously discussed. In 1946 it was shown that the slope of the tension-elongation curve observed for natural rubber diminishes more rapidly with elongation and with swelling than is predicted by equation 10, and hence also by equation 15 (23). Similar departures from theory were found for other elastomers (8,14,23). The disparity between the factors of proportionality in equations 10 and 15 representing theories for affine and phantom networks, respectively, was overshadowed by the failure of both theories to account for the relationship of stress to strain.

This circumstance led to widespread adoption of the Mooney-Rivlin relation obtained by arbitrarily appending a term proportional to the second strain invariant, $I_2 = \lambda_x^2\lambda_y^2 + \lambda_y^2\lambda_z^2 + \lambda_z^2\lambda_x^2$, to the elastic free energy. The resulting relationship of the tension to elongation is

$$f = 2C_1(\alpha - \alpha^{-2}) + 2C_2(1 - \alpha^{-3})$$

or

$$f/(\alpha - \alpha^{-2}) = 2C_1 + 2C_2/\alpha \tag{17}$$

where C_1 and C_2 are empirical constants for a given elastomer at a fixed temperature. Agreement with the observed tension-elongation relationship in simple extension is improved through use of equation 17 having the additional parameter C_2. It fails utterly in compression (or equibiaxial extension) and for biaxial strains in general (8). Even in simple elongation, departures from the linear relation prescribed by equation 17 are apparent.

With the main focus of attention on elastomers in uniaxial elongation, experiments indicate that the correction term in C_2 diminishes with dilation (swelling) and that it diminishes relative to C_1 with increase in the degree of interlinking. This was implicit in early work (23). The results shown in part in Figure 2 are particularly revealing in this connection (2). Here, the reduced nominal stress defined by

$$[f^*] \equiv (f/A^0)v_2^{\frac{1}{3}}(\alpha - \alpha^{-2})^{-1} \tag{18}$$

where A^0 is the area of the initial cross section in the reference state plotted against α^{-1} in keeping with equation 17. The volume fraction v_2 of rubber may be identified with V^0/V in equation 15b, according to which $[f^*]$ should be a constant equal to $\xi kT/V^0$ for a given network. The slopes $(2C_2)$ of the Mooney-Rivlin plots decrease with dilution, but the intercepts $(2C_1)$ remain approximately the same. Other experiments show that the intercept increases with degree of

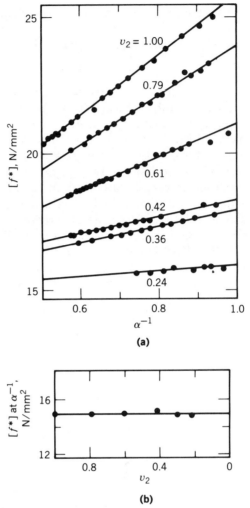

Fig. 2. Reduced nominal stress vs (**a**) α^{-1} and (**b**) v_2 (2). Courtesy of The Royal Society of Chemistry.

cross-linking (2). Thus the intercept appears to be an invariant that characterizes a given network. Observations such as these suggest that $2C_1$ may be identified with $\xi kT/V^0$ of equation 15b.

Fluctuations in Real Networks

Inasmuch as the basic premise that the stored elastic free energy resides within the chains is fully validated, observed departures from the form of the stress–strain relationship prescribed by the theories cited above implicate the connections assumed, or deduced, between the macroscopic strain and the distribution of chain vectors as the source of the discrepancy. The assumption that the latter are affine in the strain makes no allowance whatsoever for excursions of the junctions from their mean positions. Fluctuations of this nature are implicit in the molecular mobility essential to the high compliance exhibited by elastomers. It would be incorrect therefore to assume that they are suppressed altogether. On the other hand, the large, unimpeded fluctuations deduced for phantom networks may be curtailed severely by the profusion of chains in which each junction is embedded (Fig. 1). The restrictions on fluctuation considered here are operative at equilibrium. They are unrelated to the kinetic inhibitions that impede time-dependent relaxation processes.

Considerations such as these led to the suggestion that real networks may behave in a manner between the two extremes (7,24). It was suggested further that a shift in proximity to these respective extremes might be expected with strain, as phantom behavior is more closely approached with elongation or dilation (7,24). Inasmuch as the factor of proportionality to the strain function is smaller according to phantom theory than for affine theory, the observed departures from these theories might thus be explained.

Exploitation of this conjecture requires full grasp of the implications of the nonaffineness of the transformation of the distribution of chain vectors with strain (7). It is a necessary and sufficient condition for affine transformation that the neighborhood of junctions around a given junction be preserved, with distances between junctions altered in accordance with the displacement gradient λ. The environments of the junctions must therefore change with deformation in a phantom (hence, nonaffine) network, or, indeed, in any network in which the junctions undergo independent fluctuations (25,26). Since the magnitude of the fluctuations occurring in a phantom network are generally greater than the distance to the nearest spatial neighbors, drastic reshuffling of neighbors around a given junction may be required when the strain is large.

Extensive interpenetration of portions of the network, which are topologically remote in structural relation to one another, implies a maze of entanglements in which chains and junctions are inextricably intertwined. The mutual entanglement of chains and junctions confers a coherence on the real network not present in its phantom analogue comprising chains that neither preempt space nor obstruct transection of one another. This is a feature of real networks that is of foremost importance. Occurrence of the rearrangements required by phantom network theory must obviously be difficult in a real network (25–27).

The entanglements referred to here are not discrete in the sense that they engage a given chain with one of its neighbors in a unique relationship. Instead, they diffusely involve a given chain with the manifold of its neighbors. Contrary to the usual sketches of chain configurations, their trajectories do not oscillate back and forth as if guided by their time-averaged destinations (28). As follows from random-walk statistics in general, they are not self-correcting insofar as an excursion in a given direction presages correction by an opposing course. The instantaneous configuration of the chain seldom describes a path that would wind around a neighboring chain, thereby creating an entanglement similar to a crosslinkage. The diffuse entanglements prevalent in polymer networks allow extensive local rearrangements, while at the same time precluding gross displacements of neighboring members of the network.

The number of configurations accessible to a network obviously is greatly reduced by the integrity of its permanent connections and by the further constraints caused by entanglements. This reduction is inconsequential in the un-deformed network formed by interlinking randomly configured, unperturbed chains. It is the average over configuration space for an ensemble of equivalently formed networks that is relevant to the treatment of equilibrium properties. The ensemble average is unaffected by interlinking of the chains, which occurs via a random process. Upon deforming the networks thus formed, constraints from the physical integrity of the network, augmented by the effects of entanglements, contribute to the elastic free energy ΔA_{el}.

The network junctions are the members of the network most susceptible to the steric constraints imposed by the diffuse entanglements. Each of them marks the confluence of ϕ chains ($\phi \geq 3$) that encumber displacement of the junction relative to its neighbors. Although constraints obviously impinge on the chains as well, the totality of all constraints may be treated, presumably in good approximation, as if they restrict displacements of the junctions exclusively.

Network Junctions Subject to Strain-dependent Constraints

The model adopted for the purpose of giving quantitative expression to the ideas described above is shown in Figure 3. Point A represents the mean position of the chosen junction in the hypothetical phantom state of the network. The radius of the large dashed circle centered at A represents the root-mean-square fluctuation $\langle (\Delta R)^2 \rangle_{ph}^{1/2}$ around this position in the phantom state. The domain of constraints due to entanglements with surrounding real chains and to their steric requirements is represented by the smaller dashed circle centered at B and separated from A by $\bar{s}$. It is located as if the constraints were suddenly imposed at an instant during which a random excursion of the junction around its mean position A carried it to point B. (The manner in which the network was actually

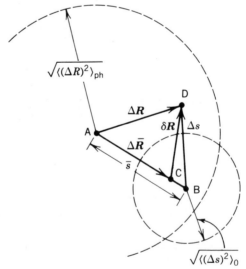

Fig. 3. Model of a network junction. See text for explanation (27). Courtesy of the American Chemical Society.

formed is irrelevant.) After the constraints have been established, the mean position of the junction is at point C removed from A by $\Delta\overline{R}$. In other words, under the combined influences of its connections with the network, ie, the phantom network forces, and of the constraints, the mean position of the junction in the unstrained, real network is at C. The instantaneous position of the junction happens to be at point D, which is outside the domain of constraints by neighbors, but inside the domain representing fluctuations of the phantom network. Both of the domain boundaries are diffuse rather than rigid; hence, the junction may wander beyond either of them, although the probability of its doing so diminishes with the distances from their centers. For simplicity and without significant sacrifice of accuracy, the action of the domain or constraint is taken to be a Gaussian function of the distance Δs of the junction from B, just as the action of the (phantom) network is a Gaussian function of ΔR.

The principal parameter κ, which characterizes the constraints, specifies the inverse ratio of the mean-square radii of the domains, that is,

$$\kappa = \langle(\Delta R)^2\rangle_{ph}/\langle(\Delta s)^2\rangle_0 \tag{19}$$

where $\langle(\Delta s)^2\rangle_0$ is the mean square of the fluctuations around B that would occur in the undeformed network if the junction is subjected only to the effect of its involvements with the surrounding chains, with constraints imposed by its connections to the network being somehow suspended. Thus κ measures the severity of the entanglement constraints relative to those of the phantom network.

Since the network is formed through random molecular processes, the instantaneous distribution of junction positions, and hence of chain vectors, must be unaffected by formation of the network. It follows that the distribution of the centers of the domains of constraint around the mean phantom positions (A) must be identical in the unstrained state with the distribution of fluctuations (ΔR) in the phantom network.

Isotropy of the network in its state of rest implies that displacement of the centers of the domains of constraint should be affine under strain. The dimensions of these domains, unlike those representing the action of the phantom network, must undergo distortion under strain. At first approximation, they may be expected to become ellipsoidal according to the macroscopic deformation gradient tensor λ; hence, the sphere represented in Figure 3 by the smaller dashed circle becomes an ellipsoid (25,27). Thus if Δx is the component of Δs along one of the principal axes of λ, on the assumption that the vectors Δs are affine in λ,

$$\langle(\Delta x)^2\rangle_\lambda/\langle(\Delta x)^2\rangle_0 = \lambda^2 \tag{20}$$

where $\lambda \equiv \lambda_x$. For $\lambda > 1$, the domain of constraint is lengthened and the severity of the constraints is diminished in this direction.

Experimental results suggest a somewhat more rapid alteration of the constraints with strain than predicted by affine deformation of the domain (25,27,28). This observation may reflect structural inhomogeneities in the network. A higher approximation is offered by (27)

$$\langle(\Delta x)^2\rangle_\lambda/\langle(\Delta x)^2\rangle_0 = \lambda^2[1 + \kappa\zeta(\lambda - 1)] \tag{21}$$

where ζ is an additional parameter. In the following development ζ is neglected, although its effect on numerical calculations will be indicated.

The primary contribution to the elastic free energy from the connectivity of the network, ie, the phantom network contribution ΔA_{ph}, is implicit in the displacement of the mean positions of the junctions in the phantom state. It is given by equation 13. The contribution ΔA_{c} from the steric constraints comprises two terms (25,27) due, respectively, to (1) alteration of the (instantaneous) distribution of the ΔR (Fig. 3) from their values in the phantom network, and (2) alteration of the distribution of displacements Δs of the junctions around the centers of their domains of constraint. The foregoing relations allow the required distributions to be formulated as functions of the strain. The free energies may then be obtained from the familiar configuration function

$$\Omega = \prod_i (\omega_i \mu_{\mathrm{J}}/\mu_{\mathrm{J},i})^{\mu_{\mathrm{J},i}} \tag{22}$$

where $\mu_{\mathrm{J},i}$ is the number of junctions at the location ΔR_i, or at Δs_i relative to the center of the domain of constraint; ω_i is the a priori probability of the state thus specified, as given by the three-dimensional Gaussian probability distribution $W(\Delta R)$ or $W_{\mathrm{c}}(\Delta s)$, the latter being ellipsoidal under strain. The contributions (1) and (2), previously discussed, follow from $-kT \ln \Omega$.

The total elastic free energy thus derived is just the sum

$$\Delta A_{\mathrm{el}} = \Delta A_{\mathrm{ph}} + \Delta A_{\mathrm{c}} \tag{23}$$

of the elastic free energy of the phantom network, ΔA_{ph} given by equation 13, and ΔA_{c} for the combined contributions (1) and (2) resulting from action of the constraints. According to the theory outlined here (26,28),

$$(kT)^{-1} \Delta A_{\mathrm{c}} = (\mu_{\mathrm{J}}/2) \sum_t \{(1 + \lambda_t^2 \kappa^{-1}) B_t - \ln [(B_t + 1)(\lambda_t^2 \kappa^{-1} B_t + 1)]\} \tag{24}$$

where

$$B_t = (\lambda_t^2 - 1)/(\lambda_t^2 \kappa^{-1} + 1)^2 \tag{25}$$

and t identifies the principal axis x, y, or z.

It will be apparent that ΔA_{c} vanishes for $\kappa \to 0$. In the opposite limit where $\kappa^{-1} \to 0$,

$$\Delta A_{\mathrm{c}} = \frac{1}{2} \mu_{\mathrm{J}} kT[I_1 - 3 - 2 \ln (V/V^0)]$$

which, when substituted in equation 23 together with equation 13, followed by replacement of $\xi + \mu_{\mathrm{J}}$ with ν according to equation 6, yields ΔA_{aff} of equation 8 (25). The present theory is therefore consistent with both affine and phantom theory at its respective limits.

The stress may be expressed similarly as the sum of contributions from the phantom network and from the entanglement constraints. In the case of simple elongation, for example, the tensile force is (25,27)

$$f = f_{\mathrm{ph}} + f_{\mathrm{c}} = f_{\mathrm{ph}}(1 + f_{\mathrm{c}}/f_{\mathrm{ph}}) \tag{26}$$

where f_{ph} is given by equation 15a or 15b. The relative contribution from the entanglement constraints is

$$f_{\mathrm{c}}/f_{\mathrm{ph}} = (\mu_{\mathrm{J}}/\xi)[\alpha K(\lambda_1^2) - \alpha^{-2} K(\lambda_2^2)](\alpha - \alpha^{-2})^{-1} \tag{27}$$

where

$$\lambda_1 = \alpha(V/V^0)^{1/3} \quad \text{and} \quad \lambda_2 = \alpha^{-1/2}(V/V^0)^{1/3},$$

and

$$K(\lambda^2) = B[\dot{B}(B + 1)^{-1} + \kappa^{-1}(\lambda^2\dot{B} + B)(\kappa\lambda^{-2} + B)^{-1}]$$

where

$$\dot{B} = \partial B/\partial\lambda^2 \tag{28}$$

In general and for perfect networks in particular, μ_J/ξ may be replaced by unity in equation 27.

Comparison of Theory with Experiments

Experimental results on elastomers in uniaxial strain are conveniently represented by plots of the reduced nominal stress $[f^*]$ (eq. 18) against α^{-1}. Results (29) covering an exceptionally wide range of extension ratio α are shown by the points in Figure 4 (28). The experiments were carried out on cross-linked polydimethylsiloxane, PDMS, without dilation, ie, with $v_2 = 1$. Those in compression, for which $\alpha^{-1} > 1$, were obtained by measuring the inflation of a sheet as a function of pressure (29). Measurements in extension were conducted on strips from the same sample. The curves have been calculated according to the theory discussed above using the parameters indicated. Use of $\zeta = 0.05$ instead of $\zeta = 0$ improves the agreement with experiment in compression $(1/\alpha > 1)$ but at the expense of agreement in extension. The divergence between theory and experiment is small compared to the range covered: fourfold in extension and sixfold in compression.

The effects (2) of swelling on the reduced force of networks of natural rubber (Fig. 2) are well represented by theory (28). Results (30) on PDMS networks

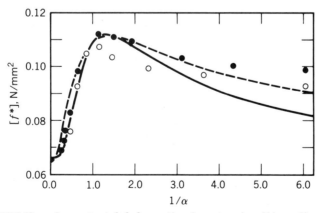

Fig. 4. PDMS under uniaxial deformation in extension $(1/\alpha < 1)$ and compression $(1/\alpha > 1)$: --------, $\kappa = 10, \zeta = 0$; ———, $\kappa = 10, \zeta = 0.05$; $\bigcirc$, measurements with increasing extension or compression; $\bullet$, measurements taken in opposite order (28).

swollen to the various degrees indicated by the volume fractions v_2 of polymer are compared in Figure 5 (30,28) with calculations according to theory for the values of κ and ζ indicated.

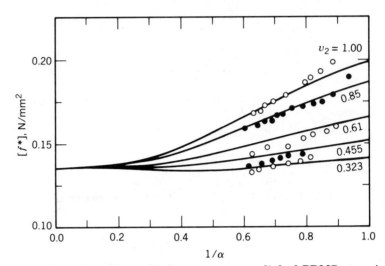

Fig. 5. Effects of swelling with benzene on cross-linked PDMS at various values of v_2. $\kappa = 6$; $\zeta = 0.12$ (28,30).

Results on tetrafunctional networks prepared by end-linking PDMS chains of different lengths, shown in Figure 6, demonstrate the effect of the degree of cross-linking (31). The curves have been calculated for values of κ chosen to be inversely proportional to the square roots of the degrees of cross-linking (ie, $\kappa \propto \xi^{-1/2}$) on the hypothesis that the constraints should be proportional to the degree of interpenetration (28) (eq. 7). All of the data for the several networks are well reproduced by the arbitrary choice of one of the κs, the others being related through the degrees of cross-linking.

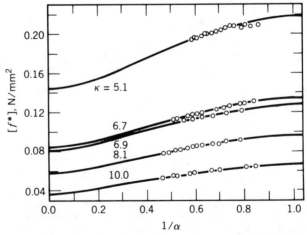

Fig. 6. End-linked PDMS at various values of κ: $v_2 = 1$; $\kappa \propto [f_{ph}^*]^{-1/2}$, $\zeta = 0$ (28,31).

Measurements of stress in biaxial extension are more definitive inasmuch as the strain is bivariate. The more complex array of data obtained from skillfully executing experiments (32) on rubber in biaxial strain long resisted rational interpretation on a molecular basis, or even in terms of the more familiar phenomenological theories (8). It is particularly significant therefore that these results are well reproduced by the theory discussed above (33,34). Thus the theory accounts for the relationship of stress to strain in elastomers virtually throughout the range accessible to experimental measurement (34). The domains of constraint postulated by the theory exert their greatest effect at small strains. Inasmuch as the domains are distorted in proportion to the principal extension ratios λ, whereas the range of the phantom fluctuations is unaffected by deformation, the relative effect of the constraints in a given direction must vary inversely with the elongation λ. The effect vanishes as λ is increased without limit. This description is oversimplified. It nevertheless explains qualitatively the attenuation of effects of the constraints at large strains or at high dilations.

The theory does not apply, of course, to elastomers in which crystallization occurs as a result of deformation (35). The theory, likewise, does not anticipate the upturn in the tension that should eventually occur at high elongation; even in the absence of crystallization, the chains approach full extension (8,36). Ordinarily, observation of the predicted upturn is precluded by rupture of the sample (35). Exceptions occur when experimental conditions do not allow a close approach to equilibrium (36) or the distribution of chain lengths is bimodal (37).

The reduced nominal stress $[f_{ph}^*]$ in the limit of high deformations or dilations emerges as the quantity that characterizes the molecular contribution of the network, as earlier experiments suggested (2,23). It obviously is essential to perform the extrapolation accurately, avoiding complications from curvature in plots of $[f^*]$ vs α^{-1}. Measurements on swollen networks offer the best procedure for this purpose.

As formally expressed, the theory takes into account only the covalent cross-linkages of the network embodied in the cycle rank ξ. Whether or not entanglements increase the effective degree of interlinking is difficult to decide on purely theoretical grounds. It is an issue best resolved by experimentation. If entanglements enhance the effective value of ξ, this enhancement should be reflected in $[f_{ph}^*]$ determined by appropriate extrapolation of experimental measurements. The value thus determined may be compared with the chemical degree of interlinking or the cycle rank.

Numerous experiments show that $[f_{ph}^*]$ obtained by extrapolation to $\alpha^{-1} = 0$, or, in some instances $[f^*]$ measured at finite extensions, to be proportional to the chemical degree of interlinking (28,31,38–43). An intercept, indicative of a threshold of entanglement cross-linkages often postulated, is not observed. The absolute magnitude of the chemical degree of interlinking is more difficult to establish with accuracy. In those instances where this objective has been achieved, the elastic and chemical degrees of interlinking are in good agreement. Included are networks of PDMS (28,31,38,40,41), poly(ethyl acrylate) (39), copoly-(isoprene–styrene) (42), and poly(cis-1,4-butadiene) (43).

Results shown in Figure 7 (28,31) illustrate the comparisons between limiting values of the reduced stress $[f_{ph}^*]$ and degrees of interlinking. Values from elasticity measurements are somewhat higher than those obtained from the chem-

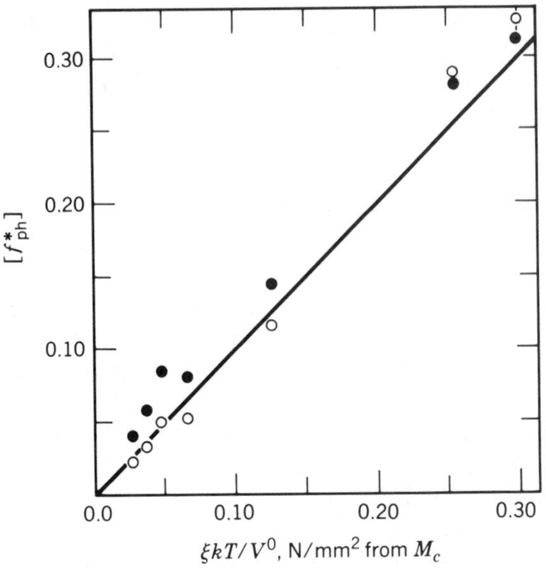

Fig. 7. Limiting values of the reduced force $[f_{ph}^*]$ for end-linked PDMS deduced from elasticity (○) and swelling (●) measurements plotted against $\xi kT/V^0$ (28,31).

ical structure, especially at low degrees of interlinking. Failure to attain ultimate elastic equilibrium may account for these departures. The results deduced from swelling equilibrium, which are not subject to this source of error, are in excellent agreement with the theoretical line based on the network structure as embodied in ξ/V^0.

The value of the parameter κ appears to be related uniquely to the degree of interpenetration Γ given by equation 7, that is (28),

$$\kappa = \text{const } \Gamma = I \langle r^2 \rangle_0^{3/2} (\mu_J/V^0) \qquad (29)$$

where I is an empirical parameter. Available data suggest that I may be the same for all tetrafunctional networks (28). If this indication is verified, it becomes possible to relate stress to strain on the basis of the degree of cross-linking which, in principle at least, is determined from the chemical constitution of the network. Only the empirical parameter ζ would then be subject to arbitrary choice. Its role in refining agreement between theory and experiment is marginal.

The molecular theory discussed here provides a comprehensive account of rubber elasticity relating the elastic equation of state to molecular constitution. This long-sought objective is achieved with a latitude of choice in only one parameter. This parameter, κ, appears to be susceptible to independent determination, approximately at least, from the cycle rank ξ that characterizes the connectivity of the network. Arbitrariness in the choice of parameters may thus be reduced to an utter minimum.

The theory also accounts for the peculiar form of the dependence of the elastic contribution to the chemical potential of the diluent in a swollen network (44). It provides the basis for a more exact treatment of strain birefringence in elastomeric networks (45,46). The theory appears to account also for the effects of functionality (47,48).

Acknowledgment

The Encyclopedia gratefully acknowledges The Society of Polymer Science, Japan for permission to print this slightly edited version of the article "Molecular Theory of Rubber Elasticity," *Polymer* **17**(1), 1 (1985).

BIBLIOGRAPHY

1. A. Ciferri, C. A. J. Hoeve, and P. J. Flory, *J. Am. Chem. Soc.* **83,** 1015 (1961).
2. G. Allen, M. J. Kirkham, J. Padget, and C. Price, *Trans. Faraday Soc.* **67,** 1278 (1971).
3. J. E. Mark, *Rubber Chem. Technol.* **46,** 593 (1973).
4. P. J. Flory, *Pure Appl. Chem., Macromol. Chem.* **8,** 1 (1972); *Rubber Chem. Technol.* **48,** 513 (1975).
5. J. S. Higgins and R. S. Stein, *J. Appl. Crystallogr.* **11,** 346 (1978).
6. P. J. Flory, *Pure Appl. Chem.* **56,** 305 (1984).
7. P. J. Flory, *Proc. R. Soc. London Ser. A* **351,** 351 (1976).
8. L. R. G. Treloar, *The Physics of Rubber Elasticity*, 3rd ed., Oxford University Press, Oxford, UK, 1975.
9. E. Guth and H. M. James, *Ind. Eng. Chem.* **33,** 624 (1941); *J. Chem. Phys.* **11,** 455 (1943).
10. F. T. Wall, *J. Chem. Phys.* **11,** 527 (1943).
11. L. R. G. Treloar, *Trans. Faraday Soc.* **39,** 241 (1943).
12. P. J. Flory and J. Rehner, Jr., *J. Chem. Phys.* **11,** 512 (1943).
13. W. Kuhn, *J. Polym. Sci.* **1,** 380 (1946).
14. P. J. Flory, *Principles of Polymer Chemistry,* Cornell University Press, Ithaca, N.Y., 1953, Chapt. 11.
15. H. M. James, *J. Chem. Phys.* **15,** 651 (1947); H. M. James and E. Guth, *ibid.,* p. 669.
16. P. J. Flory and D. Y. Yoon, *J. Chem. Phys.* **61,** 5358 (1974); P. J. Flory and V. W. C. Chang, *Macromolecules* **9,** 33 (1976).
17. P. J. Flory, *Macromolecules* **15,** 99 (1982).
18. P. J. Flory, *Contemp. Top. Polym. Sci.* **2,** 1 (1977).
19. P. J. Flory, *Trans. Faraday Soc.* **57,** 829 (1961).
20. B. E. Eichinger, *Macromolecules* **5,** 496 (1972).
21. W. W. Graessley, *Macromolecules* **8,** 865 (1975).
22. P. J. Flory, A. Ciferri, and R. Chiang, *J. Am. Chem. Soc.* **83,** 1023 (1961); R. Chiang, *J. Phys. Chem.* **70,** 2348 (1966).
23. G. Gee, *Trans. Faraday Soc.* **42,** 585 (1946).
24. G. Ronca and G. Allegra, *J. Chem. Phys.* **63,** 4990 (1975).
25. P. J. Flory, *J. Chem. Phys.* **66,** 5720 (1977).
26. P. J. Flory, *Polymer* **20,** 1317 (1979).
27. P. J. Flory and B. Erman, *Macromolecules* **15,** 800 (1982).
28. *Ibid.,* p. 806.
29. H. Pak and P. J. Flory, *J. Polym. Sci. Polym. Phys. Ed.* **17,** 1845 (1979).
30. P. J. Flory and Y. I. Tatara, *J. Polym. Sci. Polym. Phys. Ed.* **13,** 683 (1975).
31. J. E. Mark and J. L. Sullivan, *J. Chem. Phys.* **66,** 1006 (1977).
32. D. F. Jones and L. R. G. Treloar, *J. Phys. D.* **8,** 1285 (1975).
33. B. Erman, *J. Polym. Sci. Polym. Phys. Ed.* **19,** 829 (1981).
34. L. R. G. Treloar, *Br. Polym. J.* **14,** 121 (1982).
35. J. E. Mark, *Polym. Eng. Sci.* **19,** 254, 409 (1979).
36. J. Furukawa, Y. Onouchi, S. Inagaki, and H. Okamoto, *Polym. Bull.* **6,** 381 (1981).
37. M. A. Llorente, A. L. Andrade, and J. E. Mark, *Colloid Polym. Sci.* **259,** 1056 (1981); *J. Chem. Phys.* **72,** 2282 (1982); **73,** 1439 (1982); J. E. Mark and M. Y. Tang, *J. Polym. Sci. Polym. Phys. Ed.* **22,** 1849 (1984).
38. J. R. Falender, G. S. Y. Yeh, and J. E. Mark, *J. Chem. Phys.* **70,** 5324 (1979).
39. B. Erman, W. Wagner, and P. J. Flory, *Macromolecules* **13,** 1554 (1980).
40. W. Oppermann and G. Rehage, *Colloid Polym. Sci.* **259,** 1177 (1981).
41. P. J. Flory and B. Erman, *J. Polym. Sci. Polym. Phys. Ed.* **22,** 49 (1984).
42. N. P. Ning, J. E. Mark, N. Iwamoto, and B. E. Eichinger, *Macromolecules* **18,** 55 (1985).
43. R. W. Brotzman and P. J. Flory, *Macromolecules* **20** (1987).

44. R. W. Brotzman and B. E. Eichinger, *Macromolecules* **14,** 1445 (1981).
45. B. Erman and P. J. Flory, *Macromolecules* **16,** 1601 (1983).
46. *Ibid.,* p. 1607.
47. C.-Y. Jiang, L. Garrido, and J. E. Mark, *J. Polym. Sci. Polym. Phys. Ed.* **22,** 2281 (1984).
48. R. W. Brotzman, private communication.

PAUL J. FLORY
Stanford University

NEUTRON SCATTERING

Neutron scattering had its origin in 1932, the year that marked the discovery of the neutron by Chadwick. Its prehistory extends to 1924, the year of deBroglie's paper from which the wavelike properties of the neutron were recognized. The prereactor period of 1932–1942 saw the first demonstrations of Bragg scattering, neutron polarization, and estimates of the magnitude and sign of the neutron magnetic moment. The reactor period began in 1942 when the first nuclear reactor was successfully operated at the University of Chicago. The following year the Oak Ridge graphite reactor went into operation, followed in 1944 by Chicago Pile 3, a heavy-water reactor. For the first time relatively copious fluxes of the order of 10^{11}–10^{12} neutrons s^{-1} cm^{-2} were available for experimentation, and it is at these reactors that the modern phase of neutron scattering can be said to have begun. Thereafter the technique developed rapidly as core fluxes reached $\sim 5 \times 10^{13}$ neutrons s^{-1} cm^{-2} and since the early 1950s there has been intense activity in the application of neutron methods to the study of the solid and liquid state. At the present time there are three high flux reactors where core fluxes exceed 10^{15} neutrons s^{-1} cm^{-2}, at the Institut Laue-Langevin (Grenoble, France), and at the Brookhaven and Oak Ridge National Laboratories in the United States. During its initial stages the neutron technique was developed largely by physicists and most of the earlier work was in the area of solid- (or liquid-) state physics and magnetism. Since the mid-1960s, however, the technique has been increasingly applied by scientists from other disciplines, eg, chemistry, biology, and materials and polymer science. A major advance in the early 1970s was the combination of deuterium-labeling methods with small-angle neutron scattering to provide for the first time direct information on polymer-chain configurations in bulk polymers and concentrated solutions. This led to a considerable upsurge in the application of neutron scattering to polymers and further increased the number of users of neutron scattering facilities. This would not have been possible without the development of suitable spectrometers at national and international research facilities, which have been made available to outside users from the general scientific community.

A number of excellent references are available (1–4) that contain basic neutron scattering theory, though those textbooks reflect the origins of the technique and the examples are drawn largely from physics, eg, single crystals, simple liquids, monatomic gases, liquid metals, magnetic materials, etc. In view of the growing numbers of nonspecialists using neutron scattering, the need has become apparent for presentations that can provide rapid access to the method without unnecessary detail and mathematical rigor. In the field of polymer science, several reviews have been written to meet this need (5–8), and this article surveys the progress of the past two decades. During this time the unique properties of the neutron have provided a wealth of new information in all aspects of polymer structure and dynamics. The aim of this article is to aid potential users who have a general scientific background, but no specialist knowledge of scattering, to apply the technique to provide new information in areas of their own particular interests.

The neutron has a mass m of 1.0087 amu, a spin of $I = \frac{1}{2}$, and a magnetic moment of -1.91304 nuclear magnetons. It can interact with a sample through nuclear forces (nuclear scattering) or via its magnetic moment if the sample contains unpaired electrons (magnetic scattering). As polymers generally do not possess such unpaired electrons, magnetic scattering is not considered here.

Neutrons demonstrate convincingly the wave–particle duality of matter, and scattering experiments exploit both aspects of neutron behavior. A typical incident particle velocity v_0 is 750 m/s, which corresponds, via the deBroglie relation, to a wavelength λ of

$$\lambda = \frac{h}{mv_0} = 0.53 \text{ nm} \tag{1}$$

which is of the same order as the nearest-neighbor spacing between polymer molecules; h is Planck's constant.

The kinetic energy E_0 of such a neutron is given by

$$E_0 = \frac{mv_0^2}{2} = 0.003 \text{ eV}$$
$$= 4.7 \times 10^{-22} \text{ J} \tag{2}$$
$$(4.7 \times 10^{-15} \text{ erg})$$

Such energies are very much lower than electromagnetic radiation (eg, x rays) and are of the same order as the vibrational and diffusional energies of molecular systems. Exchanges of energy between the incident particle neutron and molecule give rise to inelastic scattering, which depends on the dynamics of the system studied. Although the angular dependence of the scattering of both x rays and neutrons is easily measured, the energies of molecular vibrations (~3 meV) are much lower than incident x-ray photons (~10 keV) and thus energy transfers are difficult to detect for x-ray scattering. In contrast, the energy transfers resulting from neutron scattering are easily resolved and permit the elucidation of dynamic processes. Neutrons are thus a unique probe for studying the condensed state, as they simultaneously have both the appropriate wavelength and energy to investigate the structure and dynamics of molecular systems in general, including polymers.

Theory

Energy and Momentum Transfer.   In the context of this article, scattering means the deflection of a neutron beam from its original direction by interaction with the nuclei of polymer or solvent molecules in a sample. In a scattering experiment a proportion of the incident neutrons is scattered and the remaining fraction is transmitted through the sample. The intensity of the scattered neutrons is measured as a function of the scattering angle and energy. The vector diagram shown in Figure 1 illustrates an incident neutron of wavelength λ_0 and velocity v_0 which is scattered through an angle 2θ in an inelastic process. This results in a final wavelength λ and velocity v, and the energy gained by the target (and lost by the neutron) is given by

$$\Delta E = \frac{m}{2} (v^2 - v_0^2) = \frac{\hbar^2}{2m} (k^2 - k_0^2) = \hbar\omega \tag{3}$$

where $\mathbf{k}_0$ and $\mathbf{k}$ are the initial and final wave vectors ($k = 2\pi/\lambda$). The momentum transfer is

$$\hbar\mathbf{Q} = \hbar(\mathbf{k} - \mathbf{k}_0) \tag{4}$$

$$\hbar|\mathbf{Q}| = \hbar(k^2 + k_0^2 - 2kk_0 \cos 2\theta)^{1/2} \tag{5}$$

If energy is transferred in the scattering process ($\Delta E \neq 0$), this may be regarded as a Doppler shift in the scattered wavelength due to thermal motion of the nucleus and the process is termed inelastic. If no energy change takes place ($\Delta E = \hbar\omega = 0$, $\lambda = \lambda_0$), the scattering is termed elastic and

$$|\mathbf{Q}| = \frac{4\pi}{\lambda} \sin \theta \tag{6}$$

If ΔE is small compared to the incident neutron energy ($|\Delta E| << E_0$) the scattering is termed quasi-elastic. Most neutron scattering measurements on polymers have involved neutrons scattered at small values of the momentum transfer ($Q \to 0$). This type of measurement is usually referred to as small-angle (rather than small Q) neutron scattering, though the terms are equivalent for long wavelengths ($\lambda_0 > 0.4$ nm). It is easily seen from equation 4 that for long wavelengths, $Q \to 0$ implies $k \to k_0$ and the scattering is predominantly elastic, as any neutron scattered with a large energy transfer ΔE could not satisfy both energy and momentum conservation at small Q. Coherent inelastic events with $\Delta E \neq 0$ are allowed in theory at small angles, but have extremely low probability because the cross sections for coherent inelastic scattering are orders of magnitude below the cross sections for coherent elastic scattering. For example, the cross section of a labeled (deuterated) polystyrene or polyethylene molecule with molecular weight $\overline{M}_w \sim 10^5$ in a hydrogenous matrix is ca 10^{-20} m^2 (10^8 barns) per molecule

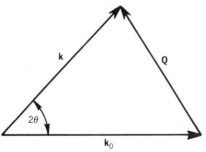

Fig. 1.   Relationship between the momentum transfer $\mathbf{Q}$, the scattered wave vector $\mathbf{k}$, and the incident wave vector $\mathbf{k}_0$ in a neutron scattering event.

or 10^{-25} m^2 (10^3 barns) per atom. This may be compared to a typical inelastic (phonon) cross section of ca 10^{-31} m^2 (10^{-3} barns) per atom. Thus although possible in principle, coherent inelastic events almost never occur in practice at small angles for the type of sample described in this article. For incoherent inelastic scattering, however, the inelastic cross sections of hydrogenous samples (eg, water) are sufficiently high that single and multiple scattering events with $\Delta E \sim E_0$ form a measurable fraction of the total scattering probability, and must be considered when this type of material is studied.

Small-angle neutron scattering (sans) experiments give information on the time-averaged structure and conformation of polymer molecules, and form the bulk of this article. Although there has been less work on quasi-elastic and inelastic processes, such experiments give valuable information on polymer dynamics, and hence the theory is described for the general case $\Delta E \neq 0$.

Scattering Length and Cross Section. Scattering theory is usually developed by considering a single atom that is fixed at the origin and hence cannot accept energy from the neutron (1,3,4). The interaction between the neutron and nucleus cannot be calculated exactly, but it is known to be short-ranged ($\sim 10^{-6}$ nm) compared to the wavelength of the neutron (~ 0.5 nm). Because of this, it is shown in standard textbooks (1,3,4,9) that the scattering can contain only zero angular momentum components. This has the important consequence that the scattering is isotropic for slow neutrons and there is no angle-dependent form factor as in the case of x rays. Thus if an incident plane wave of neutrons is described by a wave function of unit density

$$\psi_0 = e^{ik_0 z} \tag{7}$$

the scattered wave from a fixed nucleus (Fig. 2) is spherically symmetrical and of the form

$$\psi_1 = -\frac{b}{r} e^{ikr} \tag{8}$$

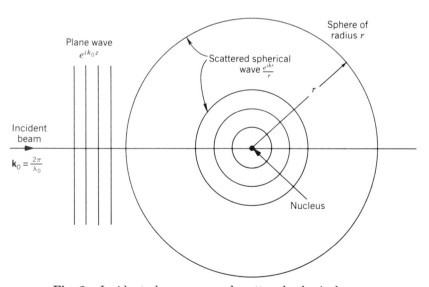

Fig. 2.   Incident plane wave and scattered spherical wave.

The quantity b has the dimensions of length and is called the scattering length. In principle it is a complex quantity, though the imaginary part is important only for nuclei with a high absorption coefficient (eg, cadmium, boron) and may be neglected for nuclei commonly found in synthetic polymers. Moreover, although the value of b is in principle dependent on the incident neutron energy E_0, the variation of energies normally encountered in neutron scattering studies is negligible ($<0.1\%$) and b may be regarded as real and a (known) constant for a given nucleus (isotope). The scattered neutrons may be envisaged as originating from a sphere of radius r centered on the nucleus, and, using the scattering length, a scattering cross section σ for the nucleus may be defined (1,3,9,10) by

$$\sigma = \frac{\text{number scattered neutrons/s}}{\text{incident neutron flux}} \tag{9}$$

The incident beam has a particle density (neutrons/unit volume) of $|e^{ik_0 z}|^2$ and hence the incident flux n_0 (ie, the number of neutrons per second per unit area) is given by

$$n_0 = v_0 |e^{ik_0 z}|^2 = v_0 \tag{10}$$

Similarly, the number of scattered neutrons passing through the surface of the sphere centered on the scattering nucleus (Fig. 2) is given by

$$n = v 4\pi r^2 \left| b \, \frac{e^{ikr}}{r} \right|^2 = 4\pi b^2 v \tag{11}$$

and hence the single-atom cross section is given by

$$\sigma = 4\pi b^2 \tag{12}$$

as the neutron velocity is unchanged after scattering from the fixed nucleus, which cannot accept energy from the neutron. It can be seen from equation 12 that σ has the dimensions of area.

To a first approximation this scattering process may be envisioned as a collision between two billiard balls, and the cross section may be regarded as the effective area that the target nucleus presents to the incident beam of neutrons for the elastic scattering process. This component of the cross section, called potential scattering (3), increases throughout the periodic table with increasing nuclear size, which is proportional to $A^{2/3}$ where A is the mass number. However, there are additional components of the cross section due to other scattering processes, eg, resonance effects due to the formation of compound nuclei (3). The net result is that the nuclear cross sections and scattering lengths vary in an apparently random manner from atom to atom and are generally treated as measured quantities rather than calculated from first principles. The cross section defined in equation 12 is usually called the bound-atom cross section, as the nucleus is considered fixed at the origin. However, where the atom is free to recoil, eg, in the gaseous state, the cross section applicable to this state is called the free-atom cross section. It may be shown (1,3,10,11) that

$$\sigma_{\text{free}} = \left(\frac{A}{A+1} \right)^2 \sigma_{\text{bound}} \tag{13}$$

The bound-atom cross section is generally relevant to polymer studies, which are virtually always conducted on samples of macroscopic dimensions in the solid or liquid state.

Coherent and Incoherent Cross Sections. The magnitude of b varies from nucleus to nucleus and is typically on the order of 10^{-12} cm (10^{-14} m). This gives rise to the usual unit for a cross section, called a barn (10^{-24} cm^2 or 10^{-28} m^2). Unlike the x-ray scattering factor f, which increases with the atomic number of the atom, there is no general trend throughout the periodic table in the values of b, which vary from isotope to isotope and from nucleus to nucleus of the same isotope if it has nonzero spin.

Because the neutron has spin $\frac{1}{2}$, it can interact with a nucleus of spin I to form one of two compound nuclei with spins ($I \pm \frac{1}{2}$), each of which has a different scattering length, b^+ or b^-, which is associated with the spin-up or spin-down states. For a given spin state J, the number of orientations is ($2J + 1$) and thus the number of possible orientations for the compound spin states of ($I + \frac{1}{2}$) and ($I - \frac{1}{2}$) are $2(I + 1)$ and $2I$, respectively. The total number of spin states is $2(2I + 1)$, and as the probabilities of each state are equal, the statistical weights are $\frac{I + 1}{2I + 1}$ and $\frac{I}{2I + 1}$, respectively.

The average (coherent) scattering length is

$$\langle b \rangle = \frac{I + 1}{2I + 1} b^+ + \frac{I}{2I + 1} b^- \tag{14}$$

where the brackets $\langle \ \rangle$ represent a thermal average over the spin-state population. A coherent cross section can be defined for each isotope by

$$\sigma_{\text{coh}} = 4\pi \langle b \rangle^2 \tag{15}$$

whereas the total-scattering cross section is given by

$$\sigma_{\text{tot}} = 4\pi \langle b^2 \rangle \tag{16}$$

The difference between the two is the incoherent cross section σ_{inc}, which is given by

$$\sigma_{\text{tot}} - \sigma_{\text{coh}} = \sigma_{\text{inc}} = 4\pi \{\langle b^2 \rangle - \langle b \rangle^2\} \tag{17}$$

If the isotope has no spin, then $\langle b^2 \rangle = \langle b \rangle^2$ as $\langle b \rangle = b$ and there is no incoherent scattering. Only the coherent-scattering cross section contains information on interference effects arising from spatial correlations of the nuclei in the system, ie, the structure of the sample. The incoherent cross section contains no information on interference effects and forms an isotropic (flat) background, which must be subtracted in sans structural investigations. It does, however, contain information on the motion of single atoms (particularly hydrogen), which may be investigated via energy analysis of the scattered beam. Although most of the atoms encountered in neutron scattering from polymers are mainly coherent scatterers (eg, carbon, oxygen), there is one important exception. In the case of hydrogen (^{1}H), the spin-up and spin-down scattering lengths have opposite sign ($b^+ = 1.080 \times 10^{-12}$ cm; $b^- = -4.737 \times 10^{-12}$ cm), and as $I = \frac{1}{2}$,

$$\sigma_{\text{coh}} = 1.76 \times 10^{-24} \text{ cm}^2 \tag{18}$$

$$\sigma_{tot} = 81.5 \times 10^{-24} \text{ cm}^2 \tag{19}$$

$$\sigma_{inc} = 79.7 \times 10^{-24} \text{ cm}^2 \tag{20}$$

According to equation 13, the free-atom cross section from ^{1}H is $\sigma_{free} = 20.36 \times 10^{-24}$ cm^2, which is close to the measured values (12,13). Cross sections and scattering lengths are given in Table 1 for atoms commonly encountered in synthetic and natural polymers. These cross sections refer to bound protons and neglect inelastic effects arising from interchange of energy with the neutron. For coherent scattering, which is a collective effect arising from the interference of scattered waves over a large correlation volume, this approximation is reasonable, especially at low Q, where recoil effects are small. However, for incoherent scattering, which depends on the uncorrelated motion of individual atoms, inelastic effects become increasingly important for long wavelength neutrons, as indicated in Figure 3, which shows the total cross section for water as a function of energy. For high E_0, the energies associated with the vibrational and translational motion of water molecules are negligible and the cross section approaches the sum of the free-atom cross sections for the hydrogen and oxygen atoms ($\approx 44 \times 10^{-24}$ cm^2). However, as $E_0 \to 0$ the scattering does not plateau at the bound-atom cross section ($\approx 167 \times 10^{-24}$ cm^2) and varies continuously with energy. This variation is due mainly to inelastic processes affecting the incoherent scattering, which is the main component of the cross section. Similar cross-sectional data are not available for hydrogen atoms in polymers, but it is known that inelastic effects are important and vary with both incident wavelength and sample temperature (14). As the cross section of a typical hydrogenous polymer molecule is largely determined by σ_{inc} for the hydrogen atoms present, the transmission of a polymer sample is also a function of the incident wavelength. The values in Table 1 may be used to calculate the RT ($T = 294$ K) transmission only for $\lambda_0 \approx 0.47$ nm (15). The effective cross section σ_{inc} changes by ca 30% for poly(methyl methacrylate) as λ_0 changes from $0.47 \to 1.0$ nm (14). Only at the former wavelength can the bound-atom cross sections given in Table 1 be used to calculate the sample transmission to ca 1% accuracy, and values calculated for $\lambda_0 \approx 1$ nm would be in error by ca 17%.

Table 1. Bound-atom Scattering Lengths and Cross Sections for Elements in Synthetic and Natural Polymers

Atom	Nucleus	Scattering length b_{coh}, 10^{-12} cm	Coherent cross section σ_{coh}, 10^{-24} cm^2 ($\sigma_{coh} = 4\pi b^2$)	Incoherent cross section σ_{inc}, 10^{-24} cm^2	Absorption cross section[a] σ_{abs}, 10^{-24} cm^2 0.475 nm	0.18 nm
hydrogen	^{1}H	−0.374	1.76	79.7	1.45	0.33
deuterium	^{2}H (D)	0.667	5.59	2.01	0.00202	0.00046
carbon	^{12}C	0.665	5.56	0	0.0145	0.0033
nitrogen	^{14}N	0.930	11.10	0	8.27	1.88
oxygen	^{16}O	0.580	4.23	0	0	0
silicon	^{28}Si	0.415	2.16	0	0.704	0.17
chlorine	Cl[b]	0.958	11.53	5.9	147.8	33.6
fluorine	^{19}F	0.566	4.03	~0	0.026	0.0098

[a] The absorption cross section σ_{abs} is a function of wavelength and for a given λ may be estimated from the above value, $\sigma_{abs}(0.475)$, by $\sigma_{abs}(\lambda) = \dfrac{\sigma_{abs}(0.475)}{0.475} \lambda$.

[b] Values are for the naturally occurring element and are an average over the mixture of isotopes.

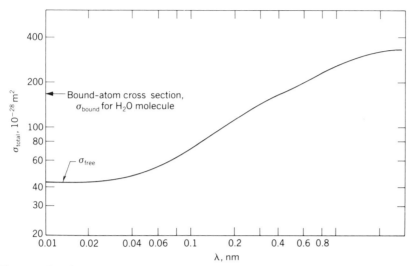

Fig. 3. Total cross section for water molecule (H_2O) vs neutron wavelength λ (nm) at ca T = 293 K (13). 1×10^{-28} m^2 = 1 barn.

It can be seen from Table 1 that there is a large difference in the coherent-scattering length between deuterium and hydrogen, and that the latter value is actually negative. This arises from a change of phase of the scattered wave and results in a marked difference in scattering (contrast) between polymer molecules synthesized with deuterium atoms or hydrogen atoms along the chain. Similarly, it may be seen that the incoherent cross section differs by more than an order of magnitude for deuterium-labeled and normal molecules containing hydrogen atoms. These scattering differences are the basis of most experiments that have been performed on polymers.

The basic experiment consists of an incident neutron beam, energy E_0, which is scattered by an assembly of sample nuclei into solid angle $d\Omega$ with energy change dE recorded by a neutron detector (Fig. 4). The double-differential-scattering cross section for unit volume of sample, $d^2\Sigma/d\Omega dE$, is defined as the number of neutrons scattered per second into a solid angle $d\Omega$ with energy change dE, divided by the incident neutron flux (neutrons per second per unit area). In this article, σ denotes the cross section of a single nucleus, and Σ denotes an assembly of nuclei, except where it denotes the standard summation sign. For such an assembly the double-differential-scattering cross section is given by standard scattering theory (4,16) as

$$\frac{d^2\Sigma}{d\Omega dE} = \frac{k}{2\pi k_0} \int_{-\infty}^{+\infty} dt \exp\left(-i\omega t\right) \left\langle \sum_{ij} b_i^* b_j F_{ij}(\mathbf{Q},t) \right\rangle$$

$$= \frac{k}{k_0} S(Q,\omega) \tag{21}$$

where

$$F_{ij}(\mathbf{Q},t) = \exp\left[-i\mathbf{Q}\cdot\mathbf{R}_i(0)\right] \exp\left[i\mathbf{Q}\cdot\mathbf{R}_j(t)\right] \tag{22}$$

In equation 21 the symbol $*$ denotes a complex conjugate and $\langle\ \rangle$ denotes a thermal average over all configurations of scatterers at position vectors $\mathbf{R}(t)$ at

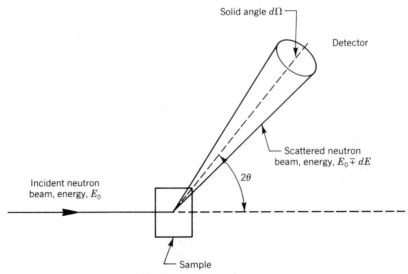

Fig. 4.   The basic scattering experiment.

time t. $S(Q,\omega)$ is called the scattering law or scattering function. Equation 21 may be separated into coherent and incoherent components of the cross section

$$\frac{d^2\Sigma^{\text{coh}}}{d\Omega dE} = \frac{k}{2\pi k_0} \int_{-\infty}^{+\infty} dt \, \exp\left(-i\omega t\right) \left\langle \sum_{ij} \langle b_i \rangle * \langle b_j \rangle F_{ij}(\mathbf{Q},t) \right\rangle \tag{23}$$

and

$$\frac{d^2\Sigma^{\text{inc}}}{d\Omega dE} = \frac{k}{2\pi k_0} \int_{-\infty}^{+\infty} dt \, \exp\left(-i\omega t\right) \left\langle \sum_{i} |b_i - \langle b \rangle_i|^2 F_{ii}(\mathbf{Q},t) \right\rangle \tag{24}$$

From equations 23 and 24 it may be seen that coherent scattering contains information on the correlations between different nuclei and hence gives information on the relative spatial arrangement of atoms in the system (eg, structure) and its time dependence. The incoherent cross section, on the other hand, contains information on correlations between the same nuclei and hence gives information on the time dependence of the motion of an individual atom (eg, vibration, diffusion, etc). The two components of the cross section may be related to time-dependent pair-correlation functions introduced by Van Hove (17), though a detailed description of this formulation is beyond the scope of this article and the reader is referred to more specialized reference works (1–4).

Equations 23 and 24 may be further simplified for model systems by treating b as a real quantity and disregarding the correlations between the scattering lengths b_i, b_j, and $F_{ij}(\mathbf{Q},t)$. For example, the scattering from polymers containing an appreciable atomic fraction of hydrogen atoms and containing no deuterated molecules is dominated by ^{1}H incoherent scattering and the cross section can be approximated by

$$\frac{d^2\Sigma}{d\Omega dE}(Q,\omega) \simeq \frac{d^2\Sigma^{\text{inc}}}{d\Omega dE} \simeq \frac{k}{k_0} \frac{N_{\text{H}}\sigma_{\text{inc}}}{4\pi} S_s(Q,\omega) \tag{25}$$

where N_H is the number of ^1H nuclei per unit volume and

$$S_s(Q,\omega) = \frac{1}{2\pi} \int_{-\infty}^{+\infty} dt \, \exp\,(-i\omega t) \sum_i \langle F_{ii}(\mathbf{Q},t)\rangle \tag{26}$$

This approximation is the basis of inelastic scattering studies of side-group motion in polymers for which coherent scattering effects form only a small correction. Similarly, small-angle neutron scattering from samples containing a fraction of deuterated chains in a matrix of normal protonated polymer results in predominantly coherent elastic scattering which is peaked at $Q = 0$. In this region ($Q < 5$ nm^{-1}) the incoherent scattering forms only a small (flat) background correction to the scattering from the labeled molecules, and the total cross section is mainly coherent. However, in the general case polymers are polyatomic systems for which the correlations of scattering lengths $b_i b_j$ and the function $F_{ij}(\mathbf{Q},t)$ cannot be neglected and hence equation 21 must remain the general starting point in the analysis (16). The separation of $d^2\Sigma/d\Omega dE$ and $S(Q,\omega)$ into coherent and incoherent contributions must be considered for the particular polymer in question before simplifying assumptions can be made (18,19).

Because the scattering law separates into coherent and incoherent components which may be further subdivided into elastic, inelastic, and quasi-elastic processes, the type of information obtainable from neutron scattering from polymers subdivides into the general categories shown in Table 2.

Small-angle Neutron Scattering

Instrumentation.   Small-angle neutron scattering is an example of predominantly coherent elastic scattering. The first applications of this technique to polymer science were made in the early 1970s in Europe, where the first small-angle scattering cameras were developed. The first sans instrument suitable for the study of polymers was built at the FRJ2 (Dido) reactor at the Kernforschungsanlage (KFA) in Jülich, FRG, in the late 1960s. This instrument pioneered the use of long-wavelength neutrons and large distances between the entrance slit, sample, and detector. The overall length of this facility is large ($\sim$40 m) and this is a direct consequence of the low source brilliance (particles per second per sterad per area) of current neutron sources, which are many orders of magnitude below that produced in conventional or synchroton x-ray sources (20). In order to compensate for this difference it is necessary to use large sample areas (1–20 cm^2), which means that the overall size of the instrument must be large in order to maintain resolution, which can be varied in the range of 2–200 nm. Other compensating factors acting in favor of neutrons include the transparency of many materials to neutrons which simplifies many experiments as a wide variety of window materials is available. Also, the sample absorption is lower than for x rays, especially for heavier elements, resulting in much thicker samples (typically 0.1–1.0 cm for polymers). The wavelength of the radiation used enters into the Q-resolution as λ^4 and this introduces another factor of several orders of magnitude in favor of the neutron technique in comparison with small-angle x-ray scattering, particularly for long-wavelength neutrons (21).

In a typical reactor, fission neutrons are produced in the core which is

Table 2. Information Obtained from Neutron-scattering Experiments on Polymers

Type of scattering	Energy change		
	Elastic, $\Delta E = 0$	Quasi-elastic, $|\Delta E| << E_0$	Inelastic, $\Delta E \neq 0$
coherent	sans chain configuration in bulk and solution, polymer compatibility, orientation mechanisms wide-angle diffraction crystal structure	Molecular dynamics in the bulk and in solution via Dopper broadening of the elastic peak	Elastic constants of crystalline polymers via phonon-dispersion curves
incoherent		Effective diffusion for segmental motion in the bulk, activation energy	Side-group vibrational frequencies and rotation barriers, intermolecular potentials

surrounded by a moderator (eg, D_2O, H_2O) and reflector (eg, Be, graphite), which reduce the neutron energy. A typical moderator and reflector at 310 K produce a Maxwellian spectrum of wavelengths peaked at ~0.1 nm (thermal neutrons). Because the λ^4 factor enters into the calculation (21) of the scattering power for a given resolution ($\Delta Q/Q$), it is highly advantageous to use long-wavelength neutrons ($\lambda > 0.4$ nm) and to increase the flux in this region. This may be accomplished by further moderating the neutrons to a lower temperature by means of a cold source containing a small volume of liquid hydrogen at $T \sim 20$ K. This source is placed near the inner end of the exit beam tube and can give flux gains of an order of magnitude at $\lambda \simeq 0.4$–1.5 nm. The FRJ-2 sans facility (21,22) was the first to employ the combination of cold neutrons and a large overall instrument size, and it was on this instrument that first sans experiments on polymers were performed. The small-angle scattering camera (D11) on the high flux reactor (HFR) at the German-French-British Institut Max von Laue–Paul Langevin (ILL) in Grenoble, incorporated many of the features of the FRJ-2 instrument, including the cold source and long (~80 m) dimensions (23). As the core flux of the Grenoble HFR is approximately an order of magnitude higher than that of the FRJ-2 reactor, the instrument has been the most versatile sans instrument since its construction in the early 1970s. Both the FRJ-2 and D11 instruments use neutron guide tubes as proposed by Maier-Leibnitz and Springer (24). These tubes are typically made of Ni-coated glass and operate by total internal reflection of neutrons. They are particularly efficient for longer wavelength neutrons as the limiting angle for total reflection increases with λ. Beam guide tubes are extensively used at the ILL and transport thermal and cold neutrons to a guide hall that contained over 20 scattering instruments in 1985. By introducing some curvature into the guides, it is possible to separate out γ-rays and fast neutrons whose wavelengths are too short to be internally reflected, and thus produce extremely low levels of background radiation in the guide hall.

In addition to the FRJ-2 and D11 instruments, a two-axis spectrometer (25) suitable for measurements in the intermediate momentum range $0.1 < Q < 1$ nm^{-1} was set up at the EL3 reactor in Saclay, France. A similar instrument (D17) was constructed at the ILL. On these instruments most of the first sans experiments were performed because there were no suitable spectrometers outside Europe at that time. The construction of such instruments in the United States began in the late 1970s at Missouri University Research Reactor (MURR), the National Bureau of Standards (NBS), and Oak Ridge National Laboratory (ORNL). In addition, a sans spectrometer (H9B) at Brookhaven National Laboratory became operational in 1981, though this instrument is designed for small-angle structural investigations on biological materials (26), and it is on the MURR (27), NBS (28), and ORNL (29) instruments that the majority of sans studies of polymers in the United States have been undertaken.

The ORNL 30-m sans instrument is operated by the National Center for Small-angle Scattering Research, which is a national user-dedicated facility that includes both sans and small-angle x-ray scattering (saxs) instrumentation (29,30). The center, sponsored by the National Science Foundation through an interagency agreement with the DOE, was established in 1978, when the construction of the 30-m facility commenced. Since becoming operational in late 1980, the 30-m sans instrument (see Fig. 5) has proved to be particularly suitable for the

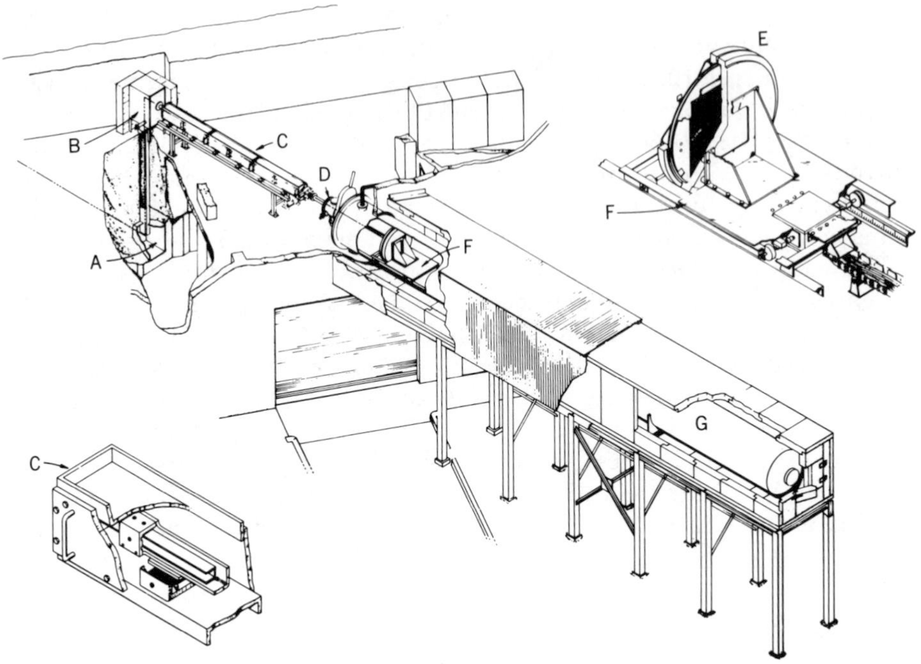

Fig. 5. Functional components of the U.S. National Small-angle Neutron Scattering Facility: A,B, graphite crystals; C, neutron guide; D, sample chamber; E, detector; F, detector carriage; G, flight path.

study of polymers, and approximately half the experiments performed have been in this area. The instrument will be described in some detail to illustrate some of the general features of sans experimentation, since it has been widely used for polymer research and it is the one with which the author is most familiar.

Monochromatization of the incident neutrons is provided by a bank of pyrolytic graphite crystals (Fig. 5A and B) to give a standard incident wavelength of $\lambda = 0.475$ nm, which can be changed to 0.238 nm by substituting graphite for the cold beryllium filter normally in position. The overall wavelength spread is $\Delta\lambda/\lambda \simeq 6\%$ and a similar mode of monochromatization is used on the MURR (27) and Saclay (25) facilities. Other instruments (22,23,28) employ rotating helical-slot velocity selectors, which have the advantage that a wider range of λ and $\Delta\lambda/\lambda$ may be covered. The incident beam may be transported by movable neutron guides (Fig. 5C) to move the apparent source to different distances (1.5–7.5 m) from the sample chamber (Fig. 5D) and hence increase the neutron flux on the sample at the expense of some loss in the resolution of the experiment. In order to minimize data acquisition times, most sans instruments are equipped with position-sensitive detectors which collect data over a wide range of angles simultaneously. On the 30-m instrument the detector (Fig. 5E) is a multiwire resistance-capacitance (RC) encoded proportional counter filled with ^{3}He-CF$_4$ with an active area of 64×64 cm^2 (31). The detector can be positioned at any distance from 1.3 to 19 m from the sample by moving the motor-driven detector carriage (Fig. 5F) along rails within the evacuated flight path (Fig. 5G) in a time period of 1–2 min. This latter feature is particularly suitable for polymer research because data can be taken over a wide Q range (0.03–6 nm^{-1}) without a time-consuming interval for demounting a fixed flight-tube assembly. With the exception of the FRJ-2 facility, this time-saving feature has not hitherto been available on other sans instruments. Other facilities employ position-sensitive counters based on the recoil products of the neutron with $^{10}BF_3$ (22,23) or scintillation counters. In general, none of the sans instruments currently available in the United States possesses the combination of modern cold-source and guide-tube technology which has made the D11 instrument at ILL so powerful. The fact that polymers are, in general, relatively strong scatterers has made it possible to accomplish significant research in the United States, though the possibility of providing truly state-of-the-art instrumentation is being actively pursued and has received significant support (32,33).

High Concentration Labeling. In experiments designed to determine the chain conformation in bulk polymer by sans, no energy discrimination is employed and the detector integrates over all energies. The scattered intensity $I(Q)$ is measured as a function of angle, and for isotropic (nonoriented) samples Q is a scalar quantity. The differential cross section is obtained by integrating equation 23 over ω, which operates on the phase factor to give a delta function in time (16). Furthermore, for the typical elements contained in polymers, the scattering lengths may be treated as real, dropping the complex conjugate. Where the element present consists predominantly of a single isotope, $\langle b \rangle$ may be replaced by b; eg, for naturally occurring carbon $\langle b \rangle = b = 0.665 \times 10^{-12}$ cm. With these simplifications equation 23 becomes

$$\frac{d\Sigma^{coh}}{d\Omega}(Q) = \frac{k}{k_0}\left\langle \sum_{ij} b_i b_j F_{ij}(\mathbf{Q},0)\right\rangle \tag{27}$$

This equation is valid for fully deuterated (>99%) or fully protonated species (eg, polymer or solvent molecules), though for partially deuterated molecules the average (real) scattering lengths $\langle b_i \rangle$, $\langle b_j \rangle$ must be used. As pointed out above, sans experiments involve measurements at small Q values. From equations 3 and 5 this implies $\lambda \simeq \lambda_0$, as the conservation of energy and momentum ensure that only neutrons scattered with $\Delta E \sim 0$ can be detected at small Q. Thus the coherent differential scattering cross section is given by

$$\frac{d\Sigma^{coh}}{d\Omega}(Q) = \left\langle \sum_{ij} b_i b_j \exp\left[i\mathbf{Q}\cdot(R_j - R_i)\right]\right\rangle \tag{28}$$

Nuclear cross sections have the dimensions of area, and as sample cross sections are normalized to unit volume, $d\Sigma(Q)/d\Omega$ has the dimensions of inverse length and is typically given in units of cm^{-1}.

For a bulk polymer sample with N molecules per unit volume of pure unlabeled component 1, a coherent scattering length of a monomer unit can be defined

$$a_1 = \sum_k b_k \tag{29}$$

where the summation runs over all the atoms in an unlabeled monomer unit (type 1). If the degree of polymerization is Z_1, the cross section may be divided into intra- and interchain components as follows:

$$\frac{d\Sigma^{coh}}{d\Omega}(Q) = Na_1^2 Z_1^2 P_1(Q) + N^2 a_1^2 Z_1^2 Q_{11}(Q) \tag{30}$$

$$= Na_1^2 Z_1^2 (P_1(Q) + NQ_{11}(Q))$$

where $P_1(Q)$ is the intrachain signal, which originates from monomer pairs belonging to the same polymer chain; it is called the form factor of the molecule ($P(0) = 1$). $Q_{11}(Q)$ is the interchain signal resulting from terms in the summation where the monomer pairs are on different chains. A similar equation may be written for a system containing a pure component 2 with only a change of subscript. If the two polymers are blended together in such a fashion that X_1 equals the mole fraction of component 1 and X_2 is the mole fraction of component 2, the cross section is given by

$$\frac{d\Sigma^{coh}}{d\Omega}(Q) = X_1 NZ_1^2 a_1^2 P_1(Q) + X_1^2 a_1^2 N^2 Z_1^2 Q_{11}(Q) + X_2 NZ_2^2 a_2^2 P_2(Q)$$

$$+ X_2^2 a_2^2 N^2 Z_2^2 Q_{22}(Q) + 2X_1 X_2 N^2 Z_1 Z_2 a_1 a_2 Q_{12}(Q) \tag{31}$$

assuming that both chains have the same number of molecules per unit volume N. In general, the intrachain functions $P_i(Q)$ and interchain functions $Q_{ij}(Q)$ are different for a blend of dissimilar components. However, in the case where the components of the blend differ only in the molecules of one component being isotopically labeled, equation 31 may be simplified as follows: assuming that deuteration of the hydrogenous molecule has a negligible effect on the monomer–monomer interactions, the average configuration of a labeled molecule does not differ from that of an unlabeled molecule and hence

$$P_1(Q) = P_2(Q) = P(Q) \tag{32}$$

and

$$Q_{11}(Q) = Q_{22}(Q) = Q_{12}(Q) = Q(Q) \tag{33}$$

If the subscript 1 is replaced by H for the unlabeled polymer (component 1) containing hydrogen atoms and similarly subscript 2 is replaced by D for its deuterated analogue, equation 31 may be simplified to

$$\frac{d\Sigma^{coh}}{d\Omega}(Q) = X_H X_D (a_H - a_D)^2 NZ^2 P(Q) + (X_H a_H + X_D a_D)^2 NZ^2 (P + NQ(Q)) \tag{34}$$

where each chain has the same degree of polymerization Z. For a pure single component without residual voids or heterogeneities (catalyst residues, impurities, stabilizers, etc), scattering is due simply to density fluctuations. In some cases (eg, amorphous polymers) density-fluctuation scattering is small (34,35,36) and, neglecting this component following a suggestion by Benoit (37), the following equation can be written:

$$P + NQ(Q) = 0 \tag{35}$$

giving the result (34)

$$\frac{d\Sigma^{coh}}{d\Omega}(Q) = X_D X_H (a_H - a_D)^2 NZ^2 P(Q) \tag{36}$$

Similar derivations were given independently in Refs. 38–40.

This shows that the form of the scattering curve in this case is governed by the single-chain form factor $P(Q)$. The mole fraction of each component modulates the scattered intensity with the maximum coherent scattering of the blend occurring at a 50:50 mixture of the two components. Thus $P(Q)$ may be obtained from the measured coherent intensity at labeling levels up to 50%. Although equation 36 is essentially the same formula derived by Von Laue (41) for random binary alloys and was implicit in some of the first expressions for the scattered intensity (42,43), the result was not appreciated in the earliest sans studies of bulk polymers and concentrated solutions. These studies relied on analogies with light and x-ray scattering, where the limit of zero concentration was required to eliminate interchain interference. The scattered signal was assumed to be proportional to C, the concentration (g/cm^3) of labeled (deuterated) polymer, and this assumption has been shown to be a reasonable approximation in this limit (5,44):

$$\frac{d\Sigma}{d\Omega}(Q) = C(a_H - a_D)^2\, \overline{M}_{wD}\, \frac{N_A}{m_D^2}\, P(Q) \qquad (37)$$

where N_A is Avogadro's number (6.023 × 10^{23}), m_D is the molecular weight of a repeat unit of the deuterated polymer, and $\overline{M}_{wD}$ is the (weight-averaged) molecular weight of a deuterated chain. It should be noted that m_D differs from the molecular weight of a repeat unit of a hydrogenous chain m_H. Similarly, the molecular weights of H and D chains with the same degree of polymerization differ by the same factor, as do the densities of H and D homopolymers. These quantities should be carefully distinguished (44) for H and D materials in order to avoid errors (up to 14% in the case of polyethylene).

The quantity $(a_H - a_D)^2$ is related to the difference in scattering power between labeled and unlabeled chains and is called the contrast factor. In general, radiation incident on a medium whose scattering power is independent of position is scattered only into the forward direction ($2\theta = 0$). For every volume element S which scatters radiation through an angle $2\theta > 0$, there is another volume element S' which scatters exactly (180°) out of phase, and therefore all scattering cancels unless the scattering power is different at S and S', ie, fluctuates from point to point in the sample. By analogy with x-ray scattering, which is caused by fluctuations in electron density, neutron scattering arises from differences in scattering length density (SLD), which is defined as the sum of coherent scattering lengths over all atoms lying in a given volume δV, divided by δV (16). For partially labeled polymer blends, the SLD is given by the coherent scattering length (eq. 29) divided by the monomer volume. The coherent cross section of a system of uniform scattering length density is zero, though fluctuations may be introduced by means of isotopic substitution, thus giving rise to a finite cross section which is proportional to $(a_H - a_D)^2$. Similarly, the contrast-variation methods, which have found wide application in biology, can sometimes be used to remove a component of the scattering by matching its SLD with the medium in which it is dispersed, thus removing the SLD fluctuation giving rise to the scattering.

The parameter used to describe the overall size of a polymer chain is the radius of gyration R_g, which may be derived from equation 36 or 37 by expanding $P(Q)$ in a power series for low $Q(Q < R_g^{-1})$ and plotting $d\Sigma^{-1}(Q)/d\Omega$ vs Q^2 (5–8,45). Alternatively, these parameters may be obtained by plotting $\ln d\Sigma(Q)/d\Omega$ vs Q^2 at low Q (45). These types of plots are conventionally referred to as Zimm and

Guinier plots, respectively, and the former is generally used for investigating polymer configurations, as it has been found to be linear over a wider Q range.

The first measurements in the bulk and concentrated solution were generally performed in the limit of low relative labeling and extrapolated to zero concentration. In this range

$$\frac{d\Sigma^{-1}}{d\Omega}(Q) = \frac{m_D^2}{C(a_H - a_D)^2\,\overline{M}_{wD}N_A}\left[1 + \frac{Q^2R_g^2}{3} + \ldots\right] \qquad (38)$$

Thus R_g and $\overline{M}_{wD}$ may be derived from the slope and $Q = 0$ intercept of such a plot. The realization that the same information could be obtained with greater accuracy at much higher levels of labeling was made (37–40) and verified (34,38–40,46–48) independently by several groups. The variation of R_g with the concentration of labeled molecules in amorphous polycarbonate (48) is shown in Figure 6; the measured values are independent of the level of labeling. Similarly, it is shown in Figure 7 that the extrapolated $Q = 0$ cross section $d\Sigma(0)/d\Omega$ is proportional to the product $X_H X_D$ for amorphous polystyrene (34), as expected from equation 36. This equation was derived on assuming equal polymerization indexes for the labeled and unlabeled chains. The effect of unequal indexes has been considered by Boué et al (49) who showed how the measured R_g and $d\Sigma(0)/d\Omega$ were perturbed by this mismatch. In this case the mole fractions X_D, X_H in equation 36 are replaced by volume fractions ϕ_D, ϕ_H (44,49). An alternative treatment of the effects of mismatch has been given in a study of samples with mismatch ratios between 0.32 and 1.94 (47).

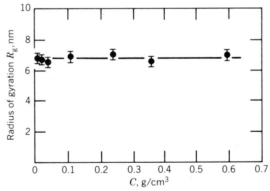

Fig. 6. Radius of gyration R_g for molecules in amorphous polycarbonate vs the concentration of labeled (deuterated) polymer. $\overline{M}_w$ (sans) = 20,100; $\overline{M}_w$ (viscometry) = 20,300; $\overline{M}_w/\overline{M}_n$ = 1.79 (48). Courtesy of *Polymer Bulletin.*

Amorphous Polymers.   The first convincing demonstration of the power of the sans technique was made in the field of bulk amorphous polymers. It is well known that there have been several theoretical approaches to the molecular conformation in these systems, based on the unperturbed Gaussian (random) coil (50,51) and the meander or bundle models (52–54), where a large fraction of the polymer molecules are envisioned in quasi-parallel arrangement. In addition, a collapsed-coil model was advanced for some systems (55). Before the development

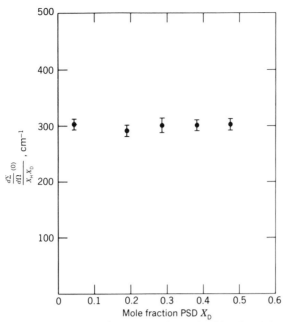

Fig. 7. Variation of $1/X_H X_D \, d\Sigma/d\Omega(0)$ with X_D for blends of deuterated (PSD) and normal (hydrogenous) polystyrene (34). Courtesy of *Polymer*.

of the sans technique there was no way of measuring the molecular conformation in bulk polymers and this led to intense debate on this issue in the literature. To the author's knowledge the first suggestion to use the contrast between deuterated and normal (hydrogenous) molecules to provide a direct determination of $P(Q)$ was made independently in at least two groups (56,57) in the late 1960s; the method was first demonstrated in principle in the early 1970s (58–64). The first quantitative sans experiment performed in the bulk polymer is shown in Figure 8, which gives the scattered intensity for various concentrations of normal hydrogenous poly(methyl methacrylate) (PMMA) dissolved in a matrix of deuterated poly(methyl methacrylate) (59–61). It may be seen from the symmetry of equation 36 that this gives the same coherent scattering as dissolving equal concentrations of deuterated polymer in a hydrogenous matrix. However, the advantage of a predominantly deuterated system is the much lower background scattering from the matrix alone ($C = 0$). Despite the fact that this background would be an order of magnitude higher for a predominantly hydrogenous system due to the higher incoherent cross section of ^{1}H (Table 1), many experiments have been performed in this manner due to the expense and scarcity of deuterated polymer. The optimum sample transmission is $\simeq 1/e \simeq 0.37$ and this implies sample thicknesses on the order of 0.1 cm and 1 cm for predominantly hydrogenous and deuterated matrices, respectively. As several alternate symbols for the scattering vector (eg, $Q = K = \mu = 4\pi/\lambda \sin \theta$) have been used in sans studies, these are shown in Figure 8 and should be taken as equivalent in these and subsequent plots. The Å unit has been widely used in sans experimentation and will be shown along with the SI unit (1 nm) to facilitate comparisons with other data in the literature. The data of Figure 8 are given in arbitrary units (counts

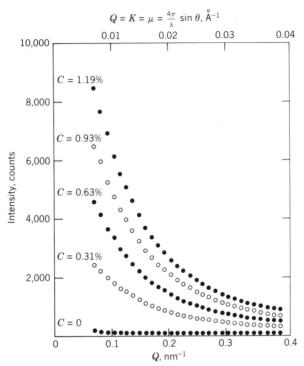

$$Q = K = \mu = \frac{4\pi}{\lambda} \sin \theta, \mathring{A}^{-1}$$

Fig. 8. Intensity vs Q for concentrations C of PMMAH dissolved in PMMAD matrix (59,60). Courtesy of Dr. Dietrich Steinkopff Verlag.

per unit time) and conversion to an absolute cross section $d\Sigma(Q)/d\Omega$, typically in units of cm^{-1}, is accomplished by multiplying by a calibration constant. Although this procedure is not essential for the measurement of the scattering dimensions (eg, R_g via equation 38), it is necessary for the measurement of $\overline{M}_w$ from the extrapolated forward scattering. Absolute calibration forms a valuable diagnostic tool for the detection of artifacts in experiments, as shown, for example, where the $\overline{M}_w$ measured by sans differs from $\overline{M}_w$ measured by other techniques (eg, chromatography) by an amount greater than the combined experimental error (43,62). The relationship between the scattering cross section and the measured count rate $I(Q)$ (counts/s) in a detector element with area Δa and counting efficiency ϵ, situated a distance r from the sample, is given by

$$\frac{d\Sigma}{d\Omega}(Q) = \frac{I(Q)}{tT} \frac{r^2}{\epsilon I_0 \, \Delta a A} \tag{39}$$

where I_0 is the intensity (counts/s·cm²) on a sample area A and thickness t with transmission T. Since the time dimension cancels in both the numerator and denominator of equation 39, the calibration constant which converts the measured intensity to an absolute cross section may be determined by comparison with a standard of known cross section run in the same scattering geometry for the same time. In order to eliminate drifts in the neutron flux on the sample, an incident-beam intensity monitor is usually employed and comparisons are made of the same number of monitor counts (65), ie, the same number of incident neutrons. Corrections are also applied for instrumental backgrounds and for the cell-by-cell variation of the detector efficiency ϵ (65). As explained earlier, the differential-scattering cross section $d\Sigma/d\Omega$ has dimensions of inverse length (cm^{-1}) and is directly analogous to the Rayleigh ratio used in light-scattering studies (61). Figure 9 shows the data of Figure 8 after conversion to absolute units and drawn in a Zimm plot with $K_1 = (a_H - a_D)^2 N_A/m_H^2$. The open circles are the result of extrapolations to $C = 0$ and $Q = 0$; these two lines intersect on the ordinate to give the inverse molecular weight $\overline{M}_{wH}$. The H and D subscripts are reversed from equations 37 and 38, which were derived for small amounts of D polymer in a hydrogenous matrix, whereas in the experiment of Figures 8 and 9, small concentrations of H molecules were dissolved in a deuterated matrix. The intensity is, of course, the same for either case for H and D chains of the same polymerization index. The molecular weight calculated from Figure 9 ($\overline{M}_{wH} \approx 220{,}000$) is in agreement with the value measured by viscometry ($\overline{M}_w \approx 250{,}000$) within the overall experimental error, indicating that the absolute calibration is reasonable for both techniques.

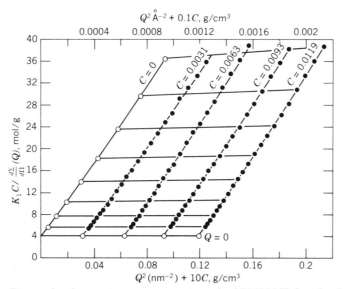

Fig. 9. Zimm plot for various concentrations C of PMMAH dissolved in PMMAD; the open circles are the result of extrapolations to $C = 0$ and $Q = 0$ (61). Courtesy of Butterworth & Co. (Publishers) Ltd.

Subsequent studies at low Q values ($Q < R_g^{-1}$) were made on a range of amorphous polymers including atactic polystyrene (63,64,66), molten polyethylene (67,68), poly(vinyl chloride) (69), polyisobutylene (70), poly(ethylene terephthalate) (71), polybutadiene (72), and a series of poly(methyl methacrylate)s of different tacticities (73). This type of experiment measures the (z-average) radius of gyration of the polymer chain R_g^z, which may be converted to the weight-average radius R_g^w if the polydispersity is known (61). According to the random-coil model, R_g^w should be proportional to $\overline{M}_w^{1/2}$, where $\overline{M}_w$ is the (weight-averaged) molecular weight, with the same constant of proportionality in the bulk as in an ideal θ-solvent. It may be seen from Table 3 that in general this prediction holds remarkably well for amorphous polymers and that there is close agreement between data on the same polymer studied independently by different groups. In the case of poly(ethylene oxide) (74), the molecular dimensions in the melt exceed those found in aqueous solutions, possibly due to the influence of salt residues on the local chain conformation. However, the dimensions in pure water and in organic θ-solvents are again close to those in the melt as found for other polymers.

Table 3. Molecular Dimensions in Bulk Amorphous Polymers

Polymer	State	$[(R_g^w)^2/\overline{M}_w]^{1/2}$, nm/g$^{1/2}$		Refs.
		Bulk	Solvent	
poly(methyl methacrylate)				
atactic	glass	0.027	0.025	59–61
	glass	0.025	0.025	73
syndiotactic	glass	0.029	0.024	
isotactic	glass	0.030	0.028	
polystyrene, atactic	glass	0.0280	0.0275	63
	glass	0.0275	0.0275	64
	melt	0.0280	0.0275	66
polyethylene	melt	0.045	0.045	67
polyethylene	melt	0.046	0.045	68
poly(vinyl chloride)	glass	0.040	0.037	69
polyisobutylene	glass	0.031	0.030	70
poly(ethylene terephthalate)	glass	0.039	0.039–0.042	71
polybutadiene	melt	0.035	0.034–0.042	72

Although these results have given impressive support to this model, they are not in themselves conclusive, since it was subsequently shown for crystalline polymers that R_g is very similar for molecules in the molten (amorphous) and solid (crystalline) states. Thus the finding that molecules exhibit the unperturbed dimensions in the molten or glassy amorphous state does not in itself rule out significant parallelism for an appreciable fraction of the molecules. In order to test how far the local molecular configuration, as opposed to the overall R_g, is described by the various models, measurements have been extended to higher values of Q. To first order the scattered intensity at a given Q is sensitive to fluctuations in the scattering length density on a distance scale $D \sim 2\pi/Q$. Thus as Q increases, the scattering is increasingly determined by the local chain configuration. This configuration may be calculated for the random-coil model using rotational isomeric statistics (51), and hence the scattered intensity may be estimated numerically and compared with the experiment. This is accomplished by measuring the scattering in the intermediate-angle range ($1 < Q < 6$ nm^{-1}), which is sensitive to the local configuration of the chain over distances $\sim$1–5 nm. Figure 10 shows the intermediate-angle neutron scattering (ians) data for molten polyethylene (68) at $T = 150°C$ compared to the rotational isomeric state (RIS) calculation of Yoon and Flory (75) and the Debye model for a coil with a Gaussian

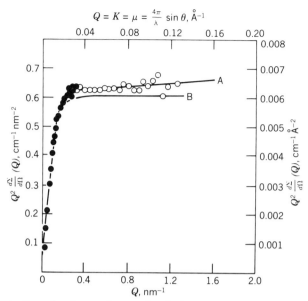

Fig. 10. Kratky plot for molten polyethylene at 150°C (68). A, Yoon and Flory rotational isomeric calculation (1976); B, Debye, Gaussian coil. $\dfrac{d\Sigma}{d\Omega}(Q) = \dfrac{2}{u^2}\dfrac{d\Sigma}{d\Omega}(0)[u - 1 + e^{-u}]$, $u = Q^2 R_g^2$. Courtesy of Butterworth & Co. (Publishers) Ltd.

distribution of chain elements (76,77). The data are plotted as $Q^2\, d\Sigma(Q)/d\Omega$ vs Q as used by Kratky (78) since this representation enhances the scattering at higher Q and facilitates comparison with different models. It may be seen from Figure 10 that $d\Sigma(Q)/d\Omega$ varies as Q^{-2} in this region, leading to a plateau in the Kratky plot, which is closely fitted by both the RIS calculation and Gaussian coil function. Similarly, the ians data for atactic polystyrene are consistent with the Debye model both in the melt and glassy states (64,66). The RIS model reflects the local conformation for the specific chain under consideration (75) and hence the agreement with the Debye model, which is based on general assumptions independent of the local chain structure, is probably fortuitous. RIS calculations for polystyrene indicate that the Kratky plot should exhibit a positive slope (75), which was not observed in previous experiments (66) that showed that the cross section varied as Q^{-2} up to $Q \simeq 4$ nm^{-1}. Subsequent experiments (79) on molecules labeled only in the main chain (D3) indicate that the Kratky plot does exhibit the expected positive slope and that the measured plateau for fully labeled (D8) molecules may result from the cancellation between this slope and the scattering due to the finite lateral dimensions ($\sim$0.5 nm) of the phenyl rings (80).

Results on PMMA (61,73) indicate that the scattering in the intermediate Q range is markedly dependent on the chain tacticity (Fig. 11). This type of measurement is particularly sensitive to corrections for the incoherent background, which is of the same order as the coherent scattering in the intermediate angle region and arises largely from hydrogen nuclei. It should be noted that the bound-atom cross sections cannot be simply used to calculate the background (Table 1), because the hydrogen incoherent cross section ($\sigma_{inc} = 79.7 \times 10^{-24}$

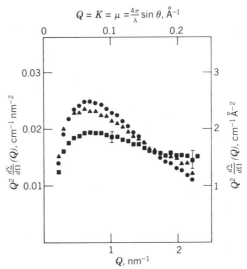

Fig. 11.   Kratky plots for 21.2 vol % PMMAH molecules in PMMAD matrix: ●, syndiotactic; ▲, atactic; ■, isotactic (73). Courtesy of the American Chemical Society.

cm^2), although widely quoted in the literature, almost never applies to real polymer systems. Methods that have been used to estimate the correction are described in Refs. 70 and 73. Comparison with the theoretical RIS calculation (81) shows that the shape of the curves as a function of tacticity is consistent with the random-coil model using rotational isomeric statistics. Similar comparisons in the intermediate Q range have also been made with polyisobutylene (70) and polycarbonate (82,83), and in each case reasonable agreement was achieved with the rotational isomeric theory. Although Pechhold (84,85) has shown that the sans and ians data on amorphous polymers can also be explained by the meander model, this has necessitated the introduction of backfolding into the model to fit the data. Schelten (86) has shown that this backfolding produces local correlations that should give rise to additional diffuse neutron scattering from a mixture of deuterated and normal polyethylene (PE) and has made a careful search for this feature. In the melt the experiments revealed no observable scattering and were inconsistent with the modified meander model. This is in clear contrast to experimental tests of the random-coil model, which has been very successful in predicting both the overall chain size and local configuration in a range of amorphous polymers, without any change other than refinements of the calculations. To the author's knowledge there are no major discrepancies between theory and the scattering experiments that might be indicative of quasi-parallel packing of polymer chains.

The successful application of the sans method to elucidate the chain structure in the amorphous state has led to a wide application of the technique in polymer science, and an exhaustive review of all the published results would be beyond the scope of this article. Some results that have had a major impact on polymer science are reviewed here with emphasis on data that have not appeared in previous reviews. The choice of the results included is to some extent personal and apology is made for the inevitable omission of significant research.

The Crystalline State. The arrangement of molecular chains within the lamellae of semicrystalline polymers has long been disputed (87–90), and sans have provided new information on this controversial subject. It is generally agreed that semicrystalline polymers exhibit a lamellar morphology both for material crystallized from the melt and from dilute solution. The thickness of lamellae is typically 10–50 nm with amorphous polymer interspersed between the crystalline regions. The molecular chains are at an angle (0–45°) to the lamellar normals and have lengths much greater than the lamellar thickness, thus traversing one or more lamellae several times. Based on considerations of density conservation at the crystal–amorphous boundary, Flory (87) demonstrated that a considerable fraction (~0.5) of chains must return to the same crystal. What is in question is whether the molecule returns with predominantly adjacent or with random re-entry to the crystallite of origin.

In general, sans experiments reveal that the radius of gyration R_g of molecules remains unchanged upon crystallization from the melt (38,44,67,68,91,92) and hence has a $\overline{M}_w^{1/2}$ dependence in both the molten and crystalline states (Table 4). This indicates that the molecules crystallize with a similar distribution of mass elements to that possessed in the melt (68) and hence is distributed over several lamellae in the crystalline state. For solution-crystallized material, however, the radius of gyration is relatively independent of molecular weight, and is usually markedly reduced from the dimensions in dilute solutions (93,94), indicating a much more compact configuration. These measurements of R_g were made at low Q and contain no information on the mutual arrangement of stems, ie, straight sections of a chain traversing a crystalline lamella. This is best examined by experiments in the intermediate-angle range, which is sensitive to the correlation of stems over distances of 1–5 nm. This type of measurement has been made for several systems (68,94–97) and compared with a variety of model calculations that simulate the chain trajectory (68,82,83,98–102). In general for crystalline materials, the approximation given in equation 35 does not apply and the second term of equation 34, resulting from density fluctuations, must be

Table 4. Comparison of Molecular Dimension in Molten and Crystallized Polymers

Polymer	Method of crystallization	$\dfrac{R_g^w}{\overline{M}_w^{1/2}}$, nm/g$^{1/2}$		Refs.
		Melt	Crystallized	
polyethylene	rapidly quenched from melt	0.045	0.045	67,68
polybutadiene,	rapidly quenched	0.047	0.047	44
hydrogenated	slowly cooled, ~1°C/min	0.048	0.047	
polypropylene	rapidly quenched	0.035	0.034	91
	isothermally crystallized at 139°C	0.035	0.038	91
	rapidly quenched from melt and subsequently annealed at 137°C	0.035	0.036	91
poly(ethylene oxide)	slowly cooled	0.042	0.052	38
polystyrene, isotactic	crystallized at 140°C for 5 h	0.026–0.028[a]	0.024–0.027	92
	crystallized at 200°C for 1 h	0.022[a]	0.024–0.029	

[a] Dimensions in the melt were not available. The values quoted are for amorphous samples (with atactic polystyrene matrices) annealed in the same way as the crystalline material (with isotactic polystyrene matrices).

included in the analysis (34,40,48) for this type of measurement. The ians data (83) for quench-crystallized polyethylene are shown in Figure 12, where the scattering function $F_n(Q)$ is defined by

$$F_n(Q) = (n + 1)Q^2 P(Q) = (n + 1)Q^2 \frac{\frac{d\Sigma}{d\Omega}(Q)}{\frac{d\Sigma}{d\Omega}(0)} \tag{40}$$

where n is the number of bonds in the chain.

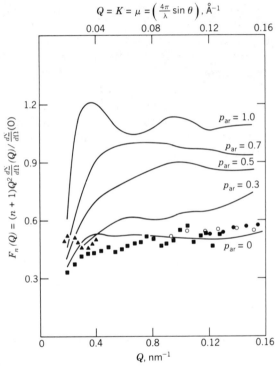

Fig. 12. Calculation of $F_n(Q)$ as a function of the probability of adjacent folding (p_{ar}) (number of bonds in chain $n = 2000$) (83). ■, Ref. 68; ○, Ref. 94; ▲, Ref. 95; ●, Ref. 96. Courtesy of IUPAC.

There is reasonable consistency between the data from several groups (62,94–96) which are compared with model calculations based on Monte Carlo statistics as a function of the probability (p_{ar}) that the stem folds adjacently along the ⟨110⟩ plane (82,83). It may be seen that the model leads to a Q^{-2} dependence for $d\Sigma(Q)/d\Omega$ and hence a plateau in the Kratky plot. The plateau levels differ by a factor ~ 2 for the extremes of random $(p_{ar} = 0)$ and adjacent $(p_{ar} = 1)$ reentry, and it was concluded (68,101) that the ians data were inconsistent with regular folding $(p_{ar} < 0.3)$. Similar model calculations have also been performed as a function of the number of stems folded adjacently in a central cluster (100). It was concluded that this model could fit the neutron data with higher probabilities ($\sim$0.7) of adjacent folding, though this involved plotting the experimental data (68) as a function of the molecular weight of the labeled chains, measured by both sans ($\overline{M}_w$ = 46,000) and chromatographic techniques ($\overline{M}_w$ = 60,000). In the opinion of the author and others (94,96), this procedure introduces unnecessary uncertainty into the height of the plateau level in $F_n(Q)$ and hence makes the comparison less precise. Equations 36 and 40 indicate that for high molecular weight material ($n > 1000$), $F_n(Q)$ is independent of molecular weight to a very good approximation since $d\Sigma(0)/d\Omega$ contains the molecular weight that is proportional to n. This parameter may therefore be canceled, thus making $F_n(Q)$ independent of molecular weight, and this would lead to a significantly lower estimate of the probability of adjacent folding or the number of stems in a central cluster. The initial comparisons of theory and experiment were made for quench-crystallized polyethylene to avoid a well-known segregation artifact which occurs in slowly cooled samples of polyethylene (43,68,94). This artifact is virtually absent for melt-crystallized polypropylene, where experiments (91) show that the plateau height in the Kratky plot is independent of the rate of cooling.

In a careful review of the scattering from both quenched and slowly cooled materials, Sadler (103) concluded that the degree of adjacency within the rows is ca 30–50% for melt-grown crystals. Although the precise definition of p_{ar} varies between different models, it seems that probabilities of this order are inconsistent with regular folding of an appreciable number (>4) of stems in one crystallographic plane. Longer sequences of adjacent stems would lead to an observable modulation of the wide-angle neutron scattering pattern ($Q > 6$ nm^{-1}), as pointed out by Stamm (104), and such modulations are not observed for melt-crystallized polyethylene (104,105). However, it was pointed out by Guttman et al (100) that a larger proportion of folds are relatively close and that the folds that are not adjacent are "near" and rarely involve stem separations greater than three nearest neighbors. This is also consistent with estimates of the distribution of distances of reentry made using RIS statistics in the interlamella amorphous regions (98) and by a recent method proposed by Fischer and co-workers for the evaluation of neutron scattering data independently of detailed structural models (106,107). The only assumption of this approach is that the molecular structure can be described as consisting of clusters of stems belonging to the same molecule in each lamella. The analysis leads to the average number of clusters per molecule N_c, the radius of gyration of the centers of stems belonging to one cluster $\langle R_{cc}^2 \rangle^{1/2}$, and the average number of tie molecules per chain. This analysis has been applied to melt-crystallized polyethylene (106), polypropylene (106), poly(ethylene oxide) (107), and poly(ethylene terephthalate) (108). Typical values of N_c are in the range of 4–13, whereas $\langle R_{cc} \rangle^{1/2}$ is on the order 1.5–6 nm. The average distance between stems $\langle a \rangle$ may be derived from $\langle R_{cc}^2 \rangle^{1/2}$ on the assumptions of a random walk or a linear arrangement of stems (regular folding). On the latter assumption the values of $\langle a \rangle$ are 2–3 times greater than the distances involved in regular folding in one crystallographic plane ($\langle a \rangle \simeq 0.5$ nm). Assuming a random walk, values of $\langle a \rangle$ are in the range 2–4 nm, which is consistent with the conclusion of near but not adjacent folding, following from the above model calculations.

Guttman et al (100) also pointed to the problem of the anomalously high densities in interfacial boundaries between crystal and amorphous regions which could result from space-filling considerations noted early by Frank (109) and

Flory (87). If the transition between crystal and amorphous regions is not abrupt and takes place over a distance of ~1 nm, space-filling anomalies can be avoided and do not invalidate the general conclusions described above (103,110,111).

As mentioned earlier, the radii of gyration of polyethylene chains in solution-grown material are much smaller than those in melt-crystallized samples and are relatively insensitive to molecular weight. The ians data also show differences with melt-crystallized samples and exhibit a peak in the Kratky plot (93,94,103,112). Based on these results Sadler and Keller (103) proposed a "superfolding" model in which the folding of the chain is not confined to a single layer, and after executing a number of folds in a given layer, the chain continues to fold in an adjoining layer. Comparisons with model calculations (98,103,112) assuming folding along a ⟨110⟩ plane have indicated that the measured intensity is too low for the folding to be all adjacent. In order to produce agreement with experiment, the adjacent stem arrangement must be "diluted" by a factor of approximately 2–3, and Yoon and Flory concluded that reproduction of the scattering function required an array of stems confined to several layers of the crystal but not densely packed in any of them. The stems of a given chain occur at small distances but seldom adjacent to one another in the growth plane (98). Alternatively, the preferred model of Sadler and co-workers (93,94,103,112) envisages a molecule distributed over approximately four neighboring ⟨110⟩ sheets. The molecule is "diluted" within each sheet, but ca 75% of folds are connecting stems at adjacent sites. Clearly some significant disagreement still persists concerning the nature of the stem "dilution" required for solution-crystallized material, and further work is in progress, which includes extending comparisons of theory and experiment to solution-crystallized polymers other than polyethylene (113). However, it seems generally agreed that the neutron data rule out the possibility that a typical molecule is regularly folded in one crystallographic plane over many stems without interruption, a model that had gained widespread support over the previous decades for both melt- and solution-crystallized material. Similarly, the extremes of random configuration have been ruled out with a large fraction of molecules usually folding in "near" reentry within a few nearest neighbors. There has thus been a considerable convergence of views to the point where the differences of interpretation are often a matter of description rather than substantial disagreement (103). Without the input of neutron data, it is very doubtful if such a convergence could have resulted on questions that have been disputed over several preceding decades.

Semidilute and Concentrated Solutions.   Small-angle neutron scattering measurements in dilute solution offer few advantages over light-scattering techniques, which permit the elucidation of chain dimensions via the electron density contrast between a macromolecule and solvent. Greater signal-to-noise may be obtained with the neutron technique since it is less sensitive to dust particles and because of the larger contrast possible with a deuterated polymer (or solvent). These factors enabled Ballard et al (114) to extend measurements to very low molecular weight polystyrenes ($\overline{M}_w \simeq$ 600–4000) in cyclohexane, where the Kratky-Porod worm chain was used to extract the persistence length. However, the main impact of sans has been in the area of semidilute and concentrated solutions, and the technique has provided a wealth of new information previously unavailable by light scattering, where intermolecular interference effects restrict measurements to dilute solutions. These effects may be overcome at higher polymer concentrations by sans measurements on systems where a fraction of the solute is isotopically labeled. As in the case of bulk polymers this type of measurement was initiated on the assumption that the labeled component should be dilute (115–120). It was subsequently demonstrated, however, that measurements may be conducted at high labeling (39,40,121–125), thus increasing the experimental signal-to-noise as in the case of bulk polymers. In parallel with the influx of new data from the sans technique, there have been considerable developments in polymer solution theory, largely stimulated by the possibility of checking predictions by scattering measurements over the whole concentration range. The theory of such solutions has been advanced by both an analytical mean-field approach (126,127) and also by renormalization group methods (128) and scaling theory (119). These theories predict the variation of R_g, the osmotic pressure π, and a screening length ξ_s introduced by Edwards (126,127). It is well established that the swelling of macromolecules in a good solvent arises from excluded-volume interactions between pairs of monomer units on a single polymer chain. As the polymer concentration increases, the excluded-volume effect is screened and diminished. In the limit of the bulk polymer, the conformation of a single chain can be described as an unperturbed random walk as originally predicted by Flory and verified by numerous sans measurements (Table 3). The screening length ξ_s may be loosely defined as the distance beyond which excluded-volume effects are absent. It varies in magnitude from the size of the polymer chain in dilute solution to the order of a monomer segment in bulk. Between them the theoretical approaches make various predictions for R_g, π, and ξ_s in the various regions of a temperature–concentration diagram introduced by Daoud and Jannink (129) as a function of the polymer concentration C and τ, where $\tau = (T - \theta)/\theta$ and θ is the θ- or Flory temperature. The predictions of these theories and their comparison with experiment have been summarized in previous review articles (5,8,130,131), which give a detailed survey; only recent experiments involving the variation of R_g and ξ_s with C are reviewed here. Previous measurements (119) indicated that for polystyrene in CS_2, R_g varies as $C^{-0.25}$ and ξ_s varies as $C^{-0.72}$ in apparent agreement with the predictions of scaling laws. Since that time, there have been reports of many violations of scaling laws (132), which should be valid in principle only in the asymptotic limit of infinite molecular weight. It has been a longstanding puzzle to understand why the scaling exponents were observed with only modest molecular weight ($\overline{M}_w = 114,000$), and this question has been reexplored by King et al (124,125) in a recent study of polystyrene in toluene. The differences between the various theoretical predictions is sometimes small and earlier measurements in the semidilute regime have been interpreted as confirming both mean-field (115) and scaling predictions (118) for the variation of ξ_s with C. For this reason, considerable efforts have been devoted (124,125) to maximizing the statistical accuracy of the experiment by careful attention to machine reproducibility and cross-checking the consistency of the data by means of redundant data sets. Use was also made of high concentration labeling methods (39,40,121–123), which have been developed since the previous measurements and permit a further increase in the signal-to-noise of the experiments. It was shown that these methods give highly reproducible results and produced data with a high degree of internal consistency between data sets taken during several experimental sessions, using different levels of polymer labeling. Results concerning the variation of ξ_s with C for several independent data sets confirmed that the exponent ($\xi_s \sim C^{-0.70}$) is close to that reported (119) and predicted by scaling theory for the semidilute regime. The variation of R_g with C is shown in Figure 13 and gives an exponent ($R_g^2 \sim C^{-0.156}$) that differs significantly from the predictions of scaling theory and with that previously found by other researchers (119). The scaling-law prediction ($R_g^2 \sim C^{-0.25}$) shown in Figure 13 falls well outside the experimental error for all independent data sets. The different interpretations that have been assigned to scattering data in the same concentration regime point to the need for a critical evaluation of the consistency of data with respect to the various theoretical predictions. Careful evaluation is needed for the effects of polydispersity (61), mismatch of labeled and unlabeled molecules (46,47,49), and inclusion of data points that fall outside the Guinier range ($QR_g > 1$). The latter effect has been considered by Ullman (133,134) who has developed numerical corrections for given ranges of QR_g. If the Guinier limit is exceeded, as is often the case in practice, R_g can be obtained by fitting $P(Q)$ to a scattering function derived from a molecular model (eg, Gaussian coil) or alternatively the measured data can be extended down to $Q = 0$ by extrapolation procedures (133,134). The former procedure has the advantage of using a much wider region of Q-space data, but has the disadvantage that the result is dependent on a model that may not accurately represent the real molecule. It is instructive to use both procedures and compare the results if the condition $QR_g > 1$ is exceeded (124,125,135).

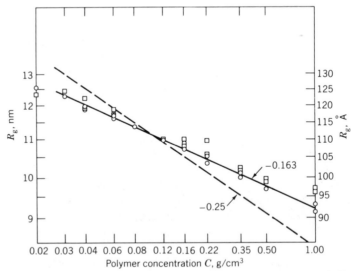

Fig. 13.   Variation of R_g with C for polystyrene in toluene. The different symbols represent results from independent data sets with different deuterium labeling (124). Courtesy of the American Chemical Society.

Although some of the conclusions of the earlier experiments may need rechecking in the light of the above factors, this should not detract from the major advances brought about by the interplay of neutron scattering and solution

theory. These include the confirmation of different regions of the temperature–concentration diagram with the demonstration of different concentration
and temperature dependence of R_g, ξ_s, etc as the boundary lines are crossed
(136,137), making this a fruitful area for further experiment and discussion.
Deuteration techniques now permit the labeling of specific sites in a molecule
(80), which will enable scattering experiments to provide even more detailed
information on the spatial arrangement of molecules in solution.

Blends and Block Copolymers.   For economic reasons it has become increasingly difficult over the past decades for the chemical industry to introduce
new commercial polymers, and hence much interest has focused on the possibility
of producing new materials by blending (alloying) existing polymers (138). Before
the application of sans, the methods used to investigate polymer compatibility
(microscopy, calorimetry, etc) could indicate macroscopic segregation but could
not demonstrate fine-grained separation or intermixing at the molecular level.
In principle, this information can be obtained by light or x-ray scattering (139),
using the natural (electron density) contrast between the species. In practice,
however, little work has been undertaken because of difficulties in preparing
samples sufficiently free of contaminants (dust, catalyst residues, etc). In general,
neutron scattering is far less sensitive to such contamination (140) in view of the
large contrast that may be obtained by deuterating a fraction of one of the species,
and the method has been widely applied to test for compatibility at the segmental
level and also to provide information on chain configuration in the blend.

Small-angle neutron scattering was first used to investigate polymer compatibility by Kirste and co-workers for both incompatible and compatible blends
(61,141,142). Most chemically dissimilar polymer pairs are incompatible, though
sans was used to demonstrate that polystyrene and poly(α-methylstyrene) form
a truly compatible mixture (143). Another important example of compatibility
is the blend of polystyrene–poly(phenylene oxide) (PPO), marketed as Noryl,
where sans studies (144,145) demonstrated that the system is compatible at the
segmental level, thus confirming earlier conclusions based on calorimetric data
(146).

These initial experiments were based on an extension to polymer blends
(147) of the Zimm analysis (148–150) developed for light scattering to give information about the molecular weight, R_g, and the second virial coefficient A_2,
which is related to the Flory interaction parameter χ. In principle, the Zimm
analysis is limited to the regime where one of the species is dilute, and the
extension to concentrated polymer mixtures has been given by Hadziioannou and
Stein (149,150) and independently by Warner et al (151) who have shown how
χ may be measured directly for homogenous mixtures. The theory was developed
(149) for a blend of two polymer species A and S, where a fraction X_D of the A
polymer chains has been replaced by D chains, leaving a fraction $X_H = 1 - X_D$
unlabeled, resulting in a three-component mixture (H, D, and S). The expression
for the differential-scattering cross section (149), which is equivalent to the Rayleigh ratio used in light scattering, is

$$\frac{d\Sigma}{d\Omega}(Q) = X_H X_D (a_H - a_D)^2 N Z^2 P(Q)$$

$$+ (X_D a_D + X_H a_H - a_S)^2 N Z^2 (P(Q) + N Q(Q)) \quad (41)$$

where a_D and a_H are the scattering lengths of the labeled and unlabeled monomers of species A, a_S is the scattering length of the S species, N is the number of A molecules per unit volume, and Z is the polymerization index of the A species; $P(Q)$ and $Q(Q)$ are the intra- and interchain functions for species A and it may be seen that the equivalent functions for the S species do not appear in the expression. The terms in equation 41 can be separated by performing combined neutron and x-ray experiments on the same sample or by performing two neutron experiments in which X_D is varied for the same overall composition of A and S species to give R_g and $\overline{M}_w$ for the labeled molecules in the blend (139,149–151). Information about the Flory interaction parameter may be obtained (149,151,152) from the second term of equation 41 by means of the random phase approximation (RPA)

$$\frac{1}{\phi Z[P + NQ(Q)]} = \frac{1}{\phi Z P^0(Q)} + \frac{1}{(1 - \phi)Z_S P_S^0(Q)} - 2\chi_{AS} \qquad (42)$$

where $P^0(Q)$ and $P_S^0(Q)$ are the intrachain functions at θ-conditions for the A and S species, Z_S is the degree of polymerization for the S species, ϕ is the volume fraction of the A species, and χ_{AS} is the Flory interaction parameter for the A and S species. The function $P + NQ(Q)$ can be obtained, for example, by performing an experiment with $X_D = 1$, where the first term of equation 41 is zero. Alternatively, two experiments may be performed for different values of X_D and equation 41 may be solved for $P + NQ(Q)$. After extrapolating to $Q = 0$, where $P^0(Q) = P_S^0(Q) = 1$, equation 42 may be solved for χ_{AS}, which may be compared with values from other techniques (149). This type of measurement has been performed for various blend systems including PS–poly(vinyl methyl ether) (149), poly(vinylidene fluoride)–PMMA (149), PS–poly(o-chlorostyrene) (153), methoxylated poly(ethylene glycol)–methoxylated poly(propylene glycol) (151), PS–polybutadiene (151), and PMMA–chlorinated polyethylene (154). In some instances, it may be shown that equations 41 and 42 reduce to a modified Zimm equation from which a second virial coefficient may be derived and related to χ (149,150,154). This approach has been used for PS–PPO blends (155), and PS–PPO, PMMA–poly(vinyl chloride), PMMA–styrene–acrylonitrile copolymers, and PS–poly(vinyl methyl ether) (PVME) blends (156).

In general, the χ parameters derived from the Zimm approach or from equation 42 are both temperature and concentration dependent. Thus the same blend can show miscibility or phase separation in different regions of the phase diagram. Herkt-Maetzky and Schelten (157) and Shibayama et al (158) have studied blends of PS–PVME that exhibit a molecular weight-dependent lower critical-solution temperature (LCST). The size of the concentration fluctuations may be characterized by a correlation length a, introduced by Ornstein and Zernicke (159). By extrapolating to the point where $a \to \infty$, the spinodal may be determined (140,157,158), although this point may not be reached in practice because of polydispersity effects (157). The critical exponents ν and γ were found to be 0.5 and 1 in both studies in accordance with the predictions of mean-field treatments, and there was reasonable agreement with cloud points determined by neutron and light scattering (158).

The above studies of polymer blends are based on the assumption that the interactions are independent of deuteration, or that the interaction parameter between labeled and unlabeled molecules of the same species χ_{HD} is zero. This assumption has also been implicit in all previous sans studies, though calculations (160) have indicated that χ_{HD} may be finite and on the order of 10^{-4}–10^{-3}. This led to the suggestion (152) that for sufficiently high molecular weights, demixing could occur in mixtures of deuterated and hydrogenous molecules of the same species, and that this could lead to the measurement of χ_{HD}. Such experiments have been performed by Bates et al (161) who examined binary mixtures of deuterium labeled and unlabeled 1,4-polybutadienes. By regarding the labeled and unlabeled molecules as different species with volume fractions ϕ_H and ϕ_D, respectively, the interaction parameter χ_{HD} between them may be estimated by fitting equation 42 to the measured scattering data. The measured cross section is shown in Figure 14 for a blend of deuterated ($\phi_D = 0.31$) and hydrogenous ($\phi_H = 0.69$) polybutadienes, and is fitted to the RPA (eq. 42) for various temperatures. The data were calibrated in absolute units (cm^{-1}) and the polymerization indexes of the H and D species ($Z_H = 960$, $Z_D = 4600$) were measured independently by chromatographic methods. The single-chain functions were represented by Debye (Gaussian) coils (76,77) with $R_g = l_K(Z/6)^{1/2}$, where $l_K = 0.69$ nm is the segment length for polybutadiene (72,162). Thus χ_{HD} is the only adjustable parameter in the fit and the RPA gives an excellent representation of the data for various blend compositions, temperatures, and chain lengths. It can be seen from Figure 14 that the extrapolated zero-Q cross section exceeds by large factors the value it would have if χ_{HD} were zero as previously assumed ($d\Sigma(0)/d\Omega \sim 100$ cm^{-1}). It was also demonstrated that the system exhibited an upper critical-solution temperature (UCST) as predicted (160). This finding is remarkable, because to the author's knowledge there was previously no example of UCST in high molecular weight polymers where LCST transitions have been universally observed. A plot of χ_{HD} is shown in Figure 15 as a function of inverse

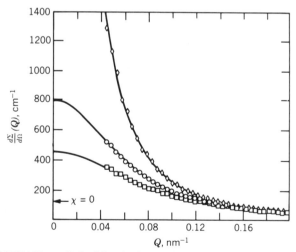

Fig. 14. $d\Sigma(Q)/d\Omega$ vs Q for blend of 69 vol % protonated and 31% deuterated 1,4-polybutadiene at the critical composition. The curves were obtained from the homogeneous mixture scattering function by adjusting χ_{HD} (one adjustable parameter): □, 361 K; ○, 274 K; ◇, 248 K (161). Courtesy of the American Physical Society.

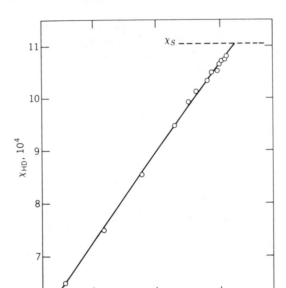

Fig. 15. Plot of χ_{HD} vs T^{-1} for blend of deuterated ($\phi_D = 0.31$) and hydrogenous ($\phi_H = 0.69$) polybutadienes (161). $\chi = \dfrac{0.326}{T} - 2.3 \times 10^{-4}$. Courtesy of the American Physical Society.

temperature (K^{-1}), which reaches the value on the spinodal at $T_S = 242.1 \pm 0.7$ K. The critical exponents exhibit the mean-field values and at room temperature ($T = 296$ K) the value of $\chi_{HD} = 8.8 \times 10^{-4}$. Similar measurements in binary mixtures of deuterated and protonated polystyrenes (163) and polydimethylsiloxanes (164) corroborate the prediction of a universal isotope effect, which has been shown to be a consequence of zero-point motion in conjunction with the anharmonicity of the interatomic potential (165). Since χ values on the order of 10^{-4}–10^{-3} are usually much smaller than the values measured between different species, the methods described above are not invalidated. However, the correction for this effect should be considered when these methods are applied to the measurement of smaller χ parameters. Similarly, the effect of a finite χ_{HD} does not invalidate the vast majority of investigations using the sans techniques, as segregation effects are important only in high molecular weight systems and have generally not been observed. In the bulk state most systems studied previously are solids at room temperature and have been exposed for only a limited time period in the liquid state, which may not be sufficient for the development of equilibrium compositional fluctuations. The polybutadiene system, with a glass-transition temperature below $-90°C$, is liquid at RT, which facilitates the attainment of equilibrium. In future the effect of a finite χ_{HD} must be carefully evaluated for liquid systems by comparing molecular weights measured by sans and independently by chromatographic techniques. This is possible, providing the sans data are taken in absolute units. Previously unexplained effects observed in such systems may be attributable to the effect of a finite χ_{HD} (166,167).

The scattering from partially labeled systems of two species has also been treated by Jahshan and Summerfield (121,122) and Koberstein (168) who derived expressions for the scattering identical to equation 41. Although this equation was derived for a homogeneous mixture (149), whereas Koberstein (168) assumed complete phase separation, all three treatments may be more general and differ only in nomenclature. The single-chain form factor $S_s(Q)$ used in Refs. 124, 125, and 168 is equivalent to $P(Q)$ used in Refs. 34, 149, 150, and 169. The function $S_T(Q)$ used in the former references is equivalent to $P + NQ(Q)$ used in the latter references and in the text of this article. Apart from a numerical constant, $S_T(Q)$ is equivalent to $S_{XR}(Q)$, which is proportional to the scattering signal observed in an x-ray experiment at small Q $(Q < R_g^{-1})$. The equations derived in Refs. 149 and 168 are equivalent apart from a factor of 4π in the definition of the Rayleigh ratio. Equation 41 or equivalent expressions (121,122) may be used to separate the domain structure from the molecular configuration within the domains for two-phase systems and several experiments have been performed along these lines. The coherent scattering from blends of normal (hydrogenated) polyethylene and polypropylene is virtually zero (169), though strong contrast may be induced by partial or complete labeling of either phase. For a fully labeled polyethylene phase $(X_D = 1)$ the first term in equation 41 vanishes and the scattering contains no information on $P(Q)$. The remaining term gives the scattering from the two-phase structure and was analyzed to provide the domain dimensions using the Debye-Bueche analysis (169–171); the mean-chord intercept lengths of both phases were in the range of 100–1000 nm. The scattering from partially labeled blends $(0 < X_D < 1)$ was analyzed to give the R_g and $\overline{M}_w$ of the labeled molecules in the polyethylene domains by solving equation 41 for two values of X_D (169). The $\overline{M}_w$ measured by sans techniques $(\overline{M}_w = 148,000)$ agreed closely with the value measured by chromatography before blending $(\overline{M}_w = 149,000)$, indicating that the labeled molecules were statistically dispersed in the domains. Similarly, the measured radius of gyration, $R_g = 13.9$ nm, was close to that measured in the homopolymer $(R_g = 16.5$ nm), as expected when the domains are large. Similar experiments to measure domain dimensions have been performed by Fernandez et al (172) on phase-separated interpenetrating polymer networks where one phase is fully labeled and also by Blundell et al (173) using the natural electron-density contrast between different species via x-ray scattering.

The state of mixing has been investigated in a range of polyolefin homo- and copolymers (174,175). Extensive segregation into separate domains is shown by 50:50 blends of high density (HDPE) and low density polyethylene (LDPE) in the solid state, with linear Debye-Bueche plots as expected for a random two-phase structure (170,171). The domain sizes calculated from these plots (~15 nm) suggest that the two components are extensively segregated into separate la-mellae, and this conclusion is consistent with the two-peak melting curves obtained for this system. Sans studies of 50:50 blends of HDPE with linear low density polyethylene (LLDPE) indicate a much smaller degree of segregation of the components and this conclusion is consistent with the single-peak melting curves obtained for such mixtures (174). Measurements on blends of polypropyl-ene and ethylene–propylene copolymers indicate extensive phase separation in copolymers with a high (> 40%) ethylene content. These systems are also phase-separated in the melt, and the domain sizes appear to increase as a function of time (175).

The attainment of optimum physical properties by blending immiscible polymers depends to a large extent on the nature of the resulting morphology, which is influenced by the blending conditions. In particular, the particle size distribution of the dispersed phase is the result of competition between breakdown and coalescence processes. Under given mixing conditions, an invariant morphology is attained that represents the steady-state condition between particle breakup in high shear regions and flow-induced coalescence of the dispersed domains. Simple models of the coalescence process predict improved efficiency with decreased shear rate and continuous-phase viscosity, though before the application of sans it was difficult to provide quantitative verification. Experiments have been designed (176) to measure the extent of coalescence during the mixing of rubber-immiscible polymers quantitatively and thereby gain understanding of its significance in the development of blend morphology. This was accomplished by preparing blends of polybutadiene (PB) and chloroprene (CP) in which PB was fully deuterated in one case and hydrogenated in the other. Equal volumes of these blends were mixed for varying durations and the loss of the initial isotopic purity in these two compositions brought about by shear-induced coalescence was determined from the magnitude of the coherent sans cross sections. The measurements showed that flow-induced coalescence occurred at a very rapid rate during mixing. Furthermore, the rheological properties of the dispersed and continuous phases, as well as the nature of the flow field used in their blending, were shown to influence the coalescence process greatly.

Extensive investigations of the morphology of block copolymers have been undertaken in the bulk and in concentrated solution. Where the two species are incompatible, microphase segregation may occur leading to mesomorphic phases with spherical, cylindrical, or lamellar morphology, depending on the block lengths, solvent, etc. Neutron scattering has proved to be particularly suitable for studying these systems because of the possibilities for varying the contrast via deuteration of one of the blocks or the solvent. The morphology of a range of polystyrene–polybutadiene (PS–PB) diblock copolymers was shown to consist of PB spheres in a PS matrix (177,178). Sans measurements on fully labeled PB block samples show well-resolved Bragg peaks originating from a macrolattice with body-centered cubic packing of the PB spheres (177,178). For partially labeled samples the theory predicts that the scattering from the phase structure should vanish when the overall scattering-length density of the PB spheres equals that of the PS matrix, leaving only the term describing the individually labeled PB molecules. This balance occurs at a labeling of 16 mol % deuterated PB blocks, where the prefactor of the second term in equation 41 is zero. To achieve this condition it should be noted that equation 41 was derived on the assumption that the H, D, and S subunits had equal volume. Where this condition does not hold, a_S must be scaled by the ratio of molar volumes of the two species (121,166,168,169). Analysis of the scattering envelopes of two samples with differing molecular weights of the PB blocks indicates that in each case the oscillations due to Bragg peaks vanish as predicted (Fig. 16). The remaining term in the scattering may be analyzed to give the radius of gyration of the PB block within the domain (166). In one case, there was a significant difference between the block $\overline{M}_w$ determined by sans and chromatography, which led to the suggestion (166) that this might result from an isotope effect and to the subsequent measurement of χ_{HD} in the polybutadiene system (161).

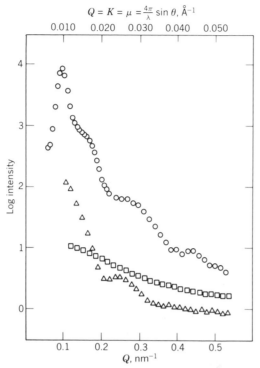

$$Q = K = \mu = \frac{4\pi}{\lambda} \sin \theta, \text{Å}^{-1}$$

Fig. 16. Sans for polystyrene–polybutadiene diblock copolymers. A matched blend of hydrogenated (H) and deuterated (D) PSH–PBD and PSH–PBH copolymers washes out the scattering due to the phase structure: ○, PSH–PBD; △, PSH–PBH; □, PSH–(16% PBD:84% PBH) (166). Courtesy of Butterworth & Co. (Publishers) Ltd.

Single-chain scattering functions in block copolymers have also been measured in polyether–polyurethanes (179) and the styrene–isoprene system (180,181). This block copolymer has been extensively studied in the spherical, cylindrical, and lamellar morphologies by small-angle x-ray scattering (181) and sans (182,183), where domain dimensions have been compared with thermodynamic theory. Block copolymers in solution have been investigated by Picot et al (184) and also by Han and Mozer (185) who have investigated the dimensions of PMMA blocks in solutions of PMMA–PS block copolymers.

Critical phenomena in copolymer melts have been studied by Bates and Hartney (186) who used the theory of Leibler (187) to determine the interaction parameter. Diblock copolymers of 1,2–1,4 polybutadiene exhibit an order–disorder critical transition, where the compositional fluctuations are manifested as a peak at $QR \sim 1$, where R is the diblock copolymer R_g. In contrast to normal blends, where critical effects are observed at $Q \to 0$, equilibrium is established very quickly, thus overcoming the diffusional limitations that have hampered previous studies of critical phenomena in the bulk. Both the interaction parameter and critical exponents were extracted from measurements of the peak intensity as a function of temperature (Fig. 17). Similar experiments were performed by Benoit et al (188) who derived general equations for the scattering from solutions of an arbitrary number of polymers and copolymers at any con-

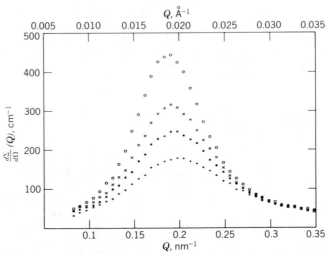

Fig. 17. $I(Q)$ vs Q for 1,2–1,4 polybutadiene block copolymer as a function of temperature: $\bigcirc$, $-7°C$; $\times$, $16°C$; $*$, $40°C$; $+$, $90°C$ (186).

centration. This treatment followed a series of recent papers (189,190) giving a new method for the evaluation of the scattering intensity of a solution of two homopolymers. A critical evaluation of the application of these methods to polymer solutions is given in Ref. 135. Quan et al (191) have studied techniques for the determination of single-chain functions of homopolymers in a block copolymer matrix. As block copolymers are frequently used as compatibilizers for blending incompatible homopolymers, these methods have important practical applications.

Deformation and Swelling. There is considerable current interest in determining the response of individual molecules in a bulk polymer to a macroscopic deformation or swelling of the sample. As shown previously, sans from partially labeled (deuterated) samples can provide direct information on the deformation of the whole molecule down to the average statistical segment. Thus it offers the possibility of checking the various models that have been developed to relate the macroscopic and microscopic properties of polymers, as for example in theories of rubber elasticity. A series of experimental and theoretical papers have appeared (192–209) with a view to testing these assumptions with respect to the deformation of individual chains in both chemically cross-linked polymers (networks) and linear polymers that contain only transitory (apparent) cross-links arising from physical entanglements.

Experiments on deformed samples are inherently more difficult than the corresponding studies of isotropic materials and some of the earlier work has proved to be inconclusive. Part of the problem arises because deformed materials produce anisotropic scattering, which cannot be radially averaged, and slices or sectors of the two-dimensional pattern must be used. This leads to inherently larger statistical errors, making it difficult to arrive at unambiguous conclusions in some experiments. Another difficulty arises with the ill-defined character of the networks resulting from broad molecular weight distributions, mismatch between the lengths of the labeled and unlabeled chains, and appreciable numbers of dangling chains, loops, or entanglements trapped between adjacent chains.

Ullman (198–201) has summarized the differences between different types of networks and pointed out that randomly cross-linked chains deform to a greater extent than end-linked chains. In order to minimize these effects and provide well-characterized model networks, many of the first experiments were carried out on polystyrene, though this polymer is a glass at room temperature and unsuitable for some experiments, as elongation must be carried out at elevated temperatures. For linear polymers, any finite time interval between stretching and quenching to the glassy state can result in stress relaxation, which changes the deformation of the molecules from its initial state. Moreover, in some of the scattering patterns of networks with labeled cross-links (192), peaks were observed that were not expected on any of the then current theories of scattering from networks. Recent theoretical developments have indicated that these peaks are due to correlation hole effects (7,207). Finally, instrumental resolution effects become a factor, as the dimensions of molecules increase in the stretch direction and can progress beyond the resolution range of the sans instrument. Generally, such effects become appreciable for $R_g > 20$ nm and corrections should be applied beyond this limit. In view of the above limitations, the conclusions of the first experiments performed should be treated with some reserve, though they proved to be valuable learning experiences and have laid the groundwork for more precise experiments in later work. The sans experiments on deformed linear polymers and networks have been reviewed (7,201) and Ullman (201) has compared recent data from networks with the theoretical models advanced for this type of system. While no general consensus has yet emerged, some tentative conclusions are beginning to be apparent.

For linear (uncross-linked) polymers, several experiments indicate that the radii of gyration perpendicular and parallel to the orientation direction transform largely in the same manner as the external macroscopic dimensions (affine deformation), provided the molecular weight is sufficiently high and the conformational relaxation processes are slow compared to the time in which the sample is deformed. Picot et al (193) studied the uniaxial deformation of hot stretched polystyrene and concluded that for low elongations with deformation ratios $\lambda_d <$ 1.7, affine deformation was the dominant mechanism. Deviations from affine behavior were found at high elongations ($\lambda_d > 3$) and an explanation was put forward in terms of a model where affine behavior holds only for distances separating effective cross-links (193). Boué et al (202) found that transverse coil radius of gyration of hot stretched linear polystyrene deforms affinely for $\lambda_d \leq$ 3, and studied its relaxation as a function of time and temperature (202). Hadziioannou et al (203) found that for extrusion-oriented high molecular weight ($\overline{M}_w \simeq 500,000$) polystyrene, the molecular deformation is nearly affine up to $\lambda_d = 10$, though for lower molecular weight material the deformation is less than affine. For linear chains the segmental orientation arises solely because of the entanglements that interconnect all chains in a macroscopic network. Chains with molecular weight less than the critical entanglement molecular weight $\overline{M}_c$, which are embedded in a matrix of longer chains, do not enter into the entanglement network and thus should not be susceptible to orientation by stretching the specimen except for transitory flow effects.

In the case of networks, chemical cross-linking does not appreciably change the molecular dimensions from those of the melt (7,72,204) and the change in dimensions introduced by swelling or stretching is much lower than predicted by affine deformation of the whole chain (chain-affine model). This is not unexpected as networks deform quite differently than linear polymers, and the chain-affine model has never been seriously advanced for chemically cross-linked materials, except perhaps where they are deformed at temperatures close to the glass-transition temperature (198). The classical theory of rubber elasticity assumes that the junction points deform affinely and it is often referred to as the junction-affine model. The phantom-network approach due to James and Guth (208) is based on the assumptions that the mean positions of the cross-link points deform affinely and that fluctuations in the cross-link points are Gaussian and independent of the applied deformation. The junction-affine model is sometimes referred to as end-to-end pulling deformation (197,206) and may be regarded as a special case of the phantom-network model (199–201), where the fluctuations in the cross-link points are suppressed. Flory has proposed a further modification of the phantom-network model in which function fluctuations are present but partially suppressed (209). The emphasis of sans experiments lies in the measurement of R_g parallel and perpendicular to the deformation axis and comparison with R_g in the underformed state. Formulas giving the variation of the deformation ratio α parallel and perpendicular to the stretch direction have been summarized as a function of λ_d (197,201,205). For swelling experiments, the deformation is isotropic, the parallel and perpendicular deformation ratios are identical, and $\lambda_d = Q_d^{1/3}$ where Q_d is the ratio of volumes in the swollen and unswollen states. The observed and calculated values of α are shown in Figure 18 for uniaxially stretched end-linked polydimethylsiloxane networks (205), which shows that for short chains the molecular deformation may be approximately described by the junction-affine (end-to-end pulling) model. However, for longer chains the deformation is lower than the prediction of even the phantom-network model. The observed molecular R_g in tetrafunctional polystyrene networks is shown in Figure 19 as a function of Q_d (206). It can be seen that the molecular

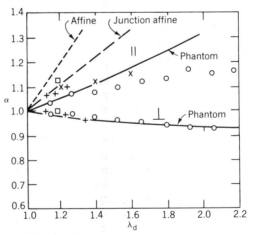

Fig. 18.   Variation of $\alpha_\parallel$ and $\alpha_\perp$ as a function of deformation ratio λ_d for tetrafunctional polydimethylsiloxane networks as a function of the (number-averaged) molecular weight ($\overline{M}_n$) compared to model predictions: □, $\overline{M}_n = 3000$; +, $\overline{M}_n = 6000$; ×, $\overline{M}_n = 10{,}000$; ○, $\overline{M}_n = 25{,}000$ (205). Courtesy of the American Chemical Society.

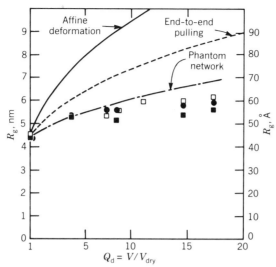

Fig. 19. Radius of gyration of the labeled chains of the network as a function of the degree of swelling. The different symbols refer to different sample–detector distances on the D11 and D17 spectrometers at the ILL: ●, D17 (2.83 m); □, D11 (5.66 m); ■, D11 (10.66 m) (206). Courtesy of the American Chemical Society.

R_g does not show any appreciable dependence on Q_d and that all models considered overestimate the variation of R_g in the high swelling range. Similar conclusions may be drawn from measurements on polystyrene networks swollen in other solvents (192), though in this case the functionality of the network was not well defined (201). In addition, Fernandez et al (72) measured a chain-expansion factor $\alpha = 1.23$ for polybutadiene networks swollen by a factor $Q_d = 4.67$ in toluene. The conclusion that may be drawn from the above experiments is that none of the models described above or envisioned in the classical theories of rubber elasticity or gel swelling can give a proper description of the effects of deformation observed at the molecular level and that new mechanisms are needed to explain the experimental results. Some alternative models and possible future developments are described in Refs. 197–201, 205, 206, and 209.

There has been much less work to investigate deformation processes in the crystalline state and only a small number of papers have appeared on this topic (103,210–212), though this promises to be a fruitful area for future research. Measurements have been performed on melt-crystallized oriented polyethylene and have been interpreted as showing changes in the molecular R_g which are approximately affine for polyethylene samples drawn near room temperature with draw ratios 5–10 (103). For polypropylene with a draw ratio of 8, the changes in R_g were found to be nonaffine and the molecules were extended by a factor ca 2 in the draw direction (210). However, these samples were drawn at 145°C and further experiments as a function of draw ratio, temperature, and molecular weight are needed before any general conclusions are possible.

It should be noted that the factor of 3 in equation 38 applies only to isotropic systems and several different radii may be defined for oriented materials (45,198,201,203,210–213). If equation 36 is expanded in the form of a Zimm equation

$$\frac{d\Sigma^{-1}}{d\Omega}(Q) = \frac{1}{X_D X_H (a_H - a_D)^2 NZ^2}[1 + \gamma R_g^2 Q^2 + \ldots] \tag{43}$$

then for an isotropic system, $\gamma = \frac{1}{3}$ would give an R_g corresponding to the root-mean-square (rms) distance of the labeled, coherently scattering elements from the center of mass. For anisotropic systems showing cylindrical symmetry about the draw direction, conventionally taken to be the z-axis, radii have been defined parallel and perpendicular to the draw direction (210). By using equation 43 for the component of Q perpendicular to the orientation axis and $\gamma = \frac{1}{2}$, the perpendicular radius gives the rms distance of scattering elements from the symmetry axis. Similarly for Q parallel to the orientation axis and $\gamma = 1$, R_g gives the rms z-distance of scattering elements from the center of mass.

An alternative definition of R_g in systems showing cylindrical symmetry about the draw direction is used in Refs. 198, 201, 203, and 212. This system of nomenclature introduces radii parallel ($R_\parallel$) and perpendicular ($R_\perp$) to the draw direction (z). Thus for Q parallel to z

$$\frac{d\Sigma^{-1}}{d\Omega}(Q_\parallel) = \frac{d\Sigma}{d\Omega}(0)\left[1 + \frac{Q_\parallel^2 R_\parallel^2}{3}\right] \tag{44}$$

and for Q perpendicular to z

$$\frac{d\Sigma^{-1}}{d\Omega}(Q_\perp) = \frac{d\Sigma^{-1}}{d\Omega}(0)\left[1 + \frac{Q_\perp^2 R_\perp^2}{3}\right] \tag{45}$$

The factor of 3 normally used for isotropic systems is retained in equations 44 and 45 for both directions perpendicular and parallel to the draw direction. Thus $R_\parallel$ and $R_\perp$ have a different definition from the parallel and perpendicular radii used in Ref. 210 and cannot be interpreted in terms of the rms distances described above. A third system of notation, which does not require the assumption of cylindrical symmetry, is used in Ref. 211. If the cross section is measured as a function of Q at angle ϕ to the z-axis, then for an intensity slice parallel to $z(\phi = 0)$

$$\frac{d\Sigma}{d\Omega}(Q,0) = \frac{d\Sigma}{d\Omega}(0)[1 + Q^2 R_z^2] \tag{46}$$

for such a slice $R_z^2 = \langle z^2 \rangle$, which is equivalent to using equation 43 with $\gamma = 1$ to give the rms z-distance of scattering elements from the center of mass. Similar equations can be written defining radii (R_x, R_y) in the x- and y-directions where $R_x^2 = \langle x^2 \rangle$ and $R_y^2 = \langle y^2 \rangle$ and the overall R_g is given by

$$R_g^2 = R_x^2 + R_y^2 + R_z^2 \tag{47}$$

Similar equations are used to define the radii of gyration, but with the subscripts $\perp$ and $\parallel$ in place of R_x, R_y, and R_z (205). In this system of notation, $R_\parallel$ and $R_\perp$ differ from the radii defined in Refs. 198 and 203 because of the difference in the factor of 3 in the Zimm equations 44 and 45.

For cylindrical symmetry $R_x^2 = R_y^2$ and for a slice perpendicular ($\phi = 90$)

to the draw direction

$$\frac{d\Sigma}{d\Omega}(Q,90) = \frac{d\Sigma^{-1}}{d\Omega}(0)[1 + Q^2R^2] \tag{48}$$

where $R^2 = (\langle x^2 \rangle + \langle y^2 \rangle)/2$. Comparison of equations 44 and 45 with equations 46 and 47 shows that these two systems of nomenclature are related by $R_z^2 = R_\parallel^2/3$ and $R_x^2 = R_y^2 = R_\perp^2/3$.

The overall R_g is given by

$$\begin{aligned} R_g^2 &= \frac{R_\parallel^2 + 2R_\perp^2}{3} \\ &= R_z^2 + 2R_y^2 \end{aligned} \tag{49}$$

In a system where the perpendicular directions ($\phi = 90$) are not equivalent, a repeat measurement on a sample rotated through 90° is needed to characterize the oriented system completely (198,201).

It is clear from the above discussion that the factor $\gamma = \frac{1}{3}$ in equation 38 cannot be simply carried over from isotropic to oriented systems as originally pointed out by Guinier (45). The different definitions of R_g that have been given in oriented systems should be carefully distinguished when comparing the data from different groups. In order to relate R_g values before and after deformation, the same definition must be used in both cases.

Other orientation studies have been undertaken to throw light on the solid-state deformation processes of individual molecules in semicrystalline polymers. It is well known that there is disagreement in the literature on the question of whether plastic deformation can be explained in terms of the melting and re-crystallization of lamellae (101) or breaking into blocks of folded chains with subsequent reformation into domains constituting a microfibrillar structure (214,215). Sans has been used to investigate the deformation process by making use of a clustering (segregation) artifact encountered by several groups (43,68,94,216). The scattering theory developed to this point is based on the assumption that the centers of gravity of the labeled molecules are randomly or statistically dispersed in the unlabeled molecules. It is known, however, that any departure from such a distribution by way of aggregation of the labeled molecules or correlations in their trajectories or positions can lead to anomalous values of $\overline{M}_w$ or R_g measured by sans (43,68,103). Extensive studies of melt-crystallized polyethylene have shown that correlated aggregates of deuterated (PED) mole-cules in hydrogenated material (PEH) are created in the melting region (125–135°C), indicating that the nonrandom distribution is caused by the difference in melting points (~6°C) between PEH and PED homopolymers. Conversely, the anomalous sans molecular weights are unaffected by annealing outside the melting region and can only be returned to their true values, as measured by chromatography, by melting and quenching the sample (43). Wignall and Wu (212) have prepared blends of 4.3 vol % PED in PEH, which show anomalously high apparent sans molecular weights and radii of gyration, resulting from correlations in the centers of gravity (clusters) of the PED molecules. These samples were subjected to solid-state deformation and Figure 20 shows Zimm plots $d\Sigma(Q)^{-1}/d\Omega$ vs Q^2 for the unoriented and oriented samples. It may be seen that $d\Sigma(0)/d\Omega$, from which the

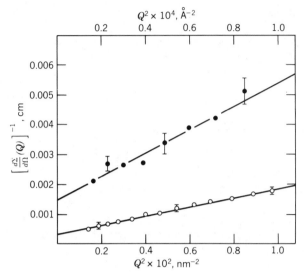

Fig. 20. $[d\Sigma^{-1}(Q)/d\Omega]$ vs Q^2 for unoriented (radial average) and oriented (horizontal slice) of 2-D scattering from 4.35 vol % PED molecules at 105°C: ●, oriented; ○, unoriented (212). Courtesy of Butterworth & Co. (Publishers) Ltd.

sans-$\overline{M}_w$ is calculated, changes by a large factor (>5) on orientation in the temperature range 20–119°C, where annealing alone is known not to affect $\overline{M}_w$. A similar reduction in the apparent sans-$\overline{M}_w$ may be achieved by melting and rapidly quenching the blend. This implies that large-scale reorganization takes place and that the level of molecular motion involved in the deformation process is on the same order of magnitude as that during the melting and crystallization process (212,217).

The exact nature of the segregation process has been the subject of debate and different models have been put forward to explain the anomalous sans scattering. Schelten et al (43) originally suggested that it might arise from the clustering together of large numbers of labeled molecules resulting in the creation of large domains of PED-rich and PED-sparse material, but later showed that the excess scattering could also arise from the enhancement and depletion of the D-monomer (CD_2) concentration in a few nearest neighbors around a given D-monomer unit involving only small perturbations in the centers of gravity of the labeled molecules (43). They introduced the term paracluster for the latter model and suggested that it was the most likely source of the excess scattering for samples quenched relatively rapidly from the melt. Other workers felt that the former (domain) model was more appropriate for melt- and solution-crystallized polyethylene (94,216), where the PED-rich and PED-sparse domains resulted from isotopic fractionation over large regions of the sample during crystallization. Irrespective of the model chosen to represent the anomalous sans-$\overline{M}_w$, a melting and recrystallization mechanism during solid-state deformation is obviously consistent with the observed reduction in forward scattering.

In order to quantify the extent that other deformation models (214) might randomize the positions of the labeled molecules and hence produce the observed reduction in $d\Sigma(0)/d\Omega$, model calculations were performed (217) based on a set of Zernicke-Prins equations (218). By an appropriate choice of structural parameters, this model can be made to include both the paracluster and domain models as special cases. After fitting it to the measured scattering, the correlation lengths describing the extent of the PED-rich domains are found to be less than the single-chain R_g of the PED molecules for both oriented and unoriented samples. Thus no large-scale segregation of these molecules takes place, and the positions of centers of gravity are in this case only slightly perturbed from a statistical distribution. This finding seems to be consistent with the results of radiation cross-linking experiments (219), and the calculations confirmed that at least a portion of the sample must undergo melting and recrystallization during deformation. Further experimental studies along these lines would be valuable along with theoretical work to explore the possibility of including a partial-melting mechanism in the microfibrillar deformation model as suggested by Peterlin (220).

Complementary saxs studies (221) on high pressure crystallized polyethylene found no evidence of a melting and recrystallization mechanism when deformed by solid-state extrusion. This type of system has been studied by Ballard et al (222) who confirmed that the molecules are largely in chain-extended form. A melting and recrystallization mechanism was originally proposed (101) for systems in which the entanglements of the melt are conserved in the solid state and subjected to large-scale irreversible deformation, and hence this prediction would not necessarily apply to chain-extended systems. Similarly, systems subjected to small reversible deformation would not be expected to undergo melting and recrystallization, and saxs studies of oriented deformation of poly(ethylene terephthalate) fibers can be interpreted in terms of a mechanical deformation scheme (223).

Scattering in the intermediate-angle range by oriented systems will clearly be of great importance, though only a limited amount of work has been undertaken to date (79,210,224). Dettenmaier et al (224) studied amorphous atactic poly(methyl methacrylate) deformed below the glass-transition temperature in a temperature range of 60–120°C for deformation ratios $\lambda_d \sim 2$–2.3. The sans data were interpreted as showing affine deformation for long chains ($\overline{M}_w > 170,000$) as generally found for other linear polymers. However, for very short chains the molecular deformation was nonaffine. Figure 21 shows the scattering function $F_n(Q)$ measured in the intermediate range for unstretched (isotropic) samples and also for an oriented sample measured perpendicular to the draw direction; the theoretical function for a Gaussian coil (F_{Gc}) is included for comparison. It was concluded that even for high molecular weight material the deformation becomes nonaffine for regions of less than 2.5 nm.

In a study of the deformation of glassy polystyrene (79), the contrast factors of the chain were varied by labeling the whole repeat unit (D8), the chain backbone (D3), or the phenyl group (D5). In the intermediate-angle range the scattering is sensitive to the finite chain cross section and the local chain rigidity. The structure factor was expressed as a thin thread function convoluted with $\exp(-Q^2 R_c^2/2)$ where R_c^2 is the mean square radius of the cross section. Comparison of the structure factors measured in the amorphous state with different contrast factors provided direct information on the finite chain cross section. Similar measurements in the deformed state (79,225) revealed an appreciable

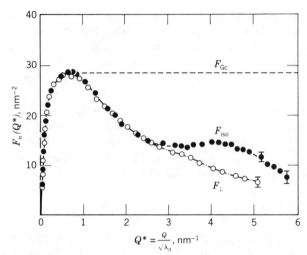

Fig. 21. Kratky plot of $F_n(Q)$ vs Q for isotropic (F_{iso}) and stretched samples of atactic poly(methyl methacrylate) measured perpendicular ($F_\perp$) to the draw direction. The Debye scattering function for a Gaussian coil (F_{Gc}) is included for comparison (224). Courtesy of the American Chemical Society.

reduction in the transverse phenyl cross section along a shear-band axis. Such measurements provide direct insight into the local deformation processes of the chain and should in the future prove to be as informative as the extensive sans studies undertaken on isotropic materials. Intermediate-angle data on atactic polystyrene deformed by extrusion-orientation processes (213) show that for $QR_g \gg 1$ the deformation is considerably less than the affine result measured at lower Q (203).

Diffusion, Relaxation, and Dynamic Studies.   The topic of the diffusion of polymer chains in the bulk polymer is of considerable interest both from the theoretical and experimental viewpoint. Following de Gennes, the diffusion on a microscopic scale is envisioned as the reptation of a chain along a tube formed by the entanglements of neighboring molecules. There had been considerable debate concerning the applicability of this concept over the distance scale ranging from a chain segment to size of the overall radius of gyration. Over the latter range, measurements had been previously performed by microdensitometry on samples of deuterated and hydrogenous polymers, which were allowed to inter-diffuse at an interface that was then sectioned and examined by ir methods. Due to the limiting thickness of sections, this method was effectively limited to molecular weights below 10^4 which needed time scales on the order of a month for measurements (226). Bartels, Crist, and Graessley (227) have used sans methods to extend the available range of diffusion coefficients that may be studied by preparing samples consisting of alternate layers of deuterated and hydrogenous polymers. As the temperature is raised and diffusion proceeds, the spatial modulations in composition decay and the scattering grows progressively, finally reaching the pattern corresponding to a uniform molecular mixture of deuterated and hydrogenous components. Analysis of the time dependence of the scattering yields the polymer diffusion coefficient D, and due to the small-distance scale probed by sans, measurements may be extended by approximately three orders of magnitude beyond the limit of ir microdensitometry (227). Initial measurements on monodisperse fractions of hydrogenated polybutadiene confirm the prediction that $D \sim M_w^{-2}$ and give activation energies in good agreement with theoretical estimates (Fig. 22). Similar experiments have been performed (228) on polyethylene, though the samples were more polydisperse and no definitive conclusions were drawn in view of the strong molecular weight dependence of D. This technique should have important future applications for understanding the mutual interdiffusion of different molecular species, one of which is deuterium-labeled.

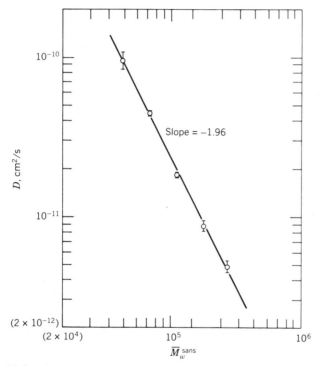

Fig. 22. Molecular weight dependence of diffusion coefficient ($T = 125°C$) for hydrogenated polybutadienes (227). Courtesy of the American Chemical Society.

The above experiment is an example of a study of a dynamic process via quasi-static techniques, where a sample is fabricated with a nonequilibrium concentration gradient, a process is allowed to proceed for a given time, and then the sample is frozen and examined by static techniques. Another example of such a measurement using sans is the investigation of transesterification kinetics in polyesters by McAlea et al (229). Mixtures of labeled and unlabeled poly(ethylene terephthalate) (PET) molecules were heat-treated for various periods, quenched to room temperature, and examined by sans. Transesterification acts to alter the lengths of the deuterated and protonated sequences due to chemical reactions between the chains. The effective block length of the deuterated segments may be measured from the extrapolated $Q = 0$ cross section and measurements show that it falls from a value corresponding to the original chain length by a factor of approximately 5 in 10 min at 270°C. To the author's knowledge such information is not available from any other technique. Similar experiments have been performed on PET by Wu et al (230) who encountered both transesterification and segregation effects and developed methods for separating them.

Quasi-static techniques have also been used (202,231) to study coil relaxation in hot-stretched polystyrene. Time–temperature superposition was used for data in the small- and intermediate-angle ranges to obtain the variation of the chain-form factor over a wide range of relaxation times, and behavior was compared with the predictions of reptation theory (202). The results were consistent with the conclusion that the polymer configuration deforms affinely down to the level of persistence length, though the time dependence of relaxation seemed to contradict current reptation theory. Further work with higher molecular weight and more monodisperse material may be necessary to test the generality of these conclusions, though this type of study seems to be very promising for the understanding of relaxation processes.

The study of real-time processes (without quenching) is just beginning, and preliminary attempts have been made (232,233) to study sans under oscillatory strain. The first experiments have also been conducted on polymer solutions under shear (233–236). Because of the nature of these experiments such studies are inherently flux limited and greatly benefit from the provision of new high intensity neutron sources (32).

Polymer Latices. One of the most important methods of polymer synthesis utilizes emulsion polymerization and in recent years much interest has focused on the preparation of latex copolymers with radial compositional gradients using a controlled monomer feed. Formation of the desired product requires that the locus of polymerization be at the aqueous interface of the growing particle (latex) and that the reaction be carried out under nearly monomer-starved conditions, ie, the rates of polymerization and monomer addition must be nearly equal. The structure of the latex markedly affects the properties of the bulk polymers and although the condition of monomer starvation is achievable, identification of the locus of polymerization cannot be investigated directly and has been the subject of considerable debate.

Grancio and Williams (237) postulated a polymer-rich spherical core surrounded by a monomer-rich shell that serves as the major locus of polymerization, giving rise to a core-shell morphology. This model, in which the first-formed polymer constitutes the core and the second-formed polymer makes up the shell, has received much support but also significant criticism. The sans technique has been used to investigate the structure of such latices by O'Reilly et al (238,239) who have studied poly(methyl methacrylate) (PMMA) latices with deuterated shells of PMMAD or PSD polymerized on the surface (238,239). The scattering from the original PMMAH latex is a Bessel function (45), exhibiting sharp maxima and minima for monodisperse particles, though in practice these sharp features are smeared by the finite experimental resolution. Accurate desmearing procedures using indirect Fourier transform methods (240,241) have been used to remove these instrumental effects and led to patterns showing the expected sharp minima with good agreement between the core radii determined by sans and independently by light scattering. Figure 23 shows the desmeared scattering

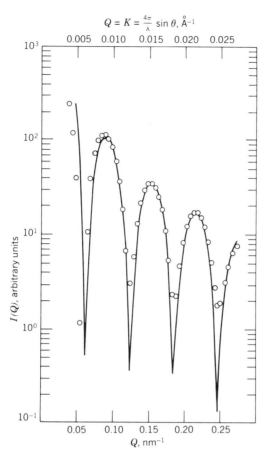

Fig. 23. Desmeared sans data (○) for PMMA latex with 3.0-nm PMMAD shell on surface (core contrast matched) compared with theoretical hollow-shell scattering (242).

pattern for a PMMAH core (radius = 49.5 nm) with a PMMAD shell (3-nm thick) polymerized on the surface (242). The sample was run in an H_2O–D_2O mixture adjusted to match the scattering-length density of the core, thus leaving only the hollow-shell scattering. The agreement between the desmeared experimental data and the theoretical hollow-shell scattering is excellent, confirming a core-shell morphology for the latex (238,239). Similar experiments on a partially deuterated PMMA shell polymerized on cores consisting of random PMMA–PS copolymers also demonstrated a core-shell structure with excellent agreement between the overall size measured by sans and transmission electron microscopy (243). In other studies (244) sans was used to measure the kinetics of swelling of polystyrene latices in styrene. The particle sizes resolved in this work (~400 nm) are close to the upper limit of resolution of currently available sans instrumentation. The experiments described above were undertaken with low volume fractions of the dispersed latices, where particle–particle interactions are minimized. At higher concentrations the mutual arrangement of the particles is reflected by a structure factor from which a latex–latex radial distribution function may be obtained. Comparison of the results of such experiments (245,246) with theoretical models (eg, hard spheres) gives information on the interparticle interactions. These techniques are relevant to a wide variety of colloidal and bio-chemical systems, and an excellent review of their application to the structure of micellar solutions is given in Ref. 16.

Information on the stabilization of nonaqueous dispersions of polymer particles by block copolymers may be obtained by sans, and the possibility of deuterating one of the block lengths allows the determination of the state of dispersion of each block in the particle or on the surface. Experiments along these lines have been performed (247) for nonaqueous dispersions of PMMA and polystyrene particles stabilized by polystyrene–polydimethylsiloxane block copolymers. The contrast variation and matching methods described above hold promise for the measurement of the R_g of molecules constrained inside latex particles with diameters smaller than the unperturbed coil dimensions, and experiments along these lines are in progress (248).

Other Applications.   Space limitations preclude a full survey of all important sans work, particularly as new papers are appearing at a rate of about four per month. Both saxs and sans have been extensively used to characterize thermoplastic materials, but have found much less application for the study of thermosets. These are highly cross-linked amorphous materials, and no detailed information on the distribution of cross-links or network homogeneity has hitherto been available. A new method has been developed by Wu and Bauer (249) who have utilized sans in conjunction with partially labeled epoxy resins. Prominent peaks are observed in the scattering patterns which reflect the intranetwork correlations. This method offers promise for the accurate characterization of network structure in thermosets.

Many structural studies have been performed on ionomers, and while there is good evidence for the aggregation of ion pairs, there is still no general agreement on the state of aggregation or the exact nature of the aggregates, which are reflected as peaks in the small-angle scattering near $Q \sim 1$ nm^{-1} (0.1 Å^{-1}). In general, saxs has had wider application than sans in this field, largely because the scattering contrast is higher due to the high electron density of the aggregates. Several systems revealing a clear peak in the saxs pattern are featureless when examined by sans, eg, poly(pentenamer sulfate) (250) and sulfonated polystyrene ionomers. However, some systems have sufficient scattering-length density contrast to be examined by sans (250–252). In water-swollen ionomers, sans offers some advantages because of the additional scattering contrast obtained with heavy water (250). Furthermore, the inclusion of a fraction of deuterated polystyrene chains permits the determination of R_g (251). For further details see Refs. 253 and 254.

A novel application of the sans technique for the determination of the partitioning of noncrystallizable comonomers in semicrystalline random copolymers has been developed (255). This method is an adaptation of an earlier procedure using saxs methods (256). In some cases the difference in scattering power between the monomer units in the copolymer is higher for sans than saxs. Where this is not the case, additional contrast may be obtained for sans experiments by deuterating one type of monomer unit. The method was successfully applied to chlorinated polyethylenes and trioxane–dioxolane copolymers with comonomer contents as low as 3%.

Liquid crystalline polymers (qv) have been widely examined by a range of techniques (eg, x rays, nmr, esr, etc), though these methods do not give direct information of chain conformation in the various phases (isotropic, smectic, nematic). The application of sans methods to these systems is just beginning (257). By partially deuterating a fraction of one of the polymers, information on the molecular R_g and the anisotropy of the coils was obtained. The study showed that large changes in conformation may occur without significantly affecting R_g and demonstrated how sans methods complement other analytical techniques.

In studies of cyclic and linear polydimethylsiloxanes in dilute solutions (258), theoretical calculations were compared to the particle scattering functions (259). The ratio of mean square R_g values for linear and cyclic molecules is close to the predicted value (2.0), though the observed Kratky plots showed some disagreement with theory and further work is in progress.

Equations 34–42 were derived on the assumption that neutrons are scattered once before being detected, but neglect the effects of multiple scattering. Goyal et al (260) have shown that this is a reasonable approximation for coherent scattering even where high concentration labeling is used. For incoherent scattering, however, this assumption is not valid and 1–2-mm samples of hydrogenous polymer or organic liquids contain an appreciable component of multiple scattering. Equation 39 assumes that the scattering is proportional to the sample thickness and transmission, though this is not valid for samples with a measurable fraction of multiple scattering. Hence a cross section that is a material (intensive) property independent of sample size and transmission cannot be calculated for this type of sample. Although incoherent effects generally constitute a small correction in the sans region, measurements in the ians region are markedly affected by such corrections. To the author's knowledge there is no exact solution to this problem, although several empirical corrections have been developed on the assumption that incoherent scattering is independent of Q for sans and ians (70,73,124,125,261).

In general, instrumental resolution (smearing) effects due to the wavelength spread ($\Delta\lambda$) of the incident beam and the finite angular range ($\Delta\theta$) over which data are collected are smaller for sans than saxs. This is because most sans experiments are performed in point geometry, whereas a significant proportion of x-ray experiments use long-slit sources, where smearing effects are much larger, particularly at small angles (262). For sans $\Delta\lambda/\lambda$ is typically 5–25% and resolution effects become appreciable for large radii ($R_g > 20$ nm) or for scatter patterns that vary sharply with angle (20,238,241,243). These effects have been discussed (20) and, for circularly symmetric scatterers, smearing effects can be treated by indirect Fourier transform methods (263). Resolution effects in anisotropic systems are discussed in Refs. 211 and 212.

Inelastic Neutron Scattering

The above examples indicate the unique information on polymer structure obtained during the past decade through the combination of sans with deuterium-labeling methods. Although there has been less work on inelastic and quasi-elastic techniques, such experiments can also give new information on polymer dynamics which is either unique or complementary to that provided by existing methods. Fortunately, several detailed reviews already exist to which reference will be made in the appropriate section (5,6,16,264–266). Apart from the quasi-static techniques that give information on translational diffusion and relaxation processes, sans gives no information on polymer dynamics because equation 23 was integrated over all ω and hence any information on the motion of molecules or segments was lost. Such information can be obtained by analysis of the energy spectrum of the scattered beam via inelastic or quasi-elastic techniques. Molecular motions range from translational (center of mass) diffusion to atomic vibrations about chemical bonds and the different types of motion give rise to a perturbation or a Doppler shift of the scattering in different energy and Q ranges. Prior to the application of neutron scattering, such motions were studied by tracer diffusion, dielectric loss measurements, and nmr relaxation methods. These techniques were used to identify the various relaxation processes, as for example, the α-, β-, γ-, and δ-transitions that occur in polymer glasses. Information on the vibration or rotation of individual side groups obtained from incoherent inelastic neutron scattering is complementary to that from nmr and optical spectroscopy (Raman, ir) as the selection rules for Raman and ir techniques do not apply to neutron spectroscopy (264,266). The thermal displacements of atoms caused by high frequency sound waves (phonons) in crystalline polymers may also be studied by inelastic scattering, though these motions are cooperative and must be studied by analysis of the coherent cross section. Conformational changes caused by segmental rotation involving several bonds are also cooperative phenomena that cause a broadening of the coherent elastic component of the scattering. These processes may be studied via energy analysis of the scattered beam, though the broadening is small and difficult to resolve within the experimental resolution of quasi-elastic techniques. The development of neutron spin-echo (nse) methods has improved the instrumental resolution by over two orders of magnitude and provides the highest resolution currently available in neutron scattering experiments. This has made it possible to observe for the first time single-chain motion in the bulk and has produced a wealth of new information on polymer dynamics which was previously unavailable by any other technique.

Side-group Motion.   Many of the first studies of side-group motion were performed on the 6H and 4H5 time-of-flight spectrometers at the Atomic Energy Research Establishment (AERE), Harwell, UK (Fig. 24) (267). The neutron beam passes through a liquid hydrogen cold source and a series of filters that remove γ-rays and fast neutrons. It is monochromatized by phased rotors (268), which select neutrons within a narrow velocity (energy) range and also pulse the beam. The incident and transmitted pulses are detected by monitors M1, M2, and M3, which give a direct measurement of the transmission and the time the pulses reach the sample. The scattered neutrons are detected in banks of BF_3 counters placed 1.1–1.9 m from the sample at fixed angles of scatter. The scattered neutrons are sorted into time channels and recorded in a multichannel analyzer. A reference timing signal is provided by a trigger pulse, generated when one of the rotors is in a preset angular position; the time τ taken by the neutron to travel a known distance is measured and gives the scattered-neutron velocity. Thus the energy transfer may be calculated along with the Q vector via equation 5. One of the first time-of-flight studies of polymers is given in Figure 25a, which shows the spectra from polyacetaldehyde and polytrifluoroacetaldehyde (269) for the

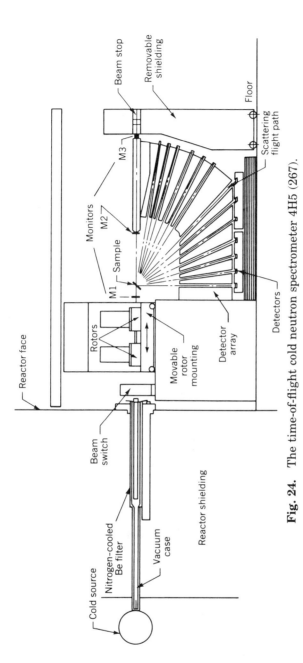

Fig. 24. The time-of-flight cold neutron spectrometer 4H5 (267).

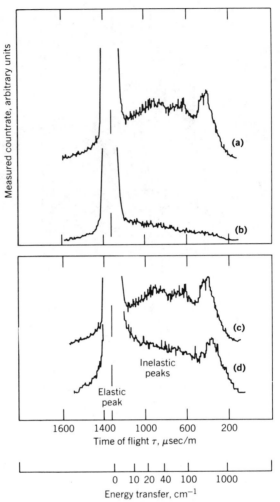

Fig. 25. Incoherent inelastic spectra taken at RT and 90° angle of scattering for (**a**) isotactic polyacetaldehyde; (**b**) polytrifluoroacetaldehyde; (**c**) isotactic polyacetalde-hyde; (**d**) atactic polyacetaldehyde (269). Courtesy of the American Institute of Physics.

counter at a scatter angle of 90°. The measured count rate is proportional to the differential scattering cross section $d^2\Sigma/d\Omega d\tau$, with respect to solid angle Ω and time of flight τ; absolute calibration may be achieved by comparison with a standard scatterer, usually vanadium (270). The cross section $d^2\Sigma/d\Omega d\tau$ is related to the differential cross section with respect to energy E via

$$\frac{d^2\Sigma}{d\Omega dE} = -\frac{\tau^3}{m}\frac{d^2\Sigma}{d\Omega d\tau} \tag{50}$$

where m is the neutron mass defined previously. The intense peak at the left-hand side of each spectrum arises from neutrons that have been elastically scat-tered with the incident neutron wavelength ($\lambda_0 = 0.52$ nm) and energy (3 meV). The inelastic peaks with energy transfers of 30, 80, and 230 cm^{-1} correspond to neutrons inelastically scattered with energy gain. Because the incident neutrons have been moderated at liquid hydrogen temperatures, they do not possess sufficient energy to excite the corresponding energy-loss processes, and the predominant inelastic process is thus energy gain in this type of spectrum. Since approximately 95% of the total scattering arises from hydrogen atoms because of their high incoherent cross section, coherent effects can be neglected to a first approximation in this type of experiment (eq. 25). Substitution of hydrogen atoms by deuterium or other atoms may be used to assign the various inelastic peaks to specific groups, as shown in the corresponding spectrum of polytrifluoroacetaldehyde. As fluorine has virtually no incoherent cross section, the effect of the substitution is the removal of the scattering contribution of the methyl group and the assignment of the three inelastic peaks in the unsubstituted spectrum to this unit. Assuming that substitution does not change the helical conformation of the chain, the peak at 230 cm^{-1} may be identified as methyl group torsion by comparison with small-molecule neutron spectra (271), and the barrier for hindered rotation of this group about the C—C axis may be calculated from the torsional frequency (269). The fact that the remaining inelastic peaks at 30 and 80 cm^{-1} in the crystalline state (Fig. 25a) are absent in the spectrum of atactic (amorphous) polyacetaldehyde as shown in Figure 25b leads to their assignment as acoustic vibrational modes. Since these modes are cooperative in nature, they are sensitive to changes in molecular configuration, and the increased disorder and reduced intermolecular forces in the amorphous (atactic) form may be expected to broaden the bands and lower their frequencies. These bands are therefore not resolved in the spectra of the atactic material, in contrast to the high frequency mode at 230 cm^{-1} which shows no change in width or frequency, supporting its assignment to a localized torsional vibration of the CH$_3$ group.

Extensive studies of the torsional rotation of the methyl group in a number of polymers have been undertaken (5,264,272–276), and the barrier heights to rotation have been calculated from the observed frequencies and compared with the energies obtained from relaxation studies (6,264). As described above, identification of a particular mode can be facilitated by labeling a particular group with deuterium, which has a much lower incoherent cross section (Table 1). The calculated potential barriers seem to be mainly dependent on the chemical structure of the repeat unit, though the stereoregularity of the side groups along the chain can cause a variation up to 50% (273). Discrepancies between the energies obtained from neutron and relaxation data have been attributed to quantum-mechanical tunneling, where a group can pass through an energy barrier greater than its kinetic energy. A recent review of tunneling via inelastic neutron scattering is given in Ref. 276.

There has been little inelastic scattering work on the motion of other side groups, though comparison of inelastic neutron scattering data of phenyl groups in polystyrene with nmr and dielectric relaxation measurements has been briefly reviewed (5). To the author's knowledge no neutron studies have yet been performed for other side groups.

Phonons and Dispersion in Crystalline Polymers.   Neutron scattering has been used extensively for the investigation of phonons in crystalline materials for several decades (277,278) and has provided much information on interatomic forces in different types of materials (metals, insulators, etc). Because phonons are quantized high frequency lattice vibrations involving the cooperative motion of many atoms, they have been predominantly studied via coherent inelastic scattering, which gives the phonon dispersion curves on single-crystal specimens. The equivalent measurements on polymers have been restricted by the lack of fully deuterated single-crystal specimens and the same limitation has applied to neutron crystallography. This technique has been widely used in biology and materials science (279) in view of its ability to locate light atoms, but there has been very little application to polymers.

Incoherent inelastic spectroscopy gives complementary information on chain dynamics and provides $\rho(\omega)$, the density of vibrational states in a given frequency interval between ω and $(\omega + d\omega)$; most of the first applications to polymers were of this type (2). The first measurements were carried out on hydrogenous specimens of polyethylene where the scattering is predominantly incoherent and is suitable for the examination of fundamental vibrational modes of low frequency. The dispersion curves have been calculated (280–282) and the system had been extensively examined by Raman and ir spectroscopy. A review of these calculations and the optical data that existed before the first neutron experiments has been given by King (285) and Allen and Higgins (264).

Initial measurements were carried out (283,284) on unoriented (isotropic) samples and a search was made for peaks as a function of energy transfer. For the time-of-flight data collected at a fixed angle, the experiment scans a region of $(Q - \omega)$ space defined by equations 3–5. Peaks occur where this scan traverses the frequency limit of a branch of the dispersion curve corresponding to a maximum in $\rho(\omega)$. Such a feature was identified at ~200 cm^{-1}, corresponding to the ν_9 acoustic mode where the dispersion surface becomes flat and gives rise to a critical peak, which can be directly compared with model predictions (285). Subsequent measurements on unoriented polyethylene (286) with substantially improved resolution revealed further peaks at 62, 94, and 123 cm^{-1}, which were in reasonable agreement with calculated frequencies (285). Measurements on samples uniaxially oriented by stretching (287) were carried out with the momentum transfer in the longitudinal direction (along the chain axis). An intense peak at 525 cm^{-1} was observed, which was absent when the momentum transfer occurred perpendicular to the chain direction; thus the peak could be assigned a frequency limit of the ν_5 longitudinal acoustic mode. Thus incoherent inelastic measurements complement the optical data and allow the observation of several modes that were absent or difficult to detect in the techniques previously used.

The scattering from materials containing appreciable numbers of hydrogen atoms is dominated by incoherent effects and hence coherent inelastic experiments must be carried out on fully deuterated or hydrogen-free materials (eg, polytetrafluoroethylene (PTFE), deuterated polyethylene, etc). Feldkamp and coworkers (288) measured the first polymer dispersion curve using oriented PED and provided the first information on the anisotropy of the force field in polyethylene from neutron scattering. Similar measurements on PTFE detected chain-axis phonons (289,290), though these studies yielded no evidence for phonons perpendicular to the chain axis. In principle, measurements of longitudinal and transverse dispersion curves would yield the crystalline force constants that could make it possible to model the observed bulk mechanical properties of polymers (291). In view of the anisotropy of the crystalline regions, several elastic constants are needed and a series of experiments to measure them have been undertaken by White and co-workers (292–295). Although some dispersion curves measured perpendicular to the chain axis have yielded elastic constants previously unavailable from other techniques (291–293), such measurements have been generally inhibited by the lack of large single-crystal specimens. High pressure annealing may be used to produce samples in largely chain-extended form (222) and this technique has been developed to produce single-crystal textured samples of PED (295). Large specimens of polyoxymethylene with even higher single-crystal perfection may be produced by radiation polymerization of single crystals of trioxane or tetroxane monomers (291,294). The resulting polymer crystals have three-dimensional order with the axes in a known relationship to those of monomer precursor crystal. Phonon measurements were undertaken at the IN2 and IN8 triple-axis spectrometers at the ILL, Grenoble, France, and also at the 10H beam hole of the Dido reactor, Harwell, UK. In this technique both monochromatization of the incident neutrons and energy analysis of the scattered beam are carried out by crystal diffraction (296). Figure 26 shows the measured phonon dispersion curves for deutero-polyoxymethylene at 298 K, where the different branches observed refer to the longitudinal acoustic and transverse acoustic modes. The displacement of the atoms is either in the direction (LA) or transverse to the direction of propagation (TA) of the phonon wave for the different branches. For phonons propagating along the c-axis, designated (00ξ) where ξ is the reduced wave vector expressed as a fraction of the appropriate lattice constant (277,278), transverse vibrations are necessarily in the basal plane. However, for phonons that propagate perpendicular to the c-axis, transverse vibrations may occur both in the basal plane or along the c-axis, and these branches must be distinguished (Fig. 26). The limiting slope of each branch gives the velocity of sound for the propagation mode from which five independent force constants were derived (294). To the author's knowledge this represents the most comprehensive analysis of

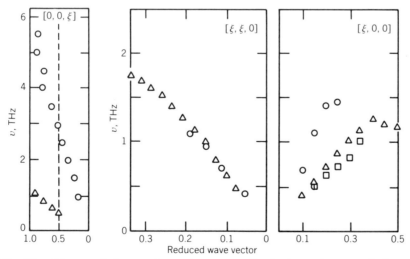

Fig. 26. Measured phonon dispersion curves for deuteropolyoxymethylene at 298 K: ○, LA; △, TA$_{basal plane}$; □, TA$_{c\text{-axis}}$ (294). Text explains LA and TA. Courtesy of Butterworth & Co. (Publishers) Ltd.

phonon spectroscopy in polymeric materials. Although the PED specimens prepared by high pressure annealing techniques (295) possessed a higher degree of perfection than any previously prepared materials, the crystalline order fell short of full single-crystal texture. The presence of ~25% amorphous material and appreciable mosaic spread (~17° full width at half maximum, FWHM) prevented a complete determination of the elastic constants, though the study did provide the most detailed information on the anisotropy of the forces in polyethylene yet available. The two independent force constants derived were used in combination with data from Raman spectroscopy to test available force models (281,282,297). The precision of the information gained from this technique is in proportion to the degree of single-crystal perfection in the samples studied. Only the wider availability of large, fully deuterated single crystals will permit the exploitation of the full potential of these methods, which have been used successfully to probe interatomic forces in other types of materials (277,278).

Quasi-elastic Scattering and Neutron Spin-echo

Quasi-elastic Methods. The inelastic spectra from hydrogen-containing materials are dominated by incoherent scattering and usually contain a prominent elastic or quasi-elastic peak centered at the energy of the incoming neutron (Fig. 25). The width of this peak is a function of the scattering angle and increases with increasing Q. The broadening arises from translational and rotational motion of individual hydrogen atoms and for simple Fick's law diffusion the intermediate scattering law is given in terms of the self-diffusion coefficient D, by (4,5,16,264,298)

$$F(Q,t) = \exp - (DQ^2t) \tag{51}$$

where

$$F(Q,t) = \left\langle \sum_{ij} b_i^* b_j F_{ij}(\mathbf{Q},t) \right\rangle \tag{52}$$

Equations 52 and 21 indicate that the scattering law $S(Q,\omega)$ is the Fourier transform of $F(Q,t)$, which is the time-correlation function as measured by photon-correlation spectroscopy (16). For simple center-of-mass diffusion (eq. 51), the scattering law $S(Q,\omega)$, which is measured by conventional neutron scattering instruments (eg, time-of-flight, triple-axis spectrometers, etc), is given by (16)

$$S(Q,\omega) = \frac{\pi DQ^2}{(\omega^2 + (DQ^2)^2)} \tag{53}$$

Thus the scattering law has a well-known Lorenzian shape with half width $\Delta\omega_{1/2} \alpha DQ^2$. Equation 53 has been used extensively to investigate translational diffusion coefficients in simple liquids, where the width of the quasi-elastic peak at half height $\Delta\omega_{1/2}$ is measured as a function of Q. After removal of the instrumental resolution effects, the diffusion coefficient may be obtained from the slope of a plot of $\Delta\omega_{1/2}$ vs Q^2 (264). An extensive review of the information available from this type of measurement has been given by Allen and Higgins (264), who have also summarized other models (eg, jump diffusion) that have been proposed in response to departures from simple diffusion in some systems.

The first application of this technique to study molecular motion in bulk polymers was made by Allen et al (264,298,299) who measured the broadening of the quasi-elastic peak from a series of dimethylsiloxanes, including both rings and chains, as a function of the degree of polymerization (n). The scattering was analyzed in terms of simple diffusion (eq. 53) and values of an effective diffusion coefficient D_{eff} were calculated from plots of $\Delta\omega_{1/2}$ vs Q^2. It was found that D_{eff} reached an asymptotic value $D_{eff} \simeq 2 \times 10^{-5}$ cm^2/s that was similar for both rings and chains. As the studies were carried out on predominantly hydrogenous material, D_{eff} refers to the motion of individual protons in the chain. The instrumental resolution of the initial time-of-flight (tof) measurements was limited ($\Delta E_{res} \simeq 0.5$ meV) and subsequent experiments were performed with higher resolution on the IN10 backscattering apparatus ($\Delta E_{res} \simeq 0.001$ meV) and high resolution tof spectrometer ($\Delta E_{res} \simeq 0.05$ meV) at the ILL, Grenoble, France. The data from each of the instruments could be superimposed on the same curve, though there was a strong departure from simple Q^{-2} dependence. The measured diffusion coefficients obtained previously resulted from approximating small sections of the curve by a Q^{-2} dependence (298). The effective diffusion coefficients obtained in this way were a function of the Q range in which they were measured and differed markedly from nmr results. Thus although no single exponent governing the Q dependence of $\Delta\omega_{1/2}$ could be resolved from these studies, analysis of the data via simple diffusion has been used to provide D_{eff} and activation energies, which served as a basis for comparisons between different polymers when measurements were made on the same instrument in the same momentum range (298).

To describe the motions other than center-of-mass diffusion or the high frequency vibrations, stretching, or rotations of individual atoms or groups about the chemical bonds, it is necessary to develop a model for the polymer that specifies the way in which cooperative disturbances are transmitted along the backbone. Such models have been developed by de Gennes (300) and Dubois-Violette (301) using earlier treatments of Rouse (302) and Zimm (303) who used a bead-and-spring model to represent the connectivity of the polymer chain. The Rouse model (302) assumes that disturbances are transmitted only along the polymer chain and has been used to describe molecular motion in the melt. In the Zimm model (303), the presence of the surrounding medium (eg, solvent molecules) introduces hydrodynamic interactions, which modify the behavior of segments of the polymer chain. This model has been used to predict molecular motion in solution. Both models led to universal predictions independent of the molecular structure and chain length in an intermediate range of distance and time, which may be explored by an appropriate choice of momentum and energy transfer in the scattering experiment. The distance scale explored is proportional to Q^{-1} and may be chosen in a range $R_g > Q^{-1} > a_s$, where a_s is a length above which the cooperative motions of the chain are independent of the structure of the repeat units. Outside this range, the effects of local bond rotation and vibration can be observed for $Q^{-1} < a_s$, whereas center-of-mass diffusion of the whole molecule is observed for $Q^{-1} > R_g$. Although there is no precise definition of a_s, it is of the same order as the statistical segment length (1–5 nm) and gives a measure of the chain stiffness. The motion of polymer chains in the range $R_g > Q^{-1} > a_s$ has been the subject of much theoretical interest and debate in recent years and the main theoretical predictions have been summarized (5,304).

In the Rouse model, the coherent scattering function $F(Q,t)$ varies as

$$F_{coh}(Q,t) \simeq \exp - (\Omega t)^{1/2} \qquad (54)$$

where Ω^{-1} is a correlation time defining the rate of motion, and varies as Q^4; Ω may be obtained from the initial time dependence of $F(Q,t)$

$$\Omega = \lim_{t \to 0} \frac{d \ln F(Q,t)}{dt} \qquad (55)$$

and is identical to the energy width of the quasi-elastic peak $\Delta\omega_{1/2}$ apart from a numerical factor. In the Zimm treatment, the $t^{1/2}$ variation in equation 54 becomes a $t^{2/3}$ dependence which predicts a Q^3 dependence for Ω and $\Delta\omega_{1/2}$. These results have been generalized by Akcasu et al (305) who have calculated the variation of Ω over the whole Q range and extended the calculations beyond the intermediate range $R_g > Q^{-1} > a_s$ covered by the Zimm and Rouse treatments. Experiments to test these predictions have been undertaken using conventional neutron scattering instruments, though these studies were limited by several factors. These instruments measure $S(Q,\omega)$ rather than $F(Q,t)$, and the different models produce predictions with only subtle differences, which are manifested in the wings of the quasi-elastic peak (7,298). Moreover, the resolution of even the best conventional instruments (eg, backscattering spectrometers) is ~0.001 meV, which is comparable to the energy widths of the peaks investigated. The model predictions had to be folded with the experimental resolution to compare theory with experiments. Furthermore, in this type of experiment, the Q dependence of the half-width has to be assumed initially in order to extract experimental values of $\Delta\omega_{1/2}$ from the best-fit model curves, and this makes the attainment of unambiguous conclusions difficult. Allen et al (306) measured a range of exponents (from 1.95 to 3.12) for solutions of polytetrahydrofuran (PTHF) in carbon disulfide, and although the data tended to favor the Zimm model, there were discrepancies between the experimental and theoretical scattering laws and no unambiguous conclusions could be drawn. The combined data from several conventional spectrometers on polydimethylsiloxane (PDMS) melts showed that the variation of $\omega_{1/2}$ with Q was consistent with the Rouse prediction, though the model underestimated the intensity in the wings of the quasi-elastic peak. Definitive experimental confirmation of the above theoretical predictions was made possible only by advances in neutron instrumentation, and in particular the development of the neutron spin-echo spectrometer. This technique is capable of resolving energy changes less than 10^{-8} eV, which represents an improvement in resolution of over two orders of magnitude beyond previously available techniques. Moreover, this instrument measures directly the intermediate scattering function (eq. 52) and thus a direct measurement was possible concerning the different model predictions for $F(Q,t)$.

Neutron Spin-echo.   The neutron spin-echo (nse) technique was conceived by Mezei (307) and the first spectrometer was built at the ILL by Mezei and Hayter. Whereas conventional scattering techniques achieve higher energy res-

olution by restricting the wavelength spread of the incident beam with a consequent loss in neutron flux, the nse spectrometer utilizes coarse monochromatization ($\Delta\lambda/\lambda \sim 10\%$), which results in a significant intensity advantage. The
neutron has both spin ($I = \frac{1}{2}$) and magnetic moment ($\mu_n = -1.913$ nuclear
magnetons), and if it is placed at an angle to a magnetic field **H**, it precesses at
the Larmor precessional frequency

$$\omega_{\mathrm{L}} = \frac{|2\mu_n H|}{\hbar} \tag{56}$$

The nse technique uses the Larmor precession as an internal clock that
monitors the scattering process of each neutron and permits a very precise determination of the energy difference between initial and final states resulting
from scattering. As only energy differences are measured, this permits the use
of a relatively wide spectrum of incident wavelengths and the resolution is limited
only by the inhomogeneity of the field in the magnets used. A particularly clear
introduction to the technique has been given by Nicholson (265), and Figure 27
illustrates the nse spectrometer. The incident (unpolarized) beam is monochromatized by a helical-slot velocity selector ($\Delta\lambda/\lambda \simeq 10\%$) and then reflected by
polarizing magnetic mirrors so that the neutron spin, which is parallel to μ_n, is
parallel to the flight path in the z-direction. Before entering a solenoidal guide
field, the plane of polarization is rotated through an angle of $\pi/2$ in a spin turn
coil (C1) so that the spins are aligned in the x-direction, thus enabling them to
precess about the first solenoidal guide field H_1. A neutron of incident wavelength

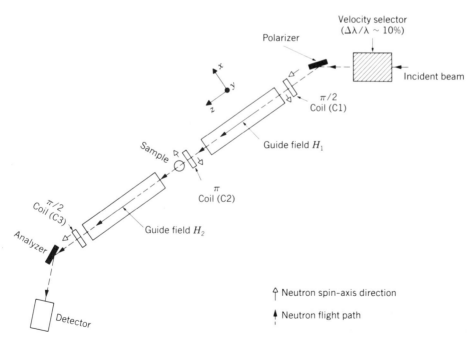

Fig. 27.   The neutron spin-echo spectrometer.

λ_0 has velocity $h/m\lambda_0$ and traverses the first guide field (length l_1) in a time

$$t_1 = \frac{l_1}{h} m\lambda_0 \tag{57}$$

Thus the total number of precessions is given by

$$N_1 = \frac{2\mu_n m\lambda_0 H_1 l_1}{h^2} \tag{58}$$

Before the sample position, a spin flipper or π turn coil (C2) is placed which has the effect of turning both the y- and z-components through an angle of π so that Lamor precession occurs in an opposite sense in the second guide field H_2. Thus the number of precessions in the second guide field, length l_2, is given by

$$N_2 = \frac{2\mu_n m\lambda H_2 l_2}{h^2} \tag{59}$$

where λ is the final wavelength after scattering. Before entering the detector, the neutrons pass through a final $\pi/2$ spin turn coil and thence to a mirror analyzer designed to transmit neutrons polarized in the z-direction. The name neutron spin-echo is taken from the spin-echo technique used in nmr that utilizes a similar sequence of spin turns (16).

If the two arms of the spectrometer are identical ($H_1 l_1 = H_2 l_2$), then for purely elastic scattering ($\lambda = \lambda_0$) the numbers of precessions in each arm are equal and in the opposite sense, thus leaving the scattered neutron in the same polarization state at the analyzer as it had at the polarizer. For quasi-elastic scattering, there is a finite energy change on scattering ($|\Delta E| \ll E_0$) and hence there are different numbers of precessions in each arm, giving rise to a nonzero net-precession angle at the analyzer. Thus the z-component of the polarization of the scattered neutrons is no longer the same as that of the incident neutrons and hence some of the signal is lost. The loss is proportional to $\cos 2\pi(N_1 - N_2)$ where

$$2\pi(N_1 - N_2) = \frac{4\pi\mu_n m H_1 l_1}{h^2} (\lambda_0 - \lambda_0) = \omega t \tag{60}$$

The wavelength difference $(\lambda_0 - \lambda_0)$ corresponds to the energy change on scattering, $\Delta E = \hbar\omega$, and as the probability of scattering with this energy transfer is $S(Q,\omega)$, the final measured polarization integrated over the wavelength spread in the incident beam $I(\lambda)$ is given by

$$P_z = \int_0^\infty I(\lambda) \, d\lambda \int S(Q,\omega) \cos \omega t \, d\omega$$

$$= \int_0^\infty I(\lambda) \, d\lambda \, F(Q,t) \tag{61}$$

Different values of t may be probed by varying the guide field H (eq. 60) for different Q values determined by the angle of scatter and incident wavelength. The signal is normalized with reference to coherent scatterer (eg, $\sim$10:1 mixture of deuterated and hydrogenous polymer) for which the scattering is predominantly elastic. Thus the normalized polarization for a given sample is given by

$$P = \frac{P_z(\text{sample})}{P_z(\text{elastic})} = \frac{\int_0^\infty I(\lambda)d\lambda\, S(Q,\omega)\cos\omega t\, d\omega}{\int_0^\infty I(\lambda)d\lambda\, S(Q,\omega)\, d\omega}$$

$$= \frac{F(Q,t)}{F(Q,0)} \tag{62}$$

The polarization thus measures directly the time dependence of the intermediate scattering law and permits a direct test of the various models for molecular motion.

In order to observe the collective motions of sections of the polymer molecules, the systems studied must be predominantly coherent scatterers, and the corrections of incoherent scattering have been developed by Hayter (308). More detailed descriptions of the technique and the principles underlying the operation of neutron polarizers and spin-flip devices are given in Refs. 16 and 308–311.

Polymer Chain Dynamics in Solution. Figure 28 shows one of the first direct measurements of $F(Q,t)$ for a solution of PDMS in deuterated benzene where the neutron polarization $P(Q,H)$ is plotted as a function of H (312). In view of the correspondence of H and t (eq. 60), this function is equivalent to $F(Q,t)$. The data are fitted to the scattering law of Dubois-Violette and de Gennes (301) calculated for the Zimm model assuming a variation of $t^{2/3}$ with time (312). These results verified unambiguously the predictions of the Zimm model and $\Delta\omega$ was

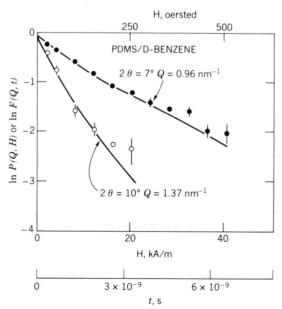

Fig. 28. Polarization $\ln P(Q,H)$ as a function of magnetic field H or $F(Q,t)$ as a function of t: solid lines (312), scattering law as calculated in Ref. 301. Courtesy of the American Physical Society.

shown to have a Q^3 variation, though the proportionality factor for the two polymers measured (PDMS and PMMA) differed from that calculated from the theory (301). Further data on polystyrene, polytetrahydrofuran, and PDMS in deuterated benzene and carbon disulfide were given by Nicholson et al (313). Unlike the PDMS data, the polystyrene and PTHF data exhibited a lower Q dependence. By fitting the data to the theory of Akcasu et al (305), it was possible to extract the step length a_s, which was found to be considerably higher for polystyrene ($a_s \simeq 4$ nm) and PTHF ($a_s \simeq 3$ nm) than for PDMS ($a_s \simeq 1$ nm). Thus some of the data for the PS and PTHF fall outside the range of $R_g > Q^{-1} > a_s$ where a Q^3 variation is expected. The correlation times in different solvents were a function of the solvent viscosity and, after normalizing to the measured step lengths, the data for all three polymers (Fig. 29) studied could be accommodated on the same theoretical curve (265,313).

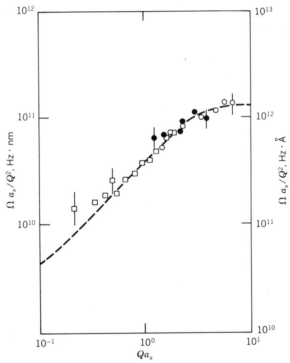

Fig. 29. Inverse correlation time for solutions of polytetrahydrofuran, polystyrene, and polydimethylsiloxane in deuterated benzene, normalized by the respective step length a_s (313). The broken line shows the theoretical curve (305): ●, polytetrahydrofuran; ○, polystyrene; □, polydimethylsiloxane. Courtesy of the American Chemical Society.

Single-chain Motion in the Melt. In Figure 30, showing nse data from PDMS melts (314), ln $P(Q,H)$ is plotted against ln H. As $P(Q,H)$ is equivalent to $F(Q,t)$ the slope of this graph on a double logarithmic scale gives the time exponent in equation 54. The nse technique gives this exponent directly and the

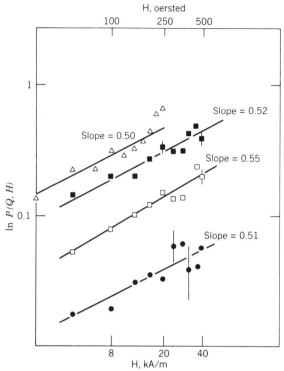

Fig. 30. Polarization decay $P(Q,H)$ obtained from 5% hydrogenous PDMS ($\overline{M}_w$ = 60,000) in deuterated PDMS ($\overline{M}_w$ = 60,000) for different Q values: △, Q = 1.32 nm^{-1}; ■, Q = 1.06 nm^{-1}; □, Q = 0.79 nm^{-1}; ●, Q = 0.53 nm^{-1} (314). Courtesy of the American Physical Society.

measured value is close to 0.5 as predicted by the Rouse model. These results have been interpreted as contradicting the theory of reptation (314), though this conclusion has been disputed (315) by the argument that the data were taken outside the Q range where reptation effects should be observed. Further debate (316) and theoretical developments in this area can be expected as more nse data appear and offer the possibility of further checking model predictions. Figure 31 shows the combined data from PDMS in solution and melt (265,317) in the form of inverse relaxation time Ω obtained by nse and backscattering techniques. The variation of Ω is consistent with the theoretical predictions of the Zimm and Rouse models for solutions and melts, respectively, and the data show how the nse technique has extended the previously available resolution from conventional neutron scattering techniques. The nse technique has also been applied to concentrated polymer solutions (312,318). Crossover effects were observed between dilute and semidilute solutions, and an exponent (-0.68) was measured for the variation of screening length ξ_s with C close to that measured by sans (119,124,125). The above examples indicate how nse has provided new information in Q and energy ranges that were hitherto inaccessible, and made possible the observation of single-chain behavior in bulk polymers and solutions. Applications to other areas of physics are given in Refs. 16, 311, and 319.

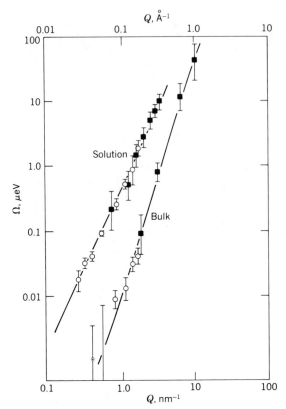

Fig. 31. Inverse correlation time for 3% hydrogenous polytetrahydrofuran in deuterated bulk polytetrahydrofuran polymer and also for a dilute solution of deuterated PTHF in CS_2: ■, backscattering spectrometer (IN10); ○, neutron spin-echo (IN11) (315).

Prospects

The above review of the application of neutron scattering to polymer science, while by no means exhaustive, is indicative of the new and valuable information that has been provided over the past two decades. The use of this method has spread far beyond experts in the field of neutron scattering and much of the work reviewed has been undertaken by nonspecialists who have applied the technique in areas of their own particular interests. This has been made possible by the development of national and international facilities that routinely provide technical assistance to a wide spectrum of outside users. It is hoped that this article encourages researchers from industrial laboratories and the general scientific community to take advantage of neutron scattering's unique potentialities for the elucidation of polymer structure and dynamics.

Virtually all the research surveyed in this article has been undertaken on reactor sources, though in the future scattering measurements at pulsed-neutron sources should be of increasing importance. At these facilities neutrons are produced by spallation, ie, the splintering of a nucleus induced by penetration of a particle (eg, proton) of suitable energy. The process creates free neutrons and is activated by an accelerator system that creates a high energy particle beam that is incident on a target and produces an intense beam of pulsed neutrons. Spallation sources have been built in the United States at Argonne and Los Alamos National Laboratories, and also in Japan at the National Laboratory for High Energy Physics (Tsukuba). The commissioning of the Spallation Neutron Source (ISIS) at the Rutherford Appleton Laboratory, UK, will provide even more intense beams than are currently available and lead to the development of a new generation of instruments. As these facilities produce pulsed rather than continuous neutron beams, the main advances should result from experiments designed to take advantage of the time structure of the beam, ie, real-time studies in experiments using periodic perturbations of the system. A particular advantage of pulsed sources is the ability to measure a wide Q range simultaneously, though experiments in a particular Q range (eg, sans) seems to be best performed at suitably designed reactor sources (320). In view of the demonstrated achievements of neutron scattering in a wide variety of fields, the construction of new or upgraded facilities is under consideration in the United States, Europe, and Japan. A review of scientific opportunities at advanced sources, including a detailed comparison between pulsed and steady-state facilities, have recently been published (320), and possible future opportunities and forecasts in the field of polymer science have been given (140). Experience has shown that conformational and morphological problems can rarely be solved by any individual technique and benefit from the combination of data from several complementary approaches. However, it seems clear from the breadth of examples surveyed here that neutron scattering will remain one of the major tools for polymer scientists in the foreseeable future.

Acknowledgment

This review is dedicated to the memory of Paul J. Flory, whose intuitive grasp of the fundamentals of polymer science led to the prediction of many of the results surveyed in this article, long before this type of measurement was conceived or undertaken.

The author wishes to acknowledge helpful suggestions and corrections by many colleagues, particularly J. B. Hayter, J. S. King, W. C. Koehler, J. M. Lefebvre, R. M. Moon, R. M. Nicklow, T. P. Russell, R. S. Stein, and R. Ullman. Thanks are also due to F. S. Bates, M. Dettenmaier, E. W. Fischer, J. S. Higgins, R. G. Kirste, J. M. O'Reilly, R. Ottewill, C. Picot, J. W. White, and W. Wu, for providing preprints of current work prior to publication. Completion of the manuscript would not have been possible without the patience of Mrs. Mary Gillespie who has coped admirably with numerous revisions and modifications.

Research sponsored by the National Science Foundation Grant #DMR-7724459 through Interagency Agreement #40-636-77 with the U.S. Department of Energy under contract DE-AC05-840R21400 with Martin Marietta Energy Systems, Inc.

BIBLIOGRAPHY

1. V. E. Turchin, *Slow Neutrons,* Israel Program for Scientific Translations, Jerusalem, 1965.
2. H. Boutin and S. Yip, *Molecular Spectroscopy with Neutrons,* MIT Press, Cambridge, Mass., 1968.
3. G. E. Bacon, *Neutron Diffraction,* Clarendon Press, Oxford, UK, 1971.
4. W. Marshall and S. W. Lovesey, *Theory of Thermal Neutron Scattering,* Clarendon Press, Oxford, UK, 1971.
5. A. Maconnachie and R. W. Richards, *Polymer* **19,** 739 (1978).
6. J. S. Higgins in G. Kostorz, ed., *Treatise on Materials Science and Technology,* Vol. 15, Academic Press, Inc., New York, 1979, p. 381.

7. L. H. Sperling, *Polym. Eng. Sci.* **24**, 1 (1984).
8. J. S. Higgins and R. S. Stein, *J. Appl. Crystallogr.* **11**, 346 (1978).
9. L. I. Schiff, *Quantum Mechanics,* McGraw-Hill Inc., New York, 1955.
10. C. G. Windsor in B. T. M. Willis, ed., *Chemical Applications of Thermal Neutron Scattering,* Oxford University Press, London, 1973.
11. W. M. Lomer and G. G. Low in P. A. Egelstaff, ed., *Thermal Neutron Scattering,* Academic Press, Inc., New York, 1965, Chapt. 1.
12. J. A. Janik and A. Kowalska in Ref. 11, Chapt. 10.
13. D. I. Garber and R. R. Kinser, eds., *Neutron Cross Sections, BNL 325,* 3rd ed., Brookhaven National Laboratory, Upton, N.Y., 1976; *BNL 325,* 2nd ed., Suppl. 2, 1964.
14. A. Maconnachie, *Polymer* **25**, 1068 (1984).
15. R. Ullman and J. S. King, private communication.
16. J. B. Hayter in V. Degiorgio and M. Corti, eds., *Proceedings of Enrico Fermic School of Physics Course XC,* Amsterdam, The Netherlands, 1985, p. 59.
17. L. Van Hove, *Phys. Rev.* **95**, 249 (1954).
18. S. W. Lovesey in S. W. Lovesey and T. Springer, eds., *Dynamics of Solid and Liquids by Neutron Scattering,* Springer-Verlag, Berlin, 1977, p. 18.
19. A. Kostorz and S. W. Lovesey in Ref. 6, p. 42.
20. J. Schelten and R. W. Hendricks, *J. Appl. Crystallogr.* **11**, 297 (1978).
21. J. Schelten in S. H. Chen, B. Chu, and R. Nossal, eds., *Scattering Techniques Applied to Supramolecular and Nonequilibrium Systems, NATO Advanced Study Series 83,* Plenum Press, New York, 1981, pp. 75–85.
22. J. Schelten, *Kerntechnik* **14**, 86 (1972).
23. K. Ibel, *J. Appl. Crystallogr.* **9**, 196 (1976).
24. H. Maier-Leibnitz and T. Springer, *J. Nucl. Energy* **17**, 217 (1963); *Annu. Rev. Nucl. Sci.* **16**, 207 (1966).
25. G. Degenkolbe and H. B. Greiff, *Kerntechnik* **15**, 437 (1973).
26. B. P. Schoenborn, D. S. Wise, and D. K. Schneider, *Trans. Amer. Crystallogr. Assoc.* **19**, 67 (1983).
27. D. F. R. Mildner, R. Berliner, O. A. Pringle, and J. S. King, *J. Appl. Crystallogr.* **14**, 370 (1981).
28. C. Glinka in J. Faber, ed., *Am. Inst. Phys. Conf. Proc.* **89**, 395 (1982).
29. W. C. Koehler, R. W. Hendricks, H. R. Child, S. P. King, J. S. Lin, and G. D. Wignall in Ref. 21, p. 35.
30. W. C. Koehler, *Physica* **137B**, 320 (1986).
31. R. K. Abele, G. W. Allin, W. T. Clay, C. E. Fowler, and M. K. Kopp, *IEEE Trans. Nucl. Sci.* **NS-28**(1), 811 (1981).
32. Major Materials Facilities Committee, National Research Council, *Major Facilities for Materials Research and Related Disciplines,* National Academy Press, Washington, D.C., 1984.
33. R. M. Moon and C. D. West, *Physica* **137B**, 347 (1986).
34. G. D. Wignall, R. W. Hendricks, W. C. Koehler, J. S. Lin, M. P. Wai, E. L. T. Thomas, and R. S. Stein, *Polymer* **22**, 886 (1981).
35. E. W. Fischer, J. H. Wendorff, M. Dettenmaier, G. Lieser, and Voigt-Martin in G. Allen and S. E. B. Petrie, eds., *Physical Structure of the Amorphous State,* Marcel Dekker, Inc., New York, 1977, p. 41.
36. D. Ullman, A. L. Renninger, G. Kritchevsky, and Vander Sander in Ref. 35, p. 153.
37. This derivation was developed by R. S. Stein (1980) following an earlier suggestion by H. Benoit (1980) and was published in Ref. 34. Similar derivations were given independently in Refs. 38–40.
38. E. W. Fischer, M. Stamm, M. Dettenmaier, and P. Herschenroeder, *Polym. Prepr. Am. Chem. Soc. Div. Polym. Chem.* **20**(1), 219 (1979).
39. C. E. Williams and co-workers, *J. Polym. Sci. Polym. Lett. Ed.* **17**, 379 (1979).
40. A. Z. Akcasu, G. C. Summerfield, S. N. Jahshan, C. C. Han, C. Y. Kim, and H. Yu, *J. Polym. Sci.* **18**, 865 (1980).
41. M. Von Laue, *Ann. Phys.* **56**, 497 (1918).
42. J. P. Cotton and co-workers, *Macromolecules* **7**, 863 (1974).
43. J. Schelten, G. D. Wignall, D. G. H. Ballard, and G. W. Longman, *Polymer* **18**, 1111 (1977).
44. B. Crist, W. Graessley, and G. D. Wignall, *Polymer* **23**, 1561 (1982).

45. A. Guinier and G. Fournet, *Small-Angle Scattering of X-Rays,* John Wiley & Sons, Inc., New York, 1955.
46. C. Tangari, G. C. Summerfield, J. S. King, R. Berliner, and D. F. R. Mildner, *Macromolecules* **13,** 1546 (1980).
47. C. Tangari, J. S. King, and G. C. Summerfield, *Macromolecules* **15,** 132 (1982).
48. W. Gawrisch, M. G. Brereton, and E. W. Fisher, *Polym. Bull.* **4,** 687 (1981).
49. F. Boué, M. Nierlich, and L. Leibler, *Polymer* **23,** 29 (1982).
50. P. J. Flory, *J. Chem. Phys.* **17,** 303 (1949); *Principles of Polymer Chemistry,* Cornell University Press, Ithaca, N.Y., 1953, p. 426.
51. P. J. Flory, *Statistical Mechanics of Chain Molecules,* Wiley-Interscience, New York, 1968, p. 34.
52. W. R. Pechhold, *Kolloid Z.* **228,** 1 (1968).
53. G. S. Y. Yeh, *Rev. Macromol. Sci.* **1,** 173 (1972).
54. G. S. Y. Yeh and P. H. Geil, *J. Macromol. Sci. Phys.* **B1**(2), 235 (1967).
55. A. Kampf, M. Hoffman, and H. Kramer, *Ber. Bunsenges* **74,** 851 (1970).
56. R. G. Kirste, *Jahresbericht 1969 des Sonderforschungsbereiches Mainz* **41,** 547 (1970).
57. G. D. Wignall, *Memo PPR G19,* Imperial Chemical Industries, Runcorn, UK, 1970.
58. J. P. Cotton, B. Farnoux, G. Jannink, J. Mons, and C. Picot, *C. R. Acad. Sci. Paris* **275,** 3C, 175 (1972).
59. R. G. Kirste, W. A. Kruse, and J. Schelten, *Makromol. Chem.* **162,** 299 (1972).
60. R. G. Kirste, W. A. Kruse, and J. Schelten, *Koll. Z. Z. Polym.* **251,** 919 (1973).
61. R. G. Kirste, W. A. Kruse, and K. Ibel, *Polymer* **16,** 120 (1975).
62. G. D. Wignall and F. S. Bates, *J. Appl. Crystallogr.* **20,** 28 (1987).
63. D. G. H. Ballard, G. D. Wignall, and J. Schelten, *Eur. Polym. J.* **9,** 965 (1973).
64. H. Benoit, J. P. Cotton, D. Decker, B. Farnoux, J. S. Higgins, G. Jannink, R. Ober, and C. Picot, *Nature* **245,** 23 (1973).
65. *National Center for Small-Angle Scattering Research Users' Guide,* Oak Ridge National Laboratory, Oak Ridge, Tenn., 1985.
66. G. D. Wignall, D. G. H. Ballard, and J. Schelten, *Eur. Polym. J.* **10,** 861 (1974).
67. G. Lieser, E. W. Fischer, and K. Ibel, *J. Polym. Sci.* **13,** 29 (1975).
68. J. Schelten, D. G. H. Ballard, G. D. Wignall, G. Longman, and W. Schmatz, *Polymer* **17,** 751 (1976).
69. P. Herschenroeder, thesis, Mainz, FRG, 1978.
70. H. Hayashi, P. J. Flory, and G. D. Wignall, *Macromolecules* **16,** 1328 (1983).
71. K. P. McAlea, J. M. Schultz, K. H. Gardner, and G. D. Wignall, *Macromolecules* **18,** 477 (1985).
72. A. M. Fernandez, L. H. Sperling, and G. D. Wignall, *Macromolecules* **19,** 2572 (1986).
73. J. M. O'Reilly, D. M. Teegarden, and G. D. Wignall, *Macromolecules* **18,** 2747 (1985).
74. J. Kugler, E. W. Fischer, M. Peuscher, and C. D. Eisenbach, *Makromol. Chem.* **184,** 2325 (1984).
75. D. Yoon and P. J. Flory, *Macromolecules* **9,** 294 (1976).
76. P. Debye, *J. Appl. Phys.* **15,** 338 (1944).
77. P. J. Flory, *Principles of Polymer Chemistry,* Cornell University Press, Ithaca, N.Y., 1969, p. 295.
78. O. Kratky, *Kolloid Z.* **182,** 7 (1962).
79. G. Coulon and co-workers, *Proceedings of the International Conference on Deformation, Yield and Fracture of Polymers,* Cambridge, UK, April 1–4, 1985.
80. M. Rawiso, R. Duplessix, and C. Picot, *Macromolecules* **20,** 630 (1987).
81. D. Y. Yoon and P. J. Flory, *Polymer* **16,** 645 (1975).
82. D. Y. Yoon and P. J. Flory, *Polym. Bull.* **4,** 692 (1981).
83. P. J. Flory, *Pure Appl. Chem.* **56,** 305 (1984).
84. R. Gennant, W. Pechhold, and H. P. Grossman, *Colloid Polym. Sci.* **225,** 285 (1977).
85. W. R. Pechhold and H. P. Grossman, *Discuss. Faraday Soc.* **68,** 58 (1979).
86. J. Schelten in *Proceedings of the U.S.-France Symposium Small-Angle X-Ray and Neutron Scattering from Polymers,* Strasbourg, France, 1980, p. 73.
87. P. J. Flory, *J. Am. Chem. Soc.* **84,** 2857 (1962).
88. E. W. Fischer and R. Lorenz, *Kolloid Z.* **189,** 97 (1963).
89. A. Keller, *Philos. Mag.* **2,** 1171 (1957); *Makromol. Chem.* **34,** 1 (1959).
90. J. D. Hoffman and J. L. Laurizen, *J. Res. Natl. Bur. Stand. Sect. A* **65A,** 297 (1961).

91. D. G. H. Ballard, P. Cheshire, G. W. Longman, and J. Schelten, *Polymer* **19**, 379 (1978).
92. J. M. Guenet, *Polymer* **22**, 313 (1981).
93. D. M. Sadler and A. Keller, *Science* **19**, 265 (1979).
94. D. M. Sadler and A. Keller, *Macromolecules* **10**, 1128 (1977).
95. G. C. Summerfield, J. S. King, and R. Ullman, *J. Appl. Crystallogr.* **11**, 548 (1978).
96. M. Stamm, E. W. Fischer, M. Dettenmaier, and P. Convert, *Discuss. Faraday Soc.* **68**, 263 (1979).
97. M. Stamm, J. Schelten, D. G. H. Ballard, *Colloid Polym. Sci.* **259**, 286 (1981).
98. D. Y. Yoon and P. J. Flory, *Discuss. Faraday Soc.* **68**, 288 (1980).
99. J. D. Hoffman, C. M. Guttman, and E. A. Dimarzio, *Discuss. Faraday Soc.* **68**, 177 (1979).
100. *Ibid.*, p. 197; *Polymer* **22**, 597 (1981).
101. P. J. Flory and D. Y. Yoon, *Nature* **272**, 226 (1977).
102. D. M. Sadler and R. Harris, *J. Polym. Sci. Polym. Phys. Ed.* **20**, 561 (1982).
103. D. Sadler in I. Hall, ed., *The Structure of Crystalline Polymers,* Elsevier Applied Science Publishers, Ltd., Barking, UK, 1983, p. 125.
104. M. Stamm, *J. Polym. Sci. Polym. Phys. Ed.* **20**, 235 (1982).
105. G. D. Wignall, L. Mandelkern, C. Edwards, and M. Glotin, *J. Polym. Sci. Polym. Phys. Ed.* **20**, 245 (1982).
106. E. W. Fischer, K. Hahn, J. Kugler, and R. Bom, *J. Polym. Sci. Polym. Phys. Ed.* **22**, 1491 (1984).
107. E. W. Fischer, *Polym. J.* **17**, 307 (1985).
108. K. P. McAlea, J. M. Schultz, K. H. Gardner, and G. D. Wignall, *J. Polym. Sci. Polym. Phys. Ed.* (in press).
109. F. C. Frank in H. R. Doremus, B. W. Roberts, and D. Tumbeill, eds., *Growth and Perfection of Crystals,* John Wiley & Sons, Inc., New York, 1958, p. 529.
110. P. J. Flory, D. Y. Yoon, and K. A. Dill, *Macromolecules* **17**, 862 (1984).
111. D. Y. Yoon and P. J. Flory, *Macromolecules* **17**, 868 (1984).
112. S. J. Spells and D. M. Sadler, *Polymer* **25**, 739 (1984).
113. D. M. Sadler, S. J. Spells, A. Keller, and G. M. Guenet, *Polymer Commun.* **25**, 290 (1984).
114. D. G. H. Ballard, M. Rayner, and J. Schelten, *Polymer* **17**, 349 (1976).
115. J. P. Cotton, B. Farnoux, and G. Jannink, *J. Chem. Phys.* **57**, 290 (1972).
116. J. P. Cotton, B. Farnoux, G. Jannink, and C. Strazielle, *J. Polym. Sci. Polym. Symp.* **42**, 981 (1973).
117. J. P. Cotton and co-workers, *Macromolecules* **7**, 863 (1974).
118. B. Farnoux and co-workers, *J. Phys. Paris Lett.* **36**, L35 (1975).
119. M. Daoud and co-workers, *Macromolecules* **8**, 805 (1975).
120. J. P. Cotton and co-workers, *J. Chem. Phys.* **65**, 1101 (1976).
121. S. N. Jahshan and G. C. Summerfield, *J. Polym. Sci. Polym. Phys. Ed.* **18**, 1859 (1980).
122. G. C. Summerfield, *J. Polym. Sci. Polym. Phys. Ed.* **19**, 1011 (1981).
123. H. Benoit, J. Koberstein, and L. Leibler, *Makromol. Chem. Suppl.* **4**, 85 (1981).
124. J. S. King, W. Boyer, G. D. Wignall, and R. Ullman, *Macromolecules* **18**, 709 (1985).
125. J. S. King, W. Boyer, G. D. Wignall, and R. Ullman in B. Sedlacek, ed., *Physical Optics of Dynamic Phenomena and Processes in Macromolecular Systems,* W. de Gruyter & Co., Berlin, New York, 1985, p. 43.
126. S. F. Edwards, *Proc. Phys. Soc. London* **85**, 613 (1965); **86**, 265 (1966).
127. S. F. Edwards, *J. Phys. Paris* **A8**, 1670 (1975).
128. J. Des Cloiseaux, *J. Phys. Paris* **36**, 281 (1975).
129. M. Daoud and G. Jannink, *J. Phys. Paris* **37**, 973 (1976).
130. R. W. Richards in J. V. Dawkins, ed., *Polymer Characterization,* Applied Science Publishers, London, 1978, p. 117.
131. B. Farnoux and G. Jannink in Ref. 21, p. 265.
132. D. W. Schaefer, *Polymer* **25**, 387 (1984).
133. R. Ullman, *J. Polym. Sci. Polym. Lett. Ed.* **22**, 521 (1983).
134. *Ibid.*, **23**, 1477 (1985).
135. R. Ullman, H. Benoit, and J. S. King, *Macromolecules* **19**, 183 (1986).
136. R. W. Richards, A. Maconnachie, and G. Allen, *Polymer* **18**, 114 (1977).
137. B. Farnoux and co-workers, *J. Phys. Paris* **39**, 77 (1978).
138. A. Manson and L. H. Sperling, *Polymer Blends and Composites,* Plenum Press, New York, 1976.
139. G. Hadziioannou, J. Gilmer, and R. S. Stein, *Polym. Bull.* **9**, 563 (1983).

140. R. S. Stein, *Neutron Scattering in the Nineties, IAEA-CN 46/42,* International Atomic Energy Agency, Vienna, 1985.

141. R. G. Kirste and B. R. Lehnen, *Makromol. Chem.* **177,** 1137 (1976).

142. W. A. Kruse, R. G. Kirste, J. Haas, J. B. Schmidt, and D. J. Stein, *Makromol. Chem.* **177,** 1145 (1976).

143. D. G. H. Ballard, M. Rayner, and J. Schelten, *Polymer* **17,** 349 (1976).

144. R. P. Kambour, R. C. Bopp, A. Maconnachie, and W. J. MacKnight, *Polymer* **21,** 133 (1980).

145. G. D. Wignall, H. R. Child, and F. Li-Aravena, *Polymer* **21,** 131 (1980).

146. J. Stoelting, F. E. Karasz, and W. J. MacKnight, *Polym. Eng. Sci.* **10,** 133 (1970).

147. B. J. Schmidt, R. G. Kirste, and J. Jelenic, *Makromol. Chem.* **181,** 1665 (1980).

148. B. H. Zimm, *J. Chem. Phys.* **16,** 157 (1948).

149. G. Hadziioannou and R. S. Stein, *Macromolecules* **17,** 567 (1984).

150. *Ibid.,* p. 1059.

151. M. Warner, J. S. Higgins, and A. J. Carter, *Macromolecules* **16,** 1931 (1983).

152. P. G. DeGennes, *Scaling Concepts in Polymer Physics,* Cornell University Press, Ithaca, N.Y., 1979, Chapt. IV.

153. C. T. Murray, J. W. Gilmer, and R. S. Stein, *Macromolecules* **18,** 996 (1985).

154. J. S. Higgins and D. J. Walsh, *Polym. Eng. Sci.* **24,** 555 (1984).

155. A. Maconnachie, R. P. Kambour, D. M. White, S. Rostani, and D. J. Walsh, *Macromolecules* **17,** 2645 (1985).

156. J. Jelenic, R. G. Kirste, R. C. Oberthur, and S. Schmitt-Strecher, *Makromol. Chem.* **185,** 129 (1984).

157. C. Herkt-Maetzky and J. Schelten, *Phys. Rev. Lett.* **51,** 896 (1983).

158. M. Shibayama, H. Yang, R. S. Stein, and C. C. Han, *Macromolecules* **18,** 2179 (1985).

159. L. F. Ornstein and F. Zernicke, *Proc. Acad. Sci. Amsterdam* **17,** 793 (1914).

160. A. B. Buckingham and H. G. E. Hentschel, *J. Polym. Sci. Polym. Phys. Ed.* **18,** 853 (1980).

161. F. S. Bates, G. D. Wignall, and W. C. Koehler, *Phys. Rev. Lett.* **55,** 2425 (1985).

162. J. Brandrup and E. H. Immergut, eds., *Polymer Handbook,* 2nd ed., John Wiley & Sons, Inc., New York, 1975.

163. F. S. Bates and G. D. Wignall, *Macromolecules* **19,** 932 (1986).

164. A. Lapp, C. Picot, and H. Benoit, *Macromolecules* **18,** 2437 (1985).

165. F. S. Bates and G. D. Wignall, *Phys. Rev. Lett.* **57,** 1429 (1986).

166. F. S. Bates, C. V. Berney, R. E. Cohen, and G. D. Wignall, *Polymer* **24,** 519 (1983).

167. A. M. Fernandez, J. M. Widmaier, L. H. Sperling, and G. D. Wignall, *Polymer* **25,** 1718 (1984).

168. J. T. Koberstein, *J. Polym. Sci. Polym. Phys. Ed.* **20,** 593 (1982).

169. G. D. Wignall, H. R. Child, and R. J. Samuels, *Polymer* **23,** 957 (1982).

170. P. Debye and A. M. Bueche, *J. Appl. Phys.* **20,** 518 (1949).

171. P. Debye, H. R. Anderson, Jr., and H. Brumberger, *J. Appl. Phys.* **28,** 679 (1957).

172. A. M. Fernandez, L. H. Sperling, and G. D. Wignall, *Multicomponent Polymer Materials, ACS Adv. Chem. Ser.* **211** (1985).

173. D. J. Blundell, G. W. Longman, G. D. Wignall, and M. Bowden, *Polymer* **15,** 33 (1974).

174. F. C. Stehling and G. D. Wignall, *Polym. Prepr. Am. Chem. Soc. Div. Polym. Chem.* **24(2),** 211 (1983).

175. D. J. Lohse, *Polym. Prepr. Am. Chem. Soc. Div. Polym. Chem.* **24(1),** 119 (1983); *Polym. Eng. Sci.* **26,** 1500 (1986).

176. C. M. Roland and G. G. A. Bohm, *J. Polymer Sci. Polym. Lett. Ed.* **22,** 79 (1984).

177. F. S. Bates, R. E. Cohen, and C. V. Berney, *Macromolecules* **15,** 589 (1982).

178. C. V. Berney, R. E. Cohen, and F. S. Bates, *Polymer* **23,** 1222 (1982).

179. J. A. Miller, S. L. Cooper, C. C. Han, and G. Pruckmayr, *Macromolecules* **17,** 1073 (1984).

180. R. W. Richards and J. L. Thomason, *Polymer* **22,** 581 (1981).

181. G. P. Hadziioannou and A. Skoulios, *Macromolecules* **15,** 258, 267 (1982).

182. R. W. Richards and J. L. Thomason, *Macromolecules* **16,** 982 (1983).

183. G. P. Hadziioannou and co-workers, *Macromolecules* **15,** 263 (1982).

184. C. Picot and co-workers, *Macromolecules* **10,** 436 (1977).

185. C. C. Han and B. Mozer, *Macromolecules* **10,** 44 (1977).

186. F. S. Bates and M. Hartney, *Macromolecules* **18,** 2478 (1985).

187. L. Leibler, *Macromolecules* **13,** 1602 (1980).

188. H. Benoit and co-workers, *Macromolecules* **18,** 986 (1985).

189. H. Benoit and M. Benmouna, *Polymer* **25,** 1059 (1984).

190. H. Benoit and M. Benmouna, *Macromolecules* **17,** 535 (1984).

191. X. S. Quan, I. Gancarz, J. T. Koberstein, and G. D. Wignall, *Bull. Am. Phys. Soc.* **30**(3), 291 (1985); *J. Polym. Sci.* (in press).

192. H. Benoit and co-workers, *J. Polym. Sci. Polym. Phys. Ed.* **14,** 2119 (1976).

193. C. Picot and co-workers, *Macromolecules* **10,** 436 (1977).

194. J. A. Hinkley, C. C. Han, B. Mozer, and H. Yu, *Macromolecules* **11,** 836 (1978).

195. S. B. Clough, A. Maconnachie, and G. Allen, *Macromolecules* **13,** 744 (1980).

196. A. Maconnachie, G. Allen, and R. W. Richards, *Polymer* **22,** 9 (1981).

197. S. Candeau, J. Bastide, and M. Delsanti, *Adv. Polym. Sci.* **44,** 27 (1982).

198. R. Ullman, *J. Chem. Phys.* **17**(1), 436 (1979).

199. R. Ullman, *Macromolecules* **15,** 582 (1982).

200. *Ibid.,* p. 1395.

201. R. Ullman in J. E. Mark and J. Lal, eds., *ACS Symp. Ser.* **193,** 1982, Chapt. 13.

202. F. Boue, M. Nierlich, G. Jannink, and R. Ball, *J. Phys. Paris* **43,** 137 (1982).

203. G. Hadziioannou, L. Wang, R. S. Stein, and R. S. Porter, *Macromolecules* **15,** 800 (1982).

204. M. Beltzung, C. Picot, P. Rempp, and J. Herz, *Macromolecules* **15,** 1594 (1982).

205. M. Beltzung, C. Picot, and J. Herz, *Macromolecules* **17,** 663 (1984).

206. J. Bastide, R. Dupplessix, C. Picot, and S. Candeau, *Macromolecules* **17,** 83 (1984).

207. L. Leibler and H. Benoit, *Polymer* **22,** 195 (1981).

208. H. M. James and E. Guth, *J. Chem. Phys.* **15,** 669 (1947); **21,** 1039 (1953).

209. P. J. Flory, *Macromolecules* **12,** 119 (1979).

210. D. G. H. Ballard, P. Cheshire, E. Janke, A. Nevin, and J. Schelten, *Polymer* **23,** 1875 (1982).

211. D. Sadler and P. J. Barham, *J. Polym. Sci. Polym. Phys. Ed.* **21,** 309 (1983).

212. G. D. Wignall and W. Wu, *Polymer Commun.* **24,** 354 (1983).

213. A. Hill, R. S. Stein, and G. D. Wignall, *Polymer* (in press).

214. A. Peterlin in H. Brumberger, ed., *Small-Angle X-ray Scattering,* Gordon and Breach, New York, 1967, p. 145.

215. G. Meinel and A. Peterlin, *J. Polym. Sci. Part A-2* **9,** 67 (1971).

216. G. C. Summerfield, J. S. King, and R. Ullman, *Macromolecules* **11,** 218 (1978).

217. W. Wu and G. D. Wignall, *Polymer* **26,** 661 (1985).

218. W. Wu, *Polymer* **24,** 43 (1983).

219. J. Schelten, A. Zinken, and D. G. H. Ballard, *Colloid Polym. Sci.* **259,** 260 (1981).

220. A. Peterlin, private communication.

221. H. H. Chuah, R. S. Porter, and J. S. Lin, *Macromolecules* **19,** 2732 (1986).

222. D. G. H. Ballard, A. Cunningham, and J. Schelten, *Polymer* **18,** 250 (1977).

223. W. Wu, H. G. Zachmann, and C. Rickel, *Polymer Commun.* **25,** 76 (1984).

224. M. Dettenmaier, A. Maconnachie, J. S. Higgins, H. H. Kausch, and P. Q. Nguyen, *Macromolecules* **19,** 773 (1986).

225. J. M. Lefebvre, B. Escaig, G. Coulon, and C. Picot, *Polymer* **26,** 1807 (1985).

226. J. Klein and B. J. Briscoe, *Proc. R. Soc. London Ser. A* **365,** 53 (1979).

227. C. R. Bartels, W. W. Graessley, and B. Crist, *J. Polym. Sci. Polym. Lett. Ed.* **21,** 495 (1983); *Macromolecules* **17,** 2702 (1984).

228. M. Stamm, *Polym. Prepr. Am. Chem. Soc. Div. Polym. Chem.* **24**(2), 380 (1983).

229. K. P. McAlea, J. M. Schultz, K. H. Gardner, and G. D. Wignall, *Polymer* **27,** 1581 (1986).

230. W. Wu, D. Wiswe, H. G. Zachmann, and K. Han, *Polymer* **26,** 655 (1985).

231. A. Maconnachie, G. Allen, and R. W. Richards, *Polymer* **22,** 1157 (1981).

232. A. R. Rennie and R. C. Oberthur, *Rev. Phys. Appl.* (in press).

233. R. C. Oberthur, *Rev. Phys. Appl.* (in press).

234. J. B. Hayter and J. Penfold, *J. Phys. Chem.* **88,** 4589 (1984).

235. J. B. Hayter, *Physica* **136B,** 269 (1986).

236. B. Ackerson and co-workers, *J. Chem. Phys.* **84,** 2344 (1986).

237. M. P. Grancio and D. J. Williams, *J. Polym. Sci. Part A-1* **8,** 2617 (1970).

238. J. M. O'Reilly, S. M. Melpolder, L. W. Fisher, V. Ramakrishnan, and G. D. Wignall, *Polym. Prepr. Am. Chem. Soc. Div. Polym. Chem.* **24,** 407 (1983).

239. L. W. Fisher, S. M. Melpolder, J. O'Reilly, and G. D. Wignall, *AICHE Symposium on Polymer Colloids—Interfacial Aspects,* Houston, Tex., 1983.

240. P. B. Moore, *J. Appl. Crystallogr.* **13,** 168 (1980).

241. O. Glatter, *J. Appl. Crystallogr.* **11,** 87 (1977).

242. L. W. Fischer, S. M. Melpolder, J. M. O'Reilly, V. Ramakrishnan, and G. D. Wignall, *J. Colloid Interface Sci.* (in press).

243. M. P. Wai, R. A. Gelman, M. G. Fatica, and G. D. Wignall, *Bull. Am. Phys. Soc.* **28,** 517 (1983); *Polymer* **28,** 918 (1987).

244. J. W. Goodwin, R. H. Ottewill, N. M. Harris, and J. Tabony, *Colloid Polym. Sci.* **253,** 78 (1980).

245. K. Alexander, D. J. Cebula, J. W. Goodwin, R. H. Ottewill, and A. Parentich, *Colloids Surf.* **7,** 233 (1983).

246. D. J. Cebula, J. W. Goodwin, R. H. Ottewill, G. Jenkin, and J. Tabony, *Colloid Polym. Sci.* **261,** 555 (1983); D. J. Cebula and co-workers, *Discuss. Faraday Soc.* **76,** 37 (1983).

247. J. S. Higgins, J. V. Dawkins, and G. Taylor, *Polymer* **21,** 627 (1980).

248. M. Linné, A. Klein, L. H. Sperling, and G. D. Wignall, *J. Macromol. Sci. Phys.* (submitted).

249. W. Wu and B. J. Bauer, *Polymer Commun.* **26,** 39 (1985).

250. T. R. Earnest, J. S. Higgins, and W. J. MacKnight, *Macromolecules* **15,** 1390 (1983).

251. T. R. Earnest, J. S. Higgins, D. L. Handlin, and W. J. MacKnight, *Macromolecules* **14,** 192 (1981).

252. W. C. Forsman, W. J. MacKnight, and J. S. Higgins, *Macromolecules* **17,** 490 (1984).

253. A. Eisenberg and M. King, *Ion Containing Polymers,* Academic Press, Inc., New York, 1978.

254. W. J. MacKnight and T. R. Earnest, Jr., *Macromol. Rev.* **16,** 41 (1981); *J. Polym. Sci.* (in press).

255. V. Kalepky, E. W. Fischer, P. Herchenroder, and J. Schelten, *J. Polym. Sci. Polym. Phys. Ed.* **17,** 2117 (1979).

256. R. J. Roe and C. Gieniewski, *Macromolecules* **6,** 212 (1973).

257. R. G. Kirste and H. G. Ohm, *Makromol. Chem.* (in press).

258. J. S. Higgins, K. Dodgson, and J. A. Semlyen, *Polymer* **20,** 553 (1979).

259. C. J. C. Edwards and co-workers, *Polymer* **25,** 365 (1984).

260. P. S. Goyal, J. S. King, and G. C. Summerfield, *Polymer* **24,** 131 (1983).

261. R. P. May, K. Ibel, and J. Haas, *J. Appl. Crystallogr.* **15,** 15 (1982).

262. G. D. Wignall, W. E. Munsil, and C. J. Pings, *J. Appl. Crystallogr.* **11,** 44 (1978).

263. V. Ramakrishnan, *J. Appl. Crystallogr.* **18,** 42 (1985).

264. G. Allen and J. S. Higgins, *Rep. Prog. Phys.* **36,** 1073 (1973).

265. L. K. Nicholson, *Contemp. Phys.* **22,** 451 (1981).

266. J. W. White in K. J. Ivin, ed., *Structural Studies of Macromolecules by Spectroscopic Methods,* Wiley-Interscience, New York, 1976, p. 41.

267. A. H. Baston and D. H. C. Harris, UKAEA (Harwell) Report R92 (1978).

268. R. M. Brugger in Ref. 11, Chapt. 2.

269. G. F. Longster and J. W. White, *J. Chem. Phys.* **48,** 5271 (1968).

270. A. H. Baston, UKAEA (Harwell) Memorandum AERE-M 2570 (1972); G. D. Wignall, UKAEA (Harwell) Memorandum, AERE-M 1928 (1967).

271. B. K. Aldred, R. C. Eden, and J. W. White, *Discuss. Faraday Soc.* **43,** 169 (1967).

272. J. S. Higgins, G. Allen, and P. N. Brier, *Polymer* **13,** 157 (1972).

273. G. Allen, C. J. Wright, and J. S. Higgins, *Polymer* **15,** 319 (1974).

274. H. Takeuchi, J. S. Higgins, A. Hill, A. Maconnachie, G. Allen, and G. C. Stirling, *Polymer* **23,** 499 (1982).

275. B. Gabrys, J. S. Higgins, K. T. Ma, and J. Roots, *Macromolecules* **17,** 560 (1984).

276. B. Gabrys, J. S. Higgins, and D. A. Young, *Polymer* **26,** 355 (1985).

277. R. Currat and R. Pynn in Ref. 6, p. 131.

278. R. M. Nicklow in Ref. 6, p. 191.

279. P. J. Brown in Ref. 6, p. 69.

280. M. Tasumi, T. Shimanouchi, and T. Miyazawa, *J. Mol. Spectr.* **9,** 261 (1962).

281. M. Tasumi and T. Shimanouchi, *J. Chem. Phys.* **43,** 1245 (1965).

282. M. Tasumi and S. Krimm, *J. Chem. Phys.* **46,** 755 (1967).

283. H. R. Danner, G. H. Safford, H. Boutin, and M. Berger, *J. Chem. Phys.* **40,** 1417 (1964).

284. W. R. Myers, G. C. Summerfield, and J. S. King, *J. Chem. Phys.* **44,** 184 (1966).

285. J. S. King in S. H. Chen and S. Yip, eds., *Spectroscopy in Biology and Chemistry,* Academic Press, Inc., New York, 1974, p. 235.

286. J. F. Twistleton and J. W. White in *Neutron Inelastic Scattering,* International Atomic Energy Agency, Vienna, 1972, p. 301.

287. W. R. Myers, G. C. Summerfield, and J. S. King, *J. Chem. Phys.* **44**, 184 (1966).
288. L. A. Feldkamp, G. Ventataraman, and J. S. King in *Inelastic Scattering of Neutrons,* International Atomic Energy Agency, Vienna, 1968, p. 159.
289. V. LaGarde, H. Prask, and S. Trevino, *Discuss. Faraday Soc.* **48**, 15, (1969).
290. L. Piseri, B. M. Powell, and G. Dolling, *J. Chem. Phys.* **58**, 158 (1973).
291. J. W. White in Ref. 266, Chapt. 4.
292. J. F. Twistleton and J. W. White, *Polymer* **13**, 40 (1972).
293. J. W. White in A. D. Jenkins, ed., *Polymer Science,* North-Holland, Amsterdam, The Netherlands, 1972, Chapt. 27.
294. M. R. Anderson, M. B. M. Harryman, D. K. Steinman, J. W. White, and R. Currat, *Polymer* **23**, 569 (1982).
295. J. F. Twistleton, J. W. White, and P. A. Reynolds, *Polymer* **23**, 578 (1982).
296. G. C. Stirling, in B. T. M. Willis, ed., *Chemical Application of Thermal Neutron Scattering,* Oxford University Press, Oxford, UK, 1973, p. 31.
297. D. W. Williams, *J. Chem. Phys.* **45**, 3770 (1966); **47**, 4680 (1967).
298. J. S. Higgins in Ref. 266, Chapt. 2.
299. G. Allen, P. N. Brier, A. G. Goodyear, and J. S. Higgins, *Faraday Discuss. Soc.* **6**, 169 (1972).
300. P. G. de Gennes, *Physics* **3**, 37 (1967).
301. E. Dubois-Violette and P. G. de Gennes, *Physics* **3**, 181 (1967).
302. P. E. Rouse, *J. Chem. Phys.* **21**, 1272 (1953).
303. B. H. Zimm, *J. Chem. Phys.* **24**, 269 (1956).
304. J. S. Higgins and J. E. Roots, *Polymer Prepr. Am. Chem. Soc. Div. Polym. Chem.* **24**(2), 241 (1983).
305. A. Z. Akcasu, M. Benmouna, and C. C. Han, *Polymer* **21**, 866 (1980).
306. G. Allen, R. Ghosh, J. S. Higgins, J. P. Cotton, B. Farnoux, G. Jannink, and G. Weill, *Chem. Phys. Lett.* **38**, 577 (1976).
307. F. Mezei, *J. Phys.* **225**, 146 (1972).
308. J. B. Hayter, *Z. Phys.* **B31**, 117 (1978).
309. J. B. Hayter, *J. Magn. Mater.* **14**, 319 (1979).
310. J. B. Hayter in Ref. 21, p. 49.
311. F. Mezei in F. Mezei, ed., *Neutron Spin Echo, Lecture Notes in Physics No. 128,* Springer-Verlag, New York, 1980.
312. D. Richter, J. B. Hayter, F. Mezei, and B. Ewen, *Phys. Rev. Lett.* **41**, 1484 (1978).
313. L. K. Nicholson, J. S. Higgins, and J. B. Hayter, *Macromolecules* **14**, 836 (1981).
314. D. Richter, A. Baumgartner, K. Binder, B. Ewen, and J. B. Hayter, *Phys. Rev. Lett.* **47**, 109 (1981).
315. J. M. Deutsch and N. D. Goldenfeld, *Phys. Rev. Lett.* **48**, 1694 (1982).
316. D. Richter, A. Baumgartner, K. Binder, B. Ewen, and J. B. Hayter, *Phys. Rev. Lett.* **48**, 1695 (1982).
317. J. S. Higgins, L. K. Nicholson, and J. B. Hayter, *Polymer* **22**, 137 (1981).
318. D. Richter, B. Ewen, and J. B. Hayter, *Phys. Rev. Lett.* **45**, 2121 (1980).
319. J. B. Hayter, G. Jannink, F. Brochard-Wyart, and P. G. de Gennes, *J. Phys. Paris Lett.* **41**, L451 (1980).
320. G. H. Lander and V. J. Emery, eds., *Scientific Opportunities with Advanced Facilities for Neutron Scattering, Argonne National Laboratory Report, CONF-8410256,* Shelter Island Workshop, New York, Oct. 1985.

G. D. WIGNALL
Oak Ridge National Laboratory

NEWTONIAN FLOW.   See VISCOELASTICITY.

NITRILE RUBBER.   See BUTADIENE POLYMERS.

NITROCELLULOSE. See CELLULOSE ESTERS, INORGANIC.

NITROETHYLENE POLYMERS. See ANIONIC POLYMERIZATION.

NITROSO POLYMERS

The term nitroso polymers has been used for polymers with the structure $-\text{N}-\text{O}-(\text{C})_m-$ as the repeating unit (nitroso rubbers). Discovery of a novel po-

lymeric system having the azo-*N,N'*-dioxide $-(\text{N}=\text{N})-$ structure in the repeating

unit, corresponding to the true dimeric form of nitroso compounds, required broadening the definition (1). Also, polymerization of nitrosobenzene and 1,4-dinitrosobenzene has been achieved and polymers with pendent nitroso groups have been prepared. This article discusses all polymers that derive from nitroso or nitroso dimer units in the chain backbone or as pendent groups. The emphasis is on the nitroso rubbers that are highly or completely fluorinated and have the nitrogen–oxygen–carbon moiety as a repeating unit.

Nitroso Rubbers

Trifluoronitrosomethane (CF_3NO) and tetrafluoroethylene (C_2F_4), in the absence of light and air, react at 20°C to give a mixture of perfluoro-2-methyl-1,2-oxazetidine (**1**) and a polymeric oil (2–7) having the form of structure (**2**):

$$\begin{array}{cc} CF_3-N-CF_2 & \left(\begin{array}{c} N-O-CF_2CF_2 \\ | \\ CF_3 \end{array}\right)_n \\ | \quad | \\ O-CF_2 \\ (\mathbf{1}) & (\mathbf{2}) \end{array}$$

The oxazetidine is the predominant product at ambient temperatures and above. However, below 0°C, a 1:1 alternating elastomeric copolymer having structure (**2**) is obtained in high yield. This nitroso rubber is nonflammable, completely amorphous, and has commendable low temperature flexibility ($T_g = -51$°C) (8) and excellent resistance to protic solvents, oxidizing agents, and ozone (9,10). Despite this combination of properties, nitroso rubbers have found only very limited and highly specialized applications because of the unavailability of an efficient vulcanization system, poor resistance to heat and electromagnetic radiation, and degradation to extremely toxic products.

The only nitroso polymers of any commercial importance so far are the copolymer of trifluoronitrosomethane and tetrafluoroethylene, and terpolymers that include small quantities of functional nitroso compounds as a third monomer, eg, 4-nitrosohexafluorobutyric acid, to provide cross-linking sites.

Monomers.   The first synthesis of CF_3NO from the reaction of fluorine with silver cyanide in the presence of silver nitrite or silver oxide was reported in 1936 (11). However, the first practical synthetic method involving uv irradiation of a mixture of perfluoroalkyl iodide and nitric oxide with mercury as a catalyst did not appear until the early 1950s (12–15). The simplest preparation of CF_3NO is the pyrolysis of trifluoroacetyl nitrite (CF_3COONO) in refluxing perhalocarbon solvents at 150–200°C (16–18). Trifluoroacetyl nitrite is prepared by the reaction of trifluoroacetic anhydride with either nitrosyl chloride at room temperature (16) or dinitrogen trioxide at 0°C (19). It can also be prepared by reaction of nitrosyl chloride with silver trifluoroacetate at -10°C (17,18). The synthesis and properties of numerous highly fluorinated nitroso monomers have been compiled (20–22). Equipment flow charts for the pilot-scale production of CF_3NO and C_2F_4 copolymers have also been given (20). Currently, CF_3NO is not manufactured in the United States.

Polymerization. CF_3NO does not homopolymerize, but reacts spontaneously upon mixing with an equimolar quantity of C_2F_4 to give a 1:1 alternating copolymer at temperatures below 0°C (23,24). Preparative methods for bulk and suspension polymerization have been reported (25). Simplicity of the system and easy work-up of the product make the bulk method preferable for the preparation of small quantities, eg, 100 g, of copolymer. An aqueous suspension system, however, is recommended for large-scale production, since this permits the dissipation of the exothermic heat of polymerization. In the suspension process, lithium bromide is used as a freezing-point depressant (23); magnesium carbonate is a suspending agent.

Most successful polymerizations have been carried out below 0°C. Different reaction temperatures and variations in monomer structure lead to different products. For example, perfluoro-2-nitroso-2-methylpropane and C_2F_4 combine above 170°C to give a polymer having the structure $-[ON(CF_2CF_2C(CF_3)_3)-(CF_2CF_2)_{11}]-$, whereas CF_3NO and 1,1-difluoroethylene react at 100–150°C over a period of 5–7 wk to yield the structure $-[N(CF_3)OCH_2CF_2N(CF_3)OCH=CF]-$ (26). Also, an elastomeric copolymer of CF_3NO and trifluoroacrylyl fluoride can be prepared at ambient temperature in 28% yield (27).

In these copolymerizations, the reactivity of the highly fluorinated nitroso compounds R—NO is independent of R, but the nature of the olefin strongly influences the polymerization rate (28). The reactivity of the olefins decreases with decreasing halogen substitution. The preparation and properties of several terpolymers, some of which include a reactive functional group containing nitroso compounds as the third monomer for eventual cross-linking, have been reported (20).

Mechanism.   Copolymerization of CF_3NO and C_2F_4 probably proceeds by a free-radical mechanism (28–32). However, a precise initiating step has not been elucidated (30).

Structure.   A copolymer of CF_3NO and C_2F_4 prepared at -78°C was shown by mass spectrometry to have a structure similar to that of (**2**) (33). However, high resolution ^{19}F nmr studies (34) on a polymer prepared at -25°C by a suspension method indicated the presence of the following type of minor irregular

repeat units: $-[ON(CF_3)CF_2CF_2N(CF_3)O]-$ and $-(O-CF_2CF_2-O)-$. The reaction variables, particularly temperature, influence the amount of irregular
orientation.

Properties.   The properties of the raw copolymer of CF_3NO and C_2F_4 are
shown in Table 1.

Table 1. Properties of Unvulcanized Copolymer of CF_3NO and C_2F_4

Property	Value		Refs.
Mark-Houwink constants[a] in	K	a	
trichlorotrifluoroethane (35°C)	2.80×10^{-4}	0.51	35
perfluorotributylamine (25°C)	8.77×10^{-5}	0.66	
solubility and solution properties	soluble in fluorocarbons		32,35
solubility parameter	5.2		23
specific gravity	1.93		23
n_D^{25}	1.3170		23
T_g, °C	-51		36
flammability	nonflammable; will not flame		37
resistance to			
chemicals	excellent except to amines		23,37
heat	decomposes violently at 270°C		38
radiation	both uv (λ 254 nm) and γ-radiation produce random scission in polymer chain		32,38
crystal structure[b]	amorphous		23

[a] $[\eta] = KM^a$.
[b] By x-ray diffraction.

Vulcanization.   Unsaturated terpolymers prepared from CF_3NO, C_2F_4, and
hexafluoro-1,3-butadiene (39) provide pendent $-CF=CF_2$ groups for cross-linking. Vulcanization (qv) of this terpolymer with several nitrogen-containing compounds, including a polymeric derivative of perfluoro-N,N'-difluoropiperazine
and perfluoro-2,5-diazahexane-2,5-dioxyl (3), has been studied (40–42).

$$CF_3-N-CF_2CF_2-N-CF_3$$

(3)

The goal was to produce a perfect nitroso rubber network by introducing cross-
links that impart stability by virtue of their chemical resemblance to the polymer
backbone. However, despite effective vulcanization, thermal stability of the cross-
linked polymers was no better than the raw polymer. Copolymers and terpolymers
containing methyl nitrosodifluoroacetate (43,44) cured with metal oxides and
epoxides gave vulcanizates for making gaskets and sealants with improved oxidation resistance. Nitroso co- and terpolymers can be processed using conventional rubber compounding and processing equipment. Several vulcanization
recipes and the properties of such vulcanizates are available (10,20,37).

Applications.   Nitroso polymers have the unique property of nonflamma-
bility in pure oxygen. For this reason, they have been used by the National
Aeronautics and Space Administration (NASA) in the Apollo spacecraft. How-

ever, this polymer has become obsolete for commercial purposes because of poor strength and stability, poor resistance to heat and radiation, and decomposition to highly toxic products. Currently, nitroso polymers are not manufactured in the United States.

Poly(azoalkylene-*N,N'*-dioxide)

Bifunctional nitrosoalkanes with suitable structural features form stable poly(azoalkylene-*N,N'*-dioxide) (poly(nitrosoalkane)) polymers, as shown below (1):

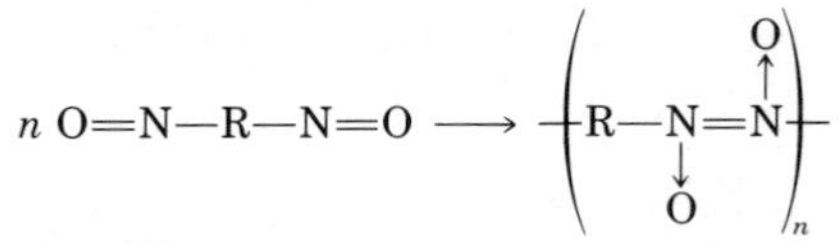

$$n \; O{=}N{-}R{-}N{=}O \; \longrightarrow \; \left(R{-}\underset{\underset{O}{\downarrow}}{\overset{\overset{O}{\uparrow}}{N{=}N}} \right)_n$$

The polymerization apparently proceeds by dimerization of nitroso groups, a process already known for the dimerization of a simple nitroso compound. Polymers with molecular weights up to 23,000 have been obtained (45). Preliminary investigations indicate that poly(nitrosoalkane)s may display liquid crystalline properties near their melt temperatures of 170–220°C (46) (see also LIQUID CRYSTALLINE POLYMERS).

Aromatic Nitroso Polymers

Various condensation polymers have been synthesized by the reaction of monomers containing nitroso groups, such as dimeric 4-nitroso-3,5-dichlorobenzoyl chloride with diols or diamines (47). These polymers have $-\underset{\underset{O}{\downarrow}}{\overset{\overset{O}{\uparrow}}{N{=}N}}-$ linkages in the chain backbone. The nitrosation of numerous addition polymers for the purpose of cross-linking by dimerization of nitroso groups has also been described (47). Such cross-links are reversible because they are thermally labile; they can be broken at elevated temperatures for easy processing and restored upon cooling to impart better physical properties to the polymer (see CROSS-LINKING, REVERSIBLE).

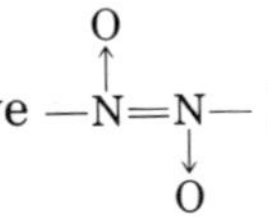

Aromatic nitroso compounds with a free para position undergo condensations in the presence of concentrated acids to give nitrosodiarylhydroxylamines. The polymerization of nitrosobenzene in dimethylformamide using aluminum chloride or sulfuric acid as catalysts has been reported (49). The degree of polymerization in the presence of aluminum chloride was 10. Nitrosobenzene does not copolymerize with styrene or α-methylstyrene in methylene chloride at 0°C with boron trifluoride etherate as a catalyst. Although novel in structure, nitrosobenzene polymers have apparently found no commercial applications.

An aromatic nitroso polymer that has attained some commercial importance is poly(*p*-dinitrosobenzene). This polymer is known to have the repeat unit $(C_6H_4N_2O_2)_n$, but data on its synthesis, structural elucidation, and molecular weight are not available in the literature. The polymer in a hydrocarbon base was previously marketed by DuPont under the trade name POLYAC. The same product is currently available from R. T. Vanderbilt Co. as VANAX-PY. Its applications include a thermoplastic can-end sealant composition (50), thermal stabilization of butadiene resin (51), and preparation of graft copolymers (52).

Introduction of pendent nitroso groups on the polymer chain has been achieved by nitrosating (53) polystyrene at the para position by acetoxymercuration followed by treatment with nitrosyl chloride. p-Nitrosated polystyrene derived from high molecular weight polystyrene shows a tendency to gel; that obtained from low molecular weight polystyrene is soluble in tetrahydrofuran and hence the nitroso groups are accessible for further chemical reactions (53).

Preparation of cation-exchange fluororesins containing $-\text{N}(\text{C}_6\text{F}_4\text{R})-\text{O}-\overset{|}{\text{C}}-\overset{|}{\text{C}}$

as the repeating unit has been reported (54). The synthesis involves the copolymerization of 2,3,5,6-tetrafluoronitrosobenzene with fluoroalkenes such as chlorotrifluoroethylene followed by chlorosulfonation and hydrolysis. The ion-exchange capacity of such resins is ca 2.26 meq/g.

Health and Safety Information

LC_{50} for C_2F_4 is 40,000 ppm/4 h. This compound has been reported (55) to detonate unpredictably, particularly if uninhibited. All operations with it should be carried out behind a suitable barricade. Nitroso rubbers are not known to be toxic, but their degradation products, ie, CF_2O, CF_3NCF_2, are extremely toxic. LD_{50} and LTV data for CF_3NO, nitroso rubbers, poly(nitrosoalkane)s, and VANAX-PY are not available in the literature. VANAX-PY, however, has been reported (56) to irritate skin and the respiratory system and should be used with adequate ventilation. The very limited toxicity data available on a few C-nitroso compounds suggest that toxicity varies with the structure of the compound. All nitroso compounds should be handled with care.

BIBLIOGRAPHY

"Nitroso Polymers" in *EPST* 1st ed., Vol. 9, pp. 322–336, by Joseph Green, Columbian Carbon Company.

1. W. L. Childress and L. G. Donaruma, *Macromolecules* **7**, 427 (1974).
2. D. A. Barr and R. N. Haszeldine, *J. Chem. Soc.* **1955**, 1881, 2532; **1960**, 1151.
3. D. A. Barr and R. N. Haszeldine, *Nature London* **175**, 991 (1955).
4. Brit. Pat. 789,254 (Jan. 15, 1958), J. B. Rose (to Imperial Chemical Industries, Ltd.).
5. D. A. Barr, R. N. Haszeldine, and C. J. Willis, *Proc. Chem. Soc.* **1959**, 230.
6. C. E. Griffen and R. N. Haszeldine, *Proc. Chem. Soc.* **1959**, 369; *J. Chem. Soc.* **1960**, 1398.
7. U.S. Pat. 3,065,214 (Nov. 20, 1962), J. B. Rose (to Imperial Chemical Industries, Ltd.).
8. G. H. Crawford, *J. Polym. Sci.* **45**, 259 (1960).
9. J. C. Montermoso, C. B. Griffis, A. Wilson, and G. H. Crawford, *Rubber Plast. Age* **42**, 514 (1961).
10. J. Green, N. B. Levine, and W. Sheehan, *Rubber Chem. Technol.* **39**, 1222 (1966).
11. O. Ruff and M. Giese, *Berichte* **69B**, 598, 684 (1936).

12. R. N. Haszeldine, *Nature London* **168,** 1028 (1951).
13. R. N. Haszeldine, *J. Chem. Soc.* **1953,** 2075.
14. J. Banus, *Nature London* **171,** 173 (1953).
15. J. Banus, *J. Chem. Soc.* **1953,** 3755.
16. J. D. Park, R. W. Rosser, and J. R. Lacher, *J. Org. Chem.* **27,** 1462 (1962).
17. C. W. Taylor, T. J. Brice, and R. L. Wear, *J. Org. Chem.* **27,** 1064 (1962).
18. R. E. Banks, R. N. Haszeldine, and M. K. McCreath, *Proc. Chem. Soc.* **1961,** 64.
19. D. E. Rice and G. H. Crawford, *J. Org. Chem.* **28,** 872 (1963).
20. M. C. Henry, C. B. Griffis, and E. C. Stump in P. Tarrant, ed., *Fluorine Chemistry Reviews,* Vol. 1, Marcel Dekker, Inc., New York, 1967, p. 1.
21. H.-C. Huang, *Hachsueh T'ung-pao* **7,** 11 (1964); *NASA Engl. Transl. N65-15669,* National Aeronautics and Space Administration, Houston, Tex., 1964.
22. I. L. Knunyants, Yu. A. Sizov, and O. V. Ukharov, *Usp. Khim.* **52,** 976, 1983.
23. C. B. Griffis and M. C. Henry, *Rubber Chem. Technol.* **39,** 481 (1966); *Rubber Plast. Age* **46,** 63 (1965).
24. U.S. Pat. 3,072,592 (Jan. 8, 1963), G. H. Crawford (to Minnesota Mining and Manufacturing Co.).
25. C. D. Padgett and E. C. Stump in W. J. Bailey, ed., *Macromolecular Syntheses,* Vol. 4, John Wiley & Sons, Inc., New York, 1972, p. 147.
26. O. V. Ukharov, Yu. A. Sizov, and I. L. Knunyants, *Zh. Vses. Khim. O. Va* **27,** 344, 346 (1982).
27. R. E. Banks, R. N. Haszeldine, M. J. Stevenson, and B. G. Willoughby, *J. Chem. Soc. C* **1969,** 2129.
28. G. H. Crawford, D. E. Rice, and B. F. Landrum, *J. Polym. Sci. Part A* **1,** 565 (1963).
29. D. A. Barr and R. N. Haszeldine, *J. Chem. Soc.* **1956,** 3416.
30. J. D. Crabtree, R. N. Haszeldine, A. J. Parker, K. Ridings, R. F. Simmons, and S. Smith, *J. Chem. Soc. Perkin Trans. 2,* 111 (1972).
31. D. T. Clark, D. Kilcast, W. J. Feast, and W. K. R. Musgrave, *J. Polym. Sci. Part A-1* **10,** 1637 (1972).
32. L. J. Fetters in L. A. Wall, ed., *Fluoropolymers,* Vol. 25 of *High Polymers,* Wiley-Interscience, New York, 1972, p. 175.
33. W. T. Flowers, R. N. Haszeldine, E. Henderson, A. K. Lee, and R. D. Sedwick, *J. Polym. Sci. Polym. Chem. Ed.* **10,** 3489 (1972).
34. D. D. Lawson and J. D. Ingham, *J. Polym. Sci. Part B* **6,** 181 (1968).
35. G. A. Morneau, P. I. Roth, and A. R. Shultz, *J. Polym. Sci.* **55,** 609 (1961).
36. G. H. Crawford, D. E. Rice, and J. C. Montermoso, *Proc. Sixth Joint Army Navy Air Force Conf. Elast. Res. Dev.* **2,** 643 (1960).
37. J. C. Montermoso, *Rubber Chem. Technol.* **34,** 1521 (1961).
38. A. R. Shultz, N. Knoll, and G. A. Morneau, *J. Polym. Sci.* **62,** 211 (1962).
39. Brit. Pat. 1,140,525 (Jan. 22, 1969), R. E. Banks and R. N. Haszeldine (to the Minister of Technology, London).
40. Ger. Offen. 2,304,712 (Aug. 9, 1973), R. E. Banks and R. N. Haszeldine (to the Secretary of State for Defense in Her Britannic Majesty's Government of the United Kingdom of Great Britain and Northern Ireland, London).
41. R. E. Banks, P. A. Carson, and R. N. Haszeldine, *J. Chem. Soc. Perkin Trans.* **1,** 1111 (1973).
42. R. E. Banks, R. N. Haszeldine, P. Mitra, T. Myerscough, and S. Smith, *J. Macromol. Sci. Chem. A* **8,** 1325 (1974).
43. U.S. Pat. 3,660,367 (May 2, 1972), N. Mayers and R. Michaels (to Thiokol Chemical Corp.).
44. U.S. Pat. 3,725,374 (Apr. 3, 1973), N. Mayers, J. Green, and R. Michaels (to Thiokol Chemical Corp.).
45. D. K. Dandge and L. G. Donaruma in B. M. Culbertson and C. U. Pittman, Jr., eds., *New Monomers and Polymers,* Vol. 25 of *Polymer Science and Technology Reviews,* Plenum Press, New York, 1984, p. 173.
46. D. K. Dandge and L. G. Donaruma, Polytechnic Institute of New York, New York, 1982, personal communication.
47. U.S. Pat. 3,872,057 (Mar. 18, 1975), J. F. Pazos (to E.I. du Pont de Nemours & Co., Inc.).
48. P. A. S. Smith, *The Chemistry of Open-Chain Organic Nitrogen Compounds,* Vol. 2, W. A. Benjamin, Inc., New York, 1965, p. 364.

49. A. Yoneda, K. Hayashi, M. Tanaka, and N. Murata, *Kobunshi Kagaku* **29,** 87 (1972).
50. U.S. Pat. 3,986,629 (Oct. 19, 1976), H. M. Singleton (to Southland Corp.).
51. U.S. Pat. 3,886,106 (May 27, 1975), D. F. Lohr, E. L. Kay, and W. R. Hausch (to the Firestone Tire & Rubber Co.).
52. Ger. Offen. 3,006,743 (Sept. 4, 1980), U. Katsuji and M. Takashi (to Sumitomo Chemical Co., Ltd.).
53. A. Yoneda, K. Sugihara, K. Hayashi, and M. Tanaka, *Kobunshi Kagaku* **30,** 180 (1973).
54. Jpn. Kokai Tokyo Koho 79 53,689 (Apr. 27, 1979) and 79 53,690 (Apr. 25, 1979), T. Kawai, K. Haraguchi, and S. Inou (to Central Glass Co., Ltd.).
55. S. Sherratt in A. Standen, ed., *Kirk-Othmer Encyclopedia of Chemical Technology*, 2nd ed., Vol. 9, Wiley-Interscience, New York, 1966, pp. 807–812.
56. *VANAX-PY, Material Safety Data Sheet*, R. T. Vanderbilt Co., Inc., Norwalk, Conn., Feb. 4, 1985.

D. K. DANDGE
New Mexico Institute of Mining and Technology

L. G. DONARUMA
University of Alabama in Huntsville

NOMENCLATURE

Nomenclature, as used in this article, refers to the naming of polymeric materials. The nomenclature of scientific communication is emphasized, although there is generally little reason for differences between scientific and other, eg, commercial, usage.

Since the publication of the first edition of this Encyclopedia, the International Union of Pure and Applied Chemistry (IUPAC) has established the Commission on Macromolecular Nomenclature, which is now the leading nomenclature body in the polymer field. The Commission is promulgating a series of rules and definitions that are placing polymer nomenclature on a much more systematic basis than had previously been the case (Table 1) (1–21). The International Standardization Organization (ISO), primarily through its Technical Committee TC/61 Plastics, and various national nomenclature bodies (such as that of the American Chemical Society) are also helping to shape the field. Recent issues of *Chemical Abstracts* are additional authoritative sources of polymer nomenclature.

At the present time, the IUPAC Commission on Macromolecular Nomenclature is developing a set of definitions for many of the basic terms dealing with polymer molecules, assemblies of polymer molecules, polymer solutions, polymer crystals, polymer melts and solids, polymerization reactions, etc. It is also extending existing nomenclature to more complicated cases, such as cross-linked polymers. When this phase of the work is completed by the late 1980s, the naming of polymers and polymer terminology will have become largely systematized and, following the IUPAC practice in other fields of chemistry, a compendium of polymer nomenclature rules will be published.

Table 1. IUPAC Publications on Polymer Nomenclature

Title	Comment	Refs.
Report on Nomenclature in the Field of Macromolecules	obsolete	1
Report on Nomenclature Dealing with Steric Regularity in High Polymers	superseded by Ref. 2	3
Revised Report on Nomenclature Dealing with Steric Regularity in High Polymers	superseded by Ref. 4	2,5
Report of the Committee on Nomenclature of the International Commission on Macromolecules	obsolete	6
Basic Definitions of Terms Relating to Polymers		7,8
List of Standard Abbreviations (Symbols) for Synthetic Polymers and Polymer Materials (1974)	superseded by Ref. 9	10
Use of Abbreviations for Names of Polymeric Substances	Recommendations 1986	9
Nomenclature of Regular Single-Strand Organic Polymers		11
Stereochemical Definitions and Notations Relating to Polymers	Provisional Recommendations 1980	12 4
Nomenclature for Regular Single-Strand and Quasi Single-Strand Inorganic and Coordination Polymers	Provisional Recommendations 1984	13 14
Note on the Terminology for Molar Masses in Polymer Science		15–17
Source-Based Nomenclature for Copolymers		18
Definitions of Terms Relating to Individual Macromolecules, Their Assemblies, and Dilute Polymer Solutions		19
Definitions of Terms Relating to Crystalline Polymers		20
A Classification of Linear Single-Strand Polymers		21

Basic Definitions

No nomenclature document is more fundamental to a given science than the definitions of basic terms used in that area. The IUPAC Commission on Macromolecular Nomenclature published a document in 1974 (8) that offers definitions of 52 terms, including polymer, constitutional unit, monomer, polymerization, regular polymer, tactic polymer, block polymer, graft polymer, monomeric unit, degree of polymerization, addition polymerization, condensation polymerization, homopolymer, copolymer, bipolymer, terpolymer, copolymerization, and many others. Both structure-based and process-based definitions are given.

Source-based Nomenclature

Traditionally, polymers have been named by attaching the prefix poly to the name of the real or assumed monomer (the "source") from which it is derived.

Thus polystyrene is the polymer made from styrene and will often be found in an index under "styrene, polymer of." When the name of the monomer consists of two or more words, parentheses should be used (1), as in poly(vinyl acetate), poly(methyl methacrylate), poly(sodium styrenesulfonate), etc. Failure to use parentheses can lead to ambiguity: polychlorostyrene can be the name of either a polychlorinated (monomeric) styrene molecule or a polymer derived from chlorostyrene; polyethylene oxide can refer to polymer (1), polymer (2), or the macrocycle (3).

$$\text{HO(CH}_2\text{CH}_2\text{O)}_n\text{H} \qquad \begin{array}{c} \text{H(CH}_2\text{CH}_2)_n \\ \diagdown \\ \diagup \quad \text{O} \\ \text{H(CH}_2\text{CH}_2)_n \end{array} \qquad \begin{array}{c} \text{CH}_2\text{(CH}_2\text{CH}_2)_n\text{CH}_2 \\ | \qquad\qquad | \\ \text{CH}_2 \text{---O---} \text{CH}_2 \end{array}$$

$$(1) \qquad\qquad\qquad (2) \qquad\qquad\qquad (3)$$

These problems are easily overcome with parentheses; names such as poly-(chloro)styrene, poly(chlorostyrene), and poly(ethylene oxide) clearly indicate the part of the name to which the prefix poly refers. The omission of parentheses is, unfortunately, quite common.

The principal deficiency of source-based nomenclature is that the chemical structure of the monomeric unit in a polymer is not identical with that of the monomer, eg, $-\text{CH}_2-\text{CHX}-$ vs $\text{CH}_2=\text{CHX}$; thus the name polymonomer is actually a misnomer. The structure of the repeating unit is also not specified in this scheme; for example, polyacrolein does not indicate whether the vinyl or the aldehyde group has polymerized (see ACROLEIN POLYMERS).

$$n\ \text{CH}_2=\text{CH}-\text{CH}=\text{O} \quad \nearrow \quad \left(\begin{array}{c} \text{CH}_2-\text{CH}- \\ | \\ \text{CH}=\text{O} \end{array}\right)_n \quad \begin{array}{l}\text{vinyl addition} \\ \text{(1,2-addition)}\end{array}$$

$$\searrow \quad \left(\begin{array}{c} \text{CH}-\text{O}- \\ | \\ \text{CH}=\text{CH}_2 \end{array}\right)_n \quad \text{carbonyl addition}$$

Different types of polymerization can take place with many other monomers, depending on the polymerization conditions. Furthermore, a name such as poly(vinyl alcohol) refers to a hypothetical source, since this polymer is obtained by hydrolysis of poly(vinyl acetate). In spite of these serious deficiencies, source-based nomenclature is still firmly entrenched in industrial literature and, to a lesser extent, in scientific communication. It originated at a time when polymer science was less developed and the structure of most polymers ill-defined. The rapid advances now being made in structural determination of polymers will gradually shift the emphasis of polymer nomenclature away from starting materials and toward the structure of the macromolecules.

Copolymers. Copolymers are polymers that are derived from more than one species of monomer (8). Because this is a process-based definition, source-based nomenclature can be easily adapted to the naming of copolymers (18). However, the arrangement of the various types of monomeric units must be specified. Seven types of arrangements have been defined and are shown in Table 2, where A, B, and C represent the names of monomers. The monomer names are linked through a connective (infix), such as -co-, to form the name of the copolymer, as in poly(styrene-co-acrylonitrile). The order of citation of the monomers is arbitrary, except for graft copolymers where the backbone monomer is named first.

An equally acceptable alternative scheme utilizes the prefix copoly followed by citation of the names of the monomers used, separated from each other by an oblique stroke. Parentheses are also needed. For example, copoly(styrene/butadiene) denotes an unspecified copolymer of styrene and butadiene. The other connectives of Table 2 are placed before such names to provide additional structural information, as in

> *stat*-copoly(styrene/butadiene)
> *ran*-copoly(ethylene/vinyl acetate)
> *alt*-copoly(styrene/maleic anhydride)
> *per*-copoly(ethylene phenylphosphonite/methyl acrylate/carbon dioxide)
> *block*-copoly(styrene/butadiene/methyl methacrylate)
> *graft*-copoly(styrene/butadiene)

It is not necessary to use parentheses to enclose vinyl acetate, maleic anhydride, methyl acrylate, etc, even though the name of each of these monomers consists of two words; the names of the polymers, as written here, are unambiguous.

The names of copolymers, derived either from the main scheme or the alternative, can be further modified to indicate various structural features. For example, the chemical nature of end groups can be specified as follows:

> α-X-ω-Y-poly(A-*alt*-B)
> α-butyl-ω-carboxy-*block*-copoly(styrene/butadiene)

Whereas subscripts placed immediately after the name of the monomer or the block designate the degree of polymerization or repetition, mass and mole fractions and molar masses, which in most cases are average quantities, are expressed by placing corresponding figures after the complete name of the copolymer. The order of citation is as for the monomeric species in the name. Unknown quantities are designated by a, b, etc. Some examples follow.

A block copolymer containing 75 mass % of polybutadiene and 25 mass % of polystyrene is

> polybutadiene-*block*-polystyrene (0.75:0.25 w) or
> *block*-copoly(butadiene/styrene) (75:25 mass %)

A graft copolymer, consisting of a polyisoprene backbone grafted with isoprene and acrylonitrile units in an unspecified arrangement, containing 85 mol % of isoprene units and 15 mol % of acrylonitrile units is

> polyisoprene-*graft*-poly(isoprene-*co*-acrylonitrile) (0.85:0.15 x) or
> *graft*-copoly[isoprene/(isoprene;acrylonitrile)] (85:15 mol %)

A graft copolymer consisting of 75 mass % of polybutadiene with a relative molecular mass of 90,000 as the backbone and 25 mass % of polystyrene in grafted chains with a relative molecular mass of 30,000 would be

> polybutadiene-*graft*-polystyrene (75:25 mass %; 90,000:30,000 M_r)

Table 2. IUPAC Nomenclature of Copolymers[a]

Type	Arrangement of monomeric units	Structure	Connective	Example
unspecified	unknown or unspecified	(A-co-B)	-co-	poly[styrene-co-(methyl methacrylate)]
statistical	obeys statistical laws	(A-stat-B)	-stat-	poly(styrene-stat-acrylonitrile-stat-butadiene)
random	obeys Bernoullian statistics	(A-ran-B)	-ran-	poly[ethylene-ran-(vinyl acetate)]
alternating	alternating sequence	(AB)$_n$	-alt-	poly[(ethylene glycol)-alt-(terephthalic acid)]
periodic	periodic with respect to at least three monomeric units	(ABC)$_n$ (ABB)$_n$ (AABB)$_n$ (ABAC)$_n$	-per-	poly[formaldehyde-per-(ethylene oxide)-per-(ethylene oxide)]
block	linear arrangement of blocks	—AAAAA—BBBBBB—	-block-[b]	polystyrene-block-polybutadiene

graft	polymeric side chain different from main chain[c]	—AAAAAAAAAAA—
		     | | | | |
		     B B B B B
			-graft-[d]	polybutadiene-graft-polystyrene

[a] Main system of the IUPAC document (18); an alternative scheme is described in the text.
[b] The connective -b- has also been used.
[c] Main chain (or backbone) is specified first in the name.
[d] The connective -g- has also been used.

195

A graft copolymer in which the polybutadiene backbone has a DP of 1700 and the polystyrene grafts have an unknown DP is named

graft-copoly(butadiene/styrene) (1700:*a* DP)

The published IUPAC copolymer document (18) should be consulted for the names of more complex copolymers, eg, those having a multiplicity of grafts or having chains radiating from a central atom (see also BLOCK COPOLYMERS; CO-POLYMERS, ALTERNATING; COPOLYMERIZATION; GRAFT COPOLYMERS).

Structure-based Nomenclature

For organic polymers that are regular, ie, have only one species of constitutional unit in a single sequential arrangement, and consist only of single strands, the IUPAC has promulgated a structure-based system of naming polymers (11). As originally devised by the Polymer Nomenclature Committee of the American Chemical Society (22), it consists of naming a polymer as poly(constitutional repeating unit), wherein the repeating unit is named as a bivalent organic radical according to the usual nomenclature rules for organic chemistry. It is important to note that in structure-based nomenclature the name of the constitutional repeating unit has no relationship to the source from which the unit was prepared. The name is simply that of the largest identifiable unit in the polymer, and locants for unsaturation, substituents, etc are dictated by the structure of the unit.

The steps involved in naming the constitutional repeating unit are (*1*) identification of the unit, taking into account the kinds of atoms in the main chain and the location of substituents; (*2*) orientation of the unit; and (*3*) naming of the unit. Examples of names for some common polymers are given in Table 3. Note that in this system parentheses are always used to enclose the repeating unit.

Structure-based nomenclature can be utilized to name polymers with great complexity, provided only that they be regular and single-stranded. Among these are polymers with constitutional repeating units which consist, themselves, of a series of smaller subunits; polymers with heteroatoms or heterocyclic ring systems in the main chain; and polymers with substituents on acyclic or cyclic subunits of constitutional repeating units. Structure-based nomenclature is also applicable to copolymers having a regular structure, regardless of the starting materials used, eg, poly(oxyethyleneoxyterephthaloyl). In principle, it should be possible to extend the existing structure-based nomenclature beyond regular, single-strand polymers to polymers that have reacted, cross-linked polymers, ladder polymers, and other more complicated systems.

Structure-based nomenclature has gained acceptance in the scientific literature, eg, *Chemical Abstracts*, because it overcomes many of the deficiencies of source-based nomenclature.

Inorganic and Coordination Polymers. The nomenclature of regular single-strand inorganic and coordination polymers (qv) is governed by the same fundamental principles as that for single-strand organic polymers (14). The name of such a polymer is that of the smallest structural repeating unit prefixed by the terms poly, *catena* (for linear chains) or other structural indicator, and designations for end groups. The structural units are named by the nomenclature rules for inorganic and coordination chemistry. Some examples are

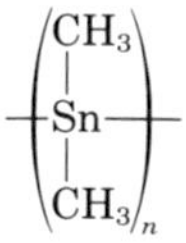

catena-poly[dimethyltin]

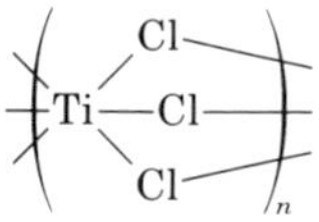

catena-poly[titanium-tri-μ-chloro]

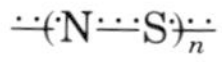

catena-poly[nitrogen-μ-thio]

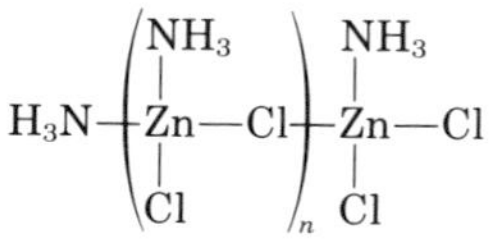

α-ammine-ω-(amminedichlorozinc)-
catena-poly[(amminechlorozinc)-μ-chloro]

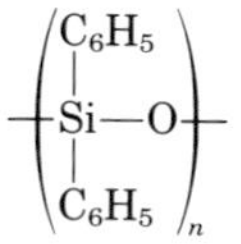

catena-poly[(diphenylsilicon)-μ-oxo]

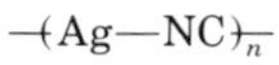

catena-poly[silver-μ-(cyano-*N*:*C*)]

Stereochemical Definitions and Notations. Structure-based nomenclature of regular polymers (4) can denote stereochemical features if the repeating unit used is the configurational unit, ie, a constitutional unit having one or more sites of defined stereoisomerism (8). Structure-based names are then derived in the usual fashion. The various stereochemical features that are possible in a polymer must be defined.

Natta and co-workers introduced the concept of tacticity, ie, the orderliness of the succession of configurational repeating units in the main chain of a polymer. For example, in poly(propylene), possible steric arrangements are shown in Fischer projections displayed horizontally:

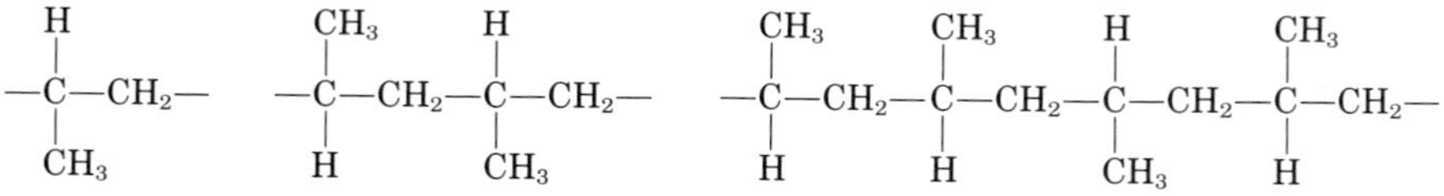

and the corresponding polymers have the following structures:

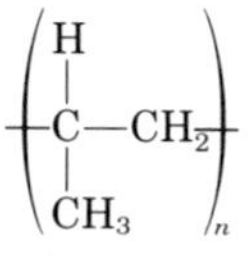

isotactic 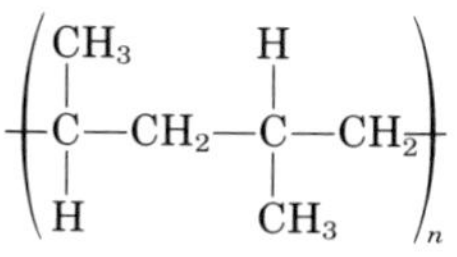   syndiotactic 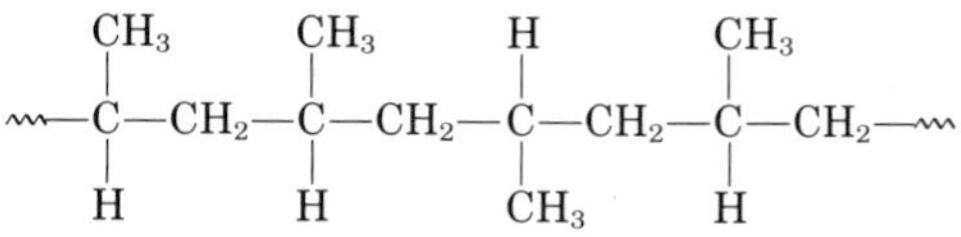   atactic

The isotactic polymer has only one species of configurational unit in a single sequential arrangement and the syndiotactic polymer shows an alternation of configurational units that are enantiomeric, whereas in the atactic polymer the

Table 3. Examples of Systematic Structure-based Names for Polymers[a]

Structure	Structure-based name	Common (source-based) name
$-(CH_2CH_2)_n-$	poly(methylene)	polyethylene
$-(CHCH_2)_n-$ $\quad CH_3$	poly(propylene)	polypropylene
$\begin{array}{c}CH_3\\ -(C-CH_2)_n-\\ CH_3\end{array}$	poly(1,1-dimethylethylene)	polyisobutylene
$-(C=CHCH_2CH_2)_n-$ $\quad CH_3$	poly(1-methyl-1-butenylene)	polyisoprene
$-(CHCH_2)_n-$ $\quad C_6H_5$	poly(1-phenylethylene)	polystyrene
$-(CHCH_2)_n-$ $\quad Cl$	poly(1-chloroethylene)	poly(vinyl chloride)
$-(CHCH_2)_n-$ $\quad CN$	poly(1-cyanoethylene)	polyacrylonitrile
$-(CHCH_2)_n-$ $\quad OOCCH_3$	poly(1-acetoxyethylene)	poly(vinyl acetate)
$-(CCH_2)_n-$ $\quad F,\ F$	poly(1,1-difluoroethylene)	poly(vinylidene fluoride)

poly(difluoromethylene)	polytetrafluoroethylene
poly[(2-propyl-1,3-dioxane-4,6-diyl)methylene]	poly(vinyl butyral)
poly[1-(methoxycarbonyl)-1-methylethylene]	poly(methyl methacrylate)
poly(oxyethylene)	poly(ethylene oxide)
poly(oxy-1,4-phenylene)	poly(phenylene oxide)
poly(oxyethyleneoxyterephthaloyl)	poly(ethylene terephthalate)
poly(iminoadipoyliminohexamethylene)	poly(hexamethylenediamine-co-adipic acid) or poly(hexamethylene adipamide)
poly[2,5-dioxotetrahydrofuran-3,4-diyl(phenylethylene)]	poly(maleic anhydride-co-styrene)

[a] Ref. 6. Courtesy of *Pure and Applied Chemistry*.

199

molecules have equal numbers of the possible configurational units in a random sequence distribution. This can be generalized as follows:

Isotactic:

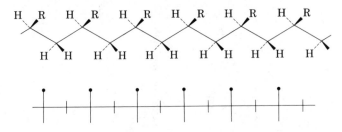

Syndiotactic:

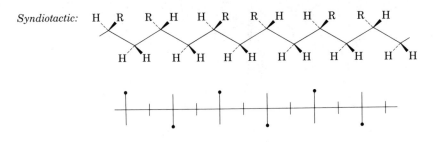

Atactic:

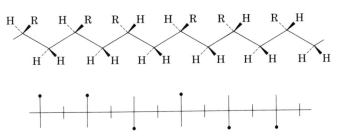

Further examples of tactic polymers are

$$\left(\!\!O-\underset{\underset{H}{|}}{\overset{\overset{CH_3}{|}}{C}}-CH_2\!\!\right)_{\!n}$$

isotactic poly(oxypropylene)

$$\left(\!\!\underset{\underset{H}{|}}{\overset{\overset{CH_3}{|}}{C}}\!\!\right)_{\!n}$$

isotactic poly(ethylidene)

$$\left(\!\!\underset{\underset{H}{|}}{\overset{\overset{CH_3\ H}{|\ |}}{C}}-\underset{\underset{CH_3}{|}}{C}\!\!\right)_{\!n}$$

syndiotactic poly(ethylidene)

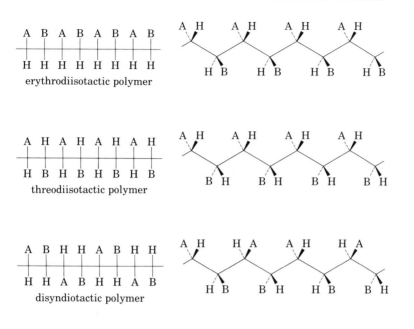

erythrodiisotactic polymer

threodiisotactic polymer

disyndiotactic polymer

isotactic poly(3-methyl-*trans*-1-butenylene) or transisotactic poly(3-methyl-1-butenylene)

diisotactic poly[*threo*-3(methoxycarbonyl)-4-methyl-*trans*-1-butenylene] or transthreodiisotactic poly[3-(methoxycarbonyl)-4-methyl-1-butenylene]

The concept of a stereoblock is illustrated in the following example of a regular poly(propylene) chain, in which the stereoblocks are denoted by ⌐‾‾⌐. The sequence of identical relative configurations of adjacent units that characterizes a stereoblock is terminated at each end of the block. The dashed line ⌐-----⌐ encloses a configurational sequence, which may or may not be identical with a stereoblock.

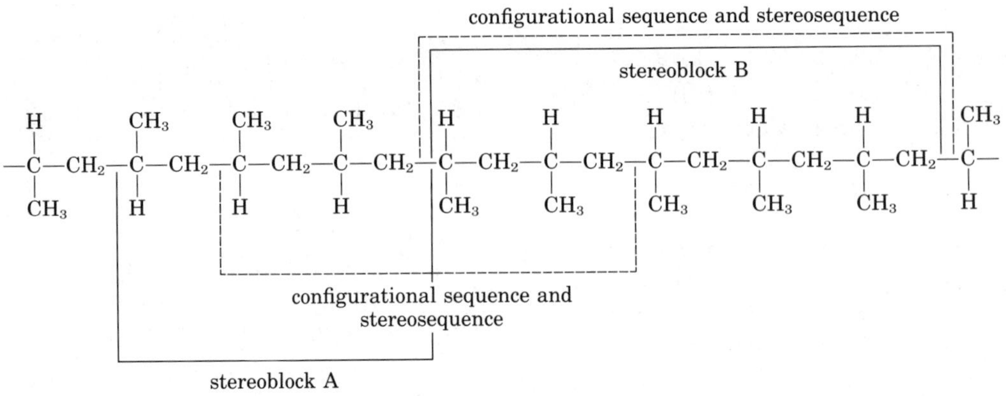

configurational sequence and stereosequence

stereoblock B

$$-\overset{\displaystyle H}{\underset{\displaystyle CH_3}{C}}-CH_2-\overset{\displaystyle CH_3}{\underset{\displaystyle H}{C}}-CH_2-\overset{\displaystyle CH_3}{\underset{\displaystyle H}{C}}-CH_2-\overset{\displaystyle CH_3}{\underset{\displaystyle H}{C}}-CH_2-\overset{\displaystyle H}{\underset{\displaystyle CH_3}{C}}-CH_2-\overset{\displaystyle H}{\underset{\displaystyle CH_3}{C}}-CH_2-\overset{\displaystyle H}{\underset{\displaystyle CH_3}{C}}-CH_2-\overset{\displaystyle H}{\underset{\displaystyle CH_3}{C}}-CH_2-\overset{\displaystyle H}{\underset{\displaystyle CH_3}{C}}-CH_2-\overset{\displaystyle CH_3}{\underset{\displaystyle H}{C}}-$$

configurational sequence and
stereosequence

stereoblock A

The published IUPAC document (4) should be consulted for more complex cases and for the notations used to designate conformations of polymer molecules (bond lengths, bond angles, torsion angles, helix sense, isomorphous and enantiomorphous structures, line repetition groups and symmetry elements, etc) as well as for the various stereochemical definitions (see also MICROSTRUCTURE; STEREOREGULAR POLYMERS).

Trade Names and Abbreviations

Because the systematic names of polymers can be cumbersome, trade names and abbreviations are frequently used as a shortcut in industrial literature and

**Table 4. List of Abbreviations from the 1986
IUPAC Recommendations**[a]

PAN	polyacrylonitrile
PCTFE	polychlorotrifluoroethylene
PEO	poly(ethylene oxide)
PETP[b]	poly(ethylene terephthalate)
PE	polyethylene
PIB	polyisobutylene
PMMA	poly(methyl methacrylate)
POM	poly(oxymethylene); polyformaldehyde
PP	polypropylene
PS	polystyrene
PTFE	polytetrafluoroethylene
PVAC	poly(vinyl acetate)
PVAL	poly(vinyl alcohol)
PVC	poly(vinyl chloride)
PVDC	poly(vinylidene dichloride)
PVDF	poly(vinylidene difluoride)
PVF	poly(vinyl fluoride)

[a] Ref. 9.
[b] The abbreviation PET is commonly used in the literature.

oral communication. For example, the simpler generic name nylon-6,6 for a polyamide, where the first number refers to the number of carbon atoms of the diamine and the second number to that of the diacid fragment, appears often in the literature rather than the systematic name poly(iminoadipoyliminohexamethylene). Useful compilations of trade names for polymers can be found in Refs. 23 and 24.

Perhaps the most widely used shortcut is the use of abbreviations for common industrial polymeric materials. The IUPAC recognizes that there may be advantages in some cases to use abbreviations, but urges that each abbreviation be fully defined the first time it appears in the text and that no abbreviation be used in titles of publications. Because there are inherent difficulties in assigning systematic and unique abbreviations to polymeric structures, only a short list has the IUPAC's official sanction (9,10) (Table 4). ISO has published a more extensive list (25), and the American Chemical Society has compiled a master list of all known abbreviations in the polymer field (26).

BIBLIOGRAPHY

"Nomenclature" in *EPST* 1st ed., Vol. 9, pp. 336–344, by Robert B. Fox, U.S. Naval Research Laboratory, and Chairman (1963–1967), Polymer Nomenclature Committee of The American Chemical Society.

1. IUPAC, *J. Polym. Sci.* **8,** 257 (1952).
2. M. L. Huggins, G. Natta, V. Desreux, and H. Mark, *Pure Appl. Chem.* **12,** 645 (1966).
3. *Ibid., J. Polym. Sci.* **56,** 153 (1962).
4. IUPAC, *Pure Appl. Chem.* **53,** 733 (1981).
5. M. L. Huggins, G. Natta, V. Desreux, and H. Mark, *Makromol. Chem.* **82,** 1 (1965).
6. M. L. Huggins, P. Corradini, V. Desreux, O. Kratky, and H. Mark, *J. Polym. Sci. Part B* **6,** 257 (1968).
7. *Appendices on Tentative Nomenclature, Symbols, Units and Standards, No. 13, IUPAC Information Bulletin*, IUPAC, Oxford, UK, 1971.
8. IUPAC, *Pure Appl. Chem.* **40,** 479 (1974).
9. IUPAC, *Pure Appl. Chem.* **59,** 691 (1987).
10. IUPAC, *Pure Appl. Chem.* **40,** 475 (1974).
11. IUPAC, *Pure Appl. Chem.* **48,** 373 (1976).
12. IUPAC, *Pure Appl. Chem.* **51,** 1101 (1979).
13. IUPAC, *Pure Appl. Chem.* **53,** 2283 (1981).
14. IUPAC, *Pure Appl. Chem.* **57,** 149 (1985).
15. IUPAC, *Makromol. Chem.* **185** (appendix to No. 1) (1984).
16. IUPAC, *J. Polym. Sci. Polym. Lett. Ed.* **22,** 57 (1984).
17. IUPAC, *J. Colloid Interface Sci.* **101,** 277 (1984).
18. IUPAC, *Pure Appl. Chem.* **57,** 1427 (1985).
19. IUPAC, *Pure Appl. Chem.,* in press.
20. IUPAC, *Pure Appl. Chem.,* in press.
21. IUPAC, *Pure Appl. Chem.,* in press.
22. Polymer Nomenclature Committee, American Chemical Society, *Macromolecules* **1,** 193 (1968).
23. M. Ash and I. Ash, *Encyclopedia of Plastics, Polymers, and Resins,* Chemical Publishing Co., Inc., New York, 1982.
24. H-G. Elias and R. A. Pethrick, eds., *Polymer Yearbook,* Harwood Academic Publishers GmbH, New York, 1984, p. 113.
25. *Plastics—Symbols and Codes—Part 1: Symbols for Basic Polymers and Their Modifications and*

for Plasticizers, International Standard ISO 1043-1984, The International Standardization Organization, New York, 1984.
26. Polymer Nomenclature Committee, American Chemical Society, *Polym. News* **9,** 101 (1983); **10** (Pt. 2), 169 (1985).

NORBERT M. BIKALES
National Science Foundation
Secretary (1978–1987), IUPAC Commission on Macromolecular Nomenclature

NONAQUEOUS DISPERSIONS. See COATINGS.

NONCOMBUSTIBLE FABRICS. See FLAMMABILITY.

NONDESTRUCTIVE TESTING. See TEST METHODS.

NON-NEWTONIAN FLOW. See VISCOELASTICITY.

NONWOVEN FABRICS

SURVEY

Nonwoven fabrics are porous, textilelike materials, usually in flat sheet form, composed primarily or entirely of fibers assembled in webs (1–3). These fabrics, also called bonded fabrics, formed fabrics, or engineered fabrics, are manufactured by processes other than spinning, weaving, or knitting. The thickness of the sheets may vary from 25 μm to several centimeters, and the weight from 10 g/m^2 to 1 kg/m^2. A sheet may resemble paper or a woven or knitted fabric in appearance and may have a unique texture or pattern. It may be as compact and crisp as paper or supple and drapable as a conventional textile; it may be resilient or limp. Its tensile properties may be barely self-sustaining or so high that it is impossible to tear, abrade, or damage the sheet by hand. The fiber components, one or several types, may be natural or synthetic, from 1–3-mm long to endless. The tensile properties may depend on frictional forces or a film-forming polymeric additive functioning as an adhesive binder. All or some of the fibers may be welded by heat or solvent. A scrim, gauze, netting, yarn, or other conventional sheet material may be added to one or both faces, or embedded within as reinforcement. The nonwoven fabric may be incorporated as a component in a composite structure.

Felted fabrics from animal hairs, eg, wool (qv), are not included even though their structure fulfills descriptions of nonwoven textile fabrics. Paper (qv), where fibers are held together by hydrogen bonding, is not considered to be a nonwoven fabric.

Textiles and nonwoven fabrics (4) are classified into three groups: fiber-web structures, netlike structures, and multiplex structures. In fiber-web structures, fibers or filaments are the fundamental units arranged into a web and bonded in such a manner that the bond-to-bond distances are greater than 50–100 times the fiber diameter. This definition clearly excludes paper, wherein bond-to-bond distances are on the order of the fiber diameter. However, this definition also excludes some nonwoven fabrics heavily impregnated with binder, where bond-to-bond distances are less than 50–100 times the fiber diameter.

Conventional textile fabrics are based on yarns and monofilaments. Fibers are parallelized and twisted by spinning to form cohesive and strong one-dimensional elements called yarns or threads. In textile fabrics, the yarns or monofilaments are interlaced in a regular repetitive design (Fig. 1). Strength and other physical properties are derived from friction between individual fibers in the yarns and between adjacent yarns. Both fiber and yarn properties of strength, resilience, and elongation are transmitted to the fabric.

In contrast to conventional textiles (5,6), the basic structure of all nonwovens is a web of fibers; thus the basic element is the single fiber. The individual fibers are arranged more or less randomly (Figs. 2 and 3). Tensile, stress–strain, and tactile properties are imparted to the web by an adhesive or other chemical and

Fig. 1.　Woven fabric of polypropylene filament yarns. Courtesy of Chicopee.

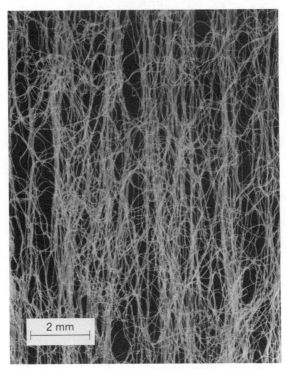

Fig. 2. Rayon card web with oriented fiber distribution. Courtesy of Chicopee.

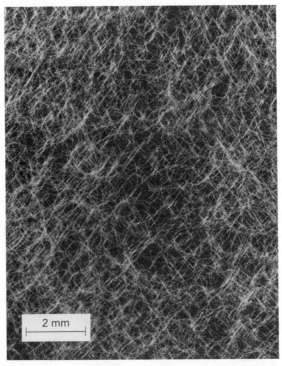

Fig. 3. Polyester air-laid web with random fiber distribution. Courtesy of Chicopee.

physical bonding, fiber-to-fiber friction created by entanglement, and reinforcement by yarns, scrims, nettings, films, foams, and additional unbonded webs.

The first patent for nonwoven textile fabrics made from staple fibers was issued in 1936. It describes an intermittently bonded nonwoven fabric made by printing an adhesive in a repetitive pattern onto a carded fiber web (7). The bonded areas provide strength, whereas the unbonded areas, where the fibers are free to move relative to each other, provide softness. In another method, patented in 1942, flexible bonded sheet is made from a random mixture of fusible and nonfusible fibers (8). A sheet held together by a system of fine-scale, random, intermittent adhesive bonds is produced when the web is subjected to conditions that soften the fusible fibers while applying the required pressure. Early nonwovens were also made from heat-drying webs saturated with water-based latices. The web-forming process was adapted from conventional carding technology, which is an intermediate step in the manufacture of yarns (spinning) from long, crimped (staple) fibers.

The technology for making nonwoven fabrics is based on the following primary elements: fibers of various lengths and diameters; a web geometrically arranged according to the method of forming and processing; the bonding within the web by adhesive, or mechanical–frictional forces created by fiber contact or entanglement; and reinforcements and composites, eg, yarns, scrims, films, nettings, and unbonded webs. Variations of one or several elements in combination allows for the enormous range of nonwoven fiber types.

Fibers

The term staple fibers was originally applied to fibers of natural origin long enough to be processed on textile machinery, but excluding endless filaments, eg, silk. Staple length was defined or characterized as approximately the longest or functionally the most important length of the natural fibers, such as cotton (qv) or wool (9). Staple fibers of natural origin have an irregular crimp, ie, a three-dimensional wavelike shape. In the present context, as applied to regenerated cellulose and synthetic fibers, staple fibers are of relatively uniform length, ca 1.3–10.2 cm, with a regular crimp. Regenerated and other extruded fibers are endless as formed. They are cut during the manufacturing process to a specified length to meet a processing or market need. Extruded fibers are also produced as continuous filaments without crimp. Woven or nonwoven fabrics made from staple fibers appear fuzzier and feel fuller than fabrics made from endless filaments. The processes for forming webs from staple fibers are different from those using continuous filaments. The products obtained from staple and filament fiber webs may differ substantially in properties.

The fibers (qv), as defined by chemical composition and as a result of their mechanical properties, determine the ultimate properties of the fabric. Web structure and bonding maximize inherent fiber characteristics.

All natural fibers that are commercially available can be formed into webs. In practice, wood pulp, which is far shorter than staple length, is the only natural fiber used in large amounts because of its high water absorbency, bulk, and low cost. Cotton and manila hemp are used in smaller amounts. Cotton has excellent inherent properties for nonwovens (10), and advances in cleaning and bleaching techniques should increase commercial consumption (11). Wool and linen are too expensive for nonwoven fabrics.

Regenerated fibers include viscose rayon (qv) and cellulose acetate (see FIBERS in CELLULOSE ESTERS, ORGANIC). Viscose rayon was the predominant fiber used for nonwoven fabrics until recently, when more stringent properties were required (12–14). Rayon is easily processed into webs and readily bonded by adhesives, notably latices, and by mechanical entanglement. It is available in regular and high wet-modulus types, but virtually only the former is used. However, it is not a strong fiber, especially when wet. Rayon is almost ideal for disposable surgical and sanitary products, filters for food and industrial solvents, wiping cloths, certain lightweight facings, and hand towels. All these applications require biodegradability, inertness to common solvents, good water absorbency, and moderate cost. In recent years, viscose-rayon production in the United States and Western Europe has been curtailed because of increasing cost of the fiber. More wood pulp is being used because of its low cost and acceptable properties. However, the shortness of the fibers (1–3 mm in common commercial grades) is a drawback in bonding, when product strength and softness are needed. Rayon has been largely replaced by polyester and polypropylene fibers.

Among the synthetic fibers, poly(ethylene terephthalate) (PET), nylon-6 and 6,6, polypropylene (PP), and poly(vinyl alcohol) fibers are employed in nonwoven fabric manufacture (15,16). The cost of polyester and polypropylene is comparable to that of viscose rayon, although their strength and resilience are superior. Facings of disposable diapers or sanitary napkins made from either of these fibers feel dry even when the absorbent, inner wood-pulp layer is saturated. As practical methods were developed to bond PET and PP fibers, they displaced rayon (qv) in lightweight facings.

Nylon, being more expensive than polyester, PP, or rayon, is used in moderate quantities, mostly in garment interlinings where it supplies strength and resilience (see POLYAMIDES, FIBERS).

Polypropylene fibers are strong, hydrophobic, and inexpensive (see OLEFIN FIBERS). Because of their low density, they give a higher fiber volume than other polymer fibers. Polypropylene cannot be bonded with latices, but is suitable for thermoplastic bonding. Staple polypropylene fibers and filaments are used in large amounts in this way. Poly(vinyl alcohol) fibers, cross-linked with formaldehyde, are manufactured in Japan. They are strong, insoluble in water, chemically resistant, and hydrophilic. Nonwovens are slightly stiffer and less absorbent, but substantially stronger than those made of rayon (17); response to latex binders is comparable. Poly(vinyl alcohol) fibers are primarily used in battery separators because of their absorbency and chemical inertness.

Cellulose acetate fibers are not expensive, but weak and difficult to bond. They are used in small amounts for decorative ribbons (3M Co.). Acrylic fibers (qv) are not used in nonwoven fabrics.

In the early stages of the nonwoven industry, the same staple fibers were used as in the conventional textile industry. Small quantities of thermoplastic fibers were used as nonwoven binders, such as plasticized cellulose acetate for fusible–nonfusible fiber mixtures (8) and Vinyon HH fibers of poly(vinyl chloride–vinyl acetate) for tea bags. More recently, the fiber industry has provided a special grade of polypropylene fibers suitable for thermal bonding as well as an extensive variety of low melting copolymer polyester fibers as thermoplastic binders. Low melting copolyamide fibers are in limited use. The processes utilizing thermoplastic fibers are called thermal bonding or thermobonding.

Manufacturers in the UK (ICI), Japan (Chisso), and the United States (BASF) have developed bicomponent fibers especially for nonwoven materials. These fibers contain a low melting and a high melting component. The former is usually the outer layer and the latter is the core; occasionally, the two components are side by side.

Length.   Natural fibers, such as wood pulp and cotton, are characterized by specific length. Commercial wood pulp fibers range in length from <1 mm to ca 3 mm. Cotton fibers typically grow to a staple length of about 28 mm.

Synthetic and regenerated fibers are made by extrusion and can be prepared in any length; they are usually furnished in precision-cut uniform length. Draw ratio is the same for fibers used for nonwoven materials and for common textiles. Staple fibers are provided in lengths of 2–6 cm, but can be cut longer. Fibers for wet processing are cut as short as 6 mm.

The length is usually chosen for convenience in the manufacturing process; when longer than 25 mm, length has little effect on product properties. Processing equipment is not designed to handle long fibers and these are often broken to a length of 25–50 mm and even shorter.

Diameter.   Regenerated and synthetic fibers are produced in a wide range of diameters for both conventional textiles and nonwoven materials. Fiber diameter is described by tex, the weight in grams of 1000 m of filament (18) or denier, the weight in grams of 9000 m; the practical metric unit is the decitex (= denier × 1.11), the weight of 10,000 m of filament. The decitex value is proportional to the square of the fiber diameter, and directly proportional to fiber polymer density (19):

$$\text{dtex} = 7.85 \times 10^{-3} \rho d^2$$

where d is the diameter, μm, and ρ is the density, g/cm^3.

The fiber diameter has a great effect on the properties of the finished fabric and must be chosen carefully. Fiber diameter influences the fabric hand, bonding properties, cover or opacity, strength and resilience, web thickness, filtering properties, and processing equipment requirements.

The diameter of a typical elliptically shaped, fine U.S. cotton fiber is around 8 μm (minor axis) and 20 μm (main axis), which is comparable to the diameter of rayon or polyester fibers of 1.65 dtex (1.5 denier) (20). Viscose-rayon fibers are roughly circular in cross section with diameters of 12–15 μm at the above decitex values; this includes the diameters of most common fibers. Finer fibers of ca 1 dtex (0.9 den) are beginning to appear on the market. Coarser fibers include heavy filaments equivalent to hundreds of decitex, as in some geofabrics.

Crimp.   Crimp refers to a three-dimensional wave along the fiber, identified as crimps per centimeter. Natural fibers such as cotton or wool have an inherent, irregular crimp. Synthetic staple fibers are crimped during manufacture by a mechanical stuffing or gear-crimping process. Rayon staple fibers are crimped mechanically or chemically.

Web-forming equipment for staple fibers in dry processing contains needles or sharp teeth to break up fiber clumps by gripping and separating the fibers. Crimp makes it possible for these needles to manipulate the fibers. In addition, it enhances necessary interfiber friction. In the finished nonwoven material, crimp imparts a third dimension and improves resilience, appearance, and feel. Fibers intended to be made into webs by wet or papermaking processes are not crimped, since this would promote fiber entanglement in the water-dispersion step. Absence of crimp contributes to the flat, dense properties typical of wet-formed nonwovens.

Cross Section.   Synthetic fibers, eg, polyester, nylon, and polypropylene, are extruded in a variety of shapes. Some noncircular shapes are better for bonding efficiency and fabric softness (21), but round cross-sectional fibers are commonly used in nonwoven fabric.

Conventional viscose-rayon fibers are approximately circular with characteristic crenulations. High wet-modulus viscose-rayon fibers do not show crenulations and are either round or bilobal. Special high water-absorbent rayon fibers have been made with a round cross section and a central lumen (22); they are not produced in significant amounts.

Spin-finish.   Spin-finish refers to the lubricant that is added in small amounts (ca 0.2 wt %) to the surface of extruded fibers to enhance processability in the mechanical operations, including spinning, by controlling friction and reducing static electricity. A fiber without finish cannot be made into a yarn or web by a dry process. The finish also affects hydrophilic or hydrophobic properties. Synthetic fibers are frequently given a hydrophilic finish to enhance wettability by water-based latex binders. For a wet-dispersion process, a synthetic fiber is given a hydrophilic finish to enhance dispersibility. Rayon fibers, however, are not given a spin finish before dispersion in water; this is called a hard finish.

Delustering.   Whiteness is achieved by dispersing a small amount (ca 1–2 wt %) of TiO_2 pigment in the dope before extrusion; this is called delustering. Delustering agents are used in synthetic fibers to reduce sheen.

Draw Ratio.   Synthetic fibers for nonwoven use are drawn during extrusion to the same extent as for general textile use except in special cases involving poly(ethylene terephthalate) and polypropylene to make them more suitable as bonding fibers.

Web Formation

All nonwoven fabrics are based on a fibrous web. Nettings, foams, scrims, and modified films have incorrectly been claimed to be nonwoven fabrics. However, some of these materials are competitive in properties and commercial applications with nonwoven fabrics.

The characteristics of the fiber web determine the physical properties of the final product. These characteristics depend largely on fiber geometry, which is determined by the mode of web formation. Fiber geometry includes the predominant fiber direction, whether oriented or random, fiber shape (straight, hooked, or curled), the extent of interfiber engagement or entanglement, crimp, and z-direction compaction (web-density control). Web characteristics are also influenced by fiber diameter, length, web weight, and chemical and mechanical properties of the polymer.

The choice of method for forming the web is determined by fiber length. Initially, the methods for forming webs from staple-length fibers were based on the textile-carding process, whereas web formation from short fibers was based on papermaking technologies. Although these technologies are still in use, new methods have been developed. For example, webs are formed from long, virtually endless filaments directly from bulk polymer; both web and fibers are produced simultaneously.

Staple Fibers

The term staple fibers denotes fibers long enough to be handled by conventional spinning equipment. These fibers are between 1.2- and 20-cm long, and even longer, but not endless. They are formed into webs by dry-processing methods. The fiber is shipped to the manufacturer in the form of bales, which are opened mechanically by machines (pickers) equipped with sharp teeth or needles to tear the tightly compacted fibers. This process is called picking. The fibers must be crimped in order to be grasped by the needles. The fibers are transferred mechanically on belts or by chutes to form heavy fiber batts, called picker laps. The picker lap is processed by carding or air-laying. This process is derived from the ancient manual methods of fiber carding, where natural staple fibers were manipulated by beds of needles.

Carding. In carding, clumps of staple fibers from the picker lap are separated mechanically into individual fibers and formed into a coherent web (23). The carding machine utilizes opposed moving beds of closely spaced needles to pull and tease the clumps apart. At its center is a large, rotating metal cylinder covered with card clothing. The card clothing is comprised of needles, wires, or fine metallic teeth embedded in a heavy cloth or in a metal foundation. Opposing moving beds of needles are wrapped on the cast-iron cylinder and a large number of narrow, cast-iron flats (in a cotton-system card), which are held on an endless belt moving over the top of the cylinder.

The needles of the two opposing surfaces must be inclined in opposite directions and must move at different speeds. The main cylinder moves faster than the flats. The clumps between the two beds of needles are separated into individual fibers, which are aligned in the machine direction as each fiber is (theoretically) held at each end by individual needles from the two beds. The fibers engage each other randomly, and, with the help of their crimp, form a coherent web at and below the surface of the needles of the clothing on the main cylinder.

The mechanism is based on the massive main cylinder. In addition to the flats, the carding machine includes means to carry the lap or batting onto the cylinder where the carding takes place. Other mechanical means remove (doff) the web from the cylinder. The doffed web is deposited onto a moving belt where it is combined with other webs; most often, the next step is bonding. Most nonwoven materials, including lightweight products, are made of webs combined from several cards to increase uniformities.

The orientation ratio of fibers in the web at the doffer of a conventional cotton-system card is ca 3:1 in the machine direction. The web is highly sensitive to stretching or drafting; as little as ca 5% significantly increases web anisotropy, ultimately to 10:1 or more by further drafting. The physical properties of nonwoven fabrics are highly dependent on fiber orientation; they are stronger in the direction parallel to the predominant fiber direction. Even saturation with binder cannot overcome the effects of fiber orientation. Card webs of rayon fibers are oriented by drafting more readily than webs of synthetic fibers, eg, polyester.

Nonwoven materials made from webs from conventional cards have high machine direction and low cross-machine direction strengths. Fabrics made from card webs and bonded by adhesives can be torn readily along the machine direction, ie, parallel to fiber orientation, but are resistant across the fiber orientation. This behavior can be used to identify nonwoven material made from oriented card webs.

In the preparation of fiber assemblies for spinning into yarns, the fibers should be oriented parallel to the machine direction. For nonwoven fabrics, it usually is desirable that fibers not be highly oriented.

Cards may be as wide as 3.5 m to produce oriented webs at 140 m/min. Modifications have been developed by Textilmaschinenfabrik and Stahlbau in Linz, Austria, to randomize fibers by air-doffing at speeds of 50–75 m/min. In some carding machines, the randomization process is mechanical by web compaction in the machine direction after doffing at speeds of over 100 m/min, but randomization is not as effective as with air-doffing.

The problem of low cross-tensile strength can be solved by cross-laying. In this process, an oriented web is laid down at or near alternating 45° angles on another oriented web of a moving belt. Cross-laying is successful with heavyweight webs but not with lightweights, because it is difficult to place them accurately enough to avoid unsightly edge lines and because speeds are limited for mechanical reasons. Cross-laying is used for preparing webs for subsequent needle bonding. Garnett cards, or modified cotton cards, can be used in this process.

Air-laying. The orientation created by carding is effectively improved by capturing fibers on a screen from an airstream (24). The apparatus is called a Rando-Webber. Starting with a lap or plied card webs, the fibers are separated by teeth or needles and introduced into an airstream. Total randomization would exclude any preferential orientation when the fibers are collected on the screen. Actually, there is a slight orientation in the machine direction which is imparted to the web by the airstream and the forward motion of the screen. Fibers deposited from an airstream appear to be more curled with a shorter effective length than fibers in a card web. Drafting the web in handling orients the fibers to a degree but less so than in card webs.

The length of fibers used in air-laying varies from about 1.9 to 6.4 cm (25). The shorter lengths allow higher production speeds and better web uniformity. Long fibers require a higher air volume, ie, a lower fiber density in the airstream, to avoid tangling. High production speeds require high air velocity which disturbs the uniform air flow and reduces web quality. Problems associated with air-laying are speed, web uniformity, and weight limitations. Because of uniformity problems, it has not been practical to make isotropic webs lighter than ca 30 g/m^2 by an air-forming process. Air-laying processes are slower than carding and hence more expensive. Nevertheless, air-laying has advantages.

In the dry-mechanical processes, the tearing action breaks fibers, although damage varies from process to process, fiber to fiber, and even day to day on a single production line. Analysis of fiber-length distribution has been greatly simplified and can be used to determine fiber breakage (26,27).

The dual-rotor process, a dry mechanical method for forming a random web, has been developed by Johnson & Johnson (28). This air-laying process produces a web containing two different fibers, eg, wood pulp and staple rayon or polyester. The fibers are fed to oppositely rotating "licker-ins" which are rotated at speeds that are optimum for fiber individualization. A licker-in is a drum or cylinder in a carding machine that removes the lap from the feed rollers. The fibers from each supply are entrained in airstreams which are impelled at high speed toward each other; mixing of dissimilar fibers can be controlled.

Short Fibers

Wet-forming. Wet processes employ very short fibers, typically wood pulp and short synthetic or regenerated fibers. Initially, webs were formed from short fibers by modified papermaking techniques, whereby the fibers were continuously dispersed in a large volume of water and caught on a moving endless wire screen. Modifications for nonwoven materials include the inclined wire collector (C. H. Dexter Division) (29) and the Rotoformer (Sandy Hill Corp.) (30).

After the web is caught on the screen, it is transferred to belts or felts and dried on heated drums. In addition to wood pulp, rayon or synthetic fibers are included; these are cut to short lengths and dispersed in the water system. The presence of the other short fibers or the absence of hydrogen bonding between pulp fibers defines the product as nonwoven material rather than paper.

Large-diameter fibers are dispersible at lengths of up to ca 1.9 cm or more. Short-diameter fibers (1.7–3.3 dtex or 1.5–3 den) are employed in lengths of 6–12 mm. Longer fibers may not disperse completely and tend to entangle and form nonuniform sheets. The C. H. Dexter Division and the International Paper Co. have developed commercial processes where staple-length fibers (1.7 dtex or 1.5 den and >2-cm long) can be dispersed in water (31). These fibers in water must be free of crimp and have either no finish, in the case of rayon, or a hydrophilic finish, in the case of synthetic fibers. It might be expected that the wet process produces random webs. However, even at the speeds of nonwoven manufacture (moderate compared to paper), fibers tend to be oriented in the machine direction. The Rotoformer is designed to control this orientation. In addition to wood pulp and regenerated and synthetic fibers, cotton linters and manila hemp can be incorporated in the web.

Dry-forming. Webs containing predominantly wood-pulp fibers can be prepared by air-forming methods. The fibers are dispersed in air and caught on a screen to form webs. In addition to wood pulp, long, fine-diameter synthetic fibers can be included. This method is used by Kroyer and Danweb in Denmark and Honshu in Japan.

Wet-formed webs are flat and papery because of hydrogen bonding between the pulp fibers, the lack of crimp of nonpulp fibers, and the powerful hydraulic forces exerted on the web during the drying stages. Short-fiber webs made by air-laying are loftier and less paperlike; hydrogen bonding is not a significant factor.

Spunbonded Webs

In the most recent techniques, fibers and web are made simultaneously directly from bulk polymer. The bulk polymer is melted, extruded, and drawn (often by triboelectric forces) to filaments that are randomized and deposited onto belts as a continuous web. They are virtually endless. The weight of the web ranges from ca 10 g/m^2 or less to much heavier weights. Web uniformity varies among manufacturers and over the weight range. Common base polymers are polypropylene, polyethylene, polyester, and nylon, but others have been used experimentally.

The spunbond process produces webs of low crimp filaments in the normal diameter range of about 1.7 dtex (1.5 den) or slightly higher. The birefringence and uniformity of diameter of these filaments are similar to standard textile fibers and filaments.

Spunbonded webs are made by DuPont, Crown Zellerbach Corp., James River Corp. of Virginia, Kimberly-Clark Corp., and Monsanto Co. in the United States, and by Freudenberg and Lurgi in the FRG (see also SPUNBONDED in this article).

Despite the large capital requirements, the high capacity results in a low cost product. Each production line is suitable for a specific polymer and a single-bonding system. Only the weight of the final material can be altered (within narrow limits) (32).

Webs are also made directly from bulk polymers by the meltblown process (33). The molten polymer is forced through very fine holes in a special die into a high velocity airstream where the polymer is formed into very fine, although irregular, filaments of indeterminate lengths. The process appears to degrade the polymer. Polypropylene as well as polyethylene, nylon, and polyesters are used. The filaments are simultaneously formed into a web where melting and resolidification, and possibly static forces, hold them together. The web consists primarily of filaments with very fine diameters which can only be characterized under a scanning electron microscope; because they are nonuniform, they cannot be described in terms of decitex or denier.

Garment insulation (Thinsulate, 3M Co.) is made by combining meltblown filaments during the web-forming process with conventional fibers. Alternatively, a meltblown web is inserted between two lightweight, spunbonded polypropylene webs (Kimberly-Clark Co.) (see also NONWOVEN FABRICS, MELTBLOWN in the Supplement).

The orientation, hence the birefringence of meltblown fibers, is low but difficult to determine because of their extremely fine diameter (<1 μm). Meltblown webs are weak and are only used where strength is not required. They may be bonded with embossed patterns by heat or ultrasonic energy and pressure.

Bonding

The bonding of fibers gives the strength to the web and influences other properties (see also ADHESION AND BONDING). Both adhesive and mechanical means are used. Mechanical bonding employs the engagement of fibers by frictional forces.

Bonding can also be achieved by chemical reaction, ie, formation of covalent bonds between binder and fibers (34,35). The chemical bonding system was based on fiber-reactive dye technology. Some large-scale latex bonding of rayon is enhanced by excess cross-linking agent in the latex, eg, N-methylolacrylamide (36). In a laboratory experiment, a binder was applied to a web of cellulose fibers and formed covalent bonds (37).

Adhesive Bonding

Adhesive binder can be applied in many different ways, in numerous kinds and amounts, in uniform, nonuniform, random, or patterned distribution. The binder may be added in the form of a synthetic water-based emulsion or dispersion, ie, a latex; a solution; a foam; thermoplastics, including fibers and powders; a monomer or oligomer for polymerization *in situ;* plastisols; or polymers that react with the fibers. Any film-forming polymeric substance that can be dissolved or dispersed, or partially or fully fused, which subsequently solidifies while in contact with the fibers, may be used. In addition, special fibers can be thermally or chemically softened or fused to give them adhesive properties.

Thermal binders are included in the original web. Otherwise, the bonding substance is added to the formed compressed web. Nonwovens used for insulation are compressed very little or not at all, and thermoplastic binder fibers are incorporated in the web as it is formed. The binder is always a film-forming polymer. For the formation of a continuous chain of bridging throughout the web, each fiber must be bonded at least two or more times to different fibers. Unbonded fibers are likely to be lost during use. Fibers that are bonded only once do not contribute to tensile properties, but may enhance bulk, softness, and absorbency.

The mechanism of adhesive bonding is not clearly understood and may not be the same for different adhesive systems. It is often assumed that bonding is directly related to shear or peel adhesion, or a combination of the two. An alternative to adhesion is the mechanical entrapment of fibers within encompassing bonds. Corona-discharge treatment of polymer film in a reactive environment greatly enhances peel adhesion of latex to the film (38). The same treatment, applied to fiber webs of the same nominal chemical composition as the film and bonded by the same latex, does not improve strength. The peel adhesion of a latex on nylon-6,6 film is improved fivefold by modifying the ζ-potential or electrophoretic mobility of the latex. However, application of the modified latex to a web of nylon-6,6 fibers does not increase the strength of the nonwoven fabric (39).

Polyester film was pretreated with 1,4-diacetyl-2,3-butanedione, an adhesive, and coated with an acrylic monomer polymerized by electron-beam radiation. The pretreatment improved the peel adhesion of the polymer to the polyester film. In a similar experiment, a web of poly(ethylene terephthalate) fibers was pretreated with diacetyl-2,3-butanedione and bonded with monomer polymerized *in situ* by electron-beam radiation. In this case tensile properties were not improved by pretreatment (40). Nonwoven fabrics have been bonded experimentally by electron-beam polymerization of monomer or oligomer *in situ* (41,42), but results do not justify the high cost of the special monomers required and the technical difficulties, although power requirements are low.

However, there are contrary data indicating that a specific adhesion can be important. For example, a lightweight nonwoven fabric composed of viscose-rayon fibers was bonded with ca 2 wt % cellulose, applied as a viscose solution, and subsequently regenerated. Microscopic examination revealed no interface between binder and fibers. In contrast, rayon-fiber fabrics require 10–15 wt % of latex binders to achieve comparable properties.

Polyolefins adhere only to themselves, whereas thermoplastic nylons and polyesters adhere effectively to other polar and semipolar fibers.

An insoluble problem associated with adhesive bonding is that strength and softness cannot be combined. Strength may be increased only by the sacrifice of suppleness and vice versa.

Latex Bonding.   Until recently, the most widely used bonding agents have been synthetic latices (43). Current latex or water-dispersed binders are usually vinyl polymers that are emulsion polymerized by free radicals (44,45); polyacrylates, polyacetates, polychlorides, polyacrylonitriles, and copolymers are included (see also LATICES). They offer many advantages, including numerous varieties, ease of handling, and low cost. They are easily modified with plasticizers or tailor-made for special needs. When combined with webs of staple fibers formed by dry processes (carding or air-laying), they are suitable for small production runs. Dry-formed staple fiber can be easily bonded with latex on inexpensive equipment; changeover to other fibers, web weights, and latices is easy.

Most latex binders include a small amount of cross-linking agent, typically *N*-methylolacrylamide (46). The advent of the cross-linking latices at the end of the 1950s led to durable, washable, wet-abrasion-resistant nonwoven fabrics. The incorporation of a cross-linking monomer does not increase the modulus of the cured binder, but improves water and solvent resistance of both binder and nonwoven fabric.

Latices are effective binders for webs made from rayon and other hydrophilic fibers. They are not nearly as effective as binders for the hydrophobic fibers, ie, polyester, polypropylene, which have mostly replaced rayon. In the case of hydrophobic fibers, fibers bonded by the latices are not as strong as the expected fiber properties. Thus high tenacity synthetic fibers do not give high strength nonwovens. The presence in latices of residual monomers, surfactants, stabilizers, and formaldehyde-containing cross-linked agents is a disadvantage (46).

The amount of binder added to the web varies from 5 to 50 wt % (usually 10–35 wt %), depending on the desired fabric properties. The higher concentrations tend to obscure fiber properties. Higher concentrations are used in garment interlinings made of nylon fibers and characterized by high resilience.

Saturation Bonding.   When a fabric is saturated or impregnated with latex, the fiber web is immersed in the latex and the excess is removed. The web can be fed directly into the bath under an idler roll and through mangle rolls. An unbonded web is not strong enough for this simple process and requires support on a carrier or, more likely, is carried between two open-mesh screens. Saturation bonding with latex is conceptually the simplest way to make a nonwoven fabric. Variations in the amount of binder, distribution throughout the web, the method of drying and curing, and the degree of compacting the web provide a great diversity in products.

A small amount of binder imparts a textilelike flexibility, ie, hand or drape, to a nonwoven fabric. The web is saturated with soft binder with low film modulus or low second-order transition temperature (47–49). This treatment ties fibers and fiber ends, giving a thin, smooth product.

Print Bonding. Fabric hand is also improved by intermittent bonding, using a binder with a printed pattern; the resultant product has areas with and without binder. The bonded areas provide the strength, whereas the unbonded areas provide bulk, flexibility, and surface texture.

In the rotogravure method, the latex is applied in a pattern to a web with an engraved roll. This method is suitable for fabrics within the range of 15–100 g/m². Print bonding is readily applicable to card webs of hydrophilic fibers, eg, rayon, cotton, poly(vinyl alcohol), and blends. Rotogravure print bonding is not applied to staple-fiber webs made by the air-laying process.

Print-pattern Design. The earliest print-pattern design includes a series of parallel lines running transverse to the machine direction, which was the predominant fiber orientation (7). This pattern maximizes cross-strength, as though each transverse binder line acted as a continuous filling yarn. A variation of this simple line pattern, ie, a pattern of horizontal, parallel wavy lines, is used extensively for bonding card webs oriented in the machine direction.

Other designs were developed to increase total energy absorption, resilience, and web-abrasion resistance. Cross-directional toughness and resilience is improved by diagonal line patterns or oval island units (50,51). Web-abrasion resistance is improved by a miniature pattern (52). Latex binders are frequently pigmented, making the pattern visible.

In the rotogravure system, the latex binder is usually printed onto a wet web. The water-based latex penetrates and migrates (diffuses). Excess migration blurs the carefully designed pattern. Migration can be controlled by coagulating the latex in place as it contacts the wet web (53). Heat-sensitive surfactants prevent binder migration to the fabric faces during drying (54,55).

Spray Bonding. Binders, particularly latices, can be applied by spraying. The binders must be of low viscosity and mechanically stable in order to withstand the shear conditions at the spray nozzle; nonplugging nozzles and an airless spray are required. The spray is usually applied to a dry web that may be composed of short or long fibers, including wood pulp alone and mixtures with staple fibers. Light spray tends to penetrate only partly into the web and is used primarily where high loft is desired in the final product. It can be applied to one or both sides. In heavy applications the binder penetrates completely, impregnating the web. Spray bonding, especially when applied to dry-formed wood pulp, may be assisted by pressure embossing. This method is used most often for bonding air-formed pulp webs.

Foam Bonding. Binder may be applied to the web as a water–air foam. This method requires less water and saves energy in the drying process. A small amount of binder can be applied uniformly, and a bulky web structure is produced. Foam may be applied to one or both sides, depending on web weight. It can be distributed and forced to penetrate by a padder (56). Binder applied as foam that collapses during drying is called froth; with some formulations, the foam structure survives the drying process.

Thermal Bonding. Softening or fusing a thermoplastic by heat while in close contact with fibers is called thermal bonding. During cooling and solidification, the polymer binds the fibers to each other. The thermoplastic material may be in the form of fibers, rods (short, high tex filaments), powder, granules, netting, film, or particles in irregular shapes (8,56,57). Bonding with thermoplastics permits a great number of variations in chemical and adhesive properties.

Melting temperature and rheological properties allow variations in processing conditions (58).

At the present time, thermal bonding is the leading process in nonwoven technology in both volume and variety of products (Fig. 4). The bonding agent is usually a heat-fusible fiber incorporated in the web, but thermoplastic powder adhesives are also used. The earliest thermoplastic fiber binders were plasticized cellulose acetate and Vinyon HH (poly(vinyl chloride–vinyl acetate)). Plasticized cellulose acetate binder fibers have disappeared from the market. Poly(vinyl chloride–vinyl acetate) fibers are still used in special products, such as tea bags, where they function as heat-seal elements rather than binders.

1 mm

Fig. 4. Thermal bonding. Low melting copolymer polyester bonds in poly(ethylene terephthalate) web-heated without pressure. Courtesy of Chicopee.

Heat-fusible textile fibers were also used as thermoplastic binders, including polypropylene, nylon-6, and poly(ethylene terephthalate) (PET). However, fusing properties of these fibers are difficult to achieve (59).

A polypropylene fiber has been developed with slightly different characteristics and extrusion draw ratio from those of standard textile fiber, providing a wider range of processing temperatures. Molten polypropylene does not adhere well to other fibers and is therefore used in high concentration (60). Polypropylene–rayon mixtures are made by a process where webs containing thermoplastic fibers are passed between opposing, heated spiral engraved rolls (61). The resultant fabric is bonded by rows of diamond-shaped embossings, which correspond to the crossovers of the lines from the two rolls.

Polypropylene fibers modified only for bonding purposes are used at 100% concentration in lightweight staple-fiber and spunbond webs for disposable diapers; this represents the largest single use for any fusible fiber and the largest volume nonwoven fabric product manufactured.

Copolymer polyester fibers have been developed specifically for thermal bonding in the United States, UK, Japan, and Switzerland (62). They are available in different ranges of melting temperatures and with different melting characteristics. They vary from highly crystalline materials with sharp melting temperatures to noncrystalline polymers that melt over a broad temperature range similar to polypropylene. Crystalline polyesters are used in blends with nonheat-sensitive fibers; noncrystalline types are intended to be used without any other polymer. Polyester fibers as a class adhere better than polypropylene to other fibers such as rayon, wood pulp, or PET. Low melting copolymer nylon fibers have found limited use as binders in clothing interlining.

Bicomponent Fibers. An entirely new class are bicomponent fibers, consisting of a low melting and high melting polymer. The two polymers may exist side by side in each fiber, but more often the fiber core is high melting and the skin low melting. Combinations include a polypropylene core–polyethylene skin, polyester core–polyethylene skin, polyester core–polypropylene skin, and nylon-6,6 core–nylon-6 skin. The low melting skin is the bonding ingredient and is ordinarily present in amounts of 33–50 wt %.

Bicomponent fibers can be used alone or in webs with a substantial proportion of nonmelting fibers such as rayon or PET. The bonds may be activated by heating without pressure by ir in an oven or by forced-air bonding. They may be bonded by a hot calender with or without an embossing pattern. The bonding action takes place wherever the molten skin of one fiber touches another. With polyethylene, the binder tends to concentrate at the bonding intersections. This phenomenon occurs much less, for example, with nylon-6. Nonwoven fabrics made from bicomponent fibers with polyethylene skin and heated without pressure are soft, porous, and bulky, and not very strong. They are used as facings for absorbent products. With nylon-6 as the bonding skin, the web is heated with simultaneous application of pressure. These fabrics can be extremely strong and tough.

Spunbond and Meltblown Processes. Other thermal-bonding fibers are manufactured by spunbond and meltblown processes. The spunbond processes, which produce isotropic webs of filaments with textilelike diameters, are virtually always based on melt extrusion. For the Reemay (DuPont) line of products, a mixture of PET and low melting polyester (10–15 wt %) is extruded. It may be bonded by overall heating to activate the low melting fibers only, or by overall heating plus point embossing to melt all the low melting fibers and most of the PET fibers in a fine pattern. DuPont also manufactures another important product designated Tyvek by a flash-extrusion process. These fibers are 100% polyethylene with irregular diameters, bonded by heat and calender pressure. Tyvek is used in products as diverse as mailing envelopes and disposable garments. These articles resemble paper and are extremely strong (63). Kimberly-Clark, on the other hand, manufactures flexible, lightweight facings by the spunbond process for disposable diapers from polypropylene fibers that are bonded by hot embossing. Melt blowing produces heat-fusible filaments with much smaller diameters than those of typical textile (staple) fibers. Melt blowing is mostly employed with polyolefin fibers. Some fusion bonding occurs during the actual web formation, but not enough to impart the desired strength properties. Strength properties are enhanced by hot embossing; ultrasonic energy with pressure can be used to generate the heat that creates a bonding pattern in the web (64). This process is used on a large scale to make wiping cloths, for example. The fine capillary structure of meltblown webs is reflected in good absorbency properties or, when treated with repellent, in good moisture-barrier properties. Meltblown fabrics have many attractive properties, but suffer from low strength.

Solution Bonding.   Bonding with organic solutions of polymers confers good strength properties, especially at low binder content (65). The problems of solvent hazards, cost, and the difficulties of handling and applying viscous but low solids solutions have prevented the widespread use of this method. However, aqueous binder solutions have specialized uses, eg, poly(vinyl alcohol) cross-linked with formaldehyde (66). Cellulose xanthate dissolved in aqueous NaOH, so-called viscose, has been used on a small scale for many years on rayon and cotton-fiber webs; viscose bonding, however, is no longer in commercial use. The products were effectively bonded with much lower binder content than is customary for latex systems; binder efficiency was excellent and specific adhesion very high.

Mechanical Bonding

Mechanical-bonding methods are based on the frictional forces between fibers. The strength–softness interdependence characteristic of adhesive bonding is absent. Increasing the frictional engagement of fibers increases the tensile properties of the fabric without affecting softness.

Although wool felt is a nonwoven fabric held together by interfiber frictional forces, it is not considered here. The oldest friction-bonding process is needle punching (also called needle felting). It is carried out on a large scale and produces durable articles (67–70). A web is carried under banks of rapidly reciprocating barbed needles pointing down. The barbs penetrate through the web and entrap bundles of fibers forcing them downward through the web, forming numerous small loci of fiber entanglement. As the needles are withdrawn, the tangled bundles stay in place. The process is slow, but useful for heavy webs. Strong supple products, such a felts, carpet backings, and civil engineering textiles (geotextiles), are made by this method.

A soft, highly absorbent, all-cotton pad can be made by exposing a web of unbonded cotton fibers to a NaOH solution. A card web of short cotton fibers is carefully floated without tension on cold aqueous NaOH solution of ca 20%. Mercerization without tension is accompanied by curling and tangling of the fibers.

Spunlace Process.   Fibers can be rearranged by fluid forces. A vigorous spray of water, passing first through a patterned stencil, then through a staple-fiber web held loosely in place by a porous backing belt, or in the reverse direction, imparts a fine regular pattern to the web (71–73). The product is trademarked Keyback, and the process is termed fiber rearranging.

After thorough washing and drying, bulky, resilient material is strong enough for use as a specialty absorbent (74).

A textureless web can be given a novel appearance and resembles woven, knitted, or embroidered fabrics. Some fiber engagement takes place. The patents describe these products as self-sustaining. However, adhesive bonding is required. Webs made by fiber rearrangement and bonded with latices are used for light-weight facings and wiping cloths.

In 1969, DuPont patented a process where fibers of a web on a supporting permeable backing are entangled by the application of multiple banks of closely spaced, fine columnar water jets at very high pressures (75). As the jets penetrate the web, they pull some fibers through. Some of the water deflects off the perme-able backing and splashes back into the web with considerable force. Fiber seg-ments are carried by the turbulent fluid and become entangled. The frictional forces resulting from the entanglement are the basis for mechanical bonding. The products are called spunlace (Fig. 5). Their manufacture requires precision equipment and presents engineering problems; production is highly energy- and capital-intensive. The products have excellent strength and softness properties, and are used for surgical and cleaning applications. They can be made from staple fibers, spunbonded webs, and common fibers. Most spunlaced products are made from rayon, polyester, or their mixtures. They can be combined by entanglement with wood pulp and made with a textured or smooth, homogenous appearance, depending on the geometry of the backing. Although tensile strengths are high, stretch recovery and bias are poor (76). Spunlaced products are manufactured by DuPont and Chicopee in the United States, Suominen in Finland, and Unicharm in Japan.

In the Unicharm process, the web is run over smooth solid rolls while under

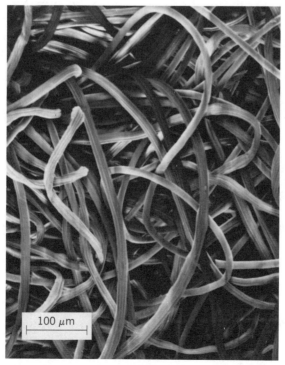

Fig. 5.   Spunlaced fabric, blend of rayon and polyester fibers, mechanically bonded by entanglement. Courtesy of Chicopee.

the impact of the water jets. It is claimed that this latter process requires less energy for entanglement.

Reinforcements and Composites

Reinforcements and composites are expected to play an important role in nonwoven fabrics (77). Reinforcements were utilized in yarn scrims (Kimberly-Clark) and plastic netting (Chicopee) within an adhesively bonded base web. The reinforcements increase the tensile force-to-break, but do not have the same stress–strain properties as the nonwoven web. Under stress, the reinforcement tended to separate. A superior fabric is manufactured by DuPont and Chicopee for use in hospital operating rooms. It consists of a web of staple polyester fibers entangled with a layer of wood pulp fibers and is finished with a fluorocarbon-containing repellent formulation. The polyester component provides the product strength, whereas the wood pulp forms an almost continuous layer with liquid barrier properties. Another composite fabric made by Chicopee is used as a sanitary absorbent. It contains an absorbent core of wood pulp and thermally sensitive bicomponent fibers, and a facing of staple bicomponent fibers activated by heat. The facing is heat- or sonic-sealed into the absorbent core.

Stitchbonded Fabrics. Laboratories supported by the governments of the GDR and Czechoslovakia have developed a group of nonwoven fabrics made by extensive overstitching or knitting of yarns into fiber webs (78). The products resemble soft woven or knitted textiles, and are used for upholstery and drapery fabrics. They are also made in small quantities in the United States. These fabrics meet the requirements of both reinforced composite fabrics and friction-bonded fabrics. Trade names include Arachne, Malivlies, Schusspol, Maliwatt, Malimo, Malipol, and Voltex.

Manufacturing Process

The following technologies are used for the manufacture of nonwoven fabrics (79):

Dry-formed, carded, and bonded. The webs are formed from staple fibers by carding or air-laying, bonded overall or in a pattern with latex or other water-borne adhesives.

Thermal-bonded. Dry-formed webs of staple fibers are bonded with fusible fibers or composed entirely of fusible fibers.

Air-laid. Wood-pulp fibers, with or without added staple fibers, are bonded with latex or similar adhesives.

Wet-formed. Short fibers, wood pulp, and short regenerated or synthetic fibers are formed into a web by processes derived from papermaking technology, followed by bonding with latex or thermal binders.

Spunbonded. Webs composed of long filaments with normal textile diameter are formed directly from bulk polymer and are usually bonded thermally.

Meltblown (microfiber). Webs of long, extremely fine-diameter fibers are formed directly from bulk polymer and are usually bonded by hot embossing processes.

Spunlaced. Dry-formed webs, usually from staple textile fibers, are mechanically entangled by multiple fine, high pressure water jets, in most cases without adhesive binder.

Needlepunched. Fibers are mechanically entangled by multiple reciprocating banks of barbed needles.

Laminated. Different layers are combined in composite or reinforced fabric by an adhesive, thermal fusion, or entanglement.

Stitchbonded. Staple-fiber webs are reinforced or mechanically entrapped by yarns stitched or knitted through the webs.

The early nonwoven materials were based on staple-fiber webs. At present, new technologies are being developed based on the direct production of fiber webs from bulk polymer (spunbonded and meltblown) or on thermal-bonding, spunlace, and lamination–reinforcement–composite processes. Nonwoven fabrics are manufactured in two steps. A web is formed from fiber or directly from polymer and bonded by chemical adhesion or mechanical friction. The web may be rearranged (on a fiber-to-fiber scale) to give a textilelike appearance. During or after the bonding, the fabric may be laminated or combined with other layers. The product may be finished for special applications, eg, impregnated with a fire retardant, a water or solvent repellent, or a liquid cleaning formulation. The equipment for the manufacturing processes is often elaborate and complicated.

The modern nonwoven industry was started after World War II and has become a significant part of the textile industry with increasing influence on the quality of daily life. The convenience of disposable diapers and sanitary protection, the medical safety of patients as a result of advanced surgical dressings, surgeons' gowns, and other hospital materials are made possible by nonwoven fabrics. Other new kinds of textiles have been created from nonwoven materials, including engineering fabrics, floppy disk liners, battery separators, upholstery fabrics, carpet backings, disposable garments, wall coverings, tea bags, cleaning materials, interlinings, and many others.

Economic Aspects and Uses

Nonwoven fabrics can be classified into disposables and durables. The former are used once and are exemplified by facings for disposable diapers, sanitary napkins, surgical supplies, and other medical applications; absorbency or repellency is the main characteristic (80). Semidurable nonwovens include wiping cloths, which require wet-abrasion characteristics and can be laundered several times. Disposable nonwovens have replaced lightweight nonwovens, knits, or paper.

Durable nonwoven fabrics include garment interlinings, fabrics used in automobiles, substrates for vinyl-coated upholstery and luggage, home furnishings such as mattress covers and draperies (stitch-through technology), geotextiles (civil engineering fabrics), floppy disk liners, indoor and outdoor carpets (needlepunched), wall coverings, garment insulation and fiber fill, architectural roofing fabrics, and suedelike composite structures.

The United States is the largest producer and consumer of nonwoven fabrics in the world, accounting for approximately 60% of world production; sales figures are given in Table 1. In 1984, production of nonwovens in the United States was

Table 1. Economic Data of the U.S. Nonwoven Industry[a], $10^{6 b}$

Category	1974	1984	Growth to 1990[c]
Disposable[d]			
diaper cover stock	75	215	360
sanitary napkins, underpads, adult diapers	35	65	
roll goods, surgical gowns, packs	33	160	285
wiping cloths, roll towels	32	145	285
Durable[d]			
interlinings	65	115	
coated fabrics	35	150	
geotextile and roofing		100	245
Total disposables	*230*	*870*	*1610*
Total durable	*355*	*690*	*1310*
Products incorporating nonwovens			
disposable baby diapers	560	2700	
sanitary napkins		975	
fabricated medical products	175	845	1630
industrial–institutional disposables	90	600	1300
Process			
staple-fiber dry-formed webs	300	300	29[e]
spunbonded	140	415	32[e]
wet-laid	70	180	11[e]
spunlaced		100	11[e]
meltblown		75	8[e]
air-laid pulp		55	5[e]
Raw material consumption			
staple fibers		440	570
polyester		150	
polypropylene		120	
rayon		120	
polymer for spunbond–meltblown		315	575

[a] From Refs. 78 and 81.
[b] Unless otherwise stated.
[c] Estimated.
[d] Not inclusive.
[e] Percentage of total.

estimated at more than 500×10^3 t. Western European production was estimated at ca 270×10^3 t. Production in Japan is less than 90×10^3 t. Nonwoven fabrics are also made in Central and South America, Australia, India, South Africa, Israel, and the Eastern-bloc countries.

BIBLIOGRAPHY

"Nonwoven Fabrics" in *EPST* 1st ed., Vol. 9, pp. 345–355, by Norbert M. Bikales, Consultant.

1. P. Coppin and co-workers, *The Definition of Nonwovens Discussed, European Disposables and Nonwovens Association Symposium,* Gothenberg, Sweden, June 6–7, 1974.
2. *Guide to Nonwoven Fabrics, INDA,* The Association of the Nonwoven Fabrics Industry, New York, 1978.
3. *ASTM D123-70, Terminology Relating to Textile Materials,* American Society for Testing and Materials, Philadelphia, 1970.

4. S. K. Batra, S. P. Hersh, R. L. Barker, D. R. Buchanan, B. S. Grysta, T. W. George, and M. H. Mohamed, *Nonwovens Ind.* **16**(9), 28 (Sept. 1985).
5. W. D. Freeston, Jr. and M. M. Platt, *Text. Res. J.* **35,** 48 (1965).
6. S. Backer and D. R. Petterson, *Text. Res. J.* **30,** 704 (1960).
7. U.S. Pat. 2,039,312 (May 5, 1936), J. H. Goldman.
8. U.S. Pat. 2,277,049 (Mar. 24, 1942), R. Reed (to The Kendall Co.).
9. E. R. Schwarz, K. R. Fox, and N. V. Wiley in H. R. Mauersberger, ed., *Matthews Textile Fibers,* 5th ed., John Wiley & Sons, Inc., New York, 1947, Chapt. XXIV.
10. A. R. Winch, *Eighth INDA Technical Symposium,* Orlando, Fla., March 19–21, 1980, p. 224.
11. A. R. Winch, *Text. Ref. J.* **50,** 64 (1980).
12. R. Remirez, *Chem. Eng.* **86**(7), 113 (Mar. 26, 1979).
13. *Chem. Week* **124**(23), 27 (June 6, 1979).
14. *Nonwovens Ind.* **15**(11), 42 (Nov. 1984).
15. A. F. Turbak and co-workers, *ACS Symp. Ser.* **58**, 40, 71 (1977).
16. D. Harrison, *Nonwovens Ind.* **17**(4), 48 (Apr. 1986).
17. U.S. Pat. 3,930,086 (Dec. 30, 1974), C. Harmon (to Johnson & Johnson).
18. A. N. J. Heyn, *Fiber Microscopy, A Textbook and Laboratory Manual,* Interscience Publishers, New York, 1954, pp. 162 and 164.
19. L. McMeekin, Chicopee, New Brunswick, N.J., June 1986, personal communication.
20. R. F. Nickerson in Ref. 9, Chapt. V, p. 178.
21. G. G. Allen and L. A. Smith, *Cellul. Chem. Technol.* **2,** 80 (1968).
22. M. J. Welch and J. A. McCombes in Ref. 10, p. 3.
23. G. R. Merrill and co-workers, *American Cotton Handbook,* 2nd ed., Textile Book Publishers, Inc., New York, 1949, Chapt. 7.
24. F. M. Buresh, *Nonwoven Fabrics,* Reinhold Publishing Co., New York, 1962.
25. G. B. Harvey, *Formed Fabric Ind.* **6**(9), 10 (1975).
26. W. L. Balls, *A Method of Measuring the Length of Cotton Hairs,* London, 1921, as cited in Ref. 9, p. 1094.
27. *Evaluation of the Motion Control Text System,* USDA Agricultural Marketing Service, Memphis, Tenn., 1974.
28. U.S. Pat. 3,740,797 (June 26, 1973), A. P. Farrington (to Johnson & Johnson).
29. U.S. Pats. 2,045,095 and 2,045,096 (June 23, 1936), F. Osborne (to C. H. Dexter Co.).
30. U.S. Pat. 2,781,699 (Feb. 19, 1957), W. J. Joslyn (to Sandy Hill Iron and Brass Works).
31. L. Kinn, *Process for Producing Long Fiber Wet-Laid Nonwovens, Fourth Annual Nonwovens Workshop,* University of Tennessee, Knoxville, Tenn., Aug. 1986.
32. P. J. Stevenson, Chicopee, New Brunswick, N.J., Sept. 1986, personal communication.
33. T. Holliday, *Nonwovens Ind.* **16**(5), 6 (May 1985).
34. G. G. Allen, G. Bullick, and A. N. Neogi, *J. Polym Sci.* **11,** 1759 (1973).
35. M. L. Miller, *Ionic Bonding in Rayon Nonwovens,* Ph.D. dissertation, University of Washington, Seattle, Wash., 1972.
36. W. F. Schlauch, *Recent Developments in Nonwoven Binder Technology Nonwoven Fabrics Forum,* Clemson University, Clemson, S.C., 1979.
37. G. G. Allan and T. Mattila, *Tappi* **53**, 1458 (1970).
38. U.S. Pat. 3,661,735 (May 9, 1972), A. Drelich (to Johnson & Johnson).
39. U.S. Pat. 3,639,327 (Feb. 1, 1972), A. Drelich and P. Condon (to Johnson & Johnson).
40. W. K. Walsh, *Radiation Processing of Textiles, Symposium,* North Carolina State University, Raleigh, N.C., May 26–27, 1976.
41. C. Houng, E. Bittencourt, J. Ennia, and W. K. Walsh, *Tappi* **59,** 98 (1976).
42. U.S. Pat. 4,146,417 (Mar. 27, 1979), A. Drelich and D. Oney (to Johnson & Johnson).
43. C. H. Kline & Co., *INDA Newsletter* **79,** 3 (Dec. 1979).
44. K. R. Barton, *Nonwovens Ind.* **10**(5), 28 (May 1979).
45. R. Nelson, *Extractables-Free Olefin Binder for Nonwovens, INDA, 11th Technical Symposium,* Baltimore, Md., Nov. 1983.
46. U.S. Pats. 4,356,229 (Oct. 26, 1982), W. DeWitt and co-workers, and 4,529,465 (July 16, 1985), W. DeWitt and R. Gill (to Rohm & Haas Co.).
47. J. W. S. Hearle and P. J. Stevenson, *Text. Res. J.* **34,** 181 (1964).
48. J. W. S. Hearle, R. I. C. Michie, and P. J. Stevenson, *Tex. Res. J.* **34,** 275 (1964).
49. R. I. C. Michie, *Text. Res. J.* **36,** 501 (1966).

50. U.S. Pat. 2,705,687 (Apr. 5, 1955), D. R. Petterson and I. S. Ness (to Chicopee Manufacturing Corp.).
51. U.S. Pat. 3,009,823 (Nov. 21, 1961), A. Drelich and V. T. Kao (to Chicopee Manufacturing Corp.).
52. U.S. Pat. 2,880,111 (Mar. 31, 1959), A. Drelich and H. W. Griswold (to Chicopee Manufacturing Corp.).
53. U.S. Pat. 4,084,033 (Apr. 11, 1978), A. Drelich (to Johnson & Johnson).
54. U.S. Pat. 3,944,690 (Apr. 6, 1976), D. Distler and co-workers (to Badische Anilin-und Soda Fabrik Aktiengesellschaft).
55. U.S. Pat. 4,119,600 (Oct. 10, 1978), R. Bakule (to Rohm & Haas Co.).
56. *Textile World,* 97 (Apr. 1985), Fleissner GmbH advertisement.
57. U.S. Pats. 2,880,112 and 2,880,113 (Mar. 31, 1959), A. Drelich (to Chicopee Manufacturing Corp.).
58. A. Drelich, *J. Coated Fabr.* **15,** 154 (Jan. 1986).
59. R. P. Moffett, *Mod. Text. Mag.* **37,** 62 (1956).
60. R. G. Mansfield, *Nonwovens Ind.* **16**(3), 26 (Mar. 1984).
61. U.S. Pat. 2,507,943 (Apr. 21, 1970), J. J. Such and A. R. Olson (to The Kendall Co.).
62. U. Wild, *Text. Asia* **XV**(7), 66 (1984).
63. U.S. Pat. 3,081,519 (Mar. 19, 1963), H. Blades and J. R. White (to E. I. du Pont de Nemours & Co., Inc.).
64. G. Flood, *Nonwovens Ind.* **17**(4), 30 (Apr. 1986).
65. A. Drelich and P. N. Britton, *Physical Combinations of Cellulose Fibers and Polymers, 149th Meeting of the American Chemical Society,* Detroit, Mich., Apr. 1965.
66. U.S. Pat. 3,253,715 (May 31, 1966), E. V. Painter and co-workers (to Johnson & Johnson).
67. P. Lennox-Kerr, ed., *Needle Felted Fabrics,* The Textile Trade Press, Manchester, UK, 1972.
68. J. H. Foster, *Nonwovens Inds.* **16**(3), 12 (Mar. 1985).
69. T. Holliday, *Nonwovens Ind.* **16**(3), 6 (Mar. 1985).
70. K. Maitre, *Nonwovens Ind.* **17**(3), 16 (Mar. 1986).
71. U.S. Pat. 3,081,514 (Mar. 19, 1963), H. Griswold (to Johnson & Johnson).
72. U.S. Pat. 2,862,251 (Dec. 2, 1958), F. Kalwaites (to Chicopee Manufacturing Corp.).
73. U.S. Pat. 3,033,721 (May 8, 1962), F. Kalwaites (to Chicopee Manufacturing Corp.).
74. U.S. Pat. 2,625,733 (Jan. 20, 1953), H. Secrist (to The Kendall Co.).
75. U.S. Pat. 3,485,706 (Dec. 23, 1969), F. J. Evans (to E. I. du Pont de Nemours & Co., Inc.).
76. M. M. Johns and L. A. Auspos, *The Measurement of the Resistance to Distanglement of Spunlaced Fabrics, Seventh INDA Technical Symposium,* New Orleans, La., Mar. 1979.
77. D. Zafiroglu, *Nonwovens Ind.* **16**(7), 20 (July 1985).
78. J. D. Singelyn, *Principles of Stitch Through Technology, Nonwoven Fabrics Forum,* Clemson University, Clemson, S.C., 1978.
79. D. K. Smith, *Nonwovens Ind.* **16**(5), 111 (May 1985).
80. D. K. Smith, "The Role of Nonwoven Materials" in P. K. Chatterjee, *Absorbency,* Elsevier, Amsterdam, The Netherlands, 1985, Chapt. IX.
81. J. Starr, *Nonwovens Ind.* **15**(12), 18 (Dec. 1984); **16**(12), 26 (Dec. 1985).

General References

F. M. Buresh, *Nonwoven Fabrics,* Reinhold Publishing Co., New York, 1962.
R. Krcma, *Manual of Nonwovens,* The Textile Trade Press, Manchester, UK, 1971.
INDA Technical Symposia, Association of the Nonwoven Fabrics Industry, New York, 1973 to present.
Annual Symposium Papers, Clemson Nonwoven Fabrics Forum, Clemson University, Clemson, S.C., 1969 to present.
Nonwovens Workshops, University of Tennessee, Knoxville, Tenn., 1983 to present.
J. Lunenschloss and W. Albrecht, *Nonwoven Bonded Fabrics,* Halsted Press, a division of John Wiley & Sons, Inc., New York, 1985.
A. T. Purdy, *Text. Prog.* **12**(4), (1983).

ARTHUR DRELICH
Chicopee

SPUNBONDED

In 1984 the production of spunbonded fabrics in the United States reached a record 135×10^3 t or approximately 25% of the total nonwoven fabrics produced (1,2). Spunbonded fabrics are distinguished from other nonwoven fabrics by a one-step manufacturing procedure, which provides a complete chemical-to-fabric route in some instances and polymer-to-fabric in others. In either instance the process integrates the spinning, laydown, consolidation, and bonding of continuous filaments to form a fabric. Commercialization dates to the early 1960s in the United States and Western Europe (3,4) and to the early 1970s in Japan (5,6) (Table 1). Many of the first plants constructed are still in operation, attesting to the usefulness of the method. New production plants are being built (3,7) to supply the growing needs of the market.

The large investment required for a spunbonded plant (20×10^6–120 $\times$ 10^6, 1986) (8–10) is offset by high productivity. Although current spunbonded production is limited to Western Europe, the United States, and Japan, considerable activity is anticipated in other areas within the next decade.

Early marketing efforts centered on the substitution of woven textile fabrics by spunbonded fabrics. This was achieved where only functionality was important. Where textilelike appearance is required, progress has been slow. Nevertheless, spunbonded fabrics are recognized as a unique class of nonwoven fabrics (11).

The area of largest growth has been disposable diaper coverstock, which accounts for approximately 50% of the U.S. coverstock market (2). Forecasts for the future growth of spunbonded fabrics continue to be favorable as consumption in both durable and disposable areas continues to grow. Growth is forecast to exceed the growth of all nonwovens, which are expected to grow at 7% per year (1). Additional growth is anticipated in geotextiles, roofing, carpet backing, and medical and durable-paper applications (2,12,13).

Fabric Characteristics

Structure.   Spunbonded fabrics are filament sheets made by an integrated process of spinning, attenuation, deposition, bonding, and winding into rolls. The fabrics are up to 5.2-m wide and usually not less than 3.0 m in order to facilitate productivity. Fiber sizes range from 0.1 to 50 dtex (0.09–45 den), although a range of 1.5–20 dtex (1.36–18 den) is most common. A combination of thickness, fiber denier, and number of fibers per unit area determines the fabric basis weight or weight per unit area, which ranges from 10 to 800 g/m^2, typically 17–180 g/m^2.

Most spunbonding processes yield a sheet with planar-isotropic properties owing to the random laydown of the fibers. Unlike woven fabrics, spunbonded sheets are nondirectional and can be cut and used without concern for stretching in the bias direction or unraveling at the edges. Nonisotropic properties are obtained by controlling the orientation of the fibers during the preparation of the web. Although it is not readily apparent, most sheets are layered or shingled structures; the number of layers increase with increasing basis weights for

Table 1. Manufacturers of Spunbonded Materials

Company	Location	Polymer	Trademark
United States			
Chemie Linz	Alabama[a]	polypropylene[b]	Polyfelt
Intertech Group	Tennessee	polypropylene	Typar
(ex-DuPont)	Tennessee	polyester	Reemay
DuPont	Virginia	polyethylene	Tyvek
Hoechst	South Carolina	polyester[b]	Trevira
Kimberly-Clark	Wisconsin	polypropylene	Evolution, Kimguard,
	Mississippi	polypropylene	Spunguard
	North Carolina	polypropylene	
	Georgia	polypropylene	
Lutravil	North Carolina	polyester	Lutradur
James River			
(ex-Crown-	Washington	polypropylene	Celestra
Zellerbach,			
ex-Monsanto)	Florida	nylon-6,6	Cerex
Wayne-Tex	Virginia[a]	polypropylene	
Western Europe			
Akzo	The Netherlands	polyester–nylon	Colback
BPB Industries	FRG	polypropylene	Corovin
(ex-Benecke)			
Chemie Linz	Austria	polypropylene[b]	Polyfelt
DuPont	Luxembourg	polypropylene	Typar
	Luxembourg[a]	polyethylene	Tyvek
Freudenberg	FRG	polyester	Lutradur
		polypropylene	Lutrasil
		nylon	Lutrabond
Holmens-Bruk	Sweden	polypropylene	
Hoechst	FRG	polyester[b]	Trevira
ICI	UK	polypropylene–polyethylene	Terram
Kimberly-Clark	UK	polypropylene	
Rhone-Poulenc	France	polyester[b]	Bidim
	Italy	polypropylene	
	France[a]	polyester	Tridim
Sodoca	France	polypropylene	Sodospun
Japan			
Asahi	Moriyama	nylon–polyester	Asahi Kasei
	Nobeoka	cupra	Bemliese
Futumara	Ohgaki	rayon	Taiko TCF
Kanebo	Hofu	polyurethane	Espansione
Mitsui	Yokkaichi	polypropylene	Tafnel, Syntex
Teijin	Iwakuni	polyester	Unisel
Toray	Shiga	polyester	Axter
Toyobo	Tsuruga	polyester	Toyobo, Volans
Unitika	Okazaki	polyester	Niace, Marix, Appeal
Other areas			
Kimberly-Clark	Australia	polypropylene	
Yuhan-Kimberly	S. Korea	polypropylene	

[a] Under construction.
[b] Needlepunched.

a given product. Fabric thickness ranges between 0.1 and 4.0 mm, typically 0.2–1.5 mm.

The method of bonding greatly affects the thickness of the sheet and other characteristics. Fiber webs bonded by calendering are thinner than needle-punched webs since calendering compresses the structure through pressure, whereas needlepunching moves fibers from the xy plane of the fabric into the z-direction (thickness).

The structure of traditional woven and knit fabrics permits the fibers to move readily within the fabric when in-plane shear forces are applied, resulting in a fabric that readily conforms in three dimensions. Calender bonding of a spun web, however, causes some of the fibers to fuse, imparting integrity to the sheet. The resulting structure has a stiffer hand or drape than traditional textile fabrics. This is due to the immobilization of the fibers in the areas of fiber-to-fiber fusion. The effect may be moderated by limiting the bonds to very small areas (points) or by entangling the fibers mechanically or hydraulically. Saturation bonding of spun webs with chemical binders such as acrylic emulsions can bond the structure throughout and give stiff sheets. This technique provides dimensional stability to certain structures, whereby the emulsion binder functions as a nonthermo-plastic component within the thermoplastic matrix.

More recent techniques include powder bonding, although this method may be more suitable for bonding nonwoven fabrics made from staple fibers (14,15).

Composition.   The method of fabric manufacture determines the sheet characteristics, whereas the polymer determines the intrinsic properties. Prop-erties such as fiber density, temperature resistance, chemical and light stability, ease of coloration, surface energies, and others are a function of the base polymer. Thus since nylon absorbs more moisture than polypropylene, spunbonded fabrics made from nylon are more water-absorbent than polypropylene fabrics.

Most spunbonded fabrics are based on isotactic polypropylene and polyester (Table 1). Small quantities are made from nylon-6,6, and increasing amounts from high density polyethylene (HDPE). The properties of fibers made from dif-ferent polymers are given in Table 2. Linear low density polyethylene (LLDPE) is increasingly used as base polymer because it gives a softer fabric (16).

Table 2. Properties of Fibers

Polymer	Breaking tenacity, N/tex[a]	Elongation, %	Specific gravity	Moisture regain[b], % at 21°C, 65% rh	Melting range, °C
polyester	0.17–0.84	12–150	1.38	0.4	248–260
nylon-6,6	0.26–0.88	15–70	1.14	4.0	248–260
polypropylene	0.22–0.48	20–100	0.91	~0.0	162–171

[a] To convert N/tex to gf/den, multiply by 11.3.
[b] Equilibrium water content.

Polypropylene.   Isotactic polypropylene is the most widely used polymer for spunbonded production. It provides the highest yield (fiber per kilogram) and covering power at the lowest cost because of its low density, which is ca 70% of the density of most polyesters. Considerable advances have been made in the manufacture of polypropylene resins and additives since the first spunbonded polypropylene fabrics were commercialized in the 1960s. Although unstabilized polypropylene is rapidly degraded by uv light, improved stabilizers permit several years of outdoor exposure before fiber properties deteriorate (17,18).

Polypropylene fibers cannot be dyed by conventional methods nor readily stained, since there are no dye-receptor sites along the molecular backbone. Some spunbonded polypropylene fabrics are colored by the addition of a pigment to the polymer melt, whereby the pigment is encased within the fiber interior. These fabrics exhibit good resistance to fading and bleeding, and the color shades are easily reproduced from lot to lot. However, large quantities of off-quality material are produced when changing the color. Delustering pigments (eg, TiO_2) are sometimes added to polypropylene.

Scrap or polypropylene fibers of inferior quality may be repelletized and then blended in small amounts with fresh polymer to produce first-grade spunbonded fabrics. This reduces cost, which is important in a highly competitive industry (see also OLEFIN FIBERS).

Polyester. Polyester is employed in a number of commercial spunbonded products and offers advantages over polypropylene, although it is more expensive. Unlike polypropylene, polyester scrap is not readily recycled in spunbonded manufacturing. Tensile strength, modulus, and heat stability of polyester fabrics are superior to those of polypropylene fabrics. Polyester fabrics are easily dyed and printed with conventional equipment (see POLYESTERS, FIBERS).

Nylon. Spunbonded fabrics are made from both nylon-6 and nylon-6,6. Nylon is highly energy-intensive and therefore more expensive than polyester or polypropylene. Nylon-6,6 spunbonded fabrics are produced at weights as low as 10 g/m^2 with excellent cover and strength. Unlike olefins and polyester fabrics, those made from nylon readily absorb water through hydrogen bonding between the amide group and water molecules (see POLYAMIDES).

Polyethylene. The properties of polyethylene fibers meltspun by traditional methods are inferior to those of polypropylene fibers. Advances in polyethylene technology may lead to the commercialization of spunbonded structures with characteristics not attainable with polypropylene. A fiber-grade polyethylene was announced in late 1986 (16).

Polymer Combinations. Some fabrics are composed of several polymers. A lower melting polymer functions as the binder, which may be a separate fiber interspersed with higher melting fibers (19), or the two polymers may be combined in one fiber type (20). In the latter case the so-called bicomponent fibers possess a lower melting component, which acts as a sheath covering a higher melting core (Fig. 1). Bicomponent fibers are also spun by extrusion of two adjacent polymers. Polyethylene, nylon-6, and polyesters modified by isophthalic acid are used as bicomponent (lower melting) elements.

Polyurethane. A new type of structure was announced in Japan with the commercialization of spunbonded fabrics based on thermoplastic urethanes (21). Although spunbonded urethane fabrics have been previously described (22), this represents the first commercial production of such fabrics. Unique properties are claimed for this product which appears to be well suited for apparel and other

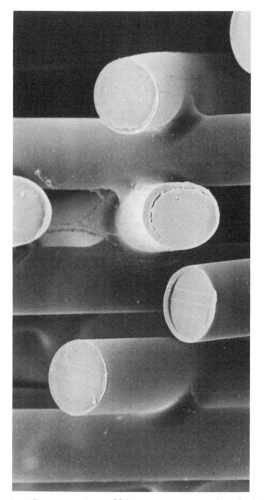

Fig. 1.   Cross section of bicomponent spunbonded fibers.

applications requiring stretch and recovery (see FIBERS, ELASTOMERIC; POLY-URETHANES).

Properties.   Spunbonded fabrics are characterized by tensile, tear, and burst strength, elongation to break, weight, thickness, porosity, and stability to heat and chemicals. These properties reflect fabric composition and structure. Comparison of generic stress–strain curves of thermally bonded and needlepunch-bonded fabrics shows that the shape of the curves is a function of the freedom of the filaments to move when the fabric is placed under stress (Fig. 2).

Some applications require special tests for sunlight, oxidation, burning resistance, moisture vapor and liquid transport, coefficient of friction, seam strength, and aesthetic properties. Most properties can be determined with standardized test procedures (nonwoven standards, International Nonwovens and Disposable Association (INDA)) (23). Typical physical properties are given in Table 3.

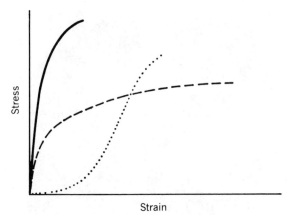

Fig. 2. Typical stress–strain curves of nonwoven fabrics. Solid = woven; dashed = thermally bonded nonwoven; and dotted = needlepunched nonwoven.

Spinning and Web Formation

Spunbonding combines fiber spinning with web formation. In the traditional methods the fiber is first spun and collected, and then converted into a fabric by a separate process. In spunbonding the bonding device is placed in line with spinning (Fig. 3). In some arrangements the web is bonded in a separate step which first appears to be less efficient. However, this arrangement is more flexible if more than one type of bonding is required for the same web.

The spinning process is similar to the production of continuous-filament yarns and utilizes similar extruder conditions for a given polymer (24). Fibers are formed as the molten polymer exits the spinnerets and is quenched by cool air. The objective of the process is to produce a wide web (eg, 3 m), and, therefore, many spinnerets are placed side by side to generate sufficient fibers across the width. The grouping of spinnerets is often called a block or bank. In commercial production two or more blocks are used in tandem in order to increase the coverage of fibers.

Before deposition on a moving belt or screen, the output of a spinneret, usually a hundred or more individual filaments, must be attenuated to orient the molecular chains of the fibers in order to increase fiber strength and decrease extensibility. This is accomplished by rapidly hauling off the plastic fibers immediately after exiting the spinneret. In practice the fibers are accelerated mechanically (25) or pneumatically (24,26,27). In most processes the fibers are pneumatically accelerated in multiple filament bundles; however, other arrangements have been described wherein a linearly aligned row (or rows) or individual filaments is pneumatically accelerated (28,29).

In traditional textile spinning some orientation of fibers is achieved by winding the filaments at a rate of ca 3200 m/min to produce partially oriented yarns (POYs) (30). The POYs can be mechanically drawn in a separate step for maximum strength. In spunbonded production filament bundles are partially oriented by pneumatic acceleration speeds of 6000 m/min or higher (27,31). Such high speeds result in partial orientation and high rates of web formation, par-

Table 3. Typical Physical Properties of Spunbonded Products

Product	Basis weight, g/m²	Thickness, mm	Tensile strength[a], N[b]	Tear strength, N[b]	Mullen burst, kPa[c]	Bonding method
Accord	69		144 MD 175 XD	36 MD 41 XD	324	point thermal
Bidim	150		495	279	1550	needlepunch
Cerex	34	0.14	182 MD[d] 116 XD	40 MD[e] 32 XD	240	chemically induced area
Colback	100	0.6	300/5 cm[f]	120[g]		area thermal (sheath core)
Corovin	75		130[f]	15[h]		point thermal
Lutradur	84	0.44	275[d] 297 XD	86[e] MD 90 XD	600[i]	copolymer area thermal
Polyfelt	137		585[d]	225[j]	1450[k]	needlepunch
Reemay	68	0.29	225[d] MD 180 XD	45[j] MD 50 XD	331[k]	copolymer area thermal
Terram	137	0.7	850[d]	250[l]	1100[m]	area thermal (sheath core)
Trevira	155		630[n] MD 495 XD	270[j] MD 248 XD	1520[k]	needlepunch
Typar	137	0.38	650[o] MD 740 XD	345[p] MD 355 XD	1210[n]	undrawn segments–area thermal
Tyvek	54	0.15	4.6 MD 5.1 XD	4.5 MD 4.5 XD		area and point thermal

[a] MD = machine direction; XD = transverse direction.
[b] To convert from N to pound-force, divide by 4.448.
[c] To convert kPa to psi, multiply by 0.145.
[d] ASTM D1117-1682.
[e] ASTM D2263.
[f] DIN 53857.
[g] DIN 53363.
[h] DIN 53356.
[i] ASTM D774.
[j] ASTM D1117.
[k] ASTM D3786.
[l] BS4303-1968.
[m] BS4768-1972.
[n] ASTM D751.
[o] ASTM D1682.
[p] ASTM D4533.

ticularly for lightweight structures (eg, 17g/m²). The formation of wide webs at high speeds is a highly efficient operation.

For many applications partial orientation sufficiently increases strength and decreases extensibility to give a functional fabric (eg, diaper coverstock). However, some applications, such as primary carpet backing, require filaments with very high tensile strength and low degree of extension. For such applications, the filaments are drawn over heated rolls with a typical draw ratio of ca 3.5:1 (25). There, the filaments are pneumatically accelerated onto a moving belt or screen. This process is slower, but gives stronger webs.

The web is formed by the pneumatic deposition of the filament bundles onto

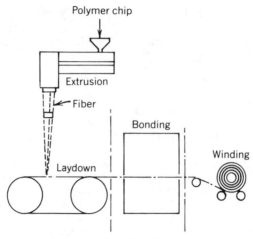

Fig. 3. Spunbond process.

the moving belt (24,31). A pneumatic gun uses high pressure air to move the filaments through a constricted area of lower pressure but higher velocity as in a Venturi tube (Fig. 4). In order for the web to achieve maximum uniformity and cover, individual filaments must be separated before reaching the belt. This is accomplished by inducing an electrostatic charge onto the bundle while under tension and before pneumatic deposition. The charge may be induced triboelectrically or by applying a high voltage charge, the former a result of rubbing the filaments against a grounded, conductive surface. The electrostatic charge on the filaments must be at least 30,000 esu/m^2 of surface area (24).

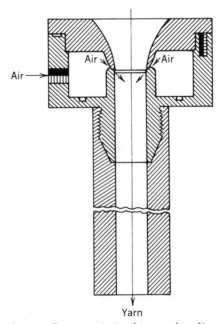

Fig. 4. Pneumatic jet for spunbonding.

The belt is usually made of conductive wire and connected to the ground. Deposition onto the belt discharges the filaments. This method is simple and reliable. Webs produced by spinning rectilinearly arranged filaments through a so-called slot die obviate such bundle-separating devices (28,29).

Filaments are also separated by mechanical or aerodynamic forces. Figure 5 illustrates a method that utilizes a rotating deflector plane to separate the filaments by depositing them in overlapping loops (32); a suction holds the fiber mass in place.

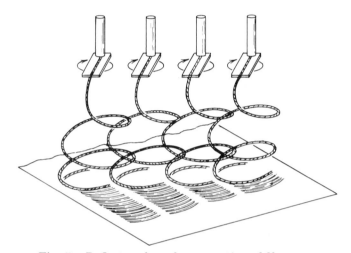

Fig. 5.   Deflector plane for separation of filaments.

For some applications, the filaments are laid down randomly without orienting the bundles with respect to the direction of the laydown belt (31). In order to achieve a particular characteristic in the final fabric, the directionality of the splayed filament is controlled by traversing the filament bundles mechanically (27,33) or aerodynamically (25,34) as they move toward the collecting belt. In the aerodynamic method, alternating pulses of air are supplied on either side of the filaments as they emerge from the pneumatic jet. By proper arrangement of the spinneret blocks and the jets, laydown can be achieved predominantly in the desired direction. The production of a web with predominantly machine and cross-machine direction filament laydown is shown in Figure 6 (25). Highly ordered patterns can be generated by oscillating filament bundles between closely spaced plates for a high degree of parallelism (34).

If the laydown belt is moving and filaments are being rapidly traversed across this direction of motion, the filaments are being deposited in a zigzag or sine-wave pattern on the surface of the moving belt. The effect of the traverse motion on the coverage and uniformity of the web has been treated mathematically (35,36). The relationships between the collecting belt speed, period of traverse, and the width of filament curtain being traversed determine the appearance of the formed web. Figure 7 illustrates the laydown for a process where the collecting belt travels a distance equal to the width of the filament curtain x during one complete period of traverse across a belt width y. If the belt speed is

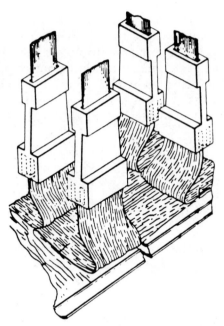

Fig. 6. Web production with predominantly machine and cross-machine direction.

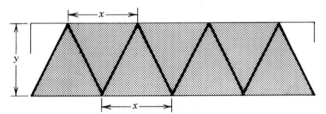

Fig. 7. Web laydown pattern.

V_b and the traverse speed is V_t, the number of layers deposited z is calculated by

$$z = \frac{xV_t}{yV_b}$$

If the traverse speed is twice the belt speed and if x and y are equal, a double coverage occurs over all areas of the belt.

Bonding

Many methods can be used to bond the fibers in the spun web. Although most procedures were developed for nonwoven staple fibers, three were adapted for continuous filaments: mechanical needling, thermal bonding, and chemical bonding. The last two may bond large regions (area bonding) or small regions (point bonding) of the web by fusion or adhesion of fibers. Point bonding results in the fusion of fibers at points with fibers in between the point bonds remaining relatively free. Other methods used with staple-fiber webs but not routinely with continuous-filament webs include stitchbonding (36,37), ultrasonic fusing (38), and hydraulic entanglement (39). The last method has the potential to produce very different continuous-filament structures, but is more complex and expensive.

Mechanical needling, also called needlepunching or needlebonding, is the simplest and least expensive method. Although it is the oldest process, it continues to be widely used. Higher production rates and flexibility have increased the sales of needlebonded fabrics, particularly in geotextiles. An excellent review of mechanical-needling technology was recently published (36).

In the needlepunching process a continuous-filament web is subjected to barbed needles that are rapidly passed through the plane of the moving spun web (Fig. 8). The needles pass in and out of the web at a rate of up to 2200 strokes per min, which can give as many as 500 penetrations per cm^2, depending on the needle density and the line speed, typically 5–25 m/min (40). This operation interlaces the fibers and bonds the structure together, relying only on the mechanical entanglement and fiber-to-fiber friction. The product tends to be more comfortable and bulky than fabrics bonded by thermal- or chemical-binder methods. Since the fibers have freedom to move over each other, the fabric is easily deformed and exhibits a low initial modulus (Fig. 2).

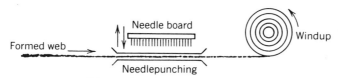

Fig. 8. Needlepunch process.

The variables in needlepunching are the needle design, punch density, and depth of punch (36). Needling produces a 100% fiber fabric without points or areas of fusion or melting. It is easily adapted to most fiber webs and requires less precise control than thermal bonding. In addition, it is the only bonding method suitable for the production of heavyweight spunbonded fabrics, eg, 800 g/m^2. It is, however, only suitable for the production of uniform fabrics ≥ 100 g/m^2, since needling tends to concentrate fibers in areas resulting in loss of visual uniformity at lower weights.

Unlike mechanical needling, both thermal- and chemical-binder bonding depend on fiber-to-fiber fusion as the means of establishing fabric integrity. The degree of fusion determines many of the fabric qualities, most notably hand or softness. Since point bonding can be accomplished with as little as 10% bonding area (ie, 90% unbonded area), such fabrics are considerably softer than area-bonded structures. Fiber mobility is retained in part or in total outside the areas of the point bonds. Thermal bonding is far more common than chemical-binder bonding and is more economical, since the latter method still requires thermal curing as the final step. Both area- and point-thermal bonding are rapid processes having line speeds in excess of 120 m/min during production of lightweight fabrics.

For area-thermal bonding, the spun web is passed through a source of heat, usually steam or hot air. Before entering the bonding area, the spun web may be consolidated by passing it under compressional restraint through a heated prebonding area which strengthens the web (19). While in the bonding area, the web is exposed to hot air or pressurized steam which causes fusion between some, but not all, fiber crossover points. Complete fusion gives a paperlike structure with little resistance to tearing. The spun web may contain small amounts, typically 5–30%, of a lower melting fiber (19) or the filaments may contain undrawn segments which melt lower than the drawn or matrix segment (33). Heterofilament structures utilize a lower melting covering (eg, sheath) to effect fusion. Both polyethylene and nylon-6 have been used as the lower melting sheath in commercial spunbonded products.

The use of steam is limited to polypropylene and polyethylene fusion, since the pressures needed to reach the temperature (eg, >200°C) required for bonding polyesters are impractical. In general, better temperature control is required for area-bonding polypropylene than for other polymers, since the temperature difference between the matrix and binder fibers can be only 3°C (33).

Area-thermal bonding is based mainly on temperature and applied to sophisticated webs containing binder fibers. Point-thermal bonding, however, utilizes both temperature and pressure to effect fiber-to-fiber fusion. It is more flexible, since lower melting fibers or segments are not required in the web. Point bonding is usually accomplished by passing a preheated, consolidated web through heated nip rolls, one of which has a raised pattern on its surface (Fig. 9). Bonding temperatures for polypropylene usually do not exceed 170°C, but pressures on the raised points are between 138 and 310 MPa (20,000–45,000 psi) (41). The bonding between the points can be controlled by adjusting the ratio of the heights of the raised points to the depth of the web (42). Typically, only 10–25% of the

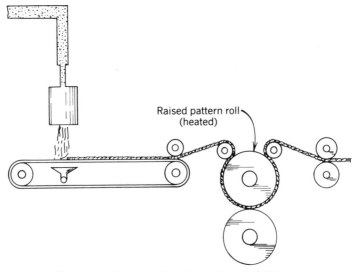

Fig. 9.   Point bonding of spunbonded fabrics.

surface available for bonding is converted to fused, compacted areas of bonding. Optimum conditions of pressure and temperature depend on many variables, including the nature of the web, line speed, and the engraved pattern. Even subtle changes can result in significant changes in the final product (43).

Since engraved-point-bonding rolls can be as wide as 5 m, the problem of maintaining uniform pressure across the width must be addressed. Small differences in pressure across the width can produce unacceptable variations. The hydraulic pressure applied at the ends of the roll results in a slight deflection, ie, less pressure is applied in the center than at the ends. This problem can be solved in various ways (43), most commonly by cambering, wherein the roll diameter decreases slightly from the center to the ends.

For spunbonded fabrics, chemical-binder bonding is used less frequently than thermal bonding; the reverse is true for staple-fiber nonwovens. Resin binders achieve special characteristics on spunbonded webs that cannot be attained thermally (44). In a typical procedure acrylic resins are applied to saturate the web; excess resin is removed by nip rolls and the wet web is passed through a drying oven to remove excess water and cure the resin which tends to concentrate at fiber–fiber junctions. Resin binders may instead be applied in discrete points in a pattern in order to immobilize fewer fibers and produce a softer fabric. However, it is difficult to control resin diffusion; the drying step is a disadvantage.

Chemical bonding with hydrogen chloride gas has been used with spun webs of nylon-6,6 to produce spunbonded nylon fabrics (45). In this unusual process the activating hydrogen chloride gas is passed over web fibers held in close contact by tension. The HCl gas ruptures the hydrogen bonds between the polymer chains and forms a complex with the amide group. Desorption of the gas reverses the process, and new hydrogen bonds are formed between polymer chains in different fibers. This method has been refined further to permit only pattern bonds to be formed, whereby fiber mobility is retained between the bonded areas, conferring a softer hand to the bonded fabric (46).

Certain generalizations apply to web bonding. If the web is highly bonded, most fibers are bonded to another fiber. The resulting structure is stiff and paperlike with high tensile and modulus properties but low tear resistance. On the other hand, if the web is only slightly bonded, fewer fiber-to-fiber bonds are present, and the structure is softer with lower tensile and modulus properties but higher resistance to tear propagation; surface-abrasion resistance is also low. Point bonding affords a greater variety of structures than area bonding because of the almost endless variety of available bonding-roll patterns. However, because of the high cost of the bonder roll, only one or two patterns are selected.

Meltblown Fabrics

The fibers of meltblown fabrics are composed of discontinuous filaments and are smaller than those of spunbonded fabrics. Although meltblown fabrics are not generally referred to as spunbonded because of discontinuous filaments, the integration of spinning, attenuation, laydown, and bonding during the production of meltblown webs describes a process traditionally defined as spunbonding. Fibers produced by melt blowing are very fine with a typical diameter of 3 μm (47,48), which is smaller by an order of magnitude than the smallest traditional spunbonded fiber. The webs are weak and easily distorted, since the fibers are extremely fine and largely unoriented. Most commercial products are made of polyester or high melt-flow polypropylene, but other thermoplastic polymers have been used.

The term meltblown is descriptive of the process used to produce these fibers. A special die uses heated, pressurized air to attenuate the molten polymer filament as it leaves the orifice of the dye or nozzle (Fig. 10). Air temperatures range from 260 to 480°C with flow rates of 1.4–7 kg/min per cm^2 of slot area (49).

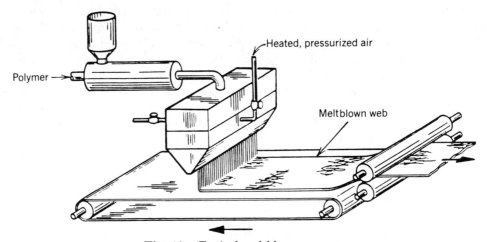

Fig. 10.   Typical meltblown process.

The rapidly moving hot air greatly attenuates the fibers as they leave the orifices, creating the subdenier size. The weak discontinuous fibers are deposited on the forming screen as a random, entangled web which may be thermally point bonded to improve strength and appearance. The web may also be deposited onto a conventional spun but not bonded web to which it is then thermally bonded. Sandwich structures have been created with the meltblown web between two conventional spunbonded webs (50). Other materials eg, cellulosics, have been blended into the meltblown filament stream to yield a meltblown structure with a unique combination of properties (51). Mixtures of meltblown fibers and crimped bulking fibers are sold as thin thermal insulation for outdoor clothing and gear (52). Meltblown technology has been adapted to produce nontraditional spunbonded fabrics (21).

The great quantity of very fine fibers in a meltblown web results in unique properties, such as very large surface areas and very small pore sizes. These materials are used for hospital gowns, sterile wrappings, incontinence devices, oil absorbers, battery separators, and special filters. Many new composite structures are expected to be designed from meltblown webs.

Flashspun Fabrics

The production of spunbonded webs by flash spinning is a radical departure from the conventional melt-spinning methods. In the latter a molten polymer is extruded through a spinneret containing about a hundred small holes to produce a fiber bundle containing as many fibers as there are holes in the spinneret; each fiber is typically 15–50 μm in diameter. The fibers within a bundle are separate from each other until the bonding operation connects some or all of the individual filaments.

In flash spinning, on the other hand, a 10–15% solution of, for example, high density polyethylene in trichlorofluoromethane or methylene chloride (53) is heated to ca 200°C and pressurized to 4500 kPa (650 psi) or more. The pressurized vessel is connected to a spinneret containing a single hole. When the pressurized solution is permitted to expand rapidly through the single hole, the low boiling solvent is instantaneously flashed off, leaving a three-dimensional film–fibril network referred to as a plexifilament. The film thickness is 4 μm or less; the three dimensionality results from the cross-linking interconnection of the subdenier fibers (53). Thus a multitude of individual but interconnected fibers are created from a single-hole spinneret.

Bubbles form rapidly as the pressurized solution is depressurized during spinning. These bubbles may grow and fracture, forming the plexifilamental network (53). Gases insoluble in the solvent may be added to the pressurized solution to promote high rates of bubble nucleation.

When a multiplicity of singular spinnerets are assembled across a width, the plexifilaments may be used to form a web that can be thermally bonded to produce a flat sheet (54). Web formation is faciliated by a baffle that deflects the stream of plexifilaments leaving the spinneret.

Unlike the subdenier fibers prepared by melt blowing, the plexifilaments from flash spinning are oriented and exhibit high tenacities (>0.1 N/tex or >1 gf/den). As a result of their very high surface areas (ca 2 m^2/g), plexifilaments scatter light and form opaque webs. In addition, the very fine fibrils give materials of exceptional softness. The webs are area or point bonded to give a paperlike or clothlike material, respectively. The paperlike sheets may be printed by conventional inks and printing equipment, whereas the point-bonded structures are very soft and are used in disposable protective clothing.

Flash spinning is the most complex and difficult method of manufacturing spunbonded fabrics because of the need to spin heated pressurized solutions under precise conditions. The result, however, is a unique structure that shows great promise. Although the process is almost 30 years old, it is used only by one company.

Test Methods

Spunbonded fabrics are characterized by standardized test procedures originally developed for textile fabrics and paper products. The Association of the Nonwoven Fabrics Industry (INDA) has published procedures (Table 4) that are

Table 4. Test Methods for Nonwoven and Spunbonded Fabrics[a]

Property	IST-INDA[b]	ASTM[c]	AATCC[d]
weight	130.0-70	D1910-64[e]	
grab strength	110.0-70	D1682-64	
tear strength	100.0-70	D1117-74	
burst strength	30.0-70	D774	
thickness	120.0-70	D1777-64[d]	
wash shrinkage		D1905-73	96-1975; 135-1973
dry-heat shrinkage		D3334-74T	
air permeability	70.0-70	D737-75	
wash-and-wear performance		D3135-72	124-1975; 88B,C-1975
fuzzing and pilling		D1375-72	
abrasion resistance	20.0-70	D1175-71	93-1974
flame retardance	50.0-71	D1230-61[f]	34-1966
color fastness			
to abrasion		D2814-73	
to cracking			8-1974
to dry cleaning			132-1973
to light			16-1974
to washing			61-1975
mildew resistance			30-1974
actinic degradation			111A,B,C,D-1975
softness	90.0-75		
seam strength		D1683-68	
water repellancy	80.9-70		127-1974
absorbency	80.0-1970	D1117-74	79-1975
stretch recovery		D2594-72; D3107-75	
drape, stiffness		D1388-64[e]	
crease recovery		D1295-67	

[a] Ref. 55.
[b] Data from *IST-INDA Standard Tests.*
[c] Data from *Book of ASTM Standards.*
[d] Data from *AATCC Technical Manual.*
[e] Reappraised 1975.
[f] Reappraised 1972.

routinely used to determine specific physical characteristics of nonwoven and spunbonded fabrics. Many tests are established for the evaluation of properties such as washability, stiffness, and softness. Advances permit the quantitative evaluation of the hand of materials for textile applications such as clothing (56).

New applications, eg, geotextiles (57), require new test methods. Aging characteristics must also be determined. Roofing applications, for example, require that the saturated fabrics retain their strength for many years in a severe environment. Testing is accelerated by heating the fabric at several different temperatures higher than the expected conditions. A curve can be constructed and the expected conditions can be extrapolated (58).

Applications

The uses for spunbonded fabrics have traditionally been grouped into durable and disposable categories, although for some applications categorization is difficult. In the early 1970s spunbonded materials were predominantly used for durable applications, such as carpet backing, furniture, bedding, and geotextiles. By 1980, however, disposable applications accounted for an increasingly large percentage, primarily because of the acceptance of lightweight (eg, 17 g/m^2), spunbonded polypropylene fabrics as a coverstock for diapers and incontinence devices (2).

Both the durable and disposable markets for spunbonded materials have approximately doubled since 1980. The main areas of growth for durable applications have been in the building and construction industries, where spunbonded materials are used in geotextiles and roofing membranes. Growth has also been achieved in primary carpet backing, particularly in automotive carpets and carpet modules, where spunbonded materials confer moldability and high dimensional stability.

With the possible exception of geotextiles, no new major markets have been established as a result of the special characteristics of spunbonded fabrics. In most cases spunbonded fabrics have replaced woven fabrics, other nonwoven fabrics (including knits), paper, or film. Spunbonded materials can be less expensive and give a better performance, which contributes to the growth of specific markets. Although several new applications appear to have been developed because of the cost–performance characteristics of a spunbonded fabric, such as the wrapping of houses to reduce air penetration, it remains to be seen if new spunbonded structures can be developed with sufficiently unusual properties to permit the development of significant new markets (1,2,59,60).

Of the four polymer types employed for spunbonding, polypropylene, polyethylene, polyester, and nylon, polyester and nylon are more expensive. This cost disadvantage could be offset by lighter weight, but in general olefin-based products have an economic advantage.

In some applications, however, higher costs are justified. In roofing membranes, a key requirement is dimensional stability to temperatures near 200°C, which is above the melting point of both polypropylene and polyethylene, but well within the performance limits of polyester.

Although polyester fibers offer other advantages, such as dyeability and higher modulus, these are not in demand in the current markets for spunbonded fabrics.

Durable Fabrics

The current markets for nonwoven fabrics in the United States and Western Europe are shown in Tables 5 and 6. Total U.S. production for 1985 is estimated to be 450×10^3 t of which approximately 25% were spunbonded. This represents nearly 8.4×10^9 m^2 of nonwoven material with growth projected at >5%/yr to 1990 (2).

Home Furnishings. Among the first applications for a spunbonded product was spunbonded polypropylene for primary carpet backing. It was introduced in the mid-1960s as a replacement for woven jute, and still enjoys a unique position in applications that require isotropic-planar properties for dimensional stability such as in printed or patterned carpets. The finer fiber allows tufting needles to penetrate with little deflection (Fig. 11). Because the spunbonded backing is

Table 5. U.S. Growth of Nonwoven Products, 1983–1987, 10⁶ $ (Point-of-use Value)ᵃ

Application	1983	1987
Consumer disposables		
baby diapers	2300	4000
adult diapers	25	300
wipes and roll towels	250	500
fabric-in-drier softeners	225	300
surgical packs	200	350
surgical gowns	60	150
surgical masks, caps, shoe covers	70	125
hospital bed underpads	80	115
medical dressings	70	150
Industrial disposables		
filtration media	300	600
wipe cloths	100	250
other	50	75
Durables		
interlining-interfacing (apparel)	125	200
coated and laminated products	100	100
home furnishings	100	180
carpet backing (primary, secondary)	125	175
geotextiles, civil engineering	NAᵇ	NAᵇ
roofing	100	200
other	100	150
Total	*4380*	*7920*

ᵃ Ref. 13; data from Maran Marketing Co., Inc. (New York).
ᵇ NA = not available.

Table 6. 1984 European Nonwoven Productionᵃ

Use	10³ t	%	10⁶ m²	%
coverstock	64.4	25.1	3242.9	48.5
medical–surgical	8.8	3.5	258.3	3.9
bed, table linen	3.5	1.4	85.5	1.3
wiping cloth	24.6	9.6	632.8	9.5
filtration	8.8	3.5	172.6	2.6
apparel interlining	15.8	6.1	303.3	4.5
civil engineering	49.2	19.1	333.9	5.0
coating substrate	15.9	6.2	228.3	3.4
furnishingsᵇ	25.1	9.7	225.7	3.4
vehicle interior	5.6	2.2	83.8	1.2
abrasives, tea bags	19.7	7.7	869.2	13.0
others	15.3	5.9	251.4	3.7
Total	*256.7*	*100.0*	*6687.7*	*100.0*

ᵃ Ref. 61.
ᵇ Wall covering, carpet underlays, and upholstery.

bonded together at many fiber junctions, it maintains clean edges after tufting, especially in small rugs where the unraveling of woven ribbon backings can be a concern. The first spunbonded carpet backing was made from polypropylene, but later products were based on polyester and polyester–nylon.

In furniture construction less expensive spunbonded fabrics have replaced

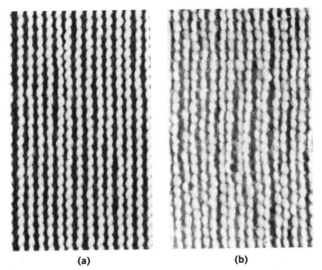

(a)   (b)

Fig. 11.   Tufted polypropylene: (a) spunbonded; (b) woven.

woven sheeting. Spunbonded fabrics are used in hidden areas requiring high strength and support in chairs, sofas, and other seating. The bottoms of chairs are often covered with dust covers or cambric cloth made of spunbonded fabrics because they do not fray, are highly porous, and give excellent cover. In addition, they are resistant to rot and mildew. In bedding spunbonded materials are used as spring insulators and wrapping, dust covers, mattress-pad covers, and quilting-cloth facing. In draperies spunbonded fabrics serve as a stitching medium.

A new application for spunbonded fabrics is as an air barrier, where the surfaces of houses are covered with a layer of spunbonded fabric followed by the external sheathing such as siding or masonry. The objective is to construct a barrier to the infiltration of air into the wall cavities and interiors of a building. Tests conducted by the NBS and the National Association of Homebuilders confirmed the effectiveness of the air barrier as a means of reducing heating costs (62). Flash-spunbonded fabrics are suitable for this application, because they resist the penetration of water and air currents, but pass moisture vapor. In winter, warm moist air from inside the house can penetrate through the wall to the outside. If the barrier material is not permeable to moisture vapor, condensation could occur inside the wall, causing moisture damage. In addition, the effective R-value of the insulation (eg, fiber glass) inside the wall cavity would be reduced by the presence of water. The combination of water and air-current resistance combined with "breathability" to moisture vapor and high tensile and tear strength is an unusual combination of properties.

Automotive.   The uses for nonwovens in automobiles have grown from a modest beginning 20 years ago to a position of significance. Needlepunched nonwoven fabrics have been used in large-area applications such as backing for vinyl seats and landau tops, whereas spunbonded fabrics have been utilized in lower volume applications such as labels, spring insulators, seat listings, and as a coated fabric for ducting. Spunbonded polyester as a tuftable carpet backing provides molding precision, improved dimensional stability, and resistance to puncture

(59). Future spunbonded automotive applications, such as headliners, depend to a large extent on the development of new structures, possibly composites, which offer performance and cost improvements over existing materials.

Roofing.   The market opportunity for roofing applications of spunbonded products is thought to exceed $185 \times 10^6 \, \text{m}^2$ for commercial buildings (flat roofing) in the United States alone. Many roofing applications were developed in Europe and slowly became accepted in the United States. Although fiber glass fabrics are predominantly used in roofing, spunbonded polyester and polypropylene have made considerable inroads (63). The main difference between glass fiber and polyester is the ability of the latter to flex and stretch without damage to the filaments. Since rooftops expand and contract with seasonal changes, polyester fabrics are less susceptible to damage from roof movement and sudden temperature fluctuations which initiate rapid dimensional changes.

Spunbonded polyester is a carrier for the bituminous waterproofing membrane. The fabric saturated with bitumen provides strength and dimensional stability. As bituminous coatings were modified with elastomeric polymers such as atactic polypropylene or sequenced butadiene–styrene (SBS), changes occurred in the installation and manufacture of membranes. Traditionally, roofs were built up by mopping hot bitumen into felts placed on the roof deck. Today, the roof membrane is manufactured under tightly controlled conditions in a factory at a distance from the site of application. The spunbonded fabric is typically saturated with modified bitumen by dipping into a tank of bitumen which is heated up to 200°C. Excess bitumen is metered off and the fabric is coated with a release material such as talc to prevent sticking on the roll. The composite is packaged into rolls approximately 1-m wide and 50-m long. The rolls are shipped to the site and applied to the flat roof surface by slowly unrolling while heating the underside to tackiness with a propane torch, enabling it to adhere to the deck. Spunbonded polyester is also used in the so-called cold-roof method for roof maintenance. A cold-applied mastic is first applied over a fiber glass base sheet, followed by more mastic, another layer of polyester, more mastic, and a final topcoat. The use of spunbonded polyester fabrics in the cold-applied process was reviewed (64).

In Europe bitumen-coated spunbonded polypropylene fabrics are widely accepted for roof linings under concrete, clay, or ceramic tile. Here, the spunbonded fabric is a critical element of the membrane, since the roof lining is fastened between rafters and depends on the strength of the spunbonded material for support. The bitumen coating waterproofs the spunbonded fabric and allows it to shed leakage between the tiles during driving snow and rain. Spunbonded fabrics coated with nonbituminous materials such as acrylics are currently being introduced in Europe (65). Roof linings represent an expanding market for spunbonded fabrics in Europe and in the southwestern United States.

Geotextiles.   Nonwoven fabrics have played an important part in the development of geotextile applications. Both needlepunch fabrics manufactured from staple fibers and spunbonded continuous-filament products have found worldwide acceptance (40). In 1985 approximately $125 \times 10^6 \, \text{m}^2$ of geotextiles were used in the United States and annual growth is forecast at 12–15% (57). Geotextile fabrics are porous to water but not to the soil fines, permitting separation or partition of soil fines from other elements. For example, in the construction of a new road, the subsoil is separated from the gravel or aggregate by the geotextile. Separation prevents the aggregate from being driven into the subsoil by the weight of vehicles; no soil fines are pumped up into the aggregate. However, water is freely transported through the fabric, permitting proper drainage without the buildup of hydrostatic pressures. Thus the road resists rutting and sustains the weight of traffic more effectively, while permitting proper drainage through the fabric (66). Drainage ditches often utilize geotextile fabrics to prevent perforated drainage pipes from becoming clogged by wrapping the pipes with a geotextile prior to installation.

Spunbonded fabrics function as filters because they are layered structures of fine fibers whose three-dimensional structure creates a torturous path. Even thin spunbonded fabrics (eg, 0.2–0.25 mm) present a challenge to the passage of soil fines and are suitable in some filtration applications. The porosity of geotextile fabrics is expressed as flux in $L/(m^2 \cdot min)$; the equivalent opening size (EOS) is a measure of the apparent size of the openings in the fabric. The flux measures the porosity to liquid water, whereas the EOS measures the porosity to solid particles of a known diameter (67). Technical literature is available from most manufacturers (68). Geotextile applications have become sufficiently important, and many universities offer courses in this subject to civil engineering students. Excellent textbooks are available for these courses (57,69,70).

Other Durable Applications. Other durable applications, such as interlinings and coating and laminating substrates, do not appear to offer much opportunity for growth for spunbonded fabrics (1,60). This situation could be reversed by the introduction of new structures or developments in the marketplace. In interlinings, for example, spunlaced nonwovens have received wide acceptance because of their outstanding drape and softness unavailable from any other fabric. Spunbonded fabrics occupy a small percentage of the coated-fabric market which is dominated by needlepunched nonwovens. The latter offer more bulk and resiliency which is required for funtionality in automotive and furniture seating.

Many filtration requirements are fulfilled by spunbonded structures and a small but technically complex market has developed during the last two decades (71). Their use in cartridge filters was recently reviewed (72).

Disposable Fabrics

The outlook for spunbonded disposable applications is bright with an 8% compounded annual growth forecast between 1984 and 1990 (73). Disposable market growth for spunbonded fabrics is forecast to exceed all other nonwoven forms with the exception of spunlaced and meltblown fabrics. Key markets are coverstock for diapers and incontinence devices, surgical gowns, sterilization wrap, envelopes, and protective clothing (Table 7).

Diapers and Incontinence Devices. The use of spunbonded fabrics as coverstock for diapers and incontinence devices has grown dramatically in the past decade and by 1984 annual consumption exceeded 3.3×10^9 m^2 in the United States and 3×10^9 m^2 in Europe (74). A coverstock is required to function essentially as a part of a one-way medium through which body fluids are transported to the absorbent layer. The laminar structure feature of the coverstock helps keep the skin of the user dry and comfortable. Although spunbonded poly-

Table 7. Disposable Nonwoven Fabric Consumption[a,b]

Use	1983[c]	1984[c]	1985[d]	1986[e]
industrial garments	126	147	171	199
feminine protection	95	103	111	118
wipes	502	585	654	735
filtration	15	15	14	14
hospital–medical	790	905	970	1024
baby diaper coverstock	1880	1906	1923	1940
adult diaper–pad	225	269	359	425
fabric softener substrate	418	460	501	546
envelopes	459	558	653	746
Total	*4510*	*4948*	*5356*	*5747*

[a] Ref. 73.
[b] In 10^6 m².
[c] Actual.
[d] Estimate.
[e] Forecast.

ester (17 g/m²) had been extensively used in diaper coverstock, it has been supplanted largely by an equivalent weight spunbonded polypropylene. A summary of U.S. disposable diaper sales history is provided in Table 8 (75).

Changes in diaper design involve not only the coverstock but the nature of the absorbent layer and the shape of the diaper itself. The use of form-fitting legs and refastenable closures has accelerated the acceptance of disposable diapers for both infants and adults. Spunbonded coverstock is also widely used in sanitary napkins and to a limited extent in tampons.

It appears that the coverstock uses of spunbonded fabrics in diapers and incontinence devices will remain unchallenged in the foreseeable future. Other nonwoven materials appear to be at a cost disadvantage compared to spunbonded polypropylene (Table 9). Any competitive threat is likely to come from advances in film technology such as improvements in perforated film currently used in some coverstock applications, particularly sanitary napkins.

Medical Markets. In medical applications many traditional materials have been replaced by high performance spunbonded materials (76). Flash-spunbonded polyethylene was the first 100% spunbonded material to find acceptance in medical uses such as disposable operating room gowns, shoe covers, and sterilizable packaging. Other spunbonded fabrics of polypropylene or nylon were utilized to increase the strength of cellulosic composites. Composite structures of spunbonded polypropylene and meltblown polypropylene fibers have been used for operating room applications; spunlaced polyester–cellulose are also widely used.

Operating gowns are worn by members of the surgical team and demand excellent fabric properties. Key requirements include breathability for comfort, low noise, resistance to fluid penetration, no linting, sterilizability, and impermeability to bacteria. Woven cotton fabrics were used for many years, but had to be reused because of their high cost, increased by laundering. A number of studies compared disposable and reusable fabrics in an attempt to correlate the effect of linting with postoperative infections. Although no correlation has been established, it was demonstrated that single-use spunbonded olefin fabrics generate significantly fewer particles than cotton fabrics (77). Other studies show that the

Table 8. The Disposable Diaper Market[a]

Year	Consumption Net births, thousands	Consumption 10⁶ units	Consumption Change, %	Consumption Penetration[b], %	Sales, 10⁶ $ Total	Sales, 10⁶ $ Average per unit	Procter & Gamble Total	Procter & Gamble Pampers	Procter & Gamble Luvs	Kimberly-Clark	Others
1980	3,566	12,200	11	53.6	1,410	0.116	70%	59%	11%	9%	21%
1982	3,662	13,900	7	59.4	1,900	0.137	66	48	18	12	22
1984	3,650	15,600	4	66.9	2,510	0.161	54	34	20	25	21
1986[c]	3,730	16,900	5	72.0	3,040	0.180	47	30	17	33	20
1988[c]	3,770	18,600	5	79.5	3,550	0.191					
1989[c]	3,780	19,530	5	83.9	3,850	0.197					

[a] Refs. 71 and 75.

[b] Reflects disposable consumption as percentage of total diaper consumption potential.

[c] Manufacturers' sales 1986–1989 are based on prevailing average daily-unit consumption continuing at 7.0 units daily or declining steadily to 6.5 units as more effective product containment features reduce daily consumption. Average-unit realizations reflect gross price before dealer discounts or allowances and weighted average sales mix (branded vs private label products). Average annual unit price inflation 1986–1989 estimated at 3%. Estimates by Goldman, Sachs & Co.

Table 9. Cost of Nonwoven Materials[a]

Material	$/kg	g/m^2	$/m^2
wet-laid long fiber	2.89	65.3	0.187
Polypropylene			
spunbonded, light	3.117	18.9	0.055
meltblown	3.173	30.9	0.097
carded			
needlepunched	3.430	137	0.405
thermal-bonded	3.740	24.3	0.090
spunbonded meltblown	5.271	51.5	0.268
Poly(ethylene terephthalate)			
spunbonded			
light	4.294	51.5	0.219
heavy	3.123	206	0.635
spunlaced pulp	4.198	72	0.300

[a] Data from John R. Starr Inc., Management Consultants (Osterville, Mass.).

rate of postoperative wound infection is reduced with the use of high barrier spunbonded olefin gowns and drapes (78).

Medical devices are often sterilized after the nonsterile device is sealed in a package (79). A part of this package, often the lid, is made from spunbonded fabric which selectively permits the sterilizing gas of ethylene oxide to pass through but not bacteria. Special high quality grades of spunbonded fabric are manufactured to ensure a barrier to bacterial penetration. Spunbonded fabric is superior to coated papers in both penetration of sterilizing gas and resistance to bacterial penetration (80). Spunbonded meltblown laminates are also widely used as a sterilizable wrap (76).

Spunbonded fabrics are used as shoe covers in the operating room. The covers are usually sewn with an elastic band at the top to allow the covers to be held snugly in place. The fabric requirements are toughness, porosity for comfort, lack of linting, and resistance to slippage. The fabric is usually treated with an anti-static coating; static could cause sparks which could damage sensitive electronic devices or result in a fire. Other medical applications for spunbonded fabrics include head covers, face masks, drapes, and other uses requiring barrier properties.

Protective Clothing.   A large growth area for spunbonded fabrics is the disposable protective clothing market. The demand for high performance disposable protective clothing has followed high technology manufacturing and environmental demands. The manufacture of particulate-sensitive electronic components, such as integrated circuits, requires clean rooms where dust generation was partly controlled through the use of nonlinting, comfortable clothing made from spunbonded fabrics. The continuous filaments of the spunbonded fabric produce no lint, and the microporous structure allows the exchange of moisture and air, increasing comfort. Since the spunbonded garment is worn over other clothing, the pore size must be sufficiently small to prevent the passage of lint and other particles through the garment and into the clean room.

The removal of asbestos from buildings has created the need for clothing that is not penetrated by small asbestos fibrids, but which is inexpensive enough to permit disposal at the end of the day. Spunbonded fabrics have demonstrated excellent resistance to asbestos penetration and particles as small as 0.5 μm (81).

Similarly, the handling of hazardous materials has prompted the need for affordable, disposable protective clothing (82). After exposure to toxic waste, pesticides, or radioactive materials, the clothing must be disposed. Garments that offer maximum protection are often made from composites of spunbonded fabrics and polyethylene coatings, or laminates of poly(vinylidene chloride) film.

Packaging.   Packaging applications for spunbonded fabrics are for the most part a speciality area in which paper products and plastic films are not satisfactory. Examples include medical sterile packaging and high performance envelopes. Many businesses prefer tear- and puncture-resistant envelopes over conventional paper products. The lighter weight of the spunbonded envelope reduces postage. Spunbonded fabrics are also used as the envelope or sleeve for floppy disks.

Other Disposable Applications.   Other disposable applications for spunbonded fabrics include bale wrap, metal-core wrap, wipes, and in-drier softeners.

Much of U.S. synthetic staple-fiber production is packaged in bales of spunbonded fabrics that have been extrusion-coated for impermeability. Synthetic fibers have been shipped all over the world in these wrappings.

Many types of spunbonded fabrics are utilized in the large wipe market. Some are treated with surfactants or other coatings to enhance the performance. Most uses benefit from low cost and low linting.

In in-drier fabric softener sheets, the spunbonded fabric is coated with a complex combination of compounds which are released into the environment of a hot clothes drier. The release of these compounds softens and perfumes the clothes, and the spunbonded sheet provides a simple and cost-effective medium to store the chemical compounds until their release in the drier.

Outlook

Spunbonded fabrics are expected to grow at a rate exceeding the rate for nonwovens (1,2), but depend on continued sales growth in existing markets. New plant construction will bring increased capacity to a level that depends on real growth to keep sales abreast with production. Since much of the new production is for construction, home furnishings, and automotive uses, a slowdown in the economy could adversely affect profitability. A sudden increase in oil prices or decrease in feedstocks availability would adversely affect profitability and growth.

No new fiber technologies are on the horizon that would change the manner in which spunbonded structures are produced. Any serious challenges to existing markets would likely come from film, foam, or advances in other nonwoven technologies.

BIBLIOGRAPHY

"Nonwoven Fabrics" in *EPST* 1st ed., Vol. 9, pp. 345–355, by Norbert M. Bikales, Consultant.

1. J. R. Starr, *Nonwovens Ind.*, **18,** 56 (Jan. 1987).
2. *Ibid.*, **15,** 18 (Dec. 1984).
3. R. G. Mansfield, *Nonwovens Ind.* **16,** 26 (Feb. 1985).
4. *Nonwovens Markets* **1,** 1 (Feb. 21, 1986).

5. *Nonwovens Ind.* **17,** 36 (Mar. 1986).
6. *Nonwovens World* **1,** 36 (May–June 1986).
7. *Text. World* **136,** 35 (Aug. 1985).
8. F. Hand, *Text. Ind.* **143,** 86 (July 1979).
9. *Eur. Chem. News* **45,** 21 (Nov. 25, 1985).
10. *Nonwovens World* **1,** 11 (Aug. 1986).
11. G. Goldstein, *Tappi J.* **67,** 44 (Oct. 1984).
12. G. Ranganayaki and co-workers, *Synth. Fibers,* 6 (Oct.–Dec. 1985).
13. L. E. Seidel, *Text. Ind.* **148,** 58 (Mar. 1984).
14. D. O'Ryan and T. Kaiser, *Nonwovens World* **1,** 106 (May–June 1986).
15. M. F. Meyer and R. L. McConnell, *Nonwovens World* **1,** 57 (Nov. 1986).
16. *Nonwovens Ind.* **17,** 166 (Oct. 1986).
17. F. Gugumus, *Polypropylene Fibers and Textiles, Third International Conference,* October 4–6, 1983, University of York, UK, the Plastics and Rubber Institute, London, p. 18.1.
18. F. Gugumus and H. Linhart, *Chem. Vlakna* **32,** 94 (1982).
19. U.S. Pat. 3,989,788 (Nov. 2, 1976), L. L. Estes, A. F. Fridrichsen, and V. S. Koshkin (to E.I. du Pont de Nemours & Co., Inc.).
20. UK Pat. 1,157,437 (July 9, 1969), B. L. Davies (to ICI, Ltd.).
21. Y. Ogawa, *Nonwovens World* **1,** 79 (May–June 1986).
22. U.S. Pat. 3,439,085 (Apr. 15, 1969), L. Hartmann (to C. Freudenberg).
23. *Guide to Nonwoven Fabrics,* INDA, New York, 1978, p. 24.
24. U.S. Pat. 3,338,992 (Aug. 29, 1967), G. A. Kinney (to E.I. du Pont de Nemours & Co., Inc.).
25. U.S. Pat. 3,991,244 (Nov. 9, 1976), S. C. Debbas (to E.I. du Pont de Nemours & Co., Inc.).
26. UK Pat. 1,436,545 (May 19, 1976), James Brock (to ICI, Ltd.).
27. U.S. Pat. 4,017,580 (Apr. 12, 1977), Jacques Barbey (to Rhone-Poulenc, Inc.).
28. U.S. Pat. 3,502,763 (Mar. 24, 1970), L. Hartmann (to C. Freudenberg).
29. U.S. Pat. 4,405,297 (Sept. 20, 1983), D. W. Appel and M. T. Morman (to Kimberly-Clark Corp.).
30. U.S. Pat. 3,771,307 (Nov. 13, 1973), Dennis G. Petrille (to E.I. du Pont de Nemours & Co., Inc.).
31. U.S. Pat. 3,692,618 (Sept. 19, 1972), O. Dorschner, F. Carduck, and C. Storkebaum (to Metallgesellschaft AG).
32. U.S. Pat. 4,163,305 (Aug. 7, 1979), V. Semjonow and J. Foedrowitz (to Hoechst AG).
33. U.S. Pat. 3,322,607 (May 30, 1967), S. L. Jung (to E.I. du Pont de Nemours & Co., Inc.).
34. UK Pat. 2,006,844 (Oct. 10, 1978), P. Ellis and R. Gibb (to ICI, Ltd.).
35. UK Pat. 1,231,066 (Oct. 4, 1968), C. H. Weightman (to ICI, Ltd.).
36. H. Külter in J. Lunenschloss and W. Albrecht, eds., *Non-Woven Bonded Fabrics,* Ellis Horwood, Ltd., Chichester, UK, p. 178.
37. D. L. Heydt, *12th Technical Symposium,* INDA, May 22–23, 1984, Washington, D.C., p. 293.
38. G. Flood, *Nonwovens Ind.* **17,** 30 (Apr. 1986).
39. D. F. Beaumont, *Nonwovens World* **1**(3), 76 (Nov. 1986).
40. E. Fehrer, *Can. Text. J.* **102,** 67 (Dec. 1985).
41. U.S. Pat. 3,855,046 (Dec. 17, 1974), P. B. Hansen and L. B. Pennings (to Kimberly-Clark Corp.).
42. U.S. Pat. 3,855,045 (Dec. 17, 1974), R. J. Brock (to Kimberly-Clark Corp.).
43. D. H. Muller and S. Barnhardt, *Nonwovens Rep. Int.,* 19 (Mar. 1986).
44. U.S. Pat. 4,125,663 (Nov. 14, 1978), P. Eckhart (to Hoechst AG).
45. U.S. Pat. 3,542,615 (Nov. 24, 1970), E. J. Dobo, D. W. Kim, and W. C. Mallonee (to Monsanto Co.).
46. U.S. Pat. 4,075,383 (Feb. 21, 1978), R. M. Anderson and co-workers (to Monsanto Co.).
47. R. G. Volkman, *Ninth Technical Symposium,* INDA, Mar. 11, 1981, Atlanta, Ga., p. 106.
48. L. C. Wadsworth and A. M. Jones, *Nonwovens Ind.* **17,** 44 (Nov. 1986).
49. U.S. Pat. 3,972,759 (Aug. 3, 1976), R. R. Buntin (to Exxon).
50. U.S. Pat. 4,374,888 (Feb. 22, 1983), S. R. Bornslaeger (to Kimberly-Clark Corp.).
51. U.S. Pat. 4,100,324 (July 11, 1978), R. A. Anderson, R. C. Sokolowski, and K. W. Ostermeier (to Kimberly-Clark Corp.).
52. U.S. Pat. 4,118,531 (Oct. 3, 1978), E. R. Hauser (to 3M Co.).
53. U.S. Pat. 3,081,519 (Mar. 19, 1963), H. Blades and J. R. White (to E.I. du Pont de Nemours & Co., Inc.).
54. U.S. Pat. 3,442,740 (May 6, 1969), J. C. David (to E.I. du Pont de Nemours & Co., Inc.).
55. *Guide to Nonwoven Fabrics,* INDA, New York.

56. S. Kawabata, *The Standardization and Analysis of Hand Evaluation*, 2nd ed., Textile Machinery Society of Japan, Osaka, Japan, 1980.

57. R. M. Koerner, *Designing with Geosynthetics*, Prentice-Hall, Inc., Englewood Cliffs, N.J., 1986, p. 15.

58. J. D. M. Wisse and S. Birkenfeld, *2nd International Conference on Geotextiles,* Las Vegas, Nev., 1982, p. 283.

59. M. Jacobsen, *Nonwovens Ind.* **17,** 18 (May 1986).

60. L. E. Seidel, *Text. Ind.* **148,** 58 (Mar. 1984).

61. *Pira Seminar,* Specialty Papers & Nonwovens, Feb. 5, 1986, Leatherhead, Surrey, UK.

62. *Tyvek Housewrap, DuPont Bulletin E-60658,* Wilmington, Del., 1984.

63. R. G. Mansfield, *Text. World* **135,** 45 (Feb. 1984).

64. D. Currier-Liftig, *Contractors Guide* **9,** 54 (June 1986).

65. *Stamisol DW F 4250 Literature, CH-8193,* Stamoid AG, Eglisau, FRG.

66. Ref. 57, p. 12.

67. Ref. 57, p. 43.

68. *Design Guideline for Typar, DuPont Bulletin E-77041,* Wilmington, Del., 1985.

69. R. M. Koerner and J. P. Welsh, *Construction and Geotechnical Engineering Using Synthetic Fibers,* John Wiley & Sons, Inc., New York, 1980.

70. R. Veldhuijzen Van Zanten, ed., *Geotextiles and Geomembranes in Civil Engineering,* A. A. Balkema, Rotterdam, The Netherlands, 1986.

71. W. Shoemaker, *Nonwovens Ind.* **15,** (Oct. 1984).

72. H. N. Sandstedt and J. J. Weisenberger, *Filtr. Sep.* (Mar.–Apr. 1985).

73. *Nonwovens Ind.* **17,** 26 (Oct. 1986).

74. *Nonwovens Markets* **1**(4) (1986).

75. *Nonwovens Markets* **1**(1) (Jan. 10, 1986).

76. J. R. Starr, *Nonwovens World* **1,** 111 (Aug. 1986).

77. S. P. Scheinberg, J. H. O'Toole, and S. K. Rudys, *Particulate and Microb. Control* (July–Aug. 1983).

78. J. A. Moylan and B. V. Kennedy, *Surg. Gynecol. Obstet.* **151,** 465 (1980).

79. *Packag. Dig.,* 37 (Aug. 1982).

80. Th. Mengen and A. Jordy, *Eighth Symposium of the Austrian Society for Hygiene, Microbiology and Preventative Medicine,* Vienna, 1984.

81. *Arthur D. Little Report C87353,* Cambridge, Mass., May 1982.

82. D. M. Lesak, *Fire Eng.* **138,** 44 (Feb. 1985).

RONALD L. SMORADA
E.I. du Pont de Nemours & Co., Inc.
Reemay, Inc.

NONWOVEN FABRICS, MELTBLOWN.   See Supplement.

NONWOVEN FABRICS, WOOD PULP.   See Supplement.

NORBORNADIENE.   See ETHYLENE-PROPYLENE ELASTOMERS.

NOVOLOID FIBERS.   See PHENOLIC RESINS; Supplement.

NOVOLACS.   See PHENOLIC RESINS.

NUCLEAR MAGNETIC RESONANCE

Nuclear magnetic resonance (nmr) spectroscopy is a method of great interest and importance for the study of polymers. The earliest studies dealt with polymers in the solid state and the spectra were of the so-called broad-band or wide-line type (1). These were almost always proton spectra and their width and detailed shape, particularly when combined with measurements of the rates of spin–lattice relaxation, could be interpreted to give information concerning the nature and frequency of molecular motion. The broadness of the resonances, however, obscured structural information.

Because of motional averaging effects, the resonance lines of spectra obtained for samples in solution are greatly narrowed and detailed information concerning the covalent structure of the polymer chains may be obtained (2). More recent developments in nmr instrumentation, particularly the use of pulsed Fourier transform methods with spectrum accumulation and magic-angle spinning, have made it possible to obtain dynamic information from solution spectra and also to obtain high resolution spectra, and therefore structural detail, directly from solids (3–5). Therefore, the barrier between two once quite separate disciplines, solid- and liquid-state nmr spectroscopy, has all but disappeared.

The Nmr Phenomenon

Magnetic Energy Levels and Transitions.   When the spin quantum number I of a nucleus is nonzero, the nucleus possesses a magnetic moment. This condition is met if the mass number and atomic number are not both even, as for carbon-12 and oxygen-16. The proton (^1_1H) has a spin I of $\frac{1}{2}$. When placed in a magnetic field of strength B_0, nuclei with nonzero I occupy quantized magnetic energy levels, called Zeeman levels, the number of which is equal to $2I + 1$ (6). The relative populations of the Zeeman levels are normally given by a Boltzmann distribution. Table 1 lists properties of magnetic nuclei of interest for polymer studies. All nuclei with spin greater than $\frac{1}{2}$ have nonspherical distributions of positive charge and therefore also possess electric quadrupole moments.

Table 1. Nuclei of Interest for Polymer Nmr Spectroscopy

Isotope	Natural abundance, %	Resonant frequency in 1-T field[a]	Spin	Magneto-gyric ratio[b], γ	Relative sensitivity[c]
^1H	99.9844	42.577	$\frac{1}{2}$	267.43	1.000
^2H (D)	0.0156	6.536	1	41.05	0.0096
^{13}C	1.108	10.705	$\frac{1}{2}$	67.24	0.0159
^{14}N	99.635	3.076	1	19.32	0.0010
^{15}N	0.365	4.315	$\frac{1}{2}$	-27.102	0.0010
^{19}F	100	40.055	$\frac{1}{2}$	251.59	0.834
^{29}Si	4.7	8.460	$\frac{1}{2}$	-53.14	0.0785
^{31}P	100	17.235	$\frac{1}{2}$	108.25	0.0664

[a] 1 tesla (T) = 10,000 gauss (G).
[b] In units of $s^{-1}T^{-1} \times 10^{-6}$.
[c] For equal numbers of nuclei at constant B_0.

The separation of Zeeman levels is given by

$$\Delta E \ = \ h\nu_0 \ = \ \mu B_0/I \tag{1}$$

For spin-$\frac{1}{2}$ nuclei, there are two energy levels for which

$$\Delta E \ = \ h\nu_0 \ = \ 2\mu B_0 \tag{2}$$

For spin-1 nuclei, there are three energy levels separated by μB_0; μ is the magnetic moment of the nucleus (Fig. 1).

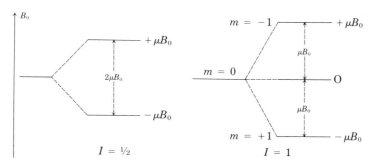

Fig. 1.   Magnetic energy levels for spin-$\frac{1}{2}$ and spin-1 nuclei.

Transitions between energy levels can be made to occur by means of a resonant radio-frequency (r-f) field B_1 of frequency ν_0. In Figure 2, the resonance frequency of the proton is plotted against six magnetic-field strengths employed in commercial spectrometers. Generation of fields above ca 2.5 T (25,000 G)

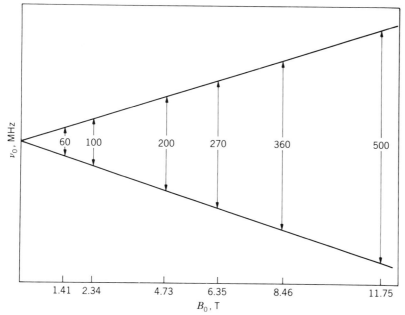

Fig. 2.   The splitting of magnetic energy levels of protons, expressed as frequency ν_0 in varied magnetic field B_0.

requires superconducting solenoid magnets. For nuclei other than protons, the resonant frequency may be obtained by multiplying the values in the third column of Table 1 by the appropriate magnitude of B_0. The resonance condition for any nucleus may also be expressed as

$$\omega_0 = \gamma B_0 \tag{3}$$

where ω_0 is the frequency in rad/s, equal to $2\pi\nu_0$, and γ is the magnetogyric ratio, equal to $2\pi\mu/Ih$. By using magnetogyric ratios, resonant frequencies are always directly proportional to γ, regardless of the spin.

An alternative way of picturing the resonance phenomenon arises from the fact that when placed in a magnetic field, a nucleus undergoes Larmor precession about the field direction at a rate given by ν_0 or ω_0 in the above equations. Transitions between energy levels occur when the frequency of the r-f field equals the Larmor precession frequency.

The Chemical Shift. Resonance occurs at slightly different frequencies for each type of proton, depending on its chemical binding and position in a molecule. This variation is caused by the cloud of electrons about each nucleus, which shields the nucleus against the magnetic field, thus requiring a slightly lower value of ν_0 to achieve resonance than for a bare proton (7,8). Protons attached to or near electronegative groups such as OH, OR, OCOR, COOR, and halogens experience a lower density of shielding electrons and resonate at higher ν_0. Protons farther removed from such groups, as in hydrocarbon chains, resonate at lower ν_0. Similar structural dependence is observed for ^{13}C nuclei. These variations are termed chemical shifts and are commonly expressed in relation to the resonance of tetramethylsilane (TMS) as the zero of reference. The total range of proton chemical shifts in organic compounds is on the order of 10 ppm, eg, ca 1 kHz in a magnetic field of 2.34 T (23,400 G) (Fig. 2). For ^{13}C nuclei and all other magnetic nuclei, the range is much greater, ie, >200 ppm; this is the principal reason for the great interest in the study of polymers by ^{13}C nmr (Fig. 3).

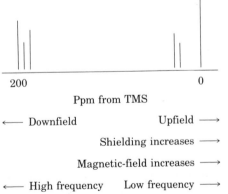

Ppm from TMS

$\longleftarrow$ Downfield   Upfield $\longrightarrow$

Shielding increases $\longrightarrow$

Magnetic-field increases $\longrightarrow$

$\longleftarrow$ High frequency   Low frequency $\longrightarrow$

Fig. 3. Schematic representation of a ^{13}C nmr spectrum, showing the directions of increasing shielding, magnetic field, and frequency.

For any nucleus, the separations of chemically shifted resonances, expressed in Hz, are proportional to B_0; when expressed in ppm, as is common, they are independent of B_0. An important advantage of high field magnets is the greater resolution of peaks and finer discrimination of structural features of polymer chains that they make possible.

The electronic screening of nuclei is actually anisotropic so that the chemical shift is a directional quantity and depends on the orientation of the molecule with respect to the magnetic-field direction. In solution, motional averaging produces an isotropic value of the chemical shift, but in solid polymers, where such motion cannot occur or is very slow, the anisotropy of the chemical shift becomes evident.

Nuclear Coupling.   Nuclei sufficiently removed from each other do not feel the effects of the others' magnetic fields; the local magnetic field at each nucleus is essentially equal to B_0 (actually slightly smaller because of electronic screening). If B_0 can be made very homogeneous over the sample, the width of the resonance lines may be as small as a few parts per 10^9 or 0.1–0.2 Hz.

Direct Dipole–Dipole Coupling.   In most substances, protons contribute to local fields and are sufficiently numerous to have a marked effect not only on proton spectra but on those of other nuclei. For example, the local field at a ^{13}C nucleus is given by

$$B_{\text{loc}} = \pm \frac{h}{4\pi} \gamma_{\text{H}} \frac{(3 \cos^2 \theta_{\text{C-H}} - 1)}{r^3_{\text{C-H}}} \qquad (4)$$

The $\pm$ sign indicates that the local field may add to or subtract from B_0, depending on whether a neighboring proton dipole is aligned with or against the direction of B_0, an almost equal probability. The angle $\theta_{\text{C-H}}$ is defined in Figure 4; r is the ^{13}C–^1H internuclear distance. If r and θ were fixed throughout the sample, this interaction would result in a splitting of the ^{13}C resonance into two equal components, the separation of which would depend on the orientation of the sample in the magnetic field. In actual polymer samples, which are glassy or microcrystalline, a summation over many values of θ and r is required, resulting in a dipolar broadening of many kilohertz, sufficient to mask all chemical-shift information. As the rate of molecular motion begins to exceed the linewidth, the resonance starts to narrow. When the reorienting C–H vectors sample all angles $\theta_{\text{C-H}}$ in a time short compared with the dipolar coupling (expressed as a fre-

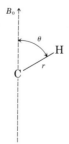

Fig. 4.   Representation of an internuclear C–H vector of length r, making an angle θ with the magnetic field.

quency), dipolar broadening is reduced to a small value because the time average of the term in parentheses in equation 4 may be replaced by a space average:

$$\int_0^\pi (3 \cos^2 \theta_{C-H} - 1) \sin \theta_{C-H} d\theta_{C-H} = 0 \tag{5}$$

The study and interpretation of dipolar-broadened spectra, particularly of polymer protons, yield valuable insights into molecular motion (1).

Indirect Nuclear Coupling. Magnetic nuclei may transmit information to each other concerning their spin states not only directly through space but also through the intervening covalent bonds. This is indirect or scalar nuclear coupling, also known as J coupling. Rapid tumbling of the molecule does not reduce this interaction to zero. If a nucleus has n sufficiently close, equivalently coupled spin-$\frac{1}{2}$ neighbors, its resonance will be split into $n + 1$ peaks, corresponding to the $n + 1$ spin states of the neighboring group of spins. Intensities are given by simple statistical considerations and are therefore proportional to the coefficients of the binomial expansion. Thus one neighboring spin splits the observed resonance to a doublet, two produce a 1:2:1 triplet, three a 1:3:3:1 quartet, and so on. The strength of the coupling is denoted by a coupling constant J and is expressed in Hz; to first order, it is independent of B_0.

The coupling of protons on adjacent saturated carbon atoms, termed vicinal coupling, varies with the dihedral angle ϕ (9,10). Trans couplings, where $\phi = 180°$, are substantially larger than gauche couplings, where $\phi = 60°$.

trans   gauche

J_{gauche} is typically 2–4 Hz and J_{trans} ranges from 8 to 13 Hz. These considerations are of great importance in studying the conformations of polymer chains. The dependence of J on ϕ is generally well described by the so-called Karplus relationship (9) (Fig. 5), although the magnitude of the coupling depends somewhat on the nature of the substituents on the bonded carbon atoms.

The chemical shifts and J couplings of nuclei in polymers do not differ from those of small molecules of analogous structure. The spectrum of ethyl orthoformate, $CH(OCH_2CH_3)_3$ (Fig. 6a), shows a single peak for the formal proton and a quartet and triplet for the CH_2 and CH_3 protons, respectively, of the ethyl group, in accordance with the rules just described. The spectrum of poly(vinyl ethyl ether) (Fig. 6b) shows corresponding resonances and, in addition, those of the main-chain β-protons appearing at 1.5 ppm and of the α-protons under the CH_2 quartet. The broadening of the lines in Figure 6b is in part due to the slower motions of the large molecule in solution and also to other structural complexities.

Nuclear coupling between ^{13}C nuclei and directly bonded protons is strong (125–250 Hz). The resulting multiplicity is often helpful in making resonance assignments but does not directly supply information concerning molecular conformation. It is usually abolished by double resonance, ie, by providing a second r-f field tuned to the proton-resonance frequency. The resulting multiplet collapse

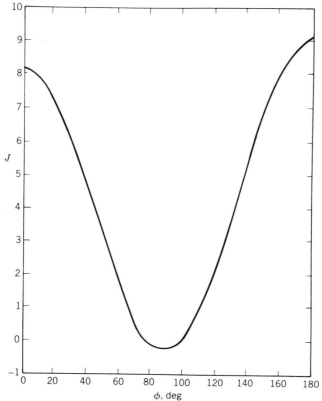

Fig. 5. J coupling of two protons, expressed in Hz, as a function of dihedral angle ϕ.

and accompanying nuclear Overhauser enhancement (NOE) give a striking improvement in signal-to-noise ratio. This is very helpful in view of the low natural abundance and small magnetogyric ratio of carbon-13 compared to the proton, both of which reduce its sensitivity to detection.

Nuclear Relaxation. Even in the highest magnetic field now employed, the separation of magnetic energy levels is small. For protons in an 11.75-T (117,500 G) field, it is only about 0.5 joule (~0.12 cal), so that the excess population of the lower state, given by a $2\mu B_0/kT$ at equilibrium, is on the order of 100 ppm (6). For all other nuclei, it is even less. Thus the net degree of nuclear polarization is very small.

Spin–Lattice Relaxation. A spin system may be readily disturbed from its equilibrium state. Before the sample is placed in the magnet, the spin states are equally populated. In the magnetic field, the spins establish a new equilibrium population, given by the Boltzmann distribution. This population may be equalized by an appropriate pulse of resonant r-f energy, and it is even possible by this means to invert the populations. The process by which the spin system attains equilibrium from a nonequilibrium state is called spin–lattice relaxation, characterized by a time T_1. Since such processes affecting the energy of the spin system affect the magnetization along the field direction, they are also sometimes

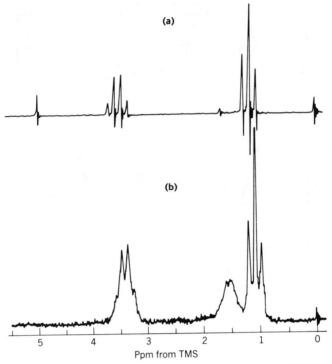

Fig. 6. The proton nmr spectrum of (**a**) ethyl orthoformate $HC(OCH_2CH_3)_3$ and (**b**) poly(vinyl ethyl ether), both observed at 60 MHz as 15% solutions in carbon tetrachloride.

called longitudinal relaxation processes. The "lattice," ie, the other molecules in the sample (a term used for both fluids and solids), serves as a heat sink in the equilibration, but there must be a link by which thermal energy may be exchanged. This link is provided by molecular motion.

For the relaxation of protons and ^{13}C nuclei in polymers, the only significant mechanism is dipole–dipole interaction with neighboring magnetic nuclei, principally protons (11,12). For nuclei of spin 1 or higher, eg, deuterium, the dominant mechanism is the interaction of the nuclear electric quadrupole with fluctuating molecular electric-field gradients (13). The motions of neighboring nuclei give rise to fluctuating magnetic fields and are characterized by a broad range of frequencies. To the extent that these motions have components at the resonant frequency of the observed nucleus ν_0, they induce spin–lattice relaxation since they provide fields equivalent to the resonant r-f field B_1. The link between the spins and the lattice is in effect a weak and inefficient one, so that spin–lattice relaxation is slow. Thus T_1 for both benzene protons and ^{13}C nuclei is ca 20 s at room temperature. For polymers in solution, it is generally shorter, but for solids it may be very long.

Spin–Spin Relaxation. Nuclear spins are not merely small static magnetic dipoles; even in a rigid solid, they are precessing about the direction of B_0. Such a precessing moment may be resolved into static and rotating components (Fig. 7). The rotating component is the right type of magnetic field to induce transitions in a neighboring nucleus precessing at the same frequency. If this spin exchange

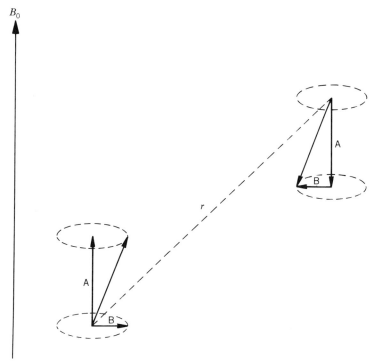

Fig. 7. Static (A) and rotating (B) components of a precessing nuclear moment.

or flip occurs, the nuclei exchange magnetic energy states with no overall change in the energy of the system but with a shortening of the life of each. This introduces an uncertainty broadening, in conformity with the Heisenberg principle, of the same form and magnitude as the broadening by the static components. Both effects are included in the quantity T_2, which is the characteristic time for spin–spin relaxation, also termed transverse relaxation, and is taken as the inverse of the linewidth:

$$T_2 = \frac{1}{\pi \delta \nu} \qquad (6)$$

Spin–spin exchange and dipolar broadening do not always occur together. For dilute nuclei such as ^{13}C, broadening by local proton dipoles is the principal effect; ^{13}C–^{13}C spin flip-flops do not contribute significantly and ^{13}C–^{1}H flip-flops cannot occur because of the disparate resonant frequencies.

Another contribution to line broadening is inhomogeneity δB_0 in the laboratory field. This may be minimized by careful shimming, ie, homogenization of the magnetic field, and is usually included in the experimental value of spin–spin relaxation, commonly termed $T_2{}^*$:

$$1/T_2{}^* = 1/T_2 + \frac{\gamma \delta B_0}{2} \qquad (7)$$

The "true" T_2, however, may be separately measured by spin-echo methods, which are discussed below.

Experimental Methods

Continuous-wave and Pulsed Fourier Transform Nmr. The heart of an nmr spectrometer is its magnet and r-f circuit. Figure 8 shows a cross section of one particular type of magnet and r-f coil combination in which the magnetic field is generated by a solenoid bathed in liquid helium so that its electrical resistance is essentially zero, thereby achieving superconductivity. Electromagnets and permanent magnets are also commonly used for nmr spectroscopy (14). The r-f coil in Figure 8 is of the Helmholtz design and is used to provide the appropriate r-f energy. A solenoid-type coil is often employed with electromagnet and permanent magnet systems.

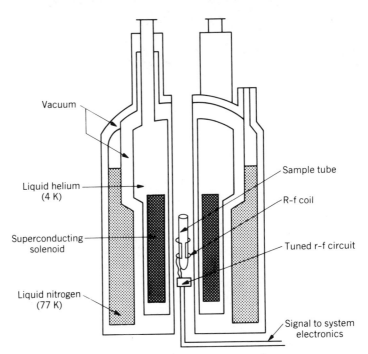

Fig. 8. Cross section of a superconducting magnet for nmr spectroscopy, showing the location of the sample. The magnet and sample tube are not drawn on the same scale. The diameter of the magnet assembly is ca 70 cm, whereas the diameter of the sample tube is ca 1 cm.

The magnetic field B_0 removes the degeneracy of the nuclear-magnetic-spin energy levels, and the electromagnetic r-f field B_1 excites transitions between these energy levels. The resonance condition occurs when the frequency of the r-f field B_1, in MHz, is exactly equal to the Larmor frequency of the nucleus under observation:

$$B_1 = \nu_0 = \gamma \frac{B_0}{2\pi} \qquad (8)$$

The actual method for achieving resonance is more complicated than it first appears. Because most molecules generally have more than one type of magnetically nonequivalent group, eg, a methyl and a methine, a single sample has multiple Larmor frequencies, perhaps fractions of a part per million (ppm) apart, corresponding to each of the chemical shifts. Any excitation method must cover all of the Larmor frequencies in the sample. The two principal methods to achieve the necessary resonant condition, continuous-wave (cw) or Fourier transform (ft), are conceptually different. In the cw method, the resonant condition is met for each nucleus by sweeping either the B_1 radio frequency or the B_0 field. Each magnetically equivalent nucleus in the sample is successively brought into resonance by the sweeping process. When a particular nucleus comes into resonance, it induces a voltage in the pickup coil. This signal, microvolts or less in magnitude, is amplified and detected directly in the frequency domain and recorded in a plot of voltage (or intensity) vs frequency.

In contrast, the ft method involves detection in the time domain and subsequent Fourier transformation into the frequency domain. A short burst or pulse of r-f signal, nominally monochromatic at ν_0, is used to excite all of the Larmor frequencies in a sample simultaneously. In terms of the energy-level diagrams of Figure 1, the effect of the r-f pulse is to equalize the populations of the nuclear-spin energy levels. Following the r-f pulse, the spins undergo a free induction decay (FID) as they reestablish their equilibrium populations. Vector diagrams are useful for visualizing the effect of the applied B_1 on the nuclear spins. In an assembly of spins, more spins are aligned along the B_0 field direction than against it; the excess population is given by the Boltzmann distribution. In a vector diagram, this is shown by drawing the net magnetic moment M_0 as a vector along the field, or z′, direction (Fig. 9a). The primes indicate that the vector-diagram frame of reference is a rotating frame, rotating at the Larmor frequency. The rotating frame simplifies the mathematical description of the spins. Earth, for example, is a rotating frame of reference. A ball thrown straight up appears to come straight down, whereas to an extraterrestrial observer, it would appear to traverse a parabolic trajectory.

An ideal r-f pulse, applied perpendicular to the field direction, equalizes the populations of the energy levels by flipping the magnetization completely into the x′y′ plane. Such a pulse is called a 90° pulse (Fig. 9b and c). The time necessary to achieve a 90° pulse is inversely related to the power of the r-f transmitter, but is generally between 3 and 30 μs. Once in the x′y′ plane, the spins strive to reestablish their initial state (Fig. 9d and e) through T_1 and T_2 relaxation processes. The T_2 process is a dephasing of the spins in the x′y′ plane, caused by spin–spin interactions. T_2 values in solids are short because there are many close-by and fixed spin–spin interactions, whereas in liquids T_2 may be as long as T_1. The spin–lattice or T_1 relaxation causes the spins to relax along the z′ direction. Spin–lattice relaxation times for polymers in solution are generally short, often less than 1 s. However, for polymers in the solid state, the T_1 can be very long, even minutes. The T_1 for carbon in gem-quality diamonds is estimated to be on the order of hours (15), as is that of ^{29}Si in silicon wafers (16). Generally, many signals must be accumulated to obtain a spectrum with an adequate signal-to-noise ratio, especially in the case of nuclei with low natural abundance, such as ^{13}C and ^{29}Si. The T_1 relaxation time determines the rate at which a sample can be pulsed and retain intensity fidelity in the spectrum.

The events in the vector diagrams in Figure 9 may also be cast into the pulse-sequence representation shown in Figure 10a, which illustrates that the

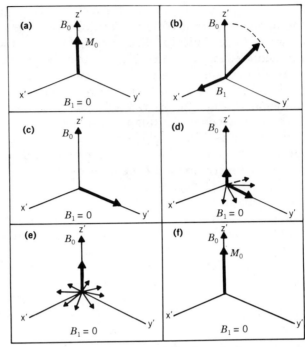

Fig. 9. Rotating-frame diagrams describing the pulsed nmr experiment. The "primed" axes indicate a rotating coordinate system. (**a**) The net magnetization M_0 is aligned along the magnetic-field direction. (**b**) and (**c**) An r-f field B_1 is applied perpendicular to B_0. The duration of B_1 is sufficient to tip the net magnetization by 90°. (**d**) and (**e**) The spins begin to relax in the x'y' plane by spin–spin (T_2) processes and in the z' direction by spin–lattice (T_1) processes. (**f**) The equilibrium magnetization is reestablished along B_0.

detected signal, or FID, is obtained as a voltage vs time. This pulse sequence also shows the delay time, the time needed to wait for T_1 relaxation before performing the pulse cycle again. Generally, many FIDs are added to produce a spectrum. Functions such as the FID can be expressed as an infinite sum of sine and cosine functions, called a Fourier series. The time-domain and frequency-domain signals are Fourier inverses, and the frequency-domain spectrum is obtained from the time domain signal by

$$F(\omega) = \int_{-\infty}^{+\infty} f(t)e^{-i\omega t} \, dt \qquad (9)$$

All nmr spectra are reported in the frequency domain, even though they may be acquired in the time domain. The chief advantage of pulsed Fourier transform techniques is in time savings. The data can be collected all at once, rather than from a slow sweep of the field. The ft method is also well suited to signal averaging, as many FIDs from a weak signal can be added prior to Fourier transformation. The ft method is compatible with digital filtering, digital-data manipulation, and signal-enhancement techniques available on minicomputers.

Figure 11 shows a block diagram of an nmr spectrometer and relates the signal acquisition part just examined to the rest of the spectrometer. The steps in obtaining an nmr signal are as follows. The sample is placed in a sample cell in the probe and the assembly is positioned so that the sample resides in the

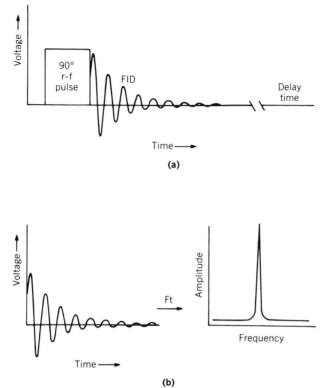

Fig. 10. (**a**) A pulse sequence, showing a single r-f pulse and the ensuing decay signal. (**b**) Fourier transformation of the time-domain FID into the frequency-domain nmr signal.

most homogeneous part of the magnetic field (Fig. 8). The operator then has the computer instruct the pulse programmer to begin the nmr experiment. The pulse programmer is a timing device that sends out precisely timed digital signals. The r-f transmitter superimposes r-f signals on these digital signals, and the r-f pulses are then amplified and sent to the sample probe. This burst of r-f energy excites the spins in the sample. This excitation produces the FID, which is amplified by the preamplifier and detected by conversion to an audio signal. These audio signals are filtered and then converted from their analog form to a digital representation using an analog-to-digital (A–D) converter. The signals are stored in digital form in the computer for further processing and eventual plotting.

Decoupling in Solids and in Solution. The smaller of the two sources of nuclear coupling is the scalar, or J coupling, that arises from indirect spin–spin coupling. Substantial simplification of ^{13}C spectra is obtained upon removing this coupling by applying an additional r-f field B_2 resonant at the frequency of the spin, the influence of which is to be removed, usually protons. This auxiliary field must be larger in frequency than the spin–spin interaction. It causes the spin to oscillate rapidly compared to the spin–spin interaction, thereby decoupling its influence. Figure 12 shows a solution-state ^{13}C nmr spectrum of dioxane with and without such scalar decoupling. The spectral simplification and increased sensitivity that decoupling produces are readily evident.

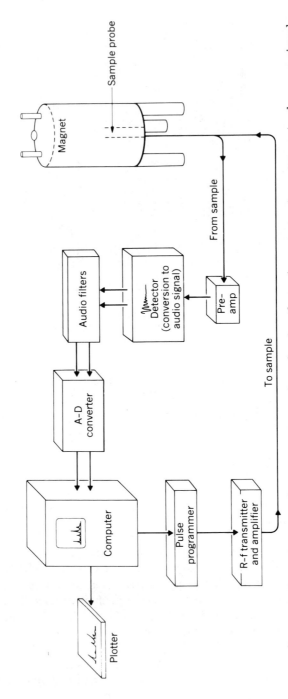

Fig. 11. Block diagram of an nmr spectrometer, showing the main components necessary to observe a signal. (The lock and decoupler are not shown.) Figure 8 shows a cross section of the magnet and illustrates the position of the sample.

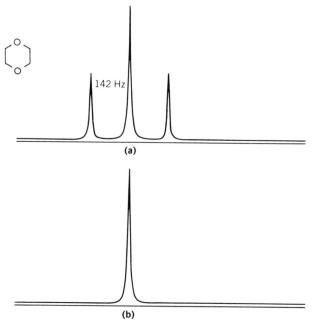

Fig. 12. ^{13}C Nmr spectra of dioxane (**a**) without proton decoupling; (**b**) with proton decoupling. J_{C-H} is 142 Hz.

Dipolar decoupling, or high power proton decoupling, is used to remove the effects of dipolar coupling in solids. It is analogous to the removal of the scalar J couplings; however, the strength in frequency of the applied auxiliary field must be greater than the dipole interaction. This means that the decoupling field must be of much greater magnitude than is normally used in liquids, which in turn requires r-f circuits with more robust components than found in a solution-state probe. Dipolar coupling and its effects in solids and in solution are summarized in Figure 13.

Figure 14**a** shows the type of spectrum that is obtained from a solid polymer sample (Hytrel copolyester) when dipolar decoupling is employed (17). Only a broad, featureless envelope is observed without dipolar decoupling. The three broad peaks from left to right can be assigned to overlapping carbonyl and aromatic resonances, OCH$_2$ groups, and CH$_2$CH$_2$ carbons. The resolution in this spectrum is inferior compared with that of this polymer in solution (Fig. 14**d**). The principal source of broadening after dipolar coupling is removed (Fig. 14**a**) is the chemical-shift anisotropy.

Chemical-shift Anisotropy Powder Patterns. The resonance frequency, or chemical shift, of a particular nucleus is related to the electronic screening about the nucleus. This screening is anisotropic and is described by the chemical-shift tensor. The chemical-shift tensor has principal values that define the magnitude of the shielding, and direction cosines that specify the orientation of the tensor with respect to the molecular framework. Because rapid molecular tumbling occurs in solution, the three-dimensionality of the electronic screening can be ignored. The single line that is observed in solution nmr, eg, Figure 12**b**, represents the isotropic average of the tensorial quantity describing the electronic

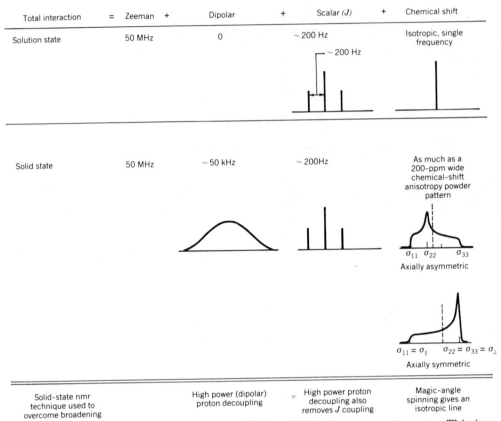

Total interaction	= Zeeman +	Dipolar +	Scalar (J) +	Chemical shift
Solution state	50 MHz	0	~ 200 Hz	Isotropic, single frequency
Solid state	50 MHz	~ 50 kHz	~ 200 Hz	As much as a 200-ppm wide chemical-shift anisotropy powder pattern
Solid-state nmr technique used to overcome broadening	High power (dipolar) proton decoupling	High power proton decoupling also removes J coupling	Magic-angle spinning gives an isotropic line

Fig. 13. Solution-state vs high resolution solid-state nmr spectroscopy. This is a specific example showing ^{13}C nuclear-spin interactions in a 4.7-T (47,000-G) field (50.3 MHz for ^{13}C).

screening. In general, the chemical-shift interaction is proportional to $\gamma \hbar \sigma B_0$, where

$$\sigma = \sigma_{11}\lambda_{11}^2 + \sigma_{22}\lambda_{22}^2 + \sigma_{33}\lambda_{33}^2 \qquad (10)$$

where σ_{ii} is the principal value of the chemical-shift tensor and λ_{ii} the direction cosine (3,18). The isotropic chemical shift, or the chemical shift that is observed in solution, is described as one-third the sum of the diagonal elements of the tensor (the sum of the diagonal elements of the tensor is called the trace):

$$\sigma_i = \frac{1}{3}(\sigma_{11} + \sigma_{22} + \sigma_{33}) \qquad (11)$$

These equations show that the chemical shift of a particular nucleus in the solid state depends on the orientation of the molecule with respect to the magnetic field. The chemical shift of a particular carbon in a single crystal (where all of the carbons have the same orientation with respect to the magnetic field) changes as the crystal is rotated in the field. In a powder, however, all crystallite orientations are present and the resultant nmr spectrum describes the chemical-shift-tensor powder pattern, or the probability of finding the various orientations with respect to the magnetic field.

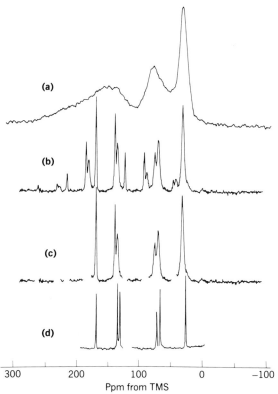

Fig. 14. ^{13}C Nmr spectra (50.3 MHz) of a Hytrel copolyester (17): (**a**) static powder pattern of the solid polymer obtained by cross-polarization and dipolar decoupling; (**b**) spectrum obtained with magic-angle spinning (mas) at 2.3 kHz (contains sidebands); (**c**) spectrum in which sidebands have been removed; and (**d**) solution-state nmr spectrum of the same polymer. Courtesy of the American Chemical Society.

Two theoretical chemical-shift-tensor powder patterns are illustrated in Figure 13. The axially asymmetric pattern is obtained when $\sigma_{11} \neq \sigma_{22} \neq \sigma_{33}$, which is the case for most solids. The axially symmetric powder pattern is obtained when either $\sigma_{11} = \sigma_{22}$ or $\sigma_{22} = \sigma_{33}$. The positions of the isotropic chemical shifts are shown as dotted lines in Figure 13. In the axially symmetric case shown, σ_{11} is the observed frequency when the principal axis system is parallel with the field direction and σ_{33} is the frequency when the principal axis system is perpendicular to the field direction.

Although the width of the chemical-shift powder pattern usually obscures detailed structural information (Fig. 14**a**), it contains considerable angular-dependent information that is particularly useful when determining the exact nature of anisotropic molecular motion. Table 2 shows the principal values for some representative chemical-shift tensors for polymers. In general, carbonyl, carboxyl, and aromatic carbons have the largest powder-pattern widths ($\sigma_{11}-\sigma_{33}$), approximately 180–250 ppm. The widths for methyl, methylene, and methine carbons are usually much smaller, on the order of 30–60 ppm.

Magic-angle Spinning.   Used in conjunction with dipolar decoupling, magic-angle spinning (mas) produces high resolution nmr spectra of solids. Mas consists

Table 2. Principal Components of ^{13}C Chemical Shieldings^a for Representative Polymers

Carbon type	Sample	σ_{11}	σ_{22}	σ_{33}	$\|\sigma_{11}-\sigma_{33}\|$, ppm	Ref.
aromatic						
protonated	poly(butylene terephthalate)				198	17
	poly(ethylene terephthalate)	226	153	15	211	19
nonprotonated	poly(butylene terephthalate)				202	17
	poly(ethylene terephthalate)	226	153	15	211	19
	graphite	251	20	20	231	20
carbonyl, ester,	poly(butylene terephthalate)				127	17
amide	poly(ethylene terephthalate)	250	122	122	128	19
	glycylglycine (amide C)	244	177	88	174	21
olefinic	*trans*-polyacetylene	218	138	22	196	22
	cis-polyacetylene	219	144	47	172	22
	trans-polybutadiene	228	116	50	178	23
CH_2	poly(butylene terephthalate)					
	—OCH_2—				60	17
	—CH_2CH_2—				14	17
	poly(ethylene terephthalate)	80	80	28	52	19
	polyethylene	50	36	12	38	24
	n-eicosane^b	50	38	17	33	25
CH_3	*n*-eicosane^b	26	22	3	23	25

^{a} Relative to TMS; $\sigma_{11} > \sigma_{22} > \sigma_{33}$; determined at ambient temperature unless specified otherwise.
^{b} Determined at $-95°C$.

of mechanically rotating the sample about an axis that forms the "magic angle" (54.7°) with respect to the external magnetic field B_0. The significance of this angle has long been recognized (26,27). Its potential to cause narrowing of nmr lines was tested experimentally in the late 1950s (26); mas was used to narrow the ^{23}Na line in NaCl and the dipolar-broadened ^{19}F lines in CaF_2. Early attempts with mas on dipolar-broadened lines were not totally successful because it is not possible to spin the sample fast enough to average out the ca 50-kHz interaction strength. However, in 1976 it was recognized that if the large dipolar broadening was removed by dipolar decoupling, the remaining broadening due to the chemical-shift anisotropy powder pattern could be removed by spinning the sample about the magic angle (28).

Rapid mechanical rotation of solid samples about an angle β with respect to the magnetic-field direction causes the direction cosines to become time-dependent in the rotor period. The expression for the chemical shift becomes

$$\sigma = ½ \sin^2 \beta(\sigma_{11} + \sigma_{22} + \sigma_{33}) + ½(3 \cos^2 \beta - 1)$$
$$\times \text{(functions of direction cosines)} + \text{(time-dependent terms)} \quad (12)$$

When β is 54.7°, the first term on the right is one-third times the sum of the diagonal elements of the chemical-shift tensor, which is precisely the expression for the isotropic chemical shift (eq. 11). The $(3 \cos^2 \beta - 1)$ term is zero at the magic angle. The final time-dependent terms in the above equation produce spinning sidebands.

Rapid sample rotation causes the angular term in the expression for the dipole–dipole interaction to become time-dependent. If it were possible to spin the sample rapidly enough at the magic angle, this interaction would also be averaged to zero, but this is generally not the case for dipolar broadening between protons and carbons. It is a possibility for nuclei that have no directly bonded protons, eg, ^{29}Si in polysilanes and ^{31}P in polyphosphazenes. The scalar, or J coupling, does not vanish with mas. However, J couplings between carbons and protons, for example, are generally not observed in solid-state nmr spectroscopy because the carbon spectra are obtained with high power decoupling, which readily removes the much weaker scalar coupling (Fig. 13). It is also possible to have dipolar coupling from adjacent quadrupolar nuclei. If this coupling is small enough, it can be effectively averaged out by mas. Often, however, the quadrupole interaction is a substantial perturbation of the Zeeman interaction, in which case the quadrupolar nucleus is quantized along an axis different from that of the spin-½ nucleus, and the optimum angle for removing the dipolar coupling is no longer the magic angle (29,30).

The effect of mas is illustrated in Figure 14. The chemical-shift-tensor powder pattern of the Hytrel copolyester of Figure 14**a** is narrowed into a series of centerbands and sidebands in Figure 14**b**. The sidebands arise because the sample is spun at 2.3 kHz, not fast enough to average completely the ca 10-kHz chemical-shift anisotropy for the carbonyl and aromatic carbons of this sample. Figure 14**c** has been edited to show only the centerbands, or isotropic chemical shifts. This spectrum compares favorably to the solution-state spectrum of the same polymer, shown in Figure 14**d**.

In the mas experiment, the solid sample is placed in a polymeric or ceramic sample holder and spun rapidly about an axis that forms the magic angle with respect to the magnetic-field direction (Fig. 15). The rate of spinning is usually greater than 2 kHz and only rarely is it possible to spin the sample faster than 5 kHz (3 kHz corresponds to 180,000 rpm). Dipolar decoupling and special pulse sequences are performed during the spinning experiment. The sample is spun on an air bearing with high velocity drive gas, usually nitrogen, from the bottom jets and stabilized with a slower stream of gas from the upper jets. There are many designs, geometries, and materials for spinners. Some of the more common materials are Delrin (acetal resin), Macor (machinable ceramic), boron nitride, Vespel, aluminum oxide (Al_2O_3), perdeuterated poly(methyl methacrylate) (PMMA), and Kel-F (polychlorotrifluoroethylene).

Cross-polarization.   In the ft method of obtaining an nmr signal, a single 90° pulse applied at the appropriate Larmor frequency produces a FID signal (Fig. 10) for both solutions and solids. However, in solids there is often limited molecular motion, leading to inefficient spin–lattice relaxation. Even in solids containing methyl groups, which generally rotate rapidly, the T_1 values for the other carbons are prohibitively long. The carbon of the methyl group cannot "communicate" its short T_1 value to other carbons in the sample by spin diffusion (mutual spin flips) because its nearest ^{13}C neighbor is on average 0.7 nm away. Even though the protons on the methyl carbon do communicate their short T_1 to other protons in the sample by diffusion to nearby spins, these other protons cannot communicate their short T_1 back to other carbons in the sample because the carbon and proton Larmor frequencies are far apart. Figure 16 shows that there is no overlap in the carbon and proton frequencies, and hence in their energies.

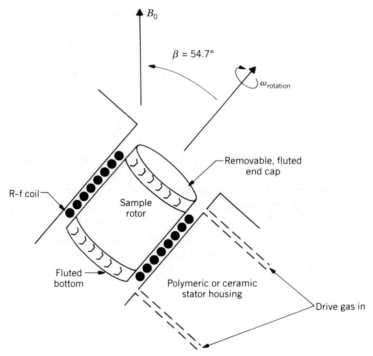

Fig. 15. A rotor (sample container) for magic-angle spinning (mas). The stator, or housing, accommodates this double air-bearing assembly.

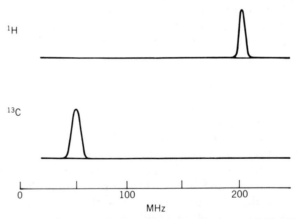

Fig. 16. Carbon and proton Larmor frequencies in a 4.7-T (47,000-G) magnetic field. There is no frequency overlap and thus no overlap in the energies.

Cross-polarization (31,32) is a technique that takes advantage of the facts that proton spin diffusion generally causes all of the protons in a solid to have the same T_1 value and that the proton T_1 is usually short compared with the carbon T_1 values. Here, cross-polarization from protons to carbons is discussed; however, these concepts apply to polarization between any abundant and rare spin pair. Cross-polarization works by effectively forcing an overlap of the proton and carbon energies in the rotating frame. It takes advantage of the uniformity of the proton spin bath, ie, the fact that all the protons can communicate efficiently with each other. Although Figure 16 illustrates that there is no energy overlap between carbons and protons in the laboratory frame, it was demonstrated in 1962 that an overlap in energies could be made to occur in the rotating frame (33). Thus energy transfer between protons and ^{13}C nuclei with disparate Larmor frequencies can be made to occur when

$$\gamma_C B_{1C} = \gamma_H B_{1H} \tag{13}$$

This equation, called the Hartmann-Hahn condition, results in a match of the rotating-frame energies for ^1H and ^{13}C. Since γ_H is four times γ_C, this match occurs when the strength of the applied carbon field B_{1C} is four times the strength of the applied proton field B_{1H}. When the proton and carbon rotating-frame energy levels match, polarization is transferred from the abundant spin (protons) to the rare spin (carbons). Because polarization is transferred from the protons to the carbons, it is the T_1 of the protons that dictates the repetition rate for signal averaging.

Vector diagrams (Fig. 17) and the pulse-sequence timing (Fig. 18) are useful for understanding this double rotating-frame experiment. Cross-polarization is called a double rotating-frame experiment because the proton frame of reference and the carbon frame of reference rotate at their respective Larmor frequencies. The goal of the experiment is to create a time dependence of the spins that is common to the two rotating frames. First, the proton and carbon spin systems

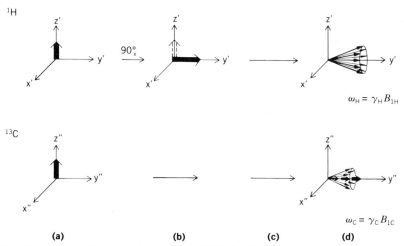

Fig. 17. Vector diagram for a ^1H and ^{13}C double rotating-frame cross-polarization experiment. The carbon frame and the proton frame are rotating at different frequencies. (a) Spin equilibration; (b) r-f pulse applied at proton Larmor frequency; (c) spin locking of protons along y' axis; r-f field applied to carbon and ^{13}C magnetization grows in direction of spin lock; (d) protons and carbons precess about the y axis.

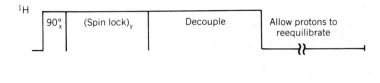

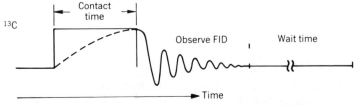

Fig. 18. The ^{1}H–^{13}C cross-polarization pulse sequence for solid-state nmr.

are allowed to equilibrate in the magnetic field (Fig. 17a). An r-f pulse (B_{1H}) is applied at the proton Larmor frequency along the x′ axis and is of sufficient duration to tip the protons by 90°, so that they are then along the y′ axis (Fig. 17b). The phase of the proton B_1 field is shifted by 90°, ie, the r-f signal is shifted by one-quarter of a wavelength, and the protons are then spin-locked along the y′ axis (34). This process is called spin locking because the proton spins are forced to precess about the y′ axis of their rotating frame for the duration of the powerful B_{1H} pulse. The protons precess about their rotating-frame y′ axis with a frequency $\omega_H = \gamma_H B_{1H}$ (Fig. 17d). Meanwhile, the carbons are put into contact with the protons by turning the carbon field B_{1C} on during the spin-lock time (Fig. 18). This causes the carbon magnetization to grow in the direction of the spin-lock field. The carbons precess about their y″ axis with a frequency $\omega_C = \gamma_C B_{1C}$ (Fig. 17d).

Figure 19 illustrates the mechanism by which the actual polarization transfer occurs. The protons are precessing about the B_{1H} field with frequency $\gamma_H B_{1H}$ and the carbons are simultaneously precessing with frequency $\gamma_C B_{1C}$. When the Hartmann-Hahn condition, $\gamma_H B_{1H} = \gamma_C B_{1C}$, is met, the z-components of both the proton and carbons must have the same time dependence. Because the z-component time dependence is common to the two spin systems, mutual spin flips can occur between the protons and the carbons. This process can be visualized as a "flow" of polarization from the abundant proton spins to the rare carbon spins.

The advantages of cross-polarization are twofold. The method circumvents the problem of the long carbon T_1 values normally found in solids. Because the carbons obtain their polarization from the protons, it is the generally shorter proton T_1 that dictates the repetition rate of the cross-polarization pulse sequence. In addition, the carbons obtain a significant enhancement in signal intensity from the cross-polarization experiment. This enhancement can be as large as the ratio of $\gamma_H:\gamma_C$, or a factor of four. Thus cross-polarization provides both a time savings and an improvement in the signal-to-noise ratio of the spectrum.

Observation of Nuclear Relaxation.   Nmr relaxation measurements on polymers are particularly useful for understanding the complex molecular motions of polymer chains, both in solution and in the solid state (11). Essentially

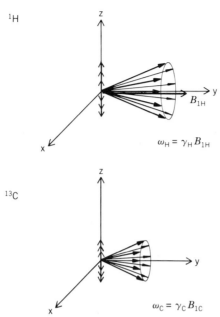

Fig. 19. A more detailed representation of Figure **17d**. The carbons are precessing with frequency $\omega_C = \gamma_C B_{1C}$, and the protons with frequency $\omega_H = \gamma_H B_{1H}$. When the Hartmann-Hahn match is achieved ($\gamma_H B_{1H} = \gamma_C B_{1C}$), the two spin systems have z-components with the same frequency dependence. Thus mutual spin flips can occur between ^{1}H and ^{13}C.

all relaxation in polymers is produced by molecular motion, which sets up fluctuating local dipolar fields. There are four primary types of nuclear relaxation: T_1, T_2, nuclear Overhauser enhancement (NOE), and $T_{1\rho}$. Their measurements are summarized in Table 3, where they are compared with other nmr methods for determining molecular motion.

Table 3. Nmr Experiments Useful for Determining Molecular Motion

Experiment	Most informative time scale, s	Comments	Refs.
T_1	10^{-6}–10^{-10}	measured by inversion–recovery or progressive saturation; useful for solutions and solids	11,35–38
NOE	10^{-8}–10^{-10}	most useful for solutions; gives information on fraction of relaxation that is dipolar	11,39
$T_{1\rho}$ (C)	10^{-2}–10^{-4}	for solids	40,41
narrowing of chemical-shift powder pattern	10^{-1}–10^{-3}	measured by off-axis spinning	42
deuterium nmr lineshapes	10^{-4}–10^{-7}	requires specific labeling of sample; for solids	13,43
T_2	10^{-5}–10^{-8}	most useful for solutions	44

The T_1, or longitudinal relaxation, describes the return of the spin system to equilibrium by coupling to its lattice. The T_2, or transverse relaxation, describes the efficiency of the spin–spin (or flip-flop) interactions. NOE measures cross-relaxation. For example, when the protons attached to a ^{13}C are irradiated, simultaneous spin flips of the carbons and protons occur, causing perturbations in the ^{13}C energy levels that are manifest as enhancements in the resonance intensities. NOE is a measure of the amount of relaxation that occurs by dipole–dipole interaction. The $T_{1\rho}$ describes the loss of spin magnetization in the rotating frame as a function of time that the spin-lock field is turned off. The theory pertaining to T_1, $T_{1\rho}$, T_2, and NOE measurements is described later.

T_1.   One of the more accurate ways to measure spin–lattice relaxation is the inversion–recovery method, illustrated in Figure 20. The sample is placed in the magnetic field and allowed to equilibrate, at which point the net magnetization M_0 is along the field direction B_0. A 180° r-f pulse tips this magnetization vector into the negative z-direction, thereby inverting the magnetization. Relaxation follows the 180° pulse and continues for a variable period τ. The return of the magnetization at time τ is "read out" by a 90° pulse. After waiting the appropriate delay time ($\sim 5 \times T_1$) for the magnetization to return to its equilibrium value M_0, the pulse sequence is repeated. In practice, data from a number of τ values are collected, and log ($M_0 - M$) is plotted as a function of τ. The T_1 is then determined from the slope of the resultant straight line ($T_1 = -1/\text{slope}$).

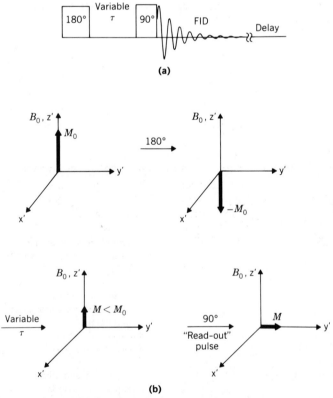

(a)

(b)

Fig. 20.   (a) Pulse-sequence and (b) vector diagrams for T_1 measurement by the inversion–recovery method.

T_2. The T_2 can be determined from the linewidth according to equation 6, or can be measured more accurately using a spin-echo method (45), which eliminates contributions to the linewidth from magnetic-field inhomogeneities (eq. 7). Figure 21 shows the pulse-sequence and vector diagrams for measurement of the T_2. The spins, which have been equilibrated in the magnetic field, are tipped into the x′y′ plane by a 90° pulse. During the variable delay time τ, the spins begin to dephase according to several broadening mechanisms, one of which is spin–spin relaxation. This causes some spins to dephase faster than others. A 180° pulse is then applied, causing the spins to refocus. The spins that had dephased the fastest have further to "travel" before refocussing at time 2τ. However, they are traveling faster in this reversed direction and arrive along the −y′ axis to form an echo at time 2τ. The dephasing caused by magnetic-field inhomogeneities is refocused, and this method is therefore a good way to separate the true T_2 from T_2^* (eq. 7).

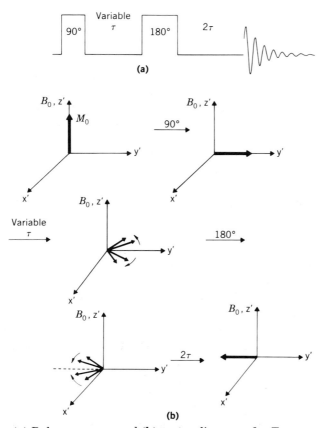

Fig. 21. (**a**) Pulse-sequence and (**b**) vector diagrams for T_2 measurement.

$T_{1\rho}$. $T_{1\rho}$ determinations in solids measure the loss of magnetization in the rotating frame as a function of the time that the spin-lock field is turned off (40,41). The modified cross-polarization pulse sequence used to make these measurements for the specific case of ^{13}C is shown in Figure 22. The pulse sequence follows that described for cross-polarization until the ^{13}C has reached its maximal

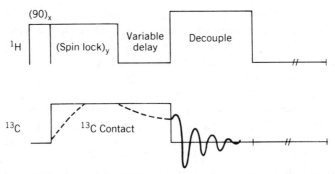

Fig. 22.   Pulse sequence for ^{13}C $T_{1\rho}$ measurements. The dotted line indicates the intensity of the ^{13}C signal.

signal intensity. At this point, the proton spin-lock field is turned off for variable delay times and the decay of the ^{13}C magnetization is monitored. Unlike the T_1, which measures motions that have characteristic frequencies in the megahertz regime, the $T_{1\rho}$ is sensitive to motions in the kilohertz regime.

Nuclear Overhauser Enhancement (NOE).   Measurement of the NOE is usually performed by obtaining the ratio of the signal intensities from two spectra, one with and one without proton irradiation. This ratio is the NOE and is equal to $1 + \eta$, where η is the NOEF, or the nuclear-Overhauser-enhancement factor. Often it is desirable to obtain an entire set of proton NOEs for a substance, which is sometimes done using two-dimensional (2D) NOE measurements.

Two-dimensional (2D Ft Nmr) Methods.   The two main types of two-dimensional nmr experiments are those that resolve chemical shifts and coupling constants and those that correlate nuclei based on interactions such as NOE or coupling. These experiments involve preparation, evolution, and detection periods (46). The correlated experiments also involve a mixing time. The mechanics of a 2D experiment are illustrated in Figure 23 for a homonuclear 2D NOE experiment.

Using the pulse sequence shown in Figure 23a, a set of *n* FIDs are obtained that differ from one another by equally spaced increments of the evolution time t_1 (Fig. 23b). The FIDs are Fourier-transformed (Fig. 23c) and the transformed spectra are then transposed by constructing new FIDs from the columns of points of the original spectra. These new FIDs (Fig. 23d) are then Fourier-transformed to provide a stacked plot (Fig. 23e). If a spin were to "communicate" with its neighbor during the mixing time, that interaction would produce an off-diagonal peak. The one-dimensional spectrum occurs on the diagonal. It is often easier to visualize the data with a contour plot which is simply a topographical map of the entire network of spin interactions (Fig. 23f).

Deuterium Quadrupole Nmr.   The deuterium nucleus (^2H) is particularly useful for probing the details of molecular motion in solid polymers (13,43,47,48). The deuterium nucleus has a spin of 1 (Table 1), thereby producing three equally spaced energy levels (Fig. 1) when placed in a magnetic field. Deuterium is a quadrupolar nucleus, and interaction of the quadrupole moment with the electric-field-gradient tensor at the nucleus produces a substantial perturbation of these three energy levels. The perturbation is actually so large in solids that the dipolar, chemical-shift, and J-coupling interactions can be ignored. The magnitude of the

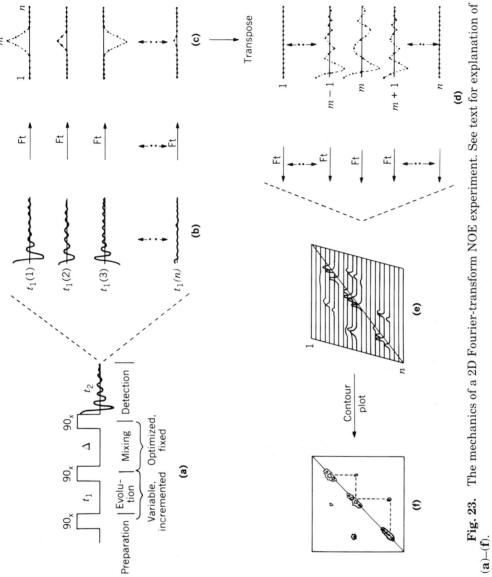

Fig. 23. The mechanics of a 2D Fourier-transform NOE experiment. See text for explanation of (a)–(f).

279

quadrupolar splitting is given by

$$\Delta \nu_{\mathrm{q}} = \frac{3e^2 qQ}{4h} (3 \cos^2 \theta - 1) \tag{14}$$

where the charge on the electron is e, eqQ is the quadrupole-coupling constant, θ is the angle that a particular carbon–deuterium bond vector makes with the external magnetic field, and h is Planck's constant.

In a powder, where all orientations of carbon–deuterium bond vectors are present, the actual spectrum is composed of two overlapping, axially symmetric powder patterns, one from each of the transitions. This pattern is often called a Pake doublet (49). These theoretical patterns are shown in Figure 24a. The left-most figure shows two axially symmetric patterns, one of which is cross-hatched. The $\perp$ and $\parallel$ frequencies occur when the C–D bonds are perpendicular and parallel to the field, respectively. The right part of Figure 24a shows the smoothed lineshape. For a static C–D bond, $\Delta \nu_{\mathrm{q}}$ or d is ca 130 kHz. Figure 24b shows the expected pattern for a rotating methyl group and Figure 24c the pattern for a two-site hop through the tetrahedral angle. Figure 24d and e illustrate the fact that solid-state deuterium nmr spectroscopy can readily distinguish aromatic rings that are flipping by 180° about their 1,4-phenylene axis from those undergoing unrestricted rotation.

Such lineshapes are sensitive to motions that have correlation times in the 10^{-4}–10^{-7}-s range. Polymers, particularly the amorphous regions in semicrystalline polymers, often have large amplitude molecular scale sub-T_g motions in this time range (see GLASS TRANSITION). For this reason, solid-state deuterium-nmr spectroscopy has been used to investigate solid polymers.

Observation of Macromolecular Structure

It is possible to overcome dipolar broadening and obtain solid-state spectra of sufficient resolution to provide structural information. Nevertheless, the narrowest lines and greatest detail are obtained from polymers in solution, where segmental motion is usually rapid enough, particularly at elevated temperatures, to provide spectra often comparable to those of small molecules (Fig. 6). Linewidth is only weakly dependent on molecular weight or the macroscopic viscosity of the solution.

Stereochemical Configuration

Nmr is particularly informative concerning stereochemical microstructure (qv), revealing details seen by no other technique.

Poly(methyl methacrylate).   Figure 25 shows the 500-MHz proton spectra of poly(methyl methacrylate) (PMMA) prepared with a free-radical initiator (A) (Fig. 25a) and with fluorenyllithium in toluene, an anionic initiator (B) (Fig. 25b) (50). The profound effect of the nature of the initiator is evident in the marked differences between these spectra. Interpretation of these spectra depends on consideration of the sequences of two monomer units or diads. There are two possible types of diads with different symmetry properties. The syndiotactic or

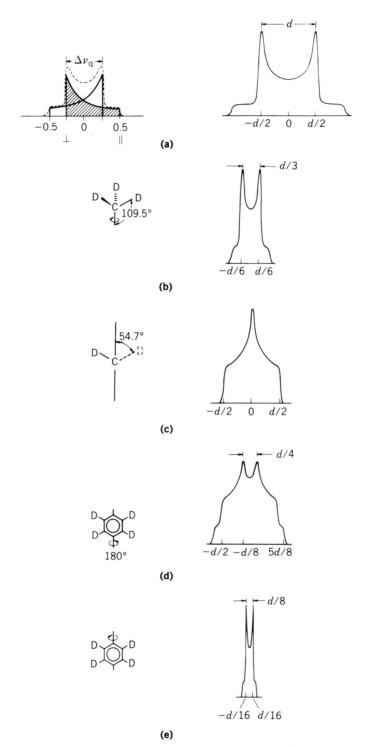

Fig. 24. Theoretical solid-state deuterium nmr powder patterns. The static pattern is shown in (**a**) $\Delta\nu_q$, or d, is ca 130 kHz. (**b**) Rotating methyl; (**c**) two-site hop; (**d**) 180° phenyl-ring flip; (**e**) free diffusion of phenyl ring.

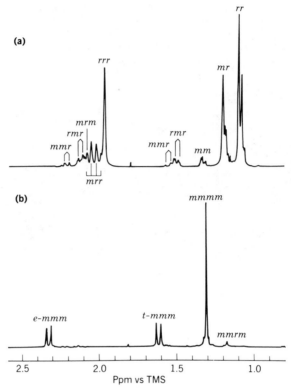

Fig. 25. 500-MHz proton nmr spectra of (**a**) syndiotactic and (**b**) isotactic PMMA (the —OCH$_3$ peaks are not shown): e and t denote erythro and threo, respectively.

racemic (r) diad has a twofold axis of symmetry, and consequently the two methylene protons are in equivalent environments on a time average over the chain conformations, which are not necessarily in the planar zigzag form shown in the projection formulas:

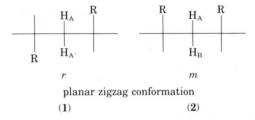

The H$_A$ and H$_A'$ protons therefore have the same chemical shift and appear as a singlet despite strong two-bond or geminal coupling between them. The isotactic or meso (m) diad has a plane of symmetry but no twofold axis, and so the two protons are nonequivalent and have different chemical shifts.

When there is no vicinal coupling to neighboring protons, as is the case in PMMA, the syndiotactic sequences should exhibit a methylene singlet, whereas the isotactic form should give two doublets, each with a spacing equal to the geminal coupling, ca 15 Hz. In Figure 25, the methylene spectrum of the anionic polymer (B) is almost exclusively a pair of doublets; quantitative assessment shows that this polymer is 95% isotactic. The methylene spectrum of the free-radical polymer (A) is more complex, but the principal resonance, at ca 1.9 ppm, is a singlet, showing that this polymer is predominantly syndiotactic but more irregular than B. This is generally the case for free-radical-initiated vinyl polymers.

Proton nmr can provide absolute stereochemical information concerning polymer chains without recourse to x-ray or other methods. Somewhat more detailed, but not absolute, information can be gained from the methyl proton resonances of PMMA near 1.2 ppm. The ester methyl resonance at ca 3.6 ppm is less sensitive to stereochemistry. In both spectra, three peaks (or, more correctly, groups of peaks) are noted, appearing in similar positions but with greatly different intensities. These correspond to the α-methyl groups in the center monomer unit of the three possible triad sequences: isotactic, syndiotactic, and heterotactic.

isotactic, *mm*   syndiotactic, *rr*

heterotactic, *mr* or *rm*

These may be more simply and appropriately designated by the *m* and *r* terminology, as indicated (51). Measurement of the relative intensities of the *mm*, *mr*, and *rr* α-methyl peaks, which appear from left to right in both spectra in this order, gives a valid statistical representation of the microstructure of each polymer.

Triad data give considerable insight into the mechanism of chain propagation and this is one of the principal uses of such information. P_m designates the probability that the polymer chain will add a monomer unit to give the same configuration as that of the last unit at its growing end, ie, that an *mm* diad will be generated. It is assumed that P_m is independent of the stereochemical configuration of the growing chain. The generation of the chain is a Bernoulli-trial process; it is like reaching into a large jar of balls marked *m* and *r* and withdrawing a ball at random. The proportion of *m* balls in the jar is P_m. Since two monomer additions are required to form a triad sequence, the probabilities of their formation are (51–53)

$$[mm] = P_m^2 \qquad (15)$$

$$[mr] = 2P_m(1 - P_m) \qquad (16)$$

$$[rr] = (1 - P_m)^2 \qquad (17)$$

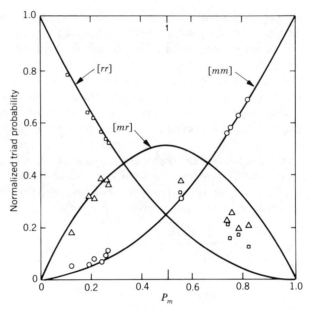

Fig. 26. The probabilities of isotactic $[mm] = P_m^2$, heterotactic $[mr] = 2P_m(1 - P_m)$, and syndiotactic $[rr] = (1 - P_m)^2$ triads as a function of P_m, the probability of m placement. The points on the left-hand side are for methyl methacrylate polymers prepared with free-radical initiators (Fig. 25a); those on the right-hand side are for polymers prepared with anionic initiators (Fig. 25b). For polymer 25a, the points for $[rr]$ have been arbitrarily placed on the $[rr]$ curve and the others fall where they may; for polymer 25b, the $[mm]$ points have been placed on the $[mm]$ curve.

The heterotactic sequence must be given double weighting because both directions, mr and rm, which are observationally indistinguishable, must be counted. A plot of these relationships is shown in Figure 26 (51). The proportion of mr (heterotactic) units rises to a maximum when P_m is 0.5, corresponding to a strictly random or atactic configuration for which $[mm]:[mr]:[rr]$ is 1:2:1. For any Bernoullian polymer, the $[mm]$, $[mr]$, and $[rr]$ sequence intensities lie on a single vertical line in Figure 26, corresponding to a single value of P_m. Spectrum **a** in Figure 25 corresponds to these simple statistics, P_m being 0.25 ± 0.01. The polymer corresponding to spectrum **b** does not. The propagation statistics in this case can be interpreted to indicate that the probability of isotactic placement depends on the stereochemical configuration of the growing chain and cannot properly be expressed by a single parameter such as P_m. Free-radical and cationic propagations always give predominantly syndiotactic chains. Anionic initiators may also do so if strongly complexing ether solvents such as dioxane or glycol dimethyl ether are employed rather than hydrocarbon solvents as in polymer B (Fig. 25b) (51,54) (see also ANIONIC POLYMERIZATION).

The fine structure in the methylene and methyl regions of Figure 25 corresponds in **a** principally to residual resonances of the stereoirregular portions of the chains; in **b** such residual resonances are less conspicuous. These residual resonances arise from sensitivity to longer stereochemical sequences than diad and triad. Table 4 shows planar zigzag projections of such sequences, together

Table 4. Diad, Triad, Tetrad, and Pentad Sequences

	α-Substituent			β-CH₂		
	Designation	Projection	Bernoullian probability	Designation	Projection	Bernoullian probability
triad	isotactic, mm (i)		P_m^2	diad meso, m		P_m
	heterotactic, mr (h)		$2P_m(1-P_m)$	racemic, r		$(1-P_m)$
	syndiotactic, rr (s)		$(1-P_m)^2$			
pentad	mmmm (isotactic)		P_m^4	tetrad mmm		P_m^3
	mmmr		$2P_m^3(1-P_m)$	mmr		$2P_m^2(1-P_m)$
	rmmr		$P_m^2(1-P_m)^2$	rmr		$P_m(1-P_m)^2$
	mmrm		$2P_m^3(1-P_m)$	mrm		$P_m^2(1-P_m)$
	mmrr		$2P_m^2(1-P_m)^2$	rrm		$2P_m(1-P_m)^2$
	rmrm (heterotactic)		$2P_m^2(1-P_m)^2$	rrr		$(1-P_m)^3$
	rmrr		$2P_m(1-P_m)^3$			
	mrrm		$P_m^2(1-P_m)^2$			
	rrrm		$2P_m(1-P_m)^3$			
	rrrr (syndiotactic)		$(1-P_m)^4$			

with their frequency of occurrence as a function of P_m, assuming Bernoullian propagation (52). The tetrads, and all "even-ads," refer to observations of β-methylene protons (or β-carbons); the "odd-ads" refer to substituents on the α-carbons (or α-carbons themselves). Resonances for tetrad sequences or higher even-ads should appear as fine structure in the diad spectra; pentad sequences or higher odd-ads should appear as fine structure on the triad resonances. The assignments to longer sequences as indicated on the spectra are based on Bernoullian probabilities in spectrum 25**a**; those in spectrum 25**b** are primarily based on **a**. The r-centered tetrads, eg, mrr, do not necessarily appear as singlets if the sequence as a whole lacks a twofold axis.

The numbers of observationally distinguishable configurational sequences, or n-ads, designated $N(n)$, obey the following relationship (52):

n	2	3	4	5	6	7	8	9
$N(n)$	2	3	6	10	20	36	72	136

or, in general,

$$N(n) = 2^{n-2} + 2^{m-1} \tag{18}$$

where $m = n/2$ if n is even and $m = (n-1)/2$ if n is odd. Discrimination of these longer sequences is unlikely beyond $n = 6$ (hexads) or $n = 7$ (heptads). The observation of such sequences permits rather searching tests of polymerization mechanisms.

Polypropylene. Polypropylene (PP) is one of the few vinyl polymers that can be prepared in both isotactic and syndiotactic forms with coordination catalysts (see INSERTION POLYMERIZATION). The proton spectra are relatively complex because of vicinal coupling between α- and β-protons and α- and methyl protons, as well as geminal methylene proton coupling in isotactic sequences (Fig. 27):

$$\underset{\alpha\beta}{-(\overset{\overset{\displaystyle CH_3}{|}}{CH}-CH_2)_n}$$

The β-protons of the syndiotactic polymer appear as a triplet at 1.03 ppm, corresponding to a single chemical shift, and J coupling to two neighboring α-protons. In the isotactic polymer, they appear as widely spaced multiplets at 1.27 ppm and 0.87 ppm, corresponding to syn and anti positions in the trans–trans conformation (56,57):

$$\text{(anti)} H \diagdown \diagup H \text{(syn)}$$
$$H \diagup \diagdown CH_3 \; H$$

Analysis of these spectra yields the following values for the vicinal main-chain couplings (in both polymers, the vicinal CH_3–H_α coupling is 5 Hz and the geminal methylene proton coupling is -13.5 Hz):

Isotactic: $J_{H_\alpha - H_{syn}}$: 6.0 Hz
$ J_{H_\alpha - H_{anti}}$: 7.0 Hz

Syndiotactic: $J_{H_\alpha - CH_2}$: 4.8, 8.3 Hz

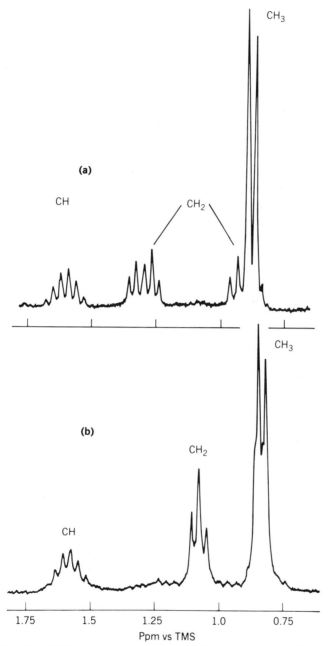

Fig. 27. The 220-MHz proton nmr spectra of (**a**) isotactic and (**b**) syndiotactic polypropylenes observed in *o*-dichlorobenzene at 165°C (55).

Such couplings strongly depend on the dihedral angle between vicinal protons (Fig. 5). In polymers, a gauche arrangement has a proton–proton coupling constant of ca 2–4 Hz and protons in a trans conformation show couplings of 8–13 Hz. A time-averaged value of J is observed owing to rapid conformational equilibration, and from this the populations of gauche and trans conformers can be

calculated (58). For PP, substantial proportions of both gauche and trans conformations are found for syndiotactic sequences, whereas isotactic chains show
a strong preference for (gt)(gt)(gt) sequences, ie, a threefold helical conformation
(see CONFORMATION AND CONFIGURATION).

The proton spectrum of atactic PP is virtually uninterpretable, being a
complex of overlapping multiplets (1). In the ^{13}C spectra of Figure 28, the multiplicity arising from ^{1}H–^{13}C J couplings has been removed by proton irradiation.
The isotactic and syndiotactic polymers give similar spectra but with readily
observable chemical-shift differences, especially for the methyl carbons. This
sensitivity to stereochemical configuration is particularly clear in the spectrum
of the atactic polymer in which the methyl resonance is split into peaks corresponding to nine of the ten possible pentad sequences (Table 4). The syndiotactic
polymer is less stereoregular than the isotactic one and the configurational statistics of both the atactic and syndiotactic polymers depart markedly from Bernoullian. This is generally the case for chains generated by coordination catalysts.

^{13}C Chemical shifts can be effectively rationalized through recognition of a

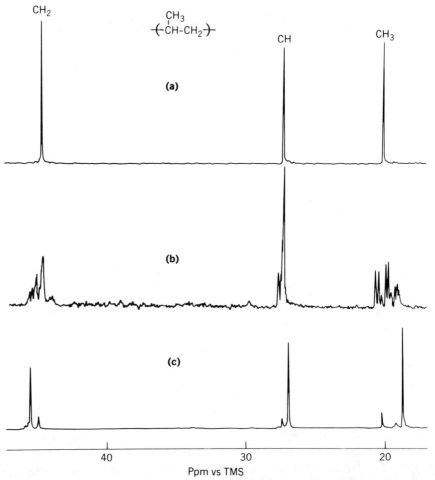

Fig. 28. The 25-MHz ^{13}C spectra of three preparations of PP observed as 20%
(wt/vol) solutions in 1,2,4-trichlorobenzene at 140°C: (**a**) isotactic; (**b**) atactic; (**c**) syndiotactic.

γ-gauche shielding effect: when two carbons separated by three bonds are gauche to each other, they shield each other by ca 5 ppm compared with the chemical shifts of the corresponding trans conformation (58). In PP, the methyl group is gauche to the methine carbon (C_α) when the chain is trans or gauche$-$, but not when it is gauche$+$ (Fig. 29). Thus the methyl group experiences differing num-

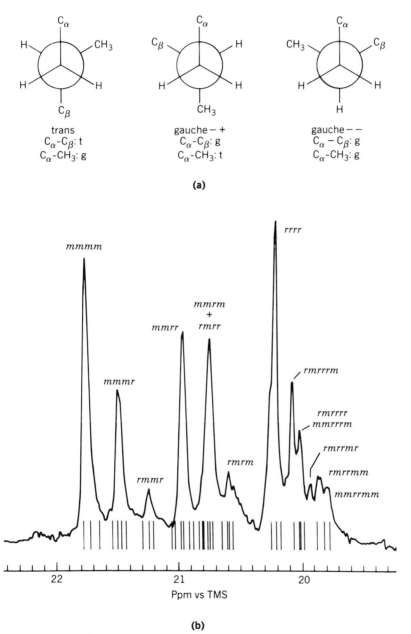

(a)

(b)

Fig. 29. (a) Conformations of a fragment of a PP chain; (b) 90-MHz ^{13}C nmr spectrum of the methyl region of atactic PP. The "stick" spectrum shows the RIS-predicted chemical shifts for the 36 heptad sequences, based on the γ-gauche effect. RIS = rotational isomeric state.

bers of gauche interactions depending on the local chain conformation. The methyl carbon chemical shift can be accurately predicted for all 36 heptad configurational sequences from the theoretically calculated populations of gauche and trans states of the main-chain bonds flanking the methine carbon in each sequence. The C_α and C_β chemical shifts may be calculated as a function of stereochemical configuration in the same way, and the treatment may be extended to other polymers.

These considerations explain the solid-state "magic-angle" ^{13}C spectra of isotactic and syndiotactic PP (59). In the crystalline state, the isotactic polymer adopts a 3_1 helical conformation composed of alternating gauche and trans bonds: (gt)(gt)(gt) (Fig. 30**a**) (61). In this structure, all CH_2, CH, and CH_3 groups occupy equivalent positions by reason of threefold symmetry and give resonances in the same relative positions as in the solution spectrum. The syndiotactic polymer, in contrast, shows a splitting of the methylene carbon resonance (Fig. 30**b**), the equal components being separated by ca 8 ppm. This is in conformity with the crystal structure, which has a repeating (gg)(tt)(gg)(tt) helical conformation with four monomer units per turn. Here, the methylene carbons reside in two different, alternating environments, which may be termed external and internal (61). The external CH_2 carbon experiences two trans interactions with carbons in the γ-positions, whereas the internal CH_2 carbon sees two gauche carbons in the γ-positions. The internal carbon should occur upfield by about two gauche interaction parameters, as appears to be the case.

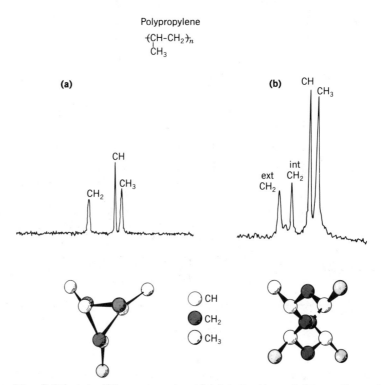

Fig. 30.   Solid-state ^{13}C nmr spectra of (**a**) isotactic and (**b**) syndiotactic PP (60); ext refers to external and int to internal.

Geometrical Isomerism

Geometrical isomerism in diene polymers may be observed and measured by nmr. In addition to the cis and trans structures found in natural isoprene polymers, formed by 1,4-addition, chain propagation of diene monomers may also proceed by incorporation of the monomer through one double bond (1,2-addition). Thus from butadiene the following isomeric chains may be produced in pure form by appropriate choice of coordination catalysts (62):

$$
\begin{array}{lll}
\text{trans-1,4} & & \text{cis-1,4}
\end{array}
$$

```
  ⋀⋀—CH₂
         \
          CH=CH                    ⋀⋀—CH₂        CH₂—⋀⋀
               \                        \        /
                CH₂—⋀⋀                   CH=CH
          trans-1,4                      cis-1,4
```

```
                                              CH₂
                                              ‖
                                              CH
                                              |
 ⋀⋀—CH—CH₂—CH—CH₂—⋀⋀        ⋀⋀—CH—CH₂—CH—CH₂—⋀⋀
      |       |                   |
      CH      CH                  CH
      ‖       ‖                   ‖
      CH₂     CH₂                 CH₂
   isotactic 1,2              syndiotactic 1,2
```

Polybutadiene. Polybutadiene formed by free-radical propagation contains all these structures (see also BUTADIENE POLYMERS). The olefinic carbons are only moderately sensitive to geometrical isomerism, but the methylene carbons are highly sensitive since they are substantially more shielded (ca 8 ppm) in the cis polymer, an effect no doubt closely related to the γ-gauche shielding effect (Fig. 31) (63,64).

Similar results are observed in the solid state. ^{13}C Spectra of *trans*-1,4-polybutadiene in the form of a compacted mass of crystals prepared from dilute heptane solution are shown in Figure 32 (23). Unlike the cis polymer, which is rubbery, the trans isomer crystallizes readily; solution-prepared material is ca 80% crystalline. The spectrum Figure 32**a** was obtained using a weak or "scalar" proton-decoupling field of sufficient power only to collapse the carbon–proton spin multiplets without affecting the direct dipole–dipole interactions with neighboring protons. Under these circumstances, only the amorphous portion of the sample is observed, mainly at the crystal surfaces, where the chains fold over and reenter the crystal and where chain motion is relatively rapid. Mas (Fig. 32**b**) causes a collapse of the residual broadening. In the spectrum Figure 32**c**, high power dipolar proton decoupling with mas allows observation of the entire sample as narrow peaks. Both the olefinic and methylene carbon resonances are split, the more shielded components being in the same positions as in **a** and **b**. The new and larger peaks are those of the crystalline stems, where there is little or no motion. The introduction of gauche and cis conformations into the chains in the amorphous regions results in an upfield shift. The strong spinning sidebands in Figure 32**c** arise only from the olefinic carbons because of their much greater chemical-shift anisotropy. The motion of the amorphous chains in **b** is sufficiently rapid and isotropic to eliminate these sidebands.

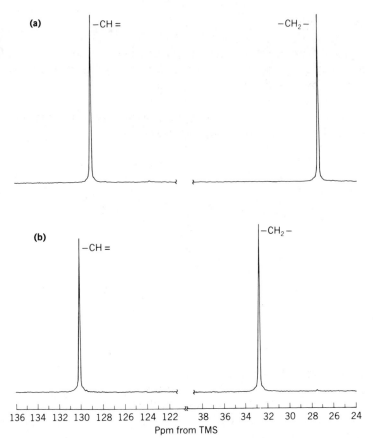

Fig. 31. 50.3-MHz ^{13}C of (**a**) *cis*-1,4-polybutadiene and (**b**) *trans*-1,4-polybutadiene observed in CDCl$_3$ at 40°C. Courtesy of F. C. Schilling.

Chains of mixed structure exhibit more complex spectra because of sequence effects (Fig. 33) (65). Although the region of olefinic carbon resonances is not fully analyzed (Fig. 33a), the olefinic carbon singlets of the pendent vinyl groups flank those of the 1,4-units. In Figure 33b, the aliphatic carbon resonances, mainly 1,4- (and 1,2-) methylene groups, appear. This part of the spectrum is shown at two values of spectral gain, 1× and 5×, to show the small resonances of the sequences containing 1,2-units. Major peak b corresponds to central methylenes in cis–cis units; the principal peak d is that of the central 1,4-unit methylene group in trans–trans and trans–cis units, not discriminated. Peaks a, c, e, and m correspond to sequences involving 1,4-units and one 1,2-unit, whereas the very small resonances f–l represent sequences containing two 1,2-units. The overall composition of the polymer is 23% cis-1,4, 58% trans-1,4, and 19% 1,2. Figure 33c is a computer simulation of Figure 33b based on the assumption of a random distribution of units in these proportions. The satisfactory fit shows that free-radical propagation in butadiene polymerization is a Bernoullian process with regard to the generation of these isomeric structures.

Polyisoprene. In a ^{13}C solution spectra of *cis*-1,4- and *trans*-1,4-polyisoprene (63,64), the olefinic carbons are relatively insensitive to isomerism at the

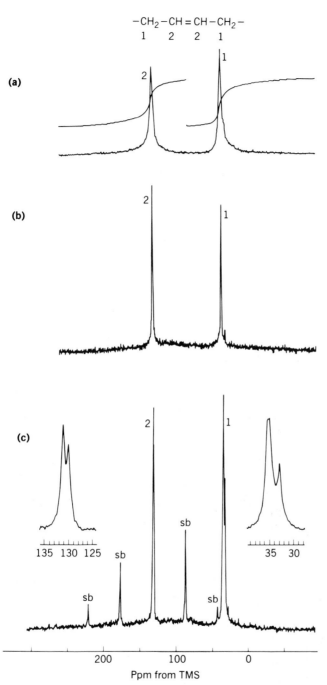

Fig. 32. 50.3-MHz ^{13}C solid-state spectra of solution-crystallized *trans*-1,4-polybutadiene (ca 80% crystalline) (23): (**a**) with low power ("scalar") proton decoupling and no spinning (the integral curves demonstrate the equal size of the resonances); (**b**) same but with mas at 2.4 kHz; (**c**) same as (**b**) but with high power proton decoupling. Insets show olefinic (left) and methylene (right) carbon resonances with expanded chemical-shift scales to exhibit crystalline stem and fold surface resonances in each region, the latter appearing more shielded; sb = sideband.

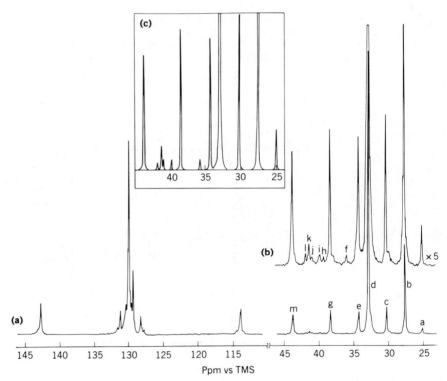

Fig. 33. The 50.3-MHz ^{13}C spectrum of free-radical polybutadiene observed at 50°C in 20 wt % solution in CDCl$_3$: (**a**) olefinic carbon spectrum; (**b**) aliphatic carbon spectrum; (**c**) computer simulation of **b** based on a random distribution of cis, trans, and 1,2-units in the proportion 23:58:19.

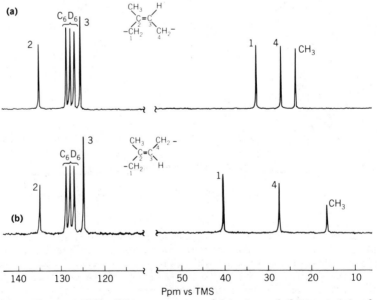

Fig. 34. The 50.3-MHz ^{13}C nmr spectra of (**a**) *cis*- and (**b**) *trans*-1,4-polyisoprene observed in 10% (wt/vol) solution in C$_6$D$_6$ at 60°C (66).

double bond, but the CH_3 and CH_2-1 carbons markedly shield each other in a cis arrangement compared with the trans (Fig. 34). The CH_2-4 carbon, cis to a carbon atom in both isomers, is less affected.

The spectrum Figure 34**a** is that of natural rubber or *Hevea brasiliensis*. The biochemical pathway to natural rubber is an enzymatic process in which isoprene as such plays no part. The polymer is highly stereoregular; no trace of the trans structure is observable. Synthetic *cis*-1,4-polyisoprene is produced commercially using lithium alkyls or Ziegler-Natta catalysts (see ISOPRENE POLYMERS). It contains 2–6% of trans units.

Regioregularity

Vinyl monomers may in principle propagate in either a head-to-tail (**3**) or a head-to-head:tail-to-tail mode (**4**):

$$-CH_2-\underset{|}{\overset{R}{CH}}-CH_2-\underset{|}{\overset{R}{CH}}-CH_2-\underset{|}{\overset{R}{CH}}-CH_2-\underset{|}{\overset{R}{CH}}-$$
(**3**)

$$-CH_2-\underset{|}{\overset{R}{CH}}-\underset{|}{\overset{R}{CH}}-CH_2-CH_2-\underset{|}{\overset{R}{CH}}-\underset{|}{\overset{R}{CH}}-CH_2-$$
(**4**)

In general, head-to-tail propagation is overwhelmingly preferred, but the other direction of addition is sometimes of considerable importance (see HEAD-TO-HEAD POLYMERS). Fluorine-substituted ethylenes are particularly subject to the generation of inverted units, presumably because fluorine atoms are relatively undemanding sterically. The physical properties of their polymers may depend strongly on the presence of such units. The presence of ^{19}F offers an additional means for detailed study, as ^{19}F chemical shifts are highly sensitive to structural variables (67).

Poly(vinyl fluoride). Poly(vinyl fluoride), a commercial plastic with high resistance to weathering (qv), has a substantial proportion of head-to-head units (Fig. 35) (68). The ^{19}F–^{1}H *J*-coupling multiplicity has been removed by proton irradiation as in ^{13}C spectroscopy. The resonance assignments, designated by capital italic letters, are as follows:

$$-\overset{A}{CH_2}-CHF-\overset{A}{CH_2}-CHF-\overset{B}{CH_2}-\overset{C}{CHF}-CHF-CH_2-CH_2-\overset{D}{CHF}-CH_2-\overset{A}{CHF}-$$

$$\longrightarrow \quad \longrightarrow \quad \longrightarrow \quad \longleftarrow \quad \longrightarrow \quad \longrightarrow$$

The stereochemical assignments are also indicated in Figure 35. The *m* and *r* designations that are not underlined represent the usual (head-to-tail) relationships between substituents in 1,3-positions (in planar zigzag projection):

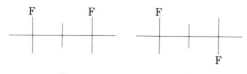

$$m \qquad\qquad\qquad r$$

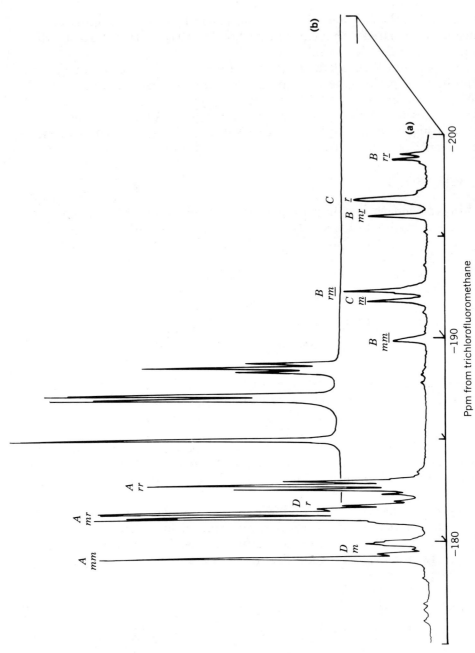

Fig. 35. The 188-MHz ^{19}F spectrum of (**a**) commercial poly(vinyl fluoride); (**b**) poly(vinyl fluoride) prepared by reductive dechlorination of poly(1-fluoro-1-chloroethylene); both spectra observed as 8% solution in N,N-dimethylformamide-d_7 at 130°C. The resonance assignments are designated by capital italic letters. Courtesy of

296

The underlined designations represent the substituents in 1,2-positions (also in planar zigzag projection):

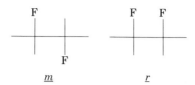

$$m \qquad\qquad r$$

For the B resonances, rm and mr are quite different structures and do not differ merely in direction as with $r\underline{m}$ and $m\underline{r}$. The fraction of inverted units is ca 11%.

The spectrum Figure 35b is that of a poly(vinyl fluoride) prepared without inversions by a special chemical route. The upfield portion of the spectrum of Figure 35a, as well as the D resonances, are absent, thus clearly identifying them as defect peaks. This material has a melting point of 210°C, compared with the normal polymer's 190°C. Both polymers are nearly atactic.

Poly(vinylidene fluoride). Poly(vinylidene fluoride) is a material of great interest because of its piezoelectric and pyroelectric properties. Nmr spectroscopy shows that it contains 3–6% of inverted units:

$$\overset{A}{-CH_2}-\overset{}{CF_2}-CH_2-\overset{C}{CF_2}-\overset{D}{CF_2}-CH_2-CH_2-\overset{B}{CF_2}-CH_2-CF_2-$$

The 188-MHz ^{19}F spectrum, shown in Figure 36, is made somewhat simpler by the absence of asymmetric carbons (67,69).

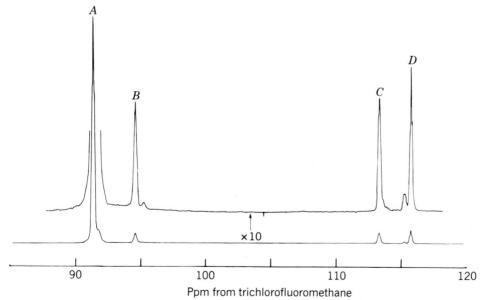

Ppm from trichlorofluoromethane

Fig. 36. The 188-MHz ^{19}F spectrum of commercial poly(vinylidene fluoride) observed in 11% solution in dimethylformamide-d_7 at 27°C. Courtesy of R. E. Cais.

Branching

Branching in vinyl polymers may be introduced deliberately by employing diene or divinyl comonomers, or branches may be produced by processes that are under less specific control and involve chain transfer (qv). Branches are of particular importance in polyethylene (PE) because their presence reduces the melting point and extent of crystallinity. Ir spectroscopy indicates that high pressure PE has unusually large numbers of methyl groups, normally expected only at chain ends. When combined with molecular weight measurements, these results indicate that there are many more ends than molecules, or, in other words, that the chains contain branches. ^{13}C Nmr can supply details of the types and distribution of these branches because of the sensitivity of ^{13}C chemical shifts to such structural variables, already seen in the discussion of PP.

In Figure 37, the resonances are labeled and described according to the scheme inset in the figure, using a large body of information obtained from model hydrocarbons and from ethylene copolymers with small proportions of olefins. The principal peak, at 30.0 ppm, not shown at its full height, corresponds to those methylene groups that are four carbons or more removed from a branch or chain

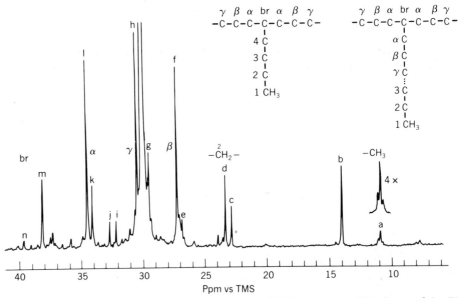

Fig. 37. The 50.3-MHz ^{13}C nmr spectrum of high pressure PE observed in 5% (wt/vol) solution in 1,2,4-trichlorobenzene at 110°C (70,71): a, CH$_3$ of an ethyl branch; b, CH$_3$ of a butyl, amyl, or longer branch; c, penultimate carbon of an amyl or longer branch; d, penultimate carbon of a butyl branch (γ to the br carbon); e, fourth carbon of an amyl branch (β to br carbon) and methylene of an ethyl branch (α to br carbon); f, main-chain β-C from an ethyl, butyl, or amyl branch and β-C of a long branch; g, third carbon of a butyl branch (β to br carbon); h, main-chain γ-C from an ethyl, butyl, or amyl branch and γ-C of a long branch; i, third carbon of a long-chain branch; j, third carbon of an amyl branch (γ to br carbon); k, fourth carbon of a butyl branch (α to br carbon) and main-chain α-C from an ethyl branch; l, main-chain α-C from a butyl or amyl branch and α-C of an amyl branch; m, branched carbon for butyl, amyl, or longer; n, branched carbon for ethyl branch; br = branched.

end. It constitutes about 80% of the spectral intensity. The C_1 carbons, ie, methyl groups, and C_2 carbons are the most shielded, branch-point carbons the least. Main-chain carbons β to the branch are more shielded, whereas those α to the branch are less shielded than unperturbed methylenes. The composition of this polyethylene is shown in Table 5. The predominant branch type is n-butyl. Both amyl and butyl branches are believed to be formed by intramolecular chain transfer or "backbiting":

$$-CH_2 \quad \overset{\cdot}{C}H_2 \longrightarrow -\overset{\cdot}{C}H \quad CH_3$$
$$(CH_2)_n \qquad\qquad (CH_2)_n$$

This reaction is evidently most probable when $n = 3$ or 4, has a low but finite probability when $n = 1$, and zero probability when $n = 0$ or 2.

Table 5. Branching in High Pressure Polyethylene[a]

Types of branch	Number of branches per 1000 backbone carbons
$-CH_3$	0.0
$-CH_2CH_3$	1.0
$-CH_2CH_2CH_3$	0.0
$-CH_2CH_2CH_2CH_3$	9.6
$-CH_2CH_2CH_2CH_2CH_3$	3.6
hexyl and longer	5.6
Total	*19.8*

[a] $\overline{M}_n = 18{,}400$; $\overline{M}_w = 129{,}000$.

The complex appearance of the ethyl branch methyl resonance at ca 11.0 ppm suggests that these branches may occur in groups or with some similar complication. The branches described in Table 5 as hexyl or longer are believed to be truly long. Their frequency of occurrence is estimated from the unique peak labeled i in Figure 37, corresponding to the third carbon from a long-chain end. Such branches are too few to influence crystallinity but are believed to affect melt rheology significantly. They are formed by transfer from the growing chain to a finished PE molecule, resulting in a free-radical site on the latter which then adds more monomer.

Copolymer Structure

Copolymers are broadly divided into three types: random, block, and graft.

Random:

Block:

Graft: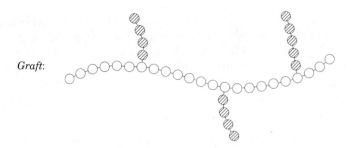

Block and graft copolymers contain relatively long sequences of one monomer bonded to similar sequences of another. Their overall composition is usually known from their method of synthesis and they do not present microstructural problems essentially different from those of homopolymers. The random type, in which two or more types of comonomer units are present in each chain, presents more complex problems. In this article, only copolymers of vinyl (or diene) monomers are discussed. Copolyesters and copolyamides are also significant, but their composition is also usually readily predictable from the ratio of comonomers employed.

   The composition of copolymers of vinyl and diene monomers are not in general the same as that of the monomer mixtures from which they are formed and cannot be deduced from the homopolymerization rates of the monomers involved (see COPOLYMERIZATION). The relationship between instantaneous copolymer composition and monomer feed composition, ie, the starting ratio of monomers, is given by the following differential equation:

$$\frac{d[M_1]}{d[M_2]} = \frac{[M_1]}{[M_2]} \frac{r_1[M_1] + [M_2]}{r_2[M_2] + [M_1]} \tag{19}$$

where M_1 and M_2 represent the two comonomers (72). The left side of this equation gives the ratio of the rates at which the two monomers enter the copolymer, which in turn must represent the composition of the copolymer being formed at any instant. The ratio $[M_1]/[M_2]$ is the mole ratio of monomers in the feed. The quantities r_1 and r_2 are the reactivity ratios, defined as the ratios of propagation rate constants.

$$r_1 = k_{11}/k_{12} \tag{20}$$

$$r_2 = k_{22}/k_{21} \tag{21}$$

Here, k_{11} is the rate constant for the addition of monomer 1 to a growing chain ending in a monomer-1 unit; k_{12} is the rate constant for the addition of monomer 2 to the growing chain ending in monomer 1; k_{21} and k_{22} are the corresponding terms for growing chains ending in a monomer-2 unit.

   Equation 19 is the copolymer equation in terms of the molar concentrations of the monomers. It is usually more convenient to express this relationship in terms of the mole fraction in both feed and copolymer. The feed mole ratio for monomer 1 is given by

$$f_1 = 1 - f_2 = \frac{[M_1]}{[M_1] + [M_2]} \tag{22}$$

The instantaneous copolymer composition is given by

$$F_1 = 1 - F_2 = \frac{d[M_1]}{d[M_1] + d[M_2]} \tag{23}$$

from which

$$F_1 = \frac{r_1 f_1^2 + f_1 f_2}{r_1 f_1^2 + 2f_1 f_2 + r_2 f_2^2} \tag{24}$$

where f_1 and f_2 represent feed mole fractions and F_1 and F_2 represent mole fractions of monomer 1 and 2 in the polymer. A parallel equation expresses F_2. These relationships deal with the instantaneous composition of the copolymer. Since the comonomers generally do not enter the polymer in the same ratio as in the feed, the latter drifts in composition as copolymerization proceeds, becoming depleted in the more reactive comonomer. As a result, the higher the monomer conversion, the more heterogeneous the product. This in no way affects the determination of the overall composition or microstructure of the product but makes it more difficult to interpret in terms of relative reactivities. It is therefore customary in fundamental studies to limit the conversion to ca 5% or less, although drifts in composition can be dealt with mathematically. In copolymer production on a practical scale, it is common practice to achieve greater structural regularity by adjusting the monomer input as the reaction proceeds. This usually means withholding the more reactive monomer.

Traditionally, the determination of reactivity ratios, which provide important information concerning the behavior of monomers and growing chains, has required the determination of the overall comonomer composition of copolymers prepared from a series of feed ratios. Elemental analysis is most commonly used. A number of computational and graphic methods are employed to do this. The theoretical treatment that predicts overall composition also predicts the frequency of occurrence of comonomer sequences. These can be readily measured by nmr (73), by which it is also possible to observe copolymer stereochemistry and the presence of anomalous units. Deviations from the simple model employed here, in which only the terminal residue of a growing chain determines its reactivity, can be detected; the effect of the penultimate unit, if any, may be clearly observed. By sequence measurements, reactivity ratios can be determined from only a single copolymer provided the feed ratio is known. It may still be desirable to observe a range of compositions to assist in resonance assignments, but it is not in principle essential.

For a random copolymerization, diad, triad, and tetrad sequences may be represented as follows, ignoring stereochemistry:

Diads: m_1m_1	m_1m_2 (or m_2m_1)	m_2m_2
Triads: $m_1m_1m_1$		$m_2m_2m_2$
$m_1m_1m_2$ (or $m_2m_1m_1$)		$m_1m_2m_2$ (or $m_2m_2m_1$)
$m_2m_1m_2$		$m_1m_2m_1$
Tetrads: $m_1m_1m_1m_1$	$m_1m_1m_2m_1(m_1m_2m_1m_1)$	$m_2m_2m_2m_2$
$m_1m_1m_1m_2(m_2m_1m_1m_1)$	$m_1m_1m_2m_2(m_2m_2m_1m_1)$	$m_2m_2m_2m_1(m_1m_2m_2m_2)$
	$m_2m_1m_2m_1(m_1m_2m_1m_2)$	
$m_2m_1m_1m_2$	$m_2m_1m_2m_2(m_2m_2m_1m_2)$	$m_1m_2m_2m_1$

The diad probabilities, ie, frequencies of occurrence, are given by

$$[m_1m_1] = F_1P_{11} \tag{25}$$

$$[m_1m_2] \text{ (or } [m_2m_1]) = 2F_1P_{12} = 2F_1(1 - P_{11}) \tag{26}$$

$$= 2F_2P_{21} = 2F_2(1 - P_{22})$$

$$[m_2m_2] = F_2P_{22} \tag{27}$$

Here, F_1 and F_2 are the overall mole fractions of m_1 and m_2, respectively.

$$[m_1m_1m_1] = F_1P_{11}^2 \tag{28}$$

$$[m_1m_1m_2] \text{ (or } [m_2m_1m_1]) = 2F_1P_{11}(1 - P_{11}) \tag{29}$$

$$[m_2m_1m_2] = F_2(1 - P_{22})(1 - P_{11}) \tag{30}$$

The quantity P_{11} expresses the conditional probability that a chain ending in m_1 will add another m_1, and P_{22} expresses the corresponding probability for m_2. P_{12} is the probability that a chain ending in m_1 will add m_2, equal to the probability that it will not add m_1, ie, $1 - P_{11}$. Corresponding to the four rate constants k_{11}, k_{12}, k_{21}, and k_{22} are the four probabilities P_{11}, P_{12}, P_{21}, and P_{22}, related by

$$P_{11} + P_{12} = 1 \tag{31}$$

$$P_{21} + P_{22} = 1 \tag{32}$$

since a growing chain has only two choices. It can be shown that P_{11} and P_{22} are given in terms of monomer-feed mole fractions and reactivity ratios by

$$P_{11} = \frac{r_1f_1}{1 - f_1(1 - r_1)} \tag{33}$$

$$P_{22} = \frac{r_2f_2}{1 - f_2(1 - r_2)} \tag{34}$$

from which

$$r_1 = \frac{(1 - f_1)[m_1m_1]}{f_1(F_1 - [m_1m_1])} \tag{35}$$

$$r_2 = \frac{(1 - f_2)[m_2m_2]}{f_2(F_2 - [m_2m_2])} \tag{36}$$

Entirely analogous relationships apply to triad and tetrad sequences.

Because a vinylidene chloride (m_1):isobutylene (m_2) copolymer has no asymmetric carbons and no vicinal J coupling (74,75), the proton nmr spectrum conveys only compositional sequence information (Fig. 38). The homopolymer of vinylidene chloride gives a single resonance for the methylene protons; the homopolymer of isobutylene (which can be prepared with cationic but not with free-radical initiators) gives singlet resonances of 3:1 intensity for the methyl and methylene protons. The spectrum of a copolymer, prepared with a free-radical initiator and containing 70 mol % vinylidene chloride, is shown in Figure 38. The methylene resonances are grouped in three chemical-shift ranges: m_1m_1-centered peaks at low field; peaks of methylene protons in m_1m_2-centered units

$$\begin{array}{cc} Cl & CH_3 \\ | & | \\ -C-CH_2-C- \\ | & | \\ Cl & CH_3 \end{array}$$

near 3 ppm; and CH_2 and CH_3 resonances of m_2-centered sequences at high field. It is evident that tetrad sequences are involved. If only diad sequences were distinguished, there would be only three methylene resonances, corresponding to m_1m_1, m_1m_2 (or m_2m_1), and m_2m_2 sequences. The upfield isobutylene peaks show considerable overlap and assignments here are less certain, but these resonances are not required for the analysis.

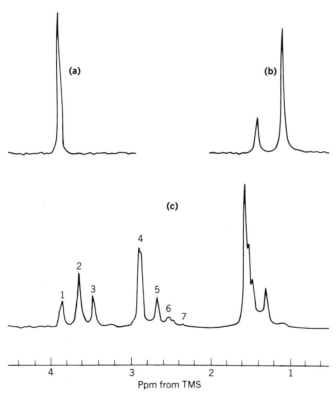

Fig. 38. The 60-MHz proton spectra of (**a**) poly(vinylidene chloride); (**b**) poly-isobutylene; (**c**) a vinylidene chloride (m_1):isobutylene (m_2) copolymer containing 70 mol % m_1 (75). Peaks are identified with monomer tetrad sequences: 1, $m_1m_1m_1m_1$; 2, $m_1m_1m_1m_2$; 3, $m_2m_1m_1m_2$; 4, $m_1m_1m_2m_1$; 5, $m_2m_1m_2m_1$; 6, $m_1m_1m_2m_2$; 7, $m_2m_1m_2m_2$. Courtesy of *Kolloid-Zeitschrift*.

From diad resonances, r_1 and r_2 may be calculated. The relative intensity of the m_1m_1 resonances centered near 3.6 ppm gives $[m_1m_1]$ as 0.426, normalized over all diad methylenes, which from the above relationships gives a value of

3.31 for r_1. Evaluation of the methylene resonance of

$$
\begin{array}{cc}
CH_3 & CH_3 \\
| & | \\
-C-CH_2-C- \\
| & | \\
CH_3 & CH_3
\end{array}
$$

cannot be readily carried out because of overlaps, even though it would most directly provide a value of r_2. Instead, note that

$$[m_1m_2] \,(+ \,[m_2m_1]) = 2F_2(1 - P_{22}) \tag{37}$$

or

$$[m_1m_2] \,(+ \,[m_2m_1]) = 2F_2 - \frac{2F_2r_2f_2}{1 - f_2(1 - r_2)} \tag{38}$$

From the group of resonances near 2.8 ppm, a value of $[m_1m_2]$ of 0.56 is obtained, from which a value of r_2 of 0.04 is calculated. Conventional analysis of this system gives

$$k_{11}/k_{12} = r_1 = 3.3 \tag{39}$$

$$k_{22}/k_{21} = r_2 = 0.05 \tag{40}$$

in agreement within experimental error. These values show that vinylidene chloride radicals prefer to add vinylidene chloride and that chains terminating in isobutylene units have only a very small tendency to add another isobutylene.

More searching tests of the propagation mechanism can be obtained from consideration of tetrad intensities. By such analysis, it is found that the relative reactivity of a growing free radical ending in vinylidene chloride depends on whether the penultimate unit is another vinylidene chloride unit or an isobutylene unit; if the latter, the chain is twice as likely to add vinylidene chloride.

Biopolymers

Nmr is particularly effective in the study of polymers of biological origin, proteins, nucleic acids, and synthetic polymers of related structure. Biopolymers even in solution commonly assume well-defined secondary structures such as helices, whereas synthetic polymers do not. The information furnished by nmr usually concerns this aspect rather than the covalent structure, which is generally known.

Polypeptides. The transition between the random-coil and α-helical conformations of synthetic polypeptides has been intensively studied by nmr (see also HELIX-COIL TRANSITIONS). In helix-supporting solvents such as chloroform, poly(γ-benzyl L-glutamate)

$$
\begin{array}{l}
O{=}C{-}O{-}CH_2C_6H_5 \\
| \\
\gamma CH_2 \\
| \\
\beta CH_2 \quad O \\
| \quad\quad \| \\
{-}(NH{-}\underset{\alpha}{CH}{-}C)_n{-}
\end{array}
$$

gives a proton spectrum in which only the side chain resonances are visible, although much broadened (Fig. 39a) (76). Upon addition of 10% trifluoroacetic acid, the peaks narrow markedly and resonances appear for NH protons near 8 ppm and α-protons near 4 ppm. Optical measurements show that the polymer is still in the α-helical form, so this transition represents the breakup of associations between helices. As the trifluoroacetic acid content is increased, new resonances appear in the NH and α-CH resonance regions. These represent random-coil segments of the polypeptide chains. They increase in intensity with increased trifluoroacetic acid content until at 20% (Fig. 39f) the polymer becomes all random coil, as indicated also by optical measurements. The behavior of the resonances at intermediate stages shows that they represent nonexchanging or slowly exchanging populations, which in turn suggests that the disrupting action of the trifluoroacetic acid, probably mainly a breakage of CO–NH hydrogen bonds within the helix, begins at the ends of the helix and works inward. When the helix-coil transition is made to take place all at once by a temperature jump, it does so at microsecond rates.

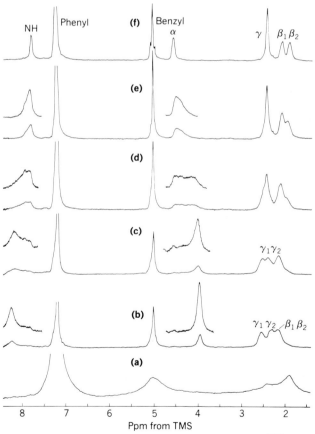

Fig. 39.   300-MHz spectra of poly(γ-benzyl L-glutamate) of $\overline{\text{DP}}$ ca 160 in CDCl$_3$ (5% wt/vol) at 32°C (76). The volume percent of trifluoroacetic acid added is (**a**) 0; (**b**) 10; (**c**) 12; (**d**) 13; (**e**) 16; (**f**) 20.

In small polypeptides, the splitting of the NH resonances gives a measure of the NH–C_aH J coupling and thus an indication of the chain conformation. In longer, helix-forming sequences, the peaks are usually too broad to permit measurement of this coupling.

Proteins. Globular proteins in the native state are composed of polypeptide chains, partly helical, which are rather tightly intertwined to form compact spheres or ellipsoids. The complete structures of more than 100 such proteins are known from x-ray diffraction studies (see PROTEINS). There is little or no free segmental motion in such structures and so the rotation of the molecule as a whole determines the linewidth. If the molecular weight is large, the correlation time is too long to permit truly high resolution spectra except perhaps for certain side-chain protons. Thus native ribonuclease (mol wt ca 13,000) gives a moderately well-resolved spectrum in aqueous solution, whereas that of bovine serum albumin (mol wt ca 65,000) is much more poorly resolved. Unfolding in 8-M urea gives a spectrum composed of narrow lines since segmental motion is possible, but of course most of the secondary structure is therefore disrupted.

The effect of molecular weight is illustrated by the 360-MHz proton spectrum of native carboxypeptidase inhibitor in D_2O solution (Fig. 40) (77), which has a molecular weight of only ca 4000. The resonances at low field (7–8 ppm) arise from aromatic and histidine ring protons (peptide NH, arginine NH, and histidine NH protons would appear at still lower field, 8.0–8.5 ppm, but have been exchanged for deuterium). The band of resonances at 4–5 ppm arises mainly from backbone α-CH protons, and is followed by a wide variety of side-chain protons and, at highest field (0.4–1.5 ppm), by methyl protons of valine, leucine, and isoleucine. The latter show small splittings (ca 6 Hz) arising from vicinal couplings.

Polynucleotides and Nucleic Acids. The nucleic acid chain is composed of pentose units connected with phosphate links, with purine and pyrimidine bases attached to the sugar units. The base–sugar–phosphate unit is known as a nucleotide unit. There are two classes of nucleic acids: those in which the sugar unit is D-ribose, called ribonucleic acids or RNA, and those in which the sugar unit is 2-deoxy-D-ribose, called deoxyribonucleic acid or DNA (see POLYNUCLEO-TIDES). A schematic segment of a DNA chain shows the bases adenine (A), thymine (T), guanine (G), and cytosine (C) attached to the sugar units:

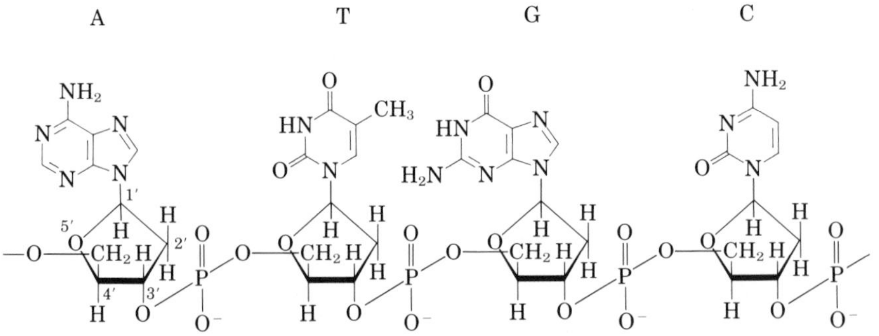

In the DNA molecule, two such chains are wound in a helical pattern about a common axis. The two chains are linked together by hydrogen bonds between

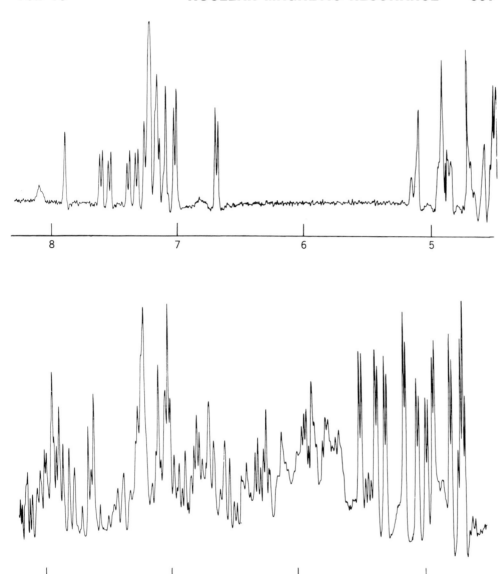

Fig. 40. 360-MHz spectrum of carboxypeptidase inhibitor (76). Courtesy of D. J. Patel.

the bases, the planes of which are perpendicular to the helix axis. Steric requirements are such that an adenine must be paired with a thymine at the same level in the other chain and a guanine with a cytosine. A synthetic double-helical polymer may be formed in which only adenine and thymine are present, arranged in an alternating sequence and giving rise to interchain hydrogen bonds; such a polymer is designated as poly(dA–dT).

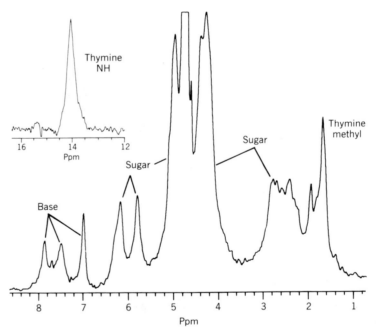

Figure 41 shows the 500-MHz proton spectrum of a helical poly(dA–dT) composed of approximately 46 base pairs (mol wt ca 32,000), observed in D_2O to avoid the large solvent resonance seen in H_2O solutions. As with proteins, the exchangeable sugar hydroxyl and NH protons have disappeared from the spectrum. The important interbase hydrogen-bond protons would appear at 14 ppm in H_2O and are shown schematically.

Fig. 41.   500-MHz proton spectrum of double-helical poly(dA–dT) having a molecular weight of ca 32,000 (78). The spectrum was obtained in D_2O and so exchangeable hydroxyl and NH proton resonances do not appear. The interbase hydrogen-bond protons would appear at ca 14 ppm in H_2O and their positions are indicated. Courtesy of *Biopolymers*.

When the double helix is opened up to the random coil, the resonances greatly narrow and undergo shifts in position; the shifts are particularly marked for the NH protons. This furnishes an important method for following and observing the details of this transition. The proton resonances are also better resolved the shorter the polynucleotide chain, and much information has been gained concerning the structure, stability, and binding of drugs by observation of small duplexes of known structure. Figure 41 represents about the limit in molecular weight at which useful nmr spectra may be obtained. At natural molecular weights, ie, 10^8–10^9, the native DNA spectra show resonances so broad as to be barely visible.

Chain Motion in Macromolecules

Measurements of nuclear relaxation furnish a powerful means for observing motions of macromolecules in two regimes: polymer solutions, in which chain motion is relatively fast, and polymers in the solid state, where motional freedom may vary from fast to essentially nil. Synthetic polymers are useful mainly for their solid-state properties, so observation of the solid state may be of greater practical importance. However, solution measurements are often more revealing of intramolecular influences, divorced from the effects of neighboring chains.

The only significant mechanism for the spin–lattice relaxation of protons and ^{13}C nuclei in polymers is dipole–dipole interaction with neighboring protons, whose motions give rise to fluctuating magnetic fields. The distribution of motional frequencies is given by a spectral-density function defined by

$$J(\omega) = \int_{-\infty}^{+\infty} G(\tau)e^{-i\omega\tau}\, d\tau \tag{41}$$

where $G(\tau)$ is the autocorrelation function of the time-dependent relation expressing the orientation of the vector between the observed nucleus and the neighboring nucleus with which it interacts:

$$G(\tau) = \langle F(t)F^*(t + \tau) \rangle \tag{42}$$

The brackets denote an ensemble average over the collection of nuclei and F represents a function expressing the position and motion of the molecule. $J(\omega)$ may be thought of as expressing the power available at frequency ω to relax the observed spins. The spectral-density and the autocorrelation function are Fourier inverses of each other in the frequency and time domains, respectively. If $G(\tau)$ decays to zero in a short time, this corresponds to a short correlation time τ_c, which means that the molecular motion is rapid and that the molecules have only a short memory of their previous state of motion.

A simple dynamic model for a polymer molecule might be a rigid sphere immersed in a viscous continuum and reoriented by small random diffusive steps. The correlation time may be thought of as the interval between these alterations in the state of motion of the molecule. For such a molecule, the loss of memory of the previous motional state is exponential with a time constant τ_c:

$$G(\tau) = e^{-(\tau/\tau_c)} \tag{43}$$

and the spectral-density function becomes

$$J(\omega) = \frac{\tau_c}{1 + \omega^2\tau_c^2} \tag{44}$$

This function is shown in Figure 42 for three values of τ_c: a long τ_c corresponding to slow motions, as for large molecules, stiff chains, or viscous solvents; a short τ_c corresponding to rapid motions of small molecules or very flexible chains; and an intermediate case for which $\tau_0\omega_0 \cong 1$; ω_0 is the resonant frequency of the observed nucleus (11). The areas under the curves are the same, ie, the molecular power available is the same; only its distribution varies with τ_c. For long correlation times, the component at ω_0 is weak. At short τ_c, the frequency spectrum is broad and so no one component, in particular that at ω_0, can be very intense. At some intermediate value, the resonant component will be a maximum. Since the rate of spin–lattice relaxation depends on the component at ω_0, it is expected to be relatively small at very long and very short τ_c, and to reach a maximum when $\tau_0\omega_0 \cong 1$. For a particular system, T_1 will then pass through a minimum when τ_c is varied, as by changing the temperature.

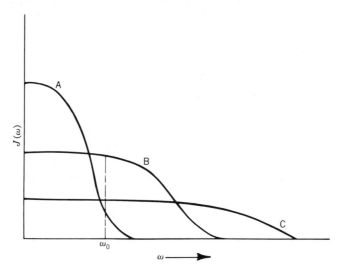

Fig. 42. Spectral-density function (motional frequency) as a function of long, medium, and short values of the correlation time τ_c: A, long τ_c, $\omega_0 > 1/\tau_c$; B, intermediate τ_c, $\omega_0 \sim 1/\tau_c$; C, short τ_c, $\omega_0 < 1/\tau_c$.

For a ^{13}C nucleus relaxed by N equivalent protons at a fixed distance r_{C-H}, it can be shown that

$$\frac{1}{NT_1} = \frac{1}{10}\frac{\gamma_H^2\gamma_C^2\hbar^2}{r_{C-H}^6}[J_0(\omega_H - \omega_C) + 3J_1\omega_C + 6J_2(\omega_H + \omega_C)] \qquad (45)$$

where J_0, J_1, and J_2 are of the form of equation 44 and prescribe the components of the motional frequency spectrum at $(\omega_H - \omega_C)$, ω_C, and $(\omega_H + \omega_C)$; ω_H and ω_C are the resonant frequencies of the proton and ^{13}C, and γ_H and γ_C are the corresponding gyromagnetic ratios. A plot of equation 45 is shown in Figure 43 for five values of the polarizing field B_0, expressed in T (10 kG). As B_0 increases, the T_1 minimum increases and moves to smaller values of τ_c.

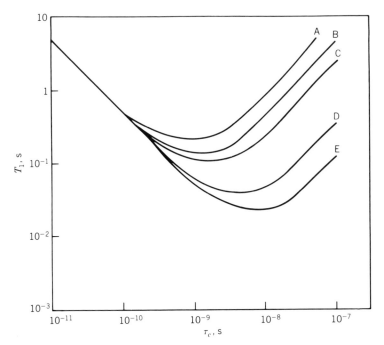

Fig. 43. Log–log plot of T_1 vs τ_c for ^{13}C nucleus relaxed by a proton (eq. 45) at varied magnetic field B_0, expressed in T: A, 11.75; B, 8.45; C, 6.34; D, 2.35; E, 1.41.

For very rapid motion, when $\omega_0\tau_c \ll 1$, all J_n become equal to τ_c (eq. 44) and equation 45 simplifies to

$$\frac{1}{NT_1} = \tau_c \cdot \frac{\gamma_H^2\gamma_C^2\hbar^2}{r_{C-H}^6} \tag{46}$$

Under these conditions of "motional narrowing," T_1 is independent of B_0, but the higher B_0, the more rapid the motion necessary to satisfy this condition. Because of the inverse sixth-power dependence on the $^{13}C-^1H$ internuclear distance, it is commonly observed for both small molecules and macromolecules that only the protons directly bonded to the observed carbon contribute significantly to T_1. Carbons without bonded protons, eg, carbonyl and quaternary carbons, exhibit relatively long T_1 values. For aliphatic carbons (CH, CH$_2$, CH$_3$), $r_{C-H} = 0.109$ nm and thus

$$\frac{1}{NT_1} = 2.03 \times 10^{10}\tau_c \tag{47}$$

or

$$\tau_c = \frac{4.92 \times 10^{-11}}{NT_1} \tag{48}$$

In the subsequent discussion, the single correlational time motional model will be employed, although it is clear that such a simple picture is not fully adequate.

Chain Motion in Solution

Most synthetic polymers in solution give fairly well-resolved proton and ^{13}C spectra, and thus must enjoy considerable segmental freedom, since rigid molecules of comparable molecular weight would give very much broader resonances. Linewidths are independent of molecular weight down to the oligomer level and very nearly independent of solution concentration and viscosity up to 25–30%. Even solutions so viscous that they barely pour from the nmr tube give quite well-resolved spectra, another indication of the dominance of local motion. Nevertheless, polymer spectra usually appear more poorly resolved than those of small-molecule counterparts. Some of this line broadening, in polymers of irregular structure, represents an envelope of chemical shifts. The true solution proton linewidth for many polymers at elevated temperatures is usually 3–5 Hz, whereas ^{13}C resonances are somewhat narrower, typically 1–2 Hz.

The motion of flexible polymers in solution may be conveniently regarded as being composed of segmental motion and overall tumbling. When the ^{13}C T_1 is measured as a function of the average number of monomer units per chain, it is found that it at first decreases and then levels off at a value of about 10^2 units. Above this relatively low molecular weight, eg, 10,000 for polystyrene (PS), overall tumbling becomes too slow to provide a significant contribution compared with segmental motion.

Table 6 lists the conditions of observation and the values of NT_1 for backbone ^{13}C nuclei of a number of synthetic polymers. Values of the correlation time τ_c are calculated from equation 48. It is found that τ_c is well on the fast-motion side of the T_1 minimum except in the case of poly(1-butene sulfone), for which T_1 decreases with increasing temperature. The following conclusions can be drawn from the data in Table 6.

Effect of Side Chains. Comparison of polyethylene with polypropylene, poly(1-butene), and poly(vinyl chloride) (PVC) shows that the presence of side chains impedes chain mobility and that the larger the side chain the slower the motion. A methyl group slows the chain by a factor of 2–3, and a chlorine is comparable; ethyl and phenyl (polystyrene) groups have a further impeding effect. The main-chain conformational barrier may be regarded as composed of a symmetrical threefold torsional potential on which is superimposed nonbonded interactions between the side groups, the latter increasing with the size of the group.

It may at first seem surprising that the presence of two α-substituents, as in polyisobutylene, poly(vinylidene chloride), and poly(methyl methacrylate), does not impede chain motion significantly more than a single substituent. This may be attributed to the fact that two substituents increase the crowding in all rotational states, including eclipsed states, without necessarily increasing the barrier between them.

Effect of Heteroatoms. The chains of polyoxyethylene (POE) are among the most flexible of those in Table 6, with a correlation time at 30°C comparable to that of PS at 100°C. Poly(propylene oxide) (PPO) shows an analogous relationship to PP. Polymethylthiirane (poly(propylene sulfide)) and polyphenylthiirane show the liberating influence of a sulfur atom; poly(styrene peroxide) shows the comparable effect of a peroxy link, —O—O—. These findings are

Table 6. ^{13}C Relaxation (T_1) Values and Approximate Correlation Times $\tau_c{}^a$ Calculated on the Isotropic Diffusion Model for Polymer Main-chain Carbons in Solution[b]

Polymer	Solvent	Temp, °C	Conc, wt %	NT_1, s[c]	τ_c, ns	Ref.
polyethylene	TCB[d]	110	33	2.52	0.019	79
	ODCB[e]	100	25	2.70	0.018	80
	ODCB	30[f]	25	1.24	0.040	80
polypropylene	ODCB	100	25	1.13	0.044	80
		30[g]	25	0.39	0.13	80
polyisobutylene	CDCl$_3$	30	5	0.30	0.16	81
poly(1-butene)	CCl$_3$CHCl$_2$	100	10	0.34[h]	0.14	82
1,4-polybutadiene						
cis	CDCl$_3$	54	20	3.00	0.016	83
trans	CDCl$_3$	54	20	2.38	0.021	83
poly(vinyl chloride)	TCB[d]	107	10	0.32	0.15	84
poly(vinylidene chloride)	HMPA-d_{18} [i]	40[j]	15	0.078[k]	0.63	85
		89	15	0.25[k]	0.20	85
poly(vinylidene fluoride)	DMF[l]	41	20	0.62	0.079	86
poly(vinyl alcohol)	DMSO	30	20	0.090	0.55	87
polyacrylonitrile	DMSO	50	20	0.15	0.33	87
polystyrene						
atactic	toluene-d_8	30	15	0.100	0.49	88
isotactic	toluene-d_8	30	15	0.090	0.55	88
poly(methyl methacrylate)						
isotactic	CDCl$_3$	38	10	0.126	0.39	89
syndiotactic	CDCl$_3$	38	10	0.080	0.62	89
atactic	DMF[l]	41	20	0.072[m]	0.81	86
poly(α-methylstyrene)	CDCl$_3$	30	10	0.100	0.49	90
polyoxymethylene	HFIP[n]	30	3	0.60	0.082	91
polyoxyethylene	C$_6$D$_6$	30	5	2.80	0.018	92
poly(propylene oxide)	CDCl$_3$		5	1.00	0.049	81
polymethylthiirane	CDCl$_3$	30	10	0.86	0.057	93
polyphenylthiirane	CDCl$_3$	25	15	0.19	0.25	94
poly(styrene peroxide)	CHCl$_3$	23	22	0.24	0.21	95
poly(1-butene sulfone)	CDCl$_3$	40	25	0.090	23[o]	96

[a] Ref. 38.
[b] Ref. 11.
[c] At 25 MHz unless otherwise noted.
[d] 1,2,4-Trichlorobenzene.
[e] o-Dichlorobenzene.
[f] Extrapolated from high temperature data with activation energy of 10.5 kJ/mol (2.51 kcal/mol) (38).
[g] Extrapolated from high temperature data with activation energy of 14.3 kJ/mol (3.42 kcal/mol) (38).
[h] At 22.6 MHz.
[i] Hexa(methyl-d_3)phosphoramide.
[j] Five other temperatures also reported.
[k] At 15 MHz.
[l] N,N'-Dimethylformamide.
[m] rrr sequences (racemic tetrads of CH$_2$ groups).
[n] Hexafluoro-2-propanol.
[o] $\overline{M}_n$ = 2020; some contribution from overall tumbling.

consistent with the known low rotational barriers at hetero atoms, but it is not so evident why polyoxymethylene (POM), with an oxygen inserted at every carbon atom, should be less flexible even than PE. This may be rationalized on the basis of the energetic preference for gauche conformers in POM

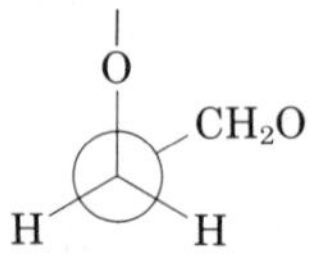

which tends to inhibit chain motions requiring a succession of both gauche and trans bonds intermixed.

Sulfone groups, in contrast to sulfur atoms alone, stiffen the chain markedly, as in poly(1-butene sulfone).

Effect of Double Bonds.   Although the double bond prevents rotation at the olefinic carbons, the reduced barriers at the allylic bonds more than compensate for this, and as a result, both *cis-* and *trans*-1,4-polybutadiene are substantially more flexible than PE.

Effect of Stereochemistry.   A dependence of chain mobility on stereochemical configuration has been observed in vinyl polymers, both for entire chains and steric sequences in atactic chains. Syndiotactic sequences commonly appear to be more restricted than isotactic. This has been reported for PMMA, PP, poly(1-butene), and PVC; for PS, the reverse is found.

Polyesters.   The effect of large groups inserted into an aliphatic chain may be comparable to that of side chains attached to them. The following T_1 values have been reported for terephthalate polyesters with aliphatic chains of varied length observed in hexafluoro-2-propanol solution at 34°C:

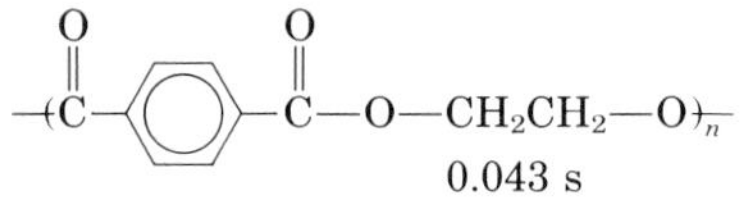

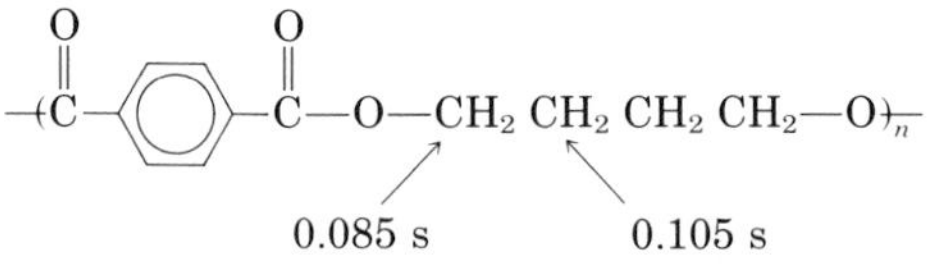

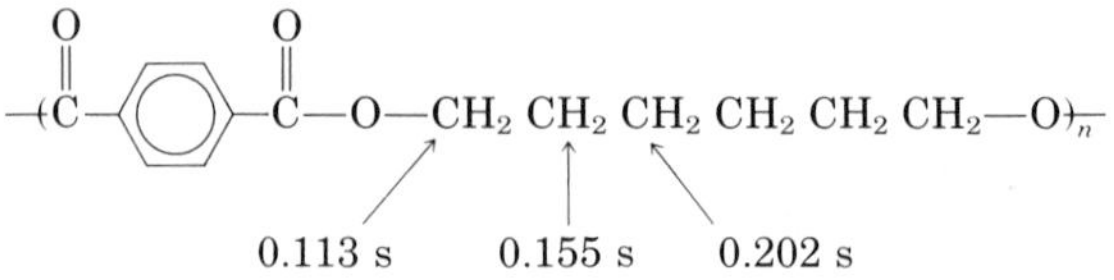

$$-\!\!\left(\!C\!-\!\bigcirc\!-\!C\!-\!O\!-\!CH_2CH_2\!-\!O\right)_{\!n}$$

0.043 s

$$-\!\!\left(\!C\!-\!\bigcirc\!-\!C\!-\!O\!-\!CH_2\,CH_2\,CH_2\,CH_2\!-\!O\right)_{\!n}$$

0.085 s   0.105 s

$$-\!\!\left(\!C\!-\!\bigcirc\!-\!C\!-\!O\!-\!CH_2\,CH_2\,CH_2\,CH_2\,CH_2\,CH_2\!-\!O\right)_{\!n}$$

0.113 s   0.155 s   0.202 s

Comparison with the values for PE in Table 6 (which must be divided by two) shows that the phenylene groups have a pronounced anchoring effect even for the six-carbon chain.

More realistic and elaborate motional models involving two or more correlation times or distributions of correlation times have been proposed and employed (11,38).

Chain Motion in the Solid State by ^{1}H Nmr

Until 1976, when cross-polarization, mas, and dipolar decoupling were combined to observe nmr spectra of solid polymers (28), ^{1}H nmr was used extensively in studies of molecular motion in solid polymers. Because polymers generally are not composed of isolated protons, the spins tend to be close enough together to produce dipolar broadening and proton spectra of solids are appropriately called wide-line spectra. The static dipolar broadening is given by equation 4. Molecular motion produces a narrowing of the static dipolar linewidth. Motion begins to affect the dipolar linewidth when

$$\frac{1}{\tau_c} \geqslant 2\pi\delta\nu \tag{49}$$

In this expression, $\delta\nu$ is the static linewidth in hertz and τ_c is the correlation time in seconds.

Calculation of the shape of the static dipolar pattern from first principles is not straightforward. However, the van Vleck second moment (97) is a method for obtaining the shape of the line from a summation over only the nearest proton neighbors in a lattice. Once the theoretical second moment is known from the van Vleck calculation method, second moments can be measured from experiments. Comparison of the theoretical and experimental second moments can provide information about molecular motions.

Polycarbonate. The polycarbonate (PC) based on 1,1-dichloro-2,2-bis(4-hydroxyphenyl)ethylene (chloral) provides a spin system that is sufficiently isolated that the dipolar doublet pattern can be observed (98):

Figure 44 shows the temperature-dependent spectra for this polymer. At very low temperatures, the polymer has a broad, almost Gaussian lineshape. The onset of line narrowing occurs at ca 200 K and the doublet shape is apparent at 260 K. A total decrease of approximately 35% is observed in the second moment as the temperature is increased. These spectra are consistent, with rotation about the 1,4-phenylene axis being the predominant mode of motion. Such a motion would not average the 2,3-phenylene proton interactions, whereas other significant main-chain motions would produce further narrowing in the spectra.

Block Copolymers. Solution nmr spectroscopy can provide information about copolymer structure, the average block length, and whether sequences are blocky, alternating, or random. However, many of the unique properties of copolymers derive from their phase-separated structure in the solid state, and solid-state proton, carbon, and deuterium nmr spectroscopies are useful for establishing the details of phase separation (see also MICROPHASE STRUCTURE). Phase separation in copolymers is monitored with proton nmr by observing the effects of

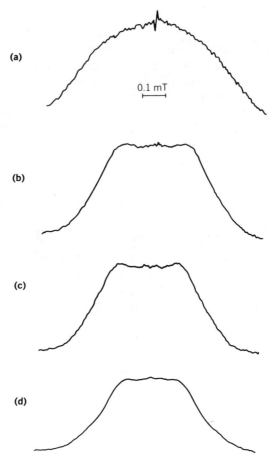

Fig. 44. Proton resonance absorption spectra for bulk chloral polycarbonate (PC) at 90 MHz as a function of temperature, and for the frozen solution in $C_2D_2Cl_4$ (98): (**a**) 153 K; (**b**) 313 K; (**c**) 373 K; (**d**) 153 K (solution). 0.1 mT = 1 G. Courtesy of *Macromolecules*.

spin diffusion, which is the process by which spins communicate with each other. For example, the protons on a rapidly rotating methyl group in a solid have a short T_1 because they are involved in motions that have spectral density in the megahertz-frequency regime. These protons can communicate their short T_1 to the entire proton spin bath through mutual spin flips with nearby protons (99). For this reason, all of the protons in a spatially homogeneous polymer have the same spin–lattice relaxation time.

    If the polymer under consideration is not spatially homogeneous, as is the case for many solid block copolymers, the process of spin diffusion can be used to measure the extent of spatial inhomogeneity. The spin-diffusion effects on T_1 and $T_{1\rho}$ are sensitive to short-range spatial inhomogeneities that occur in the range of 2–15 nm (100). The presence of two T_2 decays in the proton spectra of a solid polymer is good evidence for the existence of two distinct polymer phases whose spatial inhomogeneity spans more than several tens of nanometers. T_1 and T_2 measurements have been used to study phase separation in a polystyrene–poly(ethylene oxide) (PS–PEO) diblock copolymer (101). The PS phase is hydrophobic and glassy, whereas the PEO component is hydrophilic and semi-crystalline. The T_1 and T_2 data show individual homopolymer transitions, indicative of phase separation. However, the data also show that PS mobility increases at the PS–PEO interface above the PEO melting temperature. In addition, low temperature data suggest that PS induces localized disorder in the PEO crystalline region. Taken together, these results show that PS and PEO are incompatible, but that substantial phase mixing must occur at the PS–PEO phase boundaries (see also COMPATIBILITY).

Pulsed solid-state proton nmr has also been used to study phase separation in both polystyrene–polybutadiene–polystyrene triblock copolymers and in a series of segmented polyurethanes (102). The free induction decays for PS–PB–PS clearly show the presence of two components: the PS with its rapid decay, ie, rigid material, and the PB with its long decay, ie, the mobile component. The data for PS–PB–PS show little evidence of domain interference effects (see also STYRENE–DIENE BLOCK COPOLYMERS).

The data for PS–PB–PS, where the domains are large and well defined, can be compared to the situation encountered with the polyurethane segmented copolymers (102). The domains of the polyester soft segment and the diphenyl-methane diisocyanate hard segment were mixed by annealing, and the mixing process was followed by proton solid-state nmr spectroscopy. The data require the presence of short- and long-range degrees of mixing. The short-range degree of mixing is attributed to distances involving several molecular diameters, whereas the long-range mixing may involve chain-entanglement effects. The differences between the PS–PB–PS triblock copolymers and the segmented polyurethanes are attributed in part to the larger fraction of material that is present in the interfacial regions of the latter system. These studies further emphasize the importance of the interfacial regions in determining the ultimate properties of segmented copolymers (see also POLYURETHANES, BLOCK COPOLYMERS).

Other Polymers. Wide-line nmr results have been obtained for many polymers (1,103–105), including elastomers such as polyisobutylene and polybutadiene, and for polypropylene and poly(methyl methacrylate) (106), as well as for more rigid polymers such as polytetrafluoroethylene (PTFE) (Teflon) and poly(vinylidene fluoride) (PVDF). Generally, the second moment shows a decrease near the glass-transition temperature.

Wide-line methods can also be used to measure proton T_1 values for solid polymers (1). When these measurements are plotted as a function of temperature, the T_1 minima can usually be associated with the onset of a particular type of molecular motion, eg, methyl group rotation. Wide-line nmr can also be used to assess the degree of orientation in stretched polymers.

Chain Motion and Orientation in the Solid State by ^{13}C Nmr

Poly(ethylene terephthalate). Poly(ethylene terephthalate) (PET) fibers are employed in many commercial products, including tire cords. Fundamental questions arise concerning the degree of orientation of the polymer chains and the possible changes in orientation that may occur when the fibers undergo shrinkage due to excessive heat. From x-ray studies, it is known that the PET chain is oriented in such a way that the aromatic rings are not collinear with the fiber draw direction.

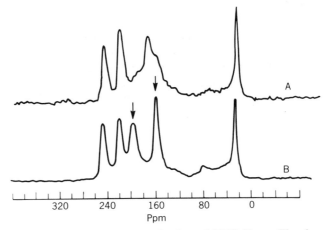

Solid-state ^{13}C nmr is useful for examining such materials because of the orientation dependence of the chemical shift (107). Without mas, perfectly oriented material should produce a single line for each magnetically distinct carbon. The chemical shifts of these lines depend on the angle that the fiber axis makes with the magnetic field. ^{13}C Nmr spectra of an oriented PET sample are shown in Figure 45. The peaks are assigned, from right to left, to C5, C3, C2, C1, and C4 (107). These spectra are obtained with the fiber draw axis parallel to the magnetic field. Spectra such as these can be interpreted to indicate overall orientational distributions in the crystalline regions.

Fig. 45. Solid-state ^{13}C nmr spectra of oriented PET fibers. The draw direction of the fibers is parallel to the magnetic field. Spectra (obtained without mas) are shown for two temperatures: A, 137°C; B, -105°C. The resonances for the ortho aromatic carbons (C2 and C3) (marked by arrows) appear separately in the low temperature spectrum (adapted from Ref. 107).

Because the ortho carbons (C2 and C3) reside in slightly different environments, their peaks are distinct at low temperature. These two peaks undergo a type of coalescence phenomenon at elevated temperatures, where it is observed that the rings undergo rapid 180° flips, thereby time-averaging the frequencies of the two peaks.

Hytrel Copolyesters.   In a solid-state carbon nmr study of molecular motion and phase mixing in a series of Hytrel copolyesters (108,109), local motion was measured at each chemically different site in the copolymer. Figure 46 shows the wide range in motional frequencies present in these copolyesters in the form of average correlation times.

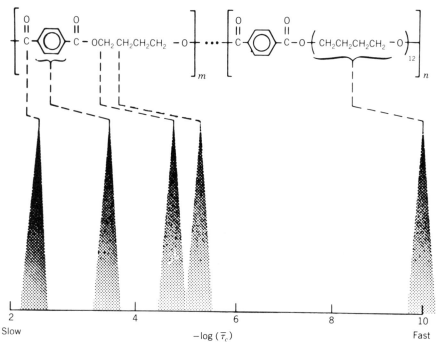

Fig. 46. Schematic representation of the different motional frequencies present in Hytrel segmented copolyesters (108).

For example, it was observed that the soft-segment carbons in a wide range of Hytrel samples give nmr spectra without requiring the solid-state nmr techniques of dipolar decoupling, cross-polarization, or mas (110), suggesting that the hard- and soft-segment carbons are phase-separated and that the soft-segment carbons are very mobile. The relaxation times (T_1 and NOE) of these carbons are independent of the mole fraction of hard segments, showing that the rates of the average local segmental motions of the soft phase are not affected by the hard segments. These solid-state relaxation times were analyzed using a broad distribution of correlation times (109), rather than the simpler single correlation time model discussed earlier. The linewidths for the soft-segment carbons are a linear function of the average hard-block length. This finding is consistent with the presence of a large interfacial area in which the angular range of the excursions of the soft-segment carbons is affected by the presence of the hard blocks.

Chemical-shift anisotropy considerations (17) and relaxation measurements (^{13}C T_1, $T_{1\rho}$) (109) were used to characterize the motions of the hard-segment carbons. The carbonyl carbons are essentially static on the nmr time scale, whereas some of the aromatic rings undergo 180° flips about the 1,4-phenylene axis. In addition, the aliphatic carbons in the segmented copolymer undergo g^+—t—g^- $\leftrightarrow g^-$—t—g^+ motions about three bonds at a rate identical to that observed for the poly(butylene terephthalate) (PBT) homopolymer.

Taken together, these results suggest that most of the hard segment is unaffected by the presence of soft-segment material.

Chain Motion in the Solid State by Deuterium Nmr

Polyurethanes.   Polyurethanes are structurally and morphologically complex materials, composed of isocyanate-derived hard segments and soft segments that are often polyether-based. It is thought that favorable hydrogen-bonding interactions in the hard segment are part of the driving force for phase separation. Deuterium nmr measurements can address the nature of phase separation, the amount of interfacial material, and the role of hard-segment material in the interface of the specifically labeled polyurethane based on 4,4′-diphenylmethane diisocyanate, chain-extended with butanediol (111,112):

$$-(C-N-\bigcirc-CH_2-\bigcirc-N-C-O-(CH_2CD_2CD_2CH_2O)_y \quad C-N-\bigcirc-CH_2-\bigcirc-N-C-O-$$

hard segment

$$CH_3$$
$$-(CH_2CH_2-O)_x-(CHCH_2-O)_z-(CH_2CH_2-O)_x-$$

soft segment

In these samples, only the central methylene carbons of the butanediol group are labeled. For the purposes of deuterium nmr spectroscopy, the soft segments and the rest of the hard segments are invisible (Fig. 47). The spectra in the center column are of the 100 wt % hard sample, ie, the homopolymer, shown at different vertical scales. Even at RT, the Pake powder pattern is substantially averaged by molecular motion. The left column of Figure 47 shows deuterium nmr spectra for the samples containing 70, 60, and 50 wt % hard segments. Because only the hard segment is labeled, the two-component nature of these spectra must arise from hard segments which reside in two different dynamical environments. Computer subtraction of appropriate amounts of the spectra in the center column from those in the left column produce the difference spectra shown in the right column. The sharp component, whose width remains constant, comprises 14, 20, and 50% of the signal intensity for the 70, 60, and 50 wt % hard samples, respectively.

The clean results from the subtraction process suggest that the central cores of the hard-segment-rich microdomains are comprised of material which behaves in a dynamical sense in a manner similar to that of the all-hard-segment material. On the basis of additional small-angle x-ray scattering (saxs) results (113), the sharp component is attributed to hard segments residing in interfacial regions. The narrow width of the solid deuterium nmr spectra for these interfacial hard segments indicates that they undergo nearly isotropic motion and suggests that there are few long-lived interurethane hydrogen-bonding interactions in this region of the copolymer.

This example shows that solid-state deuterium nmr can provide highly selective information, in this case about the interface in a segmented copolymer. Appropriate analysis of the deuterium nmr lineshapes and their temperature dependence can also yield details of both the rate and angular range of molecular motion.

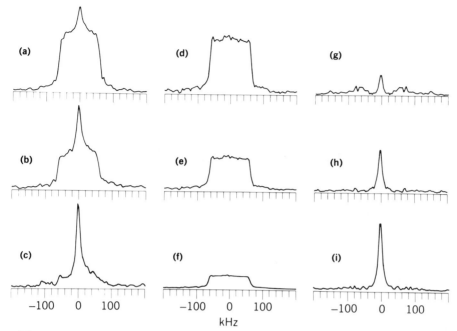

Fig. 47. Solid-state deuterium nmr spectra of hard-segment labeled polyurethanes (see text for structural formula) obtained at 55.26 MHz and 22°C (111): (**a**) 70 wt % hard; (**b**) 60 wt % hard; (**c**) 50 wt % hard; (**d**)–(**f**) 100% hard at different vertical gains; (**g**)–(**i**) subtraction of spectrum in middle column from that in left column.

Polysulfones and the Effect of Antiplasticizers. Deuterium nmr spectroscopy has been used to study molecular motion and the effect of antiplasticizers (see PLASTICIZERS) on the motion in specifically labeled polysulfones (114). Antiplasticizers produce stiffening and embrittlement of these polymers.

$$X = 95\% \ {}^1H, 5\% \ {}^2H$$

This polymer is one of a family of thermoplastic materials that is generally oxidatively stable at high temperatures and also quite resistant to acidic and basic aqueous environments. Polyarylethers have been studied by a variety of techniques. These materials exhibit a prominent secondary relaxation, the β-relaxation, at −100°C when measured at 1 Hz (115). This relaxation is enhanced by water in those polyarylethers that contain polar groups and diminished by the addition of low molecular weight antiplasticizers (see also POLYSULFONES; TRANSITIONS AND RELAXATIONS).

Figure 48 shows solid-state deuterium nmr spectra obtained for this polymer as a function of pulse-repetition rate. A 10-s delay (Fig. 48**a**) is sufficient to observe most of the signal. Different lineshapes are observed as the delay is shortened. Finally, at a very short delay (Fig. 48**c**), the pattern corresponds to that of a

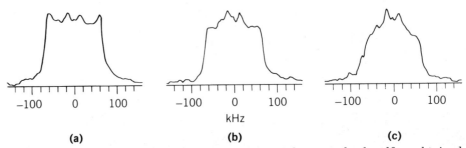

Fig. 48. Solid-state ^{2}H nmr spectra of aromatic deuterated polysulfone obtained at 23°C and 55.3 MHz, with pulse-repetition times of (**a**) 10 s; (**b**) 1 s; and (**c**) 300 ms (114).

phenylene ring undergoing rapid 180° flips (Fig. 24). These data show that there is a broad distribution of rates of phenylene-ring flips in this polymer. Some of the rings reside in environments that are nearly static, whereas others are in regions that undergo rapid (τ_c ca 10^{-7} s) ring flips. The differences are attributed to fluctuations in packing at the molecular level.

For bisphenol A polycarbonate, another strictly amorphous polymer, all of the phenyl rings undergo rapid 180° flips at or above RT (48,116), in contrast to the polysulfones, where a sizable fraction of staticlike rings are observed at RT (Fig. 48**a**). This result is expected, as the sulfone group is known to impart stiffness to the polymer chain.

Low molecular weight antiplasticizers, such as the chlorinated biphenyls, somehow restrict molecular motion and cause a decrease in the β-relaxation. The ductile mechanical properties of the polysulfone are concomitantly lost. If the phenyl-ring motions are responsible for the β-relaxation, the effect of added antiplasticizers is predicted to cause the average phenyl motions to become slower. As Figure 49 illustrates, this is the case. Therefore, phenyl flips are related with a molecular level process associated with the β-relaxation.

Interaction of Epoxy Resins with Water.   Absorption of small amounts (1–3 wt %) of water by epoxy resins produces a substantial plasticizing action, and the mechanical properties of the epoxy resin are adversely affected. Solid-state deuterium nmr studies of D_2O-exchanged epoxy resins of known water content provide information concerning the molecular details of the water–epoxy interaction (117,118). Figure 50 shows a typical quadrupole echo solid-state deuterium nmr spectrum of the exchanged epoxy resin

$$-N-CH_2-CH-CH_2-O-\bigcirc-\underset{\underset{CH_3}{|}}{\overset{\overset{CH_3}{|}}{C}}-\bigcirc-O-CH_2-CH-CH_2-N-$$

when it contains 2 wt % D_2O (Fig. 50**a**) and when this sample is dried (Fig. 50**b**). The sharp central line is attributed to the sorbed D_2O, since this peak disappears when the sample is heated or dried. This sharp peak cannot be due to free water, as it does not freeze (broaden out) at temperatures down to $-20°C$. The outer, broad part of the nmr spectrum is attributed to —OH residues that have undergone exchange with D_2O and have become —OD groups on the polymer backbone.

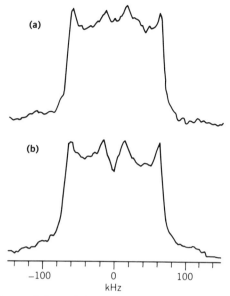

Fig. 49. Comparison of the solid-state deuterium nmr spectra of (**a**) polysulfone with 4,4′-dichlorodiphenyl sulfone added at 40 wt % with (**b**) unplasticized polysulfone (114).

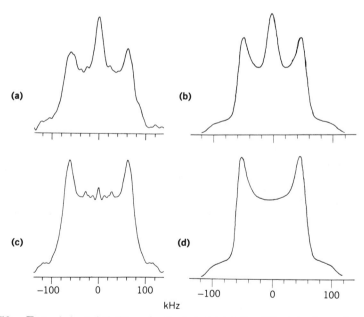

Fig. 50. Experimental (left) and calculated (right) solid-state deuterium nmr spectra of epoxy resins at 20°C (117): (**a**) and (**b**) epoxy resin containing 2 wt % D_2O; (**c**) and (**d**) deuterium exchanged and dried sample.

Spin–lattice relaxation measurements also show that there is no free water in this system (118). The T_1 of pure D_2O is 362 ms, whereas the corresponding relaxation time of the sharp signal in the epoxy resin is 12 ms. Although the water in the epoxy resin is somewhat impeded in its motion, it hops from site to site along the polymer backbone with a correlation time of ca 10^{-10} s. In this respect, the water cannot be considered tightly bound. It is also unlikely that the water disrupts the hydrogen-bonding network in the epoxy resin system, as has been proposed on the basis of other studies. This conclusion is based on the deuterium nmr data, which set an upper limit for the —OD/D_2O exchange rate of 10^3 s^{-1} and further show that the water molecules hop from site to site at least six orders of magnitude faster than this. The model that emerges from these deuterium nmr studies is one in which the water simply acts as a plasticizer in the epoxy system, just as it does in polymers that contain no exchangeable protons.

BIBLIOGRAPHY

"Nuclear Magnetic Resonance" in *EPST* 1st ed., Vol. 9, pp. 356–396, by F. A. Bovey, Bell Telephone Laboratories.

1. D. W. McCall, *Acc. Chem. Res.* **4**, 223 (1971).
2. F. A. Bovey, *High Resolution NMR of Macromolecules*, Academic Press, Inc., New York, 1972.
3. J. Schaefer and E. O. Stejskal in G. C. Levy, ed., *Topics in Carbon-13 NMR Spectroscopy*, Vol. 3, John Wiley & Sons, Inc., New York, 1979, p. 284.
4. V. J. McBrierty, *Magn. Reson. Rev.* **8**, 165 (1983).
5. R. A. Komoroski, ed., *High Resolution NMR of Synthetic Polymers in Bulk,* VCH Publishers, Inc., Deerfield Beach, Fla., 1986.
6. A. Abragam, *Principles of Nuclear Magnetism,* Oxford Press, Ltd., London, 1961.
7. F. A. Bovey, *NMR Data Tables for Organic Compounds*, Wiley-Interscience, New York, 1967.
8. Q. T. Pham, R. Petiaud, M.-F. Llauro, and H. Waton, *Proton and Carbon NMR Spectra of Polymers,* Vols. 1–3, John Wiley & Sons, Inc., New York, 1984.
9. M. Karplus, *J. Chem. Phys.* **30**, 11 (1959).
10. M. Barfield, A. M. Dean, C. J. Fallick, R. J. Spear, S. Sternhell, and P. W. Westerman, *J. Am. Chem. Soc.* **97**, 1482 (1975).
11. F. A. Bovey and L. W. Jelinski, *J. Phys. Chem.* **89**, 571 (1985).
12. J. R. Lyerla, Jr. and G. C. Levy in Ref. 3, Vol. 1, 1974, p. 79.
13. L. W. Jelinski, *Ann. Rev. Mater. Sci.* **15**, 359 (1985).
14. E. Fukushima and S. B. W. Roeder, *Experimental Pulse NMR,* Addison-Wesley Publishing Co., Inc., Reading, Mass., 1981.
15. P. M. Henrichs, Eastman Kodak Laboratories, Rochester, N.Y., 1985, personal communication.
16. T. M. Duncan, AT&T Bell Laboratories, Murray Hill, N.J., 1986, personal communication.
17. L. W. Jelinski, *Macromolecules* **14**, 1341 (1981).
18. M. Mehring, *Principles of High Resolution NMR in Solids,* 2nd ed., Springer-Verlag, New York, 1983.
19. P. D. Murphy, T. Taki, B. C. Gerstein, P. M. Henrichs, and D. J. Massa, *J. Magn. Reson.* **49**, 99 (1982).
20. T. M. Duncan, *J. Am. Chem. Soc.* **106**, 2270 (1984).
21. R. E. Stark, L. W. Jelinski, D. J. Ruben, D. A. Torchia, and R. G. Griffin, *J. Magn. Reson.* **55**, 266 (1983).
22. T. Terao, S. Maeda, T. Yamabe, K. Akagi, and H. Shirakawa, *Chem. Phys. Lett.* **103**, 347 (1984).
23. F. C. Schilling, F. A. Bovey, A. E. Tonelli, S. Tseng, and A. E. Woodward, *Macromolecules* **17**, 728 (1984).
24. J. Urbino and J. S. Waugh, *Proc. Natl. Acad. Sci. USA* **71**, 5062 (1974).
25. D. L. VanderHart, *J. Chem. Phys.* **64**, 830 (1976).
26. E. R. Andrew, A. Bradbury, and R. G. Eades, *Nature London* **183**, 1802 (1959).
27. I. J. Lowe, *Phys. Rev. Lett.* **2**, 285 (1959).
28. J. Schaefer and E. O. Stejskal, *J. Am. Chem. Soc.* **98**, 1031 (1976).
29. D. L. VanderHart, H. S. Gutowsky, and T. C. Farrar, *J. Am. Chem. Soc.* **89**, 5056 (1967).
30. J. G. Hexem, M. H. Frey, and S. J. Opella, *J. Chem. Phys.* **77**, 3847 (1982).
31. A. Pines, M. G. Gibby, and J. S. Waugh, *J. Chem. Phys.* **56**, 1776 (1972).

32. *Ibid.,* **59,** 569 (1973).
33. S. R. Hartmann and E. L. Hahn, *Phys. Rev.* **128,** 2042 (1962).
34. A. G. Redfield, *Phys. Rev.* **98,** 787 (1955).
35. G. C. Levy, R. L. Lichter, and G. L. Nelson, *Carbon-13 Nuclear Magnetic Resonance Spectroscopy,* 2nd ed., John Wiley & Sons, Inc., New York, 1980.
36. E. D. Becker, *High Resolution NMR,* 2nd ed., Academic Press, Inc., New York, 1980.
37. D. A. Torchia and A. Szabo, *J. Magn. Reson.* **49,** 107 (1982).
38. F. Heatley, *Prog. Nucl. Magn. Reson. Spectrosc.* **13,** 47 (1980).
39. J. Schaefer and D. F. S. Natusch, *Macromolecules* **5,** 416 (1972).
40. J. Schaefer, E. O. Stejskal, and R. Buchdahl, *Macromolecules* **10,** 384 (1977).
41. J. Schaefer, E. O. Stejskal, T. R. Steger, M. D. Sefcik, and R. A. McKay, *Macromolecules* **13,** 1121 (1980).
42. E. O. Stejskal, J. Schaefer, and R. A. McKay, *J. Magn. Reson.* **25,** 569 (1977).
43. H. W. Spiess, *Adv. Polym. Sci.* **66,** 23 (1985).
44. F. Lauprêtre, L. Monnerie, and J. Virlet, *Macromolecules* **17,** 1397 (1984).
45. T. C. Farrar and E. D. Becker, *Pulse and Fourier Transform NMR,* Academic Press, Inc., New York, 1971.
46. A. Bax, *Two Dimensional NMR in Liquids,* Delft University Press and D. Reidel Publishing Co., Boston, 1982.
47. L. W. Jelinski in Ref. 5, pp. 335–364.
48. H. W. Spiess, *Colloid Polym. Sci.* **261,** 193 (1983).
49. G. E. Pake, *J. Chem. Phys.* **16,** 327 (1948).
50. F. C. Schilling, F. A. Bovey, M. D. Bruch, and S. A. Kozlowski, *Macromolecules* **18,** 1418 (1985).
51. F. A. Bovey and G. V. D. Tiers, *J. Polym. Sci.* **44,** 173 (1960).
52. H. L. Frisch, C. L. Mallows, and F. A. Bovey, *J. Chem. Phys.* **45,** 1565 (1966).
53. F. A. Bovey, *Acc. Chem. Res.* **1,** 175 (1968).
54. W. Fowells, C. Schuerch, F. A. Bovey, and F. P. Hood, *J. Am. Chem. Soc.* **89,** 1396 (1967).
55. R. C. Ferguson, *Trans. N.Y. Acad. Sci.* **29,** 495 (1967).
56. F. Heatley, R. Salovey, and F. A. Bovey, *Macromolecules* **2,** 619 (1969).
57. F. Heatley and A. Zambelli, *Macromolecules* **2,** 618 (1969).
58. A. E. Tonelli and F. C. Schilling, *Acc. Chem. Res.* **14,** 233 (1981).
59. A. Bunn, E. A. Cudby, R. K. Harris, K. J. Packer, and B. J. Say, *Chem. Commun.* 15 (1981).
60. F. A. Bovey and L. W. Jelinski, *Chain Structure and Conformation of Macromolecules,* Academic Press, Inc., New York, 1982.
61. G. Natta and P. Corradini, *Nuovo Cim. Suppl. 15* **1,** 9 (1960).
62. J. Boor, Jr., *Macromol. Rev.* **2,** 115 (1967).
63. M. W. Duch and D. M. Grant, *Macromolecules* **3,** 165 (1970).
64. F. C. Schilling, AT&T Bell Laboratories, Murray Hill, N.J., 1983, personal communication.
65. Ref. 60, pp. 106–110.
66. F. A. Bovey and F. C. Schilling, AT&T Bell Laboratories, Murray Hill, N.J., 1983, personal communication.
67. R. E. Cais and N. J. A. Sloane, *Polymer* **24,** 179 (1983).
68. R. E. Cais and J. M. Kometani in J. C. Randall, ed., *NMR and Macromolecules, ACS Symp. Ser.* **247,** American Chemical Society, Washington, D.C., 1984, pp. 153–166.
69. A. E. Tonelli, F. C. Schilling, and R. E. Cais, *Macromolecules* **15,** 849 (1982).
70. F. A. Bovey, F. C. Schilling, F. L. McCrackin, and H. L. Wagner, *Macromolecules* **9,** 76 (1976).
71. Ref. 60, p. 173.
72. F. R. Mayo and F. M. Lewis, *J. Am. Chem. Soc.* **66,** 1594 (1944).
73. Ref. 60, pp. 137–154.
74. J. B. Kinsinger, T. Fischer, and C. W. Wilson III, *J. Polym. Sci. Part B* **4,** 379 (1966); **5,** 285 (1967).
75. K. H. Hellwege, U. Johnsen, and K. Kolbe, *Kolloid Z.* **214,** 45 (1966).
76. F. A. Bovey, *J. Polym. Sci. Macromol. Rev.* **9,** 1, 19 (1974).
77. D. J. Patel, AT&T Bell Laboratories, Murray Hill, N.J., 1976, personal communication.
78. R. W. Behling and D. R. Kearns, *Biopolymers* **24,** 1157 (1985).
79. F. C. Schilling, AT&T Bell Laboratories, Murray Hill, N.J., 1978, private communication, as reported by F. A. Bovey in R. H. Sarma, ed., *Stereodynamics of Molecular Systems,* Pergamon Press, Elmsford, N.Y., 1979, pp. 53–74.
80. Y. Inoue, A. Nishioka, and R. Chûjô, *Makromol. Chem.* **168,** 163 (1973).

81. F. Heatley, *Polymer* **16**, 493 (1975).
82. F. C. Schilling, R. E. Cais, and F. A. Bovey, *Macromolecules* **11**, 325 (1978).
83. W. Gronski and N. Murayama, *Makromol. Chem.* **177**, 3017 (1976).
84. F. C. Schilling, *Macromolecules* **11**, 1290 (1978).
85. K. Matsuo and W. H. Stockmayer, *Macromolecules* **14**, 544 (1981).
86. F. A. Bovey, F. C. Schilling, T. K. Kwei, and H. L. Frisch, *Macromolecules* **10**, 559 (1977).
87. Y. Inoue, A. Nishioka, and R. Chûjô, *J. Polym. Sci. Polym. Phys. Ed.* **11**, 2237 (1973).
88. W. Gronski and N. Murayama, *Makromol. Chem.* **179**, 1509 (1978).
89. J. R. Lyerla, T. T. Horikawa, and D. E. Johnson, *J. Am. Chem. Soc.* **99**, 2463 (1977).
90. F. Lauprêtre, C. Noël, and L. Monnerie, *J. Polym. Sci. Polym. Phys. Ed.* **15**, 2143 (1977).
91. G. Hermann and G. Weill, *Macromolecules* **8**, 171 (1975).
92. F. Heatley and I. Walton, *Polymer* **17**, 1019 (1976).
93. K. J. Ivin and F. Heatley, private communication, as reported in F. Heatley, *Prog. Nucl. Magn. Reson. Spectrosc.* **13**, 47 (1980).
94. R. E. Cais and F. A. Bovey, *Macromolecules* **10**, 752 (1977).
95. *Ibid.*, p. 169.
96. *Ibid.*, p. 757.
97. J. H. van Vleck, *Phys. Rev.* **74**, 1168 (1948).
98. P. T. Inglefield, A. A. Jones, R. P. Lubianez, and J. F. O'Gara, *Macromolecules* **14**, 288 (1981).
99. L. W. Jelinski and M. T. Melchior in C. R. Dybowski and R. L. Lichter, eds., *Practical NMR Spectroscopy,* Marcel Dekker, Inc., New York, 1986, pp. 253–329.
100. V. J. McBrierty, D. C. Douglass, and T. K. Kwei, *Macromolecules* **11**, 1265 (1978).
101. S. Kaplan and J. J. O'Mally, *Polym. Prepr. Am. Chem. Soc. Div. Polym. Chem.* **20**, 266 (1979).
102. R. A. Assink and G. L. Wilkes, *Polym. Eng. Sci.* **17**, 606 (1977).
103. V. J. McBrierty and D. C. Douglass, *Macromol. Rev.* **16**, 295 (1981).
104. V. J. McBrierty and D. C. Douglass, *Phys. Rep.* **63**, 61 (1980).
105. V. J. McBrierty, *Faraday Discuss. Chem. Soc.* **68**, 78 (1979).
106. J. G. Powles, B. I. Hunt, and D. J. H. Sanford, *Polymer* **5**, 505 (1964).
107. D. L. VanderHart, G. G. A. Böhm, and V. D. Mochel, *Polym. Prepr. Am. Chem. Soc. Div. Polym. Chem.* **22**, 261 (1981).
108. L. W. Jelinski, J. J. Dumais, and A. K. Engel, *Macromolecules* **16**, 403 (1983).
109. L. W. Jelinski, J. J. Dumais, P. I. Watnick, A. K. Engel, and M. D. Sefcik, *Macromolecules* **16**, 409 (1983).
110. L. W. Jelinski, F. C. Schilling, and F. A. Bovey, *Macromolecules* **14**, 581 (1981).
111. J. J. Dumais, L. W. Jelinski, L. M. Leung, I. Gancarz, A. Galambos, and J. T. Koberstein, *Macromolecules* **18**, 116 (1985).
112. A. Kintanar, L. W. Jelinski, I. Gancarz, and J. T. Koberstein, *Macromolecules* **19**, 1876 (1986).
113. J. T. Koberstein and R. S. Stein, *J. Polym. Sci. Polym. Phys. Ed.* **21**, 1439 (1983).
114. J. J. Dumais, A. L. Cholli, L. W. Jelinski, J. L. Hedrick, and J. E. McGrath, *Macromolecules* **19**, 1884 (1986).
115. L. M. Robeson, A. G. Farnham, and J. E. McGrath in R. F. Boyer and D. J. Meier, eds., *Molecular Basis of Transitions and Relaxations,* Gordon & Breach, New York, 1978, p. 405.
116. H. W. Spiess, *J. Mol. Struct.* **111**, 119 (1983).
117. L. W. Jelinski, J. J. Dumais, R. E. Stark, T. S. Ellis, and F. E. Karasz, *Macromolecules* **16**, 1019 (1983).
118. L. W. Jelinski, J. J. Dumais, A. L. Cholli, T. S. Ellis, and F. E. Karasz, *Macromolecules* **18**, 1091 (1985).

General References

Books on Nmr Spectroscopy

Refs. 2, 5, 18, 36, 46, and 60 are good general references.
C. P. Slichter, *Principles of Magnetic Resonance,* Springer-Verlag, Heidelberg, FRG, 1978.
C. A. Fyfe, *Solid State NMR for Chemists,* CFC Press, Guelph, Ontario, 1983.
M. L. Martin, J.-J. Delpuech, and G. J. Martin, *Practical NMR Spectroscopy,* Heyden and Son, Inc., Philadelphia, 1980.

Solution-state Nmr of Polymers

Consult Ref. 58 for further information on the above subject; Ref. 11 is a good review of polymer relaxation, and Ref. 38 a comprehensive review of polymer relaxation in solution.

L. W. Jelinski, *Chem. Eng. News* **62**, 26 (1984). Applications of nmr methods to synthetic polymers and biopolymers.

J. C. Randall and E. T. Hsieh, *ACS Symp. Ser.* **247**, 131 (1984). Nmr for quantitative polymer measurements.

K. J. Ivin, *Pure Appl. Chem.* **55**, 1529 (1983). Characterization of copolymers by nmr.

F. Heatley, *Macromol. Chem. London* **2**, 190 (1982). Polymer structure, conformation, and relaxation.

J. R. Ebdon, *Macromol. Chem. London* **3**, 175 (1984).

Solid-state Nmr of Polymers

Ref. 1 is a review of wide-line proton measurements on synthetic polymers; Ref. 3 reviews basic techniques of solid-state high resolution nmr; Ref. 4 is a comprehensive review of nmr of solid polymers; Refs. 13, 43, 47, and 48 are reviews of solid-state deuterium nmr of polymers; Ref. 99 presents practical and experimental aspects of high resolution solid-state nmr; and Ref. 103 reviews solid-state nmr of polymers.

F. A. Bovey
L. W. Jelinski
AT&T Bell Laboratories

NUCLEAR RADIATION. See Radiation-induced reactions.

NUCLEIC ACIDS. See Polynucleotides.

NUCLEOSIDES. See Polynucleotides.

NUCLEOTIDES. See Polynucleotides.

NUMBER-AVERAGE MOLECULAR WEIGHT. See Molecular weight determination.

NYLON. See Polyamides; Polyamides, fibers; Polyamides, plastics.

O

OFFSET PRINTING. See Decorating.

OIL-FIELD APPLICATIONS

Polymers are used extensively in the various phases of drilling, operating, and maintaining oil and gas wells, including drilling, cementing, work-over and stimulation operations, and even the abandonment of wells. Application of polymers in enhanced-oil-recovery (EOR) processes prolongs the economic lifetime of oil fields (1,2). Over 600 polymer products are marketed for these applications (3).

Both water-soluble and oil-soluble polymers are employed in oil-field technology. The principal applications of water-soluble polymers (qv) arise from their ability to alter the rheological properties of aqueous fluids used in drilling, cementing, well completion, sand control, fracturing, acidizing, and enhanced oil recovery. Cross-linked polymer systems, including water-soluble, oil-soluble, and water-dispersible systems, reduce rock permeability in a controlled manner. These systems find applications in cementing, sand-control processes, and sealing of high permeability zones. Both *in situ* polymerization and *in situ* cross-linking may be employed in sealing high permeability zones. Both organic and inorganic polymers may be used in a variety of lesser well-known applications, eg, in formation-damage control and as dispersing agents. The organic polymers may be synthesized from petrochemicals or derived from plants or microorganisms. They cover a wide range of molecular weights, chemical structures, and costs.

Oil-field Technology

Drilling Fluids. After discovery, the development of an oil or gas field begins with the drilling operation (3–5). Drilling fluids are required to cool and lubricate the drill bit, suspend formation cuttings and lift them to the surface, and control formation pressure during drilling. The circulation of drilling fluid keeps the bottom of the borehole free of formation cuttings and thus increases the drilling rate. The circulating fluid also reduces the heat of friction generated by the drill bit.

Drilling fluids are thixotropic solutions with high viscosity at low shear (during movement up and down the well bore when solids must be suspended) and low viscosity in the high shear region around the drill bit, where rapid fluid movement is essential for cooling the drill bit and removing formation cuttings. High viscosity reduces the loss of drilling fluid to the formation.

The base fluid is usually water; betonite modifies rheological properties, and salts are used to control density and stabilize the shale. Polymers find applications as water thickeners, additives, dispersants, and formation-damage control agents (1,3,6,7).

Cement Slurries and Spacers.   Casing is installed in most wells upon completion. At various stages in the drilling operation, the drill bit is removed and steel casing is lowered into the well bore. A cement slurry is pumped down the casing and up into the annular space between the casing and the formation (4,8). This operation displaces the drilling fluid. The cement sets, bonding the casing to the formation. The cement provides mechanical support for the casing and prevents the formation pressure from collapsing the pipe. Furthermore, it protects the casing from possible corrosion caused by contact with formation fluids and prevents undesirable migration of fluid from one formation to another (zone isolation). This operation is called primary cementing.

The main ingredients of typical cement slurries are given in Table 1. The type of cement used depends on the desired slurry density, formation depth, and temperature. Fluid loss from the cement slurry is a critical parameter that influences cement-set time. The resulting loss of hydrostatic pressure prevents formation fluids from entering the well bore.

Table 1. Typical Cement Slurry

Ingredient	Application
water	base fluid
cement[a]	(see text)
salts	density control
	cement-set time control
	formation-damage control
retarder[b]	cement-set time control
water-soluble polymeric	fluid-loss additive
thickner	cement-set time control
water-soluble polymer	particle dispersant
	formation-damage control
solid additive for	fluid-loss control
fluid-loss prevention	cement-set time control

[a] Usually Portland.
[b] Usually lignosulfonate.

Spacers are fluids that separate the cement slurry from the drilling fluid and remove the drilling fluid from the well bore. Spacers must be compatible with both the cement slurry and the drilling fluid. They control formation pressures and have limited fluid loss to the formation. Water-based spacers have a continuous water phase and contain polymeric additives to control pH and density; other additives control formation damage.

Incomplete displacement of drilling fluid from the annular space between the casing and the rock can lead to gaps in the cement sheath. These gaps cause undesirable fluid communication between rock zones and corrosion. So-called squeeze cementing can be used to place Portland cement, sodium silicate cement, or some other sealant in these regions. Polymer cements bond the well casing to the formation in highly corrosive environments, which would attack Portland cement, such as those found in some water and chemical-disposal wells. Waterless polymer cements (qv) are used to penetrate permafrost zones and in applications requiring superior fluid-loss control.

Completion and Work-over Fluids.   After the well has been cemented, communication between the well bore and the productive formation must be established by perforating the casing and the cement sheath. Other operations include controlling sand or water production, cleaning the well bore, and plugging perforations. During these operations, the well bore is filled with a fluid compatible with the rock formation. Well-bore fluid density may be adjusted to provide a hydrostatic pressure sufficient to prevent blowouts. Fluid-loss additives and formation-damage control additives are often employed.

Sand production from poorly consolidated formations can be a serious problem (4,9,10). Sand fills the well bore, reduces productivity, erodes down-hole and surface equipment, and creates void spaces behind the casing that may lead to formation collapse and loss of the well. Sand production may be controlled by organic polymers.

In the gravel-pack (4) and consolidated-gravel-pack sand control methods (10), a water-soluble polymer, usually a polysaccharide, is used as thickener to suspend as much as 1.8 kg of carefully sized sand particles per liter of fluid; usually, 4.8–9.2 g/L fluid is used. Suspension of sand particles is particularly important in nonvertical (deviated) well bores drilled from offshore platforms and artificial islands. A "breaker" is added for depolymerization as the sand slurry reaches the particular zone. This breaker may be an oxidizer, such as a salt of a peracid, an acid, or an enzyme. Depolymerization results in rapid loss of fluid viscosity, causing the sand to be packed against the formation. This permeable mass of sand holds the formation in place while allowing flow of oil or gas. The amount of insoluble residue formed (11) in polysaccharide depolymerization is critical, since these solids can plug flow channels in the rock and in the packed sand.

The resin coating the sand in the consolidated-sand-pack process hardens under down-hole conditions, and the sand forms a hard but permeable mass. Similar resins may be injected into the formation to coat the sand grains in the resin method of sand control. Excess resin filling the pore spaces must be flushed deeper into the formation to coat additional sand grains and leave the flow channels open when the resin hardens.

Hydraulic Fracturing.   Hydraulic fracturing increases the productivity of low permeability formations by cracking the rock to create high permeability flow channels (4,9). The fracture is generated by pumping a fluid containing suspended solids (Table 2) into the formation at a high rate and pressure. The suspended solids, called proppants, settle into the fracture and prevent sealing at the conclusion of the well treatment. Water-soluble polymers (usually cross-linked) are used to suspend the proppant and increase the solids-carrying capacity of the fracturing fluid; generally, 2–7 g of polymer per liter of fracturing fluid is used.

Table 2. Water-based Hydraulic Fracturing Fluid

Ingredient	Application
water	base fluid
potassium chloride, 1–3 wt %	formation-damage control
water-soluble polymeric thickner	modification of rheological properties[a] fluid-loss additive[b]
water-soluble polymers	formation-damage control
solid-fluid-loss additive[c]	reduction in fluid loss
liquid-fluid-loss additive	oil or oil-miscible solvent added to reduce fluid loss (often preferred over a solid-fluid-loss additive)
cross-linking agent	cross-linking of the viscosity modifier to increase solids-suspending capacity
breaker	depolymerization of the viscosity modifier to reduce viscosity at a controlled rate and time
proppant[d]	to keep fracture open

[a] Polysaccharides or their derivatives.
[b] Guar or derivatives.
[c] Silica flour or inorganic resin.
[d] Sand, resin-coated sand, or sintered bauxite.

Fluid loss from the fracture can also be critical in determining the success of the hydraulic-fracturing treatment. Rapid fluid loss can cause formation damage within the rock and reduce permeability at the formation face; a proppant barrier or bridge limits further fracture extension. Polymer-stabilized foams and fluids containing CO_2 have been developed to reduce fluid loss. Carefully sized solid particles are good fluid-loss additives, as are polymer emulsions. Commercial liquid-fluid-loss additives (dispersed oil) are often preferred when fracturing low permeability formations.

Injection is terminated when the desired amount of fluid has been pumped into the formation. This results in only partial closure of the fracture, which maintains a high permeability pathway for the fluid flow to the well bore. To aid in rapid recovery of the fracturing fluid when returning the well to production, breakers are used to depolymerize the polymer and reduce viscosity for rapid fluid recovery. The breakers are similar to those used in the gravel-packing operations described previously. The well is usually shut down for some time to permit depolymerization before starting production. In the case of incomplete depolymerization, cross-linked polymer (often referred to as gel) clogs the fracture, lowering well productivity. Depolymerization produces a solids-free fluid.

Acidizing. In an acid-fracturing treatment, an acid is injected at pressures exceeding the fracture pressure of the formation. The acid reacts with, ie, "etches," the fracture surfaces, creating a flow channel that remains open when the well is returned to production. Acid fracturing is widely used to stimulate limestone and dolomite formations. The fluid-loss characteristics of the injected fluid strongly affect fracture geometry. Hydrochloric acid solutions are commonly used in acidizing calcareous formations.

Minerals are dissolved by acidizing the matrix; this increases the diameter of flow channels near the well bore and removes solid particles plugging these flow channels (4,9,12). Blends of hydrochloric and hydrofluoric acid are used for sandstone acidizing to dissolve clays and silica.

Synthetic polymers, particularly binary or ternary acrylamide copolymers and cross-linked polymers, increase the viscosity of acid solutions (1), which in turn reduces fluid loss in fracture acidizing and acid reactivity because of lower rates of acid transfer to rock surfaces. Lower acid reactivity allows deeper penetration of the acid solution into the formation. After injecting the acid solution, the well is "turned around," the spent acid is recovered, and the well returned to production. Synthetic polymers and polysaccharides have been used to reduce friction; the former stabilize foamed acid.

Water Intrusion into the Well.   Water intrusion can be a serious problem in well drilling and production operations. Occasionally during drilling, the drilling fluid is extensively diluted by this water production from a rock zone, which can be sealed with cement. However, the rate of water intrusion may be so high that Portland cement slurries are extensively diluted and washed away. Rapidly setting sodium silicate cements can be used in these situations.

During oil and gas production, water may rise in the well to unacceptable levels. When the water enters the productive zone near the well bore, so-called water coning occurs. Polymers or a solution of organic monomer, cross-linking agent, and initiator can be injected to plug separate water-producing zones. *In situ* polymerization creates an impermeable mass which acts as a seal. *In situ* gelation of sodium silicate seals fractures through which the water flows. When water and petroleum are produced from the same zone, the water:oil ratio can be controlled by proprietary polymers to reduce rock permeability to aqueous fluids selectively while hydrocarbon permeability remains substantially unchanged (13).

Enhanced Oil Recovery.   Hydrocarbons in subterranean formations exist under pressures created by the rock overburden and generation of gases during the processes of petroleum formation. During primary recovery, these forces drive oil into the production well. As fluid is removed, these pressures decrease and oil-production rates decline. Pumps may be used to draw petroleum to the surface. Secondary recovery processes may then be applied by converting some production wells to injection wells. The geometric arrangement of injection and production wells is a complicated process. The most common secondary recovery method involves the injection of water to repressurize the reservoir. As the injected water moves through the reservoir, it displaces the oil ahead. However, the mobility of the water in the subterranean formation is usually much higher than that of the oil. The injection water flows through the formation via the most permeable rock layers. As the water does not penetrate the rock completely, substantial amounts of oil remain in the formation, as well as in the rock sections penetrated by the water, because of the high interfacial tension between the oil and the injected water.

Enhanced oil recovery (EOR) is also called tertiary oil recovery because it is often applied after primary and secondary recovery methods. Thickeners are frequently used in enhanced oil recovery (14–17). Laboratory evaluations of flood chemicals and EOR process designs are reviewed in Ref. 18. Results of numerous field-pilot projects have been reported (17,19–21). Oil production in polymer flood EOR production rose 295% between 1982 and 1984 to >1590 m^3/d (>10,000 bbl/d) (22).

The theory of the flow behavior of polymers in permeable media is discussed in detail in Ref. 23. Polymer concentrations are generally in the range 250–1000 ppm. Injection volumes are 5–25% of the reservoir pore volume. The most common types used for this application are partially hydrolyzed polyacrylamides and xanthan gum, also called biopolymer. The polymer reduces the mobility of the injected fluid by increasing solution viscosity. Polyacrylamides are extensively adsorbed on the rock surfaces, which significantly reduces permeability. Subsequently injected fluid is diverted to less permeable and less flooded portions of the rock. The resulting increase in volumetric sweep efficiency of the injected fluid increases oil recovery by an additional 2–5%. Polymer flooding recovers ca 1.2–7.7 $m^3/1000$ m^3 of rock [10–60 bbl oil/(acre·ft)], 2–10% of the oil remaining in place after secondary recovery.

Even if fresh water is used as the polymer solvent, the polymer must function as an effective viscosity modifier in the presence of salts, because the injected polymer solution is mixed with saline formation waters within the reservoir. The polymer must also be stable to the temperature and pH of the petroleum reservoir.

The economics of polymer flooding can be improved by gradual reduction of the concentration of the injected polymer, ie, by using a so-called tapered polymer slug. Initial concentrations can be as high as 2500 ppm. Operational problems include scaling, incomplete hydration of solid polymers, oxidative and bacterial degradation of polymer solutions, and stabilization of oil–water emulsions. In addition to causing polymer degradation, oxygen promotes the growth of aerobic and facultative anaerobic bacteria. Strong oxidizing agents kill or inhibit bacterial growth and remove polymer deposits causing plugging. Sulfite-type oxygen scavengers, aldehydes, halogen-type oxidizing compounds, and sulfur-containing biocides (qv) are the most common stabilizers.

Incremental oil recovery due to polymer injection is modest because the interfacial tension between the oil and injected fluid is high. This is reflected in a relatively low oil-displacement efficiency by the injected polymer solution. Polymers may be used in combination with other chemicals designed to reduce the water–oil interfacial tension. In micellar-polymer flooding, a surfactant lowers the water–oil interfacial tension, resulting in very high oil-displacement efficiency. However, the surfactants are expensive. Polymers are added to the flood water to reduce its mobility and prevent injection water bypassing the surfactant slug and the oil bank; polymers may also be added to the surfactant solution. Polymer solutions injected prior to the surfactant improve volumetric sweep efficiency of the surfactant solution and act as adsorption agents.

Micellar polymer flooding has been studied extensively in the laboratory and in a number of field trials, but the injected fluids are costly. Micellar polymer flooding has recovered 9–25 m^3 oil per 1000 m^3 rock (70–200 bbl/(acre·ft)], 15–40% of the oil remaining in place after secondary recovery.

Polymers have also been used in conjunction with caustic flooding, where the polymer reduces loss of the expensive injected chemical solution through chemical reactions and increases volumetric sweep efficiency.

The productive formation frequently consists of rock layers of varying permeability. High permeability streaks and fractures can result in rapid flow of the injected fluid from the injection well to production wells. Little of the rock comes in contact with the injected fluid and oil production is low. In addition, the ratio of water produced to oil is high, which increases the costs of pumping large volumes of water to the surface, separating the oil and disposing of the water in an environmentally acceptable manner.

Polymeric plugging agents mitigate these problems. The polymer systems are the same as those used in various well-water control processes. The most common method is the *in situ* cross-linking of polyacrylamide or xanthan gum.

Water-soluble Polymers

In oil-field applications, the most important property of a water-soluble polymer is its solution viscosity, defined as the ratio of shear stress S_s to shear rate S_r. Shear stress is the force-per-unit cross-sectional area applied to the polymer solution. Shear rate is the displacement of the polymer solution divided by the height of the solution affected by the shear stress. At 20°C, water has a viscosity of 1.0 mPa·s (= cP).

Water is a Newtonian fluid, ie, shear stress is directly proportional to shear rate. Such fluids begin to flow when only a slight force is applied. Non-Newtonian fluids do not exhibit direct proportionality between shear stress and shear rate. Since viscosity depends on the shear rate (Fig. 1), the term apparent viscosity is often used. The slope n' of each curve in Figure 1 depends on the non-Newtonian behavior of that polymer solution; for a Newtonian fluid, $n' = 1$. The greater the deviation of n' from unity (in either the positive or negative direction), the more non-Newtonian is the fluid. The shear stress ϵ (Fig. 1) is directly proportional to the apparent viscosity.

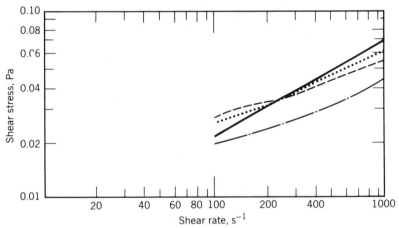

Fig. 1. Shear stress (apparent viscosity) of polysaccharides at 26°C with gel concentration of 6 g/L: ····, hydroxypropylguar; ——, hydroxyethylcellulose; ------, guar; and ------, xanthan gum.

Polymer solutions exhibit the non-Newtonian behavior known as pseudoplasticity, ie, solution viscosity decreases with increasing shear rate. At polymer concentrations used in enhanced oil recovery (<3500 ppm), both polyacrylamide and xanthan gum exist in the semidilute regime, in which polymer molecules are entangled with neighboring polymer chains (24).

Specialized experimental methods have been developed to study cross-linked polymer fluids such as those used in hydraulic fracturing (25,26); pipe viscometers and laser doppler anemometry have been employed.

Polymer adsorption can be critical in enhanced oil recovery since the injected polymer solution comes in contact with a large area of rock. During well completion or stimulation operations, polymer adsorption is less critical unless the adsorbed polymer layer causes a reduction in rock permeability (formation damage).

A recent development that holds promise for enhanced-oil-recovery, completion-fluid, and spacer-fluid applications is the use of polymer complexes. Formation of complexes in solution involving two or more different polymers increases viscosity. These complexes are disrupted by shear forces but reform at low shear (25). This behavior is different from that of cross-linked polymers, which are often irreversibly degraded by shear forces.

Ideal water-soluble polymers (qv) exhibit high apparent viscosity at low concentrations (<2500 ppm) and little permanent viscosity loss on shearing. They are stable above 120°C for long periods and resistant to oxidative, hydrolytic, and enzymatic attack. Furthermore, good material does not undergo irreversible shear degradation and does not damage the rock formation.

Polysaccharides

The chemistry and properties of polysaccharides (qv) have been discussed in a number of reviews (27–29); ^{13}C nmr spectroscopy has been used extensively in their study. The solution properties depend largely on the conformational structure in aqueous media (30).

Polysaccharides used to transport solids in sand-control and hydraulic-fracturing processes must be depolymerized while in contact with the rock; this involves cleavage of the acetal linkages of the polymer backbone (1). Polysaccharides are depolymerized by acids, enzymes, or strong oxidizing agents in catalytic processes; insoluble residues can damage the rock formation. Insoluble residues increase in the order: partially hydrolyzed polyacrylamide < hydroxyethylcellulose < hydroxypropylguar < guar (31).

Guar and Guar Derivatives. *Guar Gum.* Guar gum is a branched polysaccharide derived from the seed of the guar plant, *Cyanopsis tetragonobulus L.* It consists of galactose units connected to alternating mannose backbone units (32) (Fig. 2). The principal disadvantage of guar gum is the insoluble residue from guar bean processing, ie, between 5 and 9%; as much as 10–14% (based on the original weight of guar gum) has been reported in depolymerized guar gum

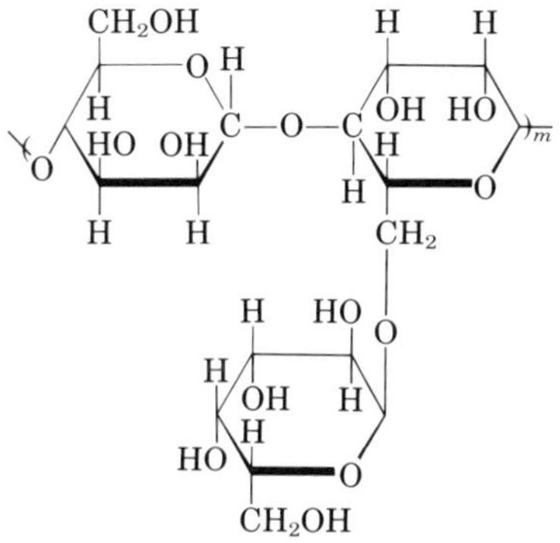

Fig. 2.   Chemical structure of guar gum; mol wt = 400–500.

solutions (32). For this reason, guar gum is unsuitable for EOR. The insoluble residue would gradually plug the injection well and reduce injection rates.

Guar gum is used in drilling fluids and cementing spacers to suspend solid particles, and in fracturing fluids to suspend sand and other proppants. It is usually cross-linked with transition-metal ions, eg, titanium, antimony, or zirconium, under acidic or basic conditions. These highly viscous cross-linked gels can transport more than 1.2 kg of propping agent per liter of fluid. In hydraulic fracturing, guar gum is used in concentrations of 2–10 g/L.

Rheological properties of guar gum solutions have been studied extensively (1). Solution viscosity as a function of polymer concentration in fresh water is shown in Figure 3. Guar gum can be hydrated in saline waters. However, compatibility with calcium and magnesium, commonly found in oil-field brines, depends on salt concentration. Guar gum does not hydrate in highly alkaline media and does not prevent cement slurry loss (33).

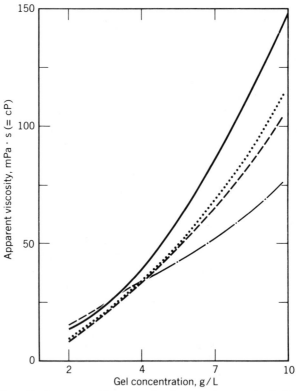

Fig. 3. Effect of polysaccharide concentration on apparent solution viscosity: ——, hydroxyethylcellulose; ······, hydroxypropylguar; -------, guar, and -·-·-·-, xanthan gum. Temperature = 26°C.

Water-swellable and slowly soluble guar gums have been used to prevent fluid loss in drilling muds, spacers, and fracturing fluids. Cross-linked guar gum gels have been used to reduce fluid loss from the well bore to the rock (so-called lost-circulation additives).

The viscosity of guar gum solutions depends on the temperature (Fig. 4). Above 79.4°C, the acetal linkage is hydrolyzed, followed by depolymerization. Additives increase the thermal stability. Although guar gum is fairly stable over a wide pH range, acids such as those used in the oil field cause rapid hydrolytic cleavage of the acetal linkage (Fig. 5). Thus guar gum is not suitable as thickener in acidizing applications, although tests using cross-linked guar to gel 3–15% acid solutions have been reported (34). Alkaline-guar-degradation rates have been studied (35). Guar gum is susceptible to attack by both cellulase- and hemicellulase-producing bacteria; enzymes may be used for depolymerization (Fig. 6).

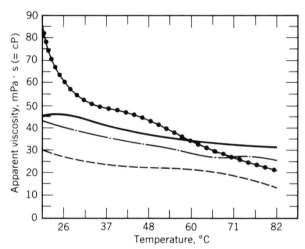

Fig. 4. Apparent viscosity of polysaccharides in aqueous 2% KCl as a function of temperature; concentration = 6 g/L. ——, Guar; ------, hydroxypropylguar; ······, hydroxyethylcellulose; and --------, xanthan gum.

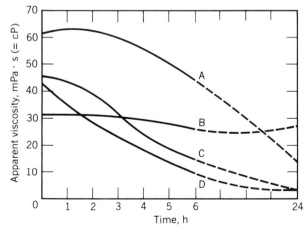

Fig. 5. Effect of 5% hydrochloric acid on apparent viscosity of polysaccharide solutions (6 g/L) in aqueous 2% KCl at 26°C: A, hydroxyethylcellulose; B, xanthan gum; C, guar; and D, hydroxypropylguar.

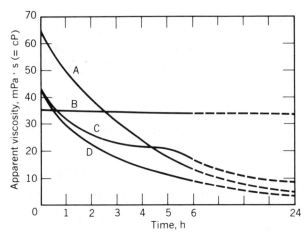

Fig. 6.   Effect of an enzyme (0.03 g/L) on apparent viscosity of polysaccharide solutions (6 g/L) in aqueous 2% KCl at 26°C: A, hydroxyethylcellulose; B, xanthan gum; C, hydroxypropylguar; and D, guar.

Guar gum hydration must be complete. Pumping undissolved solid particles down the well bore can damage the formation and affect rock permeability. Vigorous mechanical agitation and slow addition of the solid polymer greatly increases the surface area and results in rapid and complete hydration. Liquid guar concentrates or suspensions of solid guar in nonaqueous media are often used to minimize hydration problems (35). Mechanical mixing devices have been developed to improve polymer hydration (36).

Chemical treatments result in dispersion and delay of hydration. Upon treatment with borax, a complex is formed which exhibits slow hydration rates above pH 7 (2,37). The guar is dispersed in a mildly basic medium. Subsequent acidification hydrolyzes the borate group and initiates hydration. Reaction with dialdehydes such as glyoxal has also been used to reduce hydration rates (38).

Zirconium complexes, organotitanates, borates, and other compounds have been used to cross-link guar for hydraulic fracturing applications (4,9). Delayed cross-linking processes reduce pumping requirements and allow higher pumping rates (39). Breakers include strong oxidizing agents such as persulfates (1,40) and ascorbic acid salts (41). Fluid loss from cross-linked and linear guar fracturing fluids have been studied (42). The best system was a combination of diesel oil and silica flour. Guar stabilizes foamed fracturing fluids and is also used in drilling fluids. A blend of guar and calcium lignosulfonate as a drilling fluid additive has been reported (43).

Guar Gum Derivatives.   Some guar gum derivatives contain substantially lower concentrations of insoluble residue than guar gum (31,41,44). The alkaline reaction of guar gum with propylene oxide, ethylene oxide, and chloroacetic acid produces hydroxypropylguar, hydroxyethylguar, and carboxymethylguar, respectively; carboxymethylhydroxypropylguar has been produced by sequential reaction with two of the above chemical reagents (1).

The most widely used guar gum derivative is hydroxypropylguar (HPG); its principal application is in hydraulic fracturing. It is also used in drilling fluids, spacers, and to a limited degree (primarily in offshore California) in completion and work-over fluids. Cross-linked HPG reduces fluid loss from the well bore in drilling through very high permeability formations.

Cross-linking methods are similar for guar and HPG (1). Delayed cross-linking reactions occurring under low shear conditions within the rock formation provide higher and more predictable gel viscosity (45).

"Breaker" technology is similar to that used for cross-linking (40) hydroxy-propylguar solutions, which, when treated with a "breaker," contain no more than 2% insoluble residue (based on the initial weight of HPG), compared to 10–14% for untreated guar gum (31). Both formation damage and fracture permeability are reduced. Hence, HPG is highly suitable for hydraulic fracturing applications in formations with very low permeability.

Careful control of the number of moles of propylene oxide per guar unit improves solution properties. Solution viscosity as a function of HPG and of guar gum concentration in fresh water is shown in Figure 3. Hydroxypropylguar is hydrated more rapidly in cold water than guar gum and is more soluble in water-miscible solvents, eg, methanol, ethanol, and ethylene glycol. Methanol, in particular, may be used as a cosolvent in hydraulic fracturing applications where it improves the thermal stability of the gelling agent. Hydroxypropylguar is stable at higher temperatures (Fig. 4) than guar gum and more resistant to enzymatic degradation (Fig. 6).

Carboxymethylhydroxypropylguar (CMHPG) is used in fracturing applications on a modest scale (1). Cross-linking by transition-metal ions occurs at least in part via the carboxyl groups. Aluminum-promoted cross-linking is rapid via the carboxymethyl groups, whereas cross-linking by zirconium via the hydroxyl group is delayed. Above 32°C, the zirconium complex decomposes to form a species that promotes cross-linking (46,47). Such cross-linking systems reduce fluid-friction pressure and power requirements of the injection pumps.

Guar, HPG, and CMHPG form complexes with sulfonated guar and a derivatized guar containing quaternary ammonium groups (48). These complexes exhibit much higher solution viscosity than guar, HPG, and CMHPG at the same total polymer concentration and could be used in low shear spacer and completion-fluid applications. Other guar derivatives have been evaluated for oil-field applications, including an N,N-dimethylacrylamide–guar adduct for fracturing applications (49), an allyl ether–guar gum adduct proposed for acidizing and as a borehole-plugging agent (50), and a benzyltrimethylammonium chloride adduct claimed to improve viscosity characteristics (51).

Locust Bean Gum.   Locust bean gum is isolated from the fruit of the large Mediterranean evergreen legume *Ceratonia siliqua*. The ratio of mannose to galactose is 4:1 compared to 2:1 in guar gum (52). Locust bean gum is less homogeneous than guar gum with the galactose side chains distributed in blocks along the polymer backbone (32). Solution properties are similar to those of guar gum. However, during processing, heat (ca 85°C) or mechanical pregelatinization is required to achieve adequate hydration. Locust bean solution viscosity is insensitive to pH between 3 and 11 (53); solution viscosity drops rapidly above pH 11 (33). Although locust bean gum solutions are compatible with alkali-metal salts, quaternary ammonium salts cause polymer precipitation (54). Cross-linking technology is similar to that of guar gum.

Locust bean gum and xanthan gum form a complex in solution that exhibits a synergistic viscosity increase (55,56). The maximum solution viscosity of the polymer blend is observed at pH 8 with a 1:1 weight ratio of the two polymers.

Ceratonia siliqua grows slowly, which limits the availability of locust bean gum. This polymer is used in drilling fluids as a fluid-loss additive and to suspend high density solids such as barite. Locust bean gum has been proposed for high density zinc brine completion fluids (57) and for plugging high permeability zones (58). In drilling fluids, it is blended with guar gum.

Few locust bean gum derivatives have been prepared for oil-field applications. Condensation of a 1% solution (4000 mPa·s or cP) with benzyltrimethylammonium chloride produces a product with better viscosity characteristics (1% solution, 8700 mPa·s or cP) (51).

Other Gums.   Karaya gum and shiraz gum (also known as gum traganth) have had limited use as fluid-loss additives in well-stimulation fluids and in high density, brine-completion fluids (59–61) (see also GUMS, INDUSTRIAL).

Cellulose Derivatives

Cellulose (qv) or polyanhydroglucose is insoluble in water due to the very strong intermolecular and intramolecular hydrogen bonding. Substitution of organic groups on some of the glucose rings disrupts the close packing of polymer chains and reduces hydrogen bonding. These water-soluble derivatives are suitable for oil-field applications. Chemical reaction can occur at all three hydroxyl groups of the anhydroglucose unit with a maximum degree of substitution (DS) (qv) of 3.0. Molar substitution (MS) is defined as the number of moles of substituents added per anhydroglucose unit. To avoid fractional units, suppliers often report the average number of substituted hydroxyl groups and average number of moles of substituent per 10 anhydroglucose units. More than one mole of substituent may be added at a given hydroxyl group, particularly when adding epoxides (Fig. 7).

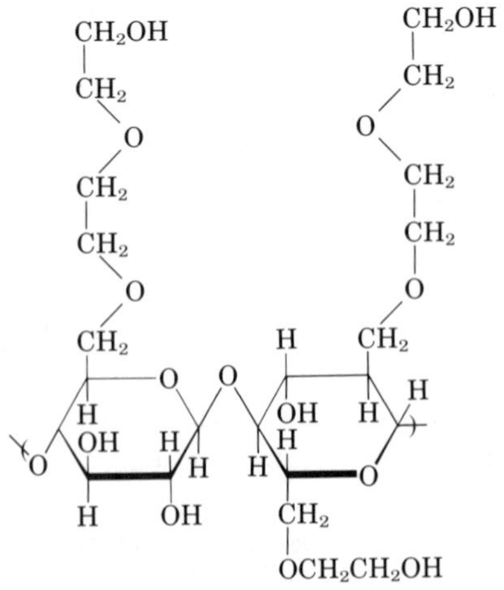

Fig. 7.   Hydroxyethylcellulose.

Carboxymethylcellulose. Carboxymethylcellulose (CMC) is mainly used in drilling fluids to suspend solids and reduce fluid loss. The *in situ* cross-linking of CMC with a multivalent metal has been proposed for plugging high permeability zones and to improve flooding efficiency (62,63). Stabilization of foamed fracturing fluids and diversion of steam foam are other uses, as are spacer fluids. The properties of blends of CMC and polyacrylamide are altered by ultrasonic treatment to increase salt tolerance for fluid-loss applications (64) (see also CELLULOSE ETHERS).

Cross-linking agents of CMC include sodium dichromate (65), chromium lignosulfonate (66), potassium permanganate (63), and trivalent aluminum (67). The degree of polymerization does not affect thermal stability. Condensation with phenol, aniline, and triethanolamine stabilizes CMC (68). Addition of diethylamine reduces bacterial and thermal degradation of CMC drilling fluids (69).

Commercial grades of CMC have a DS (qv) of 0.4–1.4; the most widely used grade has a DS of 0.7. Below a DS of 0.4, CMC polymers are not water-soluble. Solution viscosity and polymer stability in fresh water are at a maximum at pH 7–9. As a good shale stabilizer, CMC is commonly used in drilling fluids. It is compatible with monovalent metal salts. However, compatibility with divalent metal salts, such as calcium chloride, often used to increase the density of drilling fluids, is a function of DS, pH, and the mode of salt addition to the drilling fluid. This limited compatibility is due to the anionic character of the carboxyl substituent. Despite this, some use of CMC in high density, zinc brine completion fluids has been reported (57).

Hydration rate is reduced by glyoxal; this allows solid particles to disperse completely before hydration. Solids dispersion reduces "fisheye" formation.

In laboratory tests, CMC reduces permeability when injected into cores. This has limited its use to well-bore fluids designed to have low losses.

Hydroxyethylcellulose. Hydroxyethylcellulose (HEC) with a molar substitution (MS) of 1.5–2.5 is soluble in water and compatible with metal salts. It is used as a viscosity modifier for high density completion fluids containing high concentrations of salts, eg, calcium chloride, calcium bromide, zinc chloride, and zinc bromide (70). To promote particle dispersion and hydration, HEC is wetted with ethylene glycol (71) or 2-propanol (72). It is preferred for viscosity modification because of low insoluble residue even after depolymerization; effect on formation permeability is low (1,11,31). Hydroxyethylcellulose is also used in fracturing applications.

Because cross-linking HEC is difficult, its use in hydraulic fracturing is limited. It is cross-linked with Zr(IV) in cementing, completion fluids, hydraulic fracturing, and acidizing applications (73); for temporary cross-linking, glyoxal is employed (74). Polyisocyanate cross-linked HEC has been proposed for viscosity modification in enhanced oil recovery (75).

Hydroxyethylcellulose is used for viscosity modification and fluid-loss prevention in aqueous spacer fluids and cement slurries, as friction reducer in fracturing gels, and to stabilize foamed fracturing fluids. It is used as a fluid-loss additive in drilling fluids, particularly those without suspended solids which are designed to minimize formation damage due to swelling of water-sensitive clays (76). It has been suggested for polymer flooding.

Hydroxyethylcellulose is depolymerized by cleavage of the acetal linkage by acid (Fig. 5), cellulase enzymes (Fig. 6), and salts of persulfuric acid. Reversible shear thinning (Fig. 1) and temperature thinning (Fig. 4) is greater in HEC solutions than in solutions of guar and hydroxypropylguar. Complexes of HEC and sulfonated guar provide high viscosity solutions that are useful in spacers and completion fluids (48).

Addition of calcined dolomite (77), Cu(I) salts, or Cu(II) salts (78) increases HEC thermal stability. The long hydroxyethyl side chains reduce resistance to enzyme-promoted depolymerization (79). Addition of sodium hydroxide–lithium hydroxide increases resistance to enzymes (80). Treatment with glyoxal (57) and the use of HEC slurries in nonaqueous fluids (71,73) promotes particle dispersion prior to hydration.

Carboxymethylhydroxyethylcellulose. The applications of carboxymethyl-hydroxyethylcellulose (CMHEC) are based on its unique anionic–nonionic character. Commercial oil-field grades have a carboxymethyl DS of 0.3–0.4 and a hydroxyethyl MS of 0.3–0.4. These values are chosen to promote ionic characteristics, such as flocculating capability, while maintaining the nonionic characteristic of metal-salt tolerance. Carboxymethylhydroxyethylcellulose is used as a cement-set retarder and fluid-loss additive in cement slurries and aqueous spacer fluids. It is a weak-acid viscosity modifier.

An important oil-field application is cross-linking with chromium for hydraulic fracturing fluids (81). Cross-linked CMHEC is also used for temporary diversion of water (1). Zirconium(IV) cross-linked CMHEC has been used to thicken hydrochloric acid for acidizing applications (73). Cross-linked CMHEC is also used in a commercial gravel-packing process. The laboratory use of hydrated lime to gel CMHEC for a micellar-polymer-flooding process has been reported (82).

Compared to other polysaccharides, CMHEC reduces permeability of sand columns the least (83). Citric acid has been proposed as a chelating agent to delay viscosity increase and facilitate the injection of CMHEC solutions (84).

Other Cellulose Derivatives.   Other cellulose derivatives have only limited application in oil fields. Hydroxypropylcellulose (HPC) is used to thicken alcohol solutions for fracturing dry, gas-bearing formations. Hydroxypropylcellulose has been evaluated as an in-depth, matrix-plugging agent for North Sea water-injection wells. The rheological properties of HPC solutions (85) and enhanced-oil-recovery foams stabilized with HPC (86) have been reported. Fracturing fluids containing synergistic mixtures of HPC and poly(maleic anhydride-co-vinyl ether) (87) or poly(maleic anhydride-co-isobutylene) (51) have been described, but no commercial applications have been reported. Cross-linked dihydroxypropyl (MS 0.1–1.2) and hydroxyalkyl (MS 1.5–3.0) cellulose ethers have been proposed for rock-matrix plugging (88). Hydroxyethylcellulose derivatives prepared by the reaction of HEC with alkyl epoxides, such as hexadecyloxirane, have been used to emulsify oil in water (89) and may have enhanced-oil-recovery applications. Graft copolymers of cellulose and vinyl monomers, eg, acrylic acid esters and acrylonitrile, have been investigated, but no commercial uses in the oil field have been reported in a recent review (1).

Microbial Polysaccharides

Xanthan Gum.   Xanthan (qv) is the only microbial polysaccharide currently in commercial use in the oil field. It is exuded by the microorganism *Xanthamonas* *campestris* in batch-fermentation processes. It has a highly branched structure; the two D-glucose rings of the polymer backbone carry a total of three six-membered rings on a side chain (90) (Fig. 8).

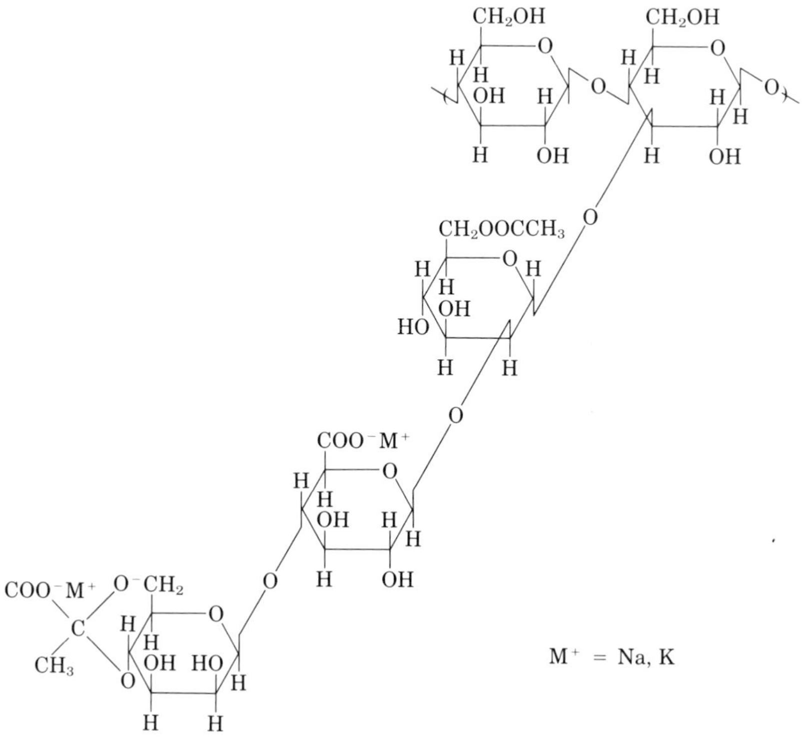

Fig. 8. Xanthan gum.

The terminal side-chain ring is fused to a five-membered pyruvate ring or to the residue formed by rupture of the pyruvate ring. The pyruvate-ring content can be controlled to some degree by the fermentation conditions (91). Pyruvate rings are present in 40–50% of the side chains (92). Rheological properties have been studied extensively (90,93) and seem to be substantially affected by the pyruvate-ring content (94). Early reports suggested the existence of double- or triple-helix structures (95); both a less-ordered and a more-ordered helical structure have been proposed (96). Highly ordered xanthan conformations have been reported to be more resistant to shear degradation (97).

Most of the xanthan gum currently used in EOR applications is in the form of a fermentation broth containing 8–15% polymer; earlier EOR grades were solids, which are still used for drilling-fluid applications. Flash drying of the fermentation broth produces a solid product with improved hydration properties (98). The fermentation broth is easily injected because of the absence of unhydrated polymer particles (microgels). Filtration of xanthan gum solutions, particularly those prepared from solid polymer, is recommended for EOR applications. Cellulase enzymes (99), polysaccharide hydrolases (100), proteases (101), caustic plus enzyme (102), methylene bis(isocyanate) (103), heat (104), and bentonite (105) treatment facilitate the filtration of xanthan solutions through diatomaceous earth filters.

The injectability of xanthan gum can be improved without adversely affecting viscosity by brief ultrasonic treatment (30 s, 60–80 MHz) of the polymer solution or by hydration of the solid polymer in low concentration aqueous boron compound solutions (105) or in solutions containing metal-ion complexing agents such as sodium citrate (106). With low viscosity, oil-continuous xanthan gum emulsions, hydration problems are minimized (107). Xanthan concentration in the dispersed aqueous phase is 9% (100). Oxidizing agents, such as hydrogen peroxide, improve the permeability of well bores plugged by injection of xanthan gum solutions (108).

Aqueous xanthan gum solutions are highly pseudoplastic and exhibit low apparent viscosity under high shear conditions (Fig. 1). This is reversible and apparent shear viscosity increases rapidly upon cessation of shear (1). The effect of xanthan gum concentration on aqueous solution viscosity is shown in Figure 3; the viscosity is maximum at pH 5.5.

The viscosity of xanthan gum solutions is less dependent on temperature than solutions of other polysaccharides used in the oil field (Fig. 4). Viscosity of xanthan gum solution was reduced substantially after 4 h at 149°C (109).

Optical-rotation measurements indicate that xanthan gum conformation is pH-dependent. Polymer conformation, probably due to the bulky side chains, results in a more stable acetal linkage compared to other polysaccharides used in the oil field. This is reflected in the stability of xanthan gum in acid solution (Fig. 5) and its resistance to enzymatic depolymerization (Fig. 6).

Xanthan gum solutions are susceptible to oxidative and bacterial degradation, and use of oxygen scavengers (23,110) and bactericides (111) is common practice. Thiourea (112), sodium dithionite (113), guanidine acetate (91), and a combination of sodium sulfite, thiourea, and 2-propanol (114) have been used as stabilizers.

Chromium and aluminum salts are used to cross-link acidic xanthan gum solutions (63,65). The active cross-linking species are trivalent metal ions; cross-linking is more efficient at low salinity. Strong oxidizing agents, eg, hydrogen peroxide, sodium hypochlorite, and persulfates, are used for depolymerization.

Xanthan gum is used in drilling fluids, completion fluids, hydraulic fracturing fluids, and enhanced oil recovery. It has been proposed as a stabilizer in enhanced-oil-recovery foams. At drilling-fluid concentrations below 2.8 g/L, xanthan gum is a more effective suspending agent than hydroxyethylcellulose, carboxymethylcellulose, and partially hydrolyzed polyacrylamide (115). Xanthan gum, stable in acid media (Fig. 5), has been used as an acid viscosity modifier. For EOR and water-control processes, enzymatic polymer broth is used; for other applications, solid polymer is used.

Xanthan gum is used in polymer-flooding and micellar-flooding applications and has been evaluated as a mobility-control agent in caustic flooding. An *in situ* cross-linking process for injection wells improves volumetric sweep efficiency. This process can be applied in waters containing as much as 16.6 wt % total dissolved solids. However, the precipitation of xanthan gum by high concentrations of divalent metal ions can be used to plug high permeability zones (116). In combination with a partially methylated melamine–formaldehyde resin, xanthan gum generates a gel *in situ* in core-flood experiments in the laboratory (117).

In the xanthan gum concentration range used in enhanced oil recovery (ca 1000 ppm), 0.5% NaCl significantly reduces viscosity (118). At higher salt concentrations, xanthan gum provides more viscosity than the same concentration of polyacrylamide. Prehydration of xanthan gum in fresh water followed by dilution with reservoir brine containing divalent metal-ion salts produces higher viscosity than direct hydration of the same xanthan gum concentration in the reservoir brine (119).

Xanthan gum polymers lacking the pyruvate side-chain ring show increased tolerance to divalent metal cations found in oil-field waters and are used to increase completion-fluid density (120). High pyruvate content xanthan gum has been evaluated for EOR applications (121). A xanthan gum containing pyruvate rings in most of the polymer repeat units and produced by a proprietary strain of *Xanthamonas campestris* shows superior tolerance to salinity and divalent metal ions (92). The pyruvate ring remains intact in buffered xanthan gum solutions heated above 100°C. These solutions exhibit improved viscosity and filterability (122).

Xanthan gum complexes with guar gum (48,123), sodium poly(styrene sulfonate) (124), polyacrylamide (125), sulfonated guar gum (48), sodium poly(vinyl sulfonate) (124), hydrolyzed sodium poly(styrene sulfonate-*co*-maleic anhydride) (48), and poly(ethylene oxide) (125) have been evaluated for use in completion-fluid and enhanced-oil-recovery applications.

Xanthan gum and polyacrylamide properties important in large-scale polymer flooding are given in Table 3. The principal advantages of xanthan gum are effectiveness as a viscosity modifier in saline waters and resistance to permanent shear degradation. Preinjection of a slug of xanthan gum to minimize interaction of subsequently injected polyacrylamide with saline injection waters has been reported (126). Xanthan adsorbs less on the formation than partially hydrolyzed polyacrylamide (127).

The higher cost of xanthan (as compared to polyacrylamides) is a disadvantage, as is the insoluble residue resulting from cellular debris. Modifications in the fermentation process and downstream treatment reduce the insoluble residue content. The maximum operating temperature for xanthan (ca 90°C) is lower than that of polyacrylamides; xanthan gum is, however, more susceptible to bacterial degradation (128).

Other Microbial Polysaccharides. An anionic graft copolymer of xanthan gum and acrylamide containing sulfomethylated groups has been reported (129). Acrylates have been grafted onto xanthan gum (130).

Scleroglucan has been evaluated as viscosity modifier for EOR (131–133). Scleroglucan exists in solution in a highly rigid conformation that is probably helical (134). Polymers synthesized by the alga *Porphyridium aeruginium* (135), the bacterium *Pseudomonas methanica* (136), and *Aerobacterium* NC1B 11883 (137) have been reported and their use in polymer flooding has been investigated. Cell-disruption techniques have been used to reduce the amount of insoluble cellular material present in *Pseudomonas* polymer solutions (138). The addition of monovalent or divalent metal ions to a solution of the polysaccharide produced by the bacterium NG1B 11870 has produced gels (139). The addition of divalent metal ions to khaya gum solutions increased viscosity (140). All these findings are relevant to oil-field applications.

The injection of bacteria and nutrients to produce water-blocking carbo-

Table 3. Properties of Water-soluble Polymers[a]

Polymer	Cost[b]	Viscosity[c], mPa·s (=cP)	Shear stability[d]	Salt tolerance[e]	Acid stability[f]	Enzyme stability[f]	Residue in broken gel[g]	Applications
guar gum	1	34	3	C	NS	NS	R	drilling fluids, spacers, friction reduction in stimulation, fracturing and lost circulation (cross-linked guar gum gels), and fluid-loss additives
hydroxypropyl guar	1	36	3	C	NS	NS	R	drilling fluids, spacers, completion and work-over fluids, friction reducer in fracturing, fracturing and lost circulation (cross-linked HPG gels), and fluid-loss additive
CMC	2	55	3	IC	NS	NS	RF	drilling fluids
HEC	2	37	3	C	NS	NS	RF	fluid-loss additive for cementing spacers, completion and work-over fluids, fracturing fluids, friction reduction in stimulation, and enhanced recovery

CMHEC	2	32	3	C	NS	NS	RF	fluid-loss additive and retarder for cementing spacers, gelling weak acids, temporary diverting agents in fracturing (cross-linked gels)
xanthan gum	4	34	1	C	MS	MS	RF	drilling, completion, fracturing, and enhanced-oil-recovery fluids
polyacrylamide, partially hydrolyzed	3	34	2	IC	S	S	RF	fluid-loss additive in cementing, drilling fluids, friction reduction in fracturing, enhanced oil recovery, and scale inhibitor
copolymer of acrylamide	3	25	2	MC	S	S	RF	fluid-loss additive in cementing, drilling fluids, friction reduction in fracturing, enhanced oil recovery, water–oil ratio reduction, improvement of injection profile (cross-linked)

[a] Taken from Ref. 1 with permission of the Society of Petroleum Engineers.
[b] The order of cost per kg of polymer, increasing from 1 to 4. The price varies with the supplier and the quantity.
[c] Solution of 4 g/L at 300 rpm.
[d] The order of shear stability, increasing from 1 to 3. Xanthan gum solution viscosity is not reduced permanently by shear.
[e] C = compatible, IC = incompatible, and MC = moderately compatible.
[f] NS = not stable, MS = moderately stable, and S = stable.
[g] R = residue present and RF = residue-free.

hydrate polymers *in situ* has been studied (141). Recombinant DNA processes may improve the polymer produced by *Xanthamonas campestris* and EOR polymers produced by proprietary strains have been prepared and evaluated (92).

A polymeric surfactant known as Emulsan and produced by *Acinebacter* RAG-1 has shown promise as a surfactant for micellar flooding (142).

Other Carbohydrate Polymers

Starches.   The glucose units of the starch backbone are joined by α-glycosidic linkages rather than by the β-glycosidic linkages present in cellulose polymers. Starch properties vary, depending on the plant source. Starch is insoluble in cold water but extensive grinding gives powders that gelatinize in cold water. The principal oil-field application is as a rheology modifier and to prevent drilling-fluid losses.

Starch (qv) is a mixture of a linear polymer, amylose, and a branched polymer, amylopectin. Solution properties of these two polymers differ and they may be separated by heating starch in water. Amylopectin dissolves readily in water to form a stable solution, whereas purified amylose exhibits limited solubility in hot water and precipitates on standing. Adsorption characteristics of partially hydrolyzed amylose and amylopectin on calcite have been studied (143).

Although starch is relatively stable to oxidative degradation in alkaline media, it is readily depolymerized by acid or enzymatic attack on the acetal linkage. Stability above 107°C is limited (144). Starch is highly susceptible to bacterial degradation. Formaldehyde and, more recently, isothiazolones have been used as biocides in starch-based drilling fluids (145).

Starch and most starch derivatives are poor viscosity modifiers. Starch solutions reduce permeability when injected into cores (4). Both these factors have limited oil-field applications. Hydroxyalkyl starches gelatinize readily in cold water. Starch, hydroxyethylstarch, hydroxypropylstarch, and carboxymethylstarch are used as fluid-loss additives (1,13). Graft copolymers of starch and vinyl monomers such as acrylamide have been proposed for polymer flooding (146,147). However, laboratory studies have indicated that although starch–acrylamide graft copolymers are good viscosity modifiers in low salinity waters, solution viscosity is permanently reduced by shear. Cationic starch–graft copolymers with cationic vinyl monomers show potential for water-control applications (148).

Dextran and Others.   Dextran (qv) is used to thicken water (149) and saturated calcium carbonate solutions (150). These fluids could have useful drilling, cementing, and completion-fluid applications. Formaldehyde is the most commonly used biocide for dextran solutions.

Pectins, pectates, and polygalacturonic acid solutions have been used to plug highly permeable rock zones. Subsequent injection of brine causes gelation of the previously injected carbohydrate polymer solution (151).

Vinyl Polymers

Polyacrylamide and Acrylamide–Acrylic Acid Copolymers.   Partially hydrolyzed polyacrylamide and copolymers of acrylamide and sodium acrylate or acrylic acid are commonly referred to as polyacrylamide in the oil-field literature. Most polyacrylamides used in the oil-field are partially hydrolyzed. Some hydrolysis ordinarily occurs after the polymerization, but hydrolysis is often incorporated as a separate reaction step. The copolymers are generally produced by emulsion copolymerization (152). Cobalt-60 irradiation initiates polymerization and produces high molecular weight polyacrylamides (153). The substitution pattern of acrylic acid groups along the polymer backbone is similar in partially hydrolyzed polyacrylamide and poly(acrylamide-*co*-acrylic acid) containing the same molar content of carboxylate groups (154). Acid hydrolysis of polyacrylamide results in blocks of carboxylate groups, whereas the more common alkaline hydrolysis gives a well-spaced distribution of carboxylate groups. This may offer a potential for developing superior flooding polymers (155).

The conformation of polyacrylamide in solution depends on the polymer charge density and the ionic strength of the solvent. Electrostatic repulsion of the anionic carboxylate groups increases results in chain elongation. The extent of polymer hydrolysis affects solution viscosity, compatibility with divalent metal salts, polymer adsorption on rock, and flow-resistance properties. In fresh water, maximum viscosity is observed at ca 35% polyacrylamide hydrolysis (156). On the other hand, the viscosity of aqueous poly(acrylamide-*co*-acrylic acid) solutions increases with increasing acrylic acid content (1) and increasing unhydrolyzed acrylamide sequence length above a value of 1.5 (157). However, maximum viscosity in a calcium-containing water occurs at 10–15% hydrolysis. Above 33–35% hydrolysis, Ca^{2+} ions cause polymer precipitation. Interaction of metal ions with the carboxylate groups reduces electrostatic repulsion of these groups and the polymer-chain contracts, reducing solution viscosity in saline waters (156) and oil recovery in core floods. The viscosity of partially hydrolyzed polyacrylamide solution is more sensitive to calcium ions than to sodium ions (158).

Chelating and sequestering agents minimize the effect of divalent (156) and multivalent (14,15) ions on the viscosity of partially hydrolyzed polyacrylamide solution. Rheological properties of polyacrylamide have been studied extensively (158,159).

Partially hydrolyzed polyacrylamide and poly(acrylamide-*co*-acrylic acid) can undergo irreversible shear degradation (14,15), even in laminar flow (160). Careful mixing and pumping are required to minimize shear degradation in field operations. The recent development of concentrated polyacrylamide emulsion formulations for enhanced oil recovery permits lower shear during the preparation of dilute solutions. Shear degradation is also reduced by *in situ* polymerization of acrylamide (161).

The chemical stability of polyacrylamide (162) is of particular importance for EOR applications. Oxygen (162), bacteria (111,162), and iron ions from metal equipment (163) reduce stability. Stabilizers include aldehydes, alcohols, sulfur compounds, nitrites, iodides, ethers (110,164), and small amounts of aromatic hydrocarbons (165). Oxygen exclusion is essential for long-term stability, particularly above 54°C.

The largest market for partially hydrolyzed polyacrylamide and poly(acrylamide-*co*-acrylic acid) is polymer flooding (14,15). At present, polyacrylamides account for most of the polymer being used in EOR field projects. With the trend toward emulsion formulations, most of the polyacrylamide being used for enhanced oil recovery is poly(acrylamide-*co*-acrylic acid). Properties of poly-acrylamides for flooding are given in Table 3. The main advantage of poly-acrylamide is low cost. Polyacrylamides have a higher temperature limit than xanthan gum and are less susceptible to bacterial degradation, but are more sensitive to shear. Thickening properties in saline waters are reduced more than those of xanthan gum. Field polymerization and copolymerization of acrylamide increase activity and give products with improved injectability and compatibility with saline formation waters (166). *In situ* polymerization may reduce costs.

Polyacrylamides tend to adsorb more on formation surfaces than xanthan gum (127,167); this adsorption is essentially irreversible. Lignosulfonates (168,169), asphaltenes (170), sodium carbonate, inorganic phosphates (171), low molecular weight poly(acrylamide-*co*-acrylic acid) (172), and aqueous pyridine (173) have been evaluated as sacrificial adsorption agents in polyacrylamide flooding. Trapping of polymer molecules, not adsorption, controls permeability behavior (174). Polyacrylamides have been used as mobility-control agents in micellar (16–20) and caustic flooding (175), although the base catalyzes polymer degradation (176). Polyacrylamides are used as adsorption reagents for nonionic glucans (177).

Cross-linked polyacrylamides have been extensively used to improve injection profiles in water (178), polymer, micellar-polymer (69), and caustic floods. They also reduce the produced water–oil ratio (179).

Trivalent metals, such as chromium or aluminum, are the most frequently used cross-linking agents (180). Chromium salts are injected together with the polyacrylamide, whereas aluminum compounds are injected alternately with polymer slugs (181). Sodium bisulfite or thiourea reduces Cr(VI) to the active cross-linking species Cr(III) (182,183); another technique employs a colloidal dispersion of Cr(III) hydroxide (184). Gradual dissolution of the chromium-containing species delays the cross-linking reaction. Kinetic studies suggest that the cross-linked structure contains two chromium atoms and two polymer molecules (180). Insoluble fillers such as sand, barite, and asbestos are added to the treatment solution to plug high permeability rock zones adjacent to well-bore applications (185).

Aluminum ion in the form of Al(III) citrate and sodium aluminate are also used as cross-linking agents (186). The alternating injection of polymer and cross-linking agent builds a layer of adsorbed polymer on flow-channel walls. The aluminum ion bonds the adsorbed polymer to injected dissolved polymer (187).

Cationic polyacrylamide derivatives, as the initially injected polymer slug, promote rapid polymer adsorption and interaction with subsequently injected partially hydrolyzed (anionic) polyacrylamide (188). Glyoxal (189) and formaldehyde (190) are used to cross-link polyacrylamide. Hypohalite salts promote rapid gelation in the presence of air (191) and increase gel stability (192). Treatments using high concentrations of linear polyacrylamide are prolonged by the addition of epichlorohydrin to the injection fluid (193). Blends of partially hydrolyzed polyacrylamide and cationic polyacrylamide–epichlorohydrin resins for *in situ* gel formation have been reported (194).

The *in situ* synthesis of a cross-linked polymer by injecting a mixture of acrylamide, N,N-methylenebis(acrylamide), and a polymerization catalyst is used in both injection and production wells to plug selected portions of the rock matrix (195). This method produces cross-linked polymers with superior chemical stability above 100°C (196). Both cross-linked polyacrylamides and linear polyacrylamides reduce the water–oil ratio after production-well treatments (179). Laboratory results indicate that acrylamide homopolymers and copolymers cross-linked with dialdehydes are poor gelling agents because of unstable cross-links (197).

Polyacrylamides in hydraulic fracturing fluids reduce friction (198). Increased carboxylate content improves polyacrylamide performance in this application. However, the presence of an acid or divalent metal cation reduces this effect. Polyacrylamide thickens the aqueous phase of hydrochloric acid emulsions used in acidizing (199) and modifies the viscosity of sulfuric acid–cement mixtures (200). Partially hydrolyzed polyacrylamides are used as flocculants in drilling fluids, where they stabilize shale (3). The inversion of an oil-continuous emulsion containing polyacrylamide in the dispersed aqueous droplets provides a shear-thickening drilling fluid designed to reduce fluid loss (201) (see also ACRYLAMIDE POLYMERS; FLOCCULATION).

Polymers of Acrylamide Derivatives.   Replacement of the acrylamide amide hydrogen atoms by other groups reduces the hydrolysis rate and increases viscosity in saline water, particularly brines containing divalent metal salts. Poly(N,N-dimethylacrylamide) and sodium poly(2-acrylamido-2-methylpropanesulfonate), which are resistant to hydrolysis under reaction conditions, promote rapid polyacrylamide hydrolysis (1). Acrylamide copolymers tend to be shear-sensitive.

Homopolymers and acrylamide copolymers of 2-acrylamido-2-methylpropanesulfonic acid salts (14,15), random copolymers of acrylamide and 2-sulfoethyl methacrylate (202), terpolymers of acrylamide, sodium acrylate, and vinyl chloride (203), terpolymers of acrylamide, sodium 2-acrylamido-2-methylpropanesulfonate, and N-vinylpyrrolidinone (204), and various proprietary acrylamide copolymers have been tested in the laboratory for use in polymer flooding.

The salt tolerance of poly(acrylamide-*co*-diacetone acrylamide), sodium poly(acrylamide-*co*-2-acrylamido-2-methylpropanesulfonate), and poly(acrylamide-*co*-2-sulfoethyl methacrylate) is higher than that of poly(acrylamide-*co*-sodium acrylate) (205). The solution behavior of acrylamide copolymers with sodium acrylate, sodium acrylamido-2-methylpropanesulfonate, and diacetone acrylamide have been studied as a function of added electrolytes, temperature, and copolymer composition and microstructure (205,206).

Starch–acrylamide (207), dextran–acrylamide (208), and lignin–acrylamide (209) graft copolymers have been proposed for enhanced oil recovery and drilling fluid. Ethoxylated, sulfonated, partially hydrolyzed polyacrylamide has been evaluated for polymer flooding. Increased oil recovery was reported in a well-injection test (210).

Copolymers of acrylamide with monomers of the general formula $MO_3SC(COOR)=CHCOOR$ have been commercialized for EOR applications (211). An effective modifier for fresh-water viscosity, this copolymer performs poorly in brines.

In an effort to improve polyacrylamide viscosity properties in saline waters, copolymers with more rigid backbones were synthesized. Partial hydrolysis gives the best viscosity properties (212); shear stability was also improved. The ternary copolymers have the general structure

$$-(CH_2CH)_x-(CH_2CH)_y-(CH_2\underset{O\diagdown N\diagdown O}{\overbrace{}})_z$$
$$\quad\quad | \quad\quad\quad\quad |$$
$$\quad CONH_2 \quad COO^-Na^+$$

Poly(N-methylacrylamide), poly(N-isopropylacrylamide), poly(N-butylacrylamide, and copolymers of N-alkylacrylamides and acrylic acid were investigated for flooding (213). Ethoxylated 2-acrylamido-2-methylpropanesulfonic acid polymers have been proposed as acid viscosity modifiers (214) and for polymer flooding (215). Ethoxylated acetone–diacetone acrylamide copolymers (216) and N,N-dimethylacrylamide copolymers with sodium styrene sulfonate, N-hydroxymethylacrylamide, and sodium 2-acrylamido-2-methylpropanesulfonate (217) have been proposed as mobility-control agents in caustic flooding. Copolymers of N,N-dimethylacrylamide, sodium-2-acrylamido-2-methylpropanesulfonate, and acrylic acid prevent fluid loss in high density brine completion fluids (218).

Acrylamide copolymers are used as friction reducers in fracturing fluids and as cement-slurry, fluid-loss additives. Sodium poly(acrylic acid) is used as a drilling-mud, fluid-loss additive and as an inhibitor of calcium sulfate scale on the formation face (3). Terpolymers of acrylamide, acrylic acid, and 2-hydroxypropyl acrylate prevent fluid loss in drilling mud (219). Interaction with Ca^{2+} present in field waters causes formation of a plugging, gelatinous mass. Copolymers of acrylamide and N-vinyl-2-pyrrolidinone modify the viscosity of drilling fluid and have been proposed for polymer flooding of deep, hot reservoirs in which thermal stability is required (220). Other N-vinylpyrrolidinone copolymers stabilize mineral fines (221) and modify brine viscosity (222).

Low molecular weight (~36,000) polymers of poly(2-acrylamido-2-methylpropanesulfonic acid) condensed with 3 mol ethylene oxide per polymer unit are thickeners for hydrofluoric–hydrochloric acid blends used in sandstone acidizing (223). Other water-soluble vinyl polymers and their oil-field applications are given in Table 4.

Sodium Silicate Gels

Sodium silicate gels are used to seal channels, vugs, and high permeability rock zones in production wells (258). In injection wells, *in situ* gelation of sodium silicate improves the injection profile (259) and stops water leakage during drilling operations by plugging highly permeable channels (4).

In the presence of divalent metal cations and strong bases or acids, sodium silicate precipitates and polymerizes to form a stiff gel. The base-promoted process is applicable between 15 and 110°C. In a basic system, divalent metal cations can cause premature gelation, but not in an acid system. The polymerization rate is a function of the concentration of sodium silicate (260). Injection wells are treated by this method (261).

A fine-structure, loose sodium silicate gel formed *in situ* is used for mobility control in enhanced oil recovery (262). Combinations of water glass and a phosgenated aniline–formaldehyde condensate are used to plug high permeability rock zones (263).

Table 4. Oil-field Applications of Vinyl Polymers

Polymer	Applications	Refs.
poly(acrylamide-co-2-acrylamido-2-methylpropanesulfonic acid)	acid viscosity modifier	197
poly(acrylamide-co-2-acrylamido-2-methylpropane sulfonic acid-co-acrylic acid)	polymer flooding	221
poly[acrylamide-co-(3-acrylamido-3-methyl)butyltrimethylammonium chloride]	micellar flood preflush	224
poly(acrylamide-co-ethenesulfonic acid)	polymer flooding	225
poly(acrylamide-co-2-hydroxypropyl acrylate-co-sodium acrylate)	drilling-fluid, lost-circulation additive	226
poly[acrylamide-co-2-(methacryloyloxyethyl) trimethylammonium methosulfate-co-sodium-2-acrylamido-2-methylpropanesulfonic acid]	drilling-fluid, lost-circulation additive	227
poly[acrylamide-co-(methacryloyloxyethyl) trimethylammonium methosulfate]	acid viscosity modifier	228
poly(dimethylaminoethyl methacrylate-co-methacrylic acid)	mineral-fines stabilizer	218
poly(acrylic acid-co-methyl acrylate)	drilling-fluid viscosity modifier	229
poly(acrylic acid-co-methacrylic acid)	drilling-fluid dispersant	230
poly(methacrylic acid-co-N-alkylacrylamide-co-alkyl methacrylate)	drilling-fluid viscosity modifier	231
poly(acrylic acid-co-N-alkylacrylamide-co-alkyl acrylate)	drilling-fluid viscosity modifier	231
methacrylic acid copolymers	lost-circulation additive for drilling fluid	232,233
sodium poly(acrylic acid-co-methallyl sulfonate)	drilling fluid, cement slurry fluid-loss additive	234
sodium poly(methacrylate-co-methallyl sulfonate)	drilling fluid, cement slurry fluid-loss additive	234
sodium poly(N,N-dimethylacrylamide-co-2-acrylamido-2-methylpropane sulfonate)	polymer flooding in saline waters	235
sodium polycrotonate	drilling-fluid dispersant	236
sodium poly(2-acrylamido-2-methylpropanesulfonate-co-methacrylamide-co-acrylate)	drilling-fluid viscosity modifier	237
poly(dimethylaminoethyl methacrylate-co-methacrylic acid)	mineral-fines stabilizer	238,239
poly(dimethylaminoethyl methacrylate-co-dimethylaminoethyl methacrylate methyl chloride salt)	high density brine, fluid-loss additive	218
poly(diethylammonium methacrylate-co-methacrylic acid)	water control	218
poly(ethyl acrylate-co-methacrylic acid)	drilling-fluid viscosity modifier	240
poly(sodium methacrylate-co-methacrylic acid) cross-linked with epichlorohydrin	water control	241
poly(methyl methacrylate-co-ethylenediamine)	corrosion inhibitor	242
poly(vinylpyrrolidinone)	preflushing for micellar flooding, acid viscosity modifier	243,244
poly(N-vinyl-2-pyrrolidinone-co-ethenesulfonic acid)	preflushing for micellar flooding	243

Table 4. (Continued)

Polymer	Applications	Refs.
sodium poly(vinylsulfonate-co-vinyl amide)	high temperature, fluid-loss additive for drilling fluids	244,245
sodium poly(2-acrylamido-2-methylpropanesulfonate-co-N-vinylacetamide) + poly(acrylic acid-co-vinyl formamide-co-vinylpyrrolidinone)	acid-gelling agent	246
poly(pyrrolidinium methanesulfonate)	polymer flooding	247
poly(sodium styrenesulfonate-co-maleic anhydride)	fluid-loss additive for drilling fluid	248
ethoxylated poly(ethenesulfonic acid-co-acrylamide)	polymer flooding	249
ethoxylated poly(vinylpyrrolidinone-co-hydroxyethyl acrylate-co-2-acrylamido-2-methylpropane sulfonic acid)	polymer flooding	250
poly(maleic anhydride-co-styrene)	EOR surfactant	251
poly(vinyl alcohol)	sacrificial adsorption agent for EOR	
	fluid-loss additive and shale stabilizer for drilling fluids	
	water control	
	sacrificial adsorption agent for EOR	252
sulfonated poly(vinyl alcohol)	sacrificial adsorption agent for EOR	252
sulfonated poly(vinyl pyrrolidinone)	sacrificial adsorption agent for EOR	252
poly(acrylic acid)	fluid-loss additive for drilling fluids	253
poly(vinyl acrylate)	fluid-loss additive for drilling fluids	254
poly(hydroxysulfonated styrene)	polymer flooding	255
olefin–aminostyrene block copolymers	polymer flooding	256
poly(vinyl alcohol) cross-linked with a dialdehyde	matrix-plugging agent for water control	257

Lignosulfonates

Lignosulfonates are employed as drilling-fluid dispersants (5) and to prolong cement-slurry-set time (13). *In situ* gelation to increase the volumetric sweep efficiency of waterfloods and EOR processes is potentially the largest volume oil-field application of lignosulfonates. They appear to offer the best potential of the cross-linked polymer systems for in-depth plugging of highly permeable channels or zones.

In situ procedures to gel aqueous 2–3 wt % lignosulfonate solutions utilize trivalent chromium ion (264,265). Good results in laboratory experiments have been reported (264).

A similar system without sodium dichromate is recommended for steam-injection wells and extremely high temperature formations (264). An acidic gas (CO_2) promotes rapid gelation (265). At 232°C, sodium lignosulfonate requires longer gelation times than ammonium lignosulfonate (264).

A modification of this system utilizes a low viscosity solution of 2–5 wt %

sodium or ammonium lignosulfonate and sodium silicate (at an SiO_2:lignosulfonate weight ratio of 0.2:1) up to 71°C (266). Field results of a lignosulfonate–silicate *in situ* gelation process have been reported (266).

Lignosulfonates stabilize carboxymethylcellulose (CMC) gels formed *in situ* by cross-linking with a trivalent metal cation. Sulfoalkylated tannins are used in a similar application (267).

Application of high concentration (1–10%) lignosulfonate solutions increases the amount of oil recovered from unconsolidated sands in laboratory experiments (168). Lignosulfonates interact synergistically with petroleum sulfonates to produce ultralow interfacial tension with crude oils (268). Ultralow interfacial tensions are required for efficient oil displacement in micellar flooding. Lignosulfonates, sulfomethylated lignosulfonates, and other lignosulfonate derivatives are used in laboratory experiments and field tests as inexpensive adsorption agents for expensive surfactants and flood polymers (168,169,269–272) (see also LIGNIN).

Cationic Organic Polymers

Applications of polymers containing quaternary ammonium groups in the repeat unit are based mainly on their adsorption characteristics and their properties as organic salts. These cationic organic polymers (COPs) adsorb rapidly and tenaciously to mineral surfaces, particularly clays in an ion-exchange process (273). The multiplicity of cationic sites on a single polymer chain causes the permanency of adsorption observed for many COPs (see POLYAMINES AND POLYQUATERNARY AMMONIUM SALTS).

The effects of clays on petroleum production have been discussed in detail (4,274). Even though they may be present at low concentrations, reservoir surface chemistry is often dominated by clays. Some clays, particularly three-layer clays, such as smectite, swell when in contact with low salinity fluids such as those employed in polymer and micellar-polymer flooding. Clay swelling reduces flow-channel dimensions and causes the release of fine particles as a result of reduced cementation; these fine particles can plug the flow channel. Fines migrate also in the absence of clay swelling (275) because of rapid rates of fluid movement. Formation damage is most severe in the immediate vicinity of the well bore, where large volumes of fluids from the entire reservoir are forced to move radially through a relatively small volume of rock.

Certain cationic organic polymers can adsorb on swelling-clay surfaces and prevent formation damage (276). These COPs are widely used as additives in drilling, completion, hydraulic fracturing, and acidizing fluids, as well as in cement slurries and in conjunction with steam-injection and micellar-polymer methods of enhanced oil recovery. Commercial COPs are proprietary. However, the patent literature indicates that copolymers of dimethylamine and epichlorohydrin (277–279) poly(diallyldimethylammonium chloride) (276,279), and polymers containing quaternary ammonium derivatives of methacrylamide and methyl methacrylate in polymer repeat units (279) effectively stabilize swelling clay. Polymers containing quaternary phosphonium groups (277) and tertiary sulfonium groups (278) in the repeat unit have been claimed for the same application.

A 0.03–0.4 wt % COP solution can substantially improve initial results and permanence of well treatments, particularly hydraulic fracturing and acidizing (273,280). Certain COPs are effective above 260°C and their use in cyclic steam-injection wells increases oil production and decreases operating costs (273).

A recent development in formation-damage control is the use of proprietary substituted polyacrylates, polyacrylamides, and polymethacrylamides containing two (281) or three (282) quaternary ammonium groups in a long flexible side chain on the polymer repeat unit. These COPs stabilize mineral-fine particles, eg, kaolinite, silica, calcite, hematite, magnetite siderite, and feldspar, minimizing formation damage even in the absence of swelling clays. Similar structures containing both quaternary ammonium groups and perfluorinated alkyl groups have also been claimed for this application (283).

As ionic species, certain COPs have the same effect on the viscosity of unhydrolyzed polyacrylamide solutions as inorganic salts. This property has been utilized to reduce the viscosity of polyacrylamide solutions temporarily, increasing injection rates and reducing shear degradation (284). In the formation, rapid COP adsorption decreases salinity, resulting in an *in situ* viscosity increase. Among the cationic organic polymers effective in this application are poly(1,1,5,5-tetramethyl-1,5-diazaundecane dibromide), poly(*N,N*-dimethyl-3,5-methylene-piperidinium chloride), and poly(methacrylamidopropyltrimethylammonium chloride) (284).

A commercial COP reduces the water–oil ratio from rock zones producing both petroleum and aqueous fluids (13). The polymer is adsorbed rapidly on the formation surfaces. Little reduction in petroleum permeability occurs, although aqueous-fluid permeability is substantially reduced.

Copolymers of acrylamide, diallyldimethylammonium chloride, and a third monomer, such as styrene or vinyl acetate, reduce water influx into production wells (285); these terpolymers are also corrosion inhibitors. The *in situ* polymerization of dialkyldimethylammonium chloride and diallyl esters of diacids produces a hydrophilic gel that acts as a seal for leaking zones (286). Branched poly(diallyldimethylammonium chloride) is a viscosity modifier in fracture acidizing (287).

Oil-soluble Polymers

Furan, Epoxy, and Phenol Resins.   The cementation of poorly consolidated rock formations by cross-linking certain organic polymers adsorbed on rock surfaces can be particularly effective in minimizing sand production from poorly consolidated formations (288). This effectiveness accounts for their continued use, despite the relatively high cost of this sand-control method.

The most commonly used plastics for this application are furan and furfuryl alcohol resins, epoxy resins, and phenol–formaldehyde resins (289,290). Sand-consolidation resins should be inert under reservoir conditions, undergo little shrinkage or occlusion of diluents during curing, strongly wet and adhere to formation surfaces, produce no by-products during polymerization and cross-linking reactions, and be flexible enough to deform in response to stress.

The properties of the most common sand-consolidation resins are given in Table 5. Furan polymers (qv) appear to be preferred for high temperature ap-

Table 5. Properties of Sand-consolidation Resins

Resin	Well closing, h	Formation temp, °C	Compressive strength of consolidated sand, MPa[a]	Permeability retention, %
phenol–formaldehyde	10–48	30–93	20.6	50–70
epoxy	2–24	10–120	34.4–48.2	67
furan	1–6	4–>204	20.6	90

[a] To convert MPa to psi, multiply by 145.

plications. At 71°C, sand consolidated by two epoxy resins (qv) and two furan resins was stable to the injection of 30×10^6 pore volumes of brine (290). A water-based furan system is the most economical of the four resins tested. Furan resins retain compressive strengths in excess of 6.9 MPa (1000 psi) after 25 days at 260°C (290).

The rock surface is cleaned and conditioned by flushing with an organic solvent to promote resin adsorption (291); silane coupling agents (qv) promote adhesion of the resin to rock (292). After injection of a low viscosity resin (in an organic solvent), a spacer fluid or "afterflush" (also an organic solvent) is injected. This displaces the resin in the sand, and none remains unadsorbed in the flow channels where it can harden and greatly reduce the formation permeability. The cross-link curing agent may be present in the resin itself (internally catalyzed) or a solution of the catalyst may be injected after a spacer fluid (externally catalyzed), contacting only resin already adsorbed on sand surfaces. The external catalyst is displaced by an organic solvent (afterflush). The amount of organic solvent required is reduced by employing aqueous slurries of epoxy resins (288).

The well is shut to allow the resin to cure. The time varies according to resin, curing agent, polymer concentration, and formation temperature (Table 5). Externally catalyzed resins are usually preferred in higher temperature formations to prevent premature hardening.

A variation of the resin-consolidation treatment can be used in older wells which have already produced sand (293). A slurry consisting of a carrier fluid, resin, catalyst, and sand is prepared in such a manner that the resin coats the sand (usually 250–420 μm or 40–60 mesh). High viscosity systems transport up to 1.8 kg of resin-coated sand per liter of fluid. The sand slurry is pumped down the well bore. The sand fills the void spaces in the formation and the well bore adjacent to the formation; the process may be internally or externally catalyzed. The hardened resin-coated sand is a permeable mass that holds the poorly consolidated formation in place. An epoxy resin-coated sand is usually employed, but furan and phenolic resins (qv) have also been used, as well as formaldehyde–furfuryl alcohol and phenol–resorcinol, copolymer-coated sands (294).

Resin-coated sands are also used as propping agents in hydraulic fracturing where high mechanical strength and self-consolidation are required; phenolic resins are preferred for this application (295). Over 4.5×10^3 metric tons of resin-coated sand have been used worldwide in a five-year period for hydraulic fracturing and sand-control applications (296).

Resins are also used as cements to complete wells in corrosive environments, which may be due to corrosive brines native to the penetrated formations or to corrosive fluids being injected. The latter include brines produced during hydrocarbon production, geothermal brines, and acids, brines, and other fluids associated with chemical production. Epoxy resin is preferred in this application in a blend with Portland cement (297) or alone (298). Although expensive, its resistance to corrosion is excellent and its effectiveness of longer than three years has been reported (298).

A coating of phenol or epoxy resin protects well tubing from paraffin deposition (299). A foamable phenol–aldehyde resin composition has been proposed as a matrix-plugging agent for water control (300); nonfoaming compositions have also been proposed (301). Nonionic water-soluble oxyalkylated phenol–formaldehyde resins are used in combination with arylsulfonate surfactants to demulsify crude oil emulsions stabilized by produced solids (302).

Low molecular weight water-insoluble urea–formaldehyde resins are used to increase oil production from cyclic steam-injection wells (303). Called thin-film spreading agents, these resins break high viscosity, water-in-oil emulsions within the formation, thus freeing the oil. Large increases in oil production from treated wells have been reported, which persist for more than one production cycle. Excellent results were obtained with such a resin in a south Texas water-flood (304).

Urea–formaldehyde resins are gelled *in situ* as sand-consolidation agents, as plugging agents for high permeability zones near injection-well bores (305), and as a matrix-plugging agent in steam-injection wells (306). The addition of 0.1–0.5 wt % polyacrylamide increases the compressive strength of urea–formaldehyde formulations (307). Formulations containing additional urea or phenol–formaldehyde resins have been used to shut off water-producing zones (308,309).

Hydrocarbon resins are used as fluid-loss additives in completion fluids (310). Polyethylene (311) and polybutadiene (312) also have been used in this application. Melamine–formaldehyde resin has been evaluated for matrix-plugging applications (313). Applications of other plastics in the oil field are given in Table 6.

Economic Aspects

Enhanced Oil Recovery. The potential-target-oil reserves for polymer flooding was given as 32.9% of the original oil in place in all U.S. oil fields (336). With advanced technology in progress, more oil fields are accessible to polymer flooding and the target oil has been estimated to be 45.5% of that originally in place. In 1985, U.S. enhanced-oil-recovery production, ca 96,160 m^3/day (604,800 bbl/day) was 6.8% of domestic crude-oil production. Polymer-flood field projects in the United States increased steadily from 1977 to 1986 partly because of increasing oil prices and partly because of oil-production tax provisions. Active U.S. polymer-flood projects increased from 14 in 1971 to 1184 in 1986 (22). However, only three of these field projects were large enough to produce more than 160 m^3/day. Between 1982 and 1985, polymer-flood EOR increased more than

Table 6. Oil-field Applications of Plastics and Other Polymers

Polymer	Application	Ref.
shredded cellophane	drilling-fluid, lost-circulation additive	7
lightly sulfonated polystyrene + lightly sulfonated isoprene–styrene copolymer[a]	oil-based, drilling-fluid viscosity improvement	314
lightly sulfonated polystyrene[a]	oil-based, drilling-fluid viscosity improvement	315
1.4–4.2% sulfonated polystyrene[a]	oil-based, drilling-fluid viscosity modifier	316
lightly sulfonated ethylene-5-ethylidene-2-norbornene copolymer[a]	oil-based, drilling-fluid viscosity modifier	317
lightly sulfonated terpolymer of ethylene, propylene, and 5-ethylidene-2-norbornene[a]	matrix-plugging agent for water control	318
terpolymer of styrene, butadiene, and Diels-Alder adducts of maleic acid esters	matrix-plugging agent for water control	319
terpolymer of styrene, butadiene, and acrylamide	oil-based, drilling-fluid viscosity modifier	317
sulfonated latex rubber	oil-based, drilling-fluid, lost-circulation additive	320
poly(vinyl acetate) emulsion	drilling-fluid thinning agent	321
poly(vinyl chloride) dispersion	drilling-fluid thinning agent	322
polydiorganosiloxane	high density drilling-fluid emulsifier	323
blend of oxyalkylated alkylphenol–aldehyde resin + polyolefin	crude oil demulsifier	324
copolymer of C_{16-18} alkylmethacrylates and dialkylaminoalkyl methacrylate	crystallization inhibitors for paraffinic crude oils	325
copolymer of ethylene and vinyl acetate	crystallization inhibitor for paraffinic crude oils	326
terpolymer of ethylene, maleic anhydride, and vinyl acetate	crystallization inhibitor for paraffinic crude oils	327
polymethylene–polyphenylamine adducts with 3-(glycidyloxy)-propyltrimethoxysilane	curing agent for epoxy-resin, sand-control agents	328
low molecular weight polyolefins[b]	viscosity modifiers for supercritical carbon dioxide for EOR	329
coumarone–indene resins	fluid-loss additive for completion fluids	330
terpene resins	fluid-loss additive for completion fluids	330
poly(ethylene oxide)	methanol viscosity modifier for hydraulic fracturing	331
sucrose ether copolymer of ethylene oxide and propylene oxide	polymeric surfactant for micellar flooding	332
copolymer of hydroxylated polybutadiene and triethylenediamine	matrix-plugging agent	333
copolymer of polyalkylene–polyamine alkylenedihalide and epichlorohydrin	fluid-loss additive for cement slurries	334
fluoroelastomers	down-hole seals	335

[a] An ionomer.

[b] Poly(1-hexene), for example.

300% to 2,432 m³/day (15,300 bbl/day) as field projects matured. In addition, 22 micellar polymer field projects were in progress in 1985 with a total production of 222 m³/day (1400 bbl/day). However, few large projects have been started in the United States since 1981; international polymer-flooding activity is also very limited. Small micellar-polymer EOR projects are in progress in the UK and France.

The only flood polymers used commercially are various grades (based on molecular weight and carboxylate group content) of polyacrylamide and xanthan gum (based primarily on pyruvate content and processing variations designed to reduce insoluble residue content). Most polyacrylamides are available in the form of a ca 30 wt % active oil-continuous emulsion. The xanthan gums are available in the form of a liquid broth, with research efforts aimed at increasing the xanthan concentration. Commercialization of a continuous fermentation process could reduce xanthan gum costs.

Use of xanthan gums in enhanced oil recovery is small compared to polyacrylamide because of higher prices and greater susceptibility to bacterial degradation. In 1984, 11,800 metric tons of polyacrylamides valued at $19,600,000 were sold for oil-field applications; sales of xanthan gum were reported at 3855 tons (337). Consumption of other EOR polymers was less (453 t, $400,000 in 1982) (338).

Other Oil-field Applications. Application areas of oil-field chemicals in order of decreasing economic importance are drilling fluids, cementing chemicals, fracturing chemicals, acidizing fluids, and enhanced oil recovery. Oil-field polymers are an important component of the chemicals used in these application areas. Total value of chemical products used in North American oil-fields in 1982 was 2.084×10^9 at the manufacturer level and 5.446×10^9 at the operator level (338). The difference is due to mark-ups by formulators and service companies.

The manufactured value of natural products sold for oil-field applications is given in Table 7. Lignosulfonate consumption in 1984 has been estimated at 59,000 t, half of the 1980–1981 consumption due to a drilling slump that began in 1982 (337). Oil-field consumption of carboxymethylcellulose in 1984 has been estimated at 22,000 t worth $63,000,000 (337). Worldwide oil-field consumption of polysaccharides (primarily cellulose derivatives, guar and guar derivatives, and xanthan gum) was 90,000 t in 1982 with a value of $150,000,000 (338). Most of the 15,800 t of hydroxypropylguar (worth $53,000,000) produced in 1983 was used in oil-field applications, predominantly hydraulic fracturing.

Because of the many different types and grades of polymers used in various oil-field applications, it is difficult to obtain data concerning the amount of individual polymers used in various applications. The sequential marketing of polymers by manufacturers, formulators, and oil-field service companies to the users (drilling contractors, service companies, and oil companies) also makes it difficult to determine the volume of polymers consumed in oil fields.

As wells are drilled deeper and deeper into hotter formations, as well-stimulation methods are increasingly applied to raise oil and gas production, and as enhanced-oil-recovery methods are applied more and more, consumption of oil-field polymers will increase. However, since oil prices are determined by political as well as economic factors, it is difficult to forecast future growth rates of polymers in the oil field.

Table 7. 1984 Estimated Manufactured Value of Oil-field Polymers[a]

Polymer	10^6 \$
lignosulfonates	40.0
guar and guar derivatives	29.4
xanthan gum	8.5
starches	8.3
carboxymethylcellulose	63.4
hydroxyethylcellulose	31.0
polyacrylates	12.8
polyacrylamides	4.0
polyamines and COPs	3.1
other polymers	3.0

[a] Non-EOR applications. Data taken from Ref. 337 with permission of Frost & Sullivan, Inc.

BIBLIOGRAPHY

1. J. Chatterji and J. K. Borchardt, *J. Pet. Technol.* **33,** 2042 (1981).
2. G. V. Chilingarian and P. Vorabutr, *Drilling and Drilling Fluids,* Elsevier Science Publishing Co., Inc., New York, 1981.
3. T. M. Muhleman, Jr., *World Oil* **202,** 35 (June 1986).
4. T. O. Allan and A. P. Roberts, *Production Operations,* Vol. 1, 2nd ed., Oil & Gas Consultants, Inc., Tulsa, Okla., 1982.
5. R. K. Clark and J. J. Nahm in M. Grayson, ed., *Kirk-Othmer Encyclopedia of Chemical Technology,* Vol. 17, 3rd ed., John Wiley & Sons, Inc., New York, 1982, pp. 143–167.
6. J. Kelly, Jr., *J. Pet. Technol.* **35,** 889 (1983).
7. J. E. Cornett, II, *World Oil,* 89 (Jan. 1984).
8. D. K. Smith, *Cementing, Monograph Vol. 4,* Society of Petroleum Engineers, Dallas, Tex., 1976.
9. T. O. Allen and A. P. Roberts, *Production Operations,* Vol. 2, 2nd ed., Oil & Gas Consultants, Inc., Tulsa, Okla., 1982.
10. G. O. Suman, *World Oil* **179,** 63 (Nov. 1974).
11. K. W. Pober, M. H. Huff, and R. K. Darlington, *J. Pet. Technol.* **35,** 2185 (1983).
12. B. B. Williams, J. L. Gidley, and R. S. Schechter, *Acidizing Fundamentals, Monograph Vol. 6,* Society of Petroleum Engineers, Dallas, Tex., 1979.
13. U.S. Pat. 4,460,627 (July 17, 1984), J. D. Weaver, L. E. Harris, and W. M. Harms (to Halliburton Co.).
14. M. T. Szabo, *J. Pet. Technol.* **31,** 553 (1979).
15. *Ibid.,* p. 561.
16. R. E. Bailey and L. B. Curtis, eds., *Enhanced Oil Recovery, National Petroleum Council Report,* National Petroleum Council, Washington, D.C., 1984.
17. M. Latil, *Enhanced Oil Recovery,* Gulf Publishing Co., Houston, Tex., 1980.
18. G. O. Goodlett, M. M. Honarpour, H. B. Carroll, and P. S. Sarathi, *Oil Gas J.* **84,** 82 (June 30, 1986).
19. F. H. Poettmann, D. C. Bond, and C. R. Hocott, *Improved Oil Recovery,* Interstate Oil Compact Commission, Oklahoma City, Okla., 1983.
20. H. K. van Poollen and co-workers, *Fundamentals of Enhanced Oil Recovery,* PennWell Publishing Co., Tulsa, Okla., 1980.
21. B. B. Sandiford in D. O. Shah and R. S. Schechter, eds., *Improved Oil Recovery by Surfactant and Polymer Flooding,* Academic Press, Inc., Orlando, Fla., 1977, pp. 487–509.
22. J. L. Leonard, *Oil Gas J.* **84,** 71 (Apr. 14, 1986).
23. B. L. Knight, S. C. Jones, and R. W. Parsons, *Soc. Pet. Eng. J.* **14,** 643 (1974) and references therein.
24. E. Unsal, J. L. Duda, and E. E. Klaus in R. T. Johansen and R. L. Berg, eds., *Chemistry of Oil Recovery, ACS Symp. Ser.* **91,** American Chemical Society, Washington, D.C., 1979, pp. 114–170.

25. R. K. Prud'homme, *Soc. Pet. Eng. J.* **24,** 431 (1984).
26. D. Guillot and A. Dunand, *Soc. Pet. Eng. J.* **25,** 39 (1985).
27. G. O. Aspinall, ed., *The Polysaccharides*, Vol. 1, Academic Press, Inc., Orlando, Fla., 1982.
28. R. L. Davidson, *Handbook of Water-soluble Gums and Resins*, McGraw-Hill Inc., New York, 1980.
29. D. S. Reid, *Dev. Ionic Polym.* **1,** 269 (1983) and Refs. therein.
30. E. R. Morris and I. T. Norton, *Stud. Phys. Theor. Chem.* **26,** 549 (1983) and references therein.
31. C. J. Githens and J. W. Burnham, *Soc. Pet. Eng. J.* **17,** 5 (1977).
32. C. W. Baker and R. L. Whistler, *Carbohydr. Res.* **45,** 237 (1975).
33. H. Deuel, J. Solms, and H. Neukom, *Chimia* **8,** 64 (1954).
34. Eur. Pat. 7,012 (Jan. 23, 1980), B. L. Swanson (to Phillips Petroleum Co.).
35. R. A. Young and L. Liss, *Cellul. Chem. Technol.* **12,** 399 (1978).
36. U.S. Pat. 4,507,470 (Mar. 26, 1985), E. T. Sortwell, M. Slovinski, and A. R. Mikkelsen (to Diatec Polymers).
37. Fr. Pat. 2,513,265 (Mar. 25, 1983), F. Bayerlein, P. P. Habereder, N. Keramaris, N. Kottmair, and M. Kuhn (to Diamalt AG).
38. U.S. Pat. 3,898,165 (Aug. 5, 1975), J. Chatterji (to Halliburton Co.).
39. D. L. Free, A. F. Frederick, and J. E. Thompson, *J. Pet. Technol.* **30,** 119 (1978).
40. U.S. Pat. 4,250,044 (Feb. 10, 1981), J. J. Hinkel (to Dow Chemical Co.).
41. Can. Pat. 1,090,112 (May 31, 1977), P. E. Clark, J. S. Underwood, and T. M. Steiner (to Dow Chemical Co.).
42. J. L. Zigrye, D. L. Whitfill, and J. A. Sievert, *J. Pet. Technol.* **37,** 315 (1985).
43. U.S. Pat. 4,140,639 (Feb. 20, 1979), J. M. Jackson (to Brinadd Co.).
44. N. R. Morrow and J. P. Heller in E. C. Donaldson, G. V. Chilingarian, and F. T. Yen, eds., *Developments in Petroleum Science*, Vol. 17A, Elsevier, Amsterdam, The Netherlands, 1985, pp. 47–74.
45. G. J. Gregory, *Oil Gas J.* **80,** 84(37), (1982).
46. Br. Pat. 2,108,112 (May 11, 1983), C. H. Kucera (to Dow Chemical Co.).
47. Eur. Pat. 92,756 (Nov. 2, 1983), G. J. Rummo (to Kay-Fries, Inc.).
48. U.S. Pat. 4,524,003 (June 18, 1985), J. K. Borchardt (to Halliburton Co.).
49. U.S. Pat. 4,094,795 (June 13, 1978), R. N. DeMartino and A. B. Conciatori (to Celanese Corp.).
50. U.S. Pat. 4,057,509 (Nov. 8, 1977), J. R. Costanza, R. N. DeMartino, and A. M. Goldstein (to Celanese Corp.).
51. Jpn. Kokai Tokkyo Koho 58,120,601 (July 18, 1983) (to Mitsubishi Acetate Co., Ltd.).
52. J. J. Gonzalez, *Macromolecules* **11,** 1074 (1978).
53. R. L. Whistler and R. J. McCreadie in R. L. Whistler, ed., *Industrial Gums*, Academic Press, Inc., Orlando, Fla., 1959, p. 370.
54. S. A. Barker, M. Stacey, and G. Zweifel, *Chem. Ind. London*, 330 (1957).
55. P. Kovacs, *Food Technol.* **27,** 26 (1973).
56. J. K. Rocks, *Food Technol.* **25,** 22 (1971).
57. U.S. Pat. 4,530,601 (Sept. 21, 1982), B. Mosier, J. L. McCrary, and K. G. Guilbeau (to Dow Chemical Co.).
58. U.S. Pat. 2,731,414 (Jan. 17, 1956), G. G. Binder, Jr., R. C. West, and K. H. Andresen (to Esso Research Co.).
59. U.S. Pat. 4,466,893 (Aug. 21, 1984), W. R. Dill (to Halliburton Co.).
60. U.S. Pat. 3,934,651 (Jan. 27, 1976), D. E. Nierode, D. M. Kehn, and K. F. Kruk (to Exxon Research and Engineering Co.).
61. K. B. Stauffer in Ref. 28, pp. 11/1–11/31.
62. U.S. Pat. 4,110,231 (Aug. 29, 1978), B. L. Swanson (to Phillips Petroleum Co.).
63. U.S. Pat. 4,110,230 (Aug. 29, 1978), J. E. Hessert and C. C. Johnston, Jr. (to Phillips Petroleum Co.).
64. X. Xu, Y. Shen, K. Chen, W. Li, X. Hu, and W. Xu, *Chenqdu Keji Daxue Xuebao* (3), 1 (1982).
65. Ger. Pat. 3,11,946 (Oct. 7, 1982), L. Brandt and A. Holst (to Hoechst AG).
66. Ger. Pat. 136,271 (June 27, 1979), B. Heyne and co-workers.
67. G. P. Thomas, *Soc. Pet. Eng. J.* **16,** 130 (1976).
68. I. B. Tomokhin, V. D. Gorodnov, and V. N. Teslenko, *Nauchn. Osn. Pererab. Nefti Gaza Neftekhim. Tezisy Dokl., Vses. Konf.*, 253 (1977).

69. R. K. Andresen, *Neft. Khoz.,* 17 (1982).

70. R. F. Scheurman, *J. Pet. Technol.* **35,** 306 (1983).

71. Brit. Pat. 2,070,611 (Sept. 9, 1981), L. D. Hoover and R. F. House (to NL Industries, Inc.).

72. Fr. Pat. 2,488,325 (Feb. 12, 1982), R. F. House and L. D. Hoover (to NL Industries, Inc.).

73. U.S. Pat. 4,324,668 (Apr. 13, 1982), L. E. Harris (to Halliburton Co.).

74. U.S. Pat. 4,210,206 (July 1, 1980), J. W. Ely (to Halliburton Co.).

75. U.S. Pat. 4,096,074 (June 20, 1978), S. Stournas (to Mobil Oil Corp.).

76. H. N. Black, H. E. Ripley, W. H. Beecroft, and L. O. Pamplin, *J. Pet. Technol.* **33,** 26 (1981).

77. U.S. Pat. 4,290,899 (Sept. 22, 1981), T. R. Malone, T. D. Foster, Jr., and S. T. Executrix (to Union Carbide Corp.).

78. Br. Pat. 2,090,308 (July 7, 1982), R. H. Rygg (to Mobil Oil Corp.).

79. M. G. Wirick, *J. Polym. Sci. Part A* **1,** 1705 (1968).

80. U.S. Pat. 4,009,329 (Feb. 27, 1977), W. C. Arney, C. A. Williams, and J. E. Glass, Jr. (to Union Carbide Corp.).

81. U.S. Pat. 4,518,040 (May 21, 1985), J. D. Middleton (to Halliburton Co.).

82. U.S. Pat. 4,321,968 (Mar. 30, 1982), E. E. Clear (to Phillips Petroleum Co.).

83. J. A. Losacano and C. M. Kim, *Fracture Conductivity Damage Due to Cross-Linked Gel Residue and Closure Stress on Propped 20/40 Mesh Sand, Paper No. SPE 14436, 60th Annual Technical Conference and Exhibition,* Society of Petroleum Engineers of AIME, Las Vegas, Nev., Sept. 22–25, 1985.

84. Ger. Pat. 2,639,620 (Mar. 9, 1978), T. J. Podlas (to Hercules, Inc.).

85. J. Schurz and H. Khatami, *Papier Darmstadt* **33**(10A), 1 (1979).

86. J. T. Patton, J. T. Patton, Jr., M. Kuntamukkala, and S. Holbrook, *Polym. Prepr. Am. Chem. Soc. Div. Polym. Chem.* **22**(2), 46 (1981).

87. U.S. Pat. 4,172,055 (Oct. 23, 1979), R. N. DeMartino (to Celanese Corp.).

88. U.S. Pat. 4,523,010 (June 11, 1985), C. Lukach, T. G. Majewicz, and A. R. Reid (to Hercules, Inc.).

89. Neth. Pat. 80 03,241 (Dec. 9, 1980), L. M. Landoll (to Hercules, Inc.).

90. D. A. Rees, *Pure Appl. Chem.* **53,** 1 (1981).

91. U.S. Pat. 4,454,620 (June 12, 1984), F. Dawans, D. Binet, N. Kohler, and D. V. Quang (to Institut Francais du Petrole).

92. J. C. Phillips, J. W. Miller, W. C. Wernau, B. E. Tate, and M. H. Auerbach, *Soc. Pet. Eng. J.* **25,** 594 (1985).

93. P. J. Whitcomb and C. W. Macosko, *J. Rheol.* **22,** 493 (1978).

94. P. A. Sandford and co-workers in P. A. Sandford and A. Laskin, eds., *Extracellular Microbial Polysaccharides, ACS Symp. Ser.* **45,** American Chemical Society, Washington, D.C., 1977, p. 192.

95. G. Holzwarth, *Biochemistry* **15,** 4333 (1976).

96. S. Paoletti, A. Cesaro, and F. Delben, *Carbohydr. Res.* **123,** 173 (1983).

97. C. S. H. Chen and E. W. Sheppard, *Polym. Eng. Sci.* **20,** 512 (1980).

98. U.S. Pat. 4,053,699 (Oct. 11, 1977), P. T. Cahalan, J. A. Peterson, and D. A. Arndt (to General Mills Chemical Co.).

99. M. Rinaudo and M. Milas, *Int. J. Biol. Macromol.* **2,** 45 (1980).

100. Eur. Pat. 145,217 (June 19, 1985), D. Beck, J. W. Miller, W. C. Wernau, and T. B. Young, III (to Pfizer, Inc.).

101. Brit. Pat. 2,065,689 (July 1, 1981), T. J. Holding and G. W. Pace (to Tate and Lyle).

102. U.S. Pat. 4,165,257 (Aug. 21, 1979), O. M. Stokke (to Conoco, Inc.).

103. Rom. Pat. 84,829 (July 30, 1984), H. Lewis and R. L. Miller (to Pfizer, Inc.).

104. U.S. Pat. 3,771,462 (Jan. 16, 1973), M. K. Abdo (to Mobil, Inc.).

105. Ger. Offen. 2,809,136 (Sept. 6, 1979), W. H. Carter, C. A. Christopher, and T. Jefferson (to Texaco Development Corp.).

106. P. A. Sandford and A. Laskin, eds., *Extracellular Microbial Polysaccharides, ACS Symp. Ser.* **45,** American Chemical Society, Washington, D.C., 1977.

107. Eur. Pat. 137,538 (Apr. 17, 1985), J. J. Bleeker, J. H. Lammers, J. B. Roest, and R. J. A. Eckert (to Shell Internationale Research Maatschappij BV).

108. W. M. Hensel, Jr., R. L. Sullivan, and R. H. Stallings, *Pet. Eng. Int.,* 155 (May 1981).

109. F. N. Deily, G. P. Lindblom, J. Patton, and W. E. Holman, *Oil Gas J.* **65,** 62 (June 26, 1967).

110. U. Grollman and W. Schnabel in *Polymer Degradation and Stability,* Applied Science Publishers, Ltd., London, 1982, pp. 353–362.
111. M. C. Cadmus and co-workers, *Appl. Environ. Microbiol.,* 5 (Aug. 1982).
112. Ger. Offen. 2,715,026 (Oct. 13, 1977), S. L. Lee (to Shell Internationale Research Maatschappij BV).
113. Eur. Pat. 106,666 (Apr. 25, 1984), C. J. Philips (to Pfizer, Inc.).
114. Can. Pat. 1,070,492 (Jan. 29, 1980), S. L. Wellington (to Shell Canada, Ltd.).
115. J. C. Salamone, S. B. Clough, A. B. Salamone, K. I. G. Reid, and D. E. Jamison, *Soc. Pet. Eng. J.* **22**, 555 (1982).
116. U.S. Pat. 3,581,824 (June 1, 1971), B. G. Hurd (to Mobil Oil Corp.).
117. U.S. Pat. 4,157,322 (June 5, 1979), G. T. Colegrove (to Merck and Co., Inc.).
118. C. S. H. Chen and E. W. Sheppard, *J. Macromol. Sci. Chem. Part A,* 239 (1979).
119. U.S. Pat. 4,104,493 (Aug. 1, 1978), W. H. Carter, C. A. Christopher, and T. Jefferson (to Texaco, Inc.).
120. Ger. Pat. 2,848,984 (May 17, 1979), W. C. Wernau (to Pfizer Corp.).
121. J. C. Phillips, J. W. Miller, W. C. Wernau, B. E. Tate, and M. H. Auerbach, *Soc. Pet. Eng. J.* **25**, 594 (1985).
122. Eur. Pat. 103,483 (Mar. 21, 1984), G. M. Holzwarth (to Exxon Research and Engineering Co.).
123. U.S. Pat. 3,919,092 (Nov. 11, 1975), C. J. Norton and D. D. Falk (to Marathon Oil Corp.).
124. U.S. Pat. 4,508,629 (Apr. 2, 1985), J. K. Borchardt (to Halliburton Co.).
125. U.S. Pat. 4,039,028 (Aug. 2, 1977), R. K. Knight (to Union Oil Co., Calif.).
126. U.S. Pat. 4,195,689 (Apr. 1, 1980), H. L. Chang (to Cities Service Co.).
127. J. G. Dominguez and G. P. Willhite, *Soc. Pet. Eng. J.* **17**, 111 (1977).
128. T. P. Castor, J. B. Edwards, and F. J. Passman in D. O. Shah, ed., *Surface Phenomena in Enhanced Oil Recovery,* American Chemical Society, Washington, D.C., 1981, pp. 773–820.
129. U.S. Pat. 4,105,605 (Aug. 8, 1978), I. W. Cottrell, R. A. Empey, and J. S. Racciato (to Merck & Co., Inc.).
130. T. Nagabhushanam and co-workers, *J. Polym. Sci. Polym. Chem. Ed.* **12**, 2953 (1974).
131. U.S. Pat. 4,457,372 (July 3, 1984), M. S. Doster, A. J. Nute, and C. A. Christopher (to Texaco, Inc.).
132. H. J. Dietzel and G. Pusch, *Soc. Pet. Eng. J.* **25**, 9 (1985).
133. M. Rinaudo and M. Vincendon, *Carbohydr. Polym.* **2**, 135 (1982).
134. N. E. Rogers in R. L. Whistler, ed., *Industrial Gums,* 2nd ed., Academic Press, Inc., Orlando, Fla., 1973, Chapt. XXII.
135. U.S. Pat. 4,079,544 (Mar. 21, 1978), J. G. Savins (to Mobil Oil Corp.).
136. U.S. Pat. 4,096,073 (June 20, 1978), G. O. Hitzman (to Phillips Petroleum Co.).
137. Eur. Pat. 138,255 (Apr. 24, 1985), J. D. Linton, M. W. Evans, and A. R. Godley (to Shell Internationale Research Maatschappij BV).
138. Eur. Pat. 135,953 (Apr. 3, 1985), P. R. Betteridge (to Shell Internationale Research Maatschappij BV).
139. Eur. Pat. 134,649 (Mar. 20, 1985), I. W. Sutherland (to National Research Development Corp.).
140. M. Aslam, G. Pass, and G. O. Phillips, *J. Chem. Res. Part S,* 320 (1978).
141. T. R. Jack and E. DiBlosio in J. E. Zajic and E. C. Donaldson, eds., *Microbes and Oil Recovery,* Bioresources Publications, El Paso, Tex., 1983, pp. 205–212.
142. Z. Zosim, S. Goldman, D. L. Gutnick, and E. Rosenberg in Ref. 141, pp. 92–99.
143. M. Rinaudo and C. Noik, *Polym. Bull. Berlin* **9**, 543 (1983).
144. P. Kovacs, *Food Technol.* **27**, 26 (1973).
145. T. Haack, D. A. Shaw, and D. E. Greenley, *Oil Gas J.* **84**, 82 (Jan. 6, 1986).
146. R. Methrotra and B. Raanby, *J. Appl. Polym. Sci.* **22**, 2991 (1978).
147. *Ibid.,* p. 3003.
148. Jpn. Kokai Tokkyo Koho 80,142,014 (Nov. 6, 1980) (to Nichiden Kagaku Co., Ltd.).
149. U.S. Pat. 3,084,122 (Apr. 2, 1963), J. D. Cypert and co-workers (to Jersey Production Research Co.).
150. U.S. Pat. 3,042,611 (July 3, 1962), J. T. Patton (to Jersey Production Research Co.).
151. U.S. Pat. 4,210,204 (July 1, 1980), C. A. Christopher, Jr. (to Texaco, Inc.).
152. U.S. Pat. 4,439,332 (Mar. 27, 1984), S. Frank, A. T. Coscia, and J. M. Schmitt (to American Cyanamid Co.).
153. Fr. Pat. 2,495,217 (June 4, 1982), J. Boutin and F. Contat (to Rhone Poulenc Industries SA).

154. J. Klein and R. Heitzman, *Makromol. Chem.* **179,** 1895 (1978).
155. F. H. Halverson and J. E. Lancaster, *Macromolecules* **18,** 1139 (1985).
156. G. Muller, J. P. Laine, and J. C. Fenyo, *J. Polym. Sci. Polym. Chem. Ed.* **17,** 659 (1979).
157. G. S. Chen, H. H. Niedlinger, and C. L. McCormick, *Prepr. Pap. Nat. Meet. Div. Pet. Chem. Am. Chem. Soc.* **29**(4), 1147 (1984).
158. J. S. Ward and F. D. Martin, *Soc. Pet. Eng. J.* **21,** 623 (1981).
159. B. B. Sandiford in D. O. Shah and R. S. Schechter, eds., *Improved Oil Recovery by Surfactant and Polymer Flooding,* Academic Press, Inc., Orlando, Fla., 1977, pp. 487–509.
160. D. McIntyre, A. L. Shih, J. Savoca, R. Seeger, and A. MacArthur, *Org. Coat. Appl. Polym. Sci. Proc.* **48,** 612 (1983).
161. U.S. Pat. 3,490,533 (Jan. 20, 1970), H. C. McLaughlin (to Halliburton Co.).
162. R. D. Shupe, *J. Pet. Technol.* **33,** 1513 (1981).
163. L. W. Holm, *Oil Gas J.* **82**(29), 82 (1984).
164. Jpn. Kokai Tokkyo Koho 65,696 (May 17, 1980) (to Nitto Chemical Industry Co., Ltd. and Sekiyu Shigen Kaihatsu KK).
165. U.S. Pat. 4,249,608 (Feb. 10, 1981), W. H. Castor (to Texaco, Inc.).
166. U.S. Pat. 4,439,334 (Mar. 27, 1984), J. K. Borchardt (to Halliburton Co.).
167. E. Pefferkorn, L. Nabzar, and A. Carroy, *J. Colloid Interface Sci.* **106,** 94 (1985).
168. B. B. Bansal, V. Hornof, and G. Neale, *Can. J. Chem. Eng.* **57,** 203 (1979).
169. J. Novosad, *J. Can. Pet. Technol.* **23,** 24 (1984).
170. U.S. Pat. 4,113,013 (Sept. 12, 1978), W. A. Ledoux, W. Schoen, and A. Kumar (to Texaco, Inc.).
171. U.S. 3,469,630 (Sept. 30, 1969), B. G. Hurd and W. R. Foster (to Mobil Oil Co.).
172. U.S. Pat. 3,804,173 (Apr. 16, 1974), R. R. Jennings (to Dow Chemical Co.).
173. U.S. Pat. 3,414,054 (Dec. 3, 1968), G. G. Bernard (to Union Oil Co., Calif.).
174. F. Friedman, *SPE Reservoir Eng.* **1,** 261 (1986).
175. K. L. Goyal, P. D. Arora, and C. V. Chilingar, *Energy Sources* **5,** 45 (1980).
176. S. Sawant and H. Morawetz, *Macromolecules* **17,** 2427 (1984).
177. U.S. Pat. 4,450,084 (May 22, 1984), M. K. Abdo (to Mobil Oil Corp.).
178. E. Malachosky and M. Herd, *Pet. Eng. Int.* **58,** 48 (June 1986).
179. L. R. Peddycoart, *Oil Gas J.* **78,** 52 (Feb. 4, 1980).
180. R. K. Prud'homme, J. T. Uhl, J. P. Poinsatte, and F. Halverson, *Soc. Pet. Eng. J.* **23,** 804 (1983).
181. U.S. Pat. 4,343,363 (Aug. 10, 1982), C. J. Norton and D. O. Falk (to Marathon Oil Co.).
182. R. K. Prud'homme and J. T. Uhl, *Kinetics of Polymer/Metal-Ion Gelation, Paper No. SPE/DOE 12640, Fourth Joint Symposium on Enhanced Oil Recovery,* Tulsa, Okla., Apr. 15–18, 1984.
183. D. J. Jordan, D. W. Green, R. E. Terry, and G. P. Wilhite, *Soc. Pet. Eng. J.* **22,** 463 (1982).
184. U.S. Pat. 3,687,200 (Aug. 29, 1972), W. G. Routson (to Dow Chemical Co.).
185. USSR Pat. 909,125 (Feb. 28, 1982), V. I. Krylov, O. N. Mironenko, R. F. Ukhanov, and S. S. Dzhangirov (to All-Union Scientific Research Institute of Bracing Wells and Drilling Mud).
186. U.S. Pat. 4,413,680 (Nov. 8, 1983), B. B. Sandiford, H. T. Dovan, and R. D. Hutchins (to Union Oil Co., Calif.).
187. G. P. Wilhite and D. J. Jordan, *Polym. Prepr. Am. Chem. Soc. Div. Polym. Chem.* **22**(2), 53 (1981).
188. B. Sloat, *Pet. Eng. Int.,* 20 (1977).
189. U.S. Pat. 4,155,405 (May 22, 1979), L. Vio (to Societe Nationale Elf Aquitaine).
190. U.S. Pat. 4,098,337 (July 4, 1978), P. A. Argabright, J. S. Rhudy, and B. L. Phillips (to Marathon Oil Co.).
191. U.S. Pat. 4,125,478 (Nov. 14, 1978), E. J. Sullivan and G. D. Jones (to Dow Chemical Co.).
192. Eur. Pat. 5,835 (Dec. 2, 1979), R. J. Pliny and T. W. Regulski (to Dow Chemical Co.).
193. USSR Pat. 1,040,118 (Sept. 7, 1983), I. A. Sidorov and co-workers (to All-Union Scientific Research Institute of Petroleum Gas).
194. U.S. Pat. 4,579,667 (Apr. 1, 1986), E. Ech and R. D. Lees (to Hercules, Inc.).
195. U.S. Pat. 3,334,689 (Oct. 8, 1967), H. C. McLaughlin (to Halliburton Co.).
196. I. Lakatos, J. Lakatos Szabo, and M. Kiss Gaspar, *Banyasz. Kohasz. Lapok. Koolaj Foldgaz* **14,** 359 (1981).
197. L. R. Norman, M. W. Conway, and J. M. Wilson, *J. Pet. Technol.* **36,** 2011 (1984).
198. Y. L. Meltzer, *Water-soluble Polymers: Technology and Applications, Chem. Process Rev.* **64,** Noyes Data Corp., Park Ridge, N.J., 1972, pp. 17–19.
199. A. M. Khasaev, M. G. Sadykhov, and Kh. G. Kurbanova, *Azerb. Neft. Khoz.,* 32 (1978).

200. USSR Pat. 700,643 (Nov. 30, 1979), N. N. Kubareva and co-workers (to Tartar Petroleum Scientific Research Institute).
201. C. L. Hamburger, Y. Tsao, B. Morrison, and E. Drake, *J. Pet. Technol.* **37,** 499 (1985).
202. C. L. McCormick, R. D. Hester, H. H. Niedlinger, and G. C. Wildman in B. Linville, ed., *BETC Prog. Rev.* **20,** 61 (1979).
203. U.S. Pat. 4,448,697 (May 15, 1984), D. R. McCoy and R. M. Gipson (to Texaco, Inc.).
204. Ger. Offen. 3,220,503 (Dec. 1, 1983), F. Engelhardt, U. Greiner, H. Schmitz, W. Gulden, and S. P. von Halasz (to Casella AG).
205. C. L. McCormick, K. R. Blackmon, and D. L. Elliott, *Prepr. Pap. Nat. Meet. Div. Pet. Chem. Am. Chem. Soc.* **29**(4), 1159 (1984).
206. C. L. McCormick, *J. Macromol. Sci. Chem. Part A* **22**, 955 (1985).
207. H. Pledger, Jr., J. J. Meister, T. E. Hogen-Esch, and G. B. Butler, *Polym. Prepr. Am. Chem. Soc. Div. Polym. Chem.* **22**(2), 72 (1981).
208. H. H. Neidlinger and C. L. McCormick, *Polym. Prepr. Am. Chem. Soc. Div. Polym. Chem.* **20**(1), 901 (1979).
209. J. J. Meister and co-workers, *Polym. Prepr. Am. Chem. Soc. Div. Polym. Chem.* **25**(1), 266 (1984).
210. U.S. Pat. 4,440,652 (Apr. 3, 1984), W. D. Hunter (to Texaco Development Corp.).
211. U.S. Pat. 4,485,224 (Nov. 27, 1984), R. A. Smith (to Goodyear Tire and Rubber Co.).
212. F. D. Martin, M. J. Hatch, M. Abouelezz, and J. C. Oxley, *Polym. Mater. Sci. Eng.* **51,** 688 (1984).
213. U.S. Pat. 4,110,232 (Aug. 29, 1979), F. C. Schwab, E. W. Sheppard, and C. S. H. Chen (to Texaco, Inc.).
214. U.S. Pat. 4,200,154 (Apr. 29, 1980), J. F. Tate (to Texaco, Inc.).
215. U.S. Pat. 4,200,151 (Apr. 29, 1980), J. F. Tate (to Texaco, Inc.).
216. U.S. Pat. 4,430,481 (Feb. 7, 1984), W. D. Hunter (to Texaco Development Corp.).
217. Eur. Pat. 94,898 (Nov. 23, 1983), K. F. Castner (to Goodyear Tire and Rubber Co.).
218. U.S. Pat. 4,554,081 (Nov. 19, 1985), J. K. Borchardt and S. P. Rao (to Halliburton Co.).
219. Fr. Pat. 2,441,049 (June 6, 1980) (to Milchem, Inc.).
220. Eur. Pat. 115,836 (Aug. 15, 1984), G. A. Stahl, I. J. Westerman, H. S. Hsigh, and A. Moradi-Araghi (to Phillips Petroleum Co.).
221. U.S. Pat. 4,536,303 (Aug. 20, 1985), J. K. Borchardt (to Halliburton Co.).
222. U.S. Pat. 4,219,429 (Aug. 26, 1980), J. C. Allen and J. F. Tate (to Texaco, Inc.).
223. U.S. Pat. 4,206,058 (June 3, 1980), J. F. Tate (to Texaco, Inc.).
224. U.S. Pat. 4,143,716 (Mar. 13, 1979), G. Kalfoglou and K. H. Fluornoy (to Texaco, Inc.).
225. U.S. Pat. 4,343,712 (Aug. 10, 1982), W. D. Hunter (to Texaco Development Corp.).
226. U.S. Pat. 4,268,400 (May 19, 1981), J. M. Lucas, A. C. Perricone, and D. P. Enright (to Milchem, Inc.).
227. Ger. Offen. 3,003,747 (Sept. 11, 1980), J. M. Lucas, D. P. Enright, and A. C. Perricone (to Milchem, Inc.).
228. U.S. Pat. 4,205,724 (June 3, 1980), L. E. Roper and B. L. Swanson (to Phillips Petroleum Co.).
229. Jpn. Kokai Tokkyo Koho 58,104,980 (June 22, 1983) (to Toa Gosei Chemical Industry Co., Ltd. and Ternite Co., Ltd.).
230. Eur. Pat. 122,789 (Oct. 24, 1984), W. M. Hann and J. Natoli (to Rohm and Haas Co.).
231. U.S. Pat. 4,423,199 (Dec. 27, 1983), C. J. Chang and T. E. Stevens (to Rohm and Haas Co.).
232. Ger. Pat. 3,418,397 (Dec. 6, 1984), M. Okada, K. Noda, and M. Shikata (to Sanyo Chemical Industries, Ltd.).
233. A. I. Pen'kov, *Neft. Khoz.* (7), 15 (1979).
234. Ger. Offen. 3,338,431 (May 2, 1985), M. Hille, W. Friede, H. Wittkus, F. Engelhardt, and U. Riegel (to Hoechst AG).
235. U.S. Pat. 4,404,411 (Sept. 13, 1983), L. Bi, M. E. Dillon, and C. Sharik (Atlantic Richfield Co.).
236. Jpn. Kokai Tokkyo Koho 59,152,983 (Aug. 31, 1984) (to Daicel Chemical Industries).
237. PCT Int. Pat. 83 02,449 (July 21, 1983), K. Uhl, J. K. Bannerman, F. Engelhardt, and A. Patel (to Cassella AG, Hoechst AG, and Dresser Industries, Inc.).
238. U.S. Pat. 4,558,741 (Dec. 17, 1985), J. K. Borchardt and B. M. Young (to Halliburton Co.).
239. I. G. Yusopov, M. G. Gataulin, A. Sh. Gazizov, and A. F. Slivchenko, *Neftepromsyl. Delo,* 27 (1981).
240. U.S. Pat. 4,301,016 (Nov. 17, 1981), D. B. Carriere and R. V. Lauzon (to NL Industries, Inc.).
241. USSR Pat. 765,497 (Sept. 23, 1980), G. N. Shvareva and co-workers.
242. U.S. Pat. 4,315,087 (Feb. 9, 1982), D. Redmore and B. T. Outlaw (to Petrolite Corp.).

243. U.S. Pat. 4,207,946 (June 17, 1980), W. C. Haltmar and E. S. Lacey (to Texaco, Inc.).

244. Eur. Pat. 44,508 (Jan. 27, 1982), F. Englehardt, H. Schmitz, J. Hax, and W. Gulden (to Cassella AG).

245. U.S. Pat. 4,440,649 (Apr. 3, 1984), R. E. Loftin and A. J. Son (to Halliburton Co.).

246. PCT Int. Pat. 82 02,052 (June 24, 1982), F. Englehardt, S. Piesch, J. Balzer, and J. C. Dawson (to Cassella AG and Dresser Industries, Inc.).

247. U.S. Pat. 4,504,622 (Mar. 12, 1985), K. D. Schmitt (to Mobil Oil Corp.).

248. B. G. Chesser and D. P. Enright, *J. Pet. Technol.* **32,** 950 (1980).

249. U.S. Pat. 4,228,019 (Oct. 14, 1980), W. D. Hunter (to Texaco Development Corp.).

250. U.S. Pat. 4,210,205 (July 1, 1980), J. C. Allen and J. F. Tate (to Texaco, Inc.).

251. U.S. Pat. 4,284,517 (Aug. 18, 1981), C. S. H. Chen and E. W. Sheppard (to Mobil Oil Corp.).

252. Brit. Pat. 2,148,356 (May 30, 1985), J. H. Clint, P. K. G. Hodgson, and E. J. Tinley (to British Petroleum Co. PLC).

253. U.S. Pat. 3,764,530 (Oct. 9, 1973), P. D. Burland, J. L. Stephenson, and E. H. Stobart (to Milchem, Inc.).

254. Jpn. Kokai Tokkyo Koho 58,219,289 (Dec. 20, 1983), (to Shimizu Construction Co., Ltd.).

255. U.S. Pat. 4,120,801 (Oct. 17, 1978), C. S. H. Chen, F. C. Schwab, and E. W. Sheppard (to Mobil Corp.).

256. U.S. Pat. 4,110,232 (Sept. 29, 1978), F. C. Schwab, E. W. Sheppard, and C. S. H. Chen (to Mobil Oil Co.).

257. U.S. Pat. 4,498,540 (1985), M. L. Marrocco (to Cities Service Oil and Gas Corp.).

258. U.S. Pat. 3,435,899 (Apr. 1, 1969), R. R. Koch, J. Ramos, and H. C. McLaughlin (to Halliburton Co.).

259. Can. Pat. 1,070,936 (Feb. 5, 1980), E. A. Elphingstone, H. C. McLaughlin, and C. W. Smith (to Halliburton Co.).

260. C. B. Hurd and H. A. Letteron, *J. Phys. Chem.* **36,** 604 (1932).

261. S. L. Downs and K. G. Manhar, *J. Pet. Technol.* **26,** 557 (1974).

262. Hung. Teljes 31,817 (May 28, 1984), Z. Heinemann, G. Milley, E. Nemeth, A. Szittar, and O. Wagner.

263. Ger. Pat. 2,908,746 (Sept. 18, 1980), F. Meyer, H. Mehesch, R. Kubens, and M. Winkelmann (to Bergwerksverband GmbH and Bayer AG).

264. B. J. Felber and D. L. Dauben, *Soc. Pet. Eng. J.* **17,** 391 (1977).

265. U.S. Pat. 4,428,429 (Jan. 31, 1984), B. J. Felber and C. A. Christopher (to Standard Oil Co., Ind.).

266. U.S. Pat. 4,275,789 (June 30, 1981), D. D. Lawrence and B. J. Felber (Standard Oil Co., Ind.).

267. U.S. Pat. 4,110,226 (Aug. 29, 1978), B. L. Swanson (to Phillips Petroleum Co.).

268. K. Manasrah, G. H. Neale, and V. Hornof, *Cellul. Chem. Technol.* **19,** 291 (1985).

269. Can. Pat. 1,111,237 (Oct. 27, 1981), G. Kalfoglou (to Texaco Development Co.).

270. U.S. Pat. 4,219,082 (Aug. 16, 1980), G. Kalfoglou (to Texaco, Inc.).

271. U.S. Pat. 4,196,177 (Apr. 8, 1980), G. Kalfoglou (to Texaco, Inc.).

272. U.S. Pat. 4,172,498 (Oct. 30, 1979), G. Kalfoglou (to Texaco, Inc.).

273. B. M. Young, H. C. McLaughlin, and J. K. Borchardt, *J. Pet. Technol.* **32,** 2121 (1980).

274. R. F. Krueger, *J. Pet. Technol.* **38,** 131 (1986).

275. K. C. Khilar and H. S. Fogler, *Soc. Pet. Eng. J.* **23,** 55 (1983).

276. U.S. Pat. 4,366,071 (Dec. 28, 1982), H. C. McLaughlin and J. D. Weaver (to Halliburton Co.).

277. U.S. Pat. 4,366,074 (Dec. 28, 1982), H. C. McLaughlin and J. D. Weaver (to Halliburton Co.).

278. U.S. Pat. 4,366,073 (Dec. 28, 1982), H. C. McLaughlin and J. D. Weaver (to Halliburton Co.).

279. U.S. Pat. 4,366,072 (Dec. 28, 1982), H. C. McLaughlin and J. D. Weaver (to Halliburton Co.).

280. H. C. McLaughlin, E. A. Elphingstone, R. E. Remington, and S. Coates, *World Oil,* 58 (May 1977).

281. U.S. Pat. 4,497,596 (Feb. 5, 1985), J. K. Borchardt and B. M. Young (to Halliburton Co.).

282. U.S. Pat. 4,536,305 (Aug. 20, 1985), J. K. Borchardt and B. M. Young (to Halliburton Co.).

283. U.S. Pat. 4,536,304 (Aug. 20, 1985), J. K. Borchardt (to Halliburton Co.).

284. J. K. Borchardt and D. L. Brown, *Prepr. Pap. Nat. Meet. Div. Pet. Chem. Am. Chem. Soc.* **29**(4), 1142 (1984).

285. U.S. Pat. 4,484,631 (Aug. 25, 1982), N. S. Sherwood, C. A. Costello, and G. F. Matz (to Calgon Corp.).

286. Ger. Pat. 159,015 (Feb. 16, 1983), G. Pacholke and co-workers.

287. U.S. Pat. 4,225,445 (Sept. 30, 1980), K. W. Dixon (to Calgon Corp.).
288. K. C. Hong and R. S. Millhone, *J. Pet. Technol.* **29**, 1657 (1977).
289. T. W. Muecke, *J. Pet. Technol.* **26**, 157 (1974).
290. R. F. Rensvold, *Soc. Pet. Eng. J.* **23**, 238 (1983).
291. W. L. Penberthy, Jr., C. M. Shaughnessy, C. Gruesbeck, and W. M. Salathiel, *J. Pet. Technol.* **30**, 845 (1978).
292. U.S. Pat. 3,404,735 (Oct. 8, 1968), B. M. Young, and K. D. Totty (to Halliburton Co.).
293. U.S. Pat. 4,259,205 (Mar. 31, 1981), J. R. Murphey (to Halliburton Co.).
294. D. R. Underdown and K. Das, *J. Pet. Technol.* **37**, 2006 (1985).
295. U.S. Pat. 4,527,627 (July 9, 1975), J. W. Graham and A. R. Sinclair (to SANTROL Products).
296. Ref. 9, p. 65.
297. Rom. Pat. 84,507 (Aug. 30, 1984), C. D. Craciun and T. Pantu (to Trustul Petrolului, Boldesti).
298. R. S. Cole and J. K. Borchardt, *Drilling,* 44 (Apr. 1985).
299. U.S. Pat. 4,389,320 (June 21, 1983), R. L. Clampitt (to Phillips Petroleum Co.).
300. U.S. Pat. 3,686,372 (Aug. 29, 1972), A. J. Whitworth, S. Y. S. Tung, and E. A. Hajto (to Borden Co.).
301. U.S. Pat. 4,299,690 (Nov. 11, 1981), B. W. Allan (to Texaco, Canada, Inc.).
302. C. M. Blair, Jr., R. E. Scribner, C. A. Stout, and J. E. Fredericksen, *Oil Gas J.* **82**, 90 (1984).
303. U.S. Pat. 4,337,828 (July 6, 1982), C. M. Blair, Jr. (to Magna Corp.).
304. C. M. Blair, Jr. and C. A. Stout, *Oil Gas J.* **83**, 55 (May 20, 1985).
305. Fr. Pat. 2,551,451 (Mar. 8, 1985), M. Soreau and D. Siegel (to Societe Francaise Hoechst SA).
306. Can. Pat. 1,187,404 (May 21, 1985), M. Navritil, J. P. Batycky, M. Sovak, and M. S. Mitchell (to Borden Co., Ltd.).
307. USSR Pat. 675,168 (July 25, 1979), N. M. Makeev and co-workers (to Perm State Scientific Research and Design Institute of the Petroleum Industry).
308. USSR Pat. 1,040,121 (Sept. 7, 1983), I. A. Sidorov and co-workers (to All-Union Scientific Research Institute of Petroleum and Gas).
309. USSR Pat. 732,494 (May 5, 1980), G. M. Shved, I. Levchenko, A. G. Storozhenko, and M. L. Sherstyanoi (to Krasnodar State Scientific Research Institute of the Petroleum Industry).
310. U.S. Pat. 4,192,753 (Mar. 11, 1980), D. S. Pye, J. P. Gallus, and P. W. Fischer (to Union Oil Co., Calif.).
311. Can. Pat. 1,050,743 (Mar. 20, 1979), J. W. Scheffel and P. W. Fischer (to Union Oil Co., Calif.).
312. Fr. Pat. 2,474,558 (July 31, 1981), C. Gadelle, J. Burger, and C. Bardon (to Institut Francais du Petrole).
313. USSR Pat. 878,904 (Nov. 7, 1981), Z. A. Balitskaya and co-workers (to Ukrainian Scientific Research Institute of the Geological Survey, Poltava).
314. U.S. Pat. 4,425,461 (Jan. 10, 1984), S. R. Turner, R. D. Lundberg, W. A. Thaler, T. O. Walker, and D. G. Peiffer (to Exxon Research and Engineering Co.).
315. Eur. Pat. 106,528 (Apr. 25, 1984), R. D. Lundberg, D. G. Peiffer, S. R. Turner, and T. O. Walker (to Exxon Research and Engineering Co.).
316. Eur. Pat. 72,245 (Feb. 16, 1983), R. D. Lundberg, D. G. Peiffer, and T. O. Walker (to Exxon Research and Engineering Co.).
317. Eur. Pat. 72,244 (Feb. 16, 1983), R. D. Lundberg, H. S. Makowski, C. P. O'Farrell, and T. O. Walker (to Exxon Research and Engineering Co.).
318. U.S. Pat. 4,183,406 (Jan. 15, 1980), R. D. Lundberg, D. E. O'Brien, H. S. Makowski, and R. R. Klein (to Exxon Research and Engineering Co.).
319. Ger. Pat. 143,451 (Aug. 20, 1980), H. Maerkert, G. Becker, R. Thiede, and K. H. Weinert (to VEB Erdgasfoerderung).
320. U.S. Pat. 4,425,462 (Jan. 10, 1984), S. R. Turner, R. D. Lundberg, and T. O. Walker (to Exxon Research and Engineering Co.).
321. USSR Pat. 692,849 (Oct. 25, 1979), M. I. Radyuk, N. E. Yushkova, and A. I. Kozubovskii (to Western Siberian Scientific Research Geological Surveying Institute for Petroleum).
322. USSR Pat. 658,161 (Apr. 25, 1979), A. I. Soshko, O. M. Dzikovskii, V. I. Shemechko, O. L. Zaionts, and V. I. Naboka (to Lvov Technical Institute).
323. Ger. Offen. 3,237,630 (July 7, 1983), P. A. Donatelli and J. W. Keil (to Dow Corning Corp.).
324. U.S. Pat. 4,175,054 (Nov. 20, 1979), W. S. Tait and R. W. Greenlee (to Petrolite Corp.).

325. Ger. Offen. 2,926,474 (Jan. 1, 1981), E. Barthell, A. Capelle, M. Chmelir, and K. Dahmen (to Chemische Fabrik Stockhausen und Cie and Servo BV).

326. N. N. Altukhova, V. G. Novikov, G. N. Yartseva, and A. D. Komarova, *Neftepromysl. Delo,* 30 (1979).

327. V. P. Vostrikova, P. V. Mikhal'kov, G. F. Tinyakov, and S. P. Didenko, *Neftepromysl. Delo,* (11), 12 (1978).

328. USSR Pat. 785,337 (Dec. 7, 1980), N. N. Altukhova and co-workers.

329. J. P. Heller, D. K. Dandge, R. Card, and L. G. Donaruma, *Soc. Pet. Eng. J.* **25,** 679 (1985).

330. U.S. Pat. 4,192,753 (Mar. 11, 1980), D. S. Pye, J. P. Gallus, and P. W. Fischer (to Union Oil Co., Calif.).

331. Brit. Pat. 2,127,070 (Apr. 4, 1984) J. Mzik.

332. Jpn. Kokai Tokkyo Koho 59 44,489 (Mar. 12, 1984) (to Teikoku Oil Co. and Kao Corp.).

333. U.S. Pat. 4,264,486 (Apr. 28, 1981), H. C. McLaughlin (to Halliburton Co.).

334. PCT Int. Pat. 85 01,935 (Aug. 8, 1984) (to Dow Chemical Co.).

335. R. D. Stevens, *Rubber World* **190**(6), 31 (1984).

336. R. E. Bailey and L. E. Curtis, *Enhanced Oil Recovery,* National Petroleum Council, Washington, D.C. (1984).

337. *U.S. Oil-field Market,* Frost & Sullivan Report, Sept. 1985.

338. *Oil-field Chemicals–North America 1983,* C. H. Kline Co.

JOHN K. BORCHARDT
Shell Development Company

OILS.

See DRYING OILS; SILICONES.

OLEATES.

See DRYING OILS.

OLEFIN–CARBON MONOXIDE COPOLYMERS

Carbon monoxide can be copolymerized with various olefins by free-radical processes; the greatest interest has been in copolymers and terpolymers with ethylene. Independent discoveries of the copolymerization of ethylene and carbon monoxide were made at Farbenfabriken Bayer (1) and at DuPont (2,3). Other olefins that have been copolymerized with carbon monoxide include acrylonitrile, tetrafluoroethylene, vinyl acetate (4), and vinyl chloride (2).

The copolymers of ethylene and carbon monoxide fall into two categories depending on composition. Those containing small amounts of carbon monoxide have properties similar to low density polyethylene, except for a greatly increased sensitivity to photodegradation. Compositions close to a 1:1 copolymer have much higher melting points and improved mechanical properties (qv). Terpolymers of ethylene, vinyl acetate, and carbon monoxide are used as permanent plasticizers (qv) for poly(vinyl chloride). Other vinyl and acrylic compounds can also be used in terpolymers with ethylene and carbon monoxide.

Properties and Preparation of the Monomers

Both ethylene and carbon monoxide are gases under ambient conditions. Their most important properties are given in Table 1. Ethylene is obtained primarily from the cracking of higher hydrocarbons, and carbon monoxide may be obtained as a by-product of steelmaking in blast furnaces or from the reaction of coal with steam, ie, the water–gas process.

Table 1. Physical Properties of the Monomers

	Ethylene	Carbon monoxide
molecular weight	28.0	28.0
boiling point, °C	− 103.7	− 191.5
melting point, °C	− 169.2	− 204.0
critical pressure, MPa	5.12	3.50
critical temperature, °C	9.9	− 140.2

Copolymerization

The free-radical copolymerization may be carried out in bulk or in inert organic solvents, eg, benzene, cyclohexane, or saturated aliphatic hydrocarbons, as water is not a suitable medium for polymerization. Since the monomers have large vapor pressures, total pressures of 20–100 MPa (200–1000 bar) are usually employed. The polymerization temperature in laboratory experiments is chosen to provide an appropriate rate of decomposition of the initiator, eg, 80°C for azobisisobutyronitrile, 100°C for benzoyl peroxide, or 135°C for di-*t*-butyl peroxide. Higher temperatures are used for continuous industrial reactors.

Because sequences of carbon monoxide do not occur, the copolymer may be considered to be a combination of —CH$_2$CH$_2$CO— and —CH$_2$CH$_2$— sequences. Thus a 1:1 copolymer is a limiting composition formed at low ethylene–carbon monoxide monomer feed ratios. An azeotropic polymerization in which the copolymer has the same ratio as the mixture of monomers from which it was made occurs at 51.5% ethylene and 48.5% carbon monoxide (5). This composition depends on the polymerization temperature.

Free-radical polymerization involving cobalt-60 γ-radiation (5) reportedly produces polymers with superior mechanical properties (6).

Strictly alternating ethylene–carbon monoxide copolymers have been made by insertion polymerization (qv) using palladium-based catalysts (7,8). These catalysts include tetrakis(triphenylphosphine)palladium, HPd(CN)$_3$ (7), and {Pd[P(C$_6$H$_5$)$_3$]$_n$(CH$_3$CN)$_{4-n}$}(BF$_4$)$_2$, where n = 1–3 (8). The conditions can be as mild as 25°C and a combined pressure of 2.07 MPa (20.7 bars).

Structure of the Copolymers

The ratio of ethylene and carbon monoxide is determined by elemental analysis. Ethylene units isolated between carbon monoxide units are 1,4-dione structures which may be determined quantitatively by nmr (9). The concentration of 1,4-diones is slightly smaller than expected for a random mixture of —CH_2CH_2CO— and —CH_2CH_2— units. Therefore, there is a slight statistical preference for sequences of ethylene units.

As in polyethylene, the chains in crystals of ethylene–carbon monoxide copolymers have a planar zigzag conformation (10,11). The keto groups are accommodated in the crystal through an expansion of the a axis of the orthorhombic unit cell. Thus at low levels of carbon monoxide, the unit cell of polyethylene is modified only slightly. At compositions close to the 1:1 alternating copolymer, the length of the crystallographic repeating unit increases from 0.254 nm (corresponding to two CH_2 groups) to 0.757 nm (corresponding to six-chain atoms). This reflects the fact that the keto groups point alternately in opposite directions.

Properties of the Copolymers

The dependence of the melting point on the percent carbon monoxide is shown in Table 2. Up to about 40% carbon monoxide, the melting point is similar to or slightly lower than that of a polyethylene made by a similar free-radical process (5). Between 40 and 50% carbon monoxide, the melting point rises sharply. The melting point of 1:1 copolymers has been reported to be as high as 244°C for products of a free-radical process and 260°C for a more perfectly alternating copolymer made with a palladium-based catalyst. In using melting-point data in this range, it is assumed that ethylene units beyond those for the 1:1 alternating composition are not included in the crystals characterized by the six-chain atom repeat (12). The heat and entropy of fusion from such calculations are compared with those for polyethylene in Table 3. The higher melting point of the 1:1 alternating ethylene–carbon monoxide copolymer is due to a smaller entropy of fusion, not a larger heat of fusion. This is attributed to more restricted rotational isomerism in the melt and possibly to the onset of motion and disorder below the melting point.

Table 2. Melting Point vs % Carbon Monoxide

% carbon monoxide	Melting point, °C
27.5	111
35.1	122
38.7	126
40.6	147
43.9	181
46.3	203
49.0	235
49.8	242

One of the most striking consequences of incorporating keto groups in polyethylene is its increased sensitivity to photodegradation by ultraviolet light. At ambient temperature, this is attributed mainly to the Norrish-II reaction, which

Table 3. Melting Parameters

	ΔH_f, kJ/mol chain atoms	ΔS_f, J/(mol·K) chain atoms
polyethylene	3.9	9.6
1:1 ethylene–carbon monoxide copolymer	2.5	4.8

involves the transfer of a proton from a γ- to an α-carbon to produce fragments terminating in a methyl ketone and a vinyl group, respectively (13) (see DEGRADATION; PHOTOCHEMISTRY).

$$-\overset{\overset{\text{O}}{\|}}{\text{C}}-\text{CH}_2-\text{CH}_2-\text{CH}_2- \xrightarrow{h\nu} -\overset{\overset{\text{O}}{\|}}{\text{C}}-\text{CH}_3 + \text{CH}_2{=}\text{CH}-$$

Copolymers having isolated keto groups are rapidly degraded in sunlight to give fragments corresponding to the chain length between keto groups. In a perfectly alternating 1:1 copolymer, there are no γ-carbon atoms and the photostability is much better.

The amount of incorporated carbon monoxide affects mechanical properties in a way similar to that for the melting point (12). There is little change up to ca 40% carbon monoxide.

Beyond that point, the strength and modulus increase, and the elongation at break decreases. Dynamic mechanical measurements on a copolymer containing 48% carbon monoxide show loss peaks at $-100°C$ and $25°C$, and possibly another near the melting point (12).

Uses

Packaging materials discarded across the countryside may produce a legislative mandate for bio- or photodegradability (14). Copolymers of ethylene with a few percent carbon monoxide can meet this demand. Whereas they remain stable when stored indoors, they become embrittled by sunlight within two weeks and break down into low molecular weight fragments.

Many plasticizers for poly(vinyl chloride) are unsatisfactory because they exude or can be extracted, but terpolymers of ethylene, vinyl acetate, and carbon monoxide are permanent plasticizers for poly(vinyl chloride) (15). They reduce PVC glass temperature and stiffness, and increase its toughness. For film, a preferred composition consists of 30–65% poly(vinyl chloride) and 35–70% of a copolymer containing 40–80% ethylene, 10–60% vinyl acetate, and 3–30% carbon monoxide (16). Other monomers, such as alkyl acrylates, may be substituted for vinyl acetate. These terpolymers can also be used in fast-curing foamable compositions (17).

BIBLIOGRAPHY

"Olefin–Carbon Monoxide Copolymers" in *EPST* 1st ed., Vol. 9, pp. 397–402, by Gustav Pieper, Farbenfabriken AG.

1. Ger. Pat. 863,711 (Aug. 27, 1941), F. Ballauf, O. Bayer, and L. Leichmann (to Farbenfabriken Bayer AG).
2. U.S. Pat. 2,495,286 (Jan. 24, 1950), M. M. Brubaker (to E. I. Du Pont de Nemours & Co., Inc.).
3. M. M. Brubaker, D. D. Coffman, and H. H. Hoehn, *J. Am. Chem. Soc.* **74**, 1509 (1952).
4. Brit. Pat. 583,172 (Dec. 11, 1946) (to Imperial Chemical Industries).
5. P. Colombo, L. E. Kukacka, J. Fontana, R. N. Chapman, and M. Steinberg, *J. Polym. Sci. Part A-1* **4**, 29 (1966).
6. M. Steinberg, R. Johnson, W. Cordel, and D. Goodman, *Radiat. Phys. Chem.* **14**, 613 (1979).
7. U.S. Pats. 3,689,460 (Sept. 5, 1972); 3,694,412 (Sept. 26, 1972); and 3,835,123 (Sept. 10, 1974), K. Nozaki (to Shell Oil Co.).
8. T. W. Lai and A. Sen, *Organometallics* **3**, 866 (1984).
9. T. K. Wu, D. W. Ovenall, and H. H. Hoehn in E. Brame, ed., *Applications of Polymer Spectroscopy*, Academic Press, Inc., Orlando, Fla., 1978, p. 19.
10. Y. Chatani, T. Takizawa, S. Murahashi, Y. Sakata, and Y. Nishimura, *J. Polym. Sci.* **55**, 811 (1961).
11. Y. Chatani, T. Takizawa, and S. Murahashi, *J. Polym. Sci.* **62**, S27 (1962).
12. H. W. Starkweather, *J. Polym. Sci. Polym. Phys. Ed.* **15**, 247 (1977).
13. G. H. Hartley and J. E. Guillet, *Macromolecules* **1**, 165 (1968).
14. Y. P. Bremer, *Polym. Plast. Technol. Eng.* **18**, 137 (1982).
15. C. F. Hammer in D. R. Paul and S. Newman, eds., *Polymer Blends*, Vol. 2, Academic Press, Inc., Orlando, Fla., 1978, p. 219.
16. U.S. Pat. 3,780,140 (Dec. 18, 1973), C. F. Hammer (to E. I. du Pont de Nemours & Co., Inc.).
17. U.S. Pat. 4,394,459 (July 19, 1983), J. Rys-Sikors (to E. I. du Pont de Nemours & Co., Inc.).

HOWARD STARKWEATHER
E. I. du Pont de Nemours & Co., Inc.

OLEFIN FIBERS

Polyolefin fibers, also called olefin fibers, are defined as "manufactured fibers in which the fiber-forming substance is any long-chain synthetic polymer of at least 85% by weight of ethylene, propylene, or other olefin units" (1). A number of crystalline polyolefins are capable of forming fibers (Table 1), but only polypropylene (PP) and, to a much lesser extent, polyethylene (PE) are of practical importance. Crystalline olefin polymers are hydrophobic and resistant to most solvents. These properties impart good environmental qualities and resistance to staining, but impede dyeing. Crystalline olefin polymers, having melting points lower than PP, have limited use in textile applications. With the exception of poly(4-methyl-1-pentene), the high cost of the higher melting polyolefins limits their applications. Furthermore, the low T_g of polyolefins may result in compressive creep. In spite of these drawbacks, polyolefin fibers are one of the fastest growing segments of the synthetic fiber market (Fig. 1). The advantages of PP fibers are low cost and light weight. Improvements in stabilization, coloration, and texturing of this versatile fiber led to substantial volume growth in the 1970s, which continues at an even higher rate owing to recent advances in fiber-forming and textile technology.

The first commercial application of polyolefin fibers was for automobile seat covers in the late 1940s. These fibers, made from low density polyethylene (LDPE) by a melt-extrusion process, lacked dimensional stability, abrasion resistance,

Table 1. Properties of Crystalline Polyolefins and Other Fiber-forming Polymers[a]

Polymer[b]	T_g, °C	T_m, °C	Softening temp, °C	Thermal-degradation temp, °C
high density polyethylene (HDPE)	−120	130	125	
i-polypropylene	−20	170	165	290
i-poly(1-butene)	−25	128		
i-poly(3-methyl-1-butene)		315		
i-poly(4-methyl-1-pentene)	18	250	244	
poly(ethylene terephthalate) (PET)	70	265	235	400
nylon-6,6	50	264	248	360

[a] Ref. 2.
[b] i = isotactic.

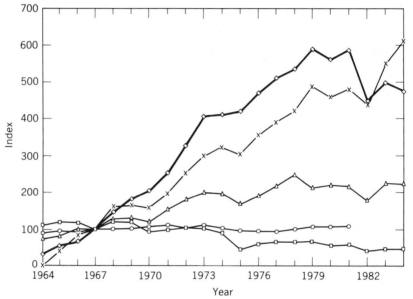

Fig. 1. Growth rates of the synthetic fiber market (1967 basis; 1967 = 100) (3): □, rayon; ○, acetate; ◇, polyester; △, nylon; x, olefin. Courtesy of *Textile Organon*.

resilience, and light stability. The olefin-fiber success story began with the introduction of high density polyethylene (HDPE) in the late 1950s. This highly crystalline, linear polyethylene gives yarns with higher tenacity than the less crystalline, branched form. Markets developed for HDPE fiber in marine rope, where water resistance and bouyancy are important. However, the low melting point, lack of resilience, and light instability limited applications.

Isotactic PP, based on the stereospecific polymerization catalysts discovered by Ziegler and Natta, was introduced commercially in the United States in 1957. The first commercial PP fibers were introduced in 1961. The first market of significance, contract carpet, was based on a three-ply, crimper-textured yarn. It competed with wool and rayon–wool blends because of its lighter weight, longer wear, and lower cost. In the mid-1960s, the discovery of improved light stabilizers led to the development of outdoor carpeting based on PP. In 1967, primary–

secondary woven carpet backing based on a film warp and fine-filament fill was introduced. Today, this application accounts for a U.S. consumption of 181×10^3 metric tons of polyolefin fiber per year. Another development in the late 1960s was spunbonded polyolefin fabrics. Spunbonded PP and PE today account for a substantial portion of the olefin-fiber market. In the early 1970s, a Taslanized-bulked continuous-filament yarn was introduced for woven, texturized upholstery. In the mid 1970s, further improvement in light stabilization of PP led to a staple product for automotive interiors and nonwoven velours for floor and wall carpet tiles. The most recent growth has been in fine-filament staple for thermally bonded nonwovens and the continued development of spunbonded, spunlaced, and meltblown nonwovens. The remarkable growth of polyolefin fibers over the last two decades (Table 2) has been sparked by advances in stabilization, production, and postprocessing technologies.

Table 2. Olefin Fiber Production[a], 10^3 t

Year	Olefin yarn	Olefin staple	Total olefin	Total noncellulosic	Olefin % of total
1965	66	15	81	998	8.1
1968	207	57	264	3229	8.2
1971	262	60	322	4293	7.5
1974	463	68	531	6212	8.5
1977	554	81	635	7312	8.7
1980	632	116	748	7874	9.5
1982	585	138	723	6442	11.2
1984	748	250	998	7473	13.4

[a] Refs. 3 and 4. Courtesy of *Textile Organon*.

Manufacture

For many years, the primary technology employed for the manufacture of PP filaments was melt spinning, similar to the methods employed for polyester and polyamide fibers. A number of texturizing operations were developed to impart characteristics of interest to carpet and upholstery manufacture. More recently, systems have been developed that are tailor-made for PP, eg, the short-spin system for low volume staple production (5).

The processes used to make fibers for nonwovens are classified as melt spinning, but there are substantial differences in these processes. Staple for thermally bonded nonwovens is produced with a much finer filament and at higher speeds than traditional carpet yarns. Spunbonded fabrics are produced in a continuous operation in which the spun yarn is laid down directly into a web and bonded. Fabric is also prepared by a meltblown process in which high velocity air is used to draw the extruded melt into fine-filament fibers which are laid down in a continuous mat. Thus the last decade has seen significant changes in both spinning and postspinning operations which have improved the manufacturing technology and broadened the market for PP fibers (see also NONWOVEN FABRICS).

The principal method of manufacturing polyolefin fibers is melt spinning.

Gel spinning has been tried in the laboratory (6–10) to produce highly oriented, high modulus fibers of PP and PE, but commercialization of these methods is still in its infancy (11,12). Meltspun fibers are commercially produced in a broad range of linear densities, ie, 0.1–10 tex (1–100 den), to fit a variety of applications, as shown in Figure 2.

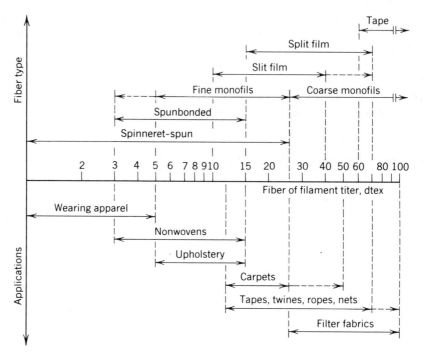

Fig. 2.   Linear density of olefin fibers in various applications (13); dtex = den.

The basic process of melt spinning is illustrated in Figure 3. The polymer resin and ingredients, primarily stabilizers, pigments, and rheological modifiers, are fed into a screw extruder, melted, and extruded through a fine-diameter spinneret. A metering pump and a mixing device are usually installed in front of the spinneret to ensure the uniform mixing and delivery rate necessary to produce uniform drawdown at high speeds. The fiber is pulled through a quench chamber by a take-up device and routed to downstream finishing operations.

Extrusion.   The extrusion (qv) of olefin fibers is largely controlled by the polymer. Polyolefin melts are strongly viscoelastic, and melt extrusion of polyolefin fibers differs from that of polyesters and polyamides. Polyolefins are manufactured in a broad range of molecular weights and ratios of weight-average to number-average molecular weights (M_w/M_n). Unlike the condensation polymers, which typically have molecular weights of 10,000–15,000 and M_w/M_n ratios of approximately 2, polyolefins have weight-average molecular weights ranging from 50,000 to 10^6 and, as polymerized, M_w/M_n ratios from 5 to 15. Further control of molecular weight and distribution is obtained by chemical or thermal degradation. The full range of molecular weights used in olefin-fiber manufacture is above 20,000, and M_w/M_n varies from 2 to 15. As molecular weight increases and

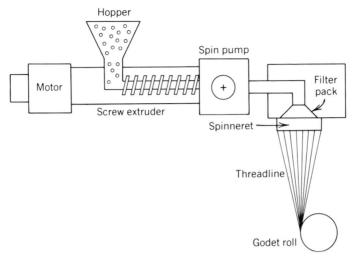

Fig. 3. Melt-spinning process.

molecular weight distribution broadens, the polymer melt becomes more pseudo-plastic (Fig. 4 and Table 3). In the sizing of extrusion equipment for olefin-fiber production, the wide range of shear viscosities and thinning effects must be considered, as these affect both power requirements and mixing efficiencies.

More important is the effect on elongational viscosity, ie, the stress–strain relationship for a uniaxial extension (see also RHEOLOGY). Fiber spinning is a uniaxial extension process, and the elongational viscosity characteristics are more important than the shear-viscosity behavior. The narrower molecular weight

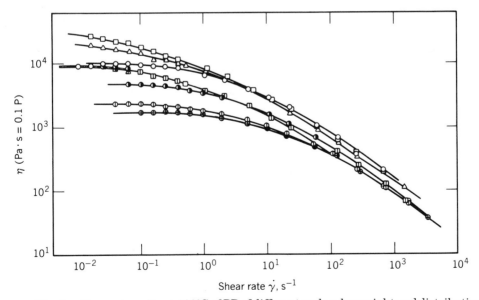

Fig. 4. Shear viscosity at 180°C of PP of different molecular weight and distribution vs shear rate (14); see Table 3 for key. Courtesy of *Polymer Engineering and Science*.

Table 3. Molecular Characterization Data for PP Samples[a,b]

Code[c]	Melt index	$M_w \times 10^{-5}$	M_w/M_n	M_z/M_w	$M_v \times 10^{-5}$
High molecular weight PP					
◯ narrow	4.2	2.84	6.4	2.59	2.40
△ regular–broad	5.0	3.03	9.0	3.57	2.42
☐ broad–regular	3.7	3.39	7.7	3.54	2.71
Middle molecular weight PP					
◑ narrow	11.6	2.32	4.7	2.81	1.92
◮ regular	12.4	2.79	7.8	4.82	2.13
⊞ broad	11.0	2.68	9.0	4.46	2.07
Low molecular weight PP					
⊕ narrow	25.0	1.79	4.6	2.47	1.52
◗ regular–narrow	23.0	2.02	6.7	3.18	1.66

[a] Ref. 14. Courtesy of *Polymer Engineering and Science.*
[b] Figs. 4–6.
[c] Narrow, regular, and broad refer to molecular weight distribution.

distributions tend to be less thinning, and elongational viscosity increases at higher extension rates (Fig. 5). This leads to higher melt orientation, which in turn is reflected in higher spun-fiber orientation, with higher tenacity and lower extensibility. In contrast, the broad molecular weight distributions tend to be more thinning and hence more prone to necking and fracture at high spinning speeds (14,15), but yield a less oriented, high elongation spun fiber. The choice of an optimum molecular weight and distribution is determined by the desired properties of the yarn and the process continuity on available equipment.

Because of the high melt viscosity of polyolefins compared with fiber-grade polyesters and nylons, normal spinning melt temperatures are 240–300°C, ie, 80–140°C above the crystalline melting point. Because of the high melt temperatures used for polyolefin-fiber spinning, thermal stabilizers such as substituted hindered phenols are added. In the presence of pigments, the melt temperature

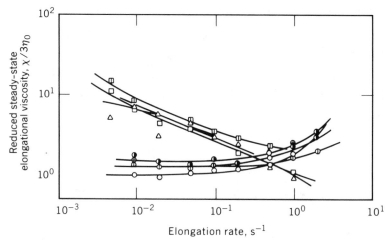

Fig. 5. Elongational viscosity at 180°C of PP of different molecular weight and distribution (14); see Table 3 for key. Courtesy of *Polymer Engineering and Science.*

must be kept below 250°C to obtain uniform color; this limits the choice of resin for colored products.

Polyolefin melts have a high degree of viscoelastic memory or elasticity. First normal stress differences of polyolefins, a rheological measure of melt elasticity, are shown in Figure 6. At a fixed molecular weight and shear rate, the first normal stress difference increases with increasing M_w/M_n ratio. Because of the high shear rate obtained in fine spinnerets, typically on the order of 10^3–10^4 s^{-1}, the filament swells (die swell) upon leaving the capillary. On a molecular scale, the residence time in the region of die swell is sufficient to allow relaxation of any shear-induced orientation. However, high die swell significantly affects the drawdown or extension rate, leading to threadline breaks. Die swell can be reduced by lower molecular weight, narrower molecular weight distribution, or higher melt temperature. Extension rate becomes particularly important as the diameter of the spun fiber is reduced.

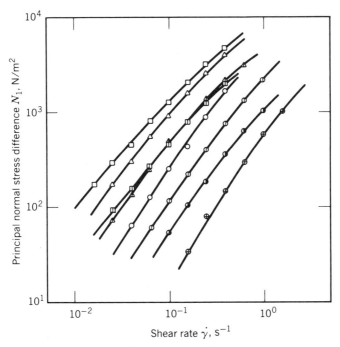

Fig. 6. First normal stress differences of PP of different molecular weight and distribution (14); see Table 3 for key. To convert N/m² to dyn/cm², multiply by 10. Courtesy of *Polymer Engineering and Science.*

Quench. PP and most other linear polyolefins crystallize more rapidly than most other crystallizable polymers. Unlike polyester, which is normally amorphous as spun, the fiber morphology is fixed in the spinning process; this limits the range of properties in subsequent drawing and annealing operations. In a low crystallinity state, sometimes called the paracrystalline or smectic form, a large degree of local order still exists. It can be reached by a rapid quench, eg, a cold-water bath. The more common commercial practice is a controlled air quench, in which the rate of cooling is controlled by the velocity and temperature of the air. During normal threadline cooling, crystallization occurs in the threadline. In-line x-ray scattering studies demonstrate that crystallization is extremely rapid, with almost the full crystalline structure being developed in fractions of a second (16).

Quench is a nonisothermal process. Attempts have been made to model the process (17–20), but the complexity of the actual system makes quench design an art. Arrangements include straight-through, outside-in, and inside-out radial patterns. The optimum configuration depends on spinneret size, hole patterns, filament size, quench-chamber dimensions, take-up rate, and desired physical properties. Process continuity and final fiber properties are governed by the temperature profile and extension rate.

Fiber spinning is an extensional process during which significant molecular orientation occurs. Under rapid crystallization, as with PP, this orientation is fixed during the spinning process. Small-angle neutron-scattering studies of quiescent PP crystallization show that the chain dimensions in both melt and crystallized forms are comparable (21). Although there may be significant relaxation of the amorphous region after spinning, the primary structure of the fiber is fixed during spinning and controls subsequent drawing and texturizing of the fiber. For fixed extrusion and take-up rates, a more rapid quench reduces the average melt-deformation temperature, increases relaxation times, and gives a more entangled melt when crystallization begins. The rapidly quenched fiber usually gives lower elongation and higher tenacity during subsequent draw (22). With a very rapid quench, the melt may not be able to relax fast enough to sustain drawdown, resulting in melt fracture. With a slow quench, the melt may totally relax, leading to ductile failure of the threadline.

Because quench is a dynamic process giving rise to nonisothermal spinning, it is difficult to model or quantify. A common measurement useful in predicting threadline behavior is fiber tension, frequently misnamed spinline stress. It is normally measured after the crystallization point in the threadline, ie, when the steady state is reached and the threadline is no longer deformed. Fiber tension increases with take-up velocity (Fig. 7) and higher molecular weight, and decreases with temperature. Crystallinity increases slightly with increasing fiber tension (Fig. 8). At low tension, the birefringence increases with increased tension, leveling off at a spinline tension of 10 MPa (1450 psi) (Fig. 9).

Take-up.   Take-up devices attenuate the spinline to the desired linear density and collect the spun yarn in a form suitable for further processing. Take-up velocities vary from 1–2 m/s for heavy monofilaments to 10–33 m/s for fine yarns. The yarn is usually wound on bobbins for subsequent processing, although it may also be stacked in cans. The most common take-up device is a godet wheel. In spunbonded processes, an aspirator is used to draw the yarn in spinning and spread it on a moving conveyor belt for subsequent bonding. In meltblown processes (Fig. 10), air jets attenuate the yarn. The product has a variety of linear densities and is taken up on moving belts.

Draw.   Polyolefin fibers are usually drawn to increase orientation and modify the physical properties of the fiber further. Linear density, necessary to control the textile properties, is more easily reduced during drawing than in spinning.

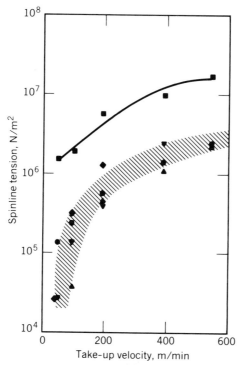

Fig. 7. Spinline tension vs take-up velocity for several PP fibers spun at 230°C (16): ■, ●, ▲, Hercules Profax PP resins of melt index 0.42, 6.6, and 12.0, respectively; ▼, ♦, Tennessee Eastman Tenite PP resins of melt index 9.0 and 2.55, respectively. To convert N/m^2 to dyn/cm^2, multiply by 10.

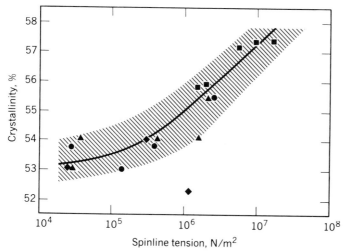

Fig. 8. Spun-yarn crystallinity vs spinline tension (16): ■, ▲, Hercules Profax PP resins of melt index 0.42 and 12.0, respectively; ♦, ●, Tennessee Eastman Tenite PP resins of melt index 9.0 and 2.55, respectively. To convert N/m^2 to dyn/cm^2, multiply by 10.

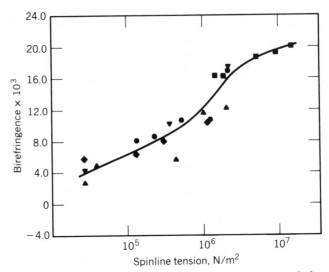

Fig. 9. Spun-yarn birefringence vs spinline tension (16); symbols are as explained in Figure 7. To convert N/m² to dyn/cm², multiply by 10.

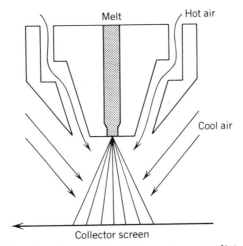

Fig. 10. A meltblown fiber spinneret (die).

The draw step can be accomplished in-line with spinning in a continuous spin–draw–texturing process (23), but is more commonly done in a second processing step. This allows simultaneous mixing of colors in a multi-ply yarn for textiles. For staple-fiber production, large tows consisting of bundles of filaments are stretched, texturized (crimped), and cut.

In secondary drawing operations, the aging properties of the spun yarn must be considered. Because PP fibers have a low T_g, the spun yarn is restructured between spinning and drawing; this is more important as the smectic content is increased (24). The aging process depends on whether the yarn is stored on bobbins, under tension, or coiled in cans with no tension on the fiber. The aging of quick-quenched (smectic) PP films is shown in Figure 11. Stored at room

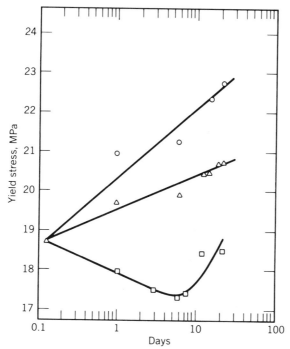

Fig. 11. Increase in yield stress of PP films upon aging at different temperatures (24): □, in ice water; △, at 25°C; ○, at 55°C. To convert MPa to psi, multiply by 145. Courtesy of *Polymer Engineering and Science.*

temperature, the increase in yield stress is 5% in 24 h. Similar data on PP spun fibers have not been published, but aging effects are similar. In addition, there are further changes in physical properties during aging of drawn fibers (25).

The crystalline structure of the spun yarn affects the draw process (26). Monoclinic yarns tend to exhibit increased tenacity and lower elongation with lower draw ratio than smectic yarns (27). They exhibit lower maximum draw ratios, undergo brittle fracture, and form microvoids at significantly lower draw temperatures, which creates a chalky appearance. Studies of the effect of spun-yarn structure on drawing behavior (26,28) show that the as-spun orientation and morphology determine fiber properties at a given draw ratio, as shown in Figure 12. However, final fiber properties can be correlated with birefringence, a measure of the average orientation (Fig. 13). Some studies show good correlation between fiber properties and amorphous orientation (Fig. 14) (29,30), but in most studies the range of spun-yarn properties is limited. Such studies suggest that the deformation during draw primarily affects the interlamellar amorphous region at low draw ratio. At higher draw ratio, the crystalline structure is substantially disrupted.

Texturing. The final step in olefin-fiber production is texturing; the method depends primarily on the application (see also TEXTURED YARN; TEXTILE PROCESSING). For carpet and upholstery, the fiber is usually bulked, a procedure in which heated fiber is deposited on a moving screen through a hot-air or steam jet. The fiber takes on a three-dimensional crimp which aids both in yarn spinning

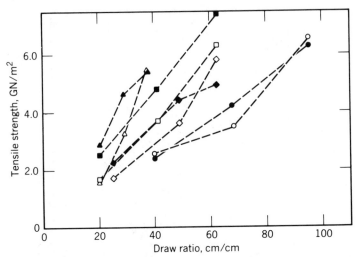

Fig. 12. Tensile strength vs draw ratio (26): ■, ▲, 0.42 melt index spun at 50 and 500 m/min, respectively; ●, ◆, 12.0 melt index spun at 200 and 500 m/min, respectively. Open symbols, cold-drawn and annealed at 140°C; filled symbols drawn at 140°C. To convert GN/m^2 to dyn/cm^2, multiply by 10^{10}.

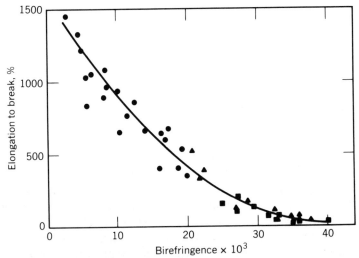

Fig. 13. Elongation to break as a function of birefringence for undrawn, hot-drawn, and cold-drawn annealed fibers (26); ●, undrawn; ▲, cold-drawn annealed at 140°C; ■, hot-drawn at 140°C.

and in developing bulk and coverage in the final fabric. "Stuffer-box" crimping imparts a two-dimensional saw-tooth crimp which is commonly found in olefin staple used in carded nonwovens and upholstery yarns.

Slit-film Fiber.   A substantial volume of olefin fiber is produced by slit-film or film-to-fiber technology (13). For producing filaments with high linear density, ie, 0.6 tex or 6 den, the production economics are more favorable than monofilament spinning (13). The fibers are used primarily for carpet backing or rope

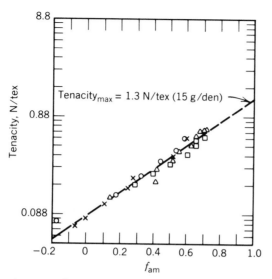

Fig. 14. Tenacity as a function of amorphous orientation for PP fibers and film (29). Film drawn at: ○, 135°C; x, 110°C; □, 90°C. △, Heat-set fiber. To convert N/tex to gf/den, multiply by 11.3.

and cordage applications. Although the processes used to make slit-film fibers are versatile and economical, their application is limited to expensive textiles.

The equipment for the slit-film fiber process is shown in Figure 15. An olefin film is usually produced and, as in melt spinning, the morphology and composition of the film determine the processing characteristics. Fibers may be produced by cutting or slitting a film, or by mechanical or chemomechanical fibrillation. In the first case, the fiber is oriented by stretching before or after the cutting process. In the latter case, film anisotropy is created by stretching before fibrillation. The

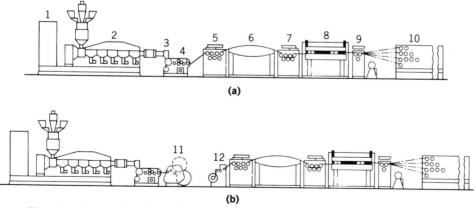

Fig. 15. Production lines for stretched film tape (13): (**a**) continuous production line for film tape; (**b**) discontinuous production lines for film and film tape. 1, Control cabinet; 2, extruder; 3, flat die; 4, chill roll; 5, septet (seven rolls); 6, hot plate; 7, septet (seven rolls); 8, heat-setting oven; 9, trio (three rolls); 10, bobbin winder; 11, film winder; 12, film-unrolling stand.

film is fibrillated mechanically by rubbing or brushing. Immiscible polymers, eg, PE or PS, may be added to PP to promote fibrillation. Many common fiber-texturing techniques such as stuffer-box, false-twist, or knife-edge treatments improve the textile characteristics of slit-film fibers.

Several more recent variations of the film-to-fiber approach result in direct conversion of film to fabric. The film may be embossed in a controlled pattern and subsequently drawn uniaxially or biaxially to produce a variety of nonwoven products (31). Addition of chemical blowing agents to the film causes fibrillation upon extrusion. Nonwovens can be formed directly from blown film with a unique radial die and control of the biaxial draw ratio (32).

Bicomponent Fibers.   PP fibers have made substantial inroads into nonwoven markets because they are easily thermally bonded. Further enhancement in thermal bonding is obtained with bicomponent fibers (33). In these fibers, two incompatible polyolefins, often PP and PE, are spun together to give a fiber with a side-by-side arrangement of the two materials. The lower melting polymer can melt and form adhesive bonds to other fibers; the higher melting component causes the fiber to retain some of its textile characteristics.

"Spurted" and Meltblown Fibers.   Unusual olefin fibers are produced by new spinning techniques. A variety of directly formed nonwovens with excellent filtration characteristics are made by meltblown processes originally developed by Exxon Corp. (34), producing very fine, ie, submicron, filaments. A simple schematic of the die is shown in Figure 10. Molten polymer is extruded from a jet with a stream of hot air which attenuates the fiber. It is quenched by cold-air jets after leaving the spinneret hole. Because the fiber cannot be wound for subsequent processing, a nonwoven web is formed directly. Mechanical integrity of the web is usually obtained by thermal bonding or needling, although other methods, such as latex bonding, can be used. Spunbonded fabrics are made commercially from PP and PE.

Pulplike olefin fibers are produced by a high pressure spurting process developed by Hercules, Inc. and Solvay, Inc. PP or PE is dissolved in volatile solvents at high temperature and pressure. After the solution is released, it expands while the solvent is volatilized, leaving a highly fluffed, pulplike product. Additives are included to modify the surface characteristics of the pulp. Uses include felted fabrics, substitution in whole or in part for wood pulp in papermaking, and replacement of asbestos in reinforcing applications (35).

High Strength Fibers.   The properties of commercial olefin fibers are far inferior to those theoretically attainable. Theoretical and actual strengths of common commercial fibers are listed in Table 4. A number of methods, including superdrawing (37), high pressure extrusion (38), spinning of liquid crystalline polymers or solutions (39), gel spinning (6–10), and hot drawing (40), produce higher strength than those given in Table 4 for commercial fibers, but these methods are tedious and uneconomical for olefin fibers. A new, high modulus commercial PE fiber with properties approaching those of aramid and graphite fibers (Table 5) is prepared by gel spinning (11). Although most of the above techniques produce substantial increases in modulus, higher tensile strengths are currently available only from gel spinning or dilute fibrillar crystal growth. Even with these techniques, the maximum strengths observed to date are only a fraction of the theoretical strengths (see HIGH MODULUS POLYMERS; ULTIMATE PROPERTIES).

Table 4. Theoretical and Actual Strengths of Commercial Fibers[a]

Polymer	Density, kg/m³	Molecular area, nm²	Theoretical strength, GPa[b]	Strength of commercial fiber, GPa[b]
polyethylene	960	0.193	31.6	0.76
nylon-6	1140	0.192	31.9	0.96
polyoxymethylene	1410	0.185	32.9	
poly(vinyl alcohol)	1280	0.228	26.7	1.08
Kevlar	1430	0.205	29.7	3.16
poly(ethylene terephthalate)	1370	0.217	28.1	1.15
polypropylene	910	0.348	17.6	0.72
poly(vinyl chloride)	1390	0.294	20.8	0.49
rayon	1500	0.346	17.7	0.69
poly(methyl methacrylate)	1190	0.667	9.2	

[a] Ref. 36. Courtesy of *Polymer Engineering and Science*.
[b] To convert GPa to psi, multiply by 145,000.

Table 5. Properties of Commercial High Strength Fibers[a]

Property	PE	Aramid	S-Glass	Graphite	Steel[b]
density, kg/m³	970	1440	2490	1730	7860
strength, GPa[c]	2.6	2.8	4.6	3.1	2.3
modulus, GPa[c]	117	113	89	227	207
elongation to break, %	3.5	2.8	5.4	1.2	1.3
filament dia, mm	0.038	0.012	0.009	0.006	0.25

[a] Ref. 12. Courtesy of TAPPI.
[b] Whiskers.
[c] To convert GPa to psi, multiply by 145,000.

Hard-elastic Fibers. Hard-elastic fibers are prepared by annealing a moderately oriented spun yarn at high temperature under tension. They are prepared from a variety of olefin polymers, acetal copolymers, and polypivalolactone (41,42). Whereas the strengths observed are comparable to those of highly drawn commercial fiber, ie, 0.52–0.61 N/tex (6–7 g/den), the recovery from elongation is substantially better. Hard-elastic fibers typically exhibit 90% recovery from 50% elongation, whereas highly drawn, high tenacity commercial fibers exhibit only 50–75% recovery from a 5% elongation. The mechanism of elastic recovery differs from the entropic models normally used to explain elastic properties. The hard-elastic fibers are thought to deform through bending of the lamellae in a stacked structure, resulting in void formations; recovery is controlled by energy considerations. Although there are potential uses in applications involving substantial deformation, eg, stretch fabrics, hard-elastic fibers are not yet used commercially (see HARD-ELASTIC BEHAVIOR in ELASTICITY).

Properties

Physical. Physical properties of commercial olefin fibers and those of other synthetic fibers are given in Table 6. Of particular interest are the lower density, allowing much lighter-weight fiber at a specified size or coverage, and the low

Table 6. Physical Properties of Commercial Fibers[a]

Polymer	Standard tenacity, GPa[b]	Breaking elongation, %	Modulus, GPa[b]	Density, kg/m³	Moisture regain at 21°C and 65% rh, %
polyester	0.37–0.73	13–40	2.1–3.7	1380	0.4
carbon	3.1	1	227	1730	
nylon	0.23–0.60	25–65	0.5–2.4	1130	4–5
rayon	0.25–0.42	8–30	0.8–5.3	1500	11–13
acetate	0.14–0.16	25–45	0.41–0.64	1320	6
acrylic	0.22–0.27	35–55	0.51–1.02	1160	1.5
olefin	0.16–0.44	20–200	0.24–3.22	910	0.01
glass	4.6	5.3–5.7	89	2490	
aramid	2.8	2.5–4.0	113	1440	4.5–7
fluorocarbon	0.18–0.74	5–140	0.18–1.48	2100	
polybenzimidazole	0.33–0.38	25–30	1.14–1.52	1430	15

[a] Ref. 43. Courtesy of *Textile World*.
[b] To convert GPa to psi, multiply by 145,000.

moisture absorption. The moderate melting temperature of PP is high enough for most applications, but low enough to permit thermal bonding more easily than in most other fibers.

Tensile. As with all polymers, tensile properties are a function of molecular weight, morphology, and testing conditions. The effects of temperature and strain on tensile properties of a typical fiber are shown in Figures 16 and 17, respectively (44). Increasing strain rate and lower temperature result in higher breaking stresses at lower elongation, consistent with the general viscoelastic behavior of polymeric materials. Under the same testing conditions, higher molecular weight at constant draw ratios results in higher tensile strength. The effect of molecular

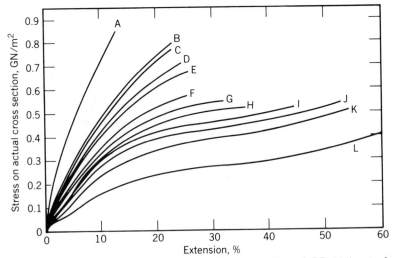

Fig. 16. Effect of temperature on tensile properties of PP (44); strain rate = 6.47×10^{-4} s⁻¹. In Kelvin: A, 90; B, 200; C, 213; D, 227; E, 243; F, 257; G, 266; H, 273; I, 278; J, 283; K, 293; L, 308 (broken at 74.8% extension, 4.84×10^3 N/m² stress). To convert GN/m² to dyn/cm², multiply by 10¹⁰.

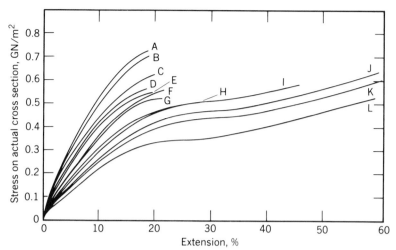

Fig. 17. Effect of strain rate on tensile properties of PP at 20°C (44): In s^{-1}: A, 4.9×10^2; B, 2.48×10^2; C, 1.26×10^2; D, 6.3×10^1; E, 3.2×10^1; F, 2.87×10^1; G, 0.230; H, 3.3×10^{-2}; I, 1.33×10^{-2}; J, 4.17×10^{-3}; K, 1.67×10^{-3}; L, 3.3×10^{-4}. To convert GN/m^2 to dyn/cm^2, multiply by 10^{10}.

weight distribution on tensile properties is complex because of the interaction with spinning conditions, but in general narrower molecular weight distributions result in higher breaking tenacity and lower elongation (27). The variation of tenacity and elongation with draw ratio for a given spun yarn correlates well with amorphous orientation (29,30). When different spun yarns are compared, neither average nor amorphous orientation completely explains these variations (26,45,46). Recent theory suggests that the number of tie molecules, both from molecules traversing the interlamellar region and especially those due to entanglements in the interlamellar region, defines the range of tensile properties achievable with draw-induced orientation (47,48). Increased entanglements (more ties) result in higher tenacity and lower elongation.

Creep, Stress Relaxation, Elastic Recovery. Olefin fibers exhibit creep and undergo stress relaxation. Because of the variety of molecular sizes and morphological states present in semicrystalline polymers, the creep and stress relaxation properties for materials such as PP cannot be represented in one curve by using time–temperature superposition principles (49). However, given the structural state of a spun yarn, curves for creep fracture (time to break under variable load) can be developed for different draw ratios (Fig. 18) (50), again indicating the importance of spun-yarn structure in a crystallizable polyolefin fiber. The same superposition can be carried out up to 110°C, where substantial reordering of polymer crystalline structure occurs (51).

High molecular weight and high orientation reduce creep. At a fixed molecular weight, the stress-relaxation modulus is higher for a highly crystalline sample prepared by slow cooling than for a smectic sample prepared by rapid quench (49). Annealing the smectic sample slightly raises the relaxation modulus, but not to the degree present in the fiber prepared by slower cooling.

Polyolefin fibers have poorer resilience than nylon; this is thought to be

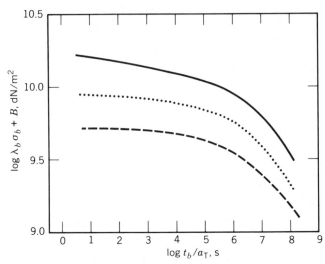

Fig. 18. Composite curve of true stress at break $\lambda_b\sigma_b$ at 40°C vs reduced time to break t_b/a_T for PP fibers of three draw ratios (50): ——, 2.7× draw, $B = 0$; ·······, 3.5× draw, $B = 0.2$; ------, 4.5× draw, $B = 0.4$. Values of B are arbitrary.

partially related to the creep properties of the polyolefins. Recovery from small-strain cyclic loading is a function of the temperature and found to be a minimum for PP at −10°C, ie, near the T_g (52). The minimum for PE is at 30°C, higher than the amorphous T_g, but probably close to the T_g of the amorphous phase in a highly oriented and crystalline PE fiber.

Chemical.   The hydrocarbon nature of olefin fibers, lacking any polarity, imparts high hydrophobicity and consequently resistance to soiling or staining by polar materials, a property important in carpet and upholstery applications. Unlike the condensation polymer fibers, eg, polyester and nylon, olefin fibers are resistant to acids and bases. At RT, PP is resistant to most organic solvents, except for some swelling in chlorinated hydrocarbon solvents. At higher temperatures, polyolefins dissolve in aromatic or chlorinated aromatic solvents, and show some solubility in high boiling hydrocarbon solvents. At high temperatures, polyolefins degrade by strong oxidizing acids.

Thermal and Oxidative Stability.   The thermal and oxidative stability of polyolefins are poor (53). They are highly sensitive to oxygen, which must be carefully controlled in all processing, particularly fiber processing. PE tends to cross-link in the presence of peroxides, whereas PP undergoes chain scission. The tertiary hydrogen in PP imparts sensitivity to oxidative degradation. Polyolefins are stabilized by hindered phenols or substituted amines (54,55). Stabilizer exudation and coloration in the final fiber create problems, more so with the amines. Preferred stabilizers are highly substituted phenols such as Cyanox 1790, Ultranox 626, and Goodrite 3114. Phosphorus compounds confer some light stability and reduce the gas yellowing resulting from hindered phenol stabilizers in the AATCC colorfastness test (56). Substituted phenols usually give satisfactory results without phosphorus synergists.

Uv Degradation.   Polyolefins are subject to light-induced degradation (57); PE is more resistant than PP. In any fiber application, stabilization against light is necessary. The stabilizer must be compatible, have low volatility, and be resistant to light and thermal degradation, ie, it must last over the lifetime of the fiber. Chemical and physical interactions with other additives must be avoided. Minimal odor and toxicity, colorlessness, resistance to gas yellowing, and low cost are additional requirements.

Stabilizers that act as uv screens or energy quenchers are usually ineffective. Because polyolefins readily form hydroperoxides, the more effective light stabilizers decompose peroxides and radical scavengers (58). Hindered amines are favored, especially polymeric amines with lower mobility and less tendency to "bloom." High molecular weight and polymeric amines limit stabilizer migration (59,60). Test results for some typical stabilizers are given in Table 7.

Table 7. Stabilization of PP Fiber by Polymeric Hindered Amine Light Stabilizers (HALS)[a,b]

HALS	Manufacturer	Carbon arc T50, h[c]	Florida T50, kJ/m^2 [d,e]
none		70	<105
Chimassorb 944	CIBA-GEIGY Corp.	300–500	293–418
Spinuvex A-36	Montedison Corp.	370–400	293
Cyasorb 3346	American Cyanamid Co.	320	418

[a] Ref. 61.
[b] Test specimens were 0.5-tex (4.5-den) filaments containing 0.25% specified HALS.
[c] Hours to 50% retention of initial tensile strength under carbon-arc exposure.
[d] kJ/m^2 to 50% retention of tensile strength; Florida under glass exposure.
[e] To convert kJ/m^2 to Langley, multiply by 2.39×10^2.

Flammability.   Flammability of polymeric materials is measured by many methods, most commonly by the oxygen-index test (ASTM D2863) or the UL 94 vertical-burn test. Most polyolefins can be made fire retardant with a stabilizer, usually a bromine-containing organic compound, and a synergist such as antimony oxide (62). However, the required loadings are usually too high for fibers to be spun. Fire-retardant PP fibers exhibit reduced light and thermal resistance. There are no commercial fire-retardant polyolefin fibers. Where applications require fire retardancy, it is usually conferred by finishes or incorporation of fire retardants in a latex, eg, in latex-bonded nonwovens.

Dyeing.   Because of their hydrophobic nature, olefin fibers are difficult to dye. Oil-soluble dispersed dyes diffuse into PP but readily "bloom" and rub off. In the first commercial dyeing of olefin fibers, nickel dyes such as UV-1084, also a light stabilizer, were used. The dyed fibers were colorfast but dull and hazy. Acid-dyeable olefin fibers have been developed by blending copolymers such as vinyltoluene-co-α-methylstyrene. A broad variety of polymeric dyesites have been blended with PP; nitrogen-containing copolymers are the most favored (63–65). A commercial acid-dyeable PP fiber is prepared with a basic amino–polyamide terpolymer (66). Dyeable blends are expensive and create problems in spinning fibers with fine filaments for apparel applications where dyeing is important. Hence, olefin fibers are usually colored by pigment blending during manufacture.

Economic Aspects

Olefin fibers are used for home furnishings and industrial applications (Table 8). The former include carpets and upholstery, and the latter rope, geotextiles (see FIBERS, ENGINEERING), and disposable and nondisposable nonwovens. Traditionally, PP fiber has had limited use in apparel because of the lower melting temperature. However, this market is increasing as manufacturers learn to take advantage of the wicking properties (moisture transport), as, for example, in lightweight sportswear (69).

Table 8. Distribution of PP Consumption[a]

Use	1967		1975		1983	
	10^3 t	%	10^3 t	%	10^3 t	%
knits	1.3	2	2.2	1	0	0
carpets	37.2	54	108.4	49	186.8	46
wovens	10.4	15	60.3	27	56.7	14
ropes	17.6	26	40.8	19	40.8	10
other	1.8	3	8.6	4	121.5	30
Total	*68.3*		*220.3*		*405.8*	

[a] Refs. 67 and 68.

In carpets, PP is used for backing (slit-film fiber) and facing. However, this application has failed to keep pace with the overall growth of the olefin market (Fig. 19). Use of olefin fibers in upholstery has grown steadily to 25% of all fiber consumed in upholstery (43), as shown in Figure 20. New growth is expected for PP fiber in carpet and wall tiles, replacing nylon.

Olefin fiber is an important material for nonwovens (70). Although the geotextile market is still small, PP is expected to be the principal fiber in such applications. The most recent use of olefin fibers is in disposable nonwoven ap-

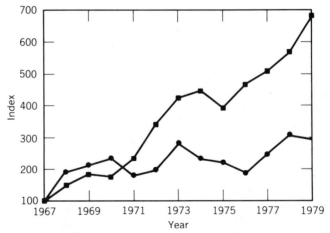

Fig. 19.   Growth of PP fiber production (■) and PP consumption in carpets (●) (3); 1967 basis (1967 = 100). Courtesy of *Textile Organon.*

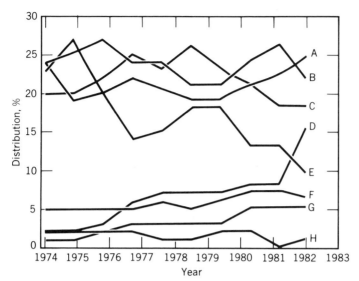

Fig. 20. Distribution of fibers in the upholstery market (43): A, olefin; B, cotton, C, nylon; D, polyester; E, rayon; F, wool; G, acrylic; H, acetate. Courtesy of *Textile World* and Phillips Fibers Corp.

plications, primarily hygienic coverstock. The two competing processes for the coverstock market are thermally bonded carded staple and spunbonded, both of which have displaced latex-bonded polyester because of improved strength, softness, and inertness. Other nonwoven applications include sanitary wipes and medical roll goods.

A special use for meltblown olefin fiber is in filtration media such as surgical masks and industrial filters (71). The high surface area associated with these ultrafine-filament fibers permits preparation of nonwoven filters with effective pore sizes as small as 0.5 μm.

Other applications, including rope, cordage, outdoor-furniture webbing, bags, and synthetic turf make up the remaining 30% of the olefin-fiber market. A rapidly growing segment is the use of spunbonded PE in packaging applications requiring high strength and low weight. Specialty olefin fibers are employed in concrete reinforcement (72,73). Hollow fibers have been tested in several filtration applications (74,75). If the economics of the high modulus olefin fibers become more favorable, substantial markets could be developed in reinforced composites such as boat hulls (12).

BIBLIOGRAPHY

"Olefin Fibers" in *EPST* 1st ed., Vol. 9, pp. 403–440, by Victor L. Erlich, Reeves Brothers, Inc.

1. The Textile Fiber Products Identification Act, Public Law 85-897, Washington, D.C., Sept. 1958.
2. D. R. Buchanan in M. Grayson, ed., *Kirk-Othmer Encyclopedia of Chemical Technology*, 3rd ed., Vol. 16, Wiley-Interscience, New York, 1981, pp. 357–385.
3. *Text. Organon* **56,** 4 (1985).

4. *Text. Organon* **46,** 3 (1975).
5. W. J. McDonald, *Fiber Producer* **11,** 38 (Aug. 1983).
6. P. Smith and P. J. Lemstra, *Makromol. Chem.* **180,** 2983 (1979).
7. P. Smith and P. J. Lemstra, *J. Mater. Sci.* **15,** 505 (1980).
8. P. Smith and P. J. Lemstra, *Polymer* **21,** 1341 (1980).
9. B. Kalb and A. J. Pennings, *Polymer* **21,** 3 (1980).
10. B. Kalb and A. J. Pennings, *J. Mater. Sci.* **15,** 2584 (1980).
11. U.S. Pat. 4,413,110 (Nov. 1, 1983), S. Kavesh and D. C. Prevorsek (to Allied Corp.).
12. R. C. Wincklhofer, "Ultra-High Strength Fibers from Polyethylene," *paper presented at Conference on Carbon and Graphite Fibers and Fabrics,* Clemson University, Clemson, S.C., Feb. 5–6, 1985, TAPPI, Atlanta, Ga., 1985.
13. H. Krassig, *J. Polym. Sci. Macromol. Rev.* **12,** 321 (1977).
14. W. Minoshima, J. L. White, and J. E. Spruiell, *Polym. Eng. Sci.* **20,** 1166 (1980).
15. O. Ishizuka, K. Murasi, K. Koyama, and K. Aoki, *Sen-i Gakkaishi* **31,** 52 (1975).
16. H. P. Nadella, H. M. Henson, J. E. Spruiell, and J. L. White, *J. Appl. Polym. Sci.* **21,** 3003 (1977).
17. S. Kase and T. Matsuo, *J. Polym. Sci. Part A* **3,** 2541 (1965).
18. S. Kase and T. Matsuo, *J. Appl. Polym. Sci.* **11,** 251 (1967).
19. C. D. Han and R. R. Lamonte, *Trans. Soc. Rheol.* **16,** 447 (1972).
20. R. R. Lamonte and C. D. Han, *J. Appl. Polym. Sci.* **16,** 3285 (1972).
21. D. G. H. Ballard and co-workers, *Polymer* **20,** 399 (1979); **23,** 1875 (1982).
22. W. C. Sheehan and T. B. Cole, *J. Appl. Polym. Sci.* **8,** 2359 (1964).
23. R. Wiedermann, *Chemiefasern Textilind.* **28/80,** 888 (1978).
24. D. M. Gezovich and P. H. Geil, *Polym. Eng. Sci.* **8,** 210 (1968).
25. C. P. Buckley and M. Habibullah, *J. Appl. Polym. Sci.* **26,** 2613 (1981).
26. H. P. Nadella, J. E. Spruiell, and J. L. White, *J. Appl. Polym. Sci.* **22,** 3121 (1978).
27. H. S. Brown, T. L. Nemzek, and C. W. Schroeder, "MWD and Its Effect on Fiber Spinning," *paper presented at the 1983 Fiber Producer Conference,* Greenville, S.C., Apr. 13, 1983, sponsored by *Fiber World,* Billiam Publishing Co., Atlanta, Ga.
28. A. Garten, D. J. Carlsson, P. Z. Sturgeon, and D. M. Wiles, *J. Polym. Sci. Polym. Phys. Ed.* **15,** 2013 (1977).
29. R. J. Samuels, *Structural Polymer Properties,* Wiley-Interscience, New York, 1974.
30. F. Geleji, L. Koczy, I. Fulop, and G. Bodor, *J. Polym. Sci. Polym. Symp.* **58,** 253 (1977).
31. U.S. Pat. 3,137,746 (June 16, 1964), D. E. Seymour and D. J. Ketteridge (to Smith and Nephews, Ltd.).
32. U.S. Pat. 4,085,175 (Apr. 19, 1978), H. W. Keuchel (to PNC Corp.).
33. *Text. Month,* 10 (Aug. 1983).
34. R. R. Buntin and D. T. Lohkamp, *Tappi* **56,** 73 (1973).
35. T. W. Rave, *Chemtech* **15,** 54 (Jan. 1985).
36. T. Ohta, *Polym. Eng. Sci.* **23,** 697 (1983).
37. M. Kamezawa, K. Yamada, and M. Takayanagi, *J. Appl. Polym. Sci.* **24,** 1227 (1979).
38. H. H. Chuah and R. S. Porter, *J. Polym. Sci. Polym. Phys. Ed.* **22,** 1353 (1984).
39. J. L. White and J. F. Fellers, *J. Appl. Polym. Sci. Appl. Polym. Symp.* **33,** 137 (1978).
40. A. F. Wills, G. Capaccio, and I. M. Ward, *J. Polym. Sci. Polym. Phys. Ed.* **18,** 493 (1980).
41. R. J. Samuels, *J. Polym. Sci. Polym. Phys. Ed.* **17,** 535 (1979).
42. S. L. Cannon, G. B. McKenna, and W. O. Statton, *J. Polym. Sci. Macromol. Rev.* **11,** 209 (1976).
43. *Text. World* **134,** 49 (Nov. 1984).
44. I. M. Hall, *J. Polym. Sci.* **54,** 505 (1961).
45. A. J. de Vries, *Pure Appl. Chem.* **53,** 1011 (1981).
46. *Ibid.,* **54,** 647 (1982).
47. D. T. Grubb, *J. Polym. Sci. Polym. Phys. Ed.* **21,** 165 (1983).
48. D. Thirion and J. F. Tassin, *J. Polym. Sci. Polym. Phys. Ed.* **21,** 2097 (1983).
49. G. Attalla, I. B. Guanella, and R. E. Cohen, *Polym. Eng. Sci.* **23,** 883 (1983).
50. A. Takaku, *J. Appl. Polym. Sci.* **26,** 3565 (1981).
51. *Ibid.,* **25,** 1861 (1980).
52. G. M. Bryant, *Text. Res. J.* **37,** 553 (1967).
53. L. Reich and S. S. Stivala, *Rev. Macromol. Chem.* **1,** 249 (1966).

54. S. Al-Malaika and G. Scott in N. S. Allen, ed., *Degradation and Stabilization of Polyolefins,* Applied Science Publishers, Ltd., London, 1983, Chapt. 6.
55. U.S. Pat. 3,639,409 (Feb. 1, 1972), K. Murayama, S. Morimura, and T. Toda (to Sankyo Co., Ltd.).
56. F. Mitterhofer, *Kunststoffe* **67,** 151 (1977).
57. D. J. Carlsson and D. M. Wiles, *J. Macromol. Sci. Rev. Macromol. Chem.* **14,** 65 (1976).
58. Ref. 54, Chapt. 7.
59. D. J. Carlsson, A. Garton, and D. M. Wiles in G. Scott, ed., *Developments in Polymer Stabilization,* Applied Science Publishers, London, 1979, p. 219.
60. F. Gugumaus in Ref. 59, p. 261.
61. L. M. Landall and A. C. Schmalz, Hercules, Inc., Oxford, Ga., 1986, private communication.
62. J. Green in M. Lewin, S. M. Atlas, and E. M. Pierce, eds., *Flame-Retardant Polymeric Materials,* Plenum Press, New York, 1982, Chapt. 1.
63. U.S. Pat. 3,873,646 (Mar. 25, 1975), H. D. Irwin (to Lubrizol Corp.).
64. U.S. Pat. 3,653,803 (Apr. 4, 1972), H. C. Frederick (to E. I. du Pont de Nemours & Co., Inc.).
65. U.S. Pat. 3,639,513 (Feb. 1, 1972), H. Masahiro and co-workers (to Mitsubishi Rayon Co., Ltd.).
66. U.S. Pat. 3,433,853 (Mar. 18, 1969), R. H. Earle, A. C. Schmalz, and C. A. Soucek (to Hercules, Inc.).
67. *Fiber Producer* **8,** 115 (Feb. 1980).
68. *Plast. World* **42,** 113 (June 1984).
69. *Am. Text.* **13**(12), 44 (Dec. 1984).
70. R. G. Mansfield, *Nonwovens Ind.* **16**(2), 28 (Feb. 1985).
71. W. Shoemaker, *Nonwovens Ind.* **15**(10), 52 (Oct. 1984).
72. D. J. Hannant, *Fiber Cements and Fiber Concretes,* John Wiley & Sons, Inc., New York, 1978.
73. *Nonwovens Rept.* **87,** 1 (July 1978).
74. A. G. Bondarenko and co-workers, *Fiber Chem.* **14,** 246 (May–June 1982).
75. *Daily News Record* **11**(84), 12 (May 4, 1981).

General References

M. Ahmed, *Polypropylene Fibers: Science and Technology,* Elsevier Science Publishing Co., Inc., New York, 1982, pp. 344–346.

L. M. Landoll
Hercules, Inc.

OLEFIN POLYMERS

This article is devoted to polymeric materials derived from higher olefins with vinyl double bonds, ie, olefins with carbon atom numbers five and higher, containing linear or branched alkyl groups as well as vinyl aromatic compounds. Poly(4-methyl-1-pentene) and polystyrene are discussed in separate articles (see 4-METHYL-1-PENTENE POLYMERS; STYRENE POLYMERS) (see also BUTENE POLYMERS; ETHYLENE POLYMERS; ISOBUTYLENE POLYMERS; PROPYLENE POLYMERS). Polymers of higher α-olefins are produced with heterogeneous Ziegler-Natta catalysts (qv). Some higher olefins can also be polymerized with cationic catalysts to materials with isomerized structures.

The first reports of the synthesis of polymers of higher α-olefins appeared simultaneously with reports of the synthesis of isotactic polypropylene (1,2). These polymers are highly isotactic and have highly regular structures with head-to-tail incorporation of monomer units in the chain (see STEREOREGULAR POLYMERS).

Monomers

The monomers of greatest interest are those available as by-products from commercial processes such as refining of petroleum products, products of thermal cracking of hydrocarbons, and ethylene oligomers. Olefins with branched alkyl groups are also produced by catalytic dehydration of the corresponding alcohols. For example, Phillips Petroleum Co. recently announced a commercial process for the synthesis of 3-methyl-1-butene from isoamyl alcohol using a high surface area, base-treated γ-alumina (3).

Careful purification, in particular removal of water and oxygenated compounds, is essential when employing Ziegler-Natta catalysts. Procedures are similar to those used for ethylene and propylene. Some physical properties of higher α-olefins are given in Table 1. These monomers are gases or liquids at ambient temperature and pressure. They are highly combustible and can form explosive mixtures with air. The primary health hazards are associated with inhalation or prolonged skin contact which can cause irritation.

Table 1. Properties of α-Olefins[a]

Monomer	Molecular weight	Bp, °C	Mp, °C	Density at 20°C, g/cm³	n_D^{20}
1-butene	56	−6.25	−185.35	0.5951	
1-pentene	70	29.96	−165.22	0.6405	1.3715
3-methyl-1-butene	70	20.05	−168.49	0.6272	1.3643
1-hexene	84	63.48	−139.83	0.6732	1.3879
3-methyl-1-pentene	84	54.17	−153.0	0.6674	1.3842
4-methyl-1-pentene	84	53.86	−153.64	0.6637	1.3827
1-heptene	98	93.64	−118.88	0.6970	1.3998
4-methyl-1-hexene	98	86.73	−141.46	0.6985	1.4000
5-methyl-1-hexene	98	85.31		0.6920	1.3967
1-octene	112	121.29	−101.72	0.7149	1.4087
5-methyl-1-heptene	112	113.3		0.7164	1.4094
vinylcyclohexane	110	128		0.8060	1.4458
styrene	104	145.16	−30.610	0.9060	1.5468
1-decene	140	170.599	−66.276	0.7408	1.4215

[a] Ref. 4.

These monomers are highly reactive from a chemical standpoint because the double bond provides the reactive site for catalytic activation and numerous radical reactions. Peroxides are readily formed by exposure to air and must be removed before polymerization.

Polymers

Physical Properties. Physical properties of isotactic polymers of higher α-olefins are given in Table 2, and crystal cell parameters in Table 3. All isotactic polymers have the helix conformation in the crystalline state (Fig. 1). Some of the polymers, in particular derivatives of α-olefins with linear alkyl groups, exist in several polymorphic modifications (Table 3). The type of the polymorph depends on crystallization conditions (8–10).

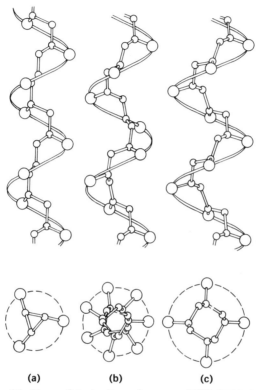

Fig. 1. Types of helices of isotactic polymers $-\!\!\!\!+\!CH_2CH(R)\!\!+\!\!\!\!_n$ of higher α-olefins ($\bigcirc$ = R) (6): (**a**) modification I, R = CH_3, C_2H_5, $CH_2CH_2CH(CH_3)_2$, C_6H_5; (**b**) modification II, R = $CH_2CH(CH_3)C_2H_5$, $CH_2CH(CH_3)_2$; (**c**) modification III, R = $CH(CH_3)_2$.

Melting points strongly depend on the substituents at the double bond. With linear alkyl substituents, the melting point of the crystalline phase rapidly decreases with increasing chain length: ca 180°C for polypropylene, 130°C for poly(1-butene), and 80°C for poly(1-pentene). However, with longer alkyl chains, crystallinity is derived from the side chains rather than from the polymer backbone (7,9–12). As a result, melting points increase up to ca 45°C for poly(1-decene), 50°C for poly(1-tetradecene), 68°C for poly(1-hexadecene), and 75°C for poly(1-octadecene). These changes are parallel to increases in the glass-transition temperature.

Polymers of olefins with branched alkyl groups generally exhibit much higher melting points. The highest values, ie, >350°C, have been reported for poly(3-methyl-1-butene) and polyvinylcyclohexane. Crystalline isotactic polymers of vinyl aromatic compounds also exhibit very high melting points. Densities of crys-

Table 2. Properties of Isotactic Polyolefins[a]

Polymer	Structural unit of polymer chain	Mp, °C	Crystal type	Space group	Density, g/cm³ Crystalline	Average	T_g, °C
poly(1-butene)[b]	$-CH_2-CH-$; C_2H_5	126–142	rhombohedral	$R\bar{3}c$	0.95	0.915	−20 to −25
poly(1-butene)[c]	$-CH_2-CH-$	122–130	tetragonal	$P\bar{4}b2$	0.96	0.90	
poly(1-pentene)[b]	$-CH_2-CH-$; $n\text{-}C_3H_7$	80	monoclinic		0.92	0.87	−40 to −55
poly(1-pentene)[c]	$-CH_2-CH-$; $i\text{-}C_3H_7$	75–80	pseudoorthorhombic		0.90		
poly(3-methyl-1-butene)[c]		350	monoclinic	$P2_1/b$	0.93	0.90	
poly(1-hexene)[d]	$-CH_2-CH-$; $n\text{-}C_4H_9$		monoclinic		0.73	0.84	−50
poly(3-methyl-1-pentene)[d]	$-CH_2-CH-$; $CH(CH_3)(C_2H_5)$	200					20–50
poly(4-methyl-1-pentene)[d]	$-CH_2-CH-$; $CH_2-CH(CH_3)_2$	235–240	tetragonal	$P\bar{4}$	0.813	0.838	20–50
poly(1-heptene)[b]	$-CH_2-CH-$; $n\text{-}C_5H_{11}$	18					
poly(4-methyl-1-hexene)[d]	$-CH_2-CH-$; $CH_2-CH(CH_3)(C_2H_5)$	188–200	tetragonal	$P\bar{4}$	0.845	0.86	−30

poly(5-methyl-1-hexene)[b]	$-CH_2-CH-$ with $(CH_2)_2$, CH, CH_3, CH_3	110–130	monoclinic	$P2_1$	0.84	0.85	
poly(1-octene)	$-CH_2-CH-$ with $n\text{-}C_6H_{13}$	~20					−65 to −45
poly(5-methyl-1-heptene)[b]	$-CH_2-CH-$ with $(CH_2)_2$, CH, CH_3, C_2H_5		tetragonal	$P\bar{4}$			
polyvinylcyclohexane[c]	$-CH_2-CH-$ (cyclohexyl)	385	tetragonal	$I4/a$	0.95	0.93	~80
polystyrene[b]	$-CH_2-CH-$ (phenyl)	213–240	rhombohedral	$R\bar{3}c$	1.12	1.08	80–100
poly(o-methylstyrene)[c]	$-CH_2-CH-$ (naphthyl)	>360	tetragonal	$I4_1cd$	1.01		
poly(m-methylstyrene)[b]		215	tetragonal	$P\bar{4}$			
poly(α-vinylnaphthalene)[c]		~360	tetragonal	$I4_1cd$	1.125		

[a] Refs. 5–7.
[b] Helix 3_1.
[c] Helix 4_1.
[d] Helix 7_2.

Table 3. Crystal Structure of Polyolefins[a]

Polymer	Crystal system	Helix type	Unit cell parameters, nm			Angles, °	Monomer unit cell
			a	b	c		
poly(1-pentene) (polymorph IA)	monoclinic	3_1	1.14	2.08	0.65	$\beta = 99.6$	12
poly(1-pentene) (polymorph IB)		3_1	2.43	1.93	0.65		
poly(1-pentene) (polymorph IIA)	pseudo-orthorhombic	4_1	1.96	1.67	0.71	$\alpha = 115.3$	16
poly(3-methyl-1-butene) (polymorph I)	monoclinic	3_1	0.95	0.85	0.68	$\alpha = 116.5$	4
poly(3-methyl-1-butene) (polymorph II)	pseudo-orthorhombic	4_1	1.92	1.72	0.66	$\alpha = 116.5$	16
poly(4-methyl-1-pentene)	tetragonal	7_2	1.86	1.86	1.38		28
poly(1-hexene)	monoclinic	7_2	2.22	0.89	1.37	$\alpha = 94.5$	14
poly(4-methyl-1-hexene)	tetragonal	7_2	1.96	1.96	1.35		28
poly(5-methyl-1-hexene)	monoclinic	3_1	1.76	1.02	0.65	$\beta = 90$	6
poly(5-methyl-1-heptene)	monoclinic	3_1	1.84	1.06	0.64	$\beta = 90$	6
polyvinylcyclohexane	tetragonal	4_1	2.20	2.20	0.65		16
polystyrene	rhombohedral	3_1	2.19	2.19	0.66	$\gamma = 120$	18
poly(o-methylstyrene)	tetragonal	4_1	1.90	1.90	0.81		16
poly(m-methylstyrene)	tetragonal	3_1	1.98	1.98	2.17		12
poly(o-fluorostyrene)	rhombohedral	3_1	2.21	2.21	0.66	$\gamma = 120$	18
poly(p-fluorostyrene)	orthorhombic	4_1	1.76	1.21	0.83		8
poly (α-vinylnaphthalene)	tetragonal	4_1	2.12	2.12	0.81		16

[a] Refs. 5–8.

talline polymers of higher α-olefins depend on the substituent. The highest reported density is for polystyrene: 1.08 g/cm^3.

Mechanical properties also depend strongly on the substituent. Polymers with linear alkyl side chains have low crystallinities under normal conditions and exhibit mechanical behavior typical for elastomers (13). On the other hand, some polymers with branched alkyl groups are highly crystalline and exhibit mechanical properties similar or superior to those of isotactic polypropylene. A good combination of mechanical and optical properties contributes to the industrial importance of poly(4-methyl-1-pentene). Poly(3-methyl-1-butene) and polyvinylcyclohexane also exhibit good mechanical characteristics. However, their high melting points, poor oxidative stability, and brittleness preclude industrial application.

All polymers of higher α-olefins exhibit good dielectric properties. Especially interesting are the dielectric characteristics of polyvinylcyclohexane: dielectric loss remains constant between -180 and 160°C, which makes it a prospective high frequency dielectric material of high thermal stability.

Chemical Properties. The chemical properties of polymers of higher olefins resemble those of polypropylene and poly(1-butene). The two tertiary carbon atoms per monomer unit in polymers derived from olefins with branched alkyl groups make them especially susceptible to oxidative degradation. The most studied example of such behavior is poly(4-methyl-1-pentene) (see 4-METHYL-1-PENTENE POLYMERS). For example, thermal destruction of poly(3-methyl-1-butene) starts at ca 300°C, whereas thermal oxidative destruction of the polymer starts at ca 100°C. The effective activation energy of this process is ca 60 kJ/mol (14.3 kcal/mol). Kinetics of thermal oxidation of isotactic poly(1-pentene) has been studied (14). The principal products of the reaction are formaldehyde, acetone, and various alkanes. Amorphous polymers of higher olefins are more reactive in the oxidation reaction than their highly crystalline analogues, mostly because of a higher diffusion rate of oxygen. The polymers can be stabilized with the same products, ie, substituted phenols, organic phosphites, as those used for stabilization of commercial isotactic polyolefins, such as polypropylene and poly(4-methyl-1-pentene).

Polymerization

Polymerization with Ziegler-Natta Catalysts. Conventional heterogeneous Ziegler-Natta catalysts (qv) are widely used for the polymerization of higher olefins under the same conditions (15–17) as those used for propylene and 1-butene (see also ETHYLENE POLYMERS; PROPYLENE POLYMERS; BUTENE POLYMERS). Catalysts most frequently used are $TiCl_4$–$Al(C_2H_5)_3$, δ-$TiCl_3$–$Al(C_2H_5)_3$, δ-$TiCl_3$–$Al(C_2H_5)_2Cl$, and supported catalysts such as $TiCl_4/MgCl_2$–$Al(C_2H_5)_3$. Polymerization of vinylcyclohexane is accompanied by extensive isomerization of the double bond with concomitant formation of ethylidene cyclohexane and ethylcyclohexenes (18). The reversed $\beta \rightarrow \alpha$ double-bond isomerization allows polymerizing olefins with internal double bonds to produce poly(α-olefins) (15,17) (see ISOMERIZATION POLYMERIZATION).

Polymers of higher α-olefins produced by polymerization with heterogeneous Ziegler-Natta catalysts have highly regular head-to-tail structures. Some head-to-head monomer addition, ie, <1%, was found only in polymers obtained with vanadium-based catalysts (15). The polymerization reactions proceed without isomerization of alkyl groups. The only exception is the polymerization of non-conjugated dienes.

The crude polymers obtained in Ziegler-Natta polymerizations can be fractionated to separate crystalline fractions of high stereoregularity (isotactic, usually insoluble in boiling n-heptane), fractions of moderate stereoregularity, and amorphous irregular fractions. For example, crude poly(4-methyl-1-pentene) obtained with δ-$TiCl_3$–$Al(i$-$C_4H_9)_3$ at 70°C contains 61% of highly crystalline polymer insoluble in boiling n-heptane, 26% of polymer soluble in boiling n-heptane but insoluble in cold n-heptane, and 13% of amorphous polymer soluble in cold n-heptane (15).

Yields of the highly crystalline fractions depend strongly on the catalytic system used. For example, yields of n-heptane-insoluble fractions of poly(3-methyl-1-butene) are 60–70% for the $TiCl_4$–$Al(C_2H_5)_3$ system but 85–95% for the $TiCl_3$–$Al(C_2H_5)_3$ system. Similarly, yields of ethyl acetate-insoluble fractions of poly(1-pentene) with the same catalyst systems are 1–3% and 30–40%, respectively (19). Yields of crystalline poly(4-methyl-1-hexene) obtained with various heterogeneous catalysts are ca 75%, and yields of crystalline poly(3,7-dimethyl-1-octene) are ca 80–84% (16). Titanium-based supported Ziegler-Natta catalysts produce the best yield of highly crystalline materials (16).

Ziegler-Natta catalysts readily copolymerize various higher α-olefins (15,20,21) with the formation of partially crystalline or amorphous products. Copolymerization of higher α-olefins with conjugated dienes is also possible.

Reactivities of higher α-olefins with Ziegler-Natta catalysts largely depend on the substituents (15,21–24). Quantitative data on reactivity of various higher α-olefins have been obtained from the analysis of reactivity ratios in copolymerization reactions of these olefins with ethylene and propylene (see also CO-POLYMERIZATION). Relative reactivities of higher α-olefins with titanium-based catalysts are given in Table 4. The strong steric effects of alkyl groups at the olefin double bond are evident. Low reactivity is exhibited by olefins with α-branched alkyl groups (3-methyl-1-butene, 3-methyl-1-pentene, vinylcyclohexane). Reactivities of α-olefins with linear alkyl groups in copolymerization with ethylene with the $TiCl_4/MgCl_2$–$Al(C_2H_5)_3$ system at 90°C are shown in Figure 2. The measure of olefin reactivity, the reactivity ratio r_2, decreases with increasing carbon number, but becomes essentially constant after 1-dodecene (22).

Table 4. Reactivities of Higher α-Olefins[a]

Monomer	Reactivity[b]
ethylene	8–20
propylene	1
1-butene	0.22–0.62
1-pentene	0.20–0.45
1-hexene	0.16–0.36
1-heptene	0.20–0.45
1-decene	0.12–0.28
1-octadecene	0.10–0.15
3-methyl-1-butene	0.024–0.06
3-methyl-1-pentene	0.048
4-methyl-1-pentene	0.15
4-methyl-1-hexene	0.10–0.17
5-methyl-1-hexene	0.34–0.50
vinylcyclohexane	0.012
styrene	0.05–0.30
vinylnaphthalene	0.07–0.10

[a] Refs. 15 and 21.
[b] Reactivity of propylene is defined as one.

Polymerization of higher α-olefins with Ziegler-Natta catalysts proceeds in a predominantly isospecific fashion (1,15,17) (see STEREOREGULAR POLYMERS). The most important factor determining isospecific growth of polyolefin chains is the steric environment of the active centers. It is caused by the conformation of the growing polymer chains and operates both in growth and initiation reactions (25).

Stereoselectivity and Stereoelectivity.   Analysis of polymerization data of higher α-olefins with asymmetric carbon atoms in the alkyl groups reveals these two phenomena. If racemic mixtures of such α-olefins are polymerized with heterogeneous Ziegler-Natta catalysts, including some supported catalysts of the

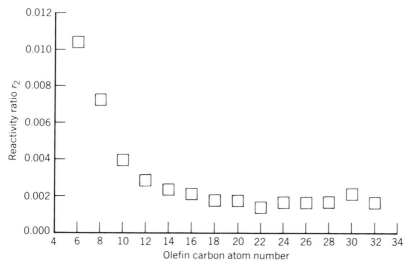

Fig. 2. Dependence of r_2 (the reactivity ratio in copolymerization with ethylene, the measure of α-olefin reactivity) on the length of linear α-olefins. Polymerization with the $TiCl_4/MgCl_2$–$Al(C_2H_5)_3$ system at 90°C (22).

type $TiCl_4/MgCl_2$–AlR_3–Lewis base (26,27), the polymers can be separated into two fractions with marked preference for incorporation of one of the olefin isomers, R or S. This stereoselection is especially strong in olefins with the asymmetric carbon atoms in the α-position to the double bond. For example, polymerization of racemic 3,7-dimethyl-1-octene with $TiCl_3$–$Al(i$-$C_4H_9)_3$ proceeds with a separation efficiency of ca 90% (28). A similar, though less strict selection, was also found in the polymerization of racemic 3-methyl-1-pentene (29).

The stereoselectivity effect in the case of supported catalysts is especially pronounced with optically active Lewis bases (27,30). It is determined by the chirality of catalytic sites which impose a particular monomer orientation before monomer addition to a polymer chain (31).

Another steric effect, stereoelectivity, was found in the case of catalysts containing asymmetric components. Such catalysts can predominantly polymerize one of the olefin stereoisomers from racemic mixtures, resulting in optically active polymers (qv). The remaining unreacted monomer is enriched with molecules of opposite chirality and acquires optical activity (15,16,32).

Cyclopolymerization of Nonconjugated Dienes. If the alkyl group in a higher olefin contains a second double bond, a special copolymerization reaction can take place, giving polymers containing cyclized monomer units. For example, 1,5-hexadiene polymerizes according to the scheme

This polymer is crystalline, with a high density (1.22 g/cm^3) and a melting point of 140–146°C (33,34). Other nonconjugated dienes produce polymers containing both cyclized and uncyclized units. Numerous examples of these reactions are described in Chapter 19 of Ref. 17 (see also CYCLOPOLYMERIZATION).

Cationic Polymerization of α-Olefins.   Higher α-olefins are also polymerized with cationic catalysts. Olefins containing linear alkyl groups, eg, 1-pentene, 1-hexene, 1-heptene, 1-octene, 1-decene, etc, are oligomerized with AlCl$_3$ or AlBr$_3$–HBr at low temperatures to oils of irregular structure (35,36). However, α-olefins with alkyl groups branched in the α-position to the double bond, eg, 3-methyl-1-butene, are readily polymerized to high molecular weight polymers at low temperatures, ie, ca -130°C, with AlCl$_3$ in solvents such as n-pentane or ethyl chloride. These polymers have highly regular structures due to a uniform isomerization polymerization (qv) (36–38):

$$RCH_2-\overset{+}{CH} \longrightarrow RCH_2CH_2-\overset{\overset{CH_3}{|}}{\underset{\underset{CH_3}{|}}{C}}{}^+ \longrightarrow RCH_2CH_2-\overset{\overset{CH_3}{|}}{\underset{\underset{CH_3}{|}}{C}}-CH_2CH_2-\overset{\overset{CH_3}{|}}{\underset{\underset{CH_3}{|}}{C}}-CH_2CH_2-\overset{\overset{CH_3}{|}}{\underset{\underset{CH_3}{|}}{C}}{}^+$$

The polymeric product of 3-methyl-1-butene typically has a viscosity-average molecular weight of 50,000. Upon increasing the polymerization temperature above -115°C, amorphous polymers containing both isomerized and non-isomerized monomer units are formed. Other Lewis acids at -130°C give either poorly crystalline polymers (AlBr$_3$, TiCl$_4$) or largely amorphous products (BF$_3$), or are unable to initiate the polymerization reaction (SnCl$_4$, VCl$_4$) (36).

The crystalline product of isomerization polymerization of 3-methyl-1-butene exists in three modifications (36,39). The most stable, the α-modification, has a tetragonal unit cell with the parameters $a = b = 1.07$ nm and $c = 0.765$ nm. The polymer exists in a planar zigzag form; it melts at 66°C and the glass-transition point is between -7 and -14°C. During the polymerization, two other crystalline modifications are formed: the β-polymorph melts at 48.5°C, and the γ-modification at 60.0°C (36).

Similar rearrangement reactions have been reported in polymerizations of 3-methyl-1-pentene (40), 3-ethyl-1-pentene (41), and 3,3-dimethyl-1-butene (42). If olefin monomers with branches in β-, γ-, or δ-positions to the double bond polymerize at low temperature with cationic catalysts, numerous isomerization reactions result in the formation of amorphous polymers. For example, 4-methyl-1-pentene produces polymer chains containing monomer units of several different structures (36,38,43–45):

$$\begin{array}{llll}
\overset{\displaystyle -CH-}{\underset{\displaystyle \underset{\displaystyle \underset{\displaystyle \overset{CH}{\diagup\diagdown}}{CH_2}}{CH_2}}{}} & -CH_2-CH- & -CH_2CH_2CH- & \overset{\displaystyle CH_3}{\underset{\displaystyle \underset{\displaystyle CH_3}{-CH_2CH_2CH_2-C-}}{}} \\
\text{1,1-enchainment} & \text{1,2-enchainment} & \text{1,3-enchainment} & \text{1,4-enchainment}
\end{array}$$

The amounts of these units depend on polymerization conditions, with 1,4-enchainments being the dominant structural type.

Processing

Processing of isotactic polymers of higher α-olefins is similar to the processing of other isotactic polymers, eg, polypropylene, poly(1-butene), and poly(4-methyl-1-pentene). Processing conditions depend on the melting points (Table 2). Antioxidants (qv) are usually required in order to prevent thermooxidative degradation.

Economic Aspects

The polymers of higher olefins discussed in this article are noncommercial, and detailed economics associated with production are not available. However, cost of production will be closely associated with monomer costs. Even-numbered-carbon α-olefin monomers with linear alkyl groups, ie, 1-hexene, 1-octene, 1-decene, etc, are produced commercially by ethylene oligomerization with $Al(C_2H_5)_3$ or homogeneous Ni-based catalysts. Current prices of these monomers are listed in Table 5.

Table 5. 1986 Prices of Linear α-Olefins

Olefin	$/kg[a]
1-butene	0.63
1-hexene	1.02
1-octene	1.02
1-decene	1.02
1-dodecene	1.00
1-tetradecene	1.01
1-hexadecene	0.98
1-octadecene	1.00
1-eicosene[b]	0.88
1-docosene[b]	0.88
1-tetracosene[b]	0.88
1-hexacosene[b]	0.86
1-octacosene[b]	0.86
1-triacontene[b]	0.86

[a] List prices, tank-car or tank-truck quantities.
[b] Available only as mixtures.

Most of the olefins with branched alkyl groups are produced as by-products of petroleum refining or are specially synthesized. Their prices are usually higher than those for linear α-olefins. For the commercial process recently announced by Phillips Petroleum Co. for 3-methyl-1-butene synthesis (3), product cost has been estimated at $6.60–11.00/kg.

Analytical and Test Methods

Analytical and test methods used to characterize polyethylene, polypropylene, poly(1-butene), and poly(4-methyl-1-pentene) are also applicable for the characterization of polymers of higher α-olefins. Stereoregularity is usually estimated by the solvent-extraction method similar to that used for isotactic polypropylene. The ^{13}C nmr method is widely used for structure characterization and conformations in the solid state (15,46–48). Several ir and Raman spectroscopic methods are used to characterize chemical structure and stereoregularity (15,49–51). Molecular weight measurements are carried out by viscosimetric methods in procedures similar to those used for poly(4-methyl-1-pentene), with the application of the Mark-Houwink equation (52). Molecular weights of various poly(α-olefins) produced from C_4–C_{18} monomers have also been measured by gpc (53).

Health and Safety Factors

Polymers of higher α-olefins are expected to be nontoxic based on structural and compositional similarities to polypropylene, poly(1-butene), and poly(4-methyl-1-pentene). The main potential health hazards are associated with residual monomer, antioxidants, and catalyst residues.

Uses

The lower molecular weight liquid oligomers of higher α-olefins have found applications as synthetic base oils in the formulation of various lubricants such as transformer oils, transmission and crankcase fluids, and the like. The true polymers of higher α-olefins, however, have found only very limited application.

Poly(1-hexene) and poly(1-octene) have been used to improve the impact resistance of polypropylene (54); poly(1-octadecene) has been oxidized with chromosulfuric acid to produce a hard oxygen-containing wax (55); poly(5-methyl-1-hexene) has been reported to function as a binder for ferric oxide in the preparation of magnetic recording tapes resistant to abrasion, heat, and solvents (56); and poly(3-methyl-1-pentene) has been used as a component in the preparation of heat-resistant compositions useful as dielectric materials (57).

BIBLIOGRAPHY

"Olefin Polymers" in *EPST* 1st ed., Vol. 9, pp. 440–459, B. S. Dyer, Imperial Chemical Industries, Ltd., and M. R. Day and W. R. Bergen, ICI America, Inc.

1. G. Natta, P. Pino, P. Corradini, F. Danusso, E. Mantica, G. Mazzanti, and G. Moraglio, *J. Am. Chem. Soc.* **77,** 1708 (1955).
2. G. Natta, *J. Polym. Sci.* **16,** 143 (1955).
3. *Chem. Eng. News* **63**(16), 27 (1985).
4. *Physical Constants of Hydrocarbons C_1 to C_{10}, ASTM Data Series Publication DS 4A*, The American Society for Testing and Materials, Philadelphia, Pa., 1971, pp. 44–50.
5. H. Tadokoro, *Structure of Crystalline Polymers,* John Wiley & Sons, Inc., New York, 1979, pp. 355–358.

6. G. Natta and F. Danusso, eds., *Stereoregular Polymers and Stereospecific Polymerizations,* Vols. 1 and 2, Pergamon Press, Ltd., Oxford, UK, 1967.

7. H. V. Boenig, *Polyolefins: Structure and Properties,* Elsevier Publishing Co., Amsterdam, 1966, p. 17.

8. M. Moser and M. Boudeulle, *J. Polym. Sci. Polym. Phys. Ed.* **16,** 971 (1978).

9. K. Blum and G. Trafara, *Makromol. Chem.* **181,** 1097 (1980).

10. G. Trafara, *Makromol. Chem.* **181,** 969 (1980).

11. J.-S. Wang, R. S. Porter, and J. R. Knox, *Polym. J.* **10,** 619 (1978).

12. G. Trafara, *Makromol. Chem. Rapid Commun.* **1,** 319 (1980).

13. M. P. Berdnikova, Y. V. Kissin, and N. M. Chirkov, *Vysokomol. Soedin.* **5,** 63 (1963).

14. S. M. Gabbay, S. S. Stivala, and L. Reich, *Polymer* **16,** 741 (1975).

15. Y. V. Kissin, *Isospecific Polymerization of Olefins with Heterogeneous Ziegler-Natta Catalysts,* Springer-Verlag, New York, 1985, pp. 12–87, 227, 228, 337.

16. P. Pino, G. Guastalla, B. Rotzinger, and R. Muelhaupt in R. P. Quirk, ed., *Transition Metal Catalyzed Polymerizations, Alkenes and Dienes,* Harwood Academic Publishers GmbH, New York, 1981, Pt. A, p. 435.

17. J. Boor, Jr., *Ziegler-Natta Catalysts and Polymerizations,* Academic Press, Inc., New York, 1979.

18. V. I. Kleiner, B. A. Krentsel, and L. L. Stotskaya, *Eur. Polym. J.* **7,** 1677 (1971).

19. G. Natta, P. Pino, E. Mantica, F. Danusso, G. Mazzanti, and M. Peraldo, *Chim. Ind. Milan* **38,** 124 (1956).

20. Y. V. Kissin, *Adv. Polym. Sci.* **15,** 91 (1974).

21. Y. V. Kissin in Ref. 16, Pt. B, p. 597.

22. Y. V. Kissin and D. L. Beach, *J. Polym. Sci. Polym. Chem. Ed.* **22,** 333 (1984).

23. P. Ammendola and A. Zambelli, *Makromol. Chem.* **185,** 2451 (1984).

24. A. Zambelli, P. Ammendola, and A. J. Sivak, *Macromolecules* **17,** 461 (1984).

25. P. Locatelli, I. Tritto, and M. C. Sacchi, *Makromol. Chem. Rapid Commun.* **5,** 495 (1984).

26. P. Pino, G. Fochi, O. Piccolo, and U. Giannini, *J. Am. Chem. Soc.* **104,** 7381 (1982).

27. P. Pino, G. Fochi, A. Oschwald, O. Piccolo, R. Muelhaupt, and U. Giannini, *Polym. Sci. Technol.* **19,** 207 (1983).

28. P. Pino, F. Ciardelli, G. Montagnolli, and O. Pieroni, *J. Polym. Sci. Part B* **5,** 307 (1967).

29. A. Zambelli, P. Ammendola, M. C. Sacchi, P. Locatelli, and G. Zannoni, *Macromolecules* **16,** 341 (1983).

30. C. Carlini and F. Ciardelli, *Chim. Ind. Milan* **63,** 486 (1981).

31. P. Corradini, G. Duerra, and V. Vincenzo, *Actas Simp. Iberoam Catal. 9th* **2,** 1631 (1984).

32. C. Carlini, R. Nocci, and F. Ciardelli, *J. Polym. Sci. Polym. Chem. Ed.* **15,** 767 (1977).

33. C. S. Marvel and J. K. Stille, *J. Am. Chem. Soc.* **80,** 1740 (1958).

34. C. S. Marvel and W. E. Garrison, *J. Am. Chem. Soc.* **81,** 4737 (1959).

35. C. M. Fontana, G. A. Kidder, and R. J. Herold, *Ind. Eng. Chem.* **44,** 1688 (1952).

36. J. P. Kennedy, *Cationic Polymerization of Olefins: A Critical Review,* John Wiley & Sons, Inc., New York, 1975.

37. J. P. Kennedy, L. S. Minckler, Jr., G. G. Wanless, and R. M. Thomas, *J. Polym. Sci. Part A* **2,** 1441 (1964).

38. W. R. Edwards and N. F. Chamberlain, *J. Polym. Sci. Part A* **1,** 2299 (1963).

39. J. P. Kennedy, J. J. Elliot, and B. Groten, *Makromol. Chem.* **77,** 26 (1964).

40. R. Bacskai, *J. Polym. Sci. Part A-1* **5,** 619 (1967).

41. J. P. Kennedy in N. M. Bikales, ed., *Encyclopedia of Polymer Science and Technology,* 1st ed., Vol. 7, Wiley-Interscience, New York, 1967, p. 754.

42. J. P. Kennedy, J. J. Elliot, and B. E. Hudson, *Makromol. Chem.* **79,** 109 (1964).

43. J. E. Goodrich and R. S. Porter, *J. Polym. Sci. Part B* **2,** 353 (1964).

44. G. G. Wanless and J. P. Kennedy, *Polymer* **6,** 111 (1965).

45. G. Ferraris, C. Corno, A. Priola, and S. Cesca, *Macromolecules* **10,** 188 (1977).

46. Y. Tanaka and H. Sato, *J. Polym. Sci. Polym. Lett. Ed.* **14,** 335 (1976).

47. D. R. Ferro and M. Ragozzi, *Macromolecules* **17,** 485 (1984).

48. M. C. Sacchi, P. Locatelli, L. Zetta, and A. Zambelli, *Macromolecules* **17,** 483 (1984).

49. Y. V. Kissin, Y. Y. Goldfarb, Y. V. Novoderzhkin, and B. A. Krentsel, *Vysokomol. Soedin. Ser. B* **18,** 167 (1976).

50. J. J. Elliot and J. P. Kennedy, *J. Polym. Sci. Polym. Chem. Ed.* **11,** 2991 (1973).

51. I. Modric, K. Holland-Moritz, K. Heinen, and D. O. Hummel, *Colloid Polym. Sci.* **251**, 913 (1973); **254**, 342 (1976).
52. P. Neuenschwander and P. Pino, *Makromol. Chem.* **181**, 737 (1980).
53. S. Mori, *J. Appl. Polym. Sci.* **18**, 2191, 2391 (1974).
54. Jpn. Pat. 72 008,370 (Mar. 10, 1972), K. Shirayama, S. Shiga, and H. Watabe (to Sumitomo Chemical Co., Ltd.).
55. Ger. Offen. 3,047,915 (July 15, 1982), K. H. Stetter (to Hoechst AG).
56. Fr. Demande 2,121,855 (Sept. 29, 1972) (to BASF).
57. Fr. Demande 2,403,360 (Apr. 13, 1979), V. M. Demidova, M. F. Utjogova, E. N. Matveeva, N. N. Gorodetskaya, and V. F. Petrova.

YURY V. KISSIN
Mobil Chemical Company

DAVID L. BEACH
Chevron Chemical Company

OLEFIN–SULFUR DIOXIDE COPOLYMERS

Olefin–sulfur dioxide copolymers [poly(olefin sulfone)s, poly(alkylene sulfone)s] (1–9) are obtained by the free-radical reaction of olefins with sulfur dioxide. In the principal class, the hydrocarbon and sulfonyl residues alternate, as in poly(propylene sulfone) (**1**), whose IUPAC systematic name is poly(sulfonyl-1-methylethylene). Such copolymers are also formed from ethylene, olefins with two or more substituents, cyclic olefins, and olefins with heteroatoms in the side chain or ring. In a second class, a nonalkyl substituent is linked directly to the olefinic carbon, and the sulfonyl groups comprise less than half the residues, eg, poly(vinyl chloride sulfone) (**2**) with $m > 1$ (typically 2).

$$-(CH_2-CH-SO_2)_n \qquad -(CH_2-CH)_m-SO_2)_n$$
$$\quad\quad\quad | \qquad\qquad\qquad\qquad |$$
$$\quad\quad\quad CH_3 \qquad\qquad\qquad\quad Cl$$
$$\quad\quad\quad (1) \qquad\qquad\qquad\qquad (2)$$

Polysulfones also form from conjugated and nonconjugated dienes and from alkynes, though not from acetylene. More than one olefinic structure may be incorporated into a single macromolecule.

A second method of preparation is by polysulfide oxidation. A tactic poly(olefin sulfone) has been prepared from a tactic polyepisulfide (10), and polysulfones with one (11), two (10,12), or more than two main-chain carbon atoms per alkyl residue are obtained (11–16). Similar polymers have been prepared from 1,4-dibromobutane and sodium dithionite (17).

A poly(phenylmethylene sulfone) (**3**) is formed by treatment of a benzenesulfonyl halide with phenyllithium (18,19), and the poly(alkylsulfonyl ester) (**4**) is formed by opening the cyclic monomer with an acidic catalyst (20).

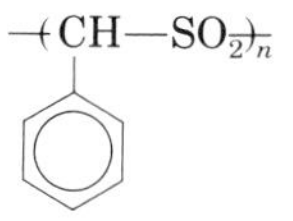

(**3**)

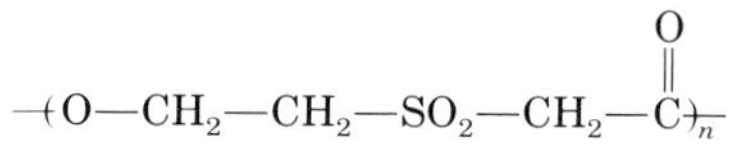

(**4**)

Divinyl sulfone reacts spontaneously with divinyl ether to give (**5**) (21) and, by step growth, combines with difunctional molecules, such as diisocyanates, diols, and dimercaptans, to give polymers, such as (**6**) and (**7**) (22,23).

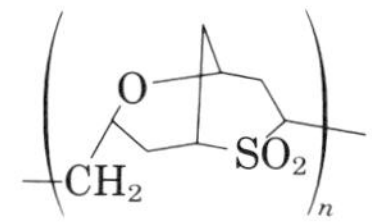

(**5**)

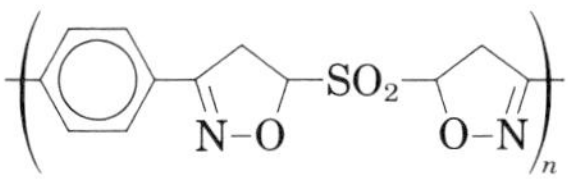

(**6**)

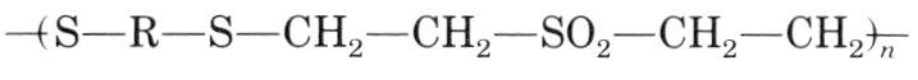

$+S-R-S-CH_2-CH_2-SO_2-CH_2-CH_2+_n$

(**7**)

Divinyl sulfone also reacts by Michael addition with diphenyl sulfonyl dichloromethane (11,24) or with organotindihydrides (25) to give (**8**) and (**9**), respectively.

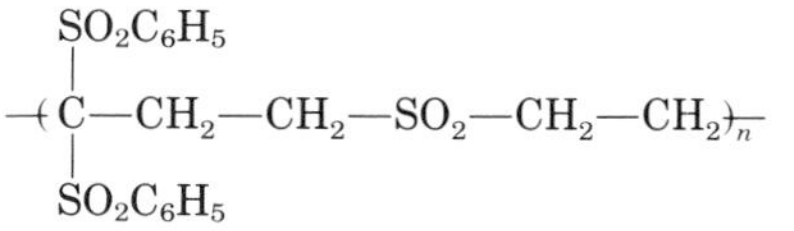

(**8**)

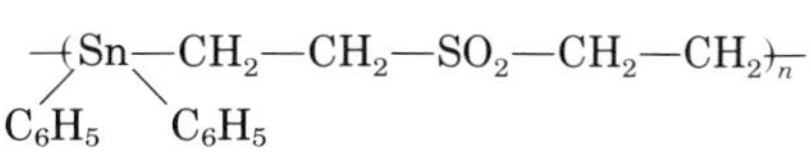

(**9**)

Polymers with sulfonyl groups in the side chain may be prepared by the anionic polymerization of a vinyl alkyl sulfone; vinyl butyl sulfone has approximately the same reactivity as acrylonitrile (26). Polymers with sulfonyl groups not in the main backbone are prepared by the reaction of 3,4-dichlorotetrahydrothiophene-1,1'-dioxide with polyamines (27) or by the addition of SO_2 to certain unsaturated polymers (28).

The term polysulfone usually refers to engineering plastics (qv), such as poly(aryl ether sulfone)s (**10**) (see POLYSULFONES).

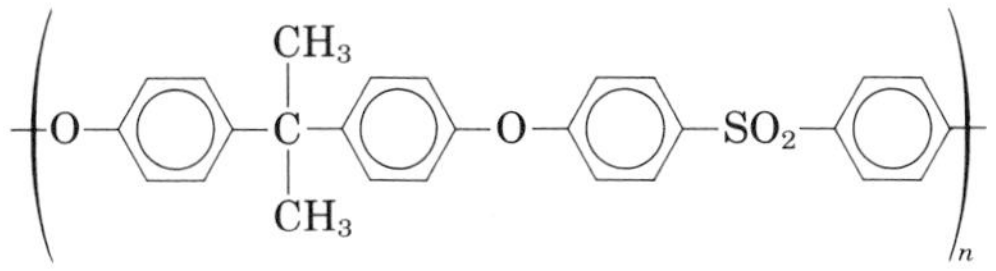

(**10**)

Olefins were first polymerized as polysulfones (28), though only much later were the substances recognized as polymeric (29–31). Poly(propene sulfone) and poly(cyclohexene sulfone) were clearly identified as alternating copolymers of the olefin and SO_2 (32).

The observation (2) that polymer formation proceeded only below a certain temperature, the ceiling temperature, led to the recognition that polymerization is thermodynamically an aggregation process (3). The head-to-tail arrangement along the polymer chain of unsymmetrical units derived from olefins was established by alkali degradation studies (33) and confirmed by ^{13}C nmr spectroscopy, which also revealed that polymers prepared with free radicals are atactic (34). The role of the olefin–sulfur dioxide charge-transfer complex in the copolymerization reaction is obscure, though it has been established that they do not initiate the reaction (35,36).

Initial dielectric studies suggested that the chains of poly(1-olefin sulfone)s were kinetically stiff random coils in solution (37,38). However, a segmental process active at high frequencies was first suggested by dynamic ^{13}C nmr studies (34,39,40) and then found to be dielectrically active (41–44). Electrostatic interactions between the sulfone dipoles create stiff sections within the chain. The shape of these sections and their size influence the equilibrium configurational properties, and their movements produce the measured dynamic features.

The readiness with which olefin–sulfur dioxide copolymers degrade to their monomers at high temperatures has prevented their application as thermoplastics, but because of sensitivity to ionizing radiation, poly(1-butene sulfone) is used as a positive resist in lithographic applications (8,45) (see also LITHOGRAPHIC RESISTS).

Monomers

Sulfur Dioxide.   The odor threshold of SO_2 (0.3–2.5 ppm (46)) is close to the recommended TLV of 2 ppm (47) and the maximum workplace concentration of 5 ppm (46). Short exposure to concentrations over 400 ppm may be lethal. The most susceptible tissues are the mucous membranes of the eye and respiratory tract where sulfurous acid may form. Prolonged low concentration exposure may result in bronchitis and pulmonary edema.

Physical properties for SO_2 are given in Table 1 (46,48–52). Density and dielectric constant are sensitive to temperature. As a highly polar liquid, SO_2 is only partially miscible with hydrocarbons, particularly at low temperatures; phase diagrams are given in Refs. 53–58. The low solubility of ethylene in SO_2 has been measured (59). The electron affinity of SO_2 (2.8 eV) is sufficient to permit the formation of charge-transfer complexes with olefins with an ionization potential near 9 eV; yellow charge-transfer bands are often referred to (60). According to the phase diagrams, 2,3-dimethyl-2-butene forms a 1:1 charge-transfer complex with SO_2 (56), isobutene forms a 2:1 complex (57), and vinyl chloride forms both a 1:1 and a 1:2 complex (58). Measurements of equilibrium constants for 1:1 complex formation over a range of temperatures for eight olefins in a hydrocarbon medium (56) gave enthalpies of complex formation between 4 and 14 kJ/mol (0.9–3.3 kcal/mol). Thus in a 10:1 mixture of SO_2 and olefin at 20°C, 40–70% of the olefin may be complexed. Complex equilibrium constants are reduced in a medium more polar than chloroform (61). Despite the formation of charge-transfer complexes, olefin–sulfur dioxide mixtures show positive deviations from Raoult's law (62,63).

Liquid sulfur dioxide solvates anions and bases, and is a common medium for inorganic and organic reactions (48–51,64) as well as for polymerizations such as that of acrylonitrile (65). It promotes olefin isomerizations (66) and the exchange of hydrogen or deuterium atoms at allylic sites in the presence of water (67). It initiates cationic (68,69) and free-radical (70,71) polymerizations and participates in the redox initiation of polysulfone formation, but inhibits the polymerization of vinyl chloride at high temperatures (72). The diffusion of SO_2 through glassy polymers has been studied (73). A polymer with pendent sulfoxide groups is selectively permeable to SO_2 (74).

Table 1. Physical Constant of Sulfur Dioxide[a,b]

Property	Value
melting point, °C	− 75.52
boiling point, °C	− 10.08
enthalpy of fusion, kJ/mol[c]	8.241
enthalpy of vaporization, kJ/mol[c]	24.94
cryoscopic constant, °C/mol	3.01
ebullioscopic constant, °C/mol	1.48
S–O distance, pm	143
O–S–O angle	119.5°
density ρ, at − 10°C, kg/m³	1540
viscosity, at 10°C, mPa·s (= cP)	0.43
dielectric constant ϵ, at − 19°C	17.4
specific conductivity, Ω/cm	3×10^{-10}
vapor pressure, kN/m²[d]	
at − 44°C	16.1
at 20°C	327

[a] Refs. 46 and 48–50.
[b] The following equations have been proposed for the liquid state:

$$\rho = 1430 - 2.486\,t - 2.63 \times 10^{-3}\,t^2$$
$$- 5.591 \times 10^{-5}\,t^3 + 8.1 \times 10^{-8}\,t^4 \text{ kg/m}^3$$
$$\epsilon = 9.512 \exp\,(-6.676 \times 10^3\,T)$$
$$\log P\,(\text{kN/m}^2) = 15.2003 - 1867.5/T$$
$$- 0.015865\,T + 0.000015574\,T^2$$

where $t = $ °C and $T = $ K.
[c] To convert J to cal, divide by 4.184.
[d] To convert kN/m² to mm Hg, multiply by 7.5.

Olefinic and Vinyl Monomers. The properties of particular olefins, vinyl, acetylenic, diene, and others, are covered elsewhere (see BUTADIENE POLYMERS; OLEFIN POLYMERS; VINYL CHLORIDE POLYMERS). Not all these monomers are capable of forming olefin–SO₂ copolymers; thermodynamic and chemical composition factors are important.

Thermodynamics of Poly(olefin sulfone) Formation

At the ceiling temperature T_c, the free energy of a mixture of SO₂ and an olefin is the same as the free energy of the 1:1 copolymer. The polymer forms upon initiation only below the T_c (2,3); above the T_c, the polymer is unstable. For a particular feed of monomers and solvent, T_c is independent of initiator and may be determined by extrapolating dilatometric or gravimetric measurements made over a range of temperatures below T_c of rates of polymerization or of molecular weight. The relationship between standard enthalpy and entropy changes of polymerization, monomer activities, and T_c is

$$T_c = \Delta H_p/[\Delta S° + R \ln (a_{SO_2}a_M)] \qquad (1)$$

This relation usually holds, regardless of the polymer solubility in the feed, but is inadequate for the polysulfones of isobutene (75) and 3-methyl-1-butene (76);

plots of ln $(a_{SO_2}a_M)$ vs $1/T$ have two branches. Perhaps for conformational reasons the solid polymer obtained from a polar, SO_2-rich feed differs from that obtained from a less polar, olefin-rich feed.

Enthalpies and entropies of polymer formation obtained with the aid of equation 1 are usually in good agreement with calorimetric values, as shown in Table 2, except for copolymers that may have a feed-related conformation.

Table 2. Heats and Entropies of Polymerization of Olefins with SO_2^a

| | ΔH_p, kJ/repeat unit[b] | | ΔS_p, |
Olefin	From equation 1	Calorimetric	J/(K·repeat unit)[b]
1-butene	86.7 ± 5.9	88.7 ± 4.2	291
1-hexadecene	80.4 ± 5.0	83.3 ± 44.2	279
cis-2-butene	87.1 ± 2.9	84.3 ± 4.2	292
isobutene[c]	78.7^d	62.4	329^d
3-methyl-1-butene[e]	92.1^d		334^d
	44.4		161^f

[a] Ref. 3.
[b] To convert J to cal, divide by 4.184.
[c] Refs. 75 and 77.
[d] SO_2-rich feed.
[e] Ref. 76.
[f] Olefin-rich feed.

Ceiling temperatures are given in Table 3 together with the values adjusted to a common reference state with the aid of equation 1. In the series of 1-olefins and 2-olefins, the T_c decreases with increasing number of carbon atoms, further substitution at the olefinic bond, and branching in the side chain, which may prevent copolymer formation because of steric hindrance. However, the highly substituted olefins 1-methylcyclopentene (83) and 1-methylcyclopropene (84) do form polysulfones, perhaps because of a favorable release of ring strain. 2-Ethyl-1-butene, 2-ethyl-1-hexene, 2,4,4-trimethyl-1-pentene, 2-methyl-2-butene, 4-methyl-2-pentene, and allyl bromide do not form polysulfones above $-80°C$; diallyl ether has a ceiling temperature above 85°C.

Near the ceiling temperature cis- and trans-2-butene yield the same polymer; rapid isomerization accompanies propagation (40,85). When sulfone copolymers of styrene or vinyl chloride are prepared, sequences of —SO_2—M—SO_2— do not form above 40 and 0°C, respectively (86,87). Similarly, the sulfone group is present in maleic anhydride copolymers with butadiene and in maleic anhydride copolymers with 1-hexene prepared below 70°C (88,89). A monomer such as allyl bromide that does not copolymerize with SO_2 even at $-80°C$ (78) may enter into a terpolymer with 1-pentene and SO_2 (90). Isobutene inhibits polysulfone formation when present in feeds with other olefins above its ceiling temperature (2).

Reactivities of Monomers

Olefins. The relative reactivities of a number of olefins in polysulfone formation have been measured by comparing the compositions of terpolymers

Table 3. Ceiling Temperatures for Olefin–Sulfur Dioxide Polymers[a]

Olefin	Observed T_c, °C	Corrected T_c, °C[b]
ethylene[c]	157	
propene	62.5, 87.9	90
1-pentene	44.5	64
1-hexene	60	60
1-hexadecene	48.5	69
isobutene	− 3.5, 6.5	4.5
2-methyl-1-pentene	− 32.5	− 34
3-methyl-1-butene	36.5	36
4,4-dimethyl-1-pentene	14.7	14
cis-2-butene	33	46
2-pentene (cis,trans)	8.5	
2-heptene (cis,trans)	− 38	
cyclopentene	73	102.5
cyclohexene	27.3	24
cycloheptene[d]	5	11
allyl alcohol	66	76
allyl ethyl ether	66.5	68
allyl formate	46	45
allyl acetate	45	45
4-pentenoic acid	66	66
vinyl acetate[e]	− 20	
chloroprene[f]	40	
cis,cis-cycloocta-1,5-diene[g]	145	

[a] From Ref. 78, except where otherwise stated.
[b] Corrected to $[SO_2][M] = 27$ mol^2/L^2, a reference state.
[c] Ref. 59.
[d] Ref. 79.
[e] Ref. 80.
[f] Ref. 81.
[g] Ref. 82.

formed from two olefins with the proportion of those two olefins in the feed in the presence of excess SO_2 (79). Well below the ceiling temperature of either olefin, relative reactivities were not temperature sensitive. At − 20°C, isobutene is more than 800 times as reactive as ethylene; the 1-olefins have an intermediate reactivity (Table 4), which suggests that the attacking radical is electrophilic (79). The relative reactivities of certain olefins with a number of dienes and acetylenes that form alternating 1:1 copolymers with SO_2 have been measured (91) (Table 4) and interpreted in terms of electron delocalization within the product hydrocarbon radical.

Vinyl Monomers. Certain vinyl monomers, such as acrylonitrile (65,92) and methyl methacrylate (93,94), homopolymerize by free-radical initiation in liquid SO_2. Acrylonitrile may copolymerize with methyl methacrylate (MMA) or α-methylstyrene in liquid SO_2 without incorporation of a sulfone group in the chain (95). When a feed mixture of a vinyl monomer such as styrene that copolymerizes with SO_2 as well as with acrylonitrile is made with one of these two monomers, the resulting copolymers contain sulfone groups, but never in a greater proportion to the styrene residues than when acrylonitrile is absent (69,96). On the other hand, terpolymers of SO_2, methyl methacrylate, and butadiene or isoprene may contain more sulfone than diene moieties (97), suggesting the presence of —SO_2—MMA—SO_2— sequences.

Table 4. Reactivities r of Monomers in Polysulfone Formation at −20°C[a]

Monomer	r^b
ethylene	0.053 (±0.001)
propene	1.0 (±0.1)
1-butene	1.3 (±0.1)
1-hexene	1.4 (±0.1)
cis-2-butene	1.6 (±0.1)
trans-2-butene	0.79 (±0.05)
isobutene	46 (±7)
cyclopentene	4.8 (±0.4)
isoprene	11.5 (±1.6)
2,3-dimethylbutadiene	15.1 (±2.6)
1,3-pentadiene	11.4 (±1.9)
1-hexyne	0.062 (±0.006)
phenylacetylene	1.9 (±0.3)
3-methyl-1-butene	1.3 (±0.1)
4,4-dimethyl-1-pentene	1.7 (±0.1)
cycloheptene	3.1 (±0.4)

[a] Depropagation promoted by ceiling temperature was negligible.
[b] Cyclohexene reference, $r = 1$. Olefin terpolymer compositions were measured by ir spectroscopy (79) and the others by ^1H nmr spectroscopy (91).

The ability of vinyl monomers to copolymerize with SO_2 can be explained in terms of the Alfrey-Price Q,e scheme (see also COPOLYMERIZATION). Radicals with electron-withdrawing substituents, and thus a positive e value, do not readily form a bond with the positively charged sulfur atom of SO_2 (5). Copolymers may not form from vinyl compounds that have Q values > 0.2 (5) because of the conjugation of the π-electron system with hetero groups, as in acrylic acid, ethyl acrylate, methacrylic acid, and others with adjacent C=O groups (98). Similar considerations may also apply to the formation of charge-transfer complexes, if these are viewed as an essential first step. However, the esr spectrum detected during the homopolymerization of methyl methacrylate in SO_2 is that of a sulfonyl radical (93).

Electron-withdrawing groups in a monomer influence the structure of the copolymers that form only if they are directly linked to the carbon of the π-bond. Thus ω-chloroolefins, including allyl chloride, form 1:1 alternating copolymers (99,100), but vinyl chloride typically gives the 2:1 copolymer (**2**) (87) and tetrafluoroethylene a random copolymer with an average 67:1 composition (101). The failure of α-methylstyrene (102,103), allyl bromide (78,100), and mono-N-alkyl- or di-N-alkyl derivatives of acrylamide to form polysulfones (103) may be due to steric crowding in the polymer as much as to mechanistic reasons.

Acetylenes and Allenes.   Although n-alkylacetylenes, such as 1-heptyne (104) and phenylacetylene (98,105), polymerize with SO_2, acetylene (106) or disubstituted acetylenes (98) do not. Terpolymers of acetylenes with SO_2 and an olefin or vinyl compound have been prepared (91,98,107,108) and reactivities measured (Table 4). Polysulfones have been prepared from difunctional molecules such as vinylacetylene (106,109–111) and methylallene (112).

Dienes. Both conjugated and nonconjugated dienes form polysulfones. Cyclic monosulfones are obtained from butadiene and isoprene with high SO_2 feeds (31,113–115) by 1,4-addition. Alternating 1:1 copolymers by 1,4-addition are also obtained with chloroprene (113,116), 2,4-dimethylbutadiene (114), 1,3-cyclohexadiene (**11**) (53,114,117–119), and cyclopentadiene (117,120) (which also forms a 2:1 copolymer at $-196°C$ (121)), but 1,2-addition may occur with 1,3-cyclooctadiene (122). The 1:1 copolymer (**12**) with bicyclo[2.2.1]-2,5-heptadiene has no sulfonyl bridge between the olefinic bonds (123) in contrast to poly(cyclopentadiene sulfone) (**13**), a 2:1 copolymer (120).

(11)   (12)   (13)

The formation of a new six-membered ring as part of the main chain is a feature of the copolymers obtained from hexa-1,5-diene (124), whereas two new six-membered rings form from *cis,cis*-1,5-cyclooctadiene (82,125,126) (**14**). The 1,5-dienes bicyclopentene and bicyclohexene may contain five-membered intramolecular rings (**15**) (127), though a six-membered ring is possible. Although diallyl ethers form polysulfones (78), 1,6-dienes, such as dicyclopentenyl ether or dicyclohexenyl ether (127), do not. In terpolymers, intramolecular cyclization of 1,4-dimethylenecyclohexane gives two new six-membered rings (128) (**16**).

(14)   (15)   (16)

However, *p*-xylylene is so reactive in homopolymerizations that the sulfone groups are irregularly placed in copolymers formed in the presence of SO_2 (**17**) (129). With *cis,trans*-1,5-cyclodecadiene, a transannular bond forms and a saturated polysulfone is obtained (**18**) by 1,4-addition, whereas *cis,cis*-1,6-cyclodecadiene gave no copolymer (130) (see also Cyclopolymerization).

(17)   (18)

Other Monomers. Carbon monoxide and maleic anhydride in the feed appear to compete with SO_2 for the sites between the olefin or diene moieties. In some patents, however, maleic anhydride is regarded as displacing the olefinic residues from the sites between the sulfone groups (131–133). Thus carbon monoxide and SO_2 form terpolymers with ethylene (134), propene, *cis,cis*-1,5-cyclooctadiene, or bicyclo[2.2.1]-2,5-heptadiene (82). Maleic anhydride forms terpolymers with SO_2 and 1-hexene (89) or butadiene (88).

Mechanism

Charge-transfer complexes (qv) are sometimes considered to play an important role in the formation of olefin–sulfur dioxide copolymers. The formation of 1:1 copolymers might also be the consequence of the radical of one monomer reacting with the other monomer (in Lewis-Mayo terms, $r_1 = r_2 = 0$) or of complicated radical dissociation processes taking place (135–137). The spontaneous initiations with cyclopentene, cyclohexene, or bicyclo[2.2.1]-2-heptene are associated with traces of peroxide impurities functioning as hydroperoxide initiators (35,36,53,123,138).

Initiation of polysulfone formation and homopolymerization of vinyl monomers in the presence of SO_2 by hydroperoxides at temperatures lower than those at which the peroxide bond spontaneously decomposes is a redox process (36,139,140). Neither di-t-butyl peroxide nor dibenzoyl peroxide initiates the copolymerization of SO_2 with pure cyclohexene, though t-butyl hydroperoxide does (36). The activation energy for the decomposition of the hydroperoxide is low and comparable with that of other redox processes; no evidence for the direct participation of the charge-transfer complex can be found (36). Strong bases decompose the hydroperoxide and suppress the copolymerization of vinyl chloride with SO_2, but polymer forms in the presence of alcohols even with the weakly nucleophilic ketones and ethers (139).

Investigations of the initiation process by esr spectroscopy indicated that the radicals generated by the reaction of SO_2 with t-butyl hydroperoxide in the presence of methanol are $H—O—\dot{S}O_2$ and t-$C_4H_9—O—\dot{S}O_2$, and in water $H—\dot{S}O_2$ (141,142). In the presence of vinyl monomers in the methanolic redox system, spectra from new species were recorded (141,142). With styrene, for example, the principal new spectral feature was attributed to $H—O—SO_2—\overset{\cdot}{Sty}$, but another radical, possibly t-$C_4H_9—O—SO_2—\overset{\cdot}{Sty}$ or t-$C_4H_9—O—\overset{\cdot}{Sty}$ was also present. Alkyl radicals were observed for α-methylstyrene, methyl methacrylate, and other monomers that do not form polysulfones. Species, such as $H—O—SO_2—CH_2—C(CH_3)_2—\dot{S}O_2$, were readily identified from olefins such as 1-hexene, cyclohexene, isobutene, and allyl alcohol, which give 1:1 copolymers with SO_2. Cyclopentene gave an allyl radical by an H-abstraction process, whereas isobutene with a low ceiling temperature gave the alkyl radical:

$$—CH_2—\underset{\underset{CH_3}{|}}{\overset{\overset{CH_3}{|}}{\underset{}{C}}}—\dot{S}O_2 \; = \; —CH_2—\underset{\underset{CH_3}{|}}{\overset{\overset{CH_3}{|}}{C}}\cdot \; + \; SO_2$$

At an electrode, ClO_4^- was thought responsible for initiating the formation of poly(butadiene sulfone), being identified by the onset of the reaction at the oxidation potential of the ClO_4^- ion (143).

The esr spectra of sulfonyl radicals have been recorded during the solution copolymerization of cyclopentene (144) and in the solid state at ambient temperature during the copolymerization with bicyclo[2.2.1]-2-heptene (145), and at low temperatures for vinyl acetate copolymerizations (146). In contrast, the esr spectrum of the alkyl radical alone was observed during the solution copolymerization of methylcyclopropene (84). On γ-radiolysis of a glassy mixture of SO_2 and 2,3-dimethylbutadiene, esr signals were observed from alkylsulfonyl, allyl, and the $SO_2^{\div}$ anion radical (146). Confirmation that the production of olefin–sulfur dioxide copolymers is a free-radical process has been provided by numerous examples where radical inhibitors have been effective (84,102,144–146) and ionic inhibitors ineffective (69,135,147). Moreover, investigations of the formation kinetics of polysulfones found values close to 0.5 for the exponent for the initiator concentration in the rate expression for the formation of poly(styrene sulfone) and poly(cis-2-butene sulfone) (7,102,135,148), indicating a mutual termination process. Though the exact mechanism is not known, it probably involves the combination of a sulfonyl and an alkyl radical (147,149). Initiator exponents greater than 0.5 have often been found (7,72,96,102,148,150,151) and attributed to partial termination by a linear process. In the particular case of poly(vinyl chloride sulfone), termination is affected by isotopic substitution at the methylene site, suggesting a rate-determining C—H scission (72).

In the formation of poly(vinyl chloride sulfone) (**2**), polymer composition depends more on temperature and less on monomer ratio or dilution (72,87,152–159). (The stoichiometry parameter m of formula (**2**) is about 1 at $-90°C$, ~5.5 at 44°C (87), and at about $-10°C$, $m = 2$.) Computer simulation (158,159) showed that Lewis-Mayo behavior ($r_s = 0$ and $r_a = 0.04 \pm 0.01$ (87)), a complex-participation model, and a penultimate-effect model were equally useful at low temperatures; the complex-participation model was applicable up to about $-18°C$. Over the whole temperature range, the observed monomer sequence distributions found by nmr spectroscopy were explained by a model that involved three reversible propagation reactions. At 0°C, —S—V—S— sequences were rare because the corresponding radical depropagates before incorporation into the chain, whereas —S—V—V—S· radicals do not depropagate significantly until much higher temperatures are reached. Similar considerations could apply to poly(styrene sulfone) (69,86,96,116,135,136,140,160–165), poly(acrylamide sulfone) (103), poly(chloroprene sulfone) (81,116,166), and other polysulfones of variable composition. The isomerization of β-deuterated-styrene during the polymerization has provided direct evidence for depropagation in that case (165).

A careful analysis of poly(3-methyl-1-butene sulfone) has shown minor and purely random deviations from the S—C ratio expected of a 1:1 copolymer (61) that typically forms from olefins. This does not imply that a 1:1 charge-transfer complex is formed during the reaction. The observation of stereospecific effects in the formation of poly(trans-2-butene sulfone) (34,167) in terpolymers containing meso units derived from trans-2-butene (168), and of a similar effect in polysulfones of partly deuterated propylene (10), may indicate that the charge-transfer complex participates. In certain terpolymerizations with acrylonitrile or n-butyl acrylate, the increase of 1-butene and sulfonyl residues upon uv irradiation has been attributed to the activation of the olefin–SO_2 charge-transfer complex (169,170).

Microstructure

Infrared spectroscopy cannot provide the same information on chain microstructure (qv) as is provided by nmr spectroscopy, but nonetheless has been very useful. The ir spectra of polysulfones display the strong bands of the asymmetric and symmetric SO_2 stretches near 1320 and 1140 cm^{-1}, as well as the other low frequency bands associated with the sulfone group identified from studies on small alkyl sulfones (171). No complete assignment has been made of the vibration spectrum of a polysulfone, but some analytical information has been obtained by partial assignments.

In the ir spectra of poly(ethylene sulfone)s (59) and poly(vinyl chloride sulfone)s (155), variations in the proportion of the sulfone groups influence the relative absorbances of certain bands. In the spectra of terpolymers of cyclic olefins with 1-olefins and SO_2, methyl bands were identified and used in the quantitative analysis of the two moieties (79). In a similar manner, a phenyl band at 700 cm^{-1} was used to determine the composition of styrene–p-isopropylstyrene terpolysulfones (96). Infrared assignments derived from polychloroprene have been used (166) to determine the proportion of the cis-1,4-, trans-1,4-, and 3,4-units in the corresponding polysulfone. The extent of dehydrochlorination of poly(vinyl chloride sulfone)s (156) and the proportion of the two hydrocarbon residues in terpolymers of isoprene, 2,3-dimethylbutadiene, and SO_2 (147) have been determined. The ir spectra of the polysulfones of cis- and $trans$-2-butenes were found to be identical (79), but the meso residues obtained when the trans olefin reacts at low temperatures can be distinguished by the absence of a band at 1220 cm^{-1} (167). The presence of a six-membered ring in the polysulfone of 1,5-hexadiene was deduced from the similarity of certain band positions to those in the spectrum of pentamethylene sulfone (124). Fourier transform ir spectroscopy has been used with x-ray photochemistry to monitor sulfone bands during degradation (99).

The ir spectra of the following polysulfones have been published: 1:1 poly(styrene sulfone) (164); 1:1 and 1:2 poly(vinyl chloride sulfone) (153); poly(butadiene sulfone) (115); hydrogenated poly(butadiene sulfone) (115); poly(tetramethylene sulfone) (14,115); poly(cis,cis-1,5-octadiene sulfone) (172); terpolymers of SO_2, CO, and propylene, 1-butene, or cis,cis-1,5-octadiene (82); and other poly(olefin sulfone)s (78). Some divergences in band assignments in the spectra of diene polysulfones have been noted (173).

The ^1H nmr shifts or spectra of a number of polysulfones give information on the chain microstructure. Those studied include the polysulfones of ethylene (174), propylene (10), 1-hexene (175), styrene (160–163), 2-butene (175), cyclopentene (175), cyclohexene (175), bicyclo[2.2.1]-2-heptene (145,175), 4-methyl-1-pentene, 5-methyl-1-hexene, and optically active 4-methyl-1-hexene (176). To aid in the assignment of main-chain proton fine structure, tactic poly(propylene sulfone) was prepared from the tactic polysulfide (10). As in the case of a number of simple alkyl sulfones (7,177), protons α with respect to the sulfone group had shifts 3–4 ppm from tetramethylsilane; the main-chain methylene protons of poly(1-olefin sulfone)s had distinctly different shifts. The fine structure of a moiety containing a chiral main-chain carbon is difficult to analyze without partial deuteration of the monomer, as shown by the studies on the polysulfones of propylene (10), styrene (161,163), vinyl chloride (157), and a number of other olefins (175). Sequence effects in the poly(ethylene sulfone)s were readily detected at 220 MHz in acetone; eg, hexamethylene sequences had shifts at 3.10, 1.87, and 1.55 ppm, whereas tetramethylene sequences had shifts at 3.10 and 2.02 ppm (174). From the 60-MHz spectrum, it could be established that the 2:1 copolymer with styrene mainly consisted of regularly ordered —Sty—Sty—SO_2— sequences (160).

The ^{1}H nmr spectra or shifts have been reported for the following conjugated dienes: 1,3-butadiene (114,115), isoprene, 2,3-dimethyl-1,3-butadiene, 1,3-cyclohexadiene, 1-methyl-1,3-butadiene (114), and chloroprene (114,116,166). Coupling constants were obtained from the spectra of the poly(chloroprene sulfone) (116) and poly(propylene sulfone) (10), indicating that in a polar solvent the main chain C—C bond of the latter polymer has the trans conformation to a large extent. It is clear from the ^{1}H spectra of the poly(1-olefin sulfone)s that the moieties are linked head-to-tail and that the dienes have predominantly 1,4-linkages. Extensive studies on poly(chloroprene sulfone)s have shown the value of ^{1}H nmr and ^{13}C nmr spectroscopy for the further elucidation of microstructure and mechanism (116).

The ^{13}C spectra of a number of soluble poly(olefin sulfone)s have been reported (34,39,40,178–180) as well as the spectra of the terpolysulfones formed from 1-butene with 2-butene (168) and 1-hexene with cyclohexene (44). As with the simple alkyl sulfones (181), which show a systematic relationship between carbon-13 shift and structure, the shifts of carbons α with respect to the sulfonyl group are well downfield. However, β-carbons experience an upfield shift effect; the magnitude depends on the rotational states of the nearby C—S bonds and the chirality of nearby carbon atoms. Fine structure in the ^{13}C spectra of β-carbons can be interpreted in these terms (34,168). The triad fine structure of the methine and methylene groups of the main chain and the diad fine structure of the first carbon of the side chain indicate that 1-olefin sulfone copolymers are nearly atactic when prepared by free-radical initiation (34,44,168). The copolymers prepared from *cis*- and *trans*-2-butenes are usually identical (78,182), but a new structure with a meso relationship of the methine carbons within a residue forms from the trans olefin below $-80°C$, according to both ^{13}C nmr (34,167,168) and ir data (167).

Preparation

Polysulfones are formed from an olefin and sulfur dioxide in the gas phase; G values (amount of chemical change per 100 eV of radiation) for initiating the process for 1-butene with high energy electrons have been reported (183). The reaction with ethylene has been initiated by γ-rays (59). At a temperature above the ceiling temperature, oily sulfinic acids are formed instead of white powders (183,184). Following uv or high energy irradiation, copolymerization has also been carried out in the solid state with isobutene (57), butadiene (185), vinyl acetate (148,186), and styrene (187) at low temperatures and bicyclo[2.2.1]-2-heptene at ambient temperatures (145).

Emulsion and suspension methods have been described (4,9,188–193); the salts of alkarylsulfonic acids are used as surfactants in aqueous SO_2 solutions. When the polymerization is homogenous, the feeds are mixtures of SO_2 with the other monomers and perhaps a solvent. Stainless steel bombs (2,170,194) or sealed tubes (2,163,194) are used as reaction vessels at elevated temperatures where the pressure may rise to >20 MPa (several hundred atmospheres) (59,195). Below the boiling points of the olefin and SO_2, a round-bottomed flask (130), an open tube (44,196), or even a thin-walled nmr tube can be used (165). Certain monomers such as *cis*-2-butene, 1-pentene, or cyclopentene may react so rapidly that the SO_2 boils off and polymer is ejected from the container (196). The sulfone co-polymers of ethylene, 1-propene, isobutene, and others precipitate in the liquid phase. Others, such as those of 1-butene, 2-butene, and cyclohexene, may be kept in solution by an excess of SO_2 (2,4,197) or by a solvent such as chloroform (61,138,151,198), benzene (166,198), or toluene (199). In the preparation of poly(cyclooctadiene sulfone) (126), tetramethylene sulfone is used as a diluent and, by promoting the intramolecular cyclization of the second double bond of the monomer (14), prevents cross-linking. To avoid branching and cross-links, the olefin should be free of peroxides (200). The cationic homopolymerization of styrene in SO_2 can be suppressed by the addition of acetone or dimethylformamide (31,69,135). The SO_2 can be introduced into the reaction vessel in the form of sodium bisulfite (201) and of the cyclic sulfone monomer (202). Continuous pro-cesses for the production of poly(olefin sulfone)s have been described (203,204).

For free-radical initiation, oxygen (2,53,54,125,138,205), ozonides (206), per-oxides (2), hydroperoxides (72,83,116,127,130,140–142), hydrogen peroxide–par-aldehyde (207,208), and ascaridole with a mineral acid (98,107,125,126,209) have been used. Ammonium nitrate (4,130) and nitrates of silver (122,151,180), lithium (2,128), and potassium (180) are effective in some copolymerizations (210,211). Silver sulfite and sulfate have been used (212) as well as cuprous chloride (213), tetraethyllead (2), magnesium perchlorate (214), and perchloric, nitrous, or nitric acids (2,215). Other initiators are amine and aniline oxides (216), nitric oxide, and nitrogen dioxide (217). Azobisisobutyronitrile (AIBN) is decomposed by heat (135,136,161,162,218) or light (135,170), but, for some applications, diphenyl-azosulfone is preferred at 0°C (161). The 2:1 complex of SO_2 with pyridine initiates the copolymerization of styrene and SO_2 in the presence of carbon tetrachloride (71).

Initiation is also effected by various forms of ionizing radiation or uv light on monomer feeds. For γ-irradiation, G values as high as 500,000 repeat units per 100 eV have been reported for a diene polysulfone (147) and, for irradiation with β-particles, the value of 500 was reported for poly(1-butene sulfone) (183). γ-Rays have been used frequently (57,59,103,147,153,157,163,164,185,194,196), occasionally sensitized by CCl_4 (72). Others have used β-particles (219) and α-particles (125). The early experiments with sunlight (28) probably depended on uv light with a wavelength of about 350 nm, which has been used by others subsequently (2,61,135,147,198,207).

Macroradicals are prepared by mechanical treatment (220), chemical treat-ment, eg, ozone (221), AIBN (222), or by irradiation (222–224), followed by the formation of graft polysulfones on polyethylene, polypropylene, polystyrene, and others.

The molecular weight of polysulfones is controlled by chain-transfer agents such as $BrCCl_3$ (162,173,180), CBr_4 (144), or thiols (144,225,226). Pyridine com-plexes with the sulfonyl radical (218) and oxygen (147) may quench the excited state of the charge-transfer complex (227). Impurities in acetone may inhibit the copolymerization of 2-methyl-1-pentene (227); the normal inhibitors of free-rad-ical polymerizations, such as benzoquinone, are also effective (31,102,145, 147,154,180). Ethylamine accelerates the copolymerization of cyclopentene (61).

If the polymer does not precipitate during formation, it may be precipitated by pouring the product into a volatile nonsolvent, such as methanol (200). Dissolution, reprecipitation, and filtration followed by vacuum drying at a moderately elevated temperature (228) reduced trapped impurities.

Properties

Solution Properties and Chain Dynamics.　Poly(methylene sulfone) and poly(propylene sulfone), where the sulfone group forms a large proportion of the repeat unit, dissolve in polar solvents, eg, concentrated sulfuric acid or dimethyl sulfoxide (11,15,31,33,229). Other polysulfones, such as those formed from 1- and cyclic olefins, dissolve in common solvents, such as acetone, chloroform, and benzene (229). The polysulfones of the 1-olefins with long alkyl groups dissolve in hydrocarbon fuel oils (131–133) and, if made with a small proportion of ionizing heteroatomic groups, cannot be extracted into water. The polysulfone of diallylamine is water soluble (230). Poly(acrylamide sulfone)s become insoluble in water with increasing sulfone content (210).

A number of poly(olefin sulfone)s have been fractionated and the solution properties measured by osmometry or light scattering and viscometry; the Mark-Houwink parameters established are given in Table 5; K_θ was obtained by the extrapolation of data from good solvents (200,232,235) and for poly(1-hexene sulfone) in θ-solvents. In the latter case, K_θ depended on solvent polarity.

Table 5. Mark-Houwink Parameters of Poly(olefin sulfone)s in Good Solvents[a]

Olefin	Solvent	Temp, °C	$K \times 10^2$	a	Ref.
1-butene	acetone	30	4.3	0.54	231
1-hexene	benzene	25	8.9	0.70	232
2-methyl-1-pentene	chloroform	20	0.59	0.81	232
cyclopentene	dioxane	25	0.53	0.76	200
cyclohexene	benzene	25	1.33	0.65	200
bicyclo[2.2.1]-2-heptene	chloroform	25	0.158	0.78	233
styrene, 2:1 copolymer	tetrahydrofuran	30	0.39	0.78	234

[a] Mark-Houwink equation: $[\eta] = KM^a$, mL/g.

Gel-permeation chromatography has been used to characterize poly(olefin sulfone)s (83,236–238). With some polymers the elution behavior was influenced by polymer–solvent and polymer–column interactions (238). Low angle laser-light scattering has also been used (239).

Examination of poly(1-olefin sulfone)s by dielectric means in a nonpolar solvent at low (3%) (37,38) and high (25%) (240) concentrations revealed a dipolar relaxation process dependent on molecular weight and associated with the rotation of the molecule. This process was absent from solutions of polysulfones of 2-olefins and cyclic olefins, such as cyclohexene (200).

Simple calculations (45,241), later supported by more detailed molecular mechanics calculations (242), demonstrated the unusual importance of electrostatic interactions between adjacent sulfone dipoles as well as side-chain steric effects as factors controlling bond conformations and the chain configuration. The strong electrostatic ordering associated with gauche main-chain C—C bonds create stiff sections that are helices in the simplest accounts (41–44). These stiff sections are linked by residues where thermal or steric effects have promoted the C—C bonds to the trans conformation and decoupled the adjacent dipoles. The polysulfone of styrene, with four carbon atoms between sulfonyl units (39,180), and the polysulfones of 2- or cyclic olefins with trans C—C bonds (241,243) are significantly more flexible than the poly(1-olefin sulfone)s (39,40). The inclusion of a small proportion of 2-olefin units within a predominantly poly(1-butene sulfone) chain does not alter the chain dynamics, according to ^{13}C nmr measurements (168).

Physical Properties.   Crystallinity has been readily detected by x-ray diffraction studies in the poly(ethylene sulfone)s of variable composition (59), in poly(butadiene sulfone) (115), and in polysulfones with three (15), four, five, or six (14) main-chain carbon atoms in the repeat unit. In these saturated polysulfones, the main chain probably adopts a planar zigzag configuration (14,15) and fibers can be drawn (14); however, no complete crystal structure has been reported. No crystallinity was detected in some poly(1-olefin sulfone)s by x-ray diffraction (244,245), but light scattering revealed crystallinity in some hot-pressed samples and films of poly(1-butene sulfone) cast from solution (246). Density and gas-permeability values suggest that the C_{18} 1-olefin copolymers may have ordered side chains (244,245).

The densities of polysulfones are given in Table 6. The density increases with increasing sulfone to hydrocarbon ratio. The glass-transition temperatures are shown in Table 7 and melt temperatures in Table 8. Examination of the polysulfones of 1-methyl- or 3-methylcyclopentene by torsional braid analysis revealed no T_g (83), but poly(1-olefin sulfone)s with long side chains have a second exothermic transition below the softening point (246). With a decreasing proportion of sulfone groups, there is a transition from brittleness, inflexibility, and hardness to elastomeric behavior (2,59,244,245). Tensile modulus is a minimum (35 N/mm^2 or 5076 psi) and elongation at the yield point is a maximum (14%) for the C_{16} olefin (244,245). Tensile strengths of the propylene and 1-butene copolymers are 2–3 times higher (2). Gaseous permeabilities are also maximum for C_{16} 1-olefin copolymers; for oxygen and carbon dioxide, the values are 2×10^{-6} and 8.5×10^{-6} mol/(m·s·Pa), respectively (244–246). The permeabilities of terpolymers support the view that long side chains are ordered in the solid state (250,251).

Table 6. Densities of Polysulfones[a]

Olefin	ρ, kg/m^3
propylene	1457
1-hexene	1220[b]
1-octadecene	990
1-butene	1245
1-hexadecene	990[b]
isobutene	1406

[a] Refs. 82, 244, and 245.
[b] The density varies smoothly with chain length between 1-hexene and 1-hexadecene (244,245).

Table 7. Glass-transition Temperatures of Polysulfones[a]

Olefin	T_g, °C
1-butene	81–95
bicyclo[2.2.1]-2-heptene	117
1-hexadecene	77
cyclopentene	82
styrene, 2:1	180–200
1-butene–cyclopentene[b]	57

[a] Refs. 225 and 246–248.
[b] 50% of each; Ref. 225 gives T_g values for other terpolymers.

Table 8. Melting Points of Polysulfones[a,b]

Olefin	T_m^b, °C
ethylene[c]	135
1-butene	150–160
1-hexene	76
cyclohexene	200–205
bicyclo[2.2.1]-2-heptene	240–290
norbornene acetate	240–245
propylene	280
isobutene	230
1-octadecene	45
cis,cis-1,5-cyclooctadiene	260
norbornene nitrile	275–280

[a] Refs. 11, 59, 82, 245, and 249.
[b] Melting is often accompanied by decomposition. For noncrystalline polymers, T_m is the temperature at which the polymer melts and leaves a trail when moved across a hot metal bar (249).
[c] 31% SO_2.

Poly(propylene sulfone) can be molded between 180 and 200°C, and poly(1-butene sulfone) between 125 and 180°C (2). Hot-pressed films have been prepared at 120°C from the polysulfones of 1-butene, cyclopentene, and bicyclo[2.2.1]-2-heptene at high pressures (140 MN/m^2 or 20,300 psi) which prevented decomposition (246). Plasticizers and stabilizers (9,249) against thermal decomposition are not available.

Stability and Degradation. The polysulfones of ethylene, propylene, butadiene, and other dienes (2,31) as well as the poly(methylene sulfone)s (14,15) dissolve in strong acids and are resistant to concentrated sulfuric acid. Poly(1,4-decalin sulfone) (**18**) (130) and poly(methylene sulfone)s (14) are resistant to alkalis, but the poly(olefin sulfone)s have the characteristic susceptibility to alkalis of β-disulfones. Liquid ammonia, aqueous sodium hydroxide, and amines degrade poly(olefin sulfone)s (33) to cyclic disulfones, eg, the copolymer of cyclohexene to a *trans,cis*-tricyclic structure (252). The thermal stability of poly(butadiene sulfone) increases by hydrogenation of the double bonds (115,202), and that of poly(1,3-cyclohexadiene sulfone) is increased by aromatization (118,119).

The thermal instability of poly(alkyl sulfone)s has been attributed to the weakness of the C—S bond (240 kJ/mol or 57.3 kcal/mol compared to 345 kJ/mol or 82.4 kcal/mol for the C—C bond) (225,253,254) and to a decomposition mechanism, depending on the presence of H atoms β to the sulfonyl group (11,199,255,256). Early experiments on poly(1-butene sulfone) revealed a tendency for the thermally promoted unzipping or depropagation reaction at high temperature to yield SO_2 and olefin; this has been confirmed by gas chromatography and mass spectroscopy or vapor phase ir studies (257) on copolymers of ethylene (253), propylene (253,258), 1-butene (253,254,258,259), 1-hexene (253, 254,258), cyclopentene (225), cyclohexene (253), 2-butene (253), bicyclo[2.2.1]-2-heptene (225), styrene (254,258,259), and certain dienes (256). Terpolymers have been similarly studied (260), eg, poly(cyclopentene-co-1-butene sulfone) (225) and poly(bicycloheptene-co-1-octadecene sulfone) (225,260), as well as random (260) and block (225) terpolymers of SO_2, an olefin, and methyl methacrylate or methacrylic acid.

Of the 1-olefin copolymers examined, the 2:1 poly(styrene sulfone) yielded large amounts of dimer (225,259) and the cis-2-butene polysulfone large amounts of the trans olefin (260). The polysulfones of 1-butene, cyclopentene, and bicycloheptene rearranged to products that included such large molecules as indene and naphthalene (260); free-radical mechanisms have been suggested.

A more recent study of the thermal decomposition of poly(1-butene sulfone) above 200°C (261) confirmed the first-order decomposition found previously (199). An initial random chain scission is followed by a rapid free-radical depropagation; the first-order activation energy was 196 (± 5) kJ/mol (46.8 kcal/mol). The more complex thermal behavior was associated with the scission of weak links. At 140°C, more olefin than SO_2 was released in contrast to the radiolytic degradation at that temperature (262).

When poly(olefin sulfone)s are exposed to high energy electrons and γ-rays, main-chain scission rather than cross-linking takes place (8), resulting in a reduction in molecular weight (237,263). Cross-linking has been noticed when the side chains, as in poly(1-hexadecene sulfone), are long and bulky (244), but, more typically, weak C—S bonds are affected in polysulfones with G scission values in vacuo of 10–12 scissions per 100 eV for γ-rays (237). The polysulfones of cyclopentene and bicyclo[2.2.1]-2-heptene are less sensitive ($G \sim 6$) (225), but the 2:1 copolymer of styrene and SO_2 has a lower value ($G = 1.1$ (264)). At low temperatures, the 1:1 copolymers lose SO_2, but, as the temperature rises toward the ceiling temperature of each copolymer, the olefin yield approaches that of SO_2 which also may rise up to a thousandfold; traces of hydrogen are detectable (228,237,262,265). The released olefins isomerize, notably poly(1-butene sulfone) and poly(3-methyl-1-butene sulfone), giving 2-butene and 3-methyl-2-butene, respectively, by hydride shifts (266). The mechanisms of the radiolytic degradations are complex and involve free-radical and cationic species (265,267–270). Esr signals of sulfonyl and alkyl radicals have been detected in polysulfones γ-irradiated at -196°C. The alkyl radicals predominate, but form sulfonyl radicals on warming (270). For polysulfones of mono-, di-, and trisubstituted olefins as well as ethylene, depropagation at high temperatures is accompanied by cationic oligomerization or homopolymerization of olefin, reducing the ratio of evolved sulfur dioxide olefin unless a cation scavenger is present (228,271,272). Poly(1-hexene sulfone) is an exception to this rule; its cationic homopolymerization appears to be unfavorable (266,273). That copolymer and the copolymers of 2-butene and of cyclohexene had enhanced depropagation rates at their individual ceiling temperatures (262). In general, the depropagation rates at ambient temperature correlate negatively with ceiling temperature so that poly(2-methyl-1-pentene sulfone), with a ceiling temperature of about $-34°C$, has a particularly high decomposition rate; its $G_{(SO_2)}$ value is 3000 (± 300) compared to values of about 100 for the other olefins examined (228,262). The degradation mechanism proposed in Ref. 8 has been confirmed in γ-irradiation studies on poly(2-methyl-1-pentene sulfone) (228). On the other hand, an initial SO_2 extrusion process might produce a polyolefin which is decomposed by further exposure (274).

Consistent with their high G scission values for γ-irradiation is the sensitivity of olefin–sulfur dioxide copolymers to high energy electrons. Doses on the order of $1-4 \times 10^{-2}$ C/m^2 reduce the molecular weight by main-chain scission, permitting solution development of patterns in thin films (45,83,247,275,276). This treatment has been commercialized for poly(1-butene sulfone) (277). Because of a low ceiling temperature, poly(2-methyl-1-pentene sulfone) films disintegrate faster than other polysulfones in a process termed vapor development. A mechanism of random scission is indicated, followed by depolymerization if the temperature is greater than 100°C, and, at lower temperatures, SO_2 loss is more rapid than that of organic residues (278,279). In contrast to the 1:1 copolymers, the 2:1 poly(styrene sulfone) is less sensitive to β-particles and no vapor development occurs; two adjacent styryl residues prevent depropagation (280). Poly(1-butene sulfone) is less susceptible to degradation by ion beams than by electrons of a similar energy (281).

In the presence of oxygen in solution, the 2:1 poly(styrene sulfone) suffers main-chain scissions when irradiated with short pulses of 265-nm light (282); poly(1-butene sulfone) behaves similarly if a sensitizer such as benzene is present in the solution. According to a vacuum uv study, the inability of complete vapor development of poly(1-butene sulfone) resists in lithography has been attributed to recombination and cross-linking reactions favored by high doses of irradiation (283).

Poly(5-hexene-2-one sulfone) (284), poly(cyclopentene sulfone) (225), and the sulfone copolymers of styrene and of other vinyl arenes (285) have been investigated as possible positive photoresists. Photosensitization of the nonaromatic polymers was enhanced by the inclusion of benzophenone (284) and CBr$_4$ (225,285); electron-beam sensitization is also enhanced (285). Lithography with x rays on poly(1-butene sulfone) has also been examined (286,287). Sensitivity to x rays and to uv light is not yet sufficient for commercial applications (8).

Details of the technical features of commercial practices in the use of polysulfones as electron-beam resists have appeared in a number of papers, reviews, and patents, which describe polymer preparation, film formation, film irradiation, and solvent treatment for film dissolution (8,225,288–301).

Uses

Thermal instability near the molding temperature has prevented olefin–sulfur dioxide copolymers from being used as bulk thermoplastics, but certain specialty uses have been developed. Expanded-foam insulation material has been obtained, but contained too much corrosive SO_2 (190,191). Gypsum and other minerals have been coated with various polysulfones to impart compatibility with polymeric hydrocarbons, such as polyethylene, to achieve a reinforcing effect (302). Certain polysulfone-based formulations, such as Stadis 450 (DuPont), have found a market as antistatic additives for hydrocarbon fuels (131–133). The biocompatibility of the poly(olefin sulfone)s (244,248) coupled with their high permeability to oxygen and to CO_2 has led to their consideration for use in the membranes of heart–lung machines (244,245) and other medical purposes (250,251). The most significant use at present exploits sensitivity to high energy radiation. Poly(1-butene sulfone) (PBS) was developed at Bell Laboratories, now known as AT&T Bell Laboratories, as a resist in the electron-beam fabrication of chromium photomasks using electron-beam exposure. It is available as a filtered solution ready for spin coating at a cost of approximately $350/L (Mead Chemical Co., Rolla, Mo., and Chisso Corp., Japan). Current interest lies in improving the resistance of polysulfones to dry etching (8,294).

BIBLIOGRAPHY

"Olefin–Sulfur Dioxide Copolymers" in *EPST* 1st ed., pp. 460–485, Niichiro Tokura, Osaka University.

1. O. Grummitt and A. Ardio, *J. Chem. Educ.* **23**, 73 (1946).
2. R. D. Snow and F. E. Frey, *Ind. Eng. Chem.* **30**, 176 (1938).
3. F. S. Dainton and K. J. Ivin, *Q. Rev. London* **12**, 61 (1958).
4. W. W. Crouch and J. E. Wicklatz, *Ind. Eng. Chem.* **47**, 160 (1955).
5. N. Tokura, *Enka Biniiru To Porima* **7**, 10 (1967).
6. E. J. Goethals, *J. Macromol. Sci. Rev. Macromol. Chem.* **2**, 73 (1968).
7. K. J. Ivin and J. B. Rose, *Adv. Macromol. Chem.* **1**, 335 (1968).
8. M. J. Bowden and J. H. O'Donnell in N. Grassie, ed., *Developments in Polymer Degradation*, Vol. 6, Elsevier Applied Science Publishers, Ltd., Barking, UK, 1985, p. 21.
9. E. M. Fettes and F. O. Davis, *High Polym.* **XIII**(III), 225 (1962).
10. K. J. Ivin and M. Navratil, *J. Polym. Sci. Part A-1* **8**, 3373 (1970).
11. E. Gipstein, E. Wellisch, and O. J. Sweeting, *J. Org. Chem.* **29**, 207 (1969).
12. J. K. Stille and J. A. Empen, *J. Polym. Sci. Part A-1* **5**, 273 (1967).
13. U.S. Pat. 2,534,366 (Dec. 19, 1950), H. D. Noether (to Celanese Corp.).
14. H. D. Noether, *Text. Res. J.* **28**, 533 (1958).
15. R. Cook, *J. Polym. Sci. Polym. Chem. Ed.* **16**, 3001 (1978).
16. V. Foldi and W. Sweeny, *Makromol. Chem.* **72**, 208 (1964).
17. E. Wellisch, E. Gipstein, and O. J. Sweeting, *J. Polym. Sci. Polym. Lett. Ed.* **2**, 35 (1964).
18. R. Fusco, S. Rossi, S. Maioranor, and G. Pagani, *Gazz. Chim. Ital.* **95**, 774 (1965).
19. N. Tokura, T. Nagai, and Y. Shirota, *J. Polym. Sci. Part C* **23**, 793 (1966).
20. K. R. Huffman and D. J. Casey, *J. Polym. Sci. Polym. Chem. Ed.* **23**, 843 (1985).
21. G. B. Butler and A. J. Sharp, *J. Polym. Sci. Polym. Lett. Ed.* **9**, 125 (1971).
22. U.S. Pat. 3,213,068 (Oct. 19, 1965), E. J. Frazza (to American Cyanamid Co.).
23. U.S. Pat. 2,505,366 (Apr. 25, 1950), D. L. Schoene (to U.S. Rubber Co.).
24. E. M. Pearce, *J. Polym. Sci.* **40**, 273 (1959).
25. F. C. Foster, *J. Am. Chem. Soc.* **74**, 2299 (1952).
26. U.S. Pat. 3,306,912 (Nov. 8, 1966), H. E. Fritz and R. P. Yunick (to Union Carbide).
27. G. J. Van Amerongen, *J. Polym. Sci.* **6**, 633 (1951).
28. Brit. Pat. 11,635 (May 11, 1914), F. E. Mathews and H. M. Elder.
29. W. Solonia, *J. Russ. Phys. Chem. Soc.* **30**, 826 (1898); *Chem. Zentre* **1**, 249 (1899).
30. Ger. Pat. 236,386 (Aug. 19, 1910) (to B.A.S.F.).
31. H. Staudinger and B. Ritzenthaller, *Ber.* **68**, 455 (1935).
32. D. S. Frederic, H. D. Cogan, and C. S. Marvel, *J. Am. Chem. Soc.* **56**, 1815 (1934).
33. C. S. Marvel and E. D. Weil, *J. Am. Chem. Soc.* **76**, 61 (1964).

34. A. H. Fawcett, F. Heatley, K. J. Ivin, C. D. Stewart, and P. Watt, *Macromolecules* **10**, 765 (1977).
35. B. Oster and R. W. Lenz, *J. Polym. Sci. Polym. Chem. Ed.* **15**, 2479 (1977).
36. G. Sartori and R. D. Lundberg, *J. Polym. Sci. Polym. Lett. Ed.* **10**, 583 (1972).
37. T. W. Bates, K. J. Ivin, and G. Williams, *Trans. Faraday Soc.* **63**, 1964 (1967).
38. *Ibid.*, p. 1976.
39. R. E. Cais and F. A. Bovey, *Macromolecules* **10**, 757 (1977).
40. W. H. Stockmayer, A. A. Jones, and T. L. Treadwell, *Macromolecules* **10**, 762 (1977).
41. A. H. Fawcett and S. Fee, *Macromolecules* **15**, 933 (1982).
42. K. Matsuo, M. L. Mansfield, and W. H. Stockmayer, *Macromolecules* **15**, 675 (1982).
43. S. Mashimo, K. Matsuo, R. H. Cole, P. Winsor, and W. H. Stockmayer, *Macromolecules* **19**, 682 (1986).
44. R. H. Cole, P. Winsor, A. H. Fawcett, and S. Fee, *Macromolecules* **20**, 157 (1987).
45. M. J. Bowden and L. F. Thompson, *Appl. Polym. Symp.* **23**, 99 (1974).
46. U. H. F. Sander, H. Fischer, U. Rothe, and R. Kola, *Sulphur, Sulphur Dioxide and Sulphuric Acid*, Verlag Chemie International, Inc., Deerfield Beach, Fla., 1984, p. 156.
47. L. Brethrick, *Hazards in the Chemical Laboratory*, 3rd ed., The Royal Society of Chemistry, London, 1981.
48. T. C. Waddington, *Nonaqueous Solvent Systems*, Academic Press, Inc., New York, 1965, p. 1253.
49. D. F. Burrow in J. J. Lagowski, ed., *The Chemistry of Nonaqueous Solvents*, Vol. 3, Academic Press, Inc., Orlando, Fla., 1970.
50. R. C. Weast, ed., *CRC Handbook of Chemistry and Physics*, 62nd ed., CRC Press, Inc., Boca Raton, Fla., 1985.
51. P. J. Elving and J. M. Markowitz, *J. Chem. Ed.* **37**, 75 (1960).
52. M. Kamoun, *J. Raman Spectrosc.* **8**, 225 (1979).
53. W. F. Seyer and E. G. King, *J. Am. Chem. Soc.* **55**, 3140 (1932).
54. W. F. Seyer and L. Hodnett, *J. Am. Chem. Soc.* **58**, 996 (1936).
55. C. S. Marvel and F. J. Glavis, *J. Am. Chem. Soc.* **60**, 2622 (1938).
56. D. Booth, F. S. Dainton, and K. J. Ivin, *Trans. Faraday Soc.* **55**, 1293 (1959).
57. M. Ito and Z. Kuri, *Kogyo Kagaku Zasshi* **69**, 2009 (1966).
58. R. E. Cais and J. H. O'Donnell, *Eur. Polym. J.* **11**, 749 (1975).
59. P. Colombo, J. Fontana, and M. Steinberg, *J. Polym. Sci. Part A-1* **6**, 3201 (1968).
60. F. de Carli, *Gazz. Chim. Ital.* **57**, 347 (1926).
61. M. Raetzsch and G. Borman, *Plaste Kautsch.* **22**, 937 (1975).
62. P. B. Ayscough, K. J. Ivin, and J. H. O'Donnell, *Trans. Faraday Soc.* **61**, 1601 (1965).
63. B. G. Brady and J. H. O'Donnell, *Trans. Faraday Soc.* **64**, 23 (1968).
64. N. Tokura, *Synthesis* **12**, 639 (1971).
65. M. Matsuda and K. Tokura, *J. Polym. Sci. Part A* **2**, 4281 (1964).
66. M. M. Rogic and D. Masilamani, *J. Am. Chem. Soc.* **99**, 5219 (1977).
67. D. Masilamani and M. M. Rogic, *J. Am. Chem. Soc.* **100**, 4634 (1978).
68. O. do Couto Filho and A. de Souza Gomes, *J. Polym. Sci. Polym. Lett. Ed.* **9**, 891 (1971).
69. N. Tokura, M. Matsuda, and K. Arakawa, *J. Polym. Sci. Part A* **2**, 3355 (1964).
70. P. Ghosh and K. F. O'Driscoll, *J. Polym. Sci. Part B* **4**, 519 (1966).
71. M. Matsuda and T. Hirayama, *Kogyo Kagaku Zasshi* **70**, 2022 (1967).
72. R. E. Cais and J. H. O'Donnell, *J. Macromol. Sci. Chem.* **A17**, 1407 (1982).
73. G. R. Ranade, R. Chandler, C. A. Plank, and W. L. S. Laukhuf, *Polym. Eng. Sci.* **25**, 164 (1985).
74. K. Imai, T. Shiomi, Y. Tezuka, and M. Satoh, *Makromol. Chem. Rapid Commun.* **6**, 413 (1985).
75. R. E. Cook, K. J. Ivin, and J. H. O'Donnell, *Trans. Faraday Soc.* **61**, 1887 (1965).
76. B. G. Brady and J. H. O'Donnell, *Trans. Faraday Soc.* **64**, 29 (1968).
77. K. J. Ivin, W. A. Keith, and H. Mackle, *Trans. Faraday Soc.* **55**, 262 (1959).
78. R. E. Cook, F. S. Dainton, and K. J. Ivin, *J. Polym. Sci.* **26**, 351 (1957).
79. J. E. Hazel and K. J. Ivin, *Trans. Faraday Soc.* **58**, 176 (1962).
80. Z. Kuri and T. Yoshimura, *J. Polym. Sci. Part B* **1**, 107 (1963).
81. F. Hrabák, J. Blazek, and J. Webr, *Makromol. Chem.* **97**, 9 (1966).
82. A. H. Frazer, *J. Polym. Sci. Part A* **3**, 3699 (1965).
83. R. J. Himics, M. Kaplan, N. V. Desai, and E. S. Poliniak, *Polym. Eng. Sci.* **17**, 406 (1977).
84. S. Iwatsuki, T. Kokubo, and Y. Yamashita, *J. Polym. Sci. Part A-1* **6**, 2411 (1968).
85. F. S. Dainton, J. Diaper, K. J. Ivin, and D. R. Sheard, *Trans. Faraday Soc.* **53**, 1269 (1957).
86. R. E. Cais, J. H. O'Donnell, and F. A. Bovey, *Macromolecules* **10**, 254 (1977).

87. R. E. Cais and J. H. O'Donnell, *J. Polym. Sci. Polym. Lett. Ed.* **15**, 659 (1977).
88. S. Iwatsuki, S. Amano, and Y. Yamashita, *Kogyo Kagaku Zasshi* **70**, 2027 (1967).
89. F. C. Cheung, Ph.D. thesis, The Queen's University of Belfast, Belfast, UK, 1985.
90. C. S. Marvel and W. H. Sharkey, *J. Org. Chem.* **9**, 113 (1944).
91. K. J. Ivin and N. A. Walker, *J. Polym. Sci. Part A-1* **9**, 2371 (1971).
92. M. Matsuda, S. Abe, and N. Tokura, *J. Polym. Sci. Part A* **2**, 3877 (1964).
93. J. J. Kearney, H. G. Clark, V. Stannett, and D. Campbell, *J. Polym. Sci. Part A-1* **9**, 1197 (1971).
94. N. Tokura, M. Matsuda, and F. Yazaki, *Makromol. Chem.* **42**, 108 (1960).
95. M. Matsuda, M. Iino, and N. Tokura, *Makromol. Chem.* **65**, 232 (1963).
96. *Ibid.*, **52**, 98 (1962).
97. J. J. Kearney, H. G. Clark, and V. Stannett, *Makromol. Chem.* **143**, 163 (1971).
98. L. L. Ryden, F. J. Glavis, and C. S. Marvel, *J. Am. Chem. Soc.* **59**, 1014 (1937).
99. M. Kaplan, *Polym. Eng. Sci.* **23**, 957 (1983).
100. M. S. Kharasch and E. Esterfeld, *J. Am. Chem. Soc.* **62**, 2559 (1940).
101. U.S. Pat. 2,411,722 (Nov. 26, 1946), J. Harman and R. M. Joyce (to E.I. du Pont de Nemours & Co., Inc.).
102. N. Tokura, M. Matsuda, and Y. Ogawa, *J. Polym. Sci. Part A* **1**, 2965 (1963).
103. R. E. Cais and G. J. Stuck, *Polymer* **19**, 179 (1978).
104. L. L. Ryden and C. S. Marvel, *J. Am. Chem. Soc.* **57**, 2311 (1935).
105. C. S. Marvel and W. W. Williams, *J. Amer. Chem. Soc.* **61**, 2710 (1939).
106. U.S. Pat. 2,225,266 (Dec. 17, 1940), F. E. Frey, R. D. Snow, and L. H. Fitch (to Phillips Petroleum Co.).
107. C. S. Marvel and F. J. Glavis, *J. Am. Chem. Soc.* **60**, 2622 (1938).
108. C. S. Marvel and W. H. Sharkey, *J. Am. Chem. Soc.* **61**, 1603 (1939).
109. A. Gulyaeva and T. Dauguleva, *Caoutch. Rubber USSR* **1**, 53 (1937); *Chem. Abstr.* **32**, 3754 (1938).
110. Y. M. Slobodin, *J. Gen. Chem. USSR* **16**, 1831 (1946).
111. Ger. Pat. 738,003 (July 1, 1943), P. Feiler (to I. G. Fabenindustrie Akt.-Ges.).
112. A. Gulyaeva and T. Dauguleva, *Caoutch. Rubber USSR* **1**, 49 (1938).
113. J. J. Kearney, V. Stannett, and H. G. Campbell, *J. Polym. Sci. Part C* **16**, 3411 (1968).
114. K. J. Ivin and N. A. Walker, *J. Polym. Sci. Polym. Lett. Ed.* **9**, 901 (1971).
115. R. S. Bauer, H. E. Lunk, and E. A. Youngman, *J. Polym. Sci. Part A-1* **8**, 1915 (1970).
116. R. E. Cais and G. J. Stuck, *Macromolecules* **13**, 415 (1980).
117. U.S. Pat. 3,476,716 (Nov. 4, 1969), L. De Vries (to Chevron Research Corp.).
118. D. J. Ballard and J. M. Keys, *Eur. Polym. J.* **11**, 565 (1975).
119. T. Yamaguchi and K. Nagai, *Kobunshi Kagaku* **28**, 129 (1971).
120. T. Yamaguchi and T. Ono, *Chem. Ind. London* **24**, 769 (1968).
121. Jpn. Pat. 7,635,000 (Mar. 27, 1976), D. Yashiro, S. Kameyama, and K. Uemoto (to Agency of Industrial Science and Technologies).
122. T. Yamaguchi and K. Nagai, *Kobunshi Kagaku* **26**, 809 (1969).
123. G. Vanhaeren and G. B. Butler, *Polym. Prepr. Am. Chem. Soc. Div. Polym. Chem.* **6**, 709 (1965).
124. J. K. Stille and D. W. Thompson, *J. Polym. Sci.* **62**, S118 (1962).
125. A. H. Frazer and W. P. O'Neill, *J. Amer. Chem. Soc.* **85**, 2613 (1963).
126. A. H. Frazer and W. P. O'Neill in E. M. Fettes, ed., *Macromolecular Synthesis*, Vol. 7, John Wiley & Sons, Inc., New York, 1979, p. 63.
127. K. Meyersen and J. Y. C. Wang, *J. Polym. Sci. Part A-1* **5**, 1827 (1967).
128. U.S. Pat. 3,331,819 (July 18, 1967), J. D. Spainhour (to Phillips Petroleum Co.).
129. L. A. Errede and J. M. Hoyt, *J. Am. Chem. Soc.* **82**, 436 (1960).
130. F. Ramp, *Polym. Prepr. Am. Chem. Soc. Div. Polym. Chem.* **7**(2), 582 (1966).
131. U.S. Pat. 3,807,977 (Apr. 30, 1974), T. E. Johnston and J. W. Matt (to E.I. du Pont de Nemours & Co., Inc.).
132. Ger. Offen. DE2 333,323 (Jan. 31, 1974), T. E. Johnston, J. W. Matt, and D. D. Johnson (to E.I. du Pont de Nemours & Co., Inc.).
133. U.S. Pat. 3,917,466 (Nov. 4, 1975), C. P. Henry (to E.I. du Pont de Nemours & Co., Inc.).
134. U.S. Pat. 2,634,254 (Apr. 7, 1953), R. D. Lipscomb (to E.I. du Pont de Nemours & Co., Inc.).
135. W. G. Barb, *Proc. R. Soc. London Ser. A* **212**, 66, 177 (1952).
136. W. G. Barb, *J. Polym. Sci.* **10**, 49 (1953).
137. C. Walling, *J. Polym. Sci.* **16**, 315 (1955).

138. M. Raetzsch and G. Borman, *Plaste Kautsch.* **20,** 600 (1973).
139. C. Mazzolini, L. Patron, A. Moretti, and M. Campanelli, *Ind. Eng. Chem. Prod. Res. Dev.* **9,** 504 (1970).
140. R. C. Schultz and A. Banihaschemi, *Makromol. Chem.* **64,** 140 (1963).
141. B. D. Flockhart, K. J. Ivin, R. C. Pink, and B. D. Sharma, *J. Chem. Soc. D,* 339 (1971).
142. B. D. Flockhart, K. J. Ivin, R. C. Pink, and B. D. Sharma in P.-O. Kinell, B. Rånby, and V. Runnström-Reio, eds., *Nobel Symp. 22,* Almqvist and Wiksell, Stockholm, Sweden, and John Wiley & Sons, Inc., New York, 1973, p. 17.
143. M. Delamar, P. C. Lacaze, B. Lemiere, J. Y. Dumousseau, and J. E. DuBois, *J. Polym. Sci. Polym. Chem. Ed.* **20,** 245 (1982).
144. Y. Yamashita, S. Iwatsuki, and K. Sakai, *Adv. Chem. Ser.* **99,** 211 (1971).
145. N. L. Zutty, C. W. Wilson, G. H. Potter, D. C. Priest, and C. J. Whitworth, *J. Polym. Sci. Part A* **3,** 2781 (1965).
146. I. L. Stoyachenko, Y. I. Shkyyarova, A. M. Kaplan, V. B. Golubev, V. P. Zubov, and V. A. Kabanov, *Vysokomol. Soedin. Ser. A* **18,** 1420 (1976).
147. J. J. Kearney, V. Stannett, and H. G. Clark, *J. Polym. Sci. Part C* **16,** 3441 (1968).
148. G. M. Bristow and F. S. Dainton, *Proc. R. Soc. London Ser. A* **229,** 509 (1955).
149. M. S. Kharasch and H. N. Friedlander, *J. Org. Chem.* **13,** 882 (1948).
150. N. Tokura and M. Matsuda, *Kogyo Kagaku Zasshi* **64,** 501 (1961).
151. F. S. Dainton and K. J. Ivin, *Proc. R. Soc. London Ser. A* **121,** 96 (1952).
152. H. Suzuki, M. Ito, and Z. Kuri, *Kogyo Kagaku Zasshi* **71,** 764 (1968).
153. C. Schneider, J. Denaxas, and D. Hummel, *J. Polym. Sci. Part C* **16,** 2203 (1967).
154. M. Matsuda and H. H. Thoi, *J. Macromol. Sci. Chem.* **11,** 1423 (1977).
155. R. E. Cais and J. H. O'Donnell, *J. Polym. Sci. Polym. Lett. Ed.* **14,** 263 (1976).
156. R. E. Cais and J. H. O'Donnell, *Makromol. Chem.* **176,** 3517 (1975).
157. R. E. Cais and J. H. O'Donnell, *Macromolecules* **9,** 279 (1976).
158. R. E. Cais, D. J. T. Hill, and J. H. O'Donnell, *J. Macromol. Sci. Chem.* **17,** 1434 (1982).
159. R. E. Cais and J. H. O'Donnell, *J. Polym. Sci. Polym. Symp.* **55,** 75 (1976).
160. M. Iino, A. Hara, and N. Tokura, *Makromol. Chem.* **98,** 81 (1966).
161. M. Iino, K. Katagiri, and M. Matsuda, *Macromolecules* **7,** 439 (1974).
162. M. Matsuda, M. Iino, T. Hirayama, and T. Miyashita, *Macromolecules* **5,** 240 (1972).
163. R. E. Cais, J. H. O'Donnell, and F. A. Bovey, *Macromolecules* **10,** 254 (1977).
164. R. C. Schulz and A. Banihaschemi, *Makromol. Chem.* **64,** 95 (1963).
165. M. Iino, H. H. Thoi, S. Shioya, and M. Matsuda, *Macromolecules* **12,** 160 (1979).
166. M. Matsuda and Y. Hara, *J. Polym. Sci. Part A-1* **10,** 837 (1972).
167. S. C. Chambers, A. H. Fawcett, S. Fee, J. H. Malone, and U. Mangar (to be submitted).
168. S. C. Chambers and A. H. Fawcett, *Macromolecules* **18,** 1710 (1985).
169. J. Furukawa, E. Kobayashi, and N. Morio, *J. Polym. Sci. Polym. Chem. Ed.* **12,** 1851 (1974).
170. J. Furukawa, E. Kobayashi, and M. Nakamura, *J. Polym. Sci. Polym. Chem. Ed.* **12,** 2789 (1974).
171. W. R. Fairheller and J. E. Katon, *Spectrochim. Acta* **20,** 1099 (1964).
172. A. H. Frazer, *J. Polym. Sci. Part A* **2,** 4031 (1964).
173. N. G. Gaylord and B. K. Patraik, *J. Polym. Sci. Polym. Chem. Ed.* **13,** 837 (1975).
174. D. W. Ovenall, R. S. Sudol, and G. A. Cabat, *J. Polym. Sci. Polym. Chem. Ed.* **11,** 233 (1973).
175. K. J. Ivin, M. Navratil, and N. A. Walker, *J. Polym. Sci. Part A-1* **10,** 701 (1972).
176. R. Bacskai, L. P. Lindeman, D. L. Ransley, and D. L. Sweeney, *J. Polym. Sci. Part A-1* **7,** 247 (1969).
177. A. H. Fawcett, Ph.D. thesis, The Queen's University of Belfast, Belfast, UK, 1968.
178. F. A. Bovey and R. E. Cais, *ACS Symp. Ser.* **103,** 1 (1979).
179. F. A. Bovey and R. E. Cais in F. Conti, ed., *Proceedings of the European Conference on NMR Macromolecules,* Lerici, Rome, 1978, p. 9.
180. F. A. Bovey in R. H. Sarma, ed., *Stereodynamics of Molecular Systems,* Pergamon Press, Oxford, UK, 1979, p. 53.
181. A. H. Fawcett, K. J. Ivin, and C. S. Stewart, *Org. Magn. Reson.* **11,** 360 (1978).
182. P. S. Skell, R. C. Woodworth, and J. H. McNamara, *J. Am. Chem. Soc.* **79,** 1253 (1957).
183. J. R. Brown and J. H. O'Donnell, *J. Polym. Sci. Part A-1* **10,** 1997 (1972).
184. F. S. Dainton and K. J. Ivin, *Trans. Faraday Soc.* **46,** 374, 382 (1950).
185. M. Ito and Z. Kuri, *Kogyo Kagaku Zasshi* **69,** 2006 (1966).
186. Z. Kuri and T. Yoshimura, *J. Polym. Sci. Polym. Lett. Ed.* **1,** 107 (1963).

187. Z. Kuri and M. Ito, *Kogyo Kagaku Zasshi* **69**, 1066 (1966).
188. U.S. Pats. 2,602,787 (July 8, 1952) and 2,593,414 (Apr. 22, 1952), W. W. Crouch (to Phillips Petroleum Co.).
189. U.S. Pat. 2,531,403 (Nov. 28, 1950), W. W. C. Rouch and L. D. Jurrens (to Phillips Petroleum Co.).
190. J. Chatelain, *Adv. Chem. Ser.* **91**, 529 (1969).
191. J. Chatelain, C. Ledous, G. Steinbach, and A. De Vries, *Ind. Chim. Belge* **32**, 423 (1967).
192. U.S. Pat. 3,904,590 (Sept. 9, 1975), A. L. Longothetis (to E.I. du Pont de Nemours & Co., Inc.).
193. Fr. Pat. 2,174,283, W. R. Moore and R. E. Erickson.
194. H. M. d'Emans, B. G. Bray, J. J. Martin, and L. C. Anderson, *Ind. Eng. Chem.* **49**, 1894 (1957).
195. U.S. Pat. 2,976,269 (Mar. 21, 1961), J. I. de Jong (to E.I. du Pont de Nemours & Co., Inc.).
196. M. A. Jobard, *J. Polym. Sci.* **29**, 275 (1958).
197. UK Pat. 528,051 (Oct. 22, 1940) (to E.I. du Pont de Nemours & Co., Inc.).
198. F. S. Dainton and K. J. Ivin, *Discuss. Faraday Soc.* **14**, 199 (1953).
199. M. A. Naylor and A. W. Anderson, *J. Am. Chem. Soc.* **76**, 3962 (1954).
200. A. H. Fawcett and K. J. Ivin, *Polymer* **13**, 439 (1972).
201. U.S. Pat. 2,943,077 (June 28, 1960), R. B. de Jong and I. M. Robinson (to E.I. du Pont de Nemours & Co., Inc.).
202. U.S. Pat. 3,576,791 (Apr. 27, 1971), H. V. Holler and E. A. Youngman (to Shell Oil Co.).
203. U.S. Pat. 2,184,295 (Dec. 26, 1939), F. E. Frey and R. D. Snow (to Phillips Petroleum Co.).
204. U.S. Pat. 2,294,027 (Aug. 25, 1942), F. E. Frey and R. D. Snow (to Phillips Petroleum Co.).
205. M. Raetzsch, *Plaste Kautsch.* **19**, 169 (1972).
206. U.S. Pat. 2,310,605 (Feb. 9, 1943), M. M. Barnett (to Freeport Sulfur Co.).
207. R. D. Snow and F. E. Frey, *J. Am. Chem. Soc.* **65**, 2417 (1943).
208. U.S. Pat. 2,169,363 (1939), C. S. Marvel and D. S. Frederick.
209. M. S. Kharasch and E. Sternfield, *J. Am. Chem. Soc.* **62**, 2559 (1940).
210. W. C. Firth and L. E. Palmer, *Macromolecules* **4**, 654 (1971).
211. K. J. Ivin, *Nature* **180**, 90 (1957).
212. U.S. Pat. 2,112,986 (Apr. 5, 1938), L. H. Fitch and F. E. Frey (to Phillips Petroleum Co.).
213. U.S. Pat. 2,258,702 (Oct. 14, 1941), F. E. Frey and L. H. Fitch (to Phillips Petroleum Co.).
214. U.S. Pat. 2,299,222 (Oct. 20, 1942), F. E. Frey, R. D. Snow, and W. A. Schulze (to Phillips Petroleum Co.).
215. U.S. Pat. 2,299,221 (Oct. 20, 1942), F. E. Frey, R. D. Snow, and W. A. Schulze (to Phillips Petroleum Co.).
216. C. S. Marvel, L. F. Audrieth, and W. H. Sharkey, *J. Am. Chem. Soc.* **64**, 1229 (1942).
217. U.S. Pat. 2,192,466 (Mar. 5, 1940), F. E. Frey and L. H. Fitch (to Phillips Petroleum Co.).
218. T. Enomoto, M. Iino, M. Matsuda, and N. Tokura, *Bull. Chem. Soc. Jpn.* **44**, 3140 (1971).
219. F. S. Dainton, K. J. Ivin, and D. R. Sherd, *Trans. Faraday Soc.* **52**, 414 (1956).
220. G. Sartori and R. D. Lundberg, *J. Polym. Sci. Polym. Chem. Ed.* **13**, 1265 (1975).
221. U.S. Pat. 3,242,231 (Mar. 22, 1966), B. Graham and W. E. Mochel (to E.I. du Pont de Nemours & Co., Inc.).
222. E. C. Eaton and K. J. Ivin, *Polymer* **6**, 339 (1965).
223. U.S. Pat. 3,170,892 (Feb. 23, 1965), W. F. Busse (to E.I. du Pont de Nemours & Co., Inc.).
224. I. Sakurada, T. Okada, T. Shioda, and Y. Matsumoto, *Mem. Jpn. Radiat. Res.* **4**, 95 (1965).
225. E. Gipstein, W. Moreau, G. Chiu, and O. U. Need, *J. Appl. Polym. Sci.* **21**, 677 (1977).
226. U.S. Pat. 3,807,977 (Apr. 30, 1974), T. E. Johnston (to E.I. du Pont de Nemours & Co., Inc.).
227. M. J. Bowden and T. Novembre, *ACS Symp. Ser.* **212**, 125 (1983).
228. T. N. Bowmer and M. J. Bowden, *ACS Symp. Ser.* **242**, 153 (1984).
229. K. J. Ivin, H. A. Ende, and G. Meyerhoff, *Polymer* **3**, 129 (1962).
230. S. Harada and M. Katayama, *Makromol. Chem.* **90**, 117 (1966).
231. J. R. Brown and J. H. O'Donnell, *J. Macromol. Sci. Chem.* **6**, 1411 (1972).
232. T. W. Bates, J. Biggins, and K. J. Ivin, *Makromol. Chem.* **87**, 180 (1965).
233. A. C. Ouano, E. Gipstein, W. Kaye, and B. Dawson, *Macromolecules* **8**, 558 (1975).
234. R. Endo, T. Manago, and M. Takeda, *Bull. Chem. Soc. Jpn.* **39**, 733 (1966).
235. T. W. Bates and K. J. Ivin, *Polymer* **8**, 263 (1967).
236. M. J. Bowden and L. F. Thompson, *J. Appl. Polym. Sci.* **19**, 905 (1975).
237. J. R. Brown and J. H. O'Donnell, *Macromolecules* **5**, 109 (1972).
238. G. N. Taylor, M. Y. Hellman, and L. E. Stillwaggon, *Chromatogr. Sci.* **19**, 237 (1981).

239. A. C. Ouano, *J. Colloid Interface Sci.* **63**, 275 (1978).
240. K. Araki, H. Aoki, and Y. Imamura, *Makromol. Chem.* **182**, 2455 (1981).
241. A. H. Fawcett and J. K. Ivin, *Polymer* **16**, 569 (1975).
242. M. Mansfield, *Macromolecules* **15**, 1587 (1982).
243. A. H. Fawcett, S. Fee, and L. Waring, *Polymer* **24**, 1571 (1983).
244. D. N. Gray, *Polym. Eng. Sci.* **17**, 719 (1977).
245. J. E. Crawford and D. N. Gray, *J. Appl. Polym. Sci.* **15**, 1881 (1971).
246. W. H. Chu, E. Gipstein, and A. C. Ouano, *J. Appl. Polym. Sci.* **21**, 1045 (1977).
247. M. J. Bowden and L. F. Thompson, *J. Electrochem. Soc.* **121**, 1620 (1974).
248. J. M. Ketteringham, W. M. Zapol, J. D. Birkett, L. L. Nelsen, A. A. Massucco, and C. Raith, *Trans. Am. Soc. Artif. Intell. Organs* **21**, 224 (1975).
249. E. H. Hill and J. R. Caldwell, *J. Polym. Sci. Part A* **2**, 1251 (1964).
250. U.S. Pat. 3,928,294 (Sept. 23, 1975), J. E. Crawford and D. E. Gray.
251. U.S. Pat. 4,423,930 (Jan. 3, 1984), D. N. Gray (to Owens-Illinois, Inc.).
252. S. Fee, Ph.D. thesis, The Queen's University of Belfast, Belfast, UK, 1985.
253. T. N. Bowmer and J. H. O'Donnell, *Polym. Degradation Stab.* **3**, 87 (1981).
254. E. Kirlan, J. K. Gillham, and E. Gipstein, *J. Appl. Polym. Sci.* **21**, 1159 (1977).
255. E. Wellish, E. Gipstein, and O. J. Sweeting, *J. Appl. Polym. Sci.* **8**, 1623 (1964).
256. H. D. Schneddemage and D. O. Hummel, *Kolloid Z. Z. Polym.* **220**, 133 (1967).
257. E. Kiran and J. K. Gillham, *J. Appl. Polym. Sci.* **20**, 931 (1976).
258. D. O. Hummel, H.-D. R. Schneddemage, and U. Pohl, *Kolloid Z. Z. Polym.* **210**, 106 (1966).
259. E. Kiran and J. K. Gillham, *Polym. Eng. Sci.* **19**, 699 (1979).
260. R. J. Gritter, M. Seegar, and E. Gipstein, *J. Polym. Sci. Polym. Chem. Ed.* **16**, 353 (1978).
261. M. J. Bowden, L. F. Thompson, W. Robinson, and M. Biolsi, *Macromolecules* **15**, 1417 (1982).
262. T. N. Bowmer and J. H. O'Donnell, *J. Polym. Sci. Polym. Chem. Ed.* **19**, 45 (1981).
263. J. R. Brown and J. H. O'Donnell, *Macromolecules* **3**, 265 (1970).
264. L. E. Stillwaggon, E. M. Dorries, L. F. Thompson, and M. J. Bowden, *Coatings Plast. Prepr.* **37**, 38 (1977).
265. T. N. Bowmer and J. H. O'Donnell, *J. Macromol. Sci.* **A17**, 243 (1982).
266. T. N. Bowmer and J. H. O'Donnell, *Polym. Bull.* **2**, 103 (1980).
267. P. B. Aysclough, K. J. Ivin, and J. H. O'Donnell, *Trans. Faraday Soc.* **61**, 1110 (1965).
268. T. N. Bowmer and J. H. O'Donnell, *Radiat. Phys. Chem.* **17**, 177 (1981).
269. D. J. T. Hill, J. H. O'Donnell, and P. J. Pomery in T. Davidson, ed., *Polymers in Electronics, ACS Symp. Ser.* **242**, American Chemical Society, Washington, D.C., 1984, p. 125.
270. T. N. Bowmer and J. H. O'Donnell, *Polymer* **22**, 71 (1981).
271. M. J. Bowden, D. L. Allara, W. I. Vroom, J. Frackoviak, L. C. Kelley, and D. R. Falcone in Ref. 269, p. 135.
272. T. N. Bowmer, J. H. O'Donnell, and P. R. Wells, *Makromol. Chem. Rapid Commun.* **1**, 1 (1980).
273. T. N. Bowmer, J. H. O'Donnell, and P. R. Wells, *Polym. Bull. Berlin* **2**, 103 (1980).
274. A. Gutierrez, J. Pacansky, and R. Kroeker, *Org. Coatings Appl. Polym. Sci. Proc.* **46**, 520 (1982).
275. M. J. Bowden and L. F. Thompson, *J. Appl. Polym. Sci.* **17**, 3211 (1973).
276. M. J. Thompson and M. J. Bowden, *J. Electrochem. Soc.* **120**, 1722 (1973).
277. M. J. Bowden, L. F. Thompson, and J. P. Ballantyne, *J. Vac. Sci. Technol.* **12**, 1294 (1975).
278. M. J. Bowden, *J. Polym. Sci. Polym. Chem. Ed.* **12**, 499 (1974).
279. M. J. Bowden and L. F. Thompson, *Polym. Eng. Sci.* **14**, 525 (1974).
280. M. J. Bowden and L. F. Thompson, *J. Electrochem. Soc.* **121**, 1620 (1974).
281. T. M. Hall, A. Wagner, and L. F. Thompson, *J. Vac. Sci. Technol.* **16**, 1889 (1979).
282. K. Horie and W. Schnabel, *Polym. Photochem.* **2**, 419 (1982).
283. P. W. Bohn, J. W. Taylor, and H. Guckel, *Anal. Chem.* **53**, 1082 (1981).
284. R. J. Himics and D. L. Ross, *Polym. Eng. Sci.* **17**, 350 (1977).
285. U.S. Pat. 3,916,036 (Oct. 28, 1975), E. Gipstein, W. M. Moreau, and O. U. Need (to I.B.M.).
286. H. Hiraoka and L. W. Welsh in Ref. 269, p. 135.
287. P. Lenzo and E. G. Spencer, *Appl. Phys. Lett.* **24**, 289 (1974).
288. M. J. Bowden and L. F. Thompson, *Polym. Eng. Sci.* **17**, 269 (1977).
289. B. Yaakobi, H. Kim, J. M. Soures, H. W. Deckman, and J. Dunsmuir, *J. Vac. Sci. Technol. Part A* **2**(2), 367 (1984).
290. M. J. Bowden, *J. Polym. Sci. Polym. Symp.* **49**, 221 (1975).
291. M. J. Bowden, *CRC Crit. Rev. Solid State Sci.* **8**, 223 (1979).

292. M. J. Bowden and J. Frackoviak, *Special Technical Publication 804*, American Society for Testing and Materials, Philadelphia, 1984.
293. S. A. Evans, J. L. Bartelt, B. J. Sloan, and G. L. Varnell, *J. Vac. Sci. Technol.* **15**, 969 (1978).
294. M. J. Bowden in L. F. Thompson, C. G. Willson, and J. M. J. Frechet, eds., *Materials for Microlithography: Radiation-Sensitive Polymers, ACS Symp. Ser.* **266**, American Chemical Society, Washington, D.C., 1984, p. 39.
295. M. J. Bowden and L. F. Thompson, *Solid State Technol.* **22**, 72 (1979).
296. L. D. Yau and L. R. Thibault, *J. Vac. Sci. Technol.* **15**, 960 (1978).
297. M. J. Bowden, *J. Appl. Polym. Sci.* **26**, 1421 (1981).
298. U.S. Pat. 3,935,331 (Jan. 27, 1976), E. S. Poliniak, H. G. Scheible, and R. J. Himics (to R.C.A. Corp.).
299. U.S. Pat. 4,097,618 (June 27, 1978), E. S. Poliniak (to R.C.A. Corp.).
300. U.S. Pat. 4,267,257 (May 12, 1981), E. S. Poliniak and N. V. Desai (to R.C.A. Corp.).
301. U.S. Pat. 4,329,410 A (May 11, 1982), D. W. Buckley (to Perkin-Elmer Corp.).
302. U.S. Pat. 3,873,492 (Mar. 25, 1975), M. Takehisa, H. Kurihara, T. Yagi, H. Wanatabe, and S. Machi (to Japan Atomic Energy Research Institute).

Allan H. Fawcett
The Queen's University of Belfast

OLIGOMERS

The International Union of Pure and Applied Chemistry (IUPAC) defines oligomer as a substance composed of molecules containing a few of one or more species of atoms or groups of atoms (constitutional units) repetitively linked to each other (1). This does not specify an absolute degree of polymerization or molecular weight that distinguishes an oligomer from a polymer, but the IUPAC definition further states that the physical properties of an oligomer vary with the addition or removal of one or a few of the constitutional units from its molecules. This structure–property definition is perhaps the most meaningful definition of an oligomer. The conversion of a monomer or a mixture of monomers into an oligomer is defined as oligomerization. This definition does not imply any constraints on the oligomer polydispersity. Therefore, although monodisperse oligomers provide more valuable information than polydisperse oligomers, the latter are still important.

The term oligomer originates from the Greek words ολιγος = few and μερος = part, and was first used in the field of synthetic polymer chemistry in the early 1950s (2), having been taken from the nomenclature of natural products, ie, oligosaccharides (3), oligopeptides (4), etc. A recent book on the history of polymer science (qv) (5) asserts that the name was first suggested by L. V. Larsen for a laboratory manual published by G. F. D'Alelio in 1943 (6). The name oligomer has been adopted by Kern (7) and Zahn (8), and their extensive work in this field led to its present widely accepted meaning.

The first attempts made by Staudinger to convince the organic chemistry community of the macromolecular nature of polymers were partially based on research performed on oligomers (9). Staudinger and Luthy (10) separated the α,ω-dimethyl ethers of polyoxymethylene (POM) with degrees of polymerization (DP) up to 14 and the α,ω-dimethyl esters of POM with DP up to 20. Although chromatographic methods were not available to demonstrate their homogeneity, these oligomers were, at that time, better identified than the homologous series of paraffins. In fact the linear and cyclic oligomers of polyethylene (PE) have been synthesized and characterized only very recently, and they will be discussed in this article. The demonstration that oligooxymethylenes exhibit the same crystal lattice as their corresponding polymers (11) was of great importance for the development of the entire field of polymer science. Carothers (12) separated and characterized a variety of oligomers by fractional crystallization of aliphatic polyesters of adipic acid. In 1939 the dimeric and trimeric cyclic compounds of ϵ-caprolactam were separated and linear oligomers from monomer to pentamer, ie, oligomers of nylon-6, were prepared (13). The linear oligomers were used as model substances for viscosimetric studies. The cyclic urethane oligomers formed from hexamethylene diisocyanate with diols were reported in 1941 (14), and systematic investigations on the cyclic (15,16) and linear oligomers (17) of nylon-6 were performed in the early 1950s.

During this same period theoretical understanding of ring-chain equilibria in condensation polymerization reactions was developed (18,19) and may be depicted

$$\wedge\wedge M_y \wedge\wedge \rightleftharpoons M_x + \wedge\wedge M_{y-x} \wedge\wedge$$

where $\wedge\wedge M_y \wedge\wedge$ denotes linear macromolecules with a degree of polymerization y and M_x denotes cyclic macromolecules with a degree of polymerization x. The following assumptions are made: chains in solution obey Gaussian statistics; all rings formed are strainless; the reactivity of all reaction sites along the chain is the same; and thermodynamic equilibrium is attained. Based on these assumptions the concentration of a macroring of a given size is governed solely by the probability of attaining a conformation that enables the closure of the macrocycle and

$$K_n = [M_n] = An^{-5/2}$$

where K_n is the equilibrium constant for the formation of the cyclic n-mer, $[M_n]$ the equilibrium concentration of the cyclic n-mer, A a proportionality coefficient, and n the degree of polymerization (qv). Therefore, for each individual macrocycle a certain thermodynamic equilibrium concentration exists which has the same physical meaning as the equilibrium monomer concentration. By increasing the ring size n the equilibrium oligomer concentration decreases according to the above equation. This theory requires that cyclization proceeds with zero enthalpy change (all rings are strainless), ie, cyclization is exclusively entropy-driven and equilibrium is attained by maximization of the entropy. Thus the equilibrium oligomer concentration, contrary to the equilibrium monomer concentration, does not depend on temperature. More recently, another method of calculating cyclic concentrations in ring-chain equilibrates has been applied to several polymeric systems (20–24) and these theories (18,19,25) have more recently been applied to cationic ring-opening polymerizations (26).

The understanding of cyclization theory has contributed tremendously to research in the field of cyclic and linear oligomers (see also MACROCYCLIC POLYMERS).

Oligomers are used to study polymerization mechanisms (27–34) and the dynamics and equilibria of cyclization (35,36), and are of enormous importance for the development of the entire field of host guest complex chemistry (37–42) or bioorganic chemistry.

The chemistry of chain-end functional oligomers is important to preparative polymer chemistry because of their use as macromonomers which are precursors for the synthesis of graft copolymers (qv), comblike polymers, and networks (qv) (43) (see MACROMERS), telechelics (see TELECHELIC POLYMERS), and reactive oligomers (45,46) (see ACETYLENE-TERMINATED PREPOLYMERS) which are of interest for the synthesis of block copolymers (qv) (46,47) and interpenetrating polymer networks (qv), etc.

Comprehensive reviews have been published on the formation of cyclic oligomers in the cationic polymerization (qv) of heterocyclic monomers (48–50), on the preparation and characterization of cyclic and linear polysiloxanes including its oligomers (51,52), as well as on the general methods for the synthesis of organic cyclic oligomers (53).

The primary reasons for studying oligomers include the use of oligomers as models for understanding polymerization mechanisms, as intermediates in preparative polymer chemistry, and as models for studying the structure–property relationships of polymers.

Models for Polymerization Mechanisms

Anionic Polymerization

Vinylic Monomers.   *Vinylpyridine Monomers.*   The oligomerization of vinyl monomers has been studied to understand the stereochemistry of polymerization and termination reactions. In particular, comparisons have been made of the stereochemistry of 2-vinylpyridine and 4-vinylpyridine polymerizations by consecutive monomer additions to a tetrahydrofuran (THF) solution of an alkali metal salt of 2-ethylpyridine or 4-ethylpyridine, followed by termination with methyl iodide. When necessary, the living anionic oligomerization products were separated by preparative liquid chromatography. Oligomers up to the heptamer have been studied. The methylation of 2-vinylpyridine oligomers is highly stereoselective, forming 95% meso product when either Li or Na is used as the counterion (54–56) due to intramolecular coordination of the alkali metal counterion by the nitrogen lone pair of the penultimate pyridine group. Coordination is possible by two different diastereomers; the diastereomer allowing syn electrophilic attack is favored. Because intramolecular coordination is not possible with 4-vinylpyridine oligomers, both methylation and monomer addition are completely nonstereoselective with equal rates of meso and racemic additions (57,58).

When the counterions are larger than lithium or sodium or when coordination is reduced by the addition of solvating agents, the stereoselectivity of 2-vinylpyridine methylation is decreased, depending on chain length (56,57). In contrast to methylation, monomer addition is only slightly stereoselective (60% isotactic), even when small counterions are used (59). Although the first monomer addition is stereoselectively trans, the trans stereoselectivity decreases with the next monomer addition (57). Equimolar mixtures of E and Z isomers are formed upon monomer addition to form the dimer and trimer as a result of equal rates of cis and trans additions (60). This may be explained by competition between coordination of the anion prior to monomer addition and coordination of the resulting anion, since the two geometries do not permit both anions to coordinate (59). Also, depolymerization side reactions resulting from E_{1cb} or E_2 eliminations occur (61). Nevertheless, the meso-prochiral conformation has been shown to be considerably more stable than the racemic-prochiral conformation by substituting methyl at the 3′ and 5′ positions of the penultimate 2-pyridine group (62).

The methylation stereoselectivity is not decreased by the presence of a methyl group α to the anion (63). However, during monomer addition the presence of an α-methyl group in the oligomerization of 2-isopropenylpyridine results in primarily syndiotactic products rather than isotactic (64), indicating steric hindrance to intramolecular chelation. The oligomerization and polymerization of 4-isopropenylpyridine also result in predominantly syndiotactic products (65); chain length, temperature, and counterion do not have any significant effect on the stereochemistry of 2- or 4-isopropenylpyridine oligomers (66).

These studies demonstrate that the geometry of the carbanion site can be an important factor in the methylene stereochemistry of anionic oligomerizations and polymerization of vinyl monomers.

Acrylic Monomers.   Because the penultimate ester group of acrylate oligomers can potentially coordinate cations, the anionic dimers of methyl, isopropyl, and *t*-butyl acrylate have been terminated with methyl iodide (67). However, it was found that coordination of the metal cation with the ester group is partially or totally absent, resulting in little stereoselectivity as compared with 2-vinylpyridine.

Vinyl Phenyl Sulfoxide.   Using the lithium salt of optically active (*R*)-ethyl phenyl sulfoxide anion as initiator, the stereochemistry of the living anionic oligomerization of racemic vinyl phenyl sulfoxide has been studied. The methylated and protonated dimers were isolated from their oligomer mixtures by preparative liquid chromatography. ^{1}H and ^{13}C nmr demonstrated that dimerization occurred with more than 98% stereoselectivity with respect to the chiral carbon, and with 92% *R/S* stereoselectivity (68). Therefore, only two of the four possible diastereomeric protonated dimers and four of the six possible methylated dimers were isolated, demonstrating coordination of the metal counterion with the sulfoxide and the stereochemical-directing ability of the chiral carbon and sulfur atoms.

The relative ratio of the diastereomeric methylated dimers strongly depended on the ratio of monomer to initiator, the optical purity of the starting materials, and the temperature (69). These results are consistent with the pro-*R* coordinated diastereomeric anions being more stable than the pro-*S*, in which case there is a kinetic preference for the pro-*R* diastereomers at low temperature, depending on reactant stoichiometry and optical purity. However, the less stable isomers are more reactive and are in greater abundance at higher temperature where isomer interconversion is more rapid.

Heterocyclic Monomers.   Most anionic ring-opening polymerizations are nucleophilic substitution reactions, eg,

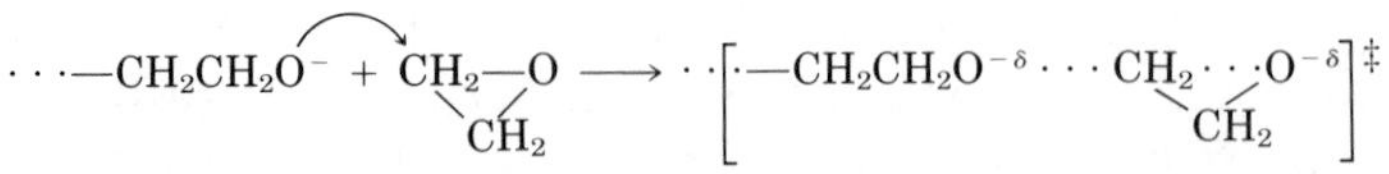

$$\cdots-CH_2CH_2O^- + \overset{\frown}{CH_2}-O \longrightarrow \cdots\left[-CH_2CH_2O^{-\delta}\cdots CH_2\cdots O^{-\delta}\right]^\ddagger$$

For some heterocyclic monomers the unique chemical structure of the growing species follows unequivocally from the monomer structure. The bond to the heteroatom is the most susceptible to chemical change. However, isomeric structures must be taken into account in the case of ambident monomers. For instance, in the polymerization of thietane, under certain conditions the carbanions and not the thiolate anions may be the growing species (70). Unsymmetrically substituted monomers can provide active species by α- or β-ring scission, eg, in the polymerization of substituted α-oxides:

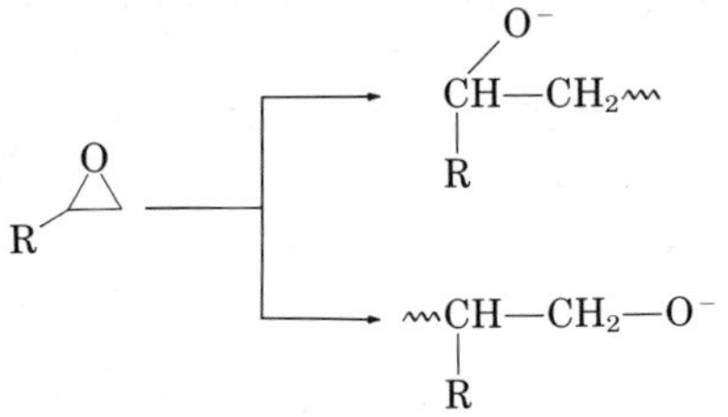

Lactones and dimers of α-hydroxycarboxylic acids can polymerize by O-acyl or O-alkyl bond scission, giving alcoholate or carboxylate anions, respectively, as the growing species. Transition from one kind of species, formed in the initiation, to the other kind, actually propagating, has also been observed (71).

A method has been developed that allows the determination of the chemical structure of active centers of a growing oligomer (72,73). This method is based on oligomeric anion-capping with $ClP(O)(OC_6H_5)_2$, followed by quantitative analysis of the resulting phosphate end groups by $^{31}P(^1H)$ nmr spectroscopy.

$$RX^-, Mt^+ + ClP(OC_6H_5)_2 \longrightarrow RXP(OC_6H_5)_2 + MtCl$$

$$\text{where } X = \text{heteroatom and } Mt = \text{metal}$$

The chemical shifts of the phosphate end groups are sensitive to the structure of the oligomeric or macromolecular substituent. Therefore, the comparison of chemical shifts of phosphate end groups with those of model compounds enables determination of the structure of the anions present (Table 1). Integration of the signals of capped active centers and of diphenyl chlorophosphate also allows the concentration of active centers to be determined.

As Table 2 shows, capping provides the possibility of determining the isomeric structures during polymerization and measuring their proportions and rates of interconversion. However, in any particular system the rate of capping must be substantially higher than the rate of interconversion.

Another method is based on the formation of different structures in the reaction of carboxylate and alcoholate anions with 2,4,6-trinitroanisole (71). Carboxylate anions react with 2,4,6-trinitroanisole to give picrate anions.

Table 1. ^{31}P Chemical Shifts of Phosphate End Groups^a

Monomer		Product of trapping	^{31}P δ, ppm	Growing species
Structure	Number			
(ethylene oxide) $CH_2\!-\!CH_2$, O	(1)	$\cdots\!-\!CH_2OP(O)(OC_6H_5)_2$	-11.5	$\cdots\!-\!CH_2O^-$
(styrene oxide) C_6H_5 $CH_2\!-\!CH$, O	(2)	$\cdots\!-\!CH_2\overset{C_6H_5}{\underset{}{C}}HOP(O)(OC_6H_5)_2$ $\cdots\!-\!CHCH_2OP(O)(OC_6H_5)_2$, C_6H_5	-13.0 -12.6	$\cdots\!-\!CH_2\overset{C_6H_5}{CHO}^-$ $\cdots\!-\!\underset{C_6H_5}{CHCH_2O}^-$
$CH_2\!-\!C\!=\!O$ $CH_2\!-\!O$	(3)	$\cdots\!-\!CH_2COP(O)(OC_6H_5)_2$ $\downarrow$ pyrophosphate	-26.0	$\cdots\!-\!CH_2C\!\!\begin{smallmatrix}O\\ \\O\end{smallmatrix}$
$CH_2\overset{CH_2}{\diagup}\!C\!=\!O$ $CH_2\underset{CH_2}{\diagdown}O$	(4)	$\cdots\!-\!(CH_2)_4OP(O)(OC_6H_5)_2$	-12.6^b	$\cdots\!-\!CH_2O^-$
$H_2C\!-\!CH_2$ $CH_2C\!=\!O$ $CH_2\underset{CH_2}{\diagdown}O$	(5)	$\cdots\!-\!(CH_2)_5OP(O)(OC_6H_5)_2$	-11.5	$\cdots\!-\!CH_2O^-$
$(CH_3)_2$ Si, O O $(CH_3)_2SiSi(CH_3)_2$, O	(6)	$\cdots\!-\!\underset{CH_3}{\overset{CH_3}{Si}}OP(O)(OC_6H_5)_2$ $\downarrow$ pyrophosphate	-26.0	$\cdots\!-\!\underset{CH_3}{\overset{CH_3}{Si}}O^-$
S $CH_2\!-\!CH$ CH_3	(7)	$\cdots\!-\!CH_2\underset{CH_3}{CH}SP(O)(OC_6H_5)_2$	$+18.9$	$\cdots\!-\!CH_2\underset{CH_3}{CHS}^-$

^{a} Ref. 72. Courtesy of *Makromolekulare Chemie Rapid Communication*.
^{b} Ref. 74.

Alcoholate anions participate in a reversible reaction leading to the formation of the Meisenheimer complex:

Table 2. ^{31}**P Nmr Determination of Concentration**^{a} **of Active Centers [AC] and Initiator [I] Converted into Phosphates**^{b}

Initiator	Monomer	$[I]_0$	$[I]$	$[AC]$	$[I] + [AC]$
$CH_3O^-K^+$	(2)	1.2×10^{-1}		1.4×10^{-1}	1.4×10^{-1}
$CH_3O^-K^+$	(5)	2.8×10^{-2}		2.4×10^{-2}	2.4×10^{-2}
$(CH_3)_3SiO^-K^+$	(6)	5.8×10^{-2}		6.0×10^{-2}	6.0×10^{-2}
$C_6H_5S^-Na^+ \cdot DBC^c$	(7)	1.5×10^{-3}	6.8×10^{-4}	8.1×10^{-4}	1.49×10^{-3}

^{a} In mol/L.
^{b} Ref. 72. Courtesy of *Makromolekulare Chemie Rapid Communication.*
^{c} DBC = dibenzo-18-crown-6 ether.

Picrate anions absorb at $\lambda = 380$ nm, whereas the Meisenheimer complex absorbs at $\lambda = 420$ and 495 nm. Thus differences in the uv spectra of polymerizing mixtures with added 2,4,6-trinitroanisole can be used to distinguish between carboxylate and alcoholate active species. When 2,4,6-trinitroanisole is used in excess, the concentration of formed picrate anions determined from uv spectra is equal to the concentration of parent carboxylate anions.

Cationic Polymerization

Heterocyclic Monomers.   Cationic ring-opening polymerizations are also nucleophilic substitution reactions, eg,

Two experiments are frequently employed to follow the cationic polymerization of heterocyclic monomers. The first is based on the polymerization of a low molar ratio of monomer and initiator directly in an nmr sample tube. This experiment requires a low rate of polymerization and allows the identification of all growing species and determination of all individual reaction step rate constants directly by ^{1}H nmr spectroscopy. Cyclic imino ethers were the first studied by this method (75) (see CYCLIC IMINO ETHERS, POLYMERIZATION); subsequently, cationic polymerization of other heterocyclic monomers has been considered (32,33,76).

Trapping experiments such as those described for the anionic polymerization of heterocyclic monomers have been designed (32,73,77) to determine the structure of the growing cations by using R_3P as a capping reagent:

Ion trapping by phosphines provides information about the fine structure of the growing species (Table 3).

Table 3. ^{31}P Chemical Shifts of Trapped Quaternary Phosphonium Salts^a

Monomer	Product of trapping	^{31}P δ, ppm	Growing species
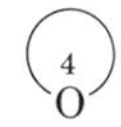 4 (O)	$\cdots$—O—$(CH_2)_3$—$\overset{+}{P}(C_6H_5)_3$	23.8	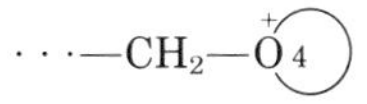 $\cdots$—CH_2—$\overset{+}{O}$ 4
 5 (O)	$\cdots$—O—$(CH_2)_4\overset{+}{P}(C_6H_5)_3$	23.4	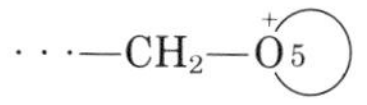 $\cdots$—CH_2—$\overset{+}{O}$ 5
 7 (O)	$\cdots$—O—$(CH_2)_6$—$\overset{+}{P}(C_6H_5)_3$	23.0	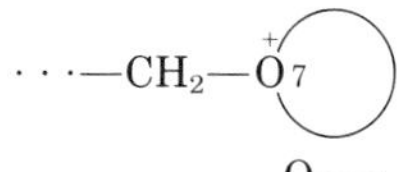 $\cdots$—CH_2—$\overset{+}{O}$ 7
C(=O)O	$\cdots$—$CH_2OCCH_2CH_2\overset{+}{P}(C_6H_5)_3$ ($\overset{\|}{O}$)	23.4	$\cdots$—CH_2O—$\overset{+}{C}$(O)
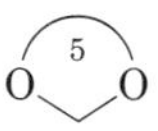 O 5 O	$\cdots$—CH_2OCH_2—$\overset{+}{P}(C_6H_5)_3$	16.7	$\cdots$—CH_2—$\overset{+}{O}CH_2$

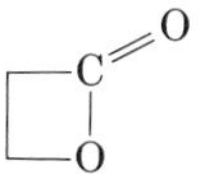

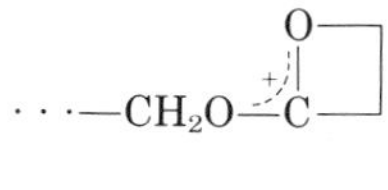

^{a} Ref. 77. Courtesy of *Makromolekulare Chemie*.

Other Ionic and Insertion Polymerizations

A novel catalytic system based on aluminum porphyrins which is particularly effective for the living ring-opening polymerization of epoxide and β-lactone
and the copolymerization of epoxide with carbon dioxide and with cyclic anhydrides has been developed (78). This initiating system has been characterized by
^{1}H nmr spectroscopy and the structure of the active species shown to be

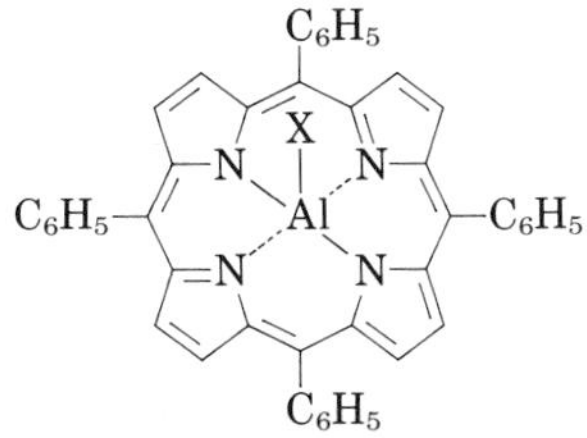

where X = OR, OCOR

R = macromolecular substituent

The active species responsible for metathesis polymerization (qv) of cyclic
olefins has also been identified by nmr spectroscopy (79,80). These species, tungsten carbenes and tungsten cyclobutane derivatives, can be simultaneously formed:

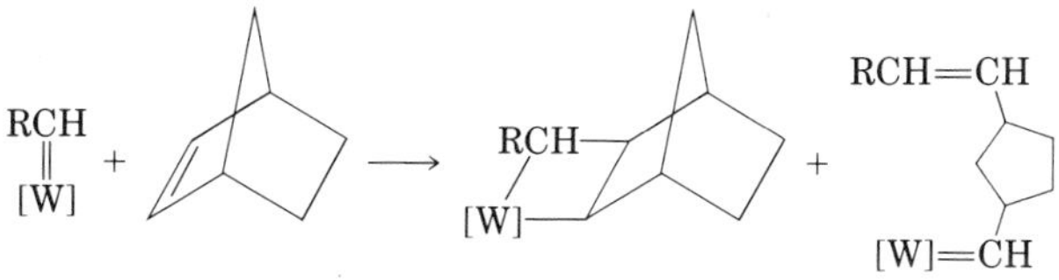

The conditions of equilibria involving these species are not yet established. The
living ring-opening polymerization of norbornene with bis(cyclopentadienyl) ti-
tanacyclobutane derivatives has also been followed by nmr spectrosocpy (81).

The mechanism of group-transfer polymerization (qv) has been studied by following the structure of the methyl methacrylate oligomers by ^{13}C nmr spectroscopy (82).

An interesting approach to the elucidation of polystyrene (PS) microstructure is based on the synthesis and characterization of individual oligomers up to pentamer and their use as nmr models for determining the stereochemical configuration of the high molecular weight PS (83).

Radical Polymerization

The most widely accepted pathway for the spontaneous generation of radicals in the thermal polymerization of styrene was proposed in 1961 and is outlined here (84,85):

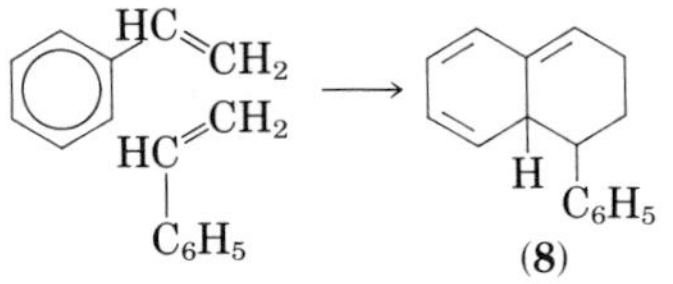

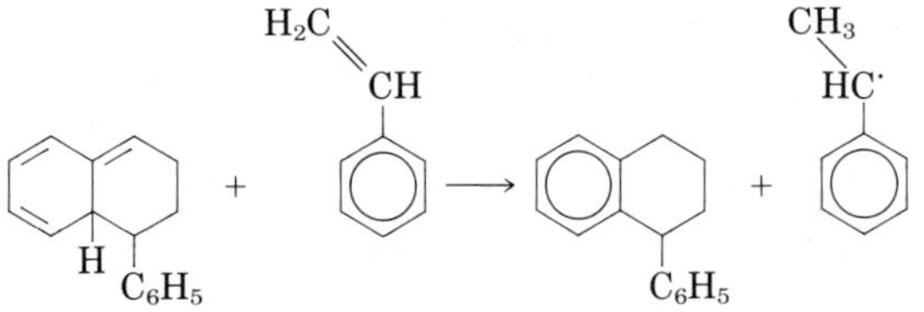

Although ample evidence for the transient formation of the Diels-Alder dimer (**8**) exists, its reaction with styrene and the identity of the radicals that initiate polymerization have remained speculative and controversial (86–89). The definitive experiment confirming this mechanism was based on the retardation of the propagation step in the thermal polymerization of styrene with $FeCl_3$ in DMF. The resulting oligomers containing between 6 and 14 monomer units were characterized by ^1H nmr spectroscopy and it was shown that radicals (**9**) and (**10**) are indeed responsible for the initiation of the polymerization (90). The identification of the oligomeric structures (**11**) and (**12**) established the identity of the initiating radicals as 1-phenylethyl (**10**) and 1-phenyl-1,2,3,4-tetrahydronaphthalenyl (**9**).

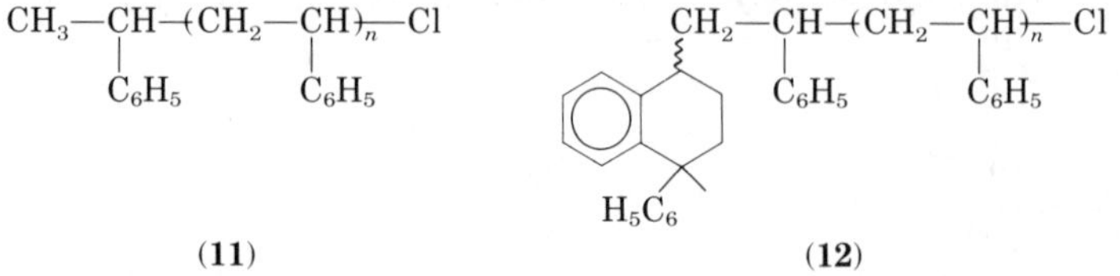

The elucidation of the nmr spectrum of the oligostyrene containing structures (**11**) and (**12**) required the use of oligomeric model compounds (**13**) and (**14**).

$$CH_3-CH-(CH_2-CH)_n-CH=CH \qquad (CH_3)_2-C-(CH_2-CH)_n-Cl$$
$$\underset{C_6H_5}{|} \qquad \underset{C_6H_5}{|} \qquad \underset{C_6H_5}{|} \qquad\qquad \underset{CN}{|} \qquad \underset{C_6H_5}{|}$$
$$(13) \qquad\qquad\qquad\qquad\qquad\qquad (14)$$

Oligomer (**13**) was prepared by the cationic oligomerization of styrene in the presence of $HClO_4$ (91) or $BF_3 \cdot O(C_2H_5)_2$ (92); structure (**14**) was made by the $FeCl_3$–DMF-retarded radical polymerization of styrene initiated by AIBN (93).

Well-defined oligomers of methyl methacrylate were used as nmr models to study the preferred conformers of poly(methyl methacrylate) (94,95). These oligomers were prepared by catalyzed chain transfer to monomer during radical polymerization of methyl methacrylate in the presence of Co(II)tetraphenyl-porphyrin (96) or cobaltoxime (97) chain-transfer agents.

Intermediates in Preparative Polymer Chemistry

Poly(ethylene oxide) Derivatives

The synthetic methods used to make functional derivatives of poly(ethylene oxide) (PEO) and their applications have been extensively reviewed (98) (see also 1,2-EPOXIDE POLYMERS). The first reported stepwise synthesis of individual PEO oligomers (99), with as many as 186 monomer units (100,101), involved William-son ether syntheses in which 2 mol of the monosodium or monopotassium salt of an appropriate oligo(oxyethylene glycol) reacted with 1 mol of the appropriate dichloride of another oligo(oxyethylene glycol). However, side products are dif-ficult to remove (102,103). Unsaturated side products result from HCl elimination; oligomers other than those desired arise from an equilibration between the so-dium salts of the original and the resultant glycols. The equilibrating side re-action can be avoided by reaction of the sodium salt of a monobenzylated glycol with the chlorine derivative of another monobenzylated glycol (104), followed by deprotection of the benzyl groups. By this method, pure and well-defined penta-, hepta-, octa-, and nonaethylene glycols have been prepared from com-mercially available n = 1–4 oligomers (n = the number of structural units). Elimination side products have been avoided by using a ditosylate of an oligo(oxyethylene glycol) with the monosodium salt of the same glycol in THF (105,106). The crude oligomeric product mixture was separated by preparative scale size-exclusion chromatography (sec) to yield oligomers n = 9, 10, 15, 20, 25, and 35 of 98–99 + % purity.

The cyclic PEO oligomers, known as crown ethers, were first prepared in 1937 (107). However, their ability to complex cationic substrates (108), which makes them useful catalysts and phase-transfer reagents, was not recognized until 1967. These materials have been reviewed (109,110). Most crown ethers are synthesized in rather high yields by template-assisted Williamson etherifi-cations, which organize the intermediate(s) in the cyclization step. However, the cyclic oligomers of PEO can also be made by oligomerization reactions of ethylene oxide, although not in high yields. Fluorine-containing metal-catalyzed cationic oligomerizations of ethylene oxide result in mixtures of cyclic (n = 3–9) and linear oligomers; 1,4-dioxane is the predominant product (111–114). The production of cyclic tetramer, pentamer, and hexamer as well as the linear oligomers is increased by the addition of alkali metal cations as templates.

Polyethylene Derivatives

Oligomers with Two Functional Chain Ends.   Telechelic polyethylene (PE) oligomers have been obtained by the selective oxidative degradation of PE single crystals at the amorphous chain folds using either fuming nitric acid or ozone (115–118) (see also TELECHELIC POLYMERS). Dicarboxylic-acid-terminated paraffins are obtained with lengths corresponding to the crystal thickness (approx. 100 carbon atoms). The dicarboxylic acid end groups can be reduced to iodides, and the iodide end groups converted to methyls (119). However, the reaction products of the nitric acid oxidation are dicarboxylic-acid-terminated paraffin chains with secondary nitro groups located in close proximity to the carboxyl end groups (120). Concentrated H_2SO_4 converts the products to keto acids in which the keto groups can be partially reduced to methylene groups by Clemmenson or Wolff-Kishner reductions. The dicarboxylic-acid-terminated paraffins have been chain-extended with 1,10-decane diol, hexamethylenediamine, or propylene glycol, and then further extended with diphenylmethane 4,4'-diisocyanate and 2,4-toluene diisocyanate (118). Multiblock copolymers of PE and polydimethylsiloxane were prepared by reaction of disilane-terminated polydimethylsiloxane with dihydroxy-terminated PE in the presence of stannous octanoate (121). The precursor PE was prepared by nitric acid oxidation (120).

Supported Reagents and Catalysts.   The use of polymers to immobilize homogeneous catalysts is of continuing interest because of the desire to combine the best features of heterogeneous and homogeneous catalysts (121–133). Insoluble polymers such as divinylbenzene-cross-linked PS have most commonly been used as the organic polymeric supports, but many other polymers have been examined (134). Soluble polymers have also been used, but have received much less attention (see CATALYSTS, POLYMER-SUPPORTED).

Ethylene oligomers with molecular weights higher than 1000 are insoluble in organic solvents at 25°C, but form solutions in a variety of solvents at 90–100°C. Based on this solubility behavior a novel concept for polymer-supported reagents and catalysts based on functionalized oligomers of ethylene has been developed (135–141). At 90–100°C these oligomers behave as soluble reagents, whereas at room temperature they become insoluble and can be separated from the reaction products by centrifugation or filtration. An additional property of ethylene oligomers is their ability to become entrapped quantitatively in PE powders by coprecipitation with high density polyethylene (HDPE) (135). Therefore, these functionalized ethylene oligomers behave either as catalysts supported on soluble polymers at high temperature or as catalysts supported on insoluble polymers at low temperature.

Two different synthetic procedures have been developed for the preparation of ethylene oligomers with functionalized chain ends. The first is based on the anionic polymerization of butadiene followed by the functionalization of growing nucleophilic chains with different electrophilic reagents and reduction of the resulting polybutadiene double bonds to the PE homologue (135). The second method is anionic oligomerization of ethylene followed by the functionalization of the nucleophilic growing chains (136–141). An example is the synthesis of diphenylphosphinated ethylene oligomers.

$$H_2C{=}CH_2 \xrightarrow[\substack{\text{tetramethylenediamine} \\ \text{hexane}}]{\text{RLi}} R{-}(CH_2CH_2)_n CH_2CH_2Li \xrightarrow{\text{ClP(C}_6\text{H}_5)_2} $$
$$R{-}(CH_2CH_2)_n CH_2CH_2P(C_6H_5)_2$$

These oligomers are used as homogeneous polymeric reagents for the synthesis of alkyl chlorides from alcohols in CCl$_4$–toluene at 90°C (136).

PE-entrapped nickel(0) catalysts have been used for the cyclooligomerization of butadiene (137,138) and a PE-bound neodimium carboxylate has been a homogeneous cocatalyst for the stereospecific polymerization of butadiene (139). A rhodium hydrogenation catalyst ligated by diphenylphosphinated ethylene oligomers is a homogeneous catalyst that effects alkene reduction of a substrate which is at the same time being oxidized by an insoluble polyvinylpyridine-bound Cr(VI) oxidant (140) (see ORGANOMETALLIC POLYMERS). A similar concept was applied to the synthesis of oligoethylenes containing either phosphonium groups or crown ether chain ends (141):

$$R{-}(CH_2CH_2)_n CH_2CH_2Li \xrightarrow{\text{ClP(C}_6\text{H}_5)_2} R{-}(CH_2CH_2)_n CH_2CH_2P(C_6H_5)_2 \xrightarrow{C_mH_{2m+1}Br}$$
$$R{-}(CH_2CH_2)_n CH_2CH_2\overset{+}{P}(C_6H_5)_2C_mH_{2m+1}Br^-$$

or

These functionalized ethylene oligomers represent a novel class of polymer-supported phase-transfer catalysts exhibiting activities comparable to those of the corresponding low molecular weight catalysts or of the insoluble PS-bound phase-transfer catalysts both in liquid–liquid or solid–liquid phase-transfer reactions (141).

Models for Structure–Property Relationships

Crystallization and Chain Folding

Polyethylene Oligomers.   One of the most extensive uses of oligomers and short polymer chains has been in studies of chain folding, which is the standard mode of polymer crystallization. That is, oligomers help to answer questions such as: at what chain length does chain folding set in? What is the length of the straight chain segments or stems? Where are the chain ends located and what is their effect on chain folding? And, what is the nature of the fold? Cyclic oligomers have been used to study the last question; linear oligomers and short polymer chains have been used to consider all four problems. A 1977 review (142) suggests that the length at which extended chains fold into crystalline lamellae depends not only on the crystallization temperature and the time at this temperature, but also on the chemical nature of the chain ends and the chain itself and on the polydispersity of the sample. For example, the interactions between highly polar hydroxyl and carboxylic acid chain ends of poly(ethylene glycol)s (PEG) (143–147) and functionalized paraffins (119,148,149), respectively, cause the chains to fold in integral fractions of the chain length such that the polar end groups are kept at the crystal surface. Therefore, varying the crystallization temperature and annealing results in discontinuous jumps in the PEG fold length, in contrast to oligomeric amides (150–153) and paraffins with noninteracting end groups. The chain-fold length of such polymers varies continuously with heat treatment. However, even noninteracting methyl end groups of the true paraffins are located at the crystal surface, such that the fold length fluctuates and the crystal surface may be uneven or rough, depending on the polydispersity of the chains (Fig. 1) (142) (see also POLYMERS, CRYSTALLINE; CRYSTALLIZATION KINETICS in the Supplement).

Besides the degradative oxidation of PE chain folds discussed previously, linear paraffins have also been obtained by extraction from polydisperse PE samples. New synthetic routes for the preparation of monodisperse n-alkanes and cyclic alkanes of all sizes have added impetus to the study of the crystallization and melting mechanisms of PE. One new method involves the oxidative coupling of terminal diacetylenes. Linear alkanes up to $C_{384}H_{770}$ and cyclic alkanes up to $C_{288}H_{576}$ have been prepared by the oxidative coupling of either 1,23-tetracosadiyne or its dimer (154). The homologous oligomers were then separated by adsorption chromatography and the triple bonds hydrogenated. Pure n-alkanes have also been synthesized by repeated Wittig couplings of a bromoaldehyde with an aldehyde-protected alkylphosphonium salt (155,156), followed by the required hydrolysis, reductions, and/or removal of bromine (157–159). Cycloalkanes $(CH_2)_n$ with n = 12–84 were prepared by metathesis reaction of cyclododecene followed by separation by preparative gpc and hydrogenation of the cyclic olefins (160–163). The cyclic olefins are obtained by a backbiting mechanism (161) similar to that which leads to cyclic compounds during the cationic ring-opening polymerization of heterocyclics (32).

The melting points of cycloalkanes and those of n-alkanes as a function of chain length are shown in Figure 2 (163). Up to about C_{30} the curve for cycloal-

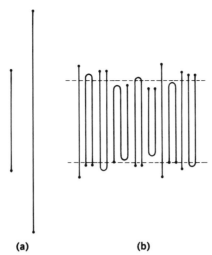

(a)   **(b)**

Fig. 1. A fold model produced by chains of lengths varying between the lengths of the two straight vertical lines on the left (**a**). If the chain ends are all to be at the crystal surface, the fold length must be uneven, as shown in (**b**). The crystallization temperature determines the mean layer thickness (dashed horizontal lines) which is observed experimentally. Different mean thicknesses are obtained by different distributions of extended, once, twice, etc, folded chains (142). Courtesy of A. Keller.

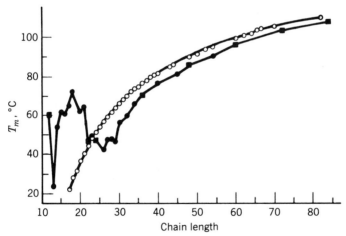

Fig. 2. Melting points of cycloalkanes (●; where ■ indicates previously compiled data) as compared with those of n-alkanes (○) as a function of number of C atoms n (163). Courtesy of *Makromolekulare Chemie*.

kanes is very irregular because of severe conformational constraints not present in n-alkanes of comparable chain length. Starting at C_{30} the curve becomes smooth, approximating a limiting value.

The melting points of the higher molecular weight cycloparaffins are lower than those of the corresponding n-alkanes, but the difference becomes smaller with increasing n. The melting points can be described by the same empirical equation as proposed for n-alkanes:

$$T_m = \frac{\Delta H}{\Delta S} = \frac{\Delta H^*(n + a)}{\Delta S^*(n + b)} = T^* \frac{n + a}{n + b}$$

	T^*, K	a	b	$Ref.$
cycloalkanes	414.3	-1.4	$+5.75$	163
n-alkanes	414.3	-1.5	$+5$	164

The use of linear and cyclic oligomers as models for polymer chain folding originates from the fact that nonpolymeric, linear molecules can crystallize in macroscopic crystals and from the observation (165) that large cyclic paraffins pack in the same way that linear paraffins pack, ie, in two parallel straight all-trans chains bridged at both ends, implying that sharp bending similar to the chain folding in PE is required (166). Thus a cyclic paraffin with the same unit cell and packing as PE was originally sought in order to reveal the exact fold configuration (166,167).

Much has been gained from studying cyclic and linear oligomers and short chains as a homologous series whose properties approach those of the corresponding polymer as chain length increases. For example, using a series of n-alkanes up to C_{384} and cyclic alkanes up to C_{288}, the length of the straight chain segments was calculated from the frequency of the longitudinal acoustic mode (LAM) observed by laser Raman spectroscopy (qv). It was found that the cycloalkanes show no additional chain folding within this range, whereas the linear alkanes fold after a critical chain length between 144 and 168 carbon atoms (154). Other studies have demonstrated by Raman spectroscopy, small-angle x-ray scattering (saxs), and differential scanning calorimetry (dsc) that chain folding occurs in n-alkanes with greater than 150 carbon atoms (157–159). In addition, chain-fold length was always an integral reciprocal of the chain length, indicative of sharp, adjacent reentry folds in which the methyl chain ends are located at the fold surface (157–159). Using computer models of chain-stem arrangements in a lattice to fit neutron scattering data (168) and to construct theoretical infrared band profiles for the CD_2 bending vibrations (169) of solution-grown mixed crystals of deuterated and normal dotriacontane and PE, data consistent with the super-folding of sheets of folded chain stems having 75% adjacent reentry and 50% dilution of a molecule along the fold plane have been obtained.

Solid-state cross-polarization–magic-angle spinning (cp–mas) ^{13}C nmr has recently allowed the conformational characterization of lamellar crystals. Comparison of the solution and solid-state ^{13}C nmr chemical shifts of cyclic alkanes with 16–80 carbon atoms indicates that the solid-state conformation is retained in solution for cyclic alkanes with up to 32 carbon atoms (170), with fast conformational interconversions of the ring occurring in the solid state as well as in solution. In the larger cyclic alkanes, however, molecular motion is frozen in the solid state, but resumes in solution, as is the case with n-alkanes (9–36 carbon atoms) and PE (170).

Another demonstration of the ability of smaller cyclic alkanes to undergo conformational interconversions in the solid state was done by correlating the thermal behavior (dsc) of cyclic alkanes (12–96 carbon atoms) with molecular mobility data obtained from temperature-dependent cp–mas ^{13}C nmr (171–173). The smaller cycloalkanes (24 carbon atoms) undergo a solid–solid transition into a mesomorphic phase of considerable disorder well below the melting temperature; cyclododecane and cyclotetradecane have basically the same conformation in all three phases. This solid–solid transition is apparently due to the onset of rotational motion, resulting in orientational disorder in the smallest rings, and conformational disorder once the cyclotetraeicosane size is reached. No solid–solid transition is observed for the medium rings (n = 36, 38), probably because of the formation of lamellar-type crystals as the ring size increases, which hinders conformational interconversions and rotational motion. The rotational behavior of the larger rings depends on the crystallization conditions. For example, solution-crystallized cyclohexanonacontane was monoclinic at RT, but melt-crystallized samples were orthorhombic. This can be explained by competition of the folds' tendency toward a minimum-energy monoclinic packing and the stems' tendency toward orthorhombic packing. The more highly strained orthorhombic modification changes completely to a higher temperature disordered phase before melting; the monoclinic modification is disordered only near the folds. The folds in PE crystals may be of this tight "orthorhombic" nature.

Linear alkanes also show increasing surface disordering as the chain length increases. Up to six carbon atoms at the chain ends near the lamellar crystalline surfaces of n-alkanes (19, 36, and 136 carbon atoms) were studied separately from the interior methylene segment by cp–mas ^{13}C nmr and compared with low molecular weight PE (174). Nonadecane melted in two steps (phase transitions) and hexatriacontane melted in three steps. There was evidence of premelting or surface melting in octahexacontahectane, and a continuous melting of PE over a wide temperature range. In all cases disordering (increasing gauche defects) started at the chain ends, but the decreasing ability for cooperative motion with increasing chain length presumably accounts for the progression from stepwise to continuous melting as a function of chain length.

Polydimethylsiloxane Oligomers.   In contrast to the cyclic alkanes, comparison of the solid-state ^{13}C and ^{29}Si nmr spectra of cyclooctamethyltetrasiloxane with its solution ^{13}C nmr spectra demonstrate that the conformational mobility of such polar oligomers is significantly determined by the molecular packing in the crystal (175), and the solid–solid transition (265 K) is only observed in solid-state experiments and not in solution. The thermal behavior (dsc) of both cyclic (5–329 structural units) and linear polydimethylsiloxanes (2–343 structural units) has been studied (176). In addition to melting and the solid–solid transition, a T_g and a cold crystallization exotherm were observed in cyclic and linear oligomers with greater than 42 and 33 structural units, respectively. Furthermore, the temperatures of these transitions were much lower than those of octamethyltetrasiloxane.

Cyclic oligomers with less than seven units and linear oligomers with less than four units are completely crystalline, showing cold crystallization and melting; the intermediate molecular weight oligomers are completely amorphous. The cyclic oligomers were prepared by a ring-chain equilibration reaction in toluene at 110°C and fractionated by vacuum fractional distillation and preparative sec. The linear oligomers were separated similarly from a commercial polydimethylsiloxane sample. Although the final molecular weight distributions were narrow

(1.01–1.23), they are actually wide when considered in relation to a lamella crystal of uniform thickness from a truly monodisperse compound. It is possible that this polydispersity accounts for the discrepancies in the two studies (175,176), since both melting and glass-transition temperatures decrease as the size of the ring increases. The molecular weight–glass-transition temperature relationship for both linear and cyclic oligomers (Figs. 3 and 4) (176) of polydimethylsiloxane has been determined. For linear oligomers T_g increases, whereas for cyclic oligomers T_g decreases with increasing molecular weight.

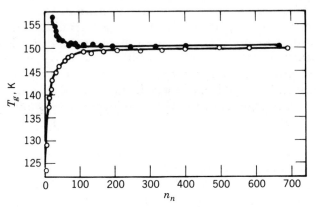

Fig. 3. T_g of the cyclic dimethylsiloxanes (●) and linear dimethylsiloxanes (○) plotted against number-average of skeletal bonds n_n (176). Courtesy of *Polymer*.

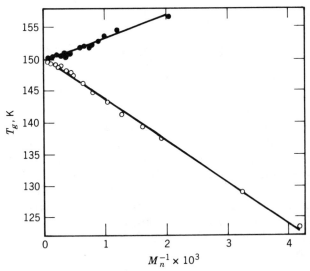

Fig. 4. T_g of the cyclic dimethylsiloxanes (●) and linear dimethylsiloxanes (○) plotted against reciprocal number-average molar mass M_n^{-1} (176). Courtesy of *Polymer*.

Poly(ethylene oxide) Oligomers. The effect of polydispersity on crystallinity was studied by comparing the monodisperse nonamer and pentadecamer of ethylene oxide with "polydisperse" (1.05–1.10) PEO (number-average degrees of polymerization from 9 to 68) (177). All samples were lamellar with a 7/2 helical structure. The monodisperse glycols were highly crystalline with their chain ends incorporated in the crystalline array, but the chain ends in the polydisperse samples were increasingly incorporated in the noncrystalline surface area as chain length increased.

The effect of the end group on the chain folding of low molecular weight PEO (ca 45 structural units) has also been determined. Low molecular weight PEO crystallized at RT in the extended chain form; end groups of seven or more atoms were required to stabilize folded chain lamellae (178). Also, it was found that the enthalpy of fusion can be used to differentiate between extended and folded chain structures, since it is essentially constant for a given structure and does not vary with the nature of the end groups.

Given that PEO is also a stacked lamellar structure (142) with alternating crystalline and amorphous layers, the ordering and disordering behavior in terms of lamella thickness and the free energies of formation from the melt of the crystalline and amorphous layers has been considered using copolymers of PEO with noncrystallizable chain ends as models with inherently alternating crystalline and amorphous regions. Short methylene chains and chains of propylene oxide satisfy the requirement of being noncrystallizable. Diblock (179), triblock (180), and multiblock (181) copolymers of PEO (40–136 repeat units) and poly(propylene oxide) (PPO) (0–12 repeat units) have been made. The diblock copolymers form extended chain crystals in which the crystalline lamellar thickness is independent of copolymer composition. The multiblock copolymers form predominantly single-fold lamellae; the triblock copolymers form single-fold lamellae when the propylene oxide chain consists of five or more units, and extended chain crystals with shorter end groups. When both the copolymer and PEO homopolymer chains are extended, the specific volume of the amorphous layer is lower than that of the supercooled melt, whereas that of single-fold copolymer and homopolymer lamellae is higher. This increase in specific volume of single-fold lamellae has been attributed to a decrease in the number of rejected molecules from the crystalline layer and a decrease in the average length of chain ends in the amorphous layer of folded chains (180). In all cases the melting temperature is low compared with that of perfectly crystalline PEO due to the positive interfacial free energy in the transition from the melt to the polycrystal, which is primarily a result of a reduction in entropy.

In the case of the multiblock copolymers the polycrystal is apparently more random than would be expected for a polycrystal with adjacent reentry, which is possible when using short propylene oxide block lengths. This was demonstrated by a low degree of crystallinity and reduced melting temperatures in comparison with PEO oligomers of the same block length. However, although the multiblock copolymers are less crystalline than the triblock copolymers, the melting points of the multiblock copolymers can be predicted to within a degree from the corresponding triblock copolymer.

Star polymers containing 1–4 oligooxyethylene branches joined by ester linkages form extended chain lamellae of the branches (182). There is an increase of lamellar spacing with increasing number of branches. The melting temperatures of the linear samples are independent of the crystallization temperature, whereas that of the branched samples show a slight increase with increasing crystallization temperature. However, this increase is much smaller than predicted (183) unless partial melting is considered, in which case the interfacial free energy must increase as branching increases.

Triblock copolymers of PEO (23–34 structural units) end-capped with methylene chains of 0–18 carbon atoms form lamella crystals in which ca 20% of the oxyethylene chain is incorporated in the amorphous layer (184). Once the methylene chain contains 26 or more carbon atoms, both blocks crystallize in a single-fold structure (185). However, the extent of crystallinity is low, indicating that the crystallization of one component precludes the crystallization of the other within the same molecule. For the folded crystals the melting temperature increases as the methylene length increases (184,185), apparently due to the high change of free energy of mixing two unlike chains upon melting and to the increased conformational restrictions of cilia. Crystallization of the entire copolymer is possible if the oligooxyethylene block is short and monodisperse, if the methylene chain ends contain 10 or more carbon atoms, and if the crystallization temperature is high (186). However, if the methylene chain contains 26 or more carbon atoms, only the methylene blocks crystallize.

The diblock methylene–ethylene oxide copolymers form both extended and single-fold chain lamellae, providing the methylene chain is long enough and the crystallization temperature low (187). However, in contrast to the triblock oligomers these chain folds are unstable and unfold to the extended form at a temperature below the melting temperature. Evidently, the methyl group at the other end of the oligooxyethylene segment is not large enough to provide end-group stabilization (178).

Although the cyclic oligomers of ethylene oxide have been prepared (111–114), they have not been used to study chain-folding mechanisms.

Poly(tetramethylene oxide) Oligomers.   As with most other oligomers the oligomers of poly(tetramethylene oxide) or polytetrahydrofuran have been used to calculate the melting temperature of the corresponding polymer of infinite chain length (419.5 K) (188) (see TETRAHYDROFURAN POLYMERS). Before living cationic polymerization methods were available to control the molecular weight of poly(tetramethylene oxide), oligomers were obtained by fractionating commercial polymer samples. Fractionation (qv) techniques continue to be developed. A unique method involves the formation of urea-inclusion complexes by isothermal extractive crystallization of poly(tetramethylene oxide)–urea complexes from methanol solutions at progressively lower crystallization temperatures (188). The narrow fractions are isolated by dissolving the urea matrix in water, then extracting the polymer from the resulting emulsion with chloroform.

Vitrification

The physical properties of polymers and oligomers vary with the degree of polymerization. Because polymeric end groups have a greater influence on properties as molecular weight decreases, PS samples have been used to test the assumption that the T_g of anionically and thermally initiated polymers coincide at high molecular weight, but differ significantly at low molecular weights where the concentration of chain ends is much higher (189). However, the same dependence of T_g on molecular weight for both types of polymers at all molecular weights was found. Since this is true even for the oligomeric samples and lower molecular weight polymers, vitrification must not be governed by free volume (see also CURING; GLASS TRANSITION). The anionically initiated PS samples were obtained commercially; the thermally initiated samples had to be fractionated from bulk polymer (190,191).

The nature of the oligomer chain ends strongly influences the T_g of poly(ether sulfone) (192–194) and poly(2,6-dimethyl-1,4-phenylene oxide) (195), and has a drastic effect on phase transitions exhibited by liquid crystalline oligomers (196).

Segmented Thermoplastic Elastomers

There is interest in tailor-making model oligomers of the "hard" and "soft" segments of thermoplastic elastomers because it is desirable to control the combined thermoplastic processing properties and elastomeric mechanical properties of the materials, especially as they are influenced by chain-length distributions (see ELASTOMERS, THERMOPLASTIC). Model urethane hard-segment oligomers from 4,4'-methylenebis(phenyl isocyanate) (MDI) and 1,4-butanediol (197,198), and from piperazine and 1,4-butanediol (198) have been prepared and react with oligo(tetramethylene oxide) soft segments to form well-defined elastomers (198). However, a transurethanification occurs between the hydrogen-bond-forming MDI urethanes at elevated temperatures, disrupting the hard segments' monodispersity and complicating their thermal behavior, which demonstrates the existence of many polymorphs (197,198). By mixing the hard-segment oligomers (198), it was concluded that the crystalline fraction of the elastomeric polymers consists of the eutectics of the short hard segments and mixed crystals of hard segments of greater than three repeat units (see also POLYURETHANES, BLOCK COPOLYMERS).

As models for the soft blocks of both polyurethanes and copoly(ether ester)s, a series of homologous α,β-dihydroxyoligo(tetramethylene oxide)s (n = 1–14), the corresponding phenyl isocyanate end-capped bis(phenyl urethane)s, and α-hydro-ω-chlorooligo(tetramethylene oxide)s have been prepared by two methods (199). The lower homologues ($n \leq 7$) were prepared by the reaction of either a diol or chloroalcohol, obtained by the cationic ring-opening of THF, with 2,3-dihydrofuran, followed by cleavage of the resulting bis- or monoacetal. The cleavage reaction results in chain extension of from 0 to 2 units and the mixture of oligomers is separated by fractional distillation. The higher homologues were prepared by a phase-transfer reaction between the appropriate α-acetal-ω-chlorooligo(tetramethylene oxide) and a diol. The acetal is then cleaved, resulting in a mixture of the mono- and bis-substituted diols, which are separated first by distillation to remove the shorter oligomers and then by preparative gpc. Although the thermal behavior of these oligomers depends on whether there is an odd or even number of repeat units and on the end group, the melting point of all extrapolate to that of pure crystalline poly(tetramethylene oxide) (200).

Oligo(tetramethylene terephthalate)s have been studied as models for the hard segments of segmented poly(ether ester)s (201–203). A series of well-defined and uniform oligomers with up to seven repeat units was prepared by a route involving protection and deprotection of butanediol with tetrahydropyran groups, and of terephthalyl chloride and benzyl groups (201,202). Again, although the end group influences the oligomers' melting behavior, the melting point of each set of oligomers extrapolated to that of pure crystalline poly(tetramethylene terephthalate) (201). In addition, the packing as determined by saxs of the oligo(tetramethylene terephthalate)s was very similar to that of the high polymer (201). Compatibility studies demonstrated that the oligoesters do not crystallize and that the melt of the lower melting component is not a good solvent for the higher melting component (201,203). The addition of oligo(tetramethylene terephthalate)s to block copoly(oxytetramethylene-*b*-tetramethylene terephthalate) induced different phase behavior in the copolymer, resulting in higher melting points in the blend than in the pure oligomer or copolymer (203). These results confirm that only part of the ester hard segments crystallize in the block copolymers, which fractionates the segments of different lengths.

The triblock copoly(ether ester)s in which the middle block was oligo-(tetramethylene oxide) ($n = 2$–6) and the end blocks were oligo(tetramethylene terephthalate) also crystallized in a lattice similar to that of the corresponding oligoester (204). They were prepared by reaction of 1 mol of the α,ω-dihydroxy-oligo(tetramethylene oxide) with 2 mol of the acid chloride of the oligo-(tetramethylene terephthalate) in which the other end was protected by a benzyl group.

Another example of well-defined segmented block copolymers is constructed of oligomethylene chains as the soft segments and the condensation products of bisphenol A with terephthalic acid as the hard segments (205,206). Hard–soft diblock copolymers and hard–soft–hard and soft–hard–soft triblock copolymers were prepared by specific stepwise chain-lengthening sequences of the appropriately protected monomers and/or telechelic oligomer blocks (205). Benzyl and *t*-butyl groups were used to protect ester and phenol functionalities, and tetrahydropyranyl groups to protect aliphatic hydroxide functionalities. Some physical investigations of these segmented block copolymers of uniform chain length and defined structure and of their constituents have been described, especially their thermal behavior with respect to phase transformations in the solid and liquid state, and thermal stability of the materials (206).

Thermal Stability of Poly(methyl methacrylate)

Poly(methyl methacrylate) (PMMA) prepared by an anionic mechanism is more thermally stable than that made by a free-radical process (207). The thermal degradation of PMMA obtained by a free-radical mechanism occurs in two distinct stages. Approximately 50% of the sample degrades between 170 and 250°C; the remainder requires temperatures well in excess of 300°C (208). It has been suggested that groups generated during the termination of the free-radical polymerization of methyl methacrylate profoundly affect the thermal stability of the resulting polymer (209). Therefore, the two separate phases of the degradation are attributed to the presence of two types of end groups, imparting different thermal stabilities. It is proposed that polymer chains having an unsaturated end group are less stable than those containing a saturated terminus (208). In the absence of transfer agents, the termination of methyl methacrylate polymerization occurs by combination and disproportionation. Disproportionation results in an unsaturated end group (**15**) and a saturated group (**16**); combination of growing radical chains yields a head-to-head linkage (**17**) within the chain.

$$-CH_2-C\!\!\begin{array}{c}CH_2\\ \diagdown\\ COOCH_3\end{array} \qquad -CH_2-\overset{CH_3}{\underset{COOCH_3}{C}}-H \qquad -CH_2-\overset{CH_3}{\underset{COOCH_3}{C}}\!\!-\!\!\overset{CH_3}{\underset{COOCH_3}{C}}-CH_2-$$

$$(\mathbf{15}) \qquad\qquad (\mathbf{16}) \qquad\qquad (\mathbf{17})$$

In order to interpret the thermal stability of PMMA, the oligomers (**18**), (**19**), and (**21**) have been synthesized as models for the polymers containing end groups or linkages formed by the different termination mechanisms (210,211):

$$CH_2=C\!\!\begin{array}{c}CH_3\\ \diagup\\ \diagdown\\ COOCH_3\end{array} \xrightarrow[\text{tetraphenylporphyrin}]{\text{cobalt}} CH_3-\overset{CH_3}{\underset{COOCH_3}{C}}\!\!-\!\!(CH_2-\overset{CH_3}{\underset{COOCH_3}{C}})_{\overline{n}}CH_2-C\!\!\begin{array}{c}CH_2\\ \diagup\\ \diagdown\\ COOCH_3\end{array} \xrightarrow{H_2}$$

$$(\mathbf{18})$$

$$CH_3-\overset{CH_3}{\underset{COOCH_3}{C}}\!\!-\!\!(CH_2-\overset{CH_3}{\underset{COOCH_3}{C}})_{\overline{n}}CH_2-\overset{CH_3}{\underset{COOCH_3}{C}}-H$$

$$(\mathbf{19})$$

$$CH_3-\overset{CH_3}{\underset{COOCH_3}{C}}-N\!\!=\!\!N-\overset{CH_3}{\underset{COOCH_3}{C}}-CH_3 + CH_2=C\!\!\begin{array}{c}CH_3\\ \diagup\\ \diagdown\\ COOCH_3\end{array} \longrightarrow CH_3-\overset{CH_3}{\underset{COOCH_3}{C}}-CH_2-\overset{CH_3}{\underset{COOCH_3}{C}}\!\!-\!\!\overset{CH_3}{\underset{COOCH_3}{C}}-CH_2-\overset{CH_3}{\underset{COOCH_3}{C}}-CH_3$$

$$(\mathbf{20}) \qquad\qquad\qquad\qquad (\mathbf{21})$$

The decomposition temperatures for the individual oligomers are 250°C for (**18**), corresponding to structure (**15**); 300°C for (**19**), which is like (**16**); and 190°C for (**21**), corresponding to (**17**). These decomposition temperatures can clearly explain the thermal behavior of radically and anionically formed PMMA. The anionically obtained polymer has a structure similar to (**16**), whereas the radically obtained polymer is a combination of (**15**)–(**17**).

Within a given sample of PMMA obtained by free-radical polymerization, longer chains are more thermally labile than their shorter counterparts (212). Most of the higher molecular weight molecules are expected to arise by combination of the growing chains and hence would contain the weak head-to-head linkage (**17**). Degradation of these chains, therefore, will be observed during the early stages of thermolysis. These results are not only of theoretical interest but also of practical importance, because even a radical polymerization can be performed in such a way that the amount of structures (**15**) and (**17**) are controlled and therefore the thermal stability of the resulting polymer designed.

Conducting Oligomers

The answer to the question of whether an interchain or intrachain process is the dominant one responsible for electrical conductivity in conducting polymers comes from the study of the electrical conductivity of polymers as a function of molecular weight (see also ELECTRICALLY CONDUCTING POLYMERS).

Poly(oxysiliconphthalocyanine)s (**23**) have been synthesized with polymerization degrees between 10 and 200 by the polycondensation of dihydroxysilicon-phthalocyanine (**22**) at 350–400°C and 1 kPa (10 mbar) (213–215).

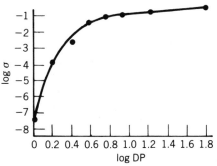

(**22**)   (**23**)

The dependence of the electrical conductivity σ on the degree of polymeri-zation of the oxidized poly(oxysiliconphthalocyanine) is shown in Figure 5. The function of the polymer chain is to bring about the desired packing of the mac-rocycles in stacks, but the actual chain length is of little consequence to the conductivity, at least in the samples of polycrystalline quality like the ones investigated (24) (see also PHTHALOCYANINE POLYMERS).

Fig. 5. RT conductivity σ in $\Omega^{-1}\cdot cm^{-1}$ as a function of degree of polymerization (214). Courtesy of *Molecular Crystals Liquid Crystals*.

Polyacetylenes [$(CH)_x$] with number-average molecular weights from 400 to ca 10^6 have been synthesized, and the crystal structure of cis and trans isomers, the thermal cis–trans isomerization, and their electrical properties have been investigated as a function of molecular weight (216,217). Very low molecular weight polyacetylene ($\overline{M}_n$ = 400–500) has the same cis crystal structure as high molecular weight polymers, but is less ordered along the c axis. It is isomerized to trans material with apparently a more compact unit cell than the high mo-lecular weight polymers. Annealing of the crystallite increases the longitudinal order during thermal isomerization. This process occurs more readily and with lower activation energy in low molecular weight polyacetylene than for polymers with high molecular weights. Isomerization of high molecular weight polymers tends to trap cis units, which can result in degradation, as evidenced by the formation of sp^3-carbon vibrations in ir spectra. This is true even for low molecular weight polyacetylene after prolonged heating (216). However, both the electrical conductivity and the thermopower coefficient of the undoped *cis*- and *trans*-polyacetylene and of the iodine-doped polymers are independent of molecular weight over a range of $\overline{M}_n$ from 400 to 870,000 (see ACETYLENE POLYMERS).

Because these experiments with both types of conducting polymers demonstrate a virtual independence of electrical conductivity upon molecular weight, interchain hopping of carriers must be the process that determines the electrical conductivity in conjugated polymers.

Thermotropic Liquid Crystalline Oligomers

Thermotropic liquid crystalline polymers (qv) containing calamitic or rod-like mesogenic units can be classified as side chain or main chain. Side-chain liquid crystalline polymers have the mesogenic units attached as side groups from the polymer backbone. The mobility of the mesogenic side groups is decoupled from the mobility of the polymer backbone through a flexible spacer (218–220). Main-chain liquid crystalline polymers have the mesogenic groups in the main chain and they are coupled either through flexible or rigid spacers (221,222).

Side-chain liquid crystalline oligomers were synthesized in order to obtain a more detailed understanding of the change in phase behavior in going from the monomer to the polymer (223,224). One study involved monodisperse oligo(methylsiloxane)s with mesogenic side chains (224).

$$m = 6; \; R = -OCH_3; \; n = \text{degree of polymerization}$$

The phase-transformation temperatures were determined as a function of the degree of polymerization DP for several different side-chain liquid crystalline polysiloxanes. For $3 < DP < 10$ a steep increase in phase-transformation temperature is observed for all systems. With growing DP the stability range of the nematic phase is more extended than the stability range of the smectic phase. Figure 6 presents an example of a polymer exhibiting both smectic and nematic mesophases. In analogy to the pressure–temperature behavior of low molar mass liquid crystals, this effect was attributed to progressively denser packing of the mesogenic side chains of the oligomers with growing degree of polymerization. This is seen macroscopically in a decrease of specific volume at the phase-transformation temperature with increasing DP. Similar experiments were performed on thermotropic main-chain liquid crystalline oligomers of polyesters (225–231), polyethers (232), and copolyethers (233).

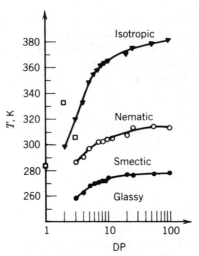

Fig. 6. T_g, T_{s-n}, and T_{n-i} transitions vs DP (224). Courtesy of *Macromolecules*.

A series of polyethers containing only bromoalkane chain ends have been made by phase-transfer-catalyzed polyetherification of 4,4′-dihydroxy-α-methylstilbene with 1,11-dibromoundecane. They are designated HMS–C-11 polyethers and can be represented

$$Br{-}(CH_2)_{11}{-}\left[O{-}\bigcirc{-}C{\overset{CH_3}{\underset{\parallel}{}}}{\underset{\overset{\parallel}{C}{-}\underset{H}{}}{}}{-}\bigcirc{-}O{-}(CH_2)_{11}{-}Br\right]_m$$

The low molecular weight polyethers are completely crystalline. On increasing their molecular weight the polymers become monotropic nematics, and at higher molecular weights enantiotropic nematics. The thermal stability of the mesophase increases with the polymer molecular weight up to about 10,000 and then remains constant (Figs. 7 and 8). Both the enthalpy and the entropy of melting and isotropization follow the same trend, ie, they increase with molecular weight, then remain constant. The phase transitions of main-chain liquid crystalline oligomers strongly depend on the nature of their chain ends (196,228,232).

The same molecular weight phase-transition dependence was observed for main-chain liquid crystalline copolymers (233).

Polymer Blends

Theoretical models have predicted that any miscible polymer blend should exhibit a lower critical solution temperature (LCST) (234–239) and that the LCST should decrease as the molecular weight of one of the two polymers increases (238,239). These LCST temperatures have been reported for a number of polymer

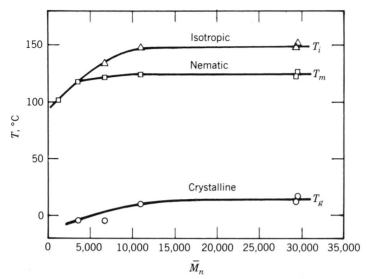

Fig. 7. The influence of number-average molecular weight ($\overline{M}_n$) on the glass-transition (T_g), melting (T_m), and isotropization (T_i) temperatures for the HMS–C-11 polyethers during the heating scan (232).

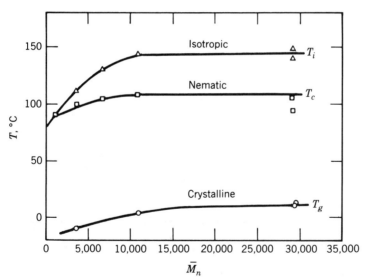

Fig. 8. The influence of number-average molecular weight ($\overline{M}_n$) on the glass-transition (T_g), crystallization (T_c), and isotropization (T_i) temperatures for the HMS–C-11 polyethers during the cooling scan (232).

mixtures (240–244). The theoretical prediction concerning the dependence of the LCST on molecular weight has been experimentally demonstrated by measuring the LCST of blends of oligomers with different molecular weights with a polymer having a constant molecular weight (240,244) (see also POLYMER BLENDS).

BIBLIOGRAPHY

"Oligomers" in *EPST* 1st ed., Vol. 9, pp. 485–506, by G. Heidemann, Textilforschungsanstalt Krefeld.

1. IUPAC Commission on Macromolecular Nomenclature, *Pure Appl. Chem.* **40**, 479 (1974).
2. G. M. van der Want and A. J. Stavermann, *Rec. Trav. Chim. Pays-Bas* **71**, 379 (1952).
3. B. Helferich, *Chem. Ber.* **63**, 989 (1930).
4. B. Helferich, *Justus Liebigs Ann. Chem.* **545**, 178 (1940).
5. H. Morawetz, *Polymers: The Origins and Growth of a Science,* John Wiley & Sons, Inc., New York, 1985, p. 6.
6. L. V. Larsen, *Chem. Eng. News* **62**, 58 (Jan. 23, 1984).
7. W. Kern, *Chem. Ztg.* **76**, 667 (1952).
8. H. Zahn, P. Rathgeber, E. Rexroth, R. Krzikalla, W. Lauer, P. Miro, H. Spoor, F. Schmidt, B. Seidel, and D. Hildebrand, *Angew. Chem.* **68**, 229 (1956).
9. H. Staudinger, *Die Hochmolekularen Organischen Verbindungen,* Springer-Verlag, Berlin, 1932.
10. H. Staudinger and M. Luthy, *Helv. Chim. Acta* **8**, 41 (1925).
11. H. Staudinger and co-workers, *Z. Phys. Chem.* **126**, 425 (1927).
12. W. H. Carothers, J. A. Arvin, and G. L. Dorough, *J. Am. Chem. Soc.* **52**, 3292 (1930).
13. P. Schlack and K. Kunz in R. Pummerer, ed., *Chemische Textilfasern, Filme und Folien,* Enke Verlag, Stuttgart, FRG, 1953, p. 629.
14. O. Bayer, *Justus Liebigs Ann. Chem.* **549**, 286 (1941).
15. P. H. Hermans, *Rec. Trav. Chim. Pays-Bas* **72**, 798 (1953).
16. P. H. Hermans, *Nature London* **177**, 126 (1956).
17. G. M. van der Want, H. Peters, and P. Inklaar, *Rec. Trav. Chim. Pays-Bas* **71**, 1221 (1952).
18. H. Jacobson and W. Stockmayer, *J. Chem. Phys.* **18**, 1600 (1950).
19. H. Jacobson, C. O. Beckmann, and W. H. Stockmayer, *J. Chem. Phys.* **18**, 1607 (1950).
20. P. J. Flory, U. W. Suter, and M. Mutter, *J. Am. Chem. Soc.* **98**, 5733 (1976).
21. *Ibid.,* p. 5740.
22. *Ibid.,* p. 5745.
23. M. Mutter, *J. Am. Chem. Soc.* **99**, 8307 (1977).
24. U. W. Suter and M. Mutter, *Makromol. Chem.* **180**, 1761 (1979).
25. J. A. Semlyen in J. A. Semlyen, ed., *Cyclic Polymers,* Elsevier Applied Science Publishers, Ltd., Barking, UK, 1986, p. 1.
26. S. Penczek, P. Kubisa, and K. Matyjaszewski, *Adv. Polym. Sci.* **68–69**, 39 (1985).
27. K. J. Ivin and T. Saegusa, eds., *Ring-Opening Polymerization,* Vols. I–III, Elsevier Applied Science Publishers, Ltd., Barking, UK, 1984.
28. J. E. McGrath, ed., "Ring-Opening Polymerization: Kinetics, Mechanisms and Synthesis," *ACS Symp. Ser.* **286**, American Chemical Society, Washington, D.C., 1985.
29. E. J. Goethals, ed., *Cationic Polymerization and Related Processes,* Academic Press, Inc., New York, 1984.
30. K. J. Ivin, *Olefin Metathesis,* Academic Press, Inc., New York, 1983.
31. V. Dragutan, A. T. Balaban, and M. Dimonie, *Olefin Metathesis and Ring-Opening Polymerization of Cyclo-Olefins,* John Wiley & Sons, Inc., New York, 1986.
32. S. Penczek, P. Kubisa, and K. Matyjaszewski, *Adv. Polym. Sci.* **37** (1980).
33. *Ibid.,* **68–69** (1985).
34. V. Percec, *ACS Symp. Ser.* **285**, 95 (1985).
35. M. A. Winnik, *Acc. Chem. Res.* **18**, 73 (1985).
36. M. A. Winnik in Ref. 25, p. 285.
37. F. Vogtle, *Top. Curr. Chem.* **98** (1981).
38. *Ibid.,* **101** (1982).
39. *Ibid.,* **121** (1984).
40. F. Vogtle and E. Weber, eds., *Host Guest Chemistry, Macrocycles, Synthesis, Structures, Applications,* Springer-Verlag, Berlin, 1985.
41. J. L. Atwood, J. E. D. Davies, and D. O. MacNicol, eds., *Inclusion Compounds,* Vol. 1, Academic Press, Inc., New York, 1984.
42. *Ibid.,* Vol. 2.
43. P. Rempp and E. Franta, *Adv. Polym. Sci.* **58**, 1 (1984).

44. F. W. Harris and H. J. Spinelli, *ACS Symp. Ser.* **282** (1985).

45. *J. Macromol. Sci. Chem.* **21**(8–9) (1984).

46. M. Morton, *Anionic Polymerization: Principles and Practices,* Academic Press, Inc., New York, 1983.

47. R. N. Young, R. P. Quirk, and L. J. Fetters, *Adv. Polym. Sci.* **56,** 1 (1984).

48. E. J. Goethals, *Pure Appl. Chem.* **48,** 335 (1976).

49. E. J. Goethals, *Adv. Polym. Sci.* **23,** 103 (1977).

50. Ref. 33, p. 35.

51. P. W. Weight and M. S. Beevers in Ref. 25, p. 85; J. A. Semlyen, *Makromol. Chem. Macromol. Symp.* **6,** 155 (1986).

52. C. J. C. Edwards and R. F. T. Stepto in Ref. 25, p. 135.

53. H. Hocker in Ref. 25, p. 197.

54. C. F. Tien and T. E. Hogen-Esch, *Macromolecules* **9,** 871 (1976).

55. C. F. Tien and T. E. Hogen-Esch, *J. Am. Chem. Soc.* **98,** 7109 (1976).

56. S. S. Huang and T. E. Hogen-Esch, *J. Polym. Sci. Polym. Chem. Ed.* **23,** 1203 (1985).

57. T. E. Hogen-Esch and W. L. Jenkins, *Macromolecules* **14,** 510 (1981).

58. C. C. Meverden and T. E. Hogen-Esch, *J. Polym. Sci. Polym. Chem. Ed.* **23,** 159 (1985).

59. S. S. Huang, C. Mathis, and T. E. Hogen-Esch, *Macromolecules* **14,** 1802 (1981).

60. C. C. Meverden and T. E. Hogen-Esch, *Makromol. Chem. Rapid Commun.* **4,** 563 (1983).

61. *Ibid.,* **5,** 749 (1984).

62. C. Mathis and T. E. Hogen-Esch, *J. Am. Chem. Soc.* **104,** 634 (1982).

63. T. E. Hogen-Esch and C. F. Tien, *Macromolecules* **13,** 207 (1980).

64. T. E. Hogen-Esch, R. A. Smith, D. Ades, and M. Fontanille, *J. Polym. Sci. Polym. Lett. Ed.* **19,** 309 (1981).

65. K. Hashimoto and T. E. Hogen-Esch, *Macromolecules* **16,** 1805 (1983).

66. *Ibid.,* p. 1809.

67. T. E. Hogen-Esch and C. F. Tien, *J. Polym. Sci. Polym. Chem. Ed.* **17,** 281 (1979).

68. M. A. Buese and T. E. Hogen-Esch, *Macromolecules* **17,** 118 (1984).

69. M. A. Buese and T. E. Hogen-Esch, *J. Am. Chem. Soc.* **107,** 4509 (1985).

70. M. Morton, R. F. Kammereck, and L. J. Fetters, *Br. Polym. J.* **3,** 120 (1971).

71. A. Hofman, S. Slomkowski, and S. Penczek, *Makromol. Chem.* **185,** 91 (1984).

72. S. Sosnowski, A. Duda, S. Slomkowski, and S. Penczek, *Makromol. Chem. Rapid Commun.* **5,** 551 (1984).

73. S. Penczek, P. Kubisa, S. Slomkowski, and K. Matyjaszewski in Ref. 28, p. 117.

74. S. Penczek, Center of Molecular and Macromolecular Studies, Polish Academy of Sciences, Lodz, Poland, 1986, personal communication.

75. S. Kobayashi and T. Saegusa in Ref. 27, Vol. II, p. 761.

76. S. Inoue and T. Aida in Ref. 27, Vol. I, p. 185.

77. K. Brzezinska, W. Chwialkowska, P. Kubisa, F. Matyjaszewski, and S. Penczek, *Makromol. Chem.* **178,** 2491 (1977).

78. S. Inoue and T. Aida in Ref. 28, p. 137.

79. J. Kress, J. A. Osborn, R. M. E. Greene, K. J. Ivin, and J. J. Rooney, *J. Chem. Soc. Chem. Commun.,* 874 (1985).

80. J. Kress, J. A. Osborn, R. M. E. Greene, K. J. Ivin, and J. J. Rooney, *5th IUPAC Symposium on Ring-Opening Polymerization,* Blois, 1986, IUPAC, Oxford, UK, 1986, p. 137; *J. Am. Chem. Soc.* **109,** 899 (1987).

81. L. R. Jilliom and R. H. Grubbs, *J. Am. Chem. Soc.* **108,** 733 (1986).

82. D. Y. Sogah and W. B. Farnham in H. Sakurai, ed., *Organosilicon and Bioorganosilicon Chemistry: Structure, Bonding, Reactivity and Synthetic Applications,* Ellis Harwood, Ltd., Chichester, UK, 1985, p. 219.

83. H. Sato and Y. Tanaka, *Macromolecules* **17,** 1964 (1984).

84. R. F. Mayo, *Polym. Prepr. Am. Chem. Soc. Div. Polym. Chem.* **2,** 55 (1961).

85. R. F. Mayo, *J. Am. Chem. Soc.* **90,** 1289 (1968).

86. W. A. Pryor and L. D. Lasswell, *Advances in Free Radical Chemistry,* Academic Press, Inc., New York, 1975, pp. 27–99.

87. W. A. Pryor, *ACS Symp. Ser.* **69,** 33 (1978).

88. W. D. Graham, J. G. Green, and W. A. Pryor, *J. Org. Chem.* **44,** 907 (1979).

89. N. J. Barr, W. I. Bengough, G. Beveridge, and G. B. Park, *Eur. Polym. J.* **14**, 245 (1978).
90. Y. K. Chong, E. Rizzardo, and D. H. Solomon, *J. Am. Chem. Soc.* **105**, 7761 (1983).
91. D. C. Pepper and P. J. Reilly, *J. Polym. Sci.* **58**, 639 (1962).
92. T. Higashimura, M. Hiza, and M. Hasegawa, *Macromolecules* **12**, 217 (1979).
93. C. H. Bamford, A. D. Jenkins, and R. Johnston, *Proc. R. Soc. London Ser. A* **239**, 214 (1959).
94. P. Cacioli, D. G. Hawthorne, S. R. Johns, D. H. Solomon, E. Rizzardo, and R. I. Willing, *J. Chem. Soc. Chem. Commun.*, 1355 (1985).
95. S. R. Johns, R. I. Willing, and D. A. Winkler, *Makromol. Chem. Rapid Commun.* **8**, 17 (1987).
96. N. S. Enikolopyan, B. R. Smirnov, G. V. Ponomarev, and I. M. Belgovskii, *J. Polym. Sci. Polym. Chem. Ed.* **19**, 879 (1981).
97. A. F. Burczyk, K. F. O'Driscoll, and G. L. Rempel, *J. Polym. Sci. Polym. Chem. Ed.* **22**, 3255 (1984).
98. J. M. Harris, *J. Macromol. Sci. Rev. Macromol. Chem. Phys.* **25**, 325 (1985).
99. S. Z. Perry and H. Hibbert, *Can. J. Res. Sect. B* **14**, 77 (1936).
100. R. Fordyce, E. L. Lovell, and H. Hibbert, *J. Am. Chem. Soc.* **61**, 1905 (1939).
101. R. Fordyce and H. Hibbert, *J. Am. Chem. Soc.* **61**, 1910 (1939).
102. P. Rempp, *Bull. Soc. Chim. Fr.*, 844 (1957).
103. K. J. Rauterkus, H. G. Schimmel, and W. Kern, *Makromol. Chem.* **50**, 166 (1961).
104. C. Coudert, M. Mpassi, G. Guillaumet, and C. Selve, *Synth. Commun.* **16**(1), 19 (1986).
105. A. Marshall, R. H. Mobbs, and C. Booth, *Eur. Polym. J.* **16**, 881 (1980).
106. H. H. Teo, R. H. Mobbs, and C. Booth, *Eur. Polym. J.* **18**, 541 (1982).
107. A. Luttringhaus and K. Zieglar, *Justus Liebigs Ann. Chem.* **528**, 155 (1937).
108. C. J. Pedersen, *J. Am. Chem. Soc.* **89**, 7017 (1967).
109. G. W. Gokel and H. D. Durst, *Synthesis,* 168 (1976).
110. G. W. Gokel and S. H. Korzeniow, *Macrocyclic Polyether Synthesis*, Springer-Verlag, Berlin, 1982.
111. J. Dale, G. Borgen, and K. Daasvatn, *Acta Chem. Scand. Ser. B* **28**, 378 (1974).
112. J. Dale, K. Daasvatn, and T. Gronneberg, *Makromol. Chem.* **178**, 873 (1977).
113. J. Dale and K. Daasvatn, *J. Chem. Soc. Chem. Commun.*, 195 (1976).
114. S. Kobayashi, T. Kobayashi, and T. Saegusa, *Polym. J.* **15**, 883 (1983).
115. D. J. Blundell, A. Keller, I. M. Ward, and I. J. Grant, *J. Polym. Sci. Part B* **4**, 781 (1966).
116. D. J. Blundell, A. Keller, and T. M. Connor, *J. Polym. Sci. Part A* **2**, 991 (1967).
117. T. Williams, D. J. Blundell, A. Keller, and I. M. Ward, *J. Polym. Sci. Part A* **2**, 1613 (1968).
118. D. G. H. Ballard and J. V. Dawkins, *Eur. Polym. J.* **9**, 211 (1973).
119. A. Keller and Y. Udagawa, *J. Polym. Sci. Part A* **2**, 221 (1972).
120. L. R. Melby, *Macromolecules* **11**, 50 (1978).
121. W. K. Basfield and J. M. G. Cowie, *Polym. Bull.* **2**, 619 (1980).
122. N. K. Mathur, C. K. Narang, and R. E. Williams, *Polymers as Aids in Organic Chemistry*, Academic Press, Inc., New York, 1980.
123. P. Hodge and D. C. Sherrington, *Polymer-Supported Reactions in Organic Synthesis*, John Wiley & Sons, Inc., New York, 1980.
124. A. Akelah and D. C. Sherrington, *Chem. Rev.* **81**, 557 (1981).
125. D. C. Bailey and S. H. Langer, *Chem. Rev.* **81**, 109 (1981).
126. A. Akelah, *Synthesis,* 413 (1981).
127. J. M. J. Frechet, *Tetrahedron* **37**, 663 (1981).
128. A. Akelah and D. C. Sherrington, *Polymer* **24**, 1369 (1983).
129. A. Akelah, *J. Mater. Sci.* **21**, 2977 (1986).
130. T. W. Ford and M. Tomoi, *Adv. Polym. Sci.* **55**, 49 (1984).
131. D. C. Sherrington, *Br. Polym. J.* **16**, 164 (1984).
132. C. U. Pittmann, Jr. in G. Wilkinson, ed., *Comprehensive Organometallic Chemistry*, Vol. 8, Pergamon Press, Inc., Elmsford, N.Y., 1982, pp. 553–611.
133. F. R. Hartley, *Supported Metal Complexes: A New Generation of Catalysts*, Reidel Publishing Co., Boston, 1985.
134. Y. Chauvin, D. Commereuc, and F. Dawans, *Prog. Polym. Sci.* **5**, 95 (1977).
135. D. E. Bergbreiter, Z. Chen, and H. Hu, *Macromolecules* **17**, 2111 (1984).
136. D. E. Bergbreiter and J. R. Blanton, *J. Chem. Soc. Chem. Commun.*, 337 (1985).
137. D. E. Bergbreiter and R. Chandran, *J. Chem. Soc. Chem. Commun.*, 1396 (1985).

138. D. E. Bergbreiter and R. Chandran, *J. Org. Chem.* **51**, 4754 (1986).

139. D. E. Bergbreiter, L. B. Chen, and R. Chandran, *Macromolecules* **18**, 1055 (1985).

140. D. E. Bergbreiter and R. Chandran, *J. Am. Chem. Soc.* **107**, 4792 (1985).

141. D. E. Bergbreiter and J. R. Blanton, *J. Org. Chem.* **50**, 5828 (1985).

142. A. Keller, main lecture presented at the symposium *La Vie,* Societe de la Vie in the Palace of Versailles, June 1977.

143. J. P. Arlie, P. Spegt, and A. Skoulios, *C. R. Acad. Sci. Paris* **260**, 5774 (1965).

144. B. Gilg, P. Spegt, and A. Skoulios, *C. R. Acad. Sci. Paris* **261**, 5482 (1965).

145. J. P. Arlie, P. Spegt, and P. Skoulios, *Makromol. Chem.* **104**, 212 (1967).

146. P. Spegt, *Makromol. Chem.* **139**, 139 (1970).

147. C. P. Buckley and A. Kovacs, *Colloid Polym. Sci.* **254**, 695 (1976).

148. A. Keller and A. O'Connor, *Polymer* **1**, 163 (1960).

149. D. M. Sadler and A. Keller, *Kolloid Z. Z. Polym.* **242**, 1081 (1970).

150. H. Zahn and W. Pieper, *Angew. Chem.* **73**, 246 (1961).

151. H. Zahn and W. Pieper, *Kolloid Z. Z. Polym.* **180**, 97 (1962).

152. W. Kern, J. Davidovits, K. J. Rautekus, and G. F. Schmidt, *Makromol. Chem.* **43**, 106 (1961).

153. F. J. Balta Calleja and A. Keller, *J. Polym. Sci. Part A* **2**, 2151 (1963).

154. K. S. Lee and G. Wegner, *Makromol. Chem. Rapid Commun.* **6**, 203 (1985).

155. O. I. Paynter, D. J. Simmonds, and M. C. Whiting, *J. Chem. Soc. Chem. Commun.*, 1165 (1982).

156. I. Bidd and M. C. Whiting, *J. Chem. Soc. Chem. Commun.*, 543 (1985).

157. G. Ungar, J. Stejny, A. Keller, I. Bidd, and M. C. Whiting, *Science* **229**, 386 (1985).

158. G. Ungar, J. Stejny, A. Keller, I. Bidd, and M. C. Whiting, poster presented at the Polymer Physics Group Conference of the Institute of Physics, London, Sept. 11–13, 1985.

159. A. Keller, *Polymer* **27**, 1835 (1986).

160. H. Hocker and R. Musch, *Makromol. Chem.* **175**, 1395 (1974).

161. H. Hocker, W. Reimann, K. Riebel, and Z. Szentivanyi, *Makromol. Chem.* **177**, 1707 (1976).

162. H. Hocker and K. Riebel, *Makromol. Chem.* **179**, 1765 (1978).

163. *Ibid.,* **178**, 3101 (1977).

164. M. G. Broadhurst, *J. Chem. Phys.* **36**, 2578 (1962).

165. A. Muller, *Helv. Chim. Acta* **16**, 155 (1933).

166. A. Keller, *Polymer* **3**, 393 (1962).

167. R. D. Burbank and A. Keller, *Bell Laboratories Technical Memorandum 60-112-114,* Murray Hill, N.J., Oct. 10, 1960.

168. S. J. Spells and D. M. Sadler, *Polymer* **25**, 739 (1984).

169. S. J. Spells, A. Keller, and D. M. Sadler, *Polymer* **25**, 749 (1984).

170. T. Yannanobe, T. Sorita, and I. Ando, *Makromol. Chem.* **186**, 2071 (1985).

171. M. Moller, W. Gronski, H. J. Cantow, and H. Hocker, *J. Am. Chem. Soc.* **106**, 5093 (1984).

172. H. Drotloff and M. Moller, *Thermochim. Acta* (in press).

173. H. Drotloff, D. Emeis, R. F. Waldron, and M. Moller, *Polymer* (in press).

174. M. Moller, H. J. Cantow, H. Drotloff, D. Emeis, K. S. Lee, and G. Wegner, *Makromol. Chem.* **187**, 1237 (1986).

175. M. Moller, *Adv. Polym. Sci.* **66**, 59 (1985).

176. S. J. Clarson, K. Dogsdon, and J. A. Semlyen, *Polymer* **26**, 930 (1985).

177. A. Marshall, R. C. Domszy, H. H. Teo, R. H. Mobbs, and C. Booth, *Eur. Polym. J.* **17**, 885 (1981).

178. C. Booth, R. C. Domszy, and Y. K. Leung, *Makromol. Chem.* **180**, 2765 (1979).

179. P. C. Ashman and C. Booth, *Polymer* **16**, 889 (1975).

180. P. C. Ashman, C. Booth, D. R. Cooper, and C. Price, *Polymer* **16**, 897 (1975).

181. P. C. Ashman and C. Booth, *Polymer* **17**, 105 (1976).

182. J. Q. G. Maclaine, P. C. Ashman, and C. Booth, *Polymer* **17**, 109 (1976).

183. P. J. Flory and A. Vrij, *J. Am. Chem. Soc.* **85**, 3548 (1963).

184. D. R. Cooper, Y. K. Leung, F. Heatley, and C. Booth, *Polymer* **19**, 309 (1978).

185. R. C. Domszy, R. H. Mobbs, Y. K. Leung, F. Heatley, and C. Booth, *Polymer* **20**, 1204 (1979).

186. R. C. Domszy and C. Booth, *Makromol. Chem.* **183**, 1051 (1982).

187. J. K. H. A. Kafaji and C. Booth, *Makromol. Chem.* **182**, 3671 (1981).

188. G. Schmidt, V. Enkelmann, U. Westphal, M. Droscher, and G. Wegner, *Colloid Polym. Sci.* **263**, 120 (1985).

189. A. Rudin and D. Burgin, *Polymer* **16**, 291 (1975).

190. T. G. Fox and P. J. Flory, *J. Polym. Sci.* **14**, 315 (1954).
191. T. G. Fox and P. J. Flory, *J. Appl. Phys.* **21**, 581 (1950).
192. V. Percec and B. C. Auman in Ref. 44, p. 91.
193. V. Percec and B. C. Auman, *Makromol. Chem.* **185**, 1867 (1984).
194. B. C. Auman, V. Percec, H. A. Schneider, W. Jishan, and H. J. Cantow, *Polymer* **28**, 119 (1987).
195. H. Nava and V. Percec, *J. Polym. Sci. Polym. Chem. Ed.* **24**, 965 (1986).
196. T. D. Shaffer and V. Percec, *Makromol. Chem. Rapid Commun.* **6**, 97 (1985).
197. Z. V. Qin, C. W. Macosko, and S. T. Wellinghoff, *Macromolecules* **18**, 553 (1985).
198. C. D. Eisenbach, H. Nefzger, M. Baumgartner, and C. Grunter, *Ber. Bunsenges. Phys. Chem.* **89**, 1190 (1985).
199. R. Bill, M. Droscher, and G. Wegner, *Makromol. Chem.* **182**, 1033 (1981).
200. *Ibid.,* **179**, 2993 (1978).
201. H. W. Hasslin, M. Droscher, and G. Wegner, *Makromol. Chem.* **179**, 1373 (1978).
202. *Ibid.,* **181**, 301 (1980).
203. H. W. Hasslin and M. Droscher, *Makromol. Chem.* **181**, 2357 (1980).
204. F. C. Schmidt and M. Droscher, *Makromol. Chem.* **184**, 2669 (1983).
205. H. Seliger, M. B. Bitar, H. Nguyen-Trong, A. Marx, R. Roberts, J. K. Kruger, and H. G. Unruh, *Makromol. Chem.* **185**, 1335 (1984).
206. J. K. Kruger, A. Marx, R. Roberts, H. G. Unruh, M. B. Bitar, H. N. Trong, and H. Seliger, *Makromol. Chem.* **185**, 1469 (1984).
207. I. C. McNeill, *Eur. Polym. J.* **4**, 21 (1968).
208. N. Grassie and H. W. Melville, *Proc. R. Soc. London Ser. A* **199**, 14 (1949).
209. N. Grassie and E. Vance, *Trans. Faraday Soc.* **49**, 184 (1953).
210. P. Cacioli, G. Moad, E. Rizzardo, A. K. Serelis, and D. H. Solomon, *Polym. Bull.* **11**, 325 (1984).
211. D. H. Solomon, P. Cacioli, and G. Moad, *Pure Appl. Chem.* **57**, 985 (1985).
212. S. L. Malhorta, L. Minh, and L. P. Blanchard, *J. Macromol. Sci. Chem.* **19**, 579 (1983).
213. E. Orthmann, V. Enkelmann, and G. Wegner, *Makromol. Chem. Rapid Commun.* **4**, 687 (1983).
214. G. Wegner, *Mol. Cryst. Liq. Cryst.* **106**, 269 (1984).
215. E. Orthmann and G. Wegner, *Makromol. Chem. Rapid Commun.* **7**, 243 (1986).
216. J. C. W. Chien and M. A. Schen, *J. Polym. Sci. Polym. Chem. Ed.* **23**, 2447 (1985).
217. M. A. Schen, F. E. Karasz, and J. C. W. Chien, *Macromol. Chem. Rapid Commun.* **5**, 217 (1984).
218. H. Finkelmann, H. Ringsdorf, and J. H. Wendorff, *Makromol. Chem.* **179**, 273 (1978).
219. H. Finkelmann, M. Happ, M. Portugal, and H. Ringsdorf, *Makromol. Chem.* **179**, 2541 (1978).
220. H. Finkelmann and G. Rehage, *Adv. Polym. Sci.* **60–61**, 99 (1984).
221. A. Roviello and A. Sirigu, *J. Polym. Sci. Polym. Lett. Ed.* **13**, 455 (1975).
222. C. K. Ober, J. I. Jin, and R. W. Lenz, *Adv. Polym. Sci.* **59**, 603 (1984).
223. S. G. Kostromin, R. V. Talroze, V. P. Shibaev, and N. A. Plate, *Makromol. Chem. Rapid Commun.* **3**, 803 (1982).
224. H. Stevens, G. Rehage, and H. Finkelmann, *Macromolecules* **17**, 851 (1984).
225. R. B. Blumstein, E. M. Stickles, and A. Blumstein, *Mol. Cryst. Liq. Cryst.* **82**, 205 (1982).
226. A. Blumstein, S. Vilasager, S. Ponrathnam, S. B. Clough, R. B. Blumstein, and G. Maret, *J. Polym. Sci. Polym. Phys. Ed.* **20**, 877 (1982).
227. A. Blumstein, *Polym. J.* **17**, 277 (1985).
228. R. B. Blumstein, E. M. Stickles, M. M. Gauthier, A. Blumstein, and F. Volino, *Macromolecules* **17**, 177 (1984).
229. A. Blumstein, O. Thomas, J. Asrar, P. Makris, S. B. Clough, and R. B. Blumstein, *J. Polym. Sci. Polym. Lett. Ed.* **22**, 13 (1984).
230. J. Majnusz, J. M. Catala, and R. W. Lenz, *Eur. Polym. J.* **19**, 1043 (1983).
231. Q. F. Zhou, X. Q. Duan, and Y. L. Liu, *Macromolecules* **19**, 247 (1986).
232. V. Percec and H. Nava, *J. Polym. Sci. Part A* **25**, 405 (1987).
233. V. Percec, H. Nava, and H. Jonsson, *J. Polym. Sci. Polym. Chem. Ed.* (in press).
234. P. J. Flory, R. A. Orwoll, and A. Vrij, *J. Am. Chem. Soc.* **86**, 3515 (1964).
235. P. J. Flory, *J. Am. Chem. Soc.* **87**, 1833 (1965).
236. B. E. Eichinger and P. J. Flory, *Trans. Faraday Soc.* **64**, 2066 (1968).
237. P. J. Flory, *Discuss. Faraday Soc.* **49**, 7 (1970).
238. L. P. McMaster, *Macromolecules* **6**, 760 (1973).
239. R. H. Lacombe and I. C. Sanchez, *J. Phys. Chem.* **80**, 2568 (1976).

240. T. Nishi and T. K. Kwei, *Polymer* **16**, 285 (1975).
241. S. Krause, *Pure Appl. Chem.* **58**, 1553 (1986).
242. J. M. Rodriguez-Parada and V. Percec, *Macromolecules* **19**, 55 (1986).
243. J. M. Rodriguez-Parada and V. Percec, *J. Polym. Sci. Part A* **24**, 579 (1986).
244. C. Pugh, J. M. Rodriguez-Parada, and V. Percec, *J. Polym. Sci. Part A* **24**, 747 (1986).

VIRGIL PERCEC
COLEEN PUGH
Case Western Reserve University

ONE-ELECTRON-INITIATED POLYMERIZATION. See RADICAL POLYMERIZATION.

OPEN-CELL FOAM. See CELLULAR MATERIALS.

OPTICAL BLEACHES. See PAPER.

OPTICAL BRIGHTENING AGENTS. See PAPER.

OPTICAL DATA STORAGE. See Supplement.

OPTICALLY ACTIVE POLYMERS

In general, the term polymer indicates synthetic macromolecular systems; accordingly, this article deals with optically active synthetic polymers only. Naturally occurring polymers, eg, biopolymers (qv), are mentioned only for comparison. Almost all biopolymers are optically active, the most conspicuous exceptions being natural rubber and gutta percha.

Only those molecules with dissymmetric or chiral structures, ie, those without inversion and reflection-symmetry elements, possess optical activity. Most applications of synthetic polymers do not require optical activity. Both low molecular weight compounds and macromolecules must obey the same symmetry rules (1) and the preparation of optically active polymers presents problems similar to those encountered in traditional organic chemistry. On the other hand, the ready availability of optically active macromolecules of natural origin encouraged the study of the chiroptical properties (optical rotary dispersion (ord) and circular dichroism (cd)) of biopolymers. These properties have been widely investigated and a large amount of information about the structure, particularly in solution, of both proteins (qv) (2,3) and nucleic acids (4,5) is available (see also POLYNUCLEOTIDES).

Polysaccharides (qv) have presented more difficulties because of the less specific conformations and the absence of chromophores absorbing in an accessible spectral region (6). The advent of vacuum cd instruments together with a large number of model compounds, as well as more advanced conformational analysis, is making the investigation of chiroptical properties very useful for this class of important optically active macromolecules.

Among model compounds employed in the analysis of chiroptical properties of biopolymers, optically active synthetic macromolecules play an important role (7,8). Some of these models, for instance, poly(α-amino acid)s for proteins, are structurally so similar that a separate discussion of chiroptical properties would not be convenient. For these reasons, this article is limited to polymers with a backbone structure different from that of naturally occurring chiral macromolecules. For additional details and examples, the reader is referred to several review papers (9–15).

Measurement of Chiroptical Properties of Polymers

When a linearly polarized light beam passes through an optically active medium, an elliptically polarized wave is generated with the main axis rotated by an angle α (optical rotation) with respect to the original polarization plane. The arc tangent of the minor to the main axis of the elliptical vibration gives ψ (the ellipticity angle) (16,17). A medium can show a nonvanishing optical rotation and ellipticity angle for chemical or physical reasons, or for both reasons.

The instruments used for measuring α and ψ, polarimeters and dichrographs, respectively, are substantially the same in the above cases, but treatment of experimental results and the type of information obtained are clearly different. In most cases, chemists are interested in obtaining indications at the molecular level, and this is better achieved by measuring chiroptical properties in dilute solutions where possible artifacts arising from orientation, aggregation, and supermolecular order, such as those occurring in liquid crystals and solid crystalline materials, are not observed. Most measurements on polymers have been performed in solution, although a few examples of measurement in the solid state are also known (18).

In solution the standard values of specific optical rotation and of the specific ellipticity angle are obtained from equations 1 and 2, respectively:

$$[\alpha]_\lambda^T = 100\alpha/lc \tag{1}$$

$$[\psi]_\lambda^T = 100\psi/lc \tag{2}$$

where λ indicates the wavelength of the incident light, T the temperature, l the cell path in dm, and c the concentration in g/dL. Equations 1 and 2 can be used for both macromolecules and low molecular weight compounds, although some differences arise when computing the corresponding molar properties. Thus in equations 3 and 4, giving the molar rotatory power (or molar optical rotation) and the molar ellipticity, respectively, M does not indicate the molecular weight of the macromolecule but that of the monomeric residue; in this way, both quantities are independent of the degree of polymerization, ie, the average number of residues per chain in the polymer.

$$[\phi]_\lambda^T = [\alpha]_\lambda^T M/100 \tag{3}$$

$$[\theta]_\lambda^T = [\psi]_\lambda^T M/100 \tag{4}$$

In a copolymer between the two monomers A and B, M is replaced by the average value $\overline{M}$, given by

$$\overline{M} = N_A M_A + N_B M_B \tag{5}$$

where N_A and N_B are the molar fractions and M_A and M_B the molecular weights of the respective monomer residues in the copolymer. It is evident that $\overline{M}$ changes with composition, and correspondingly, $[\phi]_\lambda^T$ and $[\theta]_\lambda^T$ also change. A monotonic dependence on copolymer composition, equal to that expected for the homopolymer mixture, indicates absence of significant chiral, electronic, and steric interactions between A and B units in the copolymer, or no copolymer formation. In contrast, deviation from linearity demonstrates copolymer formation and provides additional structural information (19,20).

A different treatment of experimental ellipticity data obtained from copolymers is very useful when the two comonomers contain chromophores absorbing in distinct spectral regions. In these cases the molar ellipticity in the respective region can be referred to the absorbing moiety by introducing the concentration and the molecular weight of the residue bearing the absorbing moiety into equation 5. The modified equation assumes the form given in equation 6 for the monomer A; a similar equation can clearly be written for B.

$$[\theta]_\lambda^T = \frac{\psi_A^T M_A}{l c_A} \tag{6}$$

In equation 6, M_A is the molecular weight of the residue derived from A and c_A is the corresponding concentration in g/dL (21,22). In addition to molecular weight polydispersity, macromolecules in solution are characterized by an extremely large number of conformations; thus the molar values of chiroptical properties given by equations 3, 4, and 6 are averages of many contributions coming from different structures, the relative amount of which depends on several environmental factors, such as temperature and solvent (1).

Frequently, the molar ellipticity is reported as a function of the difference between the extinction coefficients of left (ϵ_L) and right (ϵ_R) circularly polarized components of the incident light beam, as shown in equation 7:

$$\Delta\epsilon = \epsilon_L - \epsilon_R = 0.3 \times 10^{-3}[\theta] \tag{7}$$

The molar coefficient of dichroic absorption $\Delta\epsilon$ can be treated analogously to ellipticity data. Curves obtained by measuring the optical rotation at different wavelengths (ord curves) are connected with cd curves through Kronig-Kramer integral transforms (16).

Structure and Optical Activity

In order to show measurable optical activity, a polymer must consist of chiral macromolecules with one-handedness appreciably predominating. If left- and right-handed macromolecules are present in equal amounts, optically active polymers can be obtained by separating these enantiomeric macromolecules. Clearly, if chain sections with opposite chirality are included in the same macromolecule, separation is not possible.

Macromolecules in the melt and in solution assume a large number of conformations, each of which is in general dissymmetric; however, complementary arrays are equally populated and no optical activity can be observed (23). Right- or left-handed structures can predominate only if an asymmetric field makes them thermodynamically more stable. This can be achieved when asymmetric centers with a single absolute configuration are present in the main chain or the side chains (11,12,23,24). Separation of conformations with opposite chirality (resolution) is possible only if slow kinetics hinders reequilibration. This possibility is not predictable simply on the basis of linear and point symmetry analysis, but molecular rigidity and the reequilibration mechanism should be known. At present, this situation seems to be rather the exception than the rule.

Optical activity is used as an analytical tool to relate molecular structure to chiroptical properties. In many cases, these properties are very different for macromolecules and low molecular weight compounds resembling the structure of the monomeric residue. Interpretation of these differences allows development of a picture of the conformation in solution, which is difficult to obtain by other techniques. The most typical example is given by poly(α-amino acid)s and proteins for which ord features have been related to the presence of ordered secondary structures in solution (25,26). The developed theory (27) is based on a perturbative approach and is still one of the most useful when groups of electrons do not strongly interact, making it necessary to consider exchange strengths and charge transfer. The method is simple and leads to an understanding of differences in uv absorption and cd between polymers and monomeric analogues, as well as relating these differences to structure (28,29). Indeed, calculations of cd of polypeptides are now possible with good agreement between calculated values and experimental data for the α-helix (30), parallel and antiparallel β-sheet (31), polyproline (32), and disordered structures (33).

Polynucleotides require a more complex approach due to the presence in each macromolecule of four different aromatic bases (29), but results are very satisfying (34).

Similar approaches based on the application of the exciton theory (25,26) to electrostatic interactions among chromophores disposed along a helical chain have allowed calculation of cd spectra of vinyl polymers with aromatic side chains which were in excellent agreement with experimental results (35,36).

Interactions of the single chromophore with the rest of the macromolecule can also be responsible for differences in cd of polymers and monomeric analogues. This behavior is typical when a symmetric chromophore becomes dissymmetric upon insertion into a chiral molecule (37). Thus in several stereoregular polymers, optical activity is strongly affected by typical macromolecular features, such as degree of stereoregularity, conformational arrangement, and interactions with solvent, without occurrence of electronic interactions among chromophores.

Synthesis

The synthesis of optically active polymers is based mostly on the introduction in the macromolecules of stereogenic centers, such as asymmetric carbon atoms, with a single absolute configuration. A few cases are known in which it has been possible to isolate conformations of a single-handedness from macromolecules having achiral primary structure. In macromolecular chemistry, the preparation of an optically active product does not necessarily coincide with the polymerization process, which must then be followed by resolution.

Synthetic methods require an optically active agent as the monomer, as the catalyst, or as the separation support, the chiral discrimination being then of chemical origin. In some cases natural optically active molecules can be used as starting materials or asymmetric systems, such as enzymes or bacteria which are capable of chiral discrimination.

Polymerization of Optically Active Monomers. *1-Olefins.* Terminal alkenes exhibit a typical relationship between chiroptical properties and both primary and secondary structure. Closely following the discovery of Ziegler-Natta catalysts (qv) for the highly stereospecific polymerization of propylene, several optically active 1-olefins (**1**) were polymerized (**2**) in the presence of these catalysts (38–42):

$$
\begin{array}{ccc}
CH_2{=}CH & & {-}(CH_2CH){-}_n \\
| & & | \\
(CH_2)_m & \longrightarrow & (CH_2)_m \\
| & & | \\
H{-}C^*{-}CH_3 & & H{-}C^*{-}CH_3 \\
| & & | \\
R & & R \\
(\mathbf{1}) & & (\mathbf{2})
\end{array}
$$

where $m = 0, 1, 2, 3$ and $R = -C_2H_5$ or $-(CH_2)_3CH(CH_3)_2$

Since the catalyst does not produce appreciable racemization, it was assumed that the enantiomeric purity of the asymmetric carbon atoms (C^*) in the side chains of structure (**2**) was substantially the same as that in the starting monomer (**1**). Here, enantiomeric purity has the same meaning as in classical stereochemistry being given by $[(S - R)/(S + R)] \times 100$, where S and R indicate the moles of residues with S or R absolute configuration, respectively, and S is assumed to be larger than R. However, even if the catalyst does not affect the enantiomeric purity, different values of optical rotation for a given polymer are observed, depending on main-chain stereoregularity, ie, tacticity.

As a result of systematic studies, the following observations were made regarding the relationships between structure and optical rotation in structure (**2**):

1. The optical rotation increases asymptotically with increasing isotacticity (38).

2. The molar optical rotation per monomeric residue $[\phi]_\lambda^T$ (eq. 3) depends on the monomer structure and in particular on the distance of the asymmetric carbon atom from the double bond. For $m = 0$ or 1, the values of $[\phi]_\lambda^T$ are much larger in absolute value than for monomeric structural analogues. This difference is reduced with $m = 2$ and becomes practically negligible for $m = 3$ (38).

3. Polymers of structure (**2**), regardless of the values of m or the nature of R, show plain ord curves between 650 and 200 nm; extrapolation of these data by the one-term Drude equation allows evaluation of the wavelength of the lowest energy, optically active electronic transition. This is located nearly in the same position as for chiral paraffins (38,41,42). Recent cd studies in the vacuum region, ie, 200–140 nm, have shown (43) a dichroic band centered at about 160 nm which can account for the rotatory properties of poly(1-olefin)s up to the visible region, in perfect agreement with ord studies.

4. At $m = 0$ or 1, $[\phi]_\lambda^T$ increases in absolute value with increasing enantiomeric purity of the polymerized monomer, reaching the asymptotic value at ca 70% (19,44). However, a linear dependence is observed when $m = 2$ or 3 (41,44).

These experimental results have been explained on the basis of the greater conformational homogeneity in the isotactic polymer, where the monomer residue can be allowed to assume only a few conformations, all implying helical structure and high optical rotation. Good agreement between observed and calculated $[\phi]_\lambda^T$ values was indeed observed by assuming that conformations leading to one-screw sense of the helical chain were largely predominant (38). In agreement with this explanation, conformationally homogeneous chiral paraffins showed $[\phi]_\lambda^T$ of the same order of magnitude as the corresponding polymers (45). Moreover, the optical rotation in the solid state for polyolefins crystallizing in one-screw sense helices gave results very close to those measured in solution (18). The predominance of one-screw sense has been attributed to cooperative interactions among side chains with equal chirality disposed along isotactic chain sections (13,46). This agrees with the dependence of $[\phi]_\lambda^T$ on stereoregularity, side-chain enantiomeric purity, and monomer structure. It also explains the reduction in optical rotation occurring when conformationally free ethylene units are inserted in isotactic blocks of units in (2) ($m = 0$ or 1) (47,48).

Other Unsaturated Monomers.   The unsaturation in optically active monomers is usually due to a carbon–carbon double bond, eg, in vinyl and acrylic or methacrylic derivatives, whereas there are few examples of dienes, acetylene derivatives, and monomers containing multiple bonds between a carbon atom and a heteroatom. All mechanisms of chain-growth polymerization give optically active polymers, depending on type of monomer.

Vinyl ethers used as monomers are either entirely synthetic (3) (49–51) or are prepared by vinylation of naturally occurring optically active alcohols (4) (52–55).

$$\begin{array}{cc}
\text{CH}_2\!\!=\!\!\text{CH} & \text{CH}_2\!\!=\!\!\text{CH} \\
| & | \\
\text{O} & \text{O} \\
| & | \\
(\text{CH}_2)_m & \text{R*} \\
| & \\
\text{H}\!-\!\text{C*}\!-\!\text{CH}_3 & \\
| & \\
\text{R} & \\
\textbf{(3)} & \textbf{(4)}
\end{array}$$

R = C$_2$H$_5$, m = 0,1   R* = menthyl, bornyl, cholesteryl,

  C$_6$H$_5$, m = 0   or   1,2;5,6-di-O-isopropylidene-3-O-yl-α-

  glucofuranose

In the presence of both homogeneous (i-C$_4$H$_9$AlCl$_2$, BF$_3$O(C$_2$H$_5$)$_2$) and heterogeneous (Al(i-C$_4$H$_9$)$_3$–H$_2$SO$_4$, Mn/MoO$_3$–H$_2$SO$_4$) catalysts, no racemization has been observed during the polymerization of 1-methylpropyl vinyl ether ((3), R = C$_2$H$_5$ and m = 0), 2-methylbutyl vinyl ether ((3), R = C$_2$H$_5$ and m = 1), and menthyl vinyl ether (56,57). Hence, in these polymers, the side-chain enantiomeric purity must be very close to that of the starting monomer sample. In poly(1-

olefin)s, and poly(vinyl ether)s, the differences between chiroptical properties of polymers and monomeric analogues, as well as the increase of optical rotation with isotacticity, have been interpreted on the basis of helical conformations in solution with one predominant screw sense (50). The cationic polymerization of menthyl vinyl ether has been studied in various solvents and at various initiation temperatures, ie, from -72 to $30°C$; optical rotation decreases with increasing temperature and dielectric constant of solvent (55).

Optically active alkenyl vinyl ethers (**5**) have been prepared and polymerized by cationic initiators; differences in optical rotation with respect to monomeric models have been tentatively explained by considering possible asymmetric induction in the main-chain asymmetric carbon atoms (58).

<pre>
CH=CH   CH₂=CH   CH₂=CH   CH₂=C—CH₃
| | | |
C₆H₅ O   C=O   C=O   C=O
| | | |
(CH₂)ₘ   (CH₂)ₘ   O   O
| | | |
H—C*—CH₃   H—C*—CH₃   R*   R*
| |
C₂H₅   C₂H₅
(5)   (6)   (7)   (8)

   m = 0, 1, 2   m = 0, 1, 2
</pre>

Optically active alkyl vinyl ketones (**6**) have been polymerized in the presence of free-radical and anionic initiators; the latter give polymers with higher isotacticity (meso diads fraction 0.91–0.98) and higher optical rotation. Easy accessibility of the $n \rightarrow \pi^*$ electronic transition of the keto chromophore allows ord measurements in the region of the Cotton effect. Polymers from monomers (**6**) with (S) absolute configuration show a negative Cotton effect, which increases with decreasing m and increasing content of meso diads (59,60).

Acrylic (**7**) and methacrylic (**8**) esters of optically active alcohols are easily prepared with high enantiomeric purity. However, often there is more space between the backbone of the derived macromolecules and the side-chain chiral centers than in other systems. Also, the ester group imparts a high rotational freedom and the dissymmetric perturbation of the main chain occurs only with very bulky chiral side chains. Nonetheless, the easy accessibility of many optically active alcohols and the simple esterification with acrylic or methacrylic acid make monomers (**7**) and (**8**) popular; good examples include bornyl acrylate (61) and methacrylate (62), and menthyl acrylate and methacrylate (63,64). Synthetic optically active alcohols have also been used to prepare 1-methylbenzyl methacrylate (65), 2-methylbutyl methacrylate (64), and 1,3-dimethylbutyl methacrylate (66). Polymers derived from 5-oxobornyl methacrylate have been reported (67). In the case of both poly(menthyl acrylate) and poly(1-methylbenzyl methacrylate), a variation of optical rotation with tacticity has been observed, but the effect is small and the interpretation is not straightforward as in other cases (68,69). More details concerning the early study of these polymers and the relationships between chiroptical properties and tacticity and conformation are given in Refs. 10–13. Other optically active polyacrylates or polymethacrylates have been prepared by polymerization with free-radical initiators of the corresponding acryloyl and methacryloyl derivatives of carbohydrate residues (70–72).

The versatility of acrylic- or methacrylic-type functionality for the preparation of optically active polymerizable compounds is further supported by the synthesis and polymerization of several chiral amides having structure (**9**) where R = H or CH_3 (73–76):

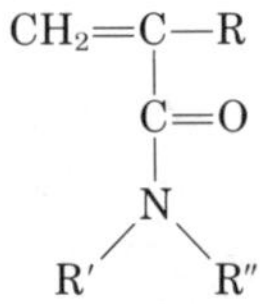

$$CH_2{=}C{-}R$$
$$|$$
$$C{=}O$$
$$|$$
$$N$$
$$R' \quad R''$$

(9)

R'	R''
H	$\overset{COOH}{\underset{(CH_2)_2}{-\overset{|}{\underset{|}{C^*}}-H}}, \quad \overset{CH_3}{\underset{COOH}{-\overset{|}{\underset{|}{C^*}}-H}}$
	$\underset{COOH}{}$
H	$-CH_2(C^*HOH)_4CH_2OH$
CH_3 or $n\text{-}C_3H_7$	$-C^*H(CH_3)(C_6H_5)$
CH_3	$-C^*H(CH_3)(C_2H_5)$

The first optically active *N*-vinyl monomers were described in the early 1960s (10,11). More recently, the cumbersome synthesis of (*S*)-3-*sec*-butyl-9-vinylcarbazole has been reported, as well as its polymerization with free-radical (AIBN) and cationic ($C_2H_5AlCl_2$ or $C_7H_7SbCl_6$) initiators (77,78). These monomers, in contrast to low molecular weight model compounds, do not show any evidence of organized conformations in solution.

Many optically active polymers have been prepared from styrene derivatives substituted with several chiral groups (**10**):

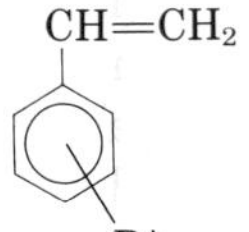

$$CH{=}CH_2$$

$$R^*$$
(10)

Most of the substituents are in the para position, eg, R* =

$$-CH_2-O-\overset{*}{C}H(CH_3)(C_2H_5) \text{ (79)}, \quad -COO-\overset{*}{C}H(CH_3)(C_2H_5) \text{ (80)},$$

$$-\overset{*}{C}H(CH_3)(C_2H_5) \text{ (81)}, \quad -\underset{O}{\overset{\downarrow}{S}}-C_6H_4-p\text{-}CH_3 \text{ (82)},$$

$$-CH_2NH-\overset{O}{\overset{\|}{C}}-\overset{OH}{\underset{OH}{\overset{|}{C}H}}\overset{}{\underset{OH}{\overset{|}{C}H}}CH-\overset{}{\underset{OH}{\overset{|}{C}H}}CH_2OH \text{ (83), and} \quad (84).$$

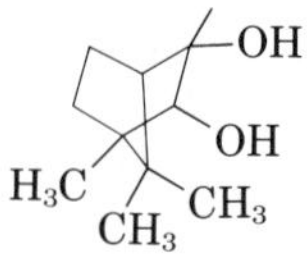

One example of ortho substitution is R* = $-CH_2-S-CH(CH_3)(C_2H_5)$ (85). This group of monomers (**10**) can also include carbazole derivatives bearing a vinyl group attached to one of the phenyl rings and a chiral group on the nitrogen atom, such as (S)-9-(2-methylbutyl)-2-vinylcarbazole and -3-vinylcarbazole (77,78).

Optically active derivatives of 1,3-dienes (**11**) have been prepared and studied (86–89); cis and trans 1,4-polymers have been obtained in the presence of Ziegler-Natta catalysts, whereas free-radical emulsion polymerization gives polymers with unsaturation in the side chains.

$$R'CH=\overset{\overset{\displaystyle R''}{|}}{C}-CH=CH_2$$

(**11**)

R'	R''
H	$-\overset{*}{C}H(CH_3)(C_2H_5)$
H	$-CH_2\overset{*}{C}H(CH_3)(C_2H_2)$
$-\overset{*}{C}H(CH_3)(C_2H_5)$	H
$-\overset{*}{C}H(CH_3)(C_2H_5)$	CH_3
$-\overset{*}{C}H(CH_3)(C_6H_5)$	H
$-CH_2\overset{*}{C}H(CH_3)(C_6H_5)$	H

The chiroptical properties of polymers from structure (**11**) are not particularly interesting, but their investigation has provided valuable information about the stereochemistry of 1,3-diene polymerization.

The (R) enantiomer of 2,3-pentadiene (a 1,3-dimethyl chiral allene) has been polymerized with an allyl nickel catalyst; when the enantiomeric purity of the monomer is above 30%, the corresponding polymer is no longer soluble in the common organic solvents (90). On the other hand, 1-alkynes (**12**) of +90% enantiomeric purity give linear soluble polymers in the presence of a catalyst obtained from iron trisacetylacetonate with triisobutylaluminum (91).

$$CH\equiv C \qquad\qquad -CH=C-\!\!-\!\!-CH=C-$$
$$\underset{(\mathbf{12})}{\overset{\displaystyle |}{(CH_2)_m}} \qquad\qquad \overset{\displaystyle |}{(CH_2)_m} \qquad \overset{\displaystyle |}{(CH_2)_m}$$

$$H_3C-\overset{\displaystyle |}{\underset{\displaystyle |}{C^*}}-H \longrightarrow \quad H_3C-\overset{\displaystyle |}{\underset{\displaystyle |}{C^*}}H \quad H_3C-\overset{\displaystyle |}{\underset{\displaystyle |}{C^*}}-H$$
$$\qquad\qquad R \qquad\qquad R \qquad\quad R$$

R = C$_2$H$_5$, m = 0, 1, 2, 3;   R = i-C$_3$H$_7$, m = 0

Because of steric repulsion between the bulky side chains in the 1,3-position, the macromolecules derived from monomers (**12**) are not planar; the cd spectra suggest the presence of an inherently chiral polyene chromophore associated with chain sections assuming one-screw sense helical arrangements (92).

A few optically active monomers containing a multiple carbon–heteroatom bond have been prepared and their polymers investigated.

Partially crystalline poly-(R)-citronellal and poly-(R)-6-methoxy-4-methylhexanal with a backbone of alternating carbon and oxygen atoms were obtained using triisobutylaluminum catalyst. These polymers show higher optical rotation than the corresponding monomers, which is attributed to the higher conformational rigidity around the asymmetric center of the side chains (93).

Optically active polymers have also been obtained by polymerizing a single enantiomer of chiral isocyanides (94–96); these macromolecules have a carbon backbone (**13**) and side chains containing nitrogen (see POLYISOCYANIDES).

$$(\!>\!\!C\!\!=\!\!N\!-\!R^*)_n$$
(**13**)

$$R^* = -\overset{*}{C}H(CH_3)COOC_2H_5, -\overset{*}{C}H(CH_3)COOCH_3, -\overset{*}{C}H[CH(CH_3)_2]COOCH_3,$$

$$-\overset{*}{C}H(C_6H_5)CH_2COOCH_3, -\overset{*}{C}H(CH_3)C_6H_5, -\overset{*}{C}H(CH_3)CH_2C_6H_5, -\overset{*}{C}H(C_2H_5)C\!\equiv\!CH,$$

$$-\overset{*}{C}H(C_2H_5)CH\!=\!CH_2, \quad or \quad -CH_2\overset{*}{C}H(CH_3)PO(C_6H_5)_2$$

In contrast, isocyanates polymerize with scission of the carbon–nitrogen double bond. Thus optically active isocyanates have been polymerized in the presence of sodium cyanide to polymers with a polyamide structure (**14**) (97,98).

$$O\!=\!C\!=\!N\!-\!R^* \longrightarrow \quad \overset{\overset{O}{\|}}{-C}\!-\!\underset{\underset{R^*}{|}}{N}\!-\!\overset{\overset{O}{\|}}{C}\!-\!\underset{\underset{R^*}{|}}{N}\!-$$
(**14**)

$$R^* = -CH_2C^*H(CH_3)(C_2H_5) \quad or$$

$$-CH_2C^*H(CH_3)(C_6H_5)$$

As observed for polyalkynes (**12**), in polyisocyanates (**14**) the steric interactions of the bulky R* groups hinder main-chain planarity; poly(n-butylisocyanate) has a helical structure (99) in the crystalline state. The dissymmetry of the backbone in polymer (**14**) is supported by a strong cd band at 253 nm assigned to the $n \to \pi^*$ electronic transition of the amide chromophore; this dichroic band has been clearly detected in the aliphatic polymer where no other interfering chromophores, such as a phenyl group, are present (98).

Copolymers of Optically Active and Achiral Unsaturated Monomers. In some copolymers made up of an optically active monomer and an achiral monomer, the nonchiral monomer units assume a predominant chirality due to insertion in the copolymer macromolecules (100). This dissymmetric effect can involve isolated units as well as sequences of the originally nonoptically active moieties. The presence of units derived from the optically active monomer is absolutely necessary to stabilize dissymmetric conformations in sequences of achiral monomer units.

The first experimental examples showing the practical realization of this concept are coisotactic copolymers of (S)-4-methyl-1-hexene with the achiral analogue 4-methyl-1-pentene (20). The optical rotation of these copolymers was higher than that of the corresponding homopolymer mixtures, indicating that 4-methyl-1-pentene units contributed to optical rotation, the contribution being of positive sign as for the (S)-4-methyl-1-hexene units. The optical rotation of the copolymer can be described by

$$[\phi]_C = N_h[\phi]_h + N_p[\phi]_p \tag{8}$$

where h and p refer to (S)-4-methyl-1-hexene and 4-methyl-1-pentene, respectively, $[\phi]$ is the molar optical rotation per monomeric residue (eq. 3), and $[\phi]_C$ is the copolymer molar optical rotation calculated by introducing the average molecular weight of the residues (eq. 5), as shown in

$$[\phi]_C = [\alpha]_C (N_h M_h + N_p M_p)/100 \tag{9}$$

M is the molecular weight of the monomeric unit and $[\alpha]_C$ the specific optical rotation (eq. 1) of the copolymer. As derived from equation 8 for different co-polymers, $[\phi]_p$ has been found to increase with increasing isotacticity and content of (S)-monomer units, reaching a limit close to that calculated for a poly(4-methyl-1-pentene) chain entirely in the left-handed helical conformation. The coisotactic copolymer macromolecules can be considered as formed by left-handed helices with randomly distributed 2-methylbutyl and isobutyl side chains (101).

Excellent proof of this chiral copolymer structure is obtained from coisotactic copolymers of optically active 1-olefins with styrene, where the only absorbing moiety above 180 nm is the phenyl group. Studies of cd in the region of the lowest energy $\pi \rightarrow \pi^*$ electronic transition of the aromatic chromophore show that it is optically active, indicating unequivocally that it occurs in a dissymmetric environment with one predominant chirality (21). The highest energy $\pi \rightarrow \pi^*$ electronic transition of the aromatic chromophore was optically active and split into a cd couplet, which was attributed to exciton-coupling interactions between aromatic side chains positioned in sequences along a one-screw sense helix (102); this interpretation has been confirmed by cd calculations (35). On the basis of this fundamental work, many other examples of conformationally induced optical activity through this copolymerization method have been reported, including cationic, anionic, and free-radical copolymerizations. Examples of random co-polymers for which evidence of induced optical activity in the achiral comonomer units has been shown are given in Table 1, including the polymerization mechanism employed.

In alternating copolymerization, similar conformational dissymmetry seems to be involved (129,130), which can be accompanied by asymmetric induction at the asymmetric main-chain carbon atoms (65,131). Examples of induced optical activity in alternating copolymers are given in Table 2.

The large and ever-increasing number of systems investigated demonstrates the usefulness of the copolymerization method described here for the simple preparation of optically active, functional polymeric materials.

Cyclic Monomers. Through ring-opening polymerization (qv), chiral cyclic monomers give rise to macromolecules with main-chain tertiary carbon atoms, asymmetric owing to the different nature of the four atoms or groups of atoms attached to them (**15**). In vinyl or acrylic polymers, each main-chain tertiary carbon atom is flanked by two methylene groups and its asymmetry depends on the different length of the two chain fragments attached to it (**16**).

$$
\begin{array}{ccc}
\underset{\substack{|\\ R}}{\overset{\substack{H\\|}}{-X-C^*-Y-X-C^*-Y-}} & \underset{\substack{|\\ H}}{\overset{\substack{R\\|}}{-X-C^*-Y-X-C^*-Y-}} & \underset{\substack{|\\ R}}{\overset{\substack{H\\|}}{-CH_2-C-CH_2-C-CH_2-C-}}
\end{array}
$$

(15a)	(15b)	
(15)		(16)

In the isotactic polymers (**15**), stereoregularity and optical activity are directly related (11,23) as chains (**15a**) and (**15b**) have no mirror-symmetry elements.

Table 1. Copolymers of Optically Active Monomers (M*) with Achiral Comonomers (M)

M*	M	Initiator or catalyst	Refs.
4-methyl-1-hexene	4-methyl-1-pentene	Ziegler-Natta	20
3,7-dimethyl-1-octene, 4-methyl-1-hexene, 5-methyl-1-heptene	styrene, o-methylstyrene, 1- and 2-vinylnaphthalene	Ziegler-Natta	21,36,102–105
4-phenyl-1-hexene	4-methyl-1-pentene	Ziegler-Natta	106,107
menthyl vinyl ether	methyl and phenyl vinyl ether	cationic	108
	N-vinylcarbazole	cationic	109
menthyl acrylate or methacrylate	styrene	free radical	110,111
	1-vinylnaphthalene	free radical	112
	6-vinylcrysene	free radical, anionic	113
	N-vinylcarbazole	free radical	114
	9-carbazolylmethylstyrene	free radical	115
	4-vinylpyridine	free radical	116
	p-acryloxybenzophenone	free radical	117
	p-vinylbenzophenone	free radical	118
	p-acryloxyazobenzene	free radical	119
	p-acryloxystilbene	free radical	120
	p-vinylstilbene	free radical	121
	p-vinyltrifluoroacetophenone	free radical	122
4-methyl-1-hexen-3-one	4-methyl-1-penten-3-one	free radical, anionic	123
1-methylbenzyl methacrylate	triphenylmethyl methacrylate	anionic	124,125
	α-phenylacrylate	anionic	126
4-[4-(2-methacryloyloxy-ethoxy)-benzoyloxy]-(1-phenylethyl-imminomethyl)-benzene	4-methoxy-4'-[4-(6-meth-acryloyloxyhexyloxy)-benzoyloxy]-biphenyl	free radical	127
vinyl p-tolyl sulfoxide	styrene	free radical	128

Table 2. Alternating Copolymers of Optically Active Monomers (M*) with Achiral Comonomers (M)

M*	M	Ref.
4-methyl-1-hexene	maleic anhydride	129
5-methyl-1-heptene	maleic anhydride	129
2,4-dimethyl-1-hexene	maleic anhydride	129
2,5-dimethyl-1-heptene	maleic anhydride	129
3,9-dimethyl-6-methyleneundecane	maleic anhydride	129
1-methylpropyl vinyl ether	maleic anhydride	132
1-methylheptyl vinyl ether	maleic anhydride	132
menthyl vinyl ether	maleic anhydride	
	dimethyl maleate	133
	dimethyl fumarate	
	styrene	134
	N-phenylmaleinimide	134
	vinylene carbonate	135
	indene	135
menthyl hydrogen maleate	styrene	136
	indene	136
menthyl hydrogen fumarate	N-vinylcarbazole	137
dimethyl fumarate	N-vinylcarbazole	137

Clearly, an equimolar mixture of (**15a**) and (**15b**) is optically inactive because of intermolecular compensation and can be considered a racemate.

Polymerization of one enantiomer of monosubstituted three-membered heterocyclic compounds has been done in the presence of several catalytic systems, giving stereospecific ring opening without or only partially affecting the asymmetric center in the monomer and the polymer enantiomeric purity. The polymers obtained have structure (**15**) with X = CH_2 and Y = O, S, or NH, corresponding to the use of propylene oxide, propylene sulfide, or 2-methylaziridine as monomers, respectively.

Potassium hydroxide, $FeCl_3$, modified zinc, or aluminum alkoxides and alkyls have been employed to obtain optically active polymers from one enantiomer of propylene oxide (138–144). The chiroptical properties of the polymer are strongly affected by the solvent (145) but not by stereoregularity (146), as occurs for polymers assuming ordered structures in solution.

Cationic polymerization of (+)-*trans*-2,3-epoxybutane yielded a meso-diisotactic polymer; the absence of optical activity derives from the alternation of equivalent asymmetric carbon atoms with opposite absolute configuration along the chain backbone (147,148).

Crystalline optically active poly(propylene sulfide) ((**15**), X = CH_2, Y = S, and R = CH_3) has been prepared with different catalytic systems (149,150).

Polymerization of enantiomers of 2-methylaziridine (151), 2-ethylaziridine (152), and 2-isobutylaziridine (153) by cationic initiators gives the corresponding optically active polymers ((**15**), X = CH_2, Y = NH, and R = CH_3,C_2H_5 and i-C_4H_9, respectively). Optically active poly(2-methylaziridine) has also been prepared by hydrolysis of poly(N-formyl-2-methylaziridine) obtained by polymerization of L-4-methyloxazoline (**17**) with dimethyl sulfate as catalyst (154,155). Cationic polymerization of N-substituted aziridines has also supplied optically active polyampholytes (qv) (156).

$$N \underset{O}{\overset{*}{\diagup}} CH_3$$

(**17**)

Several optically active polyesters (**18**) derive from the ring-opening polymerization of the corresponding lactones in the presence of cationic initiators (157–165).

$$+C\overset{O}{\overset{\|}{}}+(CH_2)_m\overset{H}{\underset{R}{\overset{|}{C^*}}}+(CH_2)_pO+_n$$

(**18**)

$m = p = 0$,	R = CH_3 or $CH(CH_3)_2$
$m = 1, p = 3$,	R = CH_3
$m = p = 2$,	R = CH_3
$m = 3, p = 1$,	R = CH_3
$m = 1, p = 0$,	R = C_2H_5

These include α-(p-substituted benzenesulfonamido)-β-lactones (166), which were polymerized with triethylamine, betaine, or butyllithium catalysts.

The properties of polyesters derived from enantiomerically pure and racemic α-disubstituted-β-propiolactones have been reported (167–170).

In addition to the cationic ring-opening polymerization of the cyclic dimer (lactide) (157–159), optically active poly(lactic acid) ((18), $m = p = 0$, R $=$ CH$_3$) has been prepared starting with L-lactic acid O-carboxyanhydride (5-methyl-dioxolan-2,4-dione) in the presence of pyridine, triethylamine, or potassium t-butoxide (171).

Ring-opening polymerization has also been employed to prepare optically active polythiolesters (19) (172–177):

$$\begin{array}{c} \text{H} \qquad \text{O} \\ | \qquad \parallel \\ +\!\!\text{S}\!-\!\!(\text{CH}_2)_m\!\!\overset{}{\text{C}}^*\!\!-\!\!(\text{CH}_2)_p\!\!\overset{}{\text{C}}\!\!+_n \\ | \\ \text{R} \end{array}$$
(19)

$m = 3, p = 1$,	R $=$ CH$_3$
$m = 2, p = 2$,	R $=$ CH$_3$
$m = 1, p = 0$,	R $=$ NHSO$_2$—⟨◯⟩—X (X $=$ H, OCH$_3$, CH$_3$, Cl, and NO$_2$)
$m = p = 0$,	R $=$ CH$_3$

Similarly, several optically active polyamides (20) have been obtained from the corresponding chiral lactams of high enantiomeric purity (178–185).

$$\begin{array}{c} \text{O} \qquad \text{H} \\ \parallel \qquad | \\ +\!\!\overset{}{\text{C}}\!\!-\!\!(\text{CH}_2)_m\!\!\overset{}{\text{C}}^*\!\!-\!\!(\text{CH}_2)_p\!\!\overset{}{\text{N}}\!\!+_n \\ | \qquad | \\ \text{R} \qquad \text{H} \end{array}$$
(20)

$m = 1, p = 3$,	R $=$ CH$_3$
$m = p = 2$,	R $=$ CH$_3$ or CH$_2$OH
$m = 0, p = 4$,	R $=$ CH$_3$
$m = 3, p = 1$,	R $=$ CH$_3$
$m = 1, p = 0$,	R $=$ CH$_3$

The optically active bicyclic oxalactam ($1R$:$5S$)-8-oxa-6-azabicyclo[3.2.1]-octan-7-one has been polymerized by ring-opening to the corresponding chiral hydrophilic polyamide (186). Using the same route, optically active polyphosphates can be prepared from enantiomers of chiral phospholanes. Thus, in the presence of triethylaluminum, optically active 4-methyl-2-oxo-2-hydro-1,3,2-dioxaphospholane gives a polymer that, by oxidation with N$_2$O$_4$, can be converted into the corresponding poly(propylene phosphate) (187).

The ring-opening polymerization of both N-carboxyanhydrides and saccharides allows the preparation of optically active poly(α-amino acid)s (188) and polysaccharides (qv) (189), respectively (see also POLYPEPTIDES).

Monomers for Step-growth Polymerization. Optically active polymers can be prepared by step-growth polymerization, starting with one enantiomer of a compound containing the two functionalities needed for chain formation. If compounds with different functionality, such as a diacid and a diamine, are polymerized, the chirality can be in either or in both (Table 3).

Table 3. Preparation of Optically Active Polyamides by Polycondensation

Diacids or derivatives	Diamine	Refs.
tartaric acid[a]	hexamethylenediamine	190
tartaric acid[a]	*o-*, *m-*, *p-*phenylenediamine	190
adipyl dichloride	lysine[a]	191
lysine[a]	lysine diketopiperazine[a]	192
succinoyl dichloride	1,2-diaminopropane[a]	193
sebacoyl dichloride	1,2-diaminopropane[a]	193
trans-1,2-cyclohexane- or	piperazine or *trans*-2,5-di-	194,195
trans-1,3-cyclohexane- or	methylpiperazine or *N,N′*-	
trans-1,2-cyclopentane-di-	dimethylethylenediamine or	
carboxylic acid[a]	2,6-diazaspiro-3,3-heptane	196
dicarbonyl dichloride of *trans*-	*trans*-diaminocyclopropane[a] or	197,198
1,2-cyclopropanedicarboxylic	*trans*-1,2-bismethylaminocyclopropane[a]	197,198
acid[a]		
2,2′-dinitro-6,6′-	*p-*, *o-*, *m-*phenylenediamine	199
dimethylbiphenyl- or 2,2′-	hexamethylenediamine	199
dichloro-6,6′-dimethylbiphenyl-	*m*-xylenediamine	199
4,4′-dicarboxylic acid[a]		
terephthaloyl dichloride	2,2′-diaminobinaphthyl-(1,1′)[a]	200
terephthalic acid	*N,N′*-diphenethylhexamethylenediamine[a]	201
pyridine 2,6-dicarboxylic acid	*N,N′*-diphenethylhexamethylenediamine[a]	201

[a] Optically active.

Because of the increasing interest in condensation polymers in applications as engineering polymers and liquid crystalline polymers (qv), synthetic approaches have been developed, increasing the number of structures available. Consequently, the number of optically active polymers prepared in this way has also been substantially increased (202).

Optically active polyamides with an asymmetric center in the backbone have been obtained either by polycondensation of a single enantiomer of a chiral diacid with a nonchiral diamine or by polycondensation of a nonchiral diacid with a single enantiomer of a chiral diamine (**21**). Some typical starting materials are listed in Table 3. As in polymers from chiral cyclic monomers, the main-chain asymmetric carbon atoms are chiral because of the four groups directly bound to them and not because of the different lengths of the two chain sections to which they are connected.

$$n\ H_2N-R'-NH_2 + n\ X-\overset{O}{\overset{\|}{C}}-R''-\overset{O}{\overset{\|}{C}}-X \longrightarrow H{-}[\overset{H}{\overset{|}{N}}-R'-\overset{H}{\overset{|}{N}}-\overset{}{\underset{O}{\overset{\|}{C}}}-R''-\overset{}{\underset{O}{\overset{\|}{C}}}]_n X + (2n-1)\ HX$$

(21)

where R′ or R″ contains an asymmetric carbon atom of a single absolute configuration

Optically active oligourethanes have been prepared by stepwise polymerization of one enantiomer of 1-isobutyl-2-acetoxyethyl isocyanate or by 1-benzyl-2-acetoxyethyl isocyanate reacting with hydroxy-terminated oligomers (203–206). An example is the preparation of carboxyl-containing polymers obtained by the polyaddition of dibenzyl L-tartrate to diisocyanates (207).

Optically active polyesters can also be prepared by polycondensation of diacids and glycols, as shown in (21) for polyamides, but the ring-opening polymerization of lactones is preferred. Another method based on polycondensation is used for chiral polyesters displaying cholesteric liquid crystal properties. Thus thermotropic liquid crystal polymers with an ester backbone (22) have been prepared from bis(4-carboxyphenyl) terephthalate and optically active propylene glycol ethers where $m = 1, 2, 3$ (208,209).

$$\left[C \!-\!\! \bigcirc \!\!-\! O \!-\! C \!-\!\! \bigcirc \!\!-\! C \!-\! O \!-\!\! \bigcirc \!\!-\! C \!-\! O \!-\! (CH_2 C^* \!-\! O)_m \right]_n$$

(22)

These polymers are characterized by cd spectra that suggest the persistence, even in solution, of a certain chiral intramolecular order of the aromatic groups (210).

Linear optically active polycondensates (23) have been prepared from a carbohydrate-based monomer, D-glucarohydrazide, and diketones (211):

$$H_2NNH \!-\! C \!-\! C \!-\! C \!-\! C \!-\! C \!-\! NHNH_2 + CH_3 \!-\! C \!-\! C \!-\! CH_3 \longrightarrow$$

$$\left[(C^*HOH)_4 \!-\! C \!-\! NH \!-\! N \!=\! C \!-\! C \!=\! N \!-\! NH \!-\! C \right]$$

(23)

The bifunctional method avoids the use of protecting groups as in the case of polycondensates from 2,3,4,5-tetra-O-acetyl-D-galactaroyl dichloride (212) or 1,6-diamino-1,6-dideoxy-2,4:3,5-di-O-methylene-D-mannitol (213). The chiroptical properties of these polymers have not yet been studied.

Polymerization of Racemic Monomers.   The preparation of racemic monomers is generally much less costly than that of pure enantiomers and they afford economical and convenient starting materials. However, optically active polymers can be obtained from racemic monomers only when the two enantiomers are polymerized at different rates and the conversion is not complete (stereoelective (214) or asymmetric-selective (215) polymerization). If the catalyst or initiator polymerizes the two enantiomers at the same rate, optically active polymers can finally be obtained by the separation of macromolecules differing in the content of monomeric units derived from the left- and right-handed monomer.

The study of the polymerization of racemic monomers can be of fundamental importance for understanding the stereochemistry of the processes (215,216). In addition to stereoelective polymerization, the degree of separation into fractions of opposite optical rotation provides information about the steric structure of active sites and stereochemical control (stereoselective polymerization) (214).

Stereoelective and stereoselective processes have been demonstrated in the polymerization of both unsaturated and cyclic racemic monomers. The former process uses optically active initiators and requires the control of conversion, whereas the latter requires a separation method according to chirality.

   Optically Active Initiators (Stereoelective Polymerization).   Polymerization of racemic propylene oxide was shown to be stereoelective by using the system diethylzinc–(+)-borneol as an optically active initiator, which gave the preferential polymerization of the enantiomer having the (R) absolute configuration (217,218). Additional examples include other epoxides or oxiranes (219–221), propylene sulfide or methylthiirane (222–224), and other thiiranes (225) (Table 4). Optically active catalysts are prepared by the reaction of dialkyl zinc or cadmium derivatives with chiral alcohols, amino acids, or glycols; the last are the most effective (225).

Table 4. Racemic Cyclic Monomers[a] Giving Stereoelective Polymerization in the Presence of Optically Active Initiators

Monomer	R[a]	Refs.
Oxides		
methyloxirane (propylene oxide)	CH_3	217–220
phenyloxirane	C_6H_5	221
t-butyloxirane	$t\text{-}C_4H_9$	225
methoxymethyloxirane	CH_2OCH_3	225
(2-methylbutoxy)methyloxirane	$CH_2OCH_2CH(CH_3)(C_2H_5)$	226
Sulfides		
methylthiirane(propylene sulfide)	CH_3	222,224
ethylthiirane	C_2H_5	225
t-butylthiirane	$t\text{-}C_4H_9$	225
isobutylthiirane	$CH_2CH(CH_3)_2$	225
methoxymethylthiirane	CH_2OCH_3	225
diethylaminomethylthiirane	$CH_2N(C_2H_5)_2$	225

[a] $CH_2\!\!-\!\!CH\!\!-\!\!R$, where X = O or S.
        $\diagdown_X\diagup$

   Preferential polymerization of one enantiomer has also been obtained with 1-olefins having the asymmetric carbon atom in positions 3 and 4, whereas no chiral discrimination is observed if the asymmetric center is further from the double bond. Several techniques have been used to enable the isotactic specific Ziegler-Natta catalysts to induce stereoelective polymerization by employing optically active metal alkyls and optically active Lewis bases and supports. However, although the occurrence of one-enantiomer-preferential polymerization was evidently proved by the presence of optical activity in both polymer and recovered nonpolymerized monomer, the enantiomeric purity was relatively low, in general not higher than 10% (226–228). Thus stereoelective polymerization has not yet assumed a preparative utility, even if it can provide interesting information about the steric requirements of the polymerization and the spatial structure of the active sites (216).

   Effective stereoelective polymerization has been achieved in the case of acrylic and methacrylic esters of racemic alcohols in the presence of catalysts such as Grignard–(−)-sparteine reagents (229,230) (Table 5). In particular, asymmetric selective polymerization of racemic 1,2-diphenylethyl methacrylate with

Table 5. Racemic Unsaturated Monomers Giving Stereoelective Polymerization in the Presence of Optically Active Initiators

Monomers	Refs.
1-Olefins	
3-methyl-1-pentene, 4-methyl-1-hexene,	227,228,231,232
3,7-dimethyl-1-octene	
Acrylates	
α-methylbenzyl acrylate	229
Methacrylates	
α-ethylbenzyl methacrylate, α-isopropylbenzyl, α-*t*-butylbenzyl,	230
s-butyl, 1-methylallyl, 2,3-epoxypropyl, 2-phenylpropyl, and	
menthyl methacrylate	
1,2-diphenylethyl methacrylate	233

C_2H_5MgBr–(–)-sparteine in toluene at $-78°C$ gave an isotactic polymer with 90% excess of the (*S*) enantiomer (233). Stereoelective polymerization of racemic menthyl methacrylate has been achieved by free-radical initiation in the presence of β-cyclodextrin (234).

   Stereoelective polymerization of vinyl ethers has not been obtained up to now by using various homogeneous and heterogeneous cationic catalysts modified with optically active agents. Alkenyl ethers appeared more promising owing to the larger steric hindrance around the double bond. Thus *cis*-1-methylpropyl propenyl ether gave, in the presence of (–)-menthoxy aluminum, a polymer with very low positive optical rotation, whereas the residual nonpolymerized monomer was converted by $BF_3O(C_2H_5)_2$ into a polymer with negative optical rotation, proving the occurrence of stereoelective polymerization. In contrast, no stereo-electivity was observed (235) for 1-menthylpropyl vinyl ether, which has no β-methyl group. However, as for olefins and vinyl ethers, optically active polymers can be prepared by polymerizing a racemic monomer in the presence of a second monomer with high enantiomeric purity. This method was first applied to 1-olefins by polymerizing racemic 3,7-dimethyl-1-octene in the presence of (*S*)-3-methyl-1-pentene of at least 90% enantiomeric purity (236,237) with isotactic-specific Ziegler-Natta catalyst. Under these conditions the two (*S*) enantiomers give rise to a random copolymer, whereas the (*R*)-3,7-dimethyl-1-octene yields a homopolymer (**24**) which is separated by solvent extraction from the more soluble (*S*) copolymer (**25**) (236,237).

$$CH_2=CH \qquad + \ CH_2=CH \qquad + \ CH_2=CH \qquad \longrightarrow$$
$$H—C^*—CH_3 \qquad CH_3—C^*—H \qquad CH_3—C^*—H$$
$$(CH_2)_3 \qquad\qquad (CH_2)_3 \qquad\qquad C_2H_5$$
$$CH(CH_3)_2 \qquad\qquad CH(CH_3)_2$$

$$\sim\!\!\sim\!CH_2—CH\sim\!\!\sim \qquad + \ \sim\!\!\sim\!CH_2—CH——CH_2—CH\sim\!\!\sim$$
$$H—C^*—CH_3 \qquad\qquad CH_3—C^*—H \quad CH_3—C^*—H$$
$$(CH_2)_3 \qquad\qquad\qquad (CH_2)_3 \qquad\quad C_2H_5$$
$$CH(CH_3)_2 \qquad\qquad\qquad CH(CH_3)_2$$
$$(R) \text{ homopolymer} \qquad\qquad (S) \text{ copolymer}$$
$$(\mathbf{24}) \qquad\qquad\qquad\qquad (\mathbf{25})$$

The same stereochemical behavior was observed by copolymerizing racemic 1-methylpropyl vinyl ether with optically active 1-phenylethyl vinyl ether in the presence of a heterogeneous catalyst obtained by reaction of aluminum alcoholates with sulfuric acid (238), and in the polymerization of racemic 1-phenylethyl vinyl ether with the pure enantiomers of N-(1-phenylethyl)maleimide by free-radical initiation (239).

Separation into Opposite Chirality Fractions. This method can be used only if the polymerization of the racemic monomer gives predominantly a mixture of homopolymers of the two enantiomers (**26**) rather than random copolymers (**27**). The process is called stereoselective polymerization (214). No separation can be expected if the polymerization of the racemic monomers leads to an alternating copolymer of the two enantiomers (240).

Polymers of racemic 1-olefins with an asymmetric carbon atom in positions 3 or 4 ((**1**), $m = 0$ and 1) have been separated into fractions with optical rotation of opposite sign by elution chromatography on isotactic poly-(S)-3-methyl-1-pentene (241,242). In the case of the most isotactic fractions of polymers derived from (**1**) ($m = 0$), the intrinsic separability was higher than 90%, strongly supporting steric control by the catalyst active centers rather than by the growing chain end (243). The possibility of separation into fractions having optical rotation of opposite sign has also been observed for other monomers. Among vinyl monomers, the polymers of racemic vinyl ethers derived from secondary alcohols bearing an asymmetric carbon atom β to the double bond ((**3**), $m = 0$) give, by elution on different optically active insoluble supports, polymer samples with positive and negative optical rotation (56). Isotactic polymers derived from racemic propylene oxide could be separated into polymers with opposite chirality because the polymer consists of enantiomeric macromolecules (**15a**) and (**15b**) (244).

Asymmetric Polymerization of Prochiral Monomers. Prochiral monomers give rise to chiral monomeric residues in spite of having a symmetry plane. Asymmetric polymerization is a process that, starting with a prochiral monomer, gives macromolecules consisting predominantly of units with a single chirality; in this case, the polymer shows appreciable optical rotation (245). An optically active agent is needed to control the stereochemistry of the polymerization process (246,247).

Examples of this type of process have been reported for substituted 1,3-dienes and cyclic monomers; 1- and 4-substituted 1,3-dienes give 1,4-polymers containing asymmetric centers in the main chain and are suitable monomers for asymmetric synthesis (245,248–255).

$$n\ \text{R'CH}=\text{CH}-\text{CH}=\text{CHR''} \longrightarrow \text{wwCH}-\text{CH}=\text{CH}-\text{CHww}$$
$$\qquad\qquad\qquad\qquad\qquad\qquad\qquad | \qquad\qquad\qquad\qquad | $$
$$\qquad\qquad\qquad\qquad\qquad\qquad\qquad \text{R'} \qquad\qquad\qquad\quad \text{R''}$$

(**28**)

R'	R''
CH$_3$	H
C$_6$H$_5$	H
COOC$_2$H$_5$	H
COOCH$_3$	C$_6$H$_5$
COOC$_4$H$_9$	CH$_3$

The optical rotation of the polymer is independent of molecular weight, indicating that asymmetric growth polymerization has taken place. Oxidative cleavage of the unsaturated polymer ((**28**), R'' = CH$_3$ or C$_6$H$_5$ and R' = COOCH$_3$) gives optically active methyl- or phenylsuccinic acid with enantiomeric purity around 6% (255). Optically active Ziegler-Natta catalysts such as titanium tetramentholate–Al(C$_2$H$_5$)$_3$ (249), anionic chiral catalysts such as butyllithium–menthyl ethyl ether (254), and free-radical initiators of monomers forming inclusion compounds in optically active matrices have been used (251). In the last method asymmetric induction of 1,3-pentadiene ((**28**), R' = CH$_3$ and R'' = H) units results in homopolymerization and copolymerization; it is called through-space asymmetric induction (256). The general validity of asymmetric inclusion polymerization for preparing optically active polymers has been further proved with 2-methyl-1,3-pentadiene (257). Additional examples are given in Ref. 258; asymmetric induction in alternating copolymerization has also been investigated (259–261).

The topochemically controlled polymerization in chiral crystals has been designated as absolute asymmetric synthesis (262). Optically active polymers can be prepared based on (2 + 2)-photocycloaddition of a racemate crystallized in one enantiomorphous single crystal (263) in the absence of any chiral agent.

Cyclic achiral monomers give rise to chiral polymers, depending on substituents and polymerization stereochemistry (23,264). In symmetrically substituted cyclic olefins (**29**), the erythrodiisotactic structure (**29a**) has a mirror glide plane and is therefore achiral, whereas the threo-diisotactic structure (**29b**) is chiral.

On the other hand, both erythro- (**30a**) and threo-diisotactic (**30b**) structures are chiral when starting with unsymmetrically disubstituted cyclic olefins.

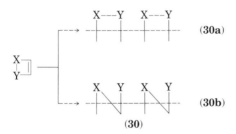

According to these theoretical considerations, polymerization of benzofuran or α- and β-naphthofuran in the presence of optically active cationic initiators results in optically active polymers (265–269).

Several vinyl polymers can have asymmetric (270) main-chain tertiary carbon atoms. By using chirally controlled polymerization processes, optical activity may also be expected in copolymers of achiral monomers because of main-chain chirality. For instance, in structure (**31**), triads AAB and ABA are asymmetric.

Polymers with measurable optical activity with one predominant configuration were obtained by asymmetric copolymerization of nonchiral monomers, such as methacrylonitrile, methyl acrylate and methacrylate, and styrene and substituted styrene derivatives with the bifunctional template monomer 3,4-O-cyclohexylidene-D-mannitol-1,2;5,6-bis-O-(4-vinylphenylboronate). After complete removal of the D-mannitol serving as template, the polymer was still optically active (271).

Polymers with Rigid Chiral Conformation. Few polymers show optical activity of purely conformational origin because, in the absence of chiral groups, right- and left-handed structures tend to reach an equimolar equilibrium. Sequences of achiral monomers units can be placed in conformations of a single handedness, but only when included in a copolymer macromolecule with optically active comonomer residues. On the other hand, homopolymers with one chiral conformation may maintain it because of backbone rigidity strongly restricting rotation around bonds. Clearly, an optically active agent is necessary either as a catalyst to build up chains of only one chiral conformation or as a separation medium to resolve the racemic mixture consisting of macromolecules with left- and right-handed conformations. Optically active poly(iminomethylene)s, more properly called poly(carbonimidoyl)s (94,272), can be prepared by polymerization of isocyanides in the presence of several catalysts.

$$n \text{ RN}{=}\text{C} \longrightarrow \begin{array}{c} {-}{(}\text{C}{)}{\overline{}_n} \\ \| \\ \text{NR} \end{array}$$

(**32**)

With bulky substituents ((**32**), R = t-butyl), the nonoptically active polymers obtained with nickel(II) catalysts can be resolved into fractions with optical rotation of negative and positive sign connected with the prevalence of stable right- and left-handed helices, respectively. This result indicates that the polymerization is stereoselective with respect to screw sense and gives rise to a racemic mixture of enantiomeric macromolecular conformers, which should correspond to rigid 4_1 one-helical structures (273).

In the case of triphenylmethyl methacrylate, optically active polymers were obtained directly with optically active catalysts. The polymer, prepared in the presence of ($-$)-sparteine–LiC$_4$H$_9$ complex in toluene, was highly isotactic and showed optical rotation at the sodium D-line as high as $+300°$ which was associated with the one-screw sense helical conformation of the macromolecules (274). This assumption was further substantiated by cd studies showing strong dichroic bands in the absorption region of the aromatic chromophore (275).

The situation is similar for polymers of chloral prepared with the optically active lithium alkoxide of cholesterol as initiator. This polymer is used as an insoluble chromatographic support for separation into fractions with optical activity of opposite sign of the equimolar mixture of the isotactic homopolymers derived from (R)- and (S)-1-methylbenzyl methacrylate (276).

Introduction of Optically Active Groups in Polymers. Optically active polymers can also be prepared by the attachment through chemical reaction of optically active groups to macromolecules. The macromolecule usually acts as a matrix and only in a few cases has the polymer become intrinsically optically active.

An obvious method is based on the neutralization of polyelectrolytes (qv) with optically active low molecular weight counterions (277). Other examples are the products obtained by reaction of poly(vinyl alcohol) with p-toluenesulfonyl-L-valine (278) and the acid chloride of ($+$)-2-methyl-6-nitrodiphenyl-2'-carboxylic acid (279). The optical rotation of the polymer depends on the extent of the reaction; this is usually incomplete and the product is a copolymer.

In spite of these limitations the method can be useful in specific applications. For example, the optically active derivative of nucleic acid nitrogenous bases grafted to linear polyethyleneimine (280) gives rise to specific steric interactions evidenced by cd spectra. The typical couplet was observed arising from dipole–

dipole electrostatic interactions between pendent heteroaromatic chromophores arranged in an ordered geometry (281). The structure of the substituted unit of the polymer is shown in (**33**), where the nitrogenous base is thymine. Addition of an optically active vinyl-substituted phenyl ester of benzoic acid (**34**) to poly-methylsiloxane gives cholesteric liquid crystalline polysiloxanes (282).

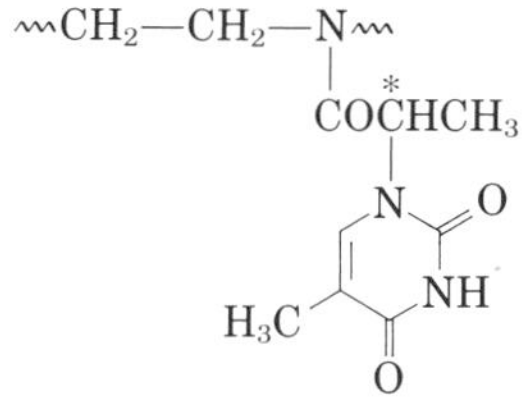

$$\text{~~CH}_2\text{—CH}_2\text{—N~~}$$

$$\overset{*}{\text{COCHCH}}_3$$

(33)

$$\text{CH}_2\text{=CH}$$

$$(\text{CH}_2)_m$$

$$\text{C=O}$$

$$\text{O—}\overset{*}{\text{CH}}(\text{CH}_3)(\text{C}_2\text{H}_5)$$

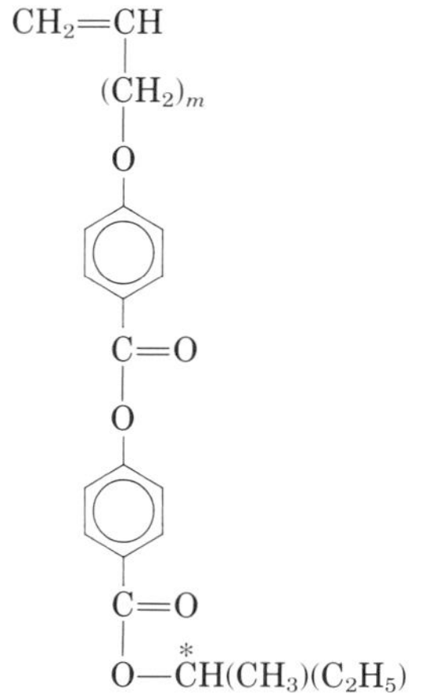

(34)

$$m = 1, 2, 3$$

Applications

Initially, optically active synthetic polymers were investigated with the aim of obtaining information on polymerization stereochemistry and macromolecular conformation (9–11,14). The former topic is related to the understanding of stereospecific polymerization, whereas the latter has contributed to the analysis of ord and cd spectra of complex naturally occurring macromolecules.

The polymerization of chiral monomers, particularly racemic mixtures, has been a very powerful tool for distinguishing between catalytic complex control and growing chain end control as responsible for the formation of stereoregular polymers (qv). The stereoelective and the stereoselective polymerizations of racemic 1-olefins clearly indicate that in Ziegler-Natta heterogeneous catalysis giving isotactic macromolecules the steric control is due to the chiral structure of the active sites (214–216).

Similar conclusions have been reached for other vinyl monomers, such as racemic vinyl ethers (56), and for the ring-opening polymerization of three-membered cyclic compounds, such as propylene oxide (215) and propylene sulfide (283). Polymerization of different D,L mixtures of α-amino acid N-carboxyanhydrides demonstrates the important role played by the chirality of the growing chain on polymerization stereochemistry (284–286). The use of chiral monomers and optically active catalysts is also helpful for the understanding of the stereochemical aspects of polymerization of dienes (248,249) and acrylic and methacrylic derivatives (234).

In conformational studies optically active polymers have given clear experimental evidence that isotactic vinyl polymers in solution can maintain the helical conformations (12,13,38) they are known to assume in the crystalline state from x-ray investigation (287,288). The stabilization of this specific conformation is due to cooperative steric interactions among monomeric residues along the stereoregular backbone, without formation of intramolecular bonds as observed in proteins where hydrogen bonding is responsible for ordered secondary structures (2,3). The understanding of this cooperative mechanism is useful in copolymers of optically active monomer with achiral comonomers. In these random copolymers, induced optical activity has been observed (100) in both isolated and sequenced residues derived from the achiral comonomer which are forced into a single chirality dissymmetric conformation by the cooperative interaction with the chiral comonomer residues. This concept has been utilized for simple preparation of optically active functional polymeric materials for a large variety of applications (124,125,128,129,289,290).

The understanding of chiroptical properties of structurally simple synthetic polymers has been helpful in analyzing ord and cd spectra of more complex biopolymers, particularly proteins.

The three standard conformations, α-helix, β-sheet, and random coil, on which the conformational analysis of polypeptides and proteins is based, are characterized by markedly different chiroptical properties (2). Model cd spectra permit analysis of proteins in terms of the α-helix, β-sheet, and random-coil conformations (291). The principal interference in such analysis is due to prosthetic groups (292) and aromatic side chains (7,108,293), which can give dichroic bands in the spectral region of the main-chain peptide chromophore whose chiroptical properties are indicative of the backbone conformation. Evidence that aromatic chromophores inserted in an ordered chiral macromolecular chain can give dichroic bands of similar ellipticity and in the same spectral region as the peptide chromophore has been obtained by studies on coisotactic copolymers of optically active 1-olefins with styrene and other vinyl aromatic monomers (108). In these copolymers, the aromatic chromophore is the sole absorbing moiety above 180 nm and shows strong cd bands between 190 and 270 nm, which have been associated with the electronic transitions of the aromatic groups disposed in the relative fixed geometry imposed by the one-screw-sense helical arrangement. Similar assistance in the cd studies of nucleic acids is given by optically active polymers containing asymmetric nucleic acid nitrogenous base derivatives as pendent groups (281,294).

In addition, optically active polymers have found application in several specialized fields as chiral reagents and catalysts for asymmetric synthesis (84,295–305); packing materials of chromatographic columns for enantiomer separation (306), eg, for the separation of racemates on helical poly(triphenylmethyl methacrylate) (307–310); and chiral materials for the preparation of cholesteric liquid crystal polymers (127,210,282,311).

Although optically active polymers have been anticipated to have unusual material properties (13), as experimentally shown by characteristic ratio values higher than those of analogous nonoptically active polymers (312), no applications in this direction have appeared yet, except possibly in the field of cholesteric liquid crystalline polymers.

BIBLIOGRAPHY

"Optically Active Polymers" in *EPST* 1st ed., Vol. 9, pp. 507–524, by Rolf C. Schulz, Universität Mainz.

1. F. Ciardelli and P. Salvadori, eds., *Fundamental Aspects and Recent Developments in ORD and CD*, Heyden & Son, Ltd., London, 1973.
2. E. R. Blout in Ref. 1, p. 352.
3. R. W. Woody, *J. Polym. Sci. Macromol. Rev.* **12**, 181 (1977).
4. J. Brahms in Ref. 1, p. 307.
5. I. Tinoco, Jr. and A. L. Williams, Jr., *Ann. Rev. Phys. Chem.* **35**, 329 (1984).
6. W. C. Johnson, Jr., *Ann. Rev. Phys. Chem.* **20**, 93 (1978).
7. E. Peggion, A. Cosani, M. Terbojevich, and M. Palumbo in E. Sélégny, ed., *Optically Active Polymers*, D. Reidel Publishing Co., Dordrecht, The Netherlands, 1979, p. 231.
8. F. Ciardelli and O. Pieroni, *Chimia* **34**, 301 (1980).
9. M. Goodman, A. Abe, and Y. L. Fan, *Macromol. Rev.* **1**, 1 (1966).
10. R. C. Schulz, *Adv. Polym. Sci.* **4**, 236 (1965).
11. P. Pino, *Adv. Polym. Sci.* **4**, 393 (1965).
12. P. Pino, F. Ciardelli, and M. Zandomeneghi, *Ann. Rev. Phys. Chem.* **21**, 561 (1970).
13. P. L. Luisi and F. Ciardelli in A. D. Jenkins and A. Ledwith, eds., *Reactivity, Mechanism and Structure in Polymer Chemistry*, John Wiley & Sons, Inc., New York, 1974, p. 471.
14. E. Sélégny, ed., *Optically Active Polymers*, D. Reidel Publishing Co., Dordrecht, The Netherlands, 1979.
15. Y. Okamoto, *Kobunshi* **32**, 191 (1983).
16. A. Moscowitz, *Adv. Chem. Phys.* **4**, 67 (1962).
17. S. F. Mason, *Qt. Rev. Chem. Soc.* **17**, 20 (1963).
18. O. Bonsignori and G. P. Lorenzi, *J. Polym. Sci. Part A-2* **8**, 1639 (1970).
19. P. Pino, F. Ciardelli, G. Montagnoli, and O. Pieroni, *J. Polym. Sci. Polym. Lett.* **5**, 307 (1967).
20. C. Carlini, F. Ciardelli, and P. Pino, *Makromol. Chem.* **119**, 244 (1968).
21. P. Pino, C. Carlini, E. Chiellini, F. Ciardelli, and P. Salvadori, *J. Am. Chem. Soc.* **90**, 5025 (1968).
22. J. L. Houben, A. Fissi, D. Bacciola, N. Rosato, O. Pieroni, and F. Ciardelli, *Int. J. Biol. Macromol.* **5**, 94 (1983).
23. M. Farina and G. Bressan in A. D. Ketley, ed., *The Stereochemistry of Macromolecules*, Vol. 3, Marcel Dekker, Inc., New York, 1968, p. 181.
24. M. Farina in F. Ciardelli, M. Farina, P. Giusti, and S. Cesca, eds., *Macromolecole: Scienza e Tecnologia*, Vol. 1, Pacini, Pisa, Italy, 1983, p. 39.
25. W. Moffitt, *J. Chem. Phys.* **25**, 467 (1956).
26. W. Moffitt and J. T. Yang, *Proc. Natl. Acad. Sci. USA* **42**, 596 (1957).
27. I. Tinoco, Jr., *Adv. Chem. Phys.* **4**, 113 (1962).
28. R. W. Woody and I. Tinoco, Jr., *J. Chem. Phys.* **46**, 4927 (1967).
29. I. Tinoco, Jr. in Ref. 14, p. 1.
30. R. W. Woody, *J. Chem. Phys.* **49**, 4797 (1968).
31. E. S. Pysh, *Proc. Natl. Acad. Sci. USA* **56**, 825 (1966).
32. E. S. Pysh, *Biopolymers* **13**, 1563 (1974).
33. E. W. Ronish and S. Krimm, *Biopolymers* **11**, 1919 (1972).
34. I. Tinoco, Jr. in Ref. 14, p. 57.
35. W. Hug, F. Ciardelli, and I. Tinoco, Jr., *J. Am. Chem. Soc.* **96**, 3407 (1974).
36. F. Ciardelli, C. Righini, M. Zandomeneghi, and W. Hug, *J. Phys. Chem.* **81**, 1948 (1977).
37. S. F. Mason, *Molecular Optical Activity and the Chiral Discriminations*, Cambridge University Press, Cambridge, UK, 1982, p. 51.
38. P. Pino, F. Ciardelli, G. P. Lorenzi, and G. Montagnoli, *Makromol. Chem.* **61**, 207 (1963).
39. P. Pino and G. P. Lorenzi, *J. Am. Chem. Soc.* **82**, 4745 (1960).
40. W. J. Bailey and E. T. Yates, *J. Org. Chem.* **25**, 1800 (1960).
41. S. Nozakura, S. Takeuchi, H. Yuki, and S. Murahashi, *Bull. Chem. Soc. Jpn.* **34**, 1673 (1961).
42. M. Goodman, K. J. Clark, M. A. Stake, and A. Abe, *Makromol. Chem.* **72**, 131 (1964).

43. C. Bertucci, P. Salvadori, F. Ciardelli, and G. Fatti, *Polym. Bull.* **13,** 469 (1985).
44. F. Ciardelli, G. Montagnoli, D. Pini, O. Pieroni, C. Carlini, and E. Benedetti, *Makromol. Chem.* **147,** 53 (1971).
45. S. Pucci, M. Aglietto, P. L. Luisi, and P. Pino in G. Chiurdoglu, ed., *Conformational Analysis: Scope and Present Limitations,* Academic Press, Inc., New York, 1971, p. 203.
46. P. L. Luisi, *Polymer* **13,** 232 (1972).
47. O. Pieroni, F. Ciardelli, and G. Stigliani, *Chim. Ind. Milan* **52,** 289 (1970).
48. F. Ciardelli, C. Carlini, E. Chiellini, P. Salvadori, L. Lardicci, R. Menicagli, and C. Bertucci in R. W. Lenz and F. Ciardelli, eds., *Preparation and Properties of Stereoregular Polymers,* D. Reidel Publishing Co., Dordrecht, The Netherlands, 1979, p. 353.
49. G. P. Lorenzi, E. Benedetti, and E. Chiellini, *Chim. Ind. Milan* **46,** 1474 (1964).
50. P. Pino, P. Salvadori, G. P. Lorenzi, E. Chiellini, L. Lardicci, G. Consiglio, O. Bonsignori, and L. Lepri, *Chim. Ind. Milan* **55,** 182 (1973).
51. G. J. Schmitt and C. Schuerch, *J. Polym. Sci.* **45,** 313 (1960).
52. D. Basagni, A. M. Liquori, and B. Pispisa, *J. Polym. Sci. Part B* **2,** 241 (1964).
53. A. M. Liquori and B. Pispisa, *J. Polym. Sci. Part B* **5,** 375 (1967).
54. A. P. Black, E. T. Dewar, and D. Rutherford, *Chem. Ind.,* 1624 (1962).
55. M. Kurokawa and Y. Minoura, *Makromol. Chem.* **181,** 707 (1980).
56. E. Chiellini, G. Montagnoli, and P. Pino, *J. Polym. Sci. Part B* **7,** 121 (1969).
57. A. M. Liquori and B. Pispisa, *Chim. Ind. Milan* **48,** 1045 (1966).
58. R. Vukovic and D. Fleš, *J. Polym. Sci. Part A-1* **13,** 49 (1975).
59. O. Pieroni, F. Ciardelli, C. Botteghi, L. Lardicci, P. Salvadori, and P. Pino, *J. Polym. Sci. Part C* **22,** 993 (1969).
60. A. Allio and P. Pino, *Helv. Chim. Acta* **57,** 616 (1974).
61. Nguyêñ-tât-Thiên, U. W. Suter, and P. Pino, *Makromol. Chem.* **184,** 2335 (1983).
62. R. C. Schulz and H. Hilpert, *Makromol. Chem.* **55,** 132 (1962).
63. R. C. Schulz, *Z. Naturforsch.* **19b,** 387 (1964).
64. E. I. Klabunovskii, Yu. I. Petrov, and M. I. Shvartsman, *Vysokomol. Soedin.* **6,** 1487 (1964).
65. N. Beredjick and C. Schuerch, *J. Am. Chem. Soc.* **78,** 2646 (1956).
66. C. L. Arcus and D. W. West, *J. Chem. Soc.,* 2699 (1959).
67. J. H. Liu, K. Kondo, and K. Takemoto, *Makromol. Chem. Rapid Commun.* **3,** 215 (1982).
68. H. Yuki, K. Ohta, K. Uno, and S. Murahashi, *J. Polym. Sci. Part A-1* **6,** 829 (1968).
69. H. Sobue, H. Matsuzaki, and S. Nakano, *J. Polym. Sci. Part A-1* **2,** 3339 (1964).
70. T. P. Bird, W. A. P. Black, E. T. Dewar, and D. Rutherford, *Chem. Ind.,* 1331 (1960).
71. S. Kimura, K. Hirai, and M. Imoto, *J. Chem. Soc. Jpn. Ind. Chem. Sect.* **65,** 688 (1962).
72. T. Ouchi and H. Chikashita, *J. Macromol. Sci. Chem.* **19,** 853 (1983).
73. R. K. Kulkarni and H. Morawetz, *J. Polym. Sci.* **54,** 491 (1961).
74. R. L. Whistler, H. P. Panzer, and H. J. Roberts, *J. Org. Chem.* **26,** 1583 (1961).
75. E. Kaiser and R. C. Schulz, *Makromol. Chem.* **81,** 273 (1968).
76. C. Braud, M. Vert, and E. Sélégny, *Makromol. Chem.* **175,** 775 (1974).
77. E. Chiellini, R. Solaro, and A. Ledwith, *Makromol. Chem.* **178,** 701 (1977).
78. *Ibid.,* **179,** 1929 (1978).
79. C. S. Marvel and C. G. Overberger, *J. Am. Chem. Soc.* **66,** 475 (1944).
80. *Ibid.,* **68,** 2106 (1946).
81. F. Ciardelli, O. Pieroni, C. Carlini, and C. Menicagli, *J. Polym. Sci. Part A-1* **10,** 809 (1972).
82. N. Kunedia, J. Shibatani, Y. Fujiwara, and M. Kinoshita, *Makromol. Chem.* **175,** 2509 (1974).
83. K. Kobayashi and H. Sumitomo, *Polym. J.* **15,** 517 (1983).
84. J. H. Liu, K. Kondo, and K. Takemoto, *Makromol. Chem.* **184,** 1547 (1983).
85. C. G. Overberger and L. C. Palmer, *J. Am. Chem. Soc.* **78,** 666 (1956).
86. E. Benedetti, F. Ciardelli, O. Pieroni, and R. Rossi, *Chim. Ind. Milan* **50,** 550 (1968).
87. Z. Janović and D. Fleš, *J. Polym. Sci. Part A-1* **9,** 1103 (1971).
88. L. Porri and D. Pini, *Chim. Ind. Milan* **55,** 196 (1973).
89. R. Radovanović-Kiprianova and D. Fleš, *J. Polym. Sci. Polym. Chem. Ed.* **13,** 1141 (1975).
90. L. Porri, R. Rossi, and G. Ingrosso, *Tetrahedron Lett.,* 1083 (1971).
91. F. Ciardelli, E. Benedetti, and O. Pieroni, *Makromol. Chem.* **103,** 1 (1967).
92. F. Ciardelli, S. Lanzillo, and O. Pieroni, *Macromolecules* **7,** 174 (1974).
93. A. Abe and M. Goodman, *J. Polym. Sci. Part A-1* **1,** 2193 (1963).

94. R. J. M. Nolte, A. J. M. van Beijnen, and W. Drenth, *J. Am. Chem. Soc.* **96**, 5932 (1974).
95. A. J. M. van Beijnen, R. J. M. Nolte, A. J. Naaktgeboren, J. W. Zwikker, W. Drenth, and A. M. F. Hezemans, *Macromolecules* **16**, 1679 (1983).
96. F. Millich and R. Sinclair, *J. Polym. Sci. Part A-1* **6**, 1417 (1968).
97. M. Goodman and S. Chen, *Macromolecules* **3**, 398 (1970).
98. *Ibid.*, **4**, 625 (1971).
99. U. Shmueli, W. Traub, and K. Rosenheck, *J. Polym. Sci. Part A-2* **7**, 515 (1969).
100. F. Ciardelli, M. Aglietto, C. Carlini, E. Chiellini, and R. Solaro, *Pure Appl. Chem.* **54**, 521 (1982).
101. F. Ciardelli and P. Salvadori, *Pure Appl. Chem.* **57**, 931 (1985).
102. F. Ciardelli, P. Salvadori, C. Carlini, and E. Chiellini, *J. Am. Chem. Soc.* **94**, 6536 (1972).
103. F. Ciardelli, P. Salvadori, C. Carlini, R. Menicagli, and L. Lardicci, *Tetrahedron Lett.*, 1779 (1975).
104. C. Carlini and E. Chiellini, *Makromol. Chem.* **176**, 519 (1975).
105. F. Ciardelli, E. Chiellini, C. Carlini, O. Pieroni, P. Salvadori, and R. Menicagli, *J. Polym. Sci. Part C* **62**, 143 (1978).
106. C. Carlini, F. Ciardelli, and D. Pini, *Makromol. Chem.* **174**, 15 (1973).
107. C. Carlini, F. Ciardelli, L. Lardicci, and R. Menicagli, *Makromol. Chem.* **174**, 27 (1973).
108. H. Yuki, K. Ohta, and N. Yajima, *Polym. J.* **1**, 164 (1970).
109. E. Chiellini, R. Solaro, A. Ledwith, and G. Galli, *Eur. Polym. J.* **16**, 875 (1980).
110. R. N. Majumdar, C. Carlini, R. Nocci, F. Ciardelli, and R. C. Schulz, *Makromol. Chem.* **177**, 3619 (1976).
111. R. N. Majumdar and C. Carlini, *Makromol. Chem.* **181**, 201 (1980).
112. R. N. Majumdar, C. Carlini, N. Rosato, and J. L. Houben, *Polymer* **21**, 941 (1980).
113. E. Chiellini, R. Solaro, and F. Ciardelli, *Makromol. Chem.* **183**, 103 (1982).
114. E. Chiellini, R. Solaro, G. Galli, and A. Ledwith, *Macromolecules* **13**, 1654 (1980).
115. E. C. Chiellini, R. Solaro, F. Ciardelli, G. Galli, and A. Ledwith, *Polym. Bull.* **2**, 577 (1980).
116. R. N. Majumdar, C. Carlini, and C. Bertucci, *Makromol. Chem.* **183**, 2047 (1982).
117. C. Carlini and F. Gurzoni, *Polymer* **24**, 101 (1983).
118. A. Altomare, C. Carlini, G. Ruggeri, and E. Taburoni, *Proceedings of the Anglo-Italian Meeting on Macromolecules,* Galzignano, Padova, Italy, Sept. 1985, Vol. II, p. 71.
119. A. Altomare, C. Carlini, F. Ciardelli, R. Solaro, and N. Rosato, *J. Polym. Sci. Polym. Chem. Ed.* **22**, 1267 (1984).
120. A. Altomare, C. Carlini, and R. Solaro, *Polymer* **23**, 1355 (1982).
121. A. Altomare, C. Carlini, M. Panattoni, and R. Solaro, *Macromolecules* **17**, 2207 (1984).
122. A. Altomare, C. Carlini, F. Ciardelli, and E. M. Pearce, *J. Polym. Sci. Polym. Chem. Ed.* **21**, 1693 (1983).
123. Ref. 61, p. 2347.
124. H. Yuki, K. Ohta, Y. Okamoto, and K. Hatada, *J. Polym. Sci. Polym. Lett. Ed.* **15**, 589 (1977).
125. *Ibid.*, **16**, 545 (1978).
126. H. Yuki, K. Ohta, K. Hatada, and H. Ishikawa, *Polym. J.* **11**, 323 (1979).
127. F. Saeva and J. Noonan, *Makromol. Chem. Rapid Commun.* **5**, 529 (1984).
128. J. E. Mulvaney and R. A. Ottaviani, *J. Polym. Sci. Part A-1* **8**, 2293 (1970).
129. M. Aglietto, C. Carlini, E. Chiellini, and F. Ciardelli, *Gazz. Chim. Ital.* **110**, 449 (1980).
130. M. Aglietto, C. Carlini, L. Crisci, and G. Ruggeri, *Gazz. Chim. Ital.* **115**, 173 (1985).
131. N. Beredjick and C. Schuerch, *J. Am. Chem. Soc.* **78**, 1933 (1958).
132. E. Chiellini, M. Marchetti, C. Villiers, C. Braud, and M. Vert, *Eur. Polym. J.* **14**, 251 (1978).
133. M. Kurokawa and Y. Minoura, *J. Polym. Sci. Polym. Chem. Ed.* **17**, 473 (1979).
134. M. Kurokawa, T. Doiuchi, H. Yamaguchi, and Y. Minoura, *J. Polym. Sci. Polym. Chem. Ed.* **16**, 129 (1978).
135. Ref. 133, p. 3297.
136. M. Kurokawa and Y. Minoura, *Makromol. Chem.* **181**, 293 (1980).
137. G. Galli, R. Solaro, E. Chiellini, and A. Ledwith, *Polymer* **22**, 1088 (1981).
138. C. C. Price, M. Osgan, R. E. Hughes, and C. Shambelan, *J. Am. Chem. Soc.* **78**, 690 (1956).
139. *Ibid.*, 3432 (1956).
140. M. Osgan and C. C. Price, *J. Polym. Sci.* **34**, 153 (1959).
141. N. S. Chu and C. C. Price, *J. Polym. Sci. Part A-1* **1**, 1105 (1963).
142. C. C. Price and R. Spector, *J. Am. Chem. Soc.* **87**, 2069 (1965).

143. T. Tsuruta, S. Inoue, and I. Tsukuma, *Makromol. Chem.* **84,** 298 (1965).
144. Y. Kumata, J. Furukawa, and T. Fueno, *Bull. Chem. Soc. Jpn.* **43,** 3663 (1970).
145. J. Furukawa in Ref. 14, p. 317.
146. E. Chiellini, P. Salvadori, M. Osgan, and P. Pino, *J. Polym. Sci. Part A-1* **8,** 1589 (1970).
147. E. J. Vandenberg, *J. Am. Chem. Soc.* **83,** 3538 (1961).
148. E. J. Vandenberg, *J. Polym. Sci. Part B* **2,** 1085 (1964).
149. N. Spassky and P. Sigwalt, *Bull. Chem. Soc. Fr.,* 4617 (1967).
150. T. Tsunetsugu, J. Furukawa, and T. Fueno, *J. Polym. Sci. Part A-1* **9,** 3541 (1971).
151. Y. Minoura, M. Takebayashi, and C. C. Price, *J. Am. Chem. Soc.* **81,** 4689 (1959).
152. K. Tsuboyama, S. Tsuboyama, and M. Yanagita, *Bull. Chem. Soc. Jpn.* **40,** 2954 (1967).
153. S. Tsuboyama, *Bull. Chem. Soc. Jpn.* **35,** 1004 (1962).
154. T. Saegusa, S. Kobayashi, and M. Ishiguro, *Macromolecules* **7,** 958 (1974).
155. J. G. Hamilton, K. J. Ivin, L. C. Kuan-Essig, and P. Watt, *Macromolecules* **9,** 67 (1976).
156. G. Smets and C. Samyn in Ref. 14, p. 331.
157. J. Kleine and H. H. Kleine, *Makromol. Chem.* **30,** 23 (1959).
158. R. C. Schulz and J. Schwaab, *Makromol. Chem.* **87,** 90 (1965).
159. R. C. Schulz and A. Guthmann, *J. Polym. Sci. Polym. Lett.* **5,** 1099 (1967).
160. Y. Iwakura, K. Iwata, S. Matsuo, and A. Tohara, *Makromol. Chem.* **122,** 275 (1969).
161. C. G. Overberger and H. Kaye, *J. Am. Chem. Soc.* **89,** 5649 (1967).
162. *Ibid.,* p. 5646.
163. J. R. Shelton, D. E. Agostini, and J. B. Lando, *J. Polym. Sci. Part A-1* **10,** 2789 (1971).
164. J. R. Shelton, J. B. Lando, and D. E. Agostini, *J. Polym. Sci. Part B* **9,** 173 (1971).
165. C. G. Overberger, G. Montaudo, T. Furukawa, and M. Goodman, *J. Polym. Sci. Part C* **31,** 33 (1970).
166. V. Jarm and D. Fleš, *J. Polym. Sci. Polym. Chem. Ed.* **15,** 1061 (1977).
167. N. Spassky, A. Leborgne, M. Reix, R. E. Prud'homme, A. Bigdeli, and R. W. Lenz, *Macromolecules* **11,** 716 (1978).
168. C. D'Hondt and R. W. Lenz, *J. Polym. Sci. Polym. Chem. Ed.* **16,** 261 (1978).
169. F. J. Carrière and C. D. Eisenbach, *Makromol. Chem.* **182,** 325 (1981).
170. D. Grenier and R. E. Prud'homme, *Macromolecules* **16,** 302 (1983).
171. H. R. Kricheldorf and J. M. Jonté, *Polym. Bull.* **9,** 276 (1983).
172. C. G. Overberger and J. K. Weise, *J. Am. Chem. Soc.* **90,** 3533 (1968).
173. *Ibid.,* p. 3525.
174. *Ibid.,* p. 3538.
175. D. Fleš and V. Tomašić, *J. Polym. Sci. Part B* **6,** 809 (1968).
176. D. Fleš, V. Tomašić, M. Samsa, D. Ahmetovic, B. Jerman, and M. Fleš, *J. Polym. Sci. Part C* **42,** 321 (1973).
177. H. G. Buhrer and H. G. Helias, *Makromol. Chem.* **41,** 140 (1970).
178. C. G. Overberger and H. Jabloner, *J. Am. Chem. Soc.* **85,** 3431 (1963).
179. D. Fleš and V. Sěke, *Chim. Ind. Fr.* **97,** 1699 (1967).
180. C. G. Overberger and G. M. Parker, *J. Polym. Sci. Part C* **22,** 387 (1968).
181. C. G. Overberger and T. Takekoshi, *Macromolecules* **1,** 1 (1968).
182. *Ibid.,* p. 7.
183. C. G. Overberger and J. H. Kozlowski, *J. Polym. Sci. Part A-1* **10,** 2265 (1972).
184. *Ibid.,* p. 2291.
185. E. Schmidt, *Angew. Makromol. Chem.* **14,** 185 (1970).
186. K. Hashimoto and H. Sumitomo, *Macromolecules* **13,** 786 (1980).
187. T. Biela, S. Penczek, S. Slomkowski, and O. Vogl, *Makromol. Chem. Rapid Commun.* **3,** 667 (1982).
188. M. Szwarc, *Adv. Polym. Sci.* **4,** 1 (1965).
189. F. J. Good, Jr. and C. Schuerch, *Macromolecules* **18,** 595 (1985).
190. Y. Minoura, S. Urayama, and Y. Noda, *J. Polym. Sci. Part A-1* **5,** 2441 (1967).
191. K. Saotome and R. C. Schulz, *Makromol. Chem.* **109,** 239 (1967).
192. V. Crescenzi, V. Giancotti, and F. Quadrifoglio, *Makromol. Chem.* **120,** 220 (1968).
193. E. Sélégny, M. Vert, and M. R. Hamoud, *J. Polym. Sci. Polym. Chem. Ed.* **12,** 85 (1974).
194. C. G. Overberger, G. Montaudo, Y. Nishimura, J. Sěbenda, and R. A. Veneski, *J. Polym. Sci. Part B* **7,** 219 (1969).

195. *Ibid.*, p. 225.
196. C. G. Overberger, Y. Okamoto and V. Bulacovschi, *Macromolecules* **8**, 31 (1975).
197. C. G. Overberger and T. Nishiyama, *J. Polym. Sci. Polym. Chem. Ed.* **19**, 331 (1981).
198. *Ibid.*, p. 349.
199. C. G. Overberger, T. Yoshimura, A. Ohnishi, and A. Gomes, *J. Polym. Sci. Part A-1* **9**, 1139 (1971).
200. R. C. Schulz and R. H. Jung, *Makromol. Chem.* **116**, 190 (1968).
201. A. P. Terentev, V. V. Dunina, and E. G. Rukhadze, *Vysokomol. Soedin. Ser. A* **9**, 599 (1967).
202. E. Sélégny and L. Merle-Aubry in Ref. 14, p. 15.
203. Y. Iwakura, K. Hayashi, and K. Iwata, *Makromol. Chem.* **104**, 46 (1967).
204. *Ibid.*, p. 56.
205. K. Iwata, Y. Iwakura, and K. Hayashi, *Makromol. Chem.* **112**, 242 (1968).
206. *Ibid.*, **116**, 250 (1968).
207. N. Kuzamoto, M. Sakamoto, T. Teshirogi, J. Komiyama, and T. Iijima, *J. Appl. Polym. Sci.* **29**, 977 (1984).
208. C. Malanga, N. Spassky, R. Menicagli, and E. Chiellini, *Polym. Bull.* **9**, 328 (1983).
209. E. Chiellini, G. Galli, C. Malanga, and N. Spassky, *Polym. Bull.* **9**, 336 (1983).
210. E. Chiellini and G. Galli, *Makromol. Chem. Rapid Commun.* **4**, 285 (1983).
211. G. A. F. Roberts and I. M. Thomas, *Makromol. Chem.* **182**, 2611 (1981).
212. M. L. Wolfrom, M. S. Toy, and A. Chaney, *J. Am. Chem. Soc.* **80**, 6328 (1958).
213. T. P. Bird, W. A. P. Black, E. T. Dewar, and T. Rutherford, *Chem. Ind.*, 1977 (1961).
214. P. Pino, F. Ciardelli, and G. Montagnoli, *J. Polym. Sci. Part C* **16**, 3265 (1968).
215. T. Tsuruta, *J. Polym. Sci. Part D* 179 (1972).
216. P. Pino, A. Oschwald, F. Ciardelli, C. Carlini, and E. Chiellini in J. C. W. Chien, ed., *Coordination Polymerization*, Academic Press, Inc., New York, 1975, p. 25.
217. T. Tsuruta, S. Inoue, N. Yoshida, and J. Furukawa, *Makromol. Chem.* **55**, 230 (1962).
218. S. Inoue, T. Tsuruta, and J. Furukawa, *Makromol. Chem.* **53**, 215 (1962).
219. T. Tsuruta, S. Inoue, and H. Koinuma, *Makromol. Chem.* **112**, 58 (1968).
220. P. H. Khanh, H. Koinuma, S. Inoue, and T. Tsuruta, *Makromol. Chem.* **134**, 253 (1970).
221. Y. Kumata, N. Asada, G. M. Parker, and J. Furukawa, *Makromol. Chem.* **136**, 291 (1970).
222. N. Spassky and P. Sigwalt, *Compt. Rend.* **265**, 624 (1967).
223. N. Spassky and P. Sigwalt, *Eur. Polym. J.* **7**, 7 (1967).
224. J. Furukawa, N. Kawabata, and A. Kato, *J. Polym. Sci. Part B* **5**, 1073 (1967).
225. N. Spassky, P. Dumas, and M. Sepulchre in Ref. 14, p. 111.
226. V. A. Ponomarenko, E. I. Klabunovskii, A. A. Il'chenko, and L. D. Tomina, *Izv. Akad. Nauk. SSSR Ser. Khim.* **4**, 923 (1970).
227. P. Pino, F. Ciardelli, and G. P. Lorenzi, *J. Am. Chem. Soc.* **85**, 3888 (1963).
228. F. Ciardelli, C. Carlini, G. Montagnoli, L. Lardicci, and P. Pino, *Chim. Ind. Milan* **50**, 860 (1968).
229. Y. Okamoto, K. Suzuki, K. Ohta, and H. Yuki, *J. Polym. Sci. Polym. Lett. Ed.* **17**, 293 (1979).
230. Y. Okamoto, K. Urakawa, and H. Yuki, *J. Polym. Sci. Polym. Chem. Ed.* **19**, 1385 (1981).
231. C. Carlini, F. Ciardelli, and R. Nocci, *J. Polym. Sci. Polym. Chem. Ed.* **15**, 767 (1977).
232. C. Carlini and F. Ciardelli, *Chim. Ind. Milan* **63**, 486 (1981).
233. Y. Okamoto, E. Yashima, K. Hatada, H. Yuki, H. Kageyama, K. Miki, and N. Kasai, *J. Polym. Sci. Polym. Chem. Ed.* **22**, 1831 (1984).
234. N. Kunieda, S. Yamane, and M. Kinoshita, *Makromol. Chem. Rapid Commun.* **6**, 305 (1985).
235. T. Higashimura and Y. Hirokawa, *J. Polym. Sci. Polym. Chem. Ed.* **15**, 1137 (1977).
236. F. Ciardelli, E. Benedetti, G. Montagnoli, L. Lucarini, and P. Pino, *Chem. Commun.*, 285 (1965).
237. F. Ciardelli, C. Carlini, and G. Montagnoli, *Macromolecules* **2**, 296 (1969).
238. E. Chiellini, *Macromolecules* **3**, 527 (1970).
239. T. Doiuchi and H. Yamaguchi, *Eur. Polym. J.* **20**, 831 (1984).
240. J. G. Hamilton, K. J. Ivin, J. J. Rooney and L. C. Waring, *Chem. Commun.*, 159 (1983).
241. P. Pino, F. Ciardelli, G. P. Lorenzi, and G. Natta, *J. Am. Chem. Soc.* **84**, 1487 (1962).
242. P. Pino, G. Montagnoli, F. Ciardelli, and E. Benedetti, *Makromol. Chem.* **93**, 158 (1966).
243. G. Montagnoli, D. Pini, A. Lucherini, F. Ciardelli, and P. Pino, *Macromolecules* **2**, 684 (1969).
244. J. Furukawa, S. Akutu, and T. Saegusa, *Makromol. Chem.* **94**, 68 (1966).
245. G. Natta, M. Farina, M. Donati, and M. Peraldo, *Chim. Ind. Milan* **42**, 1363 (1960).
246. H. L. Frisch, C. Schuerch, and M. Szwarc, *J. Polym. Sci.* **11**, 559 (1953).

247. T. Fueno and J. Furukawa, *J. Polym. Sci. Part A-1* **2**, 3681 (1964).

248. G. Natta, L. Porri, and S. Valenti, *Makromol. Chem.* **67**, 225 (1963).

249. G. Natta, L. Porri, A. Carbonaro, and G. Lugli, *Chim. Ind. Milan* **43**, 529 (1961).

250. G. Costa, P. Locatelli, and A. Zambelli, *Macromolecules* **6**, 653 (1973).

251. M. Farina, G. Audisio, and G. Natta, *J. Am. Chem. Soc.* **89**, 5071 (1967).

252. A. D. Aliev, B. A. Krentsel, and T. N. Fedotova, *Vysokomol. Soedin.* **7**, 1442 (1965).

253. A. D. Aliev and B. A. Krentsel, *Vysokomol. Soedin. Ser. A* **9**, 1464 (1967).

254. T. Tsunesugu, T. Fueno, and J. Furukawa, *J. Polym. Sci. Part A-1* **5**, 2099 (1967).

255. M. Farina, M. Modena, and W. Ghizzoni, *Rend. Accad. Naz. Lincei VIII* **32**, 91 (1962).

256. M. Farina, G. DiSilvestro, and P. Sozzani, *Makromol. Chem. Rapid Commun.* **2**, 51 (1981).

257. M. Miyata, Y. Kitahara, and K. Takemoto, *Makromol. Chem.* **184**, 1771 (1983).

258. Y. Minoura in Ref. 14, p. 159.

259. N. Kunieda, S. Yamane, H. Taguchi, and M. Kinoshita, *Makromol. Chem. Rapid Commun.* **4**, 57 (1983).

260. H. Yamaguchi, K. Sekioka, M. Kurokawa, and T. Doiuchi, *Makromol. Chem.* **184**, 455 (1983).

261. H. Yamaguchi, T. Doiuchi, K. Kawamoto, and T. Yosuzawa, *Makromol. Chem.* **185**, 2053 (1984).

262. L. Addadi, M. D. Cohen, and M. Lahav in Ref. 14, p. 183.

263. L. Addadi, J. van Mil, E. Gati, and M. Lahav, *Makromol. Chem. Suppl.* **4**, 37 (1981).

264. C. L. Arcus, *J. Chem. Soc.*, 2801 (1955).

265. G. Natta, M. Farina, M. Peraldo, and G. Bressan, *Makromol. Chem.* **43**, 68 (1961).

266. G. Natta, G. Bressan, and M. Farina, *Rend. Accad. Naz. Lincei VIII* **34**, 475 (1963).

267. M. Farina, G. Natta, and G. Bressan, *J. Polym. Sci. Part C* **4**, 141 (1964).

268. G. Bressan, M. Farina, and G. Natta, *Makromol. Chem.* **93**, 283 (1966).

269. Y. Hayakawa, T. Fueno, and J. Furukawa, *J. Polym. Sci. Part A-1* **5**, 2099 (1967).

270. G. Wulff, K. Zabrocki, and J. Hohn, *Angew. Chem. Int. Ed. Engl.* **17**, 535 (1978).

271. G. Wulff, R. Kemmerer, J. Vietmeier, and H.-G. Poll, *Nouveau J. Chem.* **6**, 681 (1982).

272. F. Millich and G. K. Baker, *Macromolecules* **2**, 122 (1969).

273. W. Drenth and R. J. M. Nolte, *Acc. Chem. Res.* **12**, 30 (1979).

274. Y. Okamoto, K. Suzuki, K. Ohta, K. Hatada, and H. Yuki, *J. Am. Chem. Soc.* **101**, 4763 (1979).

275. Y. Okamoto, K. Suzuki, and H. Yuki, *J. Polym. Sci. Polym. Chem. Ed.* **18**, 3043 (1980).

276. K. Hatada, S. Shimizu, H. Yuki, W. Harris, and O. Vogl, *Polym. Bull.* **4**, 179 (1981).

277. R. C. Schulz in Ref. 14, p. 267.

278. Y. Minoura and N. Sakota, *Nippon Kagaku Zasshi* **83**, 763 (1962).

279. R. C. Schulz and R. H. Jung, *Makromol. Chem.* **96**, 295 (1966).

280. C. G. Overberger and Y. Morishima, *J. Polym. Sci. Polym. Chem. Ed.* **18**, 1247 (1980).

281. *Ibid.*, 1267 (1980).

282. H. Finkelmann and G. Rehage, *Makromol. Chem. Rapid Commun.* **3**, 859 (1982).

283. K. J. Ivin, E. D. Lillia, P. Sigwalt, and N. Spassky, *Macromolecules* **4**, 345 (1971).

284. K. Matsuura, S. Inoue, and T. Tsuruta, *Makromol. Chem.* **85**, 284 (1965).

285. T. Tsuruta, S. Inoue, and K. Matsuura, *Biopolymers* **5**, 313 (1967).

286. S. Inoue, K. Matsuura, and T. Tsuruta, *J. Polym. Sci. Part C* **23**, 721 (1968).

287. G. Natta, *Makromol. Chem.* **35**, 93 (1960).

288. V. Petraccone, P. Ganis, P. Corradini, and G. Montagnoli, *Eur. Polym. J.* **8**, 99 (1972).

289. E. Chiellini, R. Solaro, G. Galli, and A. Ledwith, *Adv. Polym. Sci.* **62**, 143 (1984).

290. A. Altomare, C. Carlini, F. Ciardelli, M. Panattoni, R. Solaro, and J. L. Houben, *Macromolecules* **18**, 729 (1985).

291. N. Greenfield and G. D. Fasman, *Biochemistry* **8**, 4120 (1969).

292. P. M. Bayley, *Prog. Biophys. Mol. Biol.* **27**, 1 (1973).

293. P. C. Kahn in P. Colowick, ed., *Methods in Enzymology*, Vol. 61, Academic Press, Inc., New York, 1979, p. 339.

294. K. A. Brandt and C. G. Overberger, *J. Polym. Sci. Polym. Chem. Ed.* **23**, 1981 (1985).

295. P. L. Luisi in Ref. 14, p. 357.

296. F. Ciardelli, M. Aglietto, C. Carlini, S. D'Antone, G. Ruggeri, and R. Solaro in H. Benoit and P. Rempp, eds., *Macromolecules,* Pergamon Press, Ltd., Oxford, UK, 1982, p. 125.

297. E. Chiellini, R. Solaro, and S. D'Antone, *Makromol. Chem. Suppl.* **5**, 82 (1981).

298. F. Ciardelli, E. Chiellini, C. Carlini, and M. Aglietto, *Pure Appl. Chem.* **52**, 1857 (1980).

299. S. Itsuno, A. Hirao, and S. Nakahama, *Makromol. Chem. Rapid Commun.* **3**, 673 (1982).

300. T. Yamashita, E. Kagigaki, N. Takahashi, and N. Nakamura, *Makromol. Chem.* **184**, 675 (1983).

301. J. H. Liu, K. Kondo, and K. Takemoto, *Makromol. Chem.* **185**, 2125 (1984).

302. E. Chiellini, R. Solaro, and S. D'Antone in L. J. Mathias and C. Carraher, eds., *Crown Ethers and Phase Transfer Catalysis in Polymer Science,* Plenum Press, New York, 1984, p. 227.

303. D. C. Sherrington and J. Kelly in Ref. 302, p. 249.

304. Y. Kawakami and Y. Yamashita in Ref. 302, p. 263.

305. S. D'Antone, M. Penco, R. Solaro, and E. Chiellini, *Reactive Polym.* **3**, 107 (1985).

306. G. Manecke and W. Lamar in Ref. 14, p. 403.

307. H. Yuki, Y. Okamoto, and I. Okamoto, *J. Am. Chem. Soc.* **102**, 6356 (1980).

308. Y. Okamoto, E. Yashima, K. Hatada, and K. Mislow, *J. Org. Chem.* **49**, 557 (1984).

309. A. Tajiri, N. Morita, T. Asao, and M. Hatano, *Angew. Chem. Int. Ed. Engl.* **24**, 329 (1985).

310. Y. Okamoto, E. Yashima, and K. Hatada, *J. Chem. Soc. Commun.*, 1051 (1984).

311. E. Chiellini and G. Galli, *Faraday Discuss. Chem. Soc.* **79**, 241 (1985).

312. P. Neuenschwander and P. Pino, *Eur. Polym. J.* **19**, 1075 (1983).

FRANCESCO CIARDELLI
Università di Pisa

OPTICAL MICROSCOPY.   See LIGHT MICROSCOPY.

OPTICAL PROPERTIES

Plastics offer several advantages in optical applications. Manufacturing costs are low because precision injection molding produces a product that needs no finishing operations and no grinding or polishing. The ability to design mountings into moldings or to produce multiple-lens arrays reduces the assembly costs of optical products. Most plastics have a higher impact strength than silicate glass, and when they fracture, the fragments are less likely to damage the eyes. The density of plastics is less than half that of glass, reducing the weight of spectacles. Some plastics are sufficiently flexible, especially when thin, to allow formation of soft contact lenses (qv), flexible mirrors, and other products. Some plastics have a greater transmissibility than silicate glass in ir or uv parts of the spectrum.

On the other hand, certain disadvantages prevent plastics from being used in applications such as car windshields. Abrasion resistance is low. Various coatings can improve scratch resistance, but the yield stress is low. Heat tolerance is low. Even the glass-transition temperatures above 250°C are below the range of the glass transition of silicate glass (>500°C). The temperature coefficients of refractive index are high. Optical plastics are mostly homopolymers; it is not possible to adjust the refractive index or dispersion by adding metal oxides as to silicate glass. The variation in refractive index from point to point is greater than for optical glass by an order of magnitude.

The most important optical plastic is poly(methyl methacrylate) (PMMA) because of its superior weathering and scratch-resistance properties (Table 1). It is produced in two forms: a sheet that has been polymerized between glass plates with an extremely high weight-average molecular weight ($\overline{M}_w > 2 \times 10^6$) and a lower molecular weight form ($\approx 1.5 \times 10^5$) that can be injection molded (see ACRYLIC AND METHACRYLIC ESTER POLYMERS; CASTING).

Table 1. Properties of the Optical Plastics

Polymer	Abbreviation	n_D	Abbe value	Density, kg/m^3	Thermal expansion, 10^{-6}/°C	T_{max}, °C[a]
poly(methyl methacrylate)	PMMA	1.491	57.2	1190	65	90
polystyrene	PS	1.590	30.9	1100	70	80
polycarbonate	PC	1.586	29.8	1200	70	120
allyl diglycol carbonate	CR-39[b]	1.499	57.8	1320	120	100
poly(styrene-co-acrylonitrile)[c]	SAN	1.571	35.3	1070	70	95
poly(styrene-co-methyl methacrylate)[c]	NAS	1.562	34.7	1140	65	95
poly(4-methyl-1-pentene)[c]	PMP	1.466	56.4	830	117	180

[a] Maximum use temperature.
[b] Columbia Resin CR-39 (Pittsburgh Plate Glass) (1).
[c] Only small amounts are used.

The thermosetting material CR-39 can be shaped only by machining and polishing (see DIALLYL AND RELATED POLYMERS). PMP is the only semicrystalline polymer; the others are glassy. The crystalline phase of PMP is very similar in refractive index to that of the amorphous phase, which happens rarely with semicrystalline polymers (see 4-METHYL-1-PENTENE POLYMERS). The method of manufacture affects the microstructural state and the degree of homogeneity.

High molecular weight PMMA and CR-39 are isotropic and homogeneous as produced, and remain so if machined to shape and polished. The former may be thermoformed for display signs or glazing applications, or biaxially stretched to increase the toughness for use in aircraft windows. These processes produce molecular orientation (qv) in the direction of stretching and hence birefringence (see BIREFRINGENCE in the Supplement).

Injection molding (qv) is widely used to make plastic optical components from thermoplastics. With modern microprocessor control of injection pressures, speeds, and temperatures, highly reproducible parts are obtained; linear dimensions can be held to ± 8 μm for small parts (2). However, the molded part is always smaller than the steel mold. Glassy polymers have typical linear shrinkages of 0.7% in the plane of the molding; the values change with the injection pressure and the time for which the pressure is held. Shrinkage through the thickness of the molding is always greater because of the manner in which the moldings solidify. Solid skins form that restrict the in-plane shrinkage but not the through-thickness shrinkage. Ideally, moldings have a uniform wall thickness so all parts of the interior can be fed effectively during the pressure stage. With lens-shaped products, the wall thickness varies and the shrinkage is maximum at the thickest part of the lens. This causes the curvature of the molded lens to differ slightly from that of the mold; tolerances on the radius of curvature are 1–2% (2). The surface roughness of the lens should be the same as that of the highly polished stainless steel mold; alternatively, a surface texture can be etched onto the mold, imparting controlled light-diffusion properties to the article. The moldings are cooled rapidly, especially at the surfaces, and have a nonequilibrium microstructure. This occurs in two ways: the density of the glass depends on the rate at which it is cooled through the glass-transition temperature and the melt pressure at that time. The outer layers cool fastest and have the lowest densities, whereas the melt pressure may fall to zero before the interior of the lens is fully solid. Secondly, there is the molecular orientation that occurs as the melt flows into the mold. Unless the mold temperature is high, an oriented skin forms on the molding during the filling process; this is revealed by birefringence measurements.

The optical properties of plastic film used for packaging, photographic products, and other applications also depend on the method of production. Films of glassy polymers with little molecular orientation are produced by extruding the melt onto steel rolls (see FILMS, MANUFACTURE). This does not take advantage of the potential strength of semicrystalline film and is only used when dimensional change on overheating must be avoided. Most film is biaxially stretched while being cooled from the melt, eg, blown polyethylene film where the velocity and diameter of the tubular bubble are increased several-fold, or while being reheated from the glassy state, eg, poly(ethylene terephthalate) film or polypropylene film. These semicrystalline polymers have greatly increased tensile strengths in the plane of the films as a result of the biaxial orientation. Optical clarity can be improved by keeping the size of the crystalline microstructure as small as possible, by crystallization at a low temperature.

Properties

Refractive Index

The refractive index n of an isotropic material is defined in terms of the speed of light c in that material, compared with the speed of light $c_0 = 3.00 \times 10^8$ m/s in a vacuum:

$$c = c_0/n$$

The refractive index is a decreasing function of the wavelength of the light. This is a consequence of the frequency of visible light lying between that required for electronic transitions, causing absorption in the uv region, and that for atomic vibrations, which causes absorption in the ir region. The refractive index is measured at a number of standard wavelengths, such as the F, D, and C lines in atomic emission spectra (Table 2). Refractive index is usually measured with an Abbe refractometer (ASTM D542). A flat sample of the plastic is placed in contact with the refractometer prism. The method determines the critical angle for total internal reflection of sodium light (D line). An immersion fluid with a refractive index slightly higher than that of the plastic is required between the plastic and the prism to avoid spurious refraction at the air–plastic interface. However, the immersion fluid might swell the plastic surface and change its refractive index. Since the refractive index of the surface layer is being measured, the sample must be homogeneous and isotropic; most injection moldings have oriented surface layers with a nonequilibrium density.

If the Abbe refractometer is used with white light, the prisms have to be adjusted to make the colored images coincide. This then gives a parameter, the Abbe value, that is inversely proportional to D. A combination of refractive index

Table 2. Refractive Indexes and Related Properties[a]

	PMMA	PS	PC	NAS
line and wavelength, nm				
F, 486	1.497(8)	1.604(1)	1.593(4)	1.574(4)
D, 589	1.491(7)	1.590(3)	1.586(0)	1.504
C, 651	1.489(2)	1.584(9)	1.576	1.558(3)
Abbe value	57.2	30.8	34.0	35
dispersion D^b, $\times\ 10^{-2}$	1.75	3.25	2.97	2.85
linear-expansion coefficient, $\times\ 10^{-5}/°C$	6.5	6.3	6.8	6.5
dn/dT, $\times\ 10^{-5}/°C$	-8.5	-12.0	-14.3	-14.0

[a] Ref. 2.

$$^b\ D = \frac{n_F - n_C}{n_D - 1}.$$

and dispersion parameters is needed for the design of achromatic lenses. Ideally, a range of values should permit the selection of lens pairs of different plastics. This is more difficult for plastic lenses than for glass lenses, because the latter can be modified with metal oxides (Fig. 1). The metal ions dispersed throughout the network modify the refractive index. A limited number of commercial glassy plastics have suitable mechanical properties and temperature resistance. Blending any two of these produces a two-phase structure that is opaque because of light scattering. The few copolymers such as NAS and SAN that can be prepared with different compositions do not allow a wide range of refractive-index parameters.

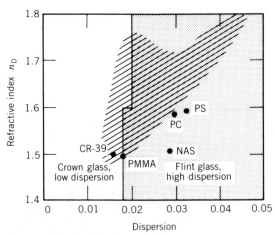

Fig. 1.   Refractive index plotted against dispersion for the principal optical plastics. The shaded area is covered by silicate glasses (3).

Effect of Chemical Structure.   Under the influence of an electric field, the electrons in the chemical bonds of a polymer chain slightly shift their average position. Both the effective electric field **E** and the induced dipole **P** are vector quantities linearly related by the bond polarizability α, which is a second-rank tensor.

$$\mathbf{P} = \alpha\mathbf{E} \tag{1}$$

Nonlinear optical phenomena, eg, the Raman effect, require $\mathbf{E}^2$ or $\mathbf{E}^3$ terms on the right-hand side of this equation, but they are outside the scope of this article.

In general, the polarizability of the chemical bonds is anisotropic, ie, it is greater along the bond than it is at a 90° angle to it. On the other hand, the random orientation distribution of the chemical bonds in a glassy polymer indicates an isotropic refractive index. The mean polarizability of a chemical bond $\bar{\alpha}$ is defined in terms of the diagonal terms of α.

$$\bar{\alpha} = \tfrac{1}{3}(\alpha_{11} + \alpha_{22} + \alpha_{33}) \tag{2}$$

The Lorentz-Lorenz formula relates the mean refractive index n to the mean polarizability of the ith kind of chemical bond and to the number N_i of such bonds per unit volume:

$$\frac{n^2 - 1}{n^2 + 1} = \frac{\Sigma\, N_i\, \bar{\alpha}_i}{3\epsilon_0} \tag{3}$$

where ϵ_0 is the permittivity of free space $= 8.85 \times 10^{-12}$ F/m. In order to achieve a high refractive index, the polymer must contain chemical bonds that have a high mean polarizability and yet do not occupy too high a volume. The refractive indexes and densities of a number of polymers are given in Table 3. In general, the refractive index increases in the following bond sequence:

$$\text{C—F, C=O, C—H,} \ \bigcirc, \ \text{C—Cl}$$

but differences in density may also affect the refractive-index ranking.

Table 3. Refractive Indexes and Densities at 25°C

Polymer	Refractive index n_D^{25}	Density, kg/m³
polytetrafluoroethylene[a]	1.3–1.4	2100–2300
polydimethylsiloxane	1.404	975
poly(vinylidene fluoride)[a]	1.42	1760
poly(4-methyl-1-pentene)	1.46	830
poly(vinyl acetate)	1.467	1197
poly(methylene oxide)[a]	1.48	1425
poly(methyl methacrylate)	1.488	1190
polyisobutylene	1.509	913
polyethylene[a]	1.51–1.54	920–960
polypropylene[a]	1.495–1.510	890–910
polybutadiene	1.515	892
cis-1,4-polyisoprene	1.519	906
polyacrylonitrile	1.518	1170
nylon-6,6[a]	1.530	1120
poly(vinyl chloride)	1.544	1406
bisphenol A polycarbonate	1.585	1200
polystyrene	1.590	1059
poly(vinylidene chloride)[a]	1.630	1876
poly(ethylene terephthalate)[a]	1.640	1380
polyether sulfone	1.65	1370
poly(p-xylylene)[a]	1.661	1120
polyimide[b]	1.78	1420

[a] Semicrystalline; values can vary with the % crystallinity.
[b] Kapton H.

If the same polymer can exist in amorphous or crystalline form, it is useful to express the right-hand side of equation 3 in terms of the density ρ of the solid. If the repeat unit or mer of the polymer has molecular mass M_m and contains n_i bonds of type i, N_i can be expressed in terms of N, the number of mers per unit volume, which in turn can be expressed in terms of Avogadro's number N_A.

$$\Sigma N_i \, \bar{\alpha}_i = N \, \Sigma \, n_i \, \bar{\alpha}_i = \frac{N_A \rho}{1000 \, M_m} \, \Sigma \, n_i \, \bar{\alpha}_i \tag{4}$$

The factor 1000 is necessary because M_m is measured in g/mol, whereas the density is measured in kg/m^3. Therefore, the refractive index increases with increasing density. The crystalline and amorphous densities of some polyolefins at 25°C are given below.

	Crystalline density, kg/m^3	Amorphous density, kg/m^3
polyethylene	1000	854
polypropylene	940	850
poly(4-methyl-1-pentene)	820	840

These polymers are always semicrystalline and the actual range of densities (Table 3) is lower than the values given above. Nevertheless, it explains the range of the refractive index of polyethylene in Table 3.

Other variables affect the refractive index indirectly by changing the density. Thus the relatively high coefficients of (linear) thermal expansion indicate that the density and hence the refractive index decrease with increasing temperature. The density of a glassy plastic increases slightly if it is cooled through the glass-transition temperature T_g under high pressure; this occurs to some extent during injection molding. Conversely, the more rapidly a glassy plastic is cooled through the T_g, the lower the density is at RT. This effect is likely to reduce the density of the outer skin of an injection molding.

Molecular orientation, whether it be of individual molecules in amorphous polymers or of molecules in crystals, can be characterized by birefringent measurements. Elastic stresses can cause the refractive index of a glassy plastic to become anisotropic (see also REFRACTOMETRY).

Transmittance

The intensity of light is reduced during passage through a plastic in two ways: first, there is a reflection at the air–plastic surfaces caused by change in refractive index, and second, there are absorption processes that occur in the plastic; light scattering also plays a role.

For light normally incident on an interface between materials 1 and 2 with refractive indexes n_1 and n_2, the reflectivity (defined as the ratio of reflected to incident-light intensity) is

$$R = \frac{(n_1 - n_2)^2}{(n_1 + n_2)^2} \tag{5}$$

This follows from the continuity of the electrical and magnetic fields at the interface (4). Thus under the optimum conditions of normal incidence from air

$(n_1 = 1)$ onto a polymer of $n_2 = 1.5$, a total of ca 8% of the light intensity is reflected at the two interfaces. When light is incident at an angle α to the interface normal (Fig. 2), the reflected light intensity depends both on α and the angle of refraction β where

$$n_1 \sin \alpha = n_2 \sin \beta \tag{6}$$

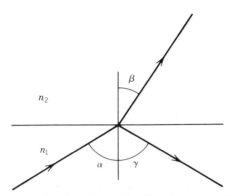

Fig. 2.   The geometry of reflection and refraction at a planar interface between media of refractive indexes n_1 and n_2. The angles α and γ are equal.

It also depends on the angle γ between the electric vector **E** of the incident light and the plane of incidence, ie, the plane containing the incident direction and the normal to the interface at the point of incidence. The total reflectivity consists of two components:

$$R = R_\| \cos^2 \gamma + R_\perp \sin^2 \gamma \tag{7}$$

where $R_\|$ is the reflectivity when **E** is parallel to the plane of incidence and $R_\perp$ when **E** is perpendicular.

$$R_\| = \frac{\tan^2 (\alpha - \beta)}{\tan^2 (\alpha + \beta)} \qquad R_\perp = \frac{\sin^2 (\alpha - \beta)}{\sin^2 (\alpha + \beta)} \tag{8}$$

Figure 3 shows the variation of the two components of reflectivity with the angle of incidence. If the incident light is unpolarized, the γ-angle has a uniform distribution of values between 0 and 90°, and average values of $\cos^2 \gamma$ and $\sin^2 \gamma$ in equation 7 are both 0.5. The reflected light is partially polarized, which explains why polarized light fringes can often be seen when a molding is viewed with obliquely reflected light.

The reflected intensity can be reduced by applying an antireflection coating to the plastic. Ideally (5), this should have a refractive index n_3, which is a geometric mean of the plastic n_2 and air $n_1 = 1$; then $n_3 = \sqrt{n_2}$, and for normal incidence it should be one-quarter wavelength thick. It is difficult to find coatings with such a low refractive index that are sufficiently durable. Magnesium fluoride can be deposited economically (2) and has a refractive index of 1.38; it reduces the reflection loss per surface to $\sim$1.6% for PMMA (Fig. 4). Consequently, in a three-element lens with six surfaces, the overall reflection losses are cut from 22 to 9%. However, since MgF$_2$ coating is extremely soft and cannot survive being

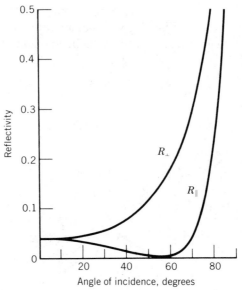

Fig. 3. Reflectivity vs angle of incidence for light with its **E** vector parallel and perpendicular to the plane of incidence, calculated for $n_1 = 1$ and $n_2 = 1.5$.

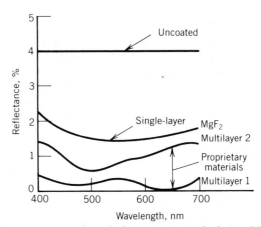

Fig. 4. Reflectance vs wavelength for an uncoated plastic MgF$_2$ and single-layer proprietary coatings (2).

cleaned with cheesecloth, such coatings can be used only in sealed systems. Figure 4 shows that multilayer proprietary coatings can reduce the reflectance per surface still further; they have better durability than MgF$_2$, but are more expensive.

Internal Absorption.   Internal absorption processes cause the transmitted intensity I to fall exponentially as the path length L increases through the plastic.

$$I = (1 - R)^2 I_0 \exp(-AL) \qquad (9)$$

where I_0 is the incident intensity and A is a constant. The transmittance T is defined as the ratio I/I_0. Figure 5 shows how the transmittance T varies with the

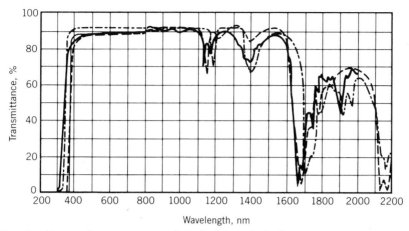

Fig. 5. Transmittance vs wavelength for optical plastics: ─, polycarbonate, 3.1-mm thick; ---, thin polystyrene; and ·─·─, acrylic, 3.1-mm thick (2).

wavelength of the light for some common optical plastics. All plastics absorb strongly in the uv region, hence the photodegradation and weathering (qv) in sunlight. Characteristic absorption bands in the ir can be used for identification. In a few cases, transmission windows in the far ir raise the possibility of cheaper alternatives to single-crystal lenses, eg, poly(4-methyl-1-pentene) has transmission peaks at 2.8, 5.0, 16, and 30 μm (6).

Plastics can be colored with inorganic pigments or organic dyes. To maintain good optical transmission, soluble dyes must be used (Fig. 6). Some fluorescent dyes absorb in one part of the visible spectrum and reemit ca 10^{-9} s later in a different part of the spectrum. Figure 6 shows the spectral analysis of the trans-

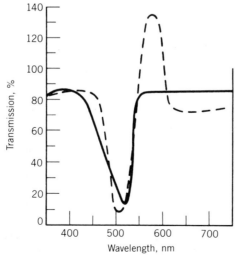

Fig. 6. Transmittance vs wavelength for a (─) yellow-dyed plastic and for a (---) fluorescent LISA sheet (7).

mitted light when such a dyed sheet is illuminated with unit intensity of white light.

Transmission becomes critical when plastics are considered for use in fiber-optics systems (see FIBERS, OPTICAL). There have been great advances in the preparation of high purity, silica-based glasses for fiber-optics communications over long distances. Plastics cannot currently compete in the telecommunication areas, but for short distances the much lower modulus of plastics and the greater failure strain indicate that high flexibility light guides can be constructed. In a 64-fiber DuPont Crofon light guide, each PMMA fiber of 250-μm dia is coated with plastic of a lower refractive index to ensure that, by total internal reflection at the interface, as much of the incident light as possible is transmitted along the fiber. The attenuation coefficient A in equation 9 has a value 0.5/m for such a light guide.

Polarizing filters are another type of absorbing material. Sheet polarizers are used in photoelastic equipment (8). If unpolarized light of unit intensity falls on a polarizing filter, the output can be considered as a plane-polarized beam of intensity K_1 with its electric vector in the 0° direction (the reference direction of the polarizer) and another plane-polarized beam of intensity K_2 with its electric vector in the 90° direction. An alternative definition of the principal transmittance K_1 is the ratio of the output to the input intensity for an incident beam plane-polarized in the 0° direction (and similarly for K_2). For Polaroid sheet polarizers, these principal transmittances are a function of wavelength (Table 4). The efficiency of a polarizer is measured with the extinction ratio K_1/K_2. The HN polarizers have a higher extinction ratio at the blue end of the visible spectrum and perform best when the amount of iodine absorbed is highest. The three grades offer different compromises between overall transmittance and extinction ratio. They cannot match the extinction ratios available from prism polarizers; specially selected calcite crystals can have extinction ratios as high as 10^7, but they are expensive and limited in size.

The transmittance of initially unpolarized light through a pair of parallel or crossed polarizers can be calculated. After the first polarizer, a beam of intensity

Table 4. Principal Transmittances of Polaroid (HN) Sheet Polarizers[a,b]

Wavelength, nm	HN 22		HN 32		HN 38	
	K_1	$10^5 K_2$	K_1	$10^5 K_2$	K_1	$10^5 K_2$
375	0.11	0.5	0.33	100	0.54	2000
400	0.21	1.0	0.47	300	0.67	4000
450	0.45	0.3	0.68	50	0.81	2000
500	0.55	0.2	0.75	5	0.86	500
550	0.48	0.2	0.70	2	0.82	70
600	0.43	0.2	0.67	2	0.79	30
650	0.47	0.2	0.70	2	0.82	30
700	0.59	0.3	0.77	3	0.86	70
750	0.69	1.0	0.84	2	0.90	400

[a] Ref. 8.
[b] H designates the polarizer type, uniaxially stretched poly(vinyl alcohol) supported by a sheet of cellulose acetate butyrate which has been treated with an iodine-containing ink; N refers to a neutral color; and the two digits to the overall percentage transmittance $T = \frac{1}{2}(K_1 + K_2)$.

$\frac{1}{2}K_1$ is polarized in the $0°$ direction, and one of intensity $\frac{1}{2}K_2$ in the $90°$ direction. If the second polarizer is also at $0°$, the intensity of the $0°$ beam is reduced to $\frac{1}{2}K_1^2$ and the $90°$ beam to $\frac{1}{2}K_2^2$. Consequently, the overall transmittance of a pair of parallel polarizers is

$$T_{\|} = \frac{1}{2}(K_1^2 + K_2^2) \tag{10}$$

If the second polarizer is at $90°$, the $0°$ beam is reduced in intensity to $\frac{1}{2}K_1K_2$, as is the $90°$ beam. Consequently, the overall transmittance of a pair of crossed polarizers is

$$T_+ = K_1K_2 \tag{11}$$

Photoelasticity

When elastic stresses in an amorphous solid cause the refractive index to become anisotropic in a reversible way, this is referred to as the photoelastic effect (9). If, however, the stresses exceed the yield stress and the refractive index changes are permanent, this is called the photoplastic effect. Two-dimensional models cut from a sheet of a photoelastic material have been widely used for stress analysis. This technique provides a visualization of high stress regions, and prototype designs can be modified before being put into production. The equipment used to load and observe the models is referred to as a polariscope. Applications of photoelasticity, particularly to crack growth and three-dimensional problems, are described in Ref. 10. Recent advances in computer-based techniques, such as finite-element analysis, have made the photoelastic technique less attractive. The more complex techniques, like the stress freezing of three-dimensional models and their subsequent slicing and analysis, require skilled technicians to prepare the models and maintain the equipment. However, two-dimensional photoelasticity experiments are relatively easy to perform and photographs of the stress patterns provide powerful visual aids. It is vital to understand the phenomena causing photoelastic fringes when examining transparent plastic moldings with polarized light, in order to avoid confusion with birefringent effects caused by permanent molecular orientation.

To explain the basic types of polariscopes, a concise method of describing the state of polarization of light is required. The model materials act as retarding plates, and the effect of these on polarized light must be analyzed to understand the information available from polariscopes.

Jones Calculus for Polarized Light. There are many ways of deriving the output of a polariscope; the most concise method uses the Jones calculus (11), where any form of polarized light is represented by a vector

$$\mathbf{a} = \begin{bmatrix} a_x \\ a_y \end{bmatrix} \tag{12}$$

in which the two quantities are the x- and y-components of the electric vector $\mathbf{E}$ of the light. The x and y axes are usually horizontal and vertical. Both a_x and a_y are complex numbers:

$$a_x = A_x e^{i(\delta_x + v)} \qquad a_y = A_y e^{i(\delta_y + v)} \tag{13}$$

where $v = \omega t + 2\pi z/\lambda$, and ω is the angular frequency and λ the wavelength of the light. The time factor e^{iv} represents the sine wave propagating along the z axis and occurs in both a_x and a_y. Usually, the Jones vector is written in a normalized form, omitting the time factor; A_x is the modulus of a_x and as such can be used to calculate the light intensity I:

$$I = A_x^2 + A_y^2 \tag{14}$$

δ_x is the argument of a_x and is used to determine the relative phase difference $\delta = \delta_x - \delta_y$ between the two components of the light. This phase difference or retardation is measured in radians; a retardation of 2π represents a one-wavelength phase shift. The initial intensity of the polarized light is normally set at unity, and the following vectors

$$\begin{bmatrix} 1 \\ 0 \end{bmatrix} \quad \begin{bmatrix} \cos\theta \\ \sin\theta \end{bmatrix} \quad \begin{bmatrix} 0 \\ 1 \end{bmatrix}$$

represent plane-polarized light with the directions of polarization horizontal, at an angle θ anticlockwise from horizontal, and vertical, respectively. The general Jones vector of equations 12 and 13 represents elliptically polarized light. On a particular xy plane, at a fixed z value, the electric vector traces out an elliptical path at an angular frequency ω. In the special case, when the phase difference $\delta = \pi/2$, the ellipse becomes a circle and the light is referred to as circularly polarized light.

Sheets of material can be categorized into two types. Polarizers are devices that split the light into two components with orthogonal electric vectors and transmit them with different intensities. Retarders or birefringent plates transmit the two components with a relative retardation δ. The fast and slow directions of a retarder indicate the relative speeds of the light components polarized in these two directions. The effect of each of these devices is summarized by a 2×2 square matrix **P** or **R**. If light with a Jones vector **a** passes through any of these devices, the emerging light has a Jones vector **a**′ given by

$$\mathbf{a'} = \mathbf{Pa} \tag{15}$$

If the light passes through a polarizer $\mathbf{P}_1$, a retarder **R**, and a polarizer $\mathbf{P}_2$ in turn, the emerging light has a Jones vector

$$\mathbf{a'} = \mathbf{P}_2\mathbf{R}\mathbf{P}_1\mathbf{a} \tag{16}$$

The sequential multiplication of the initial Jones vector by the device matrices gives the output.

The matrices for the basic forms of these devices are given in Table 5. To analyze the components of the polarized light in a new set of axes $x'y'$, which are rotated by θ anticlockwise from the xy axes, the standard transformation for vector quantities is used:

$$\mathbf{a'} = \mathbf{S}_\theta\mathbf{a} \tag{17}$$

where the matrix $\mathbf{S}_\theta$ is defined in Table 5. Consequently, the more general form of the matrices for a polarizer or analyzer at an angle θ can be found as follows: the incident light is converted into the $x'y'$ axes of the polarizer, the effect of the polarizer is calculated by multiplying by $\mathbf{P}_0$, and the output is converted back

Table 5. Jones Matrices of Polarizers and Retarders

Device	Basic device	Device at angle θ
polarizer	horizontal plane-polarizer $$\mathbf{P}_0 = \begin{bmatrix} 1 & 0 \\ 0 & 0 \end{bmatrix}$$	$$\mathbf{P}_\theta = \begin{bmatrix} c^2 & cs \\ cs & s^2 \end{bmatrix}$$
retarder	horizontal ray retarded by δ $$\mathbf{R}_{0\delta} = \begin{bmatrix} e^{i\delta} & 0 \\ 0 & 1 \end{bmatrix}$$	$$\mathbf{R}_{\theta\delta} = \begin{bmatrix} c^2 e^{i\delta} + s^2 & sc(e^{i\delta} - 1) \\ sc(e^{i\delta} - 1) & s^2 e^{i\delta} + c^2 \end{bmatrix}$$
anticlockwise rotation of the reference axis of the Jones vector by θ	$$\mathbf{S} = \begin{bmatrix} c & s \\ -s & c \end{bmatrix}^a$$	

^{a} $c \equiv \cos\theta$ and $s \equiv \sin\theta$.

into the xy axes. Therefore,

$$\mathbf{P}_\theta = \mathbf{S}_{-\theta}\mathbf{P}_0\mathbf{S}_\theta \tag{18}$$

This equation and the equivalent one for a retarder are used to calculate the device at angle θ in Table 5.

A useful special form of retarder is the quarter-wave plate whose matrix can be deduced from Table 5 as

$$\mathbf{Q}_{45} \equiv \mathbf{R}_{45,\pi/2} = \begin{bmatrix} \dfrac{i+1}{2} & \dfrac{i-1}{2} \\ \dfrac{i-1}{2} & \dfrac{i+1}{2} \end{bmatrix} = \dfrac{i+1}{2}\begin{bmatrix} 1 & i \\ i & 1 \end{bmatrix} \tag{19}$$

Plate axes are set at 45° to the horizontal or vertical plane-polarizer with which it is often combined. The effect of a quarter-wave plate on horizontally polarized light is found by using

$$\mathbf{a} = \dfrac{i+1}{2}\begin{bmatrix} 1 & i \\ i & 1 \end{bmatrix}\begin{bmatrix} 1 \\ 0 \end{bmatrix} = \dfrac{i+1}{2}\begin{bmatrix} 1 \\ i \end{bmatrix} \tag{20}$$

The components a_x' and a_y' are of equal magnitude, but differ in phase by $\pi/2$. Consequently, the light is circularly polarized, since the amplitude of $\mathbf{a}$ remains constant, but its direction rotates at the frequency of the light.

Polariscopes.   The purpose of a polariscope is to extract useful information about optical properties of the model material at every point of interest. The basic information is the orientation θ of the optic axes of the model, which is also the orientation θ of the principal stress axes, and the retardation δ, which can be used to calculate the difference between the principal stresses. The plane polariscope gives a precise value of θ, but both it and the circular polariscope can only give contour levels of δ. To obtain an exact value of δ at the point of interest, some form of compensator must be used. The Tardy compensation method, one of many, is the basis of a commercial automated polariscope, which has increased accuracy and speed of measurement.

Plane Polariscopes.   The elements of a plane polariscope are shown in Figure 7. The light passes in turn through a vertical plane-polarizer, a photoelastic model which has a fast axis at θ to the horizontal and has retardation δ, and a

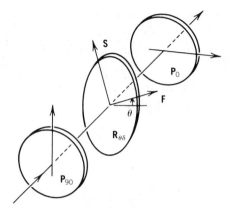

Fig. 7. Optics of a plane polariscope containing a retarder of orientation θ and retardation δ. See text for explanation of labels.

horizontal plane-polarizer. Hence, according to the Jones calculus, the output is

$$\mathbf{a}' = \mathbf{P}_0 \mathbf{R}_{\theta\delta} \mathbf{a} \tag{21}$$

where $\mathbf{a}$ is vertically polarized light. Substituting from Table 5 and carrying out the matrix multiplication produces

$$
\begin{aligned}
\mathbf{a}' &= \begin{bmatrix} 1 & 0 \\ 0 & 0 \end{bmatrix} \begin{bmatrix} c^2 e^{i\delta} + s^2 & sc(e^{i\delta} - 1) \\ sc(e^{i\delta} - 1) & s^2 e^{i\delta} + c^2 \end{bmatrix} \begin{bmatrix} 0 \\ 1 \end{bmatrix} \\
&= \begin{bmatrix} 1 & 0 \\ 0 & 0 \end{bmatrix} \begin{bmatrix} sc(e^{i\delta} - 1) \\ s^2 e^{i\delta} + c^2 \end{bmatrix} \\
&= \begin{bmatrix} (e^{i\delta} - 1) \sin\theta \cos\theta \\ 0 \end{bmatrix}
\end{aligned}
\tag{22}
$$

Hence the amplitude of a_x' is zero if

$$e^{i\delta} = 1 \qquad (\delta = 2\,n\pi) \tag{23}$$
$$\text{or} \qquad \sin 2\theta = 0 \qquad (\theta = m\pi/2)$$

where n and m are integers. The first condition defines the dark isochromatic fringes, so called because they are colored, when white, rather than monochromatic, light is used. The second condition defines the dark isoclinic fringes, which are loci of points of constant inclination of the axes of the retarder. The crossed polarizers in a plane polariscope can be rotated to positions α and $90 + \alpha$, in which case the isoclinic fringes occur for $\theta = \alpha + m\pi/2$. For every point on a photoelastic model, one isoclinic fringe passes through it. The direction of one of the retarder axes at that point is at $\theta = \alpha$ degrees to the horizontal.

Circular Polariscope. It may be somewhat confusing to have isochromatic and isoclinic fringes present at the same time. Therefore, a circular polariscope is used when only the isochromatic fringes are wanted. It differs from the plane polariscope in having quarter-wave plates at orientation $\pm 45°$ between the polarizers and the photoelastic model (Fig. 8). Hence the model is illuminated with circularly polarized light. The effect of the elements on the vertically plane-

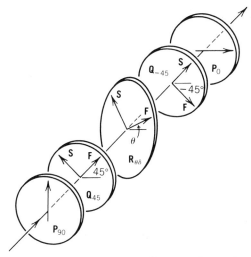

Fig. 8. Optics of a circular polariscope. See text for explanation of labels.

polarized light **a** that emerges from the first polarizer is

$$\mathbf{a'} = \mathbf{P}_0\mathbf{Q}_{-45}\mathbf{R}_{\theta\delta}\mathbf{Q}_{45}\mathbf{a} \tag{24}$$

Matrix multiplication gives the result as

$$\mathbf{a'} = \tfrac{1}{2}\begin{bmatrix} e^{i\delta} - 1 \\ 0 \end{bmatrix} \tag{25}$$

The amplitude of a_x and hence the intensity of the light is zero if

$$e^{i\delta} = 1 \quad (\delta = 2n\pi) \tag{26}$$

This is the condition for the isochromatic fringes, as before.

Both polariscopes described thus far can be used in reflection as well as in the transmission mode. All that is necessary is to have a metallic coating on the back of the model in such a way that the light passes through the model twice before entering the analyzing system (Fig. 9). This doubling of the effective thickness of the model doubles the number of isochromatic fringes, but the offset

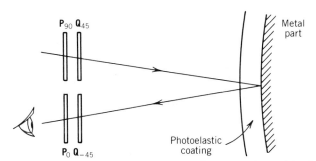

Fig. 9. Optics of a reflection polariscope used to examine a birefringent layer on the surface of a metal component. See text for explanation of labels.

of the incident-light angle introduces a slight error. The main application of a reflection polariscope is in investigations of the surface strains in real products. If the surface of the product is curved, a partially cross-linked sheet of epoxy thermoset can be contoured to the product surface and allowed to complete the cross-linking reaction. Placing a 6-mm layer of thermoset on the surface of a metal product may slightly increase the overall stiffness of the product, especially in bending. The metal has a Young's modulus that is at least 20 times greater than that of the plastic, and it is usually much thicker; thus the error so introduced is not large.

Tardy Compensator.   The Tardy compensator method is a simple adaption of the circular polariscope. It requires a knowledge of the orientation of the optic axes at the point of interest in the model. Figure 10 shows the optical arrangement; for convenience the fast axis of the model is shown as being horizontal to simplify the calculation. Circularly polarized light illuminates the model; as long as the axes of the polarizer and the quarter-wave plate are at 45° to each other, there is no need to bring them into alignment with the axes of the model. However, the fast axis of the second-quarter-wave plate must be at $-45°$ to the model fast axis. The light emerging from this quarter-wave plate is

$$\mathbf{a}' = \mathbf{Q}_{-45}\mathbf{R}_{\theta\delta}\mathbf{Q}_{45}\mathbf{a} \tag{27}$$

where $\mathbf{a}$ is vertically polarized light. Substituting the values from Table 5 and carrying out the matrix multiplication yields

$$\mathbf{a}' = \left(\frac{i+1}{2}\right)^2 \begin{bmatrix} i(e^{i\delta}-1) \\ e^{i\delta}+1 \end{bmatrix}$$

$$= \left(\frac{i+1}{2}\right)^2 e^{i\delta/2} \begin{bmatrix} \cos\left(\dfrac{\pi}{2}+\dfrac{\delta}{2}\right) \\ \sin\left(\dfrac{\pi}{2}+\dfrac{\delta}{2}\right) \end{bmatrix} \tag{28}$$

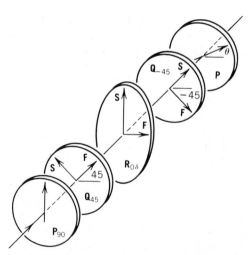

Fig. 10.   Optics of the Tardy compensator. See text for explanation of labels.

Therefore, the output is plane-polarized at an angle of $\delta/2$ anticlockwise from the y axis. The final polarizer extinguishes the light if its axis is at $\theta = \delta/2$ to the horizontal x axis. The retardation at the point of interest is

$$\delta = 2n\pi + 2\theta \tag{29}$$

where n is the fringe order of the next lowest isochromatic fringe in a circular polariscope.

Automated Polariscope. The commercial automated polariscope relies on a combination of the optics of the plane polariscope and the Tardy compensator (12). The intensity of the transmitted light is detected photoelectrically. Since it is notoriously inaccurate to find the angular position at the minimum of a sine wave, a continuously rotating polarizer system is adopted (Fig. 11). The polarizer and analyzer of the plane polariscope rotate at an angular frequency Ω. In an outer annulus of the lens system, the same rotating analyzer is used for the Tardy compensator. Equation 22 shows that the amplitude of the light from the plane polariscope varies as $\sin(\alpha - \theta)\cos(\alpha - \theta)$, where α is the angular orientation of the polarizer:

$$\alpha = \Omega t \tag{30}$$

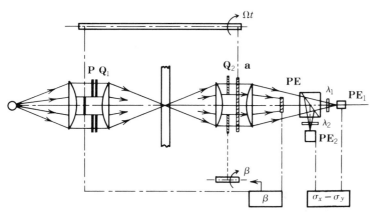

Fig. 11. Optics of the Redner automatic polariscope (12). See text for explanation of labels.

Hence the intensity of the transmitted light varies as

$$I \propto \sin^2(\alpha - \theta)\cos^2(\alpha - \theta) \tag{31}$$
$$\propto 1 - \cos 4(\Omega t - \theta)$$

ie, with angular frequency $= 4\Omega$.

If a reference voltage from a transducer connected to the rotating polarizer is compared with the output of the photoelectric cell, an accurate determination of the phase lag 4θ can be made with a phase-sensitive detector. Therefore, a value of the orientation of one of the optical axes of the model can be obtained in the range $45° > \theta > -45°$, as long as the retardation $\delta \neq 0$.

Equation 28 shows that the output intensity of the Tardy compensation system is proportional to $\sin^2 (\Omega t - \delta/2)$ and varies at an angular frequency of 2Ω. An unmodified Tardy compensator could only detect a range of retardation values of $0 < \delta < 2\pi$. To overcome this, the polariscope detects the output at two different wavelengths of light λ_1 and λ_2 for which the stress-optical coefficients are C_1 and C_2, respectively. The retardation is related to the model thickness t and the principal stress difference $\sigma_1 - \sigma_2$ by

$$\delta = 2\pi t(\sigma_1 - \sigma_2)C/\lambda \tag{32}$$

Therefore, if the output is detected photoelectrically at the wavelengths λ_1 and λ_2, and the phase difference ϕ between these two signals is detected, it is related to the principal stress difference by

$$\phi = 2\pi t(\sigma_1 - \sigma_2) \left(\frac{C_1}{\lambda_1} - \frac{C_2}{\lambda_2} \right) \tag{33}$$

A typical range of the instrument is the equivalent of five isochromatic fringes, using wavelengths $\lambda_1 = 577$ nm and $\lambda_2 = 632$ nm.

Determination of Principal Stresses.   The information obtained so far from examining a photoelastic model in a polariscope is summarized in the Mohr's circle diagram of the stress components (Fig. 12a). The radius of Mohr's circle is

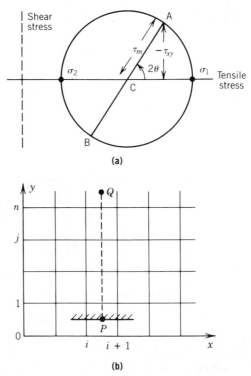

(a)

(b)

Fig. 12.   (**a**) Mohr's diagram for the stress components measured in a polariscope. The horizontal position of the center of the circle C is unknown. (**b**) Square grid of points at which the shear stress τ_{xy} is evaluated. The tensile-stress components at a point Q in the interior of a body are evaluated by a numerical integration along the path PQ from the surface at P.

known; it is the value of the maximum shear stress in any axis τ_m given by

$$\tau_m = \frac{\sigma_1 - \sigma_2}{2} \tag{34}$$

The inclination 2θ of the radius to the horizontal is given from the isoclinic angle θ. However, the position of the center of the circle representing the mean principal stress $(\sigma_1 + \sigma_2)/2$ is unknown. Until this third piece of information is found, it is not possible to evaluate the individual principal stresses σ_1 and σ_2.

There are at least three established methods of finding the individual principal stresses; two involve further experimental measurements, either the strain in the model thickness direction or the isochromatic fringe order when the model is illuminated at an oblique angle (13). The third is a numerical computation well suited to the data obtained from an automatic polariscope. The isochromatic fringe order and isoclinic angle are determined at a grid of points that have coordinates $x = is$, $y = js$, where i and j are integers and s is the separation of neighboring points (Fig. 12**b**). From these the shear stress τ_{xy} in the xy axes is calculated at every point, using

$$\tau_{xy} = -\tau_m \sin 2\theta \tag{35}$$

Figure 12**a** shows that if τ_m is positive, τ_{xy} is negative and vice versa, because 2θ always lies in the range of 0–180°. For a two-dimensional stress system, there are two conditions for stress equilibrium relating the variation with position of the stress components:

$$\frac{\partial \tau_{xy}}{\partial x} + \frac{\partial \sigma_{yy}}{\partial y} = 0, \qquad \frac{\partial \tau_{xy}}{\partial y} + \frac{\partial \sigma_{xx}}{\partial x} = 0 \tag{36}$$

Both of these can be written in finite-difference form, with the infinitesimal distance ∂x and ∂y being replaced by $\Delta x = \Delta y = s$. The first equation becomes

$$\frac{\tau_{xy}(i + 1, j) - \tau_{xy}(i,j)}{s} + \frac{\sigma_{yy}(i + \frac{1}{2}, j + \frac{1}{2}) - \sigma_{yy}(i + \frac{1}{2}, j - \frac{1}{2})}{s} = 0 \tag{37}$$

If the boundary of the model lies along the x axis at the point P, the value of σ_{yy} is zero there because of the free surface. Consequently, the value of σ_{yy} at any point Q in the interior can be obtained by summing the shear-stress differences along the path PQ:

$$\sigma_{yy}(i + \frac{1}{2}, n + \frac{1}{2}) = \sum_{j=1,n} \tau_{xy}(i,j) - \tau_{xy}(i + 1, j) \tag{38}$$

As σ_{yy} represents the horizontal coordinate of the point B on the Mohr's circle, it is clear how the individual principal stresses can be calculated. If there had been a free surface parallel to the y axis, the second of the stress-equilibrium equations (eq. 36) leads to an alternative to equation 38, evaluating σ_{xx} by summing the shear-stress differences along a horizontal path.

Model Materials. A photoelastic model material must fulfill a number of requirements. First, it should be transparent and isotropic when unstressed and should not scatter light; this eliminates semicrystalline plastics. Second, flat-sheet or three-dimensional models should be easy to prepare, ie, the models are cast from a thermoset with low shrinkage on curing. Extruded thermoplastic sheet (polycarbonate is a good example) can be used, but needs careful drying and removal of any molecular orientation by annealing at a temperature just above T_g. The sheets of material should be easily machined to shape. Third, the material should be elastic, with a linear stress–strain relation and little creep if a constant load is applied. All glassy thermoplastics, thermosets, and rubbers show a certain degree of viscoelastic behavior. The relationship between the refractive indexes n_1 and n_2 and the corresponding principal stresses σ_1 and σ_2 in the loaded model should be linear:

$$n_1 - n_2 = C(\sigma_1 - \sigma_2) \tag{39}$$

The constant C is known as the stress-optical coefficient. It is assumed the principal axes of stress and the optic axes of the material coincide. Figure 13 shows that equation 39 is only obeyed up to a certain stress for polycarbonate. Equation 39 is not in the most convenient form for calculating the principal stresses from the isochromatic fringe order n. If monochromatic light of wavelength λ passes through a model of thickness t, the refractive-index difference is given in terms of the fringe order f or in terms of the relative retardation δ by

$$n_1 - n_2 = \frac{f\lambda}{t} = \frac{\delta}{2\pi}\frac{\lambda}{t} \tag{40}$$

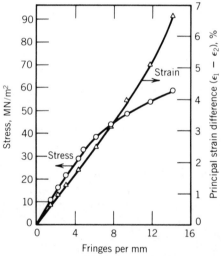

Fig. 13.   Stress and strain optical behavior of polycarbonate. To convert MN/m^2 to psi, multiply by 145 (14).

Combining equations 39 and 40 gives

$$\sigma_1 - \sigma_2 = \frac{\lambda}{C}\frac{f}{t} \tag{41}$$

where the constant λ/C is referred to as the stress-fringe value. Data on these constants for the common photoelastic model materials are given in Table 6. A high Young's modulus is considered a benefit for model materials, since the model does not distort and change the loading geometry. Similarly, a low stress-fringe

Table 6. Photoelastic and Elastic Constants of Model Materials[a]

Material	Stress-fringe value, kN/m[b]	Young's modulus E, GN/m^2 [c]	Figure of merit Q
Epoxy ERL-2774	10.2	3.27	0.32
Epoxy Epon-828	11.4	4.54	0.40
Columbia Resin CR-39	15.4	1.73	0.11
Homalite-100	23.6	3.86	0.16
polycarbonate	7.0	2.48	0.35
polyurethane rubber	0.18	0.003	0.02

[a] Ref. 15.
[b] To convert kN/m to kgf/s^2, multiply by 1000.
[c] To convert GN/m^2 to psi, multiply by 145,000.

value indicates that the model provides many fringes at a low applied stress. Consequently, a figure of merit Q, defined as the Young's modulus divided by the stress-fringe value, combines these two attributes.

Finally, a model material should be stable on exposure to the atmosphere. For some model materials, the time-edge effect can be troublesome. This is due to the slow diffusion of water into the material. The model is dry after curing and annealing, but absorbs a certain amount of water depending on the relative humidity of the atmosphere. As it can take several weeks for the water to diffuse through a 6-mm thick model, residual stresses can arise with the slightly swollen surface layers being in compression and the interior in tension. The extra retardation is only apparent at the edges of a model since in the central regions the average stress along the light path is zero. However, it is often necessary to determine the stresses at the edge of a model accurately. Figure 14 shows the development and decay of the time-edge effect in an Epon 828 model stored at constant temperature and humidity (15).

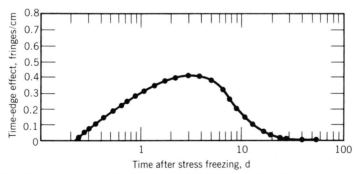

Fig. 14. The time-edge effect on Epon 828 shown as the fringe order at the surface as a function of the time of exposure to a humid atmosphere (15).

Thermoset models behave as rubbers when heated to a temperature above their glass transition. This is the basis of the frozen-stress method. The stress-fringe value of the material has to be determined at a temperature in the rubbery region; it is typically much lower than in the glassy state, as is shown by the polyurethane value in Table 6. The model is loaded in the rubbery state in an oven, which is then slowly cooled. When the model is in the glassy state, the retardation is due to molecular orientation rather than to stress, and the value does not change if the model is unloaded or sliced up if it is three-dimensional.

Isochromatic Patterns in Simple Loading Situations. The plane polariscope determines an isoclinic angle in the range of $90° > \theta > 0°$; one of the principal stress directions σ_1 lies at an angle θ anticlockwise from the horizontal x axis. At this stage it is often not known if σ_1 is larger or smaller than the other principal stress σ_2.

Many photoelastic patterns contain standard features that can be interpreted because the elastic-stress analysis is known. For example, Figure 15 shows a circular disk compressed across its diameter d. In the contact region (designated by an arrow), the stress field approximates to

$$\sigma_{rr} = \frac{2F}{\pi rt}, \qquad \sigma_{\theta\theta} = 0 \tag{42}$$

the stresses being given in polar coordinates centered on the arrow; F is the contact force and t the thickness of the model. Consequently, the isochromatics are circles, touching at the arrow, with the diameter inversely proportional to the fringe number.

A disk of photoelastic model material is often compressed diametrically in order to find the stress-optical coefficient. The maximum shear stress at the central point is the sum of the two contributions from the two opposed forces

$$\tau_m = \tfrac{1}{2}(\sigma_{rr} - \sigma_{\theta\theta}) = \frac{4F}{\pi dt} \tag{43}$$

There is no need to make observations at a series of values of F, since the value of τ can be calculated for any position on the disk. Table 7 gives the maximum shear-stress values at the coordinates x,y for a disk that has a finite contact width $2a$ with a flat plane. These values have been calculated using Westergaard stress functions (16) and the formula

$$\tau_m = \frac{2F}{\pi dt}\left(\frac{d^2}{a^2}\right)\left|\frac{iZ_1 ImZ_1}{(a^2 - Z_1^2)^{1/2}} + \frac{iZ_2 ImZ_2}{(a^2 - Z_2^2)^{1/2}} + 1\right| \tag{44}$$

where the complex number Z_1 represents the position in the reduced coordinates $(x/d, y/d)$, $Z_2 = i - Z_1$, and Im means "imaginary part of."

Table 7. Maximum Shear Stress in Units of $2F/\pi dt$

Coordinates		τ_m values for a/d		
x/d	y/d	0.01	0.02	0.04
0.00	0.10	5.52	5.41	5.03
0.00	0.14	4.14	4.10	3.95
0.00	0.18	3.38	3.36	3.29
0.00	0.22	2.91	2.90	2.86
0.00	0.26	2.60	2.59	2.56
0.00	0.30	2.38	2.38	2.36
0.00	0.40	2.08	2.08	2.07
0.00	0.50	2.00	2.00	1.99
0.10	0.50	1.78	1.77	1.77
0.20	0.50	1.25	1.25	1.25
0.30	0.50	0.69	0.69	0.70
0.40	0.50	0.27	0.27	0.27

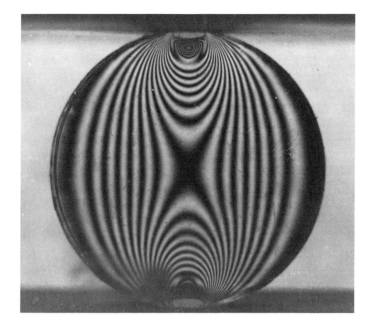

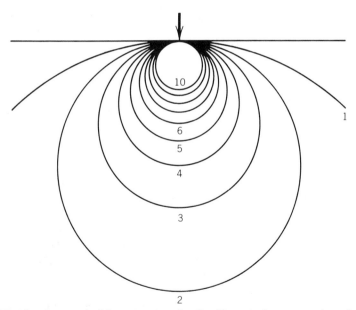

Fig. 15.   Isochromatic fringe pattern for the diametral compression of a cylinder of polystyrene. The drawing shows the theoretical form near the contact point for a line indenter on a half space.

Figure 16 shows a beam of uniform width loaded in bending. In the simple stress analysis of a beam, in which only the longitudinal tensile stresses are considered, the isochromatic fringe pattern should be straight lines between the loading points, and the total number across any section AA′ should be proportional to the bending moment.

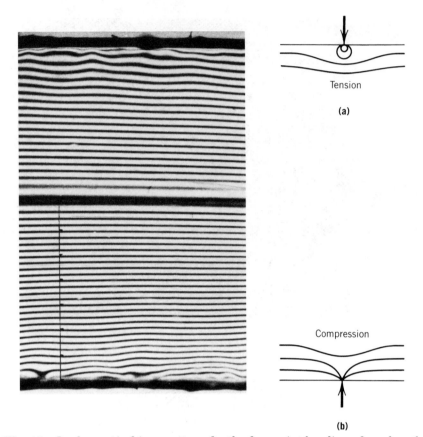

Fig. 16. Isochromatic fringe pattern for the four-point bending of a polycarbonate beam under uniform-bending moment conditions. The drawings show the effect of a surface-compressive force when the stress parallel to the surface is (**a**) tensile and (**b**) compressive.

Any surface loading of the beam disturbs this simple pattern and instead locally produces the contact force pattern of Figure 15. The effect of a contact force normal to a free surface is often useful in determining whether the principal stress difference is positive or negative at that point. If the principal stress parallel to the surface is σ_1 and the zero principal stress perpendicular to the free surface is σ_2, an additional small compressive force increases $\sigma_1 - \sigma_2$ if σ_1 is tensile and hence pushes the isochromatic fringe away from the free surface. Conversely, if the isochromatic fringes move toward the contact force, the principal stress σ_1 is compressive (Fig. 16b).

A third familiar photoelastic pattern is that at a crack tip. Figure 17 shows a compact-tension specimen, in which the crack faces are pulled open by a pair of forces F. In the region adjoining the crack tip, the isochromatic pattern consists of nested ellipses, with axes parallel to the x and y axes, and one end touching the crack tip. The equation of these ellipses in polar coordinates $r\theta$ based on the crack tip is

$$r^{1/2} = \frac{CK_1 t \sin \theta}{\lambda f \sqrt{2\pi}} \tag{45}$$

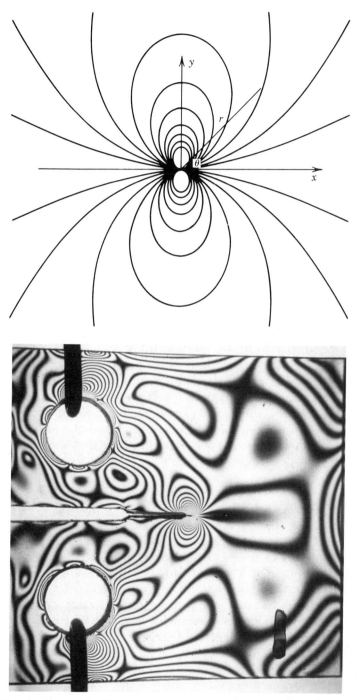

Fig. 17. Isochromatic fringe pattern for a compact-tension, fracture-mechanics specimen. The drawing shows the theoretical pattern near the tip of the crack for the crack opening mode I.

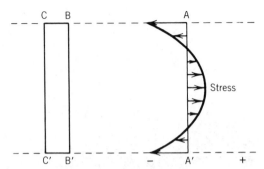

Fig. 18. Distribution of residual stresses parallel to the surface of a sheet of plastic that has been cooled rapidly during extrusion or injection molding. If the sheet is cut into a slice as shown by the lines BB′ and CC′, the residual stresses in the x-direction are reduced by at least 95%.

where f is the isochromatic fringe order. A graph of r_{max} (the value at $\theta = 90°$) vs f^{-1} should be a straight line; the stress-intensity factor K_1 can be calculated from the slope of the line. If the isochromatic pattern is found to be skew relative to the crack with r_{max} occurring for $\theta \neq 90$, it can be inferred that some mode II, crack-tip loading exists with the crack faces being sheared relative to one another.

Separation of Residual-stress Birefringence from Orientation Birefringence. When transparent glassy plastics are injection molded and the product examined in a polariscope, there may be some doubt about the origin of the isochromatic fringes. Since no external forces are applied to the article, the conventional photoelastic effect is not present. However, the stresses in the product do not have to be zero because the average tensile stress on some cross section is zero. The most common type of residual-stress distribution is shown in Figure 18; the outer layers which have solidified first are under compressive stress, balanced by the tensile stresses in the interior. Consequently, the integral

$$\int_0^t \sigma_{xx} dy = 0 \tag{46}$$

for the stresses along the line AA′. These residual stresses cause birefringence because the plastic has a nonzero, stress-optical coefficient. The other cause of birefringence that may be present is molecular orientation, giving rise to elastic stresses in the flowing melt. However, this nonequilibrium microstructure does not cause stresses in the glassy state because the molecules are no longer free to change their shapes, as the C—C bonds cannot rotate.

If the product is cut into sections with an abrasive cutoff wheel, the new free surfaces modify the residual-stress distribution. For example, if a cut is made along BB′ in Figure 18, the tensile stress σ_{xx} falls to zero along this line. If a second cut CC′ is made within a distance $t/5$ of BB′, this effectively reduces the stresses σ_{xx} in the section to zero (17). The remaining birefringence is solely due to the permanent molecular orientation. Figure 19 compares the total birefringence in a polycarbonate sheet with the molecular orientation birefringence measured in a thin slice. The difference between the two curves reveals the magnitude of the residual-stress distribution.

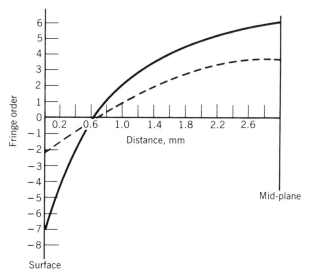

Fig. 19. Comparison of the (—) total (residual stress plus molecular orientation) fringe order with the (---) molecular orientation (thin-slice) fringe order in a rapidly cooled polycarbonate sheet.

If the birefringence of an injection molding is observed by viewing normal to a product surface (in the direction AA′ in Fig. 18), then, except near the edges of the product, the residual stress contribution is usually zero. This is the result of the condition expressed by equation 46; the total retardation is the sum of the layers through which the light passes, and the retardation in each layer is proportional to the principal stress difference. The existence of residual stresses near the edges of products can be revealed by observing if the isochromatic fringes move when a saw cut is made in this region.

Photoplasticity. In certain stress-analysis problems, limited yielding occurs in a body that otherwise remains elastic. The most common examples are the events at the tip of a crack and those that are due to concentrated surface loads as in hardness testing. As most thermoset model materials fail in a brittle manner in tension, the only suitable model materials are thermoplastics such as polycarbonate (18,19). However, many difficulties have hindered the widespread use of photoplastic modeling. The shape of the stress–strain curve after yield must be the same in the model material as in the metal that it simulates. Polycarbonate exhibits an upper-yield point in a tensile test and then cold-draws at a 20% lower stress until a strain of ∼100% is reached; this is quite unlike the behavior of any metal. Nevertheless, an isotropic polycarbonate sheet of ca 0.2-mm thickness can be used to illustrate a model of yielding at a crack tip (20). Figure 20 shows how the yielded region is clearly demarcated using a circular polariscope, because the birefringence of the polycarbonate strongly increases with the permanent strain in the yielded region. It is even possible to see contours of plastic strain close to the crack tip. However, it is not at all easy to interpret the plastic strains in a quantitative manner because of the complex relationship between birefringence, plastic strains, and stresses.

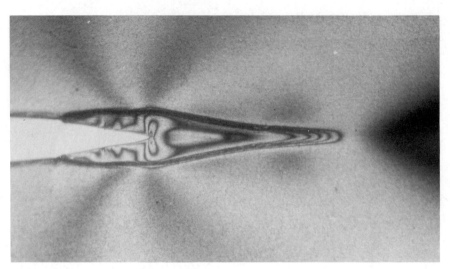

Fig. 20. Crack-tip region in 0.2-mm thick polycarbonate viewed in a circular polariscope. The nearly horizontal set of parallel fringes indicates the boundary of the yielded zone and the higher-order fringes illustrate the plastic-strain concentration at the crack tip.

Light Scattering

When a collimated beam of light passes through or is reflected from a plastic sheet or film, scattering causes that light to deviate in direction without any reduction in the total light intensity (see also LIGHT SCATTERING). The light intensity is reduced by pigments or dyes. If light is scattered through an angle exceeding 90°, the forward light intensity is reduced. Scattering as a result of surface imperfections is dealt with separately from scattering due to internal microstructural inhomogeneities, because the mechanisms for scattering are different.

Several common terms are given specific meanings when used in connection with light scattering:

Gloss refers to the reflected light intensity as a percentage of that from an ideal surface. The specular gloss is measured within a specified angular range of the ideal reflected angle r; a standard, polished, black glass of refractive index 1.567 has a specular gloss of 100%.

Haze refers to the integrated, transmitted intensity for all light that deviates by an angle $\theta > 2.5°$, as a percentage of the incident-light intensity.

Clarity is defined in terms of the image degradation of standard printed scales viewed through the plastic sheet.

Transparency refers to the transmitted intensity for all light that deviates by an angle $\theta < 2.5°$, as a percentage of the incident-light intensity. These definitions are more precise when the test equipment is specified in the appropriate national standards.

Surface Scattering of Reflected Light. A high gloss value is a more important requisite for plastic paint films than for moldings; the test methods have been developed for paints (21). Although specular gloss measurements are usually made at an angle of reflection $r = 60°$, more information on the scattering processes can be obtained when reflected intensity is measured over the range $90° > r > 0°$. Figure 21 shows the reflected-intensity variation for two black- and one white-pigmented polymer films. The maximum intensity I_S at $r = 60°$ is used to calculate the specular gloss. The intensity I_D at $r = 0°$ can be used to calculate the diffuse reflectance; it is due to high angle light scattering at the air–film interface or at pigment–polymer interfaces just below the surface. The curves differ significantly in terms of the width of peak centered at $r = 60°$; this can be characterized by the angular width of the peak at half height $W_{1/2}$. These three measures of the curves have been combined in the calculation of a gloss factor G defined by

$$G = \frac{I_S - I_D}{W_{1/2}} \tag{47}$$

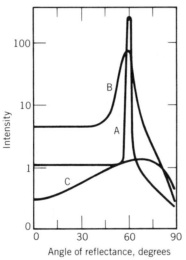

Fig. 21. Scattered-light intensity vs reflection angle for light incident at 60° on A, a white film of high gloss; B, a black film of medium gloss; and C, a black film of low gloss (21).

The gloss of paint films appears to be affected by the surface roughness, as measured by instruments such as the Talysurf (22). The average roughness R_a is defined as the mean deviation of the surface height y from the mean level $\bar{y}$ over a measuring length L.

$$R_a = \frac{1}{L} \int_0^L |y - \bar{y}| dx \tag{48}$$

The R_a value increases as the paint weathers and the pigment particles become more exposed. The relationship between the gloss factor and the average roughness can be expressed as

$$\log_{10} G = A - 3.6 R_a \tag{49}$$

where A is a constant depending on the pigment type and R_a is measured in μm.

Plastic moldings are frequently given a textured finish by etching or otherwise roughening the surface of the steel mold. This may hide molding blemishes such as weld lines and sink marks, or it may improve the appearance after a period of use when dirt has been deposited on the surface.

Surface defects are classified into three types (23). Microirregularities are defects of size below the wavelength of light that are so close together that they can only be characterized by roughness measurements such as R_a. Isolated defects of a similar kind also occur, where the light scattering is a product of an individual defect multiplied by the number of defects per unit area. Defects larger than the wavelength of light are scratches or digs (holes) caused by indentations.

There are a number of causes of microdefects. An ideal injection-molded surface replicates the surface of the steel mold, which may have been polished or given a surface texture. Surface defects occur as a result of mold filling; regular sinusoidal waves perpendicular to the flow direction occur as a result of the surface skin contracting while the mold is partially full. If glass fibers are incorporated or the melt is foamed, the surface is often very rough, but the molding is then opaque.

The total integrated scatter (TIS) is the equivalent of haze for reflected light, being the integrated intensity for all light outside the specularly reflected beam. The TIS depends on the root-mean-square surface roughness δ, the angle of reflection r, and the wavelength of the light λ (24):

$$\text{TIS} = 1 - \exp\left[-(4\pi\delta \cos r/\lambda)^2\right] \tag{50}$$

Surface Scattering of Transmitted Light.   The foregoing classification of defects applies equally well to the scattering of light transmitted through transparent plastics. Scratches and digs from use or careless cleaning of the surface are visually apparent. Figure 22 shows the surface of a motorcycle-helmet visor; scratches in the polycarbonate were made by wiping away road dirt. These scratches are 30–50 mm away from the wearer's eyes and at night produce highly distracting scattering from oncoming headlights. The U.S. military has a specification MIL-13830A for the allowable size and density of such defects; thus a no. 40 dig has a 40-μm dia, whereas a no. 30 scratch has a width of 3 μm. This does not provide a measure of the scattering power because the surface at the bottom of the scratch may be smooth or rough, depending on the cutting characteristics of the material involved. The polycarbonate scratches in Figure 22 have a rough surface.

Internal Scattering of Transmitted Light.   Light-scattering theory is an area of physics in which there is considerable activity, although the problems are rarely concerned with plastics. A variety of different theoretical approaches deal with different categories of polymer microstructure.

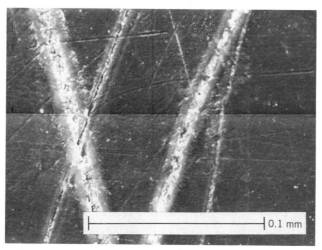

Fig. 22. Scanning electron micrograph of scratches on the outer surface of a motorcycle-helmet polycarbonate visor.

Independent Light Scattering from Spherical Particles. If the scattering centers are sufficiently separated, the total scattering is the sum of those calculated independently for each particle. Mie light-scattering theory (25) considers spherical particles of diameter d and relative refractive index $n = n_s/n_m$, where n_s is the refractive index of the sphere and n_m that of the matrix. A scattering efficiency factor K is defined as the total scattered energy divided by the energy incident on the surface of the particle. Figure 23 shows how the scattering effi-

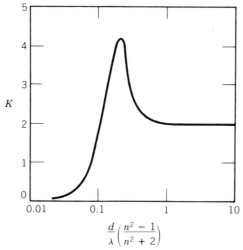

Fig. 23. Scattering coefficient K for a single sphere of relative refractive index n vs the ratio of the sphere diameter to the wavelength of the light (26).

ciency factor varies with the relative size of the sphere (24). There are three regions of interest:

When the particle diameter d is less than 10% of the wavelength of light λ, Rayleigh scattering is observed in which the efficiency factor is given by

$$K = \frac{8}{3}\pi^4 \left(\frac{d}{\lambda}\right)^4 \left(\frac{n^2 - 1}{n^2 + 2}\right)^2 \tag{51}$$

When the diameter is of the same order as the wavelength, a peak in the scattering efficiency is observed. White pigments for plastics contain particles of this size in this range.

When the diameter is large compared with the wavelength, the scattering efficiency is ~2.

The relative refractive index n affects the scattered intensity; hence TiO_2 pigment is used in white paint because the refractive index is high at 2.70. The microstructure of rubber-toughened glassy plastics is usually one of rubber spheres in a glassy matrix. In most cases, the refractive indexes of the two phases differ and the diameter of the rubber spheres is approximately 1 μm to maximize the toughness; consequently, materials such as ABS are opaque.

Refractive-index Fluctuations in Glassy Plastics.   Amorphous polymers do not contain a regular crystal lattice. It is often assumed that fluctuations in the refractive index or in the orientation of optically anisotropic molecular groups can be described by a radial correlation function $\gamma(r)$; there are local deviations η in the polarizability from its mean value $\bar{\alpha}$. If the values of the deviations are η_i and η_j at two positions i and j an r-distance apart (Fig. 24), the correlation function is defined as

$$\gamma(r) = \frac{\langle \eta_i \eta_j \rangle_r}{\langle \eta^2 \rangle_{av}} \tag{52}$$

where $\langle\ \rangle_r$ indicates that an average is taken at all pairs of points a distance r apart. The correlation function has a value 1 at $r = 0$ and is often assumed to have the exponential form

$$\gamma(r) = \exp\left(-r/a\right) \tag{53}$$

where the constant a is the correlation length. The scattering theory of Debye and Bueche (27) predicts that the scattered intensity at an angle θ to the incident beam has an intensity I given by

$$I \sim \langle \eta^2 \rangle_{av} \int_0^\infty \gamma(r) r^2 \frac{\sin Sr}{Sr}\, dr \tag{54}$$

where S is equal to $4\pi/\lambda \sin(\theta/2)$. Experimental results show that the correlation length is in the range of 100–200 nm for poly(methyl methacrylate) and polycarbonate, but much smaller for polystyrene. The root-mean-square refractive-index fluctuation is ca 2×10^{-4} for PMMA. These values indicate that the light-scattering intensity is a factor of ca 40 larger than for silicate glasses. Consequently, in fiber-optics applications, polymeric fibers are only used for transmission distances of 1–2 m, whereas silicate glass fibers are used for transmission distances of kilometers.

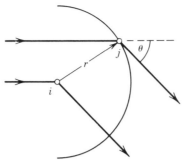

Fig. 24. Scattering caused by the correlation in polarizability deviations at points i and j, a distance r_{ij} apart.

Refractive-index and Orientation Fluctuations in Semicrystalline Polymers. The marked differences between the densities of the crystal and amorphous polyolefins have been pointed out previously. Thus the crystals that exist in the form of 10-nm thick lamellae with lateral dimensions of about 1 μm exhibit a higher refractive index than the amorphous phase. Polymer crystals are covalently bonded in the c direction, but have van der Waals interactions in the perpendicular directions. Hence they are birefringent, with, for example, $n_c - n_a = 0.031$ for the polypropylene crystal (28). Furthermore, the crystals usually form part of a spherulitic microstructure, with the polymer chain or c axes of the crystals in the tangential direction in the spherulite.

The simplest spherulite model for predicting light scattering is a sphere of radius r with a different polarizability in the radial direction α_r than in the tangential direction α_t (29). For vertically polarized incident light (Fig. 25), this predicts the intensities of the vertically polarized and horizontally polarized scattered light as

$$I_{VV} \sim \cos^4 \mu \, (\alpha_t - \alpha_r)^2 + \cdots$$
$$I_{HV} \sim \sin^2 \mu \, \cos^2 \mu \, (\alpha_t - \alpha_r)^2 \tag{55}$$

respectively, where μ is the azimuthal angle in the scattered pattern. The angular dependence of μ explains why the VV and HV patterns are two- and four-leaved, respectively. The HV component has a maximum intensity in the θ-direction at a value

$$\sin (\theta_m/2) = \frac{1.025}{\pi} \left(\frac{\lambda}{r} \right) \tag{56}$$

Consequently, the scattering maximum moves to larger θ values as the mean-spherulite radius decreases. For unpolarized incident light, the scattering pattern becomes axially symmetric with an intensity independent of the angle μ.

Transparent products can be made from semicrystalline plastics by several methods. If the article is injection molded, such as a laboratory beaker of 1-mm thickness, a polymer such as poly(4-methyl-1-pentene) must be selected in order to reduce the refractive-index difference between the crystalline and amorphous

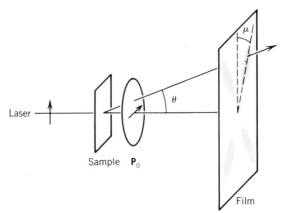

Fig. 25. Use of vertically polarized light from a laser to determine the HV scattering pattern from a spherulitic polyethylene film.

phases. If the article is thinner, such as injection blow-molded PET bottles for carbonated drinks or oriented polypropylene packaging films, larger refractive-index differences can be tolerated as long as the microstructure is suitable. This is achieved in two ways. First, crystallization is induced at a low temperature upon heating from a glassy or partly disordered state. The PET bottle preforms are cooled so rapidly when injection molded that they are glassy. On subsequent reheating to 100°C, very small crystals form. Second, the melt is stretched biaxially before crystallization occurs, resulting in a nonspherulitic microstructure and a limited range of crystal orientations (Fig. 26).

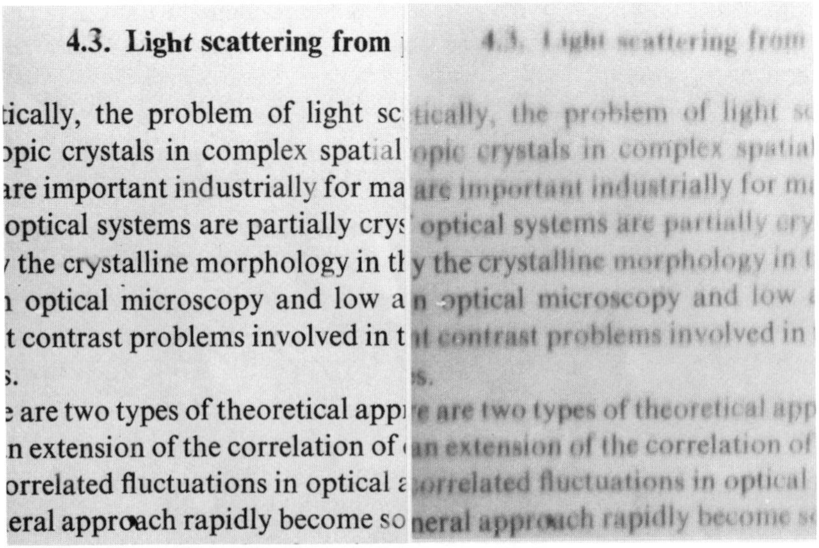

4.3. Light scattering from

tically, the problem of light sc
opic crystals in complex spatial
are important industrially for ma
optical systems are partially crys
the crystalline morphology in the
optical microscopy and low a
t contrast problems involved in the
s.

are two types of theoretical app
n extension of the correlation of
orrelated fluctuations in optical a
eral approach rapidly become so

Fig. 26. Printed text viewed through the biaxially stretched PET wall of a carbonated drink bottle 50 mm away. The right-hand photograph shows the same text viewed through a high density polyethylene bottle wall, with spherulitic crystallization and hence more light scattering.

Test Methods

The tests described are those specified by the American Society for Testing and Materials (ASTM) (30). Similar tests exist in British and other European standards.

Specular Gloss D523-80. A primary standard of highly polished black glass of refractive index 1.567 is provided with a specular gloss of 100. Working standards of ceramic tiles or ground glass are used to check the instrument adjustment in the gloss range of the panel being tested. There are strict specifications regarding the angular range of the source and the receptor. For example, the 60° receptor aperture subtends $4.4 \pm 0.1°$ in the plane of measurement and $11.7 \pm 0.2°$ perpendicular to the plane of measurement; the light must not be scattered through more than 2.2° in the plane of measurement.

Haze and Transmittance. A version of the integrating sphere hazemeter is shown in Figure 27. The sphere of diameter (200–250 mm) has a matt-white internal surface. The photocell detects the uniform intensity of scattered light inside the sphere. A circular light trap captures all the incident beam, if no specimen is present, with an annular gap of $1.3 \pm 0.1°$ between the edge of the light beam and the light trap. As the light trap subtends 8° at the specimen, the light beam at the exit port subtends 5.4° at the specimen location. Measurements of the light intensity I are made under the following conditions:

Measurement	At entrance	At exit
incident light I_1		reflectance standard or sphere rotated
total light transmitted by specimen I_2	specimen	reflectance standard or sphere rotated
light scattered by instrument I_3		light trap
light scattered by instrument and specimen I_4	specimen	light trap

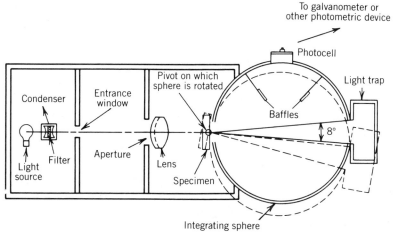

Fig. 27. Integrating-sphere hazemeter (ASTM D1003-61): —, position 1; ---, position 2.

From these measurements the total transmittance T_t is calculated as

$$T_t = I_2/I_1 \qquad (57)$$

and the diffuse transmittance T_d as

$$T_d = \frac{I_4 - I_3(I_2/I_1)}{I_1} \qquad (58)$$

In equation 58 the value of I_3 is reduced in the ratio I_2/I_1 to allow for the fact that the intensity entering the sphere is reduced by the presence of the specimen. Finally, the percent haze is calculated:

$$\% \text{ haze} = 100T_d/T_t \qquad (59)$$

The haze measurement sums up the intensity of light scattered through angles of ca 4–90°; the lower limit is imprecise because the incident beam is allowed to contain rays that deviate up to 3° from the axis of the beam.

Haze Increases as a Result of Surface Abrasion.   The most common method of abrasion testing is the Taber method D1044-82, in which two rubber wheels containing fine corundum particles rotate freely on the plastic disk being tested, which rotates at ~1 rps for 100 revolutions. The wheels with a 50-mm dia and 12.5-mm thickness are offset at ±30° from the direction of rotation of the plastic and are loaded with a force of 5 N (0.51 kgf). The abrasive action produces two sets of scratches at 60° to each other over a 10-mm wide annular track of about 80-mm dia. In British Standard AU182:1982 for plastic glazing for road vehicles, it is required that the haze value must not increase by more than 5% as a result of the abrasion test. The Taber method tracks only scattered light at relatively small angles (Fig. 28), the cross-shaped scattering pattern corresponding to the crossed sets of scratches. It is important that the angular gap around the incident light beam is maintained at 1.3° in the integrating sphere hazemeter to avoid errors of measurement. Observations of the scratched surfaces (31) show that there are scratches at spacings of ca 100 μm when the haze increases by 5%; a windscreen scratched to this extent would be unacceptable. The scratches in PMMA were mainly due to the removal of polymer by a series of brittle fractures, whereas the scratches in polycarbonate were ploughed grooves with mounds of material on either side. Coating of either material with a 5–10-μm thick layer of a polysiloxane thermoset significantly improved scratch resistance.

Clarity of Images Seen Through a Plastic Sheet.   A number of empirical tests use charts containing parallel black lines at different spacings viewed through a plastic sheet. The angular resolution of the closest spaced lines that can be resolved is a measure of the clarity. However, the resolution drops markedly as the plastic is moved away from the chart, and the specimen-to-chart distance must be specified. The physiological basis of clarity measurements is discussed in Ref. 32. Under ideal conditions, the human eye can distinguish two bright lines if their angular separation is 0.07° and if the intensity of light midway between them is <80% of the maximum intensity. Consequently, light scattered at angles between 0.07 and 2.5° may greatly affect the clarity of plastic film without contributing to the haze value.

Other optical properties are important in the transmission of information. Thus ASTM D881 considers the deviation of the line sight when an oblique sheet,

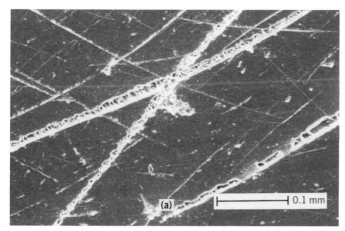 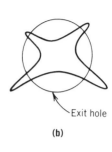

Fig. 28. (a) Surface of coated polycarbonate after 100 revolutions of a Taber abrasion meter. (b) Sketch of the light-scattering pattern from a similar specimen, compared with an exit hole that is 1.3° outside the main beam.

such as an aircraft windshield, is to be looked through. British Standard AU182: 1982 considers optical distortions in the field of view. An image of parallel lines is projected through the plastic screen onto a similarly ruled screen 4 m away. The test can detect angular deviations of 0.04°.

Applications

Glazing and Barrier Films. In a wide variety of applications, plastic sheet or film acting as a weather barrier must have high light transmission. The clarity of the sheet may be important in some cases and not so in others. In most applications the optical properties of the plastic are vital among other property requirements. The chief rival material of plastics is glass, which is not weakened by uv radiation and is not flammable. Properties of glazing plastics are given in Table 8 (33). The light-transmission values mainly reflect the number of air–plastic interfaces. The twin-walled polycarbonate extrusion typically has 0.5-mm thick walls and 0.5-mm thick ribs at ca 20-mm intervals. This hollow cellular construction gives a much better ratio of bending stiffness to weight than the solid 3-mm sheet, but at the expense of clarity. Glassy plastics are much more flexible than glass of the same thickness (the Young's moduli are 3 and 70 GN/m^2 (0.4 × 10^6 and 10 × 10^6 psi)) and need support at the edge in a different way. The traditional linseed oil putty cannot cope with movements caused by wind loads or temperature changes, and extruded plasticized PVC or rubber seals must be used. Conversely, the high elastic strains tolerated by plastics indicate that initially flat sheet can be bent into curved shapes and held in place with metal frames (34).

The impact strength of plastics is usually higher than that of glass; that of polycarbonate is excellent unless sharp cracks are present. With incorporation of a wire mesh, the plastic sheet is practically unbreakable; similarly the 0.5-

Table 8. Properties of Plastic Sheet for Outdoor Applications

Property	PMMA	PC	10-mm twin-wall PC	PVC	PVC + wire mesh
light transmission of 3-mm sheet, %	92	88	70–80	85	72
notched Izod impact strength, J/m[a]	15–20	650–950		20–500	
service temperature range, °C	− 30 to 80	− 50 to 120		− 30 to 55	− 30 to 55
weathering	good	average	average	average	average
flammability, surface spread of flame[b]	Class 3	Class 1	Class 1	Class 1	Class 1
cost relative to PVC per m²	1.5	2.5	1.3	1.0	2.4

[a] To convert J/m to ftlbf/in., divide by 53.38.
[b] BS 476:Pt 7.

mm thick walls of twin-walled polycarbonate are too thin to allow plane-strain-crack growth. The upper limiting temperature for short-term use of plastics is ca 20° below T_g (Table 8). However, slow degradation processes, such as oxidation or hydrolysis, may severely reduce the temperature that can be withstood continuously for 10 years, ie, about 65°C for polycarbonate. Weathering can produce a gradual yellowing of the plastic or the degradation of the mechanical properties to the point where brittle failure occurs under the usual service loads. PMMA has the best weathering characteristics of the sheet plastics in Table 8, but its poor flame resistance was emphasized by a number of fires and it is no longer used for the glazing of public buildings.

If the plastic sheet is manufactured by extrusion, the quality critically depends on the operation of the extrusion line (35). For instance, in order to minimize the number of foreign particles passing the filter pack, no reground material should be used. The polymer must be thoroughly dried and a degassing screw used in the extruder to minimize the number of bubbles in the sheet. Careful control of the melt temperature and the die-lip settings is necessary to minimize the thickness variation across the sheet; this can be reduced to 40 μm in a 3-mm thick sheet across a 2-m width. The surface roughness of the steel-smoothing rolls is critical in determining the surface quality of the sheet.

Horticultural Glazing.   A different set of criteria apply when plastics are used to replace glass in greenhouses. For cost reasons, film is used instead of sheet, and opaque semicrystalline film may be preferred because of its light-diffusion properties (36). For the true greenhouse effect to operate, the glazing must be transparent in the visible and near ir regions to allow photosynthesis, and yet be opaque in the far ir to reduce the loss of energy radiated by the warm soil. A 3-mm glass sheet retains 41% of the radiated energy, and plastics such as PET film can perform almost as well (33%). Losses by conduction can be relatively high with thin-plastics film; in cases where heat loss is important, double-walled polycarbonate extruded sheet can be used. It has a U value of 3.4 W/(m²·K) for a 10-mm thick sheet, compared with 5.5 W/(m²·K) for 4-mm thick glass (37).

The uv degradation of mechanical properties becomes more important for film. Polyethylene film can be used with tunnel greenhouses for only one or two seasons. An alternative is plasticized, uv-stabilized PVC reinforced with a polyester net to prevent tear propagation. More expensive films such as PET or poly(vinyl fluoride) have higher tensile strengths and better weathering characteristics, and may be economically justified for applications such as heated greenhouses.

Headlight Lenses. The incentive to use plastics to replace glass for automobile headlight lenses are twofold: as a tougher material, a plastic lens can be thinner and, as it is less dense, there is a weight saving. Furthermore, by injection molding, the lens and prism elements can be made to a higher precision than is possible by pressing glass, and a better optical performance is obtained (38). Adequate temperature resistance may be a problem. A 60-W bulb operating behind a mud-spattered lens can generate temperatures of 115°C where the light intensity on the lens is highest. Therefore, high temperature plastics such as polycarbonate must be used. Another problem is surface abrasion or spilled fuel, which may cause damage and light scattering.

In a proposed test (38) a mixture of 15% diatomaceous earth, 10% quartz powder, 30% n-heptane, 25% xylene, and 20% toluene is rubbed onto the lens for several minutes at a pressure of 20 kN/m^2 (2.9 psi); haze should not increase by more than 4%. Uncoated polycarbonate would not pass this test, and hard, solvent-resistant coatings must be developed for this application.

Aircraft Windows. A case history (39) of the failure of PMMA passenger windows in a commercial aircraft after several thousand flights is a good example of a stressed glazing application. The windows were triple glazed, with the inner pane being an unstressed scratch-resistant panel and the middle pane taking the pressure differential of 65 kN/m^2 (9.4 psi). In service the middle and outer panes crazed, reducing visibility to some extent and this should have given a warning that the design was unsatisfactory. A combination of a biaxial tensile stress due to the outward bending of the pane and some environmental agent, possibly absorbed water, caused the crazing. The solution to the problem was a biaxially stretched PMMA sheet in place of the unoriented sheet. Stretched PMMA is more resistant to crazing (qv) and crack propagation.

Electronic and Information Technology. Plastics are used increasingly in the manufacture of printed-circuit boards and other electronic devices. Although the plastic may no longer be present in the final product, it plays a vital part in the production process. An example is a dry-film photoresist system (40), which requires a photopolymer that polymerizes and cross-links when exposed to uv light. Figure 29 shows the sequence of operations in a print-and-etch method of producing a copper circuit on a circuit board. The photopolymer layer, typically 38-μm thick, is exposed to uv light and an image of the phototool is formed. The intervening PET support film transmits at least 80% of the uv light. Subsequently, the PET film is stripped off and the image developed by removing the uncross-linked polymer with a spray of trichloroethane solvent. The unwanted parts of the copper layer can be etched away before the cross-linked photoresist is stripped with a more powerful solvent such as methylene chloride (see LITHOGRAPHIC RESISTS).

Plastics such as polycarbonate are being used more and more for information

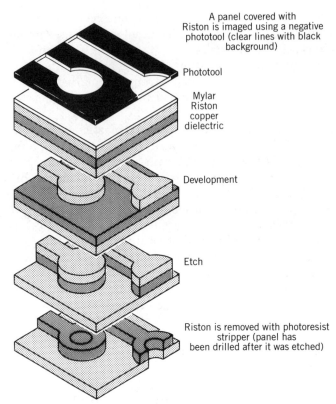

A panel covered with
Riston is imaged using a negative
phototool (clear lines with black
background)

Phototool

Mylar
Riston
copper
dielectric

Development

Etch

Riston is removed with photoresist
stripper (panel has
been drilled after it was etched)

Fig. 29. Print-and-etch sequence of operations to produce a copper circuit on a printed-circuit board using Riston dry-film photoresist (39).

storage in the form of 300-mm dia video disks or 120-mm dia audio disks (Fig. 30) (41). The plastic is employed for its ability to reproduce the surface details of a metal master using the injection-molding process. The information is stored in the form of pits of different lengths and separations. The two channels for stereosound are sampled at 44 kHz and are presented as 8-bit numbers. An additional 6 bits are added to each number to correct errors, allow synchronization, and control the laser-tracking system. A binary 1 is stored as the step at the end of a bit, and two such steps must be separated by at least three bits to avoid interference. The consequence is that in order to store 1 h of sound, the steps in level must occur in a length <0.3 μm and the spacing of the tracks is 1.7 μm. The information is then read by a small AlGaAs laser of 780-nm wavelength at a rate of 4.3 Mbits/s. A lens of numerical aperture >0.4 is used at a high magnification, giving a very limited depth of focus. Consequently, the disk must be perfectly flat. The original recording of the audio or video signal is by a photoresist process. To prepare computer data disks, a focused laser burns a hole in a light-absorbing layer on the near side of a transparent substrate; this produces holes of ca 0.6-μm dia (see OPTICAL DATA STORAGE in the Supplement).

Transparent plastic film is often used to protect switches (42) with 125-μm PET or PC membranes, which can be screen-printed with key labels. The switches usually consist of three layers: a bottom layer that is screen-printed like a printed

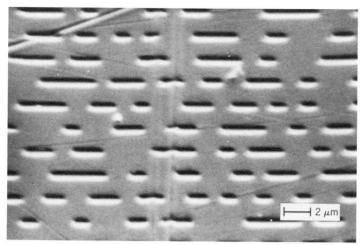

Fig. 30. Scanning electron micrograph of the surface of an audio disk.

circuit board, a middle layer that has die-cut holes beneath each key position, and a top layer that has a conducting layer beneath the key label. On pressing the membrane key, the top layer bends and makes an electrical contact between two elements of the bottom layer. The control panels are easily cleaned and the printed legends cannot be worn away.

Plastic Lenses. *Design and Mounting.* Lens design is usually carried out by computer. The lenses used in cameras and projectors often use large apertures for high light intensities, and large fields to produce a large image from a small lens. Consequently, image aberrations play an important part in lens design (43). The monochromatic aberrations can be divided into those that blur the image (spherical aberration, coma, and astigmatism) and those that deform the image (field curvature and distortion of the pincushion or barrel types). Minimization of these aberrations and of the chromatic aberrations resulting from the wavelength dependence of refractive index requires glasses with a range of refractive indexes and dispersions. Few glassy plastics provide such a range, and designs must still be based on silicate glasses.

Some single-lens types are shown in Figure 31. Since injection molding produces lenses in their finished form, the production costs of aspherical lenses (with parabolic, hyperbolic, or other surface shapes) or Fresnel prisms are the same as those of spherical-surface lenses. Since the cooling time of a molding increases as the square of the molding thickness, it is desirable to keep the lenses thin. The percent shrinkage increases in thicker regions of a lens, especially when an isolated island of melt forms in the center of a double-convex lens gated at the edge. Consequently, it is preferable to manufacture meniscus lenses in which the thickness variation is minimized (2). Figure 32a shows a four-element lens design utilizing plastic meniscus lenses. Such shapes are not made of glass because of grinding difficulties. Flat surfaces are difficult to achieve on plastic lenses because the solid skin that develops during molding has little resistance to bending. However, a slightly convex surface skin is much more stable since it must buckle before its curvature can change. Experience with the particular

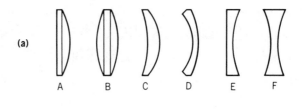

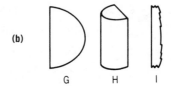

Fig. 31. (a) Spherical lenses: A, planoconvex; B, double convex; C, meniscus (concave-convex) positive; D, meniscus negative; E, planoconvave; F, double concave. (**b**) Nonspherical lenses: G, aspheric; H, cylindrical; and I, central section of a Fresnel lens (2).

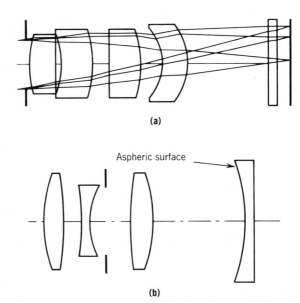

Fig. 32. (a) Four-element, plastic-lens design for an alphanumeric reader; (**b**) a field-flattening lens with an aspheric surface to improve a triplet objective. The precision requirements of the aspheric surface are reduced because the beam diameter is small at the aspheric surface near the focal plane (2).

plastic and lens type is required before lens distortions from molding shrinkage can be compensated for by modifying the mold shape (44).

With aspheric surfaces, it is often possible to reduce the number of lens elements. They are useful for field flattening in camera lenses (Fig. 32b); the image field must coincide with the photographic emulsion both at the center and at the edges of the frame. Fresnel lenses are used when a relatively thin lens must project a parallel beam of light. The prisms can be annular if a unidirectional

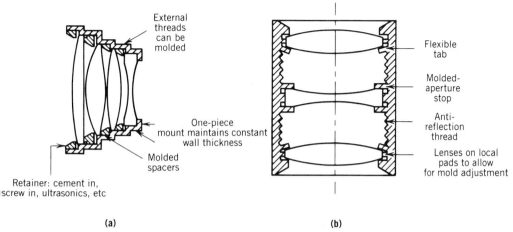

Fig. 33. (**a**) Barrel mount and (**b**) two-part clamshell mount with flexible mounting tabs (2).

beam is required, or take more complex forms if a particular angular spread is required, as in roadside warning lights or automobile headlight lenses.

Injection molding offers the advantage that the lens mounting can be molded integrally with the lens (2). Individual magnifier lenses can be mounted by an integral flange around the edge, which contains slots or elongated holes. For multiple-element lenses, a barrel mount (Fig. 33**a**) is more accurate than a two-part clamshell mount (Fig. 33**b**). The elements can be fixed in place by the standard assembly techniques, such as screw-fittings, solvent-based cements, or welding with heat or ultrasonic power. The mount must taper and have a draft angle to its sides to allow ejection from the mold. In the clamshell type, two identical halves close around the elements to form a complete cylinder (Fig. 33**b**). In order to allow for variations in the lens thicknesses, one side of each mounting must be flexible; however, this is easily achieved with thin plastic parts.

Ophthalmic Lenses. Ophthalmic plastic lenses require similar properties as glazing applications. The chief advantage of plastic is weight reduction (Table 9). The weight is proportional to the density divided by the refractive index for a lens of a given focal length. Reducing the lens weight greatly increases the comfort in wearing the eyeglasses. In addition, plastic lenses offer increased safety against breakage. In impact tests (3) for eye protectors (British Standard 2092), a spherical steel ball is fired at specified sites (Fig. 34) at a specified impact velocity. In BS 2092 there are three levels of protection corresponding to velocities

Table 9. Properties of Ophthalmic Lens Materials[a]

Material	Refractive index n_D	Density, kg/m^3	Weight of lens, g	Impact strength, J[b]
Crown glass, annealed	1.525	2.53	25.7	0.04
PMMA	1.495	1.19	12.3	
CR-39	1.498	1.32	13.6	0.4
PC, coated	1.596	1.20	11.7	22

[a] Ref. 43.
[b] To convert J to ftlbf, divide by 1.36.

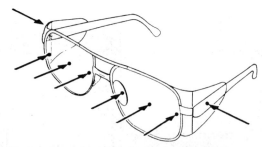

Fig. 34. Specified points of impact in tests (BS 2092) on safety eyeglasses. The lateral-impact sites only apply to the two higher grades of impact resistance (3).

of 12.2, 45.7, and 119 m/s or impact energies of 0.08, 1.29, and 8.00 J (0.06, 0.95, and 5.88 ftlbf) of a 6.5-mm dia ball. The lens must not shatter, deform toward the eye, or be knocked out of the frame. Of the materials in Table 9, only polycarbonate behaves in a ductile manner under extreme impact conditions; in the others a number of radial cracks are propagated. Crown glass, in particular, produces small sharp fragments when growing cracks divide. The main drawback of polycarbonate is lack of abrasion resistance, and if a hard coating is applied, the likelihood of brittle fracture is increased. The ductility of polycarbonate requires a special technique of cutting the lenses to the shape of the spectacle frames (45).

Polycarbonate absorbs strongly in the uv for wavelengths <390 nm; it absorbs 97% of the uv in sunlight, compared with 64% absorbed by Crown glass.

Contact lenses (qv) are invariably made from plastics. Requirements arise because the lens rests on a thin film of tears on the surface of the cornea (46). The plastic must be biocompatible with the cornea and must not contain monomer or catalyst residues that would inflame the tissue. It must be wettable by tears and have a high surface energy, a problem with some silicone lenses. For the cornea to remain transparent, a chemical reaction must occur between the nutrients in the tears and oxygen dissolved in the tear or from the air to keep the cornea partially dehydrated. Therefore, the permeability of the lens to oxygen is important. Contact-lens material is either hydrophilic or hydrophobic (Table 10) (see also HYDROGELS). Hydrophilic lenses are larger and more flexible and drape over the cornea, whereas hydrophobic lenses are smaller and more rigid. The

Table 10. Contact-lens Materials[a]

Materials	Abbreviation	Refractive index n_D	Water content, %
Hydrophilic			
poly(hydroxylethyl methacrylate)	HEMA	1.43–4.44	≤40
HEMA–poly(vinyl pyrrolidinone)	HEMA–PVP	1.43	≤72
PMMA–ethylene glycol dimethacrylic acid	PMMA–EGDMA	1.43	≤80
Hydrophobic			
PMMA		1.49	0
cellulose acetate butyrate	CAB	1.44	2
polydimethylsiloxane	PDMS	1.44	0

[a] Ref. 46.

exception is silicone rubber (PMDS), which has the typical low modulus of a
rubber; it has excellent permeability to oxygen, but is difficult to wet with tears.
The lenses are typically 0.1-mm thick and need to have their inner surface fitted
to the shape of the cornea for at least the central 5 mm. The curvature of the
cornea is measured in two principal meridians. If these values are different, large
hydrophilic lenses with a spherical outer surface cause unacceptable astigmatism.
Consequently, methods of manufacturing such lenses with aspheric shapes are
being developed. Most PMMA lenses are machined with a lathe from blanks and
then polished; the same procedure can be used for dry hydrophilic materials.

Microlenses with Distributed-index Materials. The concept of a material
with a gradient of refractive index in it is not unique to plastics (47); glasses,
silica, and silicon can be treated, but the methods of preparation may be quite
different. A cylindrical rod 150-mm long and 1–3 mm in dia can act as a lens for
light traveling along its length (Fig. 35a). It could be used for internal exami-
nations. Typically, a soft rod is produced in a high refractive-index plastic by
partial thermal polymerization of the monomer at 80°C. This rod is covered by
a lower refractive-index second monomer; diffusion of the second monomer into
the rod and unpolymerized first monomer out of the rod occurs, together with
further polymerization.

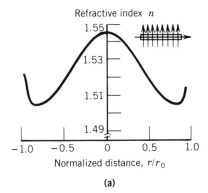

(a)

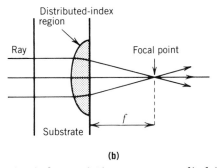

(b)

Fig. 35. (a) Refractive-index variations across a cylindrical-rod lens; (b) method
of measurement of the focal length of a planar distributed-index lens (47).

Systems include diallyl isophthalate ($n_1 = 1.57$) followed by methyl methacrylate ($n_2 = 1.49$), and CR-39 ($n_1 = 1.51$) followed by 1,1,3-trihydroperfluoropropyl methacrylate ($n_2 = 1.42$). The diffusion constants of the monomers are relatively high at low temperatures. A comparison of the measured refractive-index profile with that calculated for a monomer of constant diffusion constant D, gave a value of $D = 2.8 \times 10^{-10}$ m²/s. The refractive-index profile can be expressed as

$$n^2(r) = n^2(0)[1 - (gr)^2 + h_4(gr)^4 + h_6(gr)^6 + \cdots]$$

where r is the radial position in the fiber and $n(0)$ the refractive index at $r = 0$. The constant g in the parabolic term is known as the focusing constant. Theoretically, it has a maximum value at a time $= 0.1\ r_0^2/D$; simultaneously, the fourth-order coefficient h_4 is close to zero. To maximize the resolving power of the rod lens, the coefficient h_4 must be kept below 3 and ideally it should equal ⅔. The diffusion treatments could be modified to optimize the refractive-index profile in the cylindrical lens. A 25-mm long rod has a resolving power of 45/mm.

Another product is a planar microlens (Fig. 35**b**). When the techniques for producing single lenses are perfected, it will be feasible to produce arrays of lenses in a sheet of material that lies on top of the objects to be imaged. A sheet of polymerized diallyl isophthalate was masked with a 3.6-mm dia glass disk in such a way that on exposure to a methyl methacrylate monomer, there would be a lower concentration beneath the mask. A lens of 25-mm focal length was obtained. Smaller masks and a change in the diffusion conditions could produce smaller lenses with a range of focal lengths.

Nomenclature

A	= constant
C	= stress-optical coefficient
D	= dispersion; diffuse reflectance
E	= electric field
F	= force
G	= gloss factor
I	= intensity
K	= stress-intensity factor; scattering efficiency factor
K_1, K_2	= transmittance of polarizer
L	= length
M	= molecular weight
N	= number of bonds
N_A	= Avogadro's number
P	= polarizer matrix; dipole
Q	= quarter-wave plate matrix
R	= retarder matrix
R	= reflectivity
R_a	= roughness
S	= rotation matrix
S	= specular gloss
T	= transmittance
U	= heat-loss constant
V	= Abbe value

W = half width
Z = complex number
a = Jones vector
a = contact width
c = speed of light in material
c_0 = speed of light in vacuum
d = diameter
f = fringe order
m = maximum
m = matrix; mass
n = refractive index
r = radial distance; reflection angle
s = spacing; sphere
t = thickness, time
$\left.\begin{array}{l} x \\ y \\ z \end{array}\right\}$ = Cartesian coordinates
α = polarizability, angle
β = angle
γ = angle, correlation function
δ = retardation; surface roughness
ϵ_0 = permittivity of free space
η = deviation of polarizability
θ = angle
λ = wavelength
μ = azimuthal angle
σ = stress (principal)
τ = shear stress
Ω = angular frequency
ω = angular frequency
ρ = density

BIBLIOGRAPHY

"Optical Properties" in *EPST* 1st ed., Vol. 9, pp. 525–623, by Robert Zand, University of Michigan.

1. C. J. Parker in R. R. Shannon and J. C. Wyant, eds., *Applied Optics and Optical Engineering,* Vol. 7, Academic Press, Inc., Orlando, Fla., 1979, p. 71.
2. *The Handbook of Plastics Optics,* 2nd ed., U.S. Precision Lens, Inc., Cincinnati, Ohio, 1983.
3. *T252 Engineering Materials: An Introduction. Unit 15 Safety Spectacles,* Open University Press, Milton Keynes, UK, 1983.
4. M. Garburny, *Optical Physics,* Academic Press, Inc., Orlando, Fla., 1965.
5. A. Nussbaum and R. A. Phillips, *Contemporary Optics for Scientists and Engineers,* Prentice-Hall, Inc., Englewood Cliffs, N.J., 1976.
6. J. D. Lytle, G. W. Wilkerson, and J. G. Jaramillo, *Appl. Opt.* **18**, 1842 (1979).
7. *LISA Leaflet No. 346/83,* Bayer UK, Ltd., Newburg, UK, 1983.
8. W. A. Shurcliff, *Polarized Light,* Harvard University Press, Cambridge, Mass., 1962.
9. E. G. Coker and L. N. G. Filon, *A Treatise on Photoelasticity,* Cambridge University Press, Cambridge, UK, 1931.
10. S. A. Paipetis and G. S. Holister, eds., *Photoelasticity in Engineering Practice,* Elsevier Applied Science Publishers, Ltd., Barking, UK, 1985.
11. P. S. Theocaris and E. E. Gdoutos, *Matrix Theory of Photoelasticity,* Springer-Verlag, Berlin, 1979.

12. A. S. Redner, *Exp. Mech.* **14,** 486 (1974).
13. A. Kuske and G. Robertson, *Photoelastic Stress Analysis,* John Wiley & Sons, Inc., Chichester, UK, 1974.
14. H. E. Brinson, *Deformation and Fracture of High Polymers,* Plenum Publishing Corp., New York, 1972, p. 403.
15. R. B. Agarwal and L. W. Teufel, *Exp. Mech.* **23,** 30 (1983).
16. D. S. Dugdale and A. Ruiz, *Elasticity for Engineers,* McGraw-Hill Inc., New York, 1971.
17. N. J. Mills, *J. Mater. Sci.* **17,** 558 (1982).
18. K. Ito, *Exp. Mech.,* 373 (1962).
19. N. J. Mills, *Eng. Fract. Mech.* **6,** 537 (1974).
20. D. S. Dugdale, *J. Mech. Phys. Solids* **8,** 100 (1960).
21. B. J. Tighe in D. T. Clark and W. J. Feast, eds., *Polymer Surfaces,* John Wiley & Sons, Inc., New York, 1978, Chapt. 14.
22. *Talysurf 5 System Handbook,* Rank Taylor Hobson, Ltd., Leicester, UK, 1977.
23. J. M. Elson, H. E. Bennett, and J. M. Bennett in Ref. 1, Vol. 7, Chapt. 7.
24. H. E. Bennett and J. O. Porteus, *J. Opt. Soc. Am.* **51,** 123 (1961).
25. M. Kerker, *The Scattering of Light,* Academic Press, Inc., Orlando, Fla., 1969.
26. H. E. Bennett, *Opt. Eng.* **17,** 480 (1978).
27. P. Debye and A. M. Bueche, *J. Appl. Phys.* **20,** 518 (1949).
28. R. J. Samuels, *J. Polym. Sci. Part A* **3,** 1741 (1965).
29. R. S. Stein in R. W. Lenz and R. S. Stein, eds., *Structures and Properties of Polymeric Films,* Plenum Publishing Corp., New York, 1973, p. 1.
30. *Plastics, 1985 Annual Book of ASTM Standards,* American Society for Testing and Materials, Philadelphia, Sect. 8.
31. J. Hennig, *Kunststoffe* **71,** 103 (1981).
32. F. L. Binsbergen and J. Van Duijn, *J. Appl. Polym. Sci.* **11,** 1915 (1967).
33. *Br. Plast. Rubber,* 20 (Nov. 1983).
34. A. Sternfield, *Mod. Plast. Int.,* 36 (Mar. 1982).
35. W. Hinterkeuser, *Ind. Prod. Eng.* **8,** 107 (1984).
36. S. Hayes, *Br. Plast. Rubber,* 14 (May 1984).
37. N. J. Mills, *Plastics,* E. Arnold, London, 1986.
38. R. Ropke in *Plastics for External Car Components,* VDI-Verlag GmbH, Düsseldorf, FRG, 1980.
39. *Open University Course T351, Materials Under Stress—Units 10, 11,* Open University Press, Milton Keynes, UK, 1976.
40. *The Riston Dry Film System,* E.I. du Pont de Nemours & Co. Int. SA, Geneva, 1984.
41. G. Bouwhuis and J. J. M. Braat, "Recording and Reading Information on Optical Discs" in Ref. 1, Vol. 9, 1983.
42. *Membrane Touch Switches of DuPont Mylar,* E.I. du Pont de Nemours & Co. Int. SA, Geneva, 1984.
43. E. Hecht and A. Zajac, *Optics,* Addison-Wesley Publishing Co., Reading, Mass., 1974, Chapt. 6.
44. B. Welham, "Plastics Optical Components" in Ref. 1, Vol. 7.
45. S. Herbert, *Ind. Diamond Rev. Part 2,* 77 (1984).
46. J. W. Blaker, "Ophthalmic Optics" in Ref. 1, Vol. 9, 1983.
47. K. Iga, Y. Kokubun, and M. Oikawa, *Fundamentals of Microoptics,* Academic Press, Inc., Tokyo, 1984.

N. J. MILLS
University of Birmingham, UK

OPTICAL PROPERTIES, BIREFRINGENCE.   See Supplement.

OPTICAL ROTATORY DISPERSION.   See OPTICALLY ACTIVE POLYMERS.

ORGANOALUMINUM COMPOUNDS. See ZIEGLER-NATTA CATALYSTS.

ORGANOLITHIUM COMPOUNDS. See ANIONIC POLYMERIZATION.

ORGANOMETALLIC CATALYSTS. See ANIONIC POLYMERIZATION; COORDINATE POLYMERIZATION; ZIEGLER-NATTA CATALYSTS.

ORGANOMETALLIC POLYMERS

Macromolecules generally contain fewer than 10 different elements (C, H, N, O, S, P, halides). There are available well over 40 additional elements, many metals, which can be built into polymers. Metals often can exist in several different oxidation states. Geometries and reactivities depend on such factors as the nature and extent of substitution. Potentially then, a large number of polymer families exist in theory, limited only by intention and imagination. Organometallic polymers are those containing metals either in the backbone chains or pendent to them. The metals may be connected to the polymer by σ or π bonds to carbon, coordination bonds to elements containing free electron pairs, or, less commonly, by σ or π bonds to other elements.

Organometallic polymers can be prepared by addition, condensation, coordination, ring-opening, and other common modes of polymerization. For example, poly(η^5-vinylcyclopentadienyltricarbonylmanganese) (**1**), in which the metal is bound by π bonding to the cyclopentadienyl ring carbons, is prepared by the radical-initiated addition polymerization of (η^5-vinylcyclopentadienyl)tricarbonylmanganese (1,2).

Titanium polyether (**2**) is prepared by interfacial condensation of dicyclopentadienyltitanium dichloride with hydroquinone; the titanium atoms are part of the polymer chain (3,4).

A preformed polymer containing reactive pendent sites can react to bind a metal into the polymer structure (**3a** and **b**) (5).

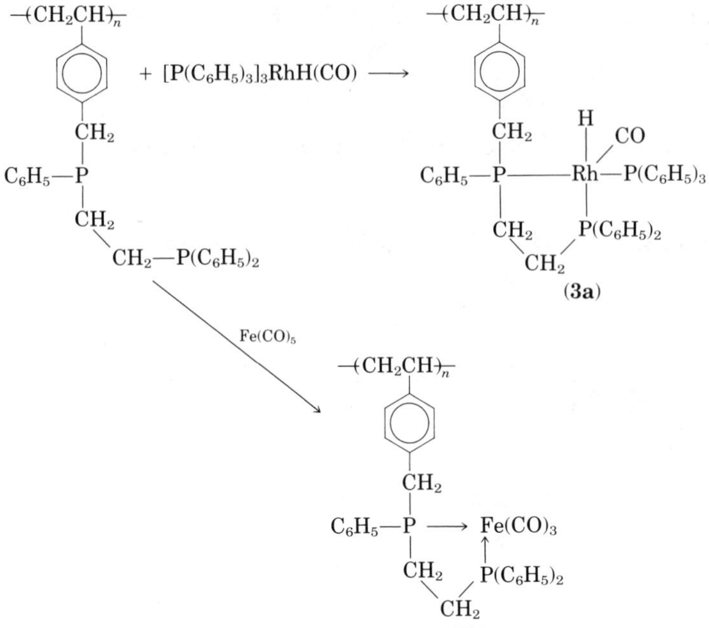

$$-(CH_2CH)_n-$$

with pendant C$_6$H$_5$—P group and $+ [P(C_6H_5)_3]_3RhH(CO) \longrightarrow$

$$\begin{array}{c} -(CH_2CH)_n- \\ CH_2 \\ C_6H_5-P \\ CH_2 \\ CH_2-P(C_6H_5)_2 \end{array}$$

$$\begin{array}{c} -(CH_2CH)_n- \\ CH_2 \quad H \quad CO \\ C_6H_5-P-Rh-P(C_6H_5)_3 \\ CH_2 \quad P(C_6H_5)_2 \\ CH_2 \end{array}$$

(3a)

Fe(CO)$_5$

$$\begin{array}{c} -(CH_2CH)_n- \\ CH_2 \\ C_6H_5-P \longrightarrow Fe(CO)_3 \\ CH_2 \quad \uparrow \quad P(C_6H_5)_2 \\ CH_2 \end{array}$$

(3b)

Preformed polymers may react with metals to give cross-linking sites which, in turn, result in elastomeric properties **(4) (6)**.

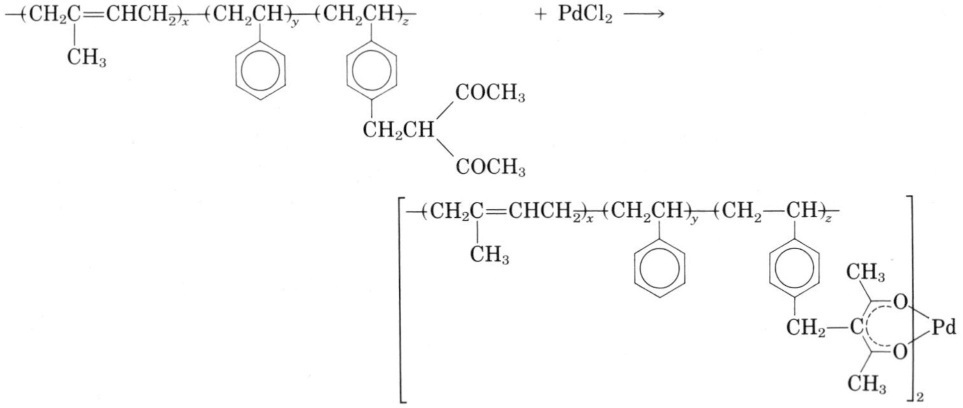

$$-(CH_2C=CHCH_2)_x-(CH_2CH)_y-(CH_2CH)_z- \quad + PdCl_2 \longrightarrow$$

with CH$_3$ and COCH$_3$ / CH$_2$CH / COCH$_3$ groups

$$\left[-(CH_2C=CHCH_2)_x-(CH_2CH)_y-(CH_2-CH)_z- \right]$$

with CH$_3$, and CH$_2$—C(CH$_3$, C=O, O)Pd, CH$_3$ group, subscript 2

(4)

Reaction of benzene-1,2,4,5-tetrathiol with Ni(OOCCH$_3$)$_2$ (or other divalent metal salts) gives the potentially conducting substance poly(benzodithiolene) **(5) (7)**.

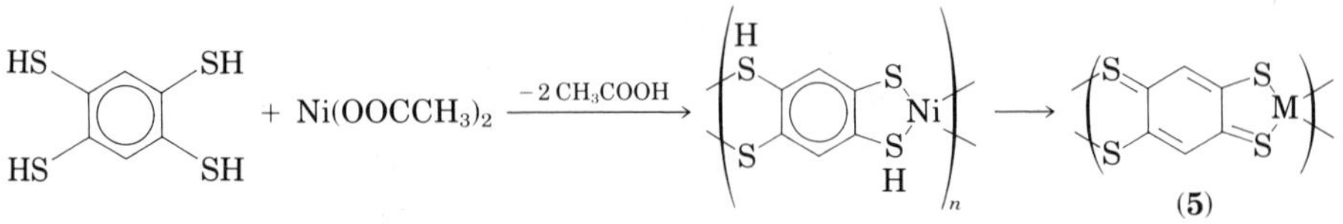

HS—(benzene ring)—SH with HS, SH $+ $ Ni(OOCCH$_3$)$_2 \xrightarrow{-2 CH_3COOH}$

$$\left(\begin{array}{c} H \\ S \quad S \\ \text{(ring)} \quad Ni \\ S \quad S \\ H \end{array} \right)_n \longrightarrow \left(\begin{array}{c} S \quad S \\ \text{(ring)} \quad M \\ S \quad S \end{array} \right)$$

(5)

This is an example of the use of a conjugated ligand as a direct bridge between metal centers. Interesting properties such as electrical conductivity, paramagnetism, ferromagnetic and/or antiferromagnetic interactions, electroactivity (allowing applications in electrode materials), and imposed mixed valency can result from these structures (see COORDINATION POLYMERS).

The Wurtz reaction on dichlorosilanes leads to polysilane (**6**), where all the polymer chain atoms are silicon (8,9). Molecular weights of 400,000 have been obtained (9), and these represent one of the few classes of polymers other than silicones (qv), phosphazenes, and $(SN)_x$ without carbon in their backbones.

$$Cl-\underset{\underset{\displaystyle R_2}{|}}{\overset{\overset{\displaystyle R_1}{|}}{Si}}-Cl + 2\,Na \longrightarrow -\!\!\left(\underset{\underset{\displaystyle R_2}{|}}{\overset{\overset{\displaystyle R_1}{|}}{Si}}-\underset{\underset{\displaystyle R_2}{|}}{\overset{\overset{\displaystyle R_1}{|}}{Si}}\right)_{\!\!n}\!\!-\ +\ 2\,NaCl$$

$$(\mathbf{6})$$

In this article organometallic polymers are defined as those containing true metals. With the exception of polysilanes, polysiloxanes and other silicon, boron, or metalloid-containing polymers are not considered as organometallic polymers here. However, a few tin-containing polymers will be mentioned. The structural varieties available within organometallic polymers are very large. Just within the subclass of coordination polymers, the structural variety is great. For example, three-dimensional networks abound and are represented by the catana-μ-(N,N'-disubstituted dithioxamido)copper complex polymer (**7**), resulting from the reaction of Cu(II) with dithiooxamides.

$$(\mathbf{7})$$

Such materials are amorphous and insoluble (10,11). Iron complexes of the dianions of oxalic acid and 2,5-hydroxyquinone, such as (**8**) and (**9**), are linear-chain and planar-network structures, respectively (12).

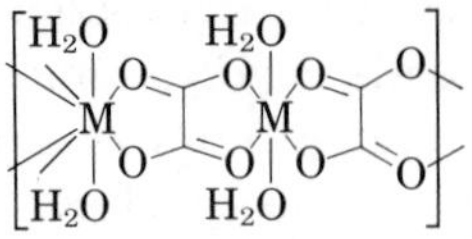

(8) 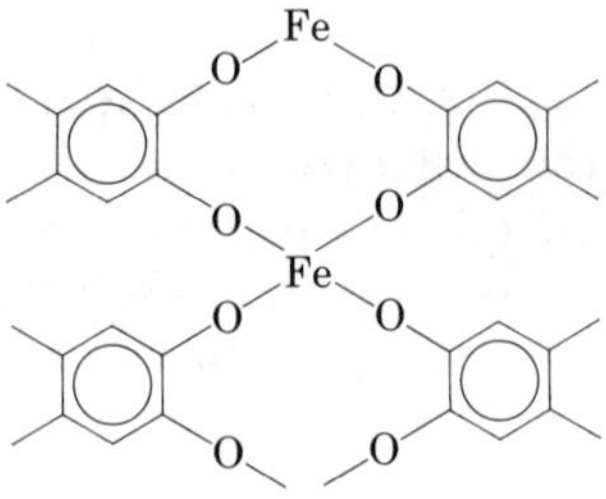   (9)

The metal phthalocyanine polymer (**10**), bridged by pyrazine groups, resembles a linear shiskabob structure (Fig. 1) (13). Reaction of ferrous phthalocyanine with pyrazine gives the dark violet solid (**10**). Other members of this group have metal and oxygen atoms alternating in the backbone (see PHTHALOCYANINE POLYMERS). 5-Phenyltetrazolate (**11**) reacts with metals from all three transition-metal series to give coordination polymers. The Ni^{2+} and Fe^{2+} adducts have been shown to have polymeric structure (**12**). These polymers give extremely viscous aqueous solutions from which threads and flexible sheets can be made (14).

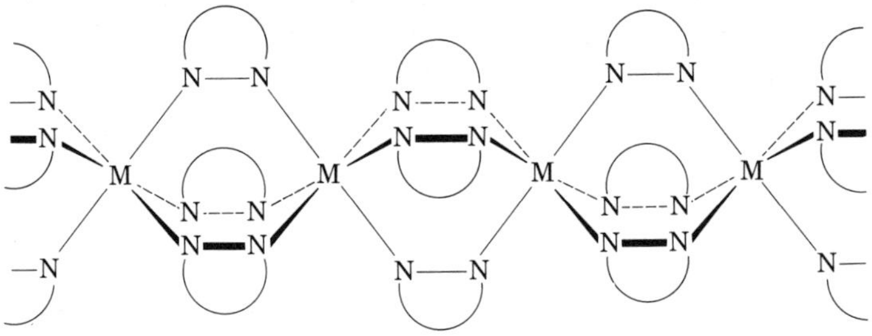

(12)

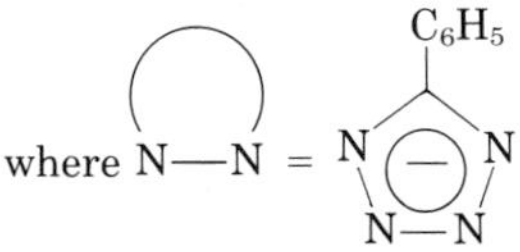

where N—N =

(11)

Classification. Organometallic polymers can be classified into condensation polymers, addition polymers, coordination polymers, and others based on their mode of preparation. For example, preformed polymers containing ligands to which metals are later complexed are considered in the coordination-polymer classes. Thus polymers (**3**) and (**4**) are coordination polymers, polymer (**1**) is an addition polymer, and (**2**) and (**6**) are condensation polymers. Polymer (**5**) is a coordination polymer. Organometallic polymers have been reviewed (15–20), as have polymer-bound metal complexes in catalysis (21–29), preparation of vinyl organometallic polymers (30), the preparation of organometallic polymers by polycondensation (31–33), conductivity of organometallic polymers (34–37), metallocene polymers (38–40), and organotin polymers for antifouling coatings (41).

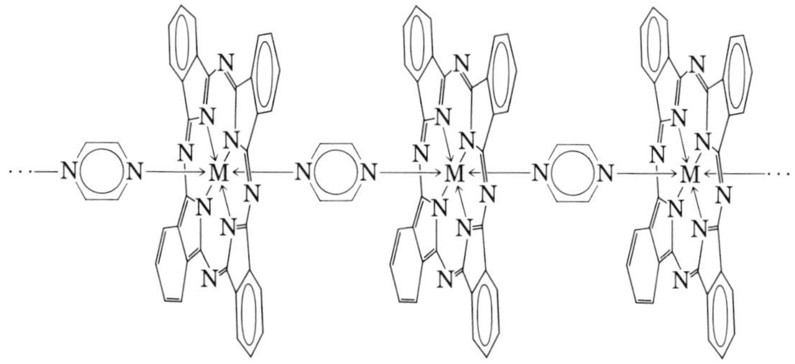

(10)

Fig. 1.   Phthalocyanine polymer.

Applications

Organometallic polymers have many current and potential uses. Poly{tris[5,5'-bis(3-acrylyl-1-propoxycarbonyl)-2,2'-bipyridine]ruthenium} is one of a series of polymerized electrochromic compounds that have been reported to go through a range of seven different colors, from orange to cherry red. This series of polymeric complexes may be useful in generating multicolor displays in electronic display panels (42). Deposited as thin films on tin oxide supports, they have been cycled more than 10^6 times through all possible oxidation states ($+2 \rightarrow -4$); 85–90% of material remains intact and a response time of a few tenths of a millisecond is maintained. Tributyltin methacrylates have found widespread use in anti-fouling paints which slowly release or ablate tributyltin hydroxide. Tin species are toxic to sliming organisms involved in biological ship-bottom fouling (40,43,44).

Commercial colorants (qv), ie, dyes or pigments, often exhibit poor migration fastness, bleeding, blooming, or plateout. Polymer-bound organometallic color-ants might reduce these difficulties. Organometallic condensation polymer dyes have been prepared by the reaction of fluorescenin and sulfonphthalein dyes with dicyclopentadienyldichlorotitanium ((**13**) and (**14**)) and organostannanes (**15**) (Fig. 2). Degrees of polymerization (DP) in excess of 100 have been obtained (45–47).

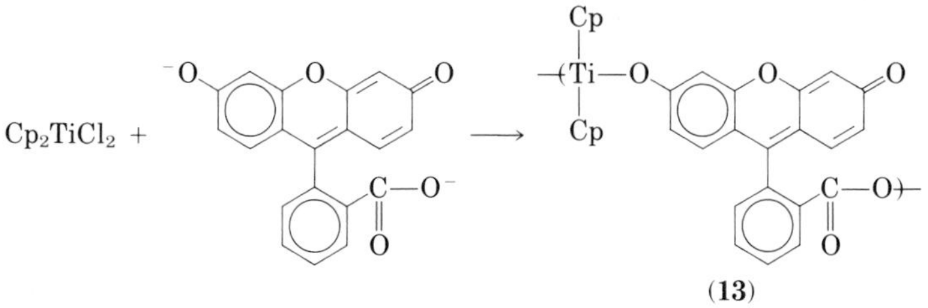

(13)

where Cp = cyclopendadienyl

Polymeric metal phosphinates having single-, double-, or triple-bridged structures (**16**)–(**18**) (Fig. 3) have been made using Al, Be, Co, Cr, Ni, Ti, and Sn (48). They form films with thermal stabilities to 450°C and the chromium(III) polyphosphinates have been used as thickening agents for silicone greases to improve their high pressure physical properties.

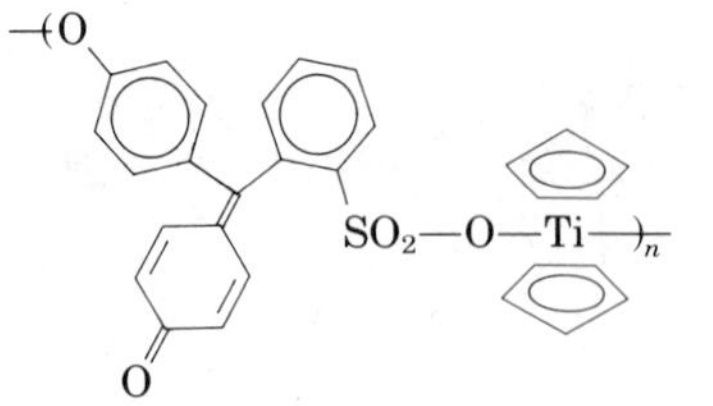

Fig. 2.   Organometallic polymer dyes.

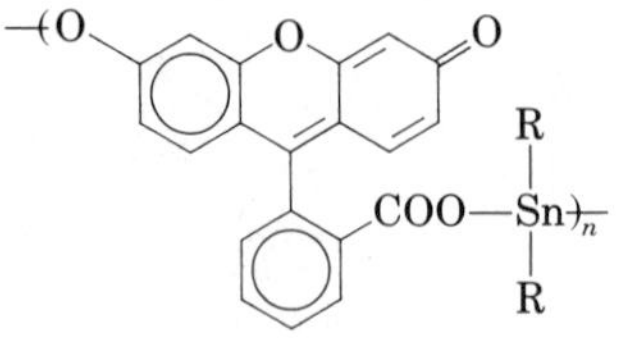

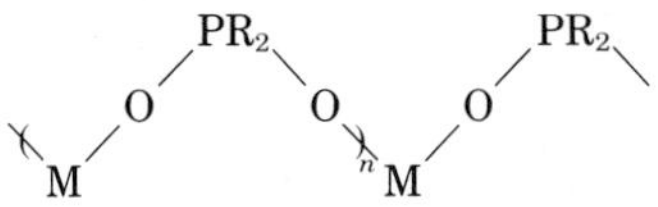

Fig. 3.   Polymeric metal phosphinates.

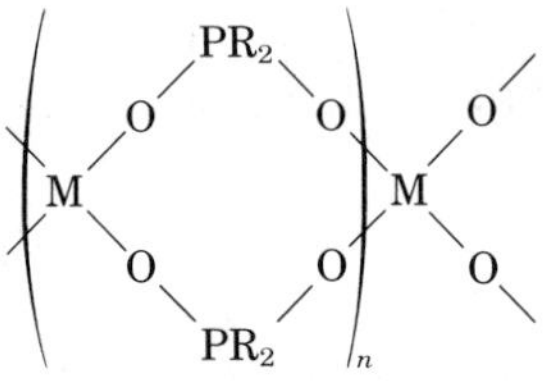

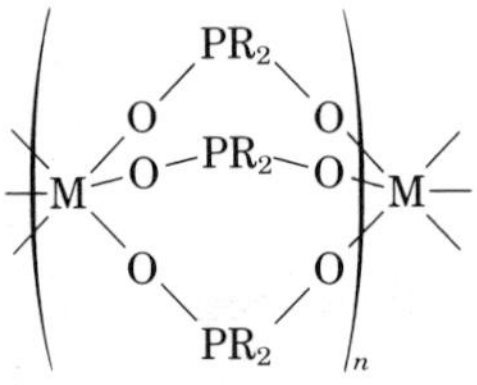

Organometallic polymers are being investigated as medicinal agents. Co-ordination of K_2PtCl_4 by poly[bis(methylamine)phosphazene] gives the platinum phosphazene polymer (**19**), which shows tumor-inhibiting activity against mouse P388 lymphocytic leukemia and in the Ehrlich ascites tumor-regression test (49).

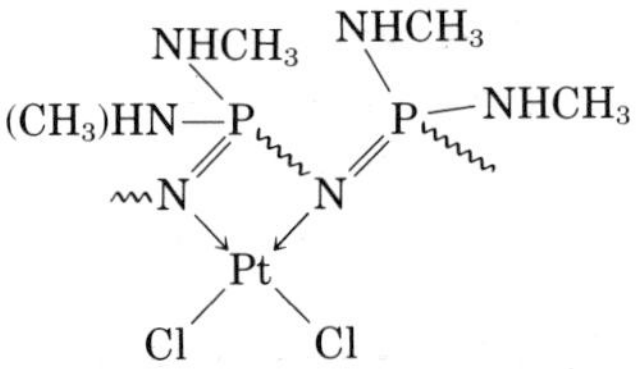

(19)

The *cis*-dichloroplatinum moiety is included in the polymer repeating units (**20**) by the reaction of K_2PtCl_4 with difunctional nitrogen compounds (50–52) such as 1,6-hexamethylenediamine, pyrimidines, purines, and hydrazines.

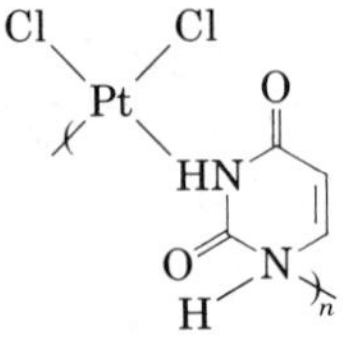

(20)

At 10–20 μg/mL they sometimes repress replication of Poliovirus I and L RNA virus, but have no activity toward L929 mouse, HeLa, and WISH cells. At concentrations above 30 μg/mL, tumoral cell growth is stopped.

Polymer matrices containing europium chelates have been used as the active species in the electronic energy-transfer processes of lasers (53,54). Europium chelates are of interest because the pump energy is absorbed by the organic chelate molecule and efficiently transferred to the europium ion (see ENERGY TRANSFER).

(21)

In one example, (21) was copolymerized with methyl methacrylate and the resulting translucent polymer showed superfluorescence when excited by pulses from xenon flashlamps (54).

One of the most vigorously studied areas of application is the use of polymer supports for homogeneous catalysts, so that fixed-catalyst beds can be used in flow processes, while the chemical selectivity inherent in the homogeneous organometallic catalyst is maintained (21–27). Other advantages of such bound catalysts include reduced reactor corrosion (associated with some homogeneous catalysts) (22); ease of product–catalyst separation; the immobilization of high concentrations of metal complex within the resin matrix, thereby retarding bimolecular reactions which deactivate the catalyst (55–57); separation of two mutually incompatible catalysts in a single-reactor vessel for multistep synthesis (58–61); and the ability to tailor the polymer-catalyst complexes to increase catalyst selectivity (62–66) and to increase reaction rate (63,66,67) (see CATALYSTS, POLYMER-SUPPORTED). Some proprietary commercial processes now are operating using polymer-supported organometallic catalysts. A polymer-supported rhodium hydroformylation process has been scaled up to pilot-plant size (68). Many patents have been issued covering the use of polymer-bound organometallic catalysts in hydrogenation (69,70), hydroformylation (23,70–75), hydroesterification of olefins (76), diene linear oligomerization (77), and others (22).

There are applications in the area of functionalized electrodes for electrocatalysis, photoelectrocatalysis, photovoltaic cells, and specialized electrode and sensor development (78–84). For example, when a [1.1]-ferrocenophane-containing polystyrene is applied as a film to the surface of a p-type semiconductor, it gives a photoelectrolysis device from which H_2 can be generated from acid at 430 mV more positive than for H_2 evolution from platinum surfaces under the same conditions (84) (see also ELECTRODES, POLYMER-MODIFIED).

Several organometallic polymers have been used as preheat shields for targets in inertial-confinement nuclear fusion (85,86). Temperatures of 50,000,000°C are needed for fusion. Small, hollow spheres containing deuterium and tritium are placed at the focal point of high intensity laser beams and subjected to 8×10^{12} W/cm (Fig. 4). The release of suprathermal electrons from the target shell causes efficiency loss in fusion due to preheating of the target. This is remedied by having a concentric shell of low average atomic-number material containing

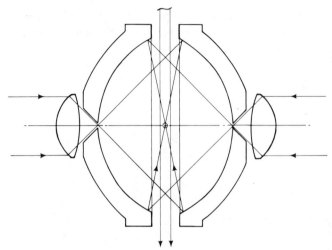

Fig. 4. Mounting of the spherical target for inertial-confinement nuclear fusion; two-sided spherical target illumination.

1–4 atom % of atomic number 50–85 uniformly distributed on the molecular level (Fig. 5). Hence, among others, polymers and copolymers of vinylruthenocene (**22**), vinylosmocene (**23**), and $(C_6H_5)_3Pb$—⟨○⟩—$CH=CH_2$ have been employed (Fig. 6).

$$⟨◇⟩—CH=CH_2 \qquad ⟨◇⟩—CH=CH_2$$
$$Ru \qquad\qquad\quad Os$$
$$⟨◇⟩ \qquad\qquad\quad ⟨◇⟩$$
$$(\mathbf{22}) \qquad\qquad\quad (\mathbf{23})$$

Other uses are mentioned in subsequent sections that discuss syntheses of specific materials.

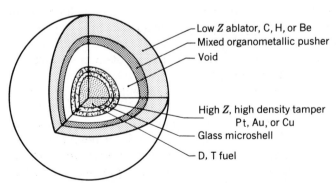

Fig. 5. One design for a multishell target for use in inertial-confinement nuclear fusion.

Fig. 6. Inside view of a hollow hemisphere made from a copolymer of $(C_6H_5)_3PbC_6H_4CH=CH_2$ and octadecyl methacrylate.

Condensation Polymers

Condensation reactions exhibit several characteristics such as expulsion of a smaller molecule (often H_2O or HX) leading to a repeat unit containing fewer atoms than the sum of the atoms in reactant molecules. The reaction site can be at a metal atom (**24**), adjacent to the metal, or further removed from the metal atom as in (**26**).

$$R_2MX_2 + H_2NRNH_2 \longrightarrow \underset{(25)}{-\!\!\left(\!\!\underset{\underset{R}{|}}{\overset{\overset{R}{|}}{M}}\!-\!\overset{\overset{H}{|}}{N}\!-\!R\!-\!\overset{\overset{H}{|}}{N}\!\right)\!\!-} + 2\ HX$$

(24)   (25)

$$\underset{(26)}{Cl-\overset{\overset{O}{||}}{C}-R-\overset{\overset{O}{||}}{C}-Cl} + HOH_2C-\!\!\underset{Fe}{\bigcirc\!\!\bigcirc}\!\!-CH_2OH \longrightarrow$$

(26)

$$\underset{(27)}{-\!\!\left(\overset{\overset{O}{||}}{C}-R-\overset{\overset{O}{||}}{C}-O-CH_2-\!\!\underset{Fe}{\bigcirc\!\!\bigcirc}\!\!-CH_2-O\right)\!\!-}$$

(27)

There is often no clear distinction between coordination and condensation reactions. Here, reactions directly involving the metal already chemically bonded to an organic moiety, such as (**24**), will be considered condensation reactions, whereas reactions involving simple metal ions and metal ion oxides (see below) undergoing chelations and electrostatic attractions will be covered in the section entitled Coordination Polymers.

$$HAR'AH + XMX \longrightarrow {+}(AR'AM){+} + 2\ HX$$

$$HARAH + MO_2{}^{2+} \longrightarrow {-}(ARAM){-}$$

Condensations at the Metal Atom

Many organometallic halides possess a high degree of covalent character and can behave as organic acid derivatives in many reactions:

$$RCOCl + H_2O \longrightarrow RCOOH + HCl$$

$${-}MCl + H_2O \longrightarrow {-}M{-}OH + HCl$$

Thus many organometallic polycondensations are extensions of organic polyesterification reactions.

$$ClCRCCl + HOR'OH \longrightarrow {+}(CRCOR'O){+} + 2\ HCl$$

$$R_2MCl_2 + HOR'OH \longrightarrow {+}(M{-}OR'O){+} + 2\ HCl$$

A number of polycondensations are based on reactions and on reactants known to be suitable for producing small molecules. Thus $(\eta^5\text{-}C_5H_5)_2ZrCl_2$ reacts with benzenethiol and 1,2-dithiolbenzene in benzene in the presence of triethylamine giving monomeric products (**28**) and (**29**). Conceptionally, it is simple to employ dithiols where bridging is not encouraged. The analogous reaction does not produce high molecular weight material, but the zirconium polythiols (**30**) can be formed using an interfacial reaction system (87).

$$\text{where Cp} = \eta^5\text{-}C_5H_5$$

Polymers can be formed by reaction of a variety of Lewis bases with organo-metallic Lewis acids such as

$$R_2TiX_2 \quad R_2SiX_2 \quad R_3As \quad R_3BiX_2$$
$$R_2ZrX_2 \quad R_2GeX_2 \quad R_3AsX_2 \quad R_2MnX_2$$
$$R_2HfX_2 \quad R_2SnX_2 \quad R_3Sb \quad RBX_2$$
$$R_2PbX_2 \quad R_3SbX_2 \quad R_2SeX_2$$

where X = halogen. The Lewis bases include diols, dithiols, diamines, amino-alcohols, hydrazines, urea, thiourea, diamides, dithioamides, dihydrazides, dithio-hydrazides, diacid salts, amino acids, and dioximes.

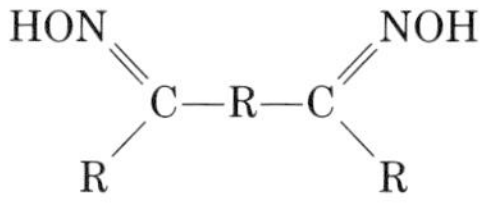

dioximes 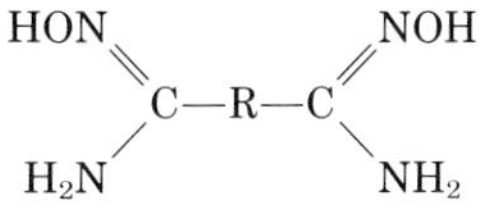   diamidoximes

An organometallic compound may also be condensed with suitable natural poly-mers, eg, dextran (qv), cellulose (qv), and wool (qv), and synthetic polymers, eg, poly(vinyl alcohol) and polyethyleneimine, that contain appropriate reactive sites (87–92).

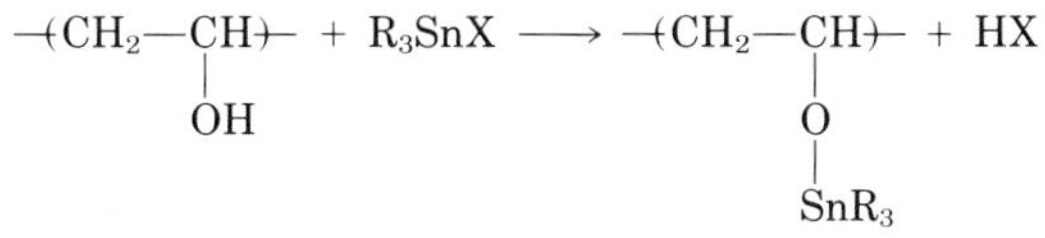

Reaction Media. Since many of the reactants and products are thermally sensitive, low temperature solution and interfacial condensation systems have typically been employed (93,94). Most solution systems are inferior to the inter-facial condensation systems. One exception is the use of aqueous solution systems employing Group IVB metallocene dihalides. Thus dicyclopentadienyltitanium dichloride (Cp_2TiCl_2) dissolves in water, forming various hydrated species con-taining the Cp_2Ti moiety. If the reaction is carried out within several minutes after the Cp_2TiCl_2 dissolves, the titanium-containing moiety reacts as though it were Cp_2Ti^{2+} and polyesters (**31**), polyethers (**32**), polythioethers (**33**), and poly-amines (**34**) are formed from reaction with diacid salts, diols, dithiols, and di-amines, respectively (93–96).

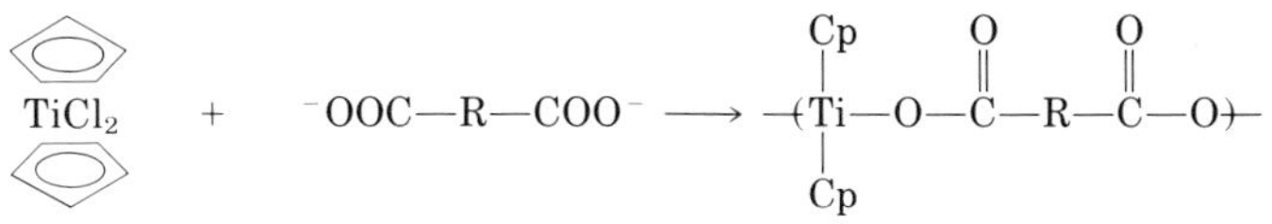

(31)

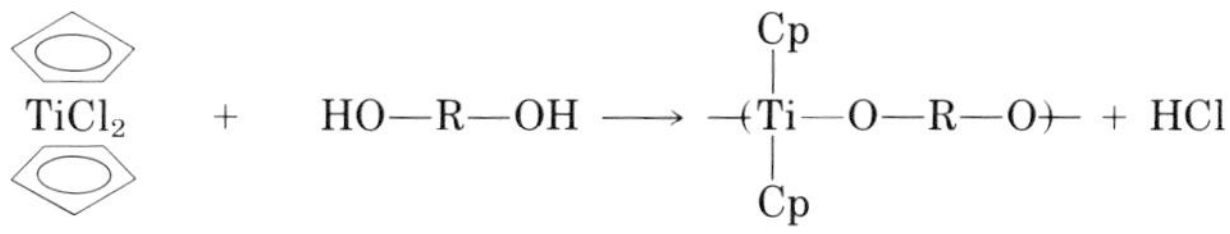

(32)

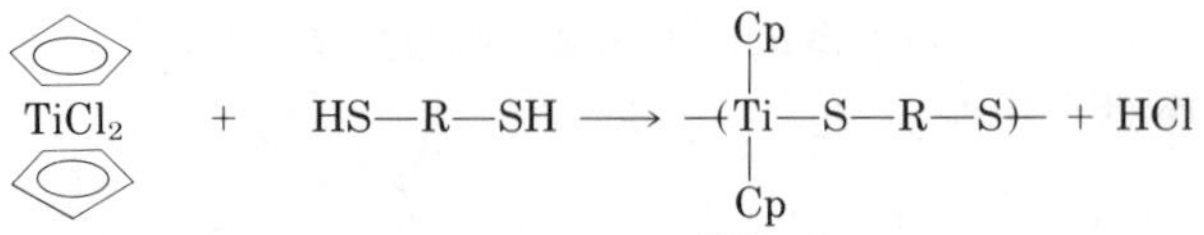

$$\text{TiCl}_2 \quad + \quad \text{HS—R—SH} \longrightarrow \text{—(Ti—S—R—S)— + HCl}$$

(33)

$$\text{TiCl}_2 \quad + \quad \text{H}_2\text{N—R—NH}_2 \longrightarrow \text{—(Ti—N—R—N)— + HCl}$$

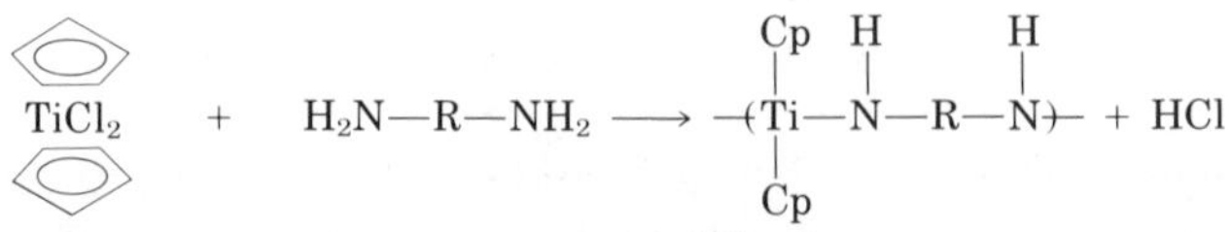

(34)

Water-based solution systems are attractive for water-stable organometallic reactants or reactants that form usable aqueous species. For reactants that are not water-soluble or that hydrolyze forming inert products, nonaqueous solution systems have been developed. Also, aqueous systems have been developed for some water-soluble monomers. R_2PbX_2 compounds are insoluble in most organic liquids and water. Polyesters have been formed initially by dissolving diacid salts in a minimum of water and then adding one to 10-fold (by volume) of DMSO in which the lead reactants are dissolved. When the two solutions are brought together, lead polyesters (**35**) form (97).

$$R_2PbX_2 + {}^-OOC—R'—COO^- \longrightarrow \text{—(Pb—OC—R'—CO)—}$$

(35)

Biological Activity.   Condensation organometallic polymers with biological activity have been made. The activity derives from the total polymer chain or through controlled release of monomeric fragments (98–103). An example is structure (**36**) obtained from metallocenes Cp_2MCl_2 and the dioxime of vitamin K where M = Ti and Zr.

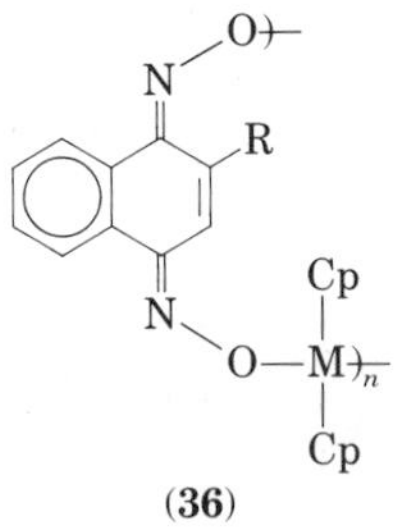

(36)

The delivery of dibutyltin moieties *in vivo* is possible using (**37**).

$$\text{—(Sn(C}_4\text{H}_9)_2\text{OCH}_2\text{CH}_2\text{O)—}$$

(37)

Use of the pyrimidine group in (**38**) encourages acceptance by a biological host and its subsequent demise through the release of the arsenic function (103).

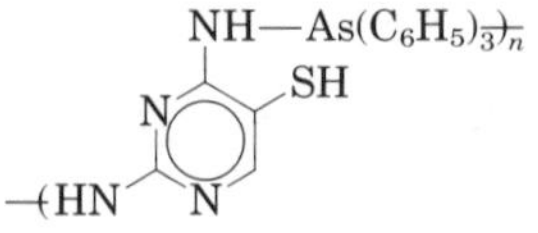

(38)

Other properties are also intriguing. A number of the polymers containing titanocene groups exhibit anomalous fiber formation reminiscent of metal whiskers. At low temperature their thermal stability is poor, but after initial weight loss below 250°C, they retain 80% of their weight to 1200°C and significant organic portions remain intact above 600°C (104).

Phthalocyanine Polymers. Shiskabob macromolecules with the stacked metal phthalocyanine structure (**39**) (M = Si, Ge) have been made (105,106) (Fig. 7). $\overline{DP} > 100$ was obtained with M = Si and $\overline{DP} > 70$ with M = Ge starting from Si(Pc)Cl$_2$ and Ge(Pc)Cl$_2$, where Pc = phthalocyanine. The dichlorides give M(Pc)(OH)$_2$, which undergoes polycondensation to give the (MO)$_n$ repeating group, and cofacial arrays of the metallophthalocyanines are achieved (Fig. 7). These polymers can be dissolved in strong acids, spun into fibers and fiber blends with Kevlar, and doped with iodine or NO$^+$X$^-$ (X = BF$_4^-$, PF$_6^-$, SbF$_6^-$) to the extent of ~36% of oxidized monomer units. Extensive spectral, structural, conductivity, and thermochemical studies of these materials have now appeared (105). Polycrystalline conductivities as high as 0.15 S/cm at 300 K have been achieved.

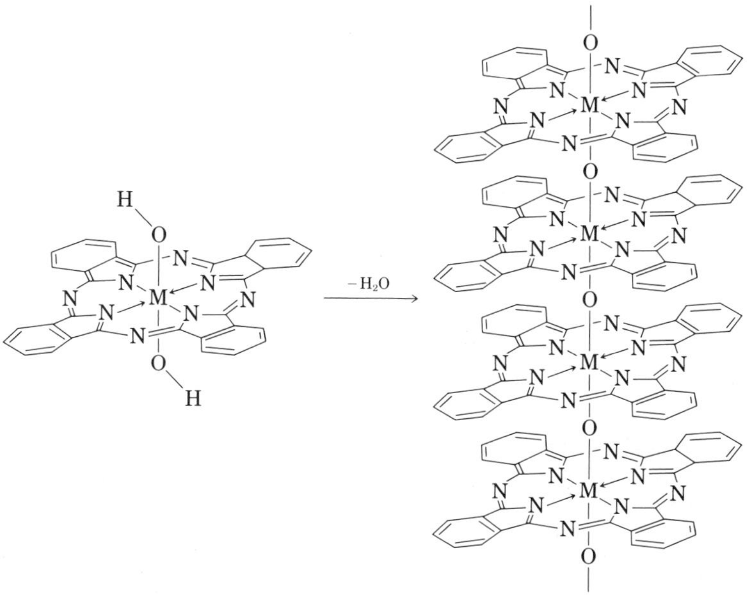

(39)

where M = Si, Ge, Sn

Fig. 7.   Formation of metallophthalocyanines.

Condensations Away From the Metal Atom

Organometallic monomers, in which the metal is not the site of condensation, have been polymerized in wide variety. In many of these cases the metal is in the backbone, however. For example, melt-phase polytransesterifications of 1,1′-bis(carbethoxy)cobalticinium hexafluorophosphate (**40**) and diols such as 1,10-decanediol or bis(hydroxymethyl)benzene give fairly high molecular weight ionic polyesters (**41**) (107). Polyamides (**42**) have been prepared by reaction of aromatic diamines with 1,1'-dicarboxycobalticenium chloride in molten antimony trichloride. The cobalticenium unit in the polymer occurs as an $SbCl_4^-$ salt, which can be isolated as the PF_6^- salt upon treatment with NH_4PF_6 (108).

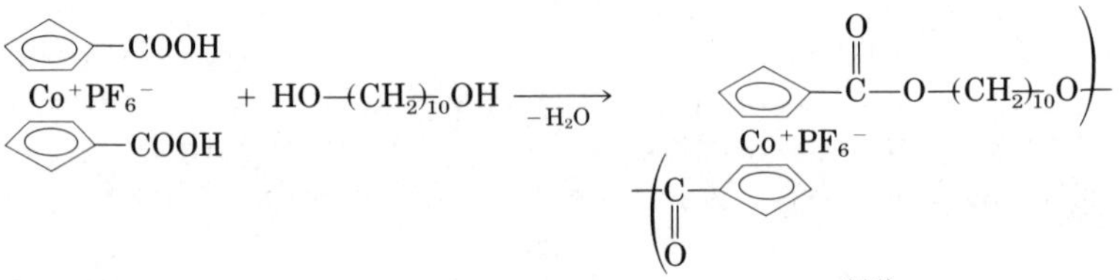

(**40**)   (**41**)

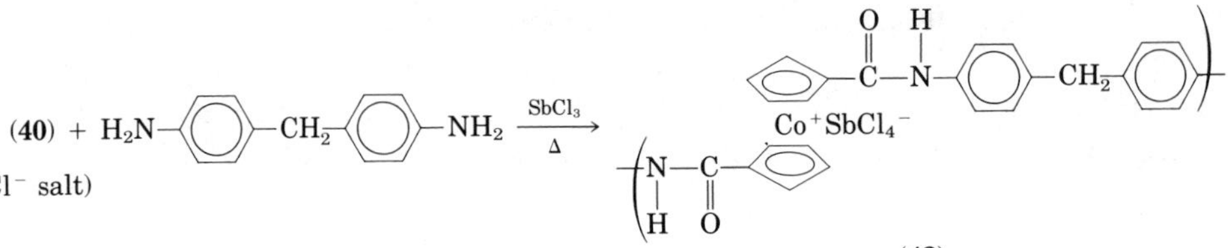

(**42**)

Ferrocene diepoxide derivatives (**43**) (109) and (**44**) (110) have been made in good yields and incorporated into epoxy resins.

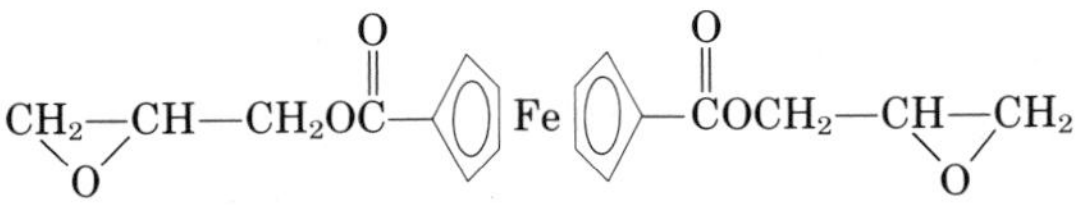

(**43**)

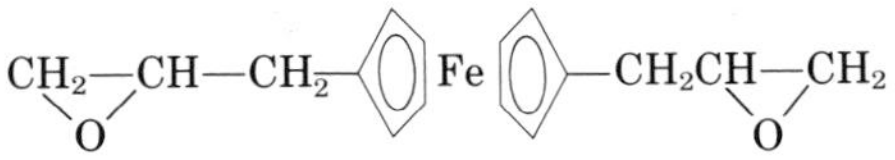

(**44**)

Many ferrocene-containing polyesters (**45**), polyamides (**46**), and polyurethanes have been reported in which classic polycondensation methods have been employed (111–113).

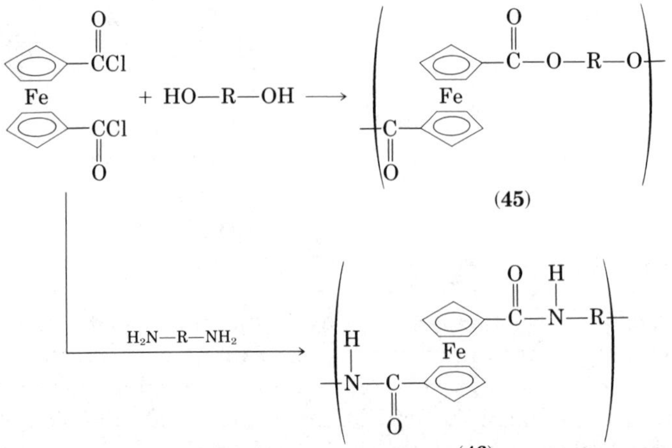

(**45**)

(**46**)

Interfacial polycondensations of 1,1′-bis(β-aminoethyl)ferrocene (**47**) with diacid chlorides and diisocyanates give new polyamides and polyureas (Fig. 8) (114). The polymerizations can be conducted at ambient temperatures in contrast to earlier high temperature polymerizations. Likewise, 1,1′-bis(β-hydroxyethyl)-ferrocene (**48**) reacts with diacid chlorides and diisocyanates to form ferrocene-containing polyesters and polyurethanes, respectively. Since the reactive groups (NH$_2$, OH) are two methylene units removed from the ferrocene nucleus, steric effects are reduced; also, the instability found in polymers of α-functional ferrocenes due to the stability of the α-ferrocenyl carbonium ion is removed. The polyamide obtained with adipoyl chloride is elastomeric and the polyureas are hard powders.

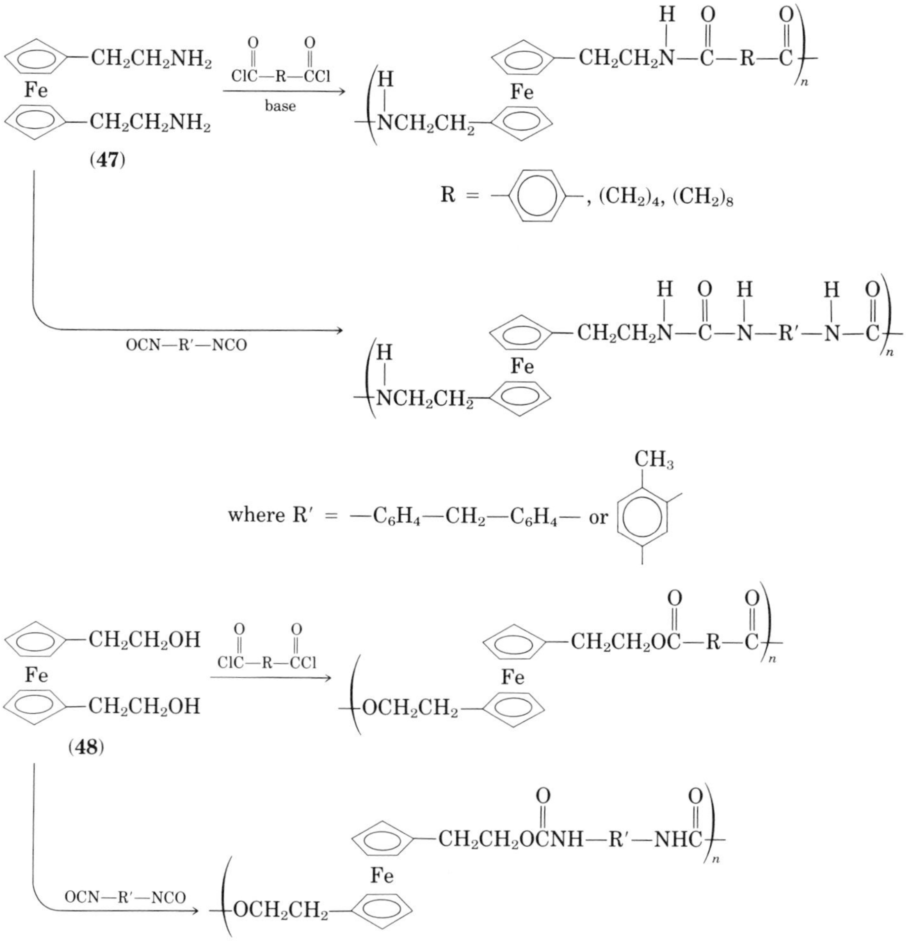

Fig. 8. Ferrocenes with functionality β to the ring.

Poly(ferrocenylphosphine oxides) (**49**) and sulfides (**50**) result from the reaction of phenyldichlorophosphonic acid or phenylphosphonothioic dichloride with ferrocene in the presence of ZnCl$_2$ (115). However, cleavage of the rings from iron also occurs and cycloalkyl-bridged ferrocene units are formed within the polymer.

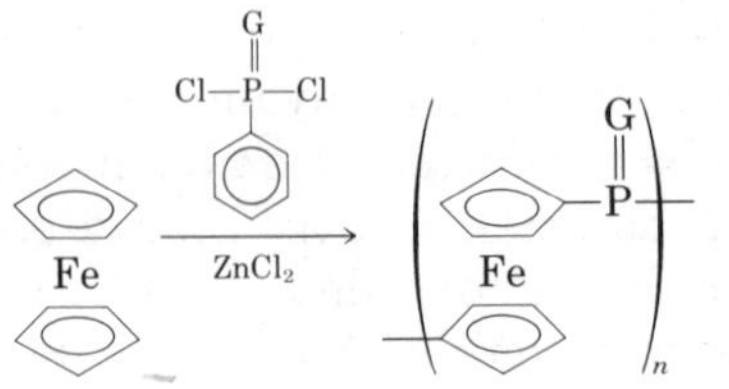

where G = O (**49**)

G = S (**50**)

A boron-containing poly(benzborimidazoline) (**51**) was synthesized from ferrocene-1,1'-diboronic acid and 3,3'-diaminobenzidine as part of an effort to produce polymers of superior thermal stability (116).

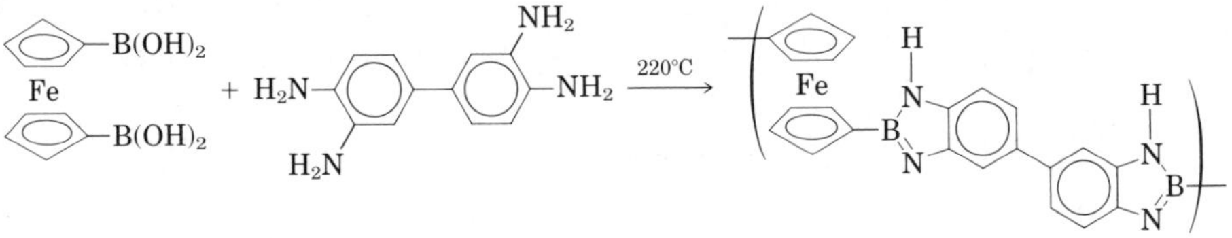

(**51**)

In other attempts to increase thermal stability while achieving processibility, silicone polymers containing ferrocene backbone groups were studied. A very early patent (117) reported that trimethylsilylalkyl derivatives of ferrocene (**52**) gave high yields of polysiloxanes (**53**) in H_2SO_4 or $CH_2Cl_2/AlCl_3$. A variety of ferrocene–silicon-containing polyamides, polyesters, polyurethanes, and polysiloxanes starting with the monomer series (**54**) have been produced (118). These polymers, by virtue of the presence of silicon, are liquids or elastomers.

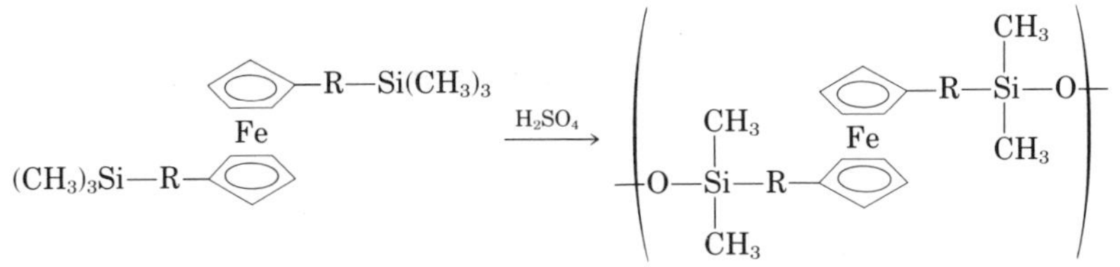

(**52**)   (**53**)

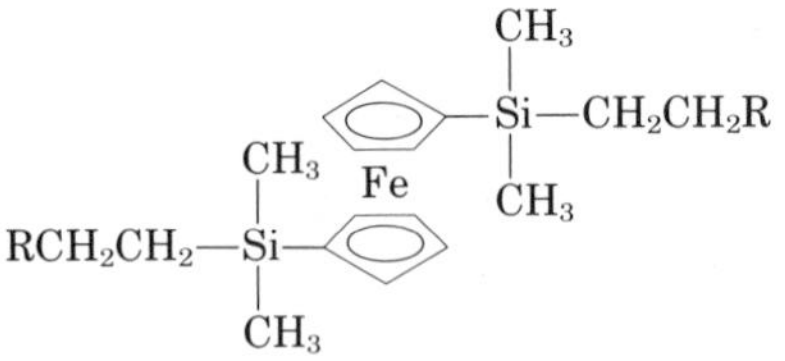

(**54**)

where R = —COOCH$_3$ —CH$_2$NH$_2$

—COOH

—CH$_2$—O—Si(CH$_3$)$_3$

—CH$_2$OH

—CH$_2$—O—CH$_2$—CH—CH$_2$

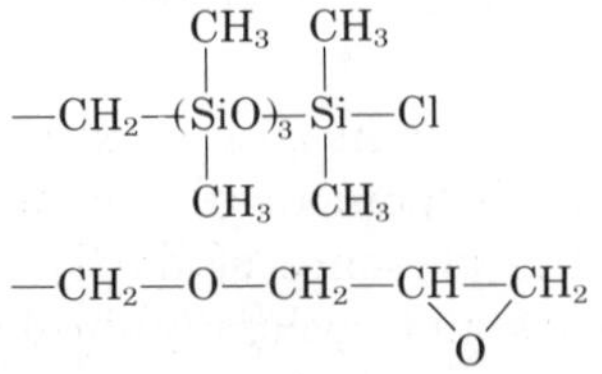

Well-characterized, ferrocene-containing siloxane polymers have been prepared via a bis(dimethylamino)silane-disilanol polycondensation (119). The monomer 1,1'-bis(dimethylaminodimethylsilyl)ferrocene (**55**) was polymerized with three aryl disilanols: dihydroxydiphenylsilane (**56**); 1,4-bis(hydroxydimethylsilyl)benzene (**57**); and 4,4'-bis(hydroxydimethylsilyl)biphenyl (**58**). Melt polymerizations at 133 Pa (1 torr) and 100°C gave the highest molecular weights (~50,000).

These oxysilane polymers, which have structures such as (**59**), are hydrolytically stable in THF–H_2O and thermally stable to over 400°C.

Coordination Polymers

Coordination polymers (qv) have long been known. Tanning leather and generating select-colored pigments depend on the coordination of metal ions. Many coordination polymers have unknown and/or irregular structures. Coordination polymers can be prepared by a number of routes; the three most common are

1. Metal coordination complexes polymerize through attached functional groups; polymer formation is a condensation or addition reaction, eg,

2. Preformed polymers containing chelating groups coordinate metal ions, eg,

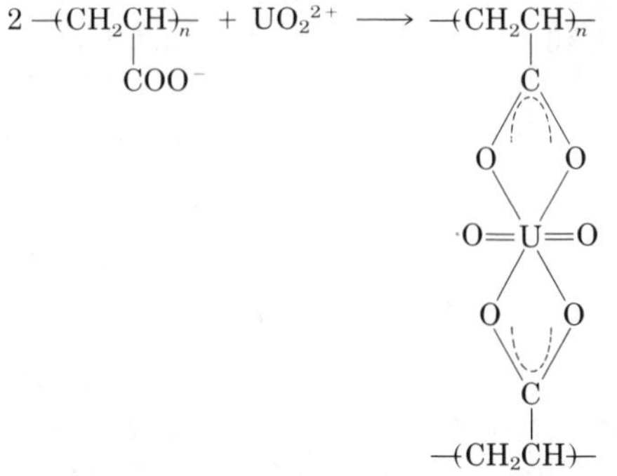

$$2 \text{-(CH}_2\text{CH)}_n + \text{UO}_2^{2+} \longrightarrow$$

3. Metal ions react with ligands capable of coordinating to two metal atoms; metal-to-donor coordination builds chains, eg,

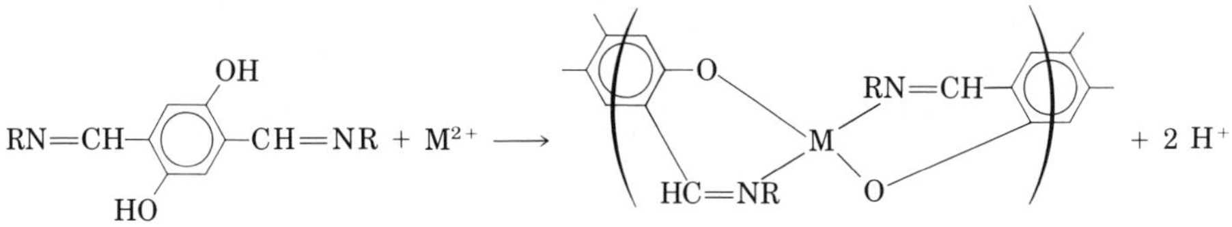

$$\text{RN}=\text{CH-} \underset{\text{HO}}{\overset{\text{OH}}{\bigcirc}} \text{-CH}=\text{NR} + \text{M}^{2+} \longrightarrow \qquad + 2\text{ H}^+$$

Ionomers are an important class of commercial polymers formed through route 2.

The classification of polymers as condensation or coordination polymers is often a subtle choice. This is illustrated by some differences in $d\pi$–$p\pi$ bonding in Group IVA systems. Silicon polyesters exist in tetrahedral geometries at the Si atom (**60**), whereas tin exists mainly with octahedral geometries in tin polyesters (**61**).

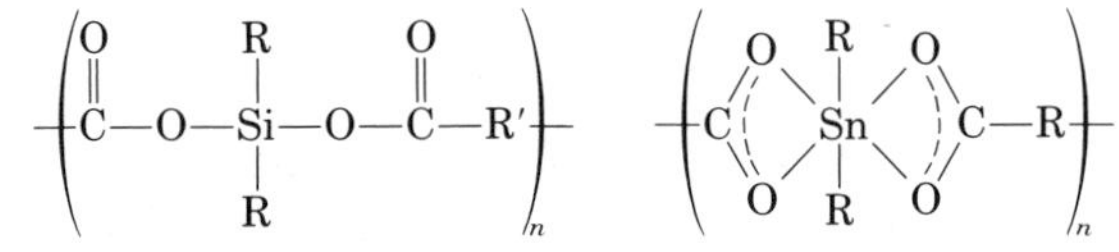

(**60**)   (**61**)

The presence or absence of particular ligands as part of the metal inner-sphere coordination is often an important contributor to polymer reactivity and solubility. Manganese(II) polyamine (**62**), which contains inner-sphere water coordination, is an example and should not be written as, or confused with, structure (**63**).

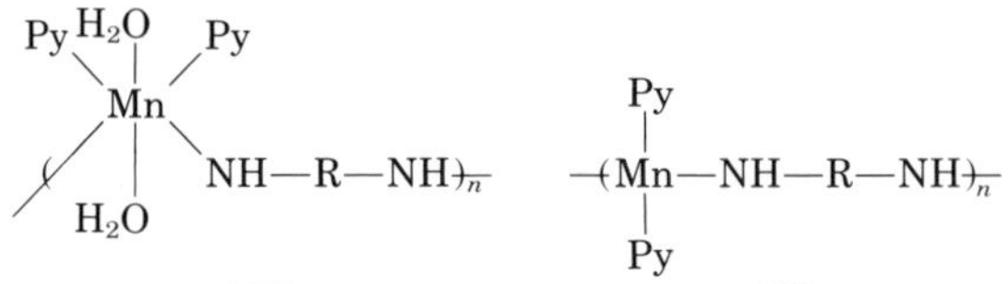

(**62**)   (**63**)

where Py = pyridine

Synthesis and characterization of coordination polymers has been supported and conducted by the U.S. Air Force in a search for materials with high thermal stabilities. Attempts to prepare stable, tractable coordination polymers simulating the exceptional thermal and/or chemical stability of model monomeric coordination compounds, such as copper ethylenediaminobisacetylacetonate(II) or copper phthalocyanine(I), have been disappointing. Processibility was also a problem. Coordination polymers were insoluble and high molecular weight species were not obtained because precipitation from solution occurred early in the polymerization. One-to-one complexes of 2,5-dihydroxy-p-benzoquinone with Cu^{2+}, Ni^{2+}, and Cd^{2+} formed polymers, and colloidal particles of the Cu^{2+} polymers (**64**) were rigid rods 100–200-nm long (120,121). Naphthazarin (**65**) and quinizarin (**66**) are other hydroxyquinones that give coordination polymers with several metals such as Co^{2+}, Cu^{2+}, Ni^{2+}, Zn^{2+}, Ba^{2+}, and Be^{2+} (122–125).

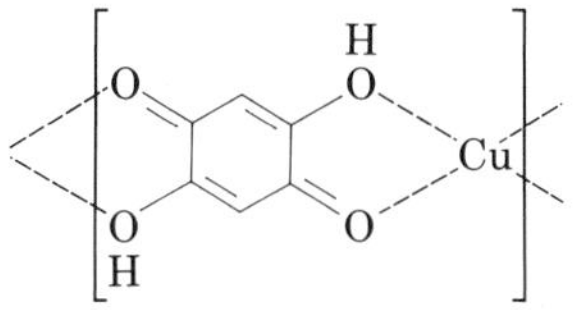

(**64**) 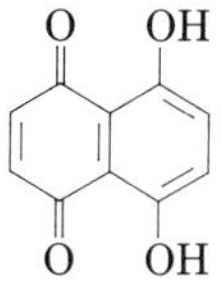   (**65**) 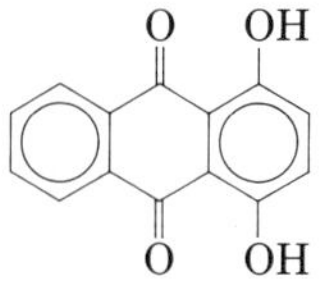   (**66**)

The degrees of polymerization in these cases probably ranged from 5 to 25 (123); similar results apply to melt polymers formed between quinizarin and metal derivatives of acetylacetone (126). Bifunctional o-hydroxyaldehydes and diamines have been used to prepare polymers through the introduction of metal ions to their Schiff bases (**67**) (127–137).

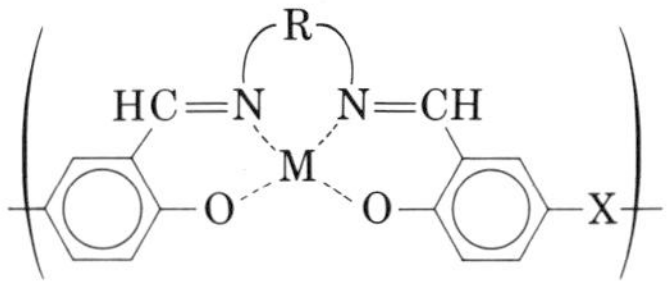

(**67**)

where X = CH_2, SO_2

These are typical ladder-type polychelates, which are highly colored, insoluble in common solvents, and weak semiconductors.

Semiconductivity. Mixed-valence, transition-metal complexes of the dianion of oxalic acid act as semiconductors. The iron(II) oxalates are linear-chain complexes as shown in (**8**). Oxidation of $Fe(C_2O_4)(H_2O)_2$ with Br_2 gives $Fe(C_2O_4)(H_2O)_{1.4}Br_{0.6}$, a semiconducting polymer with conductivity at RT of 10^{-4} S/cm. The reaction of $Fe(C_2H_2O_4)(H_2O)_2$ with I_2 results in the formation of a complex with a conductivity six orders of magnitude greater than that of the unoxidized polymer (12). The structurally related poly(metal tetrathiooxalate)s, such as (**68**), have been prepared using 1:1 stoichiometry of tetraethylammonium tetrathiooxalate and metal^{2+} salts (138), eg,

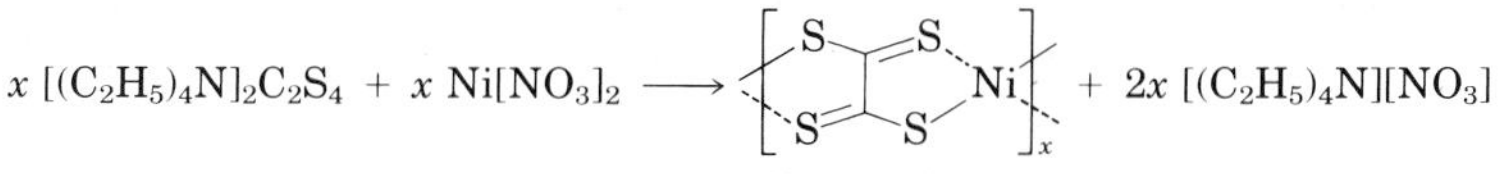

(**68**)

Conductivities of these complexes range from 5 to 20 S/cm for $[NiC_2S_4]_x$ to 1 S/cm for $[CuC_2S_4]_x$ and $[PdC_2S_4]_x$. These complexes are n-type semiconductors, as evidenced by their low negative, thermoelectric-power coefficients of ca -10 μV/K. Vapor-phase I_2 oxidation of these complexes leads to a decrease in their electrical conductivity as the concentration of mobile electrons is depleted. Tetrasodium tetrathiafulvalene tetrathiolate reacts with transition-metal salts (ML_n) to give insoluble, amorphous powders having a metal:ligand ratio of ~1:1 (139–141). These insoluble powders are believed to be oligomers containing a repeating tetrathiafulvalene–metal bisdithiolene unit (**69**).

(**69**)

The Ni(II) complex was found to have a conductivity of 30 S/cm when handled in air. Conductivities for other transition-metal derivatives were ~10^{-1} S/cm for Cu(II), ~10^{-5} S/cm for Fe(II), ~10^{-2} S/cm for Pt(II), and ~10^{-3} S/cm for Pd(II). These reactions forming the polymeric tetrathiafulvalene–metal bisdithiolenes (TTF–MBDT) were also carried out under inert conditions.

Anhydrous nickel acetylacetonate was added to a methanolic solution of sodium tetrathiolate followed by addition of a methanolic solution of $N(C_4H_9)_4$ under argon (142). An amorphous precipitate formed immediately, which had a conductivity of only 10^{-4} S/cm. This $\{[TTF–Ni(BDT)][N(C_4H_9)_4]_x\}_n$ polymer was then oxidized by adding a methanolic solution of bromine to the polymer suspension in methanol. The conductivity of the polymer rose to 20 S/cm, suggesting that a partial oxidation state, attributed to O_2 oxidation in the first case and Br_2 oxidation in the second, is necessary for high conductivity in these polymers. Metal poly(benzodithiolene)s have been prepared from divalent ions such as Co^{2+}, Ni^{2+}, Fe^{2+}, and Cu^{2+} on reaction with the benzene-1,2,4,5-tetrathiol ligand as shown in (**5**) (**7**). All of the polymers studied in this family were found to be paramagnetic conductors; conductivities ranged from 10^{-4} to 10^{-1} S/cm. The tetrathiosquarate dianion (**70**) also undergoes polychelation with Ni^{2+} or Pd^{2+} salts to give polymeric materials (**71**) (143).

(**70**)   (**71**)

These Ni- and Pd-based polymers have DPs of 10 and 25 and conductivities of 5×10^{-4} S/cm, respectively. The Pt-based polymer can be synthesized by reaction of $K_2C_4S_4$ with K_2PtCl_4 or $(C_6H_5CN)_2PtCl_2$ to give a blackish-green complex having a 1:1 ratio of $Pt:C_4S_4$ and a conductivity of 6×10^{-7} S/cm. The electronic structure of these complexes has been studied theoretically (144,145). The interchain interactions lead to an equivalency of the double and single bonds in the C_4S_4 moiety. Polymeric metal-containing hemiporphyrazines (**72**) are prepared by the reaction of metal ions with the preformed hemiporphyrazine polymer

(146–149). These dark brown-to-black metal chelates have conductivities ranging from 10^{-8} to 10^{-15} S/cm.

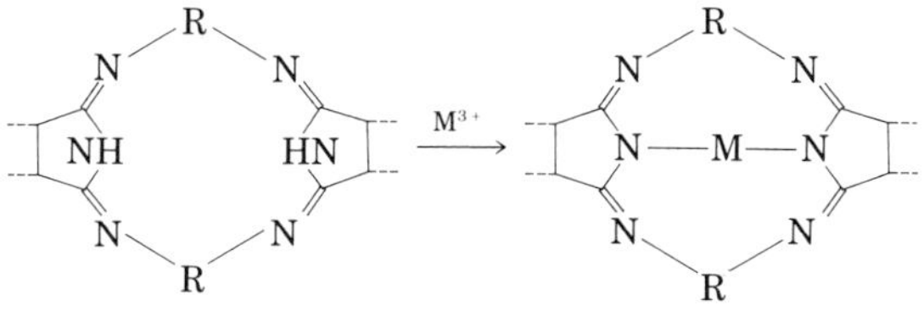

(**72**)

Polymeric tetraaza(14)annulene ligands have been prepared from both 3,3′-diaminobenzidine and 1,2,4,5-tetraaminobenzene and propynal. Their polymeric chelates, (**73**) and (**74**), were then made by reactions with metal ions (150,151).

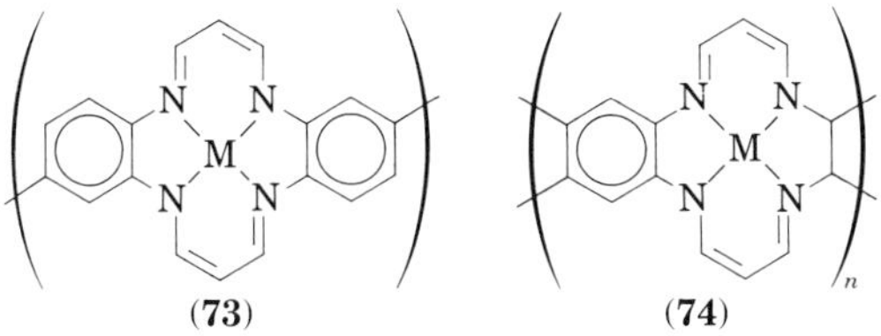

(**73**)   (**74**)

The resulting polymers were only weakly semiconducting (152). Another class of polychelates, (**75**) and (**76**), were prepared in dimethyl sulfoxide solutions by adding metal chloride salts to 4,4′-phenylenebisazodisalicyclic acid or 4,4′-(4,4′-biphenylenebisazo)salicylaldehyde–salicyclic acid and refluxing (153,154). The magnetic moments for (**75**) were 1.7×10^{-23} J/T (1.8 M_B) for the Cu^{2+}, 2.9×10^{-23} J/T (3.18 M_B) for the Ni^{2+}, and 4.2×10^{-23} J/T (4.5 M_B) for the Co^{2+} polychelates, respectively, but no useful conductivities were found. Polymers of structure (**76**) were made where M = Cr^{3+}, Mn^{2+}, Fe^{2+}, Co^{2+}, Ni^{2+}, Cu^{2+}, and Zn^{2+}.

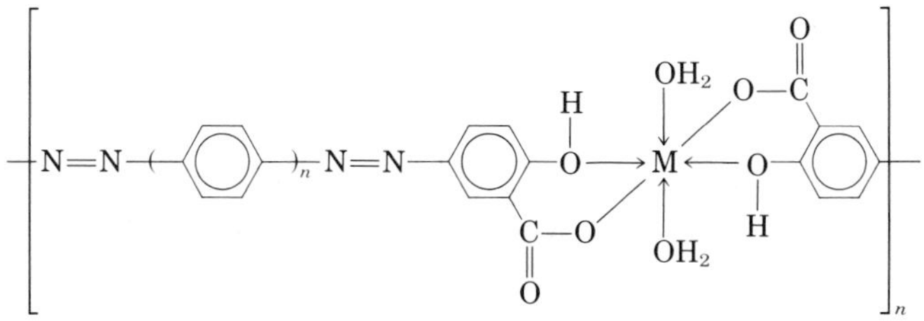

where n = 1 (**75**)

n = 2 (**76**)

Polymers (**68**)–(**71**) and (**73**)–(**75**) are formed upon chelation and their molecular weights depend on avoiding chain termination and on solubility of the specific polychelate in the solvent employed. However, the molecular weight of (**72**) depends on that of its preformed precursor polymer. Another example of metal chelation by a preformed polymer is the case of copper(II) chelates of poly(vinyl alcohol) (**77a** and **b**). Mixing poly(vinyl alcohol) with various Cu^{2+} salts followed by casting into films and doping with iodine gives cross-linked polymers with conductivities on the order of 10^{-4} S/cm (155).

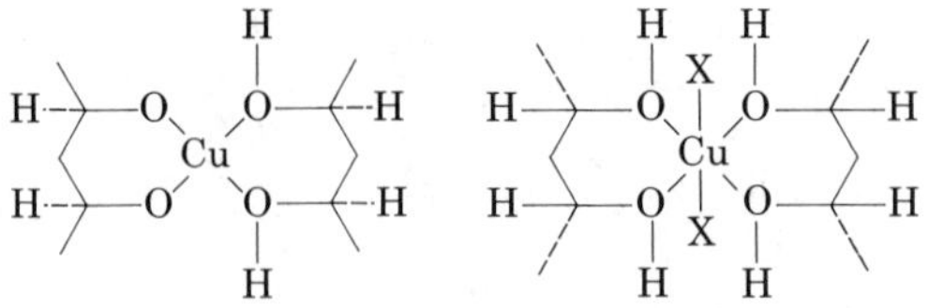

(77a)   (77b)

Preformed poly(thiosemicarbazide) (**78**) is extremely efficient in complexing Cu^{2+} ions from waste effluents from brass mills, thereby forming the polymeric copper chelate (**79**) (156). The Cu^{2+} ions are specifically complexed with high selectivity in the presence of high concentrations of other ions. Copper is tenaciously bound in polymer (**79**) and is not eluted on treatment with mineral acids, but elution with 1,4-benzoquinone releases copper in purities of 96–99% in field trials. The copper-containing polymer (**79**) can be used as a reagent for the oxidation of aldehydes to carboxylic acids and the coupling of nitroso compounds to azoxy derivatives.

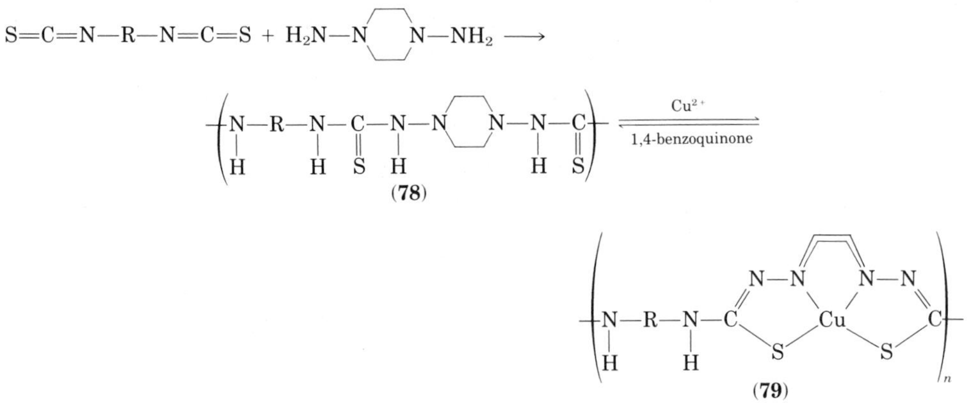

(78)

(79)

Phthalocyanine Polymers.   Shiskabob-type phthalocyanine polymers with alternating metal and pyrazine units (Fig. 1), or with alternating metal and oxygen atoms (Fig. 7) in the backbone, have been previously mentioned. In general, metal complexes of macrocyclic ligands, such as phthalocyanine, crystallize in stacks to afford linear chains that have the ability to conduct electricity along the stack axis. Phthalocyanine-like polymers with metal–metal linkages can be prepared by reaction of transition-metal salts with tetranitriles (157,158), 1,2,4,5-tetracyanobenzene (159,160), pyromellitic dianhydride (161,162), or tetracarboxylic acid derivatives (161,163). These polymers have structures generally represented by (**80a**), where the circles represent the specific coordinating macrocycle which surrounds the metal. There is direct metal–metal bonding along the stack direction. These polymers are semiconducting and exhibit pressed-pellet conductivities from 10^{-1} to 10^{-13} S/cm, depending on the metal, heterocycle, and state of doping.

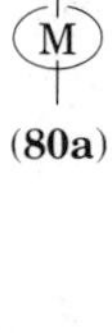

(80a)

Phthalocyanine polymers containing bridging ligands between the metal atoms (**39**) conduct electricity through their metal–ligand–metal linkages. Dehydration of phthalocyanine complexes of silicon, germanium, and tin produces face-to-face stacks of oxygen-bridged metal phthalocyanine units as shown earlier (Fig. 7) (164,165). Upon oxidation with iodine, a mixed-valence polymeric cation is formed with intercalated I_3^- ions between the chains. Conductivities range from 2×10^{-1} S/cm for $[SiPcOI_{1.40}]_n$ and 1×10^{-1} S/cm for $[GePcOI_2]_n$ to 2×10^{-4} S/cm for $[SnPcOI_{5.5}]_n$. The low conductivity of the tin polymer is attributed to a larger distance between the phthalocyanine units. Electrical conductivities on the order of 10^{-2} S/cm have been observed in undoped linear complexes with cyanide-bridging ligands, eg, $(PcCoCN)_n$. These values compare to doped metal–phthalocyanine polymers with oxygen bridges (166). The phthalocyanine polymer containing pyrazine as a bridging ligand $[Fe(Pc)(\mu\text{-pyz})]_n$ is a dark violet solid prepared from ferrous phthalocyanine and pyrazine (13). The RT conductivity of the undoped polymer was found to be on the order of 10^{-6}–10^{-7} S/cm (pressed pellet). The I_2-doped polymer showed an elevated conductivity of 10^{-1}–10^{-2} S/cm (pressed pellet).

Other Stacked Complexes.   Stacked metal-chain complexes (**80b**) without macrocyclic ligands are also known. Tetracyanoplatinate-stacked complexes resemble the metal phthalocyanine polymeric structure (**80a**), except that coordinating macrocycles do not chelate each metal atom. Instead, stacks of square-planar $Pt(CN)_4$ units are formed with anions between the stacks.

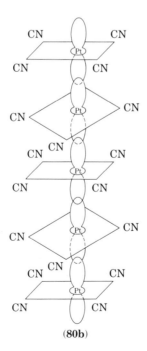

(**80b**)

The cyano groups surrounding Pt are staggered with respect to those coordinating the adjacent Pt atoms. These well-studied materials (167–169) are one-dimensional conductors with conductivities greater than 1 S/cm. For example, the electrolysis of a solution of Rb_2SO_4 and $Rb_2Pt(CN)_4$ gives reddish-brown crystals of $Rb_3Pt(CN)_4(SO_4 \cdot H_2O)$ at the electrodes.

Poly(yne)s. An unusual class of organometallic polymers is the transition-metal–poly(yne) polymers (170–175), which contain transition metals bound to carbon within the main chain and which might vaguely be considered coordination polymers. The Pt and Pd copolymers of this class have rodlike structures as indicated by (**81**) and (**82**)–(**89**) (Fig. 9). This linear structure results from the all-trans configurations at the Pt and Pd atoms. All of these polymers form nematic liquid crystals in trichloroethylene, and some of them align their main chain in a direction perpendicular to an applied magnetic field; others align parallel to the field (171,172). The nature of the transition metal within the polymer determines the magnetic properties. A typical synthesis is illustrated for polymer (**81**). Cuprous iodide reacts with *trans*-bis(tri-*n*-butylphosphine)-platinum in diethylamine under nitrogen at RT.

$$\text{HC}\equiv\text{C}-\text{C}\equiv\text{C}-\underset{\underset{\text{P(C}_4\text{H}_9)_3}{|}}{\overset{\overset{\text{P(C}_4\text{H}_9)_3}{|}}{\text{Pt}}}-\text{C}\equiv\text{C}-\text{C}\equiv\text{CH} + \text{CuI} + \text{ClPt[P(C}_4\text{H}_9)_3]_2\text{Cl} \xrightarrow{(\text{C}_2\text{H}_5)_2\text{NH}}$$

$$-\{\text{Pt[P(C}_4\text{H}_9)_3]_2-\text{C}\equiv\text{C}-\text{C}\equiv\text{C}\}_n-$$
$$(\textbf{81})$$

Evidence for the rigid-rod structures comes from large values of the Mark-Houwink exponent ($a = 1.7$ for (**81**)), independence of the intrinsic viscosity from solvent, agreement of sedimentation-equilibrium experiments with parameters derived from the ellipsoid revolution model for rigid rods, and liquid-crystal behavior of the polymers.

The reaction of butadiyne with cuprous chloride in dimethylacetamide under argon gives the three-dimensional poly(metal–yne) (**90**) (176). A similar, possibly linear polymer (**91**) is obtained by reaction of butadiyne with cupric chloride in ammonia methanol solutions. Conductivities were found to be 10^{-9} S/cm for (**90**) and (**91**). Upon iodine doping the conductivities are enhanced by ca 10 orders of magnitude to 6.3 S/cm for (**90**) and 10.2 S/cm for (**91**). Both polymers (**90**) and (**91**) display metallic behavior:

$$\text{H}-\text{C}\equiv\text{C}-\text{C}\equiv\text{C}-\text{Cu}---\overset{\overset{\displaystyle\text{H}}{|}}{\underset{\underset{\displaystyle\text{Cu}}{|}}{\overset{\overset{\displaystyle\text{C}}{\|}}{\underset{\underset{\displaystyle\text{C}}{|}}{\overset{\overset{\displaystyle\text{C}}{|}}{\underset{\underset{\displaystyle\text{C}}{\|}}{\text{C}}}}}}$$

$$---\text{Cu}-\text{C}\equiv\text{C}-\text{C}\equiv\text{C}-\text{H}$$
$$(\textbf{90})$$

$$-(\text{C}\equiv\text{C}-\text{C}\equiv\text{C}-\text{Cu})_n-$$
$$(\textbf{91})$$

Addition Organometallic Polymers

A large number of vinyl organometallic monomers have been prepared, homopolymerized, and also copolymerized with other classic vinyl monomers. The synthesis and polymerization of many organometallic vinyl monomers has

$$+(M)-C\equiv C-C\equiv C\!\!+_n \qquad\qquad \perp$$
(82)

where M = Pt, Pd

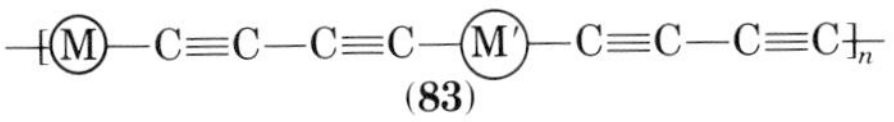

$$+(M)-C\equiv C-C\equiv C-(M')-C\equiv C-C\equiv C\!\!+_n \qquad\qquad \perp$$
(83)

where M = Pt, M' = Pd, Ni

M = Pd, M' = Ni

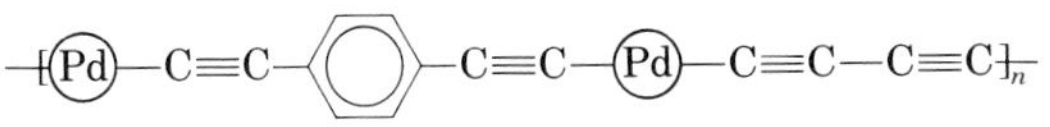

$$+(Pd)-C\equiv C-\!\!\bigcirc\!\!-C\equiv C-(Pd)-C\equiv C-C\equiv C\!\!+_n \qquad\qquad \|$$
(84)

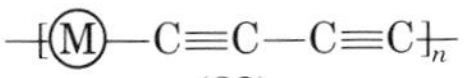

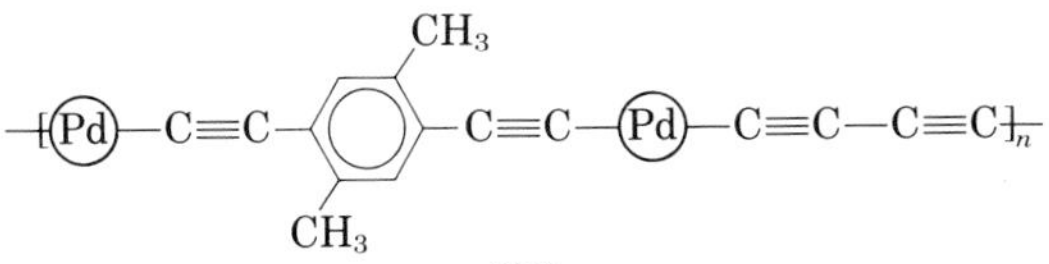

$$+(Pd)-C\equiv C-\!\!\overset{CH_3}{\underset{CH_3}{\bigcirc}}\!\!-C\equiv C-(Pd)-C\equiv C-C\equiv C\!\!+_n \qquad\qquad \|$$
(85)

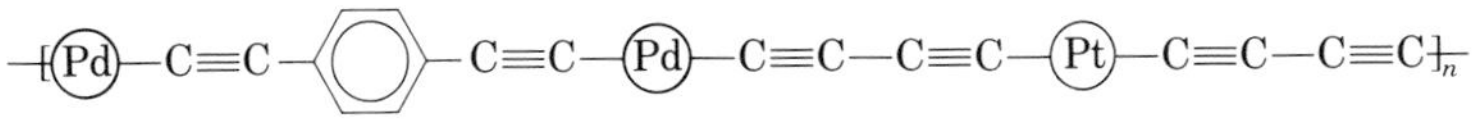

$$+(Pd)-C\equiv C-\!\!\bigcirc\!\!-C\equiv C-(Pd)-C\equiv C-C\equiv C-(Pt)-C\equiv C-C\equiv C\!\!+_n \qquad\qquad \perp$$
(86)

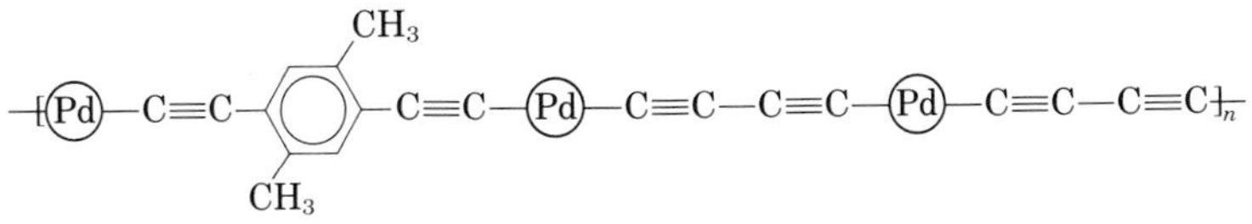

$$+(Pd)-C\equiv C-\!\!\overset{CH_3}{\underset{CH_3}{\bigcirc}}\!\!-C\equiv C-(Pd)-C\equiv C-C\equiv C-(Pd)-C\equiv C-C\equiv C\!\!+_n \qquad\qquad \perp$$
(87)

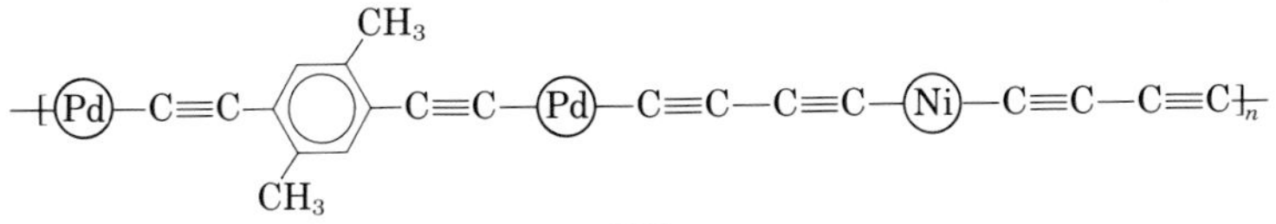

$$+(Pd)-C\equiv C-\!\!\overset{CH_3}{\underset{CH_3}{\bigcirc}}\!\!-C\equiv C-(Pd)-C\equiv C-C\equiv C-(Ni)-C\equiv C-C\equiv C\!\!+_n \qquad\qquad \perp$$
(88)

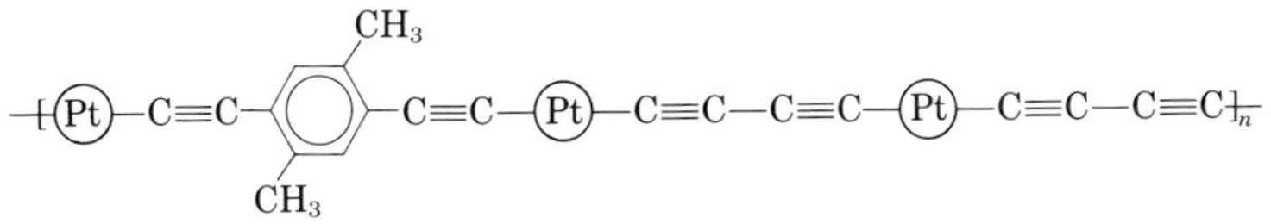

$$+(Pt)-C\equiv C-\!\!\overset{CH_3}{\underset{CH_3}{\bigcirc}}\!\!-C\equiv C-(Pt)-C\equiv C-C\equiv C-(Pt)-C\equiv C-C\equiv C\!\!+_n \qquad\qquad \|$$
(89)

Fig. 9. Rodlike polymers that align in a magnetic field. Encircled metals M = *trans*-M[P(C$_4$H$_9$)$_3$]$_2$

565

been reviewed (30,177–179). The variety of vinyl organometallic structures includes vinyl groups attached to cyclopentadienyl rings, dienes, phenyl rings, and ether or ester functions; these functions are bound to the metal. Such monomers are organometallic analogues of styrene, vinyl ethers, acrylates, and methacrylates. The vinyl groups may be directly attached to metals:

$$(CH_3)_3SiCH=CH_2 \qquad (CH_3)_3GeCH=CH_2 \qquad (CH_3)_3SnCH=CH_2$$

The first organometallic monomer that contained a transition metal was vinylferrocene (**92**), prepared in 1955 (180).

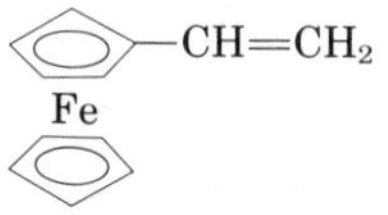

(**92**)

Its polymerization behavior has been extensively studied under radical (180–183), cationic (184), and Ziegler-Natta conditions (185); (**92**) is inert to anionic initiation (186). Vinylcymantrene (**1**) (1,187,188) and vinylcynichrodene (**93**) (189,190) have been prepared more recently and have been homopolymerized and copolymerized with a variety of electron-donating and -attracting organic comonomers, ie, styrene, methyl methacrylate, acrylonitrile, and *N*-vinylpyrrolidine (1,2,188–190).

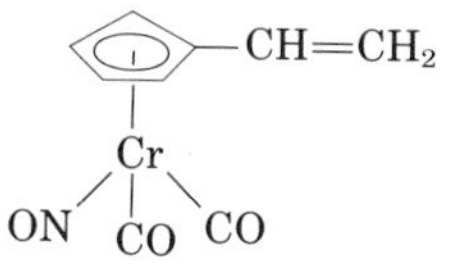

(**93**)

Preparation of Monomers

Friedel-Crafts Method. The original preparation of η^5-vinylcyclopentadienyl monomers was acetylation of the cyclopentadienyl organometallic compound by classic Friedel-Crafts chemistry, followed by standard reductions of the keto function to an alcohol and dehydration to give the vinyl function, eg, for (η^5-vinylcyclopentadienyl)tricarbonylmanganese (**1**):

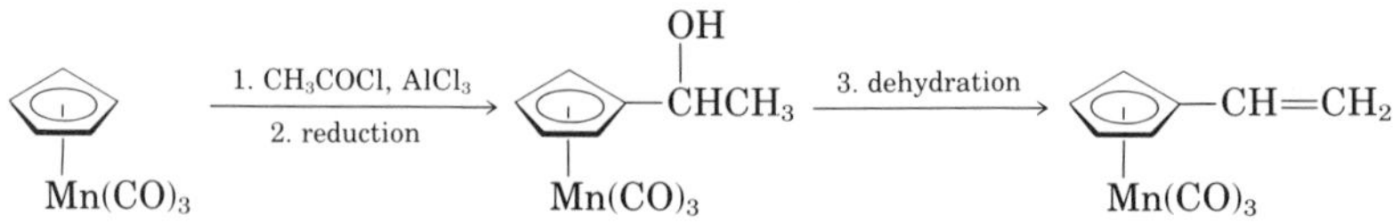

This general method has been used to make vinylferrocene (**92**) (180,191), (η^5-vinylcyclopentadienyl)dicarbonylnitrosylchromium (**93**) (189,190), vinylruthenocene (**22**) (192,193), and 3-vinylbisfulvalenediiron (**94**) (194,195).

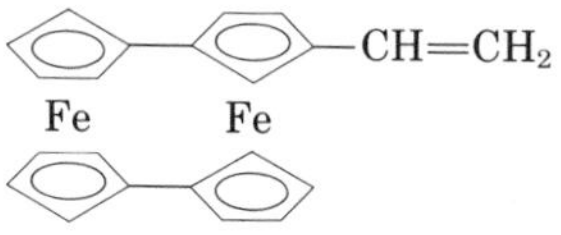

(**94**)

However, many cyclopentadienyl metal derivatives are not stable to Friedel-Crafts reaction conditions.

Wittig Method. The preparation of (η^5-vinylcyclopentadienyl)tricarbonyl-methyltungsten (**95**) was accomplished by using sodium formylcyclopentadienide (196,197), which, when refluxed with hexacarbonyltungsten in DMF, displaces three carbon monoxide molecules to give the (η^5-formylcyclopentadienyl)tri-carbonyltungsten anion. Methylation at tungsten and Wittig synthesis results in good yields (~65% overall) of the vinyl monomer (**95**). Attempts to generalize this method to other transition-metal monomers have not been successful because of low yields during the Wittig reaction.

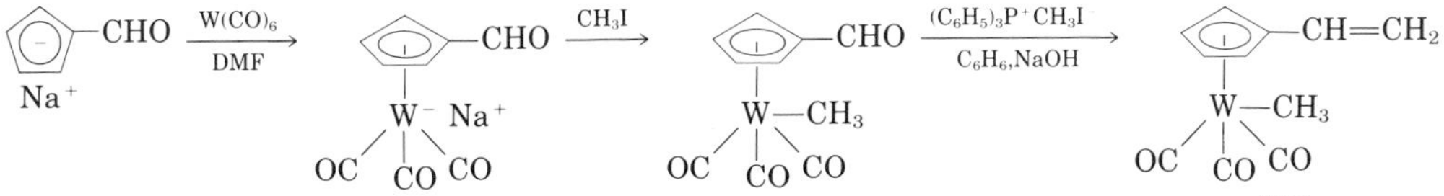

Alkenylcyclopentadienide Salts. A more general route to both η^5-vinyl- and η^5-isopropenylcyclopentadienyl monomers containing transition metals is the treatment of either 6-methylfulvene or 6,6-dimethylfulvene with lithium diiso-propylamide to give isopropenyl- (**96**) and vinylcyclopentadienide (**97**) salts, which react with a variety of organometallic compounds as summarized in Figure 10 (198,199).

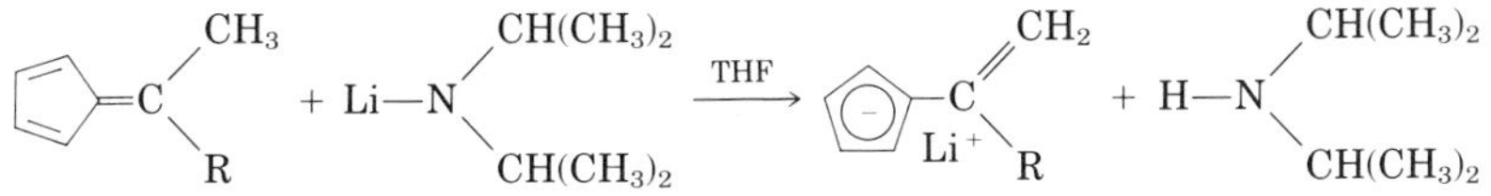

where R = CH₃ (**96**)

R = H (**97**)

For example, reaction of either (**96**) or (**97**) with either hexacarbonylmolyb-denum or tris(DMF)tricarbonyltungsten in THF, followed by nitrosylation of the intermediate metal carbonyl anions, gives the corresponding monomers (**103**) and (**104**) or (**106**) and (**107**), respectively (198–200). The monomers prepared in Figure 10 could not have been obtained by the Friedel-Crafts route. Of particular interest are the preparation of monomer (**98**), which is >50 wt % iridium, and its rhodium and cobalt analogues (**102**) and (**101**). It is clearly evident that a wide range of cyclopentadienylmetal compounds, which contain potentially polymerizable vinyl substituents, can be prepared by this new direct route involving organolithium salts (**96**) and (**97**).

Ligand Displacement. The displacement of either carbon monoxide or weakly ligated molecules from metals by π bonding organic moieties has been used to make both η^6-arene metal carbonyl monomers, such as (η^6-stryl)tricarbonyl-chromium (**108**) (201), and η^4-diene metal carbonyl monomers, such as (η^4-2,4-hexadien-1-yl acrylate)tricarbonyliron (**109**) (202).

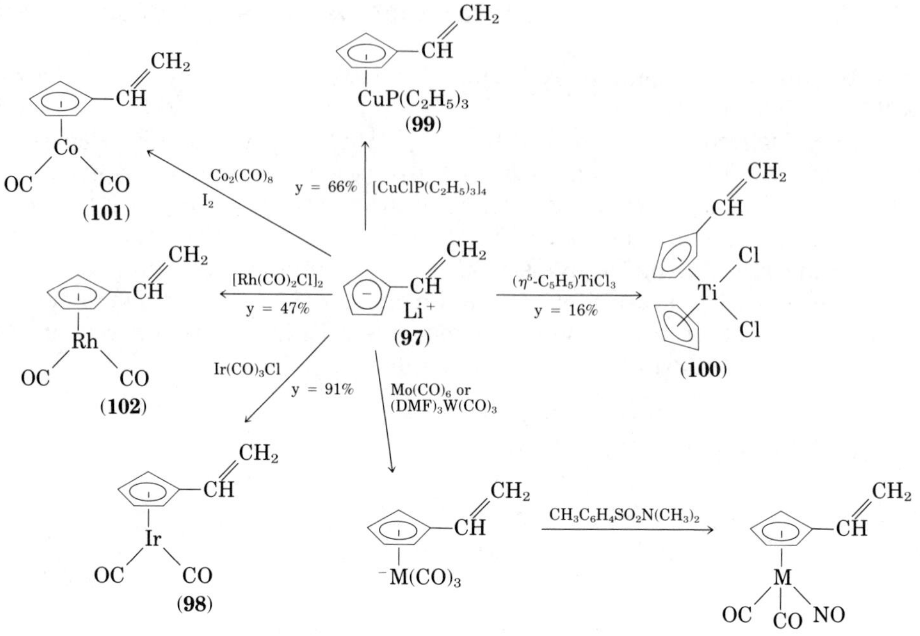

where M = Mo; y = 58% (**103**)

M = W; y = 36% (**104**)

(**a**)

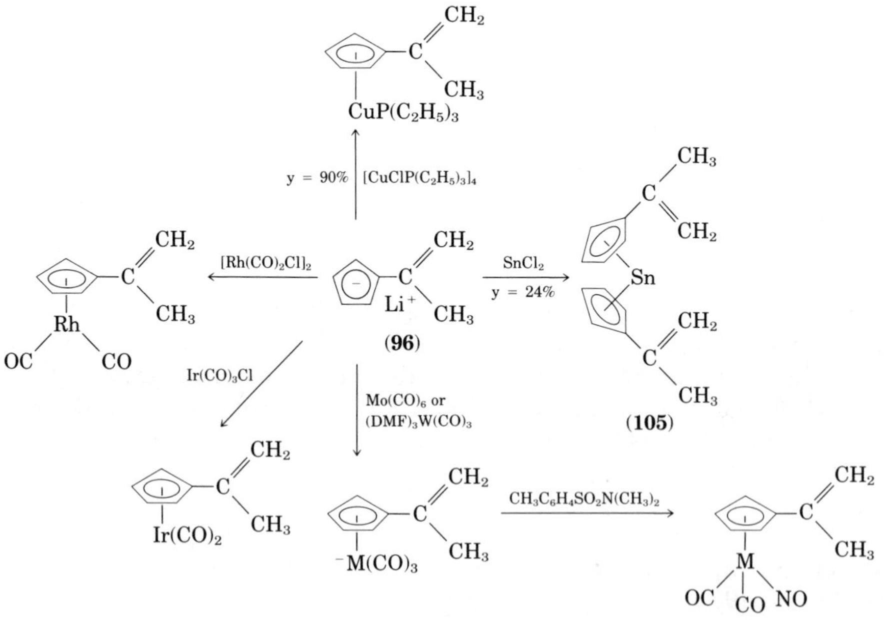

where M = Mo; y = 93% (**106**)

M = W; y = 24% (**107**)

(**b**)

Fig. 10. (**a**) Reaction of the vinylcyclopentadienide lithium salt, where y = yield; (**b**) reactions of the isopropenylcyclopentadienide lithium salt.

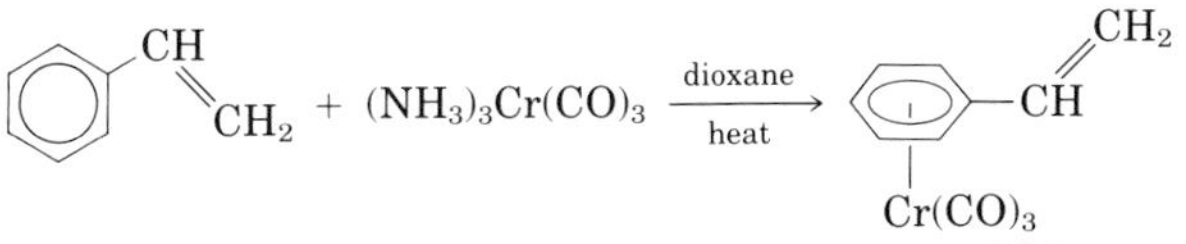

$$\text{(styrene)} + (NH_3)_3Cr(CO)_3 \xrightarrow[\text{heat}]{\text{dioxane}} \text{(arene)}-CH\!\!=\!\!CH_2 \quad Cr(CO)_3$$

(108)

$$CH_3\!\!-\!\!CH\text{(diene)}CH\!\!-\!\!CH_2OH + Fe(CO)_5 \xrightarrow{\text{heat}} CH_3\!\!-\!\!CH\cdots Fe \cdots CH\!\!-\!\!CH_2OH \xrightarrow{CH_2\!=\!CHCOCl}$$

$$CH_3\!\!-\!\!CH\cdots Fe \cdots CH\!\!-\!\!CH_2O\overset{O}{\overset{\|}{C}}CH\!\!=\!\!CH_2$$

$$OC \quad CO \quad CO$$

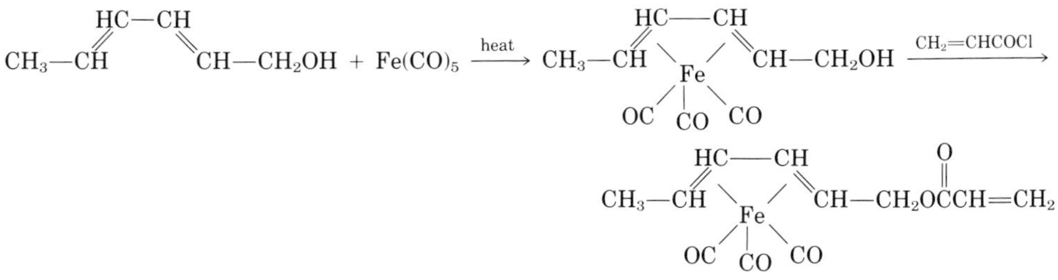

(109)

These methods have been extended to make acrylic organometallic monomers. For example, benzyl alcohol or 2-phenylethanol when refluxed in DME with $Cr(CO)_6$ give the corresponding (η^6-arene)tricarbonylchromium complexes, ie, (**110**), which can be esterified with methacryloyl chloride to give the methacrylate (**111**) (203). The synthesis and polymerization of the acrylate corresponding to (**111**) has been described in detail (204).

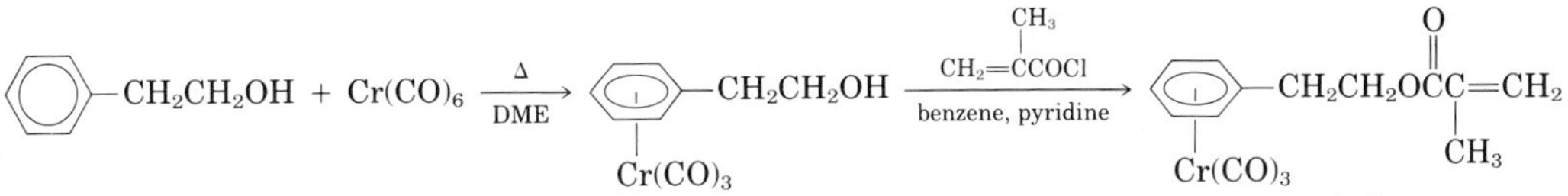

$$\text{(arene)}-CH_2CH_2OH + Cr(CO)_6 \xrightarrow[\text{DME}]{\Delta} \text{(arene)}-CH_2CH_2OH \xrightarrow[\text{benzene, pyridine}]{CH_2=CCOCl\ (CH_3)} \text{(arene)}-CH_2CH_2O\overset{O}{\overset{\|}{C}}C(CH_3)\!\!=\!\!CH_2$$

$$Cr(CO)_3 \qquad\qquad Cr(CO)_3$$

(110)   **(111)**

Acrylic and Methacrylic Esters. Acrylates or methacrylates of ferrocene (**112**) and (**113**) have been prepared from *N,N*-dimethylaminoferrocene by *N*-methylation, displacement of $(CH_3)_3N$ by hydroxide to give ferrocenylmethanol and esterification (205–208). Similar routes have been used to make other organometallic acrylates.

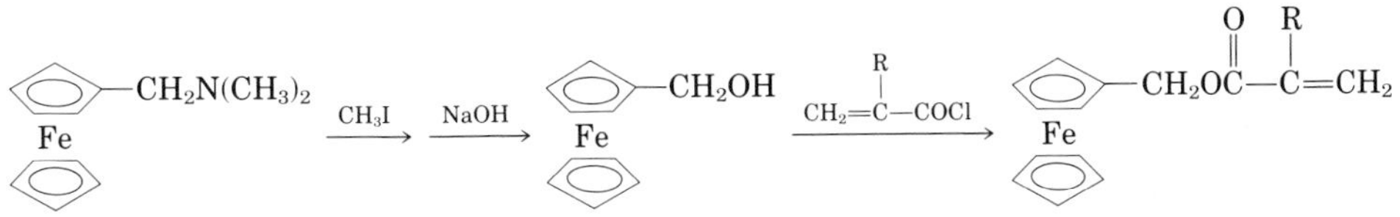

$$\text{Fc}-CH_2N(CH_3)_2 \xrightarrow[]{CH_3I\ \ NaOH} \text{Fc}-CH_2OH \xrightarrow{CH_2=C(R)-COCl} \text{Fc}-CH_2O\overset{O}{\overset{\|}{C}}\!\!-\!\!\overset{R}{C}\!\!=\!\!CH_2$$

where R = H (**112**)

R = CH_3 (**113**)

Polymerization of Vinyl Organometallic Monomers

A transition metal with its various readily available oxidation states and large steric bulk may be expected to exert unusual electronic and steric effects during polymerization. The homopolymerization of vinylferrocene (**92**) has been initiated by radical (180,181,191,209–211), cationic (184), coordination (212), and Ziegler-Natta (184) initiators.

Radical Polymerization.   Unlike the classic organic monomer, styrene, vinylferrocene undergoes oxidation at iron when peroxide initiators are employed. Thus azo initiators such as AIBN are preferred. The stability of the ferricinium ion makes ferrocene readily oxidizable by peroxides, whereas styrene would undergo polymerization. This is true of many other organometallic systems.

Unlike most vinyl monomers, the molecular weight of polyvinylferrocene does not increase with a decrease in initiator concentration (181). This is the result of the anomalously high chain-transfer constant for vinylferrocene ($C_m = 8 \times 10^{-3}$ vs 6×10^{-5} for styrene at 60°C). The rate law for vinylferrocene homopolymerization is first-order in initiator in benzene (182); thus intramolecular termination must occur.

$$R_p = 5.64 \times 10^{-4}[vinylferrocene]^{1.12}[AIBN]^{1.1} \text{ in benzene}$$

Mössbauer studies support a mechanism involving electron transfer from iron to the growing chain radical to give a zwitterion that terminates and ultimately results in a high spin Fe(III) complex (183). In dioxane the rate law is "normal," ie, first order in monomer and ca half order in AIBN, indicating a bimolecular termination mechanism (see also KINETICS OF POLYMERIZATION).

(92)

where $R_p = 5.99 \times 10^{-5}[vinylferrocene]^{0.97}[AIBN]^{0.42}$ in dioxane

Unusual homopolymerization kinetic behavior has also been observed for the radical-initiated polymerizations of (η^5-vinylcyclopentadienyl)tricarbonyl-manganese (1). In benzene, benzonitrile, and acetone, the rate was half order in AIBN concentration and three-halves order in the concentration of monomer (1). Specifically at 60°C, the rate expressions were

$$R_p = 1.303 \times 10^{-4}[M]^{1.45}[AIBN]^{0.48} \text{ in benzene}$$

$$R_p = 1.980 \times 10^{-4}[M]^{1.58}[AIBN]^{0.47} \text{ in benzonitrile}$$

$$R_p = 1.500 \times 10^{-4}[M]^{1.54}[AIBN]^{0.47} \text{ in acetone}$$

A rate expression of the form $k[M]^{1.5}[I]^{0.5}$ is derived if the initiator efficiency is low and therefore the initiator is proportional to $[M]$, ie, $f = f'[M]$. This rate equation requires that the degree of polymerization follow the expression:

$$\text{DP} = R_p/R_i = [k_p/(2f'k_tk_d)^{0.5}]([M]/[I])^{0.5}$$

Molecular weight measurements confirm that the degree of polymerization is proportional to $([M]/[I])^{0.5}$ (1).

Radical-initiated homopolymerization kinetic studies of (η^5-vinylcyclopentadienyl)methyltricarbonyltungsten (95) were carried out in benzene. The rate expression was

$$R_p = 1.13 \times 10^{-2}[M]^{0.8}[I]^{2.3}$$

$$\text{(95)}$$

The homopolymerizations were sluggish and several reinitiations were required to obtain good conversions. Chain transfer (qv) and chain termination by hydrogen abstraction from the tungsten-bound methyl group, the cyclopentadienyl ring, or the backbone methine groups were ruled out as causes of the sluggishness (197,213).

Titanium allyl and methacrylate monomers (**114**) and (**115**) give only very low molecular weight materials under benzoyl peroxide initiation and, in styrene copolymerizations, only small amounts of (**114**) and (**115**) are incorporated (214).

$$\text{(114)} \qquad \text{(115)}$$

This is in accord with a low reactivity and a high chain-transfer activity for these monomers (215). Even less reactive is (η^4-hexatrienyl)tricarbonyliron (**116**) (202).

$$\text{(116)}$$

It does not undergo either radical-initiated homopolymerization or copolymerization. Indeed, it inhibits the polymerization of both styrene and methyl acrylate. Presumably, the radical resulting from chain addition to the vinyl group of (**116**) is stable and does not undergo propagation.

The organometallic acrylates and methacrylates undergo ready radical-initiated homopolymerization and copolymerizations, and appear to behave normally. Thus, for example, homopolymerizations of 2-ferrocenylethyl acrylate (**117**) and ferrocenylethyl methacrylate (**118**) have been found to be first order in monomer and half order in initiator (216).

Mixed oxidation-state polymers have been formed from poly(ferrocenylmethyl methacrylate) and poly(vinylferrocene) (**119**) by treatment with electron acceptors such as dichlorodicyanoquinone or iodine (205,206,217–219). Mössbauer spectroscopy has been used in analyzing the fraction of ferrocene moieties oxidized to ferricenium groups (206,217). These mixed oxidation-state polymers are semiconductors via charge-hopping mechanisms.

(119)

Anionic Polymerization. The anionic initiation of vinylferrocene (**92**) does not occur (226). Furthermore, (η^5-vinylcyclopentadienyl)tricarbonylmanganese (**1**), (η^5-vinylcyclopentadienyl)dicarbonylnitrosylchromium (**93**), (η^5-vinylcyclopentadienyl)tricarbonylmethyltungsten (**95**), and (η^5-2,4-hexadien-1-yl methacrylate)tricarbonyliron all resist anionic initiation (220). The vinylcyclopentadienyl monomers resist anionic initiation because the vinyl group is exceptionally electron-rich in every case (17). Copolymerization of (**1**) with methyl acrylate, initiated by *n*-butyllithium or sodium naphthalide, yields copolymers with very low molar incorporation of (**1**).

Unlike the vinylcyclopentadienyl monomers, ferrocenylmethyl methacrylate (**113**) and acrylate (**112**) are initiated using a variety of anionic systems (220–222). The effect of the ferrocenylmethyl methacrylate/LiAlH$_4$ mole ratio on molecular weight is regular. By varying this ratio from 17 to 300, the values of $\overline{M}_n$ and $\overline{M}_w$ increase from 3,000 and 5,400 to 277,000 and 724,000, respectively (222). Under vacuum a solution of ferrocenylmethyl methacrylate polymerized using LiAlH$_4$ initiation exists as living polymers (221), allowing the use of LiAlH$_4$–tetramethylethylenediamine (TMEDA) initiation to make block copolymers of ferrocenylmethyl methacrylate (**113**) with methyl methacrylate and acrylonitrile (221) (see (**120**) and (**121**)). The poly(ferrocenylmethyl methacrylyl) anion (**122**) was unable to initiate styrene (212). Therefore, (**113**) was added to THF solution of living polystyrene, originally prepared by sodium naphthalide initiation of styrene, to give ferrocenylmethyl methacrylate–styrene block copolymers (**123**) (221).

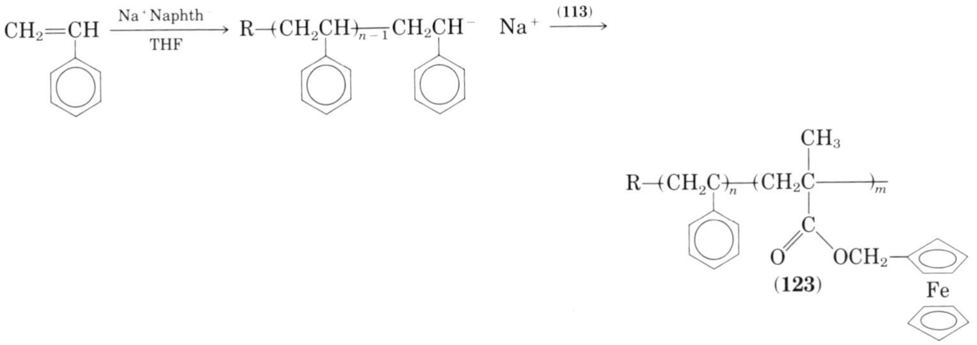

$$CH_2=CH \xrightarrow[THF]{Na \cdot Naphth} R \text{-}(CH_2CH)_{n-1}CH_2CH^- \quad Na^+ \xrightarrow{(113)}$$

(with phenyl substituents on the vinyl/repeat units)

R—(CH₂C)ₙ—(CH₂C—)ₘ structure with CH₃, phenyl, and C=O, OCH₂, Fe ferrocenyl groups

(123)

Cationic Polymerization. The electron-rich cyclopentadienyl ring in vinyl monomers (**1**), (**92**)–(**95**), (**98**)–(**100**), (**102**), and (**103**) is able to stabilize adjacent positive charge; the exceptional stability of α-ferrocenyl carbenium ions is well known. Therefore, the vinyl group in these monomers behaves as electron-rich in radical- and cationic-initiated homo- and copolymerizations and resists anionic initiation. An example is the ready cationic polymerization of 1,1'-divinylferrocene (**124**).

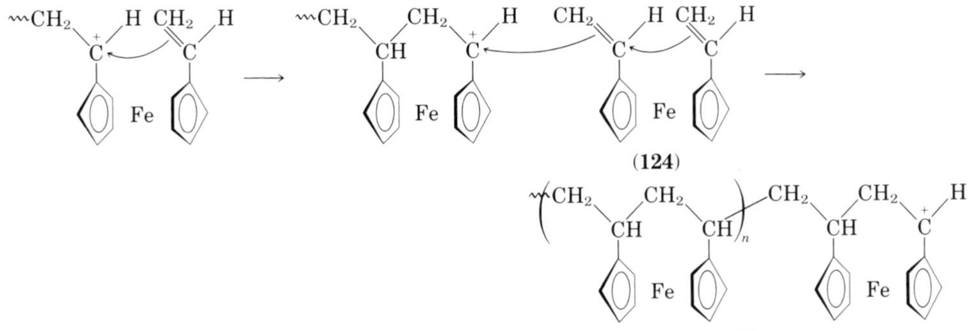

(reaction scheme showing structures **(124)** and **(125)** with ferrocene (Fe) units, CH₂, CH groups)

(124)

(125)

Molecular weights up to 35,000 have been obtained using boron trifluoroetherate initiation (223). This polymerization forms cyclolinear structures (**125**) from intramolecular electrophilic attack of the α-ferrocenylcarbenium ion on the adjacent vinyl group (193,223). Cationic initiation of some of the isopropenyl monomers shown in Figure 10**b** has been discussed (199,224,225). Although these (η^5-isopropenylcyclopentadienyl)metal monomers can be considered analogues of α-methylstyrene, (**105**) and (**106**) (Fig. 10**b**) and (**126**) and (**127**) do not readily homo- or copolymerize under a variety of cationic conditions and high molecular weight homo- or copolymers have not been achieved. The reason is unknown, but could be due to a very high ceiling temperature or perhaps to the exceptionally high stability of the tertiary α-cations.

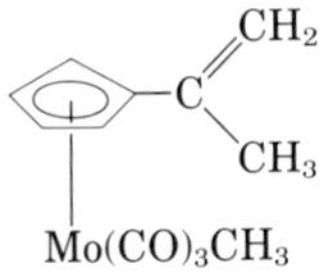

(structures **(126)** with CH₂, CH₃, Mo(CO)₃CH₃ groups and **(127)** with CH₂, CF₃, Fe groups)

(126) 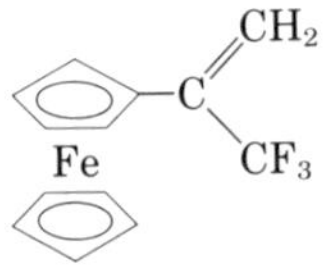 **(127)**

Most notably, in (**127**) the α-ferrocenyl carbenium ion stability is lowered by the trifluoromethyl group and the cation should be more reactive, but it does not polymerize.

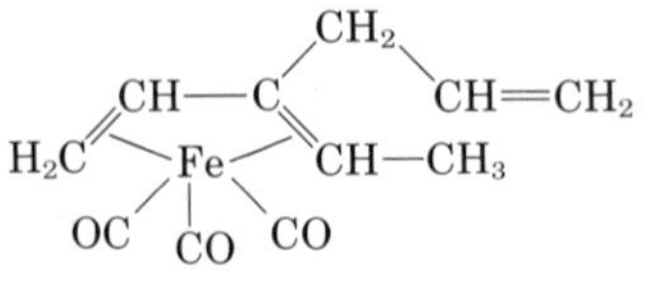

(**128**)

3-Allyl-1,4-η^4-pentadienylirontricarbonyl (**128**) gave low homopolymer yields (DP $\approx$ 20) when subjected to cationic initiation (226,227) and is inert to anionic and radical initiation. The organometallic vinyl ethers (**129**) and (**130**) give low-to-moderate yields of polymer under cationic initiation and require higher temperatures and relatively large amounts of catalyst. Molecular weights of the resulting polymers are M_n = 15,000–21,000.

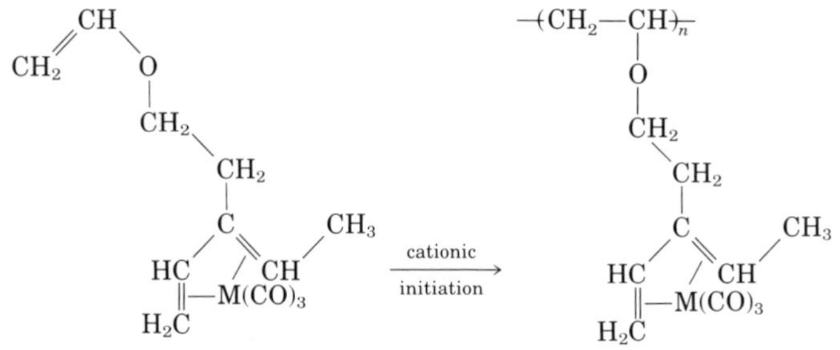

where M = Fe (**129**)

M = Ru (**130**)

Copolymerization.   The excessive electron-rich nature of vinylferrocene as a monomer is illustrated in its copolymerizations with maleic anhydride. 1:1 alternating copolymers are obtained over a wide range of M_1/M_2 feed ratios and $r_1 r_2$ = 0.003 (210). A large number of detailed copolymerization studies have been carried out between vinylferrocene and organic monomers such as styrene (193), N-vinyl-2-pyrrolidinone (211), methyl acrylate (193), methyl methacrylate (193), N-vinylcarbazole (228), and acrylonitrile (193). The relative reactivity ratios (r_1 and r_2) were obtained and from them the values of the Alfrey-Price Q and e parameters (see COPOLYMERIZATION). The value of e is a semiempirical measure of the electron richness of the vinyl group.

The best value of e for vinylferrocene is ca -2.1, compared to the e values of maleic anhydride ($+2.25$), p-nitrostyrene ($+0.39$), styrene (-0.80), p-N,N-dimethylaminostyrene (-1.37), and 1,1'-dianisylethylene (-1.96). This illustrates the exceptional electron density on the vinylferrocene vinyl group. Representative copolymerizations of monomer (**95**) are illustrated in Figure 11. The values of Q, e, and the reactivity ratios in styrene copolymerizations for several organometallic monomers are shown in Table 1 (2,199). Remarkably, the presence of different metals or the presence of electron-withdrawing carbonyl groups on metal atoms attached to the η^5-vinylcyclopentadienyl group does not significantly diminish the electron richness of the vinyl group as measured by the Alfrey-Price

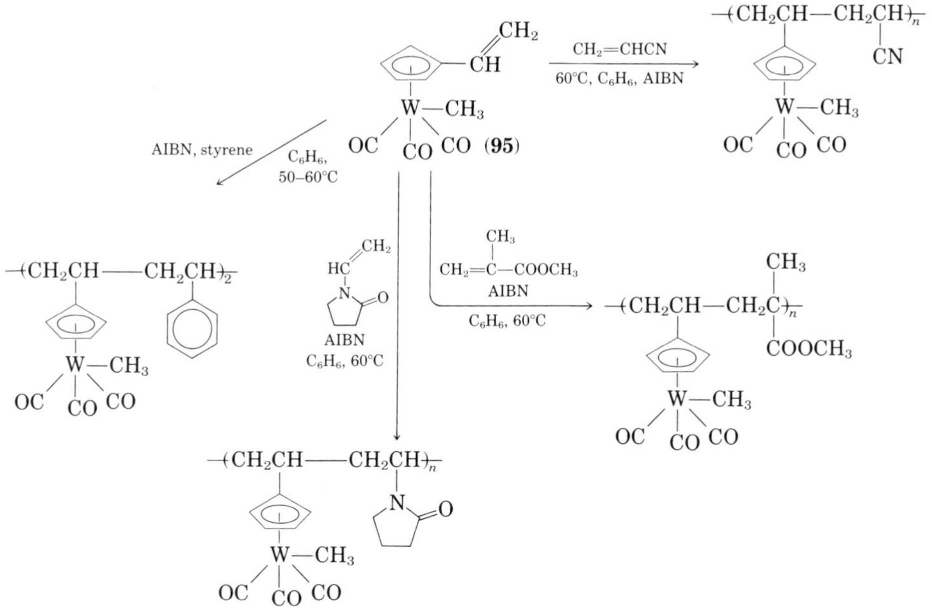

Fig. 11. Copolymerizations involving the η^5-vinylcyclopentadienyl group.

e value. Also, the powerful electron-donating effect of the p-Pd[P(C$_6$H$_5$)$_3$]$_2$Cl group on the phenyl ring in monomer (**131**) is illustrated (229) based on its e value of -1.37 (Table 1).

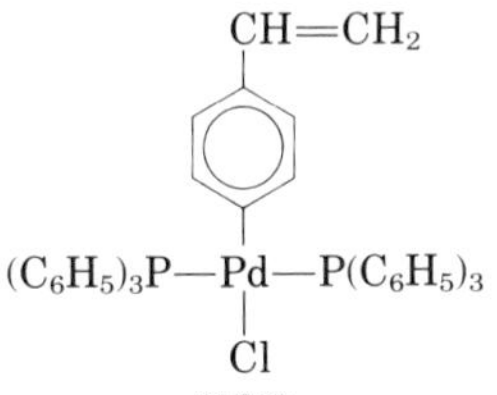

It is a significantly stronger electron donor than the p-(CH$_3$)$_2$N-group. The p-CCo$_3$(CO)$_9$ function in monomer (**132**) is also strongly electron-donating (230).

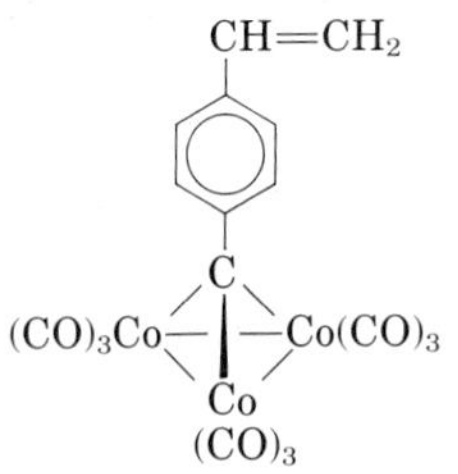

The values of Q (Table 1) indicate substantial resonance stabilization of the α-organometallic propagating radical. Thus the electron-deficient radical center is delocalized into the ring. As expected, these monomers resist homo- or copolymerization via anionic initiation.

Table 1. Alfrey-Price Q,e Values for Some Organometallic Vinyl Monomers[a]

Monomers	Structure	Reactivity ratios[b]		Q value	e value
		r_1	r_2		
Organometallics					
vinylferrocene	**92**	0.09	2.91	1.03	−2.1
(η^5-vinylcyclopentadienyl)tri-carbonylmanganese	**1**	0.098	2.50	1.1	−1.99
(η^5-vinylcyclopentadienyl)di-carbonylnitrosylchromium	**93**	0.03	0.82	3.1	−1.98
(η^5-vinylcyclopentadienyl)di-carbonylnitrosylmolybdenum	**103**	0.31	0.83	3.1	−1.98
(η^5-vinylcyclopentadienyl)tri-carbonylmethyltungsten	**95**	0.16	1.55	1.66	−1.98
(η^5-vinylcyclopentadienyl)di-carbonyliridium	**98**	0.29	1.68	4.1	−2.08
p-styrylchlorodi(triphenyl phosphine)palladium	**131**				−1.62
Nonorganometallics[c]					
1,1-di(4-*t*-butyl-phenyl)ethylene	$\left[(CH_3)_3C \text{—} \bigcirc \right]_2 C\text{=}CH_2$			1.46	−1.96
2-vinylthiophene	(S ring) CH=CH₂			2.86	−0.80
styrene	⬡—CH=CH₂			1.00	−0.80
p-nitrostyrene	$O_2N\text{—}\bigcirc\text{—}CH\text{=}CH_2$			1.63	0.39
propylene	$CH_2\text{=}CHCH_3$			0.002	−0.78
maleic anhydride	(maleic anhydride ring)			0.23	2.25
p-(*N,N*-dimethyl-amino)styrene	$(CH_3)_2N\text{—}\bigcirc\text{—}CH\text{=}CH_2$				−1.37

[a] Refs. 2 and 199.
[b] Styrene copolymerization.
[c] For comparison.

Many copolymerization studies have been conducted on organometallic monomers and a substantial number of reactivity ratios are now available for various monomer pairs in radical-initiated copolymerizations (Table 2). The r_1 values of ferrocenylmethyl acrylate (**112**) are smaller than those of 2-ferrocenylethyl acrylate (**117**) in copolymerizations with common comonomers. Similarly, the r_1 values of ferrocenylmethyl methacrylate (**113**) are smaller than the corresponding values of 2-ferrocenylethyl methacrylate (**118**) for corresponding comonomers. This shows the steric effect of the bulky ferrocene moiety. When it is removed from the reaction center, the monomer becomes more reactive. In general, the steric effects of the organometallic groups in monomers (**109**), (**112**), (**113**), (**117**), (**118**), and (**133**)–(**135**) reduce their reactivity relative to the replacement of this group with a hydrogen.

Table 2. Reactivity Ratios for Selected Copolymerizations of Vinyl Organometallic Monomers

Organometallic monomer, M_1	Comonomer, M_2	Reactivity ratios r_1	r_2	Refs.
vinylferrocene (**92**)	styrene	0.08	2.50	193
	methyl methacrylate	0.52	1.22	193
	methyl acrylate	0.82	0.62	193
	acrylonitrile	0.15	0.16	193
	N-vinylpyrolidinone	0.66	0.42	211
	N-vinylcarbazole	0.47	0.20	228
(η^5-vinylcyclopentadienyl)tricarbonyl-manganese (**1**)	styrene	0.1	2.50	2
	vinyl acetate	2.35	0.06	2
	methyl acrylate	0.19	0.47	2
	acrylonitrile	0.19	0.22	2
	vinylferrocene	0.44	0.49	2
	N-vinylpyrrolidinone	0.14	0.09	211
(η^5-vinylcyclopentadienyl)dicarbonylnitrosyl-chromium (**93**)	styrene	0.30	0.82	189,190
	N-vinylpyrrolidinone	5.3	0.08	189,190
(η^5-vinylcyclopentadienyl)tricarbonylmethyl-tungsten (**95**)	styrene	0.16	1.55	197
(η^5-vinylcyclopentadienyl)dicarbonyliridium (**98**)	styrene	0.28	0.76	198,199
(η^6-stryryl)tricarbonylchromium (**108**)	styrene	0.0	1.35	201
	methyl acrylate	0.0	0.70	201
ferrocenylmethyl acrylate (**112**)	styrene	0.02	2.3	207
	methyl acrylate	0.14	4.46	207
	methyl methacrylate	0.08	2.9	207
	vinyl acetate	1.44	0.46	207
	maleic anhydride	0.61	0.11	231
ferrocenylmethyl methacrylate (**113**)	styrene	0.03	3.7	207
	methyl acrylate	0.08	0.82	207
	methyl methacrylate	0.12	3.27	207
	vinyl acetate	1.52	0.20	207
	N-vinylpyrrolidinone	3.71	0.05	231
	acrylonitrile	0.30	0.11	231
	maleic anhydride	0.28	0.10	231
2-ferrocenylethyl acrylate (**117**)	styrene	0.41	1.06	232
	vinyl acetate	3.4	0.07	232
	methyl acrylate	0.76	0.69	232
2-ferrocenylethyl methacrylate (**118**)	styrene	0.08	0.58	232
	vinyl acetate	8.79	0.06	232
	methyl methacrylate	0.20	0.65	232
(η^6-benzylacrylate)tricarbonylchromium (**134**)	styrene	0.10	0.34	233
	methyl acrylate	0.56	0.63	233
(η^6-2-phenylethyl acrylate)tricarbonyl-chromium	styrene	0.1	0.5	234
	methyl acrylate	0.3	1.0	234
	acrylonitrile	0.6	0.2	234
(η^6-2,4-hexadien-1-yl acrylate)tricarbonyl-iron (**109**)	styrene	0.26	1.81	202
	vinyl acetate	2.0	0.05	202
	methyl acrylate	0.30	0.74	202
	acrylonitrile	0.34	0.74	202
(η^6-phenylethyl methacrylate)tricarbonyl-chromium	styrene	0.04	1.35	203
	methyl methacrylate	0.09	1.19	203
	acrylonitrile	0.07	0.79	203

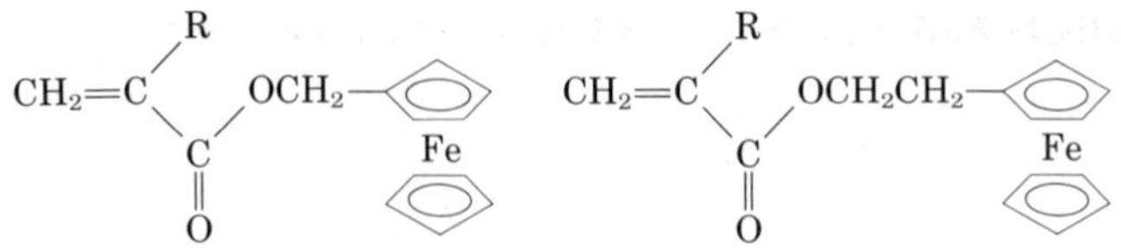

less reactive

where R = H   (112)

R = CH$_3$ (113)

more reactive

where R = H   (117)

R = CH$_3$ (118)

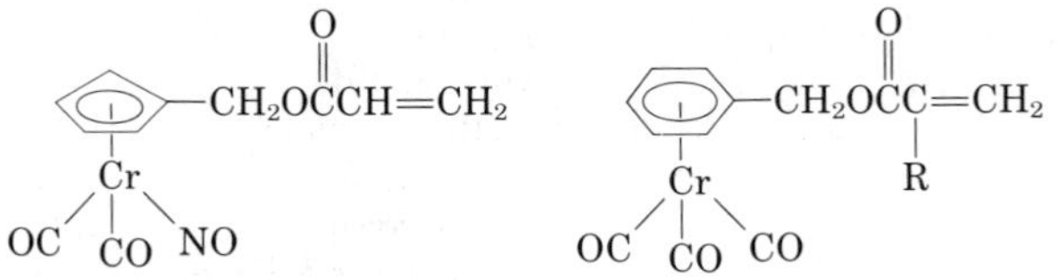

(133)

(134) R = H
(135) R = CH$_3$

Derivatization of Preformed Polymers

Catalytic Applications.   Many preformed vinyl polymers and resins have been derivatized with organometallic functions to make polymer-bound catalysts. This topic has been extensively reviewed (21–25) (see also CATALYSTS, POLYMER-SUPPORTED).

For example, polystyrene–divinylbenzene resins, modified with triphenyl-phosphine groups, have been used to prepare heterogenized nickel catalysts for use in ethylene oligomerization (235). In the equations, Ni(COD)$_2$ represents dicyclooctadienylnickel. Polymers (**136a**) and (**136b**) catalyze the linear oligomerization of olefins at rates >3 mol C$_2$H$_4$ per mol of Ni per s. The selectivity toward α-alkenes is 93–99% and the linearity >99%.

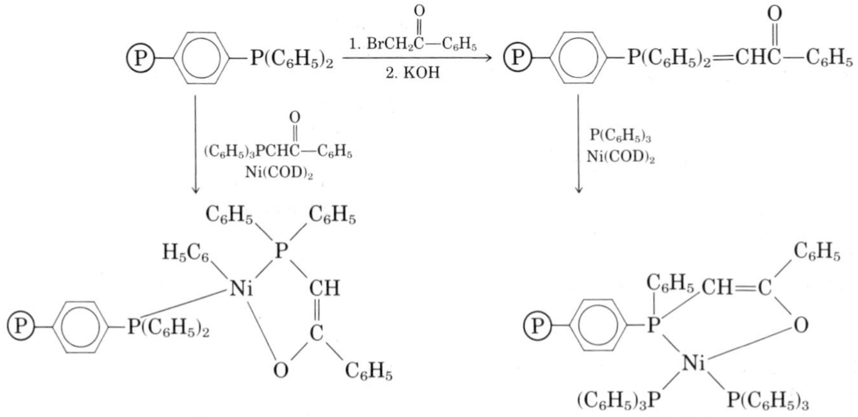

(136a)

(136b)

Polystyrene resins also react with asymmetric chelating ligands such as (−)2,3-dihydroxy-1,4-bis(diphenylphosphino)butane (DIOP). In one example, the asymmetric catalytic hydroformylation of styrene has been carried out using a polystyrene resin-bound (−)DIOP (**137**) to which PtCl$_2$ and SnCl$_2$ are attached

(236,237). The Pt atoms are chelated by the asymmetric $(-)$DIOP ligand (**138**) and therefore exist in a chiral environment. Using the dibenzophosphole analogue of DIOP bound to a polymer together with $PtCl_2$ and $SnCl_2$ gives (**139**). The highest enantiomeric excesses ever achieved in styrene hydroformylations (up to $\sim80\%$ ee) resulted when the homogeneous analogue of (**139**) was used as a catalyst.

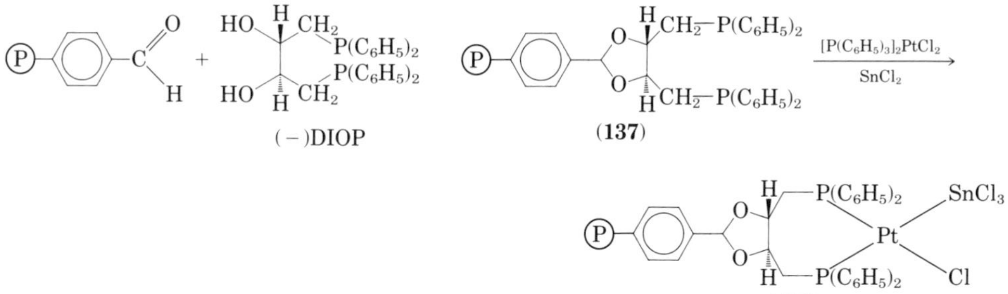

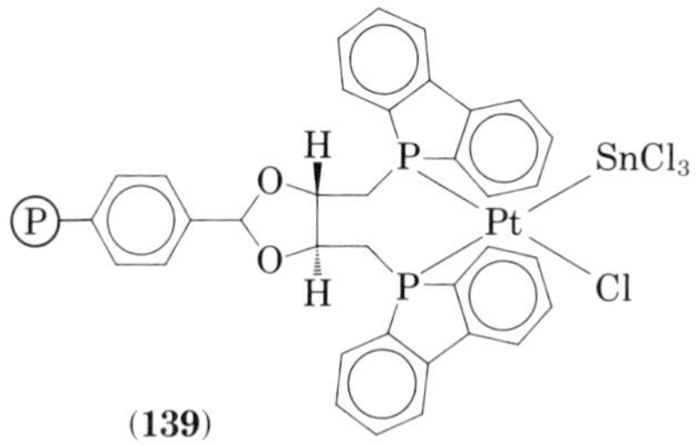

$(-)$DIOP

(**137**)

(**138**)

(**139**)

Because polystyrene is hydrophobic and many asymmetric rhodium-catalyzed hydrogenations proceed more effectively in polar or hydroxylic solvents, hydrophilic resins (**140**) have been prepared from hydroxyethyl methacrylate and ethylene dimethyacrylate, which incorporate asymmetric pyrrolidine phosphine ligands (24,238,239). The pyrrolidine phosphine ligands are built into the matrix in order to chelate rhodium in a chiral environment.

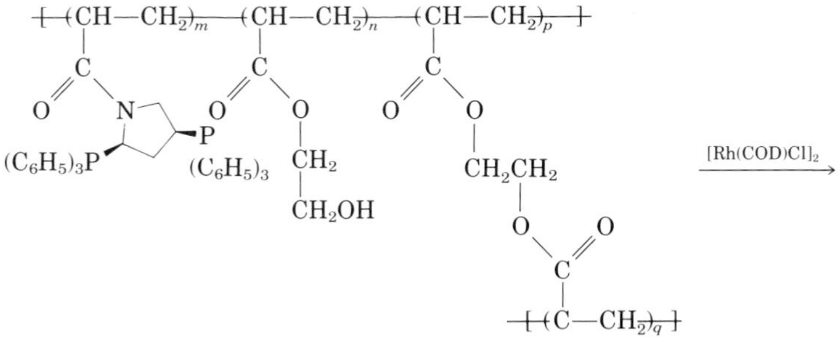

(**140**)

chiral polymer-bound rhodium catalysts

These polymer-bound chiral rhodium catalysts have been used to catalyze the hydrogenation of α-acrylamidocarboxylic acids at 5.5 MPa (800 psi) and 20°C. Enantiomeric excesses of 83–91% were achieved. The use of polymeric bipyridines

(**141**) as hydrogenation catalysts to which metals are coordinated has been studied (240–242). A wide variety of metals are readily chelated. Significant resistance to metal leaching in many catalytic reactions was observed when diaminobipyridine reacts with toluenediisocyanate to form polyureas (qv) to which metal ions are complexed.

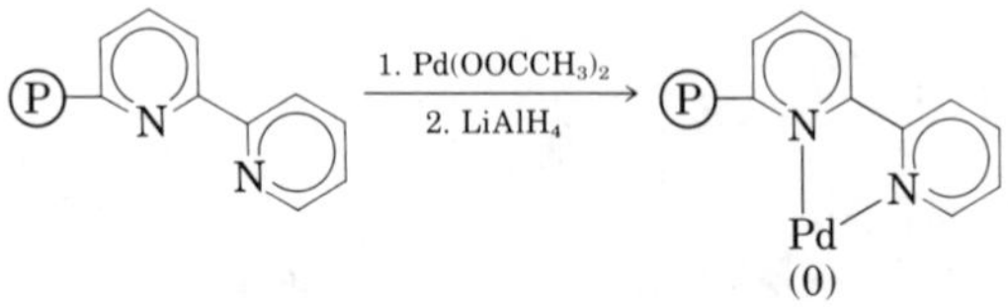

(**141**)

Polymer-supported rhodium complexes (**142**) have been used in gas-phase hydroformylations of propylene (243). H_2, CO, and $CH_2{=}CHCH_3$ are pumped through fixed-catalyst beds where the reaction takes place at the gas–solid interface. Polystyrene XAD-2 resins, functionalized at the surface with $-P(C_6H_5)_2$ groups, ligate the Rh complexes. No deactivation was observed over a 500-month period of use.

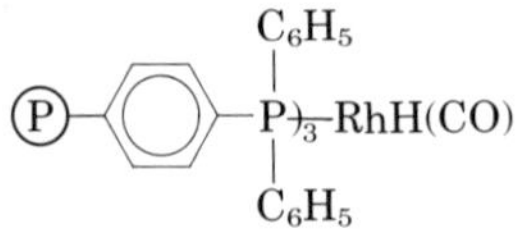

(**142**)

Selectivity of Catalysts. Steric effects of polymer matrices and the barriers they impose to diffusion of reagents to the attached catalytic site have been employed to modify the regioselectivity of homogeneous hydrogenations catalyzed by $RhCl[P(C_6H_5)_3]_3$. For example, when polystyrene–2% divinylbenzene resins, eg, (**143**), are used to anchor $RhCl[P(C_6H_5)_3]_3$, the relative rates of hydrogenation of double bonds a and b in steroids (**144**) are changed (244). The side-chain double bond b is more selectively hydrogenated with the polymer-bound catalyst because this double bond is more accessible. The selectivity factor,

$$\frac{(\%a/\%b)_{\text{reduced by polymeric catalyst}}}{(\%a/\%b)_{\text{reduced by homogeneous catalyst}}}$$

is 2–4 for various steroids in benzene. In the poorer polymer-swelling solvent, 1:1 ethanol:benzene, the selectivity factor increases to >8 as a result of increased crowding at the catalyst sites.

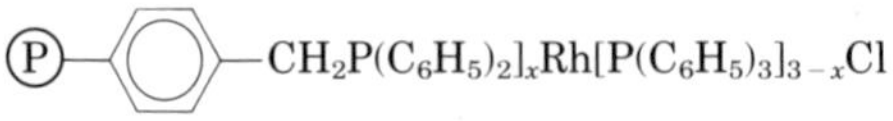

(**143**) 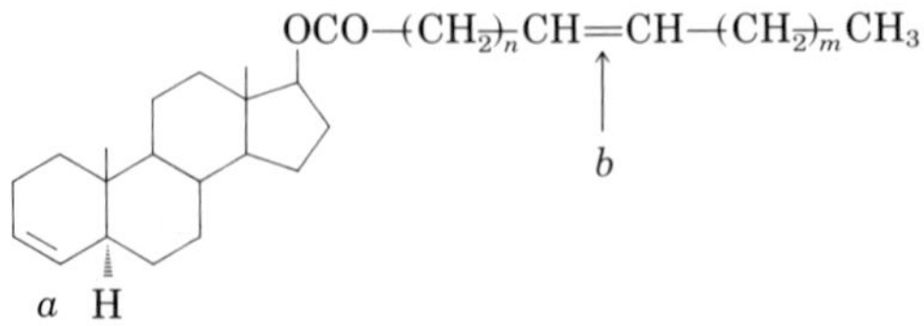   (**144**)

The use of polymeric diphenylbenzylphosphine complexes of $PdCl_2$ to catalyze the hydrogenation of linolenic acid gives faster rates than the use of homogeneous $Pd[P(C_6H_5)_3]_2$ catalysts (245,246). In the allylic amination of both *trans*- and *cis*-3-acetoxy-5-methoxycarbonyl-1-cyclohexene (**145**) with diethylamine, it was found that a mixture of cis and trans products result when using homogeneous Pd(0) catalysts. However, the application of polymer-anchored (benzyldiphenylphosphine)palladium (2% cross-linked) catalysts gave a completely stereospecific reaction to form (**146**) (247).

The use of polymer-bound palladium chloride catalysts in ethoxycarbonylation of 1-pentene gives higher normal:branched product ratios than do homogeneous catalysts at high P:Pd ratios. Also, the polymeric catalysts are more robust than their homogeneous analogues and can be used up to 160°C (76).

Polymer-bound $Rh[P(C_6H_5)_3]_3H(CO)$ gives higher normal–branched product selectivities in hydroformylation than the corresponding homogeneous systems at higher phosphine concentrations in the polymer (63,248). Also, the H_2:CO ratio used in this reaction can be increased to a higher value using the polymer-bound catalyst without loss of hydroformylation yields to competing hydrogenation.

Polymer-bound, rhodium-amine catalytic complexes have been used to convert alkenes far more selectively to alcohols under relatively mild conditions compared to their homogeneous counterparts (57).

Multistep Catalytic Reactions.   Two or more anchored catalyst centers have been used to carry out sequential multistep catalytic reactions (58–60). The products from the first catalytic reaction become reagents for the second reaction. An example is the sequential catalytic cyclooligomerization–selective hydrogenation reaction and the cyclooligomerization–hydroformylation sequences illustrated in Figure 12 where butadiene is the starting material (58,59,249). The polymers used were styrene–divinylbenzene resins functionalized with —$P(C_6H_5)_2$ groups which ligated the metal complexes.

The nickel species is a cyclooligomerization catalyst with a selectivity that varies with CO partial pressure. The ruthenium complex, bound to the same polymer particles, catalyzes the selective hydrogenation of dienes and trienes to monoenes. The rhodium complex catalyzes the selective hydroformylation of ter-

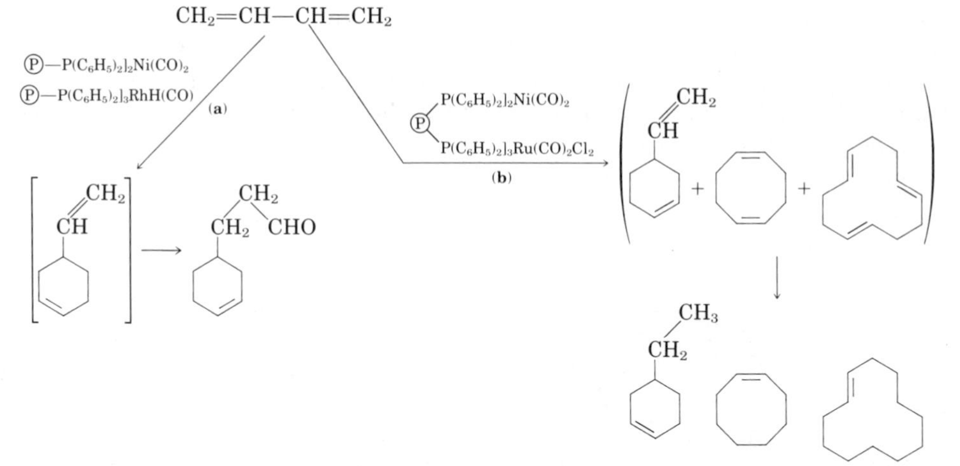

Fig. 12. Multistep catalytic reaction. (**a**) The reaction conditions are 115°C under CO for 24 h, then H_2/CO at 1.7 MPa (250 psi) and 70°C; (**b**) the reaction conditions are 100°C for 24 h, then H_2 at 1 MPa (150 psi) and 160°C.

minal olefins. Linear butadiene dimerization followed by selective hydroformylation has been carried out (250).

$$CH_2{=}CH{-}CH{=}CH_2 \xrightarrow[\text{Ⓟ}{-}P(C_6H_5)_2]_3RhH(CO)]{\text{Ⓟ}{-}P(C_6H_5)_2]_2NiBr_2\ +\ NaBH_4} [CH_3CH{=}CHCH_2CH{=}CH{-}CH{=}CH_2] \longrightarrow CH_3CH{=}CHCH_2CH{=}CH{-}CH_2CH_2CHO$$

The conditions were 100°C for 24 h, then H_2/CO at 3.4 MPa (500 psi) and 65–70°C. A three-step sequential synthesis of 2-ethylhexanal from propene and synthesis gas under a single set of conditions has been developed (61). The rhodium complex catalyzes hydroformylation (step 1) and hydrogenation (step 3), and the secondary amine catalyzes the aldol condensation (step 2).

$$CH_3{-}CH{=}CH_2 + H_2 + CO \xrightarrow[\text{step 1}]{\text{Ⓟ}(C_6H_4CH_2P(C_6H_5)_2]_2Rh(CO)Cl)(C_6H_4CH_2NHC_2H_5)} CH_3CH_2CH_2CHO \xrightarrow[\text{condensation step 2}]{\text{aldol}}$$

$$CH_3CH_2CH_2CH{=}\underset{\displaystyle CH_2CH_3}{\underset{|}{C}}{-}CHO \xrightarrow[\text{step 3}]{H_2} CH_3CH_2CH_2CH_2\underset{\displaystyle CH_2CH_3}{\underset{|}{C}H}{-}CHO$$

The polymer matrix can isolate mutually destructive catalytic sites from each other if the cross-link density is sufficiently high. In one case polymer-bound titanocene was generated and its activity studied in catalytic olefin hydrogenations. Titanocene is known to dimerize to a catalytically inactive species extremely rapidly. Comparing the hydrogenation activity as a function of titanocene-substitution levels in the polymer to a random-site statistical model illustrates that only adjacent titanocene sites dimerize; nonadjacent sites are matrix-isolated and survive to catalyze hydrogenation (67).

$(P)-C_6H_4CH_2$ — Ti (Cp)$_2$Cl$_2$ $\xrightarrow{C_4H_9Li}$ $(P)-C_6H_4CH_2$ — Ti active catalyst

adjacent sites dimerize

(P) — Ti···H···Ti — (P) \quad inactive

A method has been described that allows simultaneous reactions to be carried out with incompatible catalytic species in the same reaction vessel; the method also permits the separate recovery and isolation of both catalysts and the product (61). Linear polyethylenes (mol wt >1200), which are terminated with —P(C$_6$H$_5$)$_2$ groups, ligate rhodium to form a soluble homogeneous hydrogenation catalyst. In the same vessel a cross-linked polyvinylpyridine-bound Cr(VI) oxidant PVPCr (made from HCl and CrO$_3$) is present. When 3-cyclohexenylmethanol is heated to 100°C in xylene under H$_2$, the PVPCr oxidizes the alcohol to an aldehyde and the double bond is hydrogenated. The soluble polyethylene–rhodium catalyst is not stable to Cr(VI) reagents, but can coexist with PVPCr because the Cr(VI) sites are inside the polyvinylpyridine matrix where the soluble rhodium–polymer complex cannot diffuse. Thus the polyethylene–rhodium catalyst is physically unable to reach and react with the Cr(VI) sites.

$\text{(cyclohexenyl)}-CH_2OH \xrightarrow[\text{xylene 100°C}]{\begin{array}{c}\text{PVPCr}\\ CH_3\text{---}(CH_2CH_2)_n\text{---}CH_2P(C_6H_5)_2]_3RHCl \text{ and } H_2\end{array}} \text{(cyclohexyl)}-\overset{\overset{\displaystyle O}{\|}}{C}-H$

After the reaction is complete, the PVPCr is recovered by filtration. The hot xylene filtrate is cooled and the solubility of the polyethylene–rhodium complex drops sharply so that it precipitates. Only the product remains dissolved in the xylene.

Miscellaneous Vinyl Polymers

The use of modified vinyl polymers with covalently or ionically bound organometallic functions is of interest in the area of polymer-coated electrodes (251) (see ELECTRODES, POLYMER-MODIFIED). Polyvinylferrocene, protonated poly(4-vinylpyridine)Fe(CN)$_6^{-3}$, the sulfonated fluoropolymer Nafion modified with cobalt(II)tetraphenylporphyrin, and, more recently, Nafion containing both cobalt(II)tetraphenylporphyrin and Ru(NH$_3$)$^{3+}$ counterions and polymers of tris(4-vinyl-4'-methyl-2,2'-bipyridine)ruthenium(II) (**147**) have been used. Such polymers protect the electrodes from oxygen or from a variety of corrosive reactions that occur when carrying out electrocatalysis or electrochemistry in surrounding solutions or when promoting hole–electron separation at semiconductor surfaces (251).

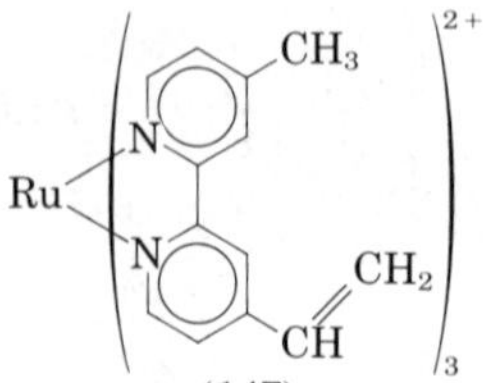

(147)

One class of vinyl backbone polymers contains metals that are directly bonded to the polymers via metal–carbon σ bonds. These types are quite rare. Usually, transition-metal–carbon bonds are subject to β-metal hydride elimination, but this is not possible when β-hydrogens are absent. The treatment of both linear and cross-linked chloromethylated polystyrenes with $Mn(CO)_5^-$, $(\eta^5\text{-}C_5H_5)Mo(CO)_3^-$, or $(\eta^5\text{-}C_5H_5)W(CO)_3^-$ results in displacement of chloride and formation of a stable metal–carbon bond within the polymers, ie, (**148**) and (**149**) (252,253).

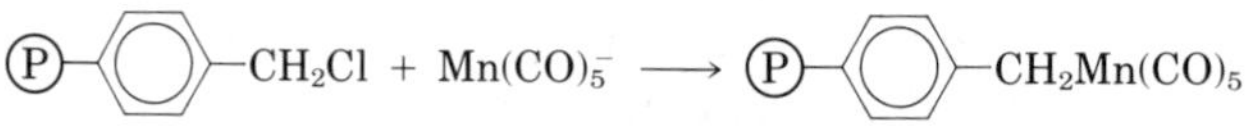

(148)

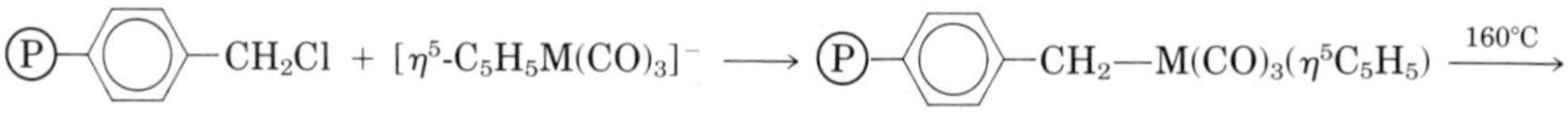

(149)

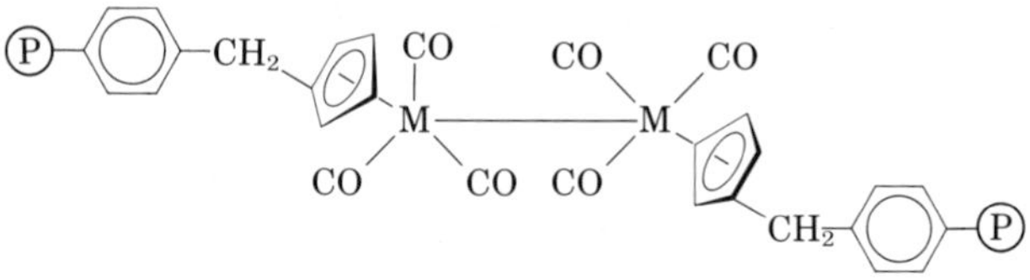

(150)

where M = W or Mo

Thermal decomposition of cross-linked resins containing these structures was carried out to see if the metal decomposition products could be dispersed within the polymer matrix. The decomposition of (**148**) released $Mn_2(CO)_{10}$ within the resin; (**149**) decomposed to produce the metal–metal dimer structure (**150**). Another approach already reviewed is the decomposition of metal carbonyls dissolved in polymers (254). This most often leads to small particles of metal dispersed in the polymer matrix. Both poly(vinyl chloride) and styrene–divinylbenzene resins have been functionalized with cationic $(\eta^5\text{-cyclopentadienyl})\text{iron}(\eta^5\text{-hexa-methylbenzene})$ functions (255–259), eg, see Figure 13. Polymer (**151**) is the starting point for the preparation of many dual-functional organometallic polymers (**152**)–(**156**), where $CpFe(CO)_2^-$, $—P(C_6H_5)_2RhCl(COD)$, $—Mo(CO)_5$, or $—P(C_6H_5)_2RhCl(CO)$ are also present. The catalytic activity of these resins and their photoactivity have been studied.

A variety of other vinylorganometallic polymers have been made. Ferrocenylacetylene (**157**) was polymerized using Ziegler-Natta, radical, and cationic

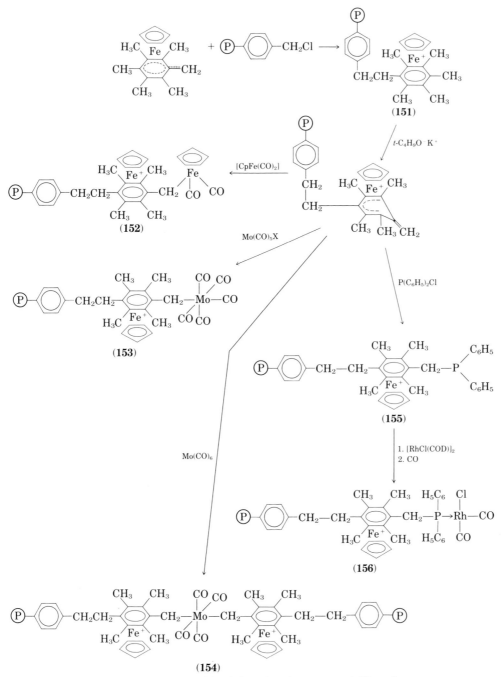

Fig. 13. Preparation of dual-functional organometallic polymers.

initiators (260–265) to give polyferrocenylacetylene (**158**). This polymer, which is shown to be completely conjugated along the backbone, is unlikely to exist totally in the structure shown by (**158**).

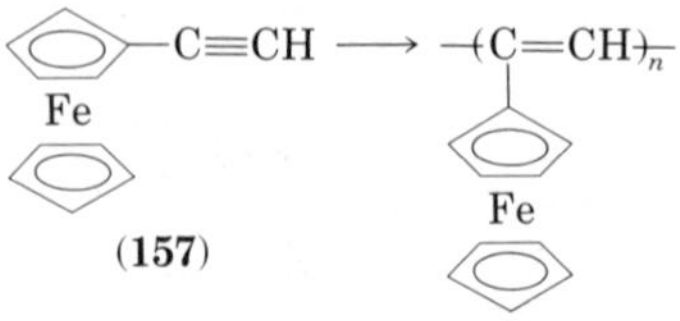

(**157**)

(**158**)

The tungsten and manganese acrylates, (**159**) and (**160**), respectively, have been prepared and copolymerized with various organic monomers (20,196)

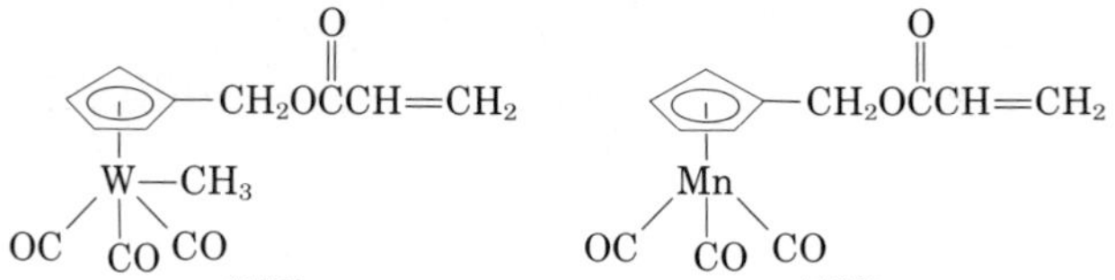

(**159**)

(**160**)

Of particular interest are polyacrylates containing the tetraphenylporphinatosilver(I), tetraphenylporphinatocobalt(II), and chloro(tetraphenylporphinato)iron(III) groups as side chains (266–270). Unlike the stacked shishkabob structures shown in (**10**), (**39**), and (**80a**), direct interaction between the porphynatometal units was not expected. Instead, as suggested by structure (**161**), it seems the pendent units could be isolated. However, uv–visible and esr spectroscopy and magnetic susceptability measurements show an antiferromagnetic interaction between metal ions in the Ag, Co, and Fe cases. It was thought that these polymers might exhibit ferromagnetic properties. This possibility exists because electron-exchange interactions between metal ions may give ferromagnetic substances (271). Thus the antiferromagnetic interaction, depicted in (**162**), may reduce the chances of observing desirable magnetic properties.

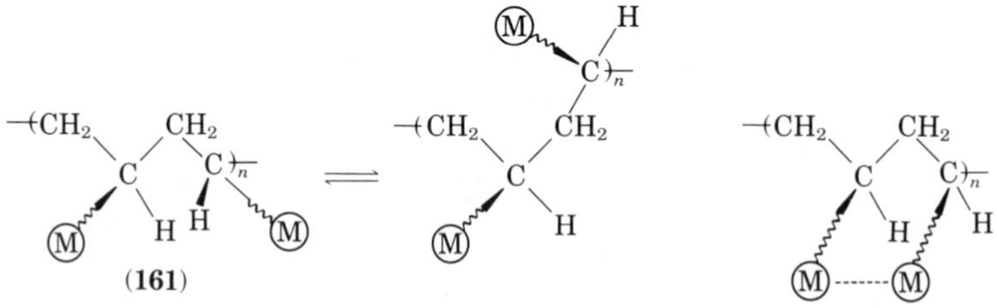

(**161**)

(**162**)

where —ᴹ represents the carboxy metal group (—COOⓂ) and

Ⓜ is a 5-(4-substituted phenyl)-10,15,20-triphenylporphinatometal group

Polysilanes

Polysilanes are a new class of organometallic polymers that have an all-silicon backbone (**6**). For 60 years, despite the success in preparing polysiloxanes, polymers with silicon backbones (polysilanes) remained elusive. Polysilane homopolymers were finally prepared by the reaction of dichlorosilanes with sodium, but the materials were insoluble, highly crystalline, and intractable. Permethylpolysilane (**163**) fits this description (272).

$$
\underset{\underset{CH_3}{|}}{\overset{\overset{CH_3}{|}}{Cl-Si-Cl}} + 2\ Na \longrightarrow \underset{\underset{CH_3}{|}}{\overset{\overset{CH_3}{|}}{+Si+_n}}
$$

(**163**)

Introduction of large organic substituent groups and copolymerization of two monomers to mix the types of substituents decreases crystallinity and increases solubility. New polysilanes can be prepared in high yields that are tractable, soluble in a variety of solvents, easily shaped, readily purified, easily molded, and castable into films (8,9,273) (see POLYSILANES AND POLYCARBOSILANES). Structures (**164**)–(**167**) fit this category (8,9). Polysilanes are formally classified as condensation polymers according to the method of preparation shown for (**163**). However, they are treated separately due to their unique all-silicon backbone.

$$
\underset{\underset{CH_3}{|}}{\overset{\overset{CH_2CH_2C_6H_5}{|}}{Cl-Si-Cl}} + Cl-\overset{\bigtriangleup}{Si}-Cl \xrightarrow[\text{refluxing toluene}]{Na} \left(\underset{\underset{CH_3}{|}}{\overset{\overset{CH_2CH_2C_6H_5}{|}}{+Si+_n}}+Si+_m\right)
$$

(**164**)

(**165**)   (**166**)   (**167**)

With the recent availability of defined high molecular weight polysilanes (values of M_w to 400,000 have been reported (9)), photochemical studies have been carried out. Exposure to 350 or 300 nm leads to rapid photolytic degradation as shown by incremental decreases in molecular weight with exposure. Copolymers with phenyl side chains directly attached to Si have strong absorptions near 330 nm resulting from a conjugative interaction between the phenyl groups and the silicon backbone which acts as a $\sigma \rightarrow \sigma^*$ or $\sigma \rightarrow \pi$ chromophore. In fact, questions about a possible conjugation along the silicon backbone still exist long after being postulated (274,275). The cyclohexylmethylsilane homopolymer (**167**) exhibits an absorption band at 326 nm (8), suggesting interaction among the chain-silicon atoms not occurring in carbon chains. This high degradation sensitivity to uv light (quantum yields from 0.97 to 0.20 have been measured for chain scission) has led IBM to investigate polysilanes in bilayer uv lithography.

Polysilanes serve as excellent reactive-ion-etching barriers for bilevel-resist applications because a protective layer of SiO_2 is formed during exposure to oxygen plasma (276) (see LITHOGRAPHIC RESISTS). The λ_{max} depends on the degree of polymerization leading to a nonlinear bleaching photochemistry. This unique "bleaching latency" is useful in contrast-enhanced lithography (277). At 3M Co., chemists have found that polysilanes have extraordinary activity as polymerization catalysts for making high molecular weight vinyl polymers, and polysilanes are being used as precursors for β-silicon carbide fibers at the Air Force Materials Laboratory (278).

BIBLIOGRAPHY

1. C. U. Pittman, Jr., C. C. Lin, and T. D. Rounsefell, *Macromolecules* **11**, 1022 (1978).
2. C. U. Pittman, Jr., G. V. Marlin, and T. D. Rounsefell, *Macromolecules* **6**, 1 (1973).
3. C. E. Carraher, Jr. and S. Bajah, *Br. Polym. J.* **15**, 9 (1974).
4. C. E. Carraher, Jr. in P. Millich and C. E. Carraher, Jr., eds., *Interfacial Synthesis,* Vol. II, Marcel Dekker, Inc., New York, 1977, Chapt. 20.
5. C. U. Pittman, Jr. and C. C. Lin, *J. Org. Chem.* **43**, 4928 (1978).
6. Y. P. Ning, J. E. Mark, N. Iwamoto, and B. E. Eichinger, *Macromolecules* **18**, 55 (1985).
7. C. W. Dirk, M. Bousseau, P. H. Barrett, F. Moraes, F. Wudl, and A. J. Heeger, *Macromolecules* **19**, 266 (1986).
8. X. H. Zhang and R. West, *J. Polym. Sci. Polym. Chem. Ed.* **22**, 159 (1984).
9. *Ibid.,* p. 225.
10. S. Kanda, *Nippon Kagaku Zasshi* **83**, 560 (1962).
11. *Ibid.,* p. 282; *Kagaku No Ryoiki Zokan* **90**, 87 (1970); *Chem. Abstr.* **73**, 88180z (1985).
12. J. T. Wrobleski and D. B. Brown, *Inorg. Chem.* **18**, 498, 2738 (1979).
13. B. N. Diel, T. Inabe, N. K. Jaggi, J. W. Lyding, O. Schneider, M. Hanack, C. R. Kannewurf, T. J. Marks, and L. H. Schwartz, *J. Am. Chem. Soc.* **106**, 3207 (1984).
14. L. Richards, I. Koufis, C. S. Chan, J. L. Richards, and C. Cotter, *Inorg. Chem.* **105**, L21 (1985).
15. J. E. Sheats, C. E. Carraher, Jr., and C. U. Pittman, Jr., eds., *Metal-containing Polymer Systems,* Plenum Publishing Corp., New York, 1985, pp. 1–523.
16. C. E. Carraher, Jr., J. E. Sheats, and C. U. Pittman, Jr., eds., *Advances in Organometallic and Inorganic Polymer Science,* Marcel Dekker, Inc., New York, 1982, pp. 1–449.
17. C. E. Carraher, Jr., J. E. Sheats, and C. U. Pittman, Jr., *Organometallic Polymers,* Academic Press, Inc., Orlando, Fla., 1978, pp. 1–346.
18. J. E. Sheats, C. U. Pittman, Jr., and C. E. Carraher, Jr., *Chem. Br.* **20**(8), 709 (1984).
19. J. E. Sheats, *J. Macromol. Sci. Chem. Part A* **16**(6), 1173, 1980.
20. C. U. Pittman, Jr., *Chem. Technol.* **1**, 416 (1971); K. A. Andrianov, *Metallorganic Polymers,* John Wiley & Sons, Inc., New York, 1965.
21. C. U. Pittman, Jr. in G. Wilkinson, F. G. A. Stone, and E. W. Abel, eds., *Comprehensive Organometallic Chemistry,* Vol. 8, Pergamon Press, Oxford, UK, 1982, Chapt. 55, pp. 553–608.
22. F. R. Hartley, *Supported Metal Complexes,* D. Reidel Publishers, Dordrecht, The Netherlands, 1985, pp. 1–318.
23. Y. Chauvin, D. Commereuc, and F. Dawans, *Prog. Polym. Sci.* **5**, 95 (1977).
24. C. U. Pittman, Jr. and G. O. Evans, *Chem. Technol.* **3**, 560 (1973).
25. C. U. Pittman, Jr. in P. Hodge and D. C. Sherrington, eds., *Polymer-supported Reactions in Organic Synthesis,* John Wiley & Sons, Inc., Chichester, UK, 1980, pp. 249–291.
26. N. Toshima, *Yuki Gosei Kagaku Kyokai Shi* **36**, 909 (1978); *Chem. Abstr.* **90**, 104033 (1979).
27. V. M. Akhmedov, A. A. Medzhidob, and A. G. Azizov, *Aserb. Khim. Zh.,* 122 (1979); *Chem. Abstr.* **92**, 11692 (1980).
28. A. L. Robinson, *Science* **194**, 1261 (1976).
29. W. T. Ford, ed., *Polymeric Reagents and Catalysts,* American Chemical Society, Washington, D.C., 1986, pp. 1–295.
30. C. U. Pittman, Jr. in E. I. Becker and M. Tsutsui, eds., *Organometallic Reactions,* Vol. 6, Marcel Dekker, Inc., New York, 1977, pp. 1–62.
31. C. E. Carraher, Jr., *Chem. Technol.,* 744 (1972).
32. C. E. Carraher, Jr., *J. Chem. Ed.* **58**(11), 921 (1981).
33. C. E. Carraher, Jr. and co-workers in C. E. Carraher, Jr. and J. Preston, eds., *Interfacial Synthesis,* Vol. III, Plenum Publishing Corp., New York, 1982, pp. 77–106.

34. B. A. Bolton in J. E. Katon, ed., *Organic Semiconducting Polymers*, Marcel Dekker, Inc., New York 1968, pp. 199–257.
35. C. Pecile, G. Zerbi, R. Bozio, and A. Girlando, eds., *Proceedings of the International Conference on the Physics and Chemistry of Low Dimensional Synthetic Metals (ICSM 84)*, Abano Terme, Italy, June 17–22, 1984.
36. A. J. Epstein and E. M. Conwell in *Mol. Cryst. Liq. Cryst.* **81,** 881 (1982).
37. J. S. Miller, ed., *Extended Linear Chain Compounds*, Vols. 1–3, Plenum Publishing Corp., New York, 1982.
38. E. W. Neuse, "Metallocene Polymers" in N. M. Bikales, ed., *Encyclopedia of Polymer Science and Technology*, Vol. 8, Wiley-Interscience, New York, 1968, p. 667.
39. E. W. Neuse and H. Rosenberg, *Metallocene Polymers*, Marcel Dekker, Inc., New York, 1970.
40. E. W. Neuse in Ref. 16, pp. 3–72.
41. R. V. Subramanian and K. N. Somasekharan in Ref. 16, pp. 73–93.
42. B. J. Spalding, *Chem. Week*, 29 (Oct. 8, 1986).
43. E. J. Dyckman and J. A. Montemarano, *Am. Paint J.* **58**(5), 66 (1973).
44. N. A. Ghanem, N. N. Messiha, N. E. Ikaldious, and A. F. Shaaban, *Eur. Polym. J.* **15,** 823 (1979).
45. C. E. Carraher, Jr., R. Schwartz, J. Schroeder, M. Schwarz, and H. M. Molloy, *Am. Chem. Soc. Div. Org. Coat. Plast. Chem. Pap.* **43,** 798 (1980).
46. C. E. Carraher, Jr., L. Tisinger, G. Solimine, M. Williams, S. Carraher, and R. Strothers, *Polym. Mater. Sci. Eng.* **55,** 469 (1986).
47. C. E. Carraher, Jr., R. Venkatachalam, T. Tiernam, M. L. Taylor, and J. A. Schroeder, *Polym. Mater. Sci. Eng.* **47,** 119 (1982).
48. B. P. Block, *Inorg. Macromol. Rev.* **1**(2), 115 (1970).
49. H. Allcock in Ref. 17, pp. 283.
50. C. E. Carraher, Jr., D. J. Giron, I. Lopez, D. R. Cerutis, and W. J. Scott, *Am. Chem. Soc. Div. Org. Coat. Plast. Chem. Pap.* **44,** 120 (1981).
51. C. E. Carraher, Jr., W. J. Scott, J. A. Schroeder, and D. J. Giron, *J. Macromol. Sci. Chem. Part A* **15**(4), 625 (1981).
52. C. E. Carraher, Jr., *Am. Chem. Soc. Div. Org. Coat. Plast. Chem. Pap.* **42,** 428 (1980).
53. N. E. Wolff and R. J. Pressley, *Appl. Phys. Lett.* **2,** 152 (1963).
54. Y. Okamoto, S. S. Wang, K. H. Zhu, E. Banks, B. Garetz, and E. K. Murphy in Ref. 15, pp. 425.
55. C. U. Pittman, Jr. and Q. Ng, *J. Organometal. Chem.* **153,** 85 (1978).
56. W. O. Haag and D. D. Whitehurst in J. W. Hightower, ed., *Catalysis*, Vol. 1, Royal Society of Chemistry, London, 1973, No. 29, p. 465.
57. R. H. Grubbs, C. Gibbons, L. C. Kroll, W. D. Bonds, Jr., and C. H. Brubaker, *J. Am. Chem. Soc.* **95,** 2374 (1973); **97,** 2128 (1975).
58. C. U. Pittman, Jr. and L. R. Smith, *J. Am. Chem. Soc.* **97,** 1749 (1975).
59. C. U. Pittman, Jr., L. R. Smith, and R. M. Hanes, *J. Am. Chem. Soc.* **97,** 1742 (1975).
60. R. F. Batchelder, B. C. Gates, and F. P. J. Kuijpers, *Preprint A40, 6th International Congress on Catalysis*, Royal Society of Chemistry, London, 1976.
61. D. E. Bergbreiter and R. Chandran, *J. Am. Chem. Soc.* **107,** 4792 (1985).
62. R. H. Grubbs, L. C. Kroll, and E. M. Sweet, *J. Macromol. Sci. Chem. Part A* **7,** 1047 (1973).
63. C. U. Pittman, Jr. and R. M. Hanes, *J. Am. Chem. Soc.* **98,** 5402 (1976).
64. C. U. Pittman, Jr. and A. Hirao, *J. Org. Chem.* **43,** 640 (1978).
65. V. A. Kabanov, V. G. Popov, V. I. Smetanyuk, and L. P. Kalinina, *Vysokomol. Soedin. Ser. B* **23,** 368 (1981).
66. S. Jacobson, W. Clements, H. Hiramoto, and C. U. Pittman, Jr., *J. Mol. Catal.* **1,** 73 (1975).
67. R. H. Grubbs, C. P. Lau, R. Cukier, and C. H. Brubaker, Jr., *J. Am. Chem. Soc.* **99,** 4517 (1977).
68. W. H. Lang, A. T. Jurewicz, W. O. Hagg, D. D. Whitehurst, and L. D. Rollman, *J. Organometal. Chem.* **134,** 85 (1977).
69. U.S. Pat. 4,313,018 (1982), N. L. Holy, W. A. Logan, and K. D. Stein (to Western Kentucky University).
70. Belg. Pat. 721,686 (1969) (to Mobil Corp.).
71. U.S. Pat. 4,098,727 (1978), W. O. Haag and D. D. Whitehurst (to Mobil Corp.).
72. A. J. Moffat, *J. Catal.* **18,** 193 (1970); **19,** 322 (1970).
73. Br. Pat. 1,277,737 and 1,295,673 (1972), K. G. Allum and R. D. Hancock (to British Petroleum).
74. U.S. Pat. 4,098,727 (1978), W. O. Haag and D. D. Whitehurst (to Mobil Corp.).

75. U.S. Pat. 4,052,461 (1977), H. B. Tinker and D. E. Morris (to Monsanto, Inc.).
76. U.S. Pat. 4,258,206 (1981), C. U. Pittman, Jr. and Q. Y. Ng (to University of Alabama).
77. U.S. Pat. 4,243,829 (1981), C. U. Pittman, Jr. and R. M. Hanes (to University of Alabama).
78. M. Kaneko and A. Yamada, *Adv. Polym. Sci.* **55**, (1983).
79. M. Kaneko and A. Yamada in Ref. 15, pp. 249–274.
80. M. S. Wrighton, ed., *Interfacial Photoprocesses: Energy Conversion and Synthesis, Adv. Chem. Ser.* **184**, American Chemical Society, Washington, D.C., 1980.
81. U.S. Pat. 4,379,740 (1983), U. T. Müller-Westerhoff and A. I. Nazzal (to IBM Corp.).
82. F. C. Anson and co-workers, *J. Am. Chem. Soc.* **106**, 59 (1984).
83. E. R. Savinova, A. I. Kokorin, A. P. Shepelin, A. V. Pashis, P. A. Zhdan, and V. N. Parmon, *J. Mol. Catal.* **32**, 149, 159 (1985).
84. T. E. Bitterwolf in Ref. 15, pp. 137–147.
85. J. E. Sheats, F. Hessel, L. Tsarouhas, K. G. Podejko, T. Porter, L. B. Kool, and R. L. Nolan, Jr. in Ref. 15, pp. 83–98.
86. J. E. Sheats, F. Hessel, L. Tsarouhas, K. G. Podejko, T. Porter, L. B. Kool, and R. L. Nolan in B. M. Culbertson and C. U. Pittman, Jr., eds., *New and Unusual Monomers and Polymers,* Plenum Publishing Corp., New York, 1983, pp. 83–98; *Polym. Mater. Sci. Eng.* **49**(2), 363 (1983); Ref. 15, pp. 83–98.
87. C. E. Carraher, Jr. and R. Nordin, *J. Appl. Polym. Sci.* **18**, 53 (1974).
88. U. Naoshima and C. E. Carraher, Jr., *Polym. Mater. Sci. Eng.* **50**, 403 (1984).
89. C. E. Carraher, Jr., T. Gehrke, D. Giron, D. R. Cerutis-Blaxall, H. M. Molloy, *J. Macromol. Sci. Chem. Part A* **17**, 1121 (1983).
90. U. Naoshima, C. E. Carraher, Jr., and G. Hess, *Polym. Mater. Sci. Eng.* **49**, 215 (1983).
91. C. E. Carraher, Jr., W. Burt, D. Giron, J. Schroeder, M. Taylor, H. Molloy, and T. Tiernon, *J. Appl. Polym. Sci.* **28**, 1919 (1983).
92. U.S. Pat. 4,312,981 (1982), C. E. Carraher, Jr., D. Giron, J. Schroeder, and C. McNeely (to Wright State University).
93. F. Millich and C. E. Carraher, Jr., eds., *Interfacial Synthesis: Technology and Applications,* Vol. II, Marcel Dekker, Inc., New York, 1978.
94. C. E. Carraher, Jr. and J. Preston, eds., *Interfacial Synthesis: Recent Advances,* Vol. III, Marcel Dekker, Inc., New York, 1981.
95. C. E. Carraher, Jr. and S. T. Bajah, *Br. Polym. J.* **14**, 42 (1973).
96. C. E. Carraher, Jr., *Makromol. Chem.* **166**, 31 (1973).
97. C. E. Carraher, Jr. and C. Deremo-Reece, *Angew. Makromol. Chem.* **65**, 95 (1972).
98. C. E. Carraher, Jr., R. S. Venkatachalam, T. O. Tiernan, and M. L. Taylor, *Appl. Polym. Sci. Proc.* **47**, 119 (1982).
99. C. E. Carraher, Jr. and C. Gebelein, eds., *Biological Activities of Polymers, ACS Symp. Ser.* **186**, American Chemical Society, Washington, D.C., 1983.
100. C. Gebelein and C. E. Carraher, Jr., eds., *Biologically Active Polymer Systems,* Plenum Publishing Corp., New York, 1984.
101. C. C. Hinckley, S. Sharif, and L. D. Russell in Ref. 15, pp. 183–195.
102. C. E. Carraher, Jr., J. Schroeder, W. Venable, C. McNeely, D. Giron, W. Woelk, and M. Feddersen in R. Seymour, ed., *Additives for Plastics,* Vol. 2, Academic Press, Inc., Orlando, Fla., 1978.
103. C. E. Carraher, Jr., W. Moon, and T. Langmorthy, *Polym. Prepr. Am. Chem. Soc. Div. Polym. Chem.* **17**, 1 (1976).
104. C. E. Carraher, Jr., *Chem. Technol.,* **741** (1972).
105. C. W. Dirk, T. Inabe, K. F. Schoch, Jr., and T. J. Marks, *J. Am. Chem. Soc.* **105**, 1539 (1983).
106. T. J. Marks and co-workers, *J. Am. Chem. Soc.* **108**, 7595 (1986).
107. C. U. Pittman, Jr., O. E. Ayers, B. Suryanarayanan, S. P. McManus, and J. E. Sheats *Makromol. Chem.* **175**, 1427 (1974).
108. E. W. Neuse in Ref. 17, p. 95.
109. S. L. Sosin, V. P. Alekseeva, M. D. Litvinova, V. V. Korshak, and A. F. Zhigach, *Vysokomol. Soedin. Ser. B* **22**(9), 703 (1976).
110. H. Watanabe, J. Motoyama, K. Hata, *Bull. Chem. Soc. Jpn.* **39**, 784 (1966).
111. C. U. Pittman, Jr., *J. Polym. Sci. Part A-1* **6**, 1687 (1968).
112. M. Okawara, Y. Takemoto, H. Kitaoka, E. Haruki, and E. Imoto, *Kogyo Kagaku Zasshi* **65**, 685 (1962).

113. H. Valot, *Acad. Sci. Ser. C* **265**(5), 403 (1966).
114. K. Gonsalves, L. Zhan-ru, and M. D. Rausch, *J. Am. Chem. Soc.* **106**, 3862 (1984).
115. C. U. Pittman, Jr., *J. Polym. Sci. Part A-1* **5**, 2927 (1967).
116. J. E. Mulvaney, J. J. Bloomfield, and C. S. Marvel, *J. Polym. Sci.* **62**, 59 (1962).
117. Fr. Pats. 1,396,271; 1,396,273; and 1,396,274 (Apr. 16, 1965) and 1,398,255 (May 7, 1965), E. V. Wilkus and A. Berger (to Compagnie Francaise Thompson-Houston).
118. Von G. Greber and M. L. Hallensleben, *Makromol. Chem.* **92**, 137 (1966).
119. W. J. Patterson, S. P. McManus, and C. U. Pittman, Jr., *J. Polym. Sci. Polym. Chem. Ed.* **12**, 837 (1974).
120. S. Kanda, *Nippon Kagaku Zasshi* **81**, 1347 (1960); *Chem. Abstr.* **55**, 22994e (1961).
121. S. Kanda and Y. Saito, *Bull. Chem. Soc. Jpn.* **30**, 192 (1957).
122. R. H. Bailes and M. Calvin, *J. Am. Chem. Soc.* **69**, 1886 (1947).
123. W. C. Drinkard, Jr. and D. N. Chakravarty, *Wright Air Development Center Technical Report No. 59–761*, Wright-Patterson AFB, Dayton, Ohio, 1960, p. 232.
124. R. S. Bottei and P. L. Gerace, *Book of Abstracts, 139th Meeting of the American Chemical Society*, St. Louis, Mo., American Chemical Society, Washington, D.C., 1961, p. 16M.
125. F. W. Knobloch and W. H. Rauscher, *J. Polym. Sci.* **38**, 261 (1959).
126. V. V. Korshak, S. V. Vinogradova, and V. S. Artemova, *Vysokomol. Soedin.* **2**, 492 (1960).
127. R. Kuhn and H. A. Staab, *Chem. Ber.* **87**, 272 (1954).
128. H. A. Goodwin and J. C. Bailor, *J. Am. Chem. Soc.* **83**, 2467 (1961).
129. G. Manecke and R. Wille, *Makromol. Chem.* **133**, 61 (1970).
130. *Ibid.*, **160**, 111 (1972).
131. W. Sawodny and M. Riederer, *Angew. Chem.* **89**, 897 (1977).
132. M. Riederer, E. Urban, and W. Sawodny, *Angew. Chem.* **89**, 898 (1977).
133. M. Riederer and W. Sawodny, *Angew. Chem.* **90**, 642 (1978).
134. M. Riederer and W. Sawodny, *J. Chem. Res. S,* 450 (1978).
135. W. Sawodny, M. Riederer, and E. Urban, *Inorg. Chim. Acta* **29**, 63 (1978).
136. F. A. Bottino and co-workers, *Inorg. Nucl. Chem. Lett.* **16**, 417 (1980).
137. M. N. Patel, S. H. Patil, and M. S. Setty, *Angew. Makromol. Chem.* **97**, 69 (1981).
138. J. R. Reynolds, F. E. Karasz, C. P. Lillya, and J. C. W. Chien, *J. Chem. Soc. Chem. Commun.*, 268 (1985).
139. N. M. Rivera, E. M. Engler, and R. R. Schumaker, *J. Chem. Soc. Chem. Commun.*, 184 (1979).
140. E. M. Engler, N. Martinez-Rivera, and R. R. Schumaker, *Am. Chem. Soc. Div. Org. Coat. Plast. Chem. Pap.* **41**, 52 (1979).
141. U.S. Pat. 4,111,857 (Sept. 5, 1978), E. M. Engler, K. H. Nichols, V. V. Patel, N. M. Rivera, and R. R. Schumaker (to IBM Corp.).
142. J. Ribas, P. Cassoux, and F. Gallais, *C. R. Acad. Sci. Paris* **293**, 64 (1981).
143. F. Götzfried, W. Beck, A. Lerf, and A. Sebald, *Angew. Chem. Int. Ed. Engl.* **18**, 463 (1979).
144. M. C. Bohm, *Phys. Status Solidi* **121**, 255 (1984).
145. M. C. Bohm, *Phys. Status Solidi Sect. B* **124**, 327 (1984).
146. D. Wöhrle, *Adv. Polym. Sci.* **10**, 35 (1972).
147. G. Manecke and D. Wöhrle, *Makromol. Chem.* **120**, 192 (1968).
148. G. Kossmehl and M. Rohde, *Makromol. Chem.* **180**, 345 (1979).
149. D. I. Packham and J. C. Haydon, *Polymer* **11**, 385 (1970).
150. R. Müller and D. Wöhrle, *Makromol. Chem.* **176**, 2775 (1975).
151. *Ibid.*, **177**, 2241 (1976).
152. *Ibid.*, **179**, 2161 (1976).
153. U. G. Deshpande and J. R. Shah, *J. Macromol. Sci. Chem. Part A* **21**, 21 (1984).
154. *Ibid.*, **23**(1), 97 (1986).
155. F. Higashi, C. S. Cho, H. Kakinoki, and O. Sumita, *J. Polym. Sci. Polym. Chem. Ed.* **17**, 313 (1979).
156. L. Donaruma, *Polym. Prepr. Am. Chem. Soc. Div. Polym. Chem.* **22**, 1 (1981).
157. G. Kossmehl and M. Rohde, *Makromol. Chem.* **178**, 715 (1977).
158. W. D. Bascom, R. L. Cottington, and T. Y. Ting, *J. Mater. Sci.* **15**, 2097 (1980).
159. C. J. Norrel and co-workers, *J. Polym. Sci. Polym. Chem. Ed.* **12**, 913 (1974).
160. W. Hanke, *Z. Chem.* **6**, 69 (1966).
161. C. S. Marvel and J. H. Rassweiler, *J. Am. Chem. Soc.* **80**, 1197 (1958).

162. L. Kreja and A. Plewka, *Electrochim. Acta* **25**, 1283 (1980).
163. A. S. Akopor, T. N. Lomova, and B. D. Berezin, *Izv. Vyssh. Uchebn. Zaved. Khim. Khim. Tekhnol.* **19**, 1177 (1976).
164. J. L. Petersen, C. S. Schramm, D. R. Stojakovic, B. M. Hoffman, and T. J. Marks, *J. Am. Chem. Soc.* **99**, 286 (1977).
165. K. F. Schock, Jr., B. R. Kundalkar, and J. T. Marks, *Am. Chem. Soc. Div. Org. Coat. Plast. Chem. Pap.* **41**, 127 (1979).
166. J. Metz and M. Hanack, *J. Am. Chem. Soc.* **105**, 828 (1983).
167. K. Krogmann and H. D. Hausen, *Z. Anorg. Allg. Chem.* **358**, 67 (1968).
168. K. Krogmann, *Angew. Chem. Int. Ed. Engl.* **8**, 35 (1969).
169. L. S. Miller and A. J. Epstein, eds., *Ann. N.Y. Acad. Sci.* **313** (1978).
170. S. Takahashi, H. Morimoto, E. Murata, S. Kataoka, K. Sonogashira, and N. Nagihara, *J. Polym. Sci. Polym. Chem. Ed.* **20**, 565 (1982).
171. S. Takahashi, Y. Takai, H. Morimoto, and K. Sonogashira, *J. Chem. Soc. Chem. Commun.*, 3 (1984).
172. S. Takahashi, Y. Takai, H. Morimoto, K. Sonogashira, and N. Hagihara, *Mol. Cryst. Liq. Cryst.* **82**, 139 (1982).
173. K. Sonogashira, S. Takahashi, and N. Hagihara, *Macromolecules* **10**, 879 (1977).
174. K. Sonogashira, Y. Fujikura, T. Yatake, N. Toyoshima, S. Takahashi, and N. Hagihara, *J. Organometal. Chem.* **145**, 101 (1978).
175. S. Takahashi, M. Kariya, T. Yatake, K. Sonogashira, and N. Hagihara, *Macromolecules* **11**(6), 1063 (1978); S. Takahashi and K. Sonogashira, *Kobunshi* **29**(5), 395 (1980); *Chem. Abstr.* **93**, 8542u (1980).
176. H. Matsuda, H. Nakanishi, and M. Kato, *J. Polym. Sci. Polym. Lett. Ed.* **22**, 107 (1984).
177. C. U. Pittman, Jr., *J. Paint Technol.* **43**(1971); *Chem. Technol.* **1**, 416 (1971).
178. C. U. Pittman, Jr., P. Grube, and R. M. Hanes, *J. Paint. Technol.* **46**, 597 (1974).
179. C. U. Pittman, Jr. in Ref. 17, pp. 1–11.
180. F. S. Arimoto and A. C. Haven, Jr., *J. Am. Chem. Soc.* **77**, 6295 (1955).
181. Y. Sasaki, L. L. Walker, E. L. Hurst, and C. U. Pittman, Jr., *J. Polym. Sci. Polym. Chem. Ed.* **11**, 1213 (1973).
182. M. H. George and G. F. Hayes, *J. Polym. Sci. Polym. Chem. Ed.* **13**, 1049 (1975).
183. *Ibid.*, **14**, 475 (1976).
184. C. Aso, T. Kunitake, and T. Nakashima, *Makromol. Chem.* **124**, 232 (1969).
185. C. R. Simionescu, *Makromol. Chem.* **163**, 59 (1973).
186. C. U. Pittman, Jr. and C. C. Lin, *J. Polym. Sci. Polym. Chem. Ed.* **17**, 271 (1979).
187. A. N. Nesmeyanov, K. N. Anisimov, N. E. Kolobova, and I. B. Zlotina, *Dokl. Akad. Nauk SSSR* **154**, 391 (1964).
188. C. U. Pittman, Jr. and T. D. Rounsefell, *Macromolecules* **9**, 937 (1976).
189. E. A. Mintz, M. D. Rausch, B. H. Edwards, J. E. Sheats, T. D. Rounsefell, and C. U. Pittman, Jr., *J. Organometal. Chem.* **137**, 199 (1977).
190. C. U. Pittman, Jr., T. D. Rounsefell, E. A. Lewis, J. E. Sheats, B. H. Edwards, M. D. Rausch, and E. A. Mintz, *Macromolecules* **11**, 560 (1978).
191. J. C. Lai, T. Rounsefell, and C. U. Pittman, Jr., *J. Polym. Sci. Part A-1* **9**, 651 (1971).
192. J. E. Sheats and T. C. Willis, *Am. Chem. Soc. Div. Org. Coat. Plast. Prepr.* **41**(2), 33 (1979).
193. J. E. Sheats and T. C. Willis, *J. Polym. Sci. Polym. Chem. Ed.* **22**, 1077 (1984).
194. C. U. Pittman, Jr. and B. Surynarayanan, *J. Am. Chem. Soc.* **96**, 7916 (1974).
195. C. U. Pittman, Jr., B. Surynarayanan, and Y. Sasaki in R. B. King, ed., *Inorganic Compounds with Unusual Properties, Adv. Chem. Ser.* **150**, American Chemical Society, Washington, D.C., 1976, pp. 46–55.
196. D. W. Macomber, M. D. Rausch, T. V. Jayaraman, R. D. Priester, and C. U. Pittman, Jr., *J. Organometal. Chem.* **205**, 353 (1981).
197. C. U. Pittman, Jr., T. V. Jayaraman, R. D. Priester, Jr., S. Spencer, M. D. Rausch, and D. Macomber, *Macromolecules* **14**, 237 (1981).
198. D. W. Macomber, W. P. Hart, M. D. Rausch, R. D. Priester, Jr., C. U. Pittman, Jr., *J. Am. Chem. Soc.* **104**, 984 (1982).
199. M. D. Rausch, D. W. Macomber, F. G. Fang, C. U. Pittman, Jr., T. V. Jayaraman, and R. D. Priester, Jr. in B. M. Culbertson and C. U. Pittman, Jr., eds., *New Monomers and Polymers*, Plenum Publishing Corp., New York, 1984, pp. 243–267.

200. D. W. Macomber and M. D. Rausch, *J. Organometal. Chem.* **250**, 311 (1983).
201. C. U. Pittman, Jr., P. L. Grube, O. A. Ayers, S. P. McManus, M. D. Rausch, and G. A. Moser, *J. Polym. Sci. Part A-1* **10**, 379 (1972).
202. C. U. Pittman, Jr., O. E. Ayers, and S. P. McManus, *J. Macromol. Sci. Chem., Part A* **7**(8), 1563 (1973).
203. C. U. Pittman, Jr., O. E. Ayers, and S. P. McManus, *Macromolecules* **7**, 737 (1974).
204. C. U. Pittman, Jr. and R. L. Voges in W. J. Bailey, ed., *Macromolecular Synthesis*, Vol. 4, John Wiley & Sons, Inc., New York, 1972, p. 175.
205. C. U. Pittman, Jr., J. C. Lai, and D. P. Vanderpool, *Macromolecules* **3**, 105 (1970).
206. C. U. Pittman, Jr., J. C. Lai, D. P. Vanderpool, M. Good, and R. Prados, *Macromolecules* **3**, 746 (1970).
207. J. C. Lai, T. D. Rounsefell, and C. U. Pittman, Jr., *Macromolecules* **4**, 155 (1971).
208. C. U. Pittman, Jr. and J. C. Lai in Ref. 204, p. 161.
209. Y. H. Chen, M. Fernandez-Refojo, and H. G. Cassidy, *J. Polym. Sci.* **40**, 433 (1959).
210. C. U. Pittman, Jr., R. L. Voges, and J. Elder, *Polym. Lett.* **9**, 191 (1971).
211. C. U. Pittman, Jr. and P. L. Grube, *J. Polym. Sci. Part A-1* **9**, 3175 (1971).
212. C. U. Pittman, Jr., *Polym. Lett.* **6**, 19 (1968).
213. C. U. Pittman, Jr., R. D. Priester, Jr., and T. V. Jayaraman, *J. Polym. Sci. Polym. Chem. Ed.* **19**, 3351 (1981).
214. V. V. Korshak, A. M. Sladkov, L. K. Luneva, and A. S. Girshovich, *Vysokomol. Soedin.* **5**, 1284 (1963).
215. R. Ralea and co-workers, *Rev. Roum. Chim.* **12**, 523 (1967).
216. C. U. Pittman, Jr., R. L. Voges, and W. R. Jones, *Macromolecules* **4**, 291 (1971).
217. C. U. Pittman, Jr., J. C. Lai, D. P. Vanderpool, M. Good, and R. Prados in C. D. Craver, ed., *Polymer Characterization: Interdisciplinary Approaches*, Plenum Publishing Corp., New York, 1971, pp. 97–124.
218. D. O. Cowan, J. Park, C. U. Pittman, Jr., Y. Sasaki, T. K. Mukherjee, and N. A. Diamond, *J. Am. Chem. Soc.* **94**, 5110 (1972).
219. C. U. Pittman, Jr., Y. Sasaki, and T. K. Mukherjee, *Chem. Lett.*, 383 (1975).
220. C. U. Pittman, Jr. and C. C. Lin, *J. Polym. Sci. Polym. Chem. Ed.* **17**, 271 (1979).
221. C. U. Pittman, Jr. and A. Hirao, *J. Polym. Sci. Polym. Chem. Ed.* **16**, 1197 (1978).
222. *Ibid.*, **15**, 1677 (1977).
223. S. L. Sosin, L. V. Jashi, B. A. Antipova, and V. V. Korshak, *Vysokomol. Soedin.* **22**(9), 699 (1970).
224. K. Gonsalves, L. Zhan-ru, R. W. Lenz, and M. D. Rausch, *J. Polym. Sci. Polym. Chem. Ed.* **23**, 1707 (1985).
225. C. U. Pittman, Jr. and M. D. Rausch, *Pure Appl. Chem.* **58**(4), 617 (1986).
226. Y. Morita, M. Yamauchi, H. Yasuda, and A. Nakamura, *Kobunshi Ronbunshu* **37**, 677 (1980).
227. H. Yasuda, I. Noda, Y. Morita, H. Nakamura, S. Miyanaga, and A. Nakamura in Ref. 15, pp. 275–290 and references therein.
228. C. U. Pittman, Jr. and P. L. Grube, *J. Appl. Polym. Sci.* **18**, 2269 (1974).
229. N. Funita and K. Sonogashira, *J. Polym. Sci. Part A-1* **12**, 2845 (1974).
230. C. U. Pittman, Jr., D. Seyferth, and S. Massad, unpublished data.
231. O. E. Ayers, S. P. McManus, and C. U. Pittman, Jr., *J. Polym. Sci. Polym. Chem. Ed.* **11**, 1201 (1973).
232. C. U. Pittman, Jr., R. L. Voges, and W. R. Jones, *Macromolecules* **4**, 298 (1971).
233. C. U. Pittman, Jr., R. L. Voges, and J. Elder, *Macromolecules* **4**, 302 (1971).
234. C. U. Pittman, Jr. and G. V. Marlin, *J. Polym. Sci. Polym. Chem. Ed.* **11**, 2753 (1973).
235. M. Peuckert and W. Keim, *J. Mol. Catal.* **22**, 289 (1984).
236. C. U. Pittman, Jr., Y. Kawabata, and L. I. Flowers, *J. Chem. Soc. Chem. Commun.*, 473 (1982).
237. G. Consiglio, P. Pino, L. I. Flowers, and C. U. Pittman, Jr., *J. Chem. Soc. Chem. Commun.*, 612 (1983).
238. G. L. Baker, S. J. Fritschel, and J. K. Stille, *Polym. Prepr. Am. Chem. Soc. Div. Polym. Chem.* **22**, 155 (1981).
239. G. L. Baker, S. J. Fritschel, and J. K. Stille, *J. Org. Chem.* **46**, 2954 (1981).
240. R. J. Card and D. C. Neckers, *J. Am. Chem. Soc.* **99**, 7733 (1977).
241. K. Zhang and D. C. Neckers, *J. Polym. Sci. Polym. Chem. Ed.* **21**, 3115 (1983).
242. S. N. Gupta and D. C. Neckers, *J. Polym. Sci. Polym. Chem. Ed.* **20**, 1609 (1982).

243. P. DeMunik and V. R. Scholten, *J. Mol. Catal.* **11**, 331 (1981).
244. R. H. Grubbs, E. M. Sweet, and S. Phisanbut in P. Rylander and H. Greenfield, eds., *Catalysis in Organic Synthesis*, Academic Press, Inc., Orlando, Fla., 1976, p. 153.
245. H. S. Bruner and J. C. Bailor, *J. Am. Oil Chem. Soc.* **49**, 533 (1972).
246. H. S. Bruner and J. C. Bailor, *Inorg. Chem.* **12**, 1465 (1973).
247. B. M. Trost and E. Keinan, *J. Am. Chem. Soc.* **100**, 7780 (1978).
248. C. U. Pittman, Jr., Q. Ng, A. Hirao, W. Honnick, and R. Hanes, *Colloq. Int. CNRS* **281**, 49 (1977).
249. S. E. Jacobson and C. U. Pittman, Jr., *J. Chem. Soc. Chem. Commun.*, 187 (1975).
250. C. U. Pittman, Jr. and L. R. Smith, *J. Am. Chem. Soc.* **97**, 341 (1975).
251. L. R. Faulkner, *Chem. Eng. News*, 28 (Feb. 27, 1984).
252. C. U. Pittman, Jr. and R. F. Felis, *J. Organometal. Chem.* **72**, 389 (1974).
253. *Ibid.*, p. 399.
254. R. Tannenbaum, E. P. Goldberg, and C. L. Flenniken in Ref. 15, pp. 303–340.
255. E. Roman, G. Valenzuela, L. Gargallo, and D. Radic, *J. Polym. Sci. Polym. Chem. Ed.* **21**, 2057 (1983).
256. D. Astruc, J. R. Hamon, G. Althoff, E. Roman, P. Batail, P. Michaud, J. P. Mariot, F. Varret, and D. Cozak, *J. Am. Chem. Soc.* **101**, 5445 (1979).
257. J. C. Green, M. R. Kelly, M. P. Payne, E. A. Seddon, D. Astruc, J. R. Hamon, and P. Michaud, *Organometallics* **2**, 211 (1983).
258. M. Rajasekharan, S. Giezynsky, J. H. Ammeter, N. Oswald, P. Michaud, J. R. Hamon, and D. Astruc, *J. Am. Chem. Soc.* **104**, 2400 (1982).
259. E. A. Roman, G. J. Valenzuela, R. O. Latorre, and J. E. Sheats in Ref. 15, pp. 165–181.
260. V. V. Korshak, L. V. Dzhashi, and S. L. Sosin, *Nuova Chim.* **49**(3), 31 (1973).
261. V. V. Korshak, T. M. Frunze, A. A. Izynee, and V. G. Samsonova, *Vysokomol. Soedin. Ser. A* **15**(3), 521 (1973).
262. C. Simionescu, T. Lixandru, I. Maxilu, and L. Tatrau, *Makromol. Chem.* **147**, 69 (1971).
263. C. U. Pittman, Jr., Y. Sasaki, and P. L. Grube, *J. Macromol. Sci. Chem.* **A8**(5), 923 (1974).
264. C. U. Pittman, Jr. and Y. Sasaki, *Chem. Lett. Jpn.*, 383 (1975).
265. T. Nakashima, T. Kunitake, and C. Aso, *Makromol. Chem.* **157**, 73 (1972).
266. M. Kamachi, H. Akimoto, and S. Nozakura, *J. Polym. Sci. Polym. Chem. Ed.* **21**, 693 (1983).
267. M. Kamachi, H. Akimoto, W. Mori, and M. Kishita, *Polym. J.* **16**, 23 (1984).
268. M. Kamachi, M. Tamaki, Y. Morishima, S. Nozakura, W. Mori, and M. Kishita, *Polym. J.* **14**, 363 (1982).
269. M. Kamachi, H. Enomoto, M. Shibaska, W. Mori, and M. Kishita, *Polym. J.* **18**, 439 (1986).
270. M. Kamachi, M. Shibaska, W. Mori, and M. Kishita, *Macromolecules* (in press).
271. O. Kahn, J. Glay, Y. Journaux, J. Jaud, and I. Morgenstern-Badarau, *J. Am. Chem. Soc.* **104**, 2165 (1982).
272. J. P. Wesson and T. C. Williams, *J. Polym. Sci. Polym. Chem. Ed.* **19**, 65 (1981).
273. R. West, L. D. David, P. I. Djurovich, H. Yu, and R. Sinclair, *Ceram. Bull.* **62**(8), 899 (1983).
274. K. M. MacKay and R. Watt, *Organometal. Chem. Rev.* **4**, 137 (1969).
275. E. Hengge, *Fortschr. Chem. Forsch.* **51**, 1 (1974).
276. D. C. Hofer, R. D. Miller, and C. G. Willson, *SPIE J* **469**, 16 (1984).
277. D. C. Hofer, R. D. Miller, C. G. Willson, and A. R. Neureuther, *SPIE J.* **469**, 108 (1984).
278. *Chem. Week*, 22, 23 (May 30, 1984).

CHARLES U. PITTMAN, JR.
Mississippi State University

CHARLES E. CARRAHER, JR.
Florida Atlantic University

JOHN R. REYNOLDS
University of Texas-Arlington

ORGANOPHOSPHORUS POLYMERS.   See Phosphorus-contain-
ing polymers.

ORGANOSOL.   See Vinyl chloride polymers.

ORIENTATION

The orientation of macromolecules in fabricated and naturally occurring polymers plays an important role in determining their performance, ranging from mechanical to optical characteristics. No laboratory for characterizing polymeric materials can be considered complete if it does not contain tools for characterizing orientation.

Historical Perspective

The Early Period.   The existence of anisotropic structure in naturally occurring organic matter such as wood must have been realized from the dawn of history. It was not until much more recently, however, that this phenomenon was quantitatively studied by scientists and associated with structural models consisting of oriented units. A satisfactory picture and representation of oriented organic polymers did not come until the twentieth century.

The study of orientation is closely associated with the investigations of the interaction of electromagnetic waves with anisotropic matter, notably mineral single crystals. In the seventeenth century, Bartholinus and Huyghens discovered and characterized the double refraction of light by Iceland spar crystal (1,2). These observations were greatly expanded by David Brewster in the second decade of the nineteenth century (3–11) who began research on relating optical behavior with the structure of matter and greatly increased the number of minerals known to exhibit birefringent or depolarizing behavior (3,4). Brewster began the study of the influence of electromagnetic radiation on oriented polymers 120 years before the acceptance of the macromolecular hypothesis and a half-century before the rise of organic chemistry. He specifically described birefringent behavior in a number of naturally occurring polymers, including flax, hemp, and cotton, a compressed rubber film, bee honeycomb, and various vegetable films (4). The observation on rubber was probably due to applied compressive forces. The influence of applied forces in inducing birefringence was subsequently specifically discovered and extensively investigated by Brewster (5–8). His initial investigations included organic substances, such as beeswax, isinglass, and calves' feet jelly (5), but were later extended to glass (6) and inorganic crystals (7). Brewster subsequently showed that many birefringent crystals absorbed light in an anisotropic manner (10,11), a phenomenon he called dichroism for a mineral that had earlier been called dichroite because it exhibited different colors in different directions. This phenomenon was noted in other minerals, including tourmaline and mica.

Brewster's efforts were organized and extended by young James Clerk Maxwell in the years 1850–1870. In an 1853 paper Maxwell (12) extended birefringence–force studies to a readily usable experimental procedure for stress analysis based on a linear birefringence–stress relationship. Maxwell used isinglass jelly models and described experiments on stretching gutta percha to produce birefringent films. The first observations of flow-induced birefringence were made by him (13) on Canada balsam solutions. Maxwell's great achievement was his electromagnetic theory of light, which he developed for isotropic and anisotropic media (14,15). He showed that birefringence phenomena resulted for materials with dielectric properties which varied with direction.

Birefringence and anisotropic swelling on cellulose membranes were described by Carl Naegeli (16–18) in Munich in the 1850s. He devised an anisotropic micellar theory in which the micelles were oriented in order to explain these observations. However these efforts received little recognition, except for Zimmermann (18). A notable exception was Ambronn in Leipzig, who in 1888 (17), inspired by Naegeli, found and described visible dichroic phenomena in cell membranes. These observations extended into the twentieth century. A. Herzog (19) investigated the complex birefringent character of cotton in a 1909 paper.

It was, however, only with the development of x-ray diffraction (20,21) in the second decade of the twentieth century that it became possible to see clearly that molecular orientation existed in polymers (22,23). Appropriately, this technique was developed at the University of Munich where Naegeli had worked. The key experiments relating to polymers were made by R. O. Herzog and his collaborators at the Kaiser Wilhelm Institute in Berlin-Dahlem (22,23) ca 1920. The diffraction studies made clear that the glucoside residues of cellulose were oriented in specific directions. This was immediately seen by Herzog and others to be consistent with and reinforced by the earlier observations of birefringence and dichroic phenomena. Herzog relates the observations to Naegeli's micellar model.

The 1930s. By the 1930s, it was well known that both x-ray diffraction and birefringence could quantitatively determine the degree of orientation in fibers. This is abundantly clear from reading Carothers and Hill's (24) 1932 study of the first meltspun synthetic fibers. The same period saw the first efforts to develop new experimental techniques for orientation measurement (18). One of the leaders of this period was J. M. Preston at the College of Technology in Manchester (UK), who in 1931 (18) sought to develop Ambronn's dichroic technique using dyestuffs in cellulosic fibers. This technique was also applied by Morey (25) to cellulosic fibers. Preston (26) also sought to use refractive-index measurements to characterize orientation. In the same period, Morey (26–28) developed a fluorescence method for measurement of orientation in cellulosic fibers involving addition of fluorescent dyestuffs to the fiber.

Papers on x-ray diffraction measurements of orientation in this period were concerned with the orientation of cellulosic chains in fibers and in cell walls (29–33). A study of cotton, wood, flax, sisal, hemp, and other cellulosic structures established the occurrence of helically oriented cellulosic materials. Naturally occurring cellulose membranes were also investigated. Simon (32) proposed the use of pole figures, ie, stereographic projection of normals to crystallographic planes, to represent the orientation of the chains of crystalline cellulose (32). The development of uniaxial and uniplanar orientation was described based on the x-ray results.

The first attempts at developing quantitative representations of uniaxial orientation were made in the 1930s by Preston (18). Two possible definitions of fraction orientation were given:

$$F_{pi} = \frac{\begin{array}{c}\text{sum of the components of particles}\\ \text{in direction of fiber axis}\end{array}}{\text{total number of particles}}$$

$$F_{pii} = \frac{\begin{array}{cc}\text{sum of components} & - & \text{sum of components}\\ \text{in direction of} & & \text{in either of other}\\ \text{fiber axis} & & \text{directions}\end{array}}{\text{total number of particles}}$$

In visible dichroism experiments, this leads to

$$F_{pi} = \frac{A_1}{A_1 + 2A_2} = \frac{D}{D + 2}$$

$$F_{pii} = \frac{A_1 - A_2}{A_1 + 2A_2} = \frac{D - 1}{D + 2}$$

where A is absorptivity and D is dichroic ratio (A_1/A_2); 1 is the fiber axis and 2 is the perpendicular direction. A later paper by Preston (34) examined the refractive indexes of cotton and cellulose fibers; the orientation was represented as

$$F_{pi} = \cos^2 \phi$$

where ϕ is the angle the polymer chain axis makes with the fiber axis. Later papers (25–28) used this representation of orientation.

In a series of papers published in 1939, P. H. Hermans and co-workers expressed uniaxial orientation (35–38) through a quantity called the Quellungsanisotropie Q, which was developed on the basis of statistical arguments.

$$Q = \frac{1}{\displaystyle\int_0^{\pi/2} J_\alpha{:}J_0 \sin^3 \phi \, d\phi} - \frac{1}{2} = \frac{1}{\sin^2 \phi} - \frac{1}{2}$$

where $J_\alpha{:}J_0$ is a statistical quantity occurring in a theory of birefringence of anisotropic lattices (39). It was shown by Hermans and Platzek (37) that

$$f_H = \frac{\Delta n}{\Delta^\circ} = \frac{2(Q - 1)}{2Q + 1}$$

which is equivalent to

$$f_H = \frac{\Delta n}{\Delta^\circ} = 1 - \tfrac{3}{2} \overline{\sin^2 \phi} \tag{1}$$

where ϕ is the angle between the chain axis and the fiber axis, and Δ° intrinsic birefringence ($\Delta^\circ = n_\parallel^\circ - n_\perp^\circ$, where $n_\parallel^\circ$ = refractive index along the chain axis and $n_\perp^\circ$ = refractive index transverse to the chain axis; $\Delta n = n_1 - n_2$, where n_1 = refractive index along the fiber axis and n_2 = refractive index transverse to the fiber axis). In a study of biaxial orientation in films carried out in the same period by Okajima and Koijumi (40–42), the three principal refractive indexes in cellulosic films were measured using an Abbe refractometer.

1940–1956. In the 1940s, Muller represented the orientation distribution function in spherical harmonics (43,44):

$$F = F_0[1 + A_1P_1 + A_2P_2 \cdots \cdots]$$

The P_js are Legendre polynomials in the mean angle between the fiber axis and the polymer chain axis. By symmetry arguments, the coefficients of the odd terms are zero. The first nonzero in P_j is

$$P_2(\phi) = \frac{3\overline{\cos^2\phi} - 1}{2}$$

which is recognizable as the Hermans orientation factor f_H of equation 1. Birefringence is analyzed in terms of the anisotropy of molecular polarizability and the Lorentz-Lorenz equation between refractive index and polarizability.

P. H. Hermans and co-workers described how the orientation factor f_H may be computed from wide-angle x-ray diffraction, and developed a simple theory of birefringence (45–47). Orientation measured by x rays and by birefringence were distinguished (46). The former represents the orientation of the crystalline regions and the latter the total orientation, including amorphous regions; crystalline orientation is invariably higher (46), ie,

$$f_H(\text{waxs}) > \frac{\Delta n}{\Delta^\circ}$$

The subject of orientation measurements by visible dichroism also received attention in this period (48–50). In 1950, a fundamental analysis of visible dichroism was published (51), showing that the dichroic effect had the same second-order form as refractive index:

$$F_{pii} = \frac{D-1}{D+2} = 1 - \tfrac{3}{2}\overline{\sin^2\phi} = f_H$$

Dichroic measurements were subsequently examined by P. H. Hermans and Heikens (52) in 1952. They showed that f_H can be expressed as

$$f_H = \frac{D-1}{D+2}\frac{D^0+2}{D^0-1} \tag{2}$$

where D^0 is the dichroic ratio at complete orientation. Hermans and Heikens compared measurements on cellulose fibers using birefringence, wide-angle x-ray diffraction results as follows:

$$f_H(\text{waxs}) > \frac{\Delta n}{\Delta^\circ} > \frac{D-1}{D+2}\frac{D^0+2}{D^0-1}$$

Depending on the value for D^0 (the dichroic ratio at complete orientation), values of f_H were the same as those determined by birefringence or x-ray diffraction.

Observations of infrared dichroism on polymeric systems were first made about 1950 using polypeptides (53,54). Fraser (55,56) developed its use to measure orientation in polymers quantitatively and establish a relationship of dichroic ratio D to the Hermans orientation factor equivalent to equation 2. The first infrared dichroism studies on polyethylene were performed in 1954 by Stein and co-workers (57,58). The paper (58) discusses the determination of uniaxial orientation in polyethylene films using wide-angle x-ray (wax) diffraction, birefringence, and infrared dichroism, and explicitly states the interrelation of the former two measurements through

$$\Delta n = X \, \Delta_c^\circ f_{Hc} + (1 - X) \, \Delta_{am}^\circ f_{Ham} + \Delta_{form}$$

where X is the crystalline fraction; f_{Hc} and f_{Ham} are Hermans orientation factors for the crystalline and amorphous regions, respectively; and Δ_c° and Δ_{am}° are intrinsic birefringences for crystalline and amorphous regions.

When multiphase systems contain anisotropic particles whose sizes are small compared to the wavelength of light and if the phases exhibit large differences in their refractive indexes, these give rise to an additional birefringence called form birefringence (voids in fibers, interfaces in incompatible polymer blends, etc). These are associated with differing mean values of the dielectric constant or refractive index associated with averaging over phase distributions.

The Modern Period Begins. In 1958, Stein defined the uniaxial orientation of crystallographic axes relative to the fiber axis as (59)

$$f_j = \frac{3 \, \overline{\cos^2 \phi_j} - 1}{2} \tag{3}$$

where j represents the crystallographic axes a, b, and c. Stein described the determination of f_a, f_b, and f_c for polyethylene using wide-angle x-ray diffraction.

In an attempt to define biaxial orientation factors, using Euler's angles of spherical coordinates, Stein defined six orientation factors for the three crystallographic axes (60). These are equation 3 plus the three expressions

$$f_\delta = 2 \, \overline{\cos^2 \delta} - 1 = \overline{\cos 2\delta} \tag{4}$$

$$f_\gamma = 2 \, \overline{\cos^2 \gamma} - 1 = \overline{\cos 2\gamma} \tag{5}$$

$$f_\sigma = 2 \, \overline{\cos^2 \sigma} - 1 = \overline{\cos 2\sigma} \tag{6}$$

where δ, γ, and σ latitudinal angles define the orientations of the **a**-, **b**-, and **c**-axis vectors. Stein described the determination of these orientation factors for polyethylene by wax diffraction, birefringence, and ir dichroism (60).

The original work was on polyethylene which possesses an orthorhombic unit cell. The problems of obtaining f_a, f_b, and f_c with nonorthorhombic monoclinic and triclinic systems was more difficult. Wilchinsky (61) described a method for using wax diffraction to determine the orientation of all crystallographic axes from available diffraction planes (62,63).

Wide-angle x-ray pole figures were applied to polypropylene (64), poly(ethylene terephthalate) (65), and polyethylene film (66,67).

The problem of biaxial orientation factors in polymer systems was considered again in 1967 (68), leading to the equations

$$F_\delta = \overline{\sin^2 \phi_{1a} \cos 2\delta} \tag{7}$$

$$F_\gamma = \overline{\sin^2 \phi_{1b} \cos 2\gamma} \tag{8}$$

$$F_\sigma = \overline{\sin^2 \phi_{1c} \cos 2\sigma} \tag{9}$$

The set of six orientation factors, equations 3 and 7–9, was related to the in-plane and out-of-plane birefringences and infrared dichroic ratios.

The use of polarized fluorescence of polymers containing dyestuffs to characterize orientation in polymers was revived in the mid-1960s (69). The method had first been used in the 1930s (28–30) and been temporarily revived in 1950 (70). The polarized fluorescence method was quantified in successive papers (71,72).

Inverse-pole-figure methods of orientation representation were introduced in 1964–1965 by Roe and Krigbaum (73,74). They expressed orientation distribution in terms of expansions of associated Legendre polynomials; this effort was expanded later by Nomura and Kawai (75,76), who also showed that fluorescence measurements of orientation correspond to mixed fourth- and second-order terms.

Other approaches to biaxial orientation proved possible. In the 1960s, Wilchinsky (77) and Stein (71) proposed that angles ϕ_{1j} and ϕ_{2j} be used between crystallographic axes and machine and transverse directions. These were developed into orientation factors by White and Spruiell (78) in 1981.

Representation of Orientation

Uniaxial Orientation.   The orientation (79) of chains in polymers is expressed in terms of the angle ϕ between the symmetry axis and the segments of the polymer chains (Fig. 1a). The distribution function $N(\phi)$ is defined to represent the distribution of segments possessing an angle ϕ with the axis. This function can be normalized:

$$2\pi \int_0^\pi N(\phi) \sin \phi \, d\phi = 1$$

According to Muller (43), the function $N(\phi)$ may be expanded in terms of a series of orthogonal polynomials, notably Legendre polynomials:

$$N(\phi) = \Sigma_\parallel a_n P_n(\cos \phi)$$

where $P_n(\cos \phi)$ are Legendre polynomials. If the assumption is made that opposite ends of polymer chains are symmetric, ie,

$$N(\phi) = N(\pi - \phi)$$

it follows that

$$a_n = 0 (n \text{ odd})$$

From the orthogonality characteristics of the Legendre polynomials,

$$\int_0^\pi P_m(\cos \phi) \, P_n(\cos \phi) \sin \phi \, d\phi = \left(\frac{2}{2n + 1}\right)\delta_{mn}$$

and

$$a_n = \left[\frac{1}{2\pi} \frac{2}{2n + 1}\right] \int_0^\pi N(\phi) \, P_n(\phi) \sin \phi \, d\phi$$

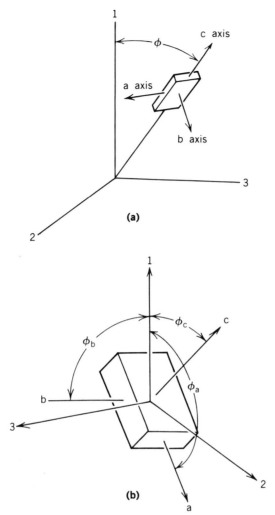

Fig. 1. (a) Uniaxial orientation; (b) angles ϕ_j between crystallographic axes and symmetry axis.

The first term in the expansion for a_n is the Hermans orientation factor (37,45,47,52):

$$f_{\mathrm{H}} = P_2(\phi) = \frac{3\,\overline{\cos^2 \phi} - 1}{2}$$

The second term is

$$P_4(\phi) = \tfrac{1}{8}[35\,\overline{\cos^4 \phi} - 30\,\overline{\cos^2 \phi} + 3]$$

The values of f_{H} of $P_2(\phi)$ range from $(+1)$ to $-\tfrac{1}{2}$ according to whether the chains are parallel to the symmetry axis,

$$\phi = 0 \qquad f_{\mathrm{H}} = P_2(\cos \phi)|_{\phi=0} = 1$$

or perpendicular to the symmetry axis,

$$\phi = \pi/2 \qquad f_{\mathrm{H}} = P_2(\cos \phi)|_{\phi = \pi/2} = -\tfrac{1}{2}$$

or characterized by isotropy,

$$\cos^2 \phi = \tfrac{1}{3} \qquad f_{\mathrm{H}} = P_2(\cos \phi) = 0$$

For a crystalline polymer, the angles ϕ_j and the distribution functions $N(\phi_j)$ between the crystallographic axes and the symmetry axes may be defined (Fig. 1b). Orientation factors for these crystallographic axes j may be defined (59).

$$f_j = \frac{3 \, \overline{\cos^2 \phi_j} - 1}{2}$$

The $\overline{\cos^2 \phi_j}$ are not independent, but are tied together through trigonometric relationships. For an orthorhombic unit cell such as polyethylene, consider the Pythagorean theorem:

$$\overline{\cos^2 \phi_a} + \overline{\cos^2 \phi_b} + \overline{\cos^2 \phi_c} = 1 \qquad (10)$$

$$f_a + f_b + f_c = 0 \qquad (11)$$

In multiphase systems, orientation distribution functions and orientation factors may be defined for each individual phase. In a semicrystalline polymer, it is possible to distinguish between f_{Hc}, the orientation factor of the crystalline phase, and f_{Ham}, the orientation factor of the amorphous regions.

A second approach to the specification of orientation in polymers is based on the anisotropy of the polarizability tensor α_{ij}:

$$f_{\mathrm{H}} = \frac{\alpha_{11} - \alpha_{22}}{\Delta\alpha_0} = \frac{\alpha_{11} - \alpha_{33}}{\Delta\alpha_0} \qquad (12)$$

where $\Delta\alpha_0$ is the value of the difference in polarizability along and perpendicular to polymer chains, and α_{11}, α_{22}, and α_{33} are the values of the polarizability tensor along "1" and perpendicular "2, 3" to the symmetry axis. The relation of f_{H} to $\overline{\cos^2 \phi}$ may be derived by coordinate transformations (43,45).

Biaxial Orientation.   Biaxial orientation in polymers can be represented in several ways, eg, by Euler's angles (60,68,75,76). This involves the angle ϕ with respect to the first symmetry axis and a second latitude. A different set of angles was proposed by Wilchinsky (77) in 1963 and later expanded (67,78) (Fig. 2). These (80–85) involve the above mentioned angle ϕ, denoted here by ϕ_1, and a second angle ϕ_2 defined from an orthogonal axis.

These representations would all seem to be equivalent; each is associated with certain advantages. The system of Euler's angles has a long history and forms the basis of the well-known set of orthogonal functions, the associated Legendre polynomials. Mathematical representation of orientation distributions are straightforward (74–76). The ϕ_1, ϕ_2 angles have the advantage of being symmetric. They yield a symmetric set of orientation factors.

Orientation factor representations have been developed for the second moments for each of these cases. In fact, two sets have been developed using Euler's angles: by Stein (60) and by Nomura, Kawai, et al (68).

Stein (1958) (60):

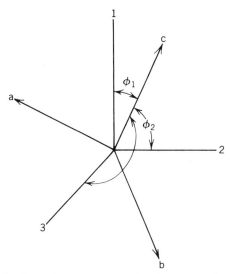

Fig. 2. Biaxial orientation representation in terms of angles ϕ_1 and ϕ_2.

$$f_\phi = \tfrac{1}{2}[3 \, \overline{\cos^2 \phi} - 1]$$

$$f_\delta = 2 \, \overline{\cos^2 \delta} - 1 = \overline{\cos 2\delta}$$

Nomura, Kawai, Kimura, and Kagiyama (1969) (68):

$$f_\phi = \tfrac{1}{2}[3 \, \overline{\cos^2 \phi} - 1] \tag{13}$$

$$f_\delta = \overline{\sin^2 \phi \cos 2\delta} \tag{14}$$

The orientation factors based on the ϕ_1 and ϕ_2 system are according to White and Spruiell (1981) (79):

$$f_1^B = 2 \, \overline{\cos^2 \phi_1} + \overline{\cos^2 \phi_2} - 1 \tag{15}$$

$$f_2^B = 2 \, \overline{\cos^2 \phi_2} + \overline{\cos^2 \phi_1} - 1 \tag{16}$$

These equations are symmetric with regard to angles ϕ_1 and ϕ_2. The orientation factors based on angles ϕ_1 and ϕ_2 are much more understandable and interpretable than those based on angles ϕ and δ. It may be seen from equations 15 and 16 that for uniaxial orientation in the 1 direction,

$$\phi_1 = 0, \; \phi_2 = \frac{\pi}{2} : f_1^B = 1 \qquad f_2^B = 0$$

and in the 2 direction,

$$\phi_1 = \frac{\pi}{2}, \; \phi_2 = 0 : f_1^B = 0 \qquad f_2^B = 1$$

For uniaxial orientation in the 3 direction,

$$\phi_1 = \frac{\pi}{2}, \; \phi_2 = \frac{\pi}{2} : f_1^B = f_2^B = -\tfrac{1}{2}$$

For random orientation,

$$\overline{\cos^2 \phi_1} = \overline{\cos^2 \phi_2} = \frac{1}{3} : f_1^B = f_2^B = 0$$

The characteristics of f_1^B and f_2^B suggest the representation in an isosceles triangle of the type shown in Figure 3.

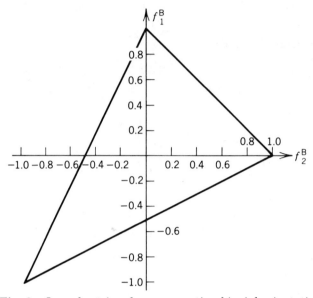

Fig. 3. Isosceles triangles representing biaxial orientation.

The previous representation of orientation may be generalized for crystalline polymers to represent all three crystallographic axes (79). They may be readily obtained from equations 15 and 16 by introducing angles between appropriate crystallographic axes and the defined laboratory axes. Equations 3–6, and 3 and 7–9, represent this for orientation factors based on Euler's angles. For orientation factors based on equations 15 and 16:

$$f_{1j}^B = 2 \overline{\cos^2 \phi_{1j}} + \overline{\cos^2 \phi_{2j}} - 1 \tag{17}$$

$$f_{2j}^B = 2 \overline{\cos^2 \phi_{2j}} + \overline{\cos^2 \phi_{1j}} - 1 \tag{18}$$

where ϕ_{1j} is the angle between the 1 laboratory axis (1 = machine direction) and the j-crystallographic axis (j = a, b, and c); ϕ_{2j} is the angle between the 2 laboratory axis (2 = transverse direction) and the j-crystallographic axis. The angles ϕ_{1j} and ϕ_{2j} are not independent of each other. For an orthorhombic unit cell, they are related by the Pythagorean theorem.

$$\overline{\cos^2 \phi_{1a}} + \overline{\cos^2 \phi_{1b}} + \overline{\cos^2 \phi_{1c}} = 1$$

$$\overline{\cos^2 \phi_{2a}} + \overline{\cos^2 \phi_{2b}} + \overline{\cos^2 \phi_{2c}} = 1$$

Simple relationships exist between the orientation factors of equations 13 and 14 and equations 15 and 16; they are equivalent (86).

Methods of Measurement

Birefringence. Optically, solids are categorized in three fundamental symmetry systems: cubic (isotropic), uniaxial, and biaxial. These systems possess one, two, or three principal refractive indexes, respectively. In other words, in cubic or isotropic systems, the optical indicatrix is a sphere ($n_1 = n_2 = n_3$); in a uniaxial system, the indicatrix is an ellipsoid of revolution ($n_1 \neq n_2 = n_3$); and, in a biaxial case (ie, orthorhombic), it is a general ellipsoid ($n_1 \neq n_2 \neq n_3$). The polymeric materials produced by processes such as compression molding, melt spinning, biaxial stretching, film blowing, and rolling possess one of these three general optical symmetries.

Birefringence is a classical method of measurement of orientation. The refractive index represents the slowing of the progress of an electromagnetic wave through a material because of interaction of the wave with polarizable molecules. This problem was first analyzed for isotropic materials (87–89) when the refractive index n was related to the molecular polarizability α through the expression

$$\frac{n^2 - 1}{n^2 + 2} = \frac{N\alpha}{3} \tag{19}$$

where N represents the number of molecules per unit volume. When polymer chains are oriented, the polarizability becomes a second-order tensor. It has been presumed that equation 19 remains valid along the principal axes of the polarizability and refractive-index ellipsoids. In oriented materials this leads to

$$\frac{n_i^2 - 1}{n_i^2 + 2} = \frac{N\alpha_i}{3}$$

Thus

$$\frac{n_i^2 - 1}{n_i^2 + 2} - \frac{n_j^2 - 1}{n_j^2 + 2} = \frac{3(n_i^2 - n_j^2)}{(n_i^2 + 2)(n_j^2 + 2)} = \frac{N(\alpha_i - \alpha_j)}{3}$$

and

$$n_i - n_j = \frac{(\overline{n^2} + 2)^2}{18\overline{n}} N(\alpha_i - \alpha_j)$$

where $\overline{n}$ represents a mean birefringence. It clearly follows that

$$\frac{\alpha_i - \alpha_j}{\alpha_{11} - \alpha_\perp} = \frac{n_i - n_j}{\Delta^\circ}$$

where

$$\Delta^\circ = \frac{(n^2 + 2)}{18\overline{n}} N(\alpha_\parallel - \alpha_\perp)$$

is the intrinsic birefringence.

The use of the previous procedure and its application of the Lorentz-Lorenz equation has been criticized (90,91). The derivation of this equation clearly rests on the assumption of isotropy. The problem for oriented chains has been analyzed using a cylindrical cavity as opposed to the spherical cavity of Lorentz (90).

The definitions of the orientation factors for uniaxial orientation lead to

$$f_H = \frac{n_1 - n_2}{\Delta^\circ} = \frac{n_1 - n_3}{\Delta^\circ} \tag{20}$$

For biaxial orientation using the orientation factors of equations 16 and 17:

$$f_1^B = \frac{n_1 - n_3}{\Delta^\circ} \qquad f_2^B = \frac{n_2 - n_3}{\Delta^\circ}$$

The developments described above presume single-phase materials. For multiple-phase materials, such as semicrystalline polymers, contributions for both phases must be introduced plus form birefringences; for uniaxial orientation (58)

$$\Delta n_{12} = X f_c \Delta_c^\circ + (1 - X) f_{am} \Delta_{am}^\circ + \Delta n_{form} \tag{21}$$

where Δn_{12} is the in-plane birefringence ($\Delta n_{12} = n_1 - n_2$, 1 = machine direction, and 2 = transverse direction); X is the crystalline fraction; f_c and f_{am} are the Hermans orientation factors for the crystalline and amorphous phases, respectively; Δ_c° and Δ_{am}° are the intrinsic birefringences of crystalline and amorphous regions; and Δn_{form} is form birefringence. For biaxially oriented systems (79)

$$\Delta n_{13} = X[f_{1c}^{Bc} \Delta_{cb}^{\circ c} + f_{1a}^{Bc} \Delta_{ab}^{\circ c}] + (1 - X) \Delta^{\circ am} f_1^{Ba} + \Delta n_{form}$$

$$\Delta n_{23} = X[f_{2c}^{Bc} \Delta_{cb}^{\circ c} + f_{2a}^{Bc} \Delta_{ab}^{\circ c}] + (1 - X) \Delta^{\circ am} f_2^{Ba} + \Delta n_{form}$$

where $\Delta_{cb}^{\circ c}$ and $\Delta_{ab}^{\circ c}$ are the cb- and ab-crystallographic direction intrinsic birefringences for the crystalline phases and $\Delta^{\circ am}$ is the intrinsic birefringence for the amorphous phase; $f_{1c}^{Bc}, f_{2c}^{Bc}, f_{1a}^{Bc}$, and f_{2a}^{Bc} are the orientation factors for the c- and a-crystallographic axes of the crystalline phase, and f_1^{Ba} and f_2^{Ba} are the orientation factors for the amorphous phase.

Retardation Techniques. Various optical systems have been used for this purpose. The most common is the polarized light microscope equipped with a compensator. Usually, the in-plane birefringence Δn_{12} (the refractive-index difference between principal axes lying in the plane normal to the axis of the microscope) is measured (Fig. 4). The monochromatic wave of light divides into two parts, polarized along the 1 and 2 principal directions which travel at different velocities v_1 and v_2:

$$R = \frac{2\pi h}{\lambda} \left(\frac{1}{v_2} - \frac{1}{v_1} \right)$$

where λ = wavelength and h = thickness.

This is sufficient for samples possessing uniaxial optical symmetry, ie, fibers and uniaxially stretched films. However, for oriented samples possessing biaxial optical symmetry (ie, orthorhombic), all three principal refractive indexes are not equal, and Δn_{13} and Δn_{23}, the out-of-plane birefringences, need to be measured (eqs. 20 and 21). An optical bench is employed (92,93), equipped with a simple goniometer, which enables the rotation of the sample about one of its main axes lying in the plane of the sample (Fig. 4).

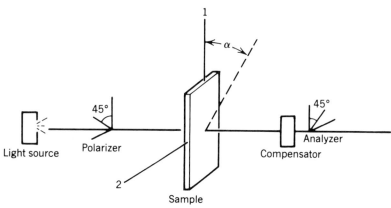

Fig. 4. Optical retardation with rotating stage to determine in-plane (Δn_{12}, $\phi = 0$) and out-of-plane (Δn_{13} or Δn_{23}, $\phi \neq 0$) birefringences. Sample tilting is about the 2 axis.

Specifically, when rotating about the 2 direction,

$$
n_3 - n_1 = \frac{\lambda}{h} \left[\frac{R(0) - R(\alpha)\left(1 - \dfrac{\sin^2 \alpha}{\bar{n}^2}\right)^{1/2}}{\sin^2 \alpha/\bar{n}} \right]
$$

where α is the angle of tilt.

Measurement of Refractive Indexes. Some of the techniques used to measure refractive indexes are the Becke line method (94), interference microscopy (95,96), Abbe refractometer (40,94,97–99), and ellipsometry (100). The Abbe refractometer (Fig. 5) can be used to determine the refractive indexes of materials belonging to all three optical symmetries. Its principle is based on the total reflection phenomenon occurring between a known and unknown isotropic media:

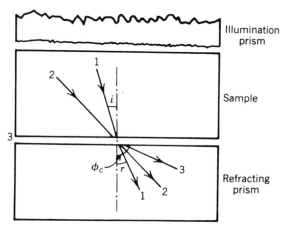

Fig. 5. Abbe refractometer technique.

$$\sin \phi_c = \frac{n}{n'}$$

where n' is the prism refractive index and n the unknown sample.

Commercial Abbe refractometers are usually calibrated to provide refractive indexes in the sodium D wavelength (589.3 nm). The Abbe refractometer is used to determine birefringence in polymer films (40–42). If the medium is not isotropic but uniaxial or biaxial, the rays entering the medium are plane-polarized in two axes of the sample traveling at different velocities determined by the refractive indexes of these directions. This results in two critical angles to be observed for a given sample orientation with respect to the prism, ie, transverse direction and normal to film surface. These can be separately determined using a polarizer on the eyepiece. This technique is simple (40,97–99), but requires a relatively smooth and flat sample surface. Its accuracy is to the fourth digit of the refractive indexes; some commercial Abbe refractometers have accuracies up to the sixth digit (95) (see REFRACTOMETRY).

Dichroism. Dichroism is a measure of anisotropy of absorption of electromagnetic waves in a medium. The absorption of the electromagnetic waves occurs when the frequency of electromagnetic waves is equal to the natural vibration frequency of electrons (101). The transition-moment vector defines this natural vibration. In the infrared wavelength range, these vibrations are associated with the molecular motions. The absorbance at a given frequency is defined as (102)

$$A = \log_{10} [I_0/I_\alpha (\mathbf{M} \cdot \mathbf{E} \cdot)^2]$$

where I_0 and I_α are the incident and transmitted intensity; $\mathbf{M}$ is the transition-moment vector associated with a certain vibration mode, and $\mathbf{E}$ is the electrical-field vector of the incoming beam.

In isotropic systems, such as gases, liquids, and randomly oriented polymeric systems, the transition-moment vector is randomly distributed in space, and thus absorbance is independent of the state of polarization of the incoming electromagnetic wave. However, when molecules are preferentially oriented, this randomness is lost, and the absorbance, no longer a scalar quantity, becomes a tensor, similar to the refractive-index tensor.

If polymer chains are preferentially oriented, the absorbance associated with a certain molecular vibration becomes anisotropic. This anisotropy is studied with a linearly polarized incoming light; the simplest representation is by dichroic ratio

$$D = \frac{A_1}{A_2}$$

where A_1 and A_2 are the absorbances when the polarization direction of the electric vector of the incoming wave is parallel "1" and perpendicular "2" to the stretching direction, respectively; D is unity for isotropic systems.

The two classes of dichroic investigations reported in the literature involve visible dichroism using dyes (17,18), and dichroism of chain segments in the infrared range (55,56,58). The latter is widely used. Visible dichroism is hampered by the unknown rules governing the orientation of absorbed dye molecules with respect to polymer chains.

In certain polymers, notably polyethylene, infrared-absorbing groups coincide with the crystallographic axes (102,103). It is thus possible to establish the orientation of the individual crystallographic axes in these materials.

Dichroism is a much more powerful technique than birefringence. The latter measures mean values over all the components and phases present, whereas dichroism is specific to particular absorption bands. It can thus be applied to different groups on the same molecule, to different segments in a copolymer (102,104), or to different components of a polymer blend (103,105).

Absorbance, like polarizability, is a second-order tensor. Thus as refractive index and birefringence lead to $P_2(\phi)$ and f_H, so do dichroic measurements.

It was shown for uniaxial orientation (18,50,52), for visible dichroism, and for infrared dichroism (56) that

$$f_H = \frac{D - 1}{D + 2} \frac{D^0 + 2}{D^0 - 1}$$

where D^0 is the dichroic ratio when the chain axes are perfectly aligned with the stretching direction; D^0 is related to the angle between the absorbing unit and the chain axis through

$$D^0 = 2 \cot^2 \alpha$$

where α represents the angles between transition-moment vector and electric vector of incoming, linearly polarized electromagnetic waves.

The application of infrared dichroism requires a knowledge of D^0 or the transition moment α in an unknown polymer. The latter is determined by identifying its value with established transition-moment angles for the same absorption band in different polymers.

When using infrared dichroism to measure biaxial orientation (68,86), absorptivities A_{11}, A_{22}, and A_{33} must be measured. In terms of the biaxial orientation factors f_1^B and f_2^B,

$$\frac{A_{11} - A_{33}}{A_{11} + A_{22} + A_{33}} = \frac{D_{13} - 1}{D_{13} + D_{23} + 1} = f_1^B \frac{D^0 - 1}{D^0 + 2}$$

$$\frac{A_{22} - A_{33}}{A_{11} + A_{22} + A_{33}} = \frac{D_{23} - 1}{D_{13} + D_{23} + 1} = f_2^B \frac{D^0 - 1}{D^0 + 2}$$

where D^0 is the value for complete uniaxial orientation. It is difficult to measure A_{11}, A_{22}, and A_{33} in a film. First the in-plane dichroic ratio is measured,

$$D_{12} = \frac{A_{11}}{A_{22}} = \frac{D_{13}}{D_{23}}$$

then the dichroic ratio D^{45} at 45°:

$$D^{45} = \frac{2A_{11}}{A_{22} + A_{33}} = \frac{2D_{13}}{D_{23} + 1}$$

Careful use of this technique requires knowledge of the refractive index at the frequency of the absorbing band. Its determination may be difficult because of anomalous dispersion in the frequency range of absorption.

Another useful dichroic technique is attenuated total reflection (atr) dichroism, which provides the surface dichroism of oriented products (106–108).

Dichroism measurements of oriented polymers are reviewed in Refs. 109 and 110.

X-ray Diffraction. The wide-angle x-ray diffraction technique (Fig. 6) is sensitive to the diffraction caused by crystalline regions. This technique is useful to obtain the orientation of a given direction of unit cell (generally a, b, and c axes).

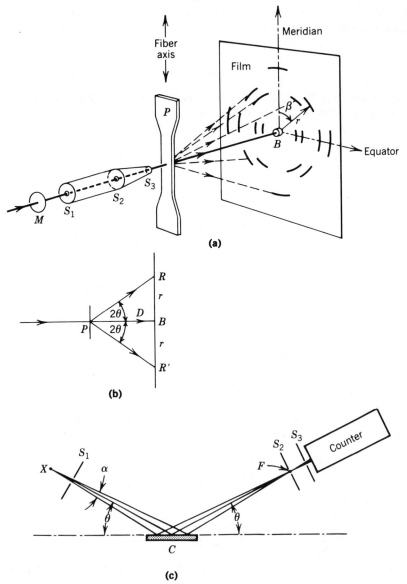

Fig. 6. Wide-angle x-ray diffraction technique. (**a**), (**b**) Flat film technique; (**c**) diffractometer technique (111). Courtesy of John Wiley & Sons, Inc.

X rays are diffracted from a specific crystallographic (hkl) plane according to Bragg's law:

$$\lambda = 2d_{hkl} \sin \theta_{hkl}$$

where λ is the wavelength of the x ray, d_{hkl} the interplanar spacing, and θ_{hkl} Bragg's angle. The c axis is the chain axis orientation, usually determined from the spatial distribution of the diffracted x rays.

In the case of uniaxial orientation, determination of orientation requires a series of hkl planes preferably perpendicular to the symmetry direction. In polyethylene, the (110), (200), and (020) reflections are strong and allow ready determination of the orientation of the a- and b-crystallographic axes. Furthermore, $\cos^2 \phi_a$ and $\cos^2 \phi_b$ may be computed directly from the angular intensity distribution of the (200) and (020) reflections, respectively. Alternatively, $\overline{\cos^2 \phi_b}$ may be obtained jointly from the (110) and (200) reflections (59). From $\cos^2 \phi_a$ and $\cos^2 \phi_b$ the Hermans-Stein orientation factors f_a and f_b of equations 10 and 11 may be determined; $\cos^2 \phi_c$ may be determined from equation 10 and f_c from equation 11.

A problem arises with many polymers, eg, polypropylene and poly(ethylene terephthalate) (PET), which lack suitable ($h00$) and ($0k0$) reflections, but exhibit monoclinic and triclinic unit cells. This problem was first attacked by Wilchinsky (61–63), who pointed out that

$$\overline{\cos^2\phi_{hkl}} = e^2 \overline{\cos^2 \phi_a} + f^2 \overline{\cos^2 \phi_b} + g^2 \overline{\cos^2 \phi_c} + 2ef \overline{\cos \phi_a \cos \phi_b}$$
$$+ 2fg \overline{\cos \phi_b \cos \phi_c} + 2ge \overline{\cos \phi_a \cos \phi_c} \quad (22)$$

where ϕ_{hkl} is the angle between a diffracting hkl plane and the reference axis. The quantities e, f, and g are geometric constants of the unit cell. In general, $\cos^2 \phi_a$, $\cos^2 \phi_b$, and $\cos^2 \phi_c$ are determined from five reflections $\cos^2 \phi_{hkl}$. On the right-hand side of equation 22, one or more terms may cancel, depending on the type of macro- and microsymmetry of the material (63,111). Specific application of this method was first made to polypropylene.

Orientation measurements for crystalline polymers are usually presented in terms of pole figures, which are stereographic projections of the diffracted intensities of normals to crystallographic planes. Originally applied to natural polymers (32), this technique has been applied to oriented synthetic polymers (64–68,83–85,111). Figure 7 shows two (010) plane, wide-angle x-ray pole figures on biaxially stretched and annealed PET films projected along two principal directions.

In general, the nth moment of the orientation distribution can be determined from wide-angle x-ray pole figures for the following relationship:

$$\overline{\cos^n \phi_j} = \frac{\int_0^{2\pi} \int_0^{2\pi} I_{hkl}(\phi_1, x_1) \cos^n \phi_1 \sin \phi_1 \, d\phi_1 \, dx_1}{\int_0^{2\pi} \int_0^{2\pi} I_{hkl}(\phi_1, x_1) \sin \phi_1 \, d\phi_1 \, dx_1}$$

where $I_{hkl}(\phi_1, x_1)$ is the diffracted intensity distribution in the pole figure.

Pole figures may be used to determine the biaxial orientation factors of crystallographic axes using equations 17 and 18 (83,84).

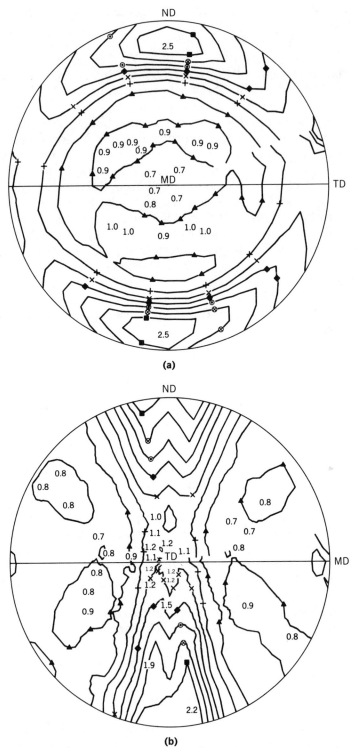

Fig. 7. Pole figures of (100) planes in stretched (4 × 1) and fixed annealed PET films. (**a**) Projection along MD; (**b**) projection along TD. MD = machine direction, TD = transverse direction, and ND = normal direction.

Other Techniques.   Almost all studies of orientation in polymer systems use birefringence, dichroism, or wide-angle x-ray diffraction, although other methods have also been developed.

Fluorescence.   The fluorescence technique is based on addition of a molecule with a fluorescing group (chromophore) to a polymer (26–28,69,70,112). It gives much more information on orientation than second-moment procedures such as birefringence and dichroism. This is due to the fact that fluorescence involves two processes: absorption of radiation and, after a delay called lifetime of excitation, a sequential emission at a different wavelength. The effect on the absorption and emission steps is to develop a fourth-moment dependence (113). More specifically, the emitted intensity I depends on

$$I \sim \cos^2 A \cos^2 B \qquad (23)$$

where A and B are the angles between the preferred axis and the axes of the absorbing and the emitting group, respectively. The angle between absorption and emission axes is generally small ($A \approx B$).

The fluorescence procedure presumes that the chromophore dye orients in the same manner as the polymer chain. This is, however, usually not the case. Studies of the literature indicate many instances where results are inconsistent with birefringence and other techniques.

Raman Scattering.   Raman scattering (114–118) involves irradiating a sample with light of one wavelength and observing radiation emitted at a different (longer) wavelength. The emitted radiation is related to the angle between the absorbing and emitting group through an expression similar to equation 23; a fourth-moment dependence again exists.

Broadline Nmr.   Broadline nmr as a technique for measuring orientation polymers (119–124) has limited application.

The nmr phenomenon is based on the fact that some nuclei possess angular momentum in addition to charge and mass. The electrical charges associated with nuclei spin due to angular momentum result in generation of a magnetic field as implied by Maxwell's equations (15,101). In a static magnetic field, the magnetic moments of spinning nuclei precess about the magnetic-field vector **H** with a frequency ω. This frequency is called the Larmor frequency. Increasing the magnetic-field strength causes increases of ω. When a second magnetic-field **H** is applied at right angles to the first field $\mathbf{H}_0$ and rotated in the plane normal to $\mathbf{H}_0$, oscillations in the angle between **M** and **H** occurs. If this rotation frequency is equal to the Larmor frequency, resonance occurs which results in absorption of power from the alternating field **H**. In dense systems like solids, the nmr absorption line is broadened due to increase of interaction of magnetic fields of neighboring pairs of, eg, protons. The width, shape, and intensity of nmr absorption spectra also depend on spin–spin and spin–lattice relaxation times.

When these pairs exhibit definite orientation as in solid materials (ie, single crystals), the nmr absorption spectra depend on the angle between the magnetic field and one of the unique axes of the sample.

Influence of Orientation on Properties

Orientation affects mechanical, optical, and electrical properties. Uniaxial orientation has a strong influence on mechanical properties. In Figure 8, tensile

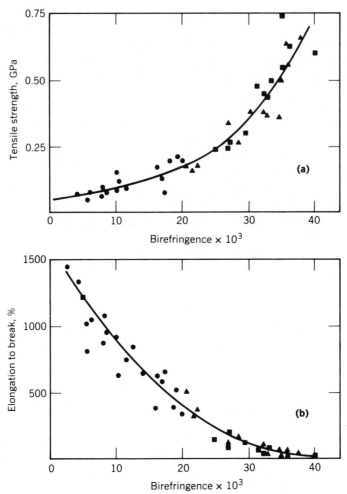

Fig. 8. (a) Tensile strength and (b) elongation to break as a function of birefringence in polypropylene fibers. ●, As spun; ▲, cold drawn and annealed at 140°C; and ■, hot drawn at 140°C. To convert GPa to dyn/cm², multiply by 10^{10}. Courtesy of *Journal of Applied Polymer Science.*

strength and elongation to break are plotted as a function of mean orientation (birefringence) in polypropylene fibers (125). Tensile strength and Young's modulus increase with uniaxial orientation. This result is typical; plots for other polyolefin fibers are similar. At low orientation, modulus and tensile strength correlate with either crystalline or amorphous orientation. After the crystallites are fully oriented, Young's modulus and tensile strength often increase further with additional stretching (126). During uniaxial stretching, birefringence continues to increase (125). From equation 21, it was concluded (127) that since f_c was constant, the key factor was amorphous orientation through f_{am}. Another quantity of importance in determining mechanical properties of oriented polymers is the chain extension as distinct from orientation. Eliminating chain folds increases modulus and tensile strength.

Uniaxially oriented samples are frequently brittle in the transverse direction, as is shown for polystyrene in Figure 9 (81). Introducing biaxial orientation

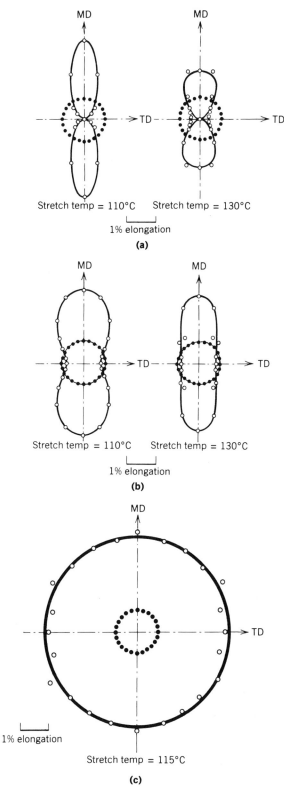

Fig. 9. Polar diagrams of elongation to break in uniaxially and biaxially stretched films. (**a**) Uniaxially stretched, free transverse direction, 3.0 × 10; (**b**) uniaxially stretched, fixed transverse direction, 3.0 × 1.0; and (**c**), simultaneous, biaxially stretched, transverse direction, 3.0 × 3.0. ND = normal direction, MD = machine direction, and TD = transverse direction. Courtesy of *Journal of Applied Polymer Science*.

produces materials with increased elongation to break in all directions in the plane of the film. Similar transverse toughening characteristics have been observed in other polymers (84).

Most extensive quantitative studies of mechanical properties have been done on three-dimensional small-strain elastic behavior. The strain tensor γ_{ij} is related to the applied stress tensor σ_{kl} through a fourth-order compliance tensor s_{ijkl}, ie, $\gamma_{ij} = s_{ijkl}\sigma_{kl}$. The compliance tensor components are related to the state of orientation; some of the s_{ijkl} components have been determined as a function of draw ratio by measuring apparent moduli cut at different angles from drawn sheets as well as by torsional studies (128–130).

BIBLIOGRAPHY

"Orientation" in *EPST* 1st ed., Vol. 9, pp. 624–648, by Zigmond W. Wilchinsky, Enjay Polymer Laboratories.

1. C. Huyghens, *Treatise on Light* (translated from the 1690 2nd ed.), Macmillan Publishing Co., New York, 1912.
2. D. Brewster, *Edinburgh Phil. J.* **1**, 289 (1819); **2**, 167 (1820); **3**, 148, 227 (1821).
3. D. Brewster, *Philos. Trans. R. Soc.* **104**, 187 (1814).
4. *Ibid.*, **105**, 29 (1815).
5. *Ibid.*, **105**, 60 (1815).
6. *Ibid.*, **106**, 156 (1816).
7. D. Brewster, *Trans. R. Soc. Edinburgh* 8, 281 (1817).
8. *Ibid.*, p. 353.
9. D. Brewster, *Philos. Trans. R. Soc.* **108**, 199 (1818).
10. *Ibid.*, **109**, 11 (1819); *Edinburgh Philos. J.* **2**, 341 (1820).
11. D. Brewster, *Edinburgh Philos. J.* **3**, 243 (1821).
12. J. C. Maxwell, *Trans. R. Soc. Edinburgh* **20**, 87 (1853).
13. J. C. Maxwell, *Proc. R. Soc.* **22**, 46 (1873).
14. J. C. Maxwell, *Philos. Trans. R. Soc.* **155**, 459 (1865).
15. J. C. Maxwell, *A Treatise on Electricity and Magnetism*, Constable, London, 1873.
16. C. Naegeli Mitt, *K. Bayer. Akad. Wiss. Sitzungsber.* (Mar. 1862); *Sitzungsber. Akad. Wiss. Munchen* **9**, 389 (1879).
17. H. Ambronn, *Ann. Phys. Chem.* **34**, 340 (1888).
18. J. M. Preston, *J. Soc. Dyers Colour.* **47**, 312 (1931).
19. A. Herzog, *Kolloid Z.* **5**, 246 (1909).
20. W. Friederich, P. Knipping, and M. Laue, *Proceedings of the Bavarian Academy of Science*, 1912, p. 303.
21. W. L. Bragg, *Proc. Camb. Philos. Soc.* **17**, 43 (1913).
22. R. O. Herzog and W. Jancke, *Ber.* **53**, 2162 (1920); R. O. Herzog, W. Jancke, and M. Polanyi, *Z. Phys.* **2**, 843 (1920).
23. R. O. Herzog, *J. Phys. Chem.* **30**, 459 (1926).
24. W. H. Carothers and J. W. Hill, *J. Am. Chem. Soc.* **54**, 1579 (1932).
25. D. R. Morey, *Text. Res. J.* **5**, 105 (1935).
26. *Ibid.*, **3**, 325 (1933).
27. *Ibid.*, **4**, 491 (1934).
28. *Ibid.*, **5**, 483 (1935).
29. G. L. Clark, *Ind. Eng. Chem.* **22**, 474 (1930).
30. W. A. Sisson and G. L. Clark, *Ind. Eng. Chem. Anal. Ed.* **5**, 296 (1933).
31. W. A. Sisson, *Ind. Eng. Chem.* **27**, 31 (1935).
32. W. A. Sisson, *J. Phys. Chem.* **40**, 343 (1936).
33. *Ibid.*, **44**, 513 (1940).
34. J. M. Preston, *Trans. Faraday Soc.* **29**, 65 (1933).

35. P. H. Hermans, O. Kratky, and P. Platzek, *Kolloid Z.* **86,** 245 (1939).
36. P. H. Hermans and P. Platzek, *Kolloid Z.* **87,** 246 (1939).
37. *Ibid.,* **88,** 68 (1939).
38. P. H. Hermans and J. de Bouys, *Kolloid Z.* **88,** 73 (1939).
39. O. Kratky and P. Platzek, *Kolloid Z.* **89,** 268 (1938).
40. S. Okajima and Y. Koizumi, *Kogyo Kagaku Zasshi* **42,** 810 (1939).
41. *Ibid.,* p. 813.
42. *Ibid.,* p. 816.
43. F. H. Muller, *Kolloid Z.* **95,** 138 (1941).
44. *Ibid.,* p. 306.
45. J. J. Hermans, P. H. Hermans, D. Vermaas, and A. Weidinger, *Rec. Trav. Chim. Pays-Bas* **65,** 427 (1946).
46. P. H. Hermans, J. J. Hermans, D. Vermaas, and A. Weidinger, *J. Polym. Sci.* **3,** 1 (1948).
47. P. H. Hermans, ed., *Contributions to the Physics of Cellulose Fibers* Elsevier, Amsterdam, The Netherlands, 1946.
48. S. Okajima and S. Kobayashi, *Nippon Kagakkai* **49,** 38 (1946).
49. S. Okajima, Y. Nakayama, and F. Adachi, *Nippon Kagakkai* **49,** 128 (1946).
50. J. M. Preston and P. C. Tsien, *J. Soc. Dyers Colour.* **62,** 368 (1946).
51. *Ibid.,* **66,** 361 (1950).
52. P. H. Hermans and D. Geikens, *Rec. Trav. Chim. Pays-Bas* **71,** 49 (1952).
53. E. J. Ambrose, A. Elliott, and Temple, *Nature London* **163,** 859 (1949).
54. E. J. Ambrose and A. Elliott, *Proc. R. Soc.* **A206,** 206 (1951).
55. R. D. B. Fraser, *J. Chem. Phys.* **21,** 1511 (1953).
56. *Ibid.,* **24,** 89 (1956).
57. R. S. Stein and M. Sutherland, *J. Chem. Phys.* **22,** 1993 (1954).
58. R. S. Stein and F. H. Norris, *J. Polym. Sci.* **21,** 381 (1956).
59. R. S. Stein, *J. Polym. Sci.* **31,** 327 (1958).
60. *Ibid.,* p. 337.
61. Z. W. Wilchinsky, *J. Appl. Phys.* **30,** 792 (1959).
62. R. Z. Sack, *J. Polym. Sci.* **59,** 543 (1961).
63. Z. W. Wilchinsky, *Adv. X Ray Anal.* **6,** 231 (1962).
64. Z. W. Wilchinsky, *J. Appl. Phys.* **31,** 1969 (1960).
65. C. J. Heffelfinger and R. L. Burton, *J. Polym. Sci.* **47,** 289 (1960).
66. P. H. Lindenmeyer and S. Lustig, *J. Appl. Polym. Sci.* **9,** 227 (1965).
67. C. R. Desper and R. S. Stein, *J. Appl. Phys.* **37,** 3990 (1966).
68. S. Nomura, H. Kawai, I. Kimura, and M. Kagiyama, *J. Polym. Sci. Part A-2* **5,** 479 (1969).
69. Y. Nishijima, Y. Onogi, and T. Asai, *J. Polym. Sci. Part C* **15,** 237 (1966).
70. J. M. Preston and Y. F. Su, *J. Soc. Chem. Ind. London* **66,** 357 (1950).
71. C. R. Desper and I. Kimura, *J. Appl. Phys.* **38,** 42225 (1967).
72. I. Kimura, M. Kagiyama, S. Nomura, and H. Kawai, *J. Polym. Sci. Part A-2* **7,** 709 (1969).
73. R. J. Roe and W. R. Krigbaum, *J. Chem. Phys.* **40,** 2603 (1964).
74. R. J. Roe, *J. Appl. Phys.* **36,** 2024 (1965).
75. H. Kawai, *Proceedings of the 5th International Rheology Congress,* Vol. 1, 1969, p. 97.
76. S. Nomura, H. Kawai, I. Kimura, and M. Kagiyama, *J. Polym. Sci. Part A-2* **8,** 383 (1970).
77. Z. W. Wilchinsky, *J. Appl. Polym. Sci.* **7,** 923 (1963).
78. J. L. White and J. E. Spruiell, *Polym. Eng. Sci.* **21,** 859 (1981).
79. D. I. Bower, *J. Polym. Sci. Polym. Phys. Ed.* **19,** 93 (1981).
80. K. J. Choi, J. L. White, and J. E. Spruiell, *J. Apppl. Polym. Sci.* **25,** 2777 (1980).
81. K. Matsumoto, J. F. Fellers, and J. L. White, *J. Appl. Polym. Sci.* **26,** 85 (1981).
82. J. L. White and A. Agarwal, *Polym. Eng. Rev.* **1,** 267 (1981).
83. K. J. Choi, J. E. Spruiell, and J. L. White, *J. Polym. Sci. Polym. Phys. Ed.* **20,** 27 (1982).
84. J. E. Flood, J. L. White, and J. F. Fellers, *J. Appl. Polym. Sci.* **27,** 2965 (1982).
85. M. Cakmak, J. E. Spruiell, and J. L. White, *Polym. Eng. Sci.* **24,** 1390 (1984).
86. J. L. White, *J. Polym. Eng.* **5,** 275 (1985).
87. H. A. Lorentz, *Ann. Phys. Chem.* **9,** 641 (1880).
88. H. A. Lorentz, *Theory of Elecrons* (translated from the 1909 2nd ed. published in Leiden, The Netherlands), Dover Publications, Inc., Mineola, N.Y., 1915.

89. L. Lorenz, *Ann. Phys. Chem.* **11,** 70 (1880).
90. R. S. Stein, *J. Polym. Sci. Part A-2* **1,** 1021 (1969).
91. S. D. Hong, C. Chang, and R. S. Stein, *J. Polym. Sci. Polym. Phys. Ed.* **13,** 1447 (1975).
92. R. S. Stein, *J. Polym. Sci.* **24,** 383 (1957).
93. B. E. Read, J. C. Duncan, and D. E. Meyer, *Polym. Test.* **143,** (1984).
94. F. D. Blass, *An Introduction to the Methods of Optical Crystallography,* Holt, Rinehart and Winston and W. B. Saunders, New York, 1967.
95. G. E. Fisher in R. Kingslate, ed., *Applied Optics and Optical Engineering,* Vol. IV, Academic Press, Inc., Orlando, Fla., 1967.
96. H. Matsuo, K. Iino, and M. Kondo, *J. Appl. Polym. Sci.* **7,** 1883 (1963).
97. G. W. Schael, *J. Appl. Polym. Sci.* **8,** 2717 (1964).
98. *Ibid.,* 903 (1968).
99. R. J. Samuels, *J. Appl. Polym. Sci.* **26,** 1383 (1981).
100. R. M. A. Azzam and N. M. Bashara, *Ellipsometry and Polarized Light,* Elsevier Science Publishing Co., Inc., New York, 1979.
101. M. Born and E. Wolf, *Principles of Optics,* Pergamon Press, Elmsford, N.Y., 1984.
102. B. E. Read and R. S. Stein, *Macromolecules* **1,** 116 (1968).
103. S. Endo, K. Min, J. L. White, and T. Kyu, *Polym. Eng. Sci.* **26,** 45 (1986).
104. G. M. Estes, R. W. Seymour, and S. L. Cooper, *Macromolecules* **4,** 452 (1971).
105. S. Onogi, T. Asada, and A. Tanaka, *J. Polym. Sci. Part A-2* **7,** 171 (1969).
106. P. A. Fluornoy and W. J. Schaffers, *Spectrochim. Acta* **22,** 5 (1966).
107. S. S. P. Sung, *Macromolecules* **14,** 591 (1981).
108. A. J. Barbetta, *Appl. Opt.* **38,** 29 (1984).
109. B. E. Read in I. M. Ward, ed., *Structure and Properties of Oriented Polymers,* Wiley-Interscience, New York, 1975.
110. B. Jasse and J. L. Koenig, *J. Macromol. Sci. Rev. Macromol. Chem.* **17**(1), 61 (1979).
111. L. E. Alexander, *X-ray Diffraction Methods in Polymer Science,* Wiley-Interscience, New York, 1969.
112. D. I. Bower in Ref. 109.
113. C. R. Desper and I. Kimura, *J. Appl. Phys.* **38,** 4225 (1967).
114. S. W. Cornell and J. Koenig, *J. Appl. Phys.* **34,** 4883 (1968).
115. R. G. Snyder, *J. Mol. Spectrosc.* **37,** 353 (1971).
116. D. I. Bower, *J. Polym. Sci. Polym. Phys. Ed.* **10,** 2135 (1972).
117. J. Purvis, D. I. Bower, and I. M. Ward, *Polymer* **14,** 398 (1973).
118. B. Jasse, R. S. Chao, and J. L. Koenig, *J. Polym. Sci. Polym. Phys. Ed.* **16,** 2157 (1973).
119. D. W. McCall and W. P. Schlichter, *J. Polym. Sci.* **26,** 171 (1951).
120. D. Hyndman and G. F. Origilio, *J. Polym. Sci.* **39,** 556 (1959); **46,** 259 (1960); *J. Appl. Phys.* **13,** 1949 (1960).
121. G. G. Olf and A. Peterlin, *J. Appl. Phys.* **35,** 3108 (1964); *Kolloid Z.* **215,** 97 (1967).
122. V. J. McBrierty and I. M. Ward, *Br. J. Appl. Phys. J. Phys. D. Ser.* **2,** 1, 1529 (1968).
123. M. Kashiwagi, M. J. Ward, and I. M. Ward, *Polymer* **12,** 697 (1971).
124. M. Kashiwagi, M. J. Folkes, J. J. Manuel, and I. M. Ward, *Polymer* **14,** 111 (1973).
125. H. P. Nadella, J. E. Spruiell, and J. L. White, *J. Appl. Polym. Sci.* **22,** 3121 (1978).
126. J. L. White, K. C. Dharod, and E. S. Clark, *J. Appl. Polym. Sci.* **18,** 2539 (1974).
127. R. J. Samuels, *Structured Polymer Properties,* Wiley-Interscience, New York, 1974.
128. G. Raumann and D. W. Saunders, *Proc. Phys. Soc. London* **77,** 1028 (1961); *Br. J. Appl. Phys.* **14,** 795 (1963).
129. D. W. Hadley and I. M. Ward in I. M. Ward, ed., *Structure and Properties of Oriented Polymers,* Wiley-Interscience, New York, 1975, Chapts. 8 and 9.
130. M. W. Darlington and D. W. Saunders in Ref. 129.

JAMES L. WHITE
MUKERREM CAKMAK
University of Akron

ORIENTATION, BIAXIAL. See ORIENTATION PROCESSES.

ORIENTATION PROCESSES

Development of Process Technology

Products with "designed-in" orientation are increasingly important in the polymer industry. They include filaments with uniaxial orientation and films and bottles with biaxial orientation. A variety of molding operations have been developed in which multiaxial orientation is built-in during the process. Uniaxial orientation results in high tensile modulus and high tensile strength in one direction, whereas equal biaxial orientation results in equal properties in all directions in the film plane. Biaxial orientation in bottles results in balancing the mechanical properties between the bottle-axis direction and the hoop direction.

Initially, processing operations were designed to produce the desired product shapes. Subsequently, they were modified to improve the properties of these products. A variety of filaments, films, and moldings were made from gutta percha and cellulose nitrate, the two principal materials of the nineteenth- and early twentieth-century plastics industry. In 1845 a patent (1) was issued to the Gutta Percha Co. for extruding filaments of gutta percha through dies and drawing them through a water bath, followed by cold drawing between slow and fast rolls. The manufacture of tubing by a similar extension process was described in a subsequent patent (2), which included compression molding of containers. A description of early inflation or blow-molding technology followed (3). Similar process technologies were developed for cellulose nitrate in the 1870s (4), including ram extrusion and compression, and injection molding (5,6). A later patent (7) describes a process in which inflatable mandrels are used to produce hollow shapes. This was later extended to blow molding (8). Similar technologies were developed for solution processing to produce fibers (9–11) and later films (12–15) from cellulosics, employing wet-bath coagulation, evaporation, and chemical reaction procedures. Tubular films were made for sausage casing (12,14,15).

These early technologies were largely concerned with the production of desirable shapes with minimal residual stresses and voids. Discussions of desirable directional characteristics first appeared in the 1920s and became explicit in the 1930s. Carothers and Hill in 1932 specifically discuss the development of uniaxial orientation in the drawing of synthetic polyester and polyamide fibers (16–18). Processes leading to uniaxially oriented ribbon sheet and film from polyamides (19) were patented by DuPont at the end of the decade. These involve drawing extrudates through a quench bath and casting film onto a chilled roll.

A 1936 patent (20) to the National Carbon Co. describes a process in which extruded tubes or rods of vinyl thermoplastics are heated above their softening temperatures and are subsequently shrunk to other shapes.

Polystyrene was commercialized by the IG Farbenindustrie in the 1930s. A Siemens patent (21) calls attention to the brittleness of polystyrene parts and the effects of elongation during processing in upgrading the mechanical performance. Polystyrene tubes were extruded, introduced into hot molds, and subsequently inflated by introducing a compressed fluid into their core. Technologies were developed in Germany to produce biaxially oriented polystyrene film, for example, by drawing over a mandrel (22) or using tentering frames. In 1930 celluloid films and sheets were produced on tentering frames (23). The patent

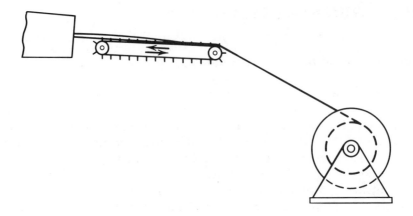

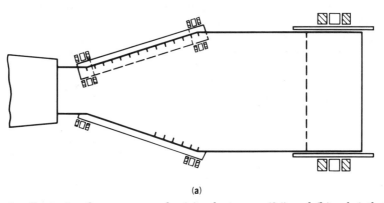

(a)

Fig. 1. Tentering-frame process for (**a**) polystyrene (24) and (**b**) poly(ethylene terephthalate) (25).

discusses shaping, but not orientation characteristics of materials. Polystyrene was developed into biaxially oriented film on a tentering frame (24) (Fig. 1a) by Norddeutsche Seekabelwerke AG.

In the 1940s the Plax Corp. (Hartford, Conn.) continued this development. A 1942 patent (26) describes shaping of quality sheets employing cellulose derivatives. Another patent (27) covers synthetic polymers and discloses equal biaxial orientation of polystyrene. Similar process technologies were applied to rubber treated with hydrochloride (28,29). In 1943 a tentering-frame process was introduced (28) by the Wingfoot Corp. (Akron, Ohio) for producing stretch film from nylon and other materials.

Tubular processes were also developed to produce biaxial orientation. The problem of unbalanced machine-direction uniaxial orientation in flexible seam-

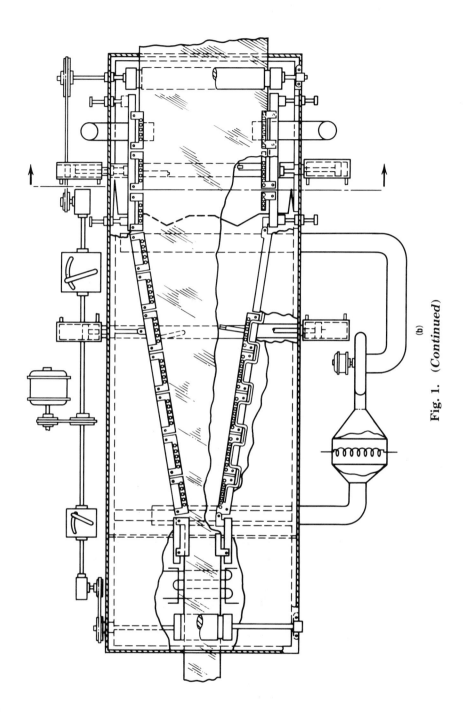

Fig. 1. (*Continued*)

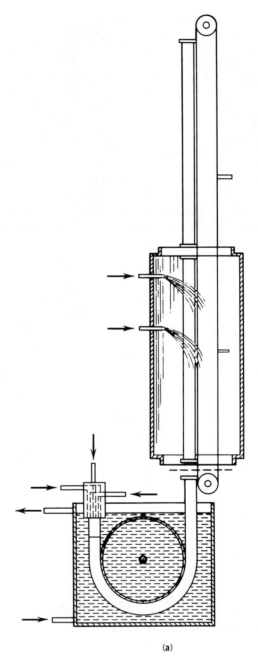

(a)

Fig. 2. Tubular-film process: (**a**) balanced single-bubble cellulosic (30); (**b**) thermoplastic films (31); and (**c**) triple bubble (32).

less cellulosic sausage casing is addressed in a 1937 patent (30) (Fig. 2a). The balanced orientation was developed by inflating the tubular extrudate emerging from a die with a high internal pressure and not subjecting it at this point to significant longitudinal tension. Film was produced from polyethylene–vinylidene chloride–vinyl chloride copolymer and other polymers using a tubular

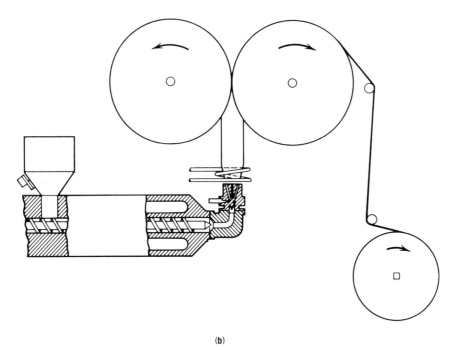

(b)

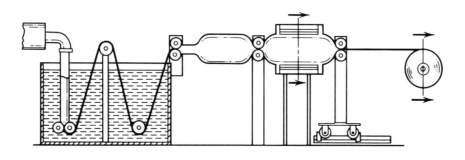

(c)

Fig. 2.   (*Continued*)

process, including air inflation (33–36). Other patents (31) (Fig. 2**b**) note that the properties of the tubular film are determined by the longitudinal drawdown ratio, degree of bubble inflation, and quench conditions (37). Vinylidene chloride–vinyl chloride copolymer was extruded as a tube and inflated in a second bubble (33–36).

In the 1950s the technologies of chilled-roll cast film, tentering frame, and tubular film were extended to biaxially oriented poly(ethylene terephthalate) (PET) film (25,38–46) (Fig. 1**b**). This involved developing biaxial orientation and controlled crystallization. Many variants on tentering-frame biaxial stretching were proposed (47–49). A continuous process for producing film oriented angularly to the machine direction uses crossed rollers (50,51).

The Dow Chemical Co. utilized the new technologies to fabricate biaxially oriented poly(vinylidene chloride) and polystyrene products. A 1954 patent (32)

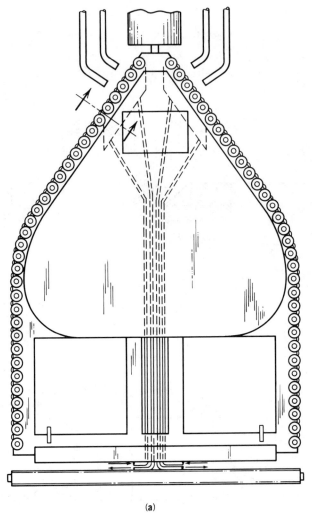

(a)

Fig. 3.   Mandrel process: (**a**) biaxially oriented polystyrene film (52); (**b**) biaxially oriented polyethylene tube (53).

(Fig. 2**c**) describes a triple-bubble technology to produce crystalline biaxially oriented poly(vinylidene chloride). Another patent (52) (Fig. 3**a**) employs a mandrel method to produce biaxially oriented polystyrene film. In a continuous process (54,55) (Fig. 4) the polymer melt is extruded and drawn out radially to produce film.

     In the 1960s attention turned to producing controlled orientation in thicker sections. Biaxially oriented thick sheet had been produced by batch versions of tentering frames (56). DuPont aimed at making biaxially oriented tubing, using a mandrel and gas pressure (53,57–60) (Fig. 3**b**). A different method of developing orientation in an extruded tube was developed at Dow (60). The outer diameter of an annular die was rotated in one direction, while the inner core or mandrel

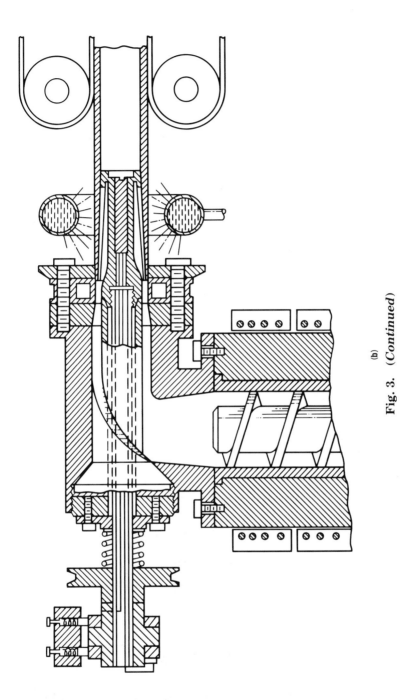

Fig. 3. *(Continued)*

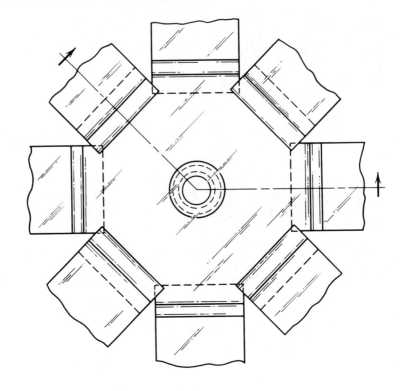

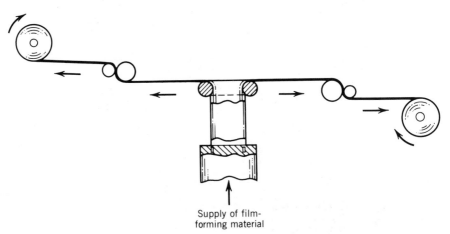

Supply of film-
forming material

Fig. 4. Radial film-stretching process (54,55).

rotated in the opposite direction (Fig. **5a**). This produced a helical flow pattern and orientation.

Efforts toward orientation processing led to new molding technology. A process for extruding, reheating, and inflating tubes produced orientation (61). In an injection-molding process the core of an annular mold rotates to produce a helically oriented molded part (62,63) (Fig. **5b**). The DuPont work on biaxially oriented poly(ethylene terephthalate) (64) involved the molding of an isotropic

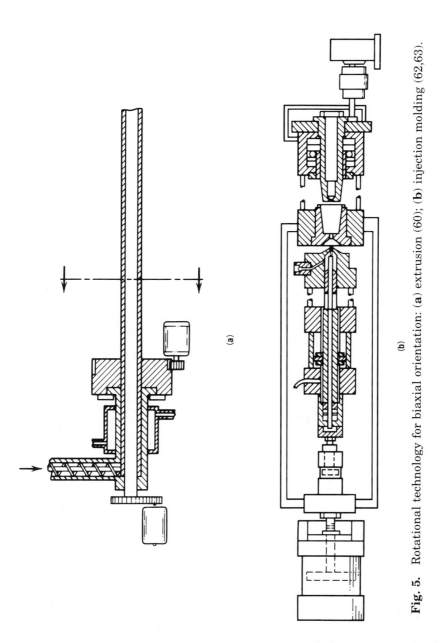

(a)

(b)

Fig. 5. Rotational technology for biaxial orientation: (**a**) extrusion (60); (**b**) injection molding (62,63).

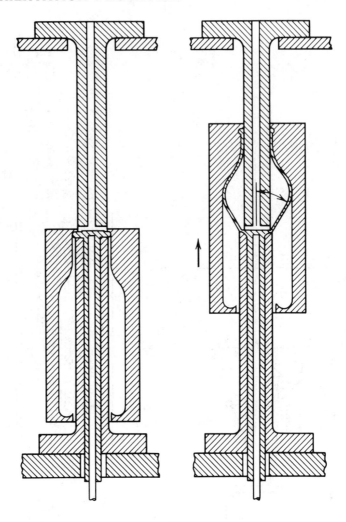

(a)

Fig. 6. Parison stretching: (**a**) a biaxially oriented bottle (64); (**b**) Corpoplast process (65).

glassy parison which was heated and inflated with air in a mold to form a biaxially oriented bottle (Fig. 6a). This gave rise to a new technology called stretch blow molding. The Corpoplast process is shown in Figure 6b.

New polymers were biaxially oriented by older technologies. Tentering-frame and double-bubble technologies were used to make a biaxially oriented polypropylene film (66–69).

Orientation and Process Conditions

In the processing of thermoplastics the degree of orientation appears to be related to the stress field at the time of solidification and the accompanying

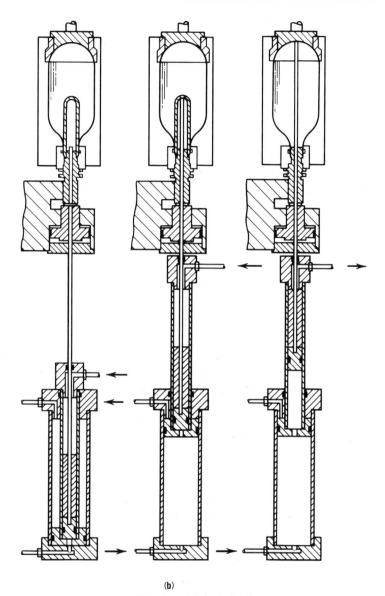

(b)

Fig. 6. (*Continued*)

thermal history. This behavior is due to the fact that flowing polymer melts consisting of flexible polymer chains obey the rheooptical law between birefringence and stress, ie, birefringence ($n_i - n_j$) is proportional to the principal stress differences ($\sigma_i - \sigma_j$):

$$n_i - n_j = C(\sigma_i - \sigma_j) \tag{1}$$

where C is the stress-optical constant. This has been verified by many research groups for different polymer melts (70–82) (see RHEOOPTICAL PROPERTIES; BIREFRINGENCE in the Supplement). Birefringence is related to the second moment of the orientation distribution in an amorphous polymer. If 1 is the machine direction, 2 the transverse direction, and 3 the thickness direction in a film, it may be shown (83,84) that

$$f_1^B = 2\,\overline{\cos^2 \phi_1} + \overline{\cos^2 \phi_2} - 1 = \frac{\Delta n_{1,3}}{\Delta^0} \tag{2}$$

$$f_2^B = 2\,\overline{\cos^2 \phi_2} + \overline{\cos^2 \phi_1} - 1 = \frac{\Delta n_{2,3}}{\Delta^0}$$

where ϕ_1 is the angle between the chain axis and the machine direction; ϕ_2 is the angle between chain axis and the transverse direction; Δ^0 is the intrinsic or maximum birefringence; and f_1^B and f_2^B are biaxial-orientation factors. From equations 1 and 2 the following expressions are derived for a flowing-melt sheet:

$$f_1^B = \frac{C}{\Delta^0}(\sigma_1 - \sigma_3) \tag{3}$$

$$f_2^B = \frac{C}{\Delta^0}(\sigma_2 - \sigma_3)$$

Both C and Δ^0 are considered to be proportional to the difference in polarizabilities along and perpendicular to the chain; C/Δ^0 appears to be a universal constant (85).

Considering the vitrification of a polymer melt under stress, it can be expected that the orientation governed by the stress field according to equation 3 is frozen-in and that equation 3 remains valid. The birefringence of vitrified glasses is the sum of this frozen-in melt-flow birefringence, plus contributions from stresses associated with nonuniform volumetric compression during solidification. This may be expressed by

$$(n_i - n_j)_g \cong C_m(\sigma_1 - \sigma_2)_m + C_s(\sigma_1 - \sigma_2)_v \tag{4}$$

where the subscript m refers to melt-phase values at the point of vitrification and g refers to the glassy state; C_g is the glassy phase stress-optical coefficient and $(\sigma_1 - \sigma_j)_v$ represents the volumetric stresses. Equation 4 is approximate as it has not been verified experimentally.

Calculations of the stress fields $(\sigma_1 - \sigma_2)_m$ and $(\sigma_1 - \sigma_2)_v$ reveal that the latter is usually larger (86,87). However, C_m of the melt state is often orders of magnitude larger than C_s of the glassy state, especially when the polymer contains phenyl groups. In polystyrene (85,88,89) (Fig. **7a**), polycarbonate (85) (Fig. **7b**), and poly(ethylene terephthalate) (90), experimental studies show that in vitrified parts

$$(n_1 - n_2) = C_m(\sigma_1 - \sigma_2)_m \tag{5}$$

However, this is not the case for poly(methyl methacrylate), where the value of C_m is closer to C_s.

In crystallizing polymer melts orientation factors are introduced, representing crystallographic axes

$$f_{1j}^B = 2\,\overline{\cos^2 \phi_{1j}} + \overline{\cos^2 \phi_{2j}} - 1 \tag{6}$$
$$f_{2j}^B = 2\,\overline{\cos^2 \phi_{2j}} + \overline{\cos^2 \phi_{1j}} - 1$$

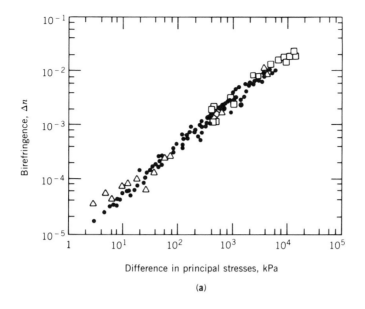

(a)

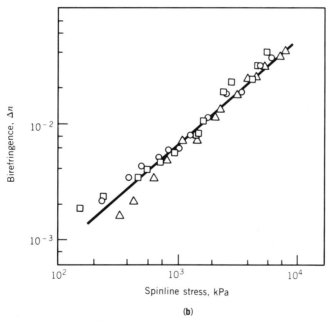

(b)

Fig. 7. Birefringence of fabricated products as a function of difference in principal stresses. (**a**) Polystyrene: □, melt spinning; ●, elongational flow; △, simple-shear flow; ◇, melt spinning (on-line measurement). (**b**) Polycarbonate, heated at 220 and 240°C; △, air cooling; ○, ice-water cooling; □, propanol cooling with dry ice. To convert kPa to dyn/cm², multiply by 10^4.

where j refers to the j-crystallographic axes (83,84). Birefringence must be considered to be the sum of contributions of crystalline and amorphous phases with the former making the main contribution.

Studies of the variation of birefringence along a fiber spinline (90,91) show that birefringence increases rapidly when crystallization occurs. This indicates that the degree of orientation of polymers crystallized under stress is much higher than that in vitrified melts.

Studies of different polymer melts crystallizing under stress in a fiber spinline demonstrated that the uniaxial crystallographic-orientation factors depend on the spinline stress rather than drawdown ratio (91–94). This was confirmed for tubular-film extrusion of polyethylene (95). The correlation of $f_{1j}^B - f_{2j}^B$ as a function of the stress field at the frost line in a polyethylene bubble is the same as that determined for a fiber spinline (92) (Fig. 8).

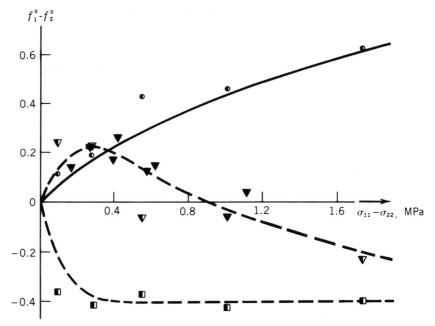

Fig. 8. Biaxial-orientation factors of fabricated film and uniaxial orientation of fibers as a function of stress field at freeze line: ▼, ▽, a axis; ◗, □, b axis; ◖, ○, c axis. Half-filled symbols indicate fibers. To convert MPa to psi, multiply by 145.

Fibers

Fibers (qv) are usually processed in such a manner as to produce uniaxial orientation along the filament axis with the chains oriented in the axial direction. This results in high Young's modulus and tensile strength, though often poor lateral properties. These are usually less important, as the primary tensions in structural members are in the axial direction.

Melt and Solution Spinning. Fiber-processing operations can be divided into a shaping step, where a polymer melt or solution is transformed into a continuous filament, and an orienting step, in which filaments are drawn in a state of uniaxial tension (see FIBERS, MANUFACTURE).

In melt spinning, orientation and crystalline morphology are governed by the stress fields (85,87,91–94), even with materials that vitrify into glasses (85,88). However, if the quenching characteristics vary significantly, glassy, paracrystalline, or truly crystalline polymers are produced. Polymers that crystallize to higher perfection can be oriented to a higher degree.

Drawing. The subsequent drawing operation elongates the polymer to several times its initial length; a neck is developed and propagated. The material in the necked region is usually much more oriented than the initial polymer. Further elongation is possible which increases the orientation. Further increases in uniaxial stretching result in little change in crystalline orientation, but produce large increases in birefringence, indicating an increase of amorphous material orientation in the semicrystalline polymer (96).

Films

Films (qv) are produced in such a manner as to yield systems with approximately equal properties in the film plane. High uniaxial orientation impairs transverse properties. Such uniaxially oriented films frequently fibrillate and are used as intermediates to produce continuous filaments. The degree of orientation and biaxiality in films varies with the process technology used (see FILMS, MANUFACTURE).

As with filaments, film production has two stages. First, the polymer melt is shaped by being forced through a slit or annular die of narrow dimension. In the second step, the desired biaxial orientation is produced.

Tubular-film Process. The most important technology for producing film is the tubular-film process (Fig. 2). The melt is extruded through a narrow annular die, and pressurized air is blown into the tube which is inflated to a larger diameter bubble. This is primarily a shaping operation to produce thin films. The stresses and drawdown in the machine and transverse directions induce some orientation of a biaxial rather than a uniaxial character. Most polyethylene films are produced in this manner.

Higher biaxial orientation may be produced in a second bubble. The film passes through an orientation and drawdown step induced by the injection of a pressure into the second bubble, which is maintained below the melting temperature.

Tentering-frame Process. In the tentering-frame process a polymer melt is cast on a quench roll which is heated to the polymer-softening temperature. The tentering frame imposes biaxial deformation at a later stage. Processing temperatures in biaxial deformation are lower than those used in tubular-film blowing. This produces films with high orientation. The tentering-frame process is used extensively for poly(ethylene terephthalate) films; additional orientation stages may be used such as annealing (heat setting) while the film is dimensionally constrained. This stabilizes the structure by increasing the crystallinity, which reduces shrinkage. Other steps may include coatings.

Molding

Traditionally, molding processes have been used to shape objects, but rarely to induce orientation. In injection molding (qv) a melt is pushed by a ram or reciprocating screw into a cold mold. The melt solidifies along the walls as it flows through the mold. The stresses developed during flow induce some orientation, which is influenced by gating, injection pressure, and mold temperature. The orientation developed is usually low.

For a high degree of orientation, a two-step process is required. First, a parison is injection molded, which may be heated above its softening temperature and reshaped. In stretch blow molding (qv), the injection-molded tubes are deformed into the desired shape, usually a bottle (Fig. 6), resulting in a higher degree of biaxial orientation.

Another procedure employs molds with rotating members (62,63); it is used to introduce hoop orientation into cups.

Other Orientation Processes

Solid-state Extrusion.   These processes involve the preparation of a solid billet and extrusion of this billet under high pressure through dies with converging entrance sections (97) or mandrel-type dies (98). The former type generally induces uniaxial orientation, whereas the latter type dies induce biaxial orientation. These processes provide end products with high chain orientation and crystallinities, but are slow (see SOLID-STATE EXTRUSION).

Gel Processing.   This process has been developed specifically for ultra high molecular weight polyethylene (mol wt $\sim$ 3–6 $\times$ 10^6) (99–101). Typical processing involves preparation of the gel from UHMWPE with decalin or paraffin as solvent. Subsequently, this gel is spun into filaments. Later stages of the operation include drawing and solvent-removal steps. The fibers obtained by this process possess very high chain orientations which result in high tensile modulus (see HIGH MODULUS POLYMERS).

BIBLIOGRAPHY

1. Brit. Pat. 10,582 (1845), R. A. Brooman.
2. Brit. Pat. 10,825 (1845), H. Bewley.
3. Brit. Pat. 11,208 (1846), C. Hancock.
4. E. C. Worden, *Nitrocellulose Industry,* Van Nostrand Reinhold Co., Inc., New York, 1914.
5. U.S. Pat. 133,229 (1872), I. S. Hyatt and J. W. Hyatt.
6. U.S. Pat. 152,232 (1874), I. S. Hyatt and J. W. Hyatt.
7. U.S. Pat. 204,228 (1878), J. W. Hyatt.
8. U.S. Pat. 237,168 (1881), W. B. Carpenter.
9. U.S. Pat. 394,559 (1888), H. de Chardonnet.
10. U.S. Pat. 531,158 (1891), H. de Chardonnet.
11. Brit. Pat. 8,700 (1892), C. F. Cross, E. J. Bevan, and C. Beadle; Brit. Pat. 1,020 (1898), C. F. Cross, E. J. Bevan, C. Beadle, and C. B. Stearn.
12. U.S. Pat. 1,163,740 (1915), W. B. Conroe.
13. U.S. Pat. 1,548,864 (1925), J. E. Brandenberger.
14. U.S. Pat. 1,601,686 (1926), W. F. Henderson.

15. U.S. Pat. 1,654,253 (1926), W. F. Henderson.
16. W. H. Carothers and J. W. Hill, *J. Am. Chem. Soc.* **54,** 1579 (1932).
17. U.S. Pat. 2,130,948 (1938), W. H. Carothers.
18. U.S. Pat. 2,137,235 (1938), W. H. Carothers.
19. U.S. Pat. 2,212,772 (1940), G. D. Graves.
20. U.S. Pat. 2,027,962 (1936), L. M. Currie.
21. U.S. Pat. 2,047,554 (1936), E. Fischer.
22. H. Horn, *Kunststoffe* **30,** 53 (1940).
23. U.S. Pat. 1,979,762 (1934), R. G. O'Kane and E. B. Derby.
24. U.S. Pat. 2,074,285 (1937), E. Studt and U. Meyer.
25. U.S. Pat. 2,770,684 (1957), F. P. Alles.
26. U.S. Pat. 2,297,645 (1942), J. Bailey.
27. U.S. Pat. 2,412,187 (1946), F. E. Wiley, R. W. Canfield, R. S. Jeslonowski, and J. Bailey.
28. U.S. Pat. 2,328,827 (1943), R. C. Martin.
29. U.S. Pat. 2,490,781 (1949), W. S. Clove.
30. U.S. Pat. 2,070,247 (1937), R. Weingard and A. Mulchinski.
31. U.S. Pat. 2,461,975 (1949), E. D. Fuller.
32. U.S. Pat. 2,688,733 (1954), J. W. McIntyre.
33. U.S. Pat. 2,409,521 (1946), R. M. Wiley.
34. U.S. Pat. 2,433,937 (1948), H. L. Tornberg.
35. U.S. Pat. 2,448,433 (1948), C. R. Irons and C. E. Sanford.
36. U.S. Pat. 2,452,080 (1949), W. T. Stephenson.
37. U.S. Pat. 2,461,976 (1949), B. H. Shenk.
38. U.S. Pat. 2,627,088 (1953), F. P. Alles and W. S. Saner.
39. U.S. Pat. 2,718,666 (1955), K. L. Knox.
40. U.S. pat. 2,755,553 (1956), D. F. Miller.
41. U.S. Pat. 2,767,435 (1956), F. P. Alles.
42. U.S. Pat. 2,784,456 (1956), T. A. Grabenstein.
43. U.S. Pat. 2,823,421 (1958), A. C. Scarlett.
44. U.S. Pat. 2,968,067 (1961), C. L. Long.
45. U.S. Pat. 3,256,379 (1966), C. J. Heffelfinger.
46. U.S. Pat. 3,257,489 (1966), C. J. Heffelfinger.
47. U.S. Pat. 2,659,931 (1953), E. V. Detmer.
48. U.S. Pat. 2,923,966 (1960), W. R. O'Tooke and E. G. Lodge.
49. U.S. Pat. 2,923,966 (1962), H. P. Koppehele.
50. U.S. Pat. 2,505,146 (1950), W. H. Ryan.
51. U.S. Pat. 2,854,697 (1958), W. H. Ryan.
52. U.S. Pats. 2,695,420 (1954), 2,718,658 (1955), and 2,770,007 (1956), M. O. Longstreth and D. W. Ryan.
53. U.S. Pat. 3,241,186 (1965), W. M. Coons.
54. U.S. Pat. 2,779,053 (1957), M. O. Longstreth and T. Alfrey.
55. U.S. Pat. 2,852,813 (1958), M. O. Longstreth and T. Alfrey.
56. U.S. Pat. 2,759,217 (1956), A. K. Peterson.
57. U.S. Pat. 2,987,765 (1961), M. T. Cichelli.
58. U.S. Pat. 2,987,767 (1961), C. E. Berry and M. T. Cichelli.
59. U.S. Pat. 3,092,874 (1963), E. L. Fallwell.
60. U.S. Pat. 3,279,501 (1966), H. J. Donald.
61. U.S. Pat. 3,288,317 (1966), F. E. Wiley.
62. U.S. Pat. 3,307,726 (1967), K. J. Cleereman.
63. U.S. Pat. 3,907,952 (1975), K. J. Cleereman.
64. U.S. Pats. 3,733,309 (1973) and 3,849,530 (1973), N. C. Wyeth and R. W. Roseveare.
65. U.S. Pat. 4,214,860 (1980), G. Kleimenhagen, O. Rosenkranz, P. Albrecht, H. Conow, H. Kofler, D. Smidth, and K. Vogel.
66. U.S. Pat. 3,302,241 (1967), R. E. Lemmon and G. E. Gould.
67. U.S. Pat. 3,311,679 (1967), E. J. Moore.
68. U.S. Pat. 3,325,575 (1967), A. G. M. Last.
69. U.S. Pat. 3,499,064 (1970), K. Tsuboshima, T. Matsuo, T. Kanon, and K. Nakamura.

70. F. D. Dexter, J. C. Miller, and W. Philippoff, *Trans. Soc. Rheol.* **5**, 193 (1961).
71. J. W. C. Adamse, H. Janeschitz-Kriegl, J. L. Den Otter, and J. L. S. Wales, *J. Polym. Sci. Part A-2* **6**, 871 (1968).
72. J. L. S. Wales, *Rheol. Acta* **8**, 38 (1969).
73. J. L. S. Wales and W. Philippoff, *Rheol. Acta* **12**, 25 (1973).
74. C. D. Han and L. H. Drexler, *J. Appl. Polym. Sci.* **17**, 2329 (1973).
75. C. D. Han, *J. Appl. Polym. Sci.* **19**, 2403 (1975); *Rheol. Acta* **14**, 173 (1975).
76. F. H. Gortmacher, M. C. Hanson, B. de Cindio, H. M. Laun, and H. Janeschitz-Kriegl, *Rheol. Acta* **15**, 256 (1976).
77. T. Matsumoto and D. C. Bogue, *J. Polym. Sci. Polym. Phys. Ed.* **15**, 1663 (1977).
78. G. V. Vinogradov, A. I. Isayev, D. A. Mustafaev, and Y. Y. Podolsky, *J. Appl. Polym. Sci.* **22**, 665 (1978).
79. M. Takahashi, T. Masuda, N. Bessho, and K. Osaki, *J. Rheol.* **29**, 517 (1980).
80. T. Arai and H. Hatta, *Nihon Reoroji Gakkaishi* **8**, 67 (1980); **8**, 110 (1980).
81. T. Arai, H. Ishikawa, and H. Hatta, *Kobunshi Ronbunshu* **38**, 29 (1981).
82. J. A. Van Aken and H. Janeschitz-Kriegl, *Rheol. Acta* **20**, 419 (1981).
83. J. L. White and J. E. Spruiell, *Polym. Eng. Sci.* **21**, 859 (1981).
84. J. L. White, *Pure Appl. Chem.* **55**, 765 (1983).
85. K. Oda, J. L. White, and E. S. Clark, *Polym. Eng. Sci.* **18**, 53 (1978).
86. A. I. Isayev, *Polym. Eng. Sci.* **23**, 271 (1983).
87. H. J. Kang and J. L. White, *Int. Polym. Proc.* **1**, 12 (1986).
88. K. J. Choi, J. L. White, and J. E. Spruiell, *J. Appl. Polym. Sci.* **25**, 2777 (1980).
89. I. Hamana, M. Matsui, and S. Kato, *Meilland Textilbr.* **4**, 382 (1969).
90. K. Katayama, T. Amano, and K. Nakamura, *Koll. Z. Z. Polym.* **226**, 125 (1968).
91. H. P. Nadella, H. M. Henson, J. E. Spruiell, and J. L. White, *J. Appl. Polym. Sci.* **21**, 3003 (1977).
92. J. R. Dees and J. E. Spruiell, *J. Appl. Polym. Sci.* **18**, 1053 (1974).
93. Y. Wang, M. Cakmak, and J. L. White, *J. Appl. Polym. Sci.* **30**, 2615 (1985).
94. J. E. Spruiell and J. L. White, *Polym. Eng. Sci.* **15**, 660 (1975).
95. K. J. Choi, J. E. Spruiell, and J. L. White, *J. Polym. Sci. Polym. Phys. Ed.* **20**, 27 (1982).
96. H. P. Nadella, J. E. Spruiell, and J. L. White, *J. Appl. Polym. Sci.* **22**, 3121 (1978).
97. J. H. Southern and R. S. Porter, *J. Appl. Polym. Sci.* **14**, 2305 (1970).
98. U.S. Pat. 4,282,237 (1981), A. R. Austen and D. V. Humpries.
99. P. Smith, P. J. Lemstra, B. Kalb and A. J. Pennings, *Polym. Bull.* **1**, 733 (1979).
100. B. Kalb and A. J. Pennings, *J. Mater. Sci.* **15**, 2584 (1980).
101. P. Smith and P. J. Lemstra, *J. Mater. Sci.* **15**, 505 (1980).

JAMES L. WHITE
MUKERREM CAKMAK
University of Akron

OSMOMETRY

The principles of a membrane osmometer are illustrated in Figure 1. At constant temperature T, if a solvent α and a solution β consisting of the solvent and a polymeric (macromolecular) solute are separated by a membrane SM, permeable to solvent only (the semipermeable membrane), then solvent α flows across SM to the solution phase β until the hydrostatic pressure in side β becomes sufficiently large to prevent the flow of solvent from side α (see also MEMBRANES). The flow of solvent from α to β can be prevented by increasing the pressure on side β from its initial value P_0 to $P_0 + \pi$, giving a pressure difference of π. This

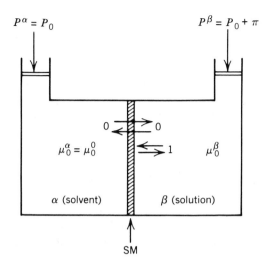

Fig. 1. Principle of a membrane osmometer. Component 0 is the solvent; component 1 is a macromolecular solute retained by the semipermeable membrane (SM), which separates the solution β from the solvent α. At constant temperature T, when the pressure difference between the two sides is π, no solvent flows from side α to β, and osmotic equilibrium is attained. At osmotic equilibrium $\mu_0 = \mu^\beta$ (1).

excess pressure π is called the osmotic pressure. The same effect can be achieved by lowering the pressure on the solvent side α from P_0 to $P_0 - \pi$. Thus as long as the pressure difference between the two sides is π, no solvent flows from α (where the solvent has a higher chemical potential) to β (where the chemical potential of the solvent is lower) (2–4). If the pressure on side β is increased beyond $P + \pi$, solvent flows from β to α; this is sometimes referred to as reverse osmosis and is being considered in the purification of brackish water (4).

Osmotic pressure is a colligative property that depends on the number of solute particles; for nonassociating solutes it provides the molecular weight M if the sample is homogeneous, or the number-average molecular weight $\overline{M}_n$ if the sample is heterogeneous. In addition, these measurements provide information about the nature of the solution, whether it is ideal or nonideal, and whether self-association occurs (2–4). It is one of the older methods for studying polymeric solutions (3). In recent years, the development of high speed membrane osmometers, in which measurements can be performed in minutes instead of hours or days, has revolutionized the technique (3,4). Osmotic-pressure measurements are generally restricted to macromolecules with a molecular weight >8000 g/mol and $<1 \times 10^6$ g/mol. For lower molecular weight solutes, vapor-pressure osmometry, cryoscopy, ebulliometry, or sedimentation equilibrium should be used (4,5) (see also MOLECULAR WEIGHT DETERMINATION).

Osmometry provides a means of obtaining the activity a_0 or the osmotic coefficient g of the solvent (2–4). For monodisperse, nonassociating solutes it gives the molecular weight M and the nonideal term B. For heterogeneous or polydisperse solutes, osmometry provides the number-average molecular weight $\overline{M}_n$ of the polymeric solute as well as the osmotic-pressure second virial coefficient B_{os}, a measure of the nonideality of the system. In fact, osmometry and other colligative methods (cryoscopy, ebulliometry, and vapor-pressure osmometry) are the only thermodynamic methods that yield an unambiguous average molecular weight $\overline{M}_n$, the number-average molecular weight. In order to obtain the weight-average molecular weight $\overline{M}_w$ from elastic light scattering or sedimentation equilibrium experiments, it must be assumed that the refractive-index increments of the polymeric components are equal. Furthermore, in sedimentation equilibrium experiments it must also be assumed that the partial specific volumes or the density increments of the polymeric components are equal. These assumptions are not necessary in osmometry (4).

When self-association occurs under ideal conditions, the number-average ($\overline{M}_{nc}$) and the weight-average ($\overline{M}_{wc}$) molecular weight and the weight fraction of monomer f_1 can be obtained; under nonideal conditions their apparent values $\overline{M}_{na}$, $\overline{M}_{wa}$, and f_a are obtained (3,4,6–8). For self-associations, there is an interrelationship between $\overline{M}_{nc}$, $\overline{M}_{wc}$, and f_1. In self-associations the molecular weights vary between M_1, the molecular weight of the monomer, and M_q, the molecular weight of the highest species present ($M_q = qM_1$, where $q = 2, 3, \ldots$). For example, in a monomer–dimer association the molecular weights vary between M_1 and $M_2 = 2M_1$. The subscript c indicates that $\overline{M}_{nc}$, $\overline{M}_{wc}$, or any other average molecular weight is concentration-dependent when self-association occurs. If the self-association is ideal, then $\overline{M}_{na} = \overline{M}_{nc}$, $\overline{M}_{wa} = \overline{M}_{wc}$, and $f_a = f_1$. For ideal self-associations it is also possible to obtain the number fraction of monomer N_c. Since the chemical equilibrium produces changes in the values of the true average molecular weights, it is customary to use the subscript c. It is also possible to study mixed associations between two solutes A and B using membrane osmometry.

Membrane osmometry (2–4,9–12) can be applied to solutions that contain homogeneous or heterogeneous, nonassociating solutes (3,4); solutions in which the solute undergoes self-association (3–8); and solutions containing two different solutes in which a mixed association occurs (3,4,13–15).

Theory

The principles of membrane osmometry can be best understood by considering the quantities noted in Figure 1 and Table 1. Initially, the chemical potential of the solvent in side α is greater than that in side β, ie, $\mu_0^\alpha(T,P_0) > \mu_0^\beta(T,P_0,X_1)$. To prevent the flow of solvent from α to β, the pressure on side β is increased from P to $P' = P_0 + \pi$; this equalizes the solvent chemical potential on side β, $\mu_0^\beta(T,P',X_1)$, to the chemical potential of the solvent in side α, $\mu_0^\alpha(T,P_0,X_0)$, where $X_0 = 1$. The chemical potential of the solvent μ_0 is a function of temperature,

Table 1. Initial and Equilibrium Properties in Osmometry

Initial conditions	Equilibrium conditions
$P^\alpha = P^\beta = P_0$	$P^\beta > P^\alpha$
$T^\alpha = T^\beta = T$	$P^\beta = P_0 + \pi = p'$
$\mu_0^\alpha > \mu_0^\beta(T,P_0,X_1)$	$T^\alpha = T^\beta$
$\mu_0^\alpha = \mu_0^0$	$\mu_0^\alpha = \mu_0^0(T,P',X_1)$

pressure, and concentration. At constant temperature and at osmotic equilibrium

$$\mu_0^\alpha = \mu_0^\beta(T, P', X_1) \tag{1}$$

$$d\mu_0^\beta = 0 \tag{2}$$

The chemical potential of the solvent is defined by

$$\mu_0 = \mu_0^0 + RT \ln a_0 = \mu_0^0 + RT \ln y_0 x_0 \tag{3}$$

where μ_0^0 is the standard-state chemical potential (for the pure solvent at temperature T), R is the gas constant, T is the absolute temperature, a_0 is the activity, y_0 is the activity coefficient, and x_0 is the mole fraction of solvent.

When $x_0 = 1$ (pure solvent), $a_0 = 1$, $y_0 = 1$, and $\mu_0 = \mu_0^0$. The solvent chemical potential can also be defined by (3,4,16)

$$\mu_0 = \mu_0^0 + gRT \ln x_0 \tag{4}$$

where g is the osmotic coefficient of the solvent. It follows that at osmotic equilibrium

$$\overline{V}_0 dP = -RT d \ln a_0 \tag{5}$$

since $\overline{V}_0$, the partial molar volume of the solvent, is given by

$$\overline{V}_0 = (\partial \mu_0 / \partial P)_T \tag{6}$$

Since the osmotic pressures encountered in macromolecular solutions are $<1.01325 \times 10^5$ N/m^2 (1 atm), it can be assumed that $\overline{V}_0$ is independent of pressure, and integration of equation 5 from P_0 to P' (and $a_0 = 1$ to $a_0 = a_0$) gives

$$\frac{\pi \overline{V}_0}{RT} = -\ln a_0 \tag{7}$$

Comparison of equations 3 and 4 indicates

$$\ln a_0 = g \ln x_0 \tag{8}$$

and

$$\frac{\pi \overline{V}_0}{RT} = -g \ln x_0 \tag{9}$$

If the solution is ideal, $a_0 = x_0$ and $g = 1$. The osmotic coefficient g is a measure of the nonideality of the solution; it is easy to show that g is also defined by

$$g = \pi_{real} / \pi_{ideal} \tag{10}$$

where π_{real} is the real osmotic pressure and π_{ideal} is the osmotic pressure the solution would have if it were ideal.

It is also possible to write the osmotic pressure equation in terms of the solute or solutes. For a two-component system

$$n_0 d\mu_0 = -n_1 d\mu_1 \quad (T, P \text{ constant}) \tag{11}$$

$$n_0 \overline{V}_0 = V - n_1 \overline{V}_1 \tag{12}$$

and thus

$$V dP = n_1 d\mu_1 \tag{13}$$

where 1 refers to the solute, n_1 is the number of moles solute, μ_1 the chemical potential of the solute, $\overline{V}_1$ the partial molar volume of the solute, and V the volume of solution. For a heterogeneous, macromolecular solution, equation 13 becomes (3,4)

$$V dP = \sum_i n_i d\mu_i \tag{14}$$

Equations 10, 13, and 14 can be used as the starting point to develop the osmotic-pressure equations under ideal or nonideal conditions for various experimental situations.

It is evident from equation 7 that osmotic-pressure experiments can give a direct measurement of the solvent activity a_0. When the mole fraction of solvent is known, the osmotic coefficient g is obtained from equation 8. With polymeric solutions the number of moles of solute introduced is usually not known and the mole-fraction scale becomes impractical. However, this approach is useful to develop the basic equations. In practice, solute concentrations are based on g per kg solvent or g per L of solution.

Nonassociating Solutes

At constant temperature for polymeric component i, whose molecular weight is M_i, it is assumed that the natural logarithm of the activity coefficient $\ln y_i$ on the c scale (g/L) can be represented by a Maclaurin's series in c_k, the concentration of polymeric component k ($k = 1, 2, \ldots, q$). Thus

$$\ln y_i = M_i \sum_{i=1}^{q} B_{ik} c_k + \ldots \tag{15}$$

$$B_{ik} = \lim_{c_k \to 0} (1/M_i)(\partial \ln y_i / \partial c_k)_{T,P,c_j \neq k} \tag{16}$$

The summation in equation 15 is carried out over all of the q polymeric solutes. When the following relations

$$\mu_i = \mu_i^0 + RT \ln y_i c_i \qquad (i = 1, 2, \ldots, q) \tag{17}$$

$$\overline{V}_i = M_i \bar{v}_i = (\partial \mu_i / \partial P)_{T,nj \neq i} \tag{18}$$

are used in equation 14, the final result is

$$\frac{\pi}{RT} = \frac{c}{M_{na}} = \frac{c}{M_n} + \frac{B_{os} c^2}{2} + \ldots \tag{19}$$

where B_{os} is the osmotic-pressure second virial coefficient and is defined by

$$B_{os} = \sum_{i=1}^{q} \sum_{k=1}^{q} f_i f_k \left(B_{ik} + \frac{\bar{v}_k}{1000 M_k} \right) \tag{20}$$

where M_i is the molecular weight of polymeric component k whose partial specific volume is $\overline{V}_i$; B_{ik} is defined by equation 16 and

$$f_j = c_j / c \qquad (j = i \text{ or } k) \tag{21}$$

is the weight fraction of component j. The quantity $\overline{M}_n$ is the number-average molecular weight and is defined by

$$\overline{M}_n = \frac{\Sigma n_i M_i}{\Sigma n_i} = \frac{c}{\Sigma(c_i/M_i)} = \frac{1}{\Sigma f_i/M_i} = \left[\int_0^{\infty} \frac{f(M)dM}{M}\right]^{-1} \tag{22}$$

Alternatively, the following expressions can be used:

$$x_0 = 1 - \Sigma x_i \tag{23}$$

$$x_i = c_i \overline{V}_0/M_i \tag{24}$$

and

$$g = 1 + B_1 c + B_2 c^2 + \ldots \tag{25}$$

in equation 10 to obtain equation 19. When the solute is homogeneous, $\overline{M}_n = M$, $\overline{M}_{na} = M_{app}$ (the apparent molecular weight), and equation 19 becomes

$$\frac{\pi}{RT} = \frac{c}{M_{app}} = \frac{c}{M} + \frac{Bc^2}{2} + \ldots \tag{26}$$

Figure 2 shows plots of π/c vs c for various proteins in 6-M guanidinium hydrochloride (GuHCl) and 0.5-M mercaptoethanol, which prevents the oxidation of the —SH groups in the proteins (17). The GuHCl solution is an excellent protein denaturant. This plot is characteristic of some nonassociating, homogeneous solutes, and the positive slope indicates that the solution is nonideal. Similar plots (17) in the absence of denaturant are horizontal or almost horizontal, indicating slight nonideality. In the case of proteins with subunits, the denaturant breaks the protein into subunits.

Figure 3 shows a plot of π/c vs c for a polystyrene sample in tetrahydrofuran at 34°C; these experiments were carried out in a Knauer high speed membrane osmometer (18). The plot is linear with a positive slope, indicating the solution is nonideal. Figure 4 shows plots of π/c vs c for polyisobutylene fractions in cyclohexane and benzene (2,10,19). These experiments were performed in a block-type membrane osmometer, and each set of measurements would take much longer than those shown in Figure 3.

In equation 19 only one nonideal term is included. The plot shown in Figure 3 and some of the plots in Figure 4 justify the omission of additional nonideal terms on the order of c^3. Sometimes solutions exhibit more nonideality and need one or more additional terms; in this case the plots of π/c vs c tend to curve upward instead of being linear (eg, plots LAA-1, LA-2, and LD-3 in Fig. 4). In some cases a linear plot of the data can be obtained using the following relation (10):

$$(\pi/c)^{1/2} = (RT/\overline{M}_n)^{1/2}\left(1 + \frac{\Gamma c}{2}\right) \tag{27}$$

Plots for ideal systems involving homogeneous (monodisperse) or heterogeneous (polydisperse) solutes may show a horizontal line; without prior knowledge about the sample, it is not known whether M or $\overline{M}_n$ is measured. Similar considerations apply to nonideal solutions. In Figure 2 M is obtained, but in Figures 3 and 4 $\overline{M}_n$ is obtained. The plots of π/c vs c in Figures 2 and 3 give inclined straight

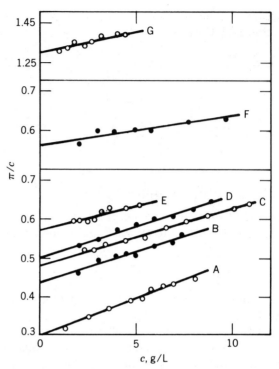

Fig. 2. Monodisperse, nonassociating solutes. Plots of π/c vs c for various proteins in 6-M guanidinium hydrochloride (GuHCl) containing 0.5-M mercaptoethanol; neither serum albumin (A) nor ovalbumin (B) have subunits, whereas the other proteins do. Alcohol dehydrogenase (D) and enolase (F) have two subunits; methemoglobin (G), lactate dehydrogenase (E), and aldolase (C) have four subunits. These data were obtained with a high speed membrane osmometer (Hewlett-Packard) (17). Courtesy of the American Chemical Society.

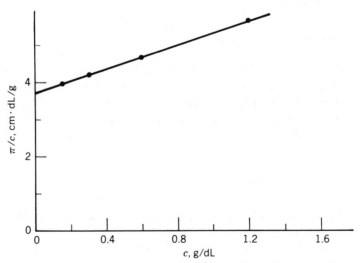

Fig. 3. Heterogeneous, nonassociating solutes. Plot of π/c vs c for a polystyrene sample in tetrahydrofuran, $\overline{M}_n = 8.05 \times 10^4$ g/mol and $B_{os} = 1.07 \times 10^{-6}$ (L·mol)/g^2 (18).

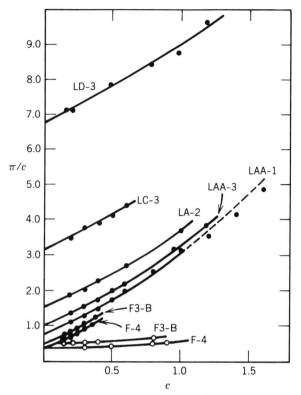

Fig. 4. Heterogeneous, nonassociating solutes. Plot of π/c vs c for polyisobutylene fractions in cyclohexane (●) and in benzene (○), both at 30°C. An older, block-type os-mometer was used (10,19). Courtesy of the American Chemical Society.

lines and so do some of the plots in Figure 4. Measurement of a different average molecular weight, such as $\overline{M}_w$, is required to identify polydispersity. With pro-teins, zone-electrophoresis experiments test for homogeneity or heterogeneity of the sample.

Self-associating Solutes

Self-associations are chemical equilibria of the type

$$nA_1 \rightleftharpoons A_n (n = 2, 3, \ldots) \tag{28}$$

$$nA_1 \rightleftharpoons qA_2 + mA_3 + \ldots \tag{29}$$

and related equilibria; here A represents the self-associating solute. Self-asso-ciations are widely encountered; many proteins, detergents, organic dyes, poly(ethylene glycol)s, and other polymers exhibit self-association (3–8). Self-associations are not analyzed in the same manner as heterogeneous or homo-geneous solutes; for the latter M or $\overline{M}_n$ is a constant. Because of the chemical equilibrium, $\overline{M}_n$ and other average molecular weights are concentration-depen-dent (Fig. 5); this is indicated by writing $\overline{M}_n$ as $\overline{M}_{nc}$ and $\overline{M}_w$ as $\overline{M}_{wc}$. Consequently,

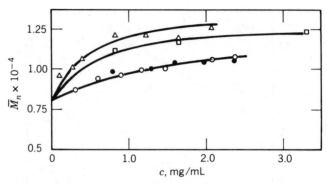

Fig. 5. Self-associations. Plot of $\overline{M}_{nc}$ vs c for Bowman-Birk soybean trypsin inhibitor in various buffers at 25°C and an ionic strength of 0.1 (22): ○, pH 7; □, pH 6; △, pH 5; ●, pH 3. Courtesy of the American Chemical Society.

the intercept of a plot of π/c vs c is RT/M_1, since at infinite dilution only monomer would be present; the limiting slope of this plot would be $RTd(1/\overline{M}_{na})/dc = RTd(1/\overline{M}_{nc})/dc + B_{os}/2$. In order to analyze self-associations, new methods must be developed. Fortunately, this is possible, since the weight-average molecular weight $\overline{M}_{wc}$ and the weight fraction of monomer f_1 can be evaluated in ideal self-associations or their apparent values $\overline{M}_{wa}$ and f_a can be obtained for nonideal self-associations.

It is assumed that the natural logarithm of the monomer's activity coefficient can be represented by

$$\ln y_i = iB^*M_1c(i = 1, 2, \ldots) \tag{30}$$

where c is the total concentration of the self-associating solute and B^* is a constant whose value depends on the temperature and the solute–solvent combination. The following relation is obtained:

$$\frac{\pi}{RT} = \frac{c}{M_{na}} = \frac{c}{M_{nc}} + \frac{Bc^2}{2} \tag{31}$$

where

$$B = B^* + \frac{\bar{v}}{1000M_1} \tag{32}$$

where $\bar{v}$ is the partial specific volume of the self-associating solute; it is assumed that $\bar{v}$ is the same for all the self-associating species. Furthermore,

$$\frac{c}{M_{nc}} = \frac{c_1}{M_1} + \frac{k_2c_1^2}{2M_1} + \frac{k_3c_1^3}{3M_1} + \cdots \tag{33}$$

(where $k_j = c_j/c_1^j, j = 2, 3, \ldots$)

$$\frac{M_1}{\overline{M}_{na}} = \frac{M_1}{\overline{M}_{nc}} + \frac{BM_1c}{2} \tag{34}$$

$$\frac{M_1}{\overline{M}_{wa}} = \frac{M_1}{\overline{M}_{wc}} + BM_1c = \frac{d}{dc}\left(\frac{cM_1}{\overline{M}_{na}}\right) = \frac{M_1}{\overline{M}_{na}} + c\frac{d}{dc}\left(\frac{M_1}{\overline{M}_{na}}\right) \tag{35}$$

and

$$\ln f_a = \int_0^c \left(\frac{M_1}{\overline{M}_{na}} - 1 \right) \frac{dc}{c} + \left(\frac{M_1}{\overline{M}_{na}} - 1 \right) = \ln f_1 + BM_1 c \tag{36}$$

These quantities can be used in various ways to analyze self-associations (3–8,20,21). In self-associations, $\overline{M}_{nc}$ and $\overline{M}_{wa}$ are interrelated but not in nonself-associating systems. The subscript c indicates that the solute of interest undergoes self-association. If a nonideal self-association is present and $BM_1 > 0$, a plot of $\overline{M}_{na}$ vs c shows a maximum, or a plot of $M_1/\overline{M}_{na}$ vs c shows a minimum. The quantity $M_1/\overline{M}_{na}$ is the osmotic coefficient.

Figure 5 shows plots of $\overline{M}_{nc}$ vs c at various pH values for solutions of the Bowman-Birk soybean trypsin inhibitor (22). These experiments were carried out at 25°C at an ionic strength of 0.1 in various buffers. The increase of $\overline{M}_{nc}$ with increasing c is characteristic of some self-associating solutes. In self-associating systems the shape of the $\overline{M}_{na}$ (or $M_1/\overline{M}_{na}$) vs c curve usually depends on the temperature, the solvent or buffer solution, the pH, and the ionic strength. For a nonideal self-association with $BM_1 > 0$, the plot of $\overline{M}_{na}$ vs c would increase with increasing c up to a maximum and then decrease with further increases in c. The corresponding plot of $M_1/\overline{M}_{na}$ vs c would show a minimum (see Ref. 4 for more details).

Self-associations are also encountered with heterogeneous (polydisperse) solutes, such as poly(ethylene glycol)s (23–25). Here the solute species at infinite dilution is referred to as a unimer.

Mixed Associations

Associations of the type

$$A + B \rightleftharpoons AB \tag{37}$$

$$\left. \begin{array}{l} nA \rightleftharpoons A_n \quad (n = 2, 3, \ldots) \\ A + B \rightleftharpoons AB \end{array} \right\} \tag{38}$$

and related equilibria between two solutes A and B are known as mixed associations (3,4,13–15). There are no interrelations between average molecular weights or their apparent values as there is with self-associations. The study of these associations is more complicated. The reactants must first be studied individually to find out whether they do self-associate, and if the solutions of A and B are ideal or nonideal. This information is used in the subsequent analysis. Since there are an infinite number of ways A and B can be combined, and since each blend would have a different value of $\overline{M}_n$ or any other average molecular weight, experiments must be carried out at β_m or β_g. Here $\beta_m = m_B^0/m_A^0$ is the ratio of the original molar concentrations of A and B, and the quantity $\beta_g = c_A^0/c_B^0$ is the ratio of the original weight (g/L) concentrations of A and B. For an ideal mixed association described by equation 37,

$$\frac{\pi}{RT} = \frac{c}{\overline{M}_n^{eq}} = \frac{c_A}{M_A} + \frac{c_B}{M_B} + \frac{kc_Ac_B}{M_{AB}} \tag{39}$$

$$\frac{c}{\overline{M}_n^0} = \frac{c_A^0}{M_A} + \frac{c_B^0}{M_B} = \frac{c_A}{M_A} + \frac{c_B}{M_A} + \frac{2kc_Ac_B}{M_{AB}} \qquad (40)$$

where $k = c_Ac_B/c_{AB}$ and

$$\Delta(c/\overline{M}_n) = \frac{c}{\overline{M}_n^0} - \frac{c}{\overline{M}_n^{eq}} = \frac{kc_Ac_B}{M_{AB}}$$

$$= \frac{k}{M_{AB}} [c_A^0 - \Delta(c/\overline{M}_n)M_A][c_B^0 - \Delta(c/\overline{M}_n)M_B] \qquad (41)$$

Here $\overline{M}_n^0$ is the value of $\overline{M}_n$ for a mixture of A and B if no association is present. For this association, $\Delta(c/\overline{M}_n)$ gives the number of moles of complex directly. In order to evaluate k, $\Delta(c/\overline{M}_n)$ is plotted vs $[c_A^0 - \Delta(c/\overline{M}_n)M_A][c_B^0 - \Delta(c/\overline{M}_n)M_B]$. If the solution is ideal and only an AB complex is formed, this plot gives a straight line whose slope is k/M_{AB}. If this model is wrong, the plot is not linear and other models must be attempted. Methods for analyzing ideal or nonideal mixed associations have been developed (4,13–15).

Sensitivity and Significance of Measurements

Osmometry is an absolute method that is fast, reliable, and simple to use. For best results, solvents and solutions must be degassed before use. Membrane osmometry is more sensitive than vapor-pressure lowering, vapor-pressure osmometry, freezing-point depression (cryoscopy), or boiling-point elevation (ebulliometry). The data shown in Table 2 clearly indicate that osmotic pressure is the most sensitive quantity detected.

Table 2. Sensitivity of Colligative Property for Studying Polymer Solutions[a]

Method	Quantity measured	Value
vapor-pressure lowering	$\dfrac{\Delta p}{p_0}$	1.8×10^{-5}
cryoscopy, °C	ΔT_f	-1.86×10^{-3}
ebulliometry, °C	ΔT_b	5.6×10^{-4}
membrane osmometry, cm	π	25.2 cm

[a] Aqueous, ideal polymer solution, $c = 10$ g/L, $M_1 = 10,000$, $T = 298.2$ K.

In a homogeneous, nonassociating sample, $\overline{M}_n = M$, the molecular weight. For self-associations or mixed associations, $\overline{M}_{nc}$ or $\overline{M}_n^{eq}$ are concentration-dependent instead of fixed quantities. They and their apparent values can be used to evaluate the equilibrium constants and the nonideal terms.

When the sample is heterogeneous and nonassociating, osmometry and other colligative property methods, which depend on the number of solute particles in solution, give $\overline{M}_n$. Elastic light scattering or sedimentation equilibrium provides $\overline{M}_w$ under circumstances described earlier, sedimentation equilibrium can also provide $\overline{M}_z$, the z-average molecular weight; $\overline{M}_n$ is defined by equation 21 and $\overline{M}_w$ and $\overline{M}_z$ are defined as follows:

$$\overline{M}_w = \frac{\Sigma n_i M_i^2}{\Sigma n_i M_i} = \frac{\Sigma c_i M_i}{c} = \Sigma f_i M_i = \int_0^\infty M f(M) dM \qquad (42)$$

$$\overline{M}_z = \frac{\Sigma n_i M_i^3}{\Sigma n_i M_i^2} = \frac{\Sigma c_i M_i^2}{\Sigma c_i M_i} = \frac{\Sigma f_i M_i^2}{\Sigma f_i M_i} = \frac{\int_0^\infty M^2 f(M) dM}{\int_0^\infty M f(M) dM} \qquad (43)$$

where n_i is the number of moles of polymeric component i whose molecular weight is M_i, c_i is the concentration of component i in g/L, c is the total solute concentration, and $f_i = c_i/c$ is the weight fraction of component i. In a homogeneous sample, $\overline{M}_n = \overline{M}_w = \overline{M}_z = M$. For a heterogeneous (polydisperse) sample, $\overline{M}_z > \overline{M}_w > \overline{M}_n$. For a continuous molecular weight distribution (MWD), the weight fraction of polymer with a molecular weight between M and $M + dM$ becomes $f(M)dM$, where $f(M)$ is the differential or frequency MWD. These molecular weights are related to the moments of the MWD; the kth moment of the MWD is defined by (26)

$$\nu_k = \sum_{i=1}^{q} f_i M_i^k = \int_0^\infty M^k f(M) dM \qquad (44)$$
$$(k = -n, \ldots, -1, 0, 1, \ldots, n)$$

These moments can also be defined for a molecular weight distribution on a number basis by using N_i, the number fraction of component i, and $\phi(M)$, the differential MWD on a number basis, in place of f_i or $f(M)dM$ in equation 44 (Table 3). The variance or second moment about the mean (σ^2) is given by

$$\sigma_w^2 = \overline{M}_w \overline{M}_z - \overline{M}_w^2 \qquad (45)$$

for a MWD on a weight basis (27) and by

$$\sigma_n^2 = \overline{M}_w \overline{M}_n - \overline{M}_n^2 \qquad (46)$$

for MWD on a number basis. $\overline{M}_n$ can be used with values of M (within the range of molecular weights in the MWD) to convert a value of $f(M)$ to $\phi(M)$ or vice

Table 3. The Moments of MWD on a Number and a Weight Basis

k	Number basis[a] $(\nu_k)_n = \int_0^\infty M^k \phi(M) dM$	Weight basis[b] $(\nu_k)_w = \int_0^\infty M^k f(M) dM$
-1	$\int_0^\infty \dfrac{\phi(M)dM}{M}$	$1/\overline{M}_n$
0	1	1
1	$\overline{M}_n$	$\overline{M}_w$
2	$\overline{M}_n \overline{M}_w$	$\overline{M}_w \overline{M}_z$

[a] $\phi(M)dM$ is the number fraction of polymer with a molecular weight between M and $M + dM$; since solvent does not appear in the equation for $(\nu_k)_n$, number fraction is used instead of mole fraction.

[b] $f(M)dM$ is the weight fraction of polymer with a molecular weight between M and $M + dM$.

versa, since

$$f(M) = \phi(M)(M/\overline{M}_n) \tag{47}$$

For a discrete MWD this becomes

$$f_i = N_i(M_i/\overline{M}_n) \tag{48}$$

The values of M or $\overline{M}_n$ obtained by osmometry agree well with those obtained by other methods, as seen in Table 4 for proteins and dextran.

Table 4. Molecular Weight Determined by Membrane Osmometry and Other Methods

Substance	Mol wt	Method	Ref.
ovalbumin	$(4.295 \pm 0.04) \times 10^4$	mo[a]	1
ovalbumin	$(4.30 \pm 0.17) \times 10^4$	se[b]	17
lysozyme	$(15.1 \pm 0.45) \times 10^3$	mo[a]	1
lysozyme	$(14.4 \pm 0.10) \times 10^3$	se[b]	28
dextran T-70[c]	4.30×10^{4d}	mo[a]	29
dextran T-70	4.10×10^{4d}	ega[e]	f
canine high density lipoprotein	2.13×10^5	mo[a]	30
canine high density lipoprotein	2.14×10^5	se[b]	31

[a] Membrane osmometry.
[b] Sedimentation equilibrium.
[c] Pharmacia Dextran T-70 Lot 7981.
[d] Number-average molecular weight.
[e] End-group analysis.
[f] These values were reported by Pharmacia Fine Chemicals, Inc. for Dextran T-70.

Modern Instrumentation

In the United States, two high speed membrane osmometers are available: the Wescan (formerly Melabs) (Wescan Instruments, Santa Clara, Calif.) and the Knauer (UIC, Inc., Joliet, Ill.). Both utilize the strain gauge (32). The Wescan has two models, one with a thermoelectric cooler capable of operation from -5 to 130°C, the other capable of operating from 20 to 130°C. Both instruments can be used with aqueous and nonaqueous solvents.

Unlike the Knauer, the Wescan instrument includes the ability to flush the solvent side of the membrane via a valving system. The manufacturer claims this allows flushing materials that have diffused through the membrane into the solvent chamber, eliminating the need to change membranes when large amounts of diffusion occur. Both osmometers have the capability of measuring osmotic pressures in excess of 80 cm of water equivalent height.

In both instruments the membrane is clamped into a stainless steel thermostatted chamber and serves as a barrier between the solvent and the solution side of the chamber. The solvent side (bottom) of the chamber is in juxtaposition with the diaphragm of the capacitance strain gauge. By means of a 10-mV, 25-cm strip-chart recorder attached to the osmometer output, the solvent transport is measured across the bottom side of the membrane in the direction of the solution which is on the topside of the membrane. When the hydrostatic pressure prevents further solvent flow, the equilibrium condition is indicated by the output chart of the recorder via the capacitance strain gauge. Any change in equilibrium such as thermal drift or diffusion of solute through the membrane is manifested by changes in the voltage output to the recorder.

The osmometer is prepared for operation by:

1. proper preconditioning and installation of the membrane and attainment of thermal equilibrium;

2. calibration of the chart-recorder span. The chart recorder is adjusted in such a way that a change in height of 10 cm represents full-scale deflection on the recorder. This procedure allows the operator to relate osmotic pressure to recorder-scale deflection;

3. adjustment of solvent zero. Solvent is introduced to the solution side and the osmometer allowed to come to equilibrium at a nonzero position. This position is shifted to zero on the chart recorder with a fine zero-adjustment potentiometer; and

4. choosing the desired sensitivity, that is, the number of cm of solvent that produces a full-scale deflection on the chart recorder.

The osmotic pressures h in cm of solvent must be converted to N/m^2 (Pa), using $P = \rho g h$, where ρ is the solvent density, which is usually expressed as g/cm^3; g is the standard acceleration of gravity whose value is 980.665 cm/s^2 (33). Thus P in $N/m^2 = \rho g h/10$ where $\rho g h$ is in cgs units (N/m = kg/s^2). To convert π/RT to moles per cubic decimeter, ie, mol/L, π/RT (in mol/L) = $\rho g h \times 10^{-4}/(8.314 RT)$. Here $R = 8.314$ J/(K·mol) and T is the absolute temperature. Since c is usually in g/L or g/dm^3 and M is usually in g/mol, the units are consistent in the relation $\pi/RT = c/M$ (or $c/\overline{M}_n$).

Membranes

For aqueous solutions, cellulose acetate membranes are usually employed, but any dialysis or ultrafiltration membrane can be used. The latter may be more retentive and take longer to reach equilibrium. The membranes should be conditioned in solvent or buffer and degassed before use while still in the solvent. For organic solvents, gel cellulose or cellophane membranes are preferred. If these membranes are stored in one solvent, they must be conditioned to a new solvent by gradual changes as described previously (4). All solvents, solutions, and membranes should be degassed before use. Wescan Instruments markets a polymeric, asymmetric, thin-film membrane, which can attain osmotic equilibrium in 5–10 min for solutes of molecular weights 15,000 or higher.

Membranes in various pore sizes are recommended for solutes with minimum molecular weights; this information is provided by the supplier and serves as a rough guide. Molecular weight determinations have been carried out on a sodium polygalacturonate sample having $\overline{M}_n = 7.73 \times 10^3$ using a cellulose acetate membrane (Schleicher and Schuell AC62) with a molecular weight cutoff of ca 15,000 (34). The $\overline{M}_n$ could be determined even though 0.05-M NaCl was used as a supporting electrolyte and there was still enough electrostatic repulsion to keep the biopolymer extended and prevent its passage through the membrane. Figure 6 shows the plot of π/c vs c for this sample. The number-average degree

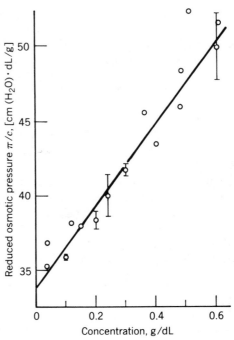

Fig. 6. Osmotic-pressure measurements of a polyelectrolyte at 35°C in a Knauer high speed membrane osmometer. Plot of π/c vs c for aqueous solutions of sodium polygalacturonate containing 0.05-M NaCl (34). Courtesy of John Wiley & Sons, Inc.

of polymerization obtained from osmometry was 39, whereas the value obtained by end-group titration was 35. These experiments were successful because the electrostatic repulsion caused the polymer to be extended.

Applications

Micelles are usually not studied by membrane osmometry. Surfactants, such as soaps, detergents, and bile salts, undergo self-association in aqueous solutions. In the case of detergents like sodium dodecyl sulfate (SDS), the micellar aggregate molecular weight, determined by light scattering, ranges from 18,500, in the absence of supporting electrolyte like NaCl, to 33,600, depending on the concentration of supporting electrolyte (35). Since the monomer molecular weight M_1 for SDS is 288.38, the monomer penetrates the semipermeable membranes, and eventually the detergent has the same concentration on both sides of the membrane. Sodium cholate, a bile salt, seems to undergo a two-equilibrium, constant, indefinite self-association in 0.154-M NaCl solutions (6); its M_1 value is 430.54. Thus this monomer penetrates virtually all available semipermeable membranes. Sodium cholate can be dialyzed in a hollow-fiber dialyzer with a 200-g cutoff, but membranes for this cutoff are not available. Although it would appear, therefore, that micelles should not be studied by membrane osmometry, such a study has been reported with SDS (36). The SDS concentration was at least twice the critical micelle concentration (CMC) in aqueous NaCl solutions on the solvent side of the membrane. The detergent was dissolved in various aqueous NaCl solutions; for any particular NaCl concentration the SDS concentration was higher than that on the solvent side of the membrane. Under these conditions the system behaves like a pseudo-nonpermeating system. Since the chemical potential gradient of the micelle across the membrane is very low, the permeation rate of the SDS monomer is also low. At 40°C the number-average degree of aggregation varied from 103 in 0.1-M NaCl to 700 in 0.8-M NaCl.

Membrane osmometry has been used together with scanning electron microscopy, solubility determination, and titration for the evaluation of poly(vinyl acetate phthalate) (PVAP) as an enteric coating excipient for pharmaceuticals; two different preparations of PVAP had $\overline{M}_n$ of 48,000 and 61,000 (37). Membrane osmometry has been applied to polysaccharides with anticoagulant properties (38). The molecular weight of the purified polysaccharide, isolated from Dragon's blood, a red resinous secretion from the fruits of *Daemonorops* species, was approximately 25,000. Another polysaccharide isolated from the pollens of *Typha augustata* (TA) had an approximate molecular weight of 30,000. Studies on the mechanism of the anticoagulant effect showed that the addition of the TA polysaccharide inhibited the release rate of fibrinopeptides by thrombin and also the aggregation of fibrin monomers.

A new method for N-deacetylation of chitosan has been proposed (39). A polymer free of N-acetyl groups and with very little decrease in molecular weight was prepared. Chitosan is said to be formed when chitin is treated with hot saturated KOH (40). Chitin (qv) is the principal structural component in the exoskeleton or shell of invertebrates. Chitin is the most abundant of the polysaccharides containing aminosugars; it is a long unbranched molecule that is constituted entirely of N-acetyl-D-glucosamine units linked by β-1,4' bonds. For chitosan (39) the degree of acetylation was determined by ir spectroscopy and conductimetry. Average molecular weights were determined by light scattering (qv) and membrane osmometry; the polydispersity ratio of $\overline{M}_w/\overline{M}_n$ was less than 2 for these polymers. This aminopolymer could be used for the investigation of chelating properties. The molecular weights of low molecular weight polymers, such as polystyrenes and a series of polyesters containing 5-fluoroacil in the main chain, have been determined by small-volume membrane osmometry and dynamic osmometry (41).

BIBLIOGRAPHY

"Osmometry" in *EPST* 1st ed., Vol. 9, pp. 659–668, by S. G. Weissberg and J. E. Brown, National Bureau of Standards.

1. B. W. McCarty, M.S. thesis (chemistry), Texas A&M University, College Station, Tex., Aug. 1984.
2. C. Tanford, *The Physical Chemistry of Macromolecules,* John Wiley & Sons, Inc., New York, 1961, pp. 210 and 238.
3. M. P. Tombs and A. R. Peacocke, *The Osmotic Pressure of Macromolecules,* Oxford University Press, London, 1974, pp. 1 and 65.
4. E. T. Adams, Jr., P. J. Wan, and E. F. Crawford, *Methods Enzymol.* **48,** 69 (1978).
5. K. E. Van Holde, G. P. Rossetti, and R. D. Dyson, *Ann. N.Y. Acad. Sci.* **162,** 279 (1969).
6. J. M. Beckerdite and E. T. Adams, Jr., *Biophys. Chem.* **21,** 103 (1985).
7. E. T. Adams, Jr., *Biochemistry* **4,** 1655 (1965).
8. E. T. Adams, Jr., L.-H. Tang, J. L. Sarquis, G. H. Barlow, and W. Norman in N. Catsimpoolas,

ed., *Physical Aspects of Protein Interactions,* Elsevier Science Publishing Co., Inc., New York, 1978, p. 1.

9. M. J. Kelly and D. W. Kupke in S. J. Leach, ed., *Physical Principles and Techniques of Protein Chemistry,* Academic Press, Inc., New York, 1973, Pt. C, p. 77.

10. P. J. Flory, *Principles of Polymer Chemistry,* Cornell University Press, Ithaca, N.Y., 1953, pp. 269 and 282.

11. J. R. Overton in A. Weissberger, ed., *Techniques of Chemistry,* Vol. 1, Wiley-Interscience, New York, 1971, Pt. V, p. 309.

12. R. D. Ulrich, *Tech. Methods Polym. Eval.* **4**(1), 9 (1975).

13. R. F. Steiner, *Biochemistry* **7**, 2201 (1968).

14. *Ibid.,* **9**, 4268 (1970).

15. B. W. Foster, R. L. Huggins, J. Robeson, and E. T. Adams, Jr., *Biophys. Chem.* **16**, 317 (1982).

16. S. Glasstone, *Textbook of Physical Chemistry,* D. Van Nostrand Co., Inc., New York, 1946, pp. 685 and 686.

17. F. J. Castellino and R. Barker, *Biochemistry* **7**, 2207 (1968).

18. J. Knutson and J. L. Armstrong, UIC, Inc., Joliet, Ill., private communication.

19. W. R. Krigbaum and P. J. Flory, *J. Am. Chem. Soc.* **75**, 1775 (1953).

20. J. Robeson, B. W. Foster, S. N. Rosenthal, E. T. Adams, Jr., and E. J. Fendler, *J. Phys. Chem.* **85**, 1254 (1981).

21. B. W. Foster, J. Robeson, N. Tagata, J. M. Beckerdite, R. L. Huggins, and E. T. Adams, Jr., *J. Phys. Chem.* **85**, 3715 (1985).

22. J. B. Harry and R. F. Steiner, *Biochemistry* **8**, 5060 (1969).

23. H.-G. Elias and R. Bareiss, *Chimia* **21**, 53 (1967).

24. K. Šolc and H.-G. Elias, *J. Polym. Sci. Polym. Phys. Ed.* **11**, 1793 (1972).

25. H.-G. Elias in M. B. Huglin, ed., *Light Scattering from Polymer Solutions,* Academic Press, Inc., New York, 1972, p. 397.

26. H. Fujita, *Foundations of Ultracentrifugal Analysis,* Wiley-Interscience, New York, 1975, pp. 335 and 354.

27. L. H. Peebles, Jr., *Molecular Weight Distributions in Polymers,* Wiley-Interscience, New York, 1971, pp. 1 and 47.

28. A. J. Sophianopoulos, C. K. Rhodes, D. N. Holcomb, and K. E. Van Holde, *J. Biol. Chem.* **237**, 1107 (1962).

29. P. J. Wan and E. T. Adams, Jr., *Biophys. Chem.* **5**, 207 (1976).

30. E. T. Adams, Jr., C. Edelstein, and A. M. Scanu, unpublished data.

31. C. Edelstein, M. Halari, and A. M. Scanu, *J. Biol. Chem.* **257**, 7189 (1982).

32. T. R. Reiff and M. Yiengst, *J. Lab. Clin. Med.* **53**, 291 (1959).

33. J. A. Dean, ed., *Lange's Handbook of Chemistry,* 11th ed., McGraw-Hill, Inc., New York, 1973, pp. 2–29.

34. M. L. Fishman, L. Pepper, and R. A. Barford, *J. Polym. Sci. Polym. Phys. Ed.* **22**, 899 (1984).

35. A. Vrij, Ph.D. dissertation, University of Utrecht, Utrecht, The Netherlands, 1959, p. 78.

36. K. U. Birdi, S. U. Dalsager, and S. Backlund, *J. Chem. Soc. Faraday Trans. 1* **76**, 2035 (1980).

37. R. U. Nesbitt, F. W. Goodhart, and R. H. Gordon, *Int. J. Pharm.* **26**, 215 (1985).

38. A. Gibbs, C. Green, and V. M. Dactor, *Thromb. Res.* **32**, 97 (1983).

39. A. Domard and M. Rinaudo, *Int. J. Biol. Macromol.* **5**, 49 (1983).

40. R. L. Whistler, *Polysaccharide Chemistry,* Academic Press, Inc., New York, 1953, pp. 395–405.

41. L. Zhang and D.-F. Zhan, *Gaofenzi Tongxun (Polym Commun.)* **8**(2), 115 (1984).

E. T. ADAMS, JR.
Texas A&M University

OUTSERT MOLDING.   See MOLDS in the Supplement.

OVALBUMIN.   See PROTEINS.

***N*-OXACRYLAMIDES.**   See Acrylamide polymers.

OXACYCLOBUTANE POLYMERS.   See Oxetane polymers.

OXAZOLINES.   See Cyclic imino ethers, polymerization.

OXETANE POLYMERS

Oxetane is a name commonly used for the unsubstituted cyclic ether with a four-membered ring and the structure (**1**):

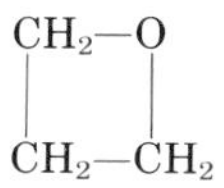

$$CH_2-O$$
$$CH_2-CH_2$$
$$(1)$$

This monomer is also known by other names including trimethylene oxide, oxacyclobutane, and 1,3-epoxypropane. The parent ether polymerizes readily as a result of ring strain associated with the small ring size. A variety of substituted oxetanes have been synthesized and many of them also polymerize readily (Table 1).

The most important chemistry needed to polymerize and understand the mechanisms of polymerization of oxetanes was observed in the late 1930s (5,6) and developed throughout the 1940s. Tetrahydrofuran (THF), the next higher member in the homologous series of cyclic ethers, was used as a model monomer (see also Tetrahydrofuran polymers). The first report of the polymerization of an oxetane was that of 3,3-bis(chloromethyl)oxetane (BCMO) (7). This was followed in 1955 by a detailed consideration of a family of 3,3-disubstituted oxetanes (8) and in 1956 by studies of the homopolymerization of oxetane itself (9,10). In each decade since then, interest in these polymerizations and in the resulting polymers has waxed and waned as monomers with different substituents became available and new applications for known polymers have been found.

During the 1950s, Hercules, Inc. developed a commercial synthesis for poly[3,3-bis(chloromethyl)oxetane] (PBCMO) and marketed the polymer for about 15 years, but it has since been withdrawn. The growing number of new publications, patents, and reports seem to indicate a resurgence of interest in PBCMO for coating and adhesive applications, especially in the USSR. Other oxetanes are being copolymerized, mainly with THF, to prepare precursors for use as the soft segment in polyurethanes (qv), polyethers, and polyamide-type elastomers. In the United States, interest presently centers on energetic polymers prepared from oxetanes in which one or more of the hydrogen atoms in the 3 position have been replaced by electron-deficient groups like $-CH_2N_3$, $-NO_2$, and $-CH_2OCH_2\overset{\displaystyle CH_3}{\underset{\displaystyle |}{C}}(NO_2)_2$. These materials are being explored as precursors to polymers that can be used as propellants and other explosives.

Table 1. Properties of Selected Oxetanes

Monomer	Structure	Abbrev	Bp, °C[a]	Mp, °C	Polymer T_m, °C	Refs.
oxetane		OX	47.8	−99	35	1,2
2-methyloxetane		2MOX	60		amorphous	1
3-methyloxetane		3MOX	67		amorphous	1
3-azidooxetane		AZOX	120–130[b]		low mol wt oil	3
3-nitrooxetane		NIOX	77[c]			3
3,3-dimethyloxetane		DMOX	80		47	1
3,3-dinitrooxetane		DNOX		70, 71	200–202	3
3,3-bis(chloromethyl)-oxetane		BCMO	91[d]	19	170	1
3,3-bis(azidomethyl)-oxetane		BAMO	liq dec 160–170		75–78	4

[a] At 101.3 kPa unless otherwise noted. To convert kPa to mm Hg, multiply by 7.5.
[b] 0.93–1.3 kPa.
[c] 0.067–0.13 kPa.
[d] 2.3 kPa.

Monomers

Properties.   More than a hundred different oxetanes have been synthesized; a few selected examples are given in Table 1 (see also Ref. 11). The parent, unsubstituted oxetane is a liquid that boils at ~48°C. Fully deuterated and fully fluorinated analogues have been prepared (1). Alkyl, cycloalkyl, bridged hydrocarbon, haloalkyl, sulfonyl chloride, ether, ester, hydroxyl, sulfide, mercapto, amino, substituted amino, carboxyl, nitrile, nitro, and azido groups can be introduced at any desired position on the ring or in side chains. Multiple substitution is common. Sometimes the functional groups are attached directly to the ring; most often there is a methylene group between the ring and the functional group. The physical properties of the resulting monomers vary widely with the substituent. Substituted oxetanes can be solids or very high boiling liquids distillable only at reduced pressure.

Many of the oxetanes are polymerizable. Quantitative conversion to polymer is often observed when polymerizations are carried out under proper conditions with suitably purified materials and appropriate initiators (12). The basicity of the monomer and ring strain are important indicators of the polymerizability of a given cyclic ether; both are high and favorable for polymerization of many oxetanes. The unsubstituted oxetane is the most basic of the unsubstituted cyclic ethers (13). It has a ring strain of 107 kJ/mol (25.6 kcal/mol), only slightly smaller than that of the readily polymerizable cyclic ether with a three-membered ring (oxirane) and more than 20 times greater than that of the cyclic ether with a six-membered ring, which does not homopolymerize (13).

Basicity and ring strain are not the only factors that determine polymerizability. Some substituted oxetanes do not polymerize under normal conditions because of the nature of their substituents. The usual mechanism by which oxetanes polymerize involves a cationic (oxonium ion) ring-opening reaction. Unless they are suitably protected, amine, hydroxyl, carboxyl, and other simple electron-donating substituents either inhibit polymerization entirely or terminate polymerization after one or two propagation reactions. Ether or formal substituents are known transfer agents and the molecular weight of any polymer formed is consequently reduced (12). Bulky substituents on the 2 and 4 positions may prevent polymerization due to steric inhibition; many electronegative substituents on the ring may prevent formation of a stable oxonium ion. However, copolymerization of such unpolymerizable oxetanes with less substituted oxetanes or with tetrahydrofuran and other monomers that polymerize by a similar mechanism may still be possible. Copolymerization of this kind is known to occur with THF (12,14).

Preparation. Syntheses of oxetanes are well established, but side reactions are common and yields are usually not more than 60% (3,15). Purification is often difficult and adds significantly to production expense.

Figure 1 summarizes the methods commonly used to form the oxetane ring (1,3). Treatment of a 3-chloropropanol or 3-chloropropyl acetate with base is by

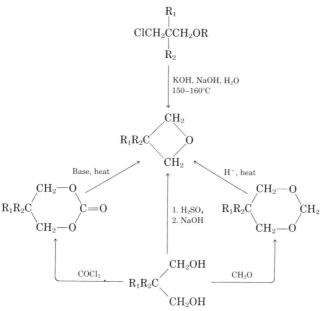

Fig. 1. Preparation of oxetanes (1,3). R_1, R_2 = H, CH_2X or other functional group.

$$R = H \text{ or } -\overset{\overset{\displaystyle O}{\|}}{C}-CH_3.$$

far the most commonly used method. 3-Bromopropanols react in the same way, but treatment of 3-iodopropanols with alkali normally gives different products such as olefins (15). Iodo- derivatives, eg, 3,3-bis(iodomethyl)oxetane, are prepared by exchanging the chlorine or bromine of the corresponding 3,3-bis(halomethyl)oxetane using NaI as the source of iodine (1). Other substituted 3,3-bis(methyl)oxetanes, eg, fluoromethyl or azidomethyl derivatives, are similarly prepared using reagents like KF (1) or NaN_3 (3), respectively. Figure 2 outlines the methods that have been used (3) to prepare a series of oxetanes with other kinds of substituents in the 3 position. The key intermediate for this series of compounds is 3-hydroxyoxetane. The hydroxyl is protected during ring closure. The protecting group can be removed with acid; the hydroxyl group is converted to its *p*-toluenesulfonate (tosylate), and the 3-(tosyl)oxetane is converted by the methods shown to the 3-azido-, 3-amino-, 3-nitro-, and 3,3-dinitrooxetane derivatives.

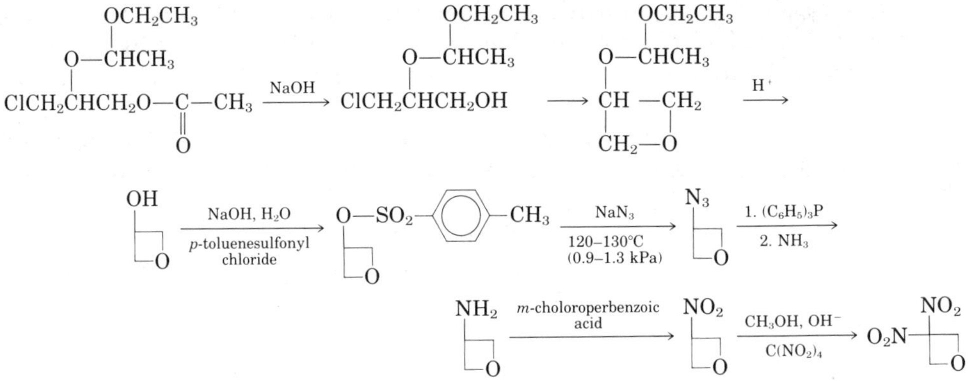

Fig. 2.   3-Hydroxyoxetane derivatives (3).

Polymerization

The ring-opening polymerization (qv) of oxetane (OX) is primarily accomplished by cationic polymerization (qv) methods. Insertion polymerization (qv) (16), anionic polymerization (qv) (17), and with certain special oxetanes, a free-radical ring-opening polymerization (18) have also been used. In certain cases, aluminum-based catalysts (18) impose a degree of microstructural control on the polymerization process that is not obtainable with a pure cationic system. The polyoxetane structure has also resulted from an anionic rearrangement polymerization. A polymer having largely the structure of poly(3-hydroxyoxetane) was obtained from the anionic polymerization of glycidol or its trimethylsilyl ether, both oxirane monomers (19).

Initiators.   The initiation of cationic polymerization requires a reagent that generates the propagating tertiary oxonium ion (**2**):

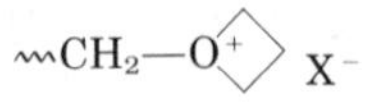

(**2**)

This can be accomplished by a variety of electrophilic reagents such as Lewis acids, preformed trialkyl oxonium ion salts, acylium ion salts, carbocationic salts, etc. Appropriate strong acids may be used, either directly or generated *in situ*, but present the complication that the secondary oxonium ion formed first (**3**)

$$HO\overset{+}{\diamondsuit}\ X^-$$

$$(3)$$

is unreactive relative to the propagating tertiary oxonium ion (**2**) (10). A counterion X^- of rather low nucleophilicity is required in order to generate a propagating oxonium ion species of adequate stability (14). Thus simple anions such as chloride or bromide are not suitable because they collapse immediately to the corresponding haloalkyl ether (or alcohol). The ions used in the earliest studies were commonly BF_4^- or BF_3OH^- which allowed for practical polymerizations of the reactive oxetane ring system, but suffered from a degree of instability evident in careful kinetic analyses. Thus more recent studies have tended to use the marginally better $SbCl_6^-$ or the superior ions PF_6^-, AsF_6^-, or SbF_6^- (14,20). The exact nature or structure of the anion formed with the alkyl aluminum–water initiators is not known, but it seems to function as a very stable ion of low nucleophilicity.

Lewis Acid Types.   The Lewis acids corresponding to the counterions just discussed are suitable initiators for oxetane polymerizations. Thus BF_3 or BF_3 etherate was used in classic studies of OX polymerization (9,10). Pure dry monomer will not polymerize with BF_3 alone (7); an active hydrogen material, generally water or an alcohol, is required and leads to a BF_3OH^- or BF_3OR^- counterion. Some of the stronger Lewis acids such as PF_5 or SbF_5 are able to self-initiate and do not require the presence of active hydrogen compounds (21).

Preformed Trialkyl Oxonium Ion Salts.   The preparation of triethyl and trimethyl tetrafluoroborate has been described (22). The corresponding hexafluorophosphate or hexafluoroantimonate salts can be prepared similarly or a method starting with a commercially available hexafluorophosphoric acid solution can be used (23). Additionally, the salts are often formed *in situ* using the Lewis acid with a so-called epoxide promoter (24). These salts are strong alkylating agents and initiate polymerization by a simple alkyl exchange reaction:

$$R_3O^+X^- + O\diamondsuit \longrightarrow R{-}O\overset{+}{\diamondsuit}\ X^- + R_2O$$

Acylium Ion Salts.   The acylium ion salts corresponding to the counterions mentioned above are also good initiators:

$$\overset{O}{\overset{\|}{RC}}{}^+ X^- + O\diamondsuit \longrightarrow \overset{O}{\overset{\|}{RC}}{-}O\overset{+}{\diamondsuit}\ X^-$$

The acyl dialkyl oxonium ion is about as reactive as a similar trialkyl ion. These salts can also be isolated and used directly, or they can be formed *in situ* from the acid chloride and a silver salt, eg, silver hexafluoroantimonate (25).

Super Acids and Their Esters.   These compounds are often excellent initiators for cationic polymerization of cyclic ethers. The fluorosulfonates and trifluoromethanesulfonates especially have been exploited. The polymerization kinetics are complicated by an ester ion equilibrium, but proper choice of a solvent with a high dielectric constant

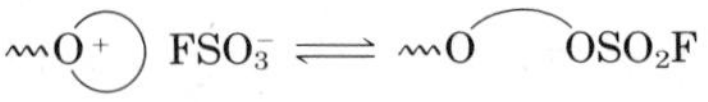

often favors the formation of the reactive ion and produces good polymerization rates (14). However, these initiators have been found to be of little value in oxetane polymerizations. For example, polymerization of BCMO initiated by ethyl trifluoromethanesulfonate proceeds extremely slowly because the above equilibrium is far to the right under polymerization conditions (26); the ester form is virtually unreactive to the cyclic ether and does not participate in the propagation reaction. Moreover, the active oxonium ion, when formed, collapses immediately to the corresponding ester. Thus a suitable mechanism for rapid polymer formation with this otherwise excellent group of initiators does not exist for BCMO and probably not for other oxetanes as well.

Aluminum Alkyl-based Initiators.   Cationic initiators can be prepared from trialkylaluminum by modification in inert solvent with 0.5–1.0 mol of water per mole of aluminum alkyl (27,28). Initiator efficiencies and rates of initiation are enhanced by the use of epoxide promoters with *in situ* trialkyl oxonium salt formation (24,28). A true cationic polymerization strictly analogous to those obtained with Lewis acid initiators is obtained, but the true nature of the counterion formed is not known and, in fact, is probably not a single species. Furthermore, it is difficult to predict the exact number of active sites that form (29,30). The alkyl aluminum coordination catalysts, developed primarily for oxirane polymerizations, can also be successfully applied to oxetanes (16). These catalysts are prepared by adding one of several coordinating modifiers to a water-modified aluminum alkyl system. The most commonly used modifier appears to be acetylacetone (2,4-pentanedione) (16,31,32).

Other Initiators.   Cationic polymerization of oxetane has been initiated electrochemically (33), by γ-rays (34), by free-radical systems (18,35), and by uv radiation (36). These methods have not achieved any great importance.

Mechanistic Considerations.   The cationic ring-opening polymerization of four-membered (and higher) cyclic ethers takes place by a nucleophilic attack of the incoming monomer oxygen on one of the two-ring, α-carbon atoms of the active cyclic oxonium ion:

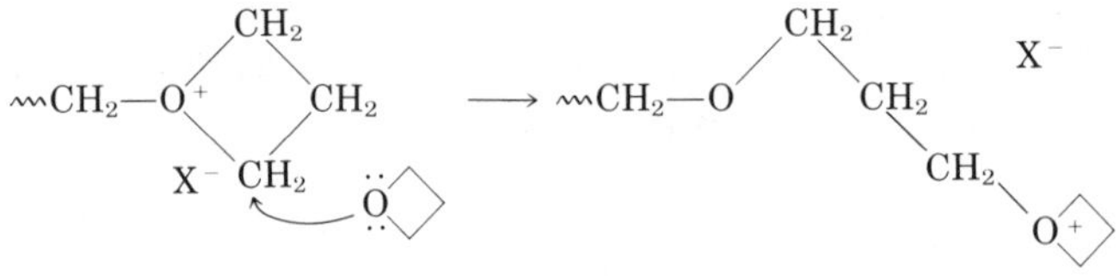

If the oxetane is unsymmetrically substituted, the microstructure of the product differs if attack occurs only at one of the two α-carbon sites or if both α-carbon atoms are subject to attack.

The monomer oxygen atom is not the only reactive oxygen atom in the system. All ether oxygen atoms in the backbone, side chains, or additives are potentially reactive. Thus polymer oxygen atoms can attack the active cyclic oxonium ion. Intramolecular attack by polymer oxygen, ie, attack of the oxonium ion end group by a polymer oxygen of the same polymer chain, can lead to either depolymerization (qv) or to formation of cyclic oligomers. Attack by the penultimate oxygen atom ion leads to depolymerization:

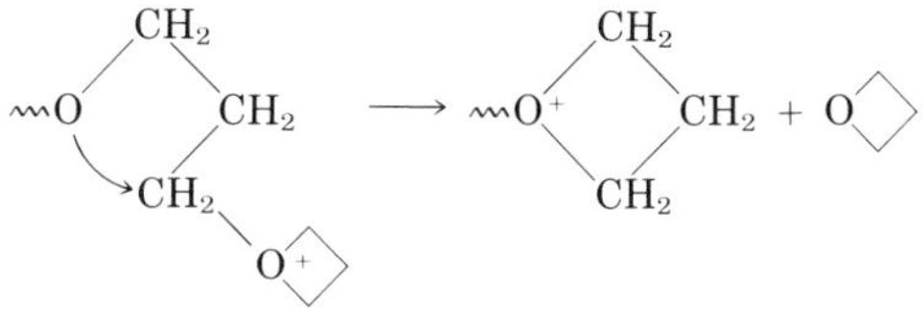

Whereas this is an important reaction for tetrahydrofuran, it is energetically unfavored for the highly ring-strained oxetane system and does not need to be considered under ordinary polymerization conditions. Attack of this same oxygen atom at either of the two-ring, α-carbon atoms leads to an eight-membered ring, which also is energetically unfavored and not found. Attack by oxygen atoms further back on the chain leads to the oxonium ion precursors of larger ring systems that are often found in substantial quantities. Intermolecular attack by polymer oxygen leads to the formation of polymeric acyclic oxonium ions:

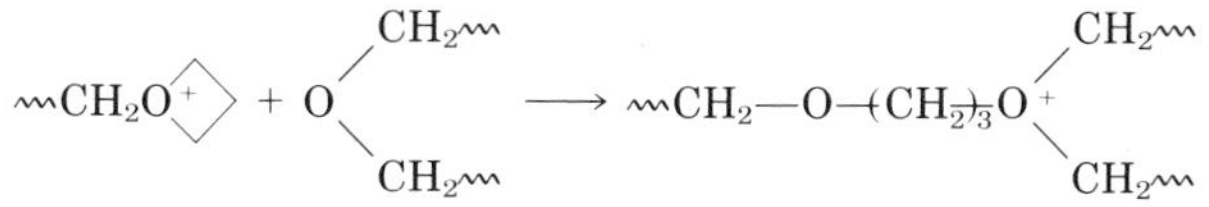

Such ions have been invoked to explain various kinetic features, eg, temporary termination as an explanation of the dramatic change from a very rapid to a slow, steady polymerization rate (37). Evidence of polymeric acyclic oxonium ions was inferred from the higher viscosity of an unterminated oxetane polymerization solution compared to the viscosity of the same solution just after amine termination (38). This effect was not observed for a DMOX polymerization where steric inhibition prevents reaction of polymer oxygen. Intermolecular reaction of polymer oxygen also accelerates broadening of the molecular weight distribution to its most probable value of two.

Kinetics. The kinetics of OX and DMOX polymerizations were first considered in 1956 (10). Subsequent analysis has shown that the kinetic expression for these polymerizations can be reduced to

$$\frac{-d[M]}{dt} = k_p[I_0][M]$$

which was thought to describe all oxetane polymerizations adequately regardless of the initiator used (39). Here only the propagation rate constant k_p is used and $[I_0]$ represents either the concentration of a fast, 100% efficient initiator or the concentration of active sites as measured directly; $[M]$ is monomer concentration. Some systems may not be this simple and rate constants obtained from actual measurements of active sites may represent average, overall rate constants encompassing several processes.

The following conclusions for OX polymerizations in methylene chloride at $-30°C$ using PF_6^- counterion have been presented (37). Initiation with the trialkyl oxonium ion salt is slow. Initiation with trityl salts is by direct addition of the trityl carbocation to oxetane and the reaction is fast. The polymerization, ie, the propagation reaction, is very fast. A competing fast reaction with polymer oxygen removes active growing sites to form dormant or "sleeping" ions by a process often referred to as temporary termination. The polymeric acyclic oxonium ion so formed is analogous in structure and reactivity to the triethyl oxonium ion and thus reacts only slowly with monomer to reinitiate the fast polymerizing active site. Such participation of polymer oxygen also accounts for the molecular weight distributions which are broader than the anticipated Poisson distribution. Furthermore, such a scheme is also consistent with the proposals that have been made to explain cyclic oligomer formation.

More recent studies with 3MOX, carried out at $-78°C$ and using acetyl hexafluorophosphate as initiator, confirm the extremely fast propagation rate (40). Near quantitative conversion to polymer in less than two minutes suggests no involvement of polymer oxygen using this monomer under these conditions. A similar observation was made when DMOX polymerization was qualitatively compared with OX polymerization, both at 25°C in methylene chloride and both initiated by triethyloxonium hexafluoroantimonate (38). The rate of polymerization of DMOX was much faster than that of OX, and a lack of the viscosity effect upon termination of the DMOX polymerization suggested little, if any, involvement with polymer oxygen in the case of DMOX. Qualitative or quantitative kinetic data bearing on this point have not yet been reported for other substituted oxetanes.

Cyclic Oligomers.   The formation of high polymer from OX monomers is often accompanied by variable amounts of low molecular weight oligomeric material, eg, the cyclic tetramer, a 16-membered tetraoxycyclic ether (9). The amount of cyclic tetramer that forms varies with polymerization conditions and with monomer structure; OX and DMOX form cyclic tetramer, whereas 2MOX and BCMO do not. Later studies showed that small amounts of a cyclic trimer can form in BCMO polymerizations (41). In more recent studies, in the case of the unsubstituted OX, both cyclic tetramer and cyclic trimer form (42). The amounts formed vary with counterion, polymerization temperature, and solvent. A more detailed study revealed that an array of cyclic oligomers results for both OX (trimer through octamer) and DMOX (tetramer through nonamer); tetramer is the major one in both cases (43). Oligomers are formed only from polymer (not by a direct reaction) and only in the presence of monomer. First, a "backbiting reaction," ie, an intramolecular reaction of polymer oxygen, forms a macrocyclic oxonium ion. Subsequent reaction with monomer at the exocyclic methylene group releases the free macrocycle and regenerates the active propagating ion (43).

The ease of macrocycle formation is very dependent on stereochemical considerations, such as the preferred conformation of polymer chains. Thus the extent of cyclic oligomer formation varies greatly with oxetane-ring substitution (see also MACROCYCLIC POLYMERS).

Microstructure.   High frequency nmr instrumentation is useful for studies of the microstructure of unsymmetrically substituted oxetanes and oxetane copolymers. Propagation occurs when monomer attacks either of the two endocyclic α-carbon atoms of the active cyclic oxonium ion. In the case of the unsubstituted OX or the 3-substituted derivatives, the molecules are symmetrical and no observable differences arise between attack at the two endocyclic α-carbon atoms.

However, for the 2-substituted oxetane, eg, 2MOX,

$$\beta \overset{CH_2}{\underset{4}{\diagup}}$$

mmO$\overset{+1}{\alpha}$ $3CH_2$

$\overset{2}{\underset{\alpha}{\diagdown}}$ CH

CH$_3$

(4)

attack at C-4 results in β-bond cleavage, opening of the O–CH$_2$ bond, whereas attack at C-2 results in α cleavage. Exclusive attack at one position to give only β (or α) cleavage results in a regular head-to-tail structure. But even occasional attack at the other position to give α (or β) cleavage results in backwards placement of the monomer unit and consequent head-to-head, tail-to-tail structural units:

$$\overset{\displaystyle CH_3}{\underset{\displaystyle |}{}}\qquad\qquad\overset{\displaystyle CH_3}{\underset{\displaystyle |}{}}$$

—CH$_2$—CH$_2$—CH—O—CH$_2$—CH$_2$—CH—O—   head-to-tail

$$\overset{\displaystyle CH_3}{\underset{\displaystyle |}{}}\quad\ \overset{\displaystyle CH_3}{\underset{\displaystyle |}{}}$$

—CH$_2$—CH$_2$—CH—O—CH—CH$_2$—CH$_2$—O—   head-to-head

$$\overset{\displaystyle CH_3}{\underset{\displaystyle |}{}}\qquad\qquad\qquad\overset{\displaystyle CH_3}{\underset{\displaystyle |}{}}$$

—CH—CH$_2$—CH$_2$—O—CH$_2$—CH$_2$—CH—O—   tail-to-tail

Furthermore, the head-to-head linkages can have either a meso or a racemic configuration of the two methyl groups.

Cationic polymerization results in both α and β cleavage. Both head-to-tail and head-to-head, tail-to-tail placement can be detected by ^{13}C nmr (20,27,44). The extent of mixed cleavage depends on polymerization temperature. Higher temperatures favor more random attack at the two sites. An alkyl aluminum catalyst system coordinated with acetylacetone results in exclusive head-to-tail placement (27,32). Either α or β cleavage occurs exclusively; this result cannot distinguish between the two modes of cleavage. In addition, the coordinate catalyst exerts steric control, and measurement of isotactic, syndiotactic, and heterotactic triad concentrations showed that syndiotactic placement was somewhat favored over the random placement found in the head-to-tail units produced by cationic catalysts. Further analysis using optically active 2MOX monomer led to pure isotactic structure with coordinate catalysis and showed that α cleavage occurs primarily with inversion of configuration of the chiral center (45,46).

In the polymerization of *cis*- and *trans*-2,4-dimethyloxetane, the meso cis isomer is converted to a polymer that has a largely racemic diisotactic structure (45,47). The racemic trans isomer yields a polymer suggesting random placement of the repeating units. Thus the direction of ring opening is influenced by the configuration of the growing oxonium ion only for the cis isomer. The anionic and insertion polymerizations of propylene oxide also include some head-to-head monomer placement (31,48).

An oxetane doubly substituted in the 2 position, the 2,2-dimethyloxetane, leads exclusively to head-to-tail structure under cationic conditions (49). Ring opening occurs exclusively by α cleavage as a result of an equilibrium between the oxonium ion and the corresponding ring-opened tertiary carbocation. It is suggested the propagation in this case is via the carbocation, whereas the oxonium ion easily undergoes transfer to monomer. In this way, the production of only low molecular weight polymer and oligomers with primary hydroxyl and isopropenyl end groups is accounted for (49).

Though symmetrically substituted, the P3MOX can, in theory, show isotactic–syndiotactic–heterotactic placement of the lone methyl along the polymer chain. It is very difficult to detect by ^{13}C nmr because the isotactic–syndiotactic–heterotactic splitting is only about 0.017 ppm (compared to 0.14 ppm for P2MOX). Though it could not be accurately assessed, the relative areas were consistent with a totally random placement of the methyl groups (40).

Polyoxetane Glycols.   Relatively low molecular weight polyethers with hydroxy end groups have been the only commercially important products resulting from THF polymerization. Thus as the next lower homologue, with increased potential for functional or modifying ring substitution, the oxetanes offer an enticing extension of this chemistry. However, as has been noted, oxetane polymerization is very fast and control of the polymerization to give low molecular weight products with end-group functionality has, until recently, not been possible. Japanese patents (50,51) describe the preparation of low molecular weight oxetane–tetrahydrofuran copolymers using methods that appear similar to those used to prepare PTHF glycols (52). The preparation of polyoxetanes and oxetane–THF copolymers substituted with energetic groups such as azido or nitro have also been described (4,53–55). The latter methods generally involve polymerization in the presence of relatively large amounts of organic diols, eg, 1,4-butanediol, or the use of preformed Lewis acid–diol mixtures as polymerization initiators. Butyllithium has been used to cleave high molecular weight polyoxetane (POX) to low molecular weight POX glycols (56). Low molecular weight POX glycols have also been prepared by ozone degradation of a high molecular weight polymer followed by lithium aluminum hydride reduction or by direct polymerization using Lewis acid–diol mixtures as initiator (57). Standard methods and recipes are then used to extend the chain of these glycols into polyurethane networks (58). POX glycols have also been used to prepare a series of block copolymers (qv) (59).

Copolymers.   Oxetane copolymerizes, often competing effectively with other oxetanes and with other cyclic ethers such as oxiranes or THF or with lactones, dioxanes, or dioxolanes (28). Copolymerization (qv) is often desired in order to modify and adjust crystallinity or crystallization rates. Symmetrical oxetanes generally crystallize readily and for some applications this is not a desirable property, eg, for elastomeric applications. Since BCMO was once commercial and its polymer is highly crystalline and high melting, its copolymerizations have been extensively studied (28). Copolymerization of BCMO with the dioxolane ring of a 1,6-anhydroglucopyranose has also been demonstrated (60). Many pairs of reactivity ratios (r_1, r_2) covering a wide variety of copolymers of oxetanes have been determined and tabulated (28,61). These values can vary substantially with conditions and need to be critically evaluated before use.

Studies of ionomers have resulted in copolymerizations of oxetane with ethyl glycidate (62) and with ethyl undecanoate (63,64). Ethyl glycidate does not copolymerize readily and only copolymers containing ca 1 mol % ethyl glycidate have been prepared. The copolymerization of DMOX with THF was studied in order to determine the T_g of various compositions (65) and the apparent r_1, r_2 values (66). The latter studies indicated that random copolymers were formed. Random copolymerization was also found for the DMOX–OX pair (67). The study suggested that OX was more reactive than DMOX toward each of the two growing ions. A study of the copolymerization of 2MOX with 2-methyltetrahydrofuran, which does not homopolymerize, showed that copolymerization of 2-methylTHF also does not occur readily. Nmr analysis indicated that both monomers entered the polymer chain in both head-to-tail and head-to-head placement (68), that is, on copolymerization both monomers behave like 2MOX (20).

Copolymerization with carbon dioxide has been demonstrated (69). Reaction with carbon dioxide to give a polycarbonate requires an alkyl tin catalyst modified with a phosphine or an amine.

Properties of Polymers

A large number of polyoxetanes have been prepared and partially characterized (1). The properties of the polymers vary greatly with the symmetry, bulk, and polarity of the substituents on the chain (Table 1). The polymers range from totally amorphous liquids to highly crystalline, high melting solids. The unsubstituted oxetane polymer has a melting temperature of 35°C, not far above ambient temperature. A single methyl substituent at either the 2 or 3 position gives a polymer with stereoisomeric possibilities. Only amorphous polymers have been reported for oxetanes with single methyl substituents (1,10). The 3,3-dimethyl derivative gives a crystalline polymer with a melting temperature of 47°C, whereas 3,3-bis(halomethyl)oxetanes melt at 135, 180, 220, and 290°C for fluoro-, chloro-, bromo-, and iodo- derivatives, respectively (1). As usual, the effect of copolymerization is to lower the melting temperature. For example, the random, high molecular weight copolymer of BCMO and THF is a tough, amorphous rubber (12).

Other crystallization parameters have been determined for some of the polymers. Numerous crystal-structure studies have been made (70–77). Isothermal crystallization rates of polyoxetane from the melt have been determined from 19 to -50°C (78,79). Similar studies have been made for PDMOX from 22 to 44°C (80,81).

The pressure dependence of the glass-transition temperature T_g of BAMO–THF copolymer has been determined by high pressure ($\leq$850 MPa or 8.5 kbar) dta and dielectric methods (82). T_g increases with increasing pressure. The results fit a modified Gibbs-DiMarzio theory. The effect of copolymer composition on T_g (65) and on dipole moments (83) has been studied over the whole composition range for DMOX–THF copolymer. The copolymers exhibit a T_g intermediate between those of the parent homopolymers.

Solubility parameters of 19.3, 16.2, and 16.2 $(J/cm^3)^{1/2}$ have been determined for POX, PDMOX, and poly(3,3,-diethyloxetane), respectively, by measuring solution viscosities (84). Heat capacities have been determined for POX and compared to those of other polyethers and polyethylene (85,86). The thermal decomposition behavior of poly[3,3-bis(ethoxymethyl)oxetane] has been examined (87).

Properties have been determined for a series of block copolymers based on poly[3,3-bis(ethoxymethyl)oxetane] and poly{[3,3-bis(methoxymethyl)oxetane]-*co*-THF}. The block copolymers had properties suggestive of a thermoplastic elastomer (59) (see ELASTOMERS, THERMOPLASTIC).

Extensive physical property measurements have been reported only for PBCMO (88,89) (Table 2). This polymer was commercially available in the United

Table 2. Typical Properties of Poly[3,3-bis(chloromethyl)oxetane]

Property	Value		Refs.
melting temperature, °C	α form, 186		90
	β form, 180		70
glass-transition temperature, °C	7–32[a]		90
density, 25°C, g/cm³			
amorphous	1.386		90
crystalline	1.47		90
compression modulus, 6.9 MPa[b]	130		91
tensile strength, MPa[b]	41.4		88,90,91
elongation at break, %	35–150		88,91
heat-distortion temperature, °C			
1.82 MPa[b]	85–93		91
0.48 MPa[b]	149		91
izod impact (notched), 228°C, J/m[c]	27		88,91
Rockwell hardness	R100		91
flexural strength, MPa[b]	34.5		88
thermal expansion, 10^{-5}/°C	8		88
volume resistivity, $\Omega\cdot$cm	1×10^{16}		88
dielectric strength, V/μm	16		88
dielectric constant, 10^6 Hz	3		88
power factor, 10^6 Hz	0.011		88
flammability	self-extinguishing		88
water absorption, % (24 h)	0.1		88
mold shrinkage, cm/cm	0.005		88
crystallographic data	α form	β form	71
crystal system	orthorhombic	monoclinic[d]	
space group	D2H-2, D2H-16	CS-1 or CS-2, CS-3	
unit cell			
a, nm	1.785[e]	0.685, 1.142	
b, nm	0.816[e]	1.142, 0.706	
c, nm	0.467–0.482	0.475, 0.482	
β, degrees		109.8, 114.5	
monomer/unit cell	4	2	
crystal density, g/cm³	1.47–1.514	1.472, 1.456	
heat of fusion, kJ/mol	23.0		
chain configuration, N*P/Q	3*1/1	3*1/1	

[a] Depends on method of determination.
[b] To convert MPa to psi, multiply by 145.
[c] To convert J/m to ftlbf/in., divide by 53.38.
[d] Two groups of data for a monoclinic cell are given in Table 2. A third group suggests an orthorhombic cell with C2V-12 space group; $a = 1.301$ nm; $b = 1.171$ nm; $c = 0.467$ nm; 4 monomers/unit cell and a crystal density of 1.447 g/cm³.
[e] One reference reverses a and b.

States from Hercules, Inc. for about 15 years under the trade name Penton, but it is not currently produced for commercial sale. It has been studied extensively in the USSR, where it is called Pentaplast, and also in Japan. The comparatively high heat-distortion temperature combined with a low water-absorption value make PBCMO suitable for articles that require sterilization. It also has excellent electrical properties, a high degree of chemical resistance, and, because of the chlorine in the backbone, is self-extinguishing. PBCMO is resistant to most solvents, specifically to ketones, water, aldehydes, esters, aromatic hydrocarbons, and weak organic bases, eg, aniline, and ammonia, weak acids, weak alkalis, and strong alkalis; it is attacked by strong acids (88). It is reported to be somewhat soluble in hot o-dichlorobenzene or cyclohexanone and in hexamethylphosphoramide at room temperature (1). It is attacked by fuming nitric acid and sulfuric acid (88). Most other crystalline oxetane polymers are also insoluble in common organic solvents (71). The amorphous and low melting oxetane polymers, in contrast, are soluble in a wide variety of organic solvents.

Processing

This discussion is limited to processing of PBCMO because this is the only oxetane polymer that has been commercialized (88). A typical commercial polymer had a number-average molecular weight of 250,000–350,000. Such PBCMO can be injection-molded in standard machines, using chrome-plated or polished hard-steel dies. As a result of its highly crystalline nature, PBCMO has a fairly low melt viscosity and easily fills molds. Molding cycles of 20 s or less are possible for parts with wall thicknesses of 1.6 mm or less. A stress-free molding with low shrinkage on cooling is obtained. The parts can be machined easily and components with exceptional dimensional stability are produced. PBCMO can also be extruded, using screw designs of the type used for nylon or polyethylene and temperatures of 182–230°C. Because of its low water absorption, predrying of the molding powder is not necessary. Coatings with a homogeneous void-free finish up to 1-mm thick can be applied to a variety of substrates (glass, steel, paper) by water or solvent-dispersion systems or by using the fluidized-bed technique of coating (88).

Economic Aspects

At the present time, polyoxetanes are not available commercially. Consequently, few oxetane monomers are available and those only in laboratory-size quantities; the monomers are expensive. The price varies with the substituent, but all are at least several dollars per gram. Typical current prices are $96/50 g of BCMO (Polysciences) and $55/25 g of oxetane (Aldrich). If large-scale commercial production is resumed for BCMO or initiated for one of the energetic monomers, lower prices will result. It is difficult to project the future economics for these potentially very useful specialty polymers.

Analytical and Test Methods

Titration.   A rapid method for determining the oxetane content of a sample by titration has been reported (92). The sample (1–2 meq) is refluxed for 3 h with 25 mL of 0.20-N pyridinium chloride in anhydrous (<0.01% water) pyridine to convert the oxetane to its 1,3-chlorohydrin. The mixture is cooled and the apparatus is rinsed with 50-mL water. Excess pyridinium chloride is titrated with standard aqueous alkali (0.15-N NaOH) using phenolphthalein indicator. A blank determination is carried out on 25 mL of the pyridinium chloride solution. The difference between the blank and the excess is a measure of the oxetane content of the sample:

$$[(B - S) \times N \times 5.806]/w = \% \text{ oxetane-ring content}$$

where B = mL of NaOH to titrate blank, S = mL of NaOH to titrate sample, N = normality of NaOH, and w = sample weight in grams.

Spectroscopy.   Ir and nmr spectroscopy are useful aids in the characterization of oxetanes and oxetane polymers. The CH frequencies of the monomer are generally observed at 2900–3000 cm^{-1}. The oxetane ring shows absorption between 960 and 980 cm^{-1} regardless of the substituents on the ring (1). An unusually high oxetane-ring ir frequency of 1000 cm^{-1} has been reported for DNOX (3). Polymer ether (COC) ir absorptions are typically near 1100 cm^{-1}. Proton nmr chemical shifts for oxetane monomer and polymer are typically ~4.0–4.8 δ (CH$_2$) and 3.5–4.7 δ (CH$_2$), respectively (3,40,44). ^{13}C nmr increasingly is being used to characterize polyoxetane microstructure (27,69).

Equivalent Weight of Hydroxyterminated Polymer.   The following procedure has been used for PAZOX (3). Approximately 0.2 g of polyoxetane, 5 mL of 1,2-dichloroethane, 2 mL of 1,1,1,3,3,3-hexamethyldisilazane, and 0.5 mL of chlorotrimethylsilane were added to a 50-mL, round-bottomed flask equipped with a magnetic stirrer, condenser, and drying tube. Volatile materials were removed at 70°C under vacuum. The residue was dissolved in 1-mL deuterochloroform and the ^{1}H nmr spectrum was recorded. The hydroxyl equivalent weight was calculated on the basis of the areas of the δ 3.3 signal (5 protons per monomer unit) and the δ 0 signal (9 protons per siloxy group). The same method was used for polymers with more complex nmr spectra by using weighed mixtures of silylated polymer and a known amount of a reference such as p-dichlorobenzene. In these cases, the equivalent weight was calculated on the basis of the reference and siloxy signals (3).

Health and Safety Factors

Very little has been reported about the toxicity and hazards of handling the oxetanes and their polymers. Both properties are influenced markedly by the substituents. Oxetane itself is a low boiling flammable organic liquid and should be handled with care like any other low boiling and flammable liquid. Details of its toxicity are unknown, although one report states that it may be narcotic in high concentrations (93). BCMO is reported to be moderately irritating to the respiratory tract with high danger of permanent injury. Animal experiments show irritant and narcotic effects (94). Because of its structure and chlorine content, BCMO can evolve toxic fumes on decomposition and is classified as dangerous in disasters (94). AZOX is one of the monomers designed to yield energetic polymer and, as such, is hazardous. Adiabatic compression tests indicate that the material is a sensitive explosive (3). AZOX can be handled safely by collecting the material directly in methylene chloride at $-78°C$ as it forms and storing it as a methylene chloride solution (3). Under no circumstances should an energetic monomer be distilled; rather purification of energetic monomers via column chromatography is suggested. Elution of BAMO with methylene chloride through a column packed with basic alumina (95) or elution of 3-nitrato-methyl-3-methyloxetane with 50/50 vol/vol chloroform/hexane through a neutral alumina packed column (96) have been recommended. Methylene chloride–hexane was used to purify DNOX by passing through silica gel (3). The polymers derived from the energetic monomers burn with a great deal of smoke. Special procedures have been developed for disposal of these energetic polymers.

Uses

Suggested uses for PBCMO include valves, bearings, precision gears, corrosion-free coatings, wedges and slot liners for starters in electric motors, films, adhesives, and even rope. Markets for this high price, high performance material were mainly in the field of chemical processing and precision injection-molded industrial parts. As a lining material PBCMO converts carbon steel tanks into vessels capable of handling high corrosive fluids at elevated temperatures (88). There are a large number of reports, especially from the USSR, of the usefulness of PBCMO in adhesive and coatings applications (97,98). There are a substantial number of recent Japanese reports, which describe the use of BCMO and other oxetanes to prepare polyglycols or copolyglycols for use in polyurethanes or polyesters or polyamide-type elastomers (51,99). PBCMO has been used in glass- and metal-filled composites (100,101). Grafting of 2- and 4-vinylpyridine to PBCMO and subsequent quaternization with methyl bromide give ion-containing membranes that demonstrate significant salt exclusion and are potentially useful as reverse-osmosis membranes (102).

The energetic polymers were specifically designed to increase the energy content of explosives and solid fuels that use polymeric binders (55). These are comparatively new materials, and are still being evaluated and compared with similar materials from less expensive starting materials. It is possible that this application is a potential growth market for polyoxetanes.

BIBLIOGRAPHY

"Oxetane Polymers" in *EPST* 1st ed., Vol. 9, pp. 668–701, by S. Okamura, H. Miyake, and N. Shimazaki, Kyoto University; pp. 702–708, by H. Boardman, Hercules, Inc.

1. E. J. Goethals, *Ind. Chim. Belge.* **30,** 559 (1965).
2. P. Luger and J. Buschman, *J. Am. Chem. Soc.* **106,** 7118 (1984).
3. K. Baum, P. J. Berkowitz, V. Grakauskas, and F. G. Archibald, *J. Org. Chem.* **48,** 2953 (1983).
4. M. B. Frankel, E. R. Wilson, D. O. Woolery, C. L. Hamermesh, and C. McArthur, *JANNAF Propulsion Committee Meeting No. AD-A103 844, CPIA Publication No. 340,* Chemical Propulsion Information Agency, Johns Hopkins University, Baltimore, Md., 1981, p. 39.

5. Ger. Pat. 741,478 (June 21, 1939), H. Meerwein (to I. G. Farbenindustrie AG).
6. H. Meerwein, D. Delfs, and H. Morshel, *Angew. Chem.* **72**, 927 (1960).
7. A. C. Farthing and R. J. W. Reynolds, *J. Polym. Sci.* **12**, 503 (1954).
8. A. C. Farthing, *J. Chem. Soc.*, 3648 (1955).
9. J. B. Rose, *J. Chem. Soc.*, 542 (1956).
10. *Ibid.*, p. 546.
11. R. S. Miller, P. A. Miller, T. N. Hall, and R. Reed, Jr., *Off. Nav. Res. U.S. Rev.*, 21 (Spring 1981).
12. M. P. Dreyfuss and P. Dreyfuss, *J. Polym. Sci. Part A-1*, **4**, 2179 (1966).
13. S. Inoue and T. Aida in K. J. Ivin and T. Saegusa, eds., *Ring-Opening Polymerization*, Vol. 1, Elsevier Science Publishing Co., Inc., New York, 1984, Chapt. 4, p. 193.
14. P. Dreyfuss, *Poly(tetrahydrofuran)*, Gordon & Breach, New York, 1982.
15. S. Searles, Jr., R. G. Nickerson, and W. K. Witsiepe, *J. Org. Chem.* **24**, 1839 (1960); S. Searles, D. G. Hummel, S. Nukina, and P. E. Throckmorton, *J. Am. Chem. Soc.* **82**, 2928 (1960).
16. E. J. Vandenberg and A. E. Robinson in E. J. Vandenberg, ed., *Polyethers, ACS Symp. Ser.* **6**, American Chemical Society, Washington, D.C., 1975, pp. 101–119.
17. T. Hiramo, S. Nakayama, and T. Tsuruta, *Makromol. Chem.* **176**, 1897 (1975).
18. N. L. Sidney, S. E. Shaffer, and W. J. Bailey, *Polym. Prepr. Am. Chem. Soc. Div. Polym. Chem.* **22**(2), 373 (1981).
19. E. J. Vandenberg, *J. Polym. Sci. Polym. Chem. Ed.* **23**, 915 (1985).
20. J. Kops, S. Hvilsted, and H. Spanggaard, *Macromolecules* **13**, 1058 (1980).
21. R. Hoene and K.-H. W. Reichert, *Makromol. Chem.* **177**, 3545 (1976).
22. H. Meerwein, *Org. Synth.* **46**, 113 (1966).
23. U.S. Pat. 3,585,227 (June 15, 1971), M. P. Dreyfuss (to B. F. Goodrich).
24. Ref. 13, pp. 27–29.
25. E. Franta, L. Reibel, J. Lehmann, and S. Penczek, *J. Polym. Sci. Polym. Symp.* **56**, 139 (1976).
26. T. Saegusa and S. Kobayashi in Ref. 16, pp. 150–168.
27. N. Oguni and Y. Hyoda, *Macromolecules* **13**, 1687 (1980).
28. P. Dreyfuss and M. P. Dreyfuss in K. C. Frisch and S. L. Reegen, eds., *Ring Opening Polymerizations*, Marcel Dekker, Inc., New York, 1969, Chapt. 2, pp. 111–158.
29. T. Saegusa and S. Matsumoto, *J. Polym. Sci. Part A-1* **6**, 1559 (1968).
30. K. Brzezinska, W. Chwialkowska, P. Kubisa, M. Matyjaszewski, and S. Penczek, *Makromol. Chem.* **178**, 2491 (1977).
31. E. J. Vandenberg, *J. Polym. Sci. Part A-1* **7**, 525 (1969).
32. J. Kops and H. Spanggaard, *Macromolecules* **15**, 1200 (1982).
33. F. Andruzzi, P. Cerrai, G. Guerra, L. Nucci, A. Pescia, and M. Tricoli, *Eur. Polym. J.* **18**, 685 (1982).
34. K. Hayashi and S. Okamura, *Makromol. Chem.* **47**, 230 (1961).
35. K. Takakura, K. Hayashi, and S. Okamura, *J. Polym. Sci. Part A-1* **4**, 1747 (1966).
36. J. V. Crivello and J. H. W. Lam, *Macromolecules* **10**, 1307 (1977).
37. P. E. Black and D. J. Worsfold, *Can. J. Chem.* **54**, 3325 (1976).
38. A. Bello, E. Perez, and J. M. G. Fatou, *Makromol. Chem.* **185**, 249 (1984).
39. T. Saegusa and S. Kobayashi, *Prog. Polym. Sci. Jpn.* **6**, 107 (1973).
40. E. Riande, J. G. De la Campa, J. Guzman, and J. De Abajo, *Macromolecules* **17**, 1431 (1984).
41. Y. Arimatsu, *J. Polym. Sci. Part A-1* **4**, 728 (1966).
42. P. Dreyfuss and M. P. Dreyfuss, *Polym. J.* **8**, 81 (1976).
43. M. Bucquoye and E. J. Goethals, *Makromol. Chem.* **179**, 1681 (1978).
44. E. Riande, J. G. De la Campa, J. Guzman, and J. De Abajo, *Macromolecules* **17**, 1891 (1984).
45. J. Kops and H. Spanggaard, *Polym. Prepr. Am. Chem. Soc. Div. Polym Chem.* **26**(1), 52 (1985).
46. A. Leborgne, N. Spassky, and J. Kops in E. J. Goethals, ed., *Cationic Polymerization and Related Processes, Proceedings of the 6th International Cationic Symposium*, Ghent, Belgium, Academic Press, Inc., London, 1984, pp. 227–236.
47. J. Kops and H. Spanggaard in Ref. 46, pp. 219–226.
48. C. C. Price, R. Spectro, and A. C. Tumolo, *J. Polym. Sci. Part A-1* **5**, 407 (1967).
49. J. Kops and H. Spanggaard, *Macromolecules* **15**, 1225 (1982).
50. Jpn. Kokai Tokkyo Koho 58/125718 [83/125,718] (July 26, 1983) (to Daicel Chemical Industries, Ltd.).

51. Ger. Offen. 3,326,178 (Oct. 18, 1984), Y. Toga, I. Okamoto, and T. Kanno (to Daicel Chemical Industries, Ltd.).
52. Ref. 13, pp. 195–205.
53. R. S. Miller, P. A. Miller, T. N. Hall, and R. Reed, Jr., *Off. Nav. Res. U.S. Res. Rev.*, 21 (Spring 1981).
54. U.S. Pat. 4,393,199 (July 12, 1983), G. E. Manser (to SRI International).
55. *Ind. Chem. News,* cover (November 1981).
56. E. J. Vandenberg, *J. Polym. Sci. Polym. Chem. Ed.* **10,** 2887 (1972).
57. S. V. Conjeevaram, R. S. Benson, and D. J. Lyman, *J. Polym. Sci. Polym. Chem. Ed.* **23,** 429 (1985).
58. Ref. 13, pp. 207–211.
59. K. E. Hardenstine, C. J. Murphy, R. B. Jones, L. H. Sperling, and G. E. Manser, *J. Appl. Polym. Sci.* **30,** 2051 (1985).
60. T. Urzu, K. Hatanaka, and K. Matsuzaki, *Makromol. Chem.* **181,** 2137 (1980).
61. G. E. Manser, R. W. Fletcher, and M. R. Knight, *Off. Nav. Res. U.S. Res. Rep. No. N00014-82-0800* (June 1985).
62. D. Tirrell, O. Vogl, T. Saegusa, S. Kobayashi, and T. Kobayashi, *Macromolecules* **13,** 1041 (1980).
63. J. Muggee and O. Vogl, *J. Polym. Sci. Polym. Chem. Ed.* **23,** 649 (1985).
64. D. A. Bansleben and O. Vogl, *J. Polym. Sci. Polym. Chem. Ed.* **23,** 673 (1985).
65. L. Garrido, E. Riande, and J. Guzman, *Makromol. Chem. Rapid Commun.* **4,** 725 (1983).
66. L. Garrido, J. Guzman, E. Riande, and J. de Abajo, *J. Polym. Sci. Polym. Chem. Ed.* **20,** 3377 (1982).
67. M. Bucquoye and E. J. Goethals, *Eur. Polym. J.* **14,** 323 (1978).
68. J. Kops and H. Spanggaard, *Macromolecules* **16,** 1544 (1983).
69. A. Baba, H. Meishou, and H. Matsuda, *Makromol. Chem. Rapid Commun.* **5,** 665 (1984).
70. P. Geil, *Polymer Single Crystals,* Wiley-Interscience, New York, 1963, p. 526.
71. R. L. Miller in J. Brandrup and E. G. Immergut, eds., *Polymer Handbook,* 2nd ed., John Wiley & Sons, Inc., New York, 1975, Chapt. III, p. 1.
72. Y. Takahashi, Y. Osaki, and H. Tadokoro, *J. Polym. Sci. Polym. Phys. Ed.* **18,** 1863 (1980).
73. *Ibid.,* **19,** 1153 (1981).
74. R. Gilardi, C. George, and J. Karle, *Off. Nav. Res. U.S. Res. Rep., No. LSM 81-1* (Oct. 15, 1981).
75. B. Moss and D. L. Dorset, *J. Polym. Sci. Polym. Phys. Ed.* **20,** 1789 (1982).
76. E. Perez, M. A. Gomez, A. Bello, and J. G. Fatou, *Colloid Polym. Sci.* **261,** 571 (1983).
77. K. E. Hardenstine, G. V. Henderson, L. H. Sperling, and C. J. Murphy, *Gov. Rep. Announce. U.S. No. AD-A148271/0/GAR* **85,** 172 (1985).
78. E. Perez, A. Bello, and J. G. Fatou, *An. Quim. Ser. A* **80,** 509 (1984); *Chem. Abstr.* **102,** 176755p (1985).
79. E. Perez, A. Bello, and J. G. Fatou, *Colloid Polym. Sci.* **262,** 605 (1984).
80. *Ibid.,* p. 913.
81. E. Perez, A. Bello, and J. G. Fatou, *Makromol. Chem.* **186,** 439 (1985).
82. K. D. Pae, C. L. Tang, and E. S. Shin, *J. Appl. Phys.* **56,** 2426 (1984).
83. L. Garrido, E. Riande, and J. Guzman, *J. Polym. Sci. Polym. Phys. Ed.* **21,** 1493 (1983).
84. E. Perez, M. A. Gomez, A. Bello, and J. G. Fatou, *J. Appl. Polym. Sci.* **27,** 3721 (1982).
85. U. Gaur and B. Wunderlich, *Polym. Prepr. Am. Chem. Soc. Div. Polym. Chem.* **20**(2), 429 (1979).
86. U. Gaur and B. Wunderlich, *J. Phys. Chem. Ref. Data* **10,** 1001 (1981).
87. R. B. Jones, C. J. Murphy, L. H. Sperling, M. Farber, S. P. Harris, and G. E. Manser, *J. Appl. Polym. Sci.* **30,** 95 (1985).
88. D. C. Miles and J. H. Briston, *Polymer Technology,* Chemical Publishing Co., Inc., New York, 1979.
89. P. Dreyfuss and M. P. Dreyfuss in M. Grayson, ed., *Kirk-Othmer Encyclopedia of Chemical Technology,* 3rd ed., Vol. 18, John Wiley & Sons, Inc., New York, 1982, pp. 645–670.
90. D. J. H. Sandiford, *J. Appl. Chem.* **8,** 188 (1958).
91. E. W. Cronin, *Mod. Plast.* **34**(6), 150 (1957).
92. R. T. Keen, *Anal. Chem.* **29,** 1041 (1957).
93. N. Irving Sax, *Dangerous Properties of Industrial Materials,* 4th ed., Van Nostrand Reinhold Co., Inc., New York, 1975, p. 1204.
94. *Ibid.,* p. 456.

95. U.S. Pat. 4,483,978 (Nov. 20, 1984), G. E. Manser (to SRI International).
96. G. E. Manser, *Nitrate Ester Polyether Glycol Prepolymers, JANNAF Propulsion Committee Meeting,* New Orleans, La., Chemical Propulsion Information Agency, Johns Hopkins University, Baltimore, Md., 1984.
97. E. N. Sokolov and A. V. Rogachev, *Vestsi Akad. Navuk B SSR Ser. Fiz. Tekh. Navuk* (2), 54 (1982).
98. N. F. Shumskii and V. I. Galkina, *Deposited Document of the Solid Propellant Rocket Static Test Panel* No. 15 Khp-D82, SPSTL, 1982.
99. Ger. Offen. 3,326,178 (Oct. 18, 1984), Y. Toga, I. Okamoto, and T. Kanno; Jpn. Pat. 83/64,801 (Apr. 13, 1983) (to Daicel Chemical Industries, Ltd.).
100. V. A. Belyi, I. M. Vertyachikh, Yu. I. Voronezhtsev, V. A. Gol'dade, and L. S. Pinchuk, *Dokl. Akad. Nauk SSSR* **275,** 639 (1984); *Chem. Abstr.* **101,** 73826n (1984).
101. N. I. Egorenkov, A. I. Kuzavkov, V. V. Evmenov, and D. G. Lin, *J. Therm. Anal.* **24,** 9 (1982).
102. E. Bittencourt, V. Stannett, J. L. Williams, and H. B. Hopfenberg, *J. Appl. Polym. Sci.* **26,** 879 (1981).

M. P. Dreyfuss
P. Dreyfuss
Michigan Molecular Institute

OXIDATION. See Antioxidants; Antiozonants; Degradation; Radiation-induced Reactions.

OXIDATION OF POLYMERS. See Degradation.

OXIDATIVE POLYMERIZATION

The polymerization of monomers by oxidative means has been known for many years. Probably the most familiar example of a commercial polymer prepared by this technique is poly(phenylene oxide) (PPO), originally synthesized at General Electric in 1959 (1).

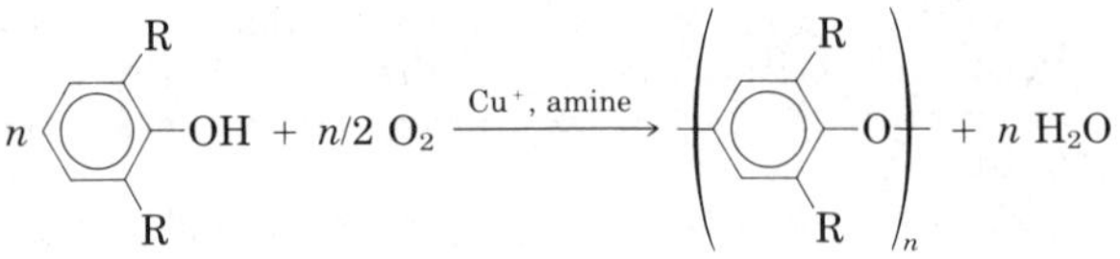

Oxidative polymerization has also been carried out successfully with aromatic amines, thiols, dialkynes, homocyclic and heterocyclic aromatic nuclei, and other miscellaneous monomers. This article focuses principally on dialkynes and aromatic moieties, with only a short discussion of other monomers. Poly(phenylene oxide)s are thoroughly discussed in a separate article (see Poly(phenylene ether)).

Dialkynes

In 1869, it was discovered that the cuprous derivative of phenylacetylene undergoes a smooth coupling in air to afford 1,4-diphenyl-1,3-butadiyne (2). Subsequently, isolation of the cuprous salt was found to be unnecessary.

In attempts to prepare polymers of acetylene via this coupling in acidic solution, however, the products vinylacetylene, divinylacetylene, and 1,5,7-octatriene-3-yne were isolated (3). There were reports of the use of cupric acetate in pyridine for coupling of terminal diynes to produce macrocycles containing α,γ-diyne units (4,5). Under different conditions (bubbling oxygen through a dioxane solution of cuprous chloride and ammonium chloride at 70°C), coupling of di-3-butynyl-sebacate produced a material, presumably short-chain polymer, that was not investigated further (4). This appears to be the first mention of a true polymeric substance obtained by oxidative polymerization of a diyne (see also DIACETYLENE POLYMERS).

Treatment of *m*-diethynylbenzene with cuprous chloride and oxygen in pyridine (6) or in tetramethylethylenediamine (TMEDA)–isopropyl alcohol (7) at room temperature yields a polymer soluble in hot chlorobenzene or nitrobenzene. The molecular weight of the polymer, as determined from end-group analysis, is ca 7000.

If rapidly heated in bulk to ca 200°C in air, nitrogen, or vacuum, it explosively evolves hydrogen and is converted to carbon. However, when the heating rate is slow, the polymer is converted to a material thermally stable to ~550°C, probably via an intramolecular cross-linking reaction to produce a polyaromatic structure (8). Under similar reaction conditions, *p*-diethynylbenzene affords a bright yellow, insoluble substance which decomposes at ca 100°C (6).

A wide variety of terminal diynes have been transformed successfully into polymers under oxidative conditions (Table 1). Copolymerization has also been carried out. For example, mixed coupling of *p*-diethynylbenzene with bis-(dimethylethynylmethoxy)methane, in the presence of cuprous chloride at 30°C, forms soluble and insoluble copolymer fractions (22). Increased relative amounts of *p*-diethynylbenzene in copolymers decreases the amount and lowers the molecular weight of the soluble fraction.

Low molecular weight individual oligomers in the series $R(C\equiv C)_n R'$ (where $n = 4$–10, 12; R,R′ = H, C_6H_5, t-C_4H_9, mesityl) have been prepared via coupling of triethylsilyl-protected terminal diynes, followed by alkaline desilylation (23–25).

Mechanism of Coupling.　The mechanism of oxidative coupling of alkynes remains unclear. The active catalyst appears to be an amine complex of a copper salt (7). Although best results have been obtained with cuprous salts, under coupling conditions, cuprous ion should be rapidly oxidized to cupric ion (26). A

Table 1. Oxidative Polymerization of Terminal Diynes[a]

R in HC≡C—R—C≡CH	Refs.
m-C$_6$H$_4$	6,7,9
p-C$_6$H$_4$	6,9
p-C$_6$H$_4$—C≡C—As—C≡C—C$_6$H$_4$-p (with C$_6$H$_5$ substituent)	9
p-CH$_2$O—C$_6$H$_4$—C(CH$_3$)(CH$_3$)—C$_6$H$_4$—OCH$_2$-p	10
p-C$_6$H$_4$—C(CH$_3$)(C$_2$H$_5$)—C$_6$H$_4$-p—	11
R′O—⟨⟩—C(CH$_3$)(CH$_3$)—⟨⟩—OR′	(R′ = C$_2$H$_5$) 12 (R′ = CH$_3$) 13
(quinoxaline structure)	14
—(CH$_2$)$_5$—, —(CH$_2$)$_6$—, —(CH$_2$)$_8$—	15[b]
—(CH$_2$)$_3$—, —(CH$_2$)$_4$—, —(CH$_2$)$_5$	16
—CH$_2$—CH(COOC$_6$H$_5$)—CH$_2$—	17
—(C(CH$_3$)(CH$_3$)—OCH$_2$—CHOH—CH$_2$)$_2$NCH$_3$	18
—C(CH$_3$)(CH$_3$)—OCH$_2$—CHOH—CH$_2$OCH$_2$—	19
—Si(C$_6$H$_5$)$_2$—, —Ge(C$_6$H$_5$)$_2$—	20
—Si(CH$_3$)$_2$OSi(CH$_3$)$_2$—	21

[a] Cu$_2$Cl$_2$ catalyst; O$_2$; pyridine, TMEDA, or mixtures as solvent.
[b] CuCl$_2$ catalyst.

three-step mechanism has been proposed for the reaction (27). However, this mechanism does not account for the observed catalysis by cuprous ion in basic media (28).

$$RC≡CH \xrightarrow{\text{base}} RC≡C^-$$

$$RC≡C^- + Cu^{2+} \longrightarrow RC≡C\cdot + Cu^+$$

$$2\,RC≡C\cdot \longrightarrow RC≡C—C≡CR$$

This effect may be attributed to formation of an alkyne–Cu$^+$ complex which facilitates ionization of the terminal alkyne (28).

$$RC{\equiv}CH + Cu^+ \longrightarrow RC{\equiv}CH$$
$$\underset{\displaystyle Cu^+}{|}$$

Uses. Diyne polymers have not attained the commercial status of poly(phenylene oxide)s. A facile cross-linking reaction brought about thermally (8,14,17,19,29) or photochemically (10,15,29) provides the main interest. This reaction may result in the formation of a network containing butatriene moieties (29), which then rearrange to give conjugated ene–yne chains (14,15,30).

Polydiynes that contain aromatic rings appear to react further under cross-linking conditions to yield graphiticlike regions of fused aromatic nuclei (8,14,31). The cross-linking is thought to be radical in nature, since it is inhibited by radical scavengers (29).

The cross-linked polymers are thermally stable and, because of the extensive conjugated unsaturated system, have been suggested for use as semiconductors and photoconductors. Copolymers of *m*- and *p*-diethynylbenzene are suitable precursors for carbon fibers (qv) (31).

Aromatic Nuclei

Benzene. Two decades ago, the facile conversion of benzene to an insoluble brown powder by treatment with aluminum chloride–cupric chloride at mild temperatures was reported (32,33).

The structure of the polymerization product was assigned that of poly(*p*-phenylene) on the basis of elemental analysis, ir spectrum, x-ray diffraction pattern, pyrolysis products, oxidative degradation, insolubility, color, and thermal stability. Polyphenylenes have been prepared by several other routes, including coupling of aryl halides via the Ullmann and Fittig reactions, free-radical arylations of aromatic substrates, and aromatization of cycloaliphatic precursors (34). Significant alternative routes have been reported (35–38) and polyphenylenes have been reviewed (34,39,40).

Polymerization Reagents and Conditions. Catalyst–oxidant systems for the conversion of benzene to polyphenylenes include aluminum chloride–cupric chloride (32,33,41), aluminum chloride–manganese dioxide (41), aluminum chloride–chloranil (41), aluminum chloride–lead dioxide (41), aluminum chloride–nitrogen dioxide (42), aluminum bromide–cupric chloride (41), antimony pentachloride–cupric chloride (41), ferric chloride (43–45), molybdenum pentachloride (46), and arsenic pentafluoride (47). The latter three function as both catalyst and oxidant. In addition, the polymerization of benzene may be effected by oxidative electrochemical techniques (48–52).

The most efficient and most widely used catalyst–oxidant combination is aluminum chloride–cupric chloride. Typically, the resulting polymer contains <2.5% Cl and has a C:H atomic ratio of ca 1.5. Substantial halogen incorporation (>10%) is obtained for the polymers prepared from $AlCl_3$ as catalyst and MnO_2, PbO_2, or chloranil as oxidant (41), whereas with NO_2 as oxidant, 1–4% nitrogen is incorporated in the product (42). The polymer from ferric chloride treatment of benzene is more highly colored and contains more chlorine than that from exposure to $AlCl_3$–$CuCl_2$. Molybdenum pentachloride produces a material similar to that derived from $AlCl_3$–$CuCl_2$.

Physical Properties and Structural Features. Poly(*p*-phenylene) (PPP) prepared from benzene–$AlCl_3$–$CuCl_2$ is a brown powder which is insoluble in all common solvents and does not possess a measurable melting or softening point (39). Subsequent discussion in this section refers to this material; it is called oxidatively coupled polymer. Investigations of the morphology by x-ray diffraction (51,52), neutron diffraction (53,54), and electron microscopy (55,56) reveal that the powder is partially crystalline and that crystallinity increases upon annealing. The material is composed of fairly short, entangled fibers of irregular widths, parts of which seem flat.

The structure of poly(*p*-phenylene) has been investigated by a number of techniques. Infrared spectral data show a dominant absorption at 803–805 cm^{-1}, indicating para substitution (33,57), with much smaller absorptions at ca 765 and 695 cm^{-1}, characteristic of monosubstituted phenyl rings (arising from end groups). Pyrolysis yields lower molecular weight oligomers in the *p*-phenylene series: terphenyl, quaterphenyl, and quinquephenyl (33). Chromic acid oxidation produces terephthalic acid and 4,4′-diphenyldicarboxylic acid (33). No isomeric materials suggestive of ortho or meta linkages have been found among the pyrolysis or oxidation products. Likewise, cross-polarization, magic-angle sample spinning ^{13}C nmr spectral data support the para-linked structure (58–62). The oxidatively coupled polymer exhibits a ^{13}C nmr spectrum virtually identical to that of poly(*p*-phenylene) prepared from organometallic coupling of 1,4-dibromobenzene, which can yield only para-linked polymer (59).

The observed insolubility of the polymer might suggest cross-linked chains (63), but consideration of the trend of decreasing solubility in the homologous *p*-phenylene series suggests that this structural feature need not be invoked. The most convincing evidence against cross-links is the solubilization of the polymer by Friedel-Crafts alkylation (64). Presumably, the alkyl groups prevent interchain π–π stacking arrangements.

The presence of quinoidal (65,66) and/or polynuclear regions (65,67,68) has been suggested to account for the dark color of the polymer (compared to sexiphenyl, which is white) and the observed esr signal. However, little direct evidence for these structural features was available until recently. Laser desorption–Fourier transform mass spectrometry (ld–ftms) data for PPP clearly indicate the presence of oligomer populations in which the phenylene chains are devoid of two hydrogens (69,70). Formation of polynuclear regions best accounts for this observation.

The extreme insolubility of poly(p-phenylene) has, until recently, precluded a direct determination of its molecular weight. Ld–ftms results indicate an average chain length of 13–14 phenylene units, with a polydispersity ratio of 1.12 (69), in good agreement with previous determinations based on vapor-pressure osmometric data for soluble alkylated PPP (64), which has a degree of polymerization of 15. Quantitative analysis of the ir spectrum of polymer prepared by oxidative coupling reveals a chain length of 12 (68). Thus this material is oligomeric, rather than truly polymeric.

Mechanism of Polymerization. The initiating species in the polymerization of benzene with aluminum chloride–cupric chloride is apparently a benzene radical cation (71,72) formed by one-electron oxidation of benzene. Propagation of the chain was earlier thought to occur via cationic (71) or radical (72) means. More recently, a novel pathway, labeled the stair-step mechanism, has been suggested (73–75).

The benzene radical cation apparently serves as the site for coordination with an appreciable number of benzene nuclei in a propagative manner. The authors' concept of this system is that the nuclei are parallel to each other, but not necessarily arranged in a highly symmetrical fashion. The radical cation is delocalized throughout the entire associated complex. When the complex incorporates a certain number of benzene nuclei, the radical cation character on the terminal rings is too small to induce further propagation, leading to subsequent covalent bond formation. The nonconjugated diene units in the resulting chain are expected to undergo isomerization in the acidic system. Facile oxidative aromatization by cupric chloride completes the reaction to give the final polymer.

This mechanistic view is strengthened by work involving solid-state synthesis of PPP from reaction of arsenic pentafluoride with *p*-terphenyl, *p*-quater-phenyl, *p*-quinquephenyl, and *p*-sexiphenyl (76,77). Close examination of the *p*-terphenyl unit cell suggests that reaction between molecules related by the *c*-axis translation is likely to be involved in the propagation steps. The precursor monomers thus are envisaged to be parallel to each other with their end carbons in close proximity.

Other Homocyclic Aromatic Hydrocarbons. Several other homocyclic aromatic molecules have been converted to polymeric materials via oxidative polymerization. Toluene (78,79) yields a polymer with a poly(*o*-phenylene) backbone and possibly some para structure upon treatment with $AlCl_3$–$CuCl_2$.

Polymerization of toluene in carbon disulfide solvent gives insoluble, infusible light-brown powder. However, in a neat system, a dark purple-brown solid was obtained with a melting point of ca 250°C and a DP of ca 7. Because of the observed color (78) and esr signal (79), the presence of polynuclear regions in the polymers has been postulated. Supporting evidence is the observation that *o*-terphenyl forms triphenylene upon treatment with aluminum chloride (80).

Biphenyl and terphenyl mainly yield sexiphenyl upon treatment with $AlCl_3$–$CuCl_2$ in *o*-dichlorobenzene at temperatures <80°C (81,82). At higher temperatures (80–155°C), soluble, processible polymers with *m*- and *p*-phenylene linkages are produced (82). Likewise, fusible polyphenylenes have been obtained from oxidative polymerization of *m*- or *p*-terphenyl in the presence of $AlCl_3$–$CuCl_2$ at 85–180°C (83). Biphenyl, *p*-terphenyl, and *p*-quaterphenyl yield conductive charge-transfer complexes (qv) of poly(*p*-phenylene) upon monomer exposure to vapors of AsF_5 (76). Anodic oxidation of biphenyl on platinum electrodes in liquid SO_2 results in formation of passive films (quaternary ammonium perchlorate as electrolyte) or conductive dendritelike deposits (quaternary ammonium tetrafluoroborate as electrolyte) (49).

Treatment of naphthalene with aluminum chloride–cupric chloride or ferric chloride–water yields a product mixture containing binaphthyls, ternaphthyls, and higher molecular weight material (79,84). The benzene-soluble fractions are comprised of chains of 3–6 naphthalene nuclei. Similar chain lengths have been obtained from reaction with aluminum chloride–nitromethane (85). The dark purple color and the concentration of radicals (esr) in the oligomers is attributed to polynuclear and/or quinoidal moieties in the chains (79,84). Interestingly, vigorous milling of naphthalene with anhydrous $MgCl_2$ produces a purple complex with a spin density of ca 10^{17}–10^{19} spins/g (86), comparable to that of the above oligomers. Naphthalene may also be oligomerized by the action of aluminum chloride alone (87). The degree of polymerization of the resulting material (ca 6) is similar to that obtained with $AlCl_3$–$CuCl_2$. However, without the oxidant, the oligomer is comprised of 1,4- or 1,2-linked dihydro units.

Fluorene (88), phenanthrene (89), and pyrene (89) principally convert to dehydrodimers and trimers when treated at room temperature with aluminum chloride–cupric chloride.

A soluble polymer with M_n of 4800 can be obtained from polymerization of *trans*-stilbene with AlCl₃–CuCl₂ in carbon disulfide solvent (90). Infrared and ¹³C and proton nmr spectra are consistent with a structure comprised of 1,4-linked phenanthrene nuclei.

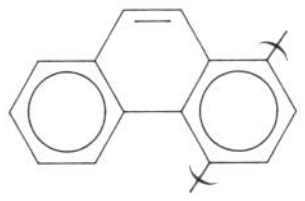

Apparently, isomerization to a cis structure and cyclization accompany the propagation.

Aryl Halides. Chlorobenzene (79,91) and fluorobenzene (91) produce oligomers with backbones comprised of *o*-phenylene units.

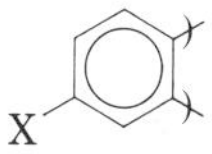

where X = Cl, F

The chlorobenzene product is red with an average DP of 10–12. The coupled product from fluorobenzene is similar. However, evidence from oxidative degradation reveals that propagation occurs to some extent by attack ortho to fluorine. 1-Chloronaphthalene is converted to oligomeric material by treatment with either FeCl₃ or AlCl₃–CuCl₂ (79).

Phenols, Thiols, Ethers, and Nitriles. Phenol is not polymerized by aluminum chloride–cupric chloride (92), presumably because of deactivation of the AlCl₃ by the phenolic hydroxyl. However, FeCl₃ successfully couples phenol, catechol, and hydroquinone (92). The phenol polymer had a molecular weight of ~1400 and spectral data (ir and nmr) were consistent with the presence of both phenolic and carbonyl groups. 2-Naphthol was coupled with aluminum chloride–nitromethane to give dimers, trimers, and polymer of narrow molecular weight distribution (93).

Dehydropolycondensation of thiophenol with MoCl₅ or WCl₆ results in the formation of a polyphenylene with retention of the SH groups (94). The powdery product softens at low temperatures and dissolves in organic solvents. Poly(*m*-phenylene disulfide) has been synthesized by oxidation of 1,3-dimercaptobenzene with dimethyl sulfoxide (95). The mild oxidant permits polymerization to occur through the thiol groups rather than through the aromatic ring.

High molecular weight linear polymers have been obtained through oxidation of di-1-naphthoxyalkanes with FeCl₃ in nitrobenzene (96).

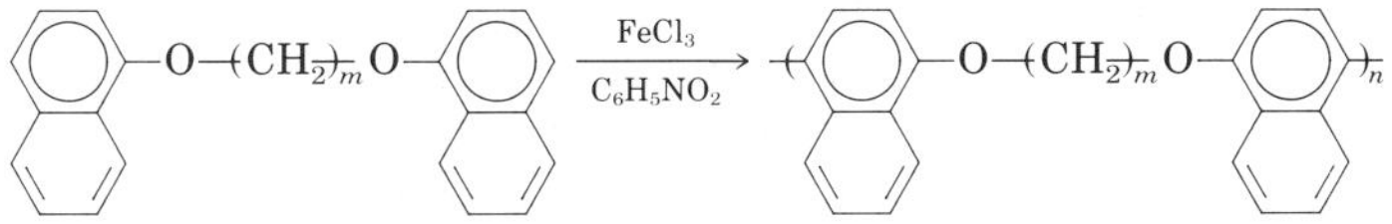

where *m* = 2–7

1,4-Dimethoxybenzene is converted to low molecular weight poly(2,5-dimethoxy-1,4-phenylene) by oxidative polymerization with AlCl₃–CuCl₂ in nitrobenzene (97). The linear polymer is soluble in concentrated sulfuric acid and fusible at ca 320°C. Similarly, high molecular weight poly(dialkoxyphenylene)s are synthesized from corresponding low molecular weight oligomers by treatment with $FeCl_3$ in nitropropane (98).

Bifunctional *p*-tolylcyanoacetic esters are oxidatively coupled with $O_2/$ $[Cu(OH)TMEDA]_2Cl_2$ in $CHCl_3$ to yield high molecular weight polymers (99). Monofunctional phenylcyanoacetic esters, under similar conditions, give oligomeric species with benzylic carbon to *p*-aromatic carbon coupling (100).

where $n = 60$–190

Amines.   Diphenylamine has been converted to a polyamine resin by ferric chloride (101,102), aluminum chloride–cupric chloride (102), or, at 200–260°C, by molecular oxygen (103). The ferric chloride product is soluble in some organic solvents and has a softening point of 80–100°C. From ir data, polymerization involves coupling of the aromatic rings; the amine groups are not involved. Polymer properties are critically dependent on reaction conditions. Physical properties of the polymer obtained from treatment with $AlCl_3$–$CuCl_2$ are not as dependent on the reaction conditions. According to spectral results (ir, uv, nmr, and esr), the polydiphenylamine from thermal oxygen oxidation contains both *p*-linked diphenylamine moieties and conjugated imino groups. The condensation is thought to proceed via H-atom abstraction from the amino groups and para positions of the diphenylamine by oxygen and radicals formed during the process. The polymer has high thermal and thermooxidative stability.

Aniline is oxidized to polyaniline (aniline black) by electrochemical (104–106) or chemical means (107–109). The polymer consists of alternating quinone–diimine and phenylene moieties.

The physical nature of the electrochemically synthesized polyaniline is dependent on the technique employed (104). Electrolysis of aniline at constant potential results in formation of an electrically passive powder. However, by cycling the potential between -0.2 and $+0.8$ V, an electroactive film of polyaniline, which strongly adheres to the electrode surface, has been prepared.

Heterocyclic Aromatic Molecules.   Relatively few reports describe the direct coupling of heteroaromatic substrates by chemical means. Dehydrocoupling of thiophene by $MoCl_5$ or WCl_6 gave products in which the original ring structure was preserved (110,111). Chlorinated monomers and dithienyl were also studied. Thiophene and 2,2′-bithienyl are converted to conductive poly(2,5-thiophenediyl) upon treatment with AsF_5 (112) or nitrosonium salts (NOSbF_6, NOPF_6, NOBF_4) (113). In a fashion analogous to the synthesis of conductive PPP from biphenyl–AsF_5 (76), the sulfur heterocycles undergo one-electron oxidation to form a radical cation, which then propagates to give the resulting polymer.

Coupling of benzothiophene is accomplished with molybdenum pentachloride or tungsten chloride (111,114). A 55–65% yield of oligomer is obtained at 343–363 K. The basic structure of the monomer is retained in the repeating units. The product is soluble in organic solvents, softens at 368–393 K, and undergoes degradation at 613 K. Oxidation of tetrahydrocarbazole by ferric chloride gives an unsymmetrical dehydrodimer product (115).

Recently, in connection with an increasing interest in conducting organic polymers, many studies of electrochemical coupling of heteroaromatic nuclei have been reported. For example, pyrrole (116–120), N-alkylpyrroles (121), thiophene (120,122,123), and furan (120) all undergo anodic oxidation to yield polymers in which the heteroaromatic nuclei are linked at the 2 and 5 positions.

where Z = NH, S or O

Polypyrrole is the most thoroughly studied product obtained from this process (119). Initially, pyrrole is oxidized to the corresponding radical cation at the electrode surface. Polymerization is thought to proceed on the surface via coupling of the radical cations at the 2 and 5 positions. Further one-electron oxidation of each repeating unit aromatizes the linear polymer. Thus two electrons per pyrrole ring must be lost for preparation of polymer. Non-nucleophilic electrolytes must be employed to avoid interference with the polymerization reaction. The material, which is conductive as formed, is obtained as a thin film typically comprised of 70% pyrrole polymer and 30% electrolyte anion by weight. With tetrafluoroborate as the anion, this weight ratio corresponds to three pyrrole moieties per anion and suggests a cationic charge of approximately 0.3 per pyrrole ring. This charge is the result of partial oxidation of the linear polymer (see also POLYPYRROLES).

Carbazole forms a film of poly(3,6-carbazolediyl) (122).

Elemental analysis of the polymer reveals that it is rich in hydrogen; thus reduced ring moieties are likely to be present. Isothianaphthene polymerizes to give polyisothionaphthene, provided the electrolyte contains nucleophilic anions such as Cl^- or Br^- (124).

Further details of electrochemical polymerization may be found in a recent review (125).

Uses. There is a wide range of potential commercial applications for PPP. Currently, the most prominent of them is the use of PPP as a conductive polymer (following doping) (63,126,127) (see ELECTRICALLY CONDUCTING POLYMERS). Although the parent polymer is an insulator, exposure to arsenic pentafluoride produces a material with a conductance of >500 S/cm, close to the value of 1200 S/cm for doped polyacetylene (see ACETYLENIC POLYMERS). Doped PPP has been investigated as a possible component of low weight, high charge-density polymer batteries (128,129).

Other uses for the polymer take advantage of the following favorable properties: thermal stability, wear resistance, and low coefficient of friction. These uses have been reviewed previously (39,40).

Polyaniline, which is currently attracting attention, is of interest because of its electrical properties and has been suggested as a possible replacement for MnO_2 in rechargeable flashlight batteries (107). Likewise, the heterocyclic polymers prepared electrochemically are primarily of interest because of their conductive nature (125).

BIBLIOGRAPHY

1. A. S. Hay, H. S. Blanchard, G. F. Endres, and J. W. Eustance, *J. Am. Chem. Soc.* **81,** 6335 (1959).
2. C. Glaser, *Chem. Ber.* **2,** 422 (1869).
3. J. A. Nieuwland, W. S. Calcott, F. B. Downing, and A. S. Carter, *J. Am. Chem. Soc.* **53,** 4197 (1931).
4. G. Eglinton and A. R. Galbraith, *J. Chem. Soc.,* 889 (1959).
5. F. Sondheimer and R. Wolovsky, *J. Am. Chem. Soc.* **84,** 260 (1962).
6. A. S. Hay, *J. Org. Chem.* **25,** 1275 (1960).
7. *Ibid.,* **27,** 3320 (1962).
8. A. E. Newkirk, A. S. Hay, and R. S. McDonald, *J. Polym. Sci. Part A-1* **2,** 2217 (1964).
9. A. S. Hay, *J. Polym. Sci. Part A-1* **7,** 1625 (1969).
10. A. S. Hay, D. A. Bolon, K. R. Leimer, and R. F. Clark, *J. Polym. Sci. Part B* **8,** 97 (1970).
11. I. L. Kotlyarevskii, A. S. Zanina, and N. M. Gusenkova, *Izv. Akad. Nauk SSSR Ser. Khim.,* 900 (1967); *Chem. Abstr.* **67,** 44145 (1967).
12. I. E. Sokolov, A. S. Zanina, and I. L. Kotlyarevskii, *Vysokomol. Soedin. Ser. B* **14,** 311 (1972); *Chem. Abstr.* **77,** 48843 (1972).
13. I. L. Kotlyarevskii, A. S. Zanina, N. M. Gusenkova, I. E. Sokolov, and E. I. Cherepov, *Vysokomol. Soedin. Ser. B* **9,** 468 (1967); *Chem. Abstr.* **67,** 82510 (1967).
14. J. J. Kane and F. E. Arnold, *Org. Coat. Plast. Chem.* **43,** 697 (1980).
15. J. B. Lando and D. R. Day, *Org. Coat. Plast. Chem.* **44,** 245 (1981).
16. U.S. Pat. 3,300,456 (Jan. 24, 1967), A. S. Hay.
17. L. A. Akopyan, E. V. Ovakimyan, and S. G. Matsoyan, *Vysokomol. Soedin. Ser. B* **14,** 752 (1972); *Chem. Abstr.* **78,** 98046 (1973).
18. L. A. Akopyan, S. B. Gevorkyan, I. S. Tsaturyan, and S. G. Matsoyan, *Arm. Khim. Zh.* **26,** 954 (1973); *Chem. Abstr.* **81,** 37864 (1974).
19. L. A. Akopyan, S. B. Gevorkyan, I. S. Tsaturyan, and S. G. Matsoyan, *Vysokomol. Soedin. Ser. B* **15,** 29 (1973); *Chem. Abstr.* **78,** 136722 (1973).
20. N. K. Lebedev, B. V. Lebedev, E. G. Kiparisova, A. M. Sladkov, and N. A. Vasneva, *Dokl. Akad. Nauk SSSR* **246,** 1405 (1979); *Chem. Abstr.* **91,** 158181 (1979).
21. D. R. Parnell and D. P. Macaione, *J. Polym. Sci. Polym. Chem. Ed.* **11,** 1107 (1973).
22. L. A. Akopyan, E. V. Ovakimyan, and S. G. Matsoyan, *Vysokomol. Soedin. Ser. B* **15,** 701 (1973); *Chem. Abstr.* **80,** 83690 (1974).
23. R. Eastmond, T. R. Johnson, and D. R. M. Walton, *Tetrahedron* **28,** 4601 (1972).
24. T. R. Johnson and D. R. M. Walton, *Tetrahedron* **28,** 5221 (1972).
25. E. Kloster-Jensen, *Angew. Chem. Int. Ed. Engl.* **11,** 438 (1972).
26. G. Eglinton and W. McRae, *Adv. Org. Chem.* **4,** 225 (1963).
27. A. L. Klebansky, I. V. Grachev, and M. Kuznetsova, *J. Gen. Chem. USSR* **27,** 3008 (1957).
28. F. Bohlmann, *Angew. Chem.* **69,** 82 (1957).

29. G. Wegner, *Makromol. Chem.* **134,** 219 (1970).
30. D. Bloor, R. J. Kennedy, and D. N. Batchelder, *J. Polym. Sci. Polym. Phys. Ed.* **17,** 1355 (1979).
31. D. M. White, *Polym. Prepr. Am. Chem. Soc. Div. Polym. Chem.* **12**(1), 155 (1971).
32. P. Kovacic and A. Kyriakis, *Tetrahedron Lett.,* 467 (1962).
33. P. Kovacic and A. Kyriakis, *J. Am. Chem. Soc.* **85,** 454 (1963).
34. P. Kovacic and F. Koch in N. M. Bikales, ed., *Encyclopedia of Polymer Science and Technology,* Vol. 11, Wiley-Interscience, New York, 1969, pp. 380–389.
35. T. Yamamoto, Y. Hayashi, and A. Yamamoto, *Bull. Chem. Soc. Jpn.* **51,** 2091 (1978).
36. H. F. VanKerckhoven, Y. K. Gilliams, and J. K. Stille, *Macromolecules* **5,** 541 (1972).
37. C. S. Marvel and G. E. Hartzell, *J. Am. Chem. Soc.* **81,** 448 (1959).
38. D. G. H. Ballard, *J. Chem. Soc. Chem. Commun.,* 954 (1983).
39. J. G. Speight, P. Kovacic, and F. Koch, *J. Macromol. Sci. Rev. Macromol. Chem.* **C5**(2), 295 (1971).
40. P. Kovacic and M. B. Jones, *Chem. Rev.,* **87** (Apr. 1987).
41. P. Kovacic and J. Oziomek, *J. Org. Chem.* **29,** 100 (1964).
42. P. Kovacic and R. J. Hopper, *J. Polym. Sci. Part A-1* **4,** 1445 (1966).
43. P. Kovacic and C. Wu, *J. Polym. Sci.* **47,** 45 (1960).
44. P. Kovacic and F. W. Koch, *J. Org. Chem.* **28,** 1864 (1963).
45. P. Kovacic, F. W. Koch, and C. E. Stephan, *J. Polym. Sci. Part A-1* **2,** 1193 (1964).
46. P. Kovacic and R. M. Lange, *J. Org. Chem.* **28,** 968 (1963).
47. M. Aldissi and R. Liepins, *J. Chem. Soc. Chem. Commun.,* 255 (1984).
48. T. Osa, A. Yildiz, and T. Kuwana, *J. Am. Chem. Soc.* **91,** 3994 (1969).
49. M. Delamare, P.-C. Lacaze, J.-Y. Dumousseau, and J.-E. Dubois, *Electrochim. Acta* **27,** 61 (1982).
50. I. Rubinstein, *J. Polym. Sci. Polym. Chem. Ed.* **21,** 3035 (1983) and references therein.
51. P. Kovacic, M. B. Feldmann, J. P. Kovacic, and J. B. Lando, *J. Appl. Polym. Sci.* **12,** 1735 (1968).
52. G. Froyer, F. Maurice, J. P. Mercier, D. Riviere, M. Le Cun, and P. Auvray, *Polymer* **22,** 992 (1981).
53. H. W. Hasslin and C. Riekel, *Synth. Met.* **5,** 37 (1982).
54. M. Stamm and J. Hocker, *J. Phys. Paris Colloq.* **44,** C3–667 (1983).
55. A. Boudet and P. Pradere, *Synth. Met.* **9,** 491 (1984).
56. P. Pradere, A. Boudet, J.-Y. Goblot, G. Froyer, and F. Maurice, *Mol. Cryst. Liq. Cryst.* **118,** 277 (1985).
57. S. I. Yaniger, D. J. Rose, W. P. McKenna, and E. M. Eyring, *Macromolecules* **17,** 2579 (1984).
58. C. E. Brown, M. B. Jones, and P. Kovacic, *J. Polym. Sci. Polym. Lett. Ed.* **18,** 653 (1980).
59. C. E. Brown, I. Khoury, M. D. Bezoari, and P. Kovacic, *J. Polym. Sci. Polym. Chem. Ed.* **20,** 1697 (1982).
60. F. Barbarin, G. Berthet, J. P. Blanc, C. Fabre, J. P. Germain, M. Hamdi, and H. Robert, *Synth. Met.* **6,** 53 (1983).
61. D. P. Murray, J. J. Dechter, and L. D. Kispert, *J. Polym. Sci. Polym. Lett. Ed.* **22,** 519 (1984).
62. J. B. Miller and C. Dybowski, *Synth. Met.* **6,** 65 (1983).
63. G. Wegner, *Angew. Chem. Int. Ed. Engl.* **20,** 361 (1981).
64. M. B. Jones, P. Kovacic, and D. Lanska, *J. Polym. Sci. Polym. Chem. Ed.* **19,** 89 (1981).
65. P. Kovacic and L-C. Hsu, *J. Polym. Sci. Part A-1* **4,** 5 (1966).
66. S. Krichene, J. P. Buisson, S. Lefrant, G. Froyer, F. Maurice, J. Y. Goblot, Y. Pelous, and C. Fabre, *Mol. Cryst. Liq. Cryst.* **118,** 301 (1985).
67. M. Nechtstein, *These d'Etat,* Grenoble, France, 1966.
68. G. Froyer, F. Maurice, P. Bernier, and P. McAndrew, *Polymer* **23,** 1103 (1982).
69. C. E. Brown, P. Kovacic, C. A. Wilkie, R. B. Cody, Jr., and J. A. Kinsinger, *J. Polym. Sci. Polym. Lett. Ed.* **23,** 453 (1985).
70. C. E. Brown, P. Kovacic, C. A. Wilkie, J. A. Kinsinger, R. E. Hein, S. I. Yaniger, R. B. Cody, Jr., *J. Polym. Sci. Polym. Chem. Ed.* **24,** 255 (1986).
71. G. G. Engstrom and P. Kovacic, *J. Polym. Sci. Polym. Chem. Ed.* **15,** 2453 (1977).
72. E. B. Mano and L. A. Alves, *J. Polym. Sci. Part A-1* **10,** 655 (1972).
73. C.-F. Hsing, I. Khoury, M. D. Bezoari, and P. Kovacic, *J. Polym. Sci. Polym. Chem. Ed.* **20,** 3313 (1982).
74. P. Kovacic and W. B. England, *J. Polym. Sci. Polym. Lett. Ed.* **19,** 359 (1981).
75. S. A. Milosevich, K. Saichek, L. Hinchey, W. B. England, and P. Kovacic, *J. Am. Chem. Soc.* **105,** 1088 (1983).

76. L. W. Shacklette, H. Eckhardt, R. R. Chance, G. G. Miller, D. M. Ivory, and R. H. Baughman, *J. Chem. Phys.* **73,** 4098 (1980).
77. T. Robinson, L. D. Kispert, and J. Joseph, *J. Chem. Phys.* **82,** 1539 (1985).
78. P. Kovacic and J. S. Ramsey, *J. Polym. Sci. Part A-1* **7,** 945 (1969).
79. C.-F. Hsing, M. B. Jones, and P. Kovacic, *J. Polym. Sci. Polym. Chem. Ed.* **19,** 973 (1981).
80. C. F. H. Allen and F. P. Pingert, *J. Am. Chem. Soc.* **64,** 1365 (1942).
81. P. Kovacic and R. M. Lange, *J. Org. Chem.* **29,** 2416 (1964).
82. C.-F. Hsing, P. Kovacic, and I. A. Khoury, *J. Polym. Sci. Polym. Chem. Ed.* **21,** 457 (1983).
83. N. Bilow and L. J. Miller, *J. Macromol. Sci. Chem.* **A3,** 501 (1969).
84. P. Kovacic and F. W. Koch, *J. Org. Chem.* **30,** 3176 (1965).
85. T. Sh. Zakirov, N. R. Bektashi, and A. V. Ragimov, *Mater. Sumgaitsko Gor. Nauchno-Tekh. Konf. Probl.,* 39 (1980); *Chem. Abstr.* **96,** 163317 (1982).
86. P. A. Holmes, D. C. W. Morley, and D. Platt, *J. Chem. Soc. Chem. Commun.,* 175 (1979).
87. H. Minato, N. Higosaki, and C. Isobe, *Bull. Chem. Soc. Jpn.* **42,** 779 (1969).
88. L. S. Wen and P. Kovacic, *Tetrahedron* **34,** 2723 (1978).
89. H. Guenther and P. Kovacic, *Synth. Commun.* **14,** 413 (1984).
90. K. P. W. Pemawansa, N.-L. Yang, and G. Odian, *Polym. Prepr. Am. Chem. Soc. Div. Polym. Chem.* **25**(1), 274 (1984).
91. P. Kovacic, J. T. Uchic, and L.-C. Hsu, *J. Polym. Sci. Part A-1* **5,** 945 (1967).
92. G. F. L. Ehlers, *IUPAC International Symposium on Macromolecular Chemistry,* Tokyo and Kyoto, Japan, Sept.–Oct. 1966, International Union of Pure and Applied Chemistry, Oxford, UK.
93. A. V. Ragimov, T. Sh. Zakirov, N. R. Bektashi, A. I. Kuzaev, and B. I. Liogon'kii, *Zh. Prikl. Khim. Leningrad* **55,** 1138 (1982); *Chem. Abstr.* **97,** 72894 (1982).
94. V. Z. Annenkova, N. I. Andreeva, V. M. Annenkova, K. A. Abzaeva, and M. G. Voronkov, *Vysokomol. Soedin. Ser. A* **26,** 854 (1984); *Chem. Abstr.* **101,** 24047 (1984).
95. C. Della Casa, P. Costa Bizzarri, and S. Nuzziello, *J. Polym. Sci. Polym. Lett. Ed.* **23,** 323 (1985).
96. R. G. Feasey, A. Turner-Jones, P. C. Daffurn, and J. L. Freeman, *Polymer* **14,** 241 (1973).
97. K. Mukai, T. Teshirogi, N. Kuramoto, and T. Kitamura, *J. Polym. Sci. Polym. Chem. Ed.* **23,** 1259 (1985).
98. Jpn. Kokai Tokkyo Koho 60 65,026 (Apr. 13, 1985) (to Kawamura Physical and Chemical Research Institute).
99. H. A. P. de Jongh, C. R. H. I. de Jongh, H. J. M. Sinnige, E. P. Magré, and W. J. Mijs, *J. Polym. Sci. Polym. Chem. Ed.* **11,** 345 (1973).
100. H. A. P. de Jongh, C. R. H. I. de Jongh, and W. J. Mijs, *J. Org. Chem.* **36,** 3160 (1971).
101. A. Bingham and B. Ellis, *J. Polym. Sci. Part A-1* **7,** 3229 (1969).
102. B. Ellis and J. V. Stevens, *J. Polym. Sci. Part A-1* **10,** 553 (1972).
103. A. A. Berlin, A. A. Ivanov, and I. I. Mirotvortsev, *J. Polym. Sci. Polym. Symp.* **40,** 175 (1973).
104. A. F. Diaz and J. A. Logan, *J. Electroanal. Chem.* **111,** 111 (1980).
105. D. M. Mohilner, R. N. Adams, and W. J. Argersinger, Jr., *J. Am. Chem. Soc.* **84,** 3618 (1962).
106. L. Dunsch, *J. Electroanal. Chem.* **61,** 61 (1975).
107. A. G. MacDiarmid, J.-C. Chiang, M. Halpern, W. S. Huang, J. R. Krawczyk, R. J. Mammone, S. L. Mu, N. L. D. Somasiri, and W. Wu, *Polym. Prepr. Am. Chem. Soc. Div. Polym. Chem.* **25**(2), 248 (1984).
108. J. Langer, *Solid State Commun.* **26,** 839 (1978).
109. A. I. Akhmedov, B. A. Tagiev, and A. V. Ragimov, *Mater. Sumgaitskoi Gor. Nauchno-Tekh. Konf. Probl.,* 41 (1980); *Chem. Abstr.* **96,** 201248 (1982).
110. M. G. Voronkov, V. Z. Annenkova, N. I. Andreeva, and V. M. Annenkova, *Vysokomol. Soedin. Ser. B* **20,** 780 (1978); *Chem. Abstr.* **90,** 55343 (1979).
111. V. Z. Annenkova, V. M. Annenkova, N. I. Andreeva, K. A. Abzaeva, and M. G. Voronkov, *Vysokomol. Soedin. Ser. B* **26,** 443 (1984); *Chem. Abstr.* **101,** 131207 (1984).
112. G. Kossmehl and G. Chatzitheodoru, *Makromol. Chem. Rapid Commun.* **2,** 551 (1981).
113. *Ibid.,* **4,** 639 (1983).
114. M. G. Voronkov, V. M. Annenkova, N. I. Andreeva, V. Z. Annenkova, and K. M. Abzaeva, *Vysokomol. Soedin. Ser. B* **24,** 409 (1982); *Chem. Abstr.* **97,** 128179 (1982).
115. I. Mester, L. Ernst, B. P. Das, B. Choudhury, and D. N. Choudhury, *Z. Naturforsch Tiel B* **39B,** 817 (1984).

116. A. F. Diaz, K. K. Kanazawa, and G. P. Gardini, *J. Chem. Soc. Chem. Commun.*, 635 (1979).
117. K. K. Kanazawa, A. F. Diaz, R. H. Geiss, W. D. Gill, H. F. Kwak, J. A. Logan, J. F. Rabolt, and G. B. Street, *J. Chem. Soc. Chem. Commun.*, 854 (1979).
118. A. F. Diaz and J. I. Castillo, *J. Chem. Soc. Chem. Commun.*, 397 (1980).
119. A. F. Diaz and K. K. Kanazawa in J. S. Miller, ed., *Extended Linear Chain Compounds*, Vol. 3, Plenum Publishing Corp., New York, 1983, p. 417.
120. G. Tourillon and F. Garnier, *J. Electroanal. Chem.* **135,** 173 (1985).
121. A. F. Diaz, J. I. Castillo, K. K. Kanazawa, J. A. Logan, M. Salmon, and O. Fajardo, *J. Electroanal. Chem.* **133,** 233 (1982).
122. J. Bargon, S. Mohmand, and R. J. Waltman, *IBM J. Res. Dev.* **27,** 330 (1983).
123. K. Kaneto, Y. Kohno, K. Yoshino, and Y. Inuishi, *J. Chem. Soc. Chem. Commun.*, 382 (1983).
124. F. Wudl, M. Kobayashi, and A. J. Heeger, *J. Org. Chem.* **49,** 3382 (1984).
125. A. F. Diaz and J. Bargon in T. Skotheim, ed., *Handbook on Conjugated Electrically Conducting Polymers,* Marcel Dekker, Inc., New York, 1986, Chapt. 3.
126. R. L. Elsenbaumer and L. Shacklette in Ref. 125, Chapt. 7.
127. R. H. Baughman, J. L. Bredas, R. R. Chance, R. L. Elsenbaumer, L. W. Shacklette, *Chem. Rev.* **82,** 209 (1982).
128. L. W. Shacklette, R. L. Elsenbaumer, R. R. Chance, J. M. Sowa, D. M. Ivory, G. G. Miller, and R. H. Baughman, *J. Chem. Soc. Chem. Commun.*, 361 (1982).
129. F. Maurice, G. Froyer, and Y. Pelous, *J. Phys. Paris* **44,** C3–587 (1983).

MARTIN B. JONES
University of North Dakota

PETER KOVACIC
University of Wisconsin-Milwaukee

OXIRANES.   See 1,2-EPOXIDE POLYMERS.

OZONE REACTIONS.   See ANTIOXIDANTS; ANTIOZONANTS.

P

PACKAGING MATERIALS

FLEXIBLE MATERIALS

Packaging protects contained products against air, water, relative humidity, odor, microorganisms, insects, light, compression, and impact. Simultaneously, packaging aids in containing desirable product constituents such as color, flavor, form, and structure.

Traditionally, ceramics and glass have been employed for packaging. Steel, aluminum, and paperboard have been used more recently (see also PAPER). Although plastics began to appear in the late nineteenth century, they were first used for packaging in the 1920s and 1930s. Cellophane, aluminum foil, and waxed papers were in common use as packaging materials before World War II. In the 1950s widespread employment of plastics for packaging began with low density polyethylene (LDPE) film for bags and in extrusion-blown form for squeeze bottles. These applications were followed by the introduction of polypropylene, nylon, polyester, and other materials.

Plastic packaging materials may be classified into rigid and flexible, with considerable overlap. The dividing line between rigid and flexible material is determined by caliper or gauge, with thicknesses <0.25 mm generally regarded as being in the flexible category. Flexible packaging is composed of both single- and multilayer structures. The latter may be further subdivided into laminated, coated, and coextruded, or combinations of these.

The U.S. Flexible Packaging Industry

In 1986 packaging in the United States was a $55–60 billion industry, growing at a real rate of 3% annually or slightly faster than the national economy. On a value basis, about 8% is represented by flexible packaging, growing at a rate of ~6%. Growth is attributable to the expansion in consumer and industrial goods and distribution, all of which require more packaging. A given volume of product requires less flexible packaging material than other packaging materials. Furthermore, because many plastics act as protection without coatings or plies, the value of goods contained by plastic flexible packaging is far higher than the value proportion in the packaging supply industry. These cost benefits have prompted the higher growth rates.

More than half of flexible packaging is used for food, ca 10% for health and beauty aids, 5% for drugs and pharmaceuticals, and the remainder for a range of product industries such as electronics, toys, hardware, tobacco, stationery, and paper goods.

Within foods, candy, bakery goods, and snack-type foods such as potato and corn chips, cookies, and crackers consume well over half of flexible packaging. Cheese, processed meat, shrink wraps, condiments, dry-drink mixes, fresh meats, and produce represent smaller applications.

More than 1000 firms participate in the flexible packaging industry, including the small number that produce polymer resins and other raw materials such as paper and aluminum. At the next level are the manufacturers that process the plastic resin into film by blowing or slot casting, a unit operation often assumed by the converter.

The processes of slitting, printing, coating, laminating, coextruding, and fabricating into preformed pouches and bags are performed by so-called converters. Almost all converters perform printing, but only about 200 laminate or coextrude.

The output of the converter, usually in roll form but sometimes in bag or pouch form, is marketed to packagers. Product manufacturers employ the materials to wrap or otherwise contain their products. Some flexible packaging, such as that for fresh meat and produce, is sold directly to retailers.

Fabrication

Flexible packaging materials are mono- or multilayer. Monolayer materials are usually plastic films that have been produced by resin melting and extrusion (see also FILMS).

Extrusion (qv) of polyethylenes and some polypropylenes through a circular die into a tubular form, which is cut and collapsed into flat film, gives blown film. Extrusion through a linear slot onto chilled rollers is called casting and is used for most polypropylenes, polyesters, and other resins.

Slot-cast films, as well as some blown films, may be further heated and stretched in the machine or in transverse directions to orient the polymer within the film and improve physical properties, such as tensile strength and low temperature resistance. Cellulosic-based materials, such as cellophane, are produced by casting, a process described below.

In coextrusion, two or more plastic melts from different extruders are combined into a single die in which the melts are joined. Coextrusion permits precise, small quantities of plastic materials to be intimately bonded to each other (see FILMS, MULTILAYER).

Free mono- and multilayer films may be adhesive- or extrusion-bonded in a laminating process. The adhesive may be water- or organic-solvent-based, in which case the solvent must be evaporated to set the adhesive. Alternatively, a temperature-dependent adhesive without solvent is heated and set by cooling. In extrusion lamination, a film of a thermoplastic (often polyethylene or a copolymer or analogue) is extruded hot to be bonded between the two film materials, which are brought together between a chilled and backup roll.

Flexible materials are printed in roll form by rotogravure or flexographic printing. In the former, a cylinder is engraved with minute depressions or wells, which accept dilute solvent-based inks by capillarity. When contacted by the packaging material, the ink is drawn from the wells to the printing surface and the solvent is evaporated by heat to set the ink. In flexographic printing, the design is elevated above the cylinder surface using rubberlike materials. Rotogravure cylinders are usually considerably more expensive than flexographic printing plates. Furthermore, the detail produced by rotogravure is finer than that produced by flexography. Thus rotogravure is used for long runs and high resolution reproduction, whereas flexography is used for shorter runs and bolder design. Both are extensively used to print flexible packaging. Coatings applied by printing or extrusion protect the printed surfaces.

A small but substantial quantity of flexible packaging is fabricated by converters into bags and pouches. Bag material is either small monolayer or large multiwall with paper as a principal substrate. Pouches are small and made from laminations. Bags contain a heat-sealed or adhesive-bonded seam running the length of the unit and a cross-seam bonded in the same fashion.

Most flexible packaging materials are sold by the converter in roll form, slit to widths useful to the packager.

Applications

Preformed bags are opened by the packager, filled with product, and closed by stitching, adhesive, or heat-sealing.

A small quantity of flexible packaging material, usually oriented polypropylene, shrink polypropylene, or polyethylene, is used to overwrap paperboard cartons. The film is wrapped around the carton and sealed by heating. Products, such as boxed chocolates, toys, cookies, and cigarettes, are overwrapped, often by a printed film.

Some flexible materials, usually containing nylon, are thermoformed, ie, heated and formed into three-dimensional shapes. Thermoforms are made from sheet rather than film. Film is used to provide a high gas-barrier, heat-sealable container for processed meat or cheese.

Large quantities of flexible packaging materials are employed in horizontal machines performing a three-step operation (forming, filling, and sealing) to enclose the contents, sometimes with a hermetic seal. The web of material is continuously unwound horizontally and folded in such a way that its longitudinal edges are in contact. Meanwhile, the product is conveyed at the same speed as the flexible packaging material into the tube. The two edges are heat-sealed and cross-heat seals are formed between the product units. This system is used for overwrapping, as well as for unit packaging for candies, cookies, and crackers.

In a variant of the horizontal form–fill–seal operation, the material, moving in a horizontal direction, is folded on itself vertically. Vertical sectors of the two faces are heat-sealed to each other to form a pouch, which may then be filled. The pouch, usually made from aluminum foil plus a plastic laminant and heat sealant, is closed by a heat seal. This type of pouch gives high protection and is used for moisture-sensitive condiments and beverage mixes.

The largest volume of flexible packaging is used in vertical form–fill–seal applications for loose, flowable products such as potato and corn chips, nuts, and coffee. The roll of flexible material is unwound over a forming collar forcing the web into a vertical, tubular shape. By heat-sealing the edges and a bottom cross-seal, an open-top tube is formed. The product is filled, the web is drawn down, and the tube is closed by another heat seal. Vertical form–fill–seal operations include water-vapor-barrier materials to package moisture-sensitive products (see also BARRIER POLYMERS).

In applications other than protective packaging, flexible materials are employed to unitize cans, bottles, cartons, or cases. Heat-shrinkable films such as LDPE are wrapped around groups of bottles or cans, sealed, and exposed to hot air to bind the contents tightly together within the film. In stretch wrapping, an extensible film cohesive to itself, eg, LLDPE, is wrapped very tightly around the contents. Surface cohesion causes the film wrap to hold to itself. As the stretched film attempts to revert to its original unstretched form, it binds the bundle more tightly.

Numerous variations and other applications are common for flexible packaging materials, eg, oxygen-permeable wraps for red meat and produce, shrinkable bags for meat and rigid tray closures.

Film Materials

The properties of film packaging materials are given in Table 1.

Cellulosics.   Cellophane, the first important packaging material, attained its peak in the 1960s before beginning a long and steady decline in the face of less expensive and functionally superior all-plastic materials.

Cellophane was developed in Europe early in this century. For many years, cellophane was virtually the only flexible packaging material available, and converting and packaging equipment was engineered for this material. Cellophane is moisture- and temperature-sensitive and has almost no barrier properties. Internal plasticizers and surface coatings, such as nitrocellulose or poly(vinylidene chloride), greatly improve performance. Cellophane is easily printed, cut, and machined on packaging equipment. However, even when coated, it does not exhibit the resistance to water-vapor transmission of most plastic films. Furthermore, it remains sensitive to climatic variations.

In 1987 cellophane was being manufactured in the United States only by one company. Less than 40,000 t, valued at ca $3.50/kg, were manufactured in or imported into the United States. A fraction of cellophane is reinforced with oriented polypropylene (see also CELLOPHANE in the Supplement).

Cellophane is derived from clean wood pulp. The cellulose is dissolved in caustic soda and converted into an alkali cellulose, which is treated with carbon disulfide to form a viscose syrup. This syrup is extruded into a sulfuric acid bath

Table 1. Properties of Packaging Materials[a,b]

	Specific gravity	Tensile strength, MPa[c]	Elongation, %	Heat-seal range, °C	Water-vapor transmission rate[d], μmol/(m²·s)[e]	Gas transmission[f], nmol/(m²·s)[g]	Use temperature, °C Maximum	Minimum
cellophane								
nitrocellulose, coated	1.44	124 (MD) 62 (TD)	15–23	93–176	5	16	depends on rh	
Saran, both sides coated	1.44	124 (MD) 62 (TD)	15–25	93–176	4.5	0.4	depends on rh	
fluorocarbon	2.2	34–69	50–400	176–204	0.25–0.5	56–120	148	−45
ionomer	0.94–0.96	21–69	350–450	87–204	13–21	1,800–3,870	71	−73
acrylonitrile	1.15	66	5	71–148	50	6.4	71–210	−45
nylon								
uncoated[h]	1.13–1.14	48–124	250–500	176–260	240–260	21	176–232	−59
Saran-coated, one side	1.13–1.14	48–124	250–500	176–260	2	4	93	−40
polycarbonate	1.2	69	92–115	204–221	97	2,060	132	−73
polyester, PET								
uncoated	1.35–1.39	172–228	120–140		13	40	204	−62
Saran-coated, one side	1.4	179–214	90–125	135–204	6[j]	3.2	coating softens at 82	−51
metallized[i]	1.35–1.39	172–228	120–140		0.3–1.4	0.32	204	−62

polyethylene								
LDPE[h]	0.910–0.925	6.9–24	225–600	121–176	12	2,000–6,700	65	−51
MDPE[h]	0.926–0.940	14–35	225–500	126–154	5–10	1,300–2,700	82–104	−51
HDPE[h]	0.941–0.965	21–52	10–500	135–154	3–6.5	260–2,000	121	−51
LLDPE[h]	0.915–0.935	24–55	400–800	121–176	12	2,000–6,700	76–82	−51
LDPE/12% EVA copolymer	0.94	21–35	300–500	93–148	39	4,100–5,200	60	−51
polypropylene								
nonoriented	0.88–0.90	21–62	400–800	162–204	5–6.5	670–3,300	121	not recommended at low temp
oriented	0.905	172–207	60–100	requires coating or additive	3–4	880	121	−51
oriented and metallized[i]	0.905	131 (MD) 220 (TD)	50–400	requires coating	1	24–80	121	−51
poly(vinyl chloride)	1.21–1.37	14–110	5–500	121–176	>40	40–12,000	93[k]	above −18
poly(vinylidene chloride)[h]	1.64–1.71	55–138	40–100	121–148	0.5–3	0.64–14	softens ca 82	above −18

[a] Ref. 1.

[b] Transparent unless otherwise stated; good resistance to grease and oils.

[c] To convert MPa to psi, multiply by 145.

[d] 2.5-μm gauge film, 38°C, 90% rh.

[e] To convert to g/(100 in.²·d), divide by 10.

[f] 25-μm gauge film, 23°C, 0% rh, 101.3 kPa.

[g] To convert to cc/(100 in.²·d), divide by 8.

[h] Transparent to translucent.

[i] Opaque.

[j] 12.5-μm gauge.

[k] Depends on plasticizer.

to coagulate the acid, decompose the viscose, and regenerate a cellulose gel, which is further processed into film.

The film is usually solvent-coated with a nitrocellulose blend or a Saran poly(vinylidene chloride) (PVDC) copolymer. The former provides heat sealability and some water-vapor barrier; PVDC imparts a much better water-vapor barrier, as well as heat sealability.

Plastics

Plastic, flexible packaging materials are thermoplastic, ie, reversibly fluid at high temperatures and solid at RT. They are predominantly based on polyethylene. In its various forms, over 1.4×10^6 t of polyethylene film are produced annually in the United States for packaging alone. In addition, large quantities of most plastic films are used for nonpackaging applications.

Plastic films may be modified by copolymerization, additives in the blend, alloying, as well as surface treatment and coating.

Low Density Polyethylene (LDPE). Derived from high pressure polymerization of ethylene gas in a continuous tubular or autoclave process, LDPE was the principal thermoplastic resin until 1980. With the development of low pressure polyethylene resins, LDPE film has lost considerable market share to the new linear low density polyethylene (LLDPE) (see ETHYLENE POLYMERS).

Although developed as an electrical insulation material in the 1930s, LDPE is used today in a wide range of applications. About 20 U.S. petrochemical companies produce resin LDPE at \$0.66–0.88/kg (1986 price), depending on grade. The film is produced by 700 or more firms and sells at \$1.30–1.54/kg or 1.25–1.87¢/m² (25-μm gauge).

Slightly cloudy, with high tensile-strength and good water-vapor-barrier properties but very high gas transmission, LDPE film is widely used in construction and agriculture, and for laundry and garbage bags and medical drapery. It is employed for shrink and stretch bundling, for dry industrial chemical bags, as drum and case liners, and to package bread, produce, paper products, diapers, toys, hardware, and food products.

The density of LDPE for film ranges from <0.92 to ca 0.93 g/cm³. The resin is mixed and melted in an extruder and converted to film by extruding through a circular die. The resulting tube is collapsed and slit into film ranging in gauge from <25 to above 75 μm. Some monolayer and coextruded polyethylene films are produced by slot-die casting. Because of its extensibility, LDPE film is usually printed on central-drum flexographic presses. Modern rotogravure presses permit rotogravure printing.

Linear Low Density Polyethylene (LLDPE). By 1985 linear low density polyethylene resin had captured more than one-quarter of the low density polyethylene film market. The resin is produced in the vapor phase under pressures about one-tenth of those required for conventional LDPE resin, which reduces manufacturing costs.

Films from these resins have 75% higher tensile and 50% higher elongation-to-break strengths, and a slightly higher but broader heat-seal initiation temperature. Impact and puncture resistance are also improved over LDPE. Water-vapor and gas-permeation properties are similar to those of LDPE films.

Linear low density polyethylene films are used in many of the same packaging applications as LDPE. The greater film extensibility permits the printing of small bags.

High Density Polyethylene (HDPE). Polyethylene resins ranging in density from 0.93 to 0.96 g/cm^3 are classified as medium density at the lower end of the scale and high density at the higher end. The characteristics of medium density polyethylene films resemble those of low density polyethylene films. High density polyethylene (HDPE) films are more translucent and stiffer, with low elongation to break; water-vapor and gas permeabilities are slightly higher than those of LDPE. Softening and melting points are high, and they are not easily sealed on flexible packaging equipment. Film is used to replace paper sacks for grocery and department stores.

High density polyethylene coextrusions with ethylene–vinyl acetate copolymer or ionomer are widely used as liners in food cartons. High density polyethylene film is produced by blown-film extrusion methods, often blowing downward because the weight of the film could cause collapse of an upwardly blowing bubble.

Polypropylene (PP). Polypropylene film is cast as unoriented (UPP) and oriented (OPP) films. The former was introduced as a cellophane replacement because of excellent clarity, gloss, and water-vapor transmission resistance. However, because of extremely poor cold-temperature resistance and a very narrow, short, heat-seal temperature range, UPP is little used for packaging.

Coextrusions of PP/PE and UPP are used to separate sliced cheese, and UPP alone is used as a transparent bag material for textile soft goods, twist wrap for candy, and other applications; less than 45,000 t of unoriented polypropylene are used for packaging. Unoriented PP is made by extrusion through a slot die; an ethylene–propylene copolymer is used to improve low temperature resistance (see PROPYLENE POLYMERS).

Oriented polypropylene film (OPP) may be classified into heat-set and non-heat-set, blown and tentered, coextruded and coated. Orientation improves the cold-temperature resistance and other physical properties, and the heat-seal properties of coatings and coextrusions. In 1985 ca 136,000 t of OPP were produced with a price of ca $3.30/kg or 1.8¢/m^2 (25-$\mu$m gauge).

Nonheat-set OPP is used as a sparkling, transparent shrink-film overwrap for cartons of candy and cosmetics. Heat-set OPP film is used to wrap bakery products, as lamination plies for potato and corn chips, and for pastas and numerous other pouch and wrap-packaging applications.

Oriented polypropylene may be manufactured by blown or slot-die extrusion processes. In the double-bubble film process, the resin is extruded through a circular die, cooled, reheated, and blown again to produce a balanced, biaxially oriented, heat-set film. Film may be blown with coextrusion of a heat-seal coating. It may also be coated with poly(vinylidene chloride) to impart water-vapor, gas-barrier, and heat-sealant properties.

In the tenter-frame process, polypropylene film is extruded through a slot die and stretched or oriented in the machine direction. The film is then reheated, gripped along its edges, and stretched outwardly while in motion to impart transverse directional orientation. Tentered OPP film usually has unbalanced orientation. To impart heat-seal properties, it may be coextruded or coated with acrylic or PVDC.

Several manufacturers have expanded the core, creating a foam structure with lower density, greater opacity, and a more paperlike feel. Another recent development is vacuum metallization to increase opacity and water-vapor-barrier properties (see METALLIZING).

Oriented polypropylene film has excellent water-vapor-barrier but poor gas-barrier properties. With its barrier clarity (or opacity in newer forms) and good heat-seal properties, it has largely replaced cellophane and glassine in packaging applications. To date, gas barriers imparted by coextrusions or coatings have been well below those required for high oxygen-barrier applications.

Poly(vinyl chloride) (PVC).   Poly(vinyl chloride) must be modified with heat stabilizers and plasticizers, which increase costs. When PVC is heated to sealing temperature, hydrochloric acid is generated, which is acrid and corrosive, a draw-back to packaging applications. Plasticized PVC film is highly transparent and soft, with a very high gas-permeation rate. The water-vapor transmission rate is relatively low. The film is so heat-sensitive that heat sealing should be avoided.

At present, PVC film is produced mostly by blown-film extrusion, although casting and calendering are employed for heavier calipers (see VINYL CHLORIDE POLYMERS).

Additives and HCl generation have led to numerous problems and proposed restrictions.

The principal packaging use of PVC film is as a gas-permeable wrap for red meat, poultry, and produce. Sparkle and transparency, combined with the ability to transmit oxygen inward to maintain red-meat color or CO_2 outward to avoid produce anaerbiosis, offer advantages in these applications.

Approximately 113,000 t of PVC film are used annually in the United States.

Polyester.   Poly(ethylene terephthalate) (PET) polyester film is used as a packaging material in limited quantities. A material with intermediate gas- and water-vapor-barrier properties, very high tensile and impact strengths, and high temperature resistance, polyester film is used for the protection of aluminum foil and boil-in-bag applications, and in nonpackaging applications for audio and video recording tapes and drafting sheet. Dust-free materials for recording require production under perfectly clean conditions, which increases costs.

In 1986 ca 22,500 t of polyester film were used for packaging at a typical price of over $4.40/kg.

Applications include use as an outer web in laminations to protect aluminum foil coated with PVDC that acts as the flat or sealing web for vacuum-packaged processed meat and cheese.

Polyester resin is a condensation product of ethylene glycol and terephthalic acid. The film is manufactured in much the same manner as OPP, ie, extruded through a slot die and biaxially oriented by stretching first in a machine and then in a transverse direction while still hot (tentering).

Polyester film cannot be heat-sealed by conventional methods and is either coextruded or coated for heat sealing.

Being heat resistant and almost inextensible, polyester film is used increas-ingly as a substrate for vacuum metallizing, a process that improves moisture-, gas-, and light-resistant properties. Vacuum-metallized polyester film is used for packaging wine (see POLYESTERS, FILMS).

Nylon.   Nylon is the designation for a family of thermoplastic polyamide materials, which in film form are moderate-oxygen barriers. The gas-barrier properties are equal to odor- and flavor-barrier properties important in food applications. Nylon films are usually tough and thermoformable, but only fair moisture barriers (see POLYAMIDES).

Less than 22,500 t of nylon are used annually in the United States for packaging, partially because of high costs. Nylon films are almost always used in lamination or coated form to ensure heat sealability and enhance barrier properties. The largest uses are as thermoforming webs for twin-web processed meat and cheese packaging under vacuum or in an inert atmosphere. Other uses include bags for red meat, boil-in-bags, and as the outer protective layer for aluminum foil in cookie packages.

Nylon polymers are formed by copolymerization of diacids with diamines (nylon-6,6) or by polymerizing amino acids (nylon-6). The digits refer to the number of carbon atoms in the monomer units. Films of both nylon-6 and -6,6 are made by slot-casting or blowing; coextrusion is used to provide heat-seal properties. The films may be oriented biaxially or uniaxially in the machine direction. Biaxially oriented nylon film is produced in Japan and now in the United States for use in coffee and soft-cookie packaging.

Poly(vinylidene chloride) (PVDC). Poly(vinylidene chloride) (PVDC) is best known in its Saran or PVC-copolymerized form. As solvent or emulsion coating, PVDC imparts high oxygen, grease, aroma, and water-vapor resistance to substrates such as cellophane, oriented polypropylene, polyester, nylon, and paperboard.

Of the common commercial resins and films, PVDC has the best water-, vapor-, and oxygen-barrier properties. High crystallinity confers resistance to the permeation of odors and flavors, as well as to grease and oil. Because of its high chlorine content, it tends to corrode processing equipment, which increases manufacturing costs. Unlike other high oxygen-barrier materials, PVDC is almost insensitive to water or water vapor (see VINYLIDENE CHLORIDE POLYMERS).

Among the more expensive films, PVDC film enjoys sales of ~4500 t annually in the United States; much larger quantities in resin form are used in coatings.

Copolymer film is produced by extrusion followed by water quenching. In-line, the film is blown, crystallized, and oriented; PVDC copolymer film is difficult to produce.

Saran film is used to wrap cheese and for vertical form–fill–seal of sausage and ground red meat. Mostly it is used as the high barrier component of laminations not containing aluminum foil. It is rarely used alone in commercial packaging because it is difficult to seal.

Polystyrene. Polystyrene packaging film has excellent clarity, stiffness, and dimensional stability. Because of high permeability to water vapor and gases, it is well suited for packaging fresh produce requiring the presence of oxygen; tear resistance is low.

Packaging applications are limited to folding-carton windows, overwraps for tomato trays, lettuce wrapping, and so on.

During extrusion, foaming polystyrene resin gives an opaque, low density sheet used for beverage-bottle labels as a water-resistant paper substitute (see STYRENE POLYMERS).

Other Films. Although less important than polyethylenes and polypropylenes, a number of other plastic films are in commercial use or development for special applications, including ethylene–vinyl acetate, ionomer, polycarbonate, ethylene–vinyl alcohol, fluorocarbon, and polyacrylonitrile.

Ethylene–vinyl acetate copolymer (EVA) forms a soft, tacky film with fairly good water-vapor-barrier but very poor gas-barrier properties. It is widely used as a low temperature initiation and broad-range, heat-sealing medium. The film also serves for lamination to other substrates for heat-sealing purposes.

Ionomers, known by their trade name Surlyn (DuPont), are ionically cross-linked thermoplastics derived from ethylene–methacrylic acid copolymers. Ionomer films are tough, extensible, and impact resistant with excellent heat-seal and hot-tack characteristics. They are extruded on other substrates. Ionomer films are used as heat-sealing and skin packaging.

Fluorocarbon films, known by their trade name Aclar (Allied Chemical), are noted for exceptionally low water-vapor permeability, about one order of magnitude better than that of PVDC. These films are very expensive and are used for pharmaceutical packaging where very long shelf life is required.

The more recent polycarbonate films are exceptionally clear and tough. High temperature resistance, and water-vapor and gas permeability are excellent. Polycarbonate films have been proposed for structural and heat-resistant plies in laminations.

Polyacrylonitrile (PAN) films have been produced from copolymers that were under FDA investigation in the 1970s and early 1980s. A material with outstanding oxygen- and CO_2-barrier properties, polyacrylonitrile has only modest water-vapor-barrier properties. It is used for processed-meat-packaging laminations where an oxygen barrier is required for vacuum packaging. In biaxially oriented form, the film is stiff and tough, but slightly yellow.

Ethylene–vinyl alcohol copolymers (EVOH) and its variants are perhaps the most widely publicized plastic packaging materials of the 1980s. Discovered in the United States and developed commercially in Japan, EVOH resins are noteworthy for very low oxygen permeability, and temperature and water sensitivity. Copolymerization with ethylene increases the water resistance of vinyl alcohol, but simultaneously increases the oxygen permeability. Thus the desired oxygen permeability must be balanced against water resistance. Encapsulation or coextrusion with low water-vapor–high gas-permeability resins such as HDPE and polypropylene solves some of these problems. Drying agents in proximity plies are also used. Biaxially oriented EVOH films have been developed with promise for use in laminations. The EVOH films are more readily fabricated, coextruded, and recycled (cost of recovering scrap is crucial) than PVDC (see VINYL ALCOHOL POLYMERS).

Coextrusions

The use of film coextrusions in packaging increased rapidly in the late 1970s and 1980s; well over 113.5×10^3 t were used in the United States in 1985. Applications are expected to increase sharply in the near future.

In coextrusion two or more thermoplastic-resin melts are extruded simultaneously from the same die. Although variants on this basic mechanism are in commercial use, the single multilayer die from multiple extruders is conceptually sound. Coextrusion permits an intimate combination of materials in precisely the quantities required to function. Incompatible plastic materials are bonded with thermoplastic adhesives.

Coextruded films may be made by extrusion-blowing or slot-casting of two or three layers (AB or ABA). Recently, five- and seven-layer blown-film equipment has been developed. Slot-casting is capable of combining up to nine layers, but most equipment is used for the simple AB, ABA, and ABD or ACBCA configurations (C is a tie layer). Cooling is better controlled in casting than blowing, and because the die is linear, the gauge uniformity is better with casting.

In the simplest combinations, oriented polypropylenes are coextruded with integral-copolymer, heat-seal layers. Low density polyethylenes are coextruded to impart toughness or slip characteristics.

In more complex combinations, HDPE and LDPE and EVA resins are coextruded to produce stiff, heat-sealable films to be used as liners in cereal, cookie, and cracker boxes. Films of EVA and white-pigmented LLDPE are used for packaging of frozen vegetables and fruits. In these applications, one layer imparts toughness, opacity, or stiffness, and another one or two layers, heat sealability.

Coextrusions of nylon with polyethylene in five layers are used for thermoforming where high gas and water-vapor barrier are required, eg, medical packaging.

BIBLIOGRAPHY

"Packaging Materials" in *EPST* 1st ed., Vol. 9, pp. 709–714, by Stanley Sacharow, Reynolds Metals Company.

1. W. C. Simms, ed., *Packaging Encyclopedia,* Cahners Publishing Co., Des Plaines, Ill., 1985.

General References

Paper Synthetics Conferences, Technical Association of the Pulp and Paper Industry, Atlanta, Ga., 1982 and 1983.
M. Bakker, ed., *The Wiley Encyclopedia of Packaging Technology,* John Wiley & Sons, Inc., New York, 1986.
C. J. Benning, *Plastic Films for Packaging,* Technomic Publishing Co., Inc., Lancaster, Pa., 1983.
A. L. Brody, *Flexible Packaging of Foods,* CRC Press, Inc., Boca Raton, Fla., 1972.
J. Eichhorn, "Films," *Mosher and Davis Industrial & Specialty Papers,* Vol. II, 1968.
R. Griffin, S. Sacharow, and A. L. Brody, *Principles of Package Development,* 2nd ed., Avi Publishing, Westport, Conn., 1985.
C. R. Oswin, *Plastic Films and Packaging,* Applied Science Publishers, Ltd., Essex, UK, 1975.
F. Paine and H. Paine, *Principles of Food Packaging,* Leonard Hill, Ltd., London, 1983.
S. Sacharow and A. L. Brody, *Packaging: An Introduction,* Harcourt Brace Jovanovich, Cleveland, Ohio, 1987.
R. S. Schotland, ed., *Proceedings of the International Conference on Coextrusion,* Vols. 1–7, Schotland Business Research, Princeton, N.J., 1981–1986.
W. C. Simms, ed., *Packaging Encyclopedia,* Cahners Publishing Co., Des Plaines, Ill., 1985.
C. M. Swalm, ed., *Chemistry of Food Packaging,* ACS Adv. Chem. Ser. **135,** American Chemical Society, Washington, D.C., 1974.

AARON L. BRODY
Schotland Business Research, Inc.

RIGID CONTAINERS

In less than 100 years, packaging has developed into a viable commercial means to maintain content integrity and has grown from heavy, crude glass, metal, wood, and paperboard into a comprehensive range of materials and structures that perform economically and effectively. Although hydrocarbon-derived polymers have been available for almost 50 years, their practical application to commercial packaging began only in the 1950s (1). Less than 5% of the GNP is invested in packaging (2).

In 1986 the sales volume of the U.S. packaging industry passed $57 billion; the real annual growth rate was above 3%, slightly higher than the average annual growth rate during this decade. Almost 7% of the total is plastic, exclusive of that used in flexible packaging and in conjunction with other materials, including less than 1% each in distribution cases and thermoformed and injection-molded pieces and 1% in closures for bottles and cans. Most rigid plastic, 4.4% of the total U.S. market, is used in bottles and jars. Jars have openings approximately the same as those of the body, whereas the neck diameter of the bottle is significantly smaller than that of the body. More than 17 billion plastic bottles and jars have been produced annually between 1983 and 1986; distribution is given in Table 1 (2,3).

Table 1. Distribution of Plastic Bottles

Application	Percent
food and beverages	49
milk	16
food	6
soft drinks	21
other beverages	6
household chemicals	17
industrial chemicals	2
automotive and marine applications	3
toiletries and cosmetics	15
medicinal and health	14
other	<1

Although rigid and semirigid plastic bottles and jars were made before 1950, production increased sharply after the mid-1960s with the replacement of glass bottles by plastic for household bleach and pasteurized milk in 3.785-L (1-gal) sizes; the development of plastic bottles for liquid detergents followed. Early plastic bottles and jars were mostly fabricated by extrusion blow molding of high density polyethylene (HDPE).

In the late 1960s and early 1970s, Monsanto, Standard Oil of Ohio (now known as SOHIO), and Borg-Warner introduced high gas-barrier polyacrylonitrile (PAN)–acrylic or styrene copolymers for carbonated-beverage bottles. However, in 1977 the FDA prohibited PAN for carbonated-beverage bottles. This was followed by the development of poly(ethylene terephthalate) (PET) bottles. The commercial success of PET bottles for carbonated beverages stimulated the production of high oxygen-barrier bottles and jars for other applications (see BARRIER POLYMERS).

Multilayer blown bottle technology, developed in Japan in 1974, allowed the incorporation of high oxygen-barrier ethylene–vinyl alcohol copolymers (EVOH) for containers for oxygen-sensitive foods. This technology has been expanded to a variety of narrow- and wide-mouth food containers, with the prospect of displacing glass packaging.

As much as 20% of extrusion blow-molded bottles are self-manufactured by dairies and liquid-detergent packagers. Perhaps more than 1000 merchant bottle blowers operate in the United States, employing extrusion and injection blow molding (2).

The U.S. plastic container industry aims to replace the almost 40 billion glass bottles and jars used annually in the domestic market by plastic. Plastic bottles, jars, and thermoformed containers are classified as semirigid, injection-molded articles as rigid.

Fabrication Processes

Compression Molding.   Compression molding is used for the production of food trays and dishes made of PET. These heat-resistant products are suitable for freezing and heating in microwave or conventional ovens. Thermosetting plastics, such as acetals and urea–formaldehyde, are compression-molded to produce glass bottle closures (4) (see COMPRESSION AND TRANSFER MOLDING).

Injection Molding.   Matched metal molds are used to produce closures, specialty packages, and bottle preforms. In conventional injection molding (qv) the plastic resin is melted in an extruder which forces a measured quantity or shot into a precision-machined mold; the nozzle of the extruder is then withdrawn.

The pressure of the extruder forces uniform distribution throughout the mold. Cooling the mold solidifies the plastic with slight shrinkage. The mold is maintained closed by mechanical or hydraulic pressure while the thermoplastic is injected and solidified. Upon opening the mold, the article is removed manually or mechanically.

Because cycle time to inject, flow, set, open, eject, and close is finite, and the face area or platen size is limited, the effective area is increased by increasing the number of mold cavities or the number of faces to several parallel replications in stack molds. In this manner the number of finished pieces per cycle may be multiplied by 10 or more.

Injection molds are constructed of precision-machined metal. Internal cooling, multiple cavities, and multifaces increase costs and create operating problems; devices for extraction or ejection of the article increase costs further. Although injection molding produces plastic pieces with accurate and precise dimensions, it is rarely used for packaging applications because it is expensive (4).

Returnable distribution cases for carbonated-beverage and milk bottles are injection-molded from high density polyethylene or polypropylene–ethylene copolymer in one- or two-cavity molds.

Crystalline and high impact polystyrene are injection-molded into small reusable packages for specialty products such as jewelry and hardware. These packages indicate a high quality article and serve as permanent receptacles.

Some HDPE and polypropylene–ethylene copolymer cups and tubs for dairy product and specialty frozen food applications are injection-molded. High and medium density polyethylenes are injection-molded as covers for metal and paperboard composite closures. Because covers and closures have a high surface area-to-depth ratio, multilevel or stack molds are used to maximize unit output per cycle. High impact polystyrene is injection-molded for refrigerated dairy product and specialty food packaging (2).

Insert injection molding is used to manufacture snap closures for dairy product cups and tubs for yogurt, ice cream, and beverage-bottle multipacks. In insert injection molding, a die-cut printed paperboard or other flat material is placed in the mold. The plastic is extruded around the insert to form a precision skeletal structure. The articles can be decorated, which is otherwise not possible (5).

Injection-molded articles can be decorated by in-mold labeling or by postmold decoration. In the former method printed film is inserted into the mold cavity before injection. The plastic forms an intimate contact with the graphic material. Postmold decoration includes hot stamping, dry offset printing, and decal or compression printing (5).

Injection molding is used for the preparation of preforms for blow molding. When the operation is performed independently of the downstream operation, it is strictly injection molding. When the injection molding is performed on the same machine or is in-line on sequential machines, it is classified according to the integral process, eg, injection blow molding.

Resins with high gas-barrier properties and very narrow melt-temperature ranges are frequently needed for shapes with narrow necks. A blow-mold parison is injected at melt temperature. The tube-shaped parison may be gently reheated to softening below melt temperature for stretching or blowing into a bottle or jar. This method combines the advantages of injection and blowing, and is used for fabricating PET into bottles.

To enhance water-vapor or gas-barrier properties, layers of different plastics may be injected together or sequentially. Multilayer injection-molded pieces may be prepared as packaging or for blowing into bottle or jar shapes.

Thermoforming. Thermoforming (qv) includes the extrusion of sheet, thicker than 0.25 mm, followed by forming the reheated sheet in an open-face mold by pressure or vacuum, or both. Sheet less than 0.25-mm thick is thermoformed in-line and filled and sealed. It is used for meats, cheeses, pharmaceuticals, and medical devices (see FLEXIBLE MATERIALS in this article).

Thermoforming operations include solid-phase pressure forming, die-cutting, and scoring sheet for plastic folding cartons.

Sheet Extrusion. Sheet for thermoforming and analogous operations is formed by extruding the melt through a slot die onto a set of polished chill rolls. The sheet is usually ca 150-cm wide. After rapid cooling, the web is coiled or cut into individual sheets.

Polystyrene, PVC, polyethylene, polypropylene, and filled polypropylene and cellulose acetate and butyrate are prepared in sheet form by extrusion. Some PVC compounds are calendered through rollers without passing through the melt phase (see CALENDERING).

Thermoformable sheet may be mono- or multilayer; the latter is produced by lamination or coextrusion. In coextrusion two or more layers are extruded through one or several closely spaced slot dies; the layers are joined by the heat of extrusion in laminar flow before entering the die. Extrudable adhesives or tie layers are inserted between noncompatible layers (see FILMS, MULTILAYER).

Multilayers are employed to incorporate high oxygen-barrier materials between structural or high water-vapor barrier plastics. Both ethylene–vinyl alcohol copolymers and poly(vinylidene chloride) are used as high oxygen-barrier interior layers with polystyrene or polypropylene as the structural layers, and polyolefin on the exterior for sealing (1).

To produce thermoformed foamed or low density plastic sheet, the resin melt is expanded with gas during the casting process. The gas may be produced by introducing a hydrocarbon gas, such as pentane, which expands by heat and is released during extrusion. This technique is employed for foamed polystyrene meat and egg trays. The hydrocarbon gas escapes into the air during extrusion, and no residue remains.

Alternatively, gas may be introduced by mixing thermally reactive chemicals such as cyanogens with the resin at the extruder. Extrusion heat initiates the reaction to release gas throughout the melt and expand it; this method is confined to nonfood applications.

Fluorocarbon gases may be injected into and mixed with the resin melt in tandem extruders. Upon casting the trapped gas expands and foams the sheet.

In steam-chest expansion the resin beads are poured into molds into which steam is injected. The steam increases the temperature close to the melting point and expands within the structure to create beads with good cushioning and insulating properties. Expanded polystyrene is widely used in this process for protective packaging of durable goods, such as electronics, and for thermal insulation for frozen food packaging.

Steam-chest foamed materials are molded directly or, more frequently, cut into the desired shapes. Thick, foamed polyolefins may be molded or cut with a hot wire.

Three-dimensional Packaging. Thermoforming is the most common method of fabricating sheet into three-dimensional packaging. In conventional thermoforming the sheet or web is heated to softening or just below the melting temperature. The softened plastic is forced by differential air pressure into an open-top mold to assume the shape of the female mold. The mold is chilled, the plastic sheet solidifies, and is removed from the mold (4).

The mold is usually prepared with orifices to permit air trapped between the sheet and the mold to escape and ensure uniform, close contact of the plastic with the mold surface. By clamping the sheet beyond the perimeter of the piece, plastic may be drawn from the peripheral areas into the mold, ensuring uniformity. Both pressure and vacuum are employed to force the softened plastic sheet into the mold; vacuum is satisfactory for noncritical finish. A shaped plug may assist the initial stretching of the sheet into the mold.

Thermoforming is used for gauges above 6 mm in some nonpackaging applications; for packaging applications the gauges are between 0.6 and 1.5 mm and may be 2.5 mm.

The material is stretched during formation, and the greater the depth of draw, the higher must be the gauge of the original sheet. Depth of draw, defined as the ratio of piece height to cross-sectional surface area, is a measure of the ability to form deep articles.

Although deep-drawn pieces are feasible, thermoforming is best suited to shallow profiles of up to 7.5 cm; deeper draws often result in thin side walls.

Any fully or partially tapered shape with an opening equal to or larger than the body can be made. Shapes can be further modified by folding, allowing the thermoforming of articles with narrow openings.

In commercial practice, packaging is produced from continuous web on intermittent-motion, thermoforming die-cutting machines. The web edge is clamped and conveyed into a heating box. If a plastic with a narrow softening temperature is used, heating is carefully controlled from top and bottom. The heated web is conveyed to the forming section where pressure or vacuum force the softened web into the mold. Polypropylene is cut within the forming mold simultaneously with forming. The mold opens and the web is conveyed to a die-cutting station.

Mold design must optimize use of the web surface by minimizing edge trim and waste. Scrap can be recovered and reused.

The unusual temperature and structural properties of polypropylene led to the development of special controlled-temperature forming processes called cold forging, billet forging, or solid-phase pressure forming (SPPF). The polypropylene sheet temperature is brought above the softening temperature but well below the melting point and subjected to very high forming pressures, three to five times those used for fabrication of polystyrene (4) (see also COLD FORMING).

Conventional thermoforming of polystyrene and PVC is the most widely used technique for packaging dairy products and for cups and trays.

Thermoforming may be integrated with filling and sealing on thermoform–fill–seal machines operated by the packager. Such systems are increasingly used for dairy products, pharmaceuticals, and medical device packaging. The base web is gripped and moved in intermittent motion through heating and forming operations. The open-top pieces are filled and a second web of material is heat-sealed to the flange of the base by heated pressure bars using the conveying belt as a backing; cutting may take place during or after sealing. Vacuum and back-gas flushing of the contents are often performed (5).

Blister packaging, widely used for packaging drugs, hardware, health and beauty aids, etc, may be performed using thermoform pieces or on thermoform–fill–seal equipment. In blister packaging the article is attached to a printed paperboard backing and heat-sealed with a transparent plastic sheet and coated aluminum foil.

Compression-molded thermoset polyester food trays can be heated in a microwave or a conventional oven. Thermoplastic polyester sheet incorporating nucleating agent may be thermoformed into heat-resistant trays. After a few seconds the plastic is partially crystallized and heat set above its T_g but below its melting point. Since partially crystallized PET tends to be brittle, PET with intrinsic viscosity >0.9 dL/g is employed with some success. Alternatively, crystallizable and amorphous PET may be coextruded and thermoformed to produce a two-layer partially crystalline and amorphous tray. The amorphous layer is sufficiently impact-resistant for distribution. In conventional heating, the amorphous layer increases its crystallinity and improves its heat resistance, a desirable attribute for the polyester trays which might otherwise soften at elevated temperatures. At freezer temperatures the amorphous layer resists impact fracturing.

Retortable trays, designed for low acid foods, may be heat-sealed and thermally sterilized up to 125°C. High oxygen-barrier properties are usually required. These trays consist of five-layer coextrusions of PVDC or EVOH as the barrier tie layers and polypropylene. The trays must be thermoformed and aged before filling and sealing to minimize heat-seal distortion due to stress. Counterpressure and temperature must be carefully controlled to avoid seal distortion. No other combinations of materials today offer any better heat resistance, oxygen and water-vapor barrier, and structural integrity.

Trays exhibiting side-by-side multiple colors are produced by thermoforming sheet in which two or more extruders feed each side of the slot die. In this manner sheet with various colors may be produced for later thermoforming.

Melt-phase thermoforming overcomes the difficulties and stresses developed in forming polypropylene sheet. As the plastic is extruded from the slot die and while it is still hot, it is transferred into the mold. To maintain thermal and mass equilibrium, the molds move continuously past the slot from which the melt is discharged. The plastic is formed in the molds. Melt-phase thermoforming has been used to produce monolayer frozen food tubs and wide-mouth snap closures. It is used for coextruding multilayer structures incorporating EVOH for hot, aseptic, and retortable foods (3,5).

In the cuspation-dilation thermoforming process developed in Australia, sheet formation is promoted by expanding blades extending into all areas and distributing the material uniformly throughout the mold. This process is claimed to provide uniform distribution of high barrier components of sheet coextrusions and laminations (3). The process also permits almost vertical side walls to cups; it is employed for the production of margarine tubs in the UK.

Scrapless forming (SPF) uses minimum heat and no trim waste is produced. Sheet is formed in the conventional manner and is cut into square blanks of size approximating the area of the final piece. The surface is lubricated and the blank is clamped and heated to T_g. Under very high pressure the blank is formed into the desired shape by so-called cold forging or billet forging. Stress-free precision pieces with no scrap are claimed; the method has not been commercialized.

Cartons. Semirigid plastic sheet can be fabricated into three-dimensional packaging by cutting, scoring, folding, locking, and adhering. Transparent PVC sheet may be die-cut, scored, folded, and laminated with adhesive into transparent folding cartons. A hot knife and bar may be required to cut and score. The cartons are used for cosmetic and jewelry packaging.

Polypropylene filled with talc or calcium carbonate may be extruded into sheet which may be printed, die-cut, scored, and folded as well as thermoformed. Adhesion may be achieved by using hot air or a flame on the contact surfaces. The package may be water-, water-vapor-, and fat-resistant with the structural integrity of paperboard. Filled polypropylene has many nonpackaging applications. Commercial packaging uses include decorated carrier devices for multiple can packs and pressure-sensitive labels without silicone-coated backing (5).

Spin Welding. In spin welding two or more thermoformed die-cut pieces are joined to produce a hollow can- or bottlelike object with a neck narrower than some part of the body. The open ends of the two thermoformed pieces are engineered in circular shape with one slightly larger than the other for an interference fit; one part is stationary while the other is rapidly rotated. Frictional heat between the two pieces at the interface increases the temperature sufficiently to weld the pieces circumferentially after rotation has ceased and ambient temperature permits setting.

Spin welding produces reverse-taper polystyrene cups and barrel-shaped cups. It has been used in Sweden to combine a hemispherical base to a conical cup to produce a plastic bottle capable of containing beer or carbonated beverages under internal pressure; both PVDC-coated PET and PAN were used.

Spin welding has also been used with high barrier coextrusions for tapered-top plastic cans to be filled with hot liquids.

Blow Molding. In blow molding (qv) a hollow shape is formed, inserted hot into a female mold, and expanded with air or gas pressure in such a manner that the skin assumes the shape of the mold.

Extrusion Blow Molding. In conventional blow molding a single extruder pushes the plastic melt through an annular die to form a tubular parison (Fig. 1). The parison is extruded into, moved into, or grasped by the open mold. The mold closes, pinching off the bottom and gripping the top. An air-blow tube is inserted through the neck opening and pressurized air is blown in to expand the

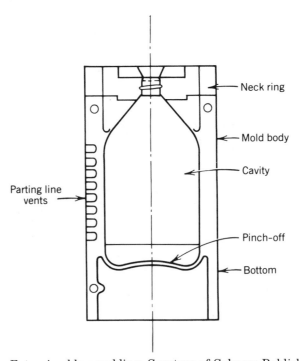

Fig. 1. Extrusion blow molding. Courtesy of Cahners Publishing.

hot, soft plastic to the chilled walls of the mold. Upon setting the parison drops or is extracted from the mold; excess flash is mechanically trimmed (1,2).

Extrusion blow molding produces narrow-neck bottles from high and low density polyethylene, PVC, and polystyrene. Bottles with adequate but not precision neck finishes are often drilled after molding. Without good parison control bottle wall thickness varies, ie, areas of greater diameter have thinner walls, whereas base, shoulder, and neck areas have thicker walls.

Extrusion-blow-molded bottles are the most widely employed for packaging of household chemicals, such as bleach, cleaners, and detergents; motor oil; automotive chemicals; and industrial chemicals in sizes up to 38–95 L.

These bottles may be fabricated with hollow handles. They may be decorated by in-mold labeling, with decals, by screen printing, or hot stamp or postmold labeling. The bottle surface must be treated with flame or corona discharge before printing to accept the decoration.

Gloss may be enhanced by two-layer coextrusion with compatible thermoplastics such as low density polyethylene on the outer surface of high density polyethylene bottles. The second layer is concentrically extruded through the annular die.

Conventional extrusion or coextrusion may be performed on vertical or horizontal rotary or shuttle die configurations. In shuttle blow molding the extruder and die are in fixed horizontal and vertical position; two or more molds shuttle into and out of position beneath the die. By reciprocating in two planes the mold may remove a parison and permit the extruder to function continuously. Since shuttle machines operate back and forth, speed is necessarily low.

Horizontal rotary machines employ multiple molds in a horizontal plane on a rotary turret. As each mold approaches the extruder die exit, it opens to accept the parison and then closes. Later, the parison is blown into bottle shape. The extruder must extrude on an intermittent basis or be intermittently withdrawn to provide a parison for each passing mold. The latter method is widely employed commercially for polyethylene liquid-detergent and bleach bottles and PVC edible-oil and beverage bottles.

Vertical rotary molds also employ multiple molds on a turret but rotate vertically. As each mold reaches the die exit, it grasps the parison and closes. Because of the vertical spacing between molds, intermittent extruder action is not required. Vertical wheels are used commercially for high volume applications such as narrow-neck motor-oil bottles.

For three or more layers, one of which is a high oxygen barrier, the material must be uniformly distributed in the body and neck. Furthermore, a mechanism is required to include the base pinch-off line. Otherwise, the mold-joining line lacks the high barrier material and permits gas transmission (3).

Because high oxygen-barrier plastics are incompatible with other thermoplastics, extrudable adhesives are inserted between the layers. Scrap can be included within the multilayer structure, provided an extrudable adhesive is incorporated.

Blow-molded, two-layer extrusion bottles are more common than three- and more-layer coextrusion-blow-molded bottles. The three-layer bottles incorporating nylon and polyolefins are made by this method for health and beauty aid applications. In Japan EVOH is incorporated as very high oxygen barrier for tomato products and mayonnaise. The technology is used in the United States and Western Europe for tomato catsup, barbecue sauce, mayonnaise, pickle relish, and other foods. Bottles fabricated from internal and external layers of polypropylene contain EVOH as the principal high oxygen-barrier material. The EVOH is protected from moisture by a desiccant–plastic blend often coextruded adjacent to the EVOH layer.

Stretch Blow Molding. Extrusion blow molding of polypropylene is difficult. Bottle orientation enhances structural strength (Fig. 2). For monolayer polypropylene bottles, the two-stage process produces a continuous extrusion of pipe which is cut into fixed parison lengths. The parisons are reheated and stretched longitudinally before circumferential blowing. Impact resistance, gloss clarity, and stiffness are improved, but barrier properties are not (3).

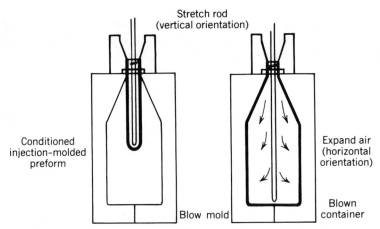

Fig. 2. Stretch biaxial-orientation blow molding used for polyester carbonated-beverage bottles. Courtesy of Cahners Publishing.

Injection Blow Molding. Materials with a very narrow melt-temperature range are formed into bottles by injection blow molding. A test-tube shape is first formed by injection. This preform is transferred to the blow-molding machine and slowly heated to uniform temperature. The heated parison is placed in the blow mold in which it is stretched to induce vertical orientation and blown to shape. Blowing induces horizontal orientation to provide circumferential or hoop structural strength. Stretching and blowing reduce wall thickness, whereas orientation improves structural strength.

Injection stretch-blow molding may be performed on a single one-stage machine in sequence (Fig. 3) or on two independent sequential machines (two-stage process, Fig. 4). Some one-stage machines involve injection molding and immediate blowing, but most involve injection, reheating, and stretch blowing (3,4).

Carbonated-beverage PET bottles are usually produced by injection stretch-blow molding. Some are produced from pipe-extrusion parisons rather than from injection-molded parisons. Pipe extrusion is claimed to be less expensive, but gives poorer quality products. Multilayer injection stretch molding has been commercialized for wide-mouth containers. The basic form is fabricated by injecting

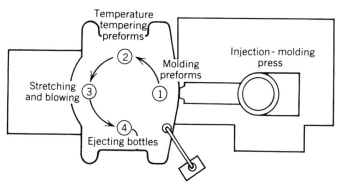

Fig. 3.   Single-stage injection stretch-blow molding: injection, thermal tempering, stretching, and blowing on one machine. Courtesy of Cahners Publishing.

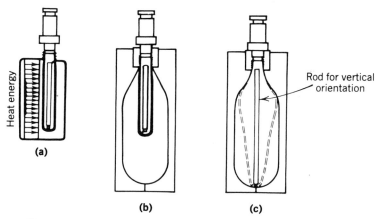

Fig. 4.   Two-step injection stretch-blow molding for polyester carbonated-beverage bottles: (**a**) reheat of injection-molded preforms; (**b**) blow mold in closed position; and (**c**) stretch blow molding. Courtesy of Cahners Publishing.

multiple layers such as polypropylene and EVOH or PET plus EVOH (and, of course, tie layers) and stretching and blowing the parison. Several high oxygen-barrier cans with plastic bodies and ends intended for metal-disk double-seam closing have been introduced. Cans containing polypropylene and EVOH are retorted after filling in order to resist temperatures up to 125°C up to 40 min. Cans fabricated with PET and EVOH are designed for hot liquids to withstand temperatures below 100°C for 1 min (3).

Heat-set Molding.   Recognition of the merits of filling hot foods into plastic packaging, followed by sealing and cooling has led to a need for high oxygen-barrier plastic containers capable of resisting temperatures up to 100°C for brief periods. Resistance to internal vacuum can be achieved by structural design. Plastics with distortion temperatures above 100°C, eg, polypropylene and high density polyethylene, may be filled with hot liquid without fear of thermal distortion, although vacuum collapse is an issue. Polyester requires physical modification to resist heat, drying to remove water, partial crystallization which may opacify the containers, and heat setting. In heat setting the container is molded and secured in the mold while at elevated temperature rather than chilled immediately. The crystalline structure of the material is thereby altered to resist moderately elevated temperatures. This technique is employed to produce PET bottles suitable for filling with hot liquids (3).

The Blow Mold–Fill–Seal System. Most extrusion and injection blow-molded bottles are produced empty, to be filled and closed. Some are produced in so-called blow mold–fill–seal operations designed for aseptic packaging. On blow mold–fill–seal machines the parison is extruded through a multilayer annular die into a sterile space containing sterile molds. The parison is blown with sterile air and immediately filled with cooled product. After filling, the bottle is closed by heat-seal fusion within the mold and removed from mold and sterile chamber through a small opening protected against contamination by the pressure of sterile air.

This method is slow because of the multiple operations on a shuttle machine. The heat of extrusion sterilizes the bottle, which is not readily achieved after molding. Blow mold–fill–seal systems are used commercially on a small scale for sterile pharmaceutical packages and some beverages (3,5).

Barrier Enhancement. Although the base plastic can provide a water-vapor barrier, a high oxygen-barrier plastic improves the barrier properties by many orders of magnitude. The coextrusion of a high oxygen-barrier plastic is not perfected. Postmold internal or external treatments have been proposed, and some are employed commercially.

Exposing the interior of high density polyethylene bottles to gaseous fluorine markedly improves chemical resistance. However, fluorine cannot be used for food containers and bottles. However, it can be used for household, automotive, and industrial chemical packaging as well as for sulfur compounds.

A high oxygen barrier, such as poly(vinylidene chloride) (PVDC), may be applied in an aqueous emulsion to the primed exterior surfaces of PET and other plastic bottles by spraying or dipping. The solvent is evaporated.

Beer PET bottles are coated with PVDC in the UK and have been used in the United States for carbonated beverages in small bottles with high surface-to-volume ratios.

Monolayer Modifications. Other methods for enhancing properties include alloying and laminar blending. In the former the resins are combined before extrusion. Incompatible resins require extrudable adhesive tie layers. Alternatively, resins are blended with extrudable adhesives, allowing monolayer designs.

In laminar blending polyolefin layers are covered with discrete platelike layers of nylon. Dispersed in laminar array, the nylon increases the resistance of polyethylene blow-molded bottles to industrial and automotive chemicals. These containers may be molded by conventional extrusion blow-molding processes (3).

Rigid and Semirigid Plastic Materials

Plastic materials for the fabrication of packaging materials by thermoforming and by injection, compression, and blow molding are usually thermoplastic. A small quantity of thermosets is employed for bottle closures; thermoset PET is employed for compression-molded food trays suitable for microwave and conventional ovens.

Thermoset Plastics. Thermoset plastics are characterized by molecular cross-linking during polymerization which is completed during the fabrication process. Polymerized or set thermoset plastics cannot be softened by heating. They are usually multicomponent systems composed of liquid monomer–polymer mixtures or partially polymerized materials. Upon heating in the mold polymerization is completed. Fillers are often incorporated.

Thermoset plastics are temperature-resistant and rigid, and resist creep and deformation under stress. The thermosetting plastics used for packaging include polyester and melamine–formaldehyde and urea–formaldehyde for bottle closures. Melamine–formaldehyde has no flavor and is hard, retains color, and has excellent resistance to oil. Phenolic resins (qv) are the preferred thermosetting packaging plastics, because of low cost, good formability, temperature resistance, and rigidity. However, they are limited to dark colors. Urea–formaldehyde is expensive, but has good flavor and chemical resistance and excellent appearance characteristics (see AMINO RESINS).

Thermoplastic Materials. High density polyethylene is the most widely used thermoplastic for fabricating rigid and semirigid packaging, followed by polyester, PVC, and polystyrene. Properties are given in Table 2.

Thermoplastics are generally noncross-linked materials that can be heated, melted, hardened, and remelted without degradation. PVC, however, is degraded.

Cellulosics. Cellulosics are included here because of their applications in semirigid packaging. The three principal cellulosic materials, acetate, butyrate, and propionate, are used primarily for thermoformable sheet for high gloss, high clarity, blister-packaging applications. Cellulosics are susceptible to changes in relative humidity, exhibit strength and toughness but limited chemical resistance, and are rarely used for food because of the potential for flavor interaction.

Cellulosic sheets are easy to thermoform. Cellulose acetate is the cheapest, but cellulose propionate has the highest tensile and impact strength (2) (see CELLULOSE ESTERS, ORGANIC).

Polyethylene. Polyethylene derived from natural gas or from naphtha cracking is characterized by toughness, excellent moisture and water resistance, fair chemical resistance, poor gas-barrier properties, and variable processability. Low density polyethylene (density 0.91–0.93 g/cm^3) produced by high pressure polymerization is flexible with high impact strength and poor heat resistance. Injection or blow molding produces translucent squeezable bottles. Low density polyethylene tends to interact with food flavors and fragrances. Linear low density polyethylene, produced at low pressures, has similar properties.

With a density of 0.93–0.94 g/cm^3, medium density polyethylene shares the moisture, liquid, and gas-barrier properties of LDPE. It is stiffer and has a higher temperature resistance. Although medium density polyethylene is a compromise between the flexibility of LDPE and the stiffness of HDPE, it is not satisfactory from a commercial standpoint (2,5,6).

High density or linear polyethylene, with a density of 0.94–0.96 g/cm^3, is a stiff but still flexible plastic with excellent water and water-vapor barrier resistance but very poor gas-barrier properties. Although readily formable by melt processes such as extrusion and injection molding, its high and narrow working temperature range precludes it from monolayer thermoforming applications. High density polyethylene is translucent and subject to flavor and aroma permeation (1,5,6). It has excellent low and modest high temperature properties up to ca

Table 2. Properties of Plastics for Rigid and Semirigid Packaging Applications

	Acrylonitrile–butadiene–styrene	Cellulosics	Nylons	Polyesters	Low density polyethylene	High density polyethylene	Polypropylene	Polystyrene	Poly(vinyl chloride)
Physical									
density, g/cm^3	1.05–1.2	1.2–1.4	1–1.4	1.37	0.91–0.93	0.95–0.96	0.9	1.04–1.07	1.3–1.5
transparency	no	yes	no	yes	no	no	no	yes	yes and no
melt temperature, °C	100–110	140–230	210–220	254–257	106–115	130–137	168	100–105	75–105
water absorption, g, in 24 h	0.2–0.6	2–7	1.3–1.9	0.08–0.09	<0.01	<0.01	0.01	0.01	0.04
Mechanical									
tensile strength, MPa^a	40.6–51.0	20.6–62.0	62.0–82.7	68.9	6.2–17.2	20–37.2	33–37.9	41.3–51.7	34.4–55.1
tensile modulus, MPa^a	1660–2200	480–1790	1170–2760	2760–4140	138–186		1100–1520	3100	2410
Izod impact, J^b	4–10.2	1.5–13.6	1.6–5.4	1.0		0.5–19	0.5–2.9	0.4	0.6–27
Rockwell hardness	75–115	85–120	120	68–78			80–102	60–75	
Thermal									
maximum use temp, °C	71–110	60–104	82–149	79	82–100	79–121	107–149	66–77	66–79
Heat-deflection temp at 180 MPa^a, °C	88–116	49–121	66–49	116	38–49	43–54	52–60	82–104	54–79
Barrier									
gas	poor	poor	excellent	good	poor	poor	poor	poor	variable
water-vapor	poor	poor	poor	good	excellent	excellent	excellent	poor	variable

^{a} To convert MPa to psi, multiply by 145.
^{b} To convert J to ftlbf, divide by 1.355.

120°C. Although high density polyethylene has good dimensional stability, it is subject to creep and softening in the presence of some chemicals.

Polyethylenes are degraded by uv light and are often blended with uv-radiation inhibitors. In some applications, however, uv absorbers are employed to accelerate degradation after use. Polyethylene, if discarded as litter, is exposed to sunlight which accelerates degradation (see also ETHYLENE POLYMERS).

Polyester. Poly(ethylene terephthalate) is preferred for carbonated-beverage bottles because of excellent gas-retention properties. Resistance to gas permeability is good, and melting points are high. The tough, clean, glossy material exhibits low creep under pressure. Chemical and oil resistance are good, as are water-vapor barrier properties. At high crystallization PET is very brittle, particularly at low temperatures.

Because of their short melting range, polyesters may be injection-molded but are not easily blow-molded or thermoformed. Blow molding must be performed on injection-molded preforms. Because of the special needs of the carbonated-beverage industry, PET resins are available in amber and green.

For bottle fabrication PET of medium intrinsic viscosity is dried and injection-molded into amorphous parisons. The parisons are heated above the T_g and stretched and blown to produce biaxial orientation; this improves tensile yield strength and impact and creep resistance.

Thermoplastic "copolyesters" (PET-G) are readily fabricated by direct extrusion blowing or by extrusion thermoforming. Copolyesters are prepared with two or more ethylene glycol molecules. They have less tendency to crystallize and are more easily processed. Barrier properties are significantly lower than those of conventional polyesters; clarity and toughness are retained (2,5,6).

Poly(vinyl chloride). Blends of poly(vinyl chloride) with large quantities of additives exhibit a wide range of properties. Hydrochloric acid is a combustion product. Because of the thermal sensitivity, organometallic tin, lead, barium, calcium, or zinc compounds are added as heat stabilizers. Only a few organotin compounds are accepted by the FDA in food applications.

Plasticizers (qv) impart flexibility, lower melt viscosity, and improve processability. Although heat stabilizers are included in almost all PVC compounds, plasticizers are omitted in compounds intended for rigid applications. For most semirigid or flexible packaging applications, heat stabilizers and plasticizers are incorporated; impact modifiers, such as ABS or EVA, enhance physical properties further. Barrier properties, structural toughness, and clarity are good; gloss and ease of processability are excellent.

Polypropylene. Although polypropylene is crystalline, commercial grades are atactic or noncrystalline. Polypropylene is sensitive to very low temperatures and has a very narrow melting range. Its moisture-barrier properties are excellent, but gas-barrier properties are poor. In molded form it is translucent to opaque. Processing and thermal properties may be improved by random or block copolymerization with ethylene comonomer (2,4,6).

Polystyrene. Polystyrene is a clear, amorphous polymer with outstanding clarity, stiffness, gloss, dimensional stability, and processability. Crystalline or unmodified polystyrene has extremely poor impact resistance which can be improved by the incorporation of styrene–butadiene, which, however, impairs the clarity. Polystyrene has very poor water-vapor-barrier properties. Residual monomer can be malodorous. The principal application of polystyrene is for structural applications. Expanded polystyrene is produced by impregnating the beads with an organic gaseous blowing agent (2,4,6).

Nylon. Nylons are not used alone in rigid and semirigid applications. Nylon-6,6 is the condensation product of a diamine and a dibasic acid; nylon-6 is formed by polymerization of polycaprolactam.

Nylons are characterized by high gas-barrier properties, toughness, good oil resistance, and excellent thermoformability. They are, however, hygroscopic and moisture-sensitive. Nylon-6,6 is used for film and nylon-6 for thermoforming applications in which gas resistance is required (2,4,6) (see POLYAMIDES, PLASTIC).

Other Resins. Acrylonitrile–butadiene–styrene is a terpolymer which may be processed with varying quantities of rubber to increase impact strength. With only modest gas- and water-vapor-barrier properties but excellent oil resistance and formability, ABS is used in thermoforming structural applications.

Polyacrylonitrile (PAN) copolymers with an acrylonitrile content above 70% have outstanding gas-barrier properties. Until 1977 PAN was the leading candidate for plastic packaging of carbonated beverages and beer, but an FDA ruling on residual acrylonitrile monomer prohibited this application. In 1986 development was continued after the monomer issue was resolved. Polyacrylonitrile has modest water-vapor-barrier properties and an inherent yellowish cast. It may be copolymerized with styrene or acrylate (see ACRYLONITRILE POLYMERS).

Polycarbonates (qv) are very tough, transparent, and heat-resistant resins. Gas- and water-vapor-barrier resistance is poor; hydrolytic breakdown occurs at high temperatures. Polycarbonate is readily thermoformed and extrusion blown (2,4,6).

BIBLIOGRAPHY

"Packaging Materials" in *EPST* 1st ed., Vol. 9, pp. 709–714, by Stanley Sacharow, The Packaging Group.

1. A. Griff, *Plastics Extrusion Technology,* Van Nostrand Reinhold Co., Inc., New York, 1968.
2. *Modern Plastics Encyclopedia,* McGraw-Hill Inc., New York, 1985.
3. *Opportunities for Barrier Plastics in Packaging,* Schotland Business Research, Inc., Princeton, N.J., 1986.
4. J. Frados, *Plastics Engineering Handbook,* Van Nostrand Reinhold Co., Inc., New York, 1976.
5. S. Sacharow and A. L. Brody, *Packaging: An Introduction,* Harcourt Brace Jovanovich, Cleveland, Ohio, 1987.
6. S. Sacharow, *Handbook of Package Materials,* AVI Publishing Co., Westport, Conn., 1976.

General References

Handbook and Buyers' Guide, Plastics Technology, New York, 1986.
J. H. Briston, *Plastic Films,* Technomic Publishing Co., Inc., Lancaster, Pa., 1983.
F. Paine, *The Packaging Media,* John Wiley & Sons, Inc., New York, 1977.

AARON L. BRODY
Schotland Business Research, Inc.

MEDICAL SUPPLIES

Changes brought about by single-use, disposable medical devices, prepackaged and sterilized by the manufacturer, include the development of new packaging materials and forms. The primary requirements for packaging of disposables are protection of the product, sterility maintenance, and ease of use. In order to satisfy these criteria, new materials such as Tyvek by DuPont and special heat-sealable coatings have been utilized. Thermoformable plastics such as polystyrene (PS), poly(vinyl chloride) (PVC), acrylics, polyesters, polypropylene (PP), and polyethylenes (PE) are used extensively in tray or blister-type packaging. Film extrusions or laminations, alone or in combination with papers or foil, meet the needs of flexible packages.

Sterilization methods, process conditions, shelf-life requirements, product resistance, and protection from environmental conditions are some of the factors to be considered in the selection or development of medical packaging materials. It is expected that the uses for polymeric materials in the packaging of medical products will continue to grow with the continued development of and concern for modern health-care practices.

Historical Development

The disposable medical-supply industry is relatively young. Traditional products used in medical treatment are made of glass, metals, or fabrics, and designed for reuse. Primary packaging was usually done in bulk form; boxes or cartons were the typical containers. Repackaging for sterilization, performed in hospitals, consisted largely of muslin or paper wrappings, tied or taped to provide a reasonable sterile barrier. Sterilization of these materials is primarily done in steam autoclaves.

Whereas these methods continue to be used today for many reusable devices, major changes in the medical-supply industry, beginning around the period of World War II, have influenced the way in which products are packaged, sterilized, and delivered to the point of use. Several factors have prompted these changes: the need for storable, sterile, ready-to-use supplies and for better infection control in hospitals; the need to reduce labor-intensive hospital functions and to produce larger quantities of medical devices for a growing market; development of lower-cost plastics to replace glass and metals and of mass sterilization methods by manufacturers; development of single-use packaging; and impending legislation for control of medical supplies.

The earliest forms of disposable device packaging consisted of paper bags or pouches, used for dressings, bandages, gloves, drapes, and several other low cost single-use supplies. As more devices were made from plastics, such as syringes, tubing, ostomy products, and clamps, they too were converted to single-use supplies and put into individual packages.

Ethylene oxide sterilization became the preferred method for presterilized devices because of its effectiveness and relative forgiveness with plastics and rubber products that cannot tolerate steam. Primary packaging was still paper, in the form of bags, pouches, or boxes with overwraps.

The introduction of plastics in disposable device packaging began with films for use in bags and pouches. A novel package developed and made by Tower Packaging in the 1960s consisted of extruded PE with a built-in linear tear feature, achieved with specially designed extrusion dies and film orientation (qv). This film, made into bags, was sufficiently permeable to permit slow sterilization with ethylene oxide (ETO) and provided easy opening with the linear tear feature (Tower-Tear Bag).

Other films (qv) were designed to be heat-sealed to paper to form pouches, which could be filled, sealed, ETO-sterilized, and maintain sterile contents for the intended shelf life of the device. A common configuration of this package has a peaked seal on one end for ease of opening and is generally called the "Chevron Pouch."

Molded products are made in many shapes, profiles, and configurations, such as stopcocks, connectors, scissors, clamps, syringes, and other devices with protruding or angular parts. Many of these were not well suited to pouch packaging. Hard, protruding plastics may wear through or puncture flexible paper packages. Therefore, plastic trays were designed to contain these products more securely.

In order to form a sterile package, heat-seal coated papers were used as lidding for trays. Plastics commonly used for medical device trays in the 1960s were cellulosic plastics, ie, cellulose acetate (CA), cellulose acetate propionate (CAP), cellulose acetate butyrate (CAB), PS, and PVC. Paper coatings were typically vinyl- or nitrocellulose-formulated to heat-seal to these plastics.

To enable the package to be ETO-sterilized, coatings were usually applied in a pattern, ie, grid or dots, to allow uncoated areas for gas transmission. The pattern coating and use of prime coats for release were also intended to minimize excessive fiber tear when lids were removed. At best these methods provided a compromise solution to meeting the needs of truly disposable device packaging.

The definite fiber orientation and two-sided physical characteristics of paper must be taken into consideration when forming a package. The peel or opening direction must be oriented with the specific face and web direction of the paper to ensure the least possible degree of fiber tear when the package is opened. Similarly, because of the limited internal strength of papers, adhesives and coatings must be designed not to exceed the paper strength in order to develop a clean-opening package. This disadvantage may be partially overcome with the use of special sizing additives to paper, which reduce the effectiveness of sealants and result in low strength seals (see PAPER).

In 1967 a new nonwoven material was introduced that has significantly influenced sterile medical device packaging. This material, a spunbonded polyolefin made from high density polyethylene (HDPE), was developed by DuPont and is marketed under the registered trade name of Tyvek. By 1969 Tyvek was being used in a limited number of device packages and within a few years was recognized as the premium packaging material in this industry (see SPUNBONDED in NONWOVEN FABRICS).

Tyvek has several characteristics that distinguish it from other synthetic or natural fiber-based flexible materials used. Among these are superior strength, wet and dry; water resistance; chemical inertness; and excellent dimensional stability. In its early commercial use Tyvek's excellent puncture- and tear-resistant qualities suggested use in protective packaging, particularly for irregularly shaped, bulky parts. Tyvek has a unique fiber structure that allows rapid transmission of gas vapors, but at the same time provides an effective barrier to microorganisms. Since 1970 Tyvek has experienced rapid and steady growth as a sterile packaging material and has set a standard of performance for other materials.

Material Characteristics

Three categories of materials are commonly used to satisfy the basic requirements of the primary package forms, ie, pouches, bags, and trays.

Flexible materials, ie, papers, Tyvek, films, composite material laminations, may be used in pouches or bags, or as lidding materials on trays. Polyolefin coextrusions are used as the thermoformed bottom part of a tray or three-dimensional package. These materials are commonly used in form–fill–seal packaging systems.

Semirigid materials, such as various plastic sheet products, typically over 0.127-mm (5 mils) thick, are used in preformed trays or may be run on form–fill–seal equipment. Semirigid tray materials may consist of PVC, PS, acrylics (XT polymer), polyesters, high impact polystyrene (HIPS), HDPE, PP, polyacrylonitrile (PAN) (Barex), polybutylene (PB) (K Resin), cellulose plastics (CA, CAP, CAB), polycarbonate (PC), or may be coextrusions of different polymers.

Sealants or adhesives are generally applied to the surface of a flexible material and enable the sealing of different packaging structures together, usually with heat and pressure. Sealants (qv) are often highly formulated adhesive coatings or may be extrusions of very adhesive polymers such as ethylene–vinyl acetate (EVA).

Flexible Material Properties

Important properties to consider include barrier properties, strength, flexibility, peelability, safety, sealability, printability, processibility, stability, and aesthetics.

Barrier Properties. A primary requirement of any sterile packaging material is that it provide a bacterial barrier. That is, any film or nonfibrous material must be pinhole-free, and fibrous or porous material must have pores below a specified size to prevent passage of microorganisms through the material.

A barrier to gases or liquids is required for packages containing liquid or volatile substances, or materials that need protection from the environment. The degree of barrier required depends on the nature of product, shelf-life requirements, and conditions to which the package will be subjected (see BARRIER POLYMERS).

Porosity is required for packages that are used in ETO sterilization or autoclaving to allow the sterilizing gases to enter and leave the package easily. Porosity is not essential for radiation sterilization.

Strength.   Materials must resist puncture, tearing, or impact, but be flexible, ie, withstand bending, folding, or wrinkling stresses common to many flexible packages. The material must be peelable, ie, the package should open cleanly without generating loose particles that might contaminate a packaged device.

Safety.   A packaging material must be clean, ie, relatively free of particulate or loose fiber matter, and should be pure, consisting of safe ingredients for direct contact with the product. It should be environmentally safe, disposable, and nonpolluting.

Processibility.   There is a requirement for sealing capability. The material must be inherently thermoplastic-coatable with sealants or capable of being sealed with another material. The material must be printable, ie, compatible with processes commonly used for package printing. In general, the material must be able to withstand converting operations, shipping, and handling.

Stability.   Packaging materials must be unaffected by the sterilization procedure used and must be stable on the shelf for the intended life of the product.

Aesthetics.   Clarity is an important consideration in evaluating appearance.

Probably no single packaging material meets all of the criteria described. In selection of materials for a specific packaging application, the most important criteria to be satisfied must be identified, then the materials that best fit these needs are selected to meet these criteria.

Tyvek.   The spunbonded material Tyvek has several attributes that make it very close to an ideal, sterile packaging material. High strength makes it a clean, nonparticulating component of many peelable packages. Tyvek is very resistant to delamination, and even if it delaminates, its continuous-fiber structure prevents the release of particulate material. The high porosity of this HDPE material makes it ideal for ETO sterilization. It has temperature limitations due to its thermoplastic nature, but may also be used in steam sterilization with suitable adhesives and well-controlled sterilization conditions (120°C).

Tyvek is not easily sealed, and in many applications, such as tray lidding, it requires a heat-seal coating to adhere to other plastics. In flexible packages with films containing EVA, plain Tyvek may be used.

The uses for Tyvek in medical-products packaging include flexible, breathable web for peelable "Chevron Pouches," lidding material for trays sterilized by ETO or radiation, cover stock for form–fill–seal packaging, lint-free inner wrap for devices, porous packages for granular or powdered dessicants, insert material, and porous vent for plastic bags. In the last case patches or strips of Tyvek may be sealed over an opening in the bag. This enables the bag to be ETO-sterilized yet remain a sterile barrier. These bags are known as header, breather, or vented bags.

Films, Laminations, and Coextrusions.   Most flexible packages for medical devices contain at least one part that is a plastic film. The most common material used in device packaging is a lamination of polyester and PE. This film provides a number of functions in a pouch: product visibility, puncture resistance, sealability, and peelability.

The materials typically used are 0.0127-mm oriented polyester film, adhesively laminated to low-to-medium density PE (0.038–0.051 mm), usually modified with EVA for better sealability. These films may be sealed to plain or coated Tyvek, plain or coated papers, or other films. Peelability depends on the type of film used or the sealant on the opposite web. These films may also be heat-seal-coated for adhesion to a wider variety of materials or to provide peelable all-film packages for barrier purposes or for radiation sterilization.

Through laminations, coextrusions, or coatings, film constructions may be tailored to perform a wide variety of specialized functions in medical packaging (Table 1).

Table 1. Laminated and Coextruded Film Constructions and Characteristics

Film layers	Properties and applications
nylon–PE	thermoformable packaging
polyester–PE–EVA	peelable, heat-seals to plain Tyvek
polyester–Surlyn[a]	wide seal range, high hot tacks, strong seals
polyester–nylon–PE	high flexibility and puncture resistance
polyester–PP	steam sterilization packaging
polyester–nylon–PP	high temperature resistance and flexibility
polyester–PVDC[b]–PE	oxygen or moisture barrier
metallized polyester–PE	light resistance and improved barrier
polyester–foil–PE	high barrier and puncture resistance
PET–foil–PET–Surlyn	high barrier and chemical resistance
PC–PE–EVA	high clarity, strength, flexibility

[a] PE copolymer ionomer resin.
[b] Poly(vinylidene chloride).

Semirigid Materials

Design.   The versatility of thermoplastics makes possible an almost unlimited variety of three-dimensional packages. Designs can include features that follow the contours of the device or hold several components in one unit. The development of the thermoformed plastic tray with peelable Tyvek lid was a key factor in the widespread use of custom procedure kits in which several components of a particular surgical procedure are contained in one tray. Other built-in features of thermoformed trays include compartments, snap-in lugs, embossed surfaces, hinges, stacking lugs, or various types of recesses.

Recesses of various types and functions can be formed into tray walls. For example, built-in hand holds provide package gripping; stacking lugs allow stable shelf storage of filled cartons; panels are used for labeling or other product information; a snap-in feature for the card base improves tray stability for dispensable items; finger spaces allow easier product removal; locators provide packaging line orientation; denesting lugs facilitate tray separation on the packaging line, particularly for plastics with high coefficients of friction; and ribs, formed into the sides and bottoms of containers, can substantially increase the strength and rigidity of a thermoformed package and in some cases enable the use of thinner-gauge materials for cost savings.

Manufacturers' logos, product trade names, or other information may be readily embossed directly into the package.

For traceability and diagnosis of process problems, individual trays in multiple arrays are commonly embossed with location numbers or lot codes relating to date of manufacture.

Although the trend in device packaging, particularly of high volume items, includes more form–fill–seal with relatively simple rectangular trays or flexible forms, there is still opportunity for innovation in the design of thermoformed packaging of specialized, multicomponent, or difficult-to-contain products.

Material Selection. Functional, economic, and aesthetic factors determine the selection of plastics for a particular application. When considering the cost of a plastic, cost per weight should be evaluated, but the density of the plastic should also be known, so its cost may be expressed in terms of area at a chosen thickness (yield). A lower-density material gives a higher yield than a higher-density plastic at the same cost/kg. Strength, heat resistance, flexibility, and formability may also be important indirect factors in cost evaluation. For example, if a particular plastic performs as well or better than another but at thinner gauges, it may result in cost savings even though its price is higher.

Other factors include ease of cutting, sealability, transparency, barrier properties, sterilizer resistance, heat resistance, stress resistance, cold-temperature impact, coefficient of friction, and compatibility with the device. Thermoformable plastics have established an important role in medical packaging, and will continue to do so, for either preformed trays or form–fill–seal packages.

Sealants and Adhesives

Sealants must provide adhesion, integrity, and ease of opening. Adhesion enables two or more components to bond. Adhesives are necessary with materials that have little or no affinity for each other. Integrity implies that the package components will hold together throughout stages of packaging, sterilization, shipping, handling, and storage. The medical device industry values the functional benefits of cleanly peelable, easy-opening packaging. This feature is primarily attributable to improved adhesives in conjunction with high strength substrates.

Sealants used in disposables packaging are usually preapplied to one or more substrate and activated by simultaneous application of heat and pressure. Other forms of sealants, such as cold-seal, pressure-sensitive, or wet adhesives, are used relatively less or not at all.

The sealant layers may be applied from different systems. Formulated coatings are usually highly specialized systems and, like many adhesives, are mixtures of various polymers, resins, and modifiers. The primary characteristics that differentiate these adhesives from others are peelability and the fact that they must also function as coatings.

Coatings (qv) may be applied from solvent systems, hot melts, or aqueous systems. The more versatile systems are based on EVA polymers and applied from aqueous or latex dispersions (see also ADHESIVE COMPOSITIONS).

Another form of sealant that is gaining in usage and functionality is extrusion coating. These coatings may be single polymers or blends, and are used in films or thermoformable plastics for form–fill–seal packaging with uncoated Tyvek.

Important characteristics of any peelable medical-packaging sealant should be considered. If the package is overall coated and intended for ETO sterilization, the coating must be porous to ETO. There must be a wide seal range, ie, the ability to run on a variety of packaging equipment with different plastics. Adhesive versatility is needed to seal to a variety of plastics and to exhibit consistent peel values. Peelability must be consistent with the strength of the other packaging materials without compromising package integrity. The sealant has to be composed of materials that will not react or degrade in the sterilization process.

Processing and Functional Properties

Typically, a thermoplastic such as PE goes through a temperature–flow cycle during which softening occurs incrementally with increasing temperature until the coating becomes molten. When the heat source is removed, the thermoplastic returns gradually to its original state, but not necessarily at the same rate as shown by its softening curve (Fig. 1). This behavior can be predicted to some extent on the basis of certain known physical properties of the material, such as melt index and molecular weight. Peelable heat-seal coatings are usually mixtures of polymers and various modifiers designed to provide a balance of functional characteristics.

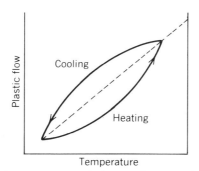

Fig. 1. A typical thermoplastic cycle.

Blocking. Thermoplastics begin to flow when they are subjected to even relatively moderate temperature increases. Therefore, a coated material may activate prematurely when it is stored in rolls under high ambient-temperature conditions, particularly if it is stored under weight. This is called blocking. Blocking can cause adhesive to transfer from one convolution of material to another, cause printed matter to offset, or render the product impossible to unwind.

Hot Tack. An adhesive's performance is characterized in terms of its tackiness rather than its plastic flow (Fig. 2). As the adhesive layer is heated, initiating plastic flow, components of the coating begin to interact and the adhesive

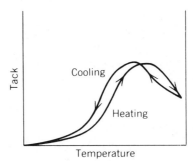

Fig. 2. Temperature–tack cycle.

becomes tacky. Tack increases with temperature up to a peak and begins to decline as the total mass becomes increasingly molten.

Measurement of hot tack aids in the determination of how effectively an adhesive will hold a package together as it cools and solidification occurs. The objective of the adhesives compounder is to make the hot-tack peak as high and as broad as formulation parameters permit. An ideal condition is one in which the hot tack is near or equal to the adhesive bond strength.

Bond Strength. A welded seal is expected to be as strong as the materials it connects. However, peelable adhesives must operate within controlled ranges of bond strength, ie, high enough to maintain package integrity and yet low enough to prevent tearing or particulate formation when the package is opened.

Characterization of bond strength appears in Figure 3. As the bond cools after sealing, its strength usually increases to a fairly well-defined range, which is generally dictated by the internal strength of the materials being connected.

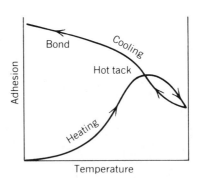

Fig. 3. Temperature–adhesion cycle.

Stress Fatigue. After the package has been securely sealed, the adhesive may continue to react, particularly under conditions encountered during ETO sterilization. The effects of elevated temperatures (49–66°C) and depressurizing usually combine to put stresses on the seal. If the adhesive bond is sufficiently softened under these conditions, it can even come apart. Stress fatigue, or creep, can be characterized by subjecting seals to fixed loads at various temperatures (see also POLYMER CREEP). Adhesive formulation plays a significant role in determining how well a particular system resists stress fatigue. However, the design of the sterilization cycle can be tailored to accommodate the plasticity of adhesives.

Cling.   This characteristic is often associated with ETO sterilization. Cling occurs when a product in contact with the adhesive coating becomes stuck to the coating. Sometimes this is due to an affinity or compatibility between the adhesive and the device, particularly if the device is composed of a soft plastic. However, cling is usually a direct function of heat-softening of the adhesive. Adhesives have been formulated to resist clinging in most cases, but this problem can be greatly alleviated by package design and controlled filling.

Figure 4 illustrates the entire range of the heat-seal characterization cycle. Such an analysis can be a useful tool in the evaluation and selection of heat-seal packaging materials.

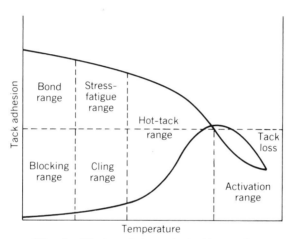

Fig. 4.   Heat-seal characterization cycle.

BIBLIOGRAPHY

General References

C. D. Marotta, "Sterile Device Packaging: An Update," *Med. Dev. Diag. Mag.* **5** (Aug. 1983).

C. D. Marotta, "Tyvek: 'Wonder' Material of Sterile Packaging," *Med. Dev. Diag. Mag.* **3** (May 1981).

A. Friedman and C. D. Marotta, "Flexible Plastics in Medical Device Packaging: Parts I and II," *Med. Dev. Diag. Mag.* **6** (Apr. and June 1984).

R. Brinkman, "Coated vs. Uncoated Webs for Thermoform, Fill and Seal," *Medical Device & Diagnostics 84 East Proceedings,* New York, May 1984, Medical Device and Diagnostics Magazine, Inc., Santa Monica, Calif., 1984.

C. D. Marotta, "Thermoformed Device Packaging: Thermoplastics for Semi-Rigid Packages," *Med. Dev. Diag. Mag.* **6** (Dec. 1984).

T. Frey, "Picking Aco-extrusion Best Suited to Needs," *Package Eng.* (Dec. 1982).

C. D. Marotta, "Adhesive Coated Packaging," *Med. Dev. Diag. Mag.* **2** (Sept. 1980).

C. D. Marotta, "Role of Heat Seal Coatings in Seal-Peel Medical Packaging," *Package Dev. Mag.* (Sept. 1971).

C. D. Marotta, "High Barrier Packaging for Wet Devices," *Med. Dev. Diag. Mag.* **4** (Jan. 1982).

"How Good is Sterile Medical Packaging," *Food Drug Packag.* (Nov. 1986).

Radiation Compatible Materials, HIMA Report 78-4.9, Health Industries Manfacturers Association, Washington, D.C., June 1980.

A. Hirsch and S. Manne, "How Radiation Affects Packaging Materials," *Packag. Dig.* **17**(1) (Jan. 1980).

Guideline for Evaluating the Safety of Materials Used in Medical Devices, HIMA Report 78-7, Health Industries Manufacturers Association, Washington, D.C., June 1978.

B. V. Harris, "Copolyester Use in Medical Packaging," *Med. Dev. Diag. Mag.* **5**(6) (June 1983).

R. C. Griffin, Jr., "Coextruded Packaging for Disposable Medical Devices," *Paper, Film, Foil Converter* **57**(1) (Jan. 1983).

A. Hirsch and S. Manne, "Factors Affecting Seal and Peel Strengths," *Med. Dev. Diag. Mag.* **4**(2) (Feb. 1982).

C. D. Marotta, "The Vital Roll that Sealants Play in Medical Device Packaging," *Packag. Technol.* **11**(6) (Dec. 1981).

Microbiological Methods for Assessment of Packaging Integrity, HIMA Report 78-4.11, Health Industries Manufacturers Association, Washington, D.C., June 1979.

C. D. Marotta, "Vented Flexible Packaging," *Med. Dev. Diag. Mag.* **3**(4) (Sept. 1981).

C. D. Marotta, "High Barrier Plastics," *Med. Dev. Diag. Mag.* **8** (June 1986).

C. D. Marotta, "Recent Developments in Autoclavable Packaging," *Med. Dev. Diag. Mag.* **8** (Aug. 1986).

C. D. Marotta, "Packaging for Radiation Sterilization," *Med. Dev. Diag. Mag.* **9** (Feb. 1987).

CARL D. MAROTTA
Innovation Technology, Inc.

PAD TRANSFER PRINTING.   See DECORATING.

PAPER

Paper consists of sheet materials that are comprised of bonded, small discrete fibers. The fibers are usually cellulosic in nature and are held together by chemical bonds which, most probably, are hydrogen bonds (see CELLULOSE). The fibers are formed into a sheet on a fine screen from a dilute water suspension. The word paper is derived from papyrus, a sheet made in ancient times by pressing together very thin strips of an Egyptian reed, *Cyperus papyrus* (1).

The early use of paper was as a writing medium, replacing clay tablets, stone, parchment, and papyrus sheets. Papyrus sheets are not considered to be paper because the individual vegetable fibers are not separated and then reformed into the sheet. Paper apparently originated in China in 105 AD and was made from flax and hemp or bark fibers of certain trees (2). The manufacture of paper from bark and bamboo spread from China to Japan where its manufacture began in ca 610. Papermaking from flax and hemp spread through Central Asia, the Middle East, and eventually to Europe. Spain and Italy were early European papermaking centers. The first European paper was made in Spain in 1150, in France by 1189, in Germany by 1320, and in the UK by 1494. Papermaking was introduced in the United States around 1700 by Rittenhouse in Philadelphia; a history of U.S. papermaking is given in Ref. 3.

Paper is made in a wide variety of types and grades to serve many functions. Writing and printing papers constitute ca 30% of the total production. The balance, except for tissue and toweling, is used primarily for packaging (see also

PACKAGING MATERIALS). Paperboard differs from paper in that it is thicker, heavier, and less flexible.

Fibrous Raw Materials

The main components used in the manufacture of paper products are listed in Table 1. More than 95% of the base material is a fibrous pulp with over 90% originating from wood (qv). Many varieties of wood are used to produce pulp, and many different manufacturing processes are involved in the conversion of wood to pulp. These range from mechanical processes, using only mechanical energy to separate the fiber from the wood matrix, to chemical processes in which the bonding material, ie, lignin (qv), is removed chemically. Many manufacturing sequences employ combinations of mechanical, chemical, and thermal methods. Pulp properties are determined by the wood type and manufacturing process, and must be matched to the needs of the final paper or board product.

Table 1. Main Components of Paper and Paperboard, wt %[a,b]

	World	North America
wood pulp		
mechanical and semichemical	20.8	21.5
unbleached kraft chemical	18.5	25.2
white chemical[c]	26.6	28.6
waste fiber	25.3	20.3
nonwood fibers	4.2	0.9
fillers and pigments	4.6	3.5

[a] Ref. 4.
[b] 1980 data.
[c] Includes unbleached sulfite.

Mechanical Pulps. In mechanical pulps the fibers are separated by mechanical energy. Because there is no chemical removal of wood components, the process results in a high pulp yield (95%). The chemical composition of mechanical pulps is similar to those of native wood, ie, they contain significant amounts of lignin and hemicellulose in addition to the basic cellulose component of the fiber. In the stone groundwood process, fibers are separated from the wood by grinding logs against revolving stone wheels. The resultant fibers are fractured and much debris is generated. These pulps are used where opacity and good printability are needed. The presence of large amounts of lignin, however, results in poor light stability, permanence, and strength. Such pulps are bleached with sequences that maintain a high pulp yield. Typical bleaching chemicals are alkaline hydrogen peroxide or sodium hydrosulfite (dithionite). Initial brightness of 80% is usually achieved.

Mechanical pulp is also produced using disk refiners to generate pulp from wood chips. Chips are passed between closely spaced revolving disks and the fiber is broken free from the wood material. Refiner mechanical pulp (RMP) is produced at atmospheric pressure from a disk refiner; this is the oldest refiner process. In the thermomechanical pulp (TMP) process, chips are steamed at ca 120°C prior to fiberization in a pressurized disk refiner. Compared to stone groundwood pulp, TMP is comprised of longer, less damaged fibers and less debris, which results in improved paper strength but some loss in opacity. Thermomechanical pulps can be used to reduce or eliminate chemical pulps in many blends. Softwood is the preferred raw material to produce optimum paper strength.

Certain chemical treatments can be employed during the TMP process to improve strength. Sodium sulfite and alkaline hydrogen peroxide have been used for chip pretreatment or posttreatment of the TMP pulp; such pulp is termed chemithermomechanical pulp (CTMP). The strength improvements of up to 50% are obtained at some sacrifice in yield (ca 85 vs 92%) and opacity. The area of CTMP production is receiving much attention as it offers an opportunity to obtain pulp with strength properties close to kraft, but with better economics due to the higher yield and lower capital costs.

Chemical Pulps.   Semichemical pulps are produced by mild chemical digestion of chips prior to reduction to pulp in a disk refiner. Yields are 70–80%, and the pulp contains a lower lignin and hemicellulose content than the wood from which it was derived. The main use for this product is in corrugated paperboard, in which advantage is taken of the stiffness resulting from the lignin and cellulose components. The base woods for semichemical pulps are usually hardwoods. Use of the neutral sulfite process is declining because of environmental problems resulting from the lack of a suitable recovery system. Newer processes, which are based on sodium carbonate and sodium sulfide, eg, the green-liquor process, or on sodium hydroxide and sodium carbonate, are replacing the neutral sulfite sequence. Waste liquor from the green-liquor process is recycled into the kraft-recovery system. Carbonate processes employ modern fluidized-bed systems to dispose of waste liquors. Compared to the neutral sulfite pulps, alkaline semichemical pulps are darker in color and have slightly higher lignin contents.

Chemical pulps have greatly reduced lignin and hemicellulose contents compared to the native wood, as these components are dissolved during chemical digestion. Because the lignin is removed, much less mechanical energy is needed to separate the fibers from the wood matrix and the resulting pulp fibers are undamaged and can produce a strong paper. Chemical pulps are used principally for strength in a variety of paper and paperboard products. When bleached, they provide a high brightness ($\sim$ 90%), light-stable product.

Kraft Process.   The principal process for producing chemical pulp is the kraft process which uses mixtures of sodium sulfide and sodium hydroxide as the pulping chemicals; yields are 46–56%. Chemical pulps in the higher yield (ca 55%) range contain about 10% lignin. They are used in products such as linerboard where strength is of prime importance. The lower yield pulps are usually bleached to remove all lignin and produce high brightness (90% +). These pulps are used where permanence and whiteness are needed in addition to strength. Bleaching technology is multistage and is based on sequential use of chlorine, hypochlorite, and chlorine dioxide (see BLEACHING OF WOOD PULP). The use of oxygen compounds, eg, gaseous oxygen, ozone, and hydrogen peroxide, is becoming more important for environmental and economic reasons. The pulping and bleaching sequences produce pulps with the brightness, yield, and strength required for a specific end use.

Acid Sulfite Process.   Small amounts of chemical pulp are manufactured by the acid sulfite process. Overall inferior product properties and the lack of an acceptable chemical recovery process has reduced production of sulfite pulps. Magnesium-, sodium-, or ammonia-based systems have largely replaced the calcium sulfite process. Sulfite pulps are used primarily in product areas that require high purity and brightness.

Other Sources.   *Reclaimed fiber* accounts for ca 20% of the total fiber used in the United States. A variety of sequences are used to disperse and clean the waste fiber. Emphasis is on mechanical screening and cleaning and only limited chemical treatment. The properties of these pulps depend largely on the raw material; strength and brightness are lower than that of a comparable virgin pulp. Contamination by plastics is a major problem.

Nonwood fibers are used in relatively small volumes (ca 4%). Examples of nonwoody pulps and products include cotton linters for writing paper and filters, bagasse for corrugated media, esparto for filter paper, or Manila hemp for tea bags. Synthetic pulps based on glass or polyolefins are also used. These pulps are relatively expensive. They are used in blends with wood pulps where they contribute a specific property to meet a specific product requirement.

Physical Properties

Most physical properties of paper depend on direction. For example, strength is greater if measured in the machine direction (MD), ie, the direction of manufacture, than in the cross-machine direction (CD). For paper made on a Fourdrinier paper machine, the ratio of the two values varies from about 1.5 to 3.5. An even greater anisotropy is observed if either of the in-plane values is compared to the out-of-plane strength.

Paper may be considered an orthotropic material, ie, one possessing three mutually perpendicular symmetry planes (5). The three principal directions are the machine, cross-machine, and thickness directions. There are several reasons for this anisotropy. Wood pulp fibers are long, slender, and, in the paper, usually ribbonlike rather than circular in cross-section. During the deposition of the slurry onto the wire, the fibers tend to align in the direction of the moving wire, and the extent to which they do this depends largely on the difference between the jet and wire speeds. The sheet tends to become stiffer and stronger in the machine direction as more fibers line up in this direction. Another important factor is the tendency for paper to become stiffer and stronger when dried under restraint, ie, prevented from shrinking during drying. Machine-made paper is dried by passing it through an array of drum dryers. The paper is thus under tension in the machine direction during this process and prevented from shrinking, making the paper stronger and stiffer in this direction.

Because the fibers are slender and often ribbonlike, they tend to be deposited on the wire in layers. There is little tendency for fibers to be oriented in the thickness direction, except for small undulations where one fiber crosses or passes beneath another. This layered structure results in different values for properties measured in the thickness direction as compared to the same properties measured in one of the in-plane directions.

The three-dimensional, orthotropic nature of paper is most easily observed in the elastic properties because it is possible to measure seven of the nine elastic moduli necessary to describe paper as an orthotropic material, using sound-wave propagation methods (5–7). It would be difficult or impossible to characterize orthotropic paper completely using the more traditional measurement techniques. The elastic properties are sensitive indicators of paper-machine operating conditions and related to end-use performance (8). Table 2 gives some of the elastic moduli for several commercial papers and laboratory sheets (handsheets).

The elastic properties of a material may be expressed in several ways. The C_{ij}s in the table are called the elastic stiffnesses and have units of stress or force per area. Paper is considered to be an orthotropic material, having elastic properties that are symmetric about each of three mutually perpendicular symmetry planes. An orthotropic material requires nine independent elastic stiffnesses to describe its elastic response to applied stresses completely. In paper the three principal axes are taken as the machine (1 or x) direction, the cross-machine (2 or y) direction, and the thickness (3 or z) direction. Thus the subscripts on the nine elastic stiffnesses refer to the directions of the applied stress and the resultant strain. For example, C_{11} is the elastic stiffness when the stress and strain are both normal to the material along the 1 (or x) axis.

Engineering constants represent the elastic response of a material to applied stresses. For orthotropic paper, nine independent elastic parameters are required.

Paper is a viscoelastic material exhibiting creep and stress relaxation phenomena (9). The mechanical and most other properties, therefore, are highly sensitive to the temperature and the moisture content of the paper. All mechanical tests should be carried out using TAPPI standard conditions of 50% rh and 23°C. The moisture content of paper, defined as the weight of water to the weight of paper and water, however, exhibits a hysteresis with relative humidity. The equilibrium moisture content vs relative humidity curve upon desorption of water lies above the curve for absorption of water. Because the differences can be as much as 1%, corresponding to an 8–10% change in some mechanical property, it is also important to specify from which direction (absorption or desorption) the equilibrium condition at 50% rh was approached. This is usually handled by preconditioning the samples to be tested at some low relative humidity (eg, 15%) and then going to the standard condition at 50%.

The weight of paper in a paper sample is the difference between the weight of the specimen and its weight after drying at 105°C until no change in weight is observed. This is specified in standard TAPPI method T412. The moisture content is then the ratio of this weight to the original weight of the specimen, expressed as a percentage.

The basis weight W is the mass in grams per square meter. In the United States, this is determined in accordance with standard method T410. Basis weight can also be expressed as kilograms of a ream of 500 sheets of a given size, but the sheet sizes are not the same for all types of paper. Typical sizes are 43.2×55.9 cm for fine papers, 61.0×91.4 cm for newsprint, and 63.5×96.5 cm for some book papers. The most common designation is pounds per 3000 square feet (1.63 g/m^2) for paper. The basis weight of board is usually expressed as pounds per thousand square feet ($\sim$ kg/204 m^2). For example, the material comprising a 69-lb linerboard weighs 69 lb/1000 ft^2 (337 g/m^2). Some typical basis weights are tissue and toweling, 16–57 g/m^2; newsprint, 49 g/m^2; grocery bag, 49–98 g/m^2; fine papers, 60–150 g/m^2; kraft linerboard, 127–439 g/m^2; and folding boxboard, 195–586 g/m^2.

Table 2. Elastic Properties[a]

Material	Apparent density ρ, kg/m³	Three-dimensional bulk stiffness, GPa[b]									Engineering constants[c], GPa[b]								
		C_{11}	C_{22}	C_{33}	C_{12}	C_{13}	C_{23}	C_{44}	C_{55}	C_{66}	E_x	E_y	E_z	ν_{xy}	ν_{xz}	ν_{yz}	G_{yz}	G_{xz}	G_{xy}
carton stock	780	8.01	3.84	0.042	1.36	0.092	0.91	0.099	0.137	2.04	7.44	3.47	0.040	0.15	0.008	0.021	0.099	0.137	2.04
linerboard	691	8.12	3.32	0.032	1.19	0.113	0.082	0.104	0.129	1.80	7.46	3.01	0.029	0.117	0.0109	0.021	0.104	0.129	1.80
corrugating medium	538									1.24	5.28	1.97		0.167					1.24
bleached-kraft softwood handsheets[d]	721	10.9	6.40	0.172				0.290	0.343	3.09	10.3	6.04		0.182			0.290	0.343	2.97
	673	16.6	2.78	0.073				0.151	0.260	2.28	16.5	2.76		0.036			0.151	0.260	2.40

[a] All tests were carried out at 50% rh, 23°C. The subscripts refer to the direction of measurement, the first normal to surface, the second in the direction of force.
[b] To convert GPa to psi, multiply by 145,000.
[c] E = Young's modulus; G = shear; ν = Poisson's ratio (dimensionless).
[d] At different densities.

The thickness or caliper is the thickness of a single sheet measured under specified conditions (TAPPI T411). It is usually expressed in micrometers or thousandths of an inch (mils). Calipers in micrometers for a number of common paper and board grades are capacitor tissue, 7.6 μm; facial tissue, 65 μm; newsprint, 85 μm; offset bond, 100 μm; linerboard, 230–640 μm; and book cover, 770–7600 μm.

The apparent density of paper is defined as basis weight divided by caliper. It is, therefore, the density of the air and fiber (and water) composite. The density of paper is usually taken to be a fundamental parameter, related to the degree of bonding between the fibers. Higher bonding levels give higher densities. For research purposes, apparent densities determined using the caliper measurement described above sometimes are not satisfactory, especially in the case of papers with rough surfaces. The surface roughness is included in the caliper measurement, leading to densities that are too low. A number of other methods for measuring caliper have been developed in recent years (10).

The tensile strength is the force per unit width parallel to the plane of the sheet that is required to produce failure in a specimen of specified width and length under specified conditions of loading (TAPPI T404 and T494). The strength of paper is also expressed in terms of breaking length, ie, the length of paper that can be supported by one end without breaking (specific strength). Breaking lengths for typical papers are from ca 2 km for newsprint to 12 km for linerboards. The values for the stronger papers compare favorably with other engineering materials. For example, breaking lengths for aluminum are ca 20–25 km.

Stretch is the extension of strain resulting from the application of a tensile load applied under specified conditions (TAPPI T404 and T494). The numerical result is usually expressed as a percentage of elongation per original length and includes the elastic and the inelastic extensibility of the paper. Stretch is greatest in the cross-machine direction, except for creped grades. It is becoming more common to evaluate the elongation as a continuous function of the applied load. The initial slope of the load-elongation curve, ie, load–width vs strain, defines the modulus of elasticity E in the machine or cross-machine directions. In this definition the units of elastic modulus are load per width.

Tensile-energy absorption (tea) or toughness is the work done on the specimen in straining it to failure (TAPPI T494). It is the area under the stress–strain curve expressed in energy units per unit area, eg, J/m^2 ($ftlbf/in.^2$). It is useful in characterizing grades where tensile strength and stretch are both important, eg, sack papers.

Ring crush is a measure of the edge-compressive strength, typically measured on linerboard and corrugating medium. Edge compression in these materials is known to be strongly related to corrugated-box performance in top-to-bottom loading. In the ring-crush test a narrow strip of paper is formed into a ring and then loaded in compression along the cylindrical axis (TAPPI T818). The ring shape minimizes buckling of the paper during the test. It is known, however, that in the case of very thin papers buckling can still occur, introducing errors. A number of other methods to minimize buckling have been studied in recent years, but the trend seems to be toward a short-span compressive test. In this method (not a TAPPI standard), a small rectangular strip of paper is clamped between two parallel jaws separated by 0.7 mm and then loaded in uniaxial compression. It is likely that this test or some similar approach will become a standard method before 1990.

The bursting strength is the hydrostatic pressure required to rupture a specimen when it is tested in a specified instrument under specified conditions. It is the pressure required to produce rupture of a circular area of the paper (30.5-mm dia) when the pressure is applied at a controlled rate (TAPPI T403). It is related to tensile strength and stretch, and used extensively throughout the industry for packaging and container grades.

Tear strength or the internal-tear resistance is the average force required to tear a single sheet of paper under standardized conditions in which a small cut is put in the specimen prior to tearing (TAPPI T414). Internal-tear resistance should be distinguished from initial or edge-tear resistance.

Stiffness is related to bending resistance. It is most commonly measured by determining the force required to produce a given deflection or by measuring the deflection produced by a given load when the paper specimen is supported rigidly at one end and the deflecting force is applied at the free end. A fundamental measure of bending stiffness is the flexural rigidity of the sample. Flexural rigidity is the product of the modulus of elasticity and the second moment of the cross section. The bending stiffness of paper varies with the cube of the thickness and directly with the modulus of elasticity (TAPPI T489 and T543).

Folding endurance refers to the number of folds a paper can withstand before failure when tested according to TAPPI T423 or T511.

Typical strength properties for some commercial and laboratory papers are given in Table 3. The MD–CD tensile ratio (anisotropy) is typically around 1.5–3.0 for most papers produced on a Fourdrinier paper machine. This ratio can be varied by changing the relative speeds of the machine wire and the fiber slurry being pumped onto the wire, which changes the extent of fiber orientation in the MD, or by changing the extent of stretching of the wet web or restraints during drying. Paper shrinks during drying, if allowed to do so, up to about 15%. If it is restrained so it cannot shrink, as it is in the machine direction, it will be stronger in this direction, but be less extensible (smaller stretch). The MD–ZD (thickness direction) anisotropy has not been extensively investigated, but published values are between 60 and 250 (11). This ratio is independent of the extent of fiber orientation in the plane of the paper, but is very sensitive to the extent of wet pressing and wet stretching and drying restraints. Paper is very weak in the thickness direction compared to the machine or cross-machine directions. Table 2 shows that it is also much less stiff in the thickness direction compared to the MD and CD.

Water resistance refers to that property of a sheet that resists passage of liquid water into or through the sheet. The tests are usually designed to simulate use conditions; consequently, there are several dozen different test methods. The more widely used are the Cobb-size (T441) and water-drop tests (T432, T492, and T819).

Water-vapor permeability refers to a specific permeability of the paper to water vapor. The two common gravimetric methods for evaluating this property are given in TAPPI T448 and T464; the tests differ in the temperature- and vapor-pressure differences which cause the permeation. The permeability is usually reported in grams of vapor permeating one square meter of paper per 24

Table 3. Strength Properties[a]

Sample	Apparent density, kg/m^3	Breaking length, km			Stretch, %		Tensile-energy absorption[c], kJ/m^2 [b]		Compressive strength[c], MPa[d]		Tear resistance, MN[e]	
		MD	CD	ZD	MD	CD	MD	CD	MD	CD	MD	CD
linerboard	721	7.22	3.71						8.28	4.83	863	912
kraft envelope	695	8.42	4.32		1.2	2.6	11.2	11.0			637	637
rag bond	687	4.9	3.47		1.8	4.7	6.29	13.2			118	226
newsprint	596	3.65	1.84		1.1	1.4	1.78	1.29				
bleached-softwood kraft handsheet[f]	721	9.11	5.28	0.066	3.2	3.7						
	673	14.3	3.15	0.056					23.1	15.1		

[a] MD = machine direction; CD = cross-machine direction; ZD = thickness direction.
[b] To convert kJ/m^2 to ftlbf/in.^{2}, divide by 2.10.
[c] Short-span test.
[d] To convert MPa to psi, multiply by 145.
[e] To convert MN to pound-force, multiply by 225,000.
[f] At different densities.

hours. Because of the unusually high affinity of cellulose for water and water vapor, water-vapor permeability generally does not correlate with permeability to other vapors and gases.

The common optical properties of paper are brightness, color, opacity, transparency, and gloss. Brightness is the reflectivity of a sheet of pulp or paper for blue light at 457 nm (TAPPI T452). The reflectivity is the reflectance for an infinitely thick sample. Color is measured by evaluating the spectral reflectivity, as given in TAPPI T442, T524, and T527. Opacity relates to that property of a sheet that prevents dark objects in contact with the back side of the sheet from being seen. It is usually evaluated by contrast ratio, which is the ratio of the diffuse reflectance of the sheet when backed by a black body to that of the sheet when backed by a white body of given absolute reflectance value (TAPPI T425). Transparency is that property of paper by which it transmits light so that objects can be seen through the paper. Transparency ratio is a measure of transparency as judged when a space separates the specimen and the object being viewed. Gloss is the ability of the surface to reflect light specularly. There are numerous definitions of gloss as it relates to appearance criteria for paper, eg, specular gloss (TAPPI T480) and low angle gloss, which is often used as a smoothness test for linerboard (TAPPI UM (useful method) 558).

Chemical Properties

The chemical composition of paper is determined by the types of fibers used and by any nonfibrous substances incorporated in or applied to the paper during the papermaking or subsequent converting operations. Paper is usually made from fibers obtained from wood pulp. The chief chemical components are cellulose, hemicellulose, and lignin. Smaller amounts of extractives and inorganics are also present. Occasionally, synthetic fibers and fibers from other plant sources are used. Paper properties that are affected directly by the chemical composition of the fibers include color, opacity, strength, permanence, and electrical properties. Development of interfiber bonding during papermaking is also strongly influenced by fiber composition. Because lignin in the fibers inhibits bonding, groundwood pulp is used in newsprint and in some book and absorbent papers which do not require a highly bonded structure. Hemicelluloses in chemical pulps contribute to bonding; therefore, pulps containing hemicelluloses are used for wrapping papers and other grades which require bonding for strength, and in glassine, which requires bonding for transparency.

In most papers, the chemical composition largely reflects nonfibrous materials that were added to the paper to achieve the desired physical, optical, or electrical properties. Examples of chemicals and resultant properties are dyes and optical brighteners to enhance appearance, resins to impart wet strength, rosin or starch size to reduce penetration of aqueous liquids, pigment coatings to provide a smooth surface for printing, mineral fillers to increase opacity, polymers applied by saturation or extrusion to impart mechanical or barrier properties, and cationic polyelectrolytes and resistive polymers used in the interior and on the surface, respectively, of papers for dielectric recording (see also PAPER ADDITIVES AND RESINS).

The performance of some papers depends on the chemical reactions of noncellulosic additives. Specialized paper coatings in which chemical reactions occur at the time of use are essential in photographic, thermal, and carbonless copy papers (see MICROENCAPSULATION). Strips of paper saturated with color-forming reagents permit rapid, inexpensive urinalyses. Phosphates or halogenated compounds are incorporated in papers to promote flame retardancy.

For acceptable performance, some grades of paper should not contain certain chemicals. Papers that are used for electrical insulation must be free of electrolytes, and papers that are used for permanent documents must be low in acidity. Reducible sulfur compounds, ie, sulfide, elemental sulfur, and thiosulfate, should not be present in papers that receive metallic coatings or in antitarnish papers that are used for wrapping polished silver or steel items. Only chemical substances approved by the FDA can be included in papers for food-contact applications.

Manufacture and Processing

Stock Preparation

Stock preparation includes the operations undertaken to prepare the pulp and additive mixture (furnish) for the papermaking process (12–15). Papermaking pulps are most conveniently handled as aqueous slurries. They can be transported, measured, subjected to desired mechanical treatments, and mixed with nonfibrous additives before delivery to the paper machine. In the case of adjacent pulping and papermaking operations, pulps are usually delivered to the paper mill in slush form directly from the pulp mill. Purchased pulps and recycled paper are typically received as dry sheets or laps, or in bales, and must be slushed before use. The objective of slushing is to separate the fibers and disperse them in water with a minimum of mechanical work in order not to alter the fiber properties. Slushing is accomplished in several types of equipment, such as the Hydrapulper illustrated in Figure 1.

Beating and Refining. These synonymous terms refer to certain mechanical actions to which pulps are subjected before being formed into a paper sheet. In commercial operation, such treatments are carried out by passsing the pulp slurry through narrow gaps between rapidly moving bar-covered surfaces. This beating or refining of pulp improves the interfiber bonding and other physical properties of the finished sheet, but also affects the dewatering behavior on the paper machine.

The impacts that pulp fibers experience as a result of bar crossings in a refiner produce physical changes. These include maceration leading to internal fibrillation and bond breaking, with accompanied fiber swelling; external fibrillation to liberate the ends of hairlike fibrils from fiber surfaces; the generation of fines of various sorts; and fiber cutting or shortening. The first two effects are the most important. Hence, refining greatly increases the wet specific surface of fibers, the swollen specific volume, and fiber flexibility, with increased affinity for water. Because of the unique cellulose–water relationship, these changes significantly enhance the ability of fibers to bond to each other when dried from a water suspension and, therefore, to form a stronger sheet.

Fig. 1. The Hydrapulper. Courtesy of Black Clawson Co.

Optical micrographs of a softwood kraft pulp before and after several periods of beating are shown in Figure 2. In Figure 3, the tensile strength, bursting strength, and tear resistance are given as functions of beating time for a softwood kraft pulp. Within the commercial range, beating generally increases tensile and bursting strengths, folding endurance, and sheet density, whereas it reduces tear resistance.

Although industrial batch-beating equipment has largely been replaced with continuous high capacity pump-through refining installations, a few batch systems remain in commercial use. Most batch equipment is a form of the Hollander beater, developed in the Netherlands in ca 1690. It consists of an elongated tub, with provision for the pulp to circulate around a dividing midfeather under the pumping action of the beater roll. The beating action occurs as the stock passes between knives or bars mounted on the roll and similar knives on a bedplate (Fig. 4**a**). Control of the load between the roll and bedplate determines the severity of the action.

The first successful continuous refiner was the Jordan, developed ca 1860 (Fig. 4**b**). The refining action is controlled by longitudinal movement of the rotating plug, regulating the pressure on the stock between the rotor and stator bars. Wide-angle refiners (Fig. 4**c**), such as the Hydrafiner (16) and the Claflin (17), were developed as modifications of the Jordan. Vanes or impellers may be incorporated into the rotating element to ensure stock transport, which flows from the narrow end to a peripheral discharge. The most recent and rapidly growing refiner design, however, is the disk refiner (Fig. 4**d**). It consists of one or more disks on a rotating shaft to provide two or more essentially flat, parallel working surfaces. The bar-covered plates mounted on the disks are typically pressed together hydraulically and guided by high precision bearing systems.

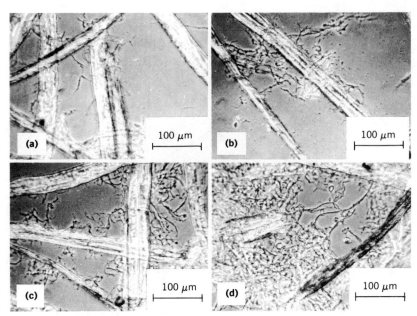

Fig. 2. Optical micrographs of a softwood sulfite pulp (**a**) before beating and (**b**)-(**d**) after beating in a Valley beater for different periods of time. Beating increases from (**b**) to (**d**). The Canadian standard freeness values (CSF; T227 OS-58) are (**a**) 700 mL; (**b**) 620 mL; (**c**) 250 mL; and (**d**) 90 mL.

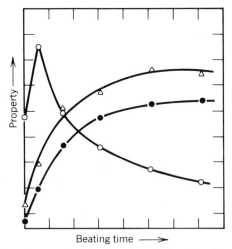

Fig. 3. Typical beating curves for a softwood kraft pulp: △, tensile strength; ●, bursting strength; ○, tear resistance.

Stock is usually fed through the center of one disk and leaves between the plates at the periphery.

The properties of the refined stock may be varied through control of operating variables. For example, high stock consistency, dull refiner bars, or low pressure or excessively large gaps between the two sets of bars may result in

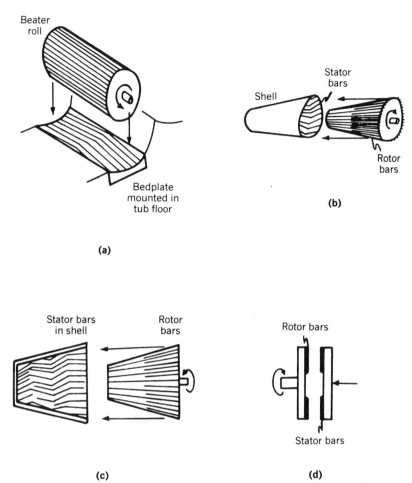

Fig. 4. Beater and refiner rotors: (**a**) cylindrical rotor (beater); (**b**) low angle conical refiner (Jordan); (**c**) wide-angle conical refiner (eg, Claflin); and (**d**) disk refiner.

mild refining action with accentuated fibrillation and swelling. Low stock consistency, sharp bars, and high pressure between elements can result in severe refining with excessive fiber-length reduction and detrimental effects on strength.

Paper-machine Operations

Continuous-sheet forming and drying came into use ca 1800; the cylinder machine and the Fourdrinier were employed. The former (Fig. 5) is especially suited to multi-ply sheets. A series of cylinders covered with wire mesh are mounted in vats containing the fiber slurry. As the cylinder revolves, water drains inward through the screen and the paper web is formed on the outside. The wet web is removed at the top of the cylinder or onto previously formed paper layers carried by a felt. The sheet with a felt then passes through press rolls for water removal and over steam-heated cylindrical dryer drums. The Fourdrinier (Fig. 6) basically consists of an endless screen belt, termed the wire, supported by

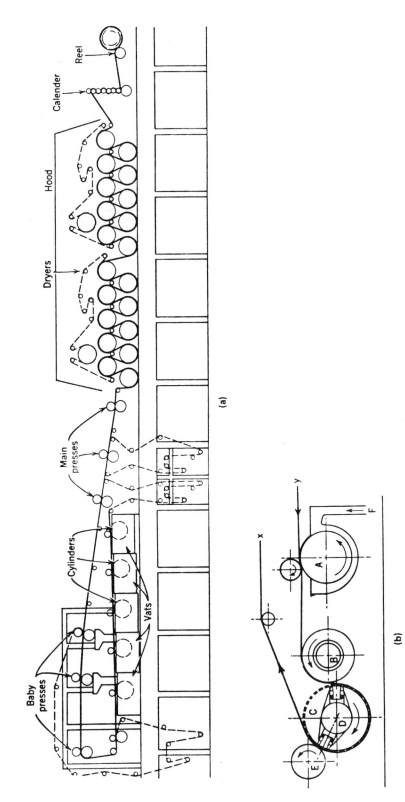

Fig. 5. (a) Cylinder paper machine. Courtesy of Beloit Corp. (b) Detail of cylinder paper machine. A, Last cylinder, taking up slurry from surrounding vat; B, extractor; C, perforated cylinder containing D, a stationary water extractor; E, press; and F, supply of slurry; x, paper on its way to the dryers; y, multiple wet-paper sheet coming from preceding cylinders (18). Courtesy of J. H. de Bussy.

734

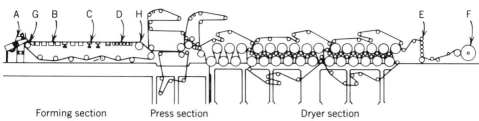

Fig. 6. Fourdrinier paper machine with A, headbox; B, Fourdrinier wet end with foil boxes; C, wet and D, dry suction boxes, pickup, and closed transfer of web through the press section and dryer section; E, calender; and F, reel; G and H are the breast roll and couch roll, respectively. Courtesy of Beloit Corp.

various devices that enhance drainage. The fiber slurry is introduced at one end through a headbox and loses water as it progresses down the wire, thereby forming the sheet. It then passes to presses and dryers as in the cylinder machine. At the end of the dryer section, many machines are equipped with a calender, a stack of loaded rolls through which the paper is threaded to produce the desired surface finish and caliper (sheet thickness), before winding on the reel.

Both of these historical machine designs are still used, according to the same principles, even though new paper machines have been designed. Cylinder machines typically employ five to seven vats for the manufacture of heavy multiply board grades. Other designs of multi-ply formers, however, are becoming more common as demands for higher production speeds increase. Even the Fourdrinier, the standard of the industry for many years, is being supplanted in some applications by certain twin-wire designs. In these machines, substantial dewatering occurs between two wires rather than on one.

Subsequent to stock preparation and dilution, the paper furnish is usually fed to the paper machine through screens or centrifugal cleaners to remove contaminants, eg, dirt and fiber bundles. The stock then enters a flowspreader, such as a tapered header, which provides a uniform stream the width of the paper machine. In all machines, except vat-type designs, the slurry then enters a headbox, whose function is to deliver to the wire a uniformly and properly dispersed flow of stock. Uniformity is a key consideration in the various paper-machine operations. Hence, the headbox must deliver a continuous, uniform flow rate of constant dilution and fiber makeup at all points across the machine. The headbox stock must be uniformly dispersed on a small scale for proper sheet formation and on a larger scale for even sheet weight. Constancy is also important in the dewatering forces on the wire, in the presses, and in the dryers. Attention to such considerations is essential for the production of a well-formed sheet, with good moisture, weight, and caliper profiles.

Headboxes. Headboxes were originally open vats extending across the width of the machine, designed to provide a constant flow of stock out the bottom opening (slice) onto the Fourdrinier wire. As requirements for speed and better dispersion increased, new designs evolved. These developments included the use of rotating perforated rolls in the box for proper turbulence and flow resistance, and the addition of an air pad at the top of the box for flow-velocity control. Air-padded headboxes of modern design are still widely used on Fourdriniers, although hydraulic headboxes are becoming increasingly common. The latter, essential on twin-wire machines, is totally filled with stock and employs channel designs to achieve the proper turbulence and dispersion. Many hydraulic headbox designs can be more appropriately called nozzles or inlets.

Filling and Loading. Materials, eg, mineral pigments for filling and loading, are added to the pulp slurry to make the papermaking furnish (see PAPER ADDITIVES AND RESINS). Pigments are used in varying amounts, depending on the paper grade, and may comprise 2–40 wt % of the final sheet (19). Fillers can improve brightness, opacity, softness, smoothness, and ink receptivity. They almost invariably reduce the degree of sizing and the strength of the sheet. The brightness, particle size, and refractive index of fillers influence the optical properties of the finished sheet. The particle size and specific gravity are important in regard to the filler retention during sheet formation. All commercial fillers are essentially insoluble in water under the conditions of use.

Kaolin or China clay is used both as a filler material and as a coating pigment. It is a low cost, naturally occurring, hydrated aluminum silicate with widespread application. The use of calcined kaolin clay as a filler pigment has been increasing. Titanium dioxide is probably the most desirable pigment for opacity improvement, particularly in fine papers, but is more expensive than other pigments; both anatase and rutile are used. Calcium carbonate is used particularly in book, cigarette papers, and other neutral-to-alkaline papers. It may be produced in the pulp mill as a by-product of causticization, or it may be obtained as ground limestone or chalk. Calcium carbonate is not used in papers that are sized in an acid furnish because of its solubility and resultant foam problems. Talc, alumina trihydrate, synthetic silicas and silicates, and barium sulfate are sometimes used as fillers. Some properties of fillers are listed in Table 4 (see FILLERS).

Table 4. Properties of Paper Fillers

Filler	Specific gravity	Refractive index	Particle size, μm	Brightness, %
kaolin clay	2.6	1.57	1–5	78–85
calcined clay	2.7	1.62	1.0	90–93
titanium dioxide				
anatase	3.9	2.55	0.3	98–99
rutile	4.2	2.7	0.3	98–99
calcium carbonate				
calcite	2.7	1.66–1.49	0.5–2	95–98
aragonite	2.9	1.68–1.53	1.0	97–99
talc	2.75	1.57	1–5	80–95
alumina trihydrate	2.4	1.57	1.0	98–101
synthetic silicates	2.1	1.55	0.01–0.5	91–100

The retention of pigments in the sheet during formation is important. Unless the white-water system is essentially closed, ie, there are no losses, high filler losses can result. So-called white water is the water drained from the sheet as it is formed. Both hydrodynamic mechanisms and colloidal or coflocculation phenomena are significant in determining filler retention (12). Polymeric retention aids are frequently used to increase titanium dioxide retention. Synthetic silicas and silicates extend the optical efficiency of titanium pigments. Talc reduces the deposition of pitchlike materials onto paper machinery.

Sizing.   Sizing additives increase sheet resistance to penetration by liquids, particularly water. Unsized or waterleaf paper freely absorbs liquids. Writing and wrapping papers are typical sized sheets in contrast to blotting paper and facial tissue which are usually unsized. Rosin, hydrocarbon, and natural waxes (qv), starches, glues, asphalt emulsions, synthetic resins, and cellulose derivatives are used as sizing agents. The agents may be added directly to the stock as beater additives to produce internal sizing, or the dry sheet may be passed through a size solution or over a roll that has been wetted with a size solution. Such sheets are said to be tub- or surface-sized.

Rosin, which is derived primarily from tall oil, is one of the most widely used sizing agents. The extracted rosin is partially saponified with caustic soda and processed to a thick paste of 70–80 wt % solids, of which as much as 30–40% is free, unsaponified rosin. Dry-rosin sizes, free-rosin sizes, completely saponified rosins, and dispersed sizes are also available. At the paper mill, the paste rosin is diluted to ca 3 wt % solids with hot water and vigorous agitation. The solution is added to the stock (0.2–3.0 wt % size based on dry fiber) usually before, but sometimes simultaneously, with one to three times as much aluminum sulfate, which precipitates the rosin on the fibers as flocculated particles. Dispersed size is used at lower concentrations. A pH of 4.5–5.5 after the addition of alum is critical. At higher pH, sodium aluminate may be used to precipitate the rosin size. The exact mechanism of sizing with rosin and alum is controversial, but good retention and distribution of the hydrophobic-size precipitates on the fiber surface are necessary.

Acid-sensitive fillers, eg, calcium carbonate, present problems in rosin-sized furnishes of pH 4.5–5.5. Much attention has been focused upon the adverse influence of acid conditions during papermaking on the deterioration of paper with time. Accordingly, production of paper under neutral–alkaline conditions is increasing. At pH 7.0–8.5, synthetic sizes are used, such as alkyl ketene dimers and alkenyl succinic anhydride.

Coloring.   Most paper and paperboards made from bleached pulps are colored by the addition of dyes and other chemicals. White papers are frequently treated with small amounts of optical brighteners to increase whiteness. Most dyes are added to the stock during preparation, although dip dyeing or dye application during calendering is also employed. Water-soluble, synthetic organic dyes are the principal paper-coloring materials. Water-insoluble dispersible pigments, eg, carbon black, vat colors, color lakes, and sulfur colors, are also used (see also DYES, MACROMOLECULAR; DYEING). The properties of basic, acid, and direct dyes are given in Table 5.

The kind of fiber and the degree to which it has been refined are important factors in paper dyeing. The undyed color of the pulp and the varying affinity for the same or different dyes, both from fiber to fiber within a pulp and between different pulps, are some of the variables that necessitate continual adjustment of dyeing techniques. Refining affects the optical properties of the pulp and, therefore, the color effect of a given dye. Generally, refining deepens the shade from a given application of a water-soluble dye, but does not change the amount of dye that is retained. Refining tends to increase the retention of pigments and

Table 5. Paper Dyes

Property	Basic dyes	Acid types	Direct dyes
cation	dye ion	Na^+, K^+, NH_4^+	Na^+
anion	Cl^-, SO_4^{2-}, NO_3^-	dye ion	dye ion
tinctorial strength	high	lower	lower
brilliancy	high	high	lower
light fastness	poor	generally good	good
acid fastness	poor	poor	variable
alkali fastness	poor	poor	variable
waterbleed fastness	generally good	generally poor	generally good
solubility	good	high	lower
affinity	strong for unbleached lignified fibers, therefore mottling occurs in mixed furnishes; no mordant necessary	none for cellulose; a mordant, eg, size and alum, is necessary	very strong for bleached or unbleached cellulose

other water-insoluble dyes, but may change the depth or hue by decreasing the pigment particle size. Many types of dyes are strongly absorbed by fillers. The two-sidedness of sheets, which results from loss of fines and filler on the wire side, contributes to two-sidedness of the color, an effect that also may result from contact with a heated dryer surface. Pigment colors tend to be concentrated on the top side of the sheet. In a complex system containing fibers, fillers, size, and dye, much colloidal activity is possible, particularly when extraneous unknown ions or particles are present. An optimum order for the addition of the filler, size, alum, and color is very difficult to establish, but addition of alum as the last usually gives the best results.

Other Beater Additives.   Beater adhesives are employed widely to enhance fiber-to-fiber bonding; cationic starches are preferred. Polyacrylamide resins and natural gums such as guar are used to some extent. Urea–formaldehyde, melamine–formaldehyde, and polyamideamine epichlorohydrin polymers provide wet strength to the finished paper sheet (see AMINO RESINS). Other natural and synthetic materials are used to alter the paper properties and influence the behavior of the system during sheet forming and drying.

Sheet Processing

Forming.   The Fourdrinier (Fig. 6) is still the most common paper machine used today. The paper between the breast roll and couch roll is dewatered by drainage through the single wire under the influence of table rolls driven by the wire, foils, and suction boxes. In the downstream nip between a table roll and the wire, a substantial "dewatering" vacuum is developed as the sheet is pulled off. As speeds increase, however, the action becomes too violent and disrupts the sheet. Hence, on many modern machines, table rolls have been replaced by foils (Fig. 7). These stationary elements, which induce a vacuum at the downstream nip, can be designed and adjusted to provide optimum drainage action. Following the foils, the wire and sheet pass over suction boxes, where more water is removed by vacuum. Most machines also employ suction in the couch roll for additional

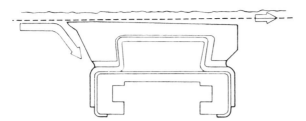

Fig. 7.   Dewatering foil. Water from the preceding foil is removed. The diverging wedge on the downstream side of the foil sucks water out of the slurry onto the wire.

water removal. In the course of dewatering on the table, the fiber content of headbox stock, typically less than 1%, is increased to the 18–20% range at the couch. The drained white water, which contains large amounts of fiber fines, filler, etc, is reused to dilute incoming stock.

Commercial Fourdrinier machines may be up to 10-m wide and over 200-m long when the press and dryer sections are included. Machine speeds depend on equipment restrictions and on limitations in water-removal rates inherent in heavy or dense grades. Heavy paperboards, glassine, and greaseproof papers may be limited to 300 m/min; linerboard and sack kraft are commonly produced at over 800 m/min. Newsprint-machine speeds were formerly limited by the wet strength of the unsupported web in the machine, especially between the forming and press sections. Incorporating a suction pickup to transfer into the first press, eliminating open draws in the press section, and attention to web handling in the dryers have increased newsprint Fourdrinier speeds over 1200 m/min.

With the advent of twin-wire machines, however, speeds in excess of 2000 m/min have become common with lightweight grades. In addition to potentially increased drainage capacity inherent in two-sided dewatering, the difference in characteristics of the two sides of the sheet is reduced, and formation, profiles, and other sheet properties are improved. This has made twin-wire technology a standard for tissue-towel and newsprint-forming sections as well as for other grades. In twin-wire formers, the headbox discharge is injected between two wires, and water is drained from the slurry by pressure rather than externally applied vacuum forces. The two wires, with the slurry between, are wrapped around a cylinder or a set of supporting bars or foils. The tension in the outer wire transmits a pressure through the slurry to the supporting structure. The pressurized slurry drains through one or both of the wires.

The Bel Baie II (Fig. 8) is an example of a twin-wire former featuring drainage through both wires. It is used extensively for newsprint, fine paper, and lighter-weight linerboards. A typical roll-type tissue former is shown in Figure 9. Drainage is single-sided and limited to low basis weights, but dewatering rates are sufficient to achieve speeds of over 2100 m/min for thin tissue.

A more recent design is the hybrid or preformer (Fig. 10). Also known as top-wire units, these forming sections employ a conventional but shortened Fourdrinier table, followed by a twin-wire section which may employ rolls or foils as dewatering elements. As in other twin-wire designs, the water expressed from the twin-wire-slurry sandwich is collected. Many of these units have been installed as retrofits to increase the capacity of conventional Fourdriniers. They

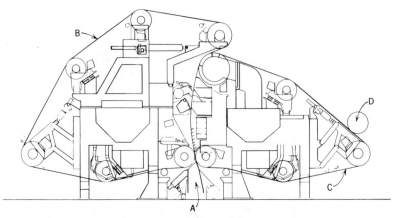

Fig. 8. The Bel Baie II twin-wire former. Stock from A, headbox is formed into sheet between B, number 1 wire and C, number 2 wire. The web is removed from wire 2 by D, a suction pickup roll. Courtesy of Beloit Corp.

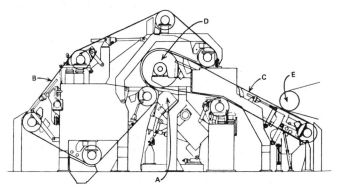

Fig. 9. A roll-type, twin-wire tissue former. Stock from A, headbox passes between B, wire number 1 and C, wire number 2 around D, a forming roll. The web is taken off the second wire by E, a suction pickup roll. Courtesy of Beloit Corp.

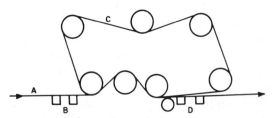

Fig. 10. Design of a top-wire unit with dewatering around rolls. A, Number 1 wire running between breast and couch rolls; B, foils; C, top wire; and D, suction boxes ahead of couch roll.

retain the advantage of formation control of the Fourdrinier table, while adding the benefits of two-sided dewatering.

As mentioned previously, the classical forming section for producing box-board and other multi-ply grades is the vat or cylinder machine. Because of the limited drainage capacities of the vats and the method of carrying the sheet, vat machines are limited to low speeds. Modern versions of the vat machine employ cast and drilled shells, suction boxes, and pressurized headboxes. These additions increase the drainage capacity and facilitate water handling, and the formers can be placed on top of the carrying felt for higher speed operation. Furthermore, certain multi-ply grades of moderate weight can be formed on Fourdriniers or twin-wire machines equipped with multilayer headboxes. Such headboxes are capable of receiving several different furnishes and depositing them at the breast roll to form a sheet of distinct plies.

Another development is the Inverform process and its more modern version, the Bel Bond shown in Figure 11. In the Inverform unit, several plies are formed on top of each other by consecutive twin-wirelike forming units above a long carrying fabric. The Inverform process is also used for lighter-weight paper grades and is capable of moderately high speeds. Other versions of board machines involve mini-Fourdriniers and twin-wires which are placed on top of a carrying fabric.

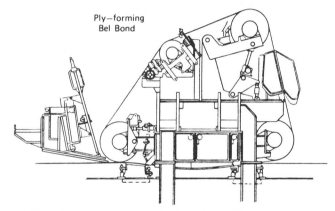

Ply—forming
Bel Bond

Fig. 11.   A Bel Bond unit of a multi-ply board machine. Courtesy of Beloit Corp.

Pressing.   The sheet leaving the forming section has typically been de-watered to near 20% solids. By passing the web through conventional rotary presses, the solids content can be increased to 35–45%, while compacting the sheet and improving properties. A given paper machine may have one or several presses, with solid, grooved, or perforated rolls, and often with suction applied through one of the roll shells. Press felts accompany the sheet through the nip, on one or both sides of the web, acting as conveyors and porous receptors of water. Double-felted presses may raise the solids content to near 48%. Under certain conditions, more water is removed with heavily loaded wide-nip presses, em-ploying soft roll covers. An expansion of this idea, the extended nip press (20), is claimed to benefit both dewatering and strength development in some grades. The nip is formed between a single roll and a lubricated belt which rides on a stationary shoe. An extended nip press may raise the solids content to >50%.

Drying.   At 35–50% solids, additional water removal by conventional mechanical means is not feasible and evaporative drying must be employed. This is a costly process and often the bottleneck of papermaking. The dryer section most commonly consists of a series of steam-heated cylinders. Alternate sides of the wet paper are exposed to the hot surface as the sheet passes from cylinder to cylinder (Fig. 6). In most cases, except for heavy board, the sheet is held close to the surface of the dryers by fabrics of controlled permeability to steam and air. Heat is transferred from the hot cylinder to the wet sheet, and water evaporates. The water vapor is removed with elaborate air systems. Most dryer sections are covered with ventilating hoods for collection and handling of the air, and heat is recovered in cold climates. The final moisture content of the dry sheet is usually 4–10%.

Other types of dryers are employed for special products or situations. The Yankee dryer is a large steam-heated cylinder, 3.7–6.1 m in diameter, which dries the sheet from one side only. It is used extensively for tissues, particularly if the sheet is creped as it leaves the dryer. Yankees also produce machine-glazed papers, where intimate contact with the polished dryer surface produces a high gloss finish on the contact side.

High velocity air drying, in which jets of hot air are directed against the sheet in a normal direction, is used in Yankee dryers and in combination with a percolation through-drying process. In the latter, hot air is sucked through the paper, thereby effecting a high heat transfer to the sheet and efficient mass transfer of water vapor from the sheet. The latter technology is commonly used for special, high quality tissue products. Infrared and other radiant drying techniques are used in special cases.

Converting

Almost all paper is converted by further treatment after manufacture. Operations include coating, printing, calendering, embossing, impregnating, saturating, laminating, and the forming of special shapes and sizes, eg, bags and boxes.

Pigment Coatings.   Pigment coatings are compositions of pigments and adhesives with small amounts of additives. They are applied to one or both sides of a paper sheet and are designed to mask or change the appearance of the base stock, improve opacity, impart a smooth and receptive surface for printing, or provide special properties for particular purposes. Pigment coatings are highly porous because the binder adhesive is insufficient to fill the void spaces among the pigment particles. This porous structure is responsible for many of the desired properties of the coated papers. For example, the high opacity of the coatings results from the scattering of light from the pigment-air and adhesive-air interfaces (see also COATINGS).

Pigment-coated printing papers must have high brightness to achieve contrast between the printed and unprinted areas. The coatings are frequently glossy. Dull-coated papers are also in demand as well as dull papers upon which glossy ink films can be printed. The coated paper should be sufficiently smooth for printing to allow full contact between the inked image area of the plate or transfer blanket and the paper surface. Requirements for smoothness decrease from gravure printing to letterpress to offset lithography. Printing smoothness involves the smoothness and compressibility of the paper, the ink, the properties of the ink-transfer and impression materials, the printing pressure, and the printing process. Consequently, smoothness is best predicted by tests that simulate the printing conditions for which the paper is intended (TAPPI UM466 and UM505).

Because only the coating layer is involved significantly in absorption of the small amount of ink fluid which is applied, tests that involve transudation of the sheet cannot be expected to provide pertinent information on ink absorbency. Comparison of ink absorbency of papers can be made according to TAPPI UM553 (see also PRINTING-INK VEHICLES).

The printing process imposes tensile stress normal to the plane of the sheet; the stress depends on the tack of the printing ink and the velocity of separation of the printing plate from the paper. The stress also tends to pick the paper, ie, remove material, unless the paper has adequate pick strength. In contrast to letterpress papers, high pick strength is required of papers to be printed by offset lithography because very tacky inks are employed. In both cases papers that are to be used for multicolor printing require higher pick strengths than those used for single-color printing because of the range of tack required in the inks. For example, gravure papers do not require high surface strength because of the low viscosity inks employed.

The pick strength is determined by tests that employ tack-graded inks or viscous test liquids at a controlled printing speed and pressure (TAPPI T499 and UM507). The pick strength of a coating depends on the pigment and the adhesive used as binder. A minimum amount of adhesive is usually required.

In offset printing, water is transferred from the blanket to the paper. If the adhesive is water-soluble, the dampened coating is more susceptible to picking at subsequent impressions. Consequently, offset papers for multicolor printing must have water-resistant coatings. To increase water resistance, insoluble compounds are added to the adhesive (TAPPI UM513 and UM468).

Application. Pigment coatings, referred to as coating colors, are normally applied to the base paper in the form of water suspensions. The total solids content, ie, pigment plus adhesive, may be 35–70%. After application, the coating must be dried. In some cases a supercalendering operation serves to smooth the surface, control surface texture, and develop a glossy finish.

Paper may be coated on equipment that is an integral part of the paper machine or separately. Many plants include both types of coating equipment and utilize each to its maximum advantage for paper and paperboard. The combination of techniques is of particular value where more than one coating must be applied to the sheet in order to obtain a product of desired quality. The paper industry uses blade, air-knife, and roll coaters (21,22).

The trailing blade coaters are frequently preferred (Fig. 12). Many variations are available, but in general the coating is smoothed and excess coating is removed by a flexible blade. The blade, supported by the paper, fills the depressions with a troweling action; thus much less coating color is applied to the high points of the sheet, resulting in a level surface. Blade coaters can be operated at

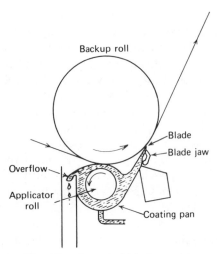

Fig. 12. Inverted-blade coater.

very high speeds (1500 m/min) using high solids coating colors of up to 70%. This coupled with their relative simplicity have made them the most widely used coaters.

The air-knife coater (Fig. 13) works on a slightly different principle. An excess of coating color is applied with a roll, and the excess is removed with a thin jet of air which acts as a doctor blade. A coating of uniform thickness is achieved, and surface contours tend to follow those of the raw stock.

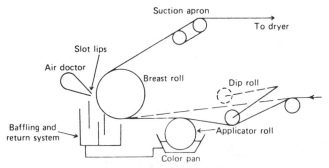

Fig. 13. Air-knife coater.

Roll coaters operate at lower speed and can apply a wider range of coating weight than blade coaters. However, film splitting takes place between the roll and paper at the exit nip which tends to leave a pattern in the coating.

In the cast-coating process, the coating is applied to a sheet and then pressed in contact with a highly polished metal drum. The coating is dried in contact with the drum to give a very smooth glossy surface, which replicates the surface smoothness of the drum.

All pigment coatings must be dried to remove the water from the coating and that which has penetrated the sheet. Drying methods include air or convection drying, contact or conduction drying, and radiant-energy drying. The speed of a coating operation is often restricted by the rate at which the water can be removed from the coating without blister formation or excessive migration of components to the hot surfaces.

Coated paper is supercalendered to improve smoothness and gloss. The paper is passed successively through the nips of a stack of alternating hard steel and soft filled rolls. The properties of the final surface are functions of the coating composition, number of rolls, pressure, temperature, and operation speed. This process also tends to densify the sheet and thus reduce porosity, opacity, and brightness.

Pigments. Pigments comprise 70–90% of the dry solids in paper coatings (23,24). In nearly all cases, the individual particles of the pigments are <5 μm in equivalent spherical dia and average less than 1 μm. Many pigments are <0.5 μm in size. These particles fill the spaces between fibers on the sheet surface and form a nearly uniform surface mat. Pigments control opacity, gloss, and the color of the raw stock. Refractive index, particle size, crystal structure, light scattering and absorption, abrasiveness, and adhesive demand are important characteristics. Minimum amounts of adhesives are used to bind the pigments; excessive amounts may affect light scattering.

Kaolin clay, a hydrated aluminosilicate, $Al_2O_3 \cdot 2SiO_2 \cdot 2H_2O$, is the most widely used coating pigment. It occurs as natural hexagonal plates with a diameter-to-thickness ratio of ca 10:1. Small amounts of impurity minerals can adversely affect pigment properties.

Commercial coating clays are fractionated by centrifugation to remove grit and large particles and to obtain an optimum range of particle sizes. Special treatments improve brightness and gloss properties, and the viscosity of the clay–water suspensions. Coating clays are produced in several grades. Brightness may vary from 75 to 90% for filler grades to >90% for special coating grades. The particles average ca 0.5 μm in equivalent spherical dia, but because of the plate form, the glossy surface is produced after calendering. The pigment produces many light-scattering surfaces and thus contributes to opacity even though its refractive index is 1.55, which is about equal to that of the cellulose fibers it must cover.

Calcined clays are produced by heating kaolin clays to remove the water of crystallization, which increases light scattering and brightness (90–93%). Calcined clays can be used as a partial replacement for more expensive TiO_2. They also improve bulkiness and ink receptivity.

The average particle size of coating-grade titanium dioxide is ca 0.3 μm. Because this size is optimum for maximum hiding power and because of its high refractive index, titanium dioxide pigment with a brightness >95% has a unique capacity to opacify and brighten coated paper. It is chemically inert and easily dispersed in water. Although both anatase and rutile crystal forms are used, rutile is preferred in coating because of its slightly higher refractive index. The amount of titanium dioxide in a coating seldom exceeds 25% by weight of the total pigment, partly because of its high cost.

Calcium carbonate from both natural sources and precipitated forms is used in coating because of its high brightness (~95–98%). The ground forms are available in many particle size ranges. The precipitated grades have a particle size of ca 0.5 μm and a very low abrasion value. Both are mixed with clays to improve brightness, opacity, printability, ink receptivity, and, in some cases, lower gloss. With the exception of specialty coatings, calcium carbonate is rarely used as the sole pigment in a coating formulation.

Talc, $3MgO \cdot 4SiO_2 \cdot H_2O$, is used as a coating pigment in Europe, and worldwide as a filler pigment, especially for pitch control. Its platelike nature, inherent softness, and brightness of 90–96% result in attractive coating properties.

Alumina trihydrate, $Al_2O_3 \cdot 3H_2O$, has the highest brightness (100%). In paper coatings, it also increases gloss and ink receptivity when used in combination with clays. Its adhesive demand is higher than that of clays. It is commonly used in specialty coatings for thermal papers and carbonless (NCR) copy paper.

Plastic pigments containing polystyrene are used in minor proportions in paper coating mainly for gloss enhancement; particle sizes ranging from 0.12 to 0.5 μm are available.

Barium sulfate pigments are used for special applications; they are marked by high whiteness, density, purity, and chemical inertness. They are ideal as bases for the sensitized layers in photographic print papers. Less important pigments include satin white for high gloss, silica for a nonskid or mat surface and ink-jet printing papers, zinc oxide for electrophotographic reproduction papers, and bentonite for carbonless papers.

Adhesives. The primary function of the adhesive in pigment coatings is to bind the pigment particles to the paper stock. The strength must be sufficient to prevent picking of the coating from the paper by tacky printing inks. The type and amount of adhesives control the characteristics of the finished paper, such as surface strength, gloss, brightness, opacity, smoothness, ink receptivity, and firmness of surface. In most coating formulations a combination of adhesives is used to maximize the properties of binder systems. The amount of adhesive in a coating formulation varies from 5 to 25% based on the pigment.

Paper-coating adhesives can be divided into water-soluble colloids, such as starch, protein and poly(vinyl alcohol), and aqueous emulsions of polymers, such as styrene–butadiene, polyacrylate, and poly(vinyl acetate).

Starch (qv) is a polymer of glucose in which the units are arranged in the linear amylose and the branched amylopectin forms. Because of the complexity and size of the structure and the high viscosity of starch in solution, starches cannot be used directly in coating formulations. They must be modified in the coating mill by thermal or enzyme conversion. Either method provides the viscosity control and degree of modification required for many applications and offers considerable cost advantages over modified starches produced by starch suppliers.

In thermal conversion, the starch is cooked continuously in a steam-injection cooker in which the thermal and mechanical shearing generated by impinging steam results in dispersion and breakdown of the starch molecules to lower molecular weights. An oxidizing agent such as ammonium persulfate (25) can be added to speed the conversion and provide stability to the starch paste. Enzyme conversion requires less energy and conversion can be stopped by applying heat.

Among the starches, corn starch is the most widely used paper-coating binder. Chemically modified starches have become less important as adhesives in paper coatings because of high costs. These include oxidized and hydroxy-ethylated starches and dextrins.

Soy protein binders are chemically modified, naturally occurring polymers derived from soybeans. These polymers have been modified to meet end-use requirements. Generally, they offer better water resistance and higher binding strength than starch and act as protective colloid and pigment dispersant. They are commonly used in board and cast coating and in some offset papers where water resistance is important.

Casein (qv), once widely used in coatings, is only rarely added to paper stock today. High price, variable quality, and availability have lead to its replacement by soy protein and alkali-swellable latex.

Poly(vinyl alcohol) is the strongest paper-coating binder available. Despite this advantage, however, high price and undesirable high shear rheology have limited its use (see VINYL ALCOHOL POLYMERS).

The aqueous emulsions of synthetic polymers can be added to the coating without special preparation as is required with water-soluble colloids. Emulsions provide properties superior to those of the natural binders, eg, low viscosity which permits high solids content and facilitates handling. The high solids allow short drying times and acceleration of coating operation. The latices are distinguished by high gloss, good response to calendering, good ink holdout, and good water resistance.

Latices (qv) based on styrene–butadiene dominate the market for synthetic paper-coating binders. A 60:40 ratio of styrene to butadiene is the most common composition. The ratio can be varied to meet specific performance requirements. The carboxylated forms are the most widely used; alkali-reactive forms are also available.

Poly(vinyl acetate) latices are mostly used on paperboard where glueability is important. Coatings containing poly(vinyl acetate) tend to be more porous and stiffer than those containing styrene–butadiene. Alkali-swellable acrylic–vinyl acetate and acrylic emulsions are also used as coating binders.

Additives. Additives control coating behavior during application or they can be used to alter the properties of the finished product (26). A single chemical additive may be used for several purposes. Some additives are essential to the production of a salable product and others may be added only to obviate problems of the coating operation.

Dispersing agents are used to make pigment slurries. The dispersant is absorbed on the pigment, causing the particles to repel one another, thereby reducing the slurry viscosity. For many years, polyphosphates were the principal dispersants. The use of polyacrylates has increased, however, because of superior long-term stability, especially at elevated temperatures. The viscosity stability of slurries containing only a polyphosphate is poor at high temperatures. Proteins, starch, and sodium carboxymethylcellulose are also used as dispersants.

Foam-control agents are commonly employed, such as adhesives that are good film formers. Most antifoam or defoamer agents are surface-active materials, such as pine oil, some alcohols, phosphate and polyglycol esters, and silicones.

Lubricants, plasticizers (qv), and flow modifiers include both soluble and insoluble soaps, sulfated oils, wax and polyethylene emulsions, amine products, and esters. Certain materials such as urea and dicyandiamide may be used to reduce the viscosity of a coating color. Lubricants improve flow, coating smoothness, finish, printability, and antidusting effects. Humectants, eg, glycerol derivatives, are used in small amounts as plasticizers; they aid in the development of the finish.

Materials used to increase the moisture resistance of a coating surface or to insolubilize the adhesive may not have film-forming properties. They may react with the hydrophilic groups in the adhesive or cross-link the polymer chains to prevent swelling with water and subsequent loss of binding strength. Water sensitivity is reduced by mixing formaldehyde donors, eg, urea– or melamine–formaldehyde resins, or glyoxal with the coating formulation. In some cases the desired results may be obtained by exposing the surface to formaldehyde or by application of a zinc, aluminum, or other metal-salt solution during calendering. Latex imparts moisture resistance.

Barrier Coatings.   In packaging applications a barrier may be needed against water, water vapor, oxygen, carbon dioxide, hydrogen sulfide, greases, fats and oils, odors, or chemicals. A water barrier can be formed by reducing the wettability of the paper surface with sizing agents. A grease or oil barrier is provided by hydrating the cellulose fibers to form a pinhole-free sheet or by coating the paper with a continuous film of a grease-resistant material. Gas or vapor barriers are formed by coating the paper with a continuous film of a suitable material.

Paraffin wax is applied in a molten form; it resists water vapor and is colorless, and free from odor, taste, or toxicity. It is applied by passing the paper through a molten bath or nip, removing the excess paraffin, and chilling. Modifiers, eg, microcrystalline wax, polyethylene, or ethylene–vinyl acetate copolymer, improve durability and film strength, raise the softening point, and increase the gloss and heat-seal strength of the coating. Polyethylene is applied by extrusion. Polyethylene coatings are more durable and flexible than wax coatings. Polymer pellets are heated rapidly with minimum air contact, and the molten material is extruded through a die and laminated immediately to the paper.

Solvent systems permit the formulation of highly sophisticated coatings comprised of a wide variety of polymers and modifiers. However, solvents are expensive and require a recovery system. Common coating resins are used, eg, cellulose derivatives, rubber derivatives, butadiene–styrene copolymers, vinyl copolymers, poly(vinylidene chloride), polyamides, polyesters, alkyds, and silicones. Emulsion or latex coatings have a high solids content at minimum viscosity. Poly(vinylidene chloride) provides excellent barrier properties.

Analytical and Test Methods

Chemical Composition.   Chemical analyses of wood pulps for papermaking include determinations of cellulose, carbohydrates, lignin, carboxyl and carbonyl groups, copper number, and viscosity (27). The carbohydrate determination involves acid hydrolysis of the pulp, preparation of volatile derivatives, and separation of the monomeric sugars by gas chromatography (TAPPI T249 and ASTM D1915); liquid chromatography, which does not require derivatization, is also employed (28). Results are used to compute percentages of cellulose and hemicellulose in the pulp. Lignin is the pulp portion that is insoluble in 72% sulfuric acid (TAPPI T222 and ASTM D1106). Lignin is determined by permanganate reduction and expressed as kappa number (TAPPI T236).

The cellulose test (TAPPI T203) and other determinations based on alkali solubility (TAPPI T212 and T235, ASTM D1109 and D1696) give the hemicellulose content of the sample and the degradation of the cellulose. Hydrolytic or oxidative pulp degradation is shown by high carboxyl (TAPPI T237 and ASTM D1926), carbonyl, and copper number values (TAPPI T430 and ASTM D919) and low viscosity. Cupriethylenediamine solutions of cellulose are used for the viscosity test (TAPPI T254); results are related to the average degree of polymerization. Molecular weight distribution may be determined by separation of cellulose tricarbanilate derivatives by gel-permeation chromatography (gpc). Monitoring the gpc column effluent with a low angle, laser-light scattering photometer provides a direct measure of cellulose molecular weight (29).

Methods for characterizing the chemistry and morphology of cellulose and lignin in pulp fibers include scanning and transmission electron microscopy (30), uv microscopy (31), ^{13}C nmr employing the cross polarization–magic angle spinning technique (32), and Raman spectroscopy (33).

Rosin size is detected by the Raspail or Lieberman-Storch test, where the paper is extracted with acidified alcohol and the ether-soluble portion of the alcohol extract is isolated (TAPPI T408 and ASTM D549). Rosin size is distinguished from residual, resinous wood components (pitch) by gas chromatography–mass spectrometry (gc–ms) of the extracts. Starch is detected by the blue color produced with application of an iodine–potassium iodide solution (TAPPI T419).

Kjeldahl nitrogen determinations are used for the urea–formaldehyde and melamine–formaldehyde wet-strength resins commonly used in paper (TAPPI T418). Melamine is determined by uv spectrophotometry (TAPPI T493). Formaldehyde-containing wet-strength resins give a red–violet color after heating the paper in a solution of chromotropic and sulfuric acids. Polyamine–polyamide epichlorohydrin wet-strength resins are determined by hydrolyzing the resin to adipic acid, which is measured by gc–ms.

Acidity or alkalinity of paper is determined by measuring the pH of a cold- or hot-water extract (TAPPI T435 and T509). Alum is the most common source of acidity.

Mineral fillers and paper-coating pigments are usually identified by x-ray diffraction analysis. The amount of filler or pigment is determined from the ash content (TAPPI T413). It may be necessary to correct for ash in the pulp and for pigment changes resulting from ashing. Elemental analysis by emission spectrography, energy-dispersive x-ray analysis, or x-ray fluorescence can aid in pigment identification and quantitation. Titanium dioxide is determined by fusion, reduction, and redox titration (TAPPI T627).

The distribution of filler through the thickness of the web is measured by removing increasingly greater amounts of paper by surface grinding and determining ash in the remaining paper. Semiquantitative data on pigment distribution are obtained from x-ray intensity maps and linescans on magnified paper cross sections. The scanning electron microscope with energy-dispersive x-ray analysis is used for these measurements.

Residual pulping chemicals in paper and paperboard are measured by the reducible sulfur test. In that determination, various forms of sulfur in the paper are reduced to hydrogen sulfide. The hydrogen sulfide reacts with p-aminodimethylaniline in the presence of ferric chloride to form methylene blue, which is measured spectrophotometrically (TAPPI T406 and ASTM D984).

Latices of synthetic resins are identified by ir spectrometry. Selective extraction with organic solvents gives purified polymer fractions for spectrometric identification. Polymer films can be identified by the multiple internal-reflectance ir technique if the film is smooth enough to permit intimate contact with the reflectance plate. Rough surfaces may be examined by diffuse-reflectance ir. These analyses are greatly facilitated by a Fourier transform ir spectrometer, with increased sensitivity and the ability to perform computerized spectral subtraction.

Volatile compounds are identified by headspace analysis with a gas chromatograph–mass spectrometer. Solvent extraction followed by gas-chromatographic analysis is used to determine paraffin wax; antioxidants, ie, butylated hydroxyanisole and butylated hydroxytoluene; and other volatile materials. Trace amounts of chlorinated organic compounds, eg, polychlorinated biphenyls, can be determined with a gas chromatograph with an electron-capture detector.

Fiber Analysis.   Paper may be composed of one or several types of fiber, eg, animal, vegetable, mineral, and synthetic. Paper is generally composed of woody vegetable fibers obtained from coniferous (softwood) and deciduous (hardwood) trees. Qualitative and quantitative methods have been developed to determine the fibrous constituents in a sheet of paper (TAPPI T401) (see also FIBERS, IDENTIFICATION). However, the recent proliferation in the number and types of pulping processes have made paper analysis much more complex (27,34).

In a common method small pieces of paper are heated to boil in 0.5% sodium hydroxide. The pieces are washed, neutralized with hydrochloric acid, and washed again, and disintegrated by vigorous shaking. A desirable fiber concentration for suspension is ca 0.05%. The suspension is examined microscopically, and the fibers are stained in order to determine the pulping process and produce contrast for the identification. The colors developed indicate the raw material and the pulping process.

Analysis of certain papers requires special treatment before they can be disintegrated. Papers containing synthetic polymers, tars, asphalt, rubber, viscose, or wet-strength resins must be analyzed individually (TAPPI T401) (27). Dyes or colors must be removed from highly colored papers before examination.

Plant-fiber identification is described in TAPPI T8 and T10. To identify synthetic fibers, it is necessary to conduct solubility and physical properties tests in addition to light-microscopy observations.

Uses

Rigid Containers

Rigid paper containers are constructed of paperboard or a combination of paperboard and paper.

Setup Boxes.   A principal requirement of the setup box is bending stiffness. Setup boxes for candy, stationery, and similar applications need not be very strong; they constitute only a small percentage of box production. Setup boxes usually are formed from a blank of single-ply, stiff paperboard or pasted boards. The board is folded along precut lines and the edges are taped with paper to hold the box together and reinforce the cut edges. A cover paper is usually glued to the outside surface of the box. Setup cartons are assembled in the manufacturer's plant before shipping.

Folding Cartons.   Folding cartons differ from setup boxes in the type of paperboard used and the method of creasing. Because of its high stiffness, the paperboard is creased by a scoring or creasing rule. The board is crushed in the area of the crease and subsequently folded along these predetermined crushed lines. Folding cartons are shipped flat and must be opened and set up for use. Examples are toothpaste boxes, cereal boxes, butter cartons, doughnut boxes, tack boxes, and milk cartons.

The type of paperboard used by the carton industry is called boxboard; categories are based on the raw material, on combinations, or on solid boxboard. Combination boxboard is made in many grades on a multicylinder paper machine using a substantial percentage of waste paper with virgin pulp. Solid boxboard is usually made on a Fourdrinier paper machine using only virgin pulp; it is bleached or coated.

Although folding cartons are made in many sizes and shapes, all are of one of three basic designs. Tray cartons are open on one face and are formed by folding a sheet of board to make the side panels; covers are glued or locked in place. Top-opening cartons are similar to tray cartons, except that one side panel is extended to serve as the top which is folded over to cover the open face; the cartons may be tucked, locked, or glued. End-opening cartons are essentially tubes, in which one or both ends are folded and sealed, locked, or tucked.

The folding carton is primarily a consumer's carton, and aesthetics are more important than strength. However, ease of assembly, absence of cracking along edges, and bending stiffness are significant.

The operations associated with the manufacture of folding cartons are printing, die cutting, and gluing. Most cartons are printed by letterpress, offset lithography, or rotogravure. Boxboard may be shipped as sheets or as a continuous web, as from a roll. The production of cartons from a continuous web is increasing. In the case of sheet stock the sheets are fed by hand or mechanically to a printing press; one or several colors may be used. After printing the sheets are fed to a die-cutting press where the carton blanks are cut and creased. The blanks are passed through a machine which folds them into tubes and glues the body closed. The formed tubes are packed flat for shipping.

Fiber Cans and Tubes.   The material used for fiber tubes and cans is a bending board. The body of a fiber can is usually of paperboard, often laminated with plastic films or metal foils, and the ends are of metal, paperboard, or plastic. The construction of the body may be spiral-wound tubes and cans, convolutely wound tubes and cans, or laminated or lap-seam cans.

Spiral-wound bodies are made on a spiral winder, which consists of a stationary cylindrical mandrel. An endless belt is looped over one end of the mandrel and a traveling saw defines the other end. The raw stock is slit into narrow rolls which are fed into the winder from either side. Each web passes over a glue roll, around the mandrel, and under the endless belt at 25–85°C. The pressure of the moving belt causes the formed tube to rotate on the mandrel, thus drawing in more material and causing the formed tube to move forward along the mandrel. The continuous spiral tube may be cut to length on the winder or rewound in multiple lengths and cut to size as a separate operation. The spirally wound tube bodies are restricted to those of circular cross section.

Convolute tubes or can bodies are made by winding two or more plies of board directly around a mandrel. The length of the tube may be the length or multiples of one can body. Convolute winding allows the formation of shapes other than round, eg, square or elliptical.

The lap-seam body is made similarly to the convolutely wound can, except that only one ply of board is used. The lap joint is glued by any one of a variety of different adhesives. Lap-seam cans can be made in various cross-sectional shapes.

The can bodies are made into cans by adding ends of metal, plastic, or paperboard. Metal ends are applied using modified metal-can-closing machines. Paperboard caps are drawn from special grades of board and may be glued to the can body.

Corrugated and Solid-fiber Boxes.   Corrugated and solid-fiber boxes are used primarily as shipping containers. Both types of containers are made from several layers of paperboard, normally referred to as combined board. Container board is the material from which the combined board is fabricated. Although both types of boxes serve the same general purpose, the handling, storage, and transportation of commodities, they differ markedly in their manufacture, structure, and performance.

Corrugated board is characterized by its cellular structure, which imparts high compressive strength at low weight. This board consists of three layers: a corrugated layer with a liner glued to both sides; a relatively thin web of paperboard, ie, the corrugating medium, passed between two fluted metal rolls to form the corrugation; and the facings or liners of high strength linerboard which are glued to the top of the flutes, thereby encasing the corrugated medium on both sides. The various steps take place sequentially: fluting of the central ply or corrugating medium, adhering the liner to one side of the fluted medium by means of aqueous adhesives, adding the other liner or facing by means of adhesives, curing the adhesive and bonding the second liner by passing the formed board over flat, steam-heated plates, and printing and scoring to define flaps and cutting the box blanks to the desired size. Corrugated board manufactured in this way is referred to as double-faced board. Double-wall corrugated board is made by combining two fluted corrugating mediums with a central liner and adding two outer-liner facings. Triple-wall corrugated board is made by combining three corrugating mediums with two inner liners and two outer facings.

The flat blanks at the end of the corrugator are passed to a printer-slotter, where they are printed with soft rubber dies and where the body scores and slots are introduced into the boards. The body scores are similar to the flap scores except that they are parallel to the flutes of the corrugating medium and thus form the vertical edges of the box when it is assembled. The slotting permits the side flaps to fold over the end flaps. The flaps form the top and bottom of the box. After the printing and slotting operations, the blanks are formed into a flat tube by taping with paper or cloth tape, stitching with metal staples, or gluing the two ends of the box. The resultant flat tubes are bundled or palletized for shipment. The final box is set up by folding and gluing, and taping or stapling the top and bottom flaps.

Construction can be varied by different weight facings and flutes. Conventional fluted rolls are designated A, B, C, and E flutes. They differ in height and in the number of flutes per length of board (Table 6).

Table 6. Flute Heights and Frequency

Flute type	Height[a] of flute[b], cm	Number flutes per meter
A	0.48	118 ± 10
B	0.24	164 ± 10
C	0.36	138 ± 10
E	0.12	308 ± 13

[a] Approximate; slight variations between manufacturers.
[b] Facing not included.

Solid-fiber combined board consists of numerous bonded plies of container board which form a solid board of high strength. It is much heavier in weight for a given thickness than corrugated board. Solid-fiber combined board is made by passing two or more webs or plies of paperboard between a number of sets of press rolls. Adhesive is applied to each ply before it passes through the press nips. In general, solid-fiber combined board is made of two to five plies, with three- and four-ply board being most common. The combined weight of the component plies, exclusive of adhesive, is 556–1758 g/m^2. In the combining operation the central plies are joined first and the outer plies are applied last. A poor grade of paperboard, eg, chipboard, is commonly used in the central plies and a strong linerboard for the outside facings or liners. This produces a combined board with high bending stiffness. Different constructions can be obtained by varying the components, number of plies, and caliper of the solid fiberboard. The subsequent operations are similar to those described for corrugated boxes.

The adhesives used in the manufacture of corrugated and solid-fiber combined board are usually starch or silicate, except where water resistance is required, in which case starch resin, modified silicate, and resin emulsions may be used.

Flexible Containers

Paper Bags. There are many types of paper bags in different shapes, styles, and number of plies, eg, single-, double-, and multiwall bags. Bag-making machines cut, fold, and glue bags from a continuous web of bleached or unbleached paper. The type depends on the requirements of the product to be packaged. The most common are the brown kraft bags, such as grocery bags and multiwall bags. Kraft bags are used for strength and are frequently of double thickness.

Single-wall bags are made in four types. Flat bags are the simplest in construction and the least expensive. They have single lengthwise seam and the bottom is folded under and glued. Satchel-bottom bags provide a flat base when filled. Square-bottom bags, eg, grocery bags, have bottoms similar to flat bags, but have bellows folds at the sides to reduce the width of the closed bag without reducing capacity. Automatic self-opening bags combine the desirable features of the other types of bags. When filled, they form a neat, squared-up package with a stable base and a center or side seam.

Multiwall Shipping Sacks.   The construction of multiwall-paper shipping sacks depends on the contents and the shipping and storage conditions. They are used primarily for the packaging of materials that need no protection against compressive forces. Their principal function is to contain the contents and protect them from contamination, and they must be able to absorb energy without rupturing (see PACKAGING MATERIALS).

Shipping sacks are made from one to six plies of high quality paper, often in combination with special coatings, laminations, or films. In a multiwall sack each ply or wall is fabricated as a tube and arranged one within the other; each layer bears its share of applied or induced stress. Better performance is obtained against shock or impact with several plies of lightweight paper than with fewer plies of heavier-weight papers. However, the latter is generally considered to be more effective against externally applied point stresses, eg, a protruding nail. The average heavy-duty, multiwall sack is constructed of a number of plies of paper with basis weights of 65–114 g/m^2. The most frequently used basis weights are 65, 81, and 98 g/m^2. Papers of heavier weight are for single- and double-ply, pasted-type shipping sacks.

The shipment of many commodities may require special barriers or treatments to resist liquids or vapors, and grease, acid, and scuffing. Special coatings are used for packing commodities, such as synthetic rubbers, asphalts, waxes, and resins, to prevent the contents from sticking to the paper.

Economic Aspects

The paper industry in the United States employs about 660,000 people in 391 pulp mills, 884 paper and board mills, and about 4,500 converting plants (35). In 1984 the United States produced nearly 36% of the total world production of pulp and almost 33% of the total world production of paper and paperboard (Table 7) (36). The per capita consumption figures for 1984 are given in Table 8 (36). Because of the relatively low consumption throughout most of the world, paper demand is growing at a relatively high rate. Since 1960 worldwide production of paper and board has grown at an average annual rate of about 6.5% per year, from a total production of 74,355 × 10^3 t in 1960 to the 1984 level of 189,911 × 10^3 t (Table 7). Over the same period of time, pulp production increased at an average annual rate of 5.6% per year.

The U.S. paper and board capacity for 1982 and 1983 and forecasts through 1987 are given in Table 9 (37). The estimated total paper and board capacity increase from 1983 to 1987 is about 1.7% per year. The production of paper and paperboard in the United States is close to capacity (paper ca 93%, paperboard ca 95%) (37). In 1983 nearly 60% of all production was for packaging or container grades.

Table 7. 1984 World Production of Pulp and Paper and Board[a], 10³ t

Continent	Pulp	Paper and paperboard
North America	70,567	76,529
United States	50,394	62,307
Canada	20,173	14,222
Europe	41,978	65,283
Asia/Australasia	19,624	37,314
Latin America	6,322	8,702
Africa	1,677	2,083
World total	140,168	189,911

[a] Ref. 36.

Table 8. 1984 Per Capita World Consumption of Paper[a]

Continent	kg
North America	279
United States	288
Canada	201
Europe	81
Asia	14
Australasia	116
Latin America	24
Africa	6
World average	41

[a] Ref. 36.

Table 9. United States Paper and Board Capacity[a], 10³ t

	1982	1983	1984[b]	1985[b]	1986[b]	1987[b]
Paper						
newsprint	5,654	5,774	5,847	5,758	5,765	5,792
printing and writing	17,809	18,633	19,770	20,462	21,087	21,953
packaging and other	6,076	6,078	5,991	5,755	5,754	5,769
tissue	5,125	5,190	5,243	5,322	5,494	5,620
Total	34,664	35,675	36,851	37,297	38,100	39,134
Paperboard						
unbleached kraft packaging	17,402	17,511	17,790	18,027	18,214	18,362
bleached packaging	4,123	4,129	4,187	4,321	4,354	4,439
semichemical	5,207	5,119	5,219	5,299	5,350	5,393
recycled	8,690	8,692	8,702	8,756	8,956	8,989
Total	35,422	35,451	35,898	36,403	36,874	37,183
Construction	3,032	2,891	2,877	2,753	2,754	2,754
Total paper and board	73,118	74,017	75,626	76,453	77,728	79,071

[a] Ref. 37.
[b] Forecast.

Environmental Issues and Mill Efficiency

Federal laws affecting the pulp and paper industry include the Clean Air Act of 1970, Clean Water Act of 1974, Resource and Recovery Act of 1976, Toxic Substances Control Act of 1977, Occupational Safety and Health Act of 1970, Federal Hazardous Substances Act, and Federal Insecticide, Fungicide, and Rodenticide Act of 1972. Recent requirements have resulted in increased lead time for expansion of basic production facilities and have affected the nature of industrial growth (38,39). Removal of the few percent of pollutants from paper-plant sources could be prohibitively expensive (40).

Initially, the concern was for general pollutants, eg, biochemical oxygen demand (BOD) and total suspended solids (TSS) in water and particulates, total reduced sulfur (TRS), and sulfur dioxide (SO_2) in air. Concern for particularly toxic chemicals, eg, chlorinated organic compounds, has increased.

Most pollutants pass from air through rain and from solids through leaching into water. Therefore, water quality is of continuing high priority to the pulp and paper industry and to the public. With the growing concern over hazardous materials, the disposal of sludges as landfill is being questioned (41) and alternative solutions must be found. Wide acceptance of incineration as a disposal method for problematic sludges will give concentrated ash, requiring special disposal methods.

Pollution Control.   Methods of pollution control in pulp and paper mills include in-mill control measures, end-of-the-pipe control measures, and monitoring and assessing environmental quality. Any change in the pulping or papermaking process directly or indirectly affects environmental quality. Environmental-control methods cannot be segregated from production technology. Since the mid-1960s, major advancements have been made in the design of pollution-abatement systems for controlling gaseous and particulate emissions in the pulp and paper industry. In the kraft pulping segment, sulfur gas and particulate emissions have been reduced within the production cycle.

Atmospheric emissions from the kraft process include gaseous and particulate matter. The principal gaseous emissions are malodorous, reduced sulfur compounds (TRS), eg, hydrogen sulfide, H_2S; methyl mercaptan, CH_3SH; dimethyl sulfide, CH_3SCH_3; dimethyl disulfide, CH_3SSCH_3; oxides of sulfur, SO_x, and of nitrogen, NO_x. In addition, most pulp-mill flue-gas streams contain appreciable amounts of water vapor. The particulate matter emissions contain primarily sodium sulfate, Na_2SO_4, and sodium carbonate, Na_2CO_3, from the recovery furnace and sodium compounds from the lime kiln and smelt tanks (42).

Odor is a serious air-pollution problem in a kraft pulp mill because the odorous gases are detectable at 1–10 ppb. The amount of odorous gases released per unit of production varies considerably between individual process units. The main sources for these include digester blow and relief gases, vacuum-washer hood and seal-tank vents, evaporation hot-well vents, the recovery furnace dissolving tanks, black-liquor oxidation tank vents, the lime kiln, and some wastewater treatment operations.

Both SO_x and NO_x are emitted in varying quantities from specific sources in the kraft chemical-recovery system. The main source of SO_2 is the recovery furnace. Smaller quantities can be released from the lime kiln and smelt-dissolving tank. Nitrogen oxides, ie, NO and NO_2, are formed by oxidation of nitrogen-containing compounds at elevated temperatures, especially when auxiliary fuels, eg, natural gas and fuel oil, are employed. When released into the atmosphere, these gases form acid rain.

Control devices for gases may include stripping, scrubbing, condensation, incineration, and adsorption. Control devices for particulate matter include cyclones, scrubbing, and electrostatic precipitators. Application of these has been quite effective in meeting the air-quality criteria.

The pulp and paper industry uses large amounts of water. In 1972, water-quality requirements had forced mills to install end-of-the-pipe secondary treatments, which are mainly biological. Means other than installation of expensive end-of-the-pipe treatments are being evaluated in terms of meeting recent requirements. Mills are recycling and reusing more process water, which results in high temperatures and increased solids in the process water, increasing corrosion slime and other deposits; chemicals are used to control these problems. Dissolved solids may impart biochemical oxygen demand (BOD), color, and toxicity to mill effluents. The suspended solids impart turbidity and long-term BOD. Directly or indirectly, all of these may affect the aquatic life.

Most of the water used in paper mills is the white water, which usually does not contain a serious amount of BOD. In most cases, receiving streams absorb the demand without being adversely affected. However, the manufacture of specialty papers may have deleterious effects. Such situations must be remedied by special mechanical or chemical treatments. White waters generally do not contain appreciable amounts of toxic materials; however, in the manufacture of asphalt-laminated paper, phenolic compounds may pass from the plant into streams. Excessive use of a slime-control agent might result in some being lost to the secondary treatment plant and river.

Modern practice is to maintain the white-water system as closed as possible, ie, as much water as is compatible with efficient machine operation is recycled. The loss of fibers and inert furnish components, particularly clay, has been greatly reduced. Fiber losses, however, still occur in the white water and a fiber-recovery system can be included in the white-water cycle based on filtration, flotation, or sedimentation. Efficient operation reduces fiber loss to $<1\%$.

In order to conform to environmental-quality guidelines, mills have installed primary and secondary treatment systems to control effluents. The primary treatment is composed of settling basins and tanks (clarifiers) which remove ca 85–100 wt % of settleable solids, such as fibers and clay. Primary sludges, which are removed by the primary clarifiers, cannot be reused by the mill in the same product. However, many mills use the sludges in lower grade products and as fertilizer and soil conditioners.

The secondary treatment usually includes a biological treatment followed by secondary clarification. Biological waste treatments include lagoons, aerated lagoons, activated sludge with air and oxygen, trickling filters, modified biological systems containing activated carbon, and combinations of these. The secondary treatments remove 90–95% BOD, most solids, most of the toxicity, but very little color. In some instances color increases after the water has been treated. Some solids from the secondary clarifiers may be recycled, but some are wasted. Biological sludges are extremely hydrophilic and are hard to dewater. They can be combined with primary sludges or other inert materials in order to improve dewatering.

As more process water is recycled to reduce overall water consumption and wastewater discharge volumes, more nonpathogenic microbial growth, ie, slime, occurs in the mill system. Slime formation prevents normal flow of stock suspensions, may make the furnish lumpy, prevents normal sheet formation, and interferes, in general, with papermaking. It has been a significant impediment to the goal of 100% closure of the water loop.

Slime problems are remedied by conditions inhibiting the growth and propagation of slime-forming organisms. The efficiency of slime control is greatly increased by ordinary mill-cleaning procedures. The application of chlorine or chloramine with or without frequent cleaning is effective in many cases. Antiseptics and disinfectants reduce or inhibit slime formation. The use of these slimicides may result in an appreciable increase in production costs. However, their use often reduces downtime caused by slime and, therefore, increases production which more than compensates for the initial cost of the slimicides.

White-water systems often contain proteolytic microorganisms that attack the machine felts and reduce their useful life. This problem may be controlled by treating the felts with a slimicide followed by cleaning with a mild acid.

Sludge Handling and Disposal.   Most waste-treatment processes generate solid wastes, which must be disposed. Pulp and paper mills generate primary sludges (containing fibers, clay filler materials, and other chemical additives) and secondary sludges largely biological in nature which are harder to handle and dewater. The disposal of sludges in landfills is being reevaluated and alternative disposal approaches are being developed (43–46).

The wide range of types of paper products results in a variety of sludges. Solid wastes result from several sources within the mill, eg, bark, sawdust, dirt, knots, pulpwood rejects, fly ash, cinders, slag, and sludges. The sludges are often disposed of in combination with residues from other sources. The pulp and paper industry generates ca 300 kg of solid waste per metric ton of finished product.

Solids content of wet sludges is 1–40 wt %. Ash content can vary from very little to over 50 wt % of the solids content. However, the solids or moisture content of a sludge is not enough for assessing the physical or engineering properties of that sludge. For example, sludges of 30–35 wt % solids from a paper recycling or deinking operation may be in a highly fluid state, whereas a low ash, high fiber pulp-mill sludge at 15–20 wt % solids may be quite dry and stable. The operating plan for a land-disposal site depends on the fluid state of the sludge (45). Operations include thickening, stabilization, conditioning, dewatering, incineration, and disposal (43).

Most sludges are disposed in landfills. Problems from this practice include possible leaching into surface and ground water, odors, methane generation, and other problems. Successful site-establishment processes are based on identifying all residue sources, including sludge, which originate from mills; characterizing the quantity and composition; assessing the storage, transportation, and materials handling that are required to transport the residues to a deposit site; and assessing the effects of various residuals on the environment of the site.

Water Quality.   Assessments of the effects of effluents on receiving streams until the mid-1970s was more subjective than objective (47,48). Since then changes in attitude within the aquatic life-science field and the regulatory system have required a restructuring of the design and foundations for effluent-impact assessments in receiving waters. For example, government regulations have proceeded from a system of stream-standards-based regulations to effluent-based regulations involving strict requirements on various pollution parameters. Pressure is being exerted to go back to the receiving water system as the ultimate test for new and more stringent discharge-control measures.

The increasing knowledge of the interrelationships between the various biological, chemical, and physical components of aquatic systems has provided significant restructuring of field assessment programs which are designed to analyze effluent impact. A single organism as an indicator of stream quality has been replaced by compositional and structural analysis. Thus total effluent effects on a broad scale can be realized. Other measured parameters are toxicity assays, algal assays, fish surveys, sediment mapping, plume mapping, sediment oxygen demand, and socioeconomic impacts.

BIBLIOGRAPHY

"Paper" in *ECT* 1st ed., Vol. 9, pp. 812–842, by H. F. Lewis, R. Shallcross, D. J. MacLaurin, T. A. Howells, W. A. Wink, B. L. Browning, I. H. Isenberg, R. C. McKee, and W. M. Van Horn, The Institute of Paper Chemistry; "Paper Coatings, Inorganic" in *ECT* 1st ed., Vol. 9, pp. 842–858, by G. Haywood, West Virginia Pulp and Paper Co.; "Paper Coatings, Organic" in *ECT* 1st ed., Vol. 9, pp. 858–867, by P. H. Yoder, Pyroxylin Products, Inc.; "Paper" in *ECT* 2nd ed., Vol. 14, pp. 494–532, by Roy P. Whitney, W. M. Van Horn, C. L. Garey, R. M. Leekley, T. A. Howells, R. C. McKee, W. A. Wink, I. H. Isenberg, and B. L. Browning, The Institute of Paper Chemistry; "Paper" in *ECT* 3rd ed., Vol. 16, pp. 768–803, by G. A. Baum, E. W. Malcolm, D. Wahren, J. W. Swanson, D. B. Easty, J. D. Litvay, H. S. Dugal, The Institute of Paper Chemistry.

1. J. N. McGovern, *Pulp Pap.* **52**(9), 112 (1978).
2. D. Hunter, *Papermaking*, renewed ed., Dover Publications, Inc., New York, 1974.
3. D. C. Smith, *History of Papermaking in the United States 1691–1969*, Lockwood Trade Journal Co., New York, 1970, p. 693.
4. *1982 Yearbook of Forest Products*, Food and Agriculture Organization of the United Nations, Rome, 1984.
5. C. C. Habeger, R. W. Mann, and G. A. Baum, *Ultrasonics* **17**, 57 (1979).
6. R. W. Mann, G. A. Baum, and C. C. Habeger, *Tappi* **63**(2), 163 (1980).
7. E. H. Fleischman, G. A. Baum, and C. C. Habeger, *Tappi* **65**(10), 115 (1982).
8. G. A. Baum, *Paper* **19**, 65 (1984).
9. P. Kolseth and A. deRuvo in R. E. Mark, ed., *Handbook of Physical and Mechanical Testing of Paper and Paperboard*, Vol. 1, Marcel Dekker, Inc., New York, 1983, p. 255ff.
10. W. A. Wink and G. A. Baum, *Tappi* **66**(9), 131 (1983).
11. G. A. Baum, K. Pers, D. R. Shepard, and T. R. Ave'Lallemant, *Tappi* **67**(5), 100 (1984).
12. K. W. Britt, *Handbook of Pulp and Paper Technology*, 2nd ed., Van Nostrand Reinhold Publishing Co., Inc., New York, 1970.
13. J. P. Casey, *Pulp and Paper Chemistry and Chemical Technology*, 3rd ed., Vol. I, Wiley-Interscience, New York, 1980.
14. J. d'A. Clark, *Pulp Technology and Treatment for Paper*, Miller Freeman Publications, Inc., San Francisco, 1978.
15. R. G. MacDonald and J. N. Franklin, *Pulp and Paper Manufacture*, 2nd ed., Vol. 3, McGraw-Hill Inc., New York, 1970.
16. U.S. Pat. 1,873,199 (Aug. 23, 1932) and 1,985,569 (Dec. 25, 1934), J. D. Haskell (to Dilts Machine Works Inc.); U.S. Pat. 1,960,753 (May 29, 1934), G. P. Prathee (to Dilts Machine Works Inc.).

17. U.S. Pat. 864,359 (Aug. 27, 1907), G. D. Claflin, Jr. (to F. N. Claflin Engineering Co.).
18. J. F. van Oss and C. J. van Oss, *Warenkennis en Technologie,* Vol V, p. 679.
19. R. W. Hagemeyer, ed., *Pigments for Paper,* TAPPI Press, Atlanta, Ga., 1984.
20. L. D. Wicks, *Tappi J.* **66**(4), 61 (1983).
21. *Blade Coating Seminar Notes,* TAPPI Press, Atlanta, Ga., 1986.
22. *Air Knife Coating Seminar Notes,* TAPPI Press, Atlanta, Ga., 1984.
23. C. L. Garey, ed., *Physical Chemistry of Pigments in Paper Coating, TAPPI Press Book No. 38;* Technical Association of the Pulp and Paper Industry, Atlanta, Ga., 1977, p. 493.
24. J. P. Casey, *Pulp and Paper,* 3rd ed., Vol. IV, Wiley-Interscience, New York, 1983, pp. 2013–2189.
25. U.S. Pat. 3,211,564 (Oct. 12, 1965), G. E. Lauterbach (to Kimberly-Clark Corp.).
26. C. G. Landes and L. Kroll, eds., *Paper Coating Additives,* TAPPI Press, Atlanta, Ga., 1978.
27. B. L. Browning, *Analysis of Paper,* 2nd ed., Marcel Dekker, Inc., New York, 1977.
28. R. C. Pettersen, V. H. Schwandt, and M. J. Effland, *J. Chromatogr. Sci.* **22,** 478 (1984).
29. J. J. Cael, D. J. Cietek, and F. J. Kolpak, *J. Appl. Polym. Sci. Appl. Polym. Symp.* **37,** 509 (1983).
30. S.-J. Kuang, S. Saka, and D. A. I. Goring, *J. Wood Chem. Technol.* **4,** 163 (1984).
31. D. A. I. Goring, "Some Recent Topics in Wood and Pulping Chemistry" in *Proceedings of the International Symposium in Wood and Pulping Chemistry,* Vol. 1, Tsukuba Science City, Japan, 1983, p. 3.
32. J. F. Haw, G. E. Maciel, and H. A. Schroeder, *Anal. Chem.* **56,** 1323 (1984).
33. R. H. Atalla, *J. Appl. Polym. Sci. Appl. Polym. Symp.* **37,** 295 (1983).
34. I. H. Isenberg, *Pulp and Paper Microscopy,* 3rd ed., The Institute of Paper Chemistry, Appleton, Wis., 1967.
35. *Lockwood's Directory of the Paper and Allied Trades,* 108th ed., Vance Publishing Corp., New York, 1984, pp. 1–8.
36. J. Pearson, P. Sutton, and H. O'Brien, eds., *Pulp Pap.* **59**(8), 55 (1985).
37. *Statistics of Paper, Paperboard, and Wood Pulp* (data through 1984), American Paper Institute, New York, 1985.
38. J. E. Huber, *Kline Guide to the Paper Industry,* 4th ed., Charles H. Kline & Co., Inc., Fairfield, N.J., 1980.
39. J. Quarles, *Federal Regulations of New Industrial Plants,* Morgan, Lewis and Brochius, Washington, D.C., Jan. 1979.
40. J. G. Strange, *The Paper Industry: A Clinical Study,* Graphic Communications Center, Inc., Appleton, Wis., 1977.
41. H. S. Dugal, "Environmental Laws and Their Impact on Mill Processes," paper presented at the *NACE Third Int. Symp. Pulp and Paper Ind. Corrosion Problems,* Atlanta, May 5–8, 1980.
42. *Environmental Pollution Control, Pulp and Paper Industry: Air, EPA 625/7-76-001,* Washington, D.C., 1976, Part I.
43. *Process and Design Manual for Sludge Treatment and Disposal, EPA 625/1-74-006,* Washington, D.C., 1974.
44. D. Marshall, *South. Pulp Pap. Manuf.* **40**(12), 19 (Dec. 1977).
45. J. J. Reinhardt and D. F. Kolberg, *Pulp Pap.* **52**(11), 128 (Oct. 1978).
46. *Pap. Trade J.* **163**(9), 34 (May 15, 1979).
47. R. Patrick and D. M. H. Martin, *Biological Surveys and Biological Monitoring in Fresh Waters,* Academy of Natural Sciences, Philadelphia, 1974.
48. J. M. Hellawell, *Biological Surveillance of Rivers,* Water Research Centre, Herts, UK, 1978.

GARY A. BAUM
EARL W. MALCOLM
JOHN D. SINKEY
JACK D. HULTMAN
DWIGHT B. EASTY
HARDEV S. DUGAL
The Institute of Paper Chemistry

PAPER ADDITIVES
AND RESINS

This article discusses the chemicals, polymers, and resins used in the papermaking and paper-coating processes. Emphasis is on polymers and resins, although there is some discussion of nonpolymeric chemicals used (see also PAPER).

Paper production is one of the largest basic industries in the world. In 1984 worldwide production of paper and paperboard totaled 190 million metric tons; consumption was 188 million metric tons. In 1984 the United States accounted for 33% of the world's paper production and consumed 35%. Annual per capita consumption of paper in the United States in 1984 was 288 kg (1), compared with the 1968 figure of 227 kg.

In 1985 sales of all paper chemicals totaled $3.8 billion. These sales are predicted to grow at a rate of 2.7%/yr to $4.2 billion in 1990 (2). The fastest growth rate is predicted for surface-treatment (3%/yr) and wet-end additives (3.2%/yr), in near agreement with another study, which predicts a growth rate of 3.5–4% for specialty paper chemicals (3). Pulping chemicals are expected to have the slowest growth rate (1.5%/yr) (2).

Since many grades of paper come into direct contact with foods, the chemicals used in papermaking often require FDA clearance (4). Many paper mills require that all chemicals used in the mill have FDA clearance even if the chemical is being used in a grade of paper that will not come in contact with food. Most mills have several machines making different grades of paper, and often the water systems are interconnected and cross-contamination from machine to machine can occur.

Additives are described in terms of their point of use in the papermaking process, because usually they are specifically designed to be used only in one location on the paper machine. Three main categories are wet-end additives, surface-treatment additives, and paper-coating additives. A distinction is made between additives designed to improve the papermaking process (processing aids) and those designed to modify the properties of the paper (functional additives).

Pulping and bleaching chemicals are discussed elsewhere (see BLEACHING OF WOOD PULP). There are numerous suppliers of the various paper chemicals, and good surveys of papermaking chemicals and the relevant suppliers are available (5,6).

Wet-end Additives

These chemicals are added to the pulp slurry prior to sheet formation and drying.

Processing Aids

Retention Aids. Paper is formed from a relatively dilute pulp slurry (~0.5% total solids). Many of the additives used in papermaking are water-insoluble and are small enough to pass through the pores of the forming wire or fabric, which are approximately equivalent to the pores of a 149-μm (100 mesh) screen. These additives are collectively called fine particles and include cellulose fines, fillers such as clay and titanium dioxide, and functional additives such as dispersions of rosin size. The fine particles generally have no built-in functionality to make them remain with cellulose fibers and thus be well retained in the final sheet of paper. Retention aids provide a mechanism for good retention in the paper product.

To retain fine particles in the forming sheet, it is necessary to convert them into species that will not pass through the pores in the wire or in the forming fiber-mat. Two mechanisms are possible: flocculation (qv) of the fine particles to a size sufficiently large to be retained in the sheet, or adsorption of the fine particles onto fibers. In water most fine particles and cellulose fibers bear an anionic surface charge. Thus most retention aids or retention systems have a cationic component to facilitate adsorption to the pulp-furnish surfaces. Retention aids may be low molecular weight materials that neutralize the anionic charge promoting a loose patchlike flocculation or high molecular weight materials that connect between fine particles and thus cause bridging flocculation.

Flocculation.   The simplest patch-type retention aid is papermaker's alum, hydrated aluminum sulfate. Alum partially neutralizes the charge on fines, causing flocculation and improved drainage and sheet formation. A major use for alum is to precipitate rosin size. Alum is often used in conjunction with another retention aid such as a high molecular weight, cationic bridging polymer.

Other patch-type retention aids are low molecular weight ($M_w = 10^3$–10^5) resins containing amine groups that provide cationic functionality. Examples are the reaction product of epichlorohydrin and the polyamide formed by adipic acid and diethylenetriamine, the condensation polymer formed from epichlorohydrin and ammonia or alkyl amines, polyethyleneimine, poly(dimethyldiallylammonium chloride) and copolymers, and water-soluble addition polymers containing the aminoacrylate functionality. These products are supplied as aqueous solutions containing 10–35% total solids.

Cationic starch is often used as a patch-type retention aid to improve the retention of mineral fillers such as clay. In recent years it has been used extensively for retaining cellulose-reactive sizes such as alkyl ketene dimer and alkenyl succinic anhydride.

Bridging polymers have molecular weights of one million or higher and most are copolymers of acrylamide and either a cationic or anionic monomer. Common cationic monomers include (meth)acryloyloxyethyltrimethylammonium salts, dimethylaminoethyl methacrylate, and dimethyldiallylammonium chloride. Acrylic acid and its salts are used as anionic monomers. Since bridging polymers have very high molecular weight, they cannot be supplied as aqueous solutions. Some bridging polymers are supplied as dry powders. They have the inherent handling disadvantage of dry materials and are generally difficult to dissolve in water. Some suppliers offer special equipment for dissolving these products (7). An alternate method of supply is as an inverted emulsion of water-swollen polymer in a nonaqueous liquid such as mineral oil (8). These products are pumpable liquids containing 20–35% active polymer solids. The emulsions are inverted in water, and the polymer dissolves to a low solids (<1%) solution just prior to use.

In some cases a combination of patch-type and bridging polymers is used for optimum retention. Thus a low molecular weight cationic polymer can be used to partially bias particles positive and form loose flocs. Addition of a high molecular weight anionic polymer can then lead to very effective retention of fine particles. A potential problem with the use of high molecular weight bridging polymers is flocculation of the fibers which can cause poor sheet formation. This problem can be partially alleviated by subjecting the stock to high shear to redisperse the flocs just prior to sheet formation.

One uncharged retention aid is poly(ethylene oxide) (PEO), a nonionic polymer. The use of high molecular weight PEO can significantly improve the first-pass retention in groundwood containing grades such as newsprint (9).

Measuring Retention. Effective retention is critical to the performance of most wet-end additives. The additives are generally either fine particles by nature or become adsorbed onto fine particles, which must subsequently be retained in the forming fiber-mat. The retention process is very complicated. Often it is difficult to measure the retention of the individual types of fine particles present on a commercial machine. Laboratory tests have been developed to simulate machine conditions and to study retention of various fine particles (10). The most commonly used equipment is the dynamic drainage jar, a modified version of which is shown in Figure 1 (11). Pulp slurry and fine-particle additives are stirred in the container, and the slurry is then allowed to drain through the fine mesh screen. The ratio of the fine particles in the drainage water to the fine particles in the original slurry is a direct measure of fine-particle retention. The test can

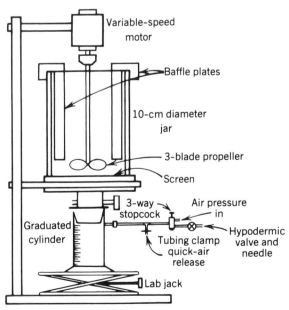

Fig. 1. Dynamic drainage jar (11).

be run with continuous stirring with no mat formation or with stoppage of the stirring just prior to drainage, which leads to the formation of a fiber mat. Studies indicate that the latter mode better simulates the papermaking process (11). Furthermore, the mechanism of retention of larger particles (>1 μm) involves flocculation followed by entrapment in the forming fiber-mat. Submicrometer particles are retained by this mechanism and by adsorption onto pulp fibers. Particle adsorption can be studied directly by using a mixture of fines-free pulp fibers and the fine-particle additive of interest (eg, reactive size) while operating the dynamic drainage jar in the continuous-stirring mode with no mat formation.

Drainage Aids. Productivity is very important to the profitability of paper machines. In many cases the critical factor determining the speed of the paper machine is the rate at which water can be removed at the wet end. This is particularly true of low freeness pulps, ie, pulps that release water slowly, such as groundwood and recycled pulp. In these cases it is often desirable to use a drainage aid, which is usually a cationic water-soluble polymer. Choice of the type of drainage aid is important. An additive that causes rapid flocculation increases initial drainage, but impedes later drainage by allowing channeling of air and loss of effective vacuum (12). Polyethyleneimine and its derivatives are effective drainage aids (13). Drainage aids are more commonly used in Europe and Japan, where use of recycled fiber is more prevalent than in the United States.

Formation Aids. Flocculation is desirable for retention of fine particles, but is often undesirable for good sheet formation. It can lead to uneven fiber distribution, which is evidenced by thick and thin spots in the sheet of paper, particularly with long fiber pulps. Flocculation can be prevented by the use of dispersants. Alternatively, water-soluble polymers such as polyacrylamide and poly(ethylene oxide), and natural gums such as guar and locust bean gum can be used to improve fiber distribution and sheet formation.

Wet-web Additives. As the sheet of paper leaves the wire, it is a fragile structure containing 65% or more water. Between the forming section and the press section, there is often an open draw which can cause web breaks. Wet-web additives improve the strength of this fragile structure. In general, the best way to improve the strength of the sheet at this point is to decrease the amount of water present. Also, sheet uniformity is very important, because uneven fiber distribution leads to thin spots which act as sources of wet-web breaks. The additives used to improve sheet formation (polyacrylamides, poly(ethylene oxide), guar, and locust bean gums) are also often used to improve wet-web strength. There is thus little need for wet-web additives on modern paper machines which use felts to support the wet web between the forming and press sections.

Defoamers. When high shear is applied to a dilute pulp slurry, foam is often a problem in the papermaking process (14). Foam is caused by surface-active impurities present in the pulp slurry itself or as components of the various papermaking additives. Another source of foam is partially hydrophobic particles, particularly in high yield and recycled pulps. Foam is detrimental to papermaking because entrained air can reduce the rate of removal of water during sheet formation and foam spots can cause thin transparent spots in the final sheet of paper. On the surface of the sheet, foam can be removed by steam or water showers. Entrained air requires the use of a defoamer. Most of the several types of defoamer available (15,16) contain a water-immiscible, surface-active material which spreads rapidly over the surface of a foam bubble. The rapid spreading applies sufficient force to rupture the foam bubble mechanically. A potential problem is that water-immiscible, surface-active species often can function as wetting agents. As a result some defoamers are very detrimental to sheet sizing (see below), and careful screening of defoamers is necessary when high levels of sizing are required (see also ANTIFOAMING AGENTS).

There are three physical forms of defoamers. Paste defoamers consist primarily of high solids (~60–70%) dispersions of fatty acids, fatty alcohols, fatty esters, and often, mineral oil. They are dispersed in water at relatively low solids content (<1%) just prior to use on the paper machine. Solid or brick defoamers have active ingredients similar to those in paste defoamers, ie, fatty acids, fatty alcohols, and fatty esters. Brick defoamers do not contain water; they contain a small amount of emulsifier and are supplied as solid, waxy bricks approximately the size of building bricks. Prior to use, brick defoamers must be emulsified in water at relatively low concentration (<1%). Automatic equipment for emulsifying the bricks is available (17). The principal component of liquid defoamers is an oil such as mineral oil or kerosene. Fatty acids and fatty alcohols are also used. Often, the active defoaming ingredient is a water-immiscible surfactant, the condensation product of a fatty acid, fatty alcohol, or fatty amide with a low molecular weight poly(ethylene oxide). These materials can also be prepared by reaction of the fatty derivative directly with ethylene oxide. A cosolvent such as 2-propanol may be used to maintain the product as a clear, homogeneous liquid. Liquid defoamers are usually emulsified in water prior to use.

Pitch Dispersants. Pitch (and "stickies") is a broad term to describe material that is water-immiscible and forms deposits at various locations on the paper machine. The source of pitch can be natural resins from the pulp slurry, or additives such as resins, asphalts, waxes, glues, and latices that are components of recycled pulps. Pitch removal requires downtime and the use of undesirable organic-solvent cleaning material. Pitch may adversely affect sheet properties by forming oily spots in the paper. Pitch can be minimized by thorough washing of the pulp, but often this is not practical. Dispersants, such as surfactants, are added to prevent precipitation of the pitch. Alternatively, an adsorbent mineral filler such as talc can be used to convert the pitch to a nontacky species that is retained in the paper (18). Recently, it has been found that low concentrations of synthetic polyolefin pulps are very effective at removing pitch and stickies (19).

Creping Aids. Water absorbency is a very important property for sanitary paper grades such as towels, napkins, facial tissue, and toilet tissue. Absorbency is enhanced by mechanically creping the sheet, by scraping the paper off a large, heated polished drum called a Yankee dryer with a doctor blade. Control of the degree of creping is important, because excessive creping can lead to large losses in sheet dry strength. The degree of creping is controlled by the degree of adhesion of the sheet to the Yankee dryer. This in turn is controlled by additives that increase adhesion (dryer adhesives) or those that decrease adhesion (release agents). Examples of adhesives include animal glues, aminopolyamide resins, and aminopolyamide–epichlorohydrin resins (see below). Release agents (qv) include silicone oils and paraffin oils. Combination products containing both adhesive and release agent are sometimes supplied as custom products for a paper machine.

Functional Additives

Dry-strength Resins

Factors Affecting Sheet Strength. Considering that it is made up of a collection of short (0.1–0.4 cm) rigid fibers, a sheet of paper has considerable strength. A bleached kraft paper sample having a basis weight of 65 g/m^2 has a tensile strength of approximately 35 N/cm (20 lbs per inch width). The exact mechanism by which paper develops dry strength is not well known. Some of the strength is due to the strength of the fibers themselves, but most of the dry strength is from bonding between the fibers. Several types of bonding have been proposed, including covalent bonds, mechanical entanglement of fibers, ionic bonds, hydrogen bonds, and van der Waals forces (20). The fact that waterleaf paper loses almost all of its strength when soaked in water suggests that the likely possibilities are ionic and hydrogen bonding. Interfiber bonding is probably due to hydrogen bonding between cellulose hydroxyls on adjacent pulp fibers.

Testing Dry Strength.   Several tests are used to measure the dry strength of paper (21). Dry tensile strength is the in-plane force required to break the sheet. Paper made on a paper machine is anisotropic due to the tension on the sheet during the papermaking process. The fibers are preferentially aligned in the machine direction due to this tension, and as a result the dry tensile strength is usually higher in the machine direction than in the cross direction. Thus it is necessary to specify direction when reporting sheet tensile strength.

Bursting strength, usually referred to as Mullen burst, is the pressure required to rupture the sheet. Since the test is nondirectional, the sheet direction is not required in reporting test results. Tear strength is usually reported as Elmendorf tear. Increases in tensile strength often result in decreases in tear strength. Internal bond is the force required to pull the sheet apart in the z-direction, that is, with force applied perpendicular to the plane of the sheet of paper. The Scott Bond Tester is one device used to measure internal bond (21).

Pulp Factors.   One of the most important factors that affect the dry strength of paper is the type of pulp used to prepare the paper. The strongest paper is produced from chemical pulps, pulps produced by the kraft, soda, or sulfite pulping processes. Within this class paper prepared from a softwood furnish is generally stronger than that from a hardwood furnish due to the longer fiber length in softwood pulps. Paper produced from the high yield pulps (groundwood, thermomechanical pulp, chemithermomechanical pulp, etc) is significantly weaker than that from chemical pulps. Chemical pulp consists essentially of cellulosic fibers devoid of lignin (qv), particularly if the pulp has been bleached. The high yield pulps contain quantities of noncellulosic material, particularly lignin, that contains no hydrogen-bonding functionality and thus cannot contribute to the normal hydrogen bonding between fibers. Recycled pulp also produces significantly weaker paper than virgin pulp, presumably due to hornification of parts of the pulp fibers during the recycling process (20).

Beating and Refining. Paper dry strength is improved by any factor that facilitates formation of hydrogen bonds between fibers. Hydration and swelling of the fiber surfaces as well as fibrillation of the fibers are important, because these factors lead to significantly higher surface area available for the formation of hydrogen bonds. Beating and refining of the pulp has a positive effect on each of these factors. A typical beating curve (Fig. 2) shows the changes in tensile or burst strength vs tear strength with increasing beating time (22). Point *A* in Figure 2 represents the strength properties of paper from defibered but unbeaten pulp. Proceeding along the curve from point *A* to beyond point *C*, the effect of increasing degrees of refining on the properties of paper from the same pulp is shown. Initially, tensile, burst, and tear increase as the pulp is hydrated and capable of producing stronger bonds. Tear strength goes through a peak and then falls off; tensile and burst continue to increase. The tear strength, which is determined by individual fiber strength and fiber-to-fiber frictional resistance, generally decreases with refining because the tearing force is concentrated into a smaller area in the stronger sheet (14). Eventually, the degree of beating becomes excessive, and tensile and burst begin to decrease as well.

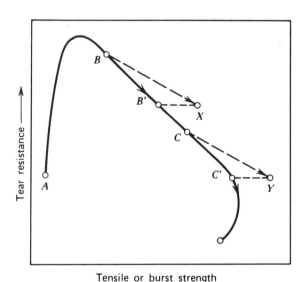

Tensile or burst strength

Fig. 2. Effects of beating and dry-strength agents on strength of paper. $B \rightarrow B'$ and $C \rightarrow C'$ represent substitution additives; $B' \rightarrow X$ and $C' \rightarrow Y$ represent fiber power-substitution additives.

Dry-strength additives can function in one of two ways. An additive that moves the sheet properties further along the existing beater curve (eg, from *B* to *B'* or *C* to *C'*) effectively accomplishes the same thing that additional refining would accomplish and as such is called a power-substitution additive. To be effective the cost of this additive must be lower than the cost of additional refining. An additive that moves sheet properties to the right (eg, from *B* to *X* or *C* to *Y*) effectively provides the properties of a stronger fiber and is called a fiber-substitution additive.

Compression Strength. Compression strength is an important strength property for corrugated board because the stacking height of corrugated boxes is dependent on the compression strength of the board (23). Board compression strength is primarily contributed by the linerboard, the outer component of the composite corrugated structure. Until recently, Mullen burst was the target strength property for linerboard, but the measurement of linerboard compression strength, using instruments such as the STFI tester, is becoming accepted as a better indicator of corrugated-board performance (24). Board compression strength can be increased by increasing the basis weight of the linerboard. However, this is costly, and efforts are under way to use chemical additives to improve compression strength. One additive system that has shown promise in model linerboard systems is a combination of carboxymethylcellulose and aminopolyamide–epichlorohydrin wet-strength resin (25). Application of these additives to the corrugating medium, the inner component of the corrugated board, is also effective in improving the compression strength of the total structure.

Natural Gums. The first dry-strength additives were naturally occurring water-soluble polymers. Waterleaf paper itself owes some of its dry strength to the soluble hemicellulose present in the pulp slurry. The Chinese were the first to use mucilages as additives in papermaking to aid in the suspension and deflocculation of the relatively long-fibered pulp used in making their paper. Later, it was discovered that these additives also improved paper-strength properties (26).

The most commonly used gums today are guar gum and locust bean gum; the active water-soluble polymer in both gums is galactomannan. The ratio of galactan to mannan units varies depending on the source of the gum (see GUMS, INDUSTRIAL). Other, less widely used gums are derived from tamarind kernels, karaya, and okra. As is true with most dry-strength additives, dry strength is decreased as the molecular weight of the gum polymer is degraded. Natural gums are similar chemically to cellulose and are capable of forming hydrogen bonds with cellulose-fiber surfaces. A 1950s study to determine the source of improved dry strength from natural gums concluded that 60% of the improvement was due to improved fiber-to-fiber bonding, 15% from increased bonding area, and 25% from improved sheet uniformity (formation) (27).

Natural gums are supplied as dry powders. They are dissolved in water at relatively low concentration (<1%) prior to use. The gum solution can be added directly to the pulp slurry or it can be sprayed onto the forming web on the Fourdrinier wire.

Starch. Historically, starches were one of the first wet-end additives used as dry-strength additives. Most of the starch (qv) used is either corn or potato starch with some lesser amounts of wheat starch and tapioca (28). Unmodified starch can be used directly, or the starch can be chemically modified by oxidation, cationization, or hydroxyethylation. The starch is usually cooked in water prior to use. Uncooked starch can also be used, but only in heavier weight grades in which the starch is heated above its gelatinization temperature prior to complete drying of the sheet. Pregelatinized starch is used in mills that do not have starch cooking equipment (29).

Because starch is a relatively inexpensive additive, fairly high amounts (2–5%) are commonly used. Excessive use of starch should be avoided to prevent machine-operating problems such as slow drainage or adverse effects on final sheet properties such as opacity. Starch improves the internal-strength properties of paper, including dry tensile, Mullen burst, and Scott bond.

Like most naturally occurring materials, many starch products, both unmodified and modified, have an anionic surface charge and have no electrostatic mechanism for retention on anionic pulp fibers. Retention is achieved either by physical adsorption promoted by a cationic material interacting between the starch and cellulosic fiber, or by heteroflocculation followed by physical entrapment in the forming mat. Each process is usually facilitated by alum as the retention aid. Cationic starches are produced by reaction of starch with a chemical having either tertiary or quaternary amine functionality, eg, diethyl(β-chloroethyl)ammonium chloride and 2,3-epoxypropyltrimethylammonium chloride (30). The reaction is usually carried out in an aqueous slurry of corn or potato starch. Cationic starches are easily adsorbed on anionic furnish components and as a result are usually used at lower concentrations than anionic starches (0.5–2%). The estimated consumption of wet-end cationic starch in 1977 was 4.77×10^7 kg (30).

Carboxymethylcellulose. Sodium carboxymethylcellulose (CMC) is the reaction product of cellulose and monochloroacetic acid. It is a water-soluble polymer available in a variety of grades with differing molecular weights and degrees of carboxymethyl substitution (31). As an internal additive for paper it can provide significant improvements in sheet dry-strength properties at relatively low levels of addition (0.05–1.0%). As an anionic polymer it requires a retention aid for good retention in the sheet. Alum can be used, although this may not lead to the optimum strengthening properties of CMC, particularly at low pH. The use of CMC in conjunction with aminopolyamide–epichlorohydrin wet-strength resins (see below) leads to synergistic improvements in both wet- and dry-strength properties of paper (32).

CMC is usually provided as a dry powder. It is dissolved in water prior to use at relatively low concentrations (<1%) because the high molecular weight results in viscous solutions at higher concentrations. An aqueous solution of a lower molecular weight CMC is also available (see also CELLULOSE ETHERS).

Acrylamide Polymers. Acrylamide-based, water-soluble polymers, introduced in the mid-1950s, were the first all-synthetic polymers that increased paper dry strength without increasing wet strength (33). The polymers are produced by the free-radical polymerization of acrylamide monomer in water. Usually, a small amount of comonomer (10 mol % or less) is also used. Molecular weight can be controlled by the polymerization process from a value of several thousand up to several million. The product is supplied either as a relatively dilute viscous solution or as a dry powder which must be dissolved in water prior to use.

A copolymer of acrylamide and acrylic acid, prepared either by direct copolymerization or by partial hydrolysis of acrylamide homopolymer, is the most widely used synthetic dry-strength additive. Alum is frequently used as the retention aid, and the anionic polymer functions well as a dry-strength additive in the presence of alum. Alternatively, the acrylamide copolymer can be used in combination with an aminopolyamide wet-strength resin. If wet strength is undesirable, a nonwet-strength cationic water-soluble polymer can also be used. Cationic copolymers of acrylamide are also available as self-retaining dry-strength additives.

Acrylamide-based dry-strength resins are very effective at improving strength properties such as dry tensile, Mullen burst, Scott bond, and wax pick. Addition is typically 0.1–1.0%, based on fiber. The primary function of the synthetic dry-strength additives is to augment the hydrogen bonding responsible for the dry strength of paper. Most dry-strength resins are chemically similar to cellulose and have hydroxyl functionality available for hydrogen bonding with cellulose-surface hydroxyls (natural gums, starch, CMC). Polyacrylamide is a good dry-strength additive because the amide functionality is excellent for hydrogen bonding.

Wet-strength Resins

Waterleaf paper has excellent dry-strength properties, but almost all of this strength is lost when the paper is wetted by water because dry strength is due primarily to hydrogen bonding (and probably some ionic bonding) and these bonds are destroyed by water, which is very effective at forming hydrogen bonds itself. Therefore, development of wet strength requires the formation of different types of bonds which are not adversely affected by exposure to water. The earliest method for developing wet strength was to fuse the cellulose fibers to one another, either by heating the paper to very high temperature or by partially solubilizing the paper with sulfuric acid (34). The latter is the method by which parchment paper is produced. The first chemical method for improving wet strength involved the use of formaldehyde, which can form methylene bridges between hydroxyl groups on the surface of cellulose fibers.

Urea–Formaldehyde Resins. The first synthetic resins used to improve wet-strength properties were the urea–formaldehyde (UF) resins, originally low molecular weight condensates of urea and formaldehyde impregnated into paper. On exposure to acid and/or alum, UF resins undergo further cross-linking and impart wet-strength properties to paper. An early improvement to UF resins was the introduction of anionic sites by reaction with sodium bisulfite or sulfur dioxide (35). This modification allowed the resins to be added to dilute pulp slurries and then retained in the paper by precipitation by alum. Later, UF resins were modified with amines such as diethylenetriamine to introduce cationic functionality that made the resins self-retaining in paper (36).

Melamine–Formaldehyde Resins. Trimethylolmelamine, the condensation product of formaldehyde and melamine (triamino-s-triazine, the cyclic trimer of cyanamid) is also used to prepare wet-strength resins. Reaction of this material with acid leads to the formation of a colloidal resin that improves paper wet strength. Since the resin contains amine functionality, it is cationic at low pH and thus can function as a self-retaining resin. Often, however, the resin is modified to introduce anionic or cationic functionality as in the case of UF resins. In general, melamine–formaldehyde resins impart higher wet and dry strength than do UF resins (see also AMINO RESINS).

Aminopolyamide–Epichlorohydrin Resins. Formaldehyde-based resins require acidic conditions for curing and the development of wet strength. The first commercially significant neutral-to-alkaline curing wet-strength resins were based on water-soluble low molecular weight aminopolyamides prepared by condensing diacids such as glutaric acid or adipic acid with polyamines such as diethylenetriamine or triethylenetetramine to form a prepolymer (37). These prepolymers contain secondary amine functionality, which is subsequently utilized in a reaction with epichlorohydrin. The resulting resin has a complex chemical structure, the unique feature of which is the azetidinium functionality (38).

The azetidinium functionality provides quaternary amine groups that impart a cationic charge to the resin that is virtually independent of pH. Thus the resin is substantive to pulp even under neutral-to-alkaline papermaking conditions. Furthermore, the electrophilic azetidinium group can either react with nucleophilic groups such as carboxyl groups on cellulose fibers or can self-cross-link with amine groups present in the resin. Either reaction leads to improved paper wet-strength properties.

Aminopolyamide–epichlorohydrin resins are significantly more expensive than formaldehyde-based resins, but the higher cost is offset by other advantages, mostly the advantage of making paper at neutral-to-alkaline pH. These include properties such as improved softness and absorbency, no tendency toward embrittlement on aging, and reduced paper-machine corrosion. Another important advantage is the elimination of formaldehyde usage.

Polyamine–Epichlorohydrin Resins.   Condensates of epichlorohydrin and polyamines such as tetraethylene pentamine were introduced as wet-strength resins in the early 1950s (39). These resins are cationic and capable of developing wet strength at neutral-to-alkaline pH, but generate lower wet strength than aminopolyamide–epichlorohydrin resins. A number of these resins are currently marketed for applications not requiring high wet strength.

A more recent development has been wet-strength resins prepared by reaction of poly(methyldiallylamine) with epichlorohydrin (40). Methyldiallyl-amine undergoes a free-radical polymerization to form a polymer containing five-membered pyrrolidine rings. Quaternization of the tertiary amines with epichlorohydrin produces the quaternary epoxide (**1**). This functionality is highly reactive with nucleophiles and probably reacts with hydroxyl groups as well as carboxyl groups on the surface of cellulose fibers. Because the functionality is so reactive, the resins are supplied in the stabilized chlorohydrin form (**2**).

As a result the resin must be converted back to the epoxide form by reaction with sodium hydroxide just prior to use. The resins are self-retaining in paper, are effective under neutral-to-alkaline papermaking conditions, and provide very high wet strength.

Glyoxal Resins.   Another class of wet-strength resin is based on the reaction product of polyacrylamide and glyoxal (41). This resin can react with
hydroxyl groups on cellulose fibers through the glyoxal functionality to form
hemiacetal bonds, resulting in improvements in paper wet strength, although
not to the same extent as aminopolyamide–epichlorohydrin resins. The base
polyacrylamide usually contains a small amount of a cationic comonomer such
as dimethyldiallylammonium chloride to make the resin substantive to pulp
fibers.

Because these resins are based on polyacrylamide, they are also effective
dry-strength additives, and dry-strength improvement is often the primary reason
for their use. The wet strength provided by glyoxal resins is temporary because
the hemiacetal bonds are subject to hydrolysis when the paper is wetted by water.
Temporary wet strength is desirable for some applications. A related temporary
wet-strength resin is dialdehyde starch, which is produced by periodate oxidation
of starch (42). Dialdehyde starch provides relatively low wet strength, and the
wet strength deteriorates rapidly when the paper is wetted by water.

Polymeric Amines.   High molecular weight polyethyleneimine (PEI) and
related polymers have been shown to impart wet strength to paper (43). However,
the wet strength is low, and PEI resins are not commercially significant wet-
strength resins.

Chitosan.   Chitosan [poly(2-aminodeoxy-1,4-glucoside)] has also been shown
to impart wet strength to paper (44). This polymer, although potentially available
in large quantities from shellfish waste, has never become a commercially significant product (see also CHITIN).

Mechanism.   With the exception of the polymeric amines, all effective wet-
strength resins undergo cross-linking reactions and some react with functionality
on the surface of cellulose pulp fibers. Two mechanisms for wet-strength development have been proposed. The protection theory suggests that the resin forms
a cross-linked network around existing fiber–fiber bonds and prevents swelling
and disruption of these bonds (45–48). The new bond theory proposes that the
wet-strength resin creates new fiber–fiber bonds that are covalent and thus are
not broken on exposure to water (49). The protection theory is supported by the
fact that a given concentration of aminopolyamide–epichlorohydrin resin imparts
wet strength that is a relatively constant percentage of the dry strength in various
pulp furnishes, even though the absolute values of wet and dry strength might
vary substantially. The new bond theory is supported by the fact that wet and
dry strength are increased by approximately the same increment with either
formaldehyde-based wet-strength resins or aminopolyamide–epichlorohydrin wet-
strength resins.

Repulping.   Repulping of paper is necessary for two reasons. During paper
manufacture broke (waste) is produced either during breaks on the paper machine
or as trim produced during the rewinding process. For the papermaking process
to be economical, this must be recycled. Also, the use of waste-paper furnishes
is increasing as the cost of virgin-pulp furnishes increases. This is particularly
true in parts of Europe and in Japan, where there are diminishing forest reserves.

Paper containing wet-strength resin is more difficult to repulp than nonwet
strength paper. Therefore, the repulpability properties of wet-strength resins
must be considered when choosing a resin. UF and MF resins can be readily
repulped under acid conditions. Alum is often used in the repulping operations

when these resins are used. Aminopolyamide–epichlorohydrin resins are not readily repulped under either acidic or basic conditions. Oxidizing agents such as sodium hypochlorite are usually required for repulping paper containing these resins. Presumably, the oxidant attacks the polyamide backbone and breaks down the resin. Oxidizing agents also function as bleaching agents, so this method of repulping is not suitable for unbleached-pulp furnishes. Paper containing poly(methyldiallylamine)–epichlorohydrin wet-strength resin is very difficult to repulp. The hydrocarbon backbone of this resin is not subject to attack by reagents commonly used in aqueous systems.

Sizing Agents

Paper is a porous structure consisting of hydrogen-bonded cellulosic fibers. Cellulose is a very hydrophilic polymer with a strong affinity for water. At 50% rh, cellulose absorbs approximately 5 wt % water. The porous–hydrophilic nature is very useful for applications requiring rapid absorption of water (eg, tissues and paper towels). However, there are a number of applications where wetting and absorption are undesirable properties for paper. Examples are printing and writing grades designed to be printed by aqueous inks, and liquid containers such as milk cartons and paper cups. In these grades the wettability of the paper is decreased through the use of chemicals called sizes or sizing agents. A paper size is defined as an additive that reduces the penetration of liquids, especially water, into paper.

To attain effective sizing it is necessary to prevent wetting of the surface of the paper. To prevent wetting the contact angle of a drop of water on the surface must be high ($>90°$) (Fig. 3). The contact angle of water on a purely

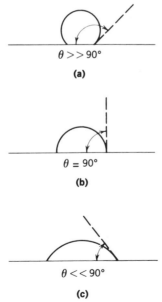

$\theta \gg 90°$

(a)

$\theta = 90°$

(b)

$\theta \ll 90°$

(c)

Fig. 3. Contact angle and wetting: (**a**) extremely limited wetting and spreading, tendency to retract, does not penetrate; (**b**) limited wetting and spreading, no tendency to penetrate; (**c**) extensive wetting and spreading, strong tendency to penetrate.

hydrophobic material such as wax is greater than 90°. For a paper size to be effective it must produce a low energy paper surface on which the initial contact angle of a drop of water is >90°. At a contact angle less than 90°, wetting, spreading, and penetration of water occur.

The rate of penetration of a liquid into a capillary pore is described by the Washburn equation and is depicted schematically in Figure 4 (50). The radius and length of the capillary pores are determined by pulp composition and papermaking conditions, and the viscosity and surface tension are determined by the nature of the liquid that will come in contact with the paper. Thus the only easily controlled variable that can change the rate of penetration of the liquid is the contact angle of the liquid on the paper surface. An effective size raises this contact angle above 90° at relatively low levels of addition.

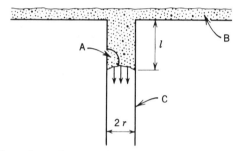

Fig. 4. Liquid flow-through capillary (Washburn equation). Time rate of penetration $= dl/dt = \frac{1}{4}(\gamma/\eta)(r/l) \cos \theta$. A, Contact angle θ between liquid and capillary wall; B, penetrating liquid, γ = surface tension, η = viscosity; C, partially filled capillary, r = radius, l = length already filled.

Size can be added to the dilute pulp slurry (internal sizing), or it can be applied from solution or dispersion to preformed paper (surface sizing). The former is the more common way to size paper. Internal sizing was first introduced in 1807 (51); it was found that the water repellency of paper could be increased by the internal addition of rosin soap and alum. Current sizing agents include rosin, fatty acids, waxes, alkyl ketene dimer, and alkenyl succinic anhydride. In 1982 the U.S. market for internal sizing agents was approximately $140 million.

Rosin Sizes.   Rosin is a natural product derived primarily from coniferous trees. There are three main types: gum rosin, which is collected as an exudate from wounds in growing trees; wood rosin, which is solvent extracted from aged wood stumps; and tall oil rosin, a by-product of the kraft pulping process. Tall oil is a mixture primarily of resin acids and fatty acids that can be skimmed off of the concentrated black liquor resulting from digestion of wood chips and filtration of the kraft pulp. Following acidification of the tall oil, rosin and fatty-acid streams are isolated by fractional distillation (see also RESINS, NATURAL).

Rosin is a complex mixture consisting primarily of several compounds called resin acids and minor amounts of neutral and disproportionated species. The structure of abietic acid, one of the predominant resin acids in rosin, is shown in Figure 5.

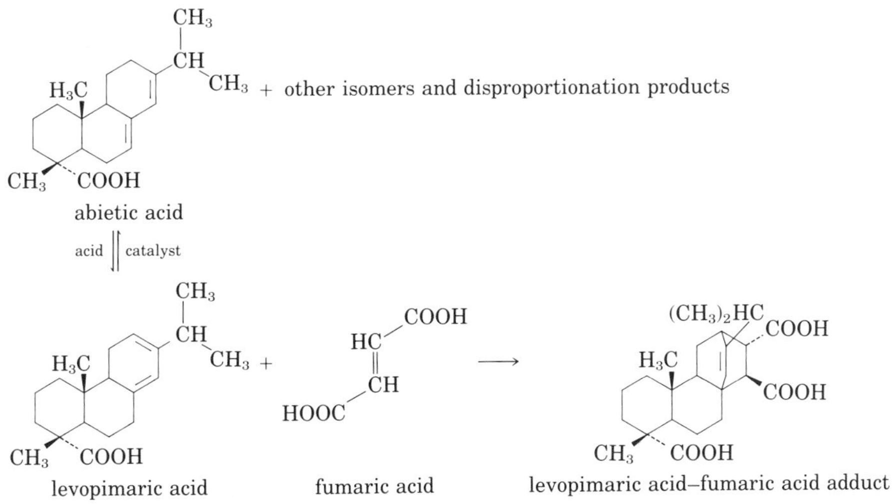

Fig. 5. Resin acids in rosin sizes.

One form of rosin size consists of the sodium or potassium soap of rosin; the degree of saponification is generally 75–100%. Unmodified rosin size is usually supplied as a viscous paste containing 60–80 wt % solids; the remainder is water. This paste must be carefully diluted in water to form a dispersion containing low solids (<1%) just prior to use. Alternatively, completely saponified rosin can be supplied as a dry powder which can be added directly to the pulp slurry without prior dilution. A third form of rosin size is a dispersion of partially saponified rosin having casein (qv) as the emulsion stabilizer. This material can be added directly to the pulp slurry.

Rosin size requires the presence of alum for sizing to develop. Rosin size and alum are added separately to the pulp slurry early enough in the process to allow good distribution. Generally, rosin size is added first followed by alum, but in some cases it is desirable to add them in the opposite order, a process known as reverse sizing.

Improvements in rosin-size efficiency were achieved with the introduction of the so-called fortified sizes which are prepared by reaction of rosin with either maleic anhydride or fumaric acid. The reaction involves initial acid isomerization of the resin acids to levopimaric acid, followed by a Diels-Alder reaction between this molecule and maleic anhydride or fumaric acid (Fig. 5). Usually, less than the theoretical amount of adduct is formed. The fortified rosin sizes are saponified and supplied either as pastes or as dispersions. Fortified sizes are used in the same way as unfortified sizes in the papermaking process.

A further improvement in efficiency was achieved with the introduction of high free-rosin dispersed sizes (52) in which the rosin is either unsaponified or lightly saponified. These products are supplied as aqueous dispersions containing 30–45% total solids. Small amounts of dispersion stabilizer such as synthetic surfactants or casein are also used. These products can be added directly to the pulp slurry, although prior dilution is sometimes desirable.

The magnitude of the improvements in efficiency are demonstrated in Table

Table 1. Efficiency of Sizing Agents

Agent	Typical usage, kg/t of paper
rosin-based	
ordinary rosin	15–20
fortified rosin	5–7.5
high free-rosin emulsion	2–4
synthetic	
alkyl ketene dimer emulsion	1–2
cationic alkyl ketene dimer emulsion	1–2
theoretical minimum	
rosin-based	0.5–1.5

1. Rosin usage has decreased from 15–20 kg/t for unmodified rosin size to 2–4 kg/t for high free-rosin dispersed size in a typical mill situation.

Because rosin sizing requires the use of alum, these sizes are only used in acid papermaking processes (pH range of 4.0–6.5). Rosin size is used in a variety of grades including printing and writing grades, bleached board, bleached and unbleached bag paper, unbleached paperboard, and cylinder board. Size concentration varies depending on the pulp furnish.

Cellulose-reactive Sizes. The first cellulose-reactive sizes, introduced in the 1950s (53,54), were based on alkyl ketene dimer (AKD) prepared by the dehydrohalogenation of fatty acid chloride.

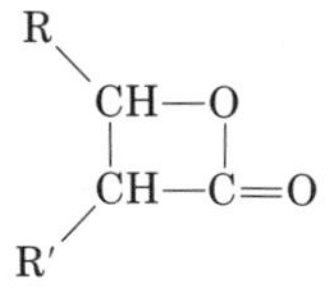

$$
\begin{array}{l}
\text{R} \\
\quad \diagdown \\
\quad \text{CH—O} \\
\quad \mid \qquad \mid \\
\quad \text{CH—C=O} \\
\quad \diagup \\
\text{R}'
\end{array}
$$

where R and R′ = C_{16}–C_{18} straight-chain hydrocarbon chains

The ketene dimer functionality is reactive toward nucleophiles and reacts with hydroxyl groups on the surface of cellulose fibers. AKD can also undergo hydrolysis to stearone, but this reaction is sufficiently slow that AKD can be supplied as an aqueous dispersion. AKD sizes are highly efficient so that low addition achieves a high degree of sizing (see Table 1).

Hydrolysis of AKD is slow as is reaction with cellulose hydroxyls. In fact, the reactivity of AKD with hydroxyl groups has been questioned (55,56). Studies on model compounds and on cellulose itself have demonstrated that AKD reacts with hydroxyl groups to form esters (57). The slow reaction can cause runnability problems on a paper machine if sizing does not develop prior to rewetting the sheet at the size press. To solve this problem cationic AKD dispersions that develop sizing more rapidly were introduced in the 1970s (58).

Other reactive sizes include alkenyl succinic anhydrides (ASA) and anhydrides of long chain fatty acids such as stearic anhydride (59,60). ASA is a commercially available size that is highly reactive with hydroxyls; consequently, it undergoes rapid hydrolysis and cannot be supplied as an aqueous emulsion. ASA is emulsified just prior to use, generally using cationic starch as a component of the emulsion stabilizer system. Stearic anhydride reacts with cellulose hydroxyls, but only half of the molecule becomes attached to cellulose. The other half forms stearic acid which does not contribute to sizing under most papermaking conditions.

As in the case of neutral-curing wet-strength resins, many of the advantages of cellulose-reactive size relate to the advantages of neutral-to-alkaline paper-making. These include higher sheet strength (which can allow the use of lower strength, lower cost pulps), reduced corrosion on the paper machine, and reduced salts in the effluent water (61). Another advantage is that neutral-to-alkaline paper does not degrade rapidly, whereas acid (alum) paper has a limited shelf life. This is important for archival papers. Alkaline-sizing agents are widely used in milk-carton board, in printing and writing paper (often in conjunction with calcium carbonate), and in gypsum board.

Wax Emulsions. Emulsions of paraffin or microcrystalline wax are also used as internal sizes for paper, usually in conjunction with rosin size (62). The emulsions usually have anionic stabilizers. Alum is added to the pulp slurry, the wax emulsion breaks, and the wax is retained by filtration in the forming paper web. Some cationic emulsions are used, and these are broken with sulfate ion. In general, wax emulsions are added to the pulp slurry after rosin and alum, often just prior to the headbox. The use of wax emulsions has been declining in favor of high efficiency rosin size and the very efficient cellulose-reactive sizes.

Fluorochemical Sizes. Several perfluorinated compounds are used as sizes against low energy penetrants such as grease and oils (63). Food packaging, paper plates, multiwall bags, and wallpaper require this type of sizing. Fluorochemical polymers and fluorinated organic acids are available, but the most commonly used products are fluorochemical phosphates. These products are usually supplied as 33% solutions in water–alcohol solvent, but some completely aqueous versions are available. They are added internally to paper and are retained through the use of cationic retention aids. Fluorochemical sizes are not always effective against penetration by high energy liquids such as water. When sizing against both grease and aqueous fluids is desired, a cellulose-reactive size is usually used along with a fluorochemical size.

Mechanism. The mechanism of internal sizing has been reviewed (64), and a general mechanism with all common sizes has been proposed (65). The general mechanism includes four steps for the development of effective sizing.

Retention. The sizing agent must be retained effectively to function well. Rosin-soap sizes are precipitated by alum and retained by electrostatic forces between the negative pulp fibers and the net positive-size precipitate. Other sizes (wax, dispersed rosin size, cellulose-reactive sizes) are supplied as dispersions or emulsified just prior to use. Cationic retention aids such as alum, cationic starch, or aminopolyamide–epichlorohydrin wet-strength resin are used to retain the size particles in the sheet.

Distribution. The size particles on the fibers must be uniformly distributed over the fiber surfaces to be effective. During drying, the size particles either melt or sinter, and can spread over adjacent-fiber surfaces. Evidence of this effect includes the fact that sheets dried at room temperature have significantly lower sizing than those dried at elevated temperatures. Low melting sizes such as wax and AKD melt and distribute readily. Rosin–alum precipitates have very high melting points and only deform and flow somewhat during the drying process.

Orientation. With the exception of wax, sizing agents are amphipathic, having both polar and nonpolar portions. To be effective these sizing agents must orient to present their nonpolar portions to the aqueous penetrant with their polar portions oriented toward the high energy cellulose surface.

Anchoring. Once oriented the size must be effectively anchored to the cellulosic fiber to prevent displacement by the aqueous penetrant. With cellulose-reactive sizes very effective anchoring is achieved by chemical reaction with the fiber surface. Anchoring with rosin–alum sizes is less well understood, but probably involves formation of a rosin–alum–cellulose complex. In the case of rosin-soap sizes the rosin–alum interaction occurs in the aqueous phase and results in the formation of size precipitate. In the case of high free-rosin dispersions the rosin–alum interaction occurs on the fiber surface during the drying process. Wax has no effective mechanism for anchoring, although attachment to the cellulose-fiber surface can be improved by pretreatment of the fiber with a cationic resin such as aminopolyamide–epichlorohydrin wet-strength resin.

Testing. There are a large number of sizing tests designed to simulate the use of the paper product. In these tests different penetrants and different test methods are used, depending on the end use. These test methods have been described in detail (64).

Fillers

Opacity is an important property for paper, particularly for grades that are printed on both sides of the sheet and in which show-through must be minimized. Newsprint has high opacity due to the use of high levels of groundwood, a high surface area, and highly light-scattering pulp. For lower surface-area pulps it is necessary to use a particulate filler to increase light scattering and opacity (66).

Mineral pigments are the most commonly used fillers (qv), including kaolin clay, calcium carbonate, titanium dioxide, silica, hydrated alumina, and talc. In the United States kaolin clay is the most widely used filler due to its low cost. Calcium carbonate is used more widely in Europe, where supply is plentiful and costs are low. Calcium carbonate can only be used at neutral-to-alkaline pH because it dissolves at low pH. The use of calcium carbonate in the United States is increasing as mills convert to alkaline-sizing systems.

With the exception of titanium dioxide the mineral fillers have refractive indexes very close to that of cellulose. Thus the increase in light scattering provided by these fillers is due solely to an increase in total surface area in the paper. One consequence of this fact is that filled paper has significantly lower strength properties than unfilled paper of the same basis weight. The mineral fillers do not bond to cellulose and reduce strength by interfering with fiber–fiber bonding, by introducing voids into the sheet, and by replacing 10–20% of the cellulose furnish. On the positive side, mineral pigments, particularly clay, are less expensive than cellulose fiber and are used to reduce costs.

Titanium dioxide has a significantly higher refractive index than cellulose.

It improves scattering due to this fact and due to increasing total surface area within the filled sheet. Titanium dioxide is quite expensive, however, and is only used in grades of paper requiring high opacity and brightness.

In recent years attempts have been made to use synthetic polymers as fillers for paper. Fillers based on polystyrene latices (67,68) and on urea–formaldehyde particles (69) have been evaluated. The low density of these fillers allows the production of lighter weight paper. The high cost relative to mineral fillers has, however, kept the synthetic fillers from developing a significant market share. A specialized additive consisting of a gas encapsulated in poly(vinylidene chloride) has also been evaluated (70). Expansion during the drying process brings about significant increases in sheet bulk with this product; however, it also is expensive.

Surface Treatments

In some cases it is not possible to add a paper chemical to the dilute pulp slurry and achieve the desired paper property. For instance, very high concentrations of some additives cannot be effectively retained internally. Other additives do not have suitable functionality for retention on pulp fibers even at low concentrations. Hence, a number of functional additives are applied to the surface of preformed paper, usually after it has been dried.

Several techniques for the surface application of chemicals to paper are available. The most common involves the use of a size press, a device consisting of two rolls forming a nip, which are flooded with a solution or dispersion of the chemical additive. Paper, usually dried below equilibrium-moisture content, is passed through the nip between the rolls to apply the additive. The paper is then dried in a second dryer section.

In tub sizing the paper is dipped in a solution or suspension of the functional additive and then redried. Spray application has also been used, usually for the application of creping aids or release agents on machines that have a Yankee dryer. Board grades are sometimes treated with solutions of functional additives at the calender stack. A more recent development has been the application of additives in the form of a foam (71). An advantage for this process is that additives can be applied to the wet sheet in the wet presses or smoothing press without breaking the sheet.

Strength Additives. The most commonly used size-press additive is starch. Various types of converted starch are used, including enzyme-converted starch, thermally converted starch, and hydroxyethylated starch. Starch is added at relatively high levels, primarily to improve sheet-strength properties. Surface starch treatment is common with printing grades. In this application the starch provides z-directional strength properties necessary for paper to survive the forces encountered during printing. In addition, starch prevents linting (transfer of small fiber fragments to printing presses or offset blankets). Starch reduces the rate of penetration of aqueous fluids by film formation and subsequent reduction of the porosity of the sheet.

Other polymers applied at the size press include poly(vinyl alcohol) and carboxymethylcellulose. These additives improve surface-strength properties and reduce sheet porosity at relatively low levels of addition. Because they are relatively expensive, they are often used in combination with starch.

Other dry-strength additives, including the natural gums, can be applied at the size press when surface strength is the desired property. Wet-strength resins are usually not added at the size press since they are efficiently retained as internal additives due to their cationic charge. An exception is wet-strength additives applied at the size press as creping aids in tissue and towel grades.

Sizing Agents.   Wax emulsions or blends of these emulsions with rosin size are applied at the size press to provide high levels of sizing. Cellulose-reactive size such as AKD size can also be applied at the size press. Surface application of AKD size can produce a slippery surface, which may be desirable or undesirable. To reduce slip AKD is sometimes applied in combination with starch or a mineral pigment. Styrene–maleic anhydride copolymers and polyurethane resins have also been used as surface sizes; they function by forming a hydrophobic film on the fiber surfaces. Fluorochemical sizes are often surface-applied at the size press, and provide grease and oil resistance for grades such as pet-food bags, labels, coupons, poultry wrap, cookie bags, snack-food bags, and reprographic papers. Chromium complexes of fatty acids are also used as surface sizes (72) to provide good water repellency and release properties. They are used in grades such as bags for ice cubes, wet vegetables, meat, and garbage, and in protective overwraps for outdoor storage.

Miscellaneous.   Creping aids and release agents are often applied at the size press for effective concentration at the sheet surface, where they are most needed when the sheet comes in contact with the Yankee dryer.

Paper that is coated on the paper machine on only one side often has curl problems due to the uneven stresses applied to the two sides of the paper during drying of the coating. This can be alleviated by applying small amounts of water to the uncoated side of the sheet at the size press. Foam application of water has been shown to be effective for applying small amounts of water for curl control (73).

Occasionally, a pigmented coating (see Paper Coatings below) is applied at the size press. These coatings are not as high quality as coatings applied by conventional coating processes, but they offer somewhat improved properties over uncoated paper.

Paper Coatings

Chemical additives may be placed onto the surface of paper in a special applicating device (ie, the coater). The paper coating is usually applied at relatively high solids compared to the surface treatments previously described. Both aqueous and nonaqueous coatings are used with both on-paper-machine and off-paper-machine application. Generally, the coating is not intended to strike deeply into the paper web.

Paper has a rough surface that is not suitable for high quality printing, since the printing elements come in contact with only a fraction of the total paper surface. For high quality printing it is necessary to provide a smooth paper surface. This is best achieved by coating the paper with a pigmented coating

(74,75). Paper coatings consist of water, a pigment (primarily kaolin clay), a binder or binders, and several other additives. An excess of the paper coating is applied to the surface of the sheet, and the excess is then removed by scraping off with a doctor blade or blowing off with an air knife. The coated paper is dried, and the result is a smooth surface suitable for high quality printing. Other properties important to the paper coating include good ink receptivity and good surface (z-directional) strength. The latter property, called pick strength, is necessary to prevent removal of the coating from the paper during the printing process.

The coating applied to paper is called a coating color. Generally, a coating color contains 40–65 wt % solids, the major component of which is pigment. Binders are used at relatively low levels, 5–30 parts by weight per 100 parts pigment. This ratio distinguishes paper coatings from paints, which are essentially filled polymer films. Other paper-coating additives include rheology modifiers, defoamers, dispersants, and lubricants.

Pigments. The most commonly used pigment for paper coatings is kaolin clay, available in a variety of grades having different particle size and degree of brightness (76). The choice of grade is dictated by the use requirements of the coated paper. Thus a premium coated paper with high gloss, good ink holdout, and printability requires a premium grade of clay having smaller particle size, higher brightness, and higher cost.

Kaolin clay particles are platelets having anionic charge on the faces and cationic charge on the edges. When clay is dispersed in water there is electrostatic interaction between these opposite charges leading to aggregation of the particles. Dispersants are used to eliminate this aggregation. Common dispersants include sodium hexametaphosphate, tetrasodium pyrophosphate, and sodium polyacrylate. The optimum concentration of dispersant for a given clay is that which gives minimum viscosity to a high solids (70%) clay slurry called a clay slip.

Calcium carbonate is rarely used as the sole pigment in a coating color, but rather in combination with kaolin clay. Calcium carbonate is produced either by grinding limestone or by a precipitation process (77). Dispersants are required to form dispersions as in the case of clay. Calcium carbonate is used more extensively in Europe, where it is readily available at relatively low cost.

Titanium dioxide is available in two crystal types: rutile and anatase. It is used as a minor component of the pigment system of a paper-coating color to impart good opacity and brightness properties (78).

A number of other mineral pigments are used in paper coatings, usually to impart a specific property (79). Amorphous silica and silicates are used as extenders for titanium dioxide. Aluminum trihydrate is used as a minor pigment component which can contribute flame retardancy to a paper coating. Barium sulfate is used almost exclusively in coatings for photographic paper. Talc is sometimes used as a pigment component to improve coating smoothness for gravure printing. Satin White, the reaction product of slaked lime and aluminum sulfate, imparts bulking properties to paper coatings. However, its dispersions are quite viscous at relatively low solids and it requires high concentrations of binder. Zinc oxide is used in coated paper for electrophotographic copying. Zinc oxide is photoconductive and thus can be used to generate a latent electrostatic image that can be made permanent through the use of a toner. There has been little growth in this type of copying due to advances and reduced costs with xerographic copying, which does not require a specialty coated paper.

Synthetic polystyrene pigments are used as components of paper coatings (80). These products are latices having an average particle size of 0.2–0.5 μm and a solids content of 48–50%. They are used as a minor pigment component (10–20 wt %) in clay-coating colors and provide improved gloss properties to paper coatings. As relatively low density pigments, they also provide lighter weight coatings.

Binders.   The earliest binders used in paper coatings were based on natural products. Casein was first used as a paper-coating binder in the 1890s (81). Casein (qv) is isolated by acid precipitation from skim milk. It is supplied as a dry powder that is dissolved in water plus alkali just prior to use and is used at about 10–20 wt %, based on the pigment. It is often cross-linked with formaldehyde or glyoxal to provide wet rub resistance to the coating. Casein was the predominant paper-coating binder until the 1950s when it was displaced by starch, soy protein isolate, and synthetic latex binders. Casein is still used as a binder, but wide fluctuations in price have put it at a disadvantage relative to the other available binders.

Starch (qv) is currently one of the largest volume paper-coating binders. It produces less viscous solutions than casein, which has advantages in the paper-coating process (29). A large variety of starch derivatives are used including oxidized starch, enzyme-converted starch, and hydroxyethylated starch. Starch is generally used at 10–20 wt %, based on the pigment.

Soy protein isolate was originally developed as a binder for paper coatings (82). It is prepared by extracting defatted soy meal with water at neutral-to-alkaline pH. The resulting solution is acidified, and the precipitate formed is isolated and dried. Soy protein isolate is dissolved in water plus alkali just prior to use and used at 8–16 wt %, based on pigment, in printing papers and 16–22% in paperboard applications. Soy protein has displaced casein in many applications. Sodium alginate has also been used as a binder–rheology modifier for paper coatings (83).

Synthetic latex binders have been a major factor in paper coatings over the past 25 years. By far the largest volume products are the styrene–butadiene latices (SBR latices) (84). SBR latices are prepared by the emulsion polymerization (qv) of styrene and butadiene using a free-radical initiator such as potassium persulfate. SBR polymers are usually about 60 wt % styrene and 40 wt % butadiene. This copolymer has a glass-transition temperature of about $-5°C$, and thus film forms readily during drying of the coating. This composition also has good strength properties resulting in good binding of the paper coating. Most SBR latexes are carboxylated, that is, a small amount (5%) of an acidic termonomer such as acrylic acid or itaconic acid is included in the emulsion polymerization. The acid groups thus introduced improve latex stability, contribute to the rheology properties of the coating color, and improve adhesion between the SBR polymer and the pigment particles. SBR latex particles are generally quite small (0.3 μm or less), which is also important for the rheological properties of the coating color. SBR latex produces coatings with a high degree of pick strength. SBR latex is usually used in combination with a natural binder such as starch or clay, although all-synthetic binders have found more widespread use in recent years (see also BUTADIENE POLYMERS; LATICES).

Poly(vinyl acetate) latices are also used widely in paper coatings (85). They are prepared by emulsion polymerization, usually with an acidic comonomer. Termonomers to modify the polymer glass-transition temperature are also often used. Poly(vinyl acetate) produces coatings of lower pick strength than SBR latices, but often with improved printability (86).

Acrylic binders are produced by the emulsion polymerization of acrylate monomers (87). The large number of acrylic monomers available allows a great deal of flexibility in designing polymer properties such as glass-transition temperature and carboxyl content (see ACRYLIC AND METHACRYLIC ESTER POLYMERS). Coatings with high pick strength can be produced. The main disadvantage of acrylic binders is their high cost relative to the other synthetic binders.

Poly(vinyl alcohol) (PVA, the water-soluble hydrolysis product of poly(vinyl acetate)) is also used as a coating binder (88). It produces coatings of very high pick strength. However, it is quite expensive and is usually used in conjunction with another binder such as SBR latex.

Rheology Modifiers. The rheological properties of paper coatings are very important during the application of the paper-coating color. The coating color must have suitable viscosity at low shear to allow pumping and pickup on the applicator roll. It must also have suitable high shear viscosity during the very high shear conditions encountered under the blade of a coater. Dilatant (shear-thickening) coatings can cause serious problems such as scratching in the paper coating.

Rheological properties are determined by a number of factors in the coating color, including total solids, water-soluble polymeric species, particle–particle interactions, and particle–dissolved polymer interactions. In the case of natural binders such as starch and protein, rheology is dictated by the coating formulation and makeup process, because the binder is a dissolved or partially dissolved polymer, and is present in relatively high concentration.

Rheology modifiers are relatively high molecular weight water-soluble polymers that provide desirable coating-color rheology. Examples are sodium carboxymethylcellulose (CMC) (88) and sodium alginate (83). Generally, polymers such as CMC produce shear-thinning coating colors that perform well under the high shear conditions typical of modern high speed coaters. In addition, these additives function as water-retention aids that prevent premature migration of coating-color components into the paper. These water-soluble polymers are often used in conjunction with natural binders such as starch. They are particularly useful in all-synthetic coating colors where they represent the only water-soluble polymeric species.

Testing. Since most paper is coated to be printed, most tests for coated paper are concerned with printability. Opacity and brightness are tested in the same way as for uncoated paper. Gloss, which is important for coated papers, is measured with a number of instruments that measure the reflectance from the coated-paper surface. Ink receptivity is also measured with a number of tests. The K&N ink test (TAPPI UM 553) is a measure of the pickup of a special ink by the coated surface. The wedge-print test uses the gloss of a printed surface at different thicknesses of ink as a measure of ink penetration into the coating. Pick strength can be measured using the IGT Printability Tester or the Prufbau Printability Tester (TAPPI T499). Ink set-off is measured on the IGT Printability

Tester by pressing an unprinted sheet against a printed sheet after various ink-drying times.

BIBLIOGRAPHY

"Paper Additives and Resins" in *EPST* 1st ed., Vol. 9, pp. 748–793, by N. R. Eldred, Union Carbide Corporation.

1. J. Pearson, D. Sutton, and H. O'Brien, *Pulp Pap.* **59,** 55 (1985).
2. *Chem. Week* **138,** 28 (May 14, 1986).
3. *Chem. Eng.* **92,** 11 (Apr. 29, 1985).
4. *Title 21, Code of Federal Regulations,* U.S. Government Printing Office, Washington, D.C., Apr. 1, 1980, Subchapt. B, Part 176, Sects. 170 and 180.
5. *Pap. Trade J.* **164**(16), 42 (Aug. 30, 1980).
6. *Survey of Paper Additives,* 1986–1987 ed., H&H Consulting Group, Flemington, N.J.
7. "Mixing Devices for Rapid Solution of Hercules (R) Water-Soluble Polymers," *Hercules Technical Information Bulletin VC-400F,* Wilmington, Del, 1985.
8. U.S. Pat. 3,206,412 (Nov. 14, 1965), (W. H. Kirkpatrick and V. L. Seale (to Nalco Chemical Co.).
9. K. W. Britt, *Tappi* **56**(10), 46 (1973).
10. R. W. Davison, *Tappi* **66**(11), 69 (1983).
11. R. H. Pelton, L. H. Allen, and H. M. Nugent, *Tappi* **64**(11), 89 (1981).
12. K. W. Britt and J. E. Unbehend, *Proceedings of the Technical Association of the Pulp and Paper Industry, 1980 Papermakers Conference,* Chicago, April 6–8, 1980, TAPPI Press, Atlanta, Ga., 1981, p. 5.
13. U.S. Pat. 3,617,440 (Nov. 2, 1971), G. W. Strother, Jr. (to The Dow Chemical Co.).
14. J. P. Casey, *Pulp and Paper Chemistry and Technology,* 3rd ed., Vols. II and III, Wiley-Interscience, New York, 1980–1981, pp. 1022–1025 and 1796–1802.
15. T. G. Rubel, *Antifoaming and Defoaming Agents, Chemical Process Review No. 60,* Noyes Data Corp., Park Ridge, N.J., 1972.
16. H. T. Kerner, *Foam Control Agents, Chemical Technology Review No. 75,* Noyes Dara Corp., Park Ridge, N.J., 1976.
17. "Paper Makers Chemicals," *Hercules Technical Bulletin No. 8518,* Wilmington, Del., Sept. 4, 1985.
18. R. L. Shelton, *Pap. Trade J.* **169**(8), 48 (1985).
19. R. W. J. McKinney and P. G. C. Currie, *Pap. Technol. Ind.* **27**(4), 182 (1986).
20. R. W. Davison in W. F. Reynolds, ed., *Dry Strength Additives,* TAPPI Press, Atlanta, Ga., 1980, pp. 1–30.
21. C. E. Brandon in J. P. Casey, ed., *Pulp and Paper, Chemistry and Chemical Technology,* 3rd ed., Vol. III, Wiley-Interscience, New York, 1981, pp. 1715–1972.
22. R. W. Davison, S. T. Putnam, R. T. Mashburn, and H. O. Ware, *Tappi* **40**(7), 499 (1957).
23. C. Fellers, "The Significance of Structure for the Compression Behavior of Paperboard" in J. A. Bristow and P. Kolseth, eds., *Paper, Structure and Properties,* Marcel Dekker, Inc., New York, 1986, pp. 281–310.
24. C. Fellers, "Edgewise Compression Strength of Paper" in R. E. Mark, ed., *Handbook of Physical and Mechanical Testing of Paper and Paperboard,* Vol. 1, Marcel Dekker, Inc., New York, 1983, pp. 349–383.
25. J. J. Becher, R. A. Stratton and G. A. Baum, *Institute of Paper Chemistry Project 3526, Prog. Rept. 2,* Appleton, Wisc., Sept. 30, 1986.
26. F. M. K. Werdouschegg in Ref. 20, pp. 67–93.
27. H. J. Leech, *Tappi* **37**(8), 343 (1954).
28. B. T. Hofreiter in Ref. 21, pp. 1475–1514.
29. M. J. Mentzer, "Starch in the Paper Industry" in R. L. Whistler, J. N. BeMiller, and E. F. Paschall, eds., *Starch, Chemistry and Technology,* Academic Press, Inc., Orlando, Fla., 1984, pp. 543–574.
30. D. S. Greif and L. A. Gaspar in Reference 20, pp. 95–117.
31. E. J. Barber, *Tappi* **44**(2), 179A (1961).
32. H. H. Espy, *TAPPI Papermakers Conference Proceedings,* TAPPI Press, Atlanta, Ga., 1983, p. 191.

33. W. F. Reynolds in Ref. 20, pp. 125–148.

34. U.S. Pat. 2,116,544 (May 10, 1938), M. O. Schur (to The Brown Co.).

35. U.S. Pat. 2,407,376 (Sept. 10, 1946), C. S. Maxwell (to American Cyanamid Co.); U.S. Pat. 2,407,599 (Sept. 10, 1946), R. W. Auten and J. L. Rainey (to Resinous Products & Chemical Co.).

36. U.S. Pat. 2,657,132 (Oct. 27, 1953), J. H. Daniel, Jr., C. G. Landes, and T. J. Suen (to American Cyanamid Co.).

37. U.S. Pats. 2,926,116 and 2,926,154 (Feb. 23, 1960), G. I. Keim (to Hercules Inc.).

38. N. A. Bates, *Tappi* **52**(6), 1162 (1969).

39. U.S. Pat. 2,595,935 (May 6, 1952), J. H. Daniel, Jr. and C. G. Landes (to American Cyanamid Co.).

40. U.S. Pats. 3,700,623 (Oct. 24, 1972) and 3,772,076 (Nov. 13, 1973), G. I. Keim (to Hercules Inc.).

41. U.S. Pat. 3,556,932 (Jan. 19, 1971), A. T. Coscia and L. L. Williams (to American Cyanamid Co.).

42. G. E. Hammerstrand, B. T. Hofreiter, D. J. Kay, and C. E. Rist, *Tappi* **46**(7), 400 (1963).

43. P. E. Trout, *Tappi* **34,** 539 (1951).

44. G. G. Allan, G. D. Crosby, J. H. Lee, M. L. Miller, and W. M. Reif, "New Bonding Systems for Paper" in *Proceedings of the Symposium on Man-Made Polymers in Papermaking,* IUPAC and EUCEPA, Helsinki, June 5–8, 1972, Finnish Paper Engineers' Association and Finnish Pulp and Paper Research Institute, Helsinki, 1973, pp. 85–96.

45. A. Jurecic, C. M. Hou, K. Sarkanen, C. P. Donofrio, and V. Stannett, *Tappi* **43**(10), 861 (1960).

46. S. J. Hazard, F. W. O'Neil, and V. Stannett, *Tappi* **44**(1), 35 (1961).

47. R. J. Kennedy, *Tappi* **45**(9), 738 (1962).

48. N. M. Fineman, *Tappi* **35**(7), 320 (1952).

49. D. J. Salley and A. F. Blockman, *Tech. Assoc. Pap.* **30,** 223 (1947); *Pap. Trade J.* **125**(1), 35 (1947).

50. E. W. Washburn, *Phys. Rev.* **17,** 273 (1921).

51. J. W. Swanson, ed., *Internal Sizing of Paper and Paperboard, TAPPI Monogr.* 33 (1971), p. 2.

52. U.S. Pat. 3,565,755 (Feb. 23, 1971), R. W. Davison (to Hercules Inc.).

53. C. A. Weissgerber and C. A. Hanford, *Tappi* **43**(12), 178A (1960).

54. J. W. Davis, W. H. Roberson, and C. A. Weissgerber, *Tappi* **39**(1), 21 (1956).

55. J. Merz, P. Rohringer, and M. Berheim, *Das Papier* **39**(5), 214 (1985).

56. J. C. Roberts and D. N. Garner, *Tappi* **68**(4), 118 (1985).

57. S. H. Nahm, *J. Wood Chem. Tech.* **6**(1), 89 (1986).

58. D. H. Dumas, *Proceedings of the Technical Association of the Pulp and Paper Industry, 1979 Tappi Papermakers Conference,* April 9–11, Boston, TAPPI Press, Atlanta, Ga., 1979, p. 67.

59. Brit. Pat. 954,526 (Apr. 8, 1964), R. J. Kulick and E. Strazdins (to American Cyanamid Co.).

60. U.S. Pat. 3,102,064 (Aug. 27, 1964), G. B. Wurtzberg, and E. D. Mazzerella (to National Starch and Chemical Corp.).

61. J. Hoppe and R. Karle, *Wochenbl. Papierfabr.* **102**(23–24), 833 (1974).

62. A. K. Plitt, *Tappi 1980 Sizing Short Course Notes,* April 16–18, 1980, Atlanta, Ga., TAPPI Press, Atlanta, Ga., 1980, p. 67.

63. C. Schwartz in Ref. 62, p. 55.

64. G. G. Spence, D. H. Dumas, J. C. Gast, S. M. Ehrhardt, and D. B. Evans, "Fatty Acids and Rosin in Paper" in R. W. Johnson and E. Fritz, eds., *Fatty Acids,* Marcel Dekker, Inc., New York, in press.

65. R. W. Davison, *Proceedings of the Technical Association of the Pulp and Paper Industry, 1986 Tappi Papermakers Conference,* April, 1986, New Orleans, La., TAPPI Press, Atlanta, Ga., 1986, p. 17.

66. Ref. 14, pp. 985–1020.

67. E. H. Rossin, *Pulp Pap.* **48**(6), 57 (1974).

68. U.S. Pat. 4,282,060 (Aug. 4, 1981), W. W. Maslanka and G. G. Spence (to Hercules Inc.).

69. P. Economou, J. F. Hardy, and J. Menashi, *Pulp Pap.* **60**(9), 109 (1976).

70. U.S. Pat. 3,556,934 (Jan. 19, 1971), F. J. Meyer (to The Dow Chemical Co.).

71. Brit. Pat. 1,551,710 (Aug. 30, 1979), B. Jenkins (to Wolvercoter Co., Ltd.).

72. J. W. Trebilcock in Ref. 62, p. 61.

73. M. C. Riddell and B. Jenkins, *Paper Technol. Ind.* **18**(6), 176 (1977).

74. W. R. Willets in R. W. Hagemeyer, ed., *Paper Coating Pigments, TAPPI Monogr.* **38,** 1 (1976).

75. F. H. Frost in R. H. Marchessault and C. Skaar, eds., *Surfaces and Coatings Related to Wood,* Syracuse University Press, Syracuse, N.Y., 1967, pp. 301–323.

76. H. H. Murray in Ref. 74, p. 69.

77. R. W. Hagemeyer in Ref. 74, p. 37.
78. W. J. McGinnis in Ref. 74, p. 191.
79. R. W. Hagemeyer, ed., *Paper Coating Pigments, TAPPI Monogr.* **38** (1976).
80. U.S. Pat. 3,949,138, E. J. Heiser (to The Dow Chemical Co.).
81. H. K. Salzberg and W. L. Marino in R. Strauss, ed., *Protein in Paper and Paperboard Coating, TAPPI Monogr.* **36,** 1 (1975).
82. R. A. Olson and P. T. Hoelderle in Ref. 81, p. 75.
83. U.S. Pat. 3,351,479 (Nov. 7, 1967), W. P. Fairchild (to Kelco Co.).
84. E. J. Heiser and F. Kaulakis in A. R. Sinclair, ed., *Synthetic Binders in Paper Coatings, TAPPI Monogr.* **37,** 22 (1975).
85. T. F. Walsh and L. A. Gaspar in Ref. 84, p. 98.
86. C. L. Parsons, *Tappi* **58**(5), 123 (1975).
87. J. J. Latimer and H. S. De Groot in Ref. 84, p. 120.
88. R. C. Jezerc and G. P. Colgan in Ref. 84, p. 64.

GAVIN G. SPENCE
Hercules Incorporated

PAPER CHROMATOGRAPHY. See CHROMATOGRAPHY.

PAPER COATINGS. See PAPER ADDITIVES AND RESINS.

PARISON. See BLOW MOLDING.

PARQUET POLYMERS. See COORDINATION POLYMERS.

PARTICLE BOARD. See COMPOSITION BOARD.

PARTING AGENTS. See RELEASE AGENTS.

PARTITION CHROMATOGRAPHY. See CHROMATOGRAPHY.

PARTS REMOVAL. See ROBOTICS.

PATENT INFORMATION

To a first approximation patents are *the* literature of technology. A patent is a legal document conferring on its owner the right to exclude others from practicing an invention, ie, making, using, or selling a particular product or carrying out a particular process. In return for this limited monopoly the inventor must fully disclose the details of the invention, and this information becomes part of the collective knowledge of humankind to serve as a spur to further invention and innovation.

The potential value of a patent monopoly more often than not causes inventors and their employers to file for patent protection, although there are some technological advances whose substance is not disclosed but is rather maintained as a trade secret. Consequently, a very substantial proportion of the published information on advances in all technologies, including the technology of polymers, is found in patent documents.

Not all of these documents are patents; many are unexamined, published patent applications, which may or may not mature into patents at some time. Up until the early 1960s most industrialized countries published only those patent applications that had been examined and deemed to be truly novel, although there have always been some countries such as Belgium that quickly issue all applications so long as they meet formal requirements. The Netherlands began a trend in 1964 for examining countries to switch to automatic publication of all patent applications approximately 18 months after their earliest file date in response to an overload on their patent examining staff. With universal publication examination can be deferred and is often avoided when an invention fails to sustain its initial promise.

The trend to early publication of all patent applications has been followed by most of the industrialized nations outside the Soviet bloc, and today the United States and Canada are the only major exceptions to the rule of early publication of all patent applications. Both still require examination for patentability and publish only those applications determined to be patentable. An exception in the United States is the publication of unexamined applications stemming from government-supported research.

Universal early publication has increased the amount of information available to the public, in comparison to the days when only patentable inventions were published. It has also accelerated the rate of dissemination of information, since nearly all patent applications are published ca 18 months after their initial filing date; patents published only after examination can take years to appear. And it has diluted the quality of the patent literature, since some of what gets published is of low quality. There is no mechanism to screen out the chaff, and the user of the patent literature must exercise considerable judgment, keeping in mind that this literature differs from the refereed technical journal literature. But with that caveat the user is reminded that patents are the chief literature of technology, containing invaluable information, much of which is not available elsewhere.

This article is primarily concerned with the information contained in patents that involve polymers, as opposed to the legal aspects of those patents. For information purposes published, unexamined applications are essentially equivalent to examined patents and, in general, the simple term patent is used here to refer to both patents and published patent applications. Readers interested in the legal aspects of patents are directed to other sources (1). Patents as information sources have also been reviewed elsewhere (2).

The number of patents that deal with polymers is large as reflected by statistics from the *Chemical Patents Index (CPI)* produced by Derwent Publications, Ltd. of London. The *CPI* is the most comprehensive abstracting and indexing system for patents with a broad definition of what constitutes a chemical patent and what constitutes a polymer-related patent. Since 1981 its polymer section, *PLASDOC,* has included nearly one-third (31.7%) of all chemical patents in the *CPI,* an annual total that reached about 45,000 new basic patents in 1985. These patents cover all aspects of polymers, from monomer preparation and purification, through polymer synthesis, compounding, and processing, to uses (Fig. 1), and this scope of information must be dealt with by information services that cover polymer patents. The range is enormous, encompassing not only chemistry and chemical engineering but also mechanical engineering, electrical engineering, physics, and other disciplines.

- Monomer preparation, purification
- Polymerization catalyst systems; initiators; modifiers
- Polymerization processes
- Polymer structure, composition
- Compounding agents, techniques
- Polymer modification reactions, processes
- Fabrication techniques, equipment
- Polymer shapes, forms
- Molecular and physical properties
- Uses

Fig. 1. Scope of polymer patent information.

Three information services are especially important in the area of polymer patents: Derwent Publications, Ltd.; Chemical Abstracts Service (CAS); and IFI/Plenum Data Co. The first two play a major part in both current awareness and retrospective information retrieval, whereas IFI/Plenum is mainly involved in retrospective retrieval. There are other information services that include polymer patent information but do not provide any special information capability regarding patents. Thus the INPADOC database produced by the International Patent Documentation Center in Vienna includes bibliographic information on all patents from virtually all countries, but has very limited subject-based retrieval capability. The *APIPAT* database produced by the American Petroleum Institute has a specialized polymer retrieval system (3), but is limited to polymers of direct interest to the petroleum industry. Neither system will be discussed here.

Current Awareness Services

The three key elements for a satisfactory current awareness publication are timeliness, scope of information coverage, and depth of information content, and a successful service must combine all three. Thus it is not sufficient to have outstanding timeliness if the information content is so thin that little information is conveyed. This is the case with the INPADOC services, which feature coverage of all subject matter and excellent promptness for most countries, but provide only bibliographic details on the patents included. The only subject-information content is the original patent title, which can be very uninformative and even misleading, and the patent classification. INPADOC products can be used very effectively for maintaining patent watch activities in selected areas and they are an outstanding source of patent family information, but of limited value for current awareness purposes.

The important current awareness products are those produced by Derwent and by CAS. Derwent has produced *PLASDOC*, its polymer patent database, since 1966. CAS produces the complete *Chemical Abstracts,* which is in general too cumbersome for use as a current awareness tool so that CAS also provides subdivisions: the Macromolecular Sections subject grouping, encompassing all of *CA* polymer coverage, and a series of *CA Selects* profiles, many of which cover polymer areas. In addition, customized selective dissemination of information (SDI) programs can be run on the computerized versions of both the Derwent and *CA* databases.

Abstracts. The heart of both the Derwent and CAS services for current awareness purposes is a professionally written abstract. Patent documents invariably include claims (or proposed claims), and generally also include an abstract provided by the applicant. However, for information purposes a claim, written in legalistic language to define a property right, often provides a very vague idea of the information content of a patent; indeed, the language is sometimes selected for obfuscatory purposes. As for applicant abstracts, these are uneven in quality and are at worst designed to provide the least information. Applicant abstracts and/or claims are available in some printed services, such as the *Official Gazette* of the U.S. Patent and Trademark Office and several competing computerized files based on U.S. patents, but these are generally inferior in information content to Derwent or *CA* abstracts. However, these applicant abstract/claim-based services are very timely, with at least a 2-month lead over Derwent and more over *CA*, so that they serve a very useful purpose despite their limited information content.

The abstracts in *CA* are high in quality, but sometimes not ideal for patents. Patents frequently take a few limited examples and from them propose far-reaching claims. The *CA* of the past generally omitted any information relating to the scope or legal aspects of a patent, describing merely what was actually done. Since about 1980 *CA* has put more emphasis on conveying claim and scope information, but it generally falls short of Derwent in this regard. *CA* has also tended in recent years to shorten its abstracts, providing less detail and thus conveying less information.

The differences between *CA* and Derwent are not surprising. *CA* covers a variety of information (journal literature, books, reports, and patents) and its audience includes the academic community as well as the industrial community. CAS management has chosen in the past not to attempt to produce products aimed specifically at patent users, but that policy is being changed. CAS has been improving its patent coverage over the past decade and is in the process of developing a new patent-oriented database.

Derwent, on the other hand, specializes in patent information for a readership of industrial scientists and engineers as well as attorneys and patent agents. Derwent abstracts have from the start reflected the information claimed in a patent. In chemically related areas, including *PLASDOC,* Derwent produces abstracts on two levels. An alerting abstract summarizes the broadest and more specific claims of a patent, describing the patent's purpose and often adding other points of interest. A documentation abstract goes beyond this to highlight what is new about the invention, to indicate, where significant, the full scope of the disclosure beyond the claims, and usually to provide detailed examples with data and drawings. Besides the abstracts, Derwent provides a highly informative rewritten title that expresses the heart of a patent's content. Original patent titles are notorious for their tendency to convey as little information as possible. Derwent rewritten titles are usually outstanding.

Derwent documentation abstracts are produced 2 weeks after their alerting abstracts and are usually the finest and most informative abstracts available. Although they are primarily produced for archival purposes, they make an outstanding alerting medium, sacrificing only 2 weeks in time with respect to the alerting abstract, but making up for it with a wealth of detail. Some of the abstracts of Japanese patents have been unsatisfactory, because of the combination of language difficulties and an enormous volume of Japanese patent publications (297,200 unexamined published applications or Kokai in 1986). Derwent documentation abstracts can often serve as surrogates for full patent documents. CAS, on the other hand, intends its abstracts to be pointers to the original documents, not surrogates. Typical differences between a *PLASDOC* documentation abstract and a *CA* abstract are shown in Figure 2, and additional examples can be found in an earlier article on this topic (4).

Timeliness. In general, Derwent also excels in timeliness. The current delay between patent issuance and the issuance of Derwent alerting abstracts is approximately 2 months for patents from most major countries and patent-issuing bodies, such as the United States, UK, FRG, the European Patent Office, and the World Intellectual Property Organization (WIPO, Patent Cooperation Treaty applications). For Japan the delay is 3 months. Derwent covers all of each week's output from any country in a single, weekly issue of its bulletins so that there is no distribution of delay times in any country. *CA,* on the other hand, has a longer delay in coverage and a distribution of lag times (Table 1). It is never certain when *CA* will cover a given patent. Thus for the five patent offices that averaged a 2-month delay with Derwent, the *CA* delay ranged from 3 to 8 months and for Japan it was 5–11 months; most items have a delay of 7–8 months.

Coverage. Subject coverage is a key aspect of both current awareness and retrospective retrieval. Derwent excels over *CA* in coverage. For example, of the first 50 patents in one *PLASDOC* issue, 23 were not covered in *CA* (4). More recently, considering the Derwent and *CA* computerized databases, of a sample of 100 items in *PLASDOC,* nearly half did not appear in *CA.* By contrast a sample of 200 patents from the Macromolecular Sections of *CA* included only four that did not appear in *PLASDOC.* Two of these were Polish patents not covered by Derwent; the other two were Japanese patents in marginally chemical areas, where Derwent coverage has some gaps. Most of the items omitted by *CA* fall into one of three categories: equipment for polymer processing, methods of poly-

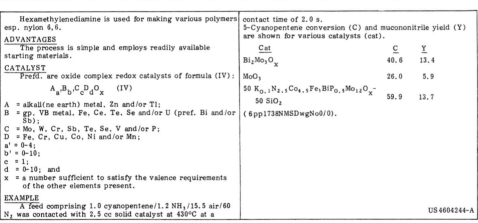

106: 33590c Adiponitrile precursors. Burrington, James D.; Grasselli, Robert K.; Kartisek, Craig T. (Standard Oil Co. (Ohio)) **U.S. US 4,604,244** (Cl. 558–320; C07C120/00), 05 Aug 1986, US Appl. 455,081, 03 Jan 1983; 6 pp. Cont.-in-part of U.S. Ser. No. 455,081, abandoned. Ammoxidn. of $H(CH_2)_aCH:CH(CH_2)_bX$ (I) (X = CN, CONH$_2$; a, b = 0–3, $a + b$ = 3) gives NCCH:CHCH:CHX (II), useful as precursors for monomers for nylon. Thus, heating 2.0 mL acrylonitrile and 0.1 equiv. AlCl$_3$ in 4 mL PhNO$_2$ with C$_3$H$_6$ at 200°/1500 psig for 4 h gave 4- and 5-hexenenitrile. Heating this mixt. (54:46 ratio) with 1.2:15.5:60 NH$_3$–air–N over Bi$_2$Mo$_3$O$_x$ at 430° for 2 s gave 13.4% mucononitrile (40.6% 5-hexenenitrile conversion).

(a)

86-225154/34   A41 E16
 STANDARD OIL CO (OHIO)
    23.04.84-US-603257   (+US-455081)   *(05.08.86)*   C07c-120   C07c-121/30
 Prepn. of mucononitrile and 5-cyano-2,4 pentadiene amide - by gas phase reaction of cyano-or amido-alkene intermediate with oxygen and ammonia in contact with solid catalyst
    C86-097154
	STAH 03.01.83
	 *US 4604-244-A
		A(1-E5, 1-E12) E(10-A15A, 10-A15B) N(1-B, 2, 3, 4-A, 4-B)
		                 A0942

Process comprises reacting a cpd. of formula (II) with oxygen and ammonia to give a prod. of formula (III) (i.e. mucononitrile or 5-cyano-2,4-pentadiene amide).

$$H-(CH_2)_a-(CH{=}CH)-(CH_2)_b \; X \quad (II)$$

$$NC-CH{=}CH-CH{=}CH-X \quad (III)$$

The novelty comprises using the reactants in the gaseous phase in contact with a solid oxidn. catalyst which catalyses ammoxidation of propylene to acrylonitrile and effecting the reaction at 250–650°C.

$$H-(CH_2)_a-(CH{=}CH)-(CH_2)_b-X \xrightarrow{O_2, NH_3}$$
$$(II)$$

$$NC-CH{=}CH-CH{=}CH-X \quad (III)$$

X = CN or CONH$_2$; and
a and b each = 0 or 1-3 such that a + b = 3.
   Also claimed is a two step process comprising
     (a) reaction of propylene with an enophile of formula CH$_2$=CH-X (I) to form cyano- or amido-pentene (IIa); and
     (b) reaction of (IIa) with O$_2$ and NH$_3$ as above to give (III).

$$CH_2{=}CH-Me \; + \; CH_2{=}CH-X \quad (I) \xrightarrow{(a)}$$

$$CH_2{=}CH-(CH_2)_3-X \quad (IIa) \xrightarrow{(b)} (III)$$

USE
   Mucononitrile may easily be converted to adiponitrile by known processes. Adiponitrile is a monomer for nylon and is also useful in mfr. of hexamethylenediamine as is (III; X = CONH$_2$).

               US4604244-A+

Hexamethylenediamine is used for making various polymers esp. nylon 6,6.

ADVANTAGES
   The process is simple and employs readily available starting materials.

CATALYST
   Prefd. are oxide complex redox catalysts of formula (IV):

$$A_{a'}B_{b'}C_cD_dO_x \quad (IV)$$

A = alkali(ne earth) metal, Zn and/or Tl;
B = gp. VB metal, Fe, Ce, Te, Se and/or U (pref. Bi and/or Sb);
C = Mo, W, Cr, Sb, Te, Se, V and/or P;
D = Fe, Cr, Cu, Co, Ni and/or Mn;
a' = 0-4;
b' = 0-10;
c = 1;
d = 0-10; and
x = a number sufficient to satisfy the valence requirements of the other elements present.

EXAMPLE
   A feed comprising 1.0 cyanopentene/1.2 NH$_3$/15.5 air/60 N$_2$ was contacted with 2.5 cc solid catalyst at 430°C at a

contact time of 2.0 s.
5-Cyanopentene conversion (C) and mucononitrile yield (Y) are shown for various catalysts (cat.).

Cat	C	Y
Bi$_2$Mo$_3$O$_x$	40.6	13.4
MoO$_3$	26.0	5.9
50 K$_{0.1}$N$_{2.5}$Co$_{4.5}$Fe$_1$BiP$_{0.5}$Mo$_{12}$O$_x$ 50 SiO$_2$	59.9	13.7

(6 pp1738NMSDwgNo0/0).

               US 4604244-A

(b)

Fig. 2. Comparison of abstracts from (**a**) *CA* and (**b**) *PLASDOC*.

---

mer processing, and uses of polymers. The roots of *CA* coverage of polymers lie in polymer chemistry and chemical engineering. Although *CA* has steadily broadened its subject scope, it still cannot match that provided by Derwent.

     Although Derwent has numerous advantages over *CA* in the area of current awareness, both information sources play an important role. For any given subject area there may be available specialized subsets of information (*CA Selects*, Derwent profiles), or a computerized SDI program may fit especially well with one

Table 1. Currency of *CA* Patent Coverage[a]

Country	Patent publication month									
	9/85	10/85	11/85	12/85	1/86	2/86	3/86	4/86	5/86	*Total*
Japan	2	1	2	13	29	0	5	0	0	*52*
FRG, United States, UK, European, and PCT[b]	0	0	0	2	2	2	2	8	4	*20*

[a] Patents in Ref. 5.
[b] Patent Cooperation Treaty.

or the other database. Comparative economics are also a consideration. Information users must consider all of these factors in deciding which alerting service(s) to use.

Retrospective Retrieval

The *Chemical Abstracts* database goes back to 1907, but material prior to 1967 is available only in printed form. Since 1967 all of *CA* is available as an online, searchable database from a number of hosts, most of whom offer special features. Both ORBIT and DIALOG offer auxiliary CAS Registry Number dictionary files; the DIALOG version has special capabilities in the identification of copolymer systems. The version offered by Questel includes the full Registry file as an adjunct, searchable by full and partial structures, plus the inclusion of molecular formula data to supplement Registry Numbers in the *CA* file. The version offered by CAS itself through STN International includes printable and searchable abstracts as well as an interface with the full Registry file. Also included in STN offerings are many pre-1965 compound listings in the Registry, and the accompanying *CAOLD* file that provides bibliographic information for these listings.

Derwent's *PLASDOC* goes back to April 1966. The *CLAIMS* files from IFI/Plenum cover a longer time span, since 1950. Both are fully computerized for online searching. Each is offered by three hosts: Derwent by ORBIT, DIALOG, and Questel; *CLAIMS* by ORBIT, DIALOG, and STN International. The availability of a database from multiple sources provides the opportunity to choose for each search the version that may include special features geared to that search, and the clustering of several patent databases on a single-host system allows the files to be used synergistically in cross-file searching.

PLASDOC's growth rate is currently about 45,000 patent cases per year. The *CA* Macromolecular Sections are growing by some 22,000 patents per year. *CLAIMS* is limited to U.S. patents, in contrast to the worldwide coverage of *CA* and *PLASDOC,* so that its collection of polymer patents is somewhat smaller and growing at a slower rate.

With databases of this size it can be a formidable problem to get thorough information retrieval without excessive clutter from irrelevant material. This situation is complicated when dealing with patent information, because patents present special problems for indexers. First, patents frequently provide information in terms of generic or Markush structures. Claims in a patent may cover polymers of a whole family of acrylate or methacrylate esters, for example, rather than a series of individual specific esters. The generic formula for an additive may encompass thousands, millions, even an infinite number of individual compounds. Also, there may be disclosures that are not covered by the claims, and some of these disclosures may not be backed up by hard data. Patents sometimes have hundreds of examples, which can cause problems from the need to index very large numbers of substances and the chance that something of importance in any one example is missed by indexers.

These problems are being solved with differing degrees of imperfection by the major database producers. A great deal of effort is meanwhile being exerted in order to improve the situation.

Chemical Abstracts. Figure 3 shows the subject-access points available in the *CA* online files. *Chemical Abstracts* has long been noted for its high quality indexing, especially the indexing of specific substances. The entirety of *CA* subject indexes is available in the online files with the ability to combine, if desired, terms from different portions of the indexing record, or alternatively, to insist that terms come from a given segment of the indexing. These index terms are supplemented by title terms and, only in the version of the *CA* file marketed by STN International, words from the abstract. The latter can be extremely useful, since at times CAS policy dictates that certain information not be indexed, useful information that can sometimes be picked up from the abstract.

- Controlled index terms
  Substances (Registry Numbers)
    Structure, substructure (Registry, DARC, dictionary files)
  General terms
- Text modification, uncontrolled language
  Linkable to index term in most online files
- Title words
- Abstract words (STN only)
- Uncontrolled keywords from *CA* issue indexes
- Patent classification (incomplete)
- *CA* section headings

Fig. 3. Subject-access points in *Chemical Abstracts.*

Patent classification information in *CA* is incomplete. First, *CA* does not include all relevant classes, such as the cross-reference classes in U.S. patents. Second, as equivalent members of a patent family issue in countries around the world, additional classes are frequently assigned by different patent offices. Such additional classes are added to the patent record in Derwent's database, but not in the *CA* file.

The CAS Registry file of chemical substances is a very extensive catalog of known chemical substances, including polymeric materials, but it is not all-inclusive and there is much information conveyed in patents that is not included in the Registry. There are many references in *CA* for which the index entries are in the form of words rather than Registry Numbers. Synthetic rubbers, for example, are frequently handled in this way. Most significantly, much of the information patents convey is in the form of generic claims and disclosures as well as unsubstantiated disclosures and claims regarding specific substances. The unsubstantiated or "paper" claim or disclosure is quite acceptable in the patent literature, although such claims may be overturned should they prove to be incorrect. At the very least they are important as prior art for future inventions. Until recently, *CA* ignored such information and only since about 1980 has it begun to index claims on specific compounds even when there are no substantiating data.

The Registry at this time lacks the capability of dealing with generic claims and disclosures. CAS is presently working on development of such a capability, but in the meantime the indexing of generics is done only in terms of words, which is inadequate.

Valuable as it is for identifying specific compounds and polymers, the CAS Registry falls short of what is needed in dealing with the compositions of copolymers. The use of words rather than Registry Numbers to index most synthetic rubbers is one example, but at least if the policy is understood, it is possible to use both words and numbers in the search strategy. It would be advantageous to users to have both names and Registry Numbers as part of the searchable record in all online *CA* files, and there may be some movement in that direction in the *CA* version provided by CAS through STN. But to date Registry Numbers tell nothing about specialized structures, such as block and graft copolymers. Thus a binary copolymer of styrene and butadiene may be a styrene–butadiene rubber (SBR); a low molecular weight unsaturated resin; a styrene graft onto polybutadiene homopolymer (which could itself have several different types of microstructure); a butadiene graft onto polystyrene homopolymer; a two-block styrene–butadiene copolymer of the A–B type; or the commercially important A–B–A styrene–butadiene–styrene thermoplastic elastomer. There can be grafts of styrene onto SBR, familiar type of impact-resistant polystyrene, or grafts of butadiene onto SBR. In short there is a wide range of possibilities, all of which actually exist. Considering also variations in molecular weight and molecular weight distribution, comonomer content and distribution, and microstructure, it becomes clear that a single Registry Number for styrene–butadiene copolymers is completely inadequate to characterize those copolymers.

Search problems are most acute for the most common monomers and their combinations, which have huge numbers of postings; the number of entries for more esoteric substances are much smaller and thereby more amenable to detailed investigation. That is, checking five or 10 references for relevancy is acceptable, but checking hundreds or even thousands of references is not. CAS has begun investigating systems for providing more discriminatory power in the registration of polymers and copolymers. At the end of 1986 it was announced that new policies will be instituted with the start of the 12th Collective Index period in 1987. Separate Registry Numbers will be assigned to random, graft, block, and alternating copolymers. This change should improve the situation in the future; however, the cost and effort needed to apply the new system to backfiles virtually assures that the backfiles will remain unenhanced, as is usually the case with improvements in database indexing and coding systems.

Whereas a better system for distinguishing among copolymer systems with different gross constitutions will be available, for numerical differentiation, ie, differing compositions, molecular weights, microstructures, or densities, the problem is more severe. The ability to apply this type of discrimination in searches of the patent literature is very much needed. It must be emphasized that this is a problem not unique to *CA*, but one that has not yet been solved by any database in this field.

Given the current status of copolymer registration, one of the most useful search capabilities is the Registry Component developed by DIALOG in its Registry dictionary files, *Chemname, Chemsis,* and *Chemzero*. When a substance is derived from other substances, as is a polymer from (co)monomer(s) or a complex from its components, DIALOG makes the individual building blocks searchable as Registry Components. Thus the Registry Numbers for all binary and higher copolymers of ethylene and propylene can be retrieved by searching the Registry Numbers of ethylene (74-85-1) and propylene (115-07-1) as Registry Components:

$$S \ RC = (74\text{-}85\text{-}1 \quad and \quad 115\text{-}07\text{-}1)$$

If desired this search can be refined further to include only terpolymers or only tetrapolymers by including the additional parameter Number of Components: $NC = 03, 04$, respectively. In practice this is a highly effective method of collecting information on families or copolymers, but it is less than comprehensive because of polymers not given Registry Numbers, such as the elastomers EPDM, SBR, or chlorinated butyl, and generic claims or disclosures in patents, as well as specific claims (pre-1980) or disclosures that are not backed by hard data.

PLASDOC.   The retrieval capabilities of *PLASDOC* are outlined in Figure 4. The primary means of searching is the *PLASDOC* code, derived from a punch card-based system originally developed by ICI Plastics and enhanced several times over the course of *PLASDOC*'s existence. A number of other searchable parameters are available, including coding terms from other sections of the Derwent database for those patents that appear in multiple sections. Search parameters based on deep coding by Derwent are available only to subscribers to Derwent services.

Each of the additional access points can be very helpful. From the middle of 1981 a registry system pinpoints some 2000 common chemicals for those patents that also appear in the deep-coded chemical sections of the *CPI*, and from 1984

- *PLASDOC* code
    Originally punch card-based
    Limited number of terms available without radical revision
- Derwent chemical code
    If patent also in *CHEMDOC*
- Manual codes
    PLASDOC, other *CPI* and *EPI* sections
- Words from augmented titles
- Abstract words
- International patent classes
- U.S. patent classes
    Indirectly by cross-file from *USCLASS, US Patents*
- Derwent classes
- Individual substance registration
    About 2000 *CHEMDOC* Registry substances, plus 750 *PLASDOC* Registry substances, 1984 +

Fig. 4.   *PLASDOC* subject-access points.

on an additional 750 substances often used in polymer systems are indexed. Searchable abstracts are also available for the entire file, an invaluable aid in pinpointing some concepts that are inadequately handled by the *PLASDOC* code. Even U.S. patent classification, not included directly in the *WPI* file, can be brought to bear in both the ORBIT and DIALOG versions of the file, by cross-file techniques involving U.S. patent files available from the same online host.

The breadth of the *PLASDOC* code matches quite closely the scope of polymer information as outlined in Figure 1. The chief shortcomings are a lack of depth and a lack of specificity, which relate in large part to the origins of the coding system. Originally, the number of search terms available was limited by the number of punch positions on a standard 80-column card. All of the codes relating to a given aspect of a polymer patent were entered on the same card, and an abstract of the patent was printed on the card. To minimize the number of cards, there was a tendency to overcode, ie, to code information that might well be entered on multiple cards all on one. Thus if a use might involve several different polymers, they might all end up on the same card and look like a mixture. Or, if several monomers were alternatives in a copolymer, all of the codes might be entered on the same card, giving the impression that terms that are merely alternatives are both essential constituents of a system.

The small number of codes allowed by the 80-column card limited the number of concepts that could be expressed and meant that highly specific and discriminating terms were less likely than more general and generic ones. Combinations of punch positions did allow some expansion of the limit, but increased retrieval noise from false coordinations. Noise was not a major problem in the earliest days of the database, when stacks of cards were fed to a card sorter to produce a small pile of prospective answers. Since Derwent's detailed documentation abstract was printed on the card, it was possible to scan quickly and pick out the real "hits," ie, relevant data.

By late 1986, however, the *PLASDOC* component of the online *WPI* files had grown to over 650,000 basic patents. The amount of noise produced by an inefficient search of this file can easily be forbidding. Only 12 mono- and diolefinic hydrocarbon monomers are searchable as polymer components in *PLASDOC*: ethylene, propylene, 1-butene, isobutylene, 4-methyl-1-pentene, styrene, vinyltoluene, α-methylstyrene, butadiene, isoprene, divinylbenzene, and piperylene. All others fall in catchall categories, including such important monomers as 1-hexene, dicyclopentadiene, ethylidenenorbornene, 1,4-hexadiene, etc.

The search capability of the code for the components of Ziegler catalysts is a grossly inadequate system that has been only slightly improved by recent enhancements. Until 1977 the only transition-metal component of Ziegler catalysts available for searching were

> transition metals and their compounds
> halides, oxyhalides
>> titanium halides, oxyhalides
>>> titanium trivalent halides
>>> other titanium halides, oxyhalides
>> other halides, oxyhalides
> oxides
> other transition metals and compounds

In 1977 terms were added to identify transition metals other than titanium. Finally, the start of compound registration in 1985 permits the identification of the more common specific catalyst components.

Derwent improved the *PLASDOC* code in 1978 by precoordinating some of the individual codes into terms known as Key Serials, but did not make it possible to link one Key Serial to another in search logic, thus preventing the identification of contextual relationships that help to minimize search noise. Then, in 1982 codes were introduced for over two dozen, specific bound-copolymer systems, a welcome advance but hardly enough. By contrast IFI's *CLAIMS-Uniterm* and *-CDB* system includes indexing terms for over 200 bound-copolymer systems. Dwarfing these numbers, DIALOG's *Chemname* file of substances with at least two postings in the CAS Registry includes some 1600 copolymer systems that include methacrylic acid as a component, and DIALOG's *Chemsis* files of single-indexed substances include nearly 5000 more. It is not necessary for Derwent to have *CA*'s incredible specificity, but certainly a system that has specific indexing terms for all commercially important polymers would be of great value to information users.

Until the introduction of indexing for 750 *PLASDOC* Registry compounds in 1984, it was not possible to search for a specific plasticizer, antioxidant, or catalyst modifier unless it was deemed to be novel and thus coded in *CHEMDOC*, Derwent's general chemical section. In searching for maleinization, a very common process of polymer modification, the searcher had to look at everything on polymer alkylation, because maleinization is not considered by Derwent to be a grafting process (in which case maleic anhydride would have been coded). In searching for matte-finished films, the coding term to search is gloss, since a single term is used to describe a property and its opposite.

Despite shortcomings the *PLASDOC* system is still an invaluable one. Many questions, especially when generic information is needed, are handled extremely well. Others are beset with excessive noise. The addition to the online file of backlog abstracts has been very helpful in ameliorating the shortcomings of the code. Derwent is to initiate a study aimed at developing improvements in the *PLASDOC* coding system. Hopefully, these will be far-reaching advances that overcome many shortcomings for the future, even though they can do little to upgrade the back files.

One more invaluable Derwent searching system must be mentioned: manual code cards. In *PLASDOC*, as in the other sections of its chemical and electrical patent databases, Derwent has established a system of codes that are used to prepare searchable sets of abstracts. Typical manual codes include

A2-A6B	transition-metal (oxy)halide polymerization catalysts
A4-G2E2	polyethylene use in film, packaging
A8-F3	phosphorus-containing flame retardants
A10-E5A	pyrolysis of waste polymer
A11-B15B	melt spinning

Scanning sets of manual code cards is still the best way of searching some types of technology; Ziegler-Natta catalyst systems are a prime example. Some manual code-card sets are dauntingly large, and manual coding is not so complete as the punch code or Key Serial coding, but this system, old-fashioned though it may be, is still an extremely valuable adjunct of *PLASDOC*.

IFI/Plenum Files.   The *CLAIMS-Uniterm file* has some special polymer capability, since it includes a number of specific indexing terms for homopolymers and bound-copolymer systems. There are over 200 of these terms, making it much easier to identify specific polymers than in *PLASDOC*. Polymers that do not have their own descriptor can be searched for by using the appropriate terms for the individual monomers, along with general terms for the type of polymer (eg, polyester). *Uniterm*'s extensive controlled vocabulary is open-ended, allowing new indexing terms to be added to the system as they become needed.

Useful as this *Uniterm* system is, its capabilities are surpassed by a considerable margin in IFI's *CLAIMS-CDB file (Comprehensive Data Base)*. Whereas *Uniterm* is available to all comers, *CDB* may only be used by subscribers who pay a substantial front-end fee. Subject-access points in the *CDB* are shown in Figure 5. With a user community considerably smaller than that of either *CA* or *PLASDOC, CDB*'s complex indexing system is relatively unfamiliar to the general public, and it is impossible to cover it in any detail within a short space. A somewhat more detailed, if still brief, description is given in an earlier publication (6).

- General index terms
    - Some with roles (present, reactant, product)
    - Numerous bound-polymer terms
    - Polymer-class terms
- Compound terms
    - All with roles
    - Associated compound Registry file for compound identification
- Fragment terms
    - For generics, uncommon compounds
    - All with roles
    - Linking of fragments for given substance
- Patent classification
    - All U.S., all IPC
- Title, abstract, and/or representative claim words
- CAS Registry Numbers, 1967–1979
- Special system of polymer roles

Fig. 5.   Access points in *CLAIMS-CDB*.

The *CDB* incorporates specific descriptors for many relatively common chemicals, including chemicals involved in polymers, a system of fragments and links to describe generic and uncommon compounds, and a system of roles. For simple chemicals the roles are the traditional ones: present, reactant, and product. For polymers there is a highly developed role system that conveys much information about not only the monomers incorporated into polymers, but even chemicals used to modify them.

The complete *CDB* role system is shown in Table 2. Table 3 shows how the role system provides an enormous amount of discrimination in describing references concerning ethylene. With some 6500 references, there are 313 in which ethylene is produced, 1155 where it is a reactant, and 285 where it is merely present. There are 384 references on the preparation of modified ethylene homopolymers and 702 on preparing modified ethylene copolymers. In all, ethylene

Table 2. Role Designation Numbers in *CLAIMS-CDB* System

	Present	Reactant	Product
compounds, general terms, fragments	10	20	30
alloys, glasses, etc	40	50	60
polyhydrocarbons, homo/co	41, 71	51, 81	61, 91
polyhydrocarbons, modified, homo/co	14, 17	15, 18	16, 19
polyethers, homo/co	42, 72	52, 82	62, 92
polyesters, homo/co	43, 73	53, 83	63, 93
polycarbonates, homo/co	44, 74	54, 84	64, 94
polyamides, homo/co	45, 75	55, 85	65, 95
polyurethanes, homo/co	46, 76	56, 86	66, 96
polysiloxanes, homo/co	47, 77	57, 87	67, 97
other regular polymers	70	80	90
unusual polymers	12	22	32
modified polymers (excluding polyhydrocarbons)	11	21	31

Table 3. Roles for Ethylene, 1982–March 1986[a]

Form of ethylene	Present	Reactant	Product
monomer	285	1155	313
in polyhydrocarbon homopolymer	2807	276	499
in polyhydrocarbon copolymer	2443	390	573
in modified polyhydrocarbon homopolymer	754	102	384
in modified polyhydrocarbon copolymer	1042	148	702
23 other roles	51	15	51

[a] Total postings: 6504.

appears in some 38 different roles (although a check of several unusual roles suggested that there might be some indexing errors). Even allowing for the possibility of some errors, the system allows pinpointing with uncanny accuracy. Thus a search for ethylene with the role 67 (polysiloxanes, product) leads to U.S. 4,312,575 with claims on glow-discharge polymerization of ethylene on polysiloxane contact lenses. There is no confusing this with a patent, eg, on a blend of polyethylene with a silicone. Similarly, a search for ethylene with the role 93 (copolyesters, product) gives U.S. 4,578,231 in which a mixture of LDPE, chlorinated PE, and an unsaturated polyester is cocured with a peroxide. Once again the system has been very informative.

The roles relate not only to monomers but also to chemical modifiers. Thus the chlorine in chlorinated polyethylene gets the role for modified polyhydrocarbons. Similarly, chlorine gets the role for polyethers in a chlorinated polyether, for polyamides in a chlorinated polyamide. Similar polymer roles are assigned to vulcanizing agents for rubbers, to hydrogen in hydrogenated polymers, and to oxygen in oxidized ones.

The *CDB* system does an excellent job of describing common polymer systems, aided substantially by the presence of the 200+ bound terms for specific homo- and copolymers. The *CDB* fragmentation system also allows it to do a remarkable job of describing complex and unusual polymer systems. Indeed, all three systems discussed here in their own ways can, in general, deal effectively with unusual polymer structures (4).

Probably the greatest shortcoming of the *CLAIMS-CDB* system is that it covers only U.S. patents. Another weakness is the inability to link indexing terms. Thus when a bound-copolymer term is lacking, it would be useful to be able to distinguish the copolymer of monomer X and the copolymer of monomer Y as the same copolymer, or two different copolymers, or X and Y as alternatives, but not comonomers in a copolymer system. Linking would also be valuable to tie together a substance to its function. Linking is an inherent part of the *CDB* because of its system of fragments and roles, and it is unfortunate that it has not been utilized more fully within the system.

Cross-file Searching. Since 1984 the availability of key patent databases on several online host systems has increased, allowing several hosts to develop clusters of patent databases. All of these hosts now feature software that enables references identified on one database to be transferred to another database and further manipulated. This type of cross-file operation requires that key patent data elements, most notably patent numbers and application and priority numbers, are in standardized form among the databases.

Table 4 shows the clusters of bibliographic patent databases currently available, or announced for the imminent future, from the hosts who offer the principal polymer patent databases. Cross-file searching has enormous potential for enhancing the value of online searches. It enables the output of several overlapping files to be combined, avoiding duplicate hits. Generally, this is done by transferring answer sets to the *WPI* database, which leads the searcher to the outstanding Derwent documentation abstracts and also provides a listing of families of equivalent patents, including English-language equivalents if available for foreign language patents. But the potential of cross-file searching goes beyond this. Thus a search on the generic *PLASDOC* database can be intersected with one on the highly specific *CA* file, enabling the searcher to pinpoint certain information. Such cross-file operations must always be done bearing in mind the different indexing conventions of different databases. In other words, when looking for a specific copolymer that might be encompassed by a generic search of *WPI*, it must be remembered that *CA* only indexes that copolymer if it is backed up by data

Table 4. Clusters of Polymer Files and Other Bibliographic Patent Databases[a]

Database	ORBIT	DIALOG	Questel	STN
WPI/L (PLASDOC)	X	X	X	
Chemical Abstracts	X	X	X	X
CLAIMS-Uniterm, -CDB	X	X		X
APIPAT	X			X[b]
US Patents	X			
JAPIO	X			
FPAT			X	
EPAT			X	
EDOC			X	
PATDPA[c]				X
CLAIMS-Reassignment, -Reexamination	X	X		X
CLAIMS-Citation		X		

[a] Hosts also offer CA Registry or Registry dictionary files, and various patent classification and search-term auxiliary files.
[b] In 1987.
[c] In German.

or if (in recent years) it is covered by a specific claim. Thus the absence of a *CA* index entry does not insure that a patent does not include information on the copolymer.

Cross-file searching is a relatively new technique and its potential is just starting to be tapped. Some of the links needed to compatibilize the files in the various file clusters have not yet been fully established, but they can be expected to be finished in the near future. Cross-file searching operations promise to become increasingly important as time goes on.

New Developments

Although this article has dwelt on the shortcomings of the various polymer patent databases, each one of them is outstandingly valuable in its own way. In combination they can answer most information needs, and the rapidly developing cross-file capabilities are improving the situation further.

Both CAS and Derwent are deeply engaged in file enhancements. CAS is committed to providing an improved patent information service as well as to the development of a Markush database, which will be able to deal with the generic chemical structure information found in patents. Derwent too is committed to developing a Markush database and is currently beginning improvements in *PLASDOC* indexing.

Derwent needs considerable improvement in the indexing of specific substances, and CAS needs to learn how to deal with generics. All of the database producers would profit by systems that describe in greater detail the aspects of polymer constitution, composition, and microstructure. There is a need for more linking of indexing terms to put information in its proper context, especially in the IFI files, but also in *PLASDOC*. And there is a real need to be able to deal with numerical information: densities within a given range, molecular weights above a given limit, etc. No database today can handle this type of information well.

The past decade has been marked by a great deal of progress in the field of patent databases, including polymer patent files. Progress will undoubtedly continue, providing still more powerful information sources for polymer patents.

BIBLIOGRAPHY

1. J. B. Gambrell, C. M. Cox, and P. E. Kreiger in M. Grayson, ed., *Kirk-Othmer Encyclopedia of Chemical Technology*, 3rd ed., Vol. 16, John Wiley & Sons, Inc., New York, 1981, pp. 851–889.
2. J. W. Lotz in Ref. 1, pp. 889–945.
3. S. M. Kaback in J. M. Bernard, ed., *Computer Handling of Generic Chemical Structures*, Gower, Aldershot, Hampshire, UK, 1984, pp. 38–48.
4. S. M. Kaback, *J. Chem. Inf. Comput. Sci.* **25**, 371 (1985).
5. *Chem. Abstr. Pat. Sect. 35* **105**(6) (Aug. 11, 1986).
6. Ref. 3, pp. 49–65.

STUART M. KABACK
Exxon Research and Engineering Company

PBT.   See POLYESTERS, THERMOPLASTIC.

PEARL POLYMERIZATION.   See SUSPENSION POLYMERIZATION.

PECTIC SUBSTANCES.   See POLYSACCHARIDES.

PEEK.   See POLYETHERETHERKETONES.

PELLETIZING

In the production of thermoplastics, the primary products frequently leave the manufacturing facility in the form of pellets. Incorporation of antioxidants, stabilizers, colors, fillers, reinforcing agents, and other additives requires that the polymer be in the melt stage to assure uniform distribution. After incorporation the melt is pelletized into uniform sizes and shapes to facilitate handling.

Polymerization products are frequently in powder form, with the raw material already compounded. Uniform pelletized form offers the following processing advantages: simpler feeding system with fewer feeders, dust-free handling, and easier cleaning between changes of feedstock; fewer unsatisfactory products because of uniformity of size and homogeneity of feed; and greater extrusion capacity and lower shipping costs as a result of higher feed-bulk density.

Offsetting these advantages may be a higher heat exposure of the product (an additional melting step), as well as the additional manufacturing costs associated with pelletizing. Molten reactor products must be pelletized.

The process involves pressure development for forced flow through a die. The extrudate is cooled to solidification and then cut into pellets, or the molten extrudate is cut as it emerges from the die and the pellets are subsequently cooled. In the latter case, both cutting and cooling may be done in air or water, or cutting may be done in air, followed by quenching in water.

Quenching solidifies the exterior shell of the pellets to prevent reagglomeration. With a water quench, enough residual heat content is left in the pellet interior to evaporate the surface moisture remaining after the excess water has been centrifuged off. The dried pellets are classified to remove both fines and agglomerates. Typical pellets are 2–4 mm maximum dimension. Diced pellets are cubic-, rectangular-, or octahedral-shaped. Strand pellets are approximately cylindrical. Die-face pellets may be cylindrical or oblate spheroids.

Equipment

Pelletizing equipment can be classified as follows:

Quenching, solidifying, and cutting	*Cutting, quenching, and solidifying die-face pelletizers*
dicers	dry-face pelletizers
strand pelletizers	water-ring pelletizers
	underwater pelletizers
	centrifugal pelletizers
	rotary-knife pelletizers

Each type of pelletizer has its function in plastics compounding, depending on the throughput rate required, the kind of compounding equipment used upstream, the material being processed, and the pellet shape required.

Compared to dicers and strand pelletizers, the more expensive die-face pelletizers utilize less space, can be more readily automated, require less supervision, and can handle higher rates. Cutting thermoplastics as a melt rather than as a solid generates fewer fines and less knife wear, but can cause more die wear if the knife must ride on the die face, as for example, with polymers like polypropylene.

Dicers. Dicers were first used in the rubber industry and subsequently applied to thermoplastics. Most poly(vinyl chloride) and acrylonitrile–butadiene–styrene copolymers are diced, as well as some other thermoplastics [1,2].

In typical operations after leaving the forming equipment (direct-strip extrusion or sheeting from a two-roll mill), the strip of polymer is quenched before entering the dicer. The strip is fed at a constant rate through nip rolls into the rotating knives operating against a stationary-bed knife. Dicers can be used with the sheet entering straight into the cutting edge or with the cutter rotor offset at a 45° angle to the entering strip. The former method uses either a notched or a ratchet-tooth knife, whereas the latter uses a stair-step knife to produce the individual pellets.

A variation of the notched-knife dicer is the slitter chopper, which first slits the polymer sheet into parallel strips that are then chopped by rotor knives acting against a bed knife. Pellet length is controlled by feed rate and cutter rotation speed.

For strip dicing, very low die pressure is required, which permits stock temperatures to remain low. In comparison to strand cutting, constant operator attention is usually not required. However, dicing systems require extensive space (30 m of water bath for a 3-mm thick strip) and noise levels are high. Frequently, the dicer is isolated in a soundproof chamber.

Strand Pelletizers. Strand pelletizers (Fig. 1) are the simplest and least expensive pelletizers. An extruder or gear pump forces the molten polymer through a row of small, round orifices. The emerging strands are conveyed through a cooling water bath with grooved rollers to keep the strands separate from each other until they are solidified. After emerging from the water bath, the strands

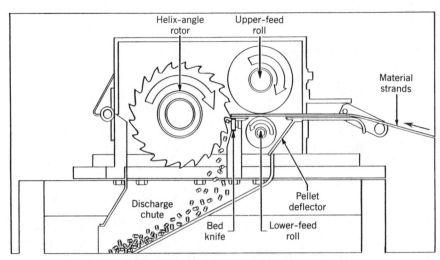

Fig. 1. Strand pelletizer. Courtesy of Conair, Inc.

may be air-dried and fed through nip rolls of a multiknife-blade rotor operating against a fixed blade. The strands, 2–4 mm in dia, are chopped into cylinders of round or slightly oval cross sections, 1–5-mm long, as determined by intake roll speed and cutter frequency.

Although strand cutting is technically feasible at very high capacities, a practical limit for many polymers is ca 2000–3000 kg/h (3). Beyond that rate, the die and bath become very wide and it becomes more difficult to maintain individual strand integrity. Frequent operator attention may be required to re-string broken or entangled strands.

Like dicing, stranding requires extensive floor space for quenching the polymer before cutting. Recent housing modifications have considerably reduced the noise associated with strand cutters.

Practically all thermoplastics can be strand-cut, except those that are very brittle or have no melt strength to enable the strand to be pulled through the bath into the cutter. In some instances, this can be overcome by moving a ribbed, endless belt in the bath at the same lineal velocity as the strands. Two belts on either side guide the strands up and down over rollers in a deep bath, thereby conserving floor space.

Die-face Pelletizers. Many thermoplastics can be pelletized by moving knives rapidly across a multiple-hole die plate. The multiblade cutter slices the polymer and hurls the pellets away from the die. The pellets are quenched in air alone with a water mist, or by subsequent immersion into a water bath; rates are up to 6000 kg/h (4).

A typical dry-face pelletizer is shown in Figure 2. The eccentrically mounted cutter may have up to six arms and may run at speeds up to 2500 rpm. Knife clearance from the die can be adjusted during operation by axial movement of the cutter spindle with respect to the die face. Band heaters bring the die up to temperature at start-up and offset heat losses during operation. Although not much heat can be transferred between the polymer and die because of the low thermal conductivity of the polymer and the short time that the polymer is in

Fig. 2.   Dry-face pelletizer. Courtesy of Buss Condux, Inc.

the lead hole and capillary of the die, it is important to maintain a uniform
temperature to provide uniform flow rate per hole. If there is a known temperature
gradient in the polymer approaching the die, it is possible to compensate for the
variation in viscosity by appropriate changes in the lengths of the capillaries
and lead holes.

Water-ring Pelletizers.   In the water-ring pelletizer, the molten strands
emerging from the die are cut with a centrally mounted cutter (Fig. 3). The
freshly cut hot pellets are thrown into a spirally rotating water stream, ~20-
mm deep, surrounding but not touching the die. The pellets are quenched as they
rotate with the water. After the pellet slurry leaves the housing, the water is
removed by screening and the pellets are dried. In some instances, a water spray
near the cutting surface may be used to deposit a skin on the molten pellets to
reduce pellet agglomeration.

The water-ring pelletizer is insensitive to process interruptions, has easier
start-up, and requires less operator attention. Although usually mounted with
the cutter axis vertical, some have a horizontal cutter axis for closer distance
between extruder and die plate.

Underwater Pelletizers.   An underwater pelletizer (Fig. 4) operates with
the die face immersed in water. The severed pellets are carried away as a slurry
in water for further cooling in transit to the dewatering, screening, and drying
equipment. The die plate may be mounted in such a way that the polymer is
extruded horizontally or downward. Underwater pelletizers are available for all
capacities up to 24,000 kg/h (5).

The critical design areas are the die plate and knife adjustment. The cutter
hub is mounted near the center of the die with all the blades continuously riding

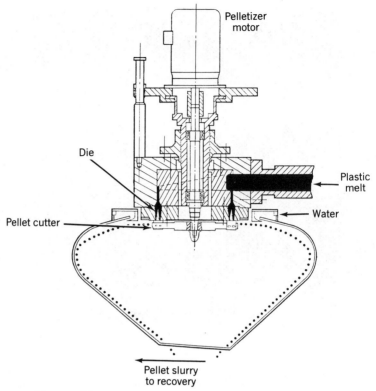

Fig. 3. Water-ring pelletizer. Courtesy of Berringer Co.

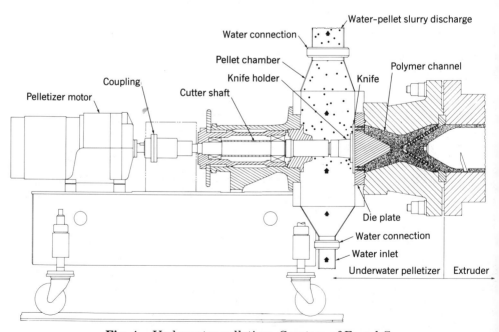

Fig. 4. Underwater pelletizer. Courtesy of Farrel Co.

on, or near, the die. The die frequently is provided with heat channels surrounding each row of holes. High temperature steam or oil heats the polymer channels to supply the large amount of heat lost to the water sink at the die face. Sometimes hardened ceramic inserts around the die holes provide insulation and better wear resistance. If there is a momentary interruption in flow, die freeze-off can be a problem, which can be overcome by adequate die-plate heating. Freeze-off is when the temperature of the polymer falls below its solidification point, thereby plugging the die hole. Start-up requires precise sequencing of polymer and water flows.

Centrifugal Pelletizers. The centrifugal pelletizer (Fig. 5) differs from all other pelletizers (6,7). The polymer melt is fed into a rotating die chamber at atmospheric pressure. Pressure to extrude the polymer through the die holes is generated by centrifugal force within the die, rather than by an extruder or a melt pump. The die holes are located in the periphery of the rotor. The emerging strands are cut by a single stationary knife, which does not run on the die as the strands are held outward by centrifugal force. The severed pellets fly off on their own momentum into an air- or water-quench system similar to that for a dry-face pelletizer. The die is heated for start-up and for offsetting heat loss to the surroundings by generation of eddy currents within the rotor as it spins between a pair of magnetic fields. In addition, the die is self-emptying upon cessation of flow.

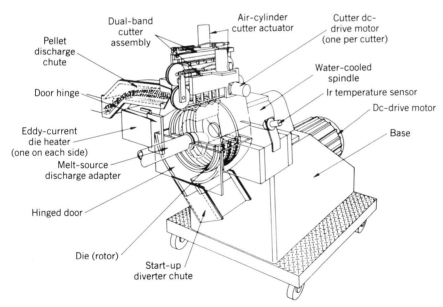

Fig. 5. Centrifugal pelletizer. Courtesy of Baker Perkins, Inc.

The centrifugal pelletizer can operate with less heat exposure since the temperature rise associated with pressure generation in an extruder or melt pump is avoided. Cutter life is extended by slow movement in the fixed position. Cutters can be changed without interruption of polymer flow. The main disadvantage of a centrifugal pelletizer is that pressure generation and cutting frequency are both dependent on rotor speed and therefore are not independent of each other.

Rotary-knife Pelletizers. The rotary-knife pelletizer (Fig. 6) operates with a die similar to a strand die, but with the rotary cutter operating in close proximity to the emerging strands. The cutter is completely enclosed. A water stream may be introduced into the rotor housing to help quench the severed pellets, with the pellets leaving the housing either coaxially with the cutter rotor or at the bottom of the housing. In addition to the simple row die shown in Figure 6, a cylindrical die can be used with several rows of die holes for flow of polymer into the cylinder for severing by a helically grooved cutter.

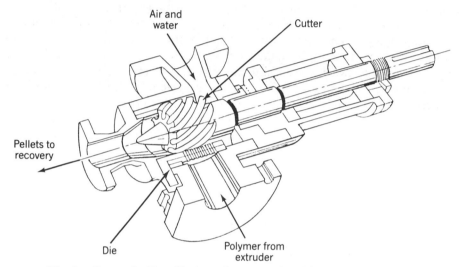

Fig. 6. Rotary-knife pelletizer. Courtesy of Welding Engineers, Inc.

Die Design. Shaping the viscous polymer melt before pelletizing requires external pressure, except for the centrifugal pelletizer or sheeting off a two-roll mill for a dicer. The viscoelastic property of many polymers causes the strand emerging from a capillary to swell in diameter, even if rapidly quenched. Thus for a given desired diameter pellet, the capillary diameter may have to be considerably smaller. Small capillaries lead to high pressure drop (up to 10 MPa or 1500 psig) through the die, and the small holes are more prone to freeze-off. Some polymers coming through a strand die possess enough melt strength so that they can be drawn down by the nip rolls to a diameter smaller than the die orifice.

The design of die plates is based on the rheological properties of the polymer, which are a function of temperature and shear rate, as well as entrance-loss correction and die-swell behavior. Dies for underwater pelletizers are subjected to high heat losses because of direct contact with water, and critical attention to heating is required to avoid freeze-off. Nonuniform die temperature is the main contributing factor to nonuniform pellet size.

Erosion within the capillaries may cause die wear, particularly with highly filled polymers. Wear is also caused by the knife passing over the die face. Irregular cutting between a rotating knife and the die face in melt cutting, or between a rotating and bed knife in stranding or dicing, can generate undesirable fines or tails on the pellets.

Pelletizer Selection. Table 1 provides general guidance for pelletizer selection. Most suppliers can demonstrate their equipment on a pilot scale of ~500 kg/h. Extremes in apparent viscosity create problems that may become acute in any pelletizer. For example, both fractional melt-index and high melt-index (>100), low density polyethylene (LDPE) are very difficult to pelletize, particularly in the same apparatus, even though an average LDPE can be pelletized in any of the described pelletizers.

Table 1. Selection Chart for Pelletizers[a]

	Pelletizer					
Polymer	Dicer	Strand	Dry-face	Water-ring	Underwater	Centrifugal
ABS copolymer	A	A	A	S	S	A
polyacetal	N	A	A	S	A	S
polyamide	A	A	N	N	S	N
polycarbonate	N	S	A	S	S	S
polyester	S	A	S	S	S	S
HDPE	S	S	A	A	A	A
LDPE	A	A	A	A	A	A
LLDPE	S	S	A	S	A	A
poly(methyl methacrylate)	N	S	S	N	A	S
polypropylene	N	S	A	N	A	A
polystyrene	N	A	A	N	A	A
poly(vinyl chloride)						
flexible	A	A	A	S	S	S
rigid	A	S	A	S	N	S
thermoplastic rubber	A	S	S	S	S	S

[a] A = acceptable; S = sometimes acceptable; and N = not acceptable.

The choice of pelletizer is also affected by considerations of total feed rate, turn-down-ratio requirements, frequency of feedstock change, operator attention, knife and die wear, and pellet shape (8).

Energy Requirements

The amount of energy required for pelletizing depends on the starting condition of the polymer, viscosity, and type of pelletizer. For example, the energy required to pelletize 4500 kg/h of a low melt-index polyethylene leaving a compounder at 200°C with an underwater pelletizer is:

Energy requirement	*kW*
extruder power to develop die-head pressure	146
heat loss from extruder	8
heat loss from die	35
underwater pelletizer drive	41
Total	*230*

Thus the specific energy required would be ca 0.05 kW·h/kg. The energy associated with quenching, drying, and classifying the pellets is additional. With a much lower viscosity, the power required for extrusion is also much lower. Strand cutters, water-ring, and dry-face pelletizers operate with less power, as the cutter rotors are turning in air rather than water. Developing pressure in a centrifugal pelletizer is independent of viscosity, and there is no heat buildup during pressure development. The specific power required for a centrifugal pelletizer in the above case would be <0.015 kW·h/kg.

Safety

With all pelletizers, there are always inherent hazards due to hot molten polymers, rotating devices, and sharp knives. Standards for the construction, care, and use of dicers and strand pelletizers are contained in ANSI B151.11 (1982). Similar standards for die-face pelletizers are under development as ANSI B151.23.

BIBLIOGRAPHY

1. H. Boulay, *Plast. Compound.*, 46 (Nov.–Dec. 1980).
2. S. H. Collins, *Plast. Compound.*, 64 (Nov.–Dec. 1981).
3. R. R. Lange, *Plast. Compound.*, 39 (Mar.–Apr. 1979).
4. J. W. Hunt, *Plast. Compound.*, 48 (July–Aug. 1978).
5. E. W. Schuler, "Advances in High Volume Pelletizing Techniques," *SPE ANTEC Technical Papers*, Vol. 21, Society of Plastic Engineers, Greenwich, Conn., 1975, pp. 477–480.
6. D. B. Todd, H. G. Karian, and J. D. Layfield, *Plast. Compound.*, 33 (July–Aug. 1983).
7. D. B. Todd and H. G. Karian, *Mod. Plast.*, 56 (Dec. 1983).
8. H. Herrmann, ed., *Granulieren von Thermoplastischen Kunststoffen*, VDI-Verlag GmbH, Dusseldorf, FRG, 1974.

General Reference

Ref. 8 is a good general reference.

D. B. TODD
Baker Perkins, Inc.

PENTADIENE POLYMERS

Polymers containing pentadiene are used as resins for adhesives and in the USSR for liquid rubbers. The monomer is usually a mixture of *cis*-1,3-pentadiene and *trans*-1,3-pentadiene, formerly called piperylene. It may also be regarded as 1-methyl-1,3-butadiene to stress its relationship to other butadiene derivatives.

After butadiene and isoprene, piperylene (1,3-pentadiene) is the most abundant diolefin hydrocarbon monomer. A similar amount of cyclopentadiene is available in the world, but although present in many refinery streams, cyclopentadiene does not share the versatile polymerizability of the other pentadienes. These monomers occur in petroleum fractions approximately in the ratio of 10 butadiene:2 isoprene:1 pentadiene (by weight). In terms of commercial utilization, however, pentadiene trails behind butadiene and isoprene, probably because of the higher boiling point and the difficulty of separating it from mixtures. More appears to be consumed as fuel. Monomer availability and price determine commercial usage more than the polymer properties. Thus 2,3-dimethylbutadiene and 2-ethylbutadiene have no commercial significance, in spite of the excellent properties of some of their polymers (see BUTADIENE DERIVATIVES, POLYMERIC).

Polymerization of pentadiene was discovered about the same time as that of isoprene. In 1881 Hofmann synthesized pentadiene from piperidine (hence the name piperylene), established its composition as C_5H_8, and noted that it polymerized on distillation. Within two years Schotten had also produced polypiperylene while trying to make piperylene. Thiele established the monomer structure and polymerized it to a rubber in 1901 (1).

Rubbery properties of the early polypentadienes were poor, and a satisfactory monomer source was lacking. The early efforts in Germany and other European countries centered more on dimethylbutadiene, isoprene, and butadiene. The extensive DuPont report on emulsion polymerization of dienes for the synthetic rubber program in World War II does not mention pentadiene (2). The U.S. team, which went to the FRG in 1946 to glean whatever it could from the unpublished reports of the German war effort, came to the conclusion that pentadiene "tended to give plastic products" (3). Both pentadiene isomers were slow in emulsion copolymerization with styrene, perhaps partly because of impurities. The emulsion polymer of *cis*-pentadiene was superior to emulsion polybutadiene in stress–strain and hysteresis properties. However, pentadiene polymers and copolymers had inferior low temperature properties (4), probably because of a higher T_g of emulsion pentadiene polymers.

With the surge in petrochemical production, which started during World War II, piperylene became as abundant as isoprene in the production of C-2 to C-5 olefins and gasoline. Piperylene occurs in waste streams containing many other hydrocarbons; the ultimate supply depends on the supply and demand of much larger commodities. Nevertheless, in the postwar period piperylene made the transition from a very minor chemical with no satisfactory means of production to a reactive diene monomer available in mixtures hardly more expensive than fuel. However, purification or separation costs are significant, depending on the application.

Products based on cationic copolymerization of pentadiene with olefinic components of the same refinery streams appeared in the 1950s. Commercial production of these resins for adhesives (qv) and tackifiers in 1986 accounts for about 65×10^6 kg, less than half of the pentadiene potentially available in the United States.

In the USSR, synthetic pentadiene rubbers (SKPs) were available in the early 1960s. Soviet production of pentadiene polymers may exceed that of the rest of the world.

Monomers

Physical properties of the monomers are given in Table 1, along with those of the most significant homologues.

Table 1. Properties of Diene Monomers[a]

Compound	Bp at 101.3 kPa[b], °C	Density at 20°C, g/mL	Refractive index
cis-1,3-pentadiene	44.086	0.6916	1.43634
trans-1,3-pentadiene	42.032	0.6771	1.43008
1,3-butadiene	−4.41	0.6211	1.4292[25]
isoprene	34.067	0.68095	1.42194

[a] From BUTADIENE DERIVATIVES, POLYMERIC.
[b] To convert kPa to mm Hg, multiply by 7.5.

Isomers and Isomerization. Unlike the other commercial diolefins, 1,3-pentadiene has two separable geometric isomers in addition to the more readily interconverted cisoid and transoid conformations that exist for all conjugated dienes (Fig. 1). Therefore, it is more precise to refer to the pentadienes than to pentadiene. Commercial pentadiene or piperylene is a mixture in which trans-1,3-pentadiene predominates, but the proportions vary. Typical cracking conditions produce an equilibrium ratio of about 65:35 trans:cis.

Other pentadiene structural isomers exist, for example, 1,4-pentadiene. They do not have the polymerization activity typical of conjugated dienes and are not within the scope of this article.

The 1,3-pentadiene isomers vary chemically and behave differently upon polymerization. Cationic initiators are relatively undiscriminating, whereas anionic polymerization tends to favor the trans isomer; Ziegler-Natta or transition-metal systems favor trans, leaving cis often untouched; Alfin polymerization favors cis; free-radical emulsion favors cis, but radiation in a channel complex is apparently effective on the trans isomer only; and alternating charge-transfer copolymerization involves both isomers, but the cis only as a result of isomerization.

Isomerization of 1,3-pentadienes is important to consider because it provides methods of purification and separation, reduces wastage when one isomer is inert, and sometimes occurs under the conditions of polymerization. Isomerization of cis- and trans-pentadienes is an equilibrium process, which can be allowed to proceed thermally, producing, for example, a trans:cis ratio of about 4:3 in 75%

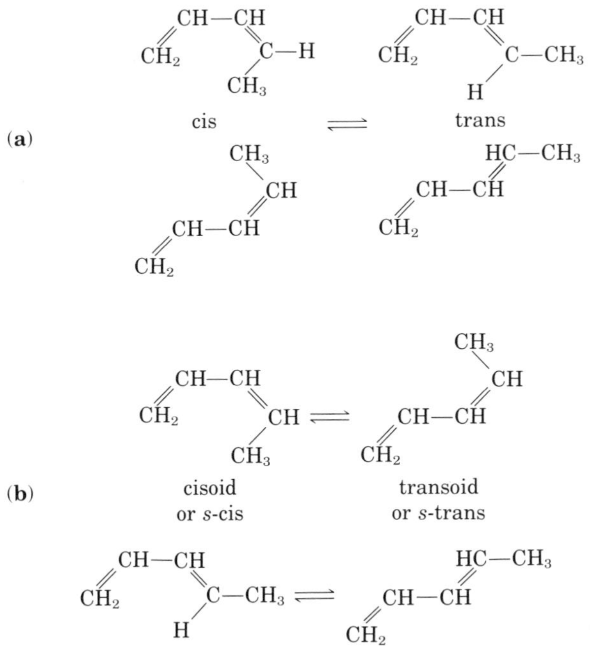

(a)

    CH—CH   CH—CH
CH₂   C—H   CH₂   C—CH₃
      CH₃   H

    cis   ⇌   trans

    CH₃   HC—CH₃
      CH   CH—CH
CH—CH   CH₂
CH₂

(b)

               CH₃
    CH—CH   CH
CH₂   CH ⇌ CH—CH
      CH₃   CH₂

    cisoid   transoid
   or s-cis   or s-trans

    CH—CH   HC—CH₃
CH₂   C—CH₃ ⇌ CH—CH
    H   CH₂

Fig. 1.   1,3-Pentadiene polymers: (**a**) geometrical interconversion by a high energy process such as protonation and deprotonation; (**b**) conformational interconversion readily by a low energy process such as bond rotation.

yield at 600°C (5). The reaction is unimolecular, probably via a cyclobutene intermediate (6). It can be accelerated by isomerization catalysts, such as iodine (5,7) or nitric oxide (8,9). The equilibrium ratio is near 5.0 at room temperature and 1.7 at 525°C. A practical consequence is that technical pentadiene contains more trans than cis, reflecting the temperature of extraction or distillation. Furthermore, cis-rich mixtures are easily converted to trans-rich by distillation over a catalyst.

Other isomerization initiators include cobalt metal or alumina-supported cobalt (10) and unmodified acidic supports such as alumina or silica, although these also promote undesired polymerization (11). Isomerization during polymerization may obscure the course of the reaction and the monomer requirements, and affect product quality. Isomerization has been detected in polymerization with trialkylaluminum–titanium tetraalkoxide (12,13) and with phosphoric acid (14,15), and in alternating copolymerization (16–18).

A very high cis fraction can be separated from mixtures predominantly cis, using cuprous ammonium chloride (19). This method apparently does not depend on isomerization per se. It is used industrially, and various improvements have been patented (20).

Photosensitized isomerization by uv irradiation has been studied for theoretical implications about the triplet state (21–23).

cis-1,3-Pentadiene may be converted to trans via the sulfone, followed by decomposition. Both isomers produce the same sulfone, which, however, yields only the trans monomer on decomposition (19). Catalysts for the sulfone formation include phenyl-β-naphthylamine and polyhydric phenols.

Pure cis can be obtained from crude mixtures by treatment with maleic anhydride. The dienophile forms insoluble adducts with trans-1,3-pentadiene and various other dienes that may be present, but not with cis-1,3-pentadiene (5).

Isomerization on silica treated with tungsten oxide and fluorine has been used, mainly to isomerize pentadienes to isoprene (24). In the presence of DMSO and KOH, 1,4-pentadiene is isomerized to a cis–trans mixture of 1,3-pentadiene (25).

Sources.   A polymer-grade pentadiene isomer mixture is obtained in steam-cracking operations while producing ethylene and propylene from petroleum feed stocks. With the increasing demand for by-products such as butadiene and aromatics, crackers are now usually managed to produce saleable pentadienes also. In addition, a C-5 fraction may be taken from the waste, which would be used for fuel after the removal of butadiene. The C-5 cut contains mainly isoprene, cyclopentene, cyclopentadiene, monoolefins, and pentadienes. Heating promotes dimerization of the cyclopentadiene, which remains in the higher boiling residue on distillation, and a more concentrated C-5 cut is obtained. Further fractionation yields an isoprene concentrate with about 50% isoprene and a piperylene concentrate with about 40% 1,3-pentadienes.

The C-5 hydrocarbons obtained per kilogram of ethylene strongly depend on the feed stock, which ranges from ethane to heavy gas–oil (26). The economics of feed stock availability and choice change with the decline of North American oil supplies and increase of imports and with market fluctuations, which may be expected to continue as oil reserves decline.

Recovery and Purification.   For cationic pentadiene–olefin copolymerization, the 40% stream (piperylene concentrate) can be employed without further enrichment. Other polymerization systems require a higher grade, which is obtained by another fractionation increasing the pentadiene concentration to as high as 95% (26). However, the boiling points of cyclopentene (44.1°C), cyclopentadiene (41°C), and 1-pentyne (39.7°C) are so close to those of the pentadienes that they cannot be removed by fractional distillation. In most polymerization systems based on transition or alkali metals, trace amounts of these particular C-5 contaminants impede the polymerization.

Highly pure pentadiene can be obtained from pentadiene–isoprene mixtures by extraction with cuprous chloride (which complexes with pentadiene) in decane (27). Methoxypropionitrile has been proposed as a liquid–liquid extraction solvent (28).

Most polymerizations require anhydrous conditions. Carbonyl and acetylenic compounds are poisons for organometallic initiators and can be removed by extraction with solutions of an aminoalcohol, such as 2-propanolamine, containing cuprous chloride (29). Supported active alkali metals, eg, sodamide or potassium hydroxide on alumina, reduce acetylenic compounds to trace amounts (30,31). Traces of cyclopentadiene inhibit organometal-catalyzed polymerizations. It can be extracted by maleic anhydride under controlled conditions in order to minimize removal of trans-pentadiene at the same time (32).

Acetylenic compounds can be polymerized and thus removed by treatment with an aluminum alkyl plus transition-metal compound of Group IVB through

VIII (33). Metals such as cobalt or titanium, which would catalyze pentadiene polymerization, are not suitable, but triethylaluminum–ferric octoate is effective (33). In the presence of cyclopentadiene, organonickel may be used in place of an iron compound (34); cyclopentadiene inactivates iron.

Sulfur compounds are largely removed by washing with aqueous copper or lead nitrate (doctor's treatment).

Gel-permeation chromatography and liquid chromatography can be used to separate *cis*- from *trans*-pentadiene and recover each isomer 99% pure. Industrially, azeotropic or extractive distillation is preferred.

Polymerization

Cationic Processes

Cationic initiators attack both 1,3-pentadiene isomers; the mixture is much less expensive than the pure isomers. Cationic polypentadienes are less useful because of low molecular weight or insolubility due to cross-linking. However, cationic copolymers of pentadiene with inexpensive branched monoolefins are soluble with reasonably controllable molecular weight and T_m. In fact, the mixtures of pentadienes and olefins occurring in refinery streams are often used only for fuel, but can be polymerized with aluminum chloride. An industry has grown up around these polymers, which are known as C-5 hydrocarbon resins, synthetic terpene resins, or synthetic polyterpene resins (26,35,36). The term petroleum resins also subsumes pentadiene resins made by cationic initiation (see also HYDROCARBON RESINS). The annual U.S. consumption of pentadienes in cationic resins is about 65×10^6 kg (to produce 110×10^6 kg of resins).

Homopolymerization of pentadiene with conventional cationic initiators produces some insoluble polymer, which is difficult to characterize. However, a low molecular weight polymer, so called oligopiperylene, is said to be used in the USSR for the manufacture of coatings (37) and as a plasticizer for polybutadiene rubber (38). Although the analysis may not characterize the industrial product, the total unsaturation is reported to be 70%, of which 73% is trans-1,4, 20% trans-1,2, and 7% cis-1,4; the structure is blocky and there is 5% reversal of the more usual head-to-tail configuration of the 1,4 units. Number-average molecular weight and oligomer unsaturation decrease with increasing polymerization temperature, and the molecular weight distribution broadens. An earlier Soviet study had shown that the cis isomer is more prone to gelation during polymerization than the trans, but the authors were able to make soluble polymers (39). Both isomers led to the same polymer, containing 62–92% of the theoretical unsaturation, of which 45–80% was trans-(1,4 + 1,2); 11–26% 1,2; up to 6% 3,4; and (with some initiators) traces of cis. The reduction in unsaturation was attributed to cyclization, and it was postulated that the rings contained a tetrasubstituted double bond.

For commercial production cationic polymerization (qv) or copolymerization of pentadienes probably involves an aluminum chloride initiator (40–45) or possibly boron trifluoride etherate (46,47). Research on such systems is continuing (48–52). Similar systems, such as titanium tetrachloride (53,54) and phosphoric acid (55), have been patented. In oligomerization of a pentadiene mixture with phosphoric acid, isomerization of the cis to the trans isomer has been observed (14,15). Modified aluminosilicate produces pentadiene polymers used for plasticizing polybutadiene (56) or coating talc (57).

A Japanese patent discloses a high polymer yield in pentane with boron trifluoride etherate modified with ethylene glycol (58). Telomerization of pentadiene with methoxymethyl chloride (59) or isobutoxymethyl chloride, cocatalyzed by zinc chloride or other cationic reagents, has been reported (60).

Pentadiene is an exceptional diene in the degree to which it hinders the cationic polymerization of isobutylene and causes chain transfer (61). It was concluded that the pentadienyl cation is a highly stabilized allyl ion, substituted on both ends.

Reactivity ratios in isobutylene copolymerization with the separated isomers have been determined: $r_{ic} = 5.0$ and $r_{it} = 2.3$ at $-55°C$; the corresponding chain transfer constants were 50 and 113, respectively (52). Most of the pentadiene units in the copolymer were trans-1,4, but about 15% were trans-1,2 (62). Polypentadiene cyclization has been studied (63).

Anionic Processes

Both 1,3-pentadiene isomers polymerize with anionic initiators (64,65), although the rates differ. With lithium or alkyllithium the trans isomer polymerizes several times faster. The difference is so significant that for practical applications a way of incorporating or using the remaining cis isomer is needed. For example, in one method alkyllithium was used first to convert more than 50% of the mixed monomer charge to polymer. The remaining monomer was recovered and polymerized in an emulsion system, in which cis polymerizes much faster (66). In another method the mixture is first polymerized with alkyllithium to high conversion, and then a cationic initiator such as aluminum chloride is added (67).

For a given polymerization rate, *trans*-pentadiene requires about three times the butyllithium concentration that isoprene requires; in copolymerization the pentadiene enters the polymer much more slowly (68).

cis-1,3-Pentadiene polymerizes more slowly than trans and also gives lower yields. The polymer has lower molecular weight, higher trans-1,4, and when made in ethers, lower 3,4 than *trans*-1,3-pentadiene gives (69,70). Vulcanizates of lithium polymer from the trans isomer have useful rubbery properties. Up to about 30%, cis is tolerable in the monomer, but adversely affects properties.

The structure is not sensitive to the particular hydrocarbon solvent used (69), but is affected by polarity changes. As in other polydienes, increasing solvent polarity increases the 1,2 units, but not as markedly as in polyisoprene and polybutadiene. Typically, 1,2 was estimated at about 10% with Li in hydrocarbons and there was no 3,4. In ethers, instead of hydrocarbon solvents, 1,2 rose to 20–30%, and 3,4 became significant, as high as 22% in THF at $-70°C$ from trans isomer (70,71).

Pentadiene has been polymerized with organolithium in the presence of dipiperidinoethane to produce a high 1,2-polypentadiene, which was hydrogenated to produce poly(1-pentene) (72). In another case, stereoregular 1,4-polypentadienes have been converted to tactic, alternating ethylene–propylene copolymers by hydrogenation (73,193).

Pentadienes have been copolymerized with cyclohexadiene in toluene with butyllithium–THF; the cis isomer is reported to have participated more than the trans isomer (74).

A method of selectively polymerizing over 90% of the pentadiene in a by-product stream employs butyllithium and THF, according to a patent (75). An example shows weight-average molecular weight near 7000. The method requires a preliminary step in which cyclopentadiene is inactivated by treatment with alkali metal. Sodium used in place of butyllithium gives a reddish-brown oil (76).

Alkali metals other than lithium give polymers of lower molecular weight, suggesting chain transfer (70). They produce more 1,2, more 3,4, and usually more trans(1,4 + 1,2). Although it can be inferred that cis(1,4 + 1,2) must have been higher in the lithium polymers, total microstructural analysis of these more complex polymers has not yet been achieved (see also ANIONIC POLYMERIZATION).

In the past, estimates of pentadiene microstructure (qv) based on nmr and ir were incomplete. Cis:trans ratios were undetermined, whether for 1,4 or 1,2. However, in 1970 it was shown that cis and trans could be measured by nmr for both 1,4 and 1,2 and that head-to-head vs head-to-tail placement for the 1,4 units could be measured by ir after hydrogenation (77); in lithium polypentadiene ca 5% head-to-head was found. Ozonolysis revealed a significant fraction of head-to-head and tail-to-tail junctions (78). Other workers found no head-to-head units by ^{13}C nmr in polymers made from either isomer with sec-butyllithium (79). Flash pyrolysis detects head-to-head units in polypentadienes (80).

Application of very high pressure to the anionic polymerization system pentadiene–butyllithium–heptane produces a small amount of 3,4 units, but does not otherwise change the structure (81).

Kinetic studies at very high lithium reveal that the average number of associated live chain ends in pentane is about 16 (82). A more recent study gives results of ca 2 at relatively low Li concentration, as for most other dienes (83). The structure of the propagating chain end is 4,1, ie, C-4 links to the previously terminal anion of the chain and C-1 becomes the new terminal anion. This is the principal conclusion of nmr studies in bulk trans-1,3-pentadiene, with a pentadiene:ethyllithium ratio of 5 or less (84). The polymer (oligomer) structure was 80–85% 1,4 (mostly but not exclusively trans) and 15–20% 1,2; there was no 3,4. It is not clear how the 1,2 units arose since 1,4 chain ends were not evident. Reactions of potassium with approximately equimolar pentadiene have also been studied (85). In the USSR quantum-chemical calculations for active centers in anionic polymerizations have been reported (86).

In the USSR anionic polypentadiene appears to have been in industrial production for about three decades. Under the designation SKP (synthetic rubber, piperylene), six grades are available, according to bulk viscosity; the midrange corresponds to a Karrer plasticity of 0.41–0.45 (87). These liquid rubbers are presumably made by sodium initiation. Soviet patent literature suggests that SKP is available as such (88,89) or as derivative products, eg, SKP-G, hydroxy-containing oligopiperylene (90); SKDP-N, on which liquid–rubber coating compositions are based (91); SKPNL, a low molecular weight piperylene rubber for an aliphatic polyester enamel (92); and ESKDP-N, epoxidized SKDP-N (93) (an analytical method for epoxy oxygen in polypiperylene is given in Ref. 94). Corrosion-resistant structural steel for SKDP-N reactors is described in Ref. 95. Some of these polymers may be made by cationic or free-radical initiation. Organosodium-initiated polydienes and liquid polymers for resins and rubber compositions have also been patented (96–98) outside the USSR.

Pentadienes are sluggish partners in anionic copolymerization with butadiene or isoprene. The reactivity ratio for isoprene is about 17 and for pentadiene about 0.06 (68). As in other diene polymerizations with lithium initiators, the rate is increased by the addition of diethyl ether, which reduces the isoprene ratio to about 5. Similar copolymers with isoprene or butadiene have been patented for liquid rubbers (99,100) for hypergolic propellant fuels. Pentadiene–cyclohexadiene copolymers have been reported (101). Pentadiene–styrene copolymers made in ether or dioxane with sodium–sodium isoproxide have been patented as colorless drying oils (102). Block copolymers with styrene, using alkyllithium initiation, have also been patented (103). Perhaps most significant are star-block copolymers intended for tire rubbers (104) or for thermoplastic elastomers in which the hard phase is polystyrene (105).

Hydroxy-terminated polypentadienes are occasionally cited in the Soviet literature, usually as components for polyesters. It is not clear in patent abstracts whether the pentadiene polymers are free radical or anionic in these cases. A Japanese patent discloses a sodium polymer, treated in THF with ethylene oxide and hydrogenated (104).

Free-radical Processes

Emulsion and Homogeneous.   Although the early assessment of emulsion polypentadienes was unfavorable, after about 1960 it was predicted that tire rubber could be made by free-radical polymerization of pentadienes (64,106,107). The change probably reflected the availability of better purification methods and the improved compounding of tire rubbers with carbon blacks, antioxidants, and accelerators.

Determination of reactivity ratios (Table 2) shows that the cis isomer enters the polymer chain faster than *trans*-pentadiene or styrene. In free-radical homopolymerization at 70°C, the relative rates are 2,3-dimethylbutadiene:isoprene: *cis*-1,3-pentadiene:*trans*-1,3-pentadiene = 25:15:3:1 (107). In another study of the higher reactivity of the cis isomer compared to the trans, conversions under very high pressure indicated that peroxy radicals prefer *cis*- to *trans*-pentadiene by a factor of 2 (108).

Table 2.   Reactivity Ratios for the Emulsion Polymerization of *cis*-1,3-Pentadiene, *trans*-1,3-Pentadiene, and Styrene[a]

Monomer	Reactivity ratio
cis-1,3-pentadiene, M_1	$r_1 = 3.6 \pm 0.7$
trans-1,3-pentadiene, M_2	$r_2 = 0.8 \pm 0.2$
cis-1,3-pentadiene, M_1	$r_1 = 1.37 \pm 0.2$
styrene, M_2	$r_2 = 0.49 \pm 0.1$
trans-1,3-pentadiene, M_1	$r_1 = 0.91 \pm 0.3$
styrene, M_2	$r_2 = 1.23 \pm 0.4$

[a] Ref. 107.

Microstructures in polypentadienes made from either isomer contained mostly trans-1,4 with some 1,2 (107). X-ray analysis of several polypentadienes, including the emulsion polymer made from trans isomer, was not helpful because the polymer was amorphous, but comparison of ir and 300-MHz nmr analyses showed that most of the 1,2 was the trans isomer (109); less than 10% of cis-1,4 and cis-1,2 were also present. The formation of cis-1,2 from trans isomer is unexpected. Some cis-1,2 was also found in syndiotactic *trans*-1,2-polypentadiene, made with Ziegler-Natta initiator from presumably all-trans isomer (110). The glass-transition temperatures of the emulsion polypentadienes were the same: $-38°C$ (107), significantly higher than those of stereoregular 1,4-polypentadienes.

Free-radical polymers made under high pressure conditions (108) were reported to contain somewhat more 1,4, less 1,2, and a few percent 3,4 (111).

A number of patents suggest that emulsion pentadiene polymers might be useful industrially, although such products do not seem to be marketed. Examples include a latex component for foam rubber (112), piperylene–styrene latices for water-emulsion paints (113), and an emulsion for adhesives (114). Piperylene–styrene resins have been made by radical polymerization and proposed for low color, fast-drying films (115). Pentadiene has been grafted by radical initiation onto styrene–acrylonitrile copolymer for high impact strength (116).

Inclusion Polypentadienes. Isotactic *trans*-1,4-polypentadiene has been made by γ-irradiation of *cis*- or *trans*-1,3-pentadiene included in solid crystalline (optically active) perhydrotriphenylene (117,118). As in other inclusion or "canal" polymers (119,120), the melting point (104°C) was higher by ca 10°C than that of the same polymer made by other means (121). Several pentadiene copolymers have been made by inclusion and analyzed by ^{13}C nmr (122); they are highly trans-1,4.

Alternating Copolymers

Maleic anhydride in the presence of a peroxide initiator or under uv radiation (18) forms an equimolar, alternating copolymer with *trans*- or *cis*-pentadiene. Acrylonitrile behaves similarly and forms an alternating copolymer with either isomer, but in the presence of ethylaluminum sesquichloride (EASC) rather than peroxide (16). The copolymers have *trans*-1,4-pentadiene units (17). These results can be attributed to the formation of a charge-transfer complex (diene donor to an acceptor, eg, maleic anhydride or acrylonitrile complexed with EASC), which polymerizes. *cis*-Pentadiene isomerizes to trans before participating in these reactions.

Alfin Polypentadienes

Both pentadiene isomers can be converted to high molecular weight polymers with Alfin initiators, such as the combination allylsodium–sodium isopropoxide–sodium chloride. Alfin polymerization of pentadiene isomers resembles free-radical rather than anionic polymerization because the cis isomer enters faster. However, it is unlike radical or anionic polymerization in producing a very high trans-1,4 polymer, with high molecular weight, independent of the initiator concentration. As with other conjugated dienes, the molecular weight can be controlled by the addition of hydrogen (123), vinyl ethers (124), 1,4-pentadiene (125,126), or 9,10-dihydroanthracene (123). The microstructure has been reported as 86% trans-1,4, 11% cis-1,2, and 3% 3,4; x-ray diffraction revealed no crystallinity (109). Copolymers with other dienes have been made (127–129).

Transition-metal Initiation

The polymerization of pentadiene with Ziegler-Natta initiators illustrates the great achievements of the era of polymer synthesis that began about 1954 (see also ZIEGLER-NATTA CATALYSTS). Several different sterically pure polymer structures were produced from diene monomers. The polymerization of pentadiene is significant because of the number and complexity of options (12,130–142) (see also INSERTION POLYMERIZATION).

Polymers of 1,3-butadiene, linear and uniform with respect to the single double bond remaining in each unit, may be cis-1,4, trans-1,4, or 1,2. Since a 1,2 unit has an asymmetric carbon, more options are possible if the polymer is uniform with respect to this asymmetry also. However, only isotactic-1,2 and syndiotactic-1,2 seem to be realistic, although more complex regularity can be conceived. Polymers of 2-substituted 1,3-butadienes introduce more options because of the possibility of 3,4 units, which are not identical to 1,2 and may be isotactic or syndiotactic. Polymers of 1-substituted 1,3-butadienes introduce more complexity because the 1,4 polymers possess an asymmetric carbon and may exist in isotactic or syndiotactic forms if they are highly stereoregular. The 1-substituted 1,3-butadiene polymers of which 1,3-pentadienes are the simplest and most accessible representatives, have one asymmetric carbon in each 1,2 polymer unit and two in each 3,4 unit. None of the theoretically possible options for stereoregular 3,4 polymers has been synthesized, probably because there is too much crowding around the 3,4 bond. Remarkably, however, four crystalline polypentadienes have been synthesized from 1,3-pentadienes: isotactic *cis*-1,4-polypentadiene, syndiotactic *cis*-1,4-polypentadiene, isotactic *trans*-1,4-polypentadiene, and syndiotactic *trans*-1,2-polypentadiene (Fig. 2). All crystalline polypentadienes have been analyzed by proton nmr (109,143) and ^{13}C nmr (144,145).

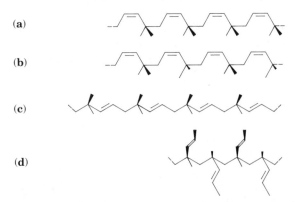

(a)

(b)

(c)

(d)

Fig. 2. Crystalline polypentadienes: (a) isotactic *cis*-1,4-polypentadiene; (b) syndiotactic *cis*-1,4-polypentadiene; (c) isotactic *trans*-1,4-polypentadiene; and (d) syndiotactic *trans*-1,2-polypentadiene.

Isotactic *trans*-1,4-Polypentadienes. Isotactic *trans*-1,4-polypentadiene was the first stereoregular polypentadiene synthesized. Treatment of a mixture of 1,3-pentadiene isomers with $(C_2H_5)_3Al$–$TiCl_3$ (violet) in heptane gave traces of this crystalline polymer in a mainly amorphous reaction product. Vanadium chloride in place of titanium chloride gives a better yield and more crystallinity (12,146–148). Both isomers polymerized, the cis giving a slightly more crystalline product; mixtures gave the amorphous products. The selective formation of crystalline trans-1,4 polymers from other diene monomers such as butadiene and isoprene with these initiators is well known, but the pentadienes are interesting because their asymmetry must also be controlled in order to produce crystalline polymer. The amorphous polypentadiene fraction was also highly trans-1,4 (146). The mode of monomer insertion has been described (132). In isotactic *trans*-1,4-polypentadiene two crystalline phases have been identified (149). The methods of analyzing pentadiene polymers have been refined (108,143–145) and rheological properties have been reported (150).

Copolymers of pentadiene with butadiene made with trialkylaluminum–vanadium trichloride are crystalline with intermediate melting points over the entire range of composition (12,151).

The initiators based on VCl_3 or violet $TiCl_3$ are heterogeneous mixtures. Similar homogeneous initiators based on alkylaluminum halides and soluble vanadium salts produced less stereoregular, amorphous *trans*-polypentadiene containing both trans-1,2 and trans-1,4 units (152,153). However, the homogeneous initiators still produced crystalline trans-1,4 polymer from butadiene. Butadiene–pentadiene copolymers produced with these initiators were crystalline at low pentadiene content; crystallinity varied with the proportion of pentadiene (12). There were other differences. Whereas the heterogeneous initiator converted both *cis*- and *trans*-pentadiene, homogeneous systems such as diethylaluminum chloride–vanadium triacetylacetonate [$V(Acac)_3$] converted only the trans to polymer, leaving the cis isomer unchanged.

The potential utility of *trans*-pentadiene as a comonomer was easily seen (152,154). It was readily available, and it could lower the crystallinity of *trans*-polybutadiene, resulting in a vulcanizable elastomer that would crystallize (and therefore strengthen) on stretching. Efforts were made in Italy to perfect and exploit the potential of the soluble vanadium system, and it was included in the survey of other rubbers potentially useful in tires (155). More information about the copolymer properties was published. A 5–10% loss in unsaturation suggested cyclization (156); tire-skid properties were superior to those of polybutadiene (157); the rheology was studied extensively (158); and vulcanizate properties were compared (159) with those of natural rubber. Copolymers, 70 butadiene–30 pentadiene, could be made with T_m and T_g very close to those of natural rubber and correspondingly excellent green strength (160). This could be very useful in a period of natural rubber scarcity.

Other initiator systems produce *trans*-1,4-polypentadiene and might be useful for butadiene–pentadiene copolymers of controlled crystallinity (160).

trans-1,4-Polypentadiene of unspecified tacticity has been produced in water with $RhCl_3$ hydrate initiator and sodium dodecylbenzene sulfonate emulsifier. The polymer contained 80% trans-1,4 when prepared from trans isomer (less from cis isomer) and 20% trans-1,2 (161). Although this is an aqueous emulsion system and the structure is like that of free-radical polypentadiene, the polymerization is considered to proceed by insertion of monomer into a π-allyl rhodium complex (161–164).

Syndiotactic *trans*-1,2-Polypentadienes. Polymer with ca 90% trans-1,2 units was prepared from *trans*-1,3-pentadiene in an aliphatic solvent using $(C_2H_5)_2AlCl–Co(Acac)_3$ (12,165,166); the cis isomer does not participate in the reaction. Even after fractionation by reprecipitation the polymer was found to be amorphous by x rays at room temperature, but crystallized on stretching. The identity period found by x rays was 0.51 nm, which identified the polymer responsible for the crystallinity as syndiotactic. However, there was evidence of some cyclization (110). Polymer made later with the intention of reproducing the crystallizable polymer was found by ^{13}C nmr to be a mixture of syndiotactic, isotactic, and atactic triads (145). Another product, made by $(C_2H_5)_2AlCl–Co(Acac)_3$ from monomer that was 70:30 trans:cis, had about 80% trans-1,2 by nmr and ir, and a glass-transition temperature of about 6°C. This polymer and another Al–Co polypentadiene with >95% trans-1,2 were sulfur-cured, and the vulcanizate properties were reported (167).

When $(C_2H_5)AlCl_2$ or $(C_2H_5)_3Al_2Cl_3$ was used in place of $(C_2H_5)_2AlCl$ in these Al–Co initiators, more of the product was cyclized (166), apparently by a cationic initiator. Amide additives minimize this tendency (168,169) and produce high trans-1,2 elastomer, which upon vulcanization retains physical properties over 10 days aging in air at 100°C.

These vulcanizates take advantage of the low stereoregularity that prevents *trans*-1,2-polypentadienes from being highly crystalline. Attempts to increase the 1,2 crystallinity have also been made. Thus the type of initiator used commercially to produce moderately crystalline syndiotactic 1,2-polybutadiene film has been applied to pentadiene in methylene chloride (170–172). A trialkylaluminum–polar compound, such as water–cobalt salt–triphenylphosphine, is used.

Syndiotactic *trans*-1,2-polypentadiene has also been prepared with iron-containing initiators. A ternary system comprising trialkylaluminum, ferric octoate $(FeOct_3)$, and an aliphatic cyano-containing ligand has been patented (173). For example, polymer that was found to be 83% syndiotactic trans-1,2 by nmr, ir, and x rays (109) was made from trans isomer in hexane with triethylaluminum–ferric octoate–butyl thiocyanate (Table 3). A similar polymer was made rapidly with azobisisobutyronitrile (AIBN) as the cyano compound with 94% conversion in 1 h. Other iron systems have produced a polymer that was probably *trans*-1,2-polypentadiene from a monomer that was mostly trans (185,186).

Syndiotactic *cis*-1,4-Polypentadienes. Syndiotactic *cis*-1,4-polypentadiene was made in an aromatic solvent using the same homogeneous $(C_2H_5)_2AlCl–Co(Acac)_3$ initiator used for the syndiotactic trans-1,2 polymer discussed previously (174,175). Natta and co-workers discovered the cis-1,4 before the trans-1,2. In their early publications they attributed the difference in polymer structure to a solvent effect, ie, aromatic vs aliphatic (165). Similar to the work on butadiene polymerization (187), water added to $(C_2H_5)_2AlCl$ forms chloroethylaluminoxane, which can be used with $Co(Acac)_3$ to polymerize trans isomer. Cobalt initiators do not polymerize cis monomer (12). The product was syndiotactic cis-1,4 polymer even when the polymerization was carried out in heptane. The aluminoxane procedure increases the polymerization rate (176).

Table 3. Stereoregular Polypentadienes Made with Transition-metal Initiators

Initiator	Polymer microstructure 1,4				Identity period, nm	T_m, °C	T_g, °C	Refs.
	Cis	Trans	Trans-1,2[a]	3,4				
Isotactic trans-1,4								
$R_3Al–VCl_3$[b]	0	98	2	0	0.48	95	−54	146–148
Syndiotactic trans-1,2								
$(C_2H_5)_2AlCl–Co(Acac)_2$	6	4	90	0	0.51		+6	165–167
$R_3Al–FeOct_3–AIBN$	12	0	83	5	0.51		+2	109,173
Syndiotactic cis-1,4								
$(C_2H_5)_2AlCl–Co(Acac)_2–$	85	0	10	5	0.84	53	−42	133,174–177
HOH								
Isotactic cis-1,4[c]								
$R_3Al–Ti(OR)_4$	85	0	10	5	0.81	44	−45	178–180
$R_3Al–CeOct_3–C_2H_5AlCl_2$	80	5	15	0	0.81		−50	109,181
$R_3Al–Nd$ stearate–$(C_2H_5)_2AlCl$	80		18	2				182
$R_3Al–FeOct_3–AIBN$[d]	99	0	1	0	0.81	50	−60	173
$R_3Al–CrNaph_3$[d]	95	0	3	2	amorphous		−59	183
$R_3Al–CrNaph_3$	75	0	21	4	amorphous		−45	184

[a] No cis has been reported.
[b] Monomer may be cis or trans.
[c] Monomer trans unless otherwise noted.
[d] Monomer cis; all others trans.

The best properties for syndiotactic cis-1,4 polymer are shown in Table 3. Reproducibility with $(C_2H_5)_2AlCl$ systems was low with cis-1,4 ranging from 50 to 80% (136). Impurities such as water may have been responsible. Replacement of $(C_2H_5)_2AlCl$ with ethylaluminum dichloride resulted in amorphous, low molecular weight, higher 1,2 polymers. Lewis bases suppress this cationic activity and provide the preferred systems for syndiotactic *cis*-1,4-polypentadiene (136). Vulcanizate properties from polymer made with an $C_2H_5AlCl_2$complex–$Co(Acac)_3$ have been reported (177).

Nickel systems also give syndiotactic *cis*-1,4-polypentadiene. Trans isomer was converted to 90–94% syndiotactic cis-1,4 polymer with a triethylaluminum–nickel octoate–BF_3 phenolate initiator. The polymers gave highly crystalline x-ray diffraction patterns, identity period 0.84 nm; the T_g was −57°C and T_m was 51°C (109). Polymer conversion, however, was only 30%, and the inherent viscosity was 1.6 dL/g or less.

Stereoregular 1,4-polypentadienes have been converted by hydrogenation (73,193) to tactic, alternating ethylene–propylene copolymers.

Isotactic *cis*-1,4-Polypentadienes. *Titanium Initiators.* Natta and co-workers announced preparation of crystalline *cis*-1,4-polypentadiene in 1963 (178,179). The initiator is trialkylaluminum–titanium tetraalkoxide with an Al:Ti ratio of 3:10. Such initiators give crystalline syndiotactic 1,2 polymer from butadiene and amorphous 3,4 polymer from isoprene (188). Thus the behavior of pentadiene seems more of a departure than with vanadium and cobalt systems. Both *cis*- and *trans*-pentadiene isomers produced some crystalline isotactic *cis*-1,4-pentadiene, although in small amounts in the case of the cis isomers (180). Products from both isomers contain amorphous polymer that could be removed by repeated fractional precipitation. The maximum cis content of the crude polymer was ca 70% from either isomer, but the yield was lower from cis.

In fact little or no cis isomer is polymerized by R_3Al–$Ti(OR)_4$ initiation. Instead, cis isomerizes to trans, which polymerizes. This was discovered by finding trans isomer in the unreacted portion of a reaction mixture of all-cis starting material (189). Later, the rate constant for isomerization was found to be highest at an Al:Ti ratio of 2.6, whereas for polymerization of trans isomer it was highest at Al:Ti = 6 (13). Another clue was provided by a polymerization in which optically active titanium menthoxide was used as the alkoxide in the initiator: the product from the cis isomer was optically active and had the same sign as the product from the trans isomer (136). However, the sign of the rotation could be controlled, and even reversed, by the choice of alkylaluminum. It seemed, therefore, that there might be more than one active complex in the initiator mixture (190).

Because of the unexpectedly different behavior with trialkylaluminum–titanium tetraalkoxide, the copolymerization of pentadiene with butadiene or isoprene was investigated. The Natta group noted that isoprene units in the copolymer were largely 3,4 and pentadiene units cis-1,4, as in the homopolymers. This was later confirmed, but more vinyl was reported in the copolymers than in the homopolymers (191). Random copolymerization seemed to be occurring with a slight tendency toward alternation. These observations suggested that the same (homogeneous) initiator produced different stereoregulation in 1,3-dienes slightly differently substituted (188). Natta and co-workers concluded that all the monomers coordinated through one double bond rather than two, but that different charge distribution accounted for the different courses followed by the different monomers.

As shown by esr analysis of homopolymerizations of butadiene, pentadiene, and isoprene with this initiator, the growing polymer in all three cases has two substituted π-allyls (from two growing chains) and one alkoxy group coordinated with Ti(III) (192). Although the polypentadiene unit is not identical to the Natta unit (188), this work supports the Natta conclusion that each diene unit coordinates with Ti via one double bond rather than two (136).

The kinetics of this soluble system have been studied (13,137,138,194). The active species is long-lived with little transfer or termination. The number of active species is extremely low compared to the number of titanium atoms. The molecular weight distribution is broad and bimodal, and there is some branching and gelation.

The vulcanization of isotactic *cis*-1,4-polypentadiene and of syndiotactic *cis*-1,4-polypentadiene have been compared (177). Both polymers had 70–75% cis-1,4 units and less than 10–15% non-1,4 units; no isomerization occurred during vulcanization. The vulcanizate made from the isotactic polymer had a higher cross-link density than that made from the syndiotactic polymer. This effect is attributed to the difference in stereoregularity.

Rare Earth Initiators. Polypentadienes with moderately high degrees of isotactic cis-1,4 stereoregularity have been made with initiators based on rare earth elements. A cerium example is shown in Table 3. Rare earth initiators polymerize only trans isomer. Molecular weights are almost a linear function of conversion, and the system has been referred to as "pseudo-living." Molecular weight can be reduced, for example, by the inclusion of dibutylaluminum hydride as the Al component or by increasing the aluminum concentration (195). Rare earth initiators appear homogeneous when prepared in the presence of the monomer, except for a slight haze in aliphatic solvents.

Cerium, but not neodymium, residues may adversely affect polymer aging. A few percent of 3,4 units have been detected in Nd polypentadiene (182), but the microstructure is substantially the same (Table 3) (131,182,196,197). Homopolymerization rates are on the order butadiene > isoprene > pentadiene (198).

Uranium Initiators. High cis-1,4 microstructure, apparently very similar to that produced by neodymium or cerium, has been reported from polymerization of 75% trans isomer with triethylaluminum–π-allyluranium–aluminum tribromide (199). Other uranium or thorium initiators have been reported (200–203).

Iron Initiators. Iron systems are the least expensive initiator systems. In addition, utilizing cis isomer, they provide the highest degree of isotactic cis-1,4 regularity, in high yield without polymer fractionation. For maximum isotactic cis-1,4 polymer, trans isomer must be minimized because it is converted to syndiotactic *trans*-1,2-polypentadiene (Table 3).

The most regular (99%) isotactic *cis*-1,4-polypentadiene reported is shown in Table 3. The initiator was R_3Al–$FeOct_3$–AIBN, and the starting monomer was 98.6% cis isomer (173). The polymer was obtained in high yield. The properties shown are those of the entire product, without the fractionation necessary to extract the prototypical Natta polymers (175,180). Further evidence of the potential utility of this polymer is the T_g within 10°C of that of natural rubber, and also a similar Mooney viscosity (ML-4 at 100°C). Iron resembles rare earth systems in appearing homogeneous and "pseudo-living." It is more tolerant of impurities than cobalt and titanium. Iron-derived polymers are highly stable in accelerated aging tests. However, iron rubbers of this type would require additional costs of monomer purification. The presence of even 5% trans isomer in the monomer raised the T_g by 5°C (173).

Chromium Initiators. Chromium produces an elastomer that is very high cis-1,4, but amorphous (Table 3) (183). Cis isomer is required for the highest cis-1,4 polymer, but trans isomer also produces high cis-1,4 (184). For example, a monomer 68% trans, 4% cis, and 27% other C-5 unsaturated compounds gives a polymer that is 75% cis-1,4 (Table 3). Isomer mixtures produce polymers with single, intermediate T_gs, all of which are in the range of practical interest for tire rubbers (204). Chromium-initiated polypentadienes, like those made with iron systems, age well. Ozone resistance of 1,4-polypentadienes is good (64).

Table 4 summarizes the polymerization behaviors of *cis*- vs *trans*-pentadienes under different initiator conditions.

Copolymers with Other Conjugated Dienes. The trans copolymers of butadiene–30 $\pm$ 5% pentadiene, made with a soluble vanadium initiator, are probably the most likely Ziegler-Natta copolymers to be of potential industrial utility.

Copolymers of butadiene–pentadiene (cis- and trans-1,4) have been made with rare earth (181) and uranium initiators (199); with these initiators and soluble vanadium only *trans*-pentadiene participates.

Iron initiators have been used to copolymerize *cis*-pentadiene–dimethylbutadiene; *trans*-pentadiene–dimethylbutadiene; *trans*-pentadiene–butadiene; *trans*-pentadiene–isoprene; and *trans*-pentadiene–*cis*-pentadiene (173). Butadiene–

Table 4. Behavior of Pentadiene Geometric Isomers with Polymerization Initiators

Catalyst	Polymerization		Polymer structure[a]; %		Relative polymerization rate		Refs.
	Cis	Trans	Cis	Trans	Cis	Trans	
cationic	+	+	mixed	mixed	high	low	
anionic, C_4H_9Li	+	+	trans-1,4; 61	trans-1,4; 54	low	moderate	64,70
free radical	+	+	trans-1,4; 70	trans-1,4; 70	moderate	low	107–109
Alfin	+	+	trans-1,4; 85	trans-1,4; 85	moderate	low	109
Transition metals							
V	+	+	iso-trans-1,4; 98	iso-trans-1,4; 98	high	high	146
Ti	–[b]	+		iso-cis-1,4; 87		low	178–180
Co, H_2O	–	+		syndio-cis-1,4; 90		moderate	174
Co	–	+		syndio-trans-1,2; 90		low	166
Ce or Nd	–	+		iso-cis-1,4; 80		moderate	109,181,182
Fe	+	+	iso-cis-1,4; 99	syndio-trans-1,2; 90	moderate	high	173
Cr	+	+	(iso-)cis-1,4; 98	cis-1,4; 75	moderate	moderate	183,184

[a] *Iso-trans* is isotactic trans stereoregular polymer; *syndio-trans* is syndiotactic trans stereoregular polymer.
[b] The cis isomer isomerized slowly to trans isomer (12,13).

dimethylbutadiene–*trans*-pentadiene have been terpolymerized. The pentadiene–dimethylbutadiene copolymers are amorphous rubbers with single T_gs, whereas the homopolymers are crystalline. The T_m for iron-initiated poly(2,3-dimethylbutadiene) is 149°C. Block copolymers seem to form from *trans*- and *cis*-pentadiene mixtures (173).

The chromium-initiated copolymers of *cis*-pentadiene–isoprene and *trans*-pentadiene–isoprene appear to be random (184,204). The pentadiene units were primarily cis-1,4; isoprene structure was about equally divided between 3,4 and 1,4.

Industrial Applications

Cationic pentadiene–olefin copolymers are used primarily in adhesives; they are a significant component (30–50% by weight) in typical formulations for pressure-sensitive adhesives (qv) (annual U.S. consumption ca 160×10^6 kg (205). Pentadiene resins improve the thermal stability of the adhesive and improve the adhesion to various substrates ranging from metal to fiber (206).

The patent literature discloses many other applications or proposed applications that employ liquid polypentadienes or liquid pentadiene copolymers after further reaction. Pentadiene polymers provide the reactive double bonds, are in conveniently miscible liquid form, lack color, and are inexpensive.

Pentadiene and its polymers can serve as vehicles for carrying reactive polar materials like maleic anhydride into polymers. Treatment of pentadiene polymers with maleic anhydride improves bonding to metal in hot melt adhesives (207), thermoset adhesives (208), and electrically insulating film (209). It increases peel strength in heat-resistant adhesives for polyester film (210). Heat-curable water-soluble adhesives for plywood manufacture can be made with liquid polypentadiene and maleic anhydride, starch, and magnesium oxide (211); water-resistant adhesives for plywood use maleic-polypentadiene adduct (212). Similarly, maleinated polypentadiene treated with cobalt and tin additives is used as an antifouling impregnant for nylon marine netting (213). Maleic anhydride–pentadiene–cyclopentadiene adducts are used in a polymerization mixture for light-curable unsaturated polyester coatings (214,215). Soldering masks can be based on a similar adduct (216). Pentadiene–dicyclopentadiene copolymer is used with maleic anhydride in a water-based coating composition (217,218).

Epoxy resins can be made from liquid polypentadienes (219–223). Epoxidation of polypentadiene in the presence of alcohols improves the pot life of polymeric formulations (224). Epoxy-oligopentadiene improves wear properties in tire-tread rubbers (93). Liquid-phase oxidation could be the source of oxidized oligopiperylene (225), and perhaps of hydroxylated piperylene, although controlled termination of anionic polymer is a more likely source of the latter. Hydroxylated polymers containing pentadiene have been reported to be adhesives (226), liquid curing agents for urethane prepolymer (227), plasticizers in melamine resin primers (228), and excellent tackifiers for tire rubbers (229). Carboxylated pentadiene oligomers (by ozonolysis of polypentadiene) have been proposed as tackifiers (230).

Hydrogenation (231), phenolation (232,233), sulfurization (234), and chlorination of polypentadiene (235) have been patented. Styrene and acrylonitrile have been grafted to polypentadiene to obtain a high impact polymer (116). Polypentadiene grafted to polyester fiber improves fiber–rubber adhesion. Terpolymers of pentadiene–acrylonitrile–methacrylic acid have been made for leather-finishing compositions (236). Various blends have been made with chlorinated PVC resin to stabilize the PVC (237), with low molecular weight acrylate for protective coatings (238), with nitrile rubber to improve mechanical properties (239), with ethylene–vinyl acetate (EVA) for resin-to-steel adhesives (240), and with concrete (241).

BIBLIOGRAPHY

1. E. G. M. Tornqvist in J. P. Kennedy and E. G. M. Tornqvist, eds., *Polymer Chemistry of Synthetic Elastomers (High Polymers 23)*, John Wiley & Sons, Inc., New York, 1968, p. 42.
2. H. W. Starkweather, P. O. Bare, A. S. Carter, F. B. Hill, V. R. Hurka, C. J. Mighton, P. A. Sanders, H. W. Walker, and M. A. Youker, *Ind. Eng. Chem.* **39**(2), 210 (1947).
3. E. R. Weidlein, *Chem. Eng. News* **24**(6), 771 (1946).
4. W. G. Taft and G. J. Tiger in G. S. Whitby, ed., *Synthetic Rubber,* John Wiley & Sons, Inc., New York, 1954, Chapt. 21, p. 686.
5. R. L. Frank, R. D. Emmick, and R. S. Johnson, *J. Am. Chem. Soc.* **69**, 2313 (1947).
6. H. M. Frey, A. M. Lamont, and R. Welsh, *J. Chem. Soc. A* **16**, 2642 (1971).
7. S. W. Benson, K. W. Egger, and D. M. Golden, *J. Am. Chem. Soc.* **87**, 468 (1965).
8. K. W. Egger and S. W. Benson, *J. Am. Chem. Soc.* **87**, 3311, 3314 (1965).
9. J. C. H. Chen and W. D. Huntsman, *J. Phys. Chem.* **75**, 430 (1971).
10. P. B. Wells and G. R. Wilson, *Disc. Faraday Soc.* **41**, 237 (1966).
11. L. Kh. Freidlin, V. Z. Sharf, and M. A. Abidov, *Neftekhimiya* **2**, 291 (1962); *Chem. Abstr.* **58**, 6679c.
12. G. Natta and L. Porri in Ref. 1, p. 597.
13. K. Boujadoux, R. Clement, J. Jozefonvicz, and J. Neel, *Eur. Polym. J.* **9**, 189 (1973).
14. V. N. Sapunov, R. I. Fedorova, P. S. Belov, N. N. Bayanova, O. Yu. Omarov, and O. A. Misbakh, *Deposited Doc. D8DEP2 VINITI,* 112 (1983); *Chem. Abstr.* **100**(15), 120443b.
15. R. I. Fedorova, S. A. Nizova, N. N. Bayanova, and O. M. Antar, *Neftekhimiya* **23**(2), 195 (1983); *Chem. Abstr.* **98**(26), 216092c.
16. N. G. Gaylord, M. Stolka, and B. K. Patnaik, *J. Macromol. Sci. Chem.* **A6**(8), 1435 (1972).
17. N. G. Gaylord, M. Nagler, and A. C. Watterson, *Eur. Polym. J.* **19**(10–11), 877 (1983); *Chem. Abstr.* **100**(14), 104145a.
18. N. G. Gaylord, R. Mehta, M. Mishra, and M. Nagler, *Polym. Prepr.* **26**(2), 180 (1985); *Chem. Abstr.* **104**(14), 110271n.
19. D. Craig, *J. Am. Chem. Soc.* **65**, 1006 (1943).
20. U.S. Pat. 3,403,196 (Sept. 24, 1968), R. B. Long and W. A. Knarr (to Esso Research & Engineering Co.).
21. G. S. Hammond, N. J. Turro, and P. A. Leermakers, *J. Phys. Chem.* **66**, 1144 (1962).
22. G. S. Hammond, J. Saltiel, A. A. Lamola, N. J. Turro, J. S. Bradshaw, D. O. Cowan, R. C. Counsell, V. Vogt, and C. Dalton, *J. Am. Chem. Soc.* **86**(16), 3197 (1964); *Chem. Abstr.* **61**, 8155b.
23. J. J. Dannenberg and J. H. Richards, *J. Am. Chem. Soc.* **87**(7), 1626 (1965); *Chem. Abstr.* **62**, 14447a.
24. V. I. Ponomarenko, B. L. Irkhin, I. M. Kolesnikov, and A. Kh. Arslanova, *Prom. Sint. Kauch.* (6), 7 (1975); *Int. Polym. Sci. Tech.* **2**(12), T85 (1975).
25. Jpn. Pat. 85 28,940 (Feb. 14, 1985) (Japan Synthetic Rubber Co., Ltd.); *Chem. Abstr.* **103**(2), 6842t.
26. B. J. Davis, "The Chemistry of C-5 Resins," *presented at the Technical Association of the Pulp and Paper Industry (TAPPI) Hot Melt Short Course,* June 1979, p. 112.

27. U.S. Pat. 3,527,831 (Sept. 8, 1970), G. C. Blytas (to Shell Oil Co.).

28. Rom. Pat. 81,274 (Apr. 30, 1983), F. Gothard, M. Cerghitescu, M. Neagoe, and N. Palibroda (to Combinatul Petrochemic, Brazil); 4 *Chem. Abstr.* **100,** 175508g.

29. U.S. Pat. 3,260,766 (July 12, 1966), W. Nudenberg and E. A. Delaney (to Texas-U. S. Chemical Co.).

30. U.S. Pats. 4,060,567 (Nov. 29, 1977) and 4,087,477 (May 2, 1978) J. J. Tazuma and A. Bergomi (to Goodyear Tire & Rubber Co.).

31. USSR Pat. 1,109,370 (Aug. 23, 1984), B. A. Saraev, V. A. Gorshkov, S. Yu. Pavlov, V. P. Bespalov, V. G. Baunova, Yu. M. Ryabov, R. Kh. Rakhimov, P. I. Kutuzov, and co-workers; *Chem. Abstr.* **101**(23), 210539n.

32. L. F. Kovrizhko, Yu. V. Bryantseva, V. I. Raevskaya, and T. P. Agarkova, *Tr. Lab. Khim. Vysokomol. Soedin.* (3), 78 (1964); *Chem. Abstr.* **65,** 12379e.

33. U.S. Pat. 3,647,913 (Mar. 7, 1972), E. Lasis (to Polymer Corp. Ltd.).

34. U.S. Pat. 4,471,153 (Sept. 11, 1984), M. C. Throckmorton (to Goodyear Tire & Rubber Co.).

35. J. A. Schlademan in D. Satas, ed., *Handbook of Pressure-sensitive Adhesive Technology,* Van Nostrand Reinhold Co., Inc., New York, 1982, Chapt. 16.

36. J. P. Kennedy and E. Marechal, *Carbocationic Polymerization,* John Wiley & Sons, Inc., New York, 1982, p. 2.

37. S. A. Egoricheva, V. A. Rozentsvet, B. I. Pantukh, M. V. Eskina, A. S. Khachaturov, and R. M. Livshits, *Lakokras. Mater. Ikh. Primen.* (1), 12 (1985); *Chem. Abstr.* **102**(20), 168329n.

38. I. I. Yukel'son, L. V. Fedorova, V. I. Raevskaya, and L. F. Polyakova, *Kauch. Rezina* (9), 11 (1977); *Chem. Abstr.* **87,** 185795a.

39. T. T. Denisova, I. A. Livshits, and E. R. Gershtein, *Vysokomol. Soedin. Ser. A* **16**(4), 880 (1974); *Polym. Sci. USSR* **16**(4), 1017 (1974).

40. Eur. Pat. Appl. 74,273 (Mar. 16, 1983), K. Mizui, M. Takeda, and T. Iwata (to Mitsui Petrochemical Industry Ltd.); *Chem. Abstr.* **98**(26), 217337y.

41. U.S. Pat. 3,577,398 (May 4, 1971), H. A. Pace and V. J. Anhorn (to Goodyear Tire & Rubber Co.).

42. Jpn. Pat. 85 118,777 (June 26, 1985) (Nippon Zeon Co. Ltd.); *Chem. Abstr.* **103**(20), 161573z.

43. Jpn. Pat. 84 164,316 (Sept. 17, 1984) (Mitsui Petrochemical Industry Ltd.); *Chem. Abstr.* **102**(12), 96630d.

44. V. P. Mardykin, B. L. Irkhin, A. V. Pavlovich, L. M. Antipin, B. N. Zorin, V. V. Zuev, and B. L. El'kin, *Zh. Prikl. Khim. Leningrad* **57**(5), 1157 (1984); *Chem. Abstr.* **102**(2), 7128f.

45. B. L. Irkhin, V. I. Ponomarenko, and K. S. Minsker, *Prom. Sint. Kauch. Nauchno Tekh. Sb* (9), 18 (1974); *Chem. Abstr.* **83,** 115074h.

46. Ger. Offen. 247, 084 (May 3, 1973), D. R. St. Cyr (to Goodyear Tire & Rubber Co.); *Chem. Abstr.* **79,** 92816c.

47. U.S. Pat. 4,189,547 (Feb. 19, 1980), R. A. Osborn and H. L. Bullard (to Goodyear Tire & Rubber Co.).

48. E. Ceausescu and co-workers, *Mater. Plast. Bucharest* **22**(2), 77 (1985); *Chem. Abstr.* **104**(12), 89067p.

49. *Ibid.,* **22**(1), 8 (1985); *Chem. Abstr.* **103**(22), 178999r.

50. A. Priola, C. Corno, and S. Cesca, *Macromolecules* **14**(3), 475 (1981).

51. *Ibid.,* **13,** 1314 (1980).

52. A. Priola, S. Cesca, G. Ferraris, and M. Bruzzone, *Makromol. Chem.* **176,** 1969 (1975).

53. USSR Pat. 1,065,433 (Jan. 7, 1984), B. I. Popov, B. I. Pantukh, G. I. Rutman, V. R. Dolidze, and G. T. Mozalevskii; *Chem. Abstr.* **100**(18), 140949z.

54. V. F. Kolbasov, D. F. Kutepov, V. I. Kul'chitskii, and N. N. Sanina, *Zh. Prikl. Khim. Leningrad* **57**(3), 631 (1984); *Chem. Abstr.* **101**(10), 73135m.

55. U.S. Pat. 4,098,982 (Mar. 8, 1977), R. T. Wojcik (to Arizona Chemical Co.); *Chem. Abstr.* **89,** 216059m.

56. V. N. Zaboristov, V. I. Anosov, M. I. Domogatskaya, and V. K. Sotnikova, *Kauch. Rezina* (6), 20 (1983); *Chem. Abstr.* **99**(10), 71962v.

57. U.S. Pats. 3,963,511 and 3,963,512 (June 15, 1976), J. D. Swift, D. G. Hawthorne, B. C. Loft, and D. H. Solomon (to Commonwealth Science & Industry Research Organization); *Chem. Abstr.* **85,** 125077s; *Chem. Abstr.* **85,** 125076r.

58. Jpn. Pat. 73 17,589 (Mar. 6, 1973), T. Gou (to Japanese Geon Co. Ltd.); *Chem. Abstr.* **79,** 19431u.

59. K. Leets, M. I. Shmidt, T. Valimae, and T. Kaal, *Zh. Org. Khim.* **20**(7), 1588 (1984); *Chem. Abstr.* **102**(3), 24084f.
60. K. Leets, M. Shmidt, T. Valimae, T. Kaal, *Eesti NSV Tead. Akad. Toim. Keem.* **34**(1), 55 (1985); *Chem. Abstr.* **103**(7), 53671w.
61. J. P. Kennedy and R. G. Squires, *J. Macromol. Sci. Chem.* **A1**(5), 861 (1967).
62. C. Corno, A. Priola, and S. Cesca, *Macromolecules* **13**, 1099 (1980).
63. M. A. Golub, *J. Polym. Sci. Polym. Lett. Ed.* **15**(6), 369 (1977).
64. Ref. 1, p. 491.
65. E. W. Duck and J. M. Locke in W. M. Saltman, ed., *The Stereo Rubbers*, John Wiley & Sons, Inc., New York, 1977, p. 205.
66. U.S. Pat. 3,147,242 (Sept. 1, 1964), R. S. Stearns (to Firestone Tire & Rubber Co.).
67. USSR Pat. 1,035,035 (Aug. 15, 1983), V. V. Moiseev, L. A. Grigor'eva, V. I. Nikulaeva, L. V. Kovtunenko, N. K. Shedogubova, S. V. Veisenberg, L. N. Mistyukova, and Z. N. Markova; *Chem. Abstr.* **99**, 195593v.
68. G. V. Rakova, A. A. Korotkov, and T.-C. Li, *Dokl. Akad. Nauk SSSR* **126**(3), 582 (1959).
69. I. A. Livshits, I. D. Afanas'ev, and E. R. Gershtein, *Kauch. Rezina* **28**(2), 4 (1967); *Chem. Abstr.* **70**, 116022h.
70. I. A. Livshits and T. T. Denisova, *Dokl. Akad. Nauk SSSR* **179**, 98 (1968); *Chem. Abstr.* **68**, 115104h.
71. F. Schue, *Bull. Soc. Chim. France,* 980 (1965).
72. A. F. Halasa, *Rubber Chem. Technol.* **54**, 634 (1981).
73. R. E. Kozulla and D. McIntyre, research presentation, Department of Polymer Science, University of Akron, Akron, Ohio, Oct. 28, 1986.
74. S. E. Radkevich, V. B. Erofeev, V. Z. Veshtort, A. Ya. Valendo, and A. G. Galeeva, *Vestsi Akad. Navuk. BSSR Khim. Navuk* (6), 71 (1985); *Chem. Abstr.* **104**(18), 149507u.
75. U.S. Pat. 4,482,771 (Nov. 13, 1984), J. W. Bozzelli, K. S. Dennis, and F. A. Donate (to Dow Chemical Co.); *Chem. Abstr.* **102**(10), 80069s.
76. U.S. Pat. 4,486,614 (Dec. 4, 1984), F. A. Donate, J. W. Bozzelli, and K. S. Dennis (Dow Chemical Co.); *Chem. Abstr.* **102**(12), 96232a.
77. J. Inomata, *Makromol. Chem.* **135**(3350), 113 (1970).
78. A. I. Jakubchik and co-workers, *Zh. Briekl. Chim.* **35**, 402 (1962); *Chem. Abstr.* **57**, 4808e.
79. O. A. Rozinova, E. R. Dolinskaya, A. S. Khachaturov, and V. A. Kormer, *Dokl. Akad. Nauk* **243**(5), 1219 (1978); *Chem. Abstr.* **90**, 138324y.
80. A. Petit and M. T. Cung in *6th Conv. Ital. Sci. Macromol.* **2**, 117 (1983); *Chem. Abstr.* **100**(26), 210708w.
81. G. Jenner, *Peint. Pigm. Vernis* **43**(3), 205 (1967); *Chem. Abstr.* **67**, 11762j.
82. F. Schue, B. Hugelin, and B. Kaempf, *Bull. Soc. Chim. France* **12**, 4673 (1967).
83. M. M. F. Al-Harrah and R. N. Young, *Polymer* **21**(1), 119 (1980).
84. M. Morton and L. A. Falvo, *Macromolecules* **6**, 190 (1973).
85. H. Yasuda, A. Yasuhara, and H. Tani, *Macromolecules* **7**(1), 145 (1974); *Chem. Abstr.* **81**, 4306z.
86. Z. M. Sabirov and co-workers, *Zh. Fiz. Khim.* **59**(5), 1136 (1985); *Chem. Abstr.* **103**(12), 88249s.
87. M. Lambert, *A Short Russian-English Dictionary of Terminology Used in the Soviet Rubber Plastics and Tyre Industries*, Maclaren & Sons, London, 1963, p. 196.
88. M. M. Mogilevich, V. B. Manerov, V. M. Troshin, and T. A. Ryazanova, *Lakokras. Mater. Ikh. Primen.* (3), 10 (1985); *Chem. Abstr.* **103**(14), 106327x.
89. V. V. Verkholantsev and T. I. Victorova, *Lakokras. Mater. Ikh. Primen.* (5), 19 (1985); *Chem. Abstr.* **104**(14), 111399x.
90. M. M. Babkina, R. A. Martynenkova, L. A. Dobrovinskii, and R. M. Livshits, *Lakokras. Mater. Ikh. Primen.* (1), 10 (1986).
91. V. S. Krasnobaeva, T. P. Kondrat'eva, I. D. Sokolova, and M. M. Mogilevich, *Lakokras. Mater. Ikh. Primen.* (5), 14 (1985); *Chem. Abstr.* **104**(14), 111393r.
92. M. M. Babkina, S. D. Yablonovskaya, R. A. Martynenkova, and co-workers, *Lakokras. Mater. Ikh. Primen.* (6), 24 (1985); *Chem. Abstr.* **104**(24), 208856g.
93. I. A. Ososhnik, V. S. Shein, and N. V. Skopintseva, *Kauch. Rezina* (10), 18 (1984); *Chem. Abstr.* **102**(4), 261095.
94. N. K. Loginova, R. A. Kuznetsova, and E. M. Fel'dblyum, *Prom. st. Sint. Kauch.* (2), 11 (1978); *Chem. Abstr.* **89**, 7366y.

95. M. D. Sokolov, B. K. Basov, V. A. Lysanov, and A. A. Zolotov, *Prom. st. Sint. Kauch.* (11), 14 (1982); *Chem. Abstr.* **98**, 90799t.

96. Jpn. Pat. 046731 (Jan. 7, 1971) (to Nippon Oil Co.); Derwent 2029416-Q.

97. Ger. Pat. 1,091,753 (Apr. 20, 1961), A. J. Parrod and G. Beinert (to Cent. Nat. Rech. Sci., Paris).

98. UK Pat. 972,246 (Oct. 14, 1964) (Phillips Petroleum Co.).

99. U.S. Pat. 2,842,926 (July 15, 1958), A. L. Ayers and C. R. Scott (to Phillips Petroleum Co.).

100. Rom. Pat. 52,000 (Dec. 24, 1969), T. Andrei, I. Negulescu, and E. Ionescu (to Inst. Petrochim. Romania); *Chem. Abstr.* **73**, 26217h.

101. USSR Pat. 998,467 (Feb. 23, 1983), B. V. Erofeev, Zh. D. Chaplanova, A. Ya. Valendo, and S. E. Radkevich; *Chem. Abstr.* **99**(6), 38919p.

102. U.S. Pat. 2,826,618 (Mar. 11, 1958), A. H. Gleason (to Esso Research & Engineering Co.).

103. U.S. Pat. 3,030,346 (Apr. 17, 1962), R. N. Cooper (to Phillips Petroleum Co.); CA 57 999f.

104. Jpn. Pat. 85 53,509 (Mar. 27, 1985). (Japan Synthetic Rubber Co. Ltd. and Bridgestone Tire Co. Ltd.); *Chem. Abstr.* **103**(4), 23633t.

105. A. G. Kharitonov, V. S. Glukhovskoi, A. N. Kondrat'ev, E. F. Mironova, and E. S. Dzhavakhadze, *Prom. st. Sint. Kauch.* (7), 13 (1983); *Chem. Abstr.* **99**(18), 141353y.

106. I. A. Livshits, S. I. Il'ina, and V. N. Reikh, *Khim. Prom.* 342 (1957); *Chem. Abstr.* **52**, 9639.

107. T. L. Hanlon, K. C. Kauffman, R. W. Kavchok, and R. G. Bauer, *Appl. Polym. Symp.* (26), 61 (1975).

108. G. Jenner, *Sc. Azione* (11), 131 (1967).

109. D. H. Beebe, C. E. Gordon, R. N. Thudium, M. C. Throckmorton, and T. L. Hanlon, *J. Polym. Sci. Polym. Chem. Ed.* **16**, 2285 (1978).

110. F. Ciampelli, M. P. Lachi, M. T. Venturi, and L. Porri, *Eur. Polym. J.* **3**, 353 (1967).

111. G. Jenner, J.-P. Doll-Robbe, and F. Schue, *Bull. Soc. Chim. France* (2), 388 (1965).

112. U.S. Pat. 3,457,201 (July 22, 1969), H. S. Smith, T. Trogdon, and T. Holt (to Dayco Corp.); *Chem. Abstr.* **71**, 71759f.

113. A. P. Rokhmistrova, E. N. Shushkina, L. V. Kosmodem'yanskii, and E. G. Lazar'yants, *Lakokras. Mater. Ikh. Primen.* (5), 23 (1966); *Chem. Abstr.* **66**, 19943u.

114. Jpn. Pat. 77 139,150 (Nov. 16, 1977), T. Kita and S. Ishibashi (to Nippon Zeon Co. Ltd.); *Chem. Abstr.* **88**, 153883g.

115. USSR Pat. 952,865 (Aug. 23, 1982), A. G. Liakumovich, T. I. Lonshchakova, P. A. Kirpichnikov, N. V. Lemaev, and co-workers; *Chem. Abstr.* **98**(10), 73305m.

116. Rom. Pat. 54,373 (June 20, 1972), C. V. Niculiu, T. F. Morel, C. Ionescu, T. Crisan, and I. Sirchis (to Inst. Petrochim.); *Chem. Abstr.* **79**, 5873c.

117. M. Farina, G. Audisio, and G. Natta, *J. Am. Chem. Soc.* **89**(19), 5071 (1967).

118. M. Farina, U. Pedretti, M. T. Gramegna, and G. Audisio, *Macromolecules* **3**(5), 475 (1970).

119. M. Miyata and K. Takemoto, *J. Polym. Sci. Symp.* (55), 279 (1976).

120. M. Miyata, Y. Kitahara, and K. Takemoto, *Polym. Bull.* **2**, 671 (1980).

121. M. Farina and G. DiSilvestro, *Makromol. Chem.* **183**, 241 (1982).

122. P. Sozzani, G. DiSilvestro, M. Grassiand, and M. Farina, *Macromolecules* **17**, 2532 (1984); **17**, 2538 (1984).

123. Neth. Pat. Appl. 6,401,978 (Aug. 30, 1965) (Shell International Research Maats NV); *Chem. Abstr.* **64**, 6837b.

124. Jpn. Pat. 73 16,058 (May 19, 1973) (Japan Synthetic Rubber Co. Ltd.).

125. Ger. Pat. 1,903,514 (Aug. 28, 1969) (Japan Synthetic Rubber Co. Ltd.); Derwent 33,920Q.

126. Belg. Pat. 722,545 (Apr. 1, 1969) (Bridgestone Tire Co., Ltd.).

127. Belg. Pat. 650,427 (Jan. 11, 1965) (National Distillers & Chemical Corp.); *Chem. Abstr.* **57**, 16830b.

128. Neth. Pat. Appl. 6,408,062 (Jan. 17, 1966) (National Distillers & Chemical Corp.); *Chem. Abstr.* **65**, 15635c.

129. Belg. Pat. 732,117 (Apr. 27, 1968) (Asahi Kasei Kogyo KK); Derwent 39,002Q.

130. L. Porri and M. C. Gallazzi, *Makromol. Chem. Rapid Commun.* **4**, 485 (1983).

131. A. Bolognesi, S. Destri, L. Porri, and F. Wang, *Makromol. Chem. Rapid Commun.* **3**, 187 (1982).

132. S. Destri, G. Gatti, and L. Porri, *Makromol. Chem. Rapid Commun.* **2**, 605 (1981).

133. M. C. Gallazzi, A. Giarrusso, and L. Porri, *Makromol. Chem. Rapid Commun.* **2**, 59 (1981).

134. L. Porri, "Structural Order in Polymers," *Proceedings of the International Symposium on Macromolecules,* Florence, Italy, 1980, p. 51.

135. S. Destri, M. C. Gallazzi, A. Giarrusso, and L. Porri, *Makromol. Chem. Rapid Commun.* **1**, 293 (1980).
136. G. Natta and L. Porri, "Elastomer Stereospecific Polymerization," *ACS Adv. Chem.* **52**, 24 (1966).
137. R. Clement, *Eur. Polym. J.* **17**, 1293 (1981).
138. *Ibid.,* p. 895.
139. G. Natta, *Science* **147**, 261 (1965).
140. W. Cooper in Ref. 65, p. 21.
141. L. Porri, in S. Carra, F. Parisi, I. Pasquon, and P. Pino, eds., *Giulio Natta: Present Significance of His Scientific Contribution,* Milan, Italy, 1982, p. 156.
142. H. J. Harwood, *Rubber Chem. Technol.* **55**(3), 769 (1982).
143. P. Aubert, J. Sledz, F. Schue, and J. Prud'homme, *Eur. Polym. J.* **16**, 361 (1980).
144. L. Zetta, G. Gatti, and G. Audisio, *Macromolecules* **11**, 763 (1978).
145. P. Aubert, J. Sledz, and F. Schue, *J. Polym. Sci. Polym. Chem. Ed.* **19**, 955 (1981).
146. G. Natta, L. Porri, P. Corradini, G. Zanini, and F. Ciampelli, *J. Polym. Sci.* **51**, 463 (1961).
147. G. Natta, L. Porri, P. Corradini, and D. Morero, *Chim. Ind.* **40**, 362 (1958).
148. U.S. Pat. 3,550,158 (Dec. 22, 1970), G. Natta, L. Porri, and G. Mazzanti (to Montecatini Edison SpA); *Chem. Abstr.* **74**, 64600f.
149. S. Brueckner, G. DiSilvestro, and W. Porzio, *Macromolecules* **19**(1), 235 (1986).
150. R. N. Shroff, D. I. Livingston, and G. S. Fielding-Russell, *Rubber Chem. Technol.* **43**(6), 1491 (1970).
151. G. Natta, L. Porri, A. Carbonaro, and G. Lugli, *Makromol. Chem.* **53**, 52 (1962); *Chem. Abstr.* **57**, 4828g.
152. L. Porri, A. Carbonaro, and F. Ciampelli, *Makromol. Chem.* **61**, 90 (1963); *Chem. Abstr.* **58**, 14263d.
153. A. Carbonaro, V. Zamboni, G. Novajra, and G. Dall'Asta, *Rubber Chem. Technol.* **46**, 1274 (1973).
154. G. Natta, *Rev. Gen. Caoutch. Plast.* **40**(5), 785 (1963); *Chem. Abstr.* **59**, 15455g.
155. C. Capitani and M. Bruzzone, *Chim. Ind. Milan* **57**, 759 (1975).
156. M. Bruzzone, A. Carbonaro, and L. Gargani, *Rubber Chem. Technol.* **51**, 907 (1978).
157. M. Bruzzone in Ref. 141, p. 67.
158. A. Carbonaro, L. Gargani, E. Sorta, and M. Bruzzone, *Proceedings of the International Rubber Conference,* Venice, Italy, 1979, p. 312; *Ind. Gomma* **24**(10), 66 (1980).
159. E. Lauretti, G. Santarelli, A. Canidio, and L. Gargani in Ref. 158, p. 322.
160. M. Bruzzone in J. E. Mark and J. Lal, eds., *ACS Symp. Ser.* **193**, 33 (1982).
161. A. A. Entezami, R. Mechin, F. Schue, A. Collet, and B. Kaempf, *Eur. Polym. J.* **13**, 193 (1977).
162. Fr. Pat. 1,488,982 (July 21, 1967), P. Teyssie, R. Dauby, and F. Dawans (to Inst. Fr. du Petrol. des Carb. et Lub.); *Chem. Abstr.* **68**, 50816v.
163. J. Zachoval, J. Krepelka, and M. Klimova, *Collect. Czech. Chem. Commun.* **37**(10), 3271 (1972); *Chem. Abstr.* **80**, 3895.
164. R. E. Rinehart, H. P. Smith, H. S. Witt, and H. Romeyn, *J. Am. Chem. Soc.* **83**, 4864 (1961); **84**, 4145 (1962).
165. G. Natta, L. Porri, and G. Sovarzi, *Eur. Polym. J.* **1**, 81 (1965).
166. L. Porri, A. diCorato, and G. Natta, *Eur. Polym. J.* **5**, 1 (1969).
167. G. S. Fielding-Russell and G. H. Smith, *Rubber Chem. Technol.* **43**(4), 771 (1970).
168. Ger. Offen. 2,015,153 (Oct. 21, 1971), P. Guenther, W. Oberkirch, F. Haas, G. Pampus, and G. Marwede (to Farb Bayer AG); *Chem. Abstr.* **76**, 60718u.
169. Belg. Pat. 764,904 (Mar. 28, 1970) (Farb Bayer AG).
170. Jpn. Pat. 70 39,273 (Dec. 10, 1970), M. Ichikawa, Y. Takeuchi, H. Kurita, and A. Kihinoki (to Japan Synthetic Rubber Co. Ltd.); *Chem. Abstr.* **75**, 77943r.
171. Jpn. Pat. 72 01,226 (Jan. 13, 1972), M. Ichikawa, Y. Takeuchi, A. Kihi, and N. Yamaguchi (to Japan Synthetic Rubber Co. Ltd.); *Chem. Abstr.* **77**, 6946z.
172. Brit. Pat. 1,158,296 (July 16, 1969) (to Japan Synthetic Rubber Co. Ltd.); *Chem. Abstr.* **71**, 82394j.
173. U.S. Pat. 4,048,418 (Sept. 13, 1977), M. C. Throckmorton (to Goodyear Tire & Rubber Co.).
174. G. Natta, L. Porri, A. Carbonaro, F. Ciampelli, and G. Allegra, *Makromol. Chem.* **51**, 229 (1962).
175. U.S. Pat. 3,301,839 (Jan. 31, 1967), G. Natta, L. Porri, A. Carbonaro, and G. Stoppa (to Montecatini Edison SpA).
176. L. Porri, A. diCorato, and G. Natta, *J. Polym. Sci. Part B* **5**(4), 321 (1967).

177. G. S. Fielding-Russell and G. H. Smith, *J. Appl. Polym. Sci.* **14**, 777 (1970).
178. U.S. Pat. 3,300,467 (Jan. 24, 1967), G. Natta, L. Porri, G. Stoppa, and A. Carbonaro (to Montecatini Edison SpA).
179. G. Natta, L. Porri, G. Stoppa, G. Allegra, and F. Ciampelli, *J. Polym. Sci. Part B* **1**, 67 (1963).
180. G. Natta, L. Porri, A. Carbonaro, and G. Stoppa, *Makromol. Chem.* **77**, 114 (1964).
181. U.S. Pat. 3,657,205 (Apr. 18, 1972), M. C. Throckmorton (to Goodyear Tire & Rubber Co.).
182. A. A. Panasenko, V. N. Odinokov, Yu. B. Monakov, L. M. Khalilov, A. S. Bezgina, V. K. Ignatyuk, and S. R. Rafikov, *Vysokomol. Soedin. Ser. B* **19**(9), 656 (1977); *Chem. Abstr.* **87**, 185152p.
183. U.S. Pat. 4,168,357 (Sept. 18, 1979), M. C. Throckmorton and C. J. Suchma (to Goodyear Tire & Rubber Co.).
184. U.S. Pat. 4,148,983 (Apr. 10, 1979), M. C. Throckmorton (to Goodyear Tire & Rubber Co.).
185. U.S. Pat. 3,936,432 (Feb. 3, 1976), M. C. Throckmorton (to Goodyear Tire & Rubber Co.).
186. U.S. Pat. 3,678,022 (July 18, 1972), J. E. Bozik, H. E. Swift, and C. Y. Wu (to Ameripol Inc.).
187. M. Gippin, *Ind. Eng. Chem. Prod. Res. Dev.* **1**, 32 (1962).
188. G. Natta, L. Porri, and A. Carbonaro, *Makromol. Chem.* **77**, 126 (1964).
189. Ref. 12, p. 646.
190. B. Costa, P. Locatelli, and A. Zambelli, *Macromolecules* **6**, 653 (1973).
191. K. Boujadoux, M. Galin, J. Jozefonvicz, and Szubarga, *Eur. Polym. J.* **10**, 1 (1974).
192. H. Hirai, K. Hiraki, I. Noguchi, T. Inoue, and S. Makishima, *J. Polym. Sci. Part A-1* **8**, 2393 (1970).
193. K. F. Elgert and W. Ritter, *Makromol. Chem.* **177**, 2021, 2781 (1976).
194. R. Clement, K. Boujadoux, J. Jozefonvicz, and G. Roques, *Eur. Polym. J.* **10**, 821 (1974).
195. M. C. Throckmorton, *Kautsch. Gummi Kunstst.* **22**(6), 293 (1969).
196. D.-M. Xie, C. G. Zhong, N.-A. Yuan, Y. F. Sun, S.-X. Xiao, and J. Ouyang, *Kao Fen Tzu T'ung Hsun* (4), 233 (1979); *Chem. Abstr.* **92**, 77008r.
197. D. Xie and C. Zhong, *Gaofenzi Tongxun* (3), 220 (1983); *Chem. Abstr.* **99**(20), 159096b.
198. Y. Jin, Y. Sun, X. Liu, X. Li, and J. Ouyang, *Fenzi Kexue Yu Huaxue Yanjiu* **4**(2), 247 (1984); *Chem. Abstr.* **101**, 131149g.
199. U.S. Pat. 3,935,175 (Jan. 27, 1976), G. Lugli, A. Mazzei, and G. Brandi (to Snam Progetti SpA).
200. U.S. Pat. 3,676,411 (July 11, 1972), M. C. Throckmorton and W. M. Saltman (to Goodyear Tire & Rubber Co.).
201. Belg. Pat. 791709 (Mar. 16, 1973) (Snam Progetti SpA).
202. Belg. Pat. 807904 (Mar. 15, 1974) (Snam Progetti SpA).
203. U.S. Pat. 4,145,497 (Mar. 20, 1979), G. Sylvester, J. Witte, and G. Marwede (to Bayer AG).
204. U.S. Pat. 4,168,374 (Sept. 18, 1979), M. C. Throckmorton and C. J. Suchma (to Goodyear Tire & Rubber Co.).
205. D. Satas in Ref. 35, p. 2.
206. C. Watson and D. Satas in Ref. 35, p. 559.
207. Jpn. Pat. 76 122,135 (Oct. 26, 1976), H. Inoue and I. Sonehara (to Nippon Zeon Co. Ltd.); *Chem. Abstr.* **86**, 172732g.
208. Jpn. Pat. 78 74,535 (July 3, 1978), M. Takahashi, K. Yanagisawa, and K. Maeshima (to Sekisui Chemical Co. Ltd.); *Chem. Abstr.* **89**, 147849e.
209. Jpn. Pat. 73 47,540 (July 6, 1973), T. Wada and Y. Mikogami (to Toshiba Silicone Co. Ltd.); *Chem. Abstr.* **80**, 4544e.
210. Jpn. Pat. 74 37,940 (Apr. 9, 1974), Y. Mikogami (to Tokyo Shibaura Electric Co. Ltd.); *Chem. Abstr.* **81**, 153952j.
211. Jpn. Pat. 74 38,938 (Apr. 11, 1974), H. Yoshimura, H. Nakagawa, A. Kimi, and I. Kogure (Nippon Zeon Co. Ltd.); *Chem. Abstr.* **81**, 92409p.
212. Jpn. Pat. 79 25,942 (Feb. 27, 1979), H. Kataoka (to Daiichi Kogyo Seiyaku Co. Ltd.); *Chem. Abstr.* **91**, 6106q.
213. Jpn. Pat. 79 20,136 (Feb. 15, 1979), K. Takahashi and H. Ishikawa (to Kanae Paint Co. Ltd.); *Chem. Abstr.* **91**, 6370w.
214. Jpn. Pat. 82 147,541 (Sept. 11, 1982) (Toshiba Chemical Products Co. Ltd.); *Chem. Abstr.* **98**(14), 108985n.
215. M. M. Babkina, S. D. Yablonovskaya, R. A. Martynenkova, and co-workers, *Lakokras. Mater. Ikh. Primen.* (6), 24 (1985); *Chem. Abstr.* **104**(24), 208856g.
216. Jpn. Pat. 82 147,296 (Sept. 11, 1982) (Toshiba Chemical Industries Co. Ltd.); *Chem. Abstr.* **98**(12), 91207k.

216. I. A. Ososhnik, V. S. Shein, and N. V. Skopintseva, *Kauch. Rezina* (10), 18 (1984); *Chem. Abstr.* **102**(4), 261095.

217. Jpn. Pat. 84 117,564 (July 6, 1984) (Nippon Zeon Co. Ltd.); *Chem. Abstr.* **102**(4), 26524y.

218. Jpn. Pat. 84 117,563 (July 6, 1984) (Nippon Zeon Co. Ltd.); *Chem. Abstr.* **102**(2), 8335h.

219. Jpn. Pat. 76 74,098 (June 26, 1976), Y. Takase, M. Kokura, and M. Ishikawa (to Asahi Denka Kogyo KK); *Chem. Abstr.* **85**, 109416w.

220. D. F. Kutepov, V. I. Kul'chitskii, V. F. Kolbasov, and V. N. Mikhailov, *Deposited Doc. VINITI,* 2842 (1975); *Chem. Abstr.* **87**, 85689a.

221. *Ibid.; Chem. Abstr.* **87**, 68957u.

222. V. F. Kolbasov, D. F. Kutepov, and V. I. Kul'chitskii, *Deposited Doc. VINITI,* 12 (1975); *Chem. Abstr.* **87**, 53985c.

223. P. T. Poluektov, T. B. Gonsovskaya, F. G. Ponomarev, and Yu. K. Gusev, *Vysokomol. Soedin. Ser. A* **15**(3), 606 (1973); *Chem. Abstr.* **79**, 6431u.

224. USSR Pat. 1,065,425 (Jan. 7, 1984), B. S. Turov, N. A. Koshel, V. V. Popova, N. S. Mineeva, O. A. Irodova, S. S. Srednev, B. K. Basov, and V. A. Lysanov; *Chem. Abstr.* **100**(18), 140948y.

225. G. Mattson and C. B. Reaves, *Oxid. Commun.* **7**(1–2), 113 (1984).

226. Ger. Offen. 2,364,886 (Aug. 8, 1974), H. Yoshimura, Y. Nakagawa, H. Yaginuma, and A. Kimi (to Nippon Zeon Co. Ltd.); *Chem. Abstr.* **82**, 31913x.

227. USSR Pat. 992,546 (Jan. 30, 1983), V. S. Shitov, A. L. Labutin, R. P Kondrat'eva, M. P. Ryazanova, G. A. Emel'yanov, V. Ya. Fraishtadt, and A. P. Vakhonin; *Chem. Abstr.* **98**, 162261m.

228. M. M. Babkina, R. A. Martynenkova, L. A. Dobrovinski, and R. M. Livshits, *Lakokras. Mater. Ikh. Primen.* (1), 10 (1986).

229. Jpn. Pat. 75 114,445 (Sept. 8, 1975), M. Nakazawa and T. Harada (to Asahi Denka Kogyo KK); *Chem. Abstr.* **84**, 45777j.

230. USSR Pat. 550,402 (Mar. 15, 1977), V. V. Beresnev, E. A. Stepanov, P. A. Kirpichnikov, N. V. Lemaev, R. G. Mirzayanov, and V. P. Kichigin; *Chem. Abstr.* **86**, 172838w.

231. USSR Pat. 975,723 (Nov. 23, 1982), A. B. Kozorez, D. K. Khasanova, A. I. Mazin, B. I. Popov, V. N. Bushuyushchii, A. L. Voitsekhovskaya, and I. I. Vol'fenzon; *Chem. Abstr.* **98**(12), 90460a.

232. Jpn. Pat. 76 92,890 (Aug. 14, 1976), T. Sato, Y. Takase, M. Kokura, and M. Ishikawa (to Asahi Denka Kogyo KK); *Chem. Abstr.* **85**, 193545t.

233. U.S. Pat. 3,431,310 (Mar. 4, 1969), B. J. Davis and L. B. Rosetti (to Reichhold Chemical Inc.); *Chem. Abstr.* **70**, 88567u.

234. Jpn. Pat. 75 110,444 (Feb. 13, 1974), N. Asada, H. Yaginuma, and K. Kita (to Nippon Zeon Co. Ltd.); *Chem. Abstr.* **84**, 45776h.

235. Yu. B. Monakov, Z. B. Shamaeva, A. A. Berg, and G. A. Tolstikov, *Dokl. Akad. Nauk SSSR* **269**(2), 408 (1983); *Chem. Abstr.* **99**(2), 6691s.

236. USSR Pat. 1,004,505 (Mar. 15, 1983), A. A. Malysheva, L. P. Kireeva, V. Ya. Klimova, S. E. Zaitseva, L. V. Kosmodem'yanskii, A. S. Doronin, V. P. Bugrov, and co-workers; *Chem. Abstr.* **99**(4), 24416k.

237. V. V. Verkholantsev and T. I. Viktorova, *Lakokras. Mater. Ikh. Primen.* (5), 19 (1985); *Chem. Abstr.* **104**(14), 111399x.

238. USSR Pat. 1,142,490 (Feb. 28, 1985), G. U. Volkov, G. V. Rudnaya, M. G. Ovchinnikova, and R. M. Livshits; *Chem. Abstr.* **102**(24), 205579s.

239. USSR Pat. 608,818 (May 30, 1978), I. I. Yukel'son, I. A. Ososhnik, and V. I. Raevskaya; *Chem. Abstr.* **89**, 111927q.

240. Jpn. Pat. 78 40,030 (Apr. 12, 1978), T. Nakagawa, T. Uchiyama, C. Miyake, Y. Nishimura, and K. Ishii (to Nippon Paint Co. Ltd.); *Chem. Abstr.* **89**, 198747c.

241. A. M. Anan'ev and co-workers, *Vestn. Belorus Un-ta Ser. 2* (1), 71 (1985); *Ref. Zh. Khim.* 1985 (Abstr. 15 M259); *Chem. Abstr.* **104**(4), 23429b.

J. NEIL HENDERSON
Consultant

M. C. THROCKMORTON
Consultant

PENULTIMATE EFFECT.   See COPOLYMERIZATION.

PEPTIDES.   See POLYPEPTIDES.

PERACIDS.   See PEROXY COMPOUNDS.

PERMANENT PRESS.   See TEXTILE RESINS.

PERMEABILITY.   See TRANSPORT PROPERTIES.

PEROXIDES, POLYMERIC.   See PEROXY COMPOUNDS.